Dangerous Properties of Industrial and Consumer Chemicals

Dangerous Properties of Industrial and Consumer Chemicals

compiled by

Nicholas P. Cheremisinoff

National Association of Safety and Health Professionals
Morganville, New Jersey

John A. King

University of Maryland
College Park, Maryland

Randi Boyko

Cycle Chem, Inc.
Elizabeth, New Jersey

MARCEL DEKKER, INC. New York · Basel · Hong Kong

Library of Congress Cataloging-in-Publication Data

Cheremisinoff, Nicholas P.
Dangerous properties of industrial and consumer chemicals / compiled by Nicholas P. Cheremisinoff, John A. King, Randi Boyko.
p. cm.
Includes indexes.
ISBN 0-8247-9183-5 (alk. paper)
1. Toxicology—Handbooks, manuals, etc. 2. Industrial toxicology—Handbooks, manuals, etc. 3. Product safety—Handbooks, manuals, etc. I. King, John Allison. II. Boyko, Randi. III. Title
RA1215.C48 1994
615.9'02--dc20 94-14912
CIP

The publisher offers discounts on this book when ordered in bulk quantities. For more information, write to Special Sales/Professional Marketing at the address below.

This book is printed on acid-free paper.

MARCEL DEKKER, INC.
270 Madison Avenue, New York, New York 10016

Current printing (last digit):
10 9 8 7 6 5 4 3 2 1

PRINTED IN THE UNITED STATES OF AMERICA

PREFACE

There are approximately 5.3 million known chemicals, of which about 55,000 are in production for domestic industrial consumption. Additionally there are hundreds of new chemicals introduced to the market place annually. Of the 55,000 commercial chemicals, a substantial number of them are thought to be present in consumer products. A major unknown is how many of these chemicals in commerce are carcinogenic in nature.

Of all the chemicals on the market, only 7,000 have been evaluated in experimental animal testing programs to assess whether they cause cancer. Thus far, about 1,000 chemicals have been found to cause tumors in animals. Although it is natural to try and extrapolate from this ratio, one must recognize that it is highly risky to generalize all chemicals. First, there is a strong bias toward testing so-called "risky chemicals," as shown by the fact that about fifty percent of the approximately 200 chemicals tested for cancer by the National Cancer Institute have been reported to be positive. Furthermore, at least several hundred additional chemicals have been reported in the literature to be mutagenic or to have reproductive effects.

One of the most extensive data bases compiled to protect consumers from chemically induced chronic hazards is that prepared by the U.S. Consumer Product Safety Commission, which has been accessed and largely used as the basis for materials in this handbook. Through their Chemical Screening Committee, the System for Tracking the Inventory of Chemicals (STIC) has been organized into a data base for managing and tracking hazardous chemicals. The editors have accessed this data base and organized the critical path data summarizing adverse chronic health effects and have highlighted those chemicals found in consumer products.

It is important to recognize that the state-of-the-art in understanding and quantifying the hazardous effects of chemical exposure, especially to consumer-type products, is poor. There are several reasons for this situation. First, adverse chronic health effects in humans often do not manifest themselves until long after the initial exposure. Secondly, the experiments used to assess the leading chronic hazard, namely cancer, in laboratory animals typically requires five years and in excess of one-half million dollars per chemical to complete. Finally, there is no single source or agency which provides comprehensive information on the chronic toxicity or even the use of chemicals in either industry or consumer products. As such, numerous publications and computerized data bases must be consulted in order to compile a single reference on the use and toxicity data necessary to define the presence of chronically hazardous chemicals in the market place.

The purpose of this handbook is to provide a general reference both for consumers and industry users. It identifies potential chronic hazards associated with chemicals used either as virgin products or as additives in consumer products. The information compiled provides preliminary toxicity data on over 1,000 chemicals.

The editors wish to acknowledge the U.S. Consumer Product Safety Commission for access and use of their System for Tracking the Inventory of Chemicals data base, and Marcel Dekker, Inc., for their fine production of this important reference.

Nicholas P. Cheremisinoff
John A. King
Randi Boyko

HOW TO USE THE HANDBOOK

The handbook provides a basic compendium of critical reviews and evaluations abstracted from various major, publicly available sources that give information on potential chronic health hazards posed by chemicals. The summaries provided for each chemical are excerpted from reference documents and reflect only the conclusions. Inclusion of a chemical in the handbook does not necessarily imply that the material produces chronic health effects, but only that published data about the chemical have been reviewed. Note also the fact that a chemical has not yet been evaluated by any of the sources cited does not mean that the chemical presents no health hazard.

One can identify a chemical reviewed in the handbook in one of two ways. Index A contains an alphabetical listing of every chemical along with the page upon which it appears. Index B contains the same listing of chemicals and their corresponding page numbers but in chronological order according to Chemical Abstract Service (CAS) numbers.

To use Index A, look up the substance name alphabetically, ignoring prefixes and numbers. Substances are listed by the chemical name or preferred synonym, rather than the IUPAC or CAS Primary name. For example, methylene chloride is listed rather than dichloromethane. One should consult the latest edition of the NIOSH Registry of Toxic Effects of Chemical Substances (RTECS) for a comprehensive listing or cross-index of synonyms or trade names.

An example of a typical abstract for a chemical, together with a brief explanation of each of the summary descriptions, follows:

[1]→ CAS NO : **62-53-3**

[2]→ CHEM NAME: **ANILINE**

→Report 1: IARC MONOGRAPH NO. 4 6/73
There is no adequate data to indicate that aniline is carcinogenic in experimental animals. At the present time, the weight of epidemiological evidence suggests that aniline is not a carcinogen for the human bladder.

→ Report 2: NTP TEST REGIMEN
The compound was found to be negative in
[3] the salmonella typhimurium mutagenicity assay.

→ Report 3: EPA RCRA HEALTH & ENV. EFFECTS PROFILE 10/80
An increased incidence of urinary bladder tumors has been noted in workers in the aniline and aniline dyestuff industries. Oral feeding studies conducted by the NCI have shown aniline to be carcinogenic in rats. Aniline is reported not to be mutagenic to six strains of salmonella; however, several aniline analogs and derivatives are mutagenic.

[4]→ Report 6: MULTI-AGENCY CROSS-CHECK
N - DHHS Annual
Y - OSHA Cand.
N - EPA Cag.

[5]→Rept'd Use: CHEMICAL INTERMEDIATE, SOLVENT, PAINT REMOVER, PRINTING INKS, SHOE BLACKS

[1]→ CAS NO. - The Chemical Abstract Service (CAS) Registry Number is a numeric designation assigned by the American Chemical Society to uniquely identify a specific chemical substance or group of substances.

[2]→ CHEM. NAME - Chemicals are listed by their most common name.

[3]→ REPORTS 1-10 - Annotated summaries of various reference documents or reports which address the chronic toxicity of a chemical.

[4]→ REPORT 6 - MULTI-AGENCY CROSS CHECK - A "Y" indicates that based on known or potential carcinogenicity, a chemical is on any of the following three lists or reports:

'DHHS Annual' is the NTP Annual Report on Carcinogens;
'OSHA CAND.' is the OSHA Candidate List for Further Review of Potential Occupational Carcinogenicity;
'EPA CAG.' is the listing of the EPA Carcinogen Assessment Group.

An 'R' indicates that the chemical has been regulated because of its carcinogenicity.

[5]→ REPT'D. USE - Uses listed are taken from standard chemical reference books, such as the Merck Index, which provide this data.

CONTENTS

Dangerous Properties of Industrial and Consumer Chemicals

I. SOURCES OF TOXICITY DATA

The following is a listing of major sources and references the reader can refer to in obtaining additional information on the toxicity of chemicals.

I. DEPARTMENT Of HEALTH AND HUMAN SERVICES (DHHS)

Annual Report on Carcinogens

Section 301(b)(4) of the Public Health Service Act, as amended by Section 262, P.L. 95622, requires that the Secretary of the DHEW (now DHHS) publish an annual report containing ". . . a list of all substances (i) which either are known to be carcinogens or which may reasonably be anticipated to be carcinogens, and (ii) to which a significant number of persons residing in the United States are exposed . . . ."

In addition, these reports are to provide information on the nature and degree of exposures to such carcinogens, and the extent to which Federal regulations are effective in reducing the risk to the public health from exposures to these substances.

II. NATIONAL TOXICOLOGY PROGRAM (NTP)

A. Carcinogenesis Testing Program

The National Toxicology Program was established in November, 1978 by the Department of HEW (now DHHS) as a department-wide effort to provide needed information to regulatory and research agencies. The DHHS agencies participating in the program are: The National

Cancer Institute (NCI), the National Institute for Environmental Health Sciences (NIEHS), the National Center for Toxicological Research (NCTR) and the National Institute for Occupational Safety and Health (NIOSH).

Within the NTP Carcinogenesis Testing Program, a cancer bioassay is a two-sex, two-species, lifetime study of experimental animals, usually rats and mice; beginning at weaning, ending 104 weeks after initiation, and using multiple dose levels of the chemical being tested. This bioassay is used to determine if a chemical causes cancer, and if it produces damaging effects on certain organ systems: liver, lung, kidney, endocrine systems, etc. The study of a single compound is expensive, costing about five hundred thousand dollars, and takes up to five years to complete.

The National Toxicology Program publishes a technical report upon completion of a bioassay and review of the results by an independent Board of Scientific Counselors.

B. Reproductive and Developmental Toxicology Program

The National Toxicology Program has a program to assess the effects of chemicals on reproductive function and development. Structural teratology testing (the testing of chemicals to determine if they produce malformations) was begun in FY79. Eight to ten chemicals are tested for teratogenic effects annually. Fetuses are examined at two different levels: 1) gross, readily apparent malformations are noted; and 2) histopathological examinations are conducted to pinpoint less readily apparent, microscopic malformations. Selected priority chemicals are also screened to determine potential reproductive hazard through germ-cell mutations.

C. Genetic Toxicology Program

The Genetic Toxicology Program tests chemicals for mutagenicity, validates existing test systems and develops new short-term test methods. The mutagenicity testing program is divided into three phases. Phase I involves Salmonella mutagenicity assays and mammalian cell cultures. Phase II includes Drosophila systems. Phase III utilizes in vivo mammalian assays. All chemicals selected for general toxicology and lifetime bioassays are tested first using the Salmonella mutagenesis screen. Chemicals which are shown to be mutagenic in Phase I assays progress to Phase II. In addition, a certain number of chemicals giving negative results in Phase I are committed also to further testing in Phase II, based primarily on known biological activity of structurally related compounds, and on estimated levels of human exposure. Chemicals which are positive mutagens in Phase II progress to Phase III.

III. WORLD HEALTH ORGANIZATION'S (WHO) INTERNATIONAL AGENCY FOR RESEARCH ON CANCER (IARC)

IARC Monographs on the Evaluation of the Carcinogenic Risk of Chemicals to Humans

In 1971 the World Health Organization's International Agency for Research on Cancer (IARC) initiated a program of evaluation of the carcinogenic risk of chemicals to humans. "The objective of the program is to elaborate and publish in the form of monographs, critical reviews of data on carcinogenicity for groups of chemicals to which humans are known to be exposed, to evaluate these data in terms of human risk . . . and to indicate where additional research efforts are needed."

The IARC Monographs are recognized as an authoritative source of information on the carcinogenicity of environmental chemicals. The first users' survey, made in 1976, indicates that the monographs are consulted routinely by various agencies in 24 countries.

IV. OCCUPATIONAL SAFETY AND HEALTH ADMINISTRATION (OSHA)

A. A List of Substances Which May Be Candidates for Further Scientific Review and Possible Identification, Classification and Regulation as Potential Occupational Carcinogens.

Publication of this list by the Occupational Safety and Health Administration is the first action taken in following the "Cancer Standard" procedures promulgated by OSHA in January, 1980.

Compilation of this list is the initial step in OSHA's screening process for suspected carcinogens. To compile this list, substances for which there was some evidence of carcinogenicity and potential occupational hazard were drawn from such lists as the NIOSH Registry of Toxic Effects of Chemical Substances (RTECS); U.S. Public Health Service, Publication No. 149; EPA's Inventory of Chemical Substances; the monographs of the International Agency for Research on Cancer; and the Annual Report of the Secretary of the U.S. Department of Health and Human Services to the President and the Congress. The substances which are found in American work places, and for which there is some evidence of carcinogenicity, are subjected to a brief scientific review of available literature to determine whether or not a substance should be subject to a more thorough review under the OSHA Cancer Policy. Based on the more detailed review, a determination may then be made as to whether a substance meets OSHA's definition of a potential occupational carcinogen.

B. Substances Regulated by the Occupational Safety and Health Administration (OSHA) on the Basis of Carcinogenicity

Since 1970, OSHA has promulgated rules regulating 18 chemicals on the basis of carcinogenicity.

V. NATIONAL INSTITUTE FOR OCCUPATIONAL SAFETY AND HEALTH (NIOSH)

A. Criteria Document for a Recommended Standard

The National Institute for Occupational Safety and Health (NIOSH) ". . . develops and periodically revises Criteria Documents that recommend standards for occupational exposure to chemical and physical hazards." The NIOSH critically evaluates all known and available medical, biological, engineering, chemical, and trade data and information that may be relevant to recommending a standard in the criteria document.

The Criteria Document provides a "valid and detailed recommendation with supporting information to be used by the Department of Labor as a basis for promulgation of an occupational health Standard."

B. Current Intelligence Bulletins

In January 1975, the National Institute for Occupational Safety and Health (NIOSH) developed a Current Intelligence Bulletin System to "...promptly review, evaluate, and supplement new information reviewed by NIOSH on occupational hazards that are either unrecognized or are greater than generally known." These hazards include, but are not limited to chemical hazards. The information is published and disseminated in the form of a "Current Intelligence Bulletin."

C. Special Hazard Reviews

The National Institute for Occupational Safety and Health (NIOSH) occasionally issues Special Hazard Reviews which "...assess, from a health standpoint, specific problems associated with a given agent or hazard, such as a potential for carcinogenic, mutagenic or teratogenic effects." The reviews also recommend appropriate control and surveillance methods. "While the reviews are not intended to supplant the more comprehensive NIOSH Criteria Documents, nor the brief NIOSH Current Intelligence Bulletins, they are prepared in such a way as to assist in the formulation of regulations."

VI. ENVIRONMENTAL PROTECTION AGENCY (EPA)

A. List of Chemicals Having Substantial Evidence of Carcinogenicity, Prepared by the Carcinogen Assessment Group (CAG)

This list was initially prepared in response to the needs of the Office of Pesticides and Toxic Substances to develop labeling regulations under section 3001 of the Resource Conservation and Recovery Act. Sources of information used in selecting chemicals as candidates for the initial list are two types: chemicals which the Carcinogen Assessment Group (CAG) previously evaluated and has determined to pose a potential risk of cancer to humans; and chemicals which CAG reviewed because one or more of three organizations (The International Agency for Research on Cancer, the National Toxicology Program Bioassay Program, and the Food and Drug Administration of the U.S. Department of Health and Human Services) concluded that these chemicals are potential human carcinogens.

B. Report of the TSCA Interagency Testing Committee to the Administrator, Environmental Protection Agency

Section 4(e) of the Toxic Substances Control Act (TSCA) (P.L. 94469) established the TSCA Interagency Testing Committee (ITC) with representation from many of the Federal research and regulatory agencies. This Committee has the continuing responsibility to identify and recommend to the Administrator of the Environmental Protection Agency, chemical substances or mixtures which should be tested to determine their hazard to human health or the environment. In the ITC review and recommendation of selected chemicals, priority attention is given to those individual or groups of chemical substances or mixtures which are known to cause or contribute to, or which are suspected of causing or contributing to cancer, gene mutations, or birth defects. The list, and reasons for making each recommendation, are required to be published in the Federal Register.

Since 1977, the ITC has published eight reports which contain a total of 46 chemical substances or categories of chemicals. One chemical has been removed from the 4(e) Priority List because EPA responded to the Committee's recommendation for testing.

C. Toxic Substances Control Act: Substantial Risk Notification

Under Section 8(e) of the Toxic Substances Control Act of 1976 (TSCA), anyone who obtains information which reasonably supports the conclusion that a substance presents a substantial risk of injury to human health or the environment must notify the Environmental Protection Agency within 15 days.

These notices are then reviewed by the Office of Pesticides and Toxic Substances. An initial evaluation of the substance is prepared containing, if appropriate, followup questions to the submitter, referrals to other agencies, and decisions to list the chemical for a Section 8 reporting rule or to undertake a formal risk assessment.

D. Toxic Substances Control Act: Premanufacture Notification (PMN)

Under Section 5(a)(1) of the Toxic Substances Control Act of 1976 (TSCA) anyone who intends to introduce into commerce a chemical substance not on the 1977 TSCA Inventory must notify the EPA at least 90 days before beginning manufacture or importation. The EPA is to use the review period to determine if the substance might present an unreasonable risk. Upon such determination, the EPA may issue an order to prohibit manufacture and may then apply to the court for an injunction to this effect. In the absence of such an order or immediate rule, the manufacturer may proceed with the plans to introduce the substance.

The EPA publishes in the Federal Register a summary of each PMN five days after its receipt.

E. List of Hazardous Constituents, Appendix VIII of the Resource Conservation and Recovery Act of 1976

Subtitle C of the Solid Waste Disposal Act, as amended by the Resource Conservation and Recovery Act of 1976 (RCRA), directs the EPA to promulgate regulations to protect human health and the environment from the improper management of hazardous wastes. Part 261 of this Act defines a hazardous waste to be a substance or mixture which may ". . . cause, or significantly contribute to, an increase in mortality or an increase in serious irreversible or incapacitating reversible illness, or pose a substantial present or potential hazard to human health or the environment when it is improperly treated, stored, transported, disposed of or otherwise managed . . ."

Appendix VIII of Part 261 is a list of toxic chemicals that have been shown in reputable scientific studies compiled by the EPA, to have toxic, carcinogenic, mutagenic, or teratogenic effects on humans or other life forms. The Appendix is used by the Agency to identify those toxic constituents which, if present in a waste, could result in the waste being placed on the hazardous waste list.

Appendix VIII presently contains 330 chemical substances and 29 classes of chemical substances. The EPA expects to add other substances to the list as additional data are developed and assessed. No time frame has been established for addition of substances to the list.

F. Lists of Hazardous Air Pollutants

Sections 108(a), lll(d), and 112(b)(i)(A) of the Clean Air Act Amendments of 1970 require that the Administrator of the EPA publish, and from time to time revise, three lists of hazardous air pollutants. As defined in the amendments, a hazardous air pollutant is an ". . . air pollutant . . . which may reasonably be anticipated to result in an increase in mortality or an increase in serious, irreversible, or incapacitating reversible, illness." The EPA Office of Air Quality Planning and Standards currently has a list of 35 chemicals under assessment for possible inclusion on one of the three lists described below.

The list required by section 108(a) is for the purpose of establishing national primary and secondary ambient air quality standards. This list includes each air pollutant, emissions of which, in the Administrator's judgement, cause or contribute to air pollution which may reasonably be anticipated to endanger public health or welfare and for which air quality criteria had not been issued before the date of enactment of the Clean Air Amendments of 1970, but for which the Administrator plans to issue air quality standards under this section. As of October 1980, six substances or groups of substances had been published on the list required by section 108(a).

Section 112(a) requires that the Administrator publish a list of hazardous air pollutants for which he intends to establish a National Emission Standard. As of October, 1980, seven substances or groups of substances had been published on the list required by this section.

Section lll(d) requires that the Administrator publish a list of each air pollutant for which air quality criteria have not been issued, or which is not included on the other two lists, but for which a standard of performance under this section would be required if the existing source were a new source. As of October 1980, 3 groups of substances had been published on the list required by this section.

G. Clean Water Act Toxic Pollutant List

Section 307(a)(1) of the Federal Water Pollution Control Act, as amended by the Clean Water Act of 1977, requires that the Administrator of the EPA publish the list of toxic pollutants listed in Table 1 of Committee Print No. 9530 of the House of Representatives Committee on Public Works and Transportation.

The Act defines toxic pollutants as "those pollutants or combination of pollutants, including disease causing agents, which after discharge and upon exposure, ingestion, inhalation, or assimilation into any organism, either directly from the environment, or indirectly by ingestion through food chains will; on the basis of information available to the Administrator, cause death, disease, behavioral malfunctions (including malfunctions in reproductions), or physical deformations, in such organisms of their offspring."

The list of 65 toxic pollutants was originally developed by a multidisciplinary task force of scientists, and forms a basis for the development of effluent limitations for categories or classes of point sources pursuant to sections of the Act.

Since the publication of the list in January 1978, several of the broad classes originally listed have been broken down into definite compounds, expanding the list to 129 substances. From time to time the Administrator of the EPA may revise the list by adding or removing pollutants. To date, no such revisions have been made.

II. TOXICOLOGICAL AND HAZARDOUS PROPERTIES DATA

CAS NO : **96-86-2**
CHEM NAME : *****ACETOPHENONE (ACP)**

REPORT 01 : HS STAFF REPORT SEP 89
ACP EXHIBITS ACUTE TOXIC EFFECTS ONLY AT HIGH DOSE LEVELS. INHALATION OF ACP ODOR IN HUMANS MAY CAUSE MALAISE, HEADACHE, EPIGASTRIC PAIN, VERTIGO, AND TENDENCY TO FALL ASLEEP. MAY CAUSE SKIN AND EYE IRRITATION, AND MILD BURN IF CONTACT IS PROLONGED AND FREQUENT. NO DATA AVAILABLE TO PERMIT JUDGEMENT CONCERNING CHRONIC (CARCINOGENIC) EFFECTS.

REPT'D USE : PERFUMES, SOLVENT PER CELLULOSE ETHERS AND ESTERS, INTERMEDIATE IN ORGANIC SYNTHESIS AND SUBSTITUTE FOR METHYLENE CHLORIDE

CAS NO : **75-07-0**
CHEM NAME : *****ACETALDEHYDE**

REPORT 01 : EPA TSCA 8(E) NOTICE 8EHQ-0383-0474
THE CELANESE CORP. PROVIDED PRELIMINARY FINDINGS FROM CHRONIC INHALATION STUDY IN WHICH RATS EXPOSED TO ACETALDEHYDE VAPOR OF 0, 750, 1500, OR 3000 PPM REVEALED HYPER- AND METAPLASTIC EPITHELIUM WITH MASSIVE KERATINIZATION IN NOSE, AND AT 3000 PPM IN LARYNX. CARCINOMA FROM RESPIRATORY EPITHELIUM FOUND IN TOP-DOSE RATS; NASAL CARCINOMA. ACETALDEHYDE CAPABLE OF INDUCING NASAL CARCINOMAS IN RATS.

REPORT 02 : NTP TECHNICAL BULLETIN 9 MAR 83
ACETALDEHYDE WAS NEGATIVE IN ONE SALMONELLA TYPHIMURIUM MUTAGENICITY ASSAY.

REPORT 03 : IARC MONOGRAPH 36 FEB 85
TESTED IN RATS BY INHALATION EXPOSURE AND IN HAMSTERS BY INHALATION EXPOSURE AND INTRATRACHEAL INSTILLATION. FOLLOWING INHALATION EXPOSURE, INCREASED INCIDENCE OF CARCINOMAS INDUCED IN NASAL MUCOSA OF RATS, AND LARYNGEAL CARCINOMAS INDUCED IN HAMSTERS. IN ANOTHER INHALATION STUDY IN HAMSTERS, USING LOWER EXPOSURE LEVEL, AND IN INTRATRACHEAL INSTILLATION STUDY, NO INCREASED INCIDENCE OF TUMORS OBSERVED. IN HAMSTERS, INHALATION ENHANCED INCIDENCE OF RESPIRATORY TRACT TUMORS PRODUCED BY INTRATRACHEAL INSTILLATION OF BENZO(A)PYRENE.

REPORT 04 : IARC MONOGRAPH 36 FEB 85
FETAL MALFORMATIONS FOUND IN MICE AND RATS TREATED WITH ACETALDEHYDE IN VIVO AND IN VITRO, AND RESORPTIONS OBSERVED IN BOTH SPECIES IN VIVO. INDUCED DNA DAMAGE AND MUTATIONS IN BACTERIA AND MUTATIONS IN YEAST MITOCHONDRIA. INDUCED CHROMOSOMAL ABERRATIONS IN PLANTS. INDUCED CHROMOSOMAL ABERRATIONS, ANEUPLOIDY, MICRONUCLEI AND SISTER CHROMATED EXCHANGES IN BONE-MARROW CELLS OBSERVED IN MICE AND HAMSTERS TREATED IN VIVO.

REPORT 06 : MULTI-AGENCY CROSS CHECK
Y HHS MAGUIRE
N OSHA CARC
N EPA CAG
Y EPA RCRA
Y EPA CLEAN AIR
N EPA CLEAN WATER

REPT'D USE : CHEMICAL INTERMEDIATE, PERFUMES, FOOD FLAVORINGS, NATURAL OCCURRENCE IN FOODS AND BOURBON

CAS NO : **60-35-5**
CHEM NAME : ***ACETAMIDE**

REPORT 01 : NTP TECHNICAL BULLETIN 3 DEC 80
THE COMPOUND WAS FOUND TO BE NEGATIVE IN THE SALMONELLA TYPHIMURIUM MUTAGENICITY ASSAY.

REPORT 02 : IARC MONOGRAPH 7 1974
ACETAMIDE IS CARCINOGENIC IN RATS FOLLOWING ORAL ADMINISTRATION, THE ONLY SPECIES AND ROUTE TESTED, PRODUCING BENIGN AND MALIGNANT LIVER TUMORS.

REPORT 03 : 20TH REPORT OF THE TSCA ITC MAY 87
ACETAMIDE SALTS, CAS NO'S 3168-72-2, 3168-73-3 AND 21429-43-6 ARE RECOMMENDED FOR TESTING WITHOUT BEING DESIGNATED FOR RESPONSE WITHIN 12 MONTHS. TESTS REQUIRED WILL BE: HEALTH

EFFECTS - SUBCHRONIC TOXICITY, ABSORPTION AND CHEMICAL DISPOSITION; CHEMICAL FATE - WATER SOLUBILITY, AEROBIC AND ANAEROBIC BIODEGRADATION, AND IDENT. OF PERSISTENT BIODEGRAD. INTERMEDIATES; ECOLOGICAL EFFECTS ACUTE TOX. TO FISH, AQUATIC INVERTEBRATES, ALGAE AND BENTHIC ORGANISMS, BIOCONCENTRATION IN FISH, CHRONIC EFFECTS ON AQUATIC AND BENTHIC FAUNA, IF LOW MG/L TOXICITY OR IF DYE DOES BIOCONCENTRATE.

REPORT 06 : MULTI-AGENCY CROSS CHECK
N HHS MAGUIRE
Y OSHA CARC
N EPA CAG
N EPA RCRA
N EPA CLEAN AIR
N EPA CLEAN WATER

REPT'D USE : SOLVENT, DYE ACCELERATOR, PLASTIC CHEMICAL INTERMEDIATE, CHEMICAL STABILIZER, DRUG PLASTICIZER, SOLDERING FLUX, LACQUER AND COSMETICS

CAS NO : **103-90-2**
CHEM NAME : ***** ACETAMINOPHEN**

REPORT 01 : NTP DRAFT TR 394 NIH PUB. NO 90-2849 NOV 90
UNDER CONDITIONS OF 2-YEAR FEED STUDIES NO EVIDENCE OF CARCINOGENIC ACTIVITY OF ACETAMINOPHEN IN MALE F344/N RATS THAT RECEIVED DIETS CONTAINING 0, 600, 3,000 OR 6,000 PPM. SOME EVIDENCE OF CARCINOGENIC ACTIVITY IN FEMALE F344/N RATS BASED ON INCREASED INCIDENCES OF MONONUCLEAR CELL LEUKEMIA. NO EVIDENCE OF CARCINOGENIC ACTIVITY IN MALE OR FEMALE B6C3F1 MICE THAT RECEIVED DIETS CONTAINING 0, 600, 3,000 OR 6,000 PPM ACETAMINOPHEN. NONEOPLASTIC LESIONS ASSOCIATED WITH EXPOSURE INCLUDED INCREASED SEVERITY OF NEPHROPATHY, INCIDENCES PARATHYROID HYPERPLASIA IN MALE RATS, SEVERITY OF NEPHROPATHY IN FEMALE RATS AND INCIDENCES OF THYROID FOLLICULAR CELL HYPERPLASIA.

REPT'D USE : ANALGESIC AND ANTIPYRETIC

CAS NO : **26446-35-5**
CHEM NAME : ***** ACETIN**

REPORT 01 : NTP TECHNICAL BULLETIN 5/8 AUG 81
ACETIN WAS POSITIVE IN THE SALMONELLA ASSAY AND INDUCED SEX-LINKED RECESSIVE LETHALS IN DROSOPHILA.

REPT'D USE : TANNING AGENT, SOLVENT, EXPLOSIVES AND FOOD ADDITIVE

CAS NO : **968-81-0**
CHEM NAME : ***** ACETOHEXAMIDE**

REPORT 01 : NCI CARC TRS 50 DHEW PUB NO NIH 78-850
UNDER CONDITIONS OF BIOASSAY ACETOHEXAMIDE WAS NOT CARCINOGENIC FOR EITHER FISCHER 344 RATS OR B6C3F1 MICE (FEEDING STUDY).

REPT'D USE : HYPOGLYCEMIC AGENT AND ANTIDIABETIC AGENT

CAS NO : **75-05-8**
CHEM NAME : *****ACETONITRILE**

REPORT 01 : EPA RCRA HEALTH AND ENV. EFFECTS PROFILE OCT 80
DEPENDING ON THE AMOUNT ABSORBED, ACETONITRILE MAY CAUSE CNS DISORDERS, AND LIVER, KIDNEY, CARDIOVASCULAR AND GASTROINTESTINAL DYSFUNCTION, REGARDLESS OF THE ROUTE OF ADMINISTRATION. THERE IS NO AVAILABLE EVIDENCE TO INDICATE THAT ACETONITRILE HAS MUTAGENIC OR CARCINOGENIC ACTIVITY. TWO STUDIES HAVE REPORTED TERATOGENIC EFFECTS IN RATS.

REPORT 02 : NIOSH CRITERIA DOCUMENT SEP 78
NIOSH RECOMMENDS A STANDARD OF 20 PPM (TWA) BASED ON LIVER, KIDNEY, RESPIRATORY, CARDIOVASCULAR, GI, AND CNS EFFECTS. THE PRESENT OSHA STANDARD IS 40 PPM (TWA).

REPORT 03 : ITC FOURTH REPORT TO EPA (44FR31866) JUN 79
THE INTERAGENCY TESTING COMMITTEE RECOMMENDS THAT THE EPA ADMINISTRATOR CONSIDER THE DEVELOPMENT OF TEST RULES TO TEST ACETONITRILE FOR CARCINOGENICITY, MUTAGENICITY, TERATOGENICITY AND OTHER CHRONIC EFFECTS; AND THAT AN EPIDEMIOLOGIC STUDY BE CONDUCTED.

REPORT 04 : NTP TECHNICAL BULLETIN 7 APR 82
ACETONITRILE WAS NEGATIVE IN THE SALMONELLA TYPHIMURIUM MUTAGENICITY ASSAY.

REPORT 05 : NTP CARC TEST PROGRAM JUL 82
IN A CURRENT NTP BIOASSAY, ACETONITRILE IS BEING ADMINISTERED BY INHALATION TO 344 RATS AND B6C3F1 MICE. THE PRECHRONIC TEST PHASE IS IN PROGRESS.

REPORT 06 : MULTI-AGENCY CROSS CHECK
N HHS MAGUIRE
N OSHA CARC
N EPA CAG
Y EPA RCRA
N EPA CLEAN AIR
N EPA CLEAN WATER

REPORT 07 : EPA RESPONSE TO ITC 48FR215 NOV 83
IN RESPONSE TO ITC 1979 DESIGNATION OF AN FOR PRIORITY TESTING UNDER TSCA 4(A) (47FR250) EPA ANNOUNCED A 1982 NEGOTIATED AGREEMENT WITH MANUFACTURERS (47FR58020) FOR HEALTH EFFECTS TESTING AND THEREFORE WILL NOT ISSUE A TEST RULE.

REPORT 08 : 47FR50942: EPA RESPONSE TO ITC NOV 83
IN RESPONSE TO ITC'S 1979 DESIGNATION OF AN FOR PRIORITY TESTING UNDER TSCA 4(E), THE NTP BEGAN TESTING AN FOR CHRONIC AND MUTAGENIC EFFECTS. MFRS OF AN PRESENTED TO EPA A PLAN FOR MUTAGENICITY AND TERATOGENICITY TESTING. EPA BELIEVES THAT THESE STEPS RESPOND TO ITC RECOMMENDATIONS EXCEPT FOR AN EPA STUDY, WHICH EPA CONCLUDES IS NOT NEEDED.

REPORT 09 : 49FR44142 NOV 84
ANNOUNCES SUBMISSIONS RECEIVED BY EPA DURING 3RD QUARTER FY 1984 FROM NEGOTIATED TEST PROGRAMS AND OTHER TESTING PROGRAMS. THESE SUBMISSIONS INCLUDE MUTAGENICITY AND TERATOGENICTY TESTING.
REPT'D USE : INDUSTRIAL SOLVENT

CAS NO : **8003-03-0**
CHEM NAME : *****P-ACETOPHENETIDIDE CAFFEINE ASPIRIN MIXTURE**

REPORT 01 : NCI CARC TRS 67 DHEW PUBNO NIH 78-1317
UNDER THE CONDITIONS OF THIS BIOASSAY EVIDENCE WAS NOT SUFFICIENT FOR THE CARCINOGENICITY OF APC IN FISCHER 344 RATS OR IN B6C3F1 MICE (FEEDING STUDY).

REPORT 02 : NTP TECHNICAL BULLETIN 5 AUG 81
P-ACETOPHENETIDIDE WAS NEGATIVE IN THE SALMONELLA TYPHIMURIUM MUTAGENICITY ASSAY IN ONE LABORATORY.
REPT'D USE : ANALGESIC AGENT AND ANTIPYRETIC/ANTIINFLAMMATION AGENT

CAS NO : **53-96-3**
CHEM NAME : *****2-ACETYLAMINE FLUORENE (2-AAF)**

REPORT 01 : ED01 STUDY CONFERENCE 1980
A STUDY USING 24,192 MICE WHICH WERE FED 2-AAF AT EIGHT DOSE LEVELS WAS CONDUCTED TO STUDY THE DOSE-RESPONSE RELATIONSHIP AT LOW DOSES. THE INCIDENCE OF BLADDER TUMORS DROPPED SHARPLY AS THE DOSE OF 2-AAF WAS REDUCED. HOWEVER, LIVER TUMORS SHOWED NEARLY A LINEAR RESPONSE OVER THE DOSE RANGE, DISPELLING ANY NOTION OF A THRESHOLD DOSE. THIS LINEAR DOSE RESPONSE FOR LIVER TUMORS ESTABLISHES THE NECESSITY OF USING LINEAR EXTRAPOLATION

FOR TUMOR RATES AT EXTREMELY LOW DOSES T0 ASSURE CONSERVATIVE RISK ASSESSMENTS.

REPORT 02 : NTP TECHNICAL BULLETIN 8 JUL 82
2-AAF WAS POSITIVE IN THE L5178Y MOUSE LYMPHOMA ASSAY.

REPORT 06 : MULTI-AGENCY CROSS CHECK
Y HHS MAGUIRE
R OSHA CARC
Y EPA CAG
Y EPA RCRA
N EPA CLEAN AIR
N EPA CLEAN WATER
REPT'D USE : LABORATORY/EXPERIMENTAL USE ONLY

CAS NO : **17433-31-7**
CHEM NAME : ***** l-ACETYL-2-PICOLINOYL HYDRAZINE (lA2PH)**

REPORT 01 : NCI BIOASSAY PROGRAM CANCER VOL. 40 1977
UNDER THE CONDITIONS OF THIS TEST, I.P. INJECTION. OF lA2PH. WAS CARCINOGENIC TO MALE AND FEMALE SWISS WEBSTERMICE, CAUSING LUNG TUMORS AND AN APPRECIABLE NUMBER OF OVARIAN AND UTERINE TUMORS. IT WAS NOT CARCINOGENIC TO SPRAGUE-DAWLEY (CD) RATS OF EITHER SEX.

REPORT 02 : NTP TECHNICAL BULLETIN 6J AN 82
THE COMPOUND WAS NEGATIVE IN THE SALMONELLA TYPHIMURIUM MUTAGENICITY ASSAY IN ONE LABORATORY.
REPT'D USE : ANTINEOPLASTIC AGENT. NO AVAILABLE PRODUCTION DATA

CAS NO : **120-66-1**
CHEM NAME : ***** ACETYL-O-TOLUIDINE**

REPORT 01 : NTP TECHNICAL BULLETIN 6 JAN 82
THE COMPOUND WAS WEAKLY POSITIVE IN THE SALMONELLA TYPHIMURIUM MUTAGENICITY ASSAY IN ONE LABORATORY.
REPT'D USE : INTERMEDIATE IN ORGANIC SYNTHESIS

CAS NO : **8048-52-0**
CHEM NAME : ***** ACRIFLAVINIUM CHLORIDE**

REPORT 01 : IARC MONOGRAPH 13 1977
NO EVALUATION CONCERNING THE CARCINOGENICITY OF ACRIFLAVINIUM CHLORIDE CAN BE MADE FROM THE ONLY LIMITED STUDY IN RATS GIVEN THE COMPOUND BY SUBCUTANEOUS INJECTION.

REPORT 02 : IARC MONOGRAPH 13 1977
ACRIFLAVINIUM CHLORIDE INDUCES POINT MUTATIONS IN E. COLI AND INDUCES FRAME-SHIFT MUTATIONS IN SALMONELLA, WITHOUT METABOLIC ACTIVATION THE CHEMICAL IS A STRONG FRAME-SHIFT MUTAGEN IN TA98 AND TA1537 WITH METABOLIC ACTIVATION. AT HIGH ORAL DOSES IN MALE MICE, IT WAS WEAKLY POSITIVE IN THE DOMINANT LETHAL TEST.

REPT'D USE : TOPICAL/URINARY ANTISEPTIC

CAS NO : **107-02-8**

CHEM NAME : ***** ACROLEIN**

REPORT 01 : IARC MONOGRAPH 19 1979
WAS TESTED IN MICE BY SKIN APPLICATION AND SUBCUTANEOUS INJECTION AND IN HAMSTERS BY INHALATION EXPOSURE. THE TWO STUDIES IN MICE SUFFER FROM SEVERAL LIMITATIONS, MAKING THEM INADEQUATE FOR EVALUATION OF CARCINOGENICITY. NO CARCINOGENIC EFFECTS WERE NOTED IN HAMSTERS.

REPORT 02 : IARC MONOGRAPH 19 FEB 79
THE PRELIMINARY AND CONFLICTING NATURE OF THE AVAILABLE DATA PRECLUDES AN EVALUATION OF THE MUTAGENICITY OF ACROLEIN. NO DATA ON THE EMBRYOTOXICITY OR TERATOGENICITY OF THIS CHEMICAL WERE AVAILABLE.

REPORT 03 : EPA AMBIENT WATER QUALITY CRITERIA OCT 80
CURRENT INFORMATION INDICATES THAT ACROLEIN IS NOT A CARCINOGEN OR COCARCINOGEN. A CRITERION, BASED ON CHRONIC TESTS RESULTS IN RATS SHOWING SIGNIFICANT CHANGES IN LIVER AND LUNG WEIGHTS, USING AN UNCERTAINTY FACTOR OF 1000, AND CORRESPONDING TO A CALCULATED ACCEPTABLE DAILY HUMAN INTAKE OF 1.09 MG, IS .321 MG/L.

REPORT 04 : EPA RCRA HEALTH AND ENV. EFFECTS PROFILE APR 80
ACROLEIN HAS NOT BEEN SHOWN TO BE A CARCINOGEN OR COCARCINOGEN IN INHALATION EXPERIMENTS. ACROLEIN IS MUTAGENIC IN SOME ASSAY SYSTEMS.

REPORT 05 : NTP TECHNICAL BULLETIN 3/8 DEC 80
ACROLEIN WAS WEAKLY POSITIVE IN THE SALMONELLA ASSAY AND DID NOT INDUCE SEX-LINKED RECESSIVE LETHALS WHEN TESTED IN DROSOPHILA. THE CHEMICAL WAS NEGATIVE IN THE MOUSE SPERM HEAD MORPHOLOGY ASSAY.

REPORT 06 : MULTI-AGENCY CROSS CHECK
N HHS MAGUIRE
N OSHA CARC
N EPA CAG
Y EPA RCRA

Y EPA CLEAN AIR
Y EPA CLEAN WATER

REPORT 07 : IARC MONOGRPH 36 FEB 85
TESTED IN MICE BY SKIN APPLICATION AND IN HAMSTERS BY INHALATION EXPOSURE. STUDY IN MICE INADEQUATE FOR AN EVALUATION OF CARCINOGENICITY. NO CARCINOGENIC EFFECT DETECTED IN HAMSTERS.

REPORT 08 : IARC MONOGRAPH 36 FEB 85
UNDER CERTAIN CONDITIONS OF DIRECT EMBRYONIC EXPOSURE, ACROLEIN TERATOGENIC TO RATS AND RABBITS. AFTER INTRAVENOUS ADMINISTRATION TO RABBITS OF MATERNALLY TOXIC DOSES, EMBRYONIC MORTALITY BUT NO SIGNIFICANT MALFORMATION SEEN IN ONE EXPERIMENT. ACROLEIN MUTAGENIC TO BACTERIA. DID NOT INDUCE DNA DAMAGE OR MUTATIONS IN FUNGI. INDUCED SISTER CHROMATID EXCHANGES IN MAMMALIAN CELLS IN VITRO. DID NOT INDUCE DOMINANT LETHAL MUTATIONS IN MICE.

REPORT 09 : IARC MONOGRAPH 36 FEB 85
GLYCIDALDEHYDE (SEE IARC, 1976) POSSIBLE METABOLITE OF ACROLEIN, TESTED IN MICE BY SKIN APPLICATION AND SUBCUTANEOUS INJECTION AND IN RATS BY SUBCUTANEOUS INJECTION. PRODUCED MALIGNANT TUMORS AT SITE OF APPLICATION IN ANIMALS OF BOTH SPECIES. GLYCIDALDEHYDE HAD INITIATING ACTIVITY IN A TW0-STAGE MOUSE-SKIN BIOASSAY. INDUCED MUTATIONS IN BACTERIA, YEAST AND MAMMALIAN CELLS IN VITRO.

REPORT 10 : EPA HEALTH ASSESSMENT DOC EPA- 600/8-86-014A DRAFT SEP 86
INADEQUATE EVIDENCE FOR MUTAGENICITY OR CHEMICAL INTERACTION WITH MAMMALIAN GERM CELLS. LIMITED EVIDENCE FOR CARCINOGENICITY IN ANIMALS CAN PRODUCE EMBRYOTOXIC AND FETOTOXIC EFFECTS AND IS TERATOGENIC UNDER CERTAIN CONDITIONS.

REPT'D USE : TEXTILE FINISHES, GLYCERIN PRODUCTION, MFG. COLLOIDAL FORMS OF METALS, PLASTICS AND PERFUMES, PRODUCTION OF ACRYLIC ACID/ESTERS AND SYNTHESIS OF METHIONINE BIOCIDE

CAS NO : **7008-42-6**
CHEM NAME : ***** ACRONYCINE**

REPORT 01 : NCI CARC TRS 49 DHEW PUBNO NIH 78-849
ACRONYCINE GIVEN BY ABDOMINAL INJECTION CAUSED A VARIETY OF CANCERS IN MALE/FEMALE RATS. IN MICE A DETERMINATION OF CARCINOGENICITY COULD NOT BE MADE BECAUSE OF LOW SURVIVAL RATES AND PROBLEMS IN TEST PROCEDURES.

REPT'D USE : EXPERIMENTAL ANTINEOPLASTIC AGENT

CAS NO : **79-06-1**
CHEM NAME : *****ACRYLAMIDE**

REPORT 01 : EPA RESPONSE TO ITC RECOMMS (45FR48510) JUL 80
IN RESPONSE TO THE ITC RECOMMENDATIONS (43FR16684) THE EPA HAS EVALUATED THE AVAILABLE HEALTH EFFECTS DATA FOR ACRYLAMIDE, PARTICULARLY ITS NEUROTOXICITY, IN VIEW OF THE EVIDENCE THAT NEUROTOXICITY IS A CONSISTENT EFFECT OF EXPOSURE OF HUMANS. THE EPA IS NOT PROPOSING TO REQUIRE FURTHER TESTING AT THIS TIME. DOW CHEMICAL PLANS TO CONDUCT ONCOGENICITY TESTING. THE EPA WILL INITIATE A PREREGULATORY ASSESSMENT OF ACRYLAMIDE BASED ON EXISTING TOXICITY DATA.

REPORT 02 : EPA TSCA 8(E) NOTICE 8EHQ-0783-0488 JUL 83
THE DOW CHEMICAL PROVIDED INTERIM RESULTS FROM CHRONIC TOXICITY/ONCOGENICITY STUDY OF FISCHER 344 RATS EXPOSED VIA DRINKING-WATER TO ACRYLAMIDE AT DOSE LEVELS OF 0, 0.01, 0.1, 0.5, OR 2.0 MG/KG/DAY. HISTOPATHOLOGIC RESULTS FOR FEMALE RATS AT THE 2.0 MG/KG/DAY DOSE LEVEL SHOWED SIGNIFICANTLY INCREASED NEOPLASMS IN BRAIN AND SPINAL CORD, MAMMARY GLAND, CLITORAL GLAND, UTERUS, ORAL CAVITY, PITUITARY GLAND, THYROID GLAND. DEGREE OF TIBIAL NERVE DEGENERATION SIGNIFICANTLY INCREASED; MORTALITY FOR FEMALE RATS INCREASED AND NUMBER WITH GROSSLY OBSERVED MASS IN MAMMARY REGION ALSO INCREASED. ANALYSIS OF MALE RAT MORTALITY SHOWED INCREASED MORTALITY AT 2.O MG/KG/DAY DOSE.

REPORT 03 : 48FR34119 JUL 83
EPA REPORTS RESULTS OF ACUTE TESTS ON AQUATIC VERTEBRATES AND INVERTEBRATES.

REPORT 04 : IARC MONOGRAPH 39, L986 JUN 85
ACRYLAMIDE SHOWED INITIATING ACTIVITY TO SKIN OF FEMALE MICE AFTER PRETREATMENT WITH 12-O-TETRADECANOYL PHORBOL 13-ACETATE. AC IN DRINKING WATER INCREASED INCIDENCE OF BENIGN OR MALIGNANT NEOPLASMS AT SEVERAL SITES IN RATS. AC NOT MUTAGENIC TO S. TYPHIMURIUM WITH OR WITHOUT METABOLIC ACTIVATION. AC INDUCED SCE'S AND CHROMOSOMAL ABERRATIONS, ANEUPLOIDY AND POLYPLOIDY FOLLOWING ORAL ADMINISTRATION TO MALE MICE. AC NOT CONSIDERED A POTENT NEUROTOXIN. NO DATA IS AVAILABLE TO EVALUATE REPRODUCTIVE, PRENATAL TOXICITY OR CARCINOGENICITY IN HUMANS. IARC SAYS SUFFICIENT EVIDENCE FOR CARCINOGENICITY OF AC IN EXPERIMENTAL ANIMALS.

REPORT 06 : MULTI-AGENCY CROSS CHECK
Y HHS MAGUIRE
N OSHA CARC

N EPA CAG
Y EPA RCRA
N EPA CLEAN AIR
N EPA CLEAN WATER

REPT'D USE : POLYMER, WATER TREATMENT CHEMICAL, PAPER, FUEL, DISHWASHER DETERGENT, GROUTS, FLOCCULANTS, CHEMICAL INTERMEDIATE AND DYES

CAS NO : **79-10-7**
CHEM NAME : ***** ACRYLIC ACID**

IARC MONOGRAPH 19 FEB 79
NO CARCINOGENICITY OR MUTAGENICITY DATA ON ACRYLIC ACID WERE AVAILABLE, ACRYLIC ACID IS EMBRYOTOXIC AND TERATOGENIC.

CAS NO : **107-13-1**
CHEM NAME : ***** ACRYLONITRILE (AN)**

REPORT 02 : NIOSH CRITERIA DOCUMENT SEP 77
NIOSH RECOMMENDS A STANDARD OF 4 PPM (TWA) TO REDUCE THE RISK OF ACRYLONITRILE-INDUCED CANCER AND PREVENT OTHER ADVERSE HEALTH EFFECTS. THE PRESENT OSHA STANDARD IS 2 PPM (TWA).

REPORT 03 : IARC MONOGRAPH 19 1979
THE DATA FROM TWO EXPERIMENTS INDICATE THAT ACRYLONITRILE IS CARCINOGENIC IN RATS, PRODUCING TUMORS OF THE FORESTOMACH, BRAIN, AND ZYMBAL GLAND. THE COMPOUND WAS ADMINISTERED ORALLY AND BY INHALATION. A PRELIMINARY EPIDEMIOLOGICAL REPORT DEMONSTRATES AN INCREASED RISK OF CANCER OF THE LUNG AND THE LARGE INTESTINE IN WORKERS EXPOSED IN CERTAIN AREAS OF A TEXTILE FIBERS PLANT. THE COMBINED EVIDENCE INDICATES THAT ACRYLONITRILE SHOULD BE REGARDED AS IF IT WERE CARCINOGENIC TO HUMANS.

REPORT 04 : DOW TOXICOLOGY RES LAB - CMA REF ANI.210XIHL DEC 80
SPRAGUE-DAWLEY RATS EXPOSED TO VAPOR CONCENTRATIONS OF 0, 20, OR 80 PPM OF ACRYLONITRILE FOR 24 MONTHS SHOWED STATISTICALLY SIGNIFICANT ADVERSE EFFECTS OF BOTH NON-TUMOROUS AND TUMOROUS NATURE. OF THE VARIOUS ORGANS WHICH SHOWED AN ONCOGENIC RESPONSE, THE CNS APPEARED MOST AFFECTED.

REPORT 05 : MONSANTO ORAL TOXICITY AND CARCINOGENICITY STUDY DEC 80
ACRYLONITRILE, ADMINISTERED AT 10, 30 AND 100 PPM IN DRINKING WATER TO F344 RATS FOR 24 MONTHS, RESULTED IN AN

INCREASED INCIDENCE OF ASTROCYTOMAS OF THE CNS AND SQUAMOUS CELL CARCINOMAS IN THE EAR CANAL, AS WELL AS MAMMARY CARCINOMAS IN FEMALES RECEIVING 100 PPM. THE INCIDENCE IN ANIMALS RECEIVING 1 AND 3 PPM WAS CONSIDERED COMPARABLE TO CONTROLS.

REPORT 06 : MULTI-AGENCY CROSS CHECK
Y HHS MAGUIRE
R OSHA CARC
Y EPA CAG
Y EPA RCRA
Y EPA CLEAN AIR
Y EPA CLEAN WATER

REPORT 07 : EPA RCRA HEALTH AND ENV. EFFECTS PROFILE OCT 80
CHRONIC EXPOSURE TO ACRYLONITRILE PRODUCES MILD LIVER DAMAGE AND FUNCTION DISORDERS OF THE CENTRAL NERVOUS SYSTEM, CARDIOVASCULAR AND HEMOPOIETIC SYSTEMS. ACRYLONITRILE HAS SHOWN MUTAGENIC ACTIVITY IN DROSOPHILA AND BACTERIA. THIS COMPOUND IS TERATOGENIC IN RATS WHETHER EXPOSURE IS BY INHALATION OR INGESTION IN DRINKING WATER. THERE ARE BOTH ANIMAL AND EPA DATA TO SUGGEST THAT ACRYLONITRILE MAY BE A HUMAN CARCINOGEN.

REPORT 08 : EPA AMBIENT WATER QUALITY CRITERIA OCT 80
ACRYLONITRILE IS SUSPECTED OF BEING A HUMAN CARCINOGEN. AMBIENT WATER CONCENTRATIONS OF .58 UG/L, .058 UG/L, AND .006 UG/L ARE ESTIMATED TO RESULT IN AN INCREASED CANCER RISK OVER A LIFETIME OF 10 -5, 10 -6, AND 10 -7 RESPECTIVELY.

REPORT 09 : NIOSH CURRENT INTELLIGENCE BULLETIN 18 DEC 76
DUPONT INFORMED NIOSH OF RESULTS OF A PRELIMINARY EPA STUDY DEMONSTRATING AN EXCESS OF CANCER AMONG WORKERS EXPOSED TO ACRYLONITRILE. NIOSH RECOMMENDS THAT AN BE HANDLED IN THE WORKPLACE AS A POTENTIAL HUMAN CARCINOGEN.

REPORT 10 : IARC MONOGRAPH 19 1979
ACRYLONITRILE, IN AQUEOUS OR GAS PHASES INDUCED REVERSE MUTATIONS IN SALMONELLA TA1530, TA1535, TA1950, TA100, TA1538, AND TA98 IN THE PRESENCE OF METABOLIC ACTIVATION. THE CHEMICAL ALSO PRODUCED REVERSE MUTATIONS IN A PLATE INCORPORATION ASSAY USING E. COLI. AN IS ALSO EMBRYOTOXIC AND TERATOGENIC.

REPORT 11 : NAS REPORT ON DIET, NUTRITION AND CANCER 1982
THERE IS NO INFORMATION ON THE HEALTH EFFECTS RESULTING FROM INGESTION OF SMALL AMOUNTS OF AN MONOMER IN THE DIET. HUMANS POTENTIALLY EXPOSED TO AN IN A SYNTHETIC FIBER PLANT WERE FOUND TO BE AT AN INCREASED RISK OF CANCER. INGESTION OR INHALATION OF LARGE AMOUNTS OF AN

ENHANCED TUMORS AT SEVERAL SITES IN RATS. THIS LIMITED EVIDENCE AND THE FINDING THAT AN IS MUTAGENIC, SUGGESTS THAT AN MAY BE CARCINOGENIC FOR HUMANS.

REPORT 12 : FINAL EPA HLTH ASSSSMENT DOC FOR ACRYLONITRILE OCT 83
AN READILY ABSORBED FOLLOWING INHALATION OR INGESTION. SYMPTOMS IN HUMANS: IRRITATION OF EYES AND MUCOSA, WEAKNESS, LABORED BREATHING, DIZZINESS, IMPAIRED JUDGEMENT, CYANOSIS, NAUSEA AND CONVULSIONS. EPA STUDIES SHOW AN HAS SIGNIFICANTLY ELEVATED RISK OF CANCERS OF LYMPH SYSTEM AND STOMACH. UNIT CANCER RISK IN AIR (1 MG/M3 FOR LIFETIME) IS 6.8X10-5.

REPORT 13 : IARC SUPPLEMENT 4 OCT 82
THERE IS LIMITED EVIDENCE FOR HUMAN CARCINOGENICITY, BUT SUFFICIENT EVIDENCE FOR CARCINOGENICITY TO ANIMALS. THERE IS ALSO SUFFICIENT EVIDENCE FOR CARCINOGENICITY TO ANIMALS IN SHORT-TERM TESTS.

REPORT 15 : EPA STST REPORT 8EHGQ-0000-0484
RESULTS OF AN INHALATION STUDY IN RATS FOR 4 WKS AT DOSE LEVELS OF 0, 7.2, 29.7, 51.7 AND 94.1 MG/M3 AND FOR 13 WKS AT DOSE LEVELS OF 0, 6.1, 22.1, AND 52.4 MG/M3 SHOWED BRONCHIO-CENTRIC INTERSTITIAL PNEUMONIA. THERE WAS NO RECOVERY IN 6 MONTHS.

REPT'D USE : CHEMICAL INTERMEDIATE IN MFR OF ACRYLIC/MODACRYLIC FIBERS, PHARMACEUTICALS DYES, ABS/SAN RESINS, NITRILE RUBBER, LATEX PAINTS, ADHESIVES AND GRAIN FUMIGANT

CAS N0 : **9003-56-9**

CHEM NAME : ***** ACRYLONITRILE-BUTADIENE-STYRENE COPOLYMER RESINS**

REPORT 01 : IARC MONOGRAPH 19 1979
NO CARCINOGENICITY DATA ON ACRYLIC AND MODACRYLIC FIBERS OR ON THE ACRYLONITRILE-BUTADIENE-STYRENE AND STYRENE-ACRYLONITRILE COPOLYMERS WERE AVAILABLE TO THE WORKING GROUP.

REPT'D USE : DRAIN WASTE, VENT PIPING, ELECTRICAL CONDUIT AND AUTOMOBILE/APPLIANCE COMPONENTS

CAS NO : **50-76-0**

CHEM NAME : ***** ACTINOMYCIN-D**

REPORT 01 : IARC MONOGRAPH 10 1976
ACTINOMYCIN-D IS CARCINOGENIC IN RATS FOLLOWING ITS INTRAPERITONEAL INJECTION: IT PRODUCED MALIGNANT MESENCHYMAL TUMORS IN THE PERITONEAL CAVITY. IN A 1979 SUPPLEMENT,

IARC DETERMINED THAT THERE IS SUFFICIENT EVIDENCE FOR THE CARCINOGENICITY OF ACTINOMYCIN-D IN ANIMALS.

REPORT 02 : NCI BIOASSAY PROGRAM - CANCER VOL. 40 1977
UNDER THE CONDITIONS OF THIS BIOASSAY, ACTINOMYCIN-D WAS CARCINOGENIC MALE AND FEMALE SPRAGUE-DAWLEY (CD) RATS BY INJECTION, CAUSING SARCOMAS IN THE PERITONEAL CAVITY AT THE SITE OF INJECTION. IT WAS NOT CARCINOGENIC TO SWISS-WEBSTER MICE OF EITHER SEX.

REPORT 03 : IARC MONOGRAPH 10 1976
ACTINOMYCIN OF UNSPECIFIED ORIGIN AND PURITY INDUCED CHROMATID BREAKS AND TRANSLOCATIONS IN SHORT TERM HUMAN PERIPHERAL LEUCOCYTE CULTURES, HUMAN EPIDERMAL FIBROBLASTS AND HELA CELLS. IN RATS, DOSES OF 50-100 UG/KG INDUCED MALFORMATIONS OF THE CNS, VISCERA AND SKELETON; IN RABBITS, EVEN SMALL DOSES OF ACTINOMYCIN-D WERE MORE EMBRYOTOXIC AND LESS TERATOGENIC.
REPT'D USE : ANTINEOPLASTIC AGENT

CAS NO : **23214-92-8**
CHEM NAME : ***** ADRIAMYCIN**

REPORT 01 : IARC MONOGRAPH 10 1976
ADRIAMYCIN WAS TESTED IN RATS BY SINGLE INTRAVENOUS INJECTION. NO EVALUATION OF THE CARCINOGENICITY OF THIS COMPOUND IS POSSIBLE ON THE BASIS OF THIS LIMITED STUDY.

REPORT 02 : IARC MONOGRAPH 10 1976
ADRIAMYCIN HC1 OF UNSPECIFIED PURITY INDUCED REVERSE MUTATIONS IN SALMONELLA. A CONCENTRATION OF .02 UG/M1 ADRIAMYCIN INDUCED A HIGH FREQUENCY OF CHROMATID AND CHROMOSOME BREAKS AND EXCHANGES IN CULTURES OF HUMAN PERIPHERAL LEUKOCYTES.

REPORT 06 : MULTI-AGENCY CROSS CHECK
Y HHS MAGUIRE
R OSHA CARC
R EPA CAG
R EPA RCRA
R EPA CLEAN AIR
R EPA CLEAN WATER
REPT'D USE : ANTIBIOTIC AND ANTINEOPLASTIC

CAS NO : **3688-53-7**
CHEM NAME : ***** AF-2(2-(2-FURYL)-3-(5-NITRO-2-FURYL)ACRYLAMIDE)**

REPORT 01 : NTP TECHNICAL BULLETIN 3 DEC 80
THE COMPOUND WAS FOUND TO BE POSITIVE IN THE SALMONELLA TYPHIMURIUM MUTAGENICITY ASSAY IN TWO LABORATORIES.

REPORT 02 : IARC MONOGRAPH 31 JUL 83
ADMINISTRATION IN DIET PRODUCED BENIGN AND MALIGNANT FORESTOMACH TUMORS IN MICE OF BOTH SEXES, BENIGN AND MALIGNANT MAMMARY TUMORS AND FORESTOMACH PAPILLOMAS IN RATS OF BOTH SEXES AND BENIGN AND MALIGNANT FORESTOMACH TUMORS IN HAMSTERS OF BOTH SEXES AND OESOPHAGEAL TUMORS IN MALE HAMSTERS. IN WEANLING MICE BY SUBCUTANEOUS INJECTION PRODUCED LUNG TUMORS. PRODUCED MULTIPLE LUNG TUMORS IN OFFSPRING OF MICE TREATED SUBCUTANEOUSLY DURING PREGNANCY.

REPORT 03 : IARC MONOGRAPH 31 JUL 83
AF-2 INDUCED DNA DAMAGE IN BACTERIA AND IN HUMAN CELLS AND MUTATIONS IN BACTERIA, FUNGI, INSECTS AND MAMMALIAN CELLS IN VIVO AND IN VITRO. CAUSED CHROMOSOMAL ANOMALIES IN MAMMALIAN CELLS, INCLUDING HUMAN CELLS. CAUSED NEOPLASTIC TRANSFORMATION IN HAMSTER CELLS. AF-2 INDUCED FETAL DEATHS IN MICE AND WAS TERATOGENIC TO SURVIVORS OF DAMS PRETREATED WITH PHENOBARBITAL.

REPT'D USE : WAS USED IN JAPAN SINCE 1965 AS A FOOD PRESERVATIVE BUT NOT USED CURRENTLY

CAS NO : **1402-68-2**
CHEM NAME : ***** AFLATOXIN**

REPORT 01 : IARC SUPPLEMENT 1 1979
AFLATOXINS ARE CARCINOGENIC IN SEVERAL SPECIES OF EXPERIMENTAL ANIMALS BY SEVERAL ROUTES OF ADMINISTRATION, PRODUCING CANCERS OF THE LIVER, COLON AND KIDNEY, EPIDEMIOLOGIC STUDIES HAVE SHOWN A POSITIVE CORRELATION BETWEEN DIETARY CONCENTRATIONS OF AFLATOXINS AND THE INCIDENCE OF LIVER CANCER. IN A 1979 SUPPLEMENT, IARC CONCLUDED THAT AFLATOXINS ARE PROBABLY CARCINOGENIC FOR HUMANS.

REPORT 02 : NAS REPORT ON DIET, NUTRITION AND CANCER 1982
AFLATOXIN IS CARCINOGENIC IN SEVERAL SPECIES OF ANIMALS AND THERE IS EVIDENCE OF A DOSE RESPONSE. A CONSISTENT BODY OF EVIDENCE, ALL BASED ON CORRELATIONAL DATA, ASSOCIATES THE CONTAMINATION OF FOODS BY AFLATOXIN WITH A HIGHER INCIDENCE OF LIVER CANCER IN PARTS OF AFRICA AND

ASIA, BUT THERE IS NO EPA EVIDENCE THAT AFLATOXIN CONTAMINATION OF FOODS IS RELATED TO CANCER RISK IN THE U.S.

REPORT 03 : IARC SUPPLEMENT 4 OCT 82
POSITIVE CORRELATION BETWEEN ESTIMATED AFLATOXIN UPTAKE IN HUMAN POPULATIONS AND THE INCIDENCE OF PRIMARY LIVER CANCER SEEN IN EPA STUDIES. NO STUDIES LINKED INCREASED RISK OF LIVER CANCER WITH AFLATOXIN INTAKE IN INDIVIDUALS. AFLATOXINS PRODUCED TUMORS IN SEVERAL SMALL ANIMALS, PLUS COLONIC AND KIDNEY CANCERS IN RATS. AFLATOXIN CAN INDUCE LIVER TUMORS IN MONKEYS, SARCOMA, ADENOCARCINOMA OF GALL BLADDER OR BILE DUCT AND PANCREATIC CARCINOMAS IN VARIOUS LAB ANIMALS. EVIDENCE FOR ACTIVITY IN SHORT-TERM TESTS WAS CONSIDERED SUFFICIENT.

REPORT 06 : MULTI-AGENCY CROSS CHECK
Y HHS MAGUIRE
N OSHA CARC
Y EPA CAG
N EPA RCRA
N EPA CLEAN AIR
N EPA CLEAN WATER
REPT'D USE : MOLD CONTAMINANT IN FOODS

CAS NO : **9002-18-0**
CHEM NAME : ***** AGAR-AGAR**

REPORT 01 : NTP BIOASSAY TRS 230 NIH PUB NO. 82-1786
UNDER THE CONDITIONS OF THIS BIOASSAY, AGAR ADMINISTERED IN THE DIET WAS NOT CARCINOGENIC FOR F344 RATS OR B6C3F1 MICE OF EITHER SEX.

REPORT 02 : FDA GRAS LISTING 21CFR182.7115
AGAR-AGAR IS GENERALLY RECOGNIZED AS SAFE FOR USE AS A FOOD STABILIZER WHEN USED IN ACCORDANCE WITH GOOD MANUFACTURING PRACTICE.
REPT'D USE : FOOD STABILIZER, GELATIN SUBSTITUTE, BACTERIAL CULTURE MEDIUM, PHOTOGRAPHIC EMULSIONS, PAPER/TEXTILE SIZING AND VETERINARY LAXATIVE

CAS NO : **2757-90-6**
CHEM NAME : *****AGARITINE(L-GLUTAMIC ACID-5-(2-(4 HYDROXYMETHYL)-PHENYLHYDRAZIDE))**

REPORT 01 : IARC MONOGRAPH 31 JUL 83
IN MICE IN DRINKING-WATER SHOWED NO INCREASE IN INCIDENCE OF TUMORS. N-ACETYL-4-(HYDROXYMETHYL) PHENYLHYDRAZINE (A STABLE DERIVATIVE OF THE HYDROLYSIS PRODUCT OF

AGARITINE) TESTED IN MICE IN DRINKING-WATER PRODUCED INCREASED INCIDENCES OF LUNG TUMORS AND OF BLOOD-VESSEL TUMORS. 4-(HYDROXYMETHYL)BENZENE-DIAZONIUM ION TETRAFLUOROBORATE (ANOTHER STABILIZED HYDROLYSIS PRODUCT OF AGARITINE) IN MICE BY SUBCUTANEOUS INJECTION, INCREASED INCIDENCE OF FIBROSARCOMAS AND OF SKIN PAPILLOMAS AND CARCINOMAS AT INJECTION SITE.

REPORT 02 : IARC MONOGRAPH 31 JUL 83
AGARITINE WEAKLY MUTAGENIC IN SALMONELLA TYPHIMURIUM.
RFPT'D USE : NATURAL SUBSTANCE FOUND IN SEVERAL MUSHROOMS OF AGARICUS SPECIES

CAS NO : **116-06-3**
CHEM NAME : ***** ALDICARB**

REPORT 01 : NCI CARC TRS 136 DHEW PUB NO NIH 79-1391
UNDER CONDITIONS OF BIOASSAY, TECHINICAL GRADE ALDICARB WAS NOT CARCINOGENIC FOR F344 RATS OR B6C3F1 MICE OF EITHER SEX (FEEDING STUDY).

REPORT 02 : NAS REPORT ON DIET, NUTRITION AND CANCER 1982
ALDICARB DOES NOT APPEAR TO BE CARCINOGENIC IN RATS OR MICE.
REPT'D USE : INSECTICIDE, NEMATOCIDE AND ACARICIDE

CAS NO : **309-00-2**
CHEM NAME : ***** ALDRIN**

REPORT 01 : NCI CARC TRS 21 DHEW PUB NO NIH 78-821
ALDRIN WAS GIVEN IN FEED TO OSBORNE-MENDEL RATS AND B6C3F1 MICE. ALDRIN WAS FOUND TO CAUSE LIVER CANCER IN MALE MICE, BUT IN FEMALE MICE INCREASES IN LIVER CANCER COULD NOT BE RELATED TO TREATMENT. IN O-M RATS, THE TUMORS COULD NOT BE CLEARLY ASSOCIATED WITH TREATMENT, BUT NOT IN HIGH-DOSE FEMALES WHEN COMPARED WITH POOLED CONTROLS. IN MALE MICE, THERE WAS A SIGNIFICANT DOSE-RELATED INCREASE IN THE INCIDENCE OF HEPATOCELLULAR CARCINOMAS WHEN COMPARED WITH EITHER MATCHED CONTROLS OR POOLED CONTROLS.

REPORT 02 : IARC MONOGRAPH 5 1974
ALDRIN HAS BEEN TESTED ONLY BY THE ORAL ROUTE IN RATS AND MICE. THE STUDY IN MICE WAS CONSIDERED TO BE INADEQUATE FOR EVALUATION. STUDIES IN RATS WERE NEGATIVE IN TWO CASES AND INADEQUATE IN TWO OTHERS.

REPORT 03 : EPA AMBIENT WATER QUALITY CRITERIA OCT 80
BOTH ALDRIN AND DIELDRIN ARE SUSPECTED OF BEING HUMAN CARCINOGENS. FOR ALDRIN, AMBIENT WATER CONCENTRATIONS OF .74 NG/L, .074 NG/L, AND .007 NG/L ARE ESTIMATED TO RESULT IN A RESPECTIVE 10 -5, 10 -6, AND 10 -7 INCREASE IN CANCER RISK OVER A LIFETIME.

REPORT 04 : EPA RCRA HEALTH AND ENV. EFFECTS PROFILE OCT 80
THE INDUCTION OF HEPATOCELLULAR CARCINOMA IN BOTH MALE AND FEMALE MICE FROM THE ADMINISTRATION OF ALDRIN LEADS TO THE CONCLUSION THAT IT IS LIKELY TO BE A HUMAN CARCINOGEN. ALDRIN HAS NOT BEEN FOUND TO BE MUTAGENIC IN SEVERAL TEST SYSTEMS ALTHOUGH IT DID INDUCE UNSCHEDULED DNA SYNTHESIS IN HUMAN FIBROBLASTS.

REPORT 06 : MULTI-AGENCY CROSS CHECK
N HHS MAGUIRE
N OSHA CARC
Y EPA CAG
Y EPA RCRA
N EPA CLEAN AIR
Y EPA CLEAN WATER
REPT'D USE : INSECTICIDE

CAS NO : **CLASS-20**
CHEM NAME : ***** ALKYL EPOXIDES**

REPORT 01 : ANPR, 49FR449 JAN 84
EPA'S RESPONSE TO ITC'S RECOMMENDATION THAT ALKYL EPOXIDES OTHER THAN ETHYLENE OXIDE, PROPYLENE OXIDE AND 1,2-BUTYLENE OXIDE, ALSO BE TESTED FOR HEALTH EFFECTS AND CHEMICAL FATE.

REPORT 02 : DECISION NOT TO TEST, 50FR36476 SEP 85
EPA IS TERMINATING ITS RULEMAKING PROCEEDINGS UNDER TSCA 4(A) TO REQUIRE TESTING FPR CERTAIN ALKYL EPOXIDES. DATA ANALYSIS REVEALS FEW PEOPLE EXPOSED TO THESE CHEMICALS, EXPOSURE LEVELS ARE LOW AND SMALL AMOUNTS ARE RELEASED TO THE ATMOSPHERE.
REPT'D USE : CHEMICAL INTERMEDIATES

CAS NO : **CLASS-02**
CHEM NAME : ***** ALKYL PHTHALATES**

REPORT 01 : 48FR34120 JUL 83
CMA THROUGH ITS PHALATE ESTERS PROGRAM PANEL IS TESTING ALKYL DIESTERS OF 1,2-BENZENEDICARBOXYLIC ACID. THE

TESTING ADDRESSES AQUATIC TOX, ENVIRONMENTAL TRANSPORT AND FATE.

REPT'D USE : PLASTICIZER

CAS NO : **107-05-1**

CHEM NAME : ***** ALLYL CHLORIDE**

REPORT 01 : NIOSH CRITERIA DOCUMENT SEP 76
NIOSH RECOMMENDS A STANDARD OF 1 PPM (TWA); 3 PPM CEILING BASED ON LIVER, KIDNEY, AND LUNG EFFECTS. THE PRESENT OSHA STANDARD IS 1 PPM (TWA).

REPORT 02 : NCI CARC TRS 73 DHEW PUB NO NIH 78-1323
UNDER THE CONDITIONS OF THIS BIOASSAY NO CONVINCING EVIDENCE WAS PRESENTED FOR THE CARCINOGENICITY OF ALLYL CHLORIDE IN OSBORNE-MENDEL RATS OF EITHER SEX. THE RESULTS ARE SUGGESTIVE THAT ALLYL CHLORIDE IS CARCINOGENIC IN MALE AND FEMALE B6C3F1 MICE SINCE THE COMPOUND, WHEN ADMINISTERED BY GAVAGE, CAUSED A LOW INCIDENCE OF NEOPLASTIC AND NON-NEOPLASTIC LESIONS OF THE FORESTOMACH.

REPORT 04 : NTP TECHNICAL BULLETIN 7 APR 82
ALLYL CHLORIDE WAS NEGATIVE IN THE RAT DOMINANT LETHAL ASSAY, THE MOUSE SPERM HEAD MORPHOLOGY ASSAY, THE DROSOPHILIA SEX-LINKED RECESSIVE LETHAL ASSAY, IN VITRO UNSCHEDULED DNA SYNTHESIS IN HUMAN FIBROBLASTS, AND THE RAT BONE MARROW CYTOLOGY ASSAY.

REPORT 05 : IARC MONOGRAPH 36 FEB 85
TESTED BY INTRAGASTRIC INCUBATION IN MICE AND RATS, BY SKIN APPLICATION IN MICE, BOTH BY REPEATED APPLICATION AND IN A TWO-STAGE ASSAY, AND BY INTRAPERITONEAL INJECTION IN MICE. FOLLOWING ORAL ADMINISTRATION TO MICE, NONSIGNIFICANT INCREASE IN INCIDENCE OF SQUAMOUS-CELL PAPILLOMAS AND CARCINOMAS OF FORESTOMACH OBSERVED; EXPERIMENT IN RATS INADEQUATE FOR EVALUATION. NO SKIN TUMOR OBSERVED IN MICE FOLLOWING REPEATED SKIN APPLICATIONS.

REPORT 06 : MULTI-AGENCY CROSS CHECK
N HHS MAGUIRE
N OSHA CARC
N EPA CAG
N EPA RCRA
Y EPA CLEAN AIR
N EPA CLEAN WATER

REPORT 07 : IARC MONOGRAPH 36 FEB 85
INHALATION EXPOSURE TO ALLYL CHLORIDE OF HIGH PURITY DID NOT INDUCE TERATOGENICITY IN RATS OR RABBITS. CAUSED DNA DAMAGE IN BACTERIA, AND WAS MUTAGENIC TO BACTERIA AND FUNGI.

REPT'D USE : ORGANIC SYNTHESIS, INTERMEDIATE IN MFR. OF GLYCEROL AND EPICHLOROHYDRIN

CAS NO : **106-92-3**

CHEM NAME : ***** ALLYL GLYCIDYL ETHER (AGE)**

REPORT 01 : NTP TECHNICAL BULLETIN 6 JUN 82
THE COMPOUND WAS POSITIVE IN THE SALMONELLA TYPHIMURIUM MUTAGENICITY ASSAY IN ONE LABORATORY.

REPORT 02 : NIOSH CRITERIA DOCUMENT JUN 78
AGE HAS SHOWN MUTAGENIC ACTIVITY IN THE SALMONELLA ASSAY, BUT MUTAGENICITY HAS NOT BEEN CONFIRMED IN OTHER TESTS. NIOSH RECOMMENDS THAT WORKER EXPOSURE BE LIMITED TO 9.6 PPM AGE, MEASURED AS A 15-MIN CEILING CONCENTRATION, TO PREVENT SYSTEMIC EFFECTS FROM INHALATION OF AGE.

REPORT 03 : ITC THIRD REPORT TO EPA (43FR50631) OCT 78
THE INTERAGENCY TESTING COMMITTEE RECOMMENDS THAT THE EPA ADMINISTRATOR CONSIDER THE DEVELOPMENT OF TEST RULES TO TEST GLYCIDOL AND SELECTED DERIVATIVES FOR CARCINOGENICITY,MUTAGENICITY,TERATOGENICITYANDREPRODUCTIVE EFFECTS.

REPORT 04 : NTP TR 376 NIH PUB. 89-2831 JAN 90
UNDER 2-YEAR INHALATION STUDIES, EQUIVOCAL EVIDENCE OF CARCINOCENICITY OF ALLYL GLYCIDYL ETHER FOR MALE OSBORNE-MENDEL RATS (AT 10 PPM), BASED ON PRESENCE OF PAPILLARY ADENOMA AND SQUAMOUS CELL CARCINOMA, AND POORLY DIFFERENTIATED ADENOCARCIOMA IN NASAL PASSAGE. NO EVIDENCE OF CARCINOGENICITY FOR FEMALE RATS. ONE PAPILLARY ADENOMA OF RESPIRATORY EPITHELIUM PRESENT IN FEMALE RAT EXPOSED AT 5 PPM. SOME EVIDENCE OF CARCINOGENICITY FOR MALE B6C3F1 MICE, BASED ON PRESENCE OF 3 ADENOMAS OF RESPIRATORY EPITHELIUM, DYSPLASIA, AND FOCAL BASAL CELL HYPERPLASIA OF RESPIRATORY EPITHELIUM IN NASAL PASSAGE OF MICE EXPOSED AT 10 PPM.

REPT'D USE : RESIN INTERMEDIATE, STABILIZER FOR CHLORINE COMPOUNDS, VINYL RESINS, RUBBER AND ADHESIVES/GLUES

CAS NO : **57-06-7**
CHEM NAME : ***** ALLYL ISOTHIOCYANATE (ATC)**

REPORT 01 : NTP TECHNICAL BULLETIN 5 AUG 81
ALLYL ISOTHIOCYANATE WAS NEGATIVE IN THE SALMONELLA TYPHIMURIUM MUTAGENICITY ASSAY IN ONE LABORATORY.

REPORT 02 : FDA GRAS LISTING 21CFR182.7115
SMALL GROUPS OF RATS FED ATC (10,000, 2500 AND 1000 PPM) FOR 26 WEEKS SHOWED NO TOXIC EFFECTS. RATS RECEIVING ATC BY GAVAGE FOR 20 DAYS SHOWED THICKENING OF NON-GLANDULAR PART OF STOMACH. MICROSCOPICALLY ACUTE AND SUBACUTE STOMACH ULCERS WERE SEEN (20 MG/KG, 50 MG/KG).

REPORT 03 : IARC MONOGRAPH 36 FEB 85
TESTED BY GASTRIC INCUBATION IN MICE AND RATS. IN MICE, NO INCREASE IN INCIDENCE OF TUMORS OBSERVED. INCREASED INCIDENCE OF EPITHELIAL HYPERPLASIA AND TRANSITIONAL-CELL PAPILLOMAS OF URINARY BLADDER OBSERVED IN MALE RATS ONLY, AND SOME SUBCUTANEOUS FIBROSARCOMAS OCCURRED IN FEMALE.

REPORT 04 : IARC MONOGRAPH 36 FEB 85
ALLYL ISOTHIOCYANATE DID NOT INDUCE DNA DAMAGE IN BACTERIA, INDUCED MUTATIONS IN BACTERIA AND INSECTS AND CHROMOSOMAL ABERRATIONS IN PLANTS. DID NOT INDUCE DOMINANT LETHAL MUTATIONS IN MICE. ALLYL ISOTHIOCYANATE HAS BEEN FOUND TO PRODUCE IRRITATION OF MUCOUS MEMBRANES AND ECZEMATOUS OR VESICULAR SKIN REACTIONS AS AN EXTERNAL ANALGESIC. NOT TERATOGENIC TO MICE, RATS, HAMSTERS OR RABBITS, BUT RESORPTIONS SEEN IN MICE AND RATS.

REPT'D USE : FLAVORING AGENT, ALCOHOL DENATURANT, MILTARY WAR GAS, DOG/CAT REPELLANT, ADHESIVES/GLUES, AEROSOLS AND EXTERNAL ANALGESIC

CAS NO : **2835-39-4**
CHEM NAME : ***** ALLYL ISOVALERATE**

REPORT 01 : NTP TECHNICAL REPORT TR 253 NIH PUBLICATION 83-2509 MAY 83
UNDER THE CONDITIONS OF THESE STUDIES, ALLYL ISOVALERATE WAS CARCINOGENIC FOR F344/N RATS AND B6C3F1 MICE, CAUSING INCREASED INCIDENCES OF HEMATOPOIETIC SYSTEM NEOPLASMS (MONONUCLEAR-CELL LEUKEMIA IN MALE RATS AND LYMPHOMA IN FEMALE MICE).

REPORT 02 : IARC MONOGRAPH 36 . FEB 85
TESTED BY INTRAGASTRIC INCUBATION IN MICE AND IN RATS. IN MICE, INDUCED PAPILLOMAS OF FORESTOMACH IN MALES AND

INCREASED INCIDENCE OF LYMPHOMAS IN FEMALES. IN RATS, INCREASED INCIDENCE OF MONONUCLEAR CELL LEUKEMIA OBSERVED IN ANIMALS OF BOTH SEXES. ALLYL ISOVALERATE WAS NOT MUTAGENIC TO BACTERIA.

REPT'D USE : FRAGRANCE AND FLAVORING AGENT

CAS NO : **107-11-7**

CHEM NAME : ***** ALLYLAMINE (2-PROPINE-1-AMINE)**

REPORT 01 : EPA 8EHQ-1089-0830 NOV 89

IN ACUTE SKIN BY HOECHST CELANESE, THREE CORROSIVITY STUDY. NEW ZEALAND RABBITS EXPOSED TO 0.5 ML (176MG/KG) OF ALLYLAMINE FOR THREE MINUTES, ONE RABBIT DIED WITHIN 24 HOURS WHILE THE REMAINING TWO SURVIVED THROUGH THE SEVENTH DAY OBSERVATION PERIOD; ALL EXHIBITED EVIDENCE OF SKIN CORROSION. THE DATA SUGGEST THAT ALLYLAMINE IS READILY ABSORBED THROUGH THE SKIN.

REPT'D USE : PHARMACEUTICAL INTERMEDIATE AND INORGANIC SYNTHESIS

CAS NO : **7429-90-5**

CHEM NAME : ***** ALUMINUM**

REPORT 01 : EPA TSCA 8 (E) EHQ-390-0915 MAR 90

ALUMINUM COMPANY OF AMERICA (ALCOA) SUBMITTED EPIDEMIOLOGY OF MORTALITY OF ALCOA WORKERS. STUDY IS PART OF SERIES OF INVESTIGATIONS AND WAS DONE, IN PART, TO CLARIFY FURTHER RELATIONSHIP BETWEEN EMPLOYMENT IN ALUMINUM REDUCTION OPERATIONS AND CANCER MORTALITY. ALCOA PREPARING COMMUNICATION PLAN TO SHARE RESULTS WITH THEIR WORKERS AND THEIR REPRESENTATIVES, AND PLANS TO UNDERTAKE MORE FOCUSED STUDY TO CLARIFY RELATIONSHIP, IF ANY, OF KIDNEY CANCER TO EMPLOYMENT.

REPT'D USE : BLDG, AUTO PARTS, ELECTRICAL LINES, PHOTOENGRAVING, MACHINERY, TUBES, PAINTS, COATINGS, PYROTECHNIC DEVICES, CATALYST, FOIL PAPER AND LIQUID FUELS.

CAS NO : **4657-93-6**

CHEM NAME : ***** 5-AMINOACENAPHTHENE**

REPORT 01 : IARC MONOGRAPH 16 1978

DATA REPORTED IN AN ABSTRACT INDICATE THAT 5-AMINOACENAPHTHENE INCREASED THE INCIDENCE OF BLADDER CARCINOMAS IN MICE FOLLOWING IMPLANTATION OF PARAFFIN WAX PELLETS CONTAINING THIS COMPOUND INTO THE BLADDER AND THAT IT PRODUCED LEUKEMIAS FOLLOWING ITS INTRAPERITONEAL INJECTION. NO EVALUATION OF THE CARCINOGENICITY OF 5-AMINOACENAPHTHENE CAN BE MADE ON THE BASIS OF THE

AVAILABLE DATA.
REPT'D USE : EXPERIMENTAL DISINFECTANT
(NO KNOWN PRODUCTION)

CAS NO : **52417-22-8**
CHEM NAME : ***** 9-AMINOACRIDINE HCL H20**

REPORT 01 : NTP TECHNICAL BULLETIN NO. 7 APR 82
THE CHEMICAL WAS POSITIVE IN THE SALMONELLA TYPHIMURIUM MUTAGENICITY ASSAY.
REPT'D USE : NO AVAILABLE PRODUCTION OR USE DATA
NO LISTING ON TSCA INVENTORY

CAS NO : **613-13-8**
CHEM NAME : ***** 2-AMINOANTHRACENE**

REPORT 01 : NTP TECHNICAL BULLETIN 3 DEC 80
THE COMPOUND WAS FOUND TO BE POSITIVE IN THE SALMONELLA TYPHIMURIUM MUTAGENICITY ASSAY BY THREE LABORATORIES.
REPT'D USE : NO USE INFORMATION AVAILABLE

CAS NO : **117-79-3**
CHEM NAME : ***** 2-AMINOANTHRAQUINONE**

REPORT 01 : NCI CARC TRS 144 DHEW PUB NO NIH 78-1399
UNDER THE CONDITIONS OF THIS BIOASSAY, DIETARY ADMINISTRATION OF 2-AMINOANTHRAQUINONE WAS CARCINOGENIC IN MALE FISCHER 344 RATS, CAUSING A COMBINATION OF HEPATOCELLULAR CARCINOMAS AND NEOPLASTIC NODULES OF THE LIVER. THE COMPOUND WAS ALSO CARCINOGENIC IN B6C3F1 MICE, CAUSING HEPATOCELLULAR CARCINOMAS IN BOTH SEXES AND MALIGNANT HEMATOPOIETIC LYMPHOMAS IN FEMALES. THERE WAS NO EVIDENCE OF CARCINOGENICITY IN FEMALE RATS.

REPORT 02 : IARC MONOGRAPH 27 1982
TECH GRADE 2-AMINOANTHRAQUINONE WAS TESTED IN ONE EXPERIMENT IN MICE AND IN ONE EXPERIMENT IN RATS BY DIETARY ADMINISTRATION. IT PRODUCED HEPATOCELLULAR CARCINOMAS IN MICE AND MALE RATS. THERE IS LIMITED EVIDENCE FOR THE CARCINOGENICITY OF THE MATERIAL TESTED--LOW PURITY TECH GRADE 2-AMINOANTHRAQUINONE. NO EVALUATION OF THE CARCINOGENICITY OF THE PURE COMPOUND CAN BE MADE.

REPORT 03 : IARC MONOGRAPH 27 1982
ALTHOUGH A CRUDE PREPARATION OF 2-AMINOANTHRAQUINONE INDUCED REVERSIONS IN SALMONELLA, THIS ACTIVITY WAS NOT

SEEN AFTER PURIFICATION OF THE COMPOUND AND WAS ATTRIBUTED TO THE PRESENCE OF 1,2-DIAMINO ANTHRAQUINONE. NO DATA WERE AVAILABLE CONCERNING EFFECTS ON REPRODUCTION AND PRENATAL TOXICITY.

REPORT 06 : MULTI-AGENCY CROSS CHECK
Y HHS MAGUIRE
Y OSHA CARC
N EPA CAG
N EPA RCRA
N EPA CLEAN AIR
N EPA CLEAN WATER

REPT'D USE : DYE INTERMEDIATE

CAS NO : **150-13-0**

CHEM NAME : *****PARA-AMINOBENZOIC ACID**

REPORT 01 : IARC MONOGRAPH 16 1978
IN THE ONLY STUDY AVAILABLE, PARA-AMINOBENZOIC ACID WAS TESTED IN MICE BY SKIN APPLICATION; NO CARCINOGENIC EFFECTS WERE OBSERVED. THE AVAILABLE DATA ARE INSUFFICIENT FOR AN EVALUATION OF THE CARCINOGENICITY OF THIS COMPOUND TO BE MADE.

REPORT 02 : IARC MONOGRAPH 16 1978
P-AMINOBENZOIC ACID WAS NOT MUTAGENIC TO EITHER E. COLI OR SALMONELLA TA98 OR TA100, WITH OR WITHOUT METABOLIC ACTIVATION.

REPORT 03 : NTP TECHNICAL BULLETIN 9 APR 83
P-AMINO BENZOIC ACID WAS NEGATIVE IN ONE SALMONELLA TYPHIMURIUM MUTAGENICITY ASSAY.

REPT'D USE : DYE INTERMEDIATE, COSMETICS, SUNSCREENING AGENT AND DRUG VETERINARY MEDICINE

CAS NO : **92-67-1**

CHEM NAME : ***** 4-AMINOBIPHENYL**

REPORT 01 : IARC MONOGRAPH 1 1972
4-AMINOBIPHENYL IS CARCINOGENIC IN THE MOUSE, RAT, RABBIT AND DOG. FOLLOWING ORAL ADMINISTRATION IT HAS PRODUCED BLADDER AND LIVER TUMORS IN MICE AND BLADDER PAPILLOMAS AND CARCINOMAS IN RABBITS AND DOGS. IN THE EPIDEMIOLGICAL STUDIES, CONFINED TO ONE SERIES OF WORKERS OCCUPATIONALLY EXPOSED TO COMMERCIAL 4-AMINOBIPHENYL, A HIGH INCIDENCE OF BLADDER CARCINOMAS WAS REPORTED. IN A 1979 SUPPLEMENT, THE IARC CONCLUDED THAT 4-ABP IS CARCINOGENIC FOR HUMANS.

REPORT 02 : NTP TECHNICAL BULLETIN 3/8 DEC 80
4-AMINOBIPHENYL WAS POSITIVE IN THE SALMONELLA ASSAY BUT DID NOT INDUCE SEX-LINKED RECESSIVE LETHALS IN DROSOPHILIA.

REPORT 06 : MULTI-AGENCY CROSS CHECK
Y HHS MAGUIRE
R OSHA CARC
Y EPA CAG
N EPA RCRA
N EPA CLEAN AIR
N EPA CLEAN WATER
REPT'D USE : RUBBER ANTIOXIDANT

CAS NO : **81-49-2**
CHEM NAME : ***** 1-AMINO-2,4-DIBROMO ANTHRAQUINONE**

REPORT 01 : NTP TECHNICAL BULLETIN 6 JAN 82
THE COMPOUND WAS POSITIVE IN THE SALMONELLA TYPHIMURIUM MUTAGENICITY ASSAY IN ONE LABORATORY.

REPORT 02 : NTP CARC TEST PROGRAM JUL 82
IN A CURRENT NTP BIOASSAY, 1-AMINO-2,4-DIBROMO ANTHRAQUINONE IS BEING ADMINISTERED IN THE DIET TO F344 RATS AND B6C3F1 MICE. THE PRECHRONIC TEST PHASE IS IN PROGRESS.
REPT'D USE : NO AVAILABLE PRODUCTION OR USE DATA

CAS NO : **88296-46-4**
CHEM NAME : *****7-AMINO-2,2-DIMETHYL-2,3-DIHYDROBENZOFURAN**

REPORT 01 : EPA RESPONSE TO ITC - 50FR29761
EPA'S RESPONSE TO ITC'S RECOMMENDATION OF PRIORITY CONSIDERATION FOR CHEMICAL FATE AND ENVIRONMENTAL IMPACT TESTING UNDER TSCA 4(A): EPA DOES NOT BELIEVE THERE IS BASIS FOR FINDING THAT SUBSTANCE MAY PRESENT AN UNREASONABLE RISK OF INJURY TO THE ENVIRONMENT NOR THAT THERE MAY BE A SUBSTANTIAL RELEASE INTO THE ENVIRONMENT.
REPT'D USE : INTERMEDIATE IN PRODUCTION OF INSECTICIDE-NEMATICIDE

CAS NO : **17026-81-2**
CHEM NAME : ***** 3-AMINO-4-ETHOXYACETANILIDE**

REPORT 01 : NCI CARC TRS 112 DHEW PUB NO NIH 78-1367
UNDER CONDITIONS OF BIOASSAY, 3-AMINO-4-ETHOXYACETANILIDE WAS CARCINOGENIC IN MALE B6C3F1 MICE CAUSING FOLLICULAR-CELL CARCINOMAS OF THE THYROID GLAND. EVIDENCE PROVIDED BY THIS BIOASSAY WAS INSUFFICIENT TO

ESTABLISH THE CARCINOGENICITY OF 3-AMINO-4-ETHOXY-ACETANILIDE IN FEMALE MICE OR IN FISCHER 344 RATS OF EITHER SEX (FEEDING STUDY).

REPORT 02 : NTP TECHNICAL BULLETIN 6 JAN 82
THE COMPOUND WAS POSITIVE IN THE SALMONELLA TYPHIMURIUM MUTAGENICITY ASSAY IN ONE LABORATORY.

REPORT 06 : MULTI-AGENCY CROSS CHECK
N HHS MAGUIRE
Y OSHA CARC
N EPA CAG
N EPA RCRA
N EPA CLEAN AIR
N EPA CLEAN WATER
REPT'D USE : DYE INTERMEDIATE

CAS NO : **132-32-1**
CHEM NAME : ***** 3-AMINO-9-ETHYLCARBAZOLE**

REPORT 01 : NCI CARC TRS 93 DHEW PUB NO NIH 78-1337
UNDER CONDITIONS OF BIOASSAY 3-AMINO-9-ETHYLCARBAZOLE (HYDROCHLORIDE) WAS CARCINOGENIC FOR THE LIVER INDUCING HEPATOCELLULAR CARCINOMAS IN FISCHER 344 RATS AND B6C3F1 MICE OF BOTH SEXES. OTHER TUMORS INDUCED IN THE RATS WERE CARCINOMAS OR PAPILLOMAS OF THE INTEGUMENTARY SYSTEM IN MALES, CARCINOMAS OF THE ZYMBAL'S GLAND OF THE EAR IN MALES AND FEMALES, AND ADENOCARCINOMAS OF THE UTERUS (FEEDING STUDY).

REPORT 06 : MULTI-AGENCY CROSS CHECK
N HHS MAGUIRE
Y OSHA CARC
N EPA CAG
N EPA RCRA
N EPA CLEAN AIR
N EPA CLEAN WATER
REP'D USE : DYE INTERMEDIATE, BIOLOGICAL STAIN AND CHEMICAL INDICATOR

CAS NO : **6109-97-3**
CHEM NAME : ***** 3-AMINO-9-ETHYLCARBAZOLE HYDROCHLORIDE**

REPORT 01 : NTP TECHNICAL BULLETIN NO. 6 JAN 82
THE CHEMICAL WAS POSITIVE IN THE SALMONELLA TYPHIMURIUM MUTAGENICITY ASSAY IN ONE LABORATORY.
REPT'D USE : NO AVAILABLE PRODUCTION OR USE DATA
NO LISTING ON TSCA INVENTORY

CAS NO : **712-68-5**
CHEM NAME : ***** 2-AMINO-5-(5-NITRO-2-FURYL)-1,3,4-THIADIAZOLE**

REPORT 01 : IARC MONOGRAPH 7 1974
2-AMINO-5-(5-NITRO-2-FURYL)-1,3,4-THIADIAZOLE IS CARCINOGENIC IN RATS FOLLOWING ORAL ADMINISTRATION, THE ONLY SPECIES AND ROUTE TESTED. IT PRODUCED MAMMARY CARCINOMAS AND FORESTOMACH PAPILLOMAS. IN A 1979 SUPPLEMENT, IARC DETERMINED THAT THERE IS SUFFICIENT EVIDENCE FOR THE CARCINOGENICITY OF 2-AMINO-5-(5-NITRO-2-FURYL)-1,3,4-THIADIAZOLE IN ANIMALS.
REPT'D USE : DRUG FOR TREATMENT OF GASTROENTERITIS AND TOPICAL ANALGESIC

CAS NO : **121-88-0**
CHEM NAME : ***** 2-AMINO-NITROPHENOL**

REPORT 01 : NTP TECHNICAL BULLETIN NO. 7 APR 82
THE CHEMICAL WAS POSITIVE IN THE SALMONELLA TYPHIMURIUM MUTAGENICITY ASSAY.

REPORT 02 : NTP TR 334 NIH PUB.88-2590 FEB 88
GAVAGE STUDIES INDICATE EVIDENCE OF CARCINOGENIC ACTIVITY FOR MALE F344/N RATS THAT RECEIVED 100 MG/KG 2-AMINO-5-NITROPHENOL SHOWN BY THE INCREASED INCIDENCE OF ACINAR CELL ADENOMAS OF THE PANCREAS. REDUCED SURVIVAL OF MALE F344/N RATS THAT RECEIVED 200 MG/KG DECREASED THE SENSITIVITY OF THIS GROUP FOR DETECTING CARCINOGENIC ACTIVITY. NO CARCINOGENIC ACTIVITY FOR FEMALE RATS. INCREASED INCIDENCES OF PREPUTIAL OR CLITORAL GLAND ADENOMAS OR CARCINOMAS (COMBINED) OCCURRED IN MALE AND FEMALE F344/N RATS ADMINISTERED 200 MG/KG 2-AMINO-5-NITRO-PHENOL.
REPT'D USE : COMPONENT OF PERMANENT HAIR DYES AND INTERMEDIATE FOR MANUFACTURE OF AZO DYE CI SOLVENT RED 8

CAS NO : **99-57-0**
CHEM NAME : ***** 2-AMINO-4-NITROPHENOL**

REPORT 01 : NTP TECHNICAL BULLETIN NO. 7 APR 82
THE CHEMICAL WAS POSITIVE IN THE SALMONELLA TYPHIMURIUM MUTAGENICITY ASSAY.

REPORT 02 : NTP TR 399 HIN 88-2595 JUN 88
UNDER 2-YR GAVAGE STUDIES, SHOWED SOME EVIDENCE OF CARCINOGENIC ACTIVITY OF 2-AMINO-4-NITROPHENOL FOR MALE F344/N RATS SHOWN BY INCREASED INCIDENCES OF RENAL CORTICAL (TUBULAR CELL) ADENOMAS. INCIDENCES OF RENAL

TUBULAR CELL HYPERPLASIA INCREASED IN MALE RATS. SURVIVAL OF MALE RATS THAT RECEIVED 2-AMINO-4-NITROPHENOL WAS REDUCED COMPARED WITH SURVIVAL OF VEHICLE CONTROL MALE RATS. NO EVIDENCE OF CARCINOGENIC ACTIVITY FOR FEMALE F344/N RATS OR FOR MALE OR FEMALE B6C3F1 MICE THAT RECEIVED 125 OR 250 MG/KG PER DAY.

REPT'D USE : CHEM. INTERMEDIATE IN MANUFACTURING OF CI MORDANT BROWN 33, CI MORDANT BROWN 1 AND USE TO DYE LEATHER, NYLON, SILK, WOOL, FUR AND HAIR DYES

CAS NO : **119-34-6**

CHEM NAME : ***** 4-AMINO-2-NITROPHENOL**

REPORT 01 : NCI CARC TRS 94 DHEW PUB NO NIH 78-1344
UNDER THE CONDITIONS OF THE BIOASSAY 4-AMINO-2-NITROPHENOL WAS CARCINOGENIC FOR MALE FISCHER 344 RATS, INDUCING TRANSITIONAL-CELL CARCINOMAS OF THE URINARY BLADDER; THE TRANSITIONAL-CELL CARCINOMAS OF THE URINARY BLADDER OBSERVED IN THREE DOSED FEMALE RATS MAY ALSO HAVE BEEN ASSOCIATED WITH ADMINISTRATION OF THE 4-AMINO-2-NITROPHENOL. THE TEST CHEMICAL WAS NOT CARCINOGENIC FOR MALE OR FEMALE B6C3F1 MICE AT THE DOSES TESTED (FEEDING STUDY).

REPORT 02 : IARC MONOGRAPH 16 1978
DATA ON 4-AMINO-2-NITROPHENOL ALONE NOT AVAILABLE SO NO EVALUATION OF ITS CARCINOGENICITY CAN BE MADE. 4-AMINO-2-NITROPHENOL INDUCED REVERSE MUTATIONS IN S. TYPHIMURIUM TA1538 IN THE ABSENCE OF LIVER ACTIVATION.

REPORT 03 : NTP TECHNICAL BULLETIN 8 JUL 82
4-AMINO-2-NITROPHENOL WAS POSITIVE IN THE L5178Y MOUSE LYMPHOMA ASSAY.

REPORT 06 : MULTI-AGENCY CROSS CHECK
N HHS MAGUIRE
Y OSHA CARC
N EPA CAG
N EPA RCRA
N EPA CLEAN AIR
N EPA CLEAN WATER

REPT'D USE : INDUSTRIAL DYE INTERMEDIATE AND HAIR COLORING COMPONENT

CAS NO : **121-66-4**
CHEM NAME : ***** 2-AMINO-5-NITROTHIAZOLE**

REPORT 01 : NCI CARC TRS 53 DHEW PUB NO NIH 78-1359
UNDER THE CONDITIONS OF THIS BIOASSAY, THE OCCURRENCE OF TUMORS OF THE HEMATOPOIETIC SYSTEM, I.E., LYMPHOMA AND GRANULOCYTIC LEUKEMIA, IN DOSED MALE FISCHER 344 RATS WAS ASSOCIATED WITH ADMINISTRATION OF 2-AMINO-5-NITROTHIAZOLE. 2-AMINO-5-NITROTHIAZOLE WAS NOT CARCINOGENIC IN FEMALE FISCHER 344 RATS OR IN MALE OR FEMALE B6C3F1 MICE (FEEDING STUDY).

REPORT 02 : IARC MONOGRAPH 31 JUL 83
ADMINISTRATION IN DIET IN FEMALE RATS INCREASED INCIDENCE OF BENIGN MAMMARY TUMORS. IN ANOTHER EXPERIMENT IN RATS INCREASED INCIDENCES OF MALIGNANT LYMPHOMAS, LYMPHOCYTIC AND UNDIFFERENTIATED LEUKEMIAS, AND GRANULOCYTIC LEUKEMIAS IN MALE RATS. RESULTS OF STUDIES IN MICE WERE NOT INDICATIVE OF A CARCINOGENIC EFFECT.

REPORT 03 : IARC MONOGRAPH 31 JUL 83
2-AMINO-4-NITROTHIAZOLE MUTAGENIC IN BACTERIA.

REPORT 06 : MULTI-AGENCY CROSS CHECK
N HHS MAGUIRE
Y OSHA CARC
N EPA CAG
N EPA RCRA
N EPA CLEAN AIR
N EPA CLEAN WATER

REPT'D USE : VETERINARY ANTIPROTOZOAL DRUG AND INTERMEDIATE IN MANUFACTURE OF DISPERSE AZO DYES

CAS NO : **95-55-6**
CHEM NAME : ***** O-AMINOPHENOL**

REPORT 01 : NTP TECHNICAL BULLETIN 3/8 DEC 80
0-AMINOPHENOL WAS POSITIVE IN THE SALMONELLA ASSAY BUT WAS NEGATIVE WHEN TESTED FOR SEX-LINKED RECESSIVE LETHALS IN DROSOPHILIA.

REPT'D USE : FUR DYE AND PHARMACEUTICAL INTERMEDIATE FOR AZOSULFUR DYES

CAS NO : **61-82-5**
CHEM NAME : ***** 3-AMINO-S-TRIAZOLE (AMITROLE)**

REPORT 01 : NTP TECHNICAL BULLETIN 8 JUL 82
AMITROLE WAS NEGATIVE IN THE L5178Y MOUSE LYMPHOMA ASSAY.

REPORT 02 : IARC MONOGRAPH 7 1974
AMITROLE INDUCED THYROID AND LIVER TUMORS IN BOTH MICE AND RATS FOLLOWING ORAL AND/OR SUBCUTANEOUS ADMINISTRATION. A SINGLE, SMALL COHORT STUDY RAISES THE SUSPICION THAT AMITROLE MAY BE CARCINOGENIC TO MAN, BUT THE FINDINGS CANNOT BE REGARDED AS CONCLUSIVE. IN A 1979 SUPPLEMENT, IARC DETERMINED THAT THERE IS SUFFICIENT EVIDENCE FOR THE CARCINOGENICITY OF AMITROLE IN ANIMALS.

REPORT 03 : IARC VOL 41 1986
SUFFICIENT EVIDENCE FOR CARCINOGENICITY OF CPD TO TEST ANIMALS BUT INADEQUATE EVIDENCE FOR CARCINOGENICITY TO HUMANS. LIMITED EVIDENCE FOR GENETIC ACTIVITY IN SHORT-TERM TESTS. ORAL ADMINISTRATION OF CPD TO MICE AND RATS INDUCED THYROID AND LIVER TUMORS. NO SKIN TUMORS SEEN AFTER TOPICAL APPLICATION ON MICE. NO CARCINOGENIC EFFECT SEEN IN HAMSTERS. A STATISTICALLY SIGNIFICANT EXCESS OF ALL TUMORS SEEN IN SUBCOHORT EXPOSED TO BOTH AMITROLE AND CHLOROPHENOXY HERBICIDES BUT NOT IN SUBCOHORT EXPOSED TO AMITROLE BUT NOT TO 2,4-D OR 2,4,5-T.

REPORT 06 : MULTI-AGENCY CROSS CHECK
Y HHS MAGUIRE
N OSHA CARC
Y EPA CAG
Y EPA RCRA
N EPA CLEAN AIR
N EPA CLEAN WATER

REPT'D USE : HERBICIDE, DYE INTERMEDIATE AND PLANT GROWTH REGULATOR

CAS NO : **2432-99-7**
CHEM NAME : ***** 11-AMINOUNDECANOIC ACID**

REPORT 01 : NTP CARC TRS 216 NIH PUB NO. 82-1772
UNDER THE CONDITIONS OF THIS BIOASSAY, 11-AMINOUNDECANOIC ACID ADMINISTERED IN THE DIET WAS CARCINOGENIC FOR MALE F344 RATS, BUT NOT FOR FEMALE RATS. NO CLEAR EVIDENCE WAS FOUND FOR CARCINOGENICITY IN B6C3F1 MICE OF EITHER SEX, ALTHOUGH THE INCREASE IN MALIGNANT LYMPHOMA IN MALE MICE MAY HAVE BEEN ASSOCIATED WITH ADMINISTRATION OF THE CHEMICAL.

REPORT 02 : IARC MONOGRAPH 39, 1986 JUN 85
11-AUA ADMINISTERED IN DIET TO MICE AND RATS LED TO INCREASED INCIDENCE OF TRANSITIONAL-CELL CARCINOMAS OF URINARY BLADDER AND NEOPLASTIC NODULES IN MALE RATS. TRANSITIONAL-CELL CARCINOMAS OF KIDNEYS AND EPIITHELIAL HYPERPLASIA OF URINARY BLADDER AND RENAL PELVIS SEEN IN MALE AND FEMALE RATS. NO CLEAR EVIDENCE OF ADVERSE EFFECTS IN MICE. DATA NOT AVAILABLE TO EVALUATE REPRODUCTIVE EFFECTS OR PRENATAL TOXICITY IN HUMANS OR ANIMALS. 11-AUA NOT MUTAGENIC TO D.MELANOGASTER. NO REPORTS OR EPA STUDIES AVAILABLE FOR EVALUATION OF CARCINOGENICITY TO HUMANS.

REPT'D USE : AUTOMOBILE PARTS, PLASTIC MONOMER FOR NYLON 11 TIRE CORDS, METAL ADHESIVES, PHOTOGRAPHIC CHEMICALS AND FUNGICIDES/BACTERICIDES

CAS NO : **7177-48-2**

CHEM NAME : ***** AMPICILLIN TRIHYDRATE**

REPORT 01 : NTP TECH REPORT NTP TR 318 NIH PUB # 87-2574 APR 87
UNDER CONDITIONS OF 2-YR GAVAGE STUDY, EQUIVOCAL EVIDENCE OF CARCINOGENICITY OF CPD FOR MALE F344/N RATS, SHOWN BY INCREASED INCIDENCE OF PHEOCHROMOCYTOMAS OF ADRENAL MEDULLA; NO EVIDENCE OF CARCINOGENICITY FOR FEMALE F344/N RATS DOSED WITH 750 OR 1500 MG/KG OR FOR MALE AND FEMALE B6C3F1 MICE DOSED WITH 1500 OR 3000 MG/KG/DAY. NONNEOPLASTIC LESIONS OF THE FORESTOMACH SEEN IN MALE RATS AND MALE AND FEMALE MICE.

REPT'D USE : ANTIBIOTIC

CAS NO : **523-50-2**

CHEM NAME : ***** ANGELICIN AND SOME SYNTHETIC DERIVATIVES**

REPORT 01 : IARC VOL 40 1986
LIMITED EVIDENCE FOR CARCINOGENICITY TO TEST ANIMALS WITH ULTRAVIOLET A RADIATION (UVA) BUT INADEQUATE EVIDENCE FOR CARCINOGENICITY TO TEST ANIMALS WITHOUT UVA. NO EVALUATION OF CARCINOGENICITY OF CPDS TO HUMANS. NO DATA AVAILABLE TO EVALUATE REPRODUCTIVE EFFECTS OR PRENATAL TOXICITY OF THE CPDS. DEGREE OF EVIDENCE FOR GENETIC ACTIVITY WAS INADEQUATE WITHOUT UVA AND SUFFICIENT WITH UVA.

REPT'D USE : RESEARCH

CAS NO : **101-05-3**
CHEM NAME : ***** ANILAZINE**

REPORT 01 : NCI CARC TRS 104 DHEW PUB NO NIH 78-1354
NO TUMORS OCCURRED IN DOSED RATS OR MICE OF EITHER SEX AT INCIDENCES THAT WERE SIGNIFICANTLY HIGHER THAN THOSE IN CORRESPONDING CONTROLS. MALE AND FEMALE RATS AND FEMALE MICE MAY HAVE BEEN ABLE TO TOLERATE HIGHER DOSES. IT IS CONCLUDED THAT UNDER THE CONDITIONS OF THIS BIOASSAY, ANILAZINE WAS NOT CARCINOGENIC FOR EITHER FISCHER 344 RATS OR B6C3F1 MICE (FEEDING STUDY).
REPT'D USE : AGRICULTURAL FUMIGANT

CAS NO : **62-53-3**
CHEM NAME : ***** ANILINE**

REPORT 01 : IARC MONOGRAPH 4 1974
THERE IS NO ADEQUATE DATA TO INDICATE THAT ANILINE IS CARCINOGENIC IN EXPERIMENTAL ANIMALS. AT THE PRESENT TIME, THE WEIGHT OF EPIDEMIOLOGICAL EVIDENCE SUGGESTS THAT ANILINE IS NOT A CARCINOGEN FOR THE HUMAN BLADDER.

REPORT 02 : RCRA HEALTH AND ENV. EFFECTS PROFILE OCT 80
AN INCREASED INCIDENCE OF URINARY BLADDER TUMORS HAS BEEN NOTED IN WORKERS IN THE ANILINE AND ANILINE DYESTUFF INDUSTRIES. ORAL FEEDING STUDIES CONDUCTED BY THE NCI HAVE SHOWN ANILINE TO BE CARCINOGENIC IN RATS. ANILINE IS REPORTED NOT TO BE MUTAGENIC TO SIX STRAINS OF SALMONELLA; HOWEVER, SEVERAL ANILINE ANALOGS AND DERIVATIVES ARE MUTAGENIC.

REPORT 03 : IARC MONOGRAPH 27 1982
ANILINE WAS INACTIVE IN BACTERIAL AND MAMMALIAN DNA REPAIR ASSAYS, IN TESTS FOR MITOTIC RECOMBINATION WITH YEAST, AND IN CELL TRANSFORMATION ASSAYS. IT DID NOT INDUCE CHROMOSOMAL ABERRATIONS IN MAMMALIAN CELLS OR IN ANIMALS. URINES FROM TREATED RATS WERE MUTAGENIC FOR SALMONELLA WITH METABOLIC ACTIVATION. ANILINE INDUCED SISTER CHROMATID EXCHANGES IN CULTURED MAMMALIAN CELLS.

REPORT 04 : ITC FOURTH REPORT TO EPA (44FR31866) JUN 79
THE INTERAGENCY TESTING COMMITTEE RECOMMENDS THAT THE EPA ADMINISTRATOR CONSIDER THE DEVELOPMENT OF TEST RULES TO TEST ANILINE, AND BROMO-, CHLORO-, AND NITROANILINES FOR CARCINOGENICITY, MUTAGENICITY, TERATOGENICITY, REPRODUCTIVE EFFECTS AND NEUROLOGICAL EFFECTS; AND THAT AN EPIDEMIOLOGIC STUDY BE CONDUCTED.

REPORT 05 : NTP TECHNICAL BULLETIN 7/8 APR 82
ANILINE WAS NEGATIVE IN THE SALMONELLA ASSAY, BUT POSITIVE IN THE L5178Y MOUSE LYMPHOMA ASSAY.

REPORT 06 : MULTI-AGENCY CROSS CHECK
N HHS MAGUIRE
Y OSHA CARC
N EPA CAG
Y EPA RCRA
N EPA CLEAN AIR
N EPA CLEAN WATER

REPORT 07 : FR 49(1) 108 ANPR JAN 84
EPA INITIATING RULEMAKING TO REQUIRE TESTING TO CHARACTERIZE HEALTH EFFECTS OF CPD. THIS IS IN RESPONSE TO ITC'S DESIGNATION OF CPD FOR TESTING.

REPT'D USE : CHEMICAL INTERMEDIATE SOLVENT, PAINT REMOVER, PRINTING INKS, SHOE BLACKS AND AUTOMOTIVE OCTANE BOOSTER

CAS NO : **142-04-1**

CHEM NAME : ***** ANILINE HYDROCHLORIDE**

REPORT 01 : NCI CARC TRS 130 DHEW PUB NO NIH 78-1385
UNDER THE CONDITIONS OF THIS BIOASSAY DIETARY ADMINISTRATION OF ANILINE HYDROCHLORIDE WAS CARCINOGENIC TO MALE AND FEMALE FISCHER 344 RATS, INDUCING HEMANGIOSARCOMAS, A COMBINATION OF FIBROSARCOMAS AND UNSPECIFIED SARCOMAS OF THE SPLEEN AND A COMBINATION OF FIBROSARCOMAS AND UNSPECIFIED SARCOMAS OF MULTIPLE BODY ORGANS. THERE WAS NO EVIDENCE OF COMPOUND-INDUCED CARCINOGENICITY IN B6C3F1 MICE OF EITHER SEX.

REPORT 02 : CIIT 24 MONTH TOXICOLOGY STUDY FEB 80
THE CIIT IS INVESTIGATING THE ABILITY OF F344 RATS TO TOLERATE ANILINE HCL AT 30, 100 AND 300 MG/KG/DAY BY DIETARY ADMINISTRATION FOR 24 MONTHS. NECROPSY OF ANIMALS SACRIFICED AT WEEKS 52 AND 78 REVEALED GROSS ALTERATIONS IN THE SPLEENS OF THE HIGH-DOSE GROUP. ALL OTHER GROSS PATHOLOGY FINDINGS APPEAR TO BE INCIDENTAL.

REPORT 03 : CIIT FINAL REPORT OF TERATOLOGY STUDY NOV 81
ANILINE HCL AT 0, 10, 30, AND 100 MG/KG/DAY WAS EVALUATED FOR TERATOGENIC EFFECTS FOLLOWING ADMINISTRATION T0 TIME-PREGNANT F344 RATS ON GESTATIONAL DAYS 7-20. ANILINE HCL WAS NOT OBSERVED TO BE TERATOGENIC, EVEN AT DOSE LEVELS WHICH PRODUCED SIGNIFICANT MATERNAL AND FETAL TOXICITY.

REPORT 04 : IARC MONOGRAPH 27 1982
ANILINE HCL WAS TESTED IN MICE BY DIETARY ADMINISTRATION, PRODUCING NO CARCINOGINIC EFFECTS. WHEN TESTED BY ORAL ADMINISTRATION IN RATS, ONE EXPERIMENT PRODUCED FIBROSARCOMAS, SARCOMAS AND HEMANGIOSARCOMAS OF THE SPLEEN OR THE PERITONEAL CAVITY. THERE IS LIMITED EVIDENCE FOR THE CARCINOGENICITY OF ANILINE HCL IN ANIMALS.

REPORT 06 : MULTI-AGENCY CROSS CHECK
N HHS MAGUIRE
Y OSHA CARC
N EPA CAG
N EPA RCRA
N EPA CLEAN AIR
N EPA CLEAN WATER

REPORT 07 : FR 49(1) 108 ANPR JAN 84
EPA IS INITIATING RULEMAKING TO REQUIRE TESTING TO CHARACTERIZE HEALTH EFFECTS OF CPD. THIS IS IN RESPONSE TO ITC'S DESIGNATION OF CPD FOR TESTING.
REPT'D USE : DYE INTERMEDIATE

CAS NO : **90-04-0**
CHEM NAME : ***** O-ANISIDINE**

REPORT 01 : IARC MONOGRAPH 27 1982
THERE IS SUFFICIENT EVIDENCE FOR THE CARCINOGENICITY OF ORTHOANISIDINE HYDROCHLORIDE IN EXPERIMENTAL ANIMALS. IN THE ABSENCE OF DATA ON HUMANS, ORTHO-ANISIDINE SHOULD BE REGARDED, FOR PRACTICAL PURPOSES, AS IF IT PRESENTED A CARCINOGENIC RISK TO HUMANS.

REPORT 06 : MULTI-AGENCY CROSS CHECK
Y HHS MAGUIRE
N OSHA CARC
N EPA CAG
N EPA RCRA
N EPA CLEANAIR
N EPA CLEAN WATER
REPT'D USE : PRODUCTION OF DYES, PHARMACEUTICALS AND TEXTILE-PROCESSING CHEMICALS

CAS NO : **134-29-2**
CHEM NAME : ***** O-ANISIDINE HYDROCHLORIDE**

REPORT 01 : NCI CARC TRS 89 DHEW PUB NO 78-1339
UNDER THE CONDITIONS OF THIS BIOASSAY O-ANISIDINE HYDROCHLORIDE WAS CARCINOGENIC FOR FISCHER 344 RATS AND B6C3F1 MICE. IT INDUCED TRANSITIONAL CELL CARCINOMAS OR PAPILLO-

MAS OF THE BLADDER IN BOTH RATS AND MICE AND IN BOTH SEXES OF EACH SPECIES, TRANSITIONAL-CELL CARCINOMAS OF THE PELVIS OR THE KIDNEY IN MALE RATS. IT ALSO INDUCED FOLLICULAR CELL TUMORS OF THE THYROID IN A FEEDING STUDY IN MALE RATS.

REPORT 02 : IARC MONOGRAPH 27 1982
O-ANISIDINE HCI WAS TESTED IN RATS AND MICE BY DIETARY ADMINISTRATION. IT WAS CARCINOGENIC IN BOTH SPECIES PRODUCING TRANSITIONAL CELL CARCINOMAS OF THE URINARY BLADDER. THERE IS SUFFICIENT EVIDENCE FOR THE CARCINOGENICITY OF O-ANISIDINE HCL IN ANIMALS.

REPORT 03 : IARC MONOGRAPH 27 1982
O-ANISIDINE HCL INDUCED REVERSE MUTATIONS IN SALMONELLA STRAINS TA153 8, TA98, AND TA100 IN THE PRESENCE OF RAT LIVER MICROSOMES. NO DATA WERE AVAILABLE CONCERNING THE EFFECT OF O-ANISIDINE HCL ON REPRODUCTION OR ITS PRENATAL TOXICITY.

REPORT 06 : MULTI-AGENCY CROSS CHECK
Y HHS MAGUIRE
N OSHA CARC
N EPA CAG
N EPA RCRA
N EPA CLEANAIR
N EPA CLEAN WATER
REPT'D USE : DYE INTERMEDIATE, NO PRODUCTION OR IMPORTATION

CAS NO : **20265-97-8**
CHEM NAME : ***** P-ANISIDINE HYDROCHLORIDE**

REPORT 01 : NCI CARC TRS 116 DHEW PUB NO NIH 78-1371
ALTHOUGH, UNDER THE CONDITIONS OF THIS BIOASSAY, THERE APPEARED TO BE AN ASSOCIATION BETWEEN CHEMICAL ADMINISTRATION AND THE INCREASED INCIDENCE OF PREPUTIAL GLAND TUMORS IN LOW DOSE MALE RATS, THE EVIDENCE WAS INSUFFICIENT TO ESTABLISH THE CARCINOGENICITY OF P-ANISIDINE HYDROCHLORIDE IN FISCHER 344 RATS. THE COMPOUND WAS NOT CARCINOGENIC IN A FEEDING STUDY IN B6C3F1 MICE.

REPORT 02 : IARC MONOGRAPH 27 1982
P-ANISIDINE HCL WAS TESTED IN MICE AND RATS BY DIETARY ADMINISTRATION. IT DID NOT PRODUCE CARCINOGENIC EFFECTS IN MICE, AND THE DATA IN RATS WERE INADEQUATE FOR EVALUATION. THE AVAILABLE DATA WERE INADEQUATE TO EVALUATE THE CARCINOGENICITY OF P-ANISIDINE HCL IN ANIMALS.

REPORT 03 : IARC MONOGRAPH 27 1982
P-ANISIDINE HCL WAS NOT MUTAGENIC IN SALMONELLA WHEN TESTED IN THE PRESENCE OF HEPATIC MICROSOMES FROM AROCHLOR-INDUCED RATS. THIS COMPOUND DID NOT INDUCE MORPHOLOGICAL TRANSFORMATIONS OF CULTURED BHK21 CELLS IN THE PRESENCE OF LIVER MICROSOMES. NO DATA WERE AVAILABLE CONCERNING THE EFFECTS OF P-ANISIDINE HCL ON REPRODUCTION OR ITS PRENATAL TOXICITY.

REPORT 06 : MULTI-AGENCY CROSS CHECK
Y HHS MAGUIRE
N OSHA CARC
N EPA CAG
N EPA RCRA
N EPA CLEANAIR
N EPA CLEAN WATER
REPT'D USE : DYE INTERMEDIATE

CAS NO : **191-26-4**
CHEM NAME : ***** ANTHANTHRENE**

REPORT 01 : IARC MONOGRAPH 32 DEC 83
IN ONE STUDY, RELATIVELY HIGH DOSES OF ANTHANTHRENE IN ACETONE PRODUCED SKIN TUMORS IN MICE; IN 3 OTHER STUDIES NO INCREASED INCIDENCE OF TUMORS WAS OBSERVED. IN ONE MOUSE-SKIN INITIATION-PROMOTION ASSAY, ANTHANTHRENE GAVE NEGATIVE RESULTS; IN ANOTHER, RESULTS WERE INCONCLUSIVE. DIRECT INJECTION OF ANTHANTHRENE INTO PULMONARY TISSUE OF RATS PRODUCED PULMONARY SQUAMOUS-CELL CARCINOMAS IN A DOSE-RELATED FASHION. THERE IS LIMITED EVIDENCE THAT ANTHRANTHRENE IS CARCINOGENIC TO EXPERIMENTAL ANIMALS.

REPORT 03 : IARC MONOGRAPH 32 DEC 83
NO DATA ON TERATOGENICITY OF ANTHANTHRENE WERE AVAILABLE. IT WAS MUTAGENIC TO SALMONELLA TYPHIMURIUM IN PRESENCE OF EXOGENOUS METABOLIC SYSTEM. THERE IS INADEQUATE EVIDENCE THAT ANTHANTHRENE IS ACTIVE IN SHORT-TERM TESTS.
REPT'D USE : COMBUSTION BY-PRODUCT

CAS NO : **120-12-7**
CHEM NAME : ***** ANTHRACENE**

REPORT 01 : NTP TECHNICAL BULLETIN 8 JUL 82
ANTHRACENE WAS POSITIVE IN THE L5178Y MOUSE LYMPHOMA ASSAY.

REPORT 02 : NTP TECHNICAL BULLETIN 9 APR 83
ANTHRACENE WAS POSITIVE IN TWO SALMONELLA TYPHIMURIUM MUTAGENICITY ASSAY.

REPORT 03 : IARC MONOGRAPH 32 DEC 83
RESULTS NOT INDICATIVE OF CARCINOGENICITY OR OF INITIATING ACTIVITY WHEN TESTED IN MICE BY SKIN APPLICATION. TESTED IN RATS BY ORAL SUBCUTANEOUS, INTRAPERITONEAL AND INTRAPULMONARY ADMINISTRATION AND IN RABBITS BY IMPLANTATION INTO BRAIN OR EYES. ORAL OR INTRAPULMONARY ADMINISTRATION PRODUCED NO EVIDENCE OF CARCINOGENICITY. SUBCUTANEOUS OR INTRAPERITONEAL ADMINISTRATTION AND IMPLANTATION IN BRAIN OR EYES INDAQEUATE FOR EVALUATION. IN SKIN APPLICATION TO MICE TOGETHER WITH EXPOSURE TO ULTRAVIOLET RADIATION CONTRADICTORY RESULTS WERE OBTAINED.

REPORT 04 : IARC MONOGRAPH 32 DEC 83
ANTHRACENE WAS NEGATIVE IN ASSAY FOR DIFFERENTIAL SURVIVAL USING DNA REPAIR-PROFICIENT/-DEFICIENT STRAINS OF BACILLUS SUBTILIS. DID NOT INDUCE MUTATIONS IN BACTERIA OR YEAST NOR UNSCHEDULED DNA SYNTHESIS OR MUTATIONS IN CULTURED MAMMALIAN CELLS.

REPT'D USE : DYE INTERMEDIATE, SMOKE SCREENS, SCINTILLATION COUNTER CRYSTALS AND COMBUSTION BY-PRODUCT

CAS NO : **118-92-3**

CHEM NAME : ***** ANTHRANILIC ACID**

REPORT 01 : NCI CARC TRS 36 DHEW PUB NO NIH 78-836
UNDER THE CONDITIONS OF THIS BIOASSAY, ANTHRANILIC ACID WAS NOT CARCINOGENIC FOR EITHER FISCHER 344 RATS OR B6C3F1 MICE (FEEDING STUDY).

REPORT 02 : IARC MONOGRAPH 16 1978
ANTHRANILIC ACID HAS BEEN TESTED IN MICE BY BLADDER IMPLANTATION. THE AVAILABLE DATA DO NOT ALLOW AN EVALUATION OF THE CARCINOGENICITY OF THIS COMPOUND TO BE MADE. AVAILABLE TERATOGENICITY DATA FOR ANTHRANALIC ACID WERE NEGATIVE. NO DATA ON THE MUTAGENICITY OF THIS CHEMICAL WERE AVAILABLE.

REPT'D USE : DYE INTERMEDIATE, MFR. OF SACCHARIN AND MOLD INHIBITOR IN SOY SAUCE

CAS NO : **84-65-1**

CHEM NAME : ***** ANTHRAQUINONE**

REPORT 01 : 50FR46090 NOV 85
EPA IS PROPOSING THAT MFRS, IMPORTERS AND PROCESSORS OF

9,10-AQ BE REQUIRED TO TEST FOR WATER SOLUBILITY, BIOCONCENTRATION AND ACUTE AQUATIC TOXICITY UNDER TSCA 4 AND THAT THEY ALSO BE REQUIRED TO SUBMIT ANNUAL REPORTS ON USAGE OF 9,10-ANTHRAQUINONE.

CAS NO : **7440-36-0**
CHEM NAME : ***** ANTIMONY**

REPORT 01 : 48FR76: EPA RESPONSE TO EPA SEP 78, JAN 86
EXPOSURE TO ANTIMONY SHALL BE CONTROLLED SO THAT EMPLOYEES ARE NOT EXPOSED TO A CONCENTRATION GREATER THAN 0.5 MG PER CUBIC METER OF AIR DETERMINED AS A TIME-WEIGHTED AVERAGE CONCENTRATION LIMIT FOR UP TO A 10-HOUR WORKSHIFT IN A 40-HOUR WORKWEEK. EXPOSURE TO ANTIMONY MAY CAUSE CARDIAC AND RESPIRATORY CHANGES AND IRRITATION OF THE SKIN AND MUCOUS MEMBRANES. SOME DATA ON ANTIMONY WORKERS RAISE THE POSSIBILITIES OF REPRODUCTIVE AND CARCINOGENIC EFFECTS; HOWEVER, THESE HAVE NOT BEEN CONFIRMED. THE PRESENT TWA OSHA STANDARD IS 0.5 MG/CU M.

REPORT 02 : EPA RCRA HEALTH AND ENV EFFECTS PROFILE OCT 80
THE ADVERSE HEALTH EFFECTS MOST COMMONLY ASSOCIATED WITH EXPOSURE TO ANTIMONY ARE PULMONARY, CARDIOVASCULAR, DERMAL, AND CERTAIN EFFECTS OF REPRODUCTION, DEVELOPMENT, AND LONGEVITY. CARDIOVASCULAR CHANGES HAVE BEEN WELL ESTABLISHED WITH EXPOSURE TO ANTIMONY AND PROBABLY REPRESENT THE MOST SERIOUS THREAT TO HUMAN HEALTH. ANTIMONY HAS NOT BEEN ASSOCIATED WITH CARCINOGENIC EFFECTS.

REPORT 03 : ITC FOURTH REPORT TO EPA (44FR31866) JUN 79
THE ITC RECOMMENDS THAT THE EPA ADMINISTRATOR CONSIDER THE DEVELOPMENT OF RULES TO TEST ANTIMONY FOR CARCINOGENICITY, MUTAGENICITY, TERATOGENICITY AND REPRODUCTIVE EFFECTS, AND OTHER CHRONIC EFFECTS; AND THAT AN EPIDEMIOLOGIC STUDY BE CONDUCTED.

REPORT 04 : 48FR716: EPA RESPONSE TO ITC JAN 83
IN RESPONSE TO ITC'S 1979 DESIGNATION OF ANTIMONY FOR PRIORITY TESTING, MFR'S SUBMITTED PROGRAM FOR MONITORING AND CONTROLLING OCCUPATIONAL EXPOSURE AND ENVIRONMENTAL RELEASE, MEDICAL SURVEILLANCE, EPA STUDIES AND HEALTH MONITORING AND TESTING TO CHARACTERIZE CHEMICAL FATE OF THE CPD. IN RESPONSE TO ITC, EPA TENTATIVELY ACCEPTED THE MFR'S PROGRAM AND IS NOT NOW INITIATING RULEMAKING UNDER TSCA4(A).

REPORT 06 : MULTI-AGENCY CROSS CHECK
N HHS MAGUIRE
N OSHA CARC
N EPA CAG
N EPA RCRA
N EPA CLEAN AIR
Y EPA CLEAN WATER

REPT'D USE : ALLOYS, ELECTROPLATING, SOLDERING FLUXES, PORCELAIN ENAMELS, FLAME RETARDANT, BATTERY GRIDS AND TYPE METAL

CAS NO : **1315-04-4**

CHEM NAME : ***** ANTIMONY SULFIDE (PENTASULFIDE)**

REPORT 01 : 48FR716: EPA RESPONSE TO ITC JAN 83
IN RESPONSE TO ITC'S DESIGNATION OF AS FOR PRIORITY TESTING, MFRS SUBMITTED PROGRAMS FOR MONITORING AND CONTROLLING OCCUPATIONAL EXPOSURE AND ENVIRONMENTAL RELEASE, MEDICAL SURVEILLANCE, EPA STUDIES AND HEALTH MONITORING AND TESTING TO CHARACTERIZE THE CHEMICAL FATE OF AS. EPA TENTATIVELY ACCEPTED THE MFR'S PROPOSED PROGRAM AND IS NOT NOW INITIATING RULEMAKING UNDER TSCA 4(A).

CAS NO : **1309-64-4**

CHEM NAME : ***** ANTIMONY TRIOXIDE**

REPORT 01 : EPA TSCA 8(E) NOTICE 8EHQ-0580-0342
ASARCO, INC. REPORTED THAT LUNG TUMORS HAVE BEEN OBSERVED IN 15/17 RATS CHRONICALLY EXPOSED VIA INHALATION TO 4.2 MG/CU M OF ANTIMONY TRIOXIDE DUST FOR ONE YEAR. SUBMITTER REPORTS THAT THE ANIMALS AT THE LOW DOSE - 1.6 MG/CU M - AND THE CONTROL ANIMALS DID NOT DEVELOP LUNG TUMORS. LIVER AND MAMMARY TUMORS WERE REPORTED TO HAVE OCCURRED IN BOTH THE EXPOSED AND CONTROL GROUPS.

REPORT 02 : 48FR716 EPA RESPONSE TO ITC
IN RESPONSE TO ITC'S 1979 DESIGNATION OF AT FOR PRIORITY TESTING, MFRS SUBMITTED PROGRAM FOR MONITORING AND CONTROLLING OCCUPATIONAL EXPOSURE AND ENVIRONMENTAL RELEASE, MEDICAL SURVEILLANCE, EPA STUDIES AND HEALTH MONITORING AND TESTING TO CHARACTERIZE CHEMICAL FATE OF AT. IN RESPONSE TO ITC, EPA TENTATIVELY ACCEPTED MFR'S PROPOSED PROGRAM AND IS NOT NOW INITIATING RULEMAKING UNDER TSCA 4(A).

CAS NO : **1345-04-6**
CHEM NAME : ***** ANTIMONY TRISULFIDE**

REPORT 01 : EPA RESPONSE TO ITC, 48FR7616 JUL 86
IN RESPONSE TO ITC 1979 DESIGNATION OF ATS FOR POSSIBLE PRIORITY TESTING, EPA HAS TENTATIVELY ACCEPTED PROGRAM PROPOSED BY THE ANTIMONY OXIDE INDUSTRY ASSOC. (AOIA) FOR MONITORING AND CONTROLLING OCCUPATIONAL EXPOSURE AND ENVIRONMENTAL RELEASES; MEDICAL SURVEILLANCE OF EXPOSED PERSONS; EPIDEMIOLOGICAL STUDIES; AND TESTS OF HEALTH EFFECTS AND CHEMICAL FATE OF ATS. THEREFORE, EPA IS NOT INITIATING RULE MAKING AT THIS TIME.
REPT'D USE : FUEL FOR DYE VAPORIZATION AND EXPLOSIVES

CAS NO : **52-46-0**
CHEM NAME : ***** APHOLATE**

REPORT 01 : IARC MONOGRAPH 9 APR 75
IN THE ONLY AVAILABLE STUDY, APHOLATE WAS INADEQUATELY TESTED BY THE ORAL ROUTE IN RATS; NO EVALUATION ON THE CARCINOGENICITY OF THIS CHEMICAL CAN BE MADE. TREATMENT OF NEUROSPORA WITH APHOLATE INDUCED AN INCREASE IN THE FREQUENCY OF RECESSIVE LETHAL MUTATIONS.
REPT'D USE : INSECT CHEMOSTERILANT

CAS NO : **147-94-4**
CHEM NAME : *****ARABINOSYLCYTOSINE**

REPORT 01 : NCI BIOASSAY PROGRAM- CANCER VOL. 40 1935-1949
UNDER THE CONDITIONS OF THE BIOASSAY, I.P. INJECTION OF ARABINOSYLCYTOSINE WAS NOT CARCINOGENIC TO SWISS-WEBSTER MICE OR SPRAGUE-DAWLEY (CD) RATS OF EITHER SEX.
REPT'D USE : ANTINEOPLASTIC/ANTIVIRAL AGENT

CAS NO : **140-57-8**
CHEM NAME : ***** ARAMITE**

REPORT 01 : IARC MONOGRAPH 5 1974
ARAMITE IS CARCINOGENIC IN THE RAT AND DOG FOLLOWING ITS ORAL ADMINISTRATION. IT PRODUCED LIVER TUMORS IN THE RAT AND CARCINOMAS OF THE GALL BLADDER AND BILIARY DUCTS IN THE DOG. ARAMITE WAS TESTED IN 4 STRAINS OF MICE BY THE ORAL ROUTE AND PRODUCED A SIGNIFICANT INCREASE OF HEPATOMAS IN MALES OF ONE STRAIN. IN A 1979 SUPPLEMENT, IARC DETERMINED THAT THERE IS SUFFICIENT EVIDENCE FOR THE CARCINOGENICITY OF ARAMITE IN ANIMALS.

REPORT 02 : IARC SUPPLEMENT 4 OCT 82
THERE IS CONSIDERED TO BE SUFFICIENT EVIDENCE OF CARCINOGENICITY OF ARAMITE IN EXPERIMENTAL ANIMALS.

REPORT 03 : 4TH ANNUAL REPORT ON CARCINOGENS (MAGUIRE REPT) 1985
ARAMITE CARCINOGENIC TO RATS AND DOGS WHEN ADMINISTERED IN DIET, PRODUCING LIVER TUMORS IN RATS, CANCER OF BILIARY DUCTS AND GALL BLADDER IN DOGS. CPD ALSO SHOWED INCREASE OF LIVER CANCER IN MALE MICE WHEN FED IN DIET. HOWEVER, DROPPED FROM THE ANNUAL REPORT BECAUSE NO U.S. RESIDENTS EXPOSED.

REPORT 06 : MULTI-AGENCY CROSS CHECK
Y HHS MAGUIRE
N OSHA CARC
Y EPA CAG
N EPA RCRA
N EPA CLEAN AIR
N EPA CLEAN WATER
REPT'D USE : ACARACIDE, MICROBICIDE AND MITICIDE

CAS NO : **10130-99-5**
CHEM NAME : ***** ARISTOLOCHIC ACID**

REPORT 01 : WHO DRUG INFO. CIRCULAR NO. 197 OCT 81
MINISTRY OF HEALTH OF GERMANY HAS INFORMED THE PHARMACEUTICALS UNIT OF THE WORLD HEALTH ORGANIZATION THAT DRUGS CONTAINING ARISTOLOCHIC ACID HAVE BEEN WITHDRAWN FROM THE MARKET. THE DECISION RESULTED FROM A DEMONSTRATION OF SIGNIFICANT CARCINOGENIC POTENTIAL IN A THREE MONTH ORAL TOXICITY STUDY IN RATS.
REPT'D USE : PHARMACEUTICAL

CAS NO : **11097-69-1**
CHEM NAME : ***** AROCLOR 1254**

REPORT 01 : NCI CARC TRS 38 DHEW PUB NO 78-838
UNDER THE CONDITIONS OF THIS BIOASSAY AROCLOR 1254 WAS NOT CARCINOGENIC IN FISCHER 344 RATS; HOWEVER, A HIGH INCIDENCE OF HEPATOCELLULAR PROFILERATIVE LESIONS IN BOTH MALE AND FEMALE RATS WAS RELATED TO ADMINISTRATION OF THE CHEMICAL. IN ADDITION, THE CARCINOMAS OF THE GASTROINTESTINAL TRACT MAY BE ASSOCIATED WITH ADMINISTRATION OF AROCLOR 1254 IN BOTH MALES AND FEMALES (FEEDING STUDY IN RATS ONLY).

REPORT 03 : IARC MONOGRAPH 18 1978
FIVE POLYCHLORINATED BIPHENYL MIXTURES HAVE BEEN TESTED

IN MICE AND/OR RATS ONLY BY ORAL ADMINISTRATION. KANECHLOR 500 AND AROCLOR 1254 ARE CARCINOGENIC IN MICE, AND AROCLOR 1260 IS CARCINOGENIC IN RATS; ALL INDUCED BENIGN AND MALIGNANT LIVER-CELL TUMORS. IN AN EXPERIMENT IN RATS OF ONLY ONE YEAR'S DURATION, KANECHLOR 500, 400, AND 300 INDUCED LIVER LESIONS DESCRIBED AS MULTIPLE HYPERPLASTIC NODULES. AN APPARENT EXCESS OF MALIGNANT MELANOMA HAS BEEN REPORTED IN WORKERS EXPOSED TO AROCLOR 1254.

REPORT 04 : IARC MONOGRAPH 18 1978
AROCHLOR 1254 DID NOT PRODUCE SIGNIFICANT CHROMOSOMAL ABNORMALITIES IN RAT BONE MARROW. THE CHEMICAL WHEN GIVEN TO RATS BY ORAL INTUBATION OR WHEN FED IN THE DIET TO RATS DID NOT CAUSE AN INCREASE IN DOMINANT LETHAL MUTATIONS. NO CHROMOSOMAL ABBERATIONS IN HUMAN LYMPHOCYTE CULTURES WERE OBSERVED WHEN TREATED WITH AROCHLOR 1254.

REPORT 06 : MULTI-AGENCY CROSS CHECK
Y HHS MAGUIRE
N OSHA CARC
N EPA CAG
N EPA RCRA
Y EPA CLEAN AIR
Y EPA CLEAN WATER

REPT'D USE : FLAME RETARDANT AND HEAT TRANSFER AGENT. NO LISTING ON TSCA INVENTORY

CAS NO : **UNKNOWN-04**
CHEM NAME : ***** C9 AROMATIC HYDROCARBONS (NAPTHA)**

REPORT 01 : EPA FYI-AX- 03388-0605 MAR 88
AMERICAN PETROLEUM INSTITUTE REPORTED ON INHALATION DEVELOPMENT TOXICITY STUDY IN FEMALE MICE GIVEN 100, 500 AND 1,500 PPM FOR 15 DAYS OF GESTATION AT 500 AND 1,500 PPM FETUSES SHOWED DEVELOPMENTAL TOXICITY BY POSTIMPLANTATION LOSS, DECREASE IN BODY WEIGHTS AND SKULL OSSIFICATION, AND INCREASED UNOSSIFIED STERNEBRAE. NO OBSERVABLE EFFECT LEVEL FOR DEVELOPMENTAL TOXICITY CONSIDERED TO BE 100 PPM, MATERIAL TOXICITY ELICITED AT 1,500 PPM.

REPT'D USE : UNKNOWN

CAS NO : **68187-58-6**
CHEM NAME : ***** AROMATIC PETROLEUM PITCH**

REPORT 01 : EPA TSCA 8(E) NOTICE 8EHQ-0780-0353
STANDARD OIL PROVIDED INTERIM RESULTS, OF A LIFETIME

MOUSE SKIN PAINTING STUDY OF TWO GRADES OF AN AROMATIC PETROLEUM PITCH, WHICH SHOW THE MIXTURE TO BE COMPARABLE IN TUMORIGENIC POTENTIAL TO THAT OF THE POSITIVE CONTROL - BENZO(A)PYRENE.

CAS NO : **98-50-0**
CHEM NAME : ***** ARSANILIC ACID**

REPORT 01 : IARC MONOGRAPH 23 1980
THERE IS INADEQUATE EVIDENCE FOR THE CARCINOGENICITY OF ARSENIC COMPOUNDS IN ANIMALS. THERE IS SUFFICIENT EVIDENCE THAT INORGANIC ARSENIC COMPOUNDS ARE SKIN AND LUNG CARCINOGENS IN HUMANS. THE EVIDENCE THAT ARSENIC COMPOUNDS CAUSE MUTATIONS AND ALLIED EFFECTS IN BACTERIA IS INCONCLUSIVE. HOWEVER, ARSENIC COMPOUNDS INDUCE CHROMOSOMAL ABERRATIONS AND MORPHOLOGICAL TRANSFORMATION IN MAMMALIAN CELLS.

REPORT 02 : NIOSH CRITERIA DOCUMENT JUN 75
NIOSH RECOMMENDS A CEILING OF 2 MICROGRAMS/CU M FOR INORGANIC ARSENIC COMPOUNDS BASED ON DERMATITIS, LUNG, AND LYMPHATIC CANCER. THE CURRENT OSHA STANDARD FOR INORGANIC ARSENIC IS 10 MICROGRAMS/CU M TWA. SEE CFR 1910.1018.

REPORT 06 : MULTI-AGENCY CROSS CHECK
Y HHS MAGUIRE
N OSHA CARC
Y EPA CAG
N EPA RCRA
N EPA CLEAN AIR
N EPA CLEAN WATER

REPT'D USE : CHEMICAL INTERMEDIATE FOR PHARMACEUTICALS, VETERINARY ANTI-INFECTANT/GROWTH PROMOTER

CAS NO : **7440-38-2**
CHEM NAME : ***** ARSENIC**

REPORT 01 : EPA AMBIENT WATER QUALITY CRITERIA OCT 80
A NUMBER OF STUDIES HAVE SHOWN THAT ARSENIC IS IMPORTANT IN THE ETIOLOGY OF HUMAN CANCERS. THERE APPEARS TO BE GENERAL AGREEMENT THAT ARSENIC IS A HUMAN CARCINOGEN. AMBIENT WATER CONCENTRATIONS OF 22 NG/L, 2.2 NG/L, AND .22 NG/L ARE ESTIMATED TO RESULT IN A RESPECTIVE 10 -5, 10 -6, AND 10 -7 INCREASE IN CANCER RISK OVER A LIFETIME.

REPORT 02 : NIOSH CURRENT INTELLIGENCE BULLETIN 14 SEP 76
NIOSH RECOMMENDATIONS FOR PERSONAL RESPIRATORY PROTECTIVE DEVICES GIVEN IN DOCUMENT.

REPORT 03 : EPA RCRA HEALTH AND ENV. EFFECTS PROFILE OCT 80
EPA STUDIES HAVE SHOWN INCREASED DEATH RATES FROM LUNG CANCER IN WORKERS EXPOSED TO ARSENIC, PROBABLY THROUGH INHALATION. OTHER HUMAN STUDIES HAVE SHOWN INCREASED SKIN CANCERS IN NON-OCCUPATIONALLY EXPOSED POPULATIONS. INCREASED INCIDENCES OF LYMPHOMAS AND HEMANGIOENDOTHELIOMAS ARE ALSO OCCASIONALLY REPORTED. ARSENICALS HAVE PRODUCED MUTAGENIC EFFECTS IN PLANTS, BACTERIA, IN VITRO LEUCOCYTE CULTURES, AND IN LYMPHOCYTES OF EXPOSED ANIMALS. TERATOGENIC EFFECTS HAVE BEEN DEMONSTRATED IN MANY ANIMAL SPECIES.

REPORT 04 : IARC MONOGRAPH 23 1980
LARGE NUMBER OF CASES OF SKIN CANCER HAVE BEEN REPORTED AMONG INDIVIDUALS EXPOSED TO INORGANIC ARSENIC THROUGH DRUGS, DRINKING WATER OR PESTICIDES. IN ONE EPIDEMIOLOGICAL STUDY, SKIN CANCER WAS POSITIVELY CORRELATED WITH HIGH ARSENIC LEVELS IN THE DRINKING WATER. THERE IS INADEQUATE EVIDENCE FOR THE CARCINOGENICITY OF ARSENIC COMPOUNDS IN ANIMALS. THERE IS SUFFICIENT EVIDENCE THAT INORGANIC ARSENIC COMPOUNDS ARE SKIN AND LUNG CARCINOGENS IN HUMANS.

REPORT 05 : IARC MONOGRAPH 23 1980
THE EVIDENCE THAT ARSENIC COMPOUNDS CAUSE MUTATIONS AND ALLIED EFFECTS IN BACTERIA IS INCONCLUSIVE. HOWEVER, ARSENIC COMPOUNDS INDUCE CHROMOSOMAL ABERRATIONS AND MORPHOLOGICAL TRANSFORMATION IN MAMMALIAN CELLS.

REPORT 06 : MULTI-AGENCY CROSS CHECK
Y HHS MAGUIRE
R OSHA CARC
Y EPA CAG
N EPA RCRA
Y EPA CLEAN AIR
Y EPA CLEAN WATER

REPORT 07 : NAS REPORT ON DIET NUTRITION AND CANCER 1982
THERE IS GOOD EVIDENCE THAT DRINKING WATER HEAVILY CONTAMINATED WITH ARSENIC INCREASES THE RISK OF SKIN CANCER IN HUMANS. HOWEVER, THE REPORTED EPA STUDIES DO NOT PROVIDE SUFFICIENT INFORMATION TO DETERMINE THE EFFECTS OF THE NORMALLY LOW LEVELS OF DIETARY ARSENIC ON CANCER RISK.

REPORT 09 : IARC SUPPLEMENT 4 OCT 82
THERE IS SUFFICIENT EVIDENCE FOR CARCINOGENICITY TO

HUMANS, BUT INADEQUATE EVIDENCE FOR CARCINOGENICITY IN ANIMALS AND LIMITED EVIDENCE FOR ACTIVITY IN SHORT-TERM TESTS.

REPORT 10 : EPA TSCA 8(E) 0688-0735 JUN 88
PENNZOIL DETERMINED NATURAL GAS SAMPLES FROM TWO PRODUCTION AREAS IN WEST VIRGINIA CONTAINED TRACE AMOUNTS OF ARSENIC, BUT BELIEVES NO SUBSTANTIAL RISK OF INJURY TO HEALTH OF THE ENVIRONMENT IS PRESENT. ALSO ARSENIC FOUND IN PRODUCED WATER FROM CERTAIN CRUDE OIL PRODUCTION WAS DETECTED IN ORGANIC METHYLATED FORM, MOSTLY IN TRIMETHYLATED SPECIES. CONCENTRATIONS RANGED FROM LESS THAN 0.3 MILLIGRAMS PER LITER (MG/L) IN WATER TAKEN FROM WELLHEADS TO 388 MG/1 IN SOME OIL STORAGE TANKS. THE WATER FROM STORAGE TANKS IS BEING DISPOSED OF AT PERMITTED DISPOSAL FACILITY.

REPT'D USE : VETERINARY MEDICINE, ALLOY ADDITIVE, ELECTRONICS, PESTICIDES AND PIGMENTS

CAS NO : **1303-28-2**

CHEM NAME : ***** ARSENIC PENTOXIDE**

REPORT 01 : IARC MONOGRAPH 23 1980
THERE IS INADEQUATE EVIDENCE FOR THE CARCINOGENICITY OF ARSENIC COMPOUNDS IN ANIMALS. THERE IS SUFFICIENT EVIDENCE THAT INORGANIC ARSENIC COMPOUNDS ARE SKIN AND LUNG CARCINOGENS IN HUMANS

REPORT 02 : NIOSH CRITERIA DOCUMENT JUN 75
NIOSH RECOMMENDS A CEILING OF 2 MICROGRAMS/CU M FOR INORGANIC ARSENIC COMPOUNDS BASED ON DERMATITIS, LUNG, AND LYMPHATIC CANCER. THE CURRENT OSHA STANDARD FOR INORGANIC ARSENIC IS 10 MICROGRAMS/CU M TWA. SEE CFR 1910.1018.

REPORT 03 : IARC MONOGRAPH 23 1980
THE EVIDENCE THAT ARSENIC COMPOUNDS CAUSE MUTATIONS AND ALLIED EFFECTS IN BACTERIA IS INCONCLUSIVE. HOWEVER, ARSENIC COMPOUNDS INDUCE CHROMOSOMAL ABERRATIONS AND MORPHOLOGICAL TRANSFORMATION IN MAMMALIAN CELLS.

REPORT 06 : MULTI-AGENCY CROSS CHECK
Y HHS MAGUIRE
N OSHA CARC
Y EPA CAG
N EPA RCRA
N EPA CLEAN AIR
N EPA CLEAN WATER

REPT'D USE : CHEMICAL INTERMEDIATE, DEFOLIANT, WOOD PRESERVATIVES, MFR. OF LINOLEUM, ELECTRICAL SEMICONDUCTORS, PYROTECHNICS AND LEATHER HIDE DEPILATORY

CAS NO : **1303-33-9**
CHEM NAME : ***** ARSENIC SULFIDE**

REPORT 01 : IARC MONOGRAPH 23 1980
THERE IS INADEQUATE EVIDENCE FOR THE CARCINOGENICITY OF ARSENIC COMPOUNDS IN ANIMALS. THERE IS SUFFICIENT EVIDENCE THAT INORGANIC ARSENIC COMPOUNDS ARE SKIN AND LUNG CARCINOGENS IN HUMANS. THE EVIDENCE THAT ARSENIC COMPOUNDS CAUSE MUTATIONS AND ALLIED EFFECTS IS INCONCLUSIVE. HOWEVER, ARSENIC COMPOUNDS INDUCE ABERRATIONS AND MORPHOLOGICAL TRANSFORMATION IN MAMMALIAN CELLS.

REPORT 02 : NIOSH CRITERIA DOCUMENT JUN 75
NIOSH RECOMMENDS A CEILING OF 2 MICROGRAMS/CU M FOR INORGANIC ARSENIC COMPOUNDS BASED ON DERMATITIS, LUNG, AND LYMPHATIC CANCER. THE CURRENT OSHA STANDARD FOR INORGANIC ARSENIC IS 10 MICROGRAMS/CU M TWA. SEE CFR 1910.1018.

REPORT 06 : MULTI-AGENCY CROSS CHECK
Y HHS MAGUIRE
N OSHA CARC
Y EPA CAG
N EPA RCRA
N EPA CLEAN AIR
N EPA CLEAN WATER
REPT'D USE : GLASS MAKING PIGMENT, LEATHER TANNING, PYROTECHNICS AND SEMICONDUCTORS

CAS NO : **1327-53-3**
CHEM NAME : ***** ARSENIC TRIOXIDE**

REPORT 01 : NIOSH CRITERIA DOCUMENT JUN 75
NIOSH RECOMMENDS A CEILING OF 2 MICROGRAMS/CU M FOR INORGANIC ARSENIC COMPOUNDS BASED ON DERMATITIS, LUNG, AND LYMPHATIC CANCER. THE CURRENT OSHA STANDARD FOR INORGANIC ARSENIC IS 10 MICROGRAMS/CU M TWA. SEE CFR 1910.1018.

REPORT 02 : IARC MONOGRAPH 23 1980
THERE IS INADEQUATE EVIDENCE FOR THE CARCINOGENICITY OF ARSENIC COMPOUNDS IN ANIMALS. THERE IS SUFFICENT EVIDENCE THAT INORGANIC ARSENIC COMPOUNDS ARE SKIN AND LUNG CARCINOGENS IN HUMANS. THE EVIDENCE THAT ARSENIC

COMPOUNDS CAUSE MUTATIONS AND ALLIED EFFECTS IN BACTERIA IS INCONCLUSIVE. HOWEVER, ARSENIC COMPOUNDS INDUCE CHROMOSOMAL ABERRATIONS AND MORPHOLOGICAL TRANSFORMATION IN MAMMALIAN CELLS.

REPORT 06 : MULTI-AGENCY CROSS CHECK
Y HHS MAGUIRE
N OSHA CARC
Y EPA CAG
N EPA RCRA
N EPA CLEAN AIR
N EPA CLEAN WATER
REPT'D USE : AGRICULTURAL PESTICIDE, DYE ACCELERATOR, DRUG CHEMICAL INTERMEDIATE AND WOOD PRESERVATIVES

CAS NO : **7784-42-1**
CHEM NAME : ***** ARSINE**

REPORT 01 : IARC MONOGRAPH 23 1980
THERE IS INADEQUATE EVIDENCE FOR THE CARCINOGENICITY OF ARSENIC COMPOUNDS IN ANIMALS. THERE IS SUFFICIENT EVIDENCE THAT INORGANIC ARSENIC COMPOUNDS ARE SKIN AND LUNG CARCINOGENS IN HUMANS. THE EVIDENCE THAT ARSENIC COMPOUNDS CAUSE MUTATIONS AND ALLIED EFFECTS IN BACTERIA IS INCONCLUSIVE. HOWEVER, ARSENIC COMPOUNDS INDUCE CHROMOSOMAL ABERRATIONS AND MORPHOLOGICAL TRANSFORMATION IN MAMMALIAN CELLS.

REPORT 02 : NIOSH CRITERIA DOCUMENT JUN 75
NOISH RECOMMENDS A CEILING OF 2 MICROGRAMS/CU M FOR INORGANIC ARSENIC COMPOUNDS BASED ON DERMATITIS, LUNG, AND LYMPHATIC CANCER. THE CURRENT OSHA STANDARD FOR INORGANIC ARSENIC IS 10 MICROGRAMS/JCU M TWA. SEE CFR 1910.1018.

REPORT 06 : MULTI-AGENCY CROSS CHECK
Y HHS MAGUIRE
N OSHA CARC
Y EPA CAG
N EPA RCRA
N EPA CLEAN AIR
N EPA CLEAN WATER
REPT'D USE : MFR. OF SEMICONDUCTOR COMPONENTS AND MILITARY WAR GAS

CAS NO : **CLASS-01**
CHEM NAME : ***** ARYL PHOSPHATES**

REPORT 01 : 48FR57452 DEC 83
EPA IS PUBLISHING THIS ANPR TO REQUIRE THE TESTING NECESSARY TO COMPLETE A PRELIMINARY ASSESSMENT OF HEALTH AND ENVIRONMENTAL EFFECTS OF ARYL PHOSPHATES. THIS IS IN REPLY TO ITC'S DESIGNATION OF ARYL PHOSPHATES FOR TESTING CONSIDERATION.
REPT'D USE : FLAME RETARDANT, PLASTICIZERS IN PRODUCTION OF POLYVINYL CHLORIDE AND FIRE RESISTANT HYRDAULIC BRAKE FLUIDS

CAS NO : **1332-21-4**
CHEM NAME : ***** ASBESTOS**

REPORT 01 : EPA AMBIENT WATER QUALITY CRITERIA OCT 80
THE DEMONSTRATED ABILITY OF ASBESTOS TO INDUCE MALIGNANT TUMORS IN DIFFERENT ANIMAL TISSUES, THE PASSAGE OF INGESTED FIBERS THROUGH THE GASTROINTESTINAL MUCOSA, AND THE EXTENSIVE HUMAN EVIDENCE FOR EXCESS PERITONEAL, GASTROINTESTINAL, AND OTHER EXTRAPULMONARY CANCER AS A RESULT OF ASBESTOS EXPOSURE SUGGESTS THAT ASBESTOS IS LIKELY TO BE A HUMAN CARCINOGEN WHEN INGESTED. WATER CONCENTRATIONS OF 300,000; 30,000 AND 3,000 FIBERS/LITER ARE ESTIMATED TO RESULT IN A RESPECTIVE 10 -5, 10 -6, AND 10 -7 INCREASE IN CANCER RISK OVER A LIFETIME.

REPORT 02 : NIOSH CURRENT INTELLIGENCE BULLETIN 5 AUG 75
REPORT OF POTENTIAL CARCINOGENIC EXPOSURE TO ASBESTOS DURING THE SERVICING OF MOTOR VEHICLE BRAKE AND CLUTCH ASSEMBLIES.

REPORT 03 : IARC MONOGRAPH 14 1977
ALL COMMERCIAL FORMS OF ASBESTOS TESTED ARE CARCINOGENIC IN MICE, RATS, HAMSTERS AND RABBITS. IN MICE, MESOTHELIOMAS WERE INDUCED FOLLOWING INTRAPERITONEAL INJECTION OF FIBERS. IN RATS, VARIOUS TYPES OF ASBESTOS PRODUCED LUNG CARCINOMAS AND MESOTHELIOMAS FOLLOWING INHALATION AND MESOTHELIOMAS AND SARCOMAS FOLLOWING INTRAPLEURAL OR INTRAPERITONEAL INJECTION. IN HUMANS, LUNG CANCER HAS BEEN ASSOCIATED WITH OCCUPATIONAL EXPOSURE TO ASBESTOS. ADDITIONALLY, MESOTHELIOMAS HAVE BEEN OBSERVED AMONG ASBESTOS WORKERS AND THEIR FAMILIES.

REPORT 04 : EPA RCRA HEALTH AND ENV. EFFECTS PROFILE OCT 80
STRONGEST EVIDENCE FOR THE CARCINOGENICITY OF INGESTED ASBESTOS IS PROVIDED BY EPIDEMIOLOGY OF POPULATIONS OCCUPATIONALLY EXPOSED TO HIGH CONCENTRATIONS OF AIRBORNE ASBESTOS DUST. INHALATION EXPOSURE TO ASBESTOS

DUST IS ACCOMPANIED BY INGESTION BECAUSE A HIGH PERCENTAGE OF INHALED FIBERS ARE REMOVED FROM THE LUNG BY MUCOCILIARY ACTION AND SUBSEQUENTLY SWALLOWED. PERITONEAL MESOTHELIOMA, OFTEN IN GREAT EXCESS AND MODEST EXCESSES OF STOMACH, ESOPHAGUS, COLON-RECTAL, AND KIDNEY CANCER HAVE BEEN LINKED TO OCCUPATIONAL EXPOSURE TO ASBESTOS.

REPORT 05 : NIOSH CRITERIA DOCUMENT 1976
NIOSH OCCUPATIONAL EXPOSURE STANDARD TO AIRBORNE ASBESTOS IN 1972 WAS A MAXIMUM OF 2 FIBERS/CC OF AIR FOR FIBERS GREATER THAN 5 MICROMETERS IN LENGTH DETERMINED AS A TWA EXPOSURE FOR AN 8-HOUR WORK DAY. ASBESTOS FIBER PEAK CONCENTRATION WAS NOT TO EXCEED 10 FIBERS/CC IN A 15 MINUTE SAMPLE. NIOSH PROPOSE A NEW EXPOSURE LIMIT IN 1976 OF 0.1 ASBESTOS FIBERS/CC DETERMINED AS A TWA EXPOSURE FOR AN 8-HOUR WORK DAY. PEAK CONCENTRATION SHOULD NOT EXCEED 0.5 FIBERS/CC (15 MIN.) PRESENT OSHA STANDARD IS 2 FIBERS/CC TWA AND 10 FIBERS/CC CEILING.

REPORT 06 : MULTI-AGENCY CROSS CHECK
Y HHS MAGUIRE
R OSHA CARC
Y EPA CAG
Y EPA RCRA
N EPA CLEAN AIR
Y EPA CLEAN WATER

REPORT 07 : NTP TR 246 NIH PUB. 90-2502 JUL 90
SHORT RANGE NOR INTERMEDIATE RANGE CHRYSOTILE ASBESTOS CARCINOGENIC INGESTED AT 1% LEVELS IN DIET BY MALE AND FEMALE SYRIAN GOLDEN HAMSTERS. INCREASES IN RATES OF ADRENAL CORTICAL ADENOMAS EXPOSED TO CHRYSOTILE NOT DIFFERENT WHEN COMPARED WITH CONTROL GROUPS. STUDIES USING IR CHRYSOTILE AND 1,2-DIMETHYLHYDRAZINE DIHYDROCHLORIDE CONSIDERED INADEQUATE BECAUSE NO INCREASE IN INTESTINAL NEOPLASIA IN DMH GROUP.

REPORT 09 : IARC MONOGRAPH 14 1977
CHRYSOTILE AND CROCIDOLITE ASBESTOS DUSTS INDUCED CHROMOSOMAL ABERRATIONS IN CULTURED CHINESE HAMSTER CELLS.

REPORT 10 : NTP TR 249 NIH PUBLICATION 84-2505 NOV 83
UNDER THE CONDITIONS OF THESE STUDIES, THE INGESTION OF AMOSITE ASBESTOS AT A LEVEL OF 1% IN THE DIET FOR THEIR LIFETIME WAS NOT TOXIC AND DID NOT CAUSE A CARCINOGENIC RESPONSE IN MALE AND FEMALE SYRIAN GOLDEN HAMSTERS.

REPORT 11 : NTP BOARD OF SCIENTIFIC COUNSELORS JUL 84
THERE WAS NO EVIDENCE OF CARCINOGENICITY IN MALE OR

FEMALE RATS EXPOSED TO SHORT RANGE CHRYSOTILE OR IN FEMALE RATS EXPOSED TO INTERMEDIATE RANGE (IR) CHRYSOTILE ASBESTOS (CAS# 12001-29-5). SOME EVIDENCE OF CARCINOGENCITY IN MALE RATS EXPOSED TO IR CHRYSOTILE ASBESTOS AS INDICATED BY AN INCREASED INDICENCE OF ADENOMATUS POLYPS IN THE LARGE INTESTINE.

REPORT 12 : NTP TR 277 NIH PUB. NO. 90-2531 MAR 90
UNDER CONDITIONS OF THIS BIOASSAY, NONFIBROUS TREMOLITE NOT TOXIC AND DID NOT CAUSE CARCINOGENIC RESPONSE WHEN INGESTED AT LEVEL OF 1% IN DIET BY MALE AND FEMALE FISCHER 344 RATS FOR THEIR LIFETIME.

REPORT 14 : NTP TR 280 NIH 89-2536 DEC 88
UNDER THE CONDITIONS OF THESE FEED STUDIES, CROCIDOLITE ASBESTOS NOT OVERTLY TOXIC AND DID NOT CAUSE CARCINOGENIC RESPONSE WHEN INGESTED AT CONCENTRATION OF 1% IN DIET BY MALE AND FEMALE F344/N RATS FOR THEIR LIFETIME.

REPT'D USE : CEMENT, FLAME RETARDANT, TEXTILE, BRAKE LININGS, BUILDING MATERIALS AND HEAT RESISTANT INSULATOR

CAS NO : **12174-11-7**
CHEM NAME : ***** ATTAPULGITE**

REPORT 01 : IARC MONOGRAPH 42 1987
LIMITED EVIDENCE FOR CARCINOGENICITY TO EXPERIMENTAL ANIMALS AND INADEQUATE EVIDENCE FOR CARCINOGENICITY TO HUMANS. SAMPLES FROM DIFFERENT DEPOSITS TESTED FOR CARCINOGENICITY BY INTRAPERITONEAL INJECTION IN RATS. SAMPLE WITH 30% FIBERS LONGER THAN 5 UM PRODUCED MESOTHELIOMAS AND SARCOMAS IN ABDOMINAL CAVITY; 1 SAMPLE WITH LESS THAN 20% OF FIBERS LONGER THAN 4 UM INDUCED MESOTHELIAL TUMORS; SAMPLES WITH < 1% OF FIBERS LONGER THAN 4 UM DID NOT CAUSE A SIGNIFICANT INCREASE IN INCIDENCE OF TUMORS.

REPORT 02 : IARC MONOGRAPH 42 1987
ATTAPULGITE DID NOT INDUCE UNSCHEDULED DNA SYNTHESIS IN PRIMARY CULTURES OF RAT HEPATOCYTES OR SISTER CHROMATID EXCHANGES IN RAT MESOTHELIAL CELLS.

REPT'D USE : PET LITTER AND SPILL CLEANUP ABSORBENT, PHARMACEUTICALS, COSMETICS, PESTICIDES, PAINTS, ADHESIVES, SEALANTS, CATALYSTS, MUD DRILLING, GELLANTS AND FERTILIZER

CAS NO : **2465-27-2**
CHEM NAME : ***** AURAMINE HYDROCHLORIDE**

REPORT 01 : EPA TSCA 8(E)-0730
IN THE...(AURAMINE HYDROCHLORIDE) STUDY, RATS WERE ADMINISTERED 0, 300, 500, 1000 OR 2000 PPM OF THE COMPOUND IN THE FEED FOR NINE MONTHS. (THE) MEAN BODY WEIGHTS AND FOOD INTAKE WERE LOWER AT ALL CONCENTRATION AND A DOSE-RELATED INCREASED ACTIVITY OF GAMMAGLUTAMYL TRANSPEPTIDASE WAS NOTED IN LIVER HOMOGENATE. HISTOPATHOLOGY INDICATE INCREASED INCIDENCES OF FOCI AND ALTERED HEPATOCYTES AMONG TREATED ANIMALS. AFTER NINE MONTHS, HYPERPLASTIC NODULES WERE SEEN IN MALES AT 1000 AND 2000 PPM AND HEPATOCELLULAR CARCINOMAS WERE OBSERVED IN THE HIGH DOSE 2000 PPM.
REPT'D USE : YELLOW DYE FOR PAPER, TEXTILES, LEATHER, ALSO AN ANTISEPTIC AND FUNGICIDE

CAS NO : **12192-57-3**
CHEM NAME : ***** AUROTHIOGLUCOSE**

REPORT 01 : IARC MONOGRAPH 13 1977
AUROTHIOGLUCOSE IS CARCINOGENIC IN MICE AFTER ITS ADMINISTRATION BY SINGLE INTRAPERITONEAL INJECTION: IT PRODUCED AN INCREASED INCIDENCE OF HEPATOMAS IN MALE MICE.
REPT'D USE : DRUG FOR TREATMENT OF RHEUMATOID ARTHRITIS

CAS NO : **320-67-2**
CHEM NAME : ***** 5-AZACYTIDINE**

REPORT 01 : NCI CARC TRS 42 DHEW PUB NO 78-842
UNDER THE CONDITIONS OF THIS BIOASSAY, THE SHORT LIFE SPAN AND SHORT DURATION OF TREATMENT OF SPRAGUE-DAWLEY RATS OF EITHER SEX AND OF MALE B6C3F1 MICE PRECLUDED EVALUATION OF THE CARCINOGENICITY OF 5-AZACYTIDINE IN THESE GROUPS; HOWEVER, THE INDUCTION OF TUMORS OF THE HEMATOPOIETIC SYSTEM IN FEMALE B6C3F1 MICE WAS ASSOCIATED WITH THE ADMINISTRATION, VIA ABDOMINAL INJECTION, OF 5-AZACYTIDINE.
REPT'D USE : EXPERIMENTAL ANTINEOPLASTIC AGENT

CAS NO : **115-02-6**
CHEM NAME : ***** AZASERINE**

REPORT 01 : IARC MONOGRAPH 10 1976
CPD IS CARCINOGENIC FOLLOWING ITS INTRAPERITONEAL INJEC-

TION IN RATS, THE ONLY SPECIES AND ROUTE TESTED, PRODUCING ADENOCARCINOMAS OF THE PANCREAS AND TUMORS OF THE KIDNEY. IN A 1979 SUPPLEMENT, IARC DETERMINED THAT THERE IS SUFFICIENT EVIDENCE FOR THE CARCINOGENICITY OF AZASERINE IN ANIMALS.

REPORT 02 : IARC MONOGRAPH 10 1976
AZASERINE WAS MUTAGENIC IN E. COLI AND SEVERAL SALMONELLA STRAINS AND HAS BEEN SHOWN TO INDUCE RECESSIVE LETHAL MUTATIONS IN DROSOPHILIA.

REPORT 06 : MULTI-AGENCY CROSS CHECK
N HHS MAGUIRE
N OSHA CARC
Y EPA CAG
Y EPA RCRA
N EPA CLEAN AIR
N EPA CLEAN WATER
REPT'D USE : EXPERIMENTAL ANTIBIOTIC/CHEMOTHERAPEUTIC AGENT

CAS NO : **446-86-6**
CHEM NAME : ***** AZATHIOPRINE**

REPORT 01 : NCI BIOASSAY PROGRAM- CANCER VOL. 40 1977
UNDER THE CONDITIONS OF THE BIOASSAY, I.P. INJECTION OF AZATHIOPRINE WAS CARCINOGENIC TO MALE AND FEMALE SWISS-WEBSTER MICE, CAUSING LYMPHOSARCOMAS AND LUNG TUMORS. IT WAS NOT CARCINOGENIC TO MALE AND FEMALE SPRAGUE-DAWLEY (CD) RATS.

REPORT 02 : IARC MONOGRAPH 26 1981
THERE IS EVIDENCE THAT AZATHIOPRINE, OFTEN COMBINED WITH PREDNISONE, IS ASSOCIATED WITH AN INCREASED INCIDENCE OF NON-HODGKINS LYMPHOMA, SQUAMOUS CELL CANCERS OF THE SKIN, HEPATO-BILIARY CARCINOMAS, MESENCHYMAL TUMORS, AND PERHAPS CERTAIN OTHER RARE NEOPLASMS. THERE IS SUFFICIENT EVIDENCE THAT AZATHIOPRINE IS CARCINOGENIC IN HUMANS.

REPORT 03 : IARC MONOGRAPH 26 1981
AZATHIOPRINE IS EMBRYOLETHAL AT DOSES NONTOXIC TO THE MOTHER AND CAN INDUCE A VARIETY OF SEVERE TERATOGENIC EFFECTS IN SEVERAL ANIMAL SPECIES.

REPORT 04 : IARC SUPPLEMENT 4 OCT 82
SUFFICIENT EVIDENCE FOR CARCINOGENICITY TO HUMANS, LIMITED EVIDENCE FOR CARCINOGENICITY TO ANIMALS AND SUFFICIENT EVIDENCE FOR ACTIVITY IN SHORT-TERM TESTS.

REPORT 06 : MULTI-AGENCY CROSS CHECK
Y HHS MAGUIRE
N OSHA CARC
N EPA CAG
N EPA RCRA
N EPA CLEANAIR
N EPA CLEAN WATER
REPT'D USE : IMMUNOSUPPRESSIVE AGENT

CAS NO : **86-50-0**
CHEM NAME : ***** AZINPHOS-METHYL**

REPORT 01 : NCI CARC TRS 69 DHEW PUB NO 78-1319
UNDER THE CONDITIONS OF THIS BIOASSAY NEOPLASMS OF THE THYROID AND PANCREATIC ISLETS SUGGEST BUT DO NOT PROVIDE SUFFICIENT EVIDENCE FOR THE CARCINOGENICITY OF AZINOPHOS-METHYL IN MALE OSBORNE-MENDEL RATS. CPD WAS NOT SHOWN TO BE CARCINOGENIC IN FEMALE OSBORNE-MENDEL RATS. AZINPHOS-METHYL WAS SHOWN TO BE NON-CARCINOGENIC IN FEMALE AND MALE OSBORNE-MENDEL RATS OR B6C3F1 MICE OF EITHER SEX IN A FEEDING STUDY.

REPORT 02 : CHEMICAL DOSSIER STAGE I
REPT'D USE : INSECTICIDE ACARICIDE

CAS NO : **151-56-4**
CHEM NAME : ***** AZIRIDINE**

REPORT 01 : IARC MONOGRAPH 9 1975
AZIRIDINE IS CARCINOGENIC IN TWO STRAINS OF MICE FOLLOWING ITS ORAL ADMINISTRATION, PRODUCING AN INCREASED INCIDENCE OF LIVER CELL AND PULMONARY TUMORS. SUBCUTANEOUS INJECTION OF SINGLE DOSES IN SUCKLING MICE PRODUCED AN INCREASED INCIDENCE OF LUNG TUMORS IN MALES. IN ONE EXPERIMENT IN RATS, SUBCUTANEOUS INJECTION IN OIL INCREASED THE INCIDENCE OF TUMORS AT THE INJECTION SITE.

REPORT 02 : IARC MONOGRAHH 9 1975
IN NEUROSPORA, AZIRIDINE PRODUCED LEAKY MUTANTS AND NON-COMPLEMENTING AND MULTILOCUS DELETIONS. IN SACCHAROMYCES, IT INDUCED MITOTIC RE-RECOMBINATION AND GENE CONVERSION. IN DROSOPHILA IT INDUCED BOTH TRANSMISSABLE TRANSLOCATIONS AND SEX-LINKED RECESSIVE LETHALS. IN SILK WORMS IT CAUSED SPECIFIC LOCUS MUTATIONS.

REPORT 06 : MULTI-AGENCY CROSS CHECK
N HHS MAGUIRE
R OSHA CARC

Y EPA CAG
N EPA RCRA
N EPA CLEAN AIR
N EPA CLEAN WATER
REPT'D USE : CHEMICAL INTERMEDIATE

CAS NO : **1072-52-2**
CHEM NAME : ***** 2-(1-AZIRIDINYL)-ETHANOL**

REPORT 01 : IARC MONOGRAPH 9 1975
2-(1-AZIRIDINYL)-ETHANOL IS CARCINOGENIC IN MICE, PRODUCING MALIGNANT TUMORS AT THE SITE OF ITS SUBCUTANEOUS INJECTION IN THE ONLY STUDY AVAILABLE.

REPORT 02 : NTP TECHNICAL BULLETIN 3 DEC 80
THE COMPOUND WAS FOUND TO BE POSITIVE IN THE SALMONELLA TYPHIMURIUM MUTAGENICITY ASSAY.

REPORT 03 : IARC MONOGRAPH 9 1975
2-(1-AZIRIDINYL) ETHANOL INDUCED SEX-LINKED RECESSIVE LETHALS IN DROSOPHILA.
REPT'D USE : CHEMICAL INTERMEDIATE POLYMER MODIFIER

CAS NO : **800-24-8**
CHEM NAME : ***** AZIRIDYL BENZOQUINONE**

REPORT 01 : IARC MONOGRAPH 9 1975
AZIRIDYL BENZOQUINONE IS CARCINOGENIC IN MICE FOLLOWING ITS INTRAPERITONEAL INJECTION, PRODUCING AN INCREASE IN THE INCIDENCE OF LUNG TUMORS.

REPORT 02 : IARC MONOGRAPH 9 1975
AZIRIDINYL BENZOQUINONE INDUCED A HIGH FREQUENCY OF CHROMOSOME ABERRATIONS IN FRESHLY-PREPARED CULTURES OF HUMAN LEUCOCYTES.
REPT'D USE : EXPERIMENTAL ANTINEOPLASTIC AGENT

CAS NO : **103-33-3**
CHEM NAME : ***** AZOBENZENE**

REPORT 01 : NCI CARC TRS 154 DHEW PUB NO 79-1710
UNDER THE CONDITIONS OF THIS BIOASSAY AZOBENZENE WAS CARCINOGENIC (SARCOMAGENIC) FOR F344 RATS, INDUCING VARIOUS TYPES OF SARCOMAS IN THE SPLEEN AND OTHER ABDOMINAL ORGANS OF BOTH MALES AND FEMALES. THE TEST CHEMICAL WAS NOT CARCINOGENIC FOR B6C3F1 MICE OF EITHER SEX (FEEDING STUDY).

REPORT 02 : IARC MONOGRAPH 8 1975
AZOBENZENE HAS BEEN REPORTED TO BE NON-MUTAGENIC WHEN TESTED IN DROSOPHILA FOR THE PRODUCTION OF X-LINKED RECESSIVE LETHALS.

REPORT 03 : IARC MONOGRAPH 8 1975
IN AN ORAL STUDY IN MICE, AZOBENZENE PRODUCED AN EXCESS OF LIVER CELL TUMORS OVER THE CONTROLS IN MALES BUT NOT IN FEMALES IN ONE OF THE TWO STRAINS USED.

REPORT 04 : NTP TECHNICAL BULLETIN 3/8 DEC 80
AZOBENZENE WAS POSITIVE IN THE SALMONELLA ASSAY, BUT NEGATIVE IN DROSOPHILA WHEN TESTED FOR SEX-LINKED RECESSIVE LETHALS.

REPORT 06 : MULTI-AGENCY CROSS CHECK
N HHS MAGUIRE
Y OSHA CARC
N EPA CAG
N EPA RCRA
N EPA CLEAN AIR
N EPA CLEAN WATER

REPT'D USE : IN SITU CONTAMINANT IN SYNTHESIS OF BENZIDINE, ACARICIDE, AND FUMIGANT

CAS NO : **123-77-3**
CHEM NAME : ***** AZODICARBONAMIDE**

REPORT 01 : NTP TECHNICAL BULLETIN NO. 3 DEC 80
AZODICARBONAMIDE WAS POSITIVE IN THE SALMONELLA TYPHIMURIUM MUTAGENICITY ASSAY IN ONE LABORATORY.

REPT'D USE : BLOWING AGENT FOR RUBBER/POLYMERIC FOAMS AND FOOD ADDITIVE

CAS NO : **225-11-6**
CHEM NAME : ***** BENZ(A)ACRIDINE**

REPORT 01 : IARC MONOGRAPH 32 DEC 83
DATA ARE INADEQUATE TO PERMIT EVALUATION OF CARCINOGENICITY OF BENZ(A)ACRIDINE IN EXPERIMENTAL ANIMALS. INADEQUATELY TESTED FOR CARCINOGENICITY IN ONE EXPERIMENT BY SKIN APPLICATION TO MICE. SINGLE REPORT OF MUTAGENICITY IN SALMONELLA TYPHIMURIUM INCONCLUSIVE.

REPT'D USE : COMBUSTION BY-PRODUCT

CAS NO : **225-51-4**
CHEM NAME : ***** BENZ(C)ACRIDINE**

REPORT 01 : IARC MONOGRAPH 3 1973
THIS COMPOUND PRODUCED SKIN TUMORS IN MICE TREATED TOPICALLY AND BLADDER TUMORS IN RATS FOLLOWING LOCAL PARAFFIN WAX PELLET IMPLANTATION.

REPORT 02 : IARC MONOGRAPH 32 DEC 83
PRODUCED LOW INCIDENCE OF SKIN TUMORS WHEN TESTED FOR CARCINOGENICITY BY SKIN APPLICATION IN ONE EXPERIMENT IN MICE. WAS INACTIVE AS AN INITIATOR IN MOUSE SKIN INITIATION-PROMOTION ASSAY. WHEN TESTED IN RATS BY BLADDER IMPLANTATION IN WAX PELLETS, AN INCREASED INCIDENCE OF BLADDER TUMORS WAS OBSERVED.

REPORT 03 : IARC MONOGRAPH 32 DEC 83
MUTAGENIC TO SALMONELLA TYPHIMURIUM IN THE PRESENCE OF AN EXOGENOUS METABOLIC SYSTEM.

REPORT 06 : MULTI-AGENCY CROSS CHECK
N HHS MAGUIRE
N OSHA CARC
Y EPA CAG
Y EPA RCRA
N EPA CLEAN AIR
N EPA CLEAN WATER
RFPT'D USE : COMBUSTION POLLUTANT

CAS NO : **98-87-3**
CHEM NAME : ***** BENZAL CHLORIDE**

REPORT 01 : EPA RCRA HEALTH AND ENV. EFFECTS PROFILE OCT 82
BENZAL CHLORIDE HAS BEEN REPORTED TO INDUCE PAPILLOMAS, CARCINOMAS, AND LEUKEMIA IN MICE. DETAILS OF THIS WORK WERE NOT AVAILABLE FOR ASSESSMENT. MUTAGENIC EFFECTS HAVE BEEN REPORTED.

REPORT 02 : IARC MONOGRAPH 29 1982
IN ONE STUDY, BENZAL CHLORIDE WAS MUTAGENIC TO BACTERIA WITH METABOLIC ACTIVATION AND WAS POSITIVE IN A REC-ASSAY WITH B. SUBTILIS. NO DATA WERE AVAILABLE TO ASSESS THE TERATOGENICITY OF THIS CHEMICAL.

REPORT 03 : IARC MONOGRAPH 29 1982
IN ONE EXPERIMENT IN WHICH BENZAL CHLORIDE TESTED BY SKIN APPLICATION IN FEMALE MICE, IT PRODUCED SQUAMOUS CELL CARCINOMAS OF THE SKIN. IN A CONCURRENT EXPERIEMNT, A LOW INCIDENCE OF PAPILLOMAS WAS OBSERVED THERE IS

LIMITED EVIDENCE THAT BC IS CARCINOGENIC IN ANIMALS.
REPT'D USE : DYE INTERMEDIATE, INTERMEDIATE IN SYNTHESIS OF BENZALDEHYDE, BENZOYL CHLORIDE

CAS NO : **100-52-7**
CHEM NAME : ***** BENZALDEHYDE**

REPORT 01 : NTP TR378 NTH PUB 90-2833 MAR 90
UNDER THE CONDITIONS OF THESE 2-YEAR GAVAGE STUDIES, THERE WAS NO EVIDENCE OF CARCINOGENIC ACTIVITY OF BENZALDEHYDE FOR MALE OR FEMALE F344/N RATS RECEIVING 200 OR 400 MG/KG PER DAY. THERE WAS SOME EVIDENCE OF CARCINOGENIC ACTIVITY OF BENZALDEHYDE FOR MALE OR FEMALE B6C3F1 MICE, AS INDICATED BY INCREASED INCIDENCES OF SQUAMOUS CELL PAPILLOMAS AND HYPERPLASIA OF THE FORESTOMACH.
REPT'D USE : FLAVORING/FRAGRANCE SYNTHESIS INTERMEDIATE, FLAVORING AGENT IN FOODS/BEVERAGES, SOLVENT AND PHARMACEUTICALS

CAS NO : **2716-12-3**
CHEM NAME : ***** BENZAMINE, 4,4- (1,4-PHENYLENE BIS (l-METHYLETHYLIDENE)) BIS (2,6-DIM**

REPORT 01 : EPA 8EHQ-1289- 0852 DEC 89
ORAL LD50 IN RATS > 5000 MG/KG DERMAL LD50 IN RATS > 2000 MG/KG TOXIC EFFECTS: REDUCED TESTES AND ENLARGED LIVER AND SPLEEN, MILD LIVER DAMAGE AND SEVERE TUBULAR DEGENERATION OF TESTES; NON-IRRITATING TO RABBIT SKIN, MILD IRRITATION TO RABBIT EYE AND NO SKIN SENSITIZING POTENTIAL IN GUINEA PIGS.
REPT'D USE : INTERMEDIATE FOR RESIN MANUFACTURE

CAS NO : **56-55-3**
CHEM NAME : ***** BENZ(A)ANTHRACENE (BA)**

REPORT 01 : IARC MONOGRAPH 3 DEC 72
BENZ(A)ANTHRACENE GIVEN BY SEVERAL ROUTES OF ADMINISTRATION HAS BEEN SHOWN TO BE CARCINOGENIC IN THE MOUSE. IT PRODUCED HEPATOMAS AND LUNG ADENOMAS FOLLOWING REPEATED ORAL ADMINISTRATION TO YOUNG MICE. THE COMPOUND ALSO PRODUCED BLADDER TUMORS IN MICE FOLLOWING IMPLANTATION. BENZ(A)ANTHRACENE IS ALSO AN INITIATOR OF SKIN CARCINOGENESIS IN MICE. IN A 1979 SUPPLEMENT, IARC DETERMINED THAT THERE IS SUFFICIENT EVIDENCE FOR THE CARCINOGENICITY OF BENZ(A)ANTHRACENE IN ANIMALS.

REPORT 02 : EPA RCRA HEALTH AND ENV. EFFECTS PROFILE OCT 80
ALTHOUGH THE PAH CLASS CONTAINS SEVERAL WELL-KNOWN POTENT CARCINOGENS BENZ(A)ANTHRACENE DISPLAYS ONLY WEAK CARCINOGENIC ACTIVITY.

REPORT 03 : EPA AMBIENT WATER QUALITY CRITERIA DEC 80
PRESENTLY AVAILABLE DATA CANNOT SUPPORT INDIVIDUAL CRITERIA FOR EACH POLYNUCLEAR AROMATIC HYDROCARBON. THE AWQC FOR CARCINOGENIC PAH COMPOUNDS IS BASED ON THE ASSUMPTION THAT EACH COMPOUND IS AS POTENT AS B(A)P AND THAT THE CARCINOGENIC EFFECT OF THE COMPOUNDS IS PROPORTIONAL TO THE SUM OF THEIR CONCENTRATIONS. THIS SUM SHOULD NOT EXCEED 28 NG/L, 2.8 NG/L, AND .28 NG/L TO KEEP THE LIFETIME CANCER RISK BELOW 10 -5, 10 -6, AND 10 -7 RESPECTIVELY.

REPORT 04 : NTP TECHNICAL BULLETIN 8 JUL 82
BENZ(A)ANTHRACENE WAS POSITIVE IN THE L5178Y MOUSE LYMPHOMA ASSAY.

REPORT 05 : NAS REPORT ON DIET NUTRITION AND CANCER 1982
ADMINISTRATION OF BA IN MINERAL OIL BY STOMACH TUBE AT INTERVALS OF 3 TO 7 DAYS INDUCED PAPILLOMAS OF THE FORESTOMACH IN MICE. BA HAS BEEN SHOWN TO BE MUTAGENIC IN THE SALMONELLA ASSAY. IT WAS ALSO GENOTOXIC IN THE HEPATOCYTE PRIMARY CULTURE-DNA REPAIR TEST.

REPORT 06 : MULTI-AGENCY CROSS CHECK
Y HHS MAGUIRE
N OSHA CARC
Y EPA CAG
Y EPA RCRA
N EPA CLEAN AIR
Y EPA CLEAN WATER

REPORT 08 : IARC MONOGRAPH 32 DEC 83
SHOWN TO BE CARCINOGENIC TO EXPERIMENTAL ANIMALS. SEE IARC MONOGRAPH 3 (1973).

REPORT 09 : IARC MONOGRAPH 32 DEC 83
MUTAGENIC TO S. TYPHIMURIUM IN PRESENCE OF AN EXOGENOUS METABOLIC-SYSTEM AND MUTAGENIC TO DROSOPHILA MELANOGASTER; ALSO MUTAGENIC TO MAMMALIAN CELLS IN-VITRO IN PRESENCE OF EXOGENOUS METABOLIC SYSTEM. POSITIVE IN ONE STUDY OF SISTER CHROMATID EXCHANGE. INDUCED UNSCHEDULED DNA SYNTHESIS IN CULTURED MAMMALIAN CELLS AND MORPHOLOGICAL TRANSFORMATION. INDUCED SISTER CHROMATID EXCHANGE IN HAMSTERS IN ONE STUDY. CONFLICTING REPORTS WERE RECEIVED ON INDUCTION OF CHROMOSOMAL ABERRATIONS IN IN-VIVO STUDIES.

REPT'D USE : COMBUSTION BY-PRODUCT AND ORGANIC SYNTHESIS. NO KNOWN COMMERCIAL USE

CAS NO : **71-43-2**
CHEM NAME : ***** BENZENE**

REPORT 01 : EPA AMBIENT WATER QUALITY CRITERIA OCT 80
BENZENE IS SUSPECTED OF BEING A HUMAN CARCINOGEN. AMBIENT WATER CONCENTRATIONS OF 6.6 UG/L, .66 UG/L; AND .066 UG/L ARE ESTIMATED TO RESULT IN A RESPECTIVE 10 -5, 10 -6, AND 10 -7 INCREASE IN CANCER RISK OVER A LIFETIME.

REPORT 02 : EPA RCRA HEALTH AND ENV. EFFECTS PROFILE OCT 80
CHRONIC EXPOSURE TO BENZENE CAUSES HEMATOLOGICAL ABNORMALITIES. BENZENE IS NOT MUTAGENIC TO BACTERIA, BUT RECENT EVIDENCE SHOWS IT TO BE CARCINOGENIC IN ANIMALS. ALSO, BENZENE HAS BEEN SHOWN TO BE LEUKEMOGENIC IN HUMANS. THERE IS SUGGESTIVE EVIDENCE THAT BENZENE MAY BE TERATOGENIC AND MAY CAUSE REDUCED FERTILITY.

REPORT 03 : NIOSH CRITERIA DOCUMENT JUL 77
BECAUSE IT IS NOT POSSIBLE AT PRESENT TO ESTABLISH A SAFE EXPOSURE LEVEL FOR A CARCINOGEN, THE NIOSH RECOMMENDATION IS TO RESTRICT EXPOSURE TO VERY LOW LEVELS THAT CAN STILL BE RELIABLY MEASURED IN THE WORKPLACE. THE NIOSH RECOMMENDATION CAN BE EXPECTED TO MATERIALLY REDUCE THE RISK OF BENZENE-INDUCED LEUKEMIA. EXPOSURE TO BENZENE CAN ALSO RESULT IN CENTRAL NERVOUS SYSTEM DEPRESSION AND SKIN IRRITATION. NIOSH RECOMMENDS A 1 PPM CEILING. THE PRESENT OSHA STANDARD IS 1 PPM TWA.

REPORT 04 : IARC SUPPLEMENT 1 1979
SEVERAL CASE REPORTS AND AN EPIDEMIOLOGIC CASE CONTROL STUDY SUGGEST A RELATIONSHIP BETWEEN BENZENE EXPOSURE AND LEUKEMIA. TWO COHORT STUDIES SHOWED AN INCREASED INCIDENCE OF ACUTE NONLYMPHOCYTIC LEUKEMIA IN WORKERS EXPOSED TO BENZENE. IN A 1979 SUPPLEMENT, THE IARC CONCLUDED THAT BENZENE IS CARCINOGENIC FOR HUMANS.

REPORT 05 : EPA REPORT ON LOW LEVEL EXPOSURE TO BENZENE SEP 78
BENZENE EXPOSURE BY INHALATION AND OTHER EXPOSURE ROUTES IS STRONGLY IMPLICATED IN THE FOLLOWING PATHOLOGICAL CONDITIONS: LEUKEMIA, ESPECIALLY ACUTE MYELOGENOUS LEUKEMIA; PANCYTOPENIA, INCLUDING APLASTIC ANEMIA; AND CHROMOSOMAL ABERRATIONS. MOST OF THE EXPOSURE STUDIES CONCERN MIDDLE-AGED TO ELDERLY MALES WHO WERE OCCUPATIONALLY EXPOSED TO BENZENE. EXPOSURE LEVELS INVOLVED DOSES IN THE RANGE OF 100-500 PPM. THERE IS NO CONVINCING

EVIDENCE THAT BENZENE CAUSES NEOPLASIAS, INCLUDING LEUKEMIA, IN ANIMALS (A COCARCINOGEN MAY BE REQUIRED).

REPORT 06 : MULTI-AGENCY CROSS CHECK
Y HHS MAGUIRE
R OSHA CARC
Y EPA CAG
Y EPA RCRA
N EPA CLEAN AIR
Y EPA CLEAN WATER

REPORT 07 : EPA TSCA 8(E) FYI-AX-1183- 0203 SUPP. NOV 83
THE AMERICAN PETROLEUM INSTITUTE REPORTS THAT IN A 90 DAYS INHALATION STUDY ON SERUM IMMUNOGLOBULIN LEVELS OF CD RATS AND CD-1 MICE EXPOSED TO BENZENE VAPOR, CONCENTRATIONS OF 10, 30, AND 300 PPM, THAT THE MICE SHOWED TRENDS TOWARDS EARLY SUPPRESSION OF BASAL SERUM LEVELS OF IMMUNOGLOBULINS FOLLOWED BY ELEVATIONS IN IGG. THE RATS SHOWED ONLY ONE CHANGE IN ONE SEX AT ONE TIME POINT WITH ONE IMMUNOGLOBULIN.

REPORT 08 : CMA DOMINANT LETHAL INHALATION STUDY DEC 80
BENZENE EXPOSURE AT DOSE LEVELS OF 1, 10 AND 30 PPM DID NOT EFFECT MALE REPRODUCTION PERFORMANCE OR HAVE A DEMONSTRABLE DOMINANT-LETHAL MUTAGENIC EFFECT. IN THE HIGH-DOSE GROUP (300 PPM) THE OBSERVED INCREASE IN MEAN NUMBER OF DEAD IMPLANTS AND MEAN MUTAGENIC RATIO WERE PRIMARILY RELATED TO THE PERFORMANCE OF A SINGLE MALE. IT IS UNCLEAR IF THESE EFFECTS WERE RELATED TO BENZENE EXPOSURE AS THE INCREASE IN GROUP VALUES FOR MEAN DEAD IMPLANTS AND MUTAGENIC RATIO WERE NOT STATISTICALLY SIGNIFICANT.

REPORT 09 : IARC MONOGRAPH 29 1982
ORAL ADMINISTRATION OF BENZENE TO RATS RESULTED IN A MEAN INCREASE IN THE INCIDENCE OF ZYMBAL GLAND CARCINOMAS. THERE IS LIMITED EVIDENCE THAT BENZENE IS CARCINOGENIC IN ANIMALS. THE RELATIONSHIP BETWEEN BENZENE EXPOSURE AND DEVELOPMENT OF ACUTE MYELOGENOUS LEUKEMIA IN HUMANS HAS BEEN ESTABLISHED IN EPA STUDIES. THERE IS SUFFICIENT EVIDENCE THAT BENZENE IS CARCINOGENIC TO HUMANS.

REPORT 10 : IARC MONOGRAPH 29 1982
BENZENE DOES NOT INDUCE SPECIFIC GENE MUTATIONS IN BACTERIAL SYSTEMS OR IN DROSOPHILA. A SINGLE REPORT SHOWED NO EVIDENCE FOR THE INDUCTION OF POINT MUTATIONS IN MAMMALIAN CELLS; HOWEVER, BENZENE INDUCED CHROMOSOMAL ABBERATIONS AND SISTER CHROMATID EXCHANGES IN MAMMALIAN CELLS IN-VIVO. EXPOSURE TO BENZENE MAY DAMAGE THE TESTES. EVIDENCE FROM MOST STUDIES IN ANIMALS

SUGGESTS THAT BENZENE IS NOT TERATOGENIC AT DOSES THAT ARE FETOTOXIC.

REPORT 11 : CMA BENZENE REPORT DEC 83
CHEM MANUFACTURERS ASSOC STUDY ON HEALTH EFFECTS OF OCCUPATIOAL EXPOSURE TO BENZENE FOUND SLIGHTLY HIGHER MORTALITY RATE FROM LYMPHATIC AND HEMATOPOIETIC CANCER AMONG EXPOSED WORKERS COMPARED TO U.S. POPULATION AND A SIGNIFICANTLY HIGHER RATE THAN IN-PLANT WORKER CONTROL GROUP. STUDY SUGGESTS POSITIVE RELATIONSHIP BETWEEN CUMULATIVE EXPOSURE AND DEATHS FROM LYMPHATIC AND HEMATOPOETIC CANCER, INCLUDING LEUKEMIA.

REPORT 12 : NTP DRAFT TECH REPORT 83-072 NIH PUB 84-2545 OCT 83
UNDER CONDITIONS OF THESE GAVAGE STUDIES, THERE WAS CLEAR EVIDENCE OF CARCINOGENICITY OF BENZENE FOR MALE F344/N RATS, FEMALE F344/N RATS, MALE B6C3F1 MICE, AND FEMALE B6C3F1 MICE.

REPORT 13 : PROPOSED RULEMAKING NAT'L DRINKING WATER STD 49FR24330 JUN 84
EPA PROPOSES TO SET A LEVEL OF ZERO FOR BENZENE IN DRINKING WATER, UNDER THE SAFE DRINKING WATER ACT.

REPORT 14 : FYI-OTS-0187-05 29 JAN 87
A STUDY OF THE LEUKEMIA RISK FROM EXPOSURE TO BENZENE INSIDE AND OUTSIDE A REFINERY FOUND NO SIGNIFICANT EXCESS OF LEUKEMIA EITHER IN BENZENE AREAS OF THE REFINERY OR INSIDE THE ENTIRE REFINERY. THIS CONCLUSION IS BASED ON A STUDY OF MORE THAN 15,000 REFINERY WORKERS OVER A 40-YEAR PERIOD.

REPORT 15 : FYI-AX-0187 -0185 JAN 87
FACE SKIN (CHEEK), SOLE OF FOOT AND DORSUM OF HAND MAY BE MORE PR

REPT'D USE : SOLVENT, CHEMICAL INTERMEDIATE AND GASOLINE ADDITIVE

CAS NO : **92-87-5**

CHEM NAME : *** **BENZIDINE**

REPORT 01 : EPA AMBIENT WATER QUALITY CRITERIA DOCUMENT OCT 80
BENZIDINE IS A PROVEN HUMAN CARCINOGEN. AMBIENT WATER CONCENTRATIONS OF 1.2 NG/L, .12 NG/L, AND .01 NG/L ARE ESTIMATED TO RESULT IN A RESPECTIVE 10 -5, 10 -6, AND 10 -7 INCREASE IN CANCER OVER A LIFETIME.

REPORT 02 : IARC MONOGRAPH 1 1972
BENZIDINE IS CARCINOGENIC IN THE MOUSE, RAT AND HAMSTER, AND POSSIBLY THE DOG. GIVEN ORALLY, IT HAS PRODUCED BLAD-

DER CARCINOMA IN THE DOG AFTER A LONG LATENT PERIOD AND LIVER TUMORS IN THE RAT AND HAMSTER. THE EPIDEMIOLOGICAL STUDIES SHOWED THAT OCCUPATIONAL EXPOSURE TO COMMERCIAL BENZIDINE ALONE WAS STRONGLY ASSOCIATED WITH BLADDER CANCER. IN A 1979 SUPPLEMENT, THE IARC CONCLUDED THAT BENZIDINE IS CARCINOGENIC FOR HUMANS.

REPORT 03 : EPA RCRA HEALTH AND ENV. EFFECTS PROFILE OCT 80
BENZIDINE IS A KNOWN CARCINOGEN AND HAS BEEN LINKED TO AN INCREASED INCIDENCE OF BLADDER CANCER IN HUMANS AND CANCERS AND TUMORS IN EXPERIMENTAL ANIMALS.

REPORT 04 : NTP CARC TEST PROGRAM JUL 82
IN A CURRENT NTP BIOASSAY, BENZIDINE IS BEING ADMINISTERED IN THE WATER TO F344 RATS. THE PRECHRONIC TEST PHASE IS IN PROGRESS.

REPORT 05 : IARC MONOGRAPH 29 1982
BENZIDINE AND ITS DIHYDROCHLORIDE HAVE BEEN SHOWN TO BE CARCINOGENIC IN MICE, RATS, HAMSTERS, AND DOGS. THERE IS SUFFICIENT EVIDENCE THAT BENZIDINE IS CARCINOGENIC IN ANIMALS. OCCUPATIONAL EXPOSURE TO BENZIDINE OR ITS 2HCL HAS BEEN STRONGLY ASSOCIATED WITH BLADDER CANCER IN NUMEROUS CASE REPORTS. THERE IS SUFFICIENT EVIDENCE THAT BENZIDINE IS CARCINOGENIC TO HUMANS.

REPORT 06 : MULTI-AGENCY CROSS CHECK
Y HHS MAGUIRE
R OSHA CARC
Y EPA CAG
Y EPA RCRA
N EPA CLEAN AIR
Y EPA CLEAN WATER

REPORT 07 : IARC MONOGRAPH 29 1982
BENZIDINE AND URINE FROM RATS FED BENZIDINE ARE MUTAGENIC TO SALMONELLA WITH METABOLIC ACTIVATION. BENZIDINE IS MUTAGENIC TO DROSOPHILA. IT INDUCES UNSCHEDULED DNA SYNTHESIS IN HELA CELLS AND IN RATS HEPATOCYTES. BENZIDINE TRANSFORMED SYRIAN HAMSTER EMBRYO CELLS AND WAS POSITIVE IN THE BHK21 CLONE 13 CELL SYSTEM. THE DATA WERE INADEQUATE TO EVALUATE THE TERATOGENICITY OF BENZIDINE.

REPT'D USE : DYE INTERMEDIATE, RUBBER HARDENER, CHEMICAL INDICATOR, SECURITY, PRINTING AND LABORATORY REAGENT

CAS NO : **CLASS-09**
CHEM NAME : ***** BENZIDINE CONGENER DYES**

REPORT 01 : NIOSH SPECIAL HAZARD REVIEW JAN 80
NIOSH RECOMMENDS BENZIDINE BASED DYES HANDLED IN THE WORKPLACE AS CARCINOGENS. PRESENTLY OSHA HAS NO STANDARD.

REPORT 02 : EPA RESPONSE TO ITC RECOMMS (46FR55004) NOV 81
IN RESPONSE TO THE ITC RECOMMENDATIONS (44FR70664) THE EPA HAS DECIDED NOT TO PURSUE TESTING PROPOSAL FOR BENZIDINE DYES AT THIS TIME BECAUSE TESTING PROGRAMS, WHICH ARE EXPECTED TO SUPPLY DATA SUFFICIENT FOR EPA NEEDS, ARE BEING PLANNED AND CONDUCTED, AND REGULATORY EFFORTS ARE BEING PURSUED BY THE OSHA AND THE CPSC.
REPT'D USE : TEXTILE/LEATHER/PAPER DYE

CAS NO : **531-85-1**
CHEM NAME : ***** BENZIDINE DIHYDROCHLORIDE**

REPORT 06 : MULTI-AGENCY CROSS CHECK
N HHS MAGUIRE
Y OSHA CARC
N EPA CAG
N EPA RCRA
N EPA CLEAN AIR
N EPA CLEAN WATER
REPT'D USE : DYE INTERMEDIATE

CAS NO : **205-99-2**
CHEM NAME : ***** BENZO(B)FLUORANTHENE**

REPORT 01 : IARC MONOGRAPH 3 1973
BENZO(B)FLUORANTHENE HAS PRODUCED SKIN TUMORS IN MICE FOLLOWING REPEATED SKIN PAINTINGS. B(B)F IS ALSO AN INITIATOR OF SKIN CARCINOGENESIS IN MICE AND PRODUCES LOCAL SARCOMAS AFTER SUBCUTANEOUS INJECTIONS. IN A 1979 SUPPLEMENT, IARC DETERMINED THAT THERE IS SUFFICIENT EVIDENCE FOR THE CARCINOGENICITY OF B(B)F IN ANIMALS.

REPORT 02 : EPA AMBIENT WATER QUALITY CRITERIA OCT 80
PRESENTLY AVAILABLE DATA CANNOT SUPPORT INDIVIDUAL CRITERIA FOR EACH POLYNUCLEAR AROMATIC HYDROCARBON. THE AWQC FOR CARCINOGENIC PAH COMPOUNDS IS BASED ON THE ASSUMPTION THAT EACH COMPOUND IS AS POTENT AS B(A)P AND THAT THE CARCINOGENIC EFFECT OF THE COMPOUNDS IS PROPORTIONAL TO THE SUM OF THEIR CONCENTRATIONS. THIS SUM SHOULD NOT EXCEED 28 NG/L, 2.8 NG/L, AND .28 NG/L TO KEEP

THE LIFETIME CANCER RISK BELOW 10 -5, 10 -6, AND 10 -7 RESPECTIVELY.

REPORT 03 : EPA RCRA HEALTH AND ENV. EFFECTS PROFILE OCT 80
B(B)F IS CARCINOGENIC TO THE SKIN OF MICE AND PRODUCES SARCOMAS WHEN INJECTED IN MICE.

REPORT 04 : IARC MONOGRAPH 32 DEC 83
SHOWN TO BE CARCINOGENIC TO EXPERIMENTAL ANIMALS. SEE IARC MONOGRAPH 3 (1973).

REPORT 05 : IARC MONOGRAPH 32 DEC 83
MUTAGENIC TO SALMONELLA TYPHIMURIUM IN PRESENCE OF AN EXOGENOUS METABOLIC SYSTEM. IN ONE AVAILABLE STUDY REPORTED TO INDUCE SISTER CHROMATID EXCHANGE BUT NOT CHROMOSOMAL ABERRATIONS IN BONE-MARROW CELLS OF HAMSTERS TREATED IN VIVO.

REPORT 06 : MULTI-AGENCY CROSS CHECK
Y HHS MAGUIRE
N OSHA CARC
Y EPA CAG
N EPA RCRA
N EPA CLEAN AIR
Y EPA CLEAN WATER

REPT'D USE : COMBUSTION POLLUTANT, RESEARCH CARCINOGEN AND FOOD CONTAMINANT

CAS NO : **203-12-3**
CHEM NAME : ***** BENZO(GHI)FLUORANTHENE**

REPORT 01 : IARC MONOGRAPH 32 DEC 83
TESTED FOR CARCINOGENICITY IN FEMALE MICE BY SKIN PAINTING; NO SKIN TUMOR OBSERVED.

REPORT 02 : IARC MONOGRAPH 32 DEC 83
BENZO(GHI)FLUORANTHENE MUTAGENIC TO S. TYPHIMURIUM IN THE PRESENCE OF AN EXOGENOUS METABOLIC SYSTEM.

REPT'D USE : COMBUSTION BY-PRODUCT

CAS NO : **205-82-3**
CHEM NAME : ***** BENZO(J)FLUORANTHENE**

REPORT 01 : IARC MONOGRAPH 3 1973
BENZO(J)FLUORANTHENE HAS ONLY BEEN TESTED IN MICE BY REPEATED SKIN PAINTING, WHEN A HIGH INCIDENCE OF SKIN CARCINOMAS WAS OBTAINED.

REPORT 02 : EPA AMBIENT WATER QUALITY CRITERIA OCT 80
AVAILABLE DATA CANNOT SUPPORT INDIVIDUAL CRITERIA FOR EACH POLY-NUCLEAR AROMATIC HYDROCARBON (PAH). THE AWQC FOR CARCINOGENIC PAH IS BASED ON THE ASSUMPTION THAT EACH COMPOUND IS AS POTENT AS B(A)P AND THAT THE CARCINOGENIC EFFECT OF THE COMPOUNDS IS PROPORTIONAL TO THE SUM OF THEIR CONCENTRATIONS. THIS SUM SHOULD NOT EXCEED 28 NG/L, 2.8 NG/L, AND .28 NG/L TO KEEP THE LIFETIME CANCER RISK BELOW 10 -5, 10 -6, AND 10 -7 RESPECTIVELY.

REPORT 03 : IARC MONOGRAPH 32 DEC 83
PRODUCED FEW BENIGN AND MALIGNANT SKIN TUMORS IN MICE WHEN TESTED BY SKIN APPLICATION. WAS ACTIVE WHEN TESTED IN MOUSE-SKIN MUTATION-PROMOTION ASSAY. SUBCUTANEOUS INJECTION OF B(K)F IN MICE PRODUCED SARCOMAS AT INJECTION SITE. PRODUCED SQUAMOUS CELL CARCINOMAS IN A DOSE-RELATED MANNER FOLLOWING DIRECT INJECTION INTO PULMONARY TISSUE.

REPORT 04 : IARC MONOGRAPH 32 DEC 83
BENZO(J)FLUORANTHENE WAS MUTAGENIC TO SALMONELLA TYPHIMURIUM IN THE PRESENCE OF AN EXOGENOUS METABOLIC SYSTEM.

REPORT 06 : MULTI-AGENCY CROSS CHECK
Y HHS MAGUIRE
N OSHA CARC
Y EPA CAG
N EPA RCRA
N EPA CLEAN AIR
N EPA CLEAN WATER

REPT'D USE : COMBUSTION POLLUTANT AND EXPERIMENTAL CARCINOGEN

CAS NO : **207-08-9**
CHEM NAME : ***** BENZO(K)FLUORANTHENE**

REPORT 01 : IARC MONOGRAPH 32 DEC 83
TESTED FOR CARCINOGENICITY IN FEMALES OF TWO STRAINS OF MICE BY SKIN APPLICATION AND PRODUCED FEW SKIN TUMORS. TESTED IN MOUSE-SKIN MUTATION-PROMOTION ASSAY AND ACTIVE AS AN INITIATOR. SUBCUTANEOUS INJECTION OF BENZO(K)FLUORANTHENE TO MICE, PRODUCED SARCOMAS AT SITE OF INJECTION. PRODUCED SQUAMOUS-CELL CARCINOMAS OF LUNG IN RATS IN DOSE-RELATED MANNER FOLLOWING DIRECT INJECTION INTO PULMONARY TISSUE.

REPORT 06 : MULTI-AGENCY CROSS CHECK
Y HHS MAGUIRE
N OSHA CARC

N EPA CAG
N EPA RCRA
N EPA CLEAN AIR
N EPA CLEAN WATER
REPT'D USE : COMBUSTION BY-PRODUCT

CAS NO : **238-84-6**
CHEM NAME : ***** BENZO(A)FLUORENE**

REPORT 01 : IARC MONOGRAPH 32 DEC 83
TESTED FOR CARCINOGENICITY IN MICE BY SKIN APPLICATION AND IN MOUSESKIN INITIATION-PROMOTION ASSAY. NEGATIVE RESULTS OBTAINED BOTH STUDIES. SUBCUTANEOUS ADMINISTRATION TO MICE, NO INJECTION-SITE TUMOR WAS OBSERVED.
REPT'D USE : COMBUSTION BY-PRODUCT

CAS NO : **243-17-4**
CHEM NAME : ***** BENZO(B)FLUORENE**

REPORT 01 : IARC MONOGRAPH 32 DEC 83
DID NOT SHOW INITIATING ACTIVITY WHEN TESTED IN MOUSE-SKIN INITIATION-PROMOTION ASSAY. THERE ARE CONFLICTING REPORTS REGARDING THE MUTAGENICITY OF BENZO(B)FLUORENE TO S. TYPHIMURIUM.
REPT'D USE : COMBUSTION BY-PRODUCT

CAS NO : **205-12-9**
CHEM NAME : ***** BENZO(C)FLUORENE**

REPORT 01 : IARC MONOGRAGH 32 DEC 83
NO INDICATION OF CARCINOGENIC OR INITIATING ACTIVITY WHEN TESTED IN MOUSE-SKIN INITIATION-PROMOTION ASSAY. THE SINGLE REPORT OF MUTAGENICITY OF BENZO(C)FLUORENE IN S. TYPHIMURIUM IS NOT CONCLUSIVE.
REPT'D USE : COMBUSTION BY-PRODUCT

CAS NO : **271-89-6**
CHEM NAME : ***** BENZOFURAN**

REPORT 01 : NTP TR370 PUB. NIH 90-2825 OCT 89
UNDER 2-YEAR GAVAGE STUDIES NO EVIDENCE OF CARCINOGENIC ACTIVITY OF BENZOFURAN FOR MALE F344/N RATS RECEIVING DOSES OF 30 OR 60 MG/KG PER DAY. SOME EVIDENCE OF CARCINOGENIC ACTIVITY FOR FEMALE F344/N RATS BASED ON TUBULAR CELL ADENOCARCINOMAS OF KIDNEY. CLEAR EVIDENCE OF

CARCINOGENIC ACTIVITY FOR MALE AND FEMALE B6C3F1 MICE BASED ALTERATION LIVER AND BRONCHIOLAR AND FORESTOMACH HYPERPLASIA.

REPT'D USE : PRODUCTION OF COUMARONE (BENZOFURAN)-INDENE HYDROCARBON RESINS USED IN RUBBER COMPOUNDS, PLASTICIZERS, ALUMINUM PAINTS AND VARNISHES, CORROSION-RESISTANT COATINGS, WATER-RESISTANT COATINGS FOR PAPER PRODUCTS AND FABRICS AND ADHESIVES.

CAS NO : **480-96-6**
CHEM NAME : ***** BENZOFURAZAN-1-OXIDE**

REPORT 01 : EPA TSCA 8(E) 8EHQ-1284-0541 S JAN 85
THE CHEMICAL TESTED POSITIVE FOR MUTAGENICITY IN BOTH AMES SALMONELLA/MICROSOME PLATE TEST (WITH ACTIVATION) AND THE MOUSE LYMPHOMA CELL FORWARD MUTATION ASSAY (WITH AND WITHOUT ACTIVATION).
REPT'D USE : LABORATORY REAGENT

CAS NO : **613-94-5**
CHEM NAME : ***** BENZOIC ACID, HYDRAZIDE**

REPORT 06 : MULTI-AGENCY CROSS CHECK
N HHS MAGUIRE
Y OSHA CARC
N EPA CAG
N EPA RCRA
N EPA CLEAN AIR
N EPA CLEAN WATER
REPT'D USE : NO USE OR PRODUCTION DATA AVAILABLE

CAS NO : **119-53-9**
CHEM NAME : ***** BENZOIN**

REPORT 01 : NTP TECHNICAL BULLETIN 5 AUG 81
BENZOIN WAS NEGATIVE IN THE SALMONELLA TYPHIMURIUM MUTAGENICITY ASSAY IN TWO LABORATORIES.

REPORT 02 : NCI/NTP CARC TRS 204 DHHS PUB NO 80-1760
BENZOIN WAS NOT CARCINOGENIC FOR F344 RATS OR B6C3F1 MICE UNDER THE TEST CONDITIONS (FEEDING STUDY).
REPT'D USE : FLAVORING AGENT, ANTISEPTIC, ORGANIC SYNTHESIS AND PHOTOPOLYMERIZATION CATALYST

CAS NO : **191-24-2**
CHEM NAME : ***** BENZO(GHI)PERYLENE**

REPORT 01 : IARC MONOGRAPH 32
NO CARCINOGENIC EFFECT OBSERVED IN SKIN APPLICATION TO FEMALE MICE. NEGATIVE RESULTS FROM THREE MOUSE-SKIN ASSASYS. NO TUMORS SEEN AT INJECTION SITE FOLLOWING SUBCUTANEOUS INJECTION. RESULTS FROM INTRAPULMONARY INJECTION IN RATS INADEQUATE FOR EVALUATION - SOME PULMONARY TUMORS OCCURRED. WHEN ADMINISTERED WITH BENZO(A)PYRENE TO SKIN OF MICE, INCREASED NUMBER OF SKIN TUMORS SEEN COMPARED TO THOSE SEEN WITH BENZO(A)PYRENE ALONE.

REPORT 02: : IARC MONOGRAPH 32 DEC 83
MUTAGENIC TO S. TYPHIMURIUM IN PRESENCE OF EXOGENOUS METABOLIC SYSTEM. NEGATIVE IN ONE STUDY OF MORPHOLOGICAL TRANSFORMATION IN MAMMALIAN CELLS.
REPT'D USE : COMBUSTION BY-PRODUCT

CAS NO : **195-19-7**
CHEM NAME : ***** BENZO(C)PHENANTHRENE**

REPORT 01 : IARC MONOGRAPH 32 DEC 83
TESTED IN MOUSE-SKIN INITIATION-PROMOTION ASSAY AND ACTIVE AS AN INITIATOR. OTHER EXPERIMENTS IN MICE BY SKIN APPLICATION AND IN MICE AND RATS BY SUBCUTANEOUS INJECTION CONSIDERED INADEQUATE FOR EVALUATION.

REPORT 02 : IARC MONOGRAPH 32 DEC 83
THE ONE STUDY EVALUATED, BENZO(C)PHENANTHRENE MUTAGENIC TO S. TYPHIMURIUM IN PRESENCE OF EXOGENOUS METABOLIC SYSTEM.
REPT'D USE : COMBUSTION BY-PRODUCT

CAS NO : **50-32-8**
CHEM NAME : ***** BENZO(A)PYRENE**

REPORT 01 : IARC MONOGRAPH 3 1973
BENZO(A)PYRENE HAS PRODUCED TUMORS IN ALL OF THE NINE SPECIES FOR WHICH DATA ARE REPORTED FOLLOWING DIFFERENT ADMINISTRATIONS INCLUDING ORAL, SKIN AND INTRATRACHEAL ROUTES. IT HAS BOTH A LOCAL AND A SYSTEMIC CARCINOGENIC EFFECT. IN A 1979 SUPPLEMENT, IARC SAYS THAT THERE IS SUFFICIENT EVIDENCE FOR THE CARCINOGENICITY OF BENZO(A)PYRENE IN ANIMALS.

REPORT 02 : NAS REPORT ON DIET, NUTRITION AND CANCER 1982
THE CARCINOGENICITY OF BAP IN VARIOUS ANIMAL SPECIES HAS BEEN WELL ESTABLISHED. BAP HAS ALSO BEEN SHOWN TO BE MUTAGENIC TO SALMONELLA, TO BE GENOTOXIC IN THE HEPATOCYTE PRIMARY CULTURE-DNA REPAIR TEST, AND TO INDUCE TRANSFORMATIONS IN SYRIAN GOLDEN HAMSTER EMBRYO CELLS.

REPORT 03 : EPA RCRA HEALTH AND ENV. EFFECTS PROFILE OCT 80
B(A)P IS AMONG THE MOST POTENT ANIMAL CARCINOGENS KNOWN AND PRODUCES TUMORS IN VIRTUALLY ALL SPECIES BY ALL ROUTES OF ADMINISTRATION. NUMEROUS EPA STUDIES SUPPORT THE BELIEF THAT CARCINOGENIC PAH, INCLUDING B(A)P, ARE ALSO HUMAN CARCINOGENS.

REPORT 04 : EPA AMBIENT WATER QUALITY CRITERIA OCT 80
PRESENTLY AVAILABLE DATA CANNOT SUPPORT INDIVIDUAL CRITERIA FOR EACH POLYNUCLEAR AROMATIC HYDROCARBON. THE AWQC FOR CARCINOGENIC PAH COMPOUNDS IS BASED ON THE ASSUMPTION THAT EACH COMPOUND IS AS POTENT AS B(A)P AND THAT THE CARCINOGENIC EFFECT OF THE COMPOUND IS PROPORTIONAL TO THE SUM OF THEIR CONCENTRATIONS. THIS SUM SHOULD NOT EXCEED 28 NG/L, 2.8 NG/L, AND .28 NG/L TO KEEP THE LIFETIME CANCER RISK BELOW 10 -5, 10 -6, AND 10 -7 RESPECTIVELY.

REPORT 05 : NTP TECHNICAL BULLETIN 3/8 DEC 80
BAP WAS POSITIVE IN THE SALMONELLA ASSAY AND THE L5178Y MOUSE LYMPHOMA ASSAY.

REPORT 06 : MULTI-AGENCY CROSS CHECK
Y HHS MAGUIRE
N OSHA CARC
Y EPA CAG
Y EPA RCRA
N EPA CLEAN AIR
Y EPA CLEAN WATER

REPORT 07 : IARC MONOGRAPH 32 DEC 83
BENZO(A)PYRENE CARCINOGENIC TO EXPERIMENTAL ANIMALS (IARC #3, 1973). EMBRYOTOXIC AND TERATOGENIC IN MICE. REDUCTION IN FERTILITY IN BOTH MALE AND FEMALE OFFSPRING OBSERVED IN MICE FOLLOWING EXPOSURE IN UTERO. METABOLIZES TO REACTIVE ELECTROPHILES CAPABLE OF BINDING COVALENTLY TO DNA. ACTIVE IN ASSAYS FOR BACTERIAL DNA REPAIR, BACTERIAPHAGE INDUCTION AND BACTERIAL MUTATION IN D. MELANOGASTER, DNA BINDING, DNA REPAIR, SISTER CHROMATID EXCHANGE, CHROMOSOMAL ABERRATIONS, POINT MUTATION AND TRANSFORMATION IN MAMMALIAN CELLS IN CULTURE; AND IN TESTS IN MAMMALS IN VIVO, INCLUDING DNA BINDING, SISTER CHROMATID EXCHANGE, CHROMOSOMAL ABERRATIONS, AND OTHER TESTS.

REPT'D USE : COMBUSTION POLLUTANT, AUTO TIRES, FURNACE BLACKS AND LABORATORY RUBBER STOPPERS

CAS NO : **192-97-2**
CHEM NAME : ***** BENZO(E)PYRENE**

REPORT 01 : IARC MONOGRAPH 3 1973
THE DATA ARE CONFINED TO TWO SKIN PAINTING EXPERIMENTS IN MICE IN WHICH BENZ(E)PYRENE EVOKED A WEAKER RESPONSE THAN EITHER B(A)P OR DB(A,H)A.

REPORT 02 : NTP TECHNICAL BULLETIN 8 JUL 82
BENZ(E)PYRENE WAS POSITIVE IN THE L5178Y MOUSE LYMPHOMA ASSAY.

REPORT 03 : IARC MONOGRAPH 32 DEC 83
TESTED FOR CARCINOGENICITY IN MICE BY SKIN APPLICATION IN TWO STUDIES AND SKIN TUMORS OBSERVED IN ONE. TESTED IN MOUSE-SKIN INITIATION PROMOTION ASSAY; PROMOTING ACTIVITY DETECTED AFTER INITIATION WITH 9,10-DIMETHYL-BENZ(A)-ANTHRACENE IN ONE STUDY; THE INITIATING ACTIVITY NOT CLEARLY EVIDENT IN OTHER STUDIES WHERE 12-O-TETRA DECANOYLPHORBOL-13-ACETATE USED AS PROMOTING AGENT. MULTIPLE INTRAPERITONEAL ADMINISTRATIONS TO NEWBORN MICE DID NOT RESULT IN SIGNIFICANT INCREASE OF TUMORS IN 2 STUDIES. IN RATS, PULMONARY INJECTION RESULTED IN ONE SQUAMOUS-CELL CARCINOMA OF LUNG AT HIGH LEVEL AND PULMONARY SARCOMA IN ONE AT MID-DOSE LEVEL.

REPORT 04 : IARC MONOGRAPH 32 DEC 83
BENZO(E)PYRENE MUTAGENIC TO SALMONELLA TYPHIMURIUM IN PRESENCE OF EXOGENOUS METABOLIC SYSTEM. DID NOT INDUCE MITOTIC RECOMBINATION IN YEAST. DID NOT INDUCE MUTATIONS OR SISTER CHROMATID EXCHANGE IN CULTURED MAMMALIAN CELLS AND NEGATIVE IN ASSAYS FOR MORPHOLOGICAL TRANSFORMATION. INDUCED UNSCHEDULED DNA SYNTHESIS IN HELA CELLS IN PRESENCE OF EXOGENOUS METABOLIC SYSTEM, BUT NOT IN PRIMARY RAT HEPATOCYTES. IN ONE AVAILABLE REPORT, DID NOT INDUCE CHROMOSOMAL ABERRATIONS IN VITRO. IN IN-VIVO STUDY, INDUCED SISTER CHROMATID EXCHANGE, BUT NOT CHROMOSOMAL ABERRATIONS IN HAMSTER BONE MARROW.
REPT'D USE : COMBUSTION BY-PRODUCT AND EXPERIMENTAL CARCINOGEN

CAS NO : **95-14-7**
CHEM NAME : ***** 1H-BENZOTRIAZOLE**

REPORT 01 : NCI CARC TRS 88 DHEW PUB NO 78-1338
IN FEMALE B6C3F1 MICE THERE WAS AN INCREASED INCIDENCE OF

ALVEOLAR/BRONCHIOLAR CARCINOMAS, SUGGESTING A POSSIBLE CARCINOGENIC EFFECT OF 1H-BENZOTRIAZOLE. IN FISCHER 344 RATS THERE WAS AN INCREASED INCIDENCE OF BRAIN TUMORS, SUGGESTING A POSSIBLE CARCINOGENIC EFFECT. HOWEVER, THERE WAS NO CONVINCING EVIDENCE THAT UNDER THE CONDITIONS OF THIS BIOASSAY 1H-BENZOTRIAZOLE WAS CARCINOGENIC IN B6C3F1 MICE OR FISCHER 344 RATS OF EITHER SEX (FEEDING STUDY).

REPT'D USE : CORROSION INHIBITOR, PHOTOGRAPHIC CHEMICAL AND HEAT TRANSFER AGENT

CAS NO : **98-07-7**

CHEM NAME : ***** BENZOTRICHLORIDE (BTC)**

REPORT 01 : EPA RCRA HEALTH AND ENV. EFFECTS PROFILE OCT 80
BENZOTRICHLORIDE HAS BEEN SHOWN TO BE MUTAGENIC IN A NUMBER OF MICROBIAL TESTS WITH AND WITHOUT METABOLIC ACTIVATION. ONE STUDY DESCRIBED THE CARCINOGENICITY IN MICE, BUT DETAILS OF THE STUDY WERE NOT AVAILABLLE FOR ASSESSMENT.

REPORT 02 : IARC MONOGRAPH 29 1982
BENZOTRICHLORIDEPRODUCEDSQUAMOUS-CELLCARCINOMASOF THE SKIN IN EACH OF THREE STUDIES BY SKIN APPLICATION TO FEMALE MICE. UPPER DIGESTIVE TRACT TUMORS WERE ALSO OBSERVED IN TWO OF THE THREE EXPERIMENTS. IN ANOTHER STUDY, BTC PRODUCED MALIGNANT TUMORS OF THE FORESTOMACH AFTER ORAL ADMINISTRATION TO FEMALE MICE. THERE IS SUFFICIENT EVIDENCE THAT BTC IS CARCINOGENIC IN MICE.

REPORT 03 : IARC MONOGRAPH 29 1982
BTC WAS MUTAGENIC TO BACTERIA WITH METABOLIC ACTIVATION, AND WAS POSITIVE IN A REC-ASSAY WITH B. SUBTILIS. NO DATA WERE AVAILABLE TO ASSESS THE TERATOGENICITY OF THIS COMPOUND.

REPORT 04 : EPA TSCA 8(E) NOTICE 8EQH-0980-0360
HOOKER SUBMITTED INFORMATION ON THE POSSIBLE CARCINOGENICITY OF BC, BTC, AND PCBTC IN MICE. THE SUBMISSION CONSISTS OF FOUR JAPANESE SCIENTIFIC ARTICLES AND AN EVALUATION PREPARED BY A HOOKER TOXICOLOGIST. SUBMITTER CONCLUDED THAT THE EXPERIMENTAL DATA INDICATE THAT BTC AND PCBTC ARE CHEMICAL CARCINOGENS.

REPORT 05 : EPA TSCA 8(E) NOTICE 8EHQ-0884-0523 AUG 84
VELSICOL CHEMICAL CORP. PROVIDED REPORTS OF A COHORT MORTALITY STUDY BETWEEN 1943 AND 1982. ITS 697 MALE EMPLOYEES AT CHLORINATION PLANT IN CHATTANOOGA BASED ON ANIMAL DATA, EPIDEMIOLOGICAL STUDIES, AND ANALYSIS BY

LENGTH OF EMPLOYMENT, IN MORTALITY STUDY DATA SUGGESTED AN ASSOCIATION BETWEEN CHLORINATION PROCESS OF TOLUENE AT CHATTANOOGA PLANT AND INCREASED RISK OF RESPIRATORY CANCER.

REPORT 06 : MULTI-AGENCY CROSS CHECK
Y HHS MAGUIRE
N OSHA CARC
N EPA CAG
Y EPA RCRA
N EPA CLEAN AIR
N EPA CLEAN WATER
REPT'D USE : DYE INTERMEDIATE

CAS NO : **98-88-4**
CHEM NAME : ***** BENZOYL CHLORIDE (BC)**

REPORT 01 : EPA TSCA 8(E) NOTICE 8EQH-0980-0360
HOOKER SUMBITTED INFORMATION ON THE POSSIBLE CARCINOGENICITY OF BC, BTC, AND PCBTC IN MICE. THE SUBMISSION CONSISTS OF FOUR JAPANESE SCIENTIFIC ARTICLES AND AN EVALUATION PREPARED BY A HOOKER TOXICOLOGIST. SUBMITTER CONCLUDED THAT THE EXPERIMENTAL DATA INDICATE THAT BTC AND PCBTC ARE CHEMICAL CARCINOGENS.

REPORT 02 : IARC MONOGRAPH 29 MAY 82
BENZOYL CHLORIDE WAS TESTED IN TWO SETS OF EXPERIMENTS BY SKIN APPLICATION TO FEMALE MICE. A FEW SKIN CARCINOMAS WERE OBSERVED IN TREATED MICE, BUT THE INCIDENCE WAS NOT STATISTICALLY SIGNIFICANT. THERE IS INADEQUATE EVIDENCE FOR THE CARCINOGENICITY OF BENZOYL CHLORIDE IN ANIMALS.

REPORT 03 : IARC MONOGRAPH 29 MAY 82
THERE IS NO CLEAR EVIDENCE THAT BENZOYL CHLORIDE IS MUTAGENIC IN BACTERIAL SYSTEMS. IT WAS NEGATIVE IN A REC-ASSAY WITH B. SUBTILIS. NO DATA WERE AVAILABLE TO ASSESS THE TERATOGENICITY OF THIS CHEMICAL.

REPORT 04 : EPA TSCA 8(E) NOTICE 8EHQ-0884-0523 AUG 84
VELSICOL CHEMICAL CORP. PROVIDED REPORTS OF A COHORT MORTALITY STUDY BETWEEN 1943 AND 1982 OF 697 MALE EMPLOYEES AT ITS CHLORINATION PLANT IN CHATTANOOGA. FOR COHORT AS A WHOLE, NO STATISTICALLY SIGNIFICANT EXCESS MORTALITY SEEN. ANALYSIS BY MAJOR CHEMICAL EXPOSURE SHOWED THAT POTENTIAL EXPOSURE TO BTC, BENZYL CHLORIDE OR BENZOYL CHLORIDE RESULTED IN SIGNIFICANTLY INCREASED RISK OF RESPIRATORY CANCER. BASED ON ANIMALS STUDIES, OTHER EPA STUDIES AND INTERNAL CONSISTENCY OF ANALYSIS BY LENGTH.

CAS NO : **94-36-0**
CHEM NAME : ***** BENZOYL PEROXIDE**

REPORT 01 : NIOSH CRITERIA DOCUMENT JUN 77
NIOSH RECOMMENDS A STANDARD OF 5 MG/CU M TWA BASED ON AIRWAY AND EYE AND SKIN IRRITATION. THE PRESENT OSHA STANDARD IS 5 MG/CU M TWA.

REPORT 02 : IARC MONOGRAPH 36 FEB 85
TESTED IN MICE AND RATS BY ORAL ADMINISTRATION IN DIET AND BY SUBCUTANEOUS ADMINISTRATION, AND IN MICE BY SKIN APPLICATION. IN THREE STUDIES BY SKIN APPLICATION IN MICE, BENZOYL PEROXIDE TESTED FOR EITHER INITIATING OR PROMOTING ACTIVITY. ALL STUDIES WERE INADEQUATE FOR AN EVALUATION OF COMPLETE CARCINOGENICITY; TWO STUDIES INDICATED THAT BENZOYL PEROXIDE HAS PROMOTING ACTIVITY IN MOUSE SKIN.

REPORT 03 : IARC MONOGRAPH 36 FEB 85
BENZOYL PEROXIDE NOT MUTAGENIC TO BACTERIA. IT DID NOT INDUCE CHROMOSOMAL ABERRATIONS IN MAMMALIAN CELLS IN VITRO AND DID NOT INDUCE DOMINANT LETHAL MUTATIONS IN MICE.
REPT'D USE : CATALYST, DRYING AGENT, PREPARATION OF PHARMACEUTICALS, DERMATOLOGICAL AGENT, BUILDING MATERIALS, POLYMER CROSS-LINKING AGENT, TEXTILE/FOOD SUBSTANCE AND BLEACHING AGENT

CAS NO : **140-11-4**
CHEM NAME : ***** BENZYL ACETATE**

REPORT 01 : NTP DRAFT TECHNICAL REPORT 1982
UNDER THE CONDITIONS OF THIS BIOASSAY BENZYL ACETATE ADMINISTERED BY GAVAGE TO FISCHER 344 RATS AT DOSE LEVELS OF 250 OR 500 MG/KG AND B63FI/N MICE AT 500 OR 1,000 MG/KG SHAVED. IT WAS CARCINOGENIC TO MALE RATS, CAUSING A STATISTICALLY SIGNIFICANT INCREASE IN ACINAR-CELL ADENOMAS IN THE PANCREAS. IT WAS NOT CARCINOGENIC TO FEMALE RATS. LIVER TUMORS (HEPATOCELLULAR ADENOMAS) IN MICE OF BOTH SEXES OCCURRED WITH STATISTICALLY SIGNIFICANT POSITIVE TRENDS.

REPORT 02 : NTP TR250, NIH PUB. #86-2506 AUG 86
IN THESE STUDIES, BA INCREASED INCIDENCE OF ACINAR-CELL ADENOMAS OF PANCREAS IN MALE F344/N RATS AND NO EVIDENCE OF CARCINOGENICITY IN FEMALE F344/N RATS. SOME EVIDENCE OF CARCINOGENICITY IN MALE AND FEMALE B6C3F1 MICE AND AN INCREASED INCIDENCE OF HEPAOTCELLULAR ADENOMAS AND SQUAMOUS CELL NEOPLASMS OF FORESTOMACH. BA DID NOT

INDUCE SCE'S OR CHROMOSOMAL ABERRATIONS. IT WAS MUTAGENIC IN CULTURED MAMMALIAN CELLS WITH ACTIVATION.

REPORT 03 : CHEMICAL HAZARD INFORMATION PROFILE SEP 82
BA CARCINOGENIC TO MALE F344/N RATS, CAUSING INCREASED INCIDENCE OF PANCREATIC ACINAR-CELL ADENOMAS FOLLOWING ORAL ADMINISTRATION IN CORN OIL. INCREASED INCIDENCE OF LIVER TUMORS (HEPATOCELLULAR ADENOMAS) IN BOTH MALE AND FEMALE RATS.

REPORT 04 : IARC VOL 40 1986
LIMITED EVIDENCE FOR CARCINOGENICITY TO EXPERIMENTAL ANIMALS. NO EVALUATION COULD BE MADE OF CARCINOGENICITY TO HUMANS. EVIDENCE IS INADEQUATE TO DETERMINE GENETIC ACTIVITY IN SHORT-TERM TESTS. BA IS ABSORBED THROUGH SKIN AND FROM GI TRACT IN TEST ANIMALS. NO CASE REPORT OE EPIDEMIOLOGICAL STUDY AVAILABLE TO WORKING GROUP FOR EVALUATION OF HUMAN CARCINOGENICITY.

REPT'D USE : SOLVENT FOR LACQUERS, RESINS, OILS, POLISHES, FRAGRANCES IN SOAPS, DETERGENTS, TOILET PREPARATIONS, FLAVORING AGENT IN FOODS AND PRINTING INKS

CAS NO : **100-51-6**
CHEM NAME : ***** BENZYL ALCOHOL**

REPORT 01 : NTP TR 343 NIH PUB #89-2599 JUN 89
UNDER CONDITIONS OF 2-YEAR GAVAGE STUDY, NO EVIDENCE OF CARCINOGENIC ACTIVITY FOR MALE OR FEMALE F344/N RATS AT 200 OR 400 MG/KG. NO EVIDENCE OF CARCINOGENIC ACTIVITY FOR MALE OR FEMALE B6C3F1 MICE AT 100 OR 200 MG/KG.

REPT'D USE : TEXTILE DYES, FOOD FLAVORINGS, BACTERIOSTAT, PERFUMES, MICROSCOPY, GELATIN, CASEIN, CELLULOSE ACETATE AND SHELLAC SOLVENT

CAS NO : **100-44-7**
CHEM NAME : ***** BENZYL CHLORIDE**

REPORT 01 : EPA RCRA HEALTH AND ENV. EFFECTS PROFILE OCT 80
BENZYL CHLORIDE HAS BEEN SHOWN TO PRODUCE CARCINOGENIC EFFECTS IN RATS FOLLOWING SC ADMINISTRATION AND IN MICE FOLLOWING IP ADMINISTRATION. WEAK MUTAGENIC ACTIVITY OF THE COMPOUND HAS BEEN DEMONSTRATED IN THE AMES SALMONELLA ASSAY AND IN E. COLI.

REPORT 02 : IARC MONOG 29 1982
BENZYL CHLORIDE IS A DIRECT ACTING MUTAGEN TO BACTERIA. IT INDUCES MITOTIC RECOMBINATION IN YEAST AND TRANSFORMS HAMSTER EMBRYO CELLS. THE AVAILABLE DATA ARE INADE-

QUATE TO ASSESS THE TERATOGENICITY OF THIS COMPOUND TO EXPERIMENTAL ANIMALS.

REPORT 03 : NTP CARC TEST PROGRAM JUL 82
IN A CURRENT NTP BIOASSAY, BENZYL CHLORIDE WAS ADMINISTERED TO F344 RATS AND B6C3F1 MICE BY GAVAGE. THE CHRONIC EXPOSURE PHASE HAS BEEN COMPLETED AND HISTOPATHOLOGY IS IN PROGRESS.

REPORT 04 : NIOSH CRITERIA DOCUMENT AUG 78
NIOSH RECOMMENDS THAT EXPOSURE TO BENZYL CHLORIDE BE CONTROLLED SO THAT NO EMPLOYEE BE EXPOSED TO A CONCENTRATION GREATER THAN 5.0 MILLIGRAMS PER CUBIC METER (MG/CU M) DETERMINED AS A CEILING CONCENTRATION DURING ANY 15-MINUTE SAMPLING PERIOD BASED ON SKIN AND EYE IRRITATIONS. THE PRESENT OSHA STANDARD IS 5 MG/CU M TWA.

REPORT 05 : IARC MONOGRAPH 29 1982
BENZYL CHLORIDE WAS TESTED IN MICE BY SKIN APPLICATION AND IN RATS BY SC INJECTION. SARCOMAS AT THE INJECTION SITE WERE NOTED IN RATS. A FEW SKIN CARCINOMAS WERE OBSERVED IN A LIMITED NUMBER OF MICE, BUT THEIR INCIDENCE WAS NOT STATISTICALLY SIGNIFICANT. THERE IS LIMITED EVIDENCE THAT BENZYL CHLORIDE IS CARCINOGENIC IN ANIMALS.

REPORT 06 : MULTI-AGENCY CROSS CHECK
N HHS MAGUIRE
N OSHA CARC
N EPA CAG
N EPA RCRA
Y EPA CLEAN AIR
N EPA CLEAN WATER

REPORT 07 : EPA TSCA 8(E) NOTICE 8EHQ-0884-0523 AUG 84
VELSICOL CHEMICAL CORP. PROVIDED REPORTS OF A COHORT MORTALITY STUDY OF BETWEEN 1943 AND 1982. ITS 697 MALE EMPLOYEES AT CHLORINATION PLANT IN CHATTANOOGA BASED ON ANIMAL DATA, EPIDEMIOLOGICAL STUDIES, AND ANALYSIS BY LENGTH OF EMPLOYMENT IN MORTALITY STUDY DATA SUGGESTED AN ASSOCIATION BETWEEN CHLORINATION PROCESS OF TOLUENE AT CHATTANOOGA PLANT AND INCREASED RISK OF RESPIRATORY CANCER.

REPT'D USE : INTERMEDIATE IN MFR OF PERFUMES, PHARMACEUTICALS, DYES AND RESINS

CAS NO : **12161-82-9**
CHEM NAME : ***** BERTRANDITE**

REPORT 01 : IARC MONOGRAPH 23 1980
ON THE BASIS OF INHALATION EXPERIMENTS IN RATS AND HAMSTERS WITH BERTRANDITE ORE, NO EVALUATION OF CARCINOGENICITY COULD BE MADE.

REPORT 06 : MULTI-AGENCY CROSS CHECK
Y HHS MAGUIRE
Y OSHA CARC
Y EPA CAG
Y EPA RCRA
Y EPA CLEAN AIR
Y EPA CLEAN WATER
REPT'D USE : MINERAL SOURCE FOR BERYLLIUM ORE

CAS NO : **7440-41-7**
CHEM NAME : ***** BERYLLIUM**

REPORT 01 : IARC MONOGRAPH 23 1980
NO DATA ON THE MUTAGENICITY, EMBRYOTOXICITY OR TERATOGENICITY OF THIS COMPOUND WERE AVAILABLE.

REPORT 02 : EPA AMBIENT WATER QUALITY CRITERIA OCT 80
BERYLLIUM WAS SHOWN TO BE CARCINOGENIC IN THREE ANIMAL SPECIES, PRODUCING CANCERS OF THE LUNG AND BONE WHEN ADMINISTERED BY INJECTION, INHALATION, OR INTRATRACHEAL INSTILLATION. INGESTION OF BERYLLIUM HAS FAILED TO PRODUCE CANCERS IN ANIMALS, POSSIBLY DUE TO ITS POOR GASTROINTESTINAL ABSORPTION. SEVERAL EPA STUDIES SUPPORT THE HYPOTHESIS THAT BERYLLIUM IS A HUMAN CARCINOGEN.

REPORT 03 : IARC MONOGRAPH 23 1980
BERYLLIUM COMPOUNDS ARE CARCINOGENIC IN THREE ANIMAL SPECIES (RAT, RABBIT, AND MONKEY). RATS EXPOSED BY INHALATION OR INTRATRACHEALLY TO VARIOUS BERYLLIUM COMPOUNDS DEVELOPED LUNG TUMORS. RABBITS EXPOSED TO SELECTED BERYLLIUM COMPOUNDS DEVELOPED OSTEOSARCOMAS FOLLOWING INTRAVENOUS AND/OR INTRAOSSEUS ADMINISTRATION. THE EPIDEMIOLOGICAL EVIDENCE THAT OCCUPATIONAL EXPOSURE TO BERYLLIUM LEADS TO INCREASED LUNG CANCER RISK IS LIMITED. IN A 1979 SUPPLEMENT, IARC CONCLUDED THAT BERYLLIUM IS CARCINOGENIC FOR HUMANS.

REPORT 04 : EPA RCRA HEALTH AND ENV. EFFECTS PROFILE OCT 80
BERYLLIUM WAS SHOWN TO BE CARCINOGENIC IN THREE ANIMAL SPECIES, PRODUCING CANCERS OF THE LUNG AND BONE WHEN ADMINISTERED BY INJECTION, INHALATION, OR INTRATRACHEAL

INSTILLATION. INGESTION OF BERYLLIUM HAS FAILED TO PRODUCE CANCERS IN ANIMALS, POSSIBLY DUE TO ITS POOR GASTROINTESTINAL ABSORPTION. SEVERAL EPA STUDIES SUPPORT THE HYPOTHESIS THAT BERYLLIUM IS A HUMAN CARCINOGEN.

REPORT 05 : NIOSH CRITERIA DOCUMENT AUG 77
NIOSH RECOMMENDS A CEILING OF 0.5 MICROGRAMS/CU M BASED ON LUNG CANCER. THE PRESENT OSHA STANDARD IS 2 MICROGRAMS/CU M TWA; 5 MICROGRAMS PER CU M CEILING.

REPORT 06 : MULTI-AGENCY CROSS CHECK
Y HHS MAGUIRE
N OSHA CARC
Y EPA CAG
N EPA RCRA
N EPA CLEAN AIR
Y EPA CLEAN WATER

REPORT 07 : IARC MONOGRAPH SUPP 4, 1982 OCT 82
THERE IS CONSIDERED TO BE SUFFICIENT EVIDENCE OF CARCINOGENICITY IN EXPERIMENTAL ANIMALS.
REPT'D USE : ELECTRONICS ALLOYS, DENTAL ALLOY, AIRCRAFT BRAKES, ROCKET FUELS, ALLOY, NEUTRON REFLECTOR AND OPTICS

CAS NO : **543-81-7**
CHEM NAME : ***** BERYLLIUM ACETATE**

REPORT 01 : IARC MONOGRAPH 23 1980
BASED ON THE INTRAOSSEUS INJECTION OF BERYLLIUM ACETATE IN RABBITS, NO EVALUATION OF CARCINOGENICITY COULD BE MADE. NO DATA ON THE MUTAGENICITY, EMBRYOTOXICITY OR TERATOGENICITY OF THIS COMPOUND WERE AVAILABLE.

REPORT 06 : MULTI-AGENCY CROSS CHECK
N HHS MAGUIRE
N OSHA CARC
N EPA CAG
Y EPA RCRA
N EPA CLEAN AIR
Y EPA CLEAN WATER
REPT'D USE : NO AVAILABLE PRODUCTION OR USE DATA. NO LISTING ON TSCA INVENTORY

CAS NO : **12770-50-2**
CHEM NAME : ***** BERYLLIUM-ALUMINUM ALLOY**

REPORT 01 : IARC MONOGRAPH 23 1980
BERYLLIUM-ALUMINUM ALLOY PRODUCED LUNG TUMORS IN RATS EXPOSED INTRA-TRACHEALLY.

REPORT 06 : MULTI-AGENCY CROSS CHECK
Y HHS MAGUIRE
N OSHA CARC
Y EPA CAG
Y EPA RCRA
N EPA CLEAN AIR
Y EPA CLEAN WATER
REPT'D USE : AIRCRAFT AND ALLOY-CASTING. NO KNOWN PRODUCTION FOR COMMERCIAL USE. NO LISTING ON TSCA INVENTORY.

CAS NO : **13106-47-3**
CHEM NAME : ***** BERYLLIUM CARBONATE**

REPORT 01 : IARC MONOGRAPH 23 1980
BERYLLIUM CARBONATE PRODUCED OSTEOSARCOMAS IN RABBITS FOLLOWING INTRAOSSEUS ADMINISTRATION. NO DATA ON THE MUTAGENICITY, EMBRYOTOXICITY OR TERATOGENICITY OF THIS COMPOUND WERE AVAILABLE.

REPORT 06 : MULTI-AGENCY CROSS CHECK
Y HHS MAGUIRE
N OSHA CARC
N EPA CAG
Y EPA RCRA
N EPA CLEAN AIR
Y EPA CLEAN WATER
REPT'D USE : LABORATORY CHEMICAL

CAS NO : **7787-47-5**
CHEM NAME : ***** BERYLLIUM CHLORIDE**

REPORT 01 : IARC MONOGRAPH 23 1980
BERYLLIUM CHLORIDE PRODUCED LUNG TUMORS IN RATS EXPOSED BY INHALATION.
REPORT 02 : IARC MONOGRAPH 23
IN THE REC ASSAY, BERYLLIUM CHLORIDE GAVE NEGATIVE RESULTS. POSITIVE RESULTS WERE OBTAINED WITH BERYLLIUM CHLORIDE IN DNA MISINCORPORATION TESTS. BERYLLIUM CHLORIDE INDUCED CHROMOSOMAL ABERRATIONS IN CULTURED PERIPHERAL LYMPHOCYTES. NO DATA ON THE EMBRYOTOXICITY

OR TERATOGENICITY OF BERYLLIUM COMPOUNDS WERE AVAILABLE.

REPORT 06 : MULTI-AGENCY CROSS CHECK
Y HHS MAGUIRE
N OSHA CARC
Y EPA CAG
Y EPA RCRA
N EPA CLEAN AIR
Y EPA CLEAN WATER

REPT'D USE : ACID CATALYST IN ORGANIC REACTIONS AND NUCLEAR REACTOR CORE MATERIAL. NO LISTING ON TSCA INVENTORY.

CAS NO : **11133-98-5**
CHEM NAME : ***** BERYLLIUM-COPPER ALLOY**

REPORT 01 : IARC MONOGRAPH 23 1980
BASED ON A SINGLE INTRATRACHEAL INSTILLATION OF BERYLLIUM-COPPER ALLOY IN RATS, NO EVALUATION OF CARCINOGENICITY COULD BE MADE.

REPORT 06 : MULTI-AGENCY CROSS CHECK
N HHS MAGUIRE
N OSHA CARC
N EPA CAG
Y EPA RCRA
N EPA CLEAN AIR
Y EPA CLEAN WATER

REPT'D USE : PEN/PENCIL CLIPS, SMALL ELECTRICAL PARTS, CIRCUIT BREAKERS AND WELDING TIPS

CAS NO : **55158-44-6**
CHEM NAME : ***** BERYLLIUM-COPPER-COBALT ALLOY**

REPORT 01 : IARC MONOGRAPH 23 1980
BASED ON A SINGLE INTRATRACHEAL INSTILLATION OF BERYLLIUM-COPPER-COBALT ALLOY IN RATS, NO EVALUATION OF CARCINOGENICITY COULD BE MADE.

REPORT 06 : MULTI-AGENCY CROSS CHECK
N HHS MAGUIRE
N OSHA CARC
N EPA CAG
Y EPA RCRA
N EPA CLEAN AIR
Y EPA CLEAN WATER

REPT'D USE : RESISTANCE WELDING, ELECTRODES AND ELECTRICAL PARTS. NO LISTING ON TSCA INVENTORY

CAS NO : **7787-49-7**
CHEM NAME : ***** BERYLLIUM FLUORIDE**

REPORT 01 : IARC MONOGRAPH 23 1980
BERYLLIUM FLUORIDE PRODUCED LUNG TUMORS IN RATS EXPOSED BY INHALATION. NO DATA ON THE MUTAGENICITY, EMBRYOTOXICITY OR TERATOGENICITY OF THIS COMPOUND WERE AVAILABLE.

REPORT 06 : MULTI-AGENCY CROSS CHECK
Y HHS MAGUIRE
N OSHA CARC
Y EPA CAG
Y EPA RCRA
N EPA CLEAN AIR
Y EPA CLEAN WATER
REPT'D USE : INTERMEDIATE IN PRODUCTION OF BERYLLIUM METALS AND GLASS MAKING

CAS NO : **13327-32-7**
CHEM NAME : ***** BERYLLIUM HYDROXIDE**

REPORT 01 : IARC MONOGRAPH 23 1980
BERYLLIUM HYDROXIDE PRODUCED LUNG TUMORS IN RATS EXPOSED INTRATRACHEALLY. NO DATA ON THE MUTAGENICITY, EMBRYOTOXICITY OR TERATOGENICITY OF THIS COMPOUND WERE AVAILABLE.

REPORT 06 : MULTI-AGENCY CROSS CHECK
Y HHS MAGUIRE
N OSHA CARC
Y EPA CAG
Y EPA RCRA
N EPA CLEAN AIR
Y EPA CLEAN WATER
REPT'D USE : INTERMEDIATE IN METAL PRODUCTION

CAS NO : **37227-61-5**
CHEM NAME : ***** BERYLLIUM-NICKEL ALLOY**

REPORT 01 : IARC MONOGRAPH 23 1980
BASED ON A SINGLE INTRATRACHEAL INSTILLATION OF BERYLLIUM-NICKEL ALLOY IN RATS, NO EVALUATION OF CARCINOGENICITY COULD BE MADE.

REPORT 06 : MULTI-AGENCY CROSS CHECK
N HHS MAGUIRE
N OSHA CARC

Y EPA CAG
Y EPA RCRA
N EPA CLEAN AIR
Y EPA CLEAN WATER

REPT'D USE : HIGH TEMPERATURE ELECTRICAL CONNECTIONS AND GLASS MAKING

CAS NO : **1304-56-9**

CHEM NAME : ***** BERYLLIUM OXIDE**

REPORT 01 : IARC MONOGRAPH 23 1980
NO DATA ON THE MUTAGENICITY, EMBRYOTOXICITY OR TERATOGENICITY OF THIS COMPOUND WERE AVAILABLE.

REPORT 02 : IARC MONOGRAPH 23 1980
BERYLLIUM OXIDE PRODUCED LUNG TUMORS IN MONKEYS FOLLOWING INTRABRONCHIAL INJECTIONS AND PRODUCED OSTEOSARCOMAS IN RABBITS FOLLOWING INTRAVENOUS AND/OR INTRAOSSEUS ADMINISTRATION. THE EPIDEMIOLOGICAL EVIDENCE THAT OCCUPATIONAL EXPOSURE TO BERYLLIUM MAY LEAD TO AN INCREASED LUNG CANCER RISK IS LIMITED. THE EXPERIMENTAL AND HUMAN DATA INDICATE THAT BERYLLIUM SHOULD BE SUSPECTED OF BEING CARCINOGENIC TO HUMANS.

REPORT 06 : MULTI-AGENCY CROSS CHECK
Y HHS MAGUIRE
N OSHA CARC
Y EPA CAG
N EPA RCRA
N EPA CLEAN AIR
N EPA CLEAN WATER

REPT'D USE : CERAMIC APPLICATIONS IN ELECTRONICS, GLASS, PLASTICS, CATALYST AND FLUORESCENT LAMPS

CAS NO : **13598-15-7**

CHEM NAME : ***** BERYLLIUM PHOSPHATE**

REPORT 01 : IARC MONOGRAPH 23 1980
BERYLLIUM PHOSPHATE PRODUCED OSTEOSARCOMAS IN RABBITS FOLLOWING INTRAVENOUS INJECTION. IN A 1979 SUPPLEMENT, IARC DETERMINED THAT THERE IS SUFFICIENT EVIDENCE FOR THE CARCINOGENICITY OF BERYLLIUM PHOSPHATE IN ANIMALS. NO DATA ON THE MUTAGENICITY, EMBRYOTOXICITY OR TERATOGENICITY OF THIS COMPOUND WERE AVAILABLE.

REPORT 06 : MULTI-AGENCY CROSS CHECK
Y HHS MAGUIRE
N OSHA CARC

Y EPA CAG
Y EPA RCRA
N EPA CLEAN AIR
Y EPA CLEAN WATER

REPT'D USE : NO AVAILABLE PRODUCTION OR USE DATA. NO LISTING ON TSCA INVENTORY.

CAS NO : **15191-85-2**
CHEM NAME : **BERYLLIUM SILICATE**

REPORT 01 : IARC MONCGRAPH 23 1980
BERYLLIUM SILICATE PRODUCED OSTEOSARCOMAS IN RABBITS FOLLOWING INTRAVENOUS INJECTION. NO DATA ON THE MUTAGENICITY, EMBRYOTOXICITY OR TERATOGENICITY OF THIS COMPOUND WERE AVAILABLE.

REPORT 06 : MULTI-AGENCY CROSS CHECK
Y HHS MAGUIRE
N OSHA CARC
N EPA CAG
Y EPA RCRA
N EPA CLEAN AIR
Y EPA CLEAN WATER

REPT'D USE : NO AVAILABLE PRODUCTION OR USE DATA. NO LISTING ON TSCA INVENTORY

CAS NO : **7787-56-6**
CHEM NAME : ***** BERYLLIUM SULFATE TETRAHYDRATE**

REPORT 01 : IARC MONOGRAPH 23 1980
BERYLLIUM SULFATE, BERYL ORE, AND BERTRANDITE PRODUCE LUNG TUMORS IN THE RAT FOLLOWING INHALATION EXPOSURE; BERYLLIUM OXIDE AND BERYLLIUM SULFATE PRODUCE TUMORS IN THE MONKEY FOLLOWING INTRABRONCHIAL IMPLANTATION OR INHALATION. IN A 1979 SUPPLEMENT, IARC DETERMINED THAT THERE IS SUFFICIENT EVIDENCE FOR THE CARCINOGENICITY OF BERYLLIUM SULFATE IN ANIMALS.

REPORT 02 : IARC MONOGRAPH 23 1980
BERYLLIUM SULPHATE WAS NOT MUTAGENIC IN VARIOUS BACTERIAL SYSTEMS. IN A REC ASSAY, BERYLLIUM SULPHATE GAVE POSITIVE RESULTS. NO DATA ON THE EMBRYOTOXICITY OR TERATOGENICITY OF BERYLLIUM COMPOUNDS WERE AVAILABLE.

REPORT 06 : MULTI-AGENCY CROSS CHECK
Y HHS MAGUIRE
N OSHA CARC
Y EPA CAG

N EPA RCRA
N EPA CLEAN AIR
N EPA CLEAN WATER
REPT'D USE : CHEMICAL INTERMEDIATE FOR BERYL AND BERTRANDITE ORES

CAS NO : **39413-47-3**
CHEM NAME : ***** BERYLLIUM ZINC SILICATE**

REPORT 01 : IARC MONOGRAPH 23 1980
ZINC BERYLLIUM SILICATE PRODUCED OSTEOSARCOMAS IN RABBITS FOLLOWING INTRAVENOUS INJECTION.

REPORT 06 : MULTI-AGENCY CROSS CHECK
N HHS MAGUIRE
N OSHA CARC
N EPA CAG
N EPA RCRA
N EPA CLEAN AIR
N EPA CLEAN WATER
REPT'D USE : PHOSPHAR IN LUMINESCENT MATERIALS. NO LISTING ON TSCA INVENTORY

CAS NO : **1302-52-9**
CHEM NAME : ***** BERYL ORE**

REPORT 01 : IARC MONOGRAPH 23 1980
BERYL ORE PRODUCED LUNG TUMORS IN RATS EXPOSED BY INHALATION. NO DATA ON THE MUTAGENICITY, EMBRYOTOXICITY OR TERATOGENICITY OF THIS COMPOUND WERE AVAILABLE.

REPORT 06 : MULTI-AGENCY CROSS CHECK
Y HHS MAGUIRE
N OSHA CARC
Y EPA CAG
Y EPA RCRA
N EPA CLEAN AIR
Y EPA CLEAN WATER
REPT'D USE : MINERAL SOURCE OF BERYLLIUM

CAS NO : **CLASS-03**
CHEM NAME : ***** BETEL-QUID AND ARECA NUT CHEWING**

REPORT 01 : IARC MONOGRAGH 37 SEP 85
SUFFICIENT EVIDENCE THAT HABIT OF CHEWING BETEL QUID CONTAINING TOBACCO OR WHEN COMBINED WITH SMOKING TOBACCO CARCINOGENIC TO HUMANS. BUT INADEQUATE EVIDENCE THAT HABIT OF CHEWING BETEL QUID WITHOUT TOBACCO

IS CARCINOGENIC TO HUMANS. LIMITED EVIDENCE THAT AQUEOUS EXTRACTS OF BETEL QUID WITH AND WITHOUT TOBACCO OR THAT ARECA NUT WITH AND WITHOUT TOBACCO IS CARCINOGENIC TO EXPERIMENTAL ANIMALS. THE DATA ARE INADEQUATE TO ALLOW EVALUATION OF CARCINOGENICITY OF BETEL LEAF OR ARECOLINE TO EXPERIMENTAL ANIMALS.

REPORY 02 : IARC MONOGRAPH 37 SEP 85
AQUEOUS EXTRACTS OF BETEL QUID WITH TOBACCO ARECAIDINE AQUEOUS EXTRACTS OF ARECA NUT AND ARECOLINE INDUCE MUTATIONS IN S. TYPHIMURIUM AND CHINESE HAMSTER V79 CELLS. AQUEOUS EXTRACTS OF BETEL LEAF ARE NOT MUTAGENIC TO SALMONELLA TYPHIMURIUM. THEY INDUCE CHROMOSOMAL ABERRATIONS IN HUMAN LYMPHOCYTES AND IN CHINESE HAMSTER OVARY CELLS. N-BUTANOL AND ETHYL ACETATE EXTRACTS OF BETEL LEAF INDUCE CHROMOSOMAL ABERRATIONS IN CHINESE HAMSTER OVARY CELLS. OTHER GENOTOXIC TESTS WERE NEGATIVE. ETHYL ACETATE AND N-BUTANOL EXTRACTS OF ARECA NUT INDUCE CHROMOSOMAL ABERRATIONS IN CHINESE HAMSTER OVARY CELLS.

REPT'D USE : CHEWING WITH AND WITHOUT TOBACCO

CAS NO : **92-52-4**

CHEM NAME : ***** BIPHENYL**

REPORT 01 : EPA FINAL TEST RULE (50FR37182) SEP 85
EPA REQUIRES MANUFACTURERS AND PROCESSORS TO TEST BIPHENYL FOR ENVIRONMENTAL EFFECTS AND CHEMICAL FATE UNDER PROVISIONS OF TSCA4(A). THIS IS IN RESPONSE TO ITC'S 10TH REPORT (MAY 25, 1986) THAT LISTS BIPHENYL FOR PRIORITY TESTING CONSIDERATION.

REPORT 02 : 52 FR 20710 BIPHENYL: TEST STDS AND RPTING REQUIREMENTS JUN 87
THIS RULE ANNOUNCES EPA'S ADOPTION OF THE STUDY PLANS AND SCHEDULE SUBMITTED BY THE BIPHENYL WORK GROUP FOR THE TESTING OF BIPHENYL FOR ENVIRONMENTAL EFFECTS AND CHEMICAL FATE. TESTS INCLUDE CHRONIC TESTING ON DAPHNIA MAGNA, EARLY LIFE STAGE TESTS ON RAINBOW TROUT, OYSTER TOXICITY AND BIOCONCENTRATION, AND AEROBIC AND ANAEROBIC BIODEGRADATION. THESE TESTS ARE REQUIRED UNDER TSCA SECT 4(A).

REPT'D USE : PRODUCTION OF DYE CARRIERS, HEAT-TRANSFER FLUIDS FOR PLASTIC AND SYNTHETIC RIBONS, FOR ALKYLATED BIPHENYL AND FUNGICIDE

CAS NO : **90-41-5**
CHEM NAME : ***** 2-BIPHENYLAMINE HYDROCHLORIDE**

REPORT 01 : NTP BOARD OF SCIENTIFIC COUNSELORS JUN 81
THE BIOASSAY REPORT CONCLUSIONS FOR 2-BIPHENYLAMINE HCL WERE REVIEWED BY THE PEER REVIEW PANEL AND APPROVED WITH MINOR MODIFICATIONS. UNDER CONDITIONS OF THIS BIOASSAY, 2-BIPHENYLAMINE HCL ADMINISTERED IN THE FEED WAS CARCINOGENIC FOR FEMALE B6C3F1 MICE INDUCING ANGIOSARCOMAS BUT WAS NOT CARCINOGENIC FOR MALE MICE (POOR SURVIVAL MAY HAVE CONTRIBUTED TO LACK OF STATISTICAL EVIDENCE). THE CHEMICAL WAS NOT CARCINOGENIC FOR F344 RATS OF EITHER SEX.

REPORT 02 : NTP TECHNICAL BULLETIN NO. 7 APR 82
2-AMINOBIPHENYL WAS POSITIVE IN THE S. TYPHIMURIUM MUTAGENICITY ASSAY IN TWO LABORATORIES, AND NEGATIVE IN A THIRD.
REPT'D USE : CHEMICAL INTERMEDIATE, DYE INTERMEDIATE AND SOLVENT

CAS NO : **7411-49-6**
CHEM NAME : ***** 3,3',4,4'-BIPHENYL-TETRAMINE**

REPORT 06 : MULTI-AGENCY CROSS CHECK
N HHS MAGUIRE
Y OSHA CARC
N EPA CAG
N EPA RCRA
N EPA CLEAN AIR
N EPA CLEAN WATER
REPT'D USE : BIOLOGICAL STAIN

CAS NO : **2168-68-5**
CHEM NAME : *****BIS(1-AZIRIDINYL)-MORPHOLINOPHOSPHINE SULFIDE**

REPORT O1 : IARC MONOGRAPH 9 1975
BIS(1-AZIRIDINYL)MORPHOLINOPHOSPHINE SULFIDE IS CARCINOGENIC IN MICE FOLLOWING ITS INTRAPERITONEAL INJECTION, PRODUCING AN INCREASE IN THE INCIDENCE OF LUNG TUMORS.
REPT'D USE : EXPERIMENTAL ANTINEOPLASTIC AND INSECT CHEMOSTERILANT

CAS NO : **154-93-8**
CHEM NAME : ***** BISCHLORDETHYL NITROSOUREA (BCNU)**

REPORT 01 : NCI BIOASSAY PROGRAM - CANCER VOL. 40 1977
UNDER THE CONDITIONS OF THE BIOASSAY, I.P. INJECTION OF BCNU WAS CARCINOGENIC TO MALE SPRAGUE-DAWLEY (CD) RATS CAUSING LUNG TUMORS, AND TO FEMALE SWISS-WEBSTER MICE INDUCING LYMPHOSARCOMAS AND LUNG TUMORS. IT WAS NOT CARCINOGENIC TO FEMALE RATS AND MALE MICE.

REPORT 02 : IARC MONOGRAPH 26 1981
BCNU IS CARCINOGENIC IN RATS, PRODUCING LUNG TUMORS AFTER IP OR IV ADMINISTRATION, AND INTRA-ABDOMINAL TUMORS AFTER IP ADMINISTRATION. TESTS IN MICE BY IP ADMINISTRATION AND SKIN APPLICATION AND IN RATS BY ORAL ADMINISTRATION COULD NOT BE EVALUATED. THERE IS SUFFICIENT EVIDENCE FOR THE CARCINOGENICITY OF BCNU IN RATS. THE DATA FROM STUDIES IN HUMANS ARE INADEQUATE TO EVALUATE THE CARCINOGENICITY ON BCNU IN MAN.

REPORT 03 : IARC MONOGRAPH 26 1981
BCNU IS EMBRYO- AND FETOLETHAL IN RATS AND RABBITS AT DOSES NONTOXIC TO THE MOTHER AND CAN INDUCE A VARIETY OF SEVERE TERATOGENIC EFFECTS IN RATS. BCNU IS MUTAGENIC IN DROSOPHILA AND MAMMALIAN CELLS. IT ALSO PRODUCES CHROMOSOMAL ABERRATIONS IN MAMMALIAN CELLS BOTH IN CELL CULTURE AND IN VIVO.

REPORT 06 : MULTI-AGENCY CROSS CHECK
Y HHS MAGUIRE
N OSHA CARC
N EPA CAG
N EPA RCRA
N EPA CLEANAIR
N EPA CLEAN WATER
REPT'D USE : CHEMOTHERAPEUTIC AGENT

CAS NO : **111-44-4**
CHEM NAME : ***** BIS(2-CHLOROETHYL)ETHER (BCEE)**

REPORT 01 : EPA AMBIENT WATER QUALITY CRITERIA OCT 80
AMBIENT WATER CONCENTRATIONS, BASED ON THE CARCINOGENICITY OF BCEE BY ORAL ADMINISTRATION TO MALE MICE, OF .3 UG/L, .03 UG/L, AND .003 UG/L ARE ESTIMATED TO RESULT IN A RESPECTIVE 10 -5, 10 -6, AND 10 -7 INCREASE IN CANCER RISK OVER A LIFETIME.

REPORT 02 : IARC MONOGRAPH 9 1975
BIS(2-CHLOROETHYL)ETHER PRODUCED AN INCREASED INCIDENCE

OF LIVER-CELL TUMORS IN MALE MICE OF TWO STRAINS FOLLOWING ITS ORAL ADMINISTRATION. ITS ADMINISTRATION BY THE SUBCUTANEOUS ROUTE IN MICE PRODUCED A LOW INCIDENCE OF SARCOMAS AT THE INJECTION SITE.

REPORT 03 : EPA RCRA HEALTH AND ENV. EFFECTS PROFILE OCT 80
ORAL ADMINISTRATION OF BCEE DID NOT PRODUCE AN INCREASE OF TUMORS IN RATS. MALE MICE SHOWED A SIGNIFICANT INCREASE IN HEPATOMAS AFTER INGESTION OF BCEE. TESTING OF BCEE IN THE AMES SALMONELLA ASSAY, IN E. COLI AND IN SACCHAROMYCES HAS SHOWN THAT THIS COMPOUND INDUCES MUTAGENIC EFFECTS.

REPORT 06 : MULTI-AGENCY CROSS CHECK
N HHS MAGUIRE
N OSHA CARC
Y EPA CAG
Y EPA RCRA
N EPA CLEAN AIR
Y EPA CLEAN WATER

REPT'D USE : PAINTS/VARNISHES, DRY CLEANING, FUMIGANT, INSECTICIDE, SOLVENT, TEXTILE, SCOURING AGENT, INTERMEDIATE IN MFR OF MORPHOLINE AND LAB REAGENT

CAS NO : **39638-32-9**
CHEM NAME : ***** BIS(2-CHLOROISOPROPYL)ETHER (BCIE)**

REPORT 01 : EPA RCRA HFALTH AND ENV. EFFECTS PROFILE OCT 80
PRELIMINARY RESULTS FROM AN NTP BIOASSAY DO NOT SHOW AN INCREASE IN TUMORS FOLLOWING ORAL ADMINISTRATION OF BCIE. BCIE HAS PRODUCED MUTAGENIC EFFECTS IN TWO BACTERIAL TEST SYSTEMS, SALMONELLA AND E.COLI, BUT HAS FAILED TO SHOW MUTAGENICITY IN ONE MAMMALIAN STUDY.

REPORT 02 : EPA AMBIENT WATER QUALITY CRITERIA OCT 80
THERE IS NO EMPIRICAL EVIDENCE THAT BCIE IS CARCINOGENIC. HOWEVER, BECAUSE OF ITS MUTAGENIC ACTIVITY AND ITS CLOSE STRUCTURAL SIMILARITY TO BCEE, THE POSSIBLE CARCINOGENICITY OF BCIE IS A MATTER OF CONCERN. A CRITERION, BASED ON NONTUMOR PATHOLOGY OF A CHRONIC STUDY IN MICE SHOWING LIVER DAMAGE, AN UNCERTAINTY FACTOR OF 10000, AND AN ACCEPTABLE DAILY HUMAN INTAKE OF 70 UG, IS 34.7 UG/L.

REPORT 06 : MULTI-AGENCY CROSS CHECK
N HHS MAGUIRE
N OSHA CARC
N EPA CAG
Y EPA RCRA
N EPA CLEAN AIR

Y EPA CLEAN WATER
REPT'D USE : INTERMEDIATE IN PREPARATION OF ION EXCHANGE RESINS AND PESTICIDES

CAS NO : **13483-18-6**
CHEM NAME : ***** 1,2-BIS(CHLOROMETHOXY)ETHANE (BCMXE)**

REPORT 01 : IARC MONOGRAPH 15 1977
1,2-BIS(CHLOROMETHOXY)ETHANE IS CARCINOGENIC IN MICE, THE ONLY ANIMAL TESTED, FOLLOWING ITS APPLICATION TO THE SKIN OR ITS SUBCUTANEOUS OR INTRAPERITONEAL ADMINISTRATION; IT PRODUCED MALIGNANT TUMORS AT THE SITES OF ADMINISTRATION.

REPORT 06 : MULTI-AGENCY CROSS CHECK
N HHS MAGUIRE
Y OSHA CARC
N EPA CAG
N EPA RCRA
N EPA CLEAN AIR
N EPA CLEAN WATER
REPT'D USE : RUBBER STABILIZER, EPOXY CURING AGENTS, TEXTILE FINISHING AGENTS AND ION EXCHANGE RESINS. NO KNOWN CURRENT PRODUCTION

CAS NO : **56894-91-8**
CHEM NAME : ***** 1,4-BIS(CHLOROMETHOXYMETHYL)-BENZENE**

REPORT 01 : IARC MONOGRAPH 15 1977
1,4-BIS(CHLOROMETHOXYMETHYL)BENZENE IS CARCINOGENIC IN MICE, THE ONLY SPECIES TESTED, FOLLOWING ITS APPLICATION TO THE SKIN OR ITS SUBCUTANEOUS ADMINISTRATION; IT PRODUCED MALIGNANT TUMORS AT THE SITES OF APPLICATION.
REPT'D USE : ION-EXCHANGE RESINS. NO LISTING ON TSCA INVENTORY

CAS NO : **542-88-1**
CHEM NAME : ***** BIS(CHLOROMETHYL) ETHER (BCME)**

REPORT 01 : EPA AMBIENT WATER QUALITY CRITERIA OCT 80
AMBIENT WATER CONCENTRATIONS, BASED ON THE CARCINOGENICITY OF BCME BY ORAL ADMINISTRATION TO MALE MICE, OF .038 NG/L, .004 NG/L, AND .0004 NG/L, ARE ESTIMATED TO RESULT IN A RESPECTIVE 10 -5, 10 -6, AND 10 -7 INCREASE IN CANCER RISK OVER A LIFETIME.

REPORT 02 : IARC MONOGRAPH 4 1974
BCME IS CARCINOGENIC TO MICE FOLLOWING INHALATION, SKIN

APPLICATION AND SUBCUTANEOUS ADMINISTRATION. IN NEWBORN MICE IT IS CARCINOGENIC AFTER A SINGLE SUBCUTANEOUS EXPOSURE. IN THE RAT IT IS CARCINOGENIC BY INHALATION AND SUBCUTANEOUS ADMINISTRATION. A HIGH INCIDENCE OF PREDOMINANTLY OAT-CELLED CARCINOMA IN A SMALL POPULATION OF LABORATORY WORKERS EXPOSED TO BCME STRONGLY SUGGESTS THAT EXPOSURE TO THIS COMPOUND CONSTITUTES A SERIOUS HUMAN LUNG CANCER HAZARD.

REPORT 03 : EPA RCRA HEALTH AND ENV. EFFECTS PROFILE OCT 80
BCME HAS BEEN SHOWN TO PRODUCE TUMORS IN ANIMALS FOLLOWING ADMINISTRATION BY SC INJECTION, INHALATION, OR DERMAL APPLICATION. EPA STUDIES OF WORKERS IN THE U.S., GERMANY, AND JAPAN WHO WERE EXPOSED TO BCME AND CMME INDICATE THAT THESE COMPOUNDS ARE HUMAN RESPIRATORY CARCINOGENS. BCME HAS PRODUCED MUTAGENIC EFFECTS IN THE AMES SALMONELLA ASSAY AND IN E. COLI.

REPORT 04 : IARC MONOGRAPHS SUPP 4, 1982 OCT 82
THERE IS CONSIDERED TO BE SUFFICIENT EVIDENCE OF CARCINOGENICITY IN EXPERIMENTAL ANIMALS.

REPORT 06 : MULTI-AGENCY CROSS CHECK
Y HHS MAGUIRE
R OSHA CARC
Y EPA CAG
N EPA RCRA
Y EPA CLEAN AIR
Y EPA CLEAN WATER

REPT'D USE : RESIN, CHEMICAL INTERMEDIATE AND RESEARCH CHEMICAL

CAS NO : **108-60-1**
CHEM NAME : ***** BIS(2-CHLORO-1-METHYLETHYL) ETHER (BCMEE)**

REPORT 01 : NCI CARC TRS 191 DHEW PUB NO 79-1747
IT IS CONCLUDED THAT UNDER THE CONDITIONS OF THIS BIOASSAY, THE TECHNICAL-GRADE TEST MATERIAL, BIS(2-CHLORO-1-METHYLETHYL) ETHER, WAS NOT CARCINOGENIC FOR F344 RATS OF EITHER SEX (STOMACH INCUBATION IN RATS ONLY).

REPORT 02 : NTP BOARD OF SCIENTIFIC COUNSELORS DEC 81
UNDER THE CONDITIONS OF THIS BIOASSAY (TEMPERATURE EXCURSIONS), TECHNICAL GRADE BCMEE WAS CARCINOGENIC FOR B6C3F1 MICE CAUSING AN INCREASED INCIDENCE OF ALVEOLAR/BRONCHIOLAR ADENOMAS IN BOTH SEXES AND HEPATOCELLULAR CARCINOMAS AND ADENOMAS IN MALES. THE OCCURRENCE OF A FEW SQUAMOUS CELL PAPILLOMAS OR CARCINOMAS OF THE STOMACH OR FORESTOMACH IN BOTH SEXES (A RARE

TUMOR IN B6C3F1 MICE) MAY HAVE BEEN RELATED TO ADMINISTRATION OF BCMEE BY GAVAGE (SEE BIOASSAY FOR CAVEATS).

REPORT 03 : IARC VOL 41 1986
LIMITED EVIDENCE FOR CARCINOGENICITY OF CPD TO TEST ANIMALS. NO EVALUATION COULD BE MADE OF THE CARCINOGENICITY OF CPD TO HUMANS. INADEQUATE EVIDENCE FOR GENETIC ACTIVITY IN SHORT-TERM TESTS. NO DATA AVAILABLE TO EVALUATE REPRODUCTIVE OR PRENATAL EFFECTS OF CPD TO TEST ANIMALS. CPD IS MUTAGENIC TO S. TYPHIMURIUM BUT DOES NOT INDUCE SEX-LINKED RECESSIVE LETHAL MUTATIONS IN D. MELANOGASTER. NO DATA AVAILABLE TO EVALUATE REPRODUCTIVE EFFECTS OR PRENATAL TOXICITY TO HUMANS. NO CASE REPORT OR EPIDEMIOLOGICAL STUDY OF THE CARCINOGENICITY OF CPD TO HUMANS.

REPORT 06 : MULTI-AGENCY CROSS CHECK
N HHS MAGUIRE
N OSHA CARC
N EPA CAG
N EPA RCRA
N EPA CLEAN AIR
Y EPA CLEAN WATER

REPT'D USE : PAINTS/VARNISH REMOVERS, CLEANING AGENTS, SPOTTING AGENTS AND DYE/CHEMICAL INTERMEDIATE. NO KNOWN CURRENT PRODUCTION.

CAS NO : **21850-44-2**

CHEM NAME : ***** BIS(2,3-DIBROMPROPYL ETHER) OF TETRABROMBISPHENOL A**

REPORT 01 : EPA TSCA 8(E) NOTICE 8EHQ-0982-0458
GREAT LAKES CHEMICAL REPORTED FINAL RESULTS FROM SEVERAL SALMONELLA ASSAYS OF THE BIS(2,3-DIBROMPROPYL ETHER) OF TETRABROMBISPHENOL A. SUBMITTER STATES THAT THE RESULTS OF A 1976 ASSAY WERE NEGATIVE, BUT THREE PURIFIED SAMPLES OF THE CHEMICAL WERE RECENTLY TESTED AND FOUND TO INDUCE BASE-PAIR SUBSTITUTION MUTATION AND TO SHOW DETOXIFICATION TO A LARGE EXTENT BY THE ADDITION OF S-9 ACTIVATION MIX.

CAS NO : **3033-62-3**

CHEM NAME : *****BIS(2-DIMETHYLAMINOETHYL)ETHER (DMAEE) NIAX CATALYST A-99**

REPORT 01 : TSCA SECT 8(E) STATUS REPORT 8EHQ-0687-0683 JUL 87
RESULTS OF AN AMBIENT TEMPERATURE INHALATION STUDY USING MALE AND FEMALE S-D RATS WERE: AT 47 PPM, ALL RATS

DIED BETWEEN 6TH AND 11TH DAYS FROM START OF EXPOSURE; AT 90 PPM, ALL RATS DIED ON 3RD OR 4TH EXPOSURE DAYS; AT 47 PPM, FOOD AND WATER CONSUMPTION SIGNIFICANTLY REDUCED; BODY WEIGHT LOSS OCCURRED IN A CONCENTRATION-RELATED FASHION; HISTOLOGICAL EXAM OF LUNGS FROM MALES (ONLY TISSUES EXAMINED UP TO THIS TIME), SHOWED CYTOPLASMIC VACUOLATION OF BRONCHIAL CELLS AT ALL DMAEE CONCENTRATIONS AND A MILD PNEUMONITIS AT 47 AND 90 PPM.

REPT'D USE : CATALYST IN MANUFACTURE OF URETHANE FOAMS.

CAS NO : **26741-53-7**

CHEM NAME : *****BIS(2,4-DI-TERT-BU TYLPHENYL)PENTAERYTHRITOL DIPHOSPHITE**

REPORT 01 : EPA TSCA 8(E)-0706

"ANIMAL STUDIES HAVE INDICATED IN ONE TEST THAT THE MATERIAL IS A NEUROTOXIN AT THE HIGH DOSE IN DOGS. THREE OTHER TESTS IN OTHER TYPES OF ANIMALS, INCLUDING A 2 YEAR STUDY IN RATS, DID NOT PRODUCE NEUROTOXICITY. THIS MATERIAL IS CONSIDERED TO BE A SUSPECT NEUROTOXIN AT HIGH LEVELS OF EXPOSURE. SYMPTOMS OF NEUROTOXICITY CAN INCLUDE WEAKNESS, STAGGERED WALK, TREMORS, OR EVEN PARALYSIS."

REPORT 02 : EPA/TSCA 8(E) HQ-0488-1287- 0706 JUN 88

FOLLOWING STUDIES CONDUCTED BY BORG-WARNER: 90-DAY ORAL TOXICITY (RATS); TWO ACUTE AND TOXICITY STUDIES (CHICKENS); TWO PRIMARY DERMAL IRRITATION STUDIES (RABBITS); HYPERSENSITIVITY STUDY (GUINEA PIGS); TWO AMES SALMONELLA TYPHIMURIUM MUTAGENICITY STUDIES; ACUTE INHALATION LC50 STUDY; ACUTE DERMAL IRRITATION TOXICITY STUDY (RABBITS); SKIN CORROSIVITY STUDY (RABBITS); ACUTE ORAL DELAYED NEMOTOXICITY STUDY (CHICKS); 8-WEEK BIOACCUMULATION STUDY (FISH); 2-YEAR AND TOXICITY STUDY (RATS); ORAL TERATOLOGY STUDY (RABBITS); 6-MONTH ORAL TOXICITY STUDY (RATS); ORAL (FEEDING) REPRODUCTION STUDY (RATS); AND ACUTE ORAL TOXICITY STUDY (RATS). NO TOXIC EFFECTS REPORTED.

REPORT 03 : EPA 8EHQ-1287- 0706 MAR 88

ULTRAMOX 626 CAUSED PROGRESSIVE FORE AND HIND LIMB PARALYSIS IN A FEMALE DOG FED AT 40MG/KG/DAY FOR 86 DAYS; MYELIN LESIONS IN SEVEN HIGH DOSE DOGS. NO EFFECT IN OTHER DOGS AT 4 OR 12 MG/KG/DAY FOR 120 DAYS. RAT ORAL LD50 5.58G/KG, RAT INHALATION LC 50 > 2MG/L, RABBIT DERMAL LD50 > 200MG/KG. NO EFFECTS IN RATS FED EITHER FOR 90 DAYS AT 100, 300, OR 1000 PPM OR FOR 2 EYARS AT 100 OR 500 PPM.

REPT'D USE : STABILIZER IN POLYMER SYSTEMS

CAS NO : **2386-90-5**
CHEM NAME : ***** BIS(2,2-EPOXYCYCLOPENTENYL) ETHER**

REPORT 01 : EPA 8E STATUS REPORT 8EHQ-1285-0577 JAN 86
CPD MAY HAVE WEAK MUTAGENIC ACTIVITY ON GERM CELLS OF SPRAGUE-DAWLEY RATS. HOWEVER, NOT SEEN AS MAJOR CONTRIBUTOR FOLLOWING ORAL EXPOSURE TO 50:50 WT.% MIXTURE WITH THE DIGLYCIDYL ETHER OF BISPHENOL A.
REPT'D USE : EPOXY RESIN HARDENER/CURING AGENT

CAS NO : **6422-86-2**
CHEM NAME : ***** BIS(2-ETHYLHEXYL)-TEREPHTHALATE**

REPORT 01 : 48FR51845 NOV 83
RESPONDING TO ITC'S DESIGNATION OF DOTP FOR HEALTH AND ENVIRONMENTAL EFFECTS TESTING, KODAK CO PRESENTED A TESTING PROGRAM AND NTP NOMINATED A VARIETY OF CHEMICALS WITH THE 2-ETHYLHEXYL MOIETY FOR TOXICITY TESTING. THEREFORE EPA IS NOT INITIATING AN ANPR AT THIS TIME.

REPORT 02 : 52FR2152 - RECE IPT OF TEST DATA JAN 87
EPA ANNOUNCES RECEIPT OF DATA FROM INDUSTRY'S VOLUNTARY TESTING PROGRAM.
REPT'D USE : PLASTICIZER FOR VINYL PRODUCTS, WEATHERSTRIPPING, OUTDOOR CARPETING, AUTOMOTIVE HOSE AND SEALANTS AND WATERSTOPS

CAS NO : **111-96-6**
CHEM NAME : ***** BIS(2-METHOXYETHYL)ETHER**

REPORT 01 : NTP TECHNICAL BULLETIN NO. 7 APR 82
THE CHEMICAL WAS FOUND TO INDUCE ANTIFERTILITY EFFECTS, WITHOUT EVIDENCE OF DOMINANT LETHAL MUTATION, IN RATS. IT WAS POSITIVE IN THE MOUSE SPERM HEAD MORPHOLOGY ASSAY, BUT WAS NEGATIVE IN VITRO UNSCHEDULED DNA SYNTHESIS IN HUMAN FIBROBLASTS AND THE RAT BONE MARROW CYTOLOGY ASSAY. RESULTS FROM THE DROSOPHILA SEX-LINKED RECESSIVE LETHAL ASSAY WERE SUGGESTIVE BUT NOT CLEARLY POSITIVE.
REPT'D USE : INDUSTRIAL SOLVENT AND REACTION MEDIUM FOR ORGANOMETALLIC SYNTHESIS

CAS NO : **80-05-7**
CHEM NAME : ***** BISPHENOL A**

REPORT 01 : NTP TECHNICAL BULLETIN 2 APR 80
THE COMPOUND WAS FOUND TO BE NEGATIVE IN THE SALMO-

NELLA TYPHIMURIUM MUTAGENICITY ASSAY IN TWO LABORATORIES.

REPORT 02 : NTP BIOASSAY DRAFT REPORT NOV 80
UNDER THE CONDITIONS OF THIS BIOASSAY, BISPHENOL A WAS NOT CARCINOGENIC FOR F334 RATS OR B6C3F1 MICE (FEEDING STUDY).

REPORT 03 : EPA FINAL TEST RULE 51FR33047 SEP 86
EPA IS ISSUING A FINAL RULE TO CONDUCT A 90-DAY INHALATION SUBCHRONIC TOXICITY STUDY. EPA IS ALSO TERMINATING ITS TEST RULE PROCESS FOR ACUTE AND CHRONIC AQUATIC TOXICITY OF BPA.

REPT'D USE : COATINGS, ADHESIVES, FILTERS, PACKAGING, NAILPOLISHES, RESIN INTERMEDIATE, FLOORINGS, FUNGICIDE AND FR CHEMICAL

CAS NO : **1675-54-3**

CHEM NAME : ***** BISPHENOL A DIGLYCIDYL ETHER**

REPORT 01 : ITC THIRD REPORT TO EPA (43FR50631) OCT 78
ITC RECOMMENDS THAT THE EPA ADMINISTRATOR CONSIDER DEVELOPMENT OF TESTS FOR CARCINOGENICITY, MUTAGENICITY, TERATOGENICITY AND REPRODUCTIVE EFFECTS OF GLYCIDOL AND SELECTED DERIVATIVES.

REPORT 06 : MULTI-AGENCY CROSS CHECK
N HHS MAGUIRE
Y OSHA CARC
N EPA CAG
N EPA RCRA
N EPA CLEAN AIR
N EPA CLEAN WATER

REPT'D USE : LAMINATE BINDER, EPOXY RESIN AND POLYESTER RESIN

CAS NO : **3064-70-8**

CHEM NAME : ***** BIS(TRICHLOROMETHYL)SULFONE**

REPORT 01 : TSCA SECT 8(E) >STATUS REPORT 83HQ-0587-0673 JUN 87
N-1386 HAN (A MIXTURE OF BIS(TRICHLOROMETHYL)SULFONE AND PETROLEUM NAPHTHA SOLVENT), PRODUCED SIGNIFICANT PULMONARY EFFECTS LEADING TO DEATH OF ALL EXPOSED RATS AFTER ONE-HOUR INHALATION EXPOSURES AT 6.96 MG/L AND 1.91 MG/L. THEREFORE THE 1-HOUR LC50 MUST BE LESS THAN 1.91 MG/L.

REPT'D USE : INTERMEDIATE IN PESTICIDE PRODUCTION

CAS NO : **8052-42-4**
CHEM NAME : ***** BITUMENS (FROM PETROLEUM REFINING)**

REPORT 01 : NIOSH CRITERIA DOCUMENT SEP 77
NIOSH RECOMMENDS A CEILING OF 5 MG/CU M ASPHALT FUMES BASED ON EYE AND RESPIRATORY IRRITATION. PRESENT OSHA STANDARD IS 2.5 MG/CU M TWA.

REPORT 02 : IARC MONOGRAPH 35 JAN 85
THERE IS SUFFICIENT EVIDENCE FOR CARCINOGENICITY OF EXTRACTS OF STEAM-REFINED BITUMENS, AIR-REFINED BITUMENS AND POOLED MIXTURES OF STEAM-AND AIR-REFINED BITUMENS IN EXPERIMENTAL ANIMALS. ALSO LIMITED EVIDENCE FOR CARCINOGENICITY OF UNDILUTED STEAM-REFINED BITUMENS AND FOR CRACKING-RESIDUE BITUMENS IN ANIMALS; AND INADEQUATE EVIDENCE FOR CARCINOGENICITY OF UNDILUTED AIR-REFINED BITUMENS. IS REASONABLE, FOR PRACTICAL PURPOSES, TO REGARD THESE CHEMICALS AS IF THEY PRESENTED A CARCINOGENIC RISK TO HUMANS.

REPORT 03 : IARC MONOGRAPH 35 JAN 85
BOTH AN EXTRACT OF ROAD-SURFACING BITUMEN AND ITS EMISSIONS WERE MUTAGENIC TO SALMONELLA TYPHIMURIUM. ROAD-SURFACING BITUMEN, INCLUDING EMISSIONS, PRODUCED MUTATIONS IN PROKARYOTES.

REPORT 06 : MULTI-AGENCY CROSS CHECK
N HHS MAGUIRE
Y OSHA CARC
N EPA CAG
N EPA RCRA
N EPA CLEAN AIR
N EPA CLEAN WATER

REPT'D USE : ROAD/AIRFIELD PAVING, ROOFING, FLOORING, METAL PROTECTION AND INDUSTRIAL PAINTS

CAS NO : **11056-06-7**
CHEM NAME : ***** BLEOMYCIN**

REPORT 01 : IARC MONOGRAPH 26 MAY 81
THE AVAILABLE DATA FROM STUDIES IN EXPERIMENTAL ANIMALS AND IN HUMANS WERE INADEQUATE TO EVALUATE THE CARCINOGENICITY OF BLEOMYCINS TO MAN.

REPORT 02 : IARC MONOGRAPH 26 MAY 81
NO DATA WERE AVAILABLE TO EVALUATE THE TERATOGENIC POTENTIAL OF BLEOMYCIN. BLEOMYCIN IS MUTAGENIC IN DROSOPHILA. IT INDUCES CHROMOSOMAL CHANGES AND INCREASES IN SISTER CHROMATID EXCHANGES IN VARIOUS MAMMALIAN

CELLS IN CULTURE. THE DRUG ALSO INDUCED NEOPLASTIC TRANSFORMATION IN A MOUSE CELL LINE.

REPT'D USE : ANTINEOPLASTIC AGENT

CAS NO : **CLASS-36**

CHEM NAME : ***** BOOT AND SHOE MANUFACTURE AND REPAIR**

REPORT 01 : IARC MONOGRAPH 25 SEP 81

EMPOLYMENT IN BOOT AND SHOE INDUSTRY IS CAUSALLY ASSOCIATED WITH THE DEVELOPMENT OF NASAL ADENOCARCINOMAS. MOST LIKELY EXPOSURE TO LEATHER DUST PLAYS A ROLE. EVIDENCE OF INCREASED RISK OF BLADDER CANCER ASSOCIATED WITH EMPLOYMENT IN LEATHER INDUSTRY. OCCURRENCE OF LEUKEMIA AND BENZENE IS WELL DOCUMENTED. HYPOTHESIS-GENERATING SURVEYS SUGGESTED ASSOCIATIONS BETWEEN BOOT AND SHOE MANUFACTURE AND REPAIR AND CANCER OF LUNG, ORAL CAVITY AND PHARYNX AND STOMACH. THESE FINDINGS CANNOT BE EVALUATED.

CAS NO : **7637-07-2**

CHEM NAME : ***** BORON TRIFLUORIDE**

REPORT 01 : NIOSH CRITERIA DOCUMENT DEC 76

NIOSH DOES NOT RECOMMEND A STANDARD ALTHOUGH RESPIRATORY SYSTEM EFFECTS HAVE BEEN ASSOCIATED WITH EXPOSURE. THE PRESENT OSHA STANDARD IS 1 PPM CEILING.

REPT'D USE : MAGNESIUM CASTING, NUCLEAR APPLICATIONS, FUMIGANT/INSECTICIDE AND INTERMEDIATE FOR BORANES

CAS NO : **UNKNOWN-31**

CHEM NAME : ***** BRACKEN FERN**

REPORT 01 : IARC VOL 40 1986

SUFFICIENT EVIDENCE FOR CARCINOGENICITY OF CPD IN TEST ANIMALS. LIMITED EVIDENCE FOR DERIVATIVE OF CPD IN EXPERIMENTAL ANIMALS. (PATAQUILOSIDE) AND INADEQUATE EVIDENCE FOR CARCINOGENICITY IN TEST ANIMALS IN ANOTHER DERIVATIVE, SHIKIMIC ACID. THERE IS INADEQUATE EVIDENCE FOR CARCINOGENICITY OF CPD TO HUMANS.

REPT'D USE : FOOD

CAS NO : **UNKNOWN-18**
CHEM NAME : *****BRILIANT RED HISALONIC**

REPORT 01 : CPSC MONOGRAPH ON FIBER REACTIVE DYES DEC 81
TREATED RATS EXHIBITED DECREASED WEIGHT GAIN AND A STATISTICALLY SIGNIFICANT DECREASE IN SERUM PROTEINS AND MORPHOLOGICAL ALTERATIONS OF THE LIVER AND KIDNEY. A MORPHOLOGICAL MODIFICATION OF WHITE BLOOD CELLS WAS ALSO OBSERVED.
REPT'D USE : DYE. NO AVAILABLE PRODUCTION INFORMATION

CAS NO : **105-36-2 SCREEN CODE: 3 CAI: NO**
CHEM NAME : ***** BROMOACETIC ACID, ETHYL ESTER**

REPORT 06 : MULTI-AGENCY CROSS CHECK
N HHS MAGUIRE
Y OSHA CARC
N EPA CAG
N EPA RCRA
N EPA CLEAN AIR
N EPA CLEAN WATER
REPT'D USE : CHEMICAL INTERMEDIATE IN ORGANIC SYNTHESIS

CAS NO : **75-27-4**
CHEM NAME : ***** BROMODICHLOROETHANE**

REPORT 01 : EPA RCRA HEALTH AND ENV. EFFECTS PROFILE OCT 80
DICHLOROBROMOMETHANE GAVE POSITIVE RESULTS IN MUTAGENICITY TESTS WITH SALMONELLA.

REPORT 02 : EPA AMBIENT WATER QUALITY CRITERIA OCT 80
IT IS NOT PRESENTLY POSSIBLE TO DERIVE A VALID WATER QUALITY CRITERION FOR BROMOCHLOROMETHANE. HOWEVER, FOR THE MAXIMUM PROTECTION OF HUMAN HEALTH FROM THE POTENTIAL CARCINOGENIC EFFECTS FROM BROMOCHLOROMETHANE, CONCENTRATIONS OF 1.9 UG/L, .19 UG/L AND .019 UG/L ARE ESTIMATED TO RESULT IN AN INCREMENTAL INCREASE OF CANCER RISK OVER LIFETIME OF 10 -5, 10 -6, AND 10 -7 RESPECTIVELY.

REPORT 03 : NTP DRAFT BIOASSAY TR REPORT AUG 86
UNDER CONDITIONS OF 2-YEAR GAVAGE STUDIES, CLEAR EVIDENCE OF CARCINOGENICITY OF BROMODICHLOROMETHANE FOR MALE AND FEMALE F344/N RATS AND B6C3F1 MICE, AS SHOWN BY INCREASED INCIDENCE OF TUBULAR CELL ADENOMAS AND ADENOCARCINOMAS IN KIDNEY AND ADENOMATOUS POLYPS AND ADENOCARCINOMAS OF LARGE INTESTINE IN RATS; INCREASED INCIDENCE OF TUBULAR CELL ADENOMAS AND ADENOCARCINOMAS IN KIDNEY OF MALE MICE; AND INCREASED INCIDENCE OF

HEPATOCELLULAR ADENOMAS AND CARCINOMAS IN FEMALE MICE.

REPORT 04 : NTP TECHNICAL BULLETIN 7 APR 82
BROMODICHLOROMETHANE WAS NEGATIVE IN THE SALMONELLA TYPHIMURIUM MUTAGENICITY ASSAY.

REPORT 06 : MULTI-AGENCY CROSS CHECK
Y HHS MAGUIRE
N OSHA CARC
N EPA CAG
N EPA RCRA
N EPA CLEAN AIR
Y EPA CLEAN WATER
REPT'D USE : CHEMICAL INTERMEDIATE, FR CHEMICAL AND SOLVENT

CAS NO : **1817-73-8**
CHEM NAME : ***** 2-BROMO-4,6-DINITROANLINE**

REPORT 01 : NTP TECHNICAL BULLETIN 9 APR 83
2-BROMO-4,6-DINITROANLINE WAS POSITIVE IN ONE SALMONELLA TYPHIMURIUM MUTAGENICITY ASSAY.

CAS NO : **74-96-4**
CHEM NAME : ***** BROMOETHANE**

REPORT 01 : NTP TR 363 NIH 90-2818 OCT 89
2-YEAR INHALATION STUDIES SHOWED SOME EVIDENCE OF CARCINOGENIC ACTIVITY OF BROMOETHANE FOR MALE F344/N RATS, INDICATED BY MARGINALLY INCREASED INCIDENCE OF PHEOCHROMOCYTOMAS OF THE ADRENAL GLAND AND NEOPLASMS OF THE BRAIN AND LUNG. FOR FEMALE F344/N RATS, THERE WAS EQUIVOCAL EVIDENCE OF CARCINOGENIC ACTIVITY, INDICATED BY MARGINALLY INCREASED INCIDENCES OF NEOPLASMS OF THE BRAIN AND LUNG. FOR MALE B6C3F1 MICE, SHOWED EQUIVOCAL EVIDENCE OF CARCINOGENIC ACTIVITY, BASED ON MARGINALLY INCREASED INCIDENCES OF NEOPLASMS OF THE LUNG. THERE WAS CLEAR EVIDENCE OF CARCINOGENIC ACTIVITY FOR FEMALE B6C3F1 MICE, AS INDICATED BY NEOPLASMS OF THE UTERUS.
REPT'D USE : ALKYLATING AGENT IN ORGANIC SYNTHESIS, MANUFACTURE OF PHARMACEUTICALS, ETHYLATION OF GASOLINE, FUMIGANT, REFRIGERANT AND SOLVENT CHEMICAL INTERMEDIATE

CAS NO : **74-83-9**
CHEM NAME : *****BROMOMETHANE (BM) (METHYL BROMIDE)**

REPORT 01 : NTP TECHNICAL BULLETIN NO. 3/7 DEC 80
TERATOLOGY ASSAYS SHOW THAT INHALATION OF METHYL BROMIDE IN RATS AND RABBITS WAS NOT ASSOCIATED WITH PRENATAL DEATH OR INCREASED MALFORMATIONS AT DOSES TOXIC TO THE MOTHER. THE CHEMICAL WAS NEGATIVE IN THE RAT DOMINANT LETHAL ASSAY, THE MOUSE SPERM HEAD MORPHOLOGY ASSAY, THE DROSOPHILA SEX-LINKED RECESSIVE LETHAL ASSAY, THE IN VITRO UNSCHEDULED DNA SYNTHESIS IN HUMAN FIBROBLASTS, AND THE RAT BONE MARROW CYTOLOGY ASSAY.

REPORT 02 : EPA AMBIENT WATER QUALITY CRITERIA OCT 80
BROMOMETHANE HAS NOT BEEN ADEQUATELY TESTED FOR CARCINOGENICITY. A CRITERION, BASED ON THE ACGIH TLV OF 20 PPM FOR KNOWN SYSTEMIC TOXICITY IN HUMANS, AND A SAFETY FACTOR OF 100, IS 1.39 MG/L. FOR THE MAXIMUM PROTECTION OF HUMAN HEALTH FROM THE POTENTIAL CARCINOGENIC EFFECTS FROM BROMOMETHANE, CONCENTRATIONS OF 1.9 UG/L, .19 UG/L AND .019 UG/L ARE ESTIMATED TO RESULT IN AN INCREMENTAL INCREASE OF CANCER RISK OVER THE LIFETIME OF 10 -5, 10 -6, AND 10 -7, RESPECTIVELY.

REPORT 03 : EPA RCRA HEALTH AND ENV. EFFECTS PROFILE OCT 80
THERE IS NO INFORMATION ON THE CHRONIC TOXICITY, CARCINOGENICITY, OR TERATOGENICITY OF BROMOMETHANE. BROMOMETHANE HAS BEEN SHOWN TO BE MUTAGENIC IN THE AMES SALMONELLA ASSAY.

REPORT 04 : NIOSH CURRENT INTELLIGENCE BULLETIN 43 SEP 84
THE MONOHALOMETHANES (METHYL CHLORIDE, METHYL BROMIDE, AND METHYL IODIDE) ARE DIRECT-ACTING MUTAGENS IN AMES ASSAY; AND IN STUDIES IN RATS OR MICE BY VARIOUS ROUTES OF ADMINISTRATION, HAVE DEMONSTRATED ABILITY TO PRODUCE CANCER. METHYL CHLORIDE PRODUCED TERATOGENIC EFFECT (HEART DEFECTS) IN OFFSPRING OF PREGNANT MICE EXPOSED BY INHALATION AT 500 AND 750 PPM. NIOSH RECOMMENDS THEY BE CONSIDERED POTENTIAL OCCUPATIONAL CARCINOGENS AND THAT METHYL CHLORIDE BE CONSIDERED A POTENTIAL OCCUPATIONAL TERATOGEN.

REPORT 06 : MULTI-AGENCY CROSS CHECK
N HHS MAGUIRE
N OSHA CARC
N EPA CAG
Y EPA RCRA
N EPA CLEAN AIR
Y EPA CLEAN WATER

REPORT 07 : IARC VOL 41 1986
LIMITED EVIDENCE FOR CARCINOGENICITY OF BM TO TEST ANIMALS BUT INADEQUATE EVIDENCE FOR CARCINOGENICITY OF BM TO HUMANS. SUFFICIENT EVIDENCE OF GENETIC ACTIVITY IN SHORT-TERM TESTS. NO DATA AVAILABLE TO EVALUATE REPRODUCTIVE EFFECTS OR PRENATAL TOXICITY TO HUMANS. BM TESTED IN RATS BY GAVAGE CAUSED AN INCREASED INCIDENCE OF SQUAMOUS CELL CARCINOMAS OF FORESTOMACH IN EACH SEX. BM MUTAGENIC TO PLANTS AND BACTERIA. NO DATA AVAILABLE TO EVALUATE REPRODUCTIVE EFFECTS IN ANIMALS. CPD DID NOT CAUSE PRENATAL TOXIC EFFECTS IN RATS OR RABBITS EXPOSED BY INHALATION.

REPORT 08 : NTP DRAFT TR 385 NIH PUB. 91-2840 NOV 90
UNDER 2-YEAR INHALATION STUDIES, METHYL BROMIDE CAUSED DEGENERATIVE CHANGES IN CEREBELLUM AND CEREBRUM, MYOCARDIAL DEGENERATION AND CARDIOMYOPATHY, STERNAL DYSPLASIA, AND OLFACTORY EPITHELIAL NECROSIS AND METAPLASIA. NO EVIDENCE OF CARCINOGENIC ACTIVITY IN MALE AND FEMALE B6C3F1 MICE. TOXIC EFFECTS PERSISTED ALTHOUGH EXPOSURE IN 100 PPM GROUP TERMINATED AFTER 20 WEEKS.

REPT'D USE : FUMIGANT, CROP DISINFECTANT, VEGETABLE OIL EXTRACTION AND CHEMICAL INTERMEDIATE

CAS NO : **16238-56-5**

CHEM NAME : ***** 7-BROMOMETHYL-12-METHYL-BENZ(A)ANTHRACENE**

REPORT 01 : NTP TECHNICAL BULLETIN 8 JUL 82
THE CHEMICAL WAS POSITIVE IN THE L5178Y MOUSE LYMPHOMA ASSAY.

CAS NO : **106-99-0**

CHEM NAME : ***** 1,3-BUTADIENE**

REPORT 01 : EPA TSCA 8(E) NOTICE 8EHQ-0382-0441
DOW CHEMICAL PROVIDED RESULTS OF AN INHALATION TERATOLOGY STUDY OF 1,3-BUTADIENE IN S-D RATS. EXPOSURE OF RATS TO 8000 PPM OF 1,3-BUTADIENE DURING DAYS 6-15 OF GESTATION RESULTED IN A SIGNIFICANT INCREASE IN MAJOR SKELETAL ABNORMALITIES. THERE WAS NO EVIDENCE OF TERATOGENICITY AT EXPOSURE LEVELS OF 200 AND 1000 PPM.

REPORT 02 : NTP CARC TEST PROGRAM JUL 82
IN A CURRENT NTP BIOASSAY, 1,3-BUTADIENE IS BEING ADMINISTERED BY INHALATION TO B6C3F1 MICE. THE CHRONIC EXPOSURE PHASE IS IN PROGRESS.

REPORT 03 : EPA TSCA 8(E) NOTICE 8EHQ-1080-0370 OCT 80
EXXON CHEMICAL AMERICAS SUBMITTED PRELIMINARY RESULTS FROM A LONG TERM INHALATION STUDY IN SPRAGUE-DAWLEY RATS WHICH INDICATE AN INCREASE IN TOTAL MAMMARY TUMORS AT BOTH 8000 AND 1000 PPM OVER THAT OBSERVED IN THE CONTROLS. MALE RATS SHOWED IN INCREASE IN INCIDENCE OF LEYDIG CELL TUMORS (BENIGN) OF THE TESTES. THERE IS ALSO A SUGGESTION OF AN INCREASE IN TUMORS OF SEVERAL OTHER ORGAN SYSTEMS.

REPORT 04 : NIH PUB 84-2544 NTP TECHNICAL REPORT 288 AUG 84
CLEAR EVIDENCE OF CARCINOGENICITY IN MALE AND FEMALE B6C3F1 MICE EXPOSED TO 625 OR 1,250PPM OF 1,3-BUTADIENE SHOWN BY INCREASED INCIDENCES AND EARLY INDUCTION OF HEMANGIOSARCOMAS OF HEART, MALIGNANT LYMPHOMAS, ALVEOLAR/BRONCHIOLAR ADENOMAS AND CARCINOMAS, AND PAPILLOMAS OF STOMACH IN MALES AND FEMALES AND OF ACINAR CELL CARCINOMAS OF OVARY AND HEPATOCELLULAR ADENOMAS AND ADENOMAS OR CARCINOMAS (COMBINED) IN FEMALES. 1,3-BUTADIENE ASSOCIATED WITH NONNEOPLASTIC LESIONS IN RESPIRATORY EPITHELIUM, LIVER NECROSIS AND TESTICULAR OR OVARIAN ATROPHY.

REPORT 05 : IARC MONOGRAPH 39, 1986 JUN 85
1,3-BD CARCINOGENIC TO BOTH SEXES BY INHALATION. NO DATA AVAILABLE TO EVALUATE REPRODUCTIVE EFFECTS OR PRENATAL TOXICITY IN HUMANS OR ANIMALS. HUMAN STUDIES GIVE EVIDENCE OF ASSOCIATION BETWEEN EXCESS LEUKEMIA AND LYMPHOMAS AND EXPOSURE TO 1,3-BD AND SOLVENTS SUCH AS BENZENE AND MONOMERS SUCH AS STYRENE. IARC: SUFFICIENT EVIDENCE FOR CARCINOGENICITY OF 1,3-BD IN ANIMALS; INADEQUATE EVIDENCE FOR CARCINOGENICITY IN HUMANS.

REPORT 06 : MULTI-AGENCY CROSS CHECK
Y HHS MAGUIRE
N OSHA CARC
N EPA CAG
N EPA RCRA
N EPA CLEAN AIR
N EPA CLEAN WATER

REPORT 07 : EPA TSCA 8EHQ-0 484-0511 MAY 84
WEAKLY POSITIVE IN IN-VITRO PRIMARY RAT HEPATOCYTE DNA REPAIR TEST, WEAKLY POSITIVE IN AN IN-VIVO MOUSE MICRONUCLEUS TEST, NEGATIVE IN A CHINESE HAMSTER OVARY CELL MUTAGENICITY TEST AND NEGATIVE IN AN IN-VITRO BALB/C MOUSE CELL TRANSFORMATION TEST.

REPT'D USE : CHEMICAL INTERMEDIATE, RUBBER/RESIN MONOMER AND FLAME RETARDANT

CAS NO : **55-98-1**
CHEM NAME : ***** 1,4-BUTANEDIOL DIMETHANESULPHONATE**

REPORT 01 : IARC MONOGRAPH 4 1974
ADMINISTRATION OF 1,4-BUTANEDIOL DIMETHANESULPHONATE TO THE MOUSE BY INTRAPERITONEAL INJECTION AND TO THE RAT BY ORAL ADMINISTRATION DID NOT SIGNIFICANTLY INCREASE THE INCIDENCE OF TUMORS. INTRAVENOUS ADMINISTRATION IN THE MOUSE SIGNIFICANTLY INCREASED THE INCIDENCE OF THYMIC LYMPHOMAS AND OVARIAN TUMORS. ALTHOUGH THERE IS EVIDENCE THAT HISTOLOGICAL AND CYTOLOGICAL CHANGES ARE ASSOCIATED WITH MYLERAN THERAPY, THERE IS NO FIRM EVIDENCE OF AN INCREASED CANCER RISK AMONG THOSE TREATED.

REPORT 02 : NTP TECHNICAL BULLETIN 9 APR 83
MYLERAN WAS POSITIVE IN ONE SALMONELLA TYPHIMURIUM MUTAGENICITY ASSAY.

REPORT 06 : MULTI-AGENCY CROSS CHECK NOV 86
Y HHS MAGUIRE
N OSHA CARC
N EPA CAG
N EPA RCRA
N EPA CLEANAIR
N EPA CLEAN WATER

REPT'D USE : CHEMOTHERAPEUTIC AGENT

CAS NO : **111-76-2**
CHEM NAME : ***** 2-BUTOXYETHANOL (EGMBE)**

REPORT 01 : EPA TSCA 8(E) 8EHQ-0483-0475 APR 83
THE EASTMAN KODAK COMPANY SUBMITTED AN INTEREM REPORT ON THE INHALATION TERATOLOGY POTENTIAL OF ETHYLENE GLYCOL MONOBUTYL ETHER (EGMBE) IN THE RAT. PREGNANT FISCHER 344 RATS WERE EXPOSED TO EGMBE VAPOR CONCENTRATIONS OF 0, 100, 200 OR 300 PPM FOR 6 HRS/DAY ON DAYS 6-15 OF GESTATION. THE ONLY SIGNIFICANT RESULTS WERE FETOTOXIC EFFECTS SUCH AS CIRCULATORY ABNORMALITES AND A HIGHER INCIDENCE OF EMBRYO AND FETAL LETHALITY.

REPORT 02 : EPA TSCA 8(E) 8EHQ-0584-0517 JUN 84
SOLVENT-CUTBACK RUST PREVENTIVE (STODDARD SOLVENT - 90.9 VOL%/ 2-BUTOXYETHANOL - 2.5 VOL%) CARCINOGENIC IN SKIN PAINTING STUDY: INDUCED SQUAMOUS CELL CARCINOMAS IN 12 % OF C3H/HEJ MICE TESTED.

REPT'D USE : SOLVENT, CLEANERS, PAINTS, FLAME RETARDANTS, METAL POLISH, SWIMMING POOL CHEMICAL, AEROSOLS, RESINS AND WAX

CAS NO : **124-17-4**
CHEM NAME : ***** 2-(2-BUTOXYETHOXY) ETHYL ACETATE (DGBA)**

REPORT 01 : 49FR45606 - EPA RESPONSE TO ITC NOV 84
IN RESPONSE TO ITC'S DESIGNATION OF DGBA FOR PRIORITY TESTING FOR REPPRODUCTIVE EFFECTS, SUBCHRONIC TOXICITY AND TOXICOKINETIC STUDIES, EPA IS ISSUING AN ANPR TO EXPAND SCOPE TO INCLUDE DIETHYLENE GLYCOL BUTYL ETHER (DGBE), TO DEFINE PROPOSED TESTING AND TO SEEK PUBLIC COMMENT.

REPORT 02 : EPA PROPOSING TEST RULE, 51FR27880 AUG 86
EPA IS PROPOSING THAT MFRS AND PROCESSORS BE REQUIRED TO PERFORM HEALTH EFFECTS TESTING FOR SUBCHRONIC TOXICITY, WITH EMPHASIS ON REPRODUCTIVE; HEMATOLOGICAL; LIVER AND KIDNEY EFFECTS; AND NEUROLOGICAL/BEHAVIORAL EFFECTS; DEVELOPMENTAL NEUROTOXICITY; PHARMACOKINETIC MUTAGENICITY; AND ONCOGENICITY. EPA ALSO PROPOSING DERMAL ABSORPTION TESTING OF DGBA. EPA USING EXPOSURE TO DGBA IN LATEX PAINT AS THE BASIS FOR SECT 4(A)(l) FINDING.
REPT'D USE : SOLVENT AND CARRIER IN INKS AND INDUSTRIAL COATINGS, LATEX PAINTS, BRAKE FLUIDS, METAL CLEANERS, STAMP PAD INKS AND FLOOR CLEANERS

CAS NO : **123-86-4**
CHEM NAME : ***** N-BUTYL ACETATE**

REPORT 01 : EPA STAT REPORT 8EHQ-0387-0659 APR 87
5 MALE AND 5 FEMALE WISTAR RATS WERE EXPOSED TO AN AEROSOL VAPOR OF CPD AT CONCENTRATIONS OF 0.8,2.2 AND 5.2 MG/L IN A HEAD-ONLY DYNAMIC INHALATION SYSTEM. THE MASS MEDIAN AERODYNAMIC DIAMETER (MMAD) AT THE HIGH DOSE LEVEL WAS 1.01 UM (STD DEV. = 3.30 UM). MORTALITY RATE WAS 6/10, 10/10 AND 10/10, RESPECTIVELY. NO SEX RELATED EFFECT SEEN. BASED ON CONCENTRATION OF CPD IN BREATHING ZONE, THE LC50 ESTIMATED TO BE 0.74 MG/L OR 160 PPM.
REPT'D USE : SOLVENT AND GASOLINE ADDITIVE

CAS NO : **141-32-2**
CHEM NAME : ***** N-BUTYL ACRYLATE**

REPORT 01 : NTP TECHNICAL BULLETIN NO. 3 DEC 80
TERATOLOGY ASSAYS SHOW THAT ORAL ADMINISTRATION OF N-BUTYL ACRYLATE CAUSES INCREASED MALFORMATIONS AT DOSES NOT TOXIC TO THE MOTHER.

REPORT 02 : IARC MONOGRAPH 39, 1986 JUN 85
NO TREATMENT RELATED TUMORS SEEN IN MICE TESTED FOR

CARCINOGENICITY VIA SKIN APPLICATIONS. NO NEOPLASTIC EFFECTS SEEN IN RATS EXPOSED TO N-BA VIA INHALATION FOR 2 YEARS. INHALATION OF MATERNALLY-TOXIC CONCENTRATIONS OF N-BA PRODUCED EMBRYOTOXICITY, NO SIGNS OF FETOTOXICITY OR TERATOGENICITY. N-BA NOT MUTAGENIC TO S.TYPHIMURIUM WITHOUT ACTIVATION. NO DATA AVAILABLE TO EVALUATE HUMAN REPRODUCTIVE EFECTS OR PRENATAL TOXICITY OF N-BA. NO REPORTS AVAILABLE TO EVALUATE CARCINOGENICITY OF N-BA TO HUMANS. IARC CONCLUDES THAT INADEQUATE EVIDENCE TO EVALUATE THE CARCINOGENICITY OF N-BA IN HUMANS.

REPT'D USE : PAINTS, PLASTIC MONOMER, LEATHER, FINISHES AND FR CHEMICAL

CAS NO : **75-65-0**

CHEM NAME : ***** T-BUTYL ALCOHOL (T-BUTANOL)**

REPORT 01 : SAFETY ASSESS'T REPT COSMETICS INGREDIENT REVIEW PANEL JUL 87

DATA INSUFFICIENT TO SUPPORT SAFETY OF CPD IN COSMETICS. RATS AND MICE GIVEN CPD IN DRINKING WATER FOR 90 DAYS DEVELOPED GROSS LESIONS IN URINARY TRACT. ORAL ADMINISTRATION TO PREGNANT MICE RESULTED IN POOR INITIAL BEHAVIOR OF PUPS WITH RECOVERY LATER. MUTAGENIC TO CULTURED HUMAN AND CHINESE HAMSTER OVARY HYBRID CELLS. RESULTS OF SKIN TESTS ON MICE BELIEVED TO SHOW CPD INACTIVE AS COMPLETE CARCINOGEN OR TUMOR PROMOTER. CAN CAUSE DERMATITIS IN HUMANS. NIOSH REPORTS 8000 PPM DANGEROUS TO HEALTH AND LIFE. OSHA AND ACGIH EXPOSURE STANDARD AT LTV-TWA OF 100 PPM.

REPT'D USE : SOLVENT IN COSMETICS, DENATURANT FOR ETHANOL, PAINT REMOVER AND FLAVORING AGENT

CAS NO : **25013-16-5**

CHEM NAME : ***** BUTYLATED HYDROXYANISOLE (BHA)**

REPORT 01 : IARC VOL 40 1986

THERE IS SUFFICIENT EVIDENCE FOR CARCIONOGENICITY OF BHA TO TEST ANIMALS, BUT NO DATA AVAILABLE TO EVALUATE CARCINOGENICITY OF BHA TO HUMANS. BHA NOT MUTAGENIC TO S. TYPHIMURIUM, D. MELANOGASTER OR TO CHINESE HAMSTER CELLS IN VITRO. DID NOT CAUSE CHROMOSOMAL DEFECTS IN D. MELANOGASTER OR CULTURED CHINESE HAMSTER CELLS. BHA OFTEN MODIFIED DNA DAMAGING MUTAGENIC AND CLASTOGENIC ACTIVITIES OF KNOWN MUTAGENS AND CARCINOGENS. BHA ALSO REDUCED ACTIVITY OF INDIRECTLY ACTING MUTAGENS AND CARCINOGENS.

REPORT 06 : MULTI-AGENCY CROSS CHECK
Y HHS MAGUIRE
N OSHA CARC
N EPA CAG
N EPA RCRA
N EPA CLEAN AIR
N EPA CLEAN WATER

REPT'D USE : ANTIOXIDANT AND PRESERVATIVE IN FOODS AND COSMETICS

CAS NO : **128-37-0**
CHEM NAME : ***** BUTYLATED HYDROXYTOLUENE (BHT)**

REPORT 01 : NCI CARC IRS 150 DHEW PUB NO 79-1706
IT IS CONCLUDED THAT UNDER THE CONDITIONS OF THIS BIOASSAY BHT WAS NOT CARCINOGENIC FOR F344 RATS OR B6C3F1 MICE (FEEDING STUDY).

REPORT 02 : FDA GRAS LISTING 21CF4182.3173
BHT IS GENERALLY RECOGNIZED AS SAFE FOR USE AS A CHEMICAL PRESERVATIVE IN FOODS WHEN THE TOTAL CONTENT OF ANTIOXIDANTS IS NOT OVER .02 PERCENT OF FAT OR OIL CONTENT, PROVIDED THE SUBSTANCE IS USED IN ACCORDANCE WITH GOOD MANUFACTURING PRACTICE.

REPORT 03 : NTP TECHNICAL BULLETIN 6 JAN 82
THE COMPOUND WAS NEGATIVE IN THE SALMONELLA TYPHIMURIUM MUTAGENICITY ASSAY IN ONE LABORATORY.

REPORT 04 : NAS REPORT ON DIET, NUTRITION AND CANCER 1982
AT LEAST ONE ADEQUATE BIOASSAY TO TEST THE CARCINOGENICITY OF BHT PROVIDES NO CLEAR EVIDENCE OF CARCINOGENICITY UNDER THE CONDITIONS OF THE TEST. TWO OTHER STUDIES PROVIDE EVIDENCE THAT BHT HAS A TUMOR PROMOTING EFFECT, ESPECIALLY FOR URETHANE AND 2-AAF. ON THE OTHER HAND, LARGE AMOUNTS OF BHT HAVE BEEN SHOWN TO INHIBIT NEOPLASIA INDUCED BY A NUMBER OF CHEMICALS.

REPORT 05 : EPA FYI-OTS -1183-0275 NOV 83
THE CHEMICAL MANUFACTURERS ASSOCIATION REPORTED PRELIMINARY RESULTS OF A LONG-TERM TOXICITY STUDY ON BHT IN RATS. MALES AND FEMALES WERE EXPOSED FOR 13 WEEKS AT 0, 25, 100 AND 500 MG/KG, METES, AND OFFSPRING EXPOSED FOR 144 WEEKS AT 0, 25, 100, AND 250 MG/KG. BASED ON STATISTICALLY SIGNIFICANT POSITIVE TREND FOR HEPATOCELLULAR CARCINOMAS AND INCREASE OF HEPATOCELLULAR ADENOMAS BHT IS CARCENOGENIC IN MALE RATS. ALSO STATISTICALLY SIGNIFICANT POSITIVE TREND OF HEPATOCELLULAR ADENOMAS DEMONSTRATED FOR FEMALE RATS.

REPORT 07 : IARC VOL 40 1986
LIMITED EVIDENCE FOR THE CARCINOGENICITY OF BHT IN TEST ANIMALS, BUT NOT ABLE TO EVALUATE CARCINOGENICITY IN HUMANS. EVIDENCE INADEQUATE TO EVALUATE GENETIC ACTIVITY IN SHORT-TERM TESTS. BHT DID NOT INDUCE DNA DAMAGE IN B. SUBTILIS OR MUTATION IN S. TYPHIMURIUM. DID NOT INDUCE CHROMOSOMAL ABERRATIONS IN PLANTS OR MUTATION OR CHROMOSOMAL ABERRATIONS IN D. MELANOGASTER. IT INDUCED SPERM ABNORMALITIES IN MICE. IT OFTEN MODIFIED DNA-DAMAGING, MUTAGENIC AND CLASTOGENIC ACTIVITIES OF KNOWN MUTAGENS OR CARCINOGENS. BHT USUALLY REDUCED ACTIVITY OF INDIRECTLY-ACTING MUTAGENS OR CARCINOGENS.

REPT'D USE : FOOD PRESERVATIVE, ADHESIVES, CLEANING AGENTS, CHEMICAL STABILIZER AND ANTIOXIDANT

CAS NO : **939-97-9**

CHEM NAME : ***** P-TERT-BUTYLBENZALDEHYDE (P-TBB)**

REPORT 01 : PROPOSED HEALTH AND SAFETY DATA REPORTING
50FR46309 NOV 85
RULE WOULD REQUIRE MFRS, IMPORTERS OR PROCESSORS TO PROVIDE EPA WITH GENERAL EXPOSURE-RELATED DATA. EPA ALSO PROPOSING TO ADD P-TBB TO LIST OF CHEMICALS SUBJECT TO SECT 8(D) REPORTING REQUIREMENTS.

REPORT 02 : PROPOSED RULE MAKING 50FR46309 NOV 85
P-TBB HAS CAUSED MODERATELY ADVERSE REPRODUCTIVE EFFECTS IN LAB ANIMALS, BUT EPA LACKS DATA ON MUTAGENICITY, CARCINOGENICITY AND TERATOGENICITY OF P-TBB AND ON ITS ENVIRONMENTAL EFFECTS.

REPORT 03 : EPA FED. REG. NOTICE 51 FR 17336 MAY 86
UNDER TSCA 8(A) AND 8(D) SECTIONS EPA ESTABLISHES REPORTING AND RECORDKEEPING REQUIREMENTS FOR P-TBBA, P-TBT, AND P-TBB BECAUSE OF POSSIBLE HUMAN REPRODUCTIVE EFFECTS.

REPT'D USE : INTERMEDIATE IN MFR OF PHARMACEUTICALS, MODIFIER/REGULATOR IN ALKYD RESINS, SOLVENT AND CHEMICAL INTERMEDIATE

CAS NO : **98-73-7**

CHEM NAME : ***** P-TERT-BUTYLBENZOIC ACID (P-TBBA)**

REPORT 01 : EPA - 51FR17336 MAY 86
ESTABLISHES REPORTING AND RECORD-KEEPING REQUIREMENTS UNDER TSCA SECT 4(A) FOR REQUIRED TESTS.

REPORT 02 : EPA TSCA 0388-0726 APR 88
HOLCHST CELANESE REPORTED ON SUBCHRONIC FERTILITY STUDY IN MOLE WISTAR RATS PRIOR TO MATING AT DOSES OF 20, 100, OR 500 PPM IN FOOD. AT 100 PPM FERTILITY OF ONE MOLE NOTHING WAS IMPAIRED AND AT 500 PPM LED ALL MOLES TO A SLOWING DOWN OF WEIGHT GAIN AND INTERFERENCE WITH FERTILITY, ALL OF WHICH WAS REVERSIBLE, EXCEPT AT 500 PPM WEIGHT OF TESTICLES WERE STILL REDUCED AND HISTOLOGICAL DAMAGE COULD BE DETECTED IN SOME TUBULUS SECTIONS OF THE GEMINAL EPITHELUIM - ITS NO EFFECT LEVEL WAS DETERMINED TO BE ABOUT 20 PPM.

REPORT 03 : EPA FEDERAL REGISTER NOTICE 51FR17336 MAY 86
UNDER TSCA 8(A) AND 8(D) SECTIONS REPORTING AND RECORDKEEPING REQUIREMENTS PER P-TBBA, P-TBT, AND P-TBB BECAUSE OF POSSIBLE HUMAN REPRODUCTIVE EFFECTS.

REPORT 04 : EPA FEDERAL REGISTER NOTICE 51FR17336 MAY 86
UNDER TSCA 8 (A) AND 8 (D) SECTIONS EPA ESTABLISHES REPORTING AND RECORDKEEPING REQUIREMENTS FOR P-TBBA, P-TBT, AND P-TBB BECAUSE OF POSSIBLE HUMAN REPRODUCTIVE EFFECTS.

REPT'D USE : CHEMICAL INTERMEDIATE

CAS NO : **85-68-7**
CHEM NAME : ***** BUTYL BENZYL PHTHALATE**

REPORT 01 : IARC MONOGRAPH 29 1982
BBP WAS TESTED IN MICE AND FEMALE RATS BY ORAL ADMINISTRATION AND IN MALE MICE BY IP INJECTION. A SOMEWHAT HIGHER INCIDENCE OF MONOCYTIC LEUKEMIAS WAS OBSERVED IN FEMALE RATS. THE AVAILABLE STUDIES ARE ADEQUATE TO EVALUATE THE CARCINOGENICITY OF BBP IN ANIMALS.

REPORT 02 : IARC MONOGRAPH 29 1982
BBP WAS NOT MUTAGENIC FOR E. COLI, B. SUBTILIS, SALMONELLA, OR S. CEREVISIATE OR IN THE MOUSE LYMPHOMA ASSAY. NO DATA WERE AVAILABLE TO ASSESS THE TERATOGENICITY OF THIS CHEMICAL.

REPORT 03 : EPA RESPONSE TO ITC RECOMMS (46FR53775) OCT 81
IN RESPONSE TO THE ITC RECOMMENDATIONS (42FR55026), EPA HAS RECEIVED A PROPOSAL FROM THE CMA IN WHICH THE PHTHALATE ESTER INDUSTRY HAS AGREED TO TEST BENZYL PHTHALATE FOR ENVIRONMENTAL AND HEALTH EFFECTS OF CONCERN TO ITC AND EPA. EPA HAS DECIDED NOT TO PROPOSE ADDITIONAL TESTS AT THIS TIME.

REPORT 04 : NTP CARC TRS 213 NIH PUB. NO. 82-1769
UNDER THE CONDITIONS OF THIS BIOASSAY, BUTYL BENZYL PHTHALATE, ADMINISTERED IN THE FEED, WAS PROBABLY CARCINOGENIC FOR FEMALE F344/N RATS, CAUSING AN INCREASED INCIDENCE OF LEUKEMIA. THE MALE F344/N RAT STUDY WAS CONSIDERED INADEQUATE FOR EVALUATION DUE TO COMPOUND-RELATED TOXICITY AND EARLY MORTALITY. BBP WAS NOT CARCINOGENIC FOR B6C3F1 MICE OF EITHER SEX.

REPORT 05 : NTP TECHNICAL BULLETIN NO. 1, 4 AND 5 DEC 80, APR 81
BBP WAS NEGATIVE IN THE SALMONELLA TYPHIMURIUM MUTAGENICITY IN FIVE ASSAYS.

REPORT 06 : MULTI-AGENCY CROSS CHECK
N HHS MAGUIRE
N OSHA CARC
N EPA CAG
N EPA RCRA
N EPA CLEAN AIR
Y EPA CLEAN WATER

REPORT 07 : 50FR36446 SEP 85
EPA PROPOSES TO REQUIRE ENVIRONMENTAL FATE AND EFFECTS TESTS FOR BBP, INCLUDING ACUTE AND CHRONIC TOXICITY TESTS IN FRESH AND SALT WATER ORGANISMS, BIOCONCENTRATION TESTING IN THE EASTERN OYSTER AND TESTING FATE OF BBP IN UNDISTURBED SEDIMENTS.

REPT'D USE : PLASTICIZER FOR VINYL FLOORING, AUTO UPHOLSTERY, ADHESIVES, CAULKING COMPOUNDS AND AIR CONDITIONER FILTERS

CAS NO : **109-69-3**

CHEM NAME : ***** N-BUTYL CHLORIDE**

REPORT 01 : NTP TR 312 NIH #86-2568 APR 86 080786
UNDER CONDITIONS OF 2 YEAR GAVAGE STUDIES NO EVIDENCE OF CARCINOGENICITY OF N-BUTYL CHLORIDE FOR MALE AND FEMALE F344/N RATS AT DAILY DOSES OF 60 OR 120 MG/KG, FOR MALE B6C3F1 MICE AT DOSES OF 250, 500, OR 1,000 MG/KG, OR FOR FEMALE B6C3F1 MICE AT DOSES OF 250 OR 500 MG/KG. CHEMICAL-INDUCED TOXICITY IN HIGH DOSE RATS (PRIMARILY FEMALES) REDUCED SENSITIVITY OF STUDY FOR DETERMINING CARCINOGENICITY.

REPT'D USE : SOLVENT, ALKYLATING AGENT, PRODUCTION OF TIN STABILIZERS AND VETERINARY DRUG

CAS NO : **2426-08-6**
CHEM NAME : ***** BUTYL GLYCIDYL ETHER (BGE)**

REPORT 01 : EPA TSCA 8(E) NOTICE 8EHQ-0678-0213
SHELL OIL PROVIDED RESULTS OF A BATTERY OF MUTAGENICITY TESTS CONDUCTED ON BUTYL GLYCIDYL ETHER. BGE IS POSITIVE IN THE SALMONELLA ASSAY AND WHERE SHOWN IN THE DOMINANT LETHAL TEST. BGE WAS NEGATIVE IN A HOST MEDIATED ASSAY AND A MICRONUCLEUS TEST.

REPORT 02 : NIOSH CRITERIA DOCUMENT JUN 78
BGE WAS MUTAGENIC IN MICROBIAL AND MAMMALIAN TEST SYSTEMS. IT PRODUCED A SIGNIFICANT INCREASE IN THE NUMBER OF FETAL DEATHS IN THE DOMINANT LETHAL TEST WHEN APPLIED TO THE SKIN OF MALE MICE. NIOSH RECOMMENDS THAT WORKER EXPOSURE TO BGE BE LIMITED TO 4.4 PPM, MEASURED AS A 15 IN CEILING CONCENTRATION BASED ON EYE AND SKIN IRRITATION AND OBSERVED MUTAGENIC ACTIVITY.

REPORT 03 : EPA TSCA 8(E) 0584-0518S MAY 84
A COMPANY PROVIDED RESULTS FROM SEVERAL IN VITRO GENOTOXICITY STUDIES OF A PROPRIETARY MIXTURE OF EPOXY RESINS THAT SHOWED THAT IT PRODUCED A POSITIVE RESPONSE WITH AND WITHOUT METABOLIC ACTIVATION IN THE AMES TEST, THE MOUSE LYMPHOMA TEST AND THE SISTER CHROMATID EXCHANGE TEST.
REPT'D USE : EPOXY RESIN REACTIVE DILUENT, STABILIZER FOR PVC RESINS, CHLORINATED PARAFFINS AND HALOGENTAED PRODUCTS AND SURFACTANTS

CAS NO : **75-91-2**
CHEM NAME : ***** T-BUTYL HYDROPEROXIDE (TBHP)**

REPORT 01 : NTP TECHNICAL BULLETIN 6/8 JAN 82
T-BUTYL HYDROPEROXIDE WAS POSITIVE IN THE SALMONELLA ASSAY, BUT NEGATIVE IN DROSOPHILA WHEN TESTED FOR SEX-LINKED RECESSIVE LETHALS.

REPORT 02 : EPA TSCA 8(E) NOTICE 8EHQ-1081-0415 OCT 81
ATLANTIC-RICHFIELD REPORTS THAT TBHP WAS FOUND TO BE MUTAGENIC, BOTH WITH AND WITHOUT METABOLIC ACTIVATION, IN SALMONELLA AND IN A MOUSE LYMPHOMA CELL MUTATION ASSAY. TBHP WAS LESS ACTIVE IN THE PRESENCE OF ACTIVATION ENZYME AND WAS NEGATIVE IN A CELL TRANSFORMATION ASSAY.
REPT'D USE : CHEMICAL INTERMEDIATE, CURING AGENT AND REACTION INITIATOR

CAS NO : **25213-39-2**
CHEM NAME : ***** 1-BUTYLMETACRYLATE/STYRENE RANDOM COPOLYMER**

REPORT 01 : TSCA SECT 8(E) STATUS REPORT 8EHQ-0487-0668 JUN 87
XEROX SUBMITTED CHRONIC STUDY FINDINGS OF XEROX 9000-TYPE PHOTOCOPYING TONER ADMINISTERED AT 1.0, 4.0 AND 16.0 MG/M3 BY INHALATION TO FISCHER 344 RATS. NO UNUSUAL HISTOPATHOLOGICAL FINDINGS EXCEPT THE 16 MG/M3 GROUP SHOWED INCREASE IN COLLAGEN, POSSIBLY RESULT OF ARTIFACT ASSOCIATED WITH THICKLY CUT HISTOLOGIC SECTIONS, AND INCREASED LUNG WEIGHT, RETENTION OF TEST MATERIAL, AND RETARDATION OF ALVEOLAR MACROPHAGE MEDIATED CLEARANCE. TONER MATERIAL USED ENRICHED TEN-FOLD IN RESPIRABLE SIZE PARTICLES WITH RESPECT TO COMMERCIALLY AVAILABLE TONER MATERIAL.

REPORT 02 : EPA TSCA 8(E) 0487-0668 JUN 87
XEROX PROVIDED REPORT FROM SUBCHRONIC INHALATION STUDY OF XEROX 9000-TYPE PHOTOCOPYING TONER IN FISCHER 344 RATS. ACGIH RESPIRABLE FRACTION OF THE TONER ENRICHED TO 35%. EXPOSURE CONCENTRATIONS USED WERE 0, 1.0, 4.0, 16.0 AND 64.0 MG/M3, INCREASE IN LUNG WEIGHT OBSERVED AT 16 AND 64 MG/M3 DOSE LEVELS. AT 64 MG/M3, FOOD COMSUMPTION IN BOTH MALE AND FEMALE RATS SLIGHTLY DECREASED, BUT BODY WEIGHT NOT AFFECTED. DOSE-RELATED INCREASE IN PARTICLELADEN ALVEOLAR MACROPHAGES. THICKENING OF ALVEOLAR WALLS OBSERVED IN HIGH EXPOSURE GROUPS. MFTD MAXIMUM LUNG BURDEN FOR WHICH MACROPHAGE MEDIATED LUNG CLEARANCE NOT SIGNIFICANTLY IMPAIRED EXCEEDED 64 MG/M3 EXPOSURE LEVEL.

REPORT 03 : EPA TSCA 8(E) 1289-0668 DEC 89
ABSENCE OF SYSTEMIC OR UPPER RESPIRATORY SYSTEM TOXICITY, COUPLED WITH CHANGES IN LUNGS OR LYMPH NODES IN CHRONIC TONER INHALATION STUDY IN RATS SHOWED RESPIRATORY SYSTEM TARGET ORGAN AND PULMONARY CHANGES ATTRIBUTED TO "LUNG OVERLOADING", A GENERIC RESPONSE OF RESPIRATORY SYSTEM TO EXCESSIVE AMOUNTS OF DUST RETAINED IN LUNGS FOR PROLONGED INTERVAL. EXPOSURE AT HIGH LEVEL RESULTED IN EXCEEDING LUNG OVERLOADING WITH CHRONIC AND PERSISTENT INFLAMMATORY RESPONSE IN RAT LUNG. EXPOSURE AT LOW LEVEL RESULTED IN BELOW LUNG OVERLOADING TRANSIENT AND REVERSIBLE INFLAMMATORY RESPONSE IN RAT LUNG.
REPT'D USE : XEROGRAPHIC TONER

CAS NO : **98-51-1**
CHEM NAME : ***** P-TERT-BUTYLTOLUENE (P-TBT)**

REPORT 02 : EPA FEDERAL REGISTER NOTICE 51FR17336 MAY 86
UNDER TSCA 8 (A) AND 8 (D) SECTIONS EPA ESTABLISHES REPORTING AND RECORDKEEPING REQUIREMENTS FOR P-TBBA, P-TBT, AND P-TBB BECAUSE OF POSSIBLE HUMAN REPRODUCTIVE EFFECTS.
REPT'D USE : CHEMICAL INTERMEDIATE

CAS NO : **3068-88-0**
CHEM NAME : ***** BETA-BUTYROLACTONE**

REPORT 01 : IARC MONOGRAPH 11 1976
B-BUTYROLACTONE IS CARCINOGENIC IN MICE BY SKIN APPLICATION AND BY SUBCUTANEOUS INJECTION AND IN RATS BY ORAL ADMINISTRATION AND BY SUBCUTANEOUS INJECTION. IT PRODUCED TUMORS AT THE SITE OF ADMINISTRATION IN BOTH SPECIES. IN A 1979 SUPPLEMENT, IARC DETERMINED THAT THERE IS SUFFICIENT EVIDENCE FOR THE CARCINOGENICITY OF B-BUTYROLACTONE IN ANIMALS.
REPT'D USE : INTERMEDIATE FOR INDUSTRIAL CHEMICALS

CAS NO : **96-48-0**
CHEM NAME : ***** GAMMA-BUTYROLACTONE**

REPORT 01 : NTP TECHNICAL REPORT NO. 7 APR 82
4-BUTYROLACTONE WAS NEGATIVE IN THE SALMONELLA TYPHIMURIUM MUTAGENICITY ASSAY.

REPORT 02 : IARC MONOGRAPH 11 1976
GAMMA-BUTYROLACTONE WAS TESTED IN MICE BY ORAL ADMINISTRATION, SUBCUTANEOUS INJECTION AND SKIN APPLICATION AND IN RATS BY ORAL AND SUBCUTANEOUS ADMINISTRATION. NO CARCINOGENIC EFFECTS WERE OBSERVED.

REPORT 03 : HS STAFF REPORT AUG 89
GBL NOT EXPECTED TO CAUSE SIGNIFICANT OBSERVABLE ACUTE TOXIC EFFECTS OR ADVERSE REPRODUCTIVE EFFECTS FOLLOWING LOW LEVEL INFREQUENT EXPOSURES. GAVE POSITIVE RESPONSE IN REC-ASSAY, AND BHK-21 CELL TRANSFORMATION TEST. CHRONIC TOXICITY INFORMATION DOES NOT INDICATE CARCINOGENIC IN RATS AND MICE; FINAL CONCLUSIONS MUST WAIT AUDIT AND REVIEW OF NTP CANCER BIOASSAY (GAVAGE).
REPT'D USE : AEROSOLS, PAINT REMOVERS, SOLVENT FOR POLYMER RESINS AND SUBSTITUTE FOR METHYLENE CHLORIDE

CAS NO : **7440-43-9**
CHEM NAME : ***** CADMIUM**

REPORT 01 : EPA AMBIENT WATER QUALITY CRITERIA OCT 80
THE CASE FOR CADMIUM AS A CARCINOGEN IS NOT PERSUASIVE WHEN THE EXISTING DATA ARE CRITICALLY REVIEWED; HOWEVER, THE WEIGHT OF EVIDENCE FOR ONCOGENIC POTENTIAL OF CADMIUM IS SUFFICIENT TO BE QUALITATIVELY SUGGESTIVE. IT IS NOT RECOMMENDED THAT CADMIUM BE CONSIDERED A SUSPECT HUMAN CARCINOGEN FOR PURPOSES OF CALCULATING A WATER QUALITY CRITERION. A CRITERION, BASED ON ESTABLISHED CHRONIC RENAL EFFECTS, OF 10 UG/L IS JUDGED TO PROVIDE AMPLE PROTECTION OF HUMAN HEALTH.

REPORT 02 : NIOSH CRITERIA DOCUMENT AUG 76
NIOSH RECOMMENDS A STANDARD OF 40 MICROGRAMS/CU M (TWA) TO PROTECT AGAINST PULMONARY DISEASE AND RENAL TOXICITY; A STANDARD BASED ON TERATOGENICITY AND CARCINOGENICITY WAS NOT RECOMMENDED DUE TO UNCERTAINTIES IN DATA. THE PRESENT OSHA STANDARD IS 0.1 MG/CU M (TWA). ICITY, TERATOGENICITY, AND MUTAGENICITY OF CADMIUM COMPOUNDS.

REPORT 03 : NAS REPORT ON DIET, NUTRITION AND CANCER 1982
DATA FROM ONE LABORATORY EXPERIMENT SUGGEST THAT CADMIUM GIVEN IN DRINKING WATER IS NOT CARCINOGENIC IN MICE, WHEREAS IM AND SC INJECTIONS OF CADMIUM SALTS INDUCE CANCER IN RATS AND MICE. THE RESULTS OF THREE EPA STUDIES SUGGEST THAT INGESTION OF CADMIUM IN FOOD OR DRINKING WATER IS ASSOCIATED WITH INCREASED CANCER RISK, BUT ANOTHER STUDY DID NOT CONFIRM THESE FINDINGS. OCCUPATIONAL EXPOSURE TO CADMIUM HAS BEEN ASSOCIATED WITH AN INCREASE IN THE RISK OF RENAL AND PROSTATE CANCER. THE EVIDENCE FROM EPA AND LAB STUDIES DOES NOT PERMIT FIRM CONCLUSIONS ABOUT THE EFFECTS OF DIETARY EXPOSURE TO CADMIUM.

REPORT 05 : EPA RCRA HEALTH AND ENV. EFFECTS PROFILE OCT 80
EPA STUDIES INDICATE THAT CADMIUM EXPOSURE MAY INCREASE THE MORTALITY LEVEL FOR CANCER OF THE PROSTRATE. LONG TERM FEEDING AND INHALATION STUDIES IN ANIMALS HAVE NOT PRODUCED TUMORS, WHILE IV ADMINISTRATION OF CADMIUM HAS PRODUCED ONLY INJECTION SITE TUMORS. MUTAGENIC EFFECTS OF CADMIUM EXPOSURE HAVE BEEN SEEN IN ANIMAL STUDIES, BACTERIAL SYSTEMS, IN VITRO TESTS, AND IN CHROMOSOMES OF OCCUPATIONALLY EXPOSED WORKERS.

REPORT 06 : MULTI-AGENCY CROSS CHECK
Y HHS MAGUIRE
N OSHA CARC
Y EPA CAG

N EPA RCRA
N EPA CLEAN AIR
Y EPA CLEAN WATER

REPORT 07 : IARC MONOGRAPH 11 1976
LOCAL TUMORS WERE ALSO PRODUCED IN RATS BY I.M. INJECTION OF CADMIUM POWDER AND CADMIUM SULFIDE. IN A 1979 SUPPLEMENT, IARC CONCLUDED THAT CADMIUM AND CERTAIN CADMIUM COMPOUNDS ARE PROBABLY CARCINOGENIC FOR HUMANS.
REPT'D USE : ELECTROPLATING, HEAT STABILIZER IN PLASTICS, BATTERIES AND ALLOYS

CAS NO : **10108-64-2**
CHEM NAME : ***** CADMIUM CHLORIDE**

REPORT 01 : IARC MONOGRAPH 11 1976
SINGLE OR REPEATED SC INJECTIONS OF SEVERAL INORGANIC CADMIUM COMPOUNDS (CADMIUM CHLORIDE, SULFATE, SULFIDE, AND OXIDE) RESULTED IN DEVELOPMENT OF INJECTION-SITE SARCOMAS IN THE RAT. INTERSTITIAL CELL TUMORS OF THE TESTIS WERE FOUND IN RATS AND MICE GIVEN SC INJECTIONS OF SOLUBLE CADMIUM SALTS (CADMIUM SULFATE AND CADMIUM CHLORIDE) IN A 1979 SUPPLEMENT, IARC DETERMINED THAT THERE IS SUFFICIENT EVIDENCE OF THE CARCINOGENICITY OF CADMIUM CHLORIDE IN ANIMALS.

REPORT 02 : NTP TECHNICAL BULLETIN NO. 6 JAN 82
THE COMPOUND WAS NEGATIVE IN THE SALMONELLA TYPHIMURIUM MUTAGENICITY ASSAY IN ONE LABORATORY.

REPORT 03 : IARC MONOGRAPH 11 1976
TERATOGENIC EFFECTS WERE PRODUCED BY DAILY ORAL ADMINISTRATION OF CADMIUM CHLORIDE TO RATS. SOME TERATOGENIC EFFECTS MAY BE EXPLAINED BY THE INTERFERENCE OF CADMIUM IN MATERNAL ZINC METABOLISM.

REPORT 04 : IARC MONOGRAPHS SUPP 4, 1982 OCT 82
THERE IS CONSIDERED TO BE SUFFICIENT EVIDENCE OF CARCINOGENICITY IN EXPERIMENTAL ANIMALS.

REPORT 06 : MULTI-AGENCY CROSS CHECK
Y HHS MAGUIRE
N OSHA CARC
Y EPA CAG
N EPA RCRA
N EPA CLEAN AIR
N EPA CLEAN WATER
REPT'D USE : TEXTILE DYEING, MIRRORS, PHOTOGRAPHIC CHEMICALS, PESTICIDE, PRINTING, ELECTROPLATING AND CADMIUM YELLOW

CAS NO : **1306-19-0**
CHEM NAME : ***** CADMIUM OXIDE**

REPORT 01 : NTP TECHNICAL BULLETIN 7 APR 82
CADMIUM OXIDE WAS NEGATIVE IN THE SALMONELLA TYPHIMURIUM MUTAGENICITY ASSAY.

REPORT 02 : IARC MONOGRAPH 11 1976
SINGLE OR REPEATED S.C. INJECTIONS OF SEVERAL INORGANIC CADMIUM COMPOUNDS (CADMIUM CHLORIDE, SULFATE, SULFIDE, AND OXIDE) RESULT IN THE DEVELOPMENT OF INJECTION-SITE SARCOMAS IN THE RAT. TWO STUDIES SUGGEST THAT OCCUPATIONAL EXPOSURE TO CADMIUM OXIDE MAY INCREASE THE RISK OF PROSTATE CANCER IN MAN, BUT THE SIZE OF THE GROUPS EXAMINED WAS TOO SMALL TO ALLOW DEFINITE CONCLUSIONS TO BE DRAWN. IN A 1979 SUPPLEMENT, IARC DETERMINED THAT THERE IS SUFFICIENT EVIDENCE FOR THE CARCINOGENICITY OF CADMIUM OXIDE IN ANIMALS.

REPORT 03 : IARC MONOGRAPHS SUPP 4, 1982 OCT 83
THERE IS CONSIDERED TO BE SUFFICIENT EVIDENCE OF CARCINOGENICITY IN EXPERIMENTAL ANIMALS.

REPORT 04 : EPA 8(E) STATUS REPORT 8EHQ-0886-0618 S AUG 86
LUNG TUMORS HAVE BEEN SEEN IN RATS FOLLOWING INHALATION EXPOSURE TO CADMIUM OXIDE FOR 20 HRS/DAY, 7 DAYS/WK FOR 18 MONTHS.

REPORT 06 : MULTI-AGENCY CROSS CHECK
Y HHS MAGUIRE
N OSHA CARC
Y EPA CAG
N EPA RCRA
N EPA CLEAN AIR
N EPA CLEAN WATER

REPT'D USE : BATTERY ELECTRODES, NEMATOCIDE, CATALYST, CHEMICAL INTERMEDIATE FOR CD SALTS, MFR OF SILVER ALLOYS, GLASS, CERAMIC AND GLAZES; ELECTROPLATING AND RUBBER AND PLASTIC ADDITIVE

CAS NO : **10124-36-4**
CHEM NAME : ***** CADMIUM SULFATE**

REPORT 01 : IARC MONOGRAPHS SUPP 4, 1982 OCT 82
THERE IS CONSIDERED TO BE SUFFICIENT EVIDENCE OF CARCINOGENICITY IN EXPERIMENTAL ANIMALS.

REPORT 02 : IARC MONOGRAPH 11 1976
SINGLE OR REPEATED SC INJECTIONS OF SEVERAL INORGANIC

CADMIUM COMPOUNDS (CADMIUM CHLORIDE, SULFATE, SULFIDE, AND OXIDE) RESULT IN DEVELOPMENT OF INJECTION-SITE SARCOMAS IN THE RAT. INTERSTITIAL CELL TUMORS OF THE TESTES WERE FOUND IN RATS AND MICE GIVEN SC INJECTIONS OF SOLUBLE CADMIUM SULFATE AND CADMIUM CHLORIDE. IN A 1979 SUPPLEMENT, IARC DETERMINED THAT THERE IS SUFFICIENT EVIDENCE FOR THE CARCINOGENICITY OF CADMIUM SULFATE IN ANIMALS.

REPORT 03 : EPA 8(E) STATUS REPORT 8EHQ- 0886-0619 S AUG 86
LUNG TUMORS HAVE BEEN SEEN IN RATS FOLLOWING EXPOSURE TO CADMIUM SULFATE. INHALATION REGIMEN INVOLVED EXPOSURES OF ABOUT 20 HRS/DAY, 7 DAYS/WK FOR 18 MONTHS.

REPORT 06 : MULTI-AGENCY CROSS CHECK
Y HHS MAGUIRE
N OSHA CARC
Y EPA CAG
N EPA RCRA
N EPA CLEAN AIR
N EPA CLEAN WATER

REPT'D USE : INTERMEDIATE IN MFR OF PIGMENTS, PLASTIC STABILIZERS, NEMATOCIDE CATALYST, PHOTOELECTRICAL CELLS, ELECTROPLATING AND FLUORESCENT SCREENS

CAS NO : **1306-23-6**
CHEM NAME : ***** CADMIUM SULFIDE**

REPORT 01 : IARC MONOGRAPH 11 1976
SINGLE OR REPEATED SC INJECTIONS OF SEVERAL INORGANIC CADMIUM COMPOUNDS (CADMIUM CHLORIDE, SULFATE, SULFIDE, AND OXIDE) RESULT IN DEVELOPMENT OF INJECTION-SITE SARCOMAS IN THE RAT. LOCAL TUMORS WERE ALSO PRODUCED IN RATS BY IM INJECTION OF CADMIUM POWDER AND CADMIUM SULFIDE. IN A 1979 SUPPLEMENT, IARC DETERMINED THAT THERE IS SUFFICIENT EVIDENCE FOR THE CARCINOGENICITY OF CADMIUM SULFIDE IN ANIMALS.

REPORT 02 : IARC MONOGRAPH 11 1976
IN GOLDEN HAMSTERS, I.V. INJECTION OF CADMIUM SULFIDE PRODUCED CLEFT LIPS AND OTHER FACIAL DEFECTS. TERATOGENIC EFFECTS DUE TO INJECTED CADMIUM HAVE ALSO BEEN SEEN IN RATS.

REPORT 03 : EPA 8(E) STATUS REPORT 8EHQ- 0886-0618 S AUG 86
LUNG TUMORS SEEN IN RATS EXPOSED FOR 20 HRS/DAY, 7 DAYS/WK FOR 18 MONTHS TO CADMIUM SULFIDE BY INHALATION.

REPORT 06 : MULTI-AGENCY CROSS CHECK
Y HHS MAGUIRE
N OSHA CARC
Y EPA CAG
N EPA RCRA
N EPA CLEAN AIR
N EPA CLEAN WATER

REPT'D USE : PIGMENT, PHOSPHORS, PRINTING INKS, ARTIST'S COLORS, CERAMIC GLAZES, SCINTILLATION COUNTERS, SHAMPOOS, PYROTECHNICS AND EPOXY RESINS

CAS NO : **58-08-2**
CHEM NAME : ***** CAFFEINE**

REPORT 03 : IARC MONOGRAPHS 51 1991
CAFFEINE TESTED FOR CARCINOGENICITY IN FIVE STUDIES IN RATS BY ORAL ADMINISTRATION. IN TWO STUDIES, NO SIGNIFICANT DIFFERENCE IN INCIDENCE OF TUMORS AT ANY SITE. OTHER THREE STUDIES FOUND INADEQUATE FOR EVALUATION.

REPT'D USE : COFFEE, TEA, MATE, SOFT DRINKS AND PHARMACEUTICAL PREPARATIONS

CAS NO : **7778-44-1**
CHEM NAME : ***** CALCIUM ARSENATE**

REPORT 01 : IARC MONOGRAPH 23 1980
THERE IS INADEQUATE EVIDENCE FOR THE CARCINOGENICITY OF ARSENIC COMPOUNDS IN ANIMALS. THERE IS SUFFICIENT EVIDENCE THAT INORGANIC ARSENIC COMPOUNDS ARE SKIN AND LUNG CARCINOGENS IN HUMANS. THE EVIDENCE THAT ARSENIC COMPOUNDS CAUSE MUTATIONS AND ALLIED EFFECTS IN BACTERIA IS INCONCLUSIVE. HOWEVER, ARSENIC COMPOUNDS INDUCE CHROMOSOMAL ABERRATIONS AND MORPHOLOGICAL TRANSFORMATION IN MAMMALIAN CELLS.

REPORT 02 : NIOSH CRITERIA DOCUMENT JUN 75
NIOSH RECOMMENDS A CEILING OF 2 MICROGRAMS/CU M FOR INORGANIC ARSENIC COMPOUNDS BASED ON DERMATITIS, LUNG, AND LYMPHATIC CANCER. THE CURRENT OSHA STANDARD FOR INORGANIC ARSENIC IS 10 MICROGRAMS/CU M TWA. SEE CFR 1910.1018.

REPORT 06 : MULTI-AGENCY CROSS CHECK
Y HHS MAGUIRE
N OSHA CARC
Y EPA CAG
N EPA RCRA

N EPA CLEAN AIR
N EPA CLEAN WATER
REPT'D USE : INSECTICIDE, PESTICIDE, HERBICIDE AND MOLLUSICIDE

CAS NO : **8012-75-7**
CHEM NAME : ***** CALCIUM CHROMATE**

REPORT 01 : EPA TSCA 8(E) NOTICE 8EHQ-1182-0463
DIAMOND SHAMROCK CORP PROVIDED A TRANSLATION OF A GERMAN SUMMARY OF A 30-MONTH STUDY OF SODIUM BICHROMATE AND CALCIUM CHROMATE ADMINISTERED AT VARYING DOSES VIA INTERTRACHEAL INSTILLATION TO S-D RATS. THE SUMMARY STATED THAT "A CARCINOGENIC EFFECT WAS ONLY PRODUCED BY BOTH CHROMATES IN THE LUNG" AND THAT "THERE WAS NO DIFFERENCE IN INTENSITY OF EFFECT BETWEEN THE SODIUM DICHROMATE AND THE DISSOLVED CALCIUM CHROMATE."

REPORT 06 : MULTI-AGENCY CROSS CHECK
Y HHS MAGUIRE
N OSHA CARC
N EPA CAG
N EPA RCRA
N EPA CLEAN AIR
N EPA CLEAN WATER
REPT'D USE : PIGMENT AND HIGH TEMPERATURE APPLICATIONS

CAS NO : **156-62-7**
CHEM NAME : ***** CALCIUM CYANAMIDE**

REPORT 01 : NCI CARC TRS 163 DHEW PUB NO 79-1719
UNDER THE CONDITIONS OF THIS BIOASSAY, THE TEST FORMULATION OF CALCIUM CYANAMIDE WAS NOT CARCINOGENIC FOR F344 RATS OR B6C3F1 MICE OF EITHER SEX (FEEDING STUDY).

REPORT 02 : NTP TECHNICAL BULLETIN 7/8 APR 82, JUL 82
CALCIUM CYANAMID WAS WEAKLY POSITIVE IN THE SALMONELLA ASSAY, AND NEGATIVE IN DROSOPHILA WHEN TESTED FOR SEX-LINKED RECESSIVE LETHALS.
REPT'D USE : INTERMEDIATE FOR MFR OF MELAMINE, UREA, THIOUREN, PESTICIDE, COTTON DEFOLIANT AND HERBICIDE

CAS NO : **139-06-0**
CHEM NAME : ***** CALCIUM CYCLAMATE**

REPORT 01 : IARC MONOGRAPH 22 1980
CALCIUM CYCLAMATE HAS BEEN TESTED BY ORAL ADMINISTRATION IN ONE TWO-GENERATION EXPERIMENT IN RATS; NO DIFFER-

ENCE IN TUMOR INCIDENCE WAS SEEN BETWEEN TREATED AND CONTROL ANIMALS.

REPORT 02 : IARC MONOGRAPH 22 1980
THE LIMITED NUMBER OF MUTAGENICITY STUDIES PUBLISHED GIVE NO EVIDENCE THAT CYCLAMATES CAUSE POINT MUTATIONS. CYCLAMATES DO CAUSE CHROMOSOME DAMAGE. THERE IS NO EVIDENCE THAT CYCLAMATES ARE TERATOGENIC.

REPT'D USE : SYNTHETIC SWEETENER

CAS NO : **UNKNOWN-35**

CHEM NAME : ***** CALCIUM DIALKYLTHIOPHOSPHATES**

REPORT 01 : MICROBIOLOGICAL ASSOC'S FINAL REPORT TO CMA OCT 84
UNDER CONDITIONS OF MOUSE LYMPHOMA MUTAGENESIS ASSAY USING BABL/3T3 MOUSE EMBRYO CELLS, CDDP GAVE A NEGATIVE RESPONSE WITH METABOLIC ACTIVATION. CDDP GAVE A NEGATIVE RESPONSE IN THE L5178 TK+/-MOUSE LYMPHOMA MUTAGENESIS ASSAY WITH METABOLIC ACTIVATION.

CAS NO : **61789-36-4**

CHEM NAME : ***** CALCIUM NAPHTHENATE**

REPORT 01 : TSCA SECT 8(E) 8EHQ-0587-0675 JUN 87
FEMALE MICE WERE TREATED FOR UP TO 2 YRS WITH EPIDERMAL APPLICATIONS OF 0.05 ML OF MIXTURE OF OIL AND CPD, UNDILUTED OIL OR A 37.5 UG/ML SOLUTION OF BENZO(A)PYRENE (B(A)P). 84% OF FEMALE MICE IN B(A)P GROUP DEVELOPED CUTANEOUS TUMORS AT TREATED SITES. NO EPIDERMAL OR DERMAL TUMORS SEEN IN UNTREATED NEGATIVE CONTROLS. 16% OF MICE DEVELOPED TUMORS OF TREATED SITE WHEN MINERAL OIL AND CPD APPLIED. APPLICATION OF CARRIER ALONE DID NOT GIVE RISE TO TUMORS. AS A RESULT THE LAB DETERMINED THAT MIXTURE OF MINERAL OIL AND CPD IS A CUTANEOUS CARCINOGEN IN MICE. STCF MICE USED IN TESTS.

REPT'D USE : NOT GIVEN

CAS NO : **23209-59-8**

CHEM NAME : **CALCIUM SODIUM METAPHOSPHATE FIBERS**

REPORT 01 : EPA TSCA 8(E) STATUS REPORT 8EHQ-0386-0619 AUG 86
CSM FIBERS IN GELATIN PLEDGETS SURGICALLY IMPLANTED IN PLEURAL CAVITY OF RATS SHOWED NO FIBROSARCOMAS AT 6, 12, OR 18 MONTHS. 3 ANIMALS SACRIFICED AT 24 MONTHS HAD SARCOMAS. THIS APPROXIMATES HISTORICAL BACKGROUND. INHALATION STUDY UNDERWAY; NO SARCOMAS OBSERVED AFTER

3 MONTHS. CSM FIBERS APPEAR TO BIODEGRADE IN CULTURED LUNG CELLS.
REPT'D USE : ASBESTOS SUBSTITUTE

CAS NO : **61345-53-7**
CHEM NAME : **CALIBAN (F/R-P 44)**

REPORT 01 : NOV 87
HS STAFF REFERRAL FOR TOXICITY DATA INDICATE CHEMICAL IS A NONTOXIC FLAME RETARDANT.
REPT'D USE : TEXTILE FLAME RETARDANT

CAS NO : **76-22-2**
CHEM NAME : ***** CAMPHOR**

CAS NO : **56-25-7**
CHEM NAME : ***** CANTHARIDIN**

REPORT 01 : IARC MONOGRAPH 10 1976
CANTHARIDIN WAS TESTED ONLY BY SKIN APPLICATION IN MICE; IT PRODUCED AN INCREASED INCIDENCE OF SKIN PAPILLOMAS AND A LOW INCIDENCE OF SKIN CARCINOMAS.
REPT'D USE : VETERINARY VESICANT AND WART REMOVAL

CAS NO : **105-60-2**
CHEM NAME : ***** CAPROLACTAM**

REPORT 01 : NTP BIOASSAY TRS 214 NIH PUB NO 81-1770
UNDER THE CONDITIONS OF THIS BIOASSAY CAPROLACTAM ADMINISTERED IN THE FEED WAS NOT CARCINOGENIC FOR F344 RATS OR B6C3F1 MICE OF EITHER SEX.

REPORT 02 : IARC MONOGRAPH 19 1979
NO DATA ON THE CARCINOGENICITY OR MUTAGENICITY OF CAPROLACTAM WERE AVAILABLE TO THE WORKING GROUP.

REPORT 03 : IARC MONOGRAPH 39, 1986 JUN 85
C TESTED FOR CARCINOGENICITY IN MICE AND RATS VIA DIET. NO CARCINOGENIC ACTIVITY SEEN. CPD HAS NOT BEEN ADEQUATELY EVALUATED FOR PRENATAL OR REPRODUCTIVE TOXICITY IN HUMANS OR ANIMALS. CPD GAVE NEGATIVE RESULTS IN A WIDE RANGE OF IN-VITRO SHORT-TERM TESTS. BORDERLINE POSITIVE RESULTS WERE GOTTEN IN TESTS FOR GENE CONVERSION IN YEAST AND MORPHOLOGICAL TRANSFORMATION IN CULTURED MAMMALIAN CELLS. IT PRODUCED SOMATIC-CELL MUTATIONS IN D.

MELANOGASTER AND IT MAY HAVE INDUCED POINT MUTATIONS IN YEAST AND CHROMOSOMAL ABERRATIONS IN HUMAN CELLS. NO CASE REPORTS OR EPA STUDIES AVAILABLE TO EVALUATE CARCINOGENICITY IN HUMANS. NO EVIDENCE OF CARCINOGENICITY TO ANIMALS.

REPT'D USE : PLASTIC MONOMER FOR NYLON 6, TEXTILES, CARPETS, HOSIERY, FOOD PACKING AND AUTO PARTS

CAS NO : **133-06-2**

CHEM NAME : ***** CAPTAN**

REPORT 01 : IARC MONOGRAPH 30 JAN 83
CAPTAN WAS TESTED FOR CARCINOGENICITY IN MICE AND RATS BY ADMINISTRATION IN DIET. IT WAS CARCINOGENIC TO ONE STRAIN OF MICE, INDUCING DUODENAL TUMORS (ADENOCARCINOMA AND ADENOMATOUS POLYP). NO EVIDENCE OF CARCINOGENICITY WAS FOUND IN RATS.

REPORT 02 : IARC MONOGRAPH 30 JAN 83
CAPTAN HAS SHOWN LITTLE, IF ANY, EMBRYOTOXIC OR TERATOGENIC POTENTIAL IN MICE OR RATS AT MATERNALLY TOLERATED DOSES. RESULTS OBTAINED IN TESTS WITH RABBITS AND HAMSTERS WERE CONSIDERED INCONCLUSIVE.

REPORT 03 : IARC MONOGRAPHI 30 JAN 83
CAPTAN MUTAGENIC TO BACTERIA AND YEAST. BOTH POSITIVE AND NEGATIVE RESULTS OBTAINED IN HOST-MEDIATED ASSAY IN MICE. WEAK OR NEGATIVE EFFECTS OBSERVED IN DROSOPHILA MELANOGASTER. INDUCED CHROMOSOMAL ABERRATIONS, SISTER CHROMATID EXCHANGE AND MUTATIONS AT SEVERAL LOCI, BUT NOT UNSCHEDULED DNA SYNTHESIS, IN CULTURED MAMMALIAN CELLS. NO INCREASE IN MICRONUCLEATED ERYTHROCYTES OR CHROMOSOMAL ABERRATIONS DETECTED IN TREATED MICE OR RATS; POSITIVE RESULTS OBTAINED IN DOMINANT LETHAL TESTS NOT CONFIRMED BY OTHER STUDIES. SUFFICIENT EVIDENCE TO ESTABLISH MUTAGENICITY IN CELLULAR SYSTEMS, BUT DATA INSUFFICIENT TO ESTABLISH MUTAGENICITY IN MAMMALS.

REPT'D USE : FUNGICIDE, BACTERIOSTAT, ADHESIVES AND GLUES

CAS NO : **63-25-2**

CHEM NAME : ***** CARBARYL**

REPORT 01 : NIOSH CRITERIA DOCUMENT SEP 76
NIOSH RECOMMENDS A STANDARD OF 5 MG/CO M TWA BASED ON CNS AND REPRODUCTIVE SYSTEM EFFECTS. THE PRESENT OSHA STANDARD IS 5 MG/CU M TWA.

REPORT 02 : IARC MONOGRAPH 12 1976
IN ONE STUDY IN RATS WITH RELATIVELY LOW SURVIVAL RATES, CARBARYL PRODUCED SARCOMAS FOLLOWING ITS ORAL BUT NOT AFTER ITS SUBCUTANEOUS ADMINISTRATION. IN ONE STUDY IN MICE BY ORAL ADMINISTRATION, NO CARCINOGENIC EFFECT WAS OBSERVED. THE AVAILABLE DATA DOES NOT ALLOW AN EVALUATION OF THE CARCINOGENICITY OF CARBARYL TO BE MADE.

REPORT 03 : NAS REPORT ON DIET, NUTRITION, AND CANCER 1982
THE DATA FOR CARBARYL ARE INCONCLUSIVE AND DO NOT PERMIT AN ASSESSMENT OF CARCINOGENICITY.

REPORT 04 : IARC MONOGRAPH 12 1976
NO REVERSE MUTATIONS WERE OBSERVED IN E. COLI, AND NO MITOTIC GENE CONVERSION WAS SEEN IN SACCHAROMYCES. IN THE ABSENCE OF METABOLIC ACTIVATION, CARBARYL WAS NOT MUTAGENIC TO SALMONELLA TA100, TA98, TA1535 AND TA1537. CARBARYL CAUSED AN INCREASED NUMBER OF RECESSIVE LETHALS IN DROSOPHILA. NO INCREASE IN DOMINANT LETHALITY COULD BE ESTABLISHED IN MICE.

REPT'D USE : INSECTICIDE, AEROSOLS, ACARICIDE AND MOLLUSCICIDE

CAS NO : **86-74-8**
CHEM NAME : ***** CARBAZOLE**

REPORT 01 : IARC MONOGRAPH 32 DEC 83
TESTED FOR CARCINOGENICITY IN MICE BY ADMINISTRATION IN DIET, BY SKIN APPLICATION AND BY SUBCUTANEOUS INJECTION. IN STUDY BY ORAL ADMINISTRATION, DOSE-DEPENDENT INCREASE IN INCIDENCE OF LIVER NEO-PLASTIC NODULES AND HEPATOCELLULAR CARCINOMAS OBSERVED. PAPILLOMAS AND CARCINOMAS OF FORESTOMACH OCCURRED IN ANIMALS RECEIVING HIGH-DOSE LEVEL. OTHER STUDIES IN MICE CONSIDERED INADEQUATE FOR EVALUATION. CARBAZOLE WAS NOT MUTAGENIC TO SALMONELLA TYPHIMURIUM.

REPT'D USE : DYE INTERMEDIATE, PHOTOGRAPHIC PLATES AND REAGENT. NO KNOWN PRODUCTION. COMBUSTION BY-PRODUCT

CAS NO : **20073-24-9**
CHEM NAME : ***** 3-CARBETHOXYPSORALEN**

REPORT 01 : IARC VOL 40 1986
NO EVIDENCE OF CARCINOGENICITY OF CPD IN COMBINATION WITH ULTRAVIOLET A (UVA) RADIATION TO MOUSE SKIN. INADEQUATE EVIDENCE FOR CARCINOGENICITY TO TEST ANIMALS IN ABSENCE OF UVA. CARCINOGENICITY TO HUMANS COULD NOT BE EVALUATED.

REPT'D USE : RESEARCH

CAS NO : **1333-86-4**
CHEM NAME : ***** CARBON BLACKS**

REPORT 01 : NIOSH CRITERIA DOCUMENT SEP 78
NIOSH RECOMMENDS A STANDARD OF 3.5 MG/CU M TWA BASED ON LUNG, HEART, SKIN EFFECTS AND CARCINOGENICITY. THE PRESENT OSHA STANDARD IS 3.5 MG/CU M TWA.

REPORT 02 : EPA TSCA 8(E) STATUS REPORT 8EHQ-0183-0468 JAN 83
PHILLIPS PETROLEUM PROVIDED FINAL REPORTS FROM TWO IN-VITRO MUTAGENICITY ASSAYS OF CARBON BLACK FEEDSTOCK. THE FEEDSTOCK DISPLAYED MUTAGENIC ACTIVITY WITH EXOGENOUS METABOLIC ACTIVATION IN AMES/SALMONELLA TYPHIMURIUM ASSAY AND A MOUSE LYMPHOMA FORWARD MUTATION ASSAY; WITHOUT ACTIVATION THE CARBON BLACK FEEDSTOCK WAS REPORTED TO BE NEGATIVE IN BOTH ASSAYS. A MIXTURE OF LIGHT, INTERMEDIATE, AND HEAVY CATALYTIC CRACKED PETROLEUM DISTILLATES (CAS NO.'S: 64741-59-9, 64741-60-2, AND 64741-61-3, RESPECTIVELY).

REPORT 03 : IARC MONOGRAPH 33 APR 84
AVAILABLE DATA ON CARBON BLACKS ARE INADEQUATE TO PERMIT EVALUATION OF CARCINOGENICITY TO ANIMALS BUT SUFFICIENT EVIDENCE THAT SOLVENT (BENZENE) EXTRACTS OF CARBON BLACKS ARE CARCINOGENIC IN ANIMALS FROM FEEDING STUDIES IN MICE AND SKIN PAINTING IN MICE AND MONKEYS. MATERIAL EXTRACTED FROM CARBON BLACKS CARCINOGENIC IN MICE PRODUCING TUMORS AFTER SUBCUTANEOUS INJECTION. A CARBON BLACK CONTAINING CARCINOGENIC POLYNUCLEAR AROMATIC COMPOUNDS PRODUCED LOCAL SARCOMAS WHEN INJECTED SUBCUTANEOUSLY IN TRICAPRYLIN (IN MICE).

REPORT 04 : IARC MONOGRAPH 33 APR 84
IN ONE MUTATION TEST IN PROKARYOTES ONE SAMPLE OF CARBON BLACK PARTICLES WAS POSITIVE IN OTHER STUDIES SOME TREATED CARBON BLACKS WERE MUTAGENIC, OTHERS WERE NOT.

REPORT 05 : IARC MONOGRAPH 33 APR 84
THERE WAS SUFFICIENT EVIDENCE FOR EXCESS OCCURRENCE IN RUBBER WORKERS EXPOSED TO CARBON BLACK DUSTS CANCER OF BLADDER, STOMACH AND LUNG AND OF LEUKEMIA; AND LIMITED EVIDENCE OF CANCER OF SKIN, COLOR, PROSTATE, AND OF LYMPHOMA.

REPT'D USE : RUBBER PIGMENT REINFORCING AGENT, PRINTING INKS, PAINTS, CLEANING AGENTS, AEROSOLS, SPACE DEODORIZERS, PHOTOGRAPHIC, CHEMICALS ADHESIVES/GLUES, BATTERIES, TONER CATALYST AND CARBON PAPER

CAS NO : **75-15-0**
CHEM NAME : ***** CARBON DISULFIDE**

REPORT 01 : NTP TECHNICAL BULLETIN 2/3 APR 80, DEC 80
THE COMPOUND WAS FOUND TO BE NEGATIVE IN THE SALMONELLA TYPHIMURIUM MUTAGENICITY ASSAY IN TWO LABORATORIES. TERATOLOGY ASSAYS SHOWED THAT INHALATION OF CARBON DISULFIDE BY RATS OR RABBITS WAS NOT ASSOCIATED WITH PRENATAL DEATH OR INCREASED MALFORMATIONS.

REPORT 02 : NIOSH CRITERIA DOCUMENT MAY 77
NIOSH RECOMMENDS A STANDARD OF 1 PPM (TWA) BASED ON CARDIAC, CNS, AND REPRODUCTIVE SYSTEM EFFECTS. THE PRESENT OSHA STANDARD IS 20 PPM TWA.

REPORT 03 : EPA RCRA HEALTH AND ENV. EFFECTS PROFILE OCT 80
WHEN RATS INHALED CARBON DISULFIDE AT 10 MG/CU M, ABNORMALITIES OF GENITOURINARY AND SKELETAL SYSTEMS WERE NOTED. IN ADDITION, DISTURBANCES OF OSSIFICATION AND BLOOD FORMATION AND DYSTROPHIC CHANGES IN LIVER AND KIDNEY WERE NOTED.

REPORT 04 : NTP CARC TEST PROGRAM JUL 82
IN A CURRENT NTP BIOASSAY, CARBON DISULFIDE IS BEING ADMINISTERED INHALATION TO F344 RATS. THE PRECHRONIC TEST PHASE IS IN PROGRESS.
REPT'D USE : SOLVENT IN MFR OF REGENERATED CELLULOSE, CARBON TETRACHLORIDE, NEOPRENE CEMENT, RUBBER VULCANIZATION AGENT AND GRAIN FUMIGANT

CAS NO : **UNKNOWN-08**
CHEM NAME : ***** CARBON FIBERS**

REPORT 01 : EPA TSCA 8(E) NOTICE 8EHQ-1281-0423 JAN 82
UNION CARBIDE PROVIDED RESULTS FROM A CHRONIC SKIN PAINTING STUDY IN MICE, OF TWO TYPES OF CARBON FIBERS. THE CARBON FIBERS WERE PULVERIZED AND SUSPENDED IN BENZENE FOR APPLICATION. A VARIETY OF HISTOLOGICALLY VERIFIED SARCOMAS, CARCINOMAS, AND PAPILLOMAS WERE FOUND AT THE APPLICATION SITES OF MICE TREATED WITH BOTH TYPES OF CARBON FIBERS. THERE WERE NO TUMORS OBSERVED FOR THE BENZENE CONTROL GROUP.
REPT'D USE : INDUSTRIAL PROCESSING

CAS NO : **630-08-0**
CHEM NAME : ***** CARBON MONOXIDE**

REPORT 01 : CARBON MONOXIDE STUDY, SEATTLE WASHINGTON 1977
IN 1977, A 20-DAY STUDY INVOLVING 36 OUTDOOR SITES, FIVE INDOOR SITES AND TWO PEDESTRIAN WALKING ROUTES WAS CONDUCTED IN DOWNTOWN SEATTLE DURING OCTOBER-NOVEMBER WHICH IS DURING PART OF THE SEASON WHEN HIGH CARBON LEVELS FREQUENTLY OCCUR. NO VALUES EXCEEDED THE 9.0 PPM EIGHT-HOUR STANDARD AT SOME STUDY SITES ON CERTAIN DAYS.

REPORT 02 : NIOSH CRITERIA DOCUMENT AUG 72
NIOSH RECOMMENDS THAT OCCUPATIONAL EXPOSURE TO CARBON MONOXIDE BE CONTROLLED SO THAT NO WORKER BE EXPOSED TO A CONCENTRATION OF NO MORE THAT 35 PPM DETERMINED AS A TIME-WEIGHTED AVERAGE (TWA) EXPOSURE FOR AN 8-HOUR WORK DAY. NO LEVEL OF CO TO WHICH WORKERS ARE EXPOSED SHALL EXCEED A CEILING CONCENTRATION OF 200 PPM BASED ON CARDIAC EFFECTS. THE PRESENT OSHA STANDARD IS 50 PPM (TWA).

REPORT 03 : REPORT OF NATIONAL ACADEMY OF SCIENCES 1977
DOCUMENT SUMMARIZES THE CARBON MONOXIDE LITERATURE RELATED TO EFFECTS ON MAN AND HIS ENVIRONMENT FOR THE CONSIDERATION OF THE EPA IN UPDATING THE INFORMATION IN THE AIR QUALITY CRITERIA FOR CO. REPORT EMPHASIZES BIOLOGICAL EFFECTS OF CO ON MAN, DATA CONCERNING THE IDENTIFICATION OF SUSCEPTIBLE POPULATIONS, AND EVIDENCE IMPLICATING CO AS A CAUSAL FACTOR IN DISEASE.

REPT'D USE : CHEMICAL INTERMEDIATE, COMBUSTION POLLUTANT AND METALLURGICAL REDUCING AGENT

CAS NO : **56-23-5**
CHEM NAME : ***** CARBON TETRACHLORIDE**

REPORT 01 : CFR NOTATION 16CFR1500.17(A) (2) MAY 68
THE CPSC HAS DECLARED, UNDER THE AUTHORITY OF THE FEDERAL HAZARDOUS SUBSTANCES ACT, CARBON TETRACHLORIDE AND MIXTURES CONTAINING IT TO BE BANNED HAZARDOUS SUBSTANCES. THIS BAN EXCLUDES UNAVOIDABLE MANUFACTURING RESIDUES OF CARBON TETRACHLORIDE IN OTHER CHEMICALS THAT DO NOT, DURING USE, RESULT IN AN AIR CONCENTRATION OF CARBON TETRACHLORIDE GREATER THAN 10 PPM.BAN ORIGINALLY PROMULGATED MAY 1968.

REPORT 02 : IARC MONOGRAPH 20 1979
CARBON TETRACHLORIDE IS FETOTOXIC. IN PLATE TESTS WITH SALMONELLA TA100, TA1535 AND TA1538, CARBON TETRACHLORIDE

WAS NON-MUTAGENIC BOTH WITH AND WITHOUT METABOLIC ACTIVATION.

REPORT 03 : EPA AMBIENT WATER QUALITY CRITERIA OCT 80
THERE IS SUFFICIENT EVIDENCE TO CONCLUDE THAT CARBON TET IS A CARCINOGEN IN ANIMALS AND, WITH APPROPRIATE ASSUMPTIONS, IT IS INTERPRETED TO BE A SUSPECT HUMAN CARCINOGEN. AMBIENT WATER CONCENTRATIONS OF 4.0 UG/L, .4 UG/L, AND .04 UG/L ARE ESTIMATED TO RESULT IN A RESPECTIVE 10 -5, 10 -6, AND 10 -7 INCREASE IN CANCER RISK OVER A LIFETIME.

REPORT 04 : NIOSH CRITERIA DOCUMENT JUN 76
NIOSH RECOMMENDS THAT WORKERS NOT BE EXPOSED TO CARBON TETRACHLORIDE IN EXCESS OF 2 PPM (12.6 MG/CU M) CEILING BASED ON LIVER CANCER. THE PRESENT OSHA STANDARD IS 10 PPM TWA.

REPORT 06 : MULTI-AGENCY CROSS CHECK
Y HHS MAGUIRE
N OSHA CARC
Y EPA CAG
Y EPA RCRA
Y EPA CLEAN AIR
Y EPA CLEAN WATER

REPORT 07 : EPA RCRA HEALTH AND ENV. EFFECTS PROFILE OCT 80
TOXICOLOGIC DATA FOR NON-HUMAN MAMMALS ARE EXTENSIVE AND SHOW THAT CARBON TETRACHLORIDE CAUSES LIVER AND KIDNEY DAMAGE, BIOCHEMICAL CHANGES IN LIVER FUNCTION, AND NEUROLOGICAL DAMAGE. CARBON TETRACHLORIDE HAS BEEN FOUND TO INDUCE LIVER CANCER IN RATS AND MICE. MUTAGENIC EFFECTS HAVE NOT BEEN OBSERVED AND TERATOGENIC EFFECTS HAVE NOT BEEN CONCLUSIVELY DEMONSTRATED.

REPORT 08 : EPA NATIONAL DRINKING WATER STANDARDS
ANPR 49FR24330 JUN 84
EPA PROPOSES A ZERO LEVEL FOR CT IN DRINKING WATER UNDER SAFE DRINKING WATER ACT. THIS RECOMMENDED MAXIMUM CONTAMINANT LEVEL WILL BE FOLLOWED, AFTER COMMENT PERIOD, BY A MAXIMUM CONTAMINANT LEVEL. AS CLOSE TO THE RMCL AS FEASIBLE, BASED ON TECHNOLOGIES, COSTS AND OTHER FACTORS.

REPORT 09 : EPA NAT'L DRINKING WATER REGS - VSOC'S. 49FR24330 JUN 84
EPA PROPOSES ESTABLISHING NATIONAL DRINKING WATER REG'S FOR VOLATILE SYNTHETIC ORGANIC CPDS AT THE "NO KNOWN EFFECT" OR "NO ANTICIPATED EFFECT" LEVEL, THE "RECOMMENDED MAXIMUM CONTAMINANT LEVELS" (RMCLS). THESE NON-ENFORCEABLE HEALTH GOALS WILL BE SET AT ZERO EFFECT LEVEL PENDING SETTING OF MAX. CONTAMINANT LEVELS FOLLOWING COMMENT PERIOD.

REPT'D USE : CHEMICAL INTERMEDIATE, SOLVENT, PESTICIDE CHEMICAL ADDITIVE, FLAME RETARDANT, CLEANING AGENT AND FIRE EXTINGUISHERS

CAS NO : **77-65-6**
CHEM NAME : ***** CARBROMAL**

REPORT 01 : NCI CARC TRS 173 DHEW PUB NO 79-1729
UNDER THE CONDITIONS OF THIS BIOASSAY, DIETARY ADMINISTRATION OF CARBROMAL WAS NOT CARCINOGENIC IN FISCHER 344 RATS OR B6C3F1 MICE.

REPORT 02 : NTP TECHNICAL BULLETIN 5 AUG 81
CARBROMAL WAS NEGATIVE IN THE SALMONELLA TYPHIMURIUM MUTAGENICITY ASSAY IN TWO LABORATORIES.
REPT'D USE : MILD SEDATIVE AND HYPNOTIC AGENT

CAS NO : **UNKNOWN-24**
CHEM NAME : ***** CARPENTRY AND JOINERY**

REPORT 01 : IARC MONOGRAPH 25 FEB 81
EPIDEMIOLOGIC DATA ARE NOT SUFFICIENT TO MAKE DEFINITIVE ASSESSMENT OF CARCINOGENIC RISK OF EMPLOYMENT AS A CARPENTER OR JOINER. MANY STUDIES RAISE THE POSSIBILITY OF INCREASED RISK OF HODGKINS DISEASE. THERE IS CONFLICTING EVIDENCE ABOUT ASSOCIATION BETWEEN NASAL ADENO-CARCINOMAS AND WORK AS A CARPENTER. EVIDENCE SUGGESTING INCREASED RISKS OF LUNG, BLADDER, AND STOMACH CANCER COMES FROM LARGE POPULATION-BASED OCCUPATIONAL MORTALITY STATISTICAL STUDIES AND IS INADEQUATE FOR AN EVALUATION OF RISKS OF THESE TUMORS DEVELOPING.

CAS NO : **9000-07-1**
CHEM NAME : ***** CARRAGEENANS**

REPORT 01 : IARC MONOGRAPH 10 1976
ONE TYPE OF CARRAGEENAN WAS TESTED IN RATS BY SUBCUTANEOUS INJECTION AND PRODUCED LOCAL SARCOMAS. IN MICE AND RATS ADMINISTERED FOOD-GRADE NATIVE CARRAGEENAN ORALLY, THE INCIDENCE OF TUMORS WAS GREATER THAN THAT IN CONTROLS; HOWEVER, THIS NEGATIVE EXPERIMENT IS INADEQUATE IN TERMS OF THE NUMBER OF ANIMALS USED.

REPORT 02 : IARC MONOGRAPH 31 JUL 83
NATIVE (UNDEGRADED) CARRAGEENAN TESTED IN RATS AND HAMSTERS IN DIET; NO EVIDENCE OF CARCINOGENICITY FOUND. IN FEMALE RATS TREATED WITH AZOXYMETHANE OR

N-NITROSOMETHYLUREA TOGETHER IN DIET, GREATER INCIDENCE OF COLORECTAL CANCERS OBSERVED THAN WITH TREATMENT BY THESE TWO CHEMICALS ALONE. DEGRADED CARRAGEENAN TESTED IN RATS IN DIET IN DRINKING-WATER AND BY ORAL INTUBATION, INDUCED COLORECTAL CANCERS.

REPORT 03 : IARC MONOGRAPH 31 JUL 83
DEGRADED CARRAGEENAN NOT MUTAGENIC IN BACTERIA OR IN MAMMALIAN CELLS IN VITRO.

REPT'D USE : FOOD, GELLING AGENT, EMULSIFIER, VISCOSITY BUILDER, CLARIFYING AGENT IN BEER, AEROSOLS AND SPACE DEODORIZERS

CAS NO : **2244-16-8**

CHEM NAME : ***** D-CARVONE**

REPORT 01 : NTP TR 381 NIH PUB. 90-2836 FEB 90
UNDER 2-YEAR GAVAGE STUDIES NO EVIDENCE OF CARCINOGENIC ACTIVITY OF D-CARVONE FOR MALE OR FEMALE B6C3F1 MICE ADMINISTERED D-CARVONE AT 375 OR 750 MG/KG, 5 DAYS PER WEEK FOR 2 YEARS.

REPT'D USE : FLAVORING AND FRAGRANCE AGENT IN FOODS, PERFUMES, LOTIONS AND DETERGENTS

CAS NO : **120-80-9**

CHEM NAME : ***** CATECHOL**

REPORT 01 : IARC MONOGRAPH 15 1977
NO DATA ON THE EMBRYOTOXICITY OR TERATOGENICITY OF CATECHOL WERE AVAILABLE. CATECHOL INDUCED CHROMOSOME ABERRATIONS IN A. CEPA.

REPORT 02 : NTP TECHNICAL BULLETIN 2 APR 80
CATECHOL WAS NEGATIVE IN THE SALMONELLA TYPHIMURIUM MUTAGENICITY ASSAY IN ONE LABORATORY.

REPORT 03 : IARC MONOGRAPH 15 1977
IN SKIN PAINTING STUDIES IN MICE, CATECHOL INCREASED THE CARCINOGENIC EFFECTS OF BENZO(A)PYRENE ON THE SKIN; THE AVAILABLE DATA DO NOT ALLOW AN EVALUATION OF THE CARCINOGENICITY OF THIS COMPOUND.

REPORT 04 : NTP TEST REGIMEN JUL 82
IN A CURRENT NTP BIOASSAY, CATECHOL IS BEING ADMINISTERED BY GAVAGE TO F344 RATS AND B6C3F1 MIDE. THE PRECHRONIC TESTING PHASE BEGAN IN MAY 80.

REPT'D USE : POLYMERIZATION INHIBITOR, ANTIOXIDANT, TOPICAL ANTISEPTIC, PHOTOGRAPHIC, CHEMICAL INTERMEDIATE IN MFR OF PHARMACEUTICALS AND PESTICIDES

CAS NO : **302-17-0**
CHEM NAME : ***** CHLORAL HYDRATE**

REPORT 01 : NTP TECHNICAL BULLETIN 3 DEC 80
THE COMPOUND WAS FOUND TO BE POSITIVE IN THE SALMONELLA TYPHIMURIUM MUTAGENICITY ASSAY IN ONE LABORATORY.
REPT'D USE : SEDATIVE, LINIMENTS AND MANUFACTURE OF DDT

CAS NO : **133-90-4**
CHEM NAME : ***** CHLORAMBEN**

REPORT 01 : NCI CARC TRS 25 DHEW PUB NO 77-82
UNDER THE CONDITIONS OF THIS BIOASSAY, THERE WERE NO TUMORS IN OSBORNE-MENDEL RATS THAT WERE SIGNIFICANTLY RELATED TO ADMINISTRATION OF THE CHEMICAL. IN B6C3F1 FEMALE MICE, CHLORAMBEN WAS CARCINOGENIC, PRODUCING HEPATOCELLULAR CARCINOMAS IN TREATED ANIMALS (FEEDING STUDY).

REPORT 03 : NTP TECHNICAL BULLETIN 6 JAN 82
THE COMPOUND WAS POSITIVE IN THE SALMONELLA TYPHIMURIUM MUTAGENICITY ASSAY IN ONE LABORATORY.

REPORT 06 : MULTI-AGENCY CROSS CHECK
N HHS MAGUIRE
Y OSHA CARC
N EPA CAG
N EPA RCRA
N EPA CLEAN AIR
N EPA CLEAN WATER
REPT'D USE : HERBICIDE

CAS NO : **305-03-3**
CHEM NAME : ***** CHLORAMBUCIL**

REPORT 01 : IARC MONOGRAPH 26 1981
CHLORAMBUCIL HAS BEEN TESTED IN MICE AND RATS BY IP INJECTION. IT PRODUCED LUNG TUMORS IN MICE AND PROBABLY PRODUCED TUMORS OF THE HAEMATOPOIETIC SYSTEM AND OVARY IN MICE AND HAEMATOPOIETIC TUMORS IN MALE RATS. THERE IS LIMITED EVIDENCE FOR THE CARCINOGENICITY OF CHLORAMBUCIL IN MICE AND RATS. THERE IS LIMITED EVIDENCE THAT IT IS CARCINOGENIC IN HUMANS.

REPORT 02 : IARC MONOGRAPH 26 1981
CHLORAMBUCIL CAN INDUCE TERATOGENIC EFFECTS IN SEVERAL ANIMAL SPECIES AND EMBRYOLETHALITY AT DOSES NONTOXIC TO THE MOTHER. CHLORAMBUCIL INDUCED POINT MUTATIONS IN E.

COLI, MITOTIC GENE CONVERSION IN THE DIPLOID STRAIN D4 OF SACCHAROMYCES, AND A DOSE DEPENDENT INCREASE IN MUTATIONS IN S.CEREVISIAE. IT HAS ALSO BEEN SHOWN TO INDUCE CHROMOSOMAL ABERRATIONS IN HUMAN LYMPHOCYTES IN CULTURE.

REPORT 03 : NCI BIOASSAY PROGRAM - CANCER VOL. 40 1977
UNDER THE CONDITIONS OF THIS BIOASSAY, I.P. INJECTION OF CHLORAMBUCIL WAS CARCINOGENIC TO MALE AND FEMALE SWISS-WEBSTER MICE, CAUSING LUNG TUMORS AND A SMALL NUMBER OF LYMPHOSARCOMAS. FEMALE MICE ALSO HAD OVARIAN TUMORS. IT WAS NOT CARCINOGENIC TO SPRAGUE-DAWLEY (CD) FEMALE RATS BUT IT WAS CARCINOGENIC TO MALE RATS, CAUSING LYMPHOMAS AND A FEW LYMPHOSARCOMAS, MYELOGENOUS LEUKEMIAS AND RETICULUM CELL SARCOMAS.

REPORT 06 : MULTI-AGENCY CROSS CHECK
Y HHS MAGUIRE
N OSHA CARC
Y EPA CAG
Y EPA RCRA
N EPA CLEAN AIR
N EPA CLEAN WATER

REPT'D USE : ANTINEOPLASTIC AGENT AND INSECT CHEMOSTERILANT

CAS NO : **10599-90-3**

CHEM NAME : ***** CHLORAMINE**

REPORT 01 : NTP DRAFT TR 392 NIH PUB. 91-2847 NOV 90
UNDER 2-YEAR STUDIES NO EVIDENCE OF CARCINOGENIC ACTIVITY OF CHLORAMINATED DRINKING WATER FOR MALE F344XN RATS. EQUIVOCAL EVIDENCE OF CARCINOGENIC ACTIVITY OF CHLORAMINATED DRINKING WATER FOR FEMALE F344/N RATS INDICATED BY INCREASE IN INCIDENCE OF MONONUCLEAR CELL LEUKEMIA. NO EVIDENCE OF CARCINOGENIC ACTIVITY OF CHLORAMINATED DRINKING WATER FOR MALE OR FEMALE B6C3F1 MICE.

REPT'D USE : DISINFECTANT AND ORGANIC SYNTHESIS REAGENT

CAS NO : **56-75-7**

CHEM NAME : ***** CHLORAMPHENICOL**

REPORT 01 : IARC SUPPLEMENT 1 1979
NO DATA WERE AVAILABLE ON THE CARCINOGENICITY OF CHLORAMPHENICOL IN EXPERIMENTAL ANIMALS. CASE REPORTS HAVE DESCRIBED LEUKEMIA IN SOME PATIENTS FOLLOWING CHLORAMPHENICOL-INDUCED APLASTIC ANEMIA.

REPORT 02 : NTP CARC TEST PROGRAM JUL 82
IN A CURRENT NTP BIOASSAY, CHLORAMPHENICOL IS BEING ADMINISTERED IN THE FEED TO F344 RATS AND B6C3F1 MICE. THE PRECHRONIC TEST PHASE IS IN PROGRESS.

REPORT 03 : IARC MONOGRAPH 10 1976
NO SIGNIFICANT INCREASE IN THE NUMBER OF CHROMOSOMAL ABERRATIONS WAS INDUCED BY CHLORAMPHENICOL IN THE RAT IN-VIVO IN THE BONE MARROW CYTOGENIC TEST OR IN CULTURES OF HUMAN PERIPHERAL LYMPHOCYTES. A PREPARATION OF UNSPECIFIED ORIGIN AND PURITY INJECTED IP INTO HYBRID C3HF MALE MICE PRODUCED NO EVIDENCE OF DOMINANT LETHAL MUTATIONS.
REPT'D USE : ANTIBIOTIC ANTIFUNGAL AND ANTIRICKETTSIAL AGENT

CAS NO : **57-74-9**
CHEM NAME : ***** CHLORDANE**

REPORT 01 : NCI CARC TRS 8 DHEW PUB NO 77-808 1977
UNDER THE CONDITIONS OF THIS BIOASSAY CHLORDANE IS CARCINOGENIC FOR THE LIVER IN MICE (FEEDING STUDY).

REPORT 02 : IARC SUPPLEMENT 1 1979
CHLORDANE AND HEPTACHLOR ARE CONSIDERED TOGETHER BECAUSE THEY ARE STRUCTURALLY SIMILAR AND OFTEN CONTAMINATED WITH EACH OTHER. CHLORDANE AND HEPTACHLOR ARE CARCINOGENIC IN MICE, PRODUCING LIVER TUMORS FOLLOWING ORAL ADMINISTRATION.

REPORT 03 : EPA AMBIENT WATER QUALITY CRITERIA OCT 80
CHLORDANE IS SUSPECTED OF BEING A HUMAN CARCINOGEN. AMBIENT WATER CONCENTRATIONS OF 4.6 NG/L, .46 NG/L, AND .046 NG/L ARE EXPECTED TO RESULT IN A RESPECTIVE 10 -5, 10 -6, AND 10 -7 INCREASE IN CANCER RISK OVER A LIFETIME.

REPORT 04 : EPA RCRA HEALTH AND ENV. EFFECTS PROFILE OCT 80
PURE CHLORDANE, WHICH IS A CIS/TRANS MIXTURE OF ISOMERS, INDUCES LIVER CANCER IN MICE AND IS MUTAGENIC IN SOME ASSAYS. CHLORDANE HAS NOT BEEN SHOWN TO BE TERATOGENIC. LITTLE INFORMATION IS AVAILABLE ON CHRONIC MAMMALIAN TOXICITY.

REPORT 05 : IARC MONOGRAPH 20 1979
ANALYTICAL GRADE CHLORDANE WAS TESTED IN ONE EXPERIMENT IN MICE AND IN ONE IN RATS BY ORAL ADMINISTRATION. IT PRODUCED HEPATOCELLULAR CAR CARCINOMAS IN MICE OF BOTH SEXES; IN RATS, THE RESULTS WERE INCONCLUSIVE. THERE IS SUFFICIENT EVIDENCE THAT CHLORDANE IS CARCINOGENIC IN MICE. A REPORT OF A NUMBER OF CASES OF CANCER IN HUMANS

WAS ALSO AVAILABLE, BUT THESE DATA DO NOT ALLOW AN EVALUATION OF THE CARCINOGENICITY OF CHLORDANE TO HUMANS TO BE MADE.

REPORT 06 : MULTI-AGENCY CROSS CHECK
N HHS MAGUIRE
N OSHA CARC
Y EPA CAG
Y EPA RCRA
N EPA CLEAN AIR
Y EPA CLEAN WATER

REPORT 07 : IARC MONOGRAPH 20 1979
PURE CHLORDANE WAS NOT MUTAGENIC IN ANY STRAIN OF SALMONELLA TESTED. IT INDUCED GENE CONVERSIONS IN SACCHAROMYCES AND INDUCED MUTATIONS IN MAMMALIAN CELLS IN CULTURE. THE CHEMICAL WAS NEGATIVE IN DOMINANT LETHAL TESTS IN MICE.

REPT'D USE : INSECTICIDE

CAS NO : **143-50-0**
CHEM NAME : ***** CHLORDECONE**

REPORT 01 : NAS REPORT ON DIET, NUTRITION, AND CANCER 1982
ON THE BASIS OF STUDIES IN ANIMALS, AND IN THE ABSENCE OF ADEQUATE DATA FROM EPA STUDIES, IT APPEARS THAT CHLORDECONE PRESENTS A CARCINOGENIC RISK TO HUMANS. HOWEVER, IT IS REASONABLE TO ASSUME THAT AMOUNTS PRESENT IN THE AVERAGE US DIET DO NOT MAKE A MAJOR CONTRIBUTION TO THE OVERALL RISK OF CANCER FOR HUMANS.

REPORT 02 : NCI CARCINOGENESIS BIOASSAY JAN 76
UNDER CONDITIONS OF THIS BIOASSAY, DIETARY ADMINISTRATION OF TECHGRADE CHLORDECONE TO OSBORNE-MENDEL RATS AND B6C3F1 MICE RESULTED IN A SIGNIFICANT INCIDENCE OF HEPATOCELLULAR CARCINOMAS OF HIGH-DOSE LEVEL RATS AND OF MICE AT BOTH DOSE LEVELS.

REPORT 03 : IARC MONOGRAPH 20 1979
CHLORDECONE WAS TESTED IN ONE EXPERIMENT IN MICE AND IN TWO IN RATS BY ORAL ADMINISTRATION: IT PRODUCED HEPATOCELLULAR CARCINOMAS IN MALES AND FEMALES OF BOTH SPECIES. CHLORDECONE IMPAIRS FERTILITY AND IS FETOTOXIC. IN A 1979 SUPPLEMENT IARC DETERMINED THAT THERE IS SUFFICIENT EVIDENCE FOR THE CARCINOGENICITY OF CHLORDECONE IN ANIMALS.

REPORT 04 : NIOSH CRITERIA DOCUMENT JAN 76
NIOSH RECOMMENDS A STANDARD OF 1 UG/CU M CEILING BASED

ON NERVOUS SYSTEM EFFECTS AND LIVER CANCER. THERE IS PRESENTLY NO OSHA STANDARD.

REPORT 05 : NTP TECHNICAL BULLETIN 7 APR 82
CHLORDECONE WAS NEGATIVE IN THE SALMONELLA TYPHIMURIUM MUTAGENICITY ASSAY.

REPORT 06 : MULTI-AGENCY CROSS CHECK
Y HHS MAGUIRE
N OSHA CARC
Y EPA CAG
Y EPA RCRA
N EPA CLEAN AIR
N EPA CLEAN WATER
REPT'D USE : INSECTICIDE BANNED BY EPA IN 1977

CAS NO : **6164-98-3**
CHEM NAME : ***** CHLORDIMEFORM**

REPORT 01 : IARC MONOGRAPH 30 JAN 83
PARA-CHLORO-ORTHO-TOLUIDINE (METABOLITE OF CHLORDIMEFORM) TESTED FOR CARCINOGENICITY IN MICE AND RATS BY ORAL ADMINISTRATION IN DIET. CARCINOGENIC IN TWO STRAINS OF MICE, PRODUCING HAEMANGIOSARCOMAS, STUDIES IN RATS WERE NOT INDICATIVE OF A CARCINOGENIC EFFECT, BUT CERTAIN LIMITATIONS IN THEIR DESIGN WERE NOTED.

REPORT 02 : IARC MONOGRAPH 30 JAN 83
CHLORDIMEFORM NEGATIVE IN TESTS FOR DNA DAMAGE OR MUTAGENICITY IN SEVERAL CELLULAR SYSTEMS. NO DATA AVAILABLE, HOWEVER, WITH REGARD TO ITS MUTAGENICITY IN MAMMALS, AND NO OVERALL EVALUATION OF MUTAGENICITY OF CHLORDIMEFORM COULD BE MADE.
REPT'D USE : INSECTICIDE

CAS NO : **115-28-6**
CHEM NAME : ***** CHLORENDIC ACID**

REPORT 01 : NTP TECH REPORT NTP TR 304, NIH PUB # 87-2560 APR 87
CPD NOT MUTAGENIC IN STRAINS TA100, TA98, OR TA1537 OF S. TYPHIMURIUM WITH AND WITHOUT METABOLIC ACTIVATION. UNDER CONDITIONS OF 2-YR FEED STUDIES, CLEAR EVIDENCE OF CARCINOGENICITY FOR MALE AND FEMALE F344/N RATS, AND FOR FEMALE B6C3F1 MICE. NO EVIDENCE OF CARCINOGENICITY OF CPD FOR FEMALE B6C3F1 GIVEN CPD IN DIET AT CONCENTRATIONS OF 620 OR 1250 PPM FOR 103 WEEKS.

REPORT 06 : MULTI-AGENCY CROSS CHECK
Y HHS MAGUIRE
N OSHA CARC
N EPA CAG
N EPA RCRA
N EPA CLEAN AIR
N EPA CLEAN WATER

REPT'D USE : REACTIVE FLAME RETARDANT, CHEMICAL INTERMEDIATE, EPOXY RESIN HARDENER AND LUBRICANT

CAS NO : **CLASS-11**
CHEM NAME : ***** CHLORINATED NAPHTHALENES (CNS)**

REPORT 01 : EPA FEDERAL REGISTER NOTICE 49FR 33619 AUG 84
UNDER TSCA 8(A) EPA PUBLISHED RULE REQUIRING MANUFACTURERS AND IMPORTERS OF CNS TO NOTIFY EPA OF MANUFACTURE OR IMPORTATION BECAUSE OF THEIR POTENTIAL FOR ONCOGENICITY, LIVER DEGENERATION, HYPERKERATOSIS, DERMATITIS AND OTHER HEALTH EFFECTS.

REPORT 02 : TSCA SECT 8(E) STATUS REPORT 8EHQ-0687-0680 JUL 87
CPD FOUND TO BE HEPATOTOXIC AT DIETARY LEVELS OF 0.3% AND 1.0%. SLIGHTLY IRRITATING TO EYES AND SKIN AND MAY CAUSE THYROID DYSFUNCTION. CPD HAS ORAL LD50 IN MALE AND FEMALE RATS OF APPROX. 1650 MG/KG AND A DERMAL LD50 IN GUINEA PIGS OF GREATER THAN 2000 MG/KG.

REPORT 06 : MULTI-AGENCY CROSS CHECK AUG 84
N HHS MAGUIRE
N OSHA CARC
N EPA CAG
N EPA RCRA
N EPA CLEAN AIR
Y EPA CLEAN WATER

REPT'D USE : CABLE INSULATION, WOOD PRESERVATIVES, AUTOMOTIVE CAPACITORS, ENGINE OILS, CUTTING AND GRINDING FLUIDS AND DYE PRODUCTION. NO DOMESTIC PRODUCTION, BUT IMPORTED FOR R AND D PURPOSES

CAS NO : **CLASS-44**
CHEM NAME : ***** CHLORINATED PARAFFIN**

REPORT 01 : EPA TSCA 8(E) NOTICE 8EHQ-0884-0527 AUG 84
NTERNATIONAL CIILORINATED PARAFFIN PRODUCERS TESTING CONSORTIUM SUBMITTED SUMMARIZED PRELIMINARY FINDINGS FROM STUDY OF "C-14/17, 52% CHLORINATED, CHLORINATED PARAFFIN ADMINISTERED IN FEED TO PREGNANT RATS PRIOR TO AND DURING GESTATION AND DURING LACTATION. AT HIGHEST

DOSE TESTED 6250 PPM, THERE WAS SUBCUTANEOUS BLEEDING AND DEATH IN RAT PUPS, BUT DAMS APPARENTLY WERE NOT AFFECTED.

REPORT 02 : EPA TSCA 8(E) NOTICE 8EHQ-0983-0491S SEP 83
INTERNATIONAL CHLORINATED PARAFFIN PRODUCERS TESTING CONSORTIUM SUBMITTED RESULTS FROM ACUTE AND CHRONIC TOXICITY STUDIES OF CHLORINATED PARAFFIN IN DAPHNIA MAGNA, CHIRONOMOUS TENTANS (MIDGE), IN MUSSELS, RAINBOW TROUT, AND MARINE ALGA. ADVERSE BIOLOGIC ACTICITY WAS LIMITED TO A SHORT CHAIN LENGTH HIGHLY CHLORINATED PARAFFIN; NON-BIODEGRADABILITY WAS FOUND; AND BIOCONCENTRATION HAD BEEN OBSERVED IN TROUT AND MUSSELS.

REPORT 03 : NTP TR 305 NIH PUB 86-2561 MAY 86
UNDER CONDITIONS OF 2 YEAR GAVAGE STUDIES NO EVIDENCE OF CARCINOGENICITY CHLORINATED PARAFFINS (C23, 43% CHLORINE) FOR MALE F344/N RATS GIVEN 1,875 OR 3,750 MG/KG PER DAY. EQUIVOCAL EVIDENCE OF CARCINOGENICITY FOR FEMALE F344/N RATS AS SHOWN BY AN INCREASED INCIDENCE OF ADRENAL GLAND MEDULLARY PHEOCHROMOCYTOMAS. CLEAR EVIDENCE OF CARCINOGENICITY FOR MALE B6C3F1 MICE SHOWN BY INCREASE IN INCIDENCE OF MALIGNANT LYMPHOMAS. EQUIVOCAL EVIDENCE OF CARCINOGENICITY FOR FEMALE B6C3F1 MICE SHOWN BY MARGINAL INCREASE IN INCIDENCE OF HEPATOCELLULAR NEOPLASMS.

REPORT 04 : NTP TR 308 PUB 86-2481 1986
CHLORINATED PARAFFIN C12, 60% CHLORINE IN F344/N RATS AND B6C3F1 MICE (GAVAGE STUDIES) SHOWED CLEAR EVIDENCE OF CARCINOGENICITY BOTH MALES AND FEMALES OF BOTH SPECIES AT DOSES OF 125 AND 250 MG/KG IN THE MICE AND 312 AND 625 MG/KG IN THE RAT. IT IS POSSIBLY CARCINOGENIC TO HUMANS.

REPORT 05 : EPS FEDERAL REGISTER NOTICE VOL 48 NO 145 JUL 83
EPA REPORTED INTERNATIONAL CHLORINATED PARAFFIN PRODUCERS TESTING CONSORTIUM SUBMITTED TERATOLOGY STUDY IN WHICH RABBITS ORALLY GIVEN A 58% CHLORINATED SHORT CHAIN PARAFFIN AT 10, 30 AND 100 MG/KG. NO MALFORMATIONS ABOVE THE CONTROL GROUP INDUCED, BUT SOME INDICATION OF EMBRYOTOXICTY (WHOLE LITTER RESORPTIONS) OBSERVED AT HIGH DOSE.

REPORT 06 : MULTI-AGENCY CROSS CHECK
Y HHS MAGUIRE
N OSHA CARC
N EPA CAG
N EPA RCRA
N EPA CLEAN AIR
N EPA CLEAN WATER

REPORT 07 : EPA FEDERAL REGISTER NOTICE VOL 49 NO 29 FEB 84
EPA REPORTED ON CONSORTIUM STUDY IN SHEEPSHEAD MINNOW LARVAE EXPOSED TO SHORT-CHAIN 58% CHLORINATED PARAFFIN FOR 32 DAYS SHOWED DIFFERENCE IN LENGTH AND WEIGHT COMPARED TO CONTROL. MUSSELS EXPOSED TO THIS PARAFFIN IN SEAWATER FOR 12 WEEKS AT 9.3 UG/LITER, SHOWED REDUCTION IN GROWTH RATE (BUT NOT AT 2.3 UG/L). IN BIOCONCENTRATION STUDY MUSSELS EXPOSED TO 15 UG/LITER OF THIS PARAFFIN FOR 84 DAYS SHOWED TOXIC EFFECTS IN ACUTE AND TOXICITY STUDIES ON SHRIMP AND RAINBOW TROUT NO TOXIC EFFECTS OBSERVED.

REPORT 08 : EPA FEDERAL REGISTER NOTICE VOL 49 NO 29 FEB 84
EPA REPORTED ON CONSORTIUM STUDY THAT SHOWED RATS ORALLY DOSED WITH SHORT-CHAIN 58% CHLORINATED PARAFFIN AT 250, 750, OR 2,000 MG/KG/DAY FOR 5 DAYS WAS NOT MUTAGENIC. DOMINANT LETHAL MUTATION AFFECTING POSTMEIOTIC STAGE OF SPERMATOGENESIS DETECTED IN POSTIVE CONTROL GROUP MALES OCCURED. IN 14-DAY DEITARY RANGE-FINDING STUDY, RATS EXPOSED TO 300, 1,000, AND 3,000 MG/KG/DAY, SHOWED MARKED REDUCTION IN BODY WEIGHT AND FOOD CONSUMPTION AT 3,000 MG DOSE. INCREASE IN LIVER WEIGHTS AND SEVERITY OF HEPATOCELLULAR HYPERTROPHY SEEN AT ALL DOSE LEVELS. MYOCARDIAL ATROPHY NOTED AT 3,000 MG DOSE LEVEL.

REPORT 09 : EPA FEDERAL REGISTER NOTICE VOL 49 NO 145 JULY 84
EPA REPORTED THAT IN A 13 WEEKS TOXICITY STUDY IN RATS WITH SHORT-CHAIN 58% CHLORINATED PARAFFIN AND LONG-CHAIN 70% CHLORINATED PARAFFIN NO TERATOGENIC EFFECTS OBSERVED

REPORT 10 : EPA FEDERAL REGISTER NOTICE VOL 48 NO 228 NOV 83
EPA REPORTED CHLORINATED PARAFFIN PRODUCERS CONSORTIUM SUBMITTED TERATOLOGY STUDY IN RABBITS WITH LONG-CHAIN 70% CHLORINATED PARAFFIN SHOWING NO DIFFERENCES FROM CONTROL IN FETAL MALFORMATIONS AT THREE DOSE LEVELS, HIGHEST AT 1,000 MG/KG/DAY. IN ORAL RANGE-FINDING STUDY IN RATS WITH SHORT-CHAIN 58% CHLORINATED PARAFFIN NO SIGNIFICANT CHANGES IN BODY WEIGHT AT 1,000 1,500, OR 2,000 MG/KG/DAY. IN 60-DAY TOXICITY EXPERIMENT WITH SHORT-CHAIN 58% CHLORINATED PARAFFIN SIGNIFICANT TOXICITY TO MUSSELS, BUT NOT TO RAINBOW TROUT. THE LC50 0.074 MG/LITER.

REPT'D USE : FIRE RETARDANT IN PLASTICS, TEXTILES, PAINTS, INSULATION COVERINGS PLASTICIZER IN PLASTICS, PAINTS, RUBBER, ADHESIVES, CAULK, POLYETHYLENE, SEALANTS, DETERGENTS AND LUBRICANT ADDITIVES

CAS NO : **61788-33-8**
CHEM NAME : ***** CHLORINATED TERPHENYL**

REPORT 01 : 49FR11181 MAR 84
RULE REQUIRES THAT MFRS OR IMPORTERS NOTIFY EPA OF ANY MFR OR IMPORTATION OF CT, A CHEMICAL RECOMMENDED FOR TESTING FOR HEALTH AND ENVIRONMENTAL IMPACTS BY ITC.

REPORT 06 : MULTI-AGENCY CROSS CHECK
N HHS MAGUIRE
Y OSHA CARC
N EPA CAG
N EPA RCRA
N EPA CLEAN AIR
N EPA CLEAN WATER
REPT'D USE : NO USE DATA AVAILABLE

CAS NO : **56802-99-4**
CHEM NAME : ***** CHLORINATED TRISODIUM PHOSPHATE**

REPORT 01 : NTP DRAFT REPORT TR 294 OCT 84
THERE WAS NO EVIDENCE OF CARCINOGENICITY FOR EITHER MALE OR FEMALE B6C3F1 MICE GIVEN CHLORINATED TRISODIUM PHOSPHATE BY GAVAGE IN WATER FOR 103 WEEKS AT DOSES OF 500 OR 1,000 MG. THE STUDIES IN F344/N RATS WERE CONSIDERED TO BE INADEQUATE STUDIES OF CARCINOGENICITY BECAUSE THE EXPERIMENTS WERE TERMINATED AT 35 WEEKS.
REPT'D USE : DISHWASHING COMPOUNDS, BLEACHES, CLEANERS, SPOT REMOVER AND BACTERIOCIDE

CAS NO : **7782-50-5**
CHEM NAME : ***** CHLORINE**

REPORT 01 : NTP DRAFT TR 392 NIH PUB. 91-2847 NOV 90
UNDER 2-YEAR STUDIES NO EVIDENCE OF CARCINOGENIC ACTIVITY OF CHLORINATED DRINKING WATER FOR MALE F344/N RATS. EQUIVOCAL EVIDENCE OF CARCINOGENIC ACTIVITY OF CHLORINATED DRINKING WATER FOR FEMALE F344/N RATS, INDICATED BY INCREASE IN INCIDENCE OF MONONUCLEAR CELL LEUKEMIA. NO EVIDENCE OF CARCINOGENIC ACTIVITY OF CHLORINATED DRINKING WATER FOR MALE OR FEMALE B6C3F1 MICE.
REPT'D USE : DISINFECTANT, BLEACHING, PLASTIC AND RESIN PRODUCTION, INTERMEDIATE IN PRODUCTION OF ETHYLENE GLYCOL ANTIFREEZE AND ETHYLENE CHLORIDE, REFRIGERANTS, PHARMACEUTICALS, COSMETICS, METAL EXTRACTIONS AND WAR GAS

CAS NO : **302-22-7**
CHEM NAME : ***** CHLORMADINONE ACETATE**

REPORT 01 : IARC MONOGRAPH 21 1979
CHLORMADINONE ACETATE WAS TESTED IN MICE, RATS AND DOGS BY ORAL ADMINISTRATION. WHEN GIVEN ALONE TO DOGS, CHLORMADINONE ACETATE PRODUCED MAMMARY TUMORS, WHEN GIVEN TO MICE IN COMBINATION WITH MESTRANOL, IT INCREASED THE INCIDENCE OF PITUITARY TUMORS IN ANIMALS OF BOTH SEXES.

REPORT 02 : IARC MONOGRAPH 21 1979
NO CHROMOSOMAL EFFECTS WERE OBSERVED WHEN CHLORMADINONE ACETATE WAS ADDED TO CULTURES OF LYMPHOCYTES GROWN FROM THE BLOOD OF HEALTHY WOMEN. CHLORMADINONE ACETATE HAS BEEN REPORTED TO BE EMBRYOLETHAL AND TERATOGENIC WHEN GIVEN DURING THE ORGANOGENESIS STAGE IN SOME ANIMAL SPECIES.
REPT'D USE : VETERINARY HORMONE

CAS NO : **494-03-1**
CHEM NAME : ***** CHLORNAPHAZINE**

REPORT 01 : IARC MONOGRAPH 4 1974
N,N-BIS(2-CHLOROETHYL)-2-NAPHTHYLAMINE IS CARCINOGENIC TO THE MOUSE LUNG BY THE INTRAPERITONEAL ROUTE AND HAS A LOCAL CARCINOGENIC EFFECT IN RATS FOLLOWING SUBCUTANEOUS INJECTION. THIS DRUG HAS BEEN ADMINISTERED TO MAN WITH 32P SODIUM PHOSPHATE FOR THE TREATMENT OF POLYCYTHAEMIA AND OF NEOPLASIAS OF THE HAEMOPOIETIC SYSTEM IN DOSES OF UP TO 400 MG/DAY. FOLLOW-UP STUDIES HAVE SHOWN THAT UNDER THESE CONDITIONS, THE DRUG IS CARCINOGENIC, PRODUCING BLADDER TUMORS AFTER ADMINISTRATION OF TOTAL DOSES AS LOW AS 4 G.

REPORT 02 : 6TH ANNUAL REPORT ON CARCINOGENS 1991
NO U.S. RESIDENTS EXPOSED, SO DROPPED FROM ANNUAL REPORT. LAST APPEARED IN 4TH ANNUAL REPORT.

REPORT 06 : MULTI-AGENCY CROSS CHECK
Y HHS MAGUIRE
Y OSHA CARC
Y EPA CAG
Y EPA RCRA
N EPA CLEAN AIR
N EPA CLEAN WATER
REPT'D USE : EXPERIMENTAL CHEMOTHERAPEUTIC AGENT

CAS NO : **107-20-0**
CHEM NAME : ***** CHLOROACETALDEHYDE**

REPORT 01 : EPA RCRA HEALTH AND ENV. EFFECTS PROFILE OCT 80
CARCINOGENIC EFFECTS WERE OBSERVED IN FEMALE ICR HA SWISS MICE FOLLOWING ADMINISTRATION OF CHLOROACETALDEHYDE VIA DERMAL APPLICATION OR SC INJECTION. MUTAGENIC EFFECTS, VARYING FROM WEAK TO STRONG, HAVE BEEN REPORTED IN THE YEASTS SCHIZOSACCHAROMYCES AND SACCHAROMYCES, AND IN CERTAIN SALMONELLA STRAINS. THERE IS NO EVIDENCE IN THE AVAILABLE LITERATURE TO INDICATE THAT CHLOROACETALDEHYDE PRODUCES TERATOGENIC EFFECTS.

REPORT 02 : 8EHQ-0387-0660 EPA STATUS REPORT APR 87
MALE AND FEMALE RATS EXPOSED FOR ONE HOUR TO CPD AT CONCENTRATIONS BETWEEN 0.14 AND 8.47 G/M3 IN AIR. THE 1-HOUR LC50 FOR COMBINED MALES AND FEMALES WAS CALCULATED TO BE BETWEEN 0.65 AND 0.79 G/M3. ANIMALS WHICH DIED SOON AFTER EXPOSURE SHOWED SIGNS OF EDEMA AND ATELECSTASIS IN LUNGS, USUALLY ACCOMPANIED BY A HYDRO-THORAX. THE SUBMITTER SAID THAT THE DATA SHOW A SIGNIFICANT RISK OF INJURY BY INHALATION.

REPORT 03 : TSCA SECT 8(E) STATUS REPORT 8EHQ-0387-0660 APR 87
ACUTE INHALATION TOXICITY OF CA STUDIED IN MALE AND FEMALE RATS WITH A ONE TIME 1 HOUR EXPOSURE TO CA IN AIR AT A CONCENTRATION BETWEEN 0.14 AND 8.47 G/M3. THE 1 HR LC50 COMBINED FOR MALES AND FEMALES WAS BETWEEN 0.65 AND 0.78 GM/M3 IN AIR. ANIMALS DYING SOON AFTER EXPOSURE SHOWED EVIDENCE OF EDEMA AND ATELECTASIS IN LUNGS WHEN AUTOPSIED. THESE WERE CONSIDERED SIGNS OF LUNG FUNCTION IMPAIRMENT.
REPT'D USE : FUNGICIDE, INTERMEDIATE IN MFR OF 2-AMINOTHIAZOLE AND TREE BARK REMOVAL

CAS NO : **78-95-5**
CHEM NAME : ***** CHLOROACETONE**

REPORT 01 : 8EHQ-0387-0660 EPA STATUS REPORT APR 87
MALE AND FEMALE RATS EXPOSED FOR ONE HOUR TO CPD AT 0.5 AND 7.9 G/M3 IN AIR. THE ONE HOUR LC50 FOR COMBINED MALES AND FEMALES WAS CALCULATED TO BE 1.9 G/M3. THE 1-HR LC50 FOR FEMALES WAS CALCULATED TO BE 2.7 G/MG3 AND FOR THE MALES, 1.2 G/M3. ANIMALS WHICH DIED SOON AFTER EXPOSURE SHOWED SIGNS OF EDEMA IN LUNGS ACCOMPANIED BY A HYDRO-THORAX. THE SUBMITTER STATED THAT THE DATA SHOW A SIGNIFICANT RISK OF INJURY BY INHALATION.

REPORT 02 : TSCA SECT 8(E) STATUS REPORT 8EHQ-0387-0660 APR 87
ACUTE INHALATION TOXICITY OF CHLOROACETONE (CA) STUDIED IN MALE AND FEMALE RATS BY EXPOSING THEM TO ATMOSPHERES CONTAINING CA AT CONCENTRATIONS BETWEEN 0.5 AND 7.9 G/M3 IN AIR. THE 1-HR LC50 FOR BOTH SEXES WAS 1.9 G/M3 (95% C.I.=1.6 AND 2.2 G/M3). LC50'S BY SEX WERE 2.7 G/M3 (95% C.I.=2.5-2.9 G/M3) FOR FEMALES AND 1.2 G/M3 (95% C.I.=1.1-1.3 G/M3) FOR MALES. AUTOPSY FINDINGS OF LUNG EDEMA CONSIDERED SIGNS OF LUNG FUNCTION IMPAIRMENT CAUSED BY INHALATION OF CA. DATA BELIEVED TO SHOW SIGNIFICANT RISK OF INJURY BY INHALATION. RATS WERE EXPOSED ONE TIME TO TEST ATMOSPHERES.
REPT'D USE : CHEMICAL INTERMEDIATE

CAS NO : **532-27-4**
CHEM NAME : ***** 2-CHLOROACETOPHENONE**

REPORT 01 : NTP TR379 NIH PUB NO. 90-2834 MAR 90
UNDER 2-YEAR INHALATION STUDIES NO EVIDENCE OF CARCINOGENIC ACTIVITY OF 2-CHLOROACETOPHENONE FOR MALE RATS EXPOSED TO 1 OR 2 MG/M3. EQUIVOCAL EVIDENCE OF CARCINOGENIC ACTIVITY FOR FEMALE F344/N RATS BASED ON MARGINAL INCREASE IN FIBROADENOMAS OF MAMMARY GLAND. NO EVIDENCE OF CARCINOGENIC ACTIVITY FOR MALE OR FEMALE B6C3F1 MICE EXPOSED TO 2 OR 4 MGMM3 2-CHLOROACETOPHENONE.
REPT'D USE : TEAR GAS

CAS NO : **140-49-8**
CHEM NAME : ***** 4'-(CHLOROACETYL)-ACETANILIDE**

REPORT 01 : NCI CARC TRS 177 DHEW PUB NO 79-1733
UNDER THE CONDITIONS OF THIS BIOASSAY, 4-(CHLOROACETYL)-ACETANILIDE WAS NOT CARCINOGENIC WHEN ADMINISTERED IN THE DIET TO FISCHER 344 RATS OR B6C3F1 MICE OF EITHER SEX.
REPT'D USE : DYE INTERMEDIATE AND PHARMACEUTICAL INTERMEDIATE

CAS NO : **106-47-8**
CHEM NAME : ***** P-CHLOROANILINE**

REPORT 01 : NCI CARC TRS 189 DHEW PUB NO 79-1745
THE FINDING OF SMALL NUMBER OF FIBROMAS AND SARCOMAS IN THE SPLEENS OF MALE RATS WAS CONSIDERED STRONGLY SUGGESTIVE OF CARCINOGENICITY BECAUSE OF THE RARITY OF THESE TUMORS IN THE SPLEENS OF CONTROL RATS. HEMANGIOMATOUS TUMORS IN DOSED MICE MAY ALSO HAVE BEEN ASSOCIATED WITH ADMINISTRATION OF P-CHLOROANILINE.

HOWEVER, IT IS CONCLUDED THAT UNDER THE CONDITIONS OF THIS BIOASSAY, SUFFICIENT EVIDENCE WAS NOT FOUND TO ESTABLISH THE CARCINOGENICITY OF P-CHLOROANILINE FOR FISCHER 344 RATS OR B6C3F1 MICE (FEEDING STUDY).

REPORT 02 : NTP CARC TEST PROGRAM JUL 82
IN A CURRENT NTP BIOASSAY, P-CHLOROANILINE IS BEING ADMINISTERED BY GAVAGE TO F344 RATS AND B6C3F1 MICE. THE CHRONIC EXPOSURE PHASE IS IN PROGRESS.

REPORT 03 : NTP TECHNICAL BULLETIN 5/8 AUG 81 JUL 82
P-CHLOROANILINE WAS NEGATIVE IN THE SALMONELLA ASSAY, BUT POSITIVE IN THE L5178Y MOUSE LYMPHOMA ASSAY.

REPORT 04 : ITC FOURTH REPORT TO EPA (44FR31866) JUN 79
THE INTERAGENCY TESTING COMMITTEE RECOMMENDS THAT THE EPA ADMINISTRATOR CONSIDER THE DEVELOPMENT OF TEST RULES TO TEST ANILINE, AND BROMO-, CHLORO- AND NITROANILINES FOR CARCINOGENICITY, MUTAGENICITY, TERATOGENICITY/REPRODUCTIVE EFFECTS AND NEUROLOGICAL EFFECTS, AND THAT AN EPIDEMIOLOGIC STUDY BE CONDUCTED.

REPORT 05 : 49(1) FR 108 ANPR JAN 84
EPA PUBLISHING ANPR TO INITAITE RULEMAKING TO REQUIRE TESTING TO CHARACTERIZE HEALTH AND ENVIRONMENTAL EFFECTS OF THE ANILINES. THIS IS IN RESPONSE TO ITC'S DESIGNATION OF ANILINES FOR TESTING IN ITS 4TH REPORT.

REPT'D USE : CHEMICAL INTERMEDIATE FOR DYES, AGRICULTURALS AND PHARMACEUTICALS

CAS NO : **108-90-7**
CHEM NAME : ***** CHLOROBENZENE**

REPORT 01 : EPA RCRA HEALTH AND ENV. EFFECTS PROFILE OCT 80
CHRONIC EXPOSURE TO CHLOROBENZENE APPEARS TO CAUSE A VARIETY OF PATHOLOGIES UNDER DIFFERENT EXPERIMANTAL REGIMENS; HOWEVER, THE LIVER AND KIDNEY APPEAR TO BE AFFECTED IN A NUMBER OF SPECIES. THERE HAVE BEEN NO STUDIES CONDUCTED TO EVALUATE THE MUTAGENIC, TERATOGENIC, OR CARCINOGENIC POTENTIAL OF CHLOROBENZENE.

REPORT 02 : NTP TR 261 NIH PUB 86-2517 OCT 85
UNDER CONDITIONS OF THESE STUDIES, CHLOROBENZENE ADMINISTRATION BY GAVAGE INCREASED OCCURRENCE OF NEOPLASTIC NODULES OF LIVER IN HIGH DOSE (120 MG/KG/DAY MALE F344/N RATS) PROVIDING SOME BUT NOT CLEAR EVIDENCE OF CARCINOGENICITY IN MALE RATS. CARCINOGENIC EFFECTS NOT OBSERRVED IN FEMALE F344/N RATS OR IN MALE OR FEMALE B6C3F1 MICE.

REPORT 03 : CMA INHALATION TERATOLOGY STUDY JAN 82
THE U.S. CHLOROBENZENE PRODUCERS, UNDER THE AUSPICES OF THE CMA, EVALUATED THE EMBRYOTOXIC AND TERATOGENIC POTENTIAL OF MCB VAPORS IN RATS AND RABBITS. INHALATION OF MCB VAPORS FOR SIX HOURS/DAY DURING THE PERIOD OF MAJOR ORGANOGENESIS WAS NOT EMBRYOTOXIC OR TERATOGENIC IN RATS AND RABBITS EVEN AT A MATERNALLY TOXIC EXPOSURE CONCENTRATION OF 590 PPM.

REPORT 04 : EPA RESPONSE TO ITC RECOMMS (45FR48524) JUL 80
IN RESPONSE TO ITC RECOMMENDATIONS (42FR55026) THE EPA PROPOSES THAT MANUFACTURERS AND PROCESSORS OF CHLORINATED BENZENES, EXCEPT HEXACHLOROBENZENE, CONDUCT TESTING FOR TERATOGENICITY AND REPRODUCTIVE EFFECTS AND OTHER CHRONIC EFFECTS. EPA WILL CONDUCT MUTAGENICITY TESTS AND DECIDE ON FURTHER TIER TESTING AS APPROPRIATE. IT IS NOT REASONABLE TO REQUIRE AN EPIDEMIOLOGIC STUDY FOR THESE CHEMICALS.

REPORT 05 : 51 CFR 130 JUL 86
EPA IS ISSUING A FINAL RULE UNDER TSCA (4) REQUIRING MFRS AND PROCESSORS OF MCB TO CONDUCT REPRODUCTIVE TESTS ACCORDING TO TSCA HEALTH EFFECTS TESTING GUIDELINES IN 40 CFR 798. THIS RULE IS PROMULGATED FOR JUDICIAL REVIEW AT 1300 HRS, JULY 22, 1986 AND BECOMES EFFECTIVE ON AUGUST 21, 1986.

REPORT 06 : MULTI-AGENCY CROSS CHECK
N HHS MAGUIRE
N OSHA CARC
N EPA CAG
Y EPA RCRA
Y EPA CLEAN AIR
Y EPA CLEAN WATER

REPORT 07 : 52FR2152 - RECE IPT OF TEST DATA JAN 87
EPA ANNOUNCES RECEIPT OF TEST DATA FROM INDUSTRY'S VOLUNTARY TESTING PROGRAM.

REPT'D USE : SOLVENT PAINTS, DRY CLEANING SOLUTIONS, ADHESIVES, POLISHES/WAXES, INSECTICIDES, DYESTUFFS, INTERMEDIATE IN SYNTHESIS OF PHENOL, DDT AND ANILINE

CAS NO : **510-15-6**
CHEM NAME : ***** CHLOROBENZILATE**

REPORT 01 : NCI CARC TRS 75 DHEW PUB NO 78-1325
UNDER THE CONDITIONS OF THIS BIOASSAY ORALLY ADMINISTERED CHLOROBENZILATE WAS CARCINOGENIC IN MALE AND FEMALE B6C3F1 MICE, CAUSING AN INCREASED INCIDENCE OF

HEPATOCELLULAR CARCINOMAS. THE RESULTS DO NOT, HOWEVER, PROVIDE SUFFICIENT EVIDENCE FOR THE CARCINOGENICITY OF CHLOROBENZILATE IN OSBORNE-MENDEL RATS (FEEDING STUDY).

REPORT 02 : NTP TECHNICAL BULLETIN 3 DEC 80
THE COMPOUND WAS FOUND TO BE NEGATIVE IN THE SALMONELLA TYPHIMURIUM MUTAGENICITY ASSAY IN ONE LABORATORY.

REPORT 03 : IARC MONOGRAPH 5 1974
CHLOROBENZILATE WAS TESTED IN A PRELIMINARY STUDY BY THE ORAL ROUTE IN TWO STRAINS OF MICE AND PRODUCED AN INCREASED INCIDENCE OF HEPATOMAS IN MALES OF BOTH STRAINS.

REPORT 04 : NTP TECHNICAL BULLETIN 5 AUG 81
CHLOROBENZILATE WAS NEGATIVE IN THE S. TYPHIMURIUM MUTAGENICITY ASSAY IN TWO LABORATORIES.

REPORT 05 : IARC MONOGRAPH JAN 83
CHLOROBENZILATE TESTED FOR CARCINOGENICITY BY ADMINISTRATION IN DIET IN THREE STAINS OF MICE AND IN TWO STRAINS OF RATS; INDUCED HEPATOCELLULAR CARCINOMAS IN BOTH SEXES OF MICE OF ONE STRAIN AND IN MALES OF TWO OTHER STRAINS. DATA ON RATS WERE INADQUATE FOR EVALUATION.

REPORT 06 : MULTI-AGENCY CROSS CHECK
N HHS MAGUIRE
N OSHA CARC
Y EPA CAG
Y EPA RCRA
N EPA CLEAN AIR
N EPA CLEAN WATER

REPORT 07 : IARC MONOGRAPH JAN 83
CHLOROBENZILATE NOT MUTAGENIC TO SALMONELLA TYPHIMURIUM WITH OR WITHOUT EXOGENOUS METABOLIC ACTIVATION. DATA FROM STUDIES ON OTHER ORGANISMS CONSIDERED INSUFFICIENT FOR EVALUATION; NO OVERALL EVALUATION OF MUTAGENICITY OF CHLOROBENZILATE COULD BE MADE.

REPT'D USE : ACARICIDE, MITICIDE AND SYNERGIST FOR DDT

CAS NO : **5216-25-1**
CHEM NAME : ***** P-CHLOROBENZOTRICHLORIDE (PCBTC)**

REPORT 01 : EPA TSCA 8(E) NOTICE 8EQH-0980-0360
HOOKER SUBMITTED INFORMATION ON THE POSSIBLE CARCINOGENICITY OF BC, BTC, AND PCBTC IN MICE. THE SUBMISSION CONSISTS OF FOUR JAPANESE SCIENTIFIC ARTICLES AND AN EVALUATION PREPARED BY A HOOKER TOXICOLOGIST. SUBMITTER

CONCLUDED THAT THE EXPERIMENTAL DATA INDICATE THAT BTC AND PCBTC ARE CHEMICAL CARCINOGENS.

REPORT 02 : 48 FR 53160 NOV 83
OCCIDENTAL CHEMICAL CORP PROPOSED MULTI-LEVEL TESTING METHOD FOR PCBTC. INCLUDED IS A BASE SET OF SUBCHRONIC EFFECTS, METABOLISM, MUTAGENICITY AND CELL TRANSFORMATION.

REPT'D USE : HERBICIDE MANUFACTURE

CAS NO : **98-56-6**

CHEM NAME : ***** 4-CHLOROBENZOTRIFLUORIDE**

REPORT 01 : ITC REPORT TO EPA (47FR5456) FEB 82
4-CBTF RECOMMENDED FOR TESTING FOR CHRONIC HEALTH EFFECTS, FOR BIOCONCENTRATION IN FISH AND FOR CHEMICAL FATE. NEGOTIATED TESTING AGREEMENT BETWEEN EPA, HOOKER CHEMICAL AND OCCIDENTAL CHEMICAL SIGNED (48FR32730) ON JULY 18, 1983.

REPORT 02 : EPA RESPONSE TO ITC (50FR42216) OCT 85
EPA WILL NOT REQUIRE FURTHER TESTING FOR HEALTH EFFECTS, ENVIRONMENTAL EFFECTS AND CHEMICAL FATE BECAUSE DATA RECEIVED ARE ADEQUATE. THERE IS NO EVIDENCE OF MUTAGENICITY AND ACUTE TOXICITY IS WELL DEFINED. DATA SHOW NO SUBSTANTIAL EVIRONMENTAL EXPOSURE OR UNREASONABLE RISK TO ENVIRONMENT OR HUMAN HEALTH.

REPT'D USE : CHEMICAL INTERMEDIATE IN PRODUCTION OF HERBICIDES, DYES AND DRUGS

CAS NO : **124-48-1**

CHEM NAME : ***** CHLORODIBROMOMETHANE**

REPORT 01 : EPA RCRA HEALTH AND ENV. EFFECTS PROFILE OCT 80
DIBROMOCHLOROMETHANE GAVE POSITIVE RESULTS IN MUTAGENICITY TESTS WITH SALMONELLA.

REPORT 03 : NTP TECHNICAL BULLETIN 5 AUG 81
CHLORODIBROMOMETHANE WAS NEGATIVE IN THE SALMONELLA TYPHIMURIUM ASSAY IN ONE LABORATORY.

REPORT 06 : MULTI-AGENCY CROSS CHECK
N HHS MAGUIRE
N OSHA CARC
N EPA CAG
Y EPA RCRA
N EPA CLEAN AIR
Y EPA CLEAN WATER

REPORT 07 : NTP TECHNICAL REPORT NO. 282 NIH PUB 85-25 -38 AUG 85
NO EVIDENCE OF CARCINOGENICITY IN MALE OR FEMALE F344/N RATS RECEIVING 40 OR 80 MG/KG CHLORODIBROMOMETHANE BY GAVAGE. FATTY METAMORPHOSIS AND GROUND-GLASS CYTOPLASMIC LIVER CHANGES IN MALE AND FEMALE RATS. EQUIVOCAL EVIDENCE OF CARCINOGENICITY FOR MALE B6C3F1 MICE (INCREASED INCIDENCE OF HEPATOCELLULAR CARCINOMAS). SOME EVIDENCE OF CARCINOGENICITY FOR FEMALE B6C3F1 MICE (INCREASED INCIDENCE OF HEPATOCELLULAR ADENOMAS AND INCREASED COMBINED INCIDENCE OF HEPATOCELLULAR ADENOMAS OR CARCINOMAS).

REPT'D USE : INTERMEDIATE IN MR OF FIRE EXTINGUISHING AGENTS, AEROSOLS, PROPELLANTS, PESTICIDES AND REFRIGERANT

CAS NO : **75-45-6**

CHEM NAME : ***** CHLORODIFLUOROMETHANE**

REPORT 01 : IARC 41 1986
LIMITED EVIDENCE FOR CARCINOGENICITY OF CPD TO TEST ANIMALS AND INADEQUATE EVIDENCE FOR CARCINOGENICITY TO HUMANS. INADEQUATE EVIDENCE FOR ACTIVITY IN SHORT-TERM TESTS FOR GENETIC ACTIVITY. TESTED IN MICE BY GAVAGE IN RATS AND BY INHALATION IN MICE AND RATS. NO INCREASE IN TUMOR INCIDENCE AFTER ORAL DOSING. INHALATION STUDY WAS INCONCLUSIVE FOR MALE MICE AND NEGATIVE FOR FEMALES. INHALATION AT HIGH DOSES LED TO INCREASED INCIDENCE OF FIBROSARCOMA AND ZYMBAL-GLAND TUMORS IN MALES. CPD IS TERATOGENIC FOR RATS, BUT NO PRENATAL TOXICITY IN RABBITS FOLLOWING INHALATION TEST. NO DATA TO EVALUATE REPRODUCTIVE EFFECTS OR PRENATAL TOXICITY TO HUMANS.

REPT'D USE : REFRIGERANT, CHEMICAL INTERMEDIATE BLOWING AGENT FOR PLASTICS AND SOLVENT

CAS NO : **75-00-3**

CHEM NAME : ***** CHLOROETHANE**

REPORT 01 : NIOSH CURRENT INTELLIGENCE BULLETIN 27 AUG 78
AS OF JULY 1978, NO NCI BIOASSAY TESTS WERE PLANNED. OSHA EXPOSURE STANDARD IS 1000 PPM FOR WORKERS. ALL CHLOROETHANE COMPOUNDS ARE KNOWN TO CAUSE CNS DEPRESSION IN LABORATORY ANIMALS. CHLOROETHANES ARE GENERALLY IRRITATING TO THE EYES AND SKIN. ANIMAL STUDIES REVEAL KIDNEY DAMAGE AND FATTY CHANGES IN LIVER, KIDNEY, AND HEART. MONOCHLOROETHANE SHOULD BE TREATED IN THE WORKPLACE WITH CAUTION AND MONITORED FOR CARCINOGENIC EFFECTS IN HUMANS BECAUSE OF ITS RELATION TO FOUR CARCINOGENIC CHLOROETHANES.

REPORT 02 : NTP TR346 NIH PUB. 90-2801 OCT 89
2-YEAR INHALATION STUDIES SHOWED EQUIVOCAL EVIDENCE OF CARCINOGENIC ACTIVITY OF CHLOROETHANE FOR MALE F344/N RATS INDICATED BY BENIGN AND MALIGNANT EPITHELIAL NEOPLASMS OF THE SKIN. FOR FEMALE F344/N RATS EQUIVOCAL EVIDENCE OF CARCINOGENIC ACTIVITY INDICATED BY THREE UNCOMMON MALIGNANT ASTROCYTOMAS OF THE BRAIN IN THE EXPOSED GROUP. THE STUDY IN MALE B6C3F1 MICE CONSIDERED TO BE INADEQUATE STUDY OF CARCINOGENICITY BECAUSE OF REDUCED SURVIVAL IN EXPOSED GROUP. HOWEVER, THERE WAS INCREASED INCIDENCE OF ALVEOLAR/BRONCHIOLAR NEOPLASMS OF LUNG. CLEAR EVIDENCE OF CARCINOGENIC ACTIVITY FOR FEMALE B6C3F1 MICE INDICATED BY CARCINOMAS OF UTERUS.

REPORT 03 : EPA RCRA HEALTH AND ENV. EFFECTS PROFILE OCT 80
THERE IS NO AVAILABLE EVIDENCE WHICH INDICATES THAT MONOCHLOROETHANE PRODUCES CARCINOGENIC, MUTAGENIC, OR TERATOGENIC EFFECTS. SYMPTOMS PRODUCED BY HUMAN POISONING WITH MICE INCLUDE CNS DEPRESSION, RESPIRATORY FAILURE AND CARDIAC ARRHYTHMIAS.

REPORT 04 : NTP TECHNICAL BULLETIN 5 AUG 81
CHLOROETHENE WAS NEGATIVE IN THE SALMONELLA TYPHIMURIUM MUTAGENICITY ASSAY IN ONE LABORATORY.

REPORT 06 : MULTI-AGENCY CROSS CHECK
N HHS MAGUIRE
N OSHA CARC
N EPA CAG
N EPA RCRA
N EPA CLEAN AIR
Y EPA CLEAN WATER

REPORT 07 : EPA TSCA 8(E) 0188-0713 JAN 88
DOW CHEMICAL REPORTED ON A 2-YEAR MOUSE INHALATION STUDY BY NTP THAT SHOWED AT THE 15,000 PPM DOSE ADENOCARCINOMAS OF THE UTERUS OBSERVED.

REPT'D USE : CHEMICAL INTERMEDIATE FOR TOPICAL ANESTHETICS, GASOLINE ANTI-KNOCK, DYES, PHARMACEUTICALS, SOLVENT FOR PHOSPHORUS, SULFUR, FATS, OILS, WAXES; INDUSTRIAL REFRIGERANT; SKIN MEDICINE, CAPTIVE USE ALKYLATING AGENT, MANUFACTURE OF TETRAETHYL LEAD AND ETHYLCELLULOSE; ORGANIS SYNTHESIS, SULFUR, OILS, RESINS AND WAXES

CAS NO : **107-07-3**
CHEM NAME : ***** 2-CHLOROETHANOL (ETHYLENE CHLOROHYDRIN)**

REPORT 01 : NTP TECHNICAL REPORT 275 NIH PUB 86-2531 NOV 85
UNDER CONDITIONS OF THESE 2 YEAR DERMAL STUDIES NO EVI-

DENCE OF CARCINOGENICITY OF 2-CHLOROETHANOL FOR MALE AND FEMALE F344/N RATS GIVEN 50 OR 100 MG/KG PER DAY OR FOR MALE AND FEMALE SWISS CD-1 MICE GIVEN 7.5 OR 15 MG PER ANIMAL PER DAY.

REPORT 02 : NTP TECHNICAL BULLETIN 3 DEC 80
THE COMPOUND WAS FOUND TO BE POSITIVE IN THE SALMONELLA TYPHIMURIUM MUTAGENICITY ASSAY IN TWO LABORATORIES.

REPORT 03 : EPA 8EHQ-1187- 698 NOV 87
INFORMATION FROM UNION CARBIDE ON AN EPA STUDY OF WORKERS EMPLOYED BETWEEN 1940 AND 1946 SHOW POSITIVE TRENDS IN PANCREATIC CANCER AND LEUKEMIA MORTALITY.

REPT'D USE : INTERMEDIATE IN MFR OF INSECTICIDES, SOLVENT IN FABRIC DYEING AND INDUSTRIAL SOLVENT

CAS NO : **999-81-5**

CHEM NAME : ***** (2-CHLOROETHYL)TRIMETHYLAMMONIUM CHLORIDE**

REPORT 01 : NCI CARC TRS 158 DHEW PUB NO 79-1714
UNDER THE CONDITIONS OF THIS BIOASSAY, (2-CHLOROETHYL)TRIMETHYLAMMONIUM CHLORIDE WAS NOT CARCINOGENIC FOR F344 RATS OR B6C3F1 MICE OF EITHER SEX (FEEDING STUDY).

REPORT 02 : NTP TECHNICAL BULLETIN 6 JAN 82
THE CHEMICAL WAS NEGATIVE IN THE SALMONELLA TYPHIMURIUM MUTAGENICITY ASSAY IN ONE LABORATORY.

REPT'D USE : PLANT GROWTH REGULATOR

CAS NO : **13909-09-6**

CHEM NAME : ***** 1-(2-CHLOROETHYL)-3-(4-METHYLCYCLOHEXYL)-1-NITROSOUREA**

REPORT 01 : NCI BIOASSAY PROGRAM - CANCER VOL. 40 1977
UNDER THE CONDITIONS OF THE BIOASSAY, I.P. INJECTION OF METHYL CCNU WAS CARCINOGENIC TO MALE SPRAGUE-DAWLEY (CD) RATS, CAUSING SUBCUTANEOUS FIBROMAS, PERITONEAL SARCOMAS, LUNG TUMORS AND MAMMARY TUMORS. IT WAS NOT CARCINOGENIC TO FEMALE RATS AND MALE AND FEMALE SWISS-WEBSTER MICE.

REPORT 06 : MULTI-AGENCY CROSS CHECK
Y HHS MAGUIRE
N OSHA CARC
N EPA CAG
N EPA RCRA
N EPA CLEAN AIR

N EPA CLEAN WATER

REPT'D USE : NO AVAILABLE PRODUCTION OR USE DATA; EXPERIMENTAL TUMORIGEN

CAS NO : **UNKNOWN-19**

CHEM NAME : ***** CHLOROFLUOROCARBON 22**

REPT'D USE : PROPELLANTS AND AEROSOLS

CAS NO : **CLASS-08**

CHEM NAME : ***** CHLOROFLUOROCARBONS**

REPORT 01 : 16CFR1401 FEDERAL REGISTER NOTICE VOL 42P.42780 AUG 77
THE COMMISSION HAS FOUND THAT FULLY HALOGENATED CHLOROFLUOROALKANES USED AS PROPELLANTS IN SELF-PRESSURIZED CONTAINERS MAY POSE A RISK OF DEPLETION OF OZONE IN THE STRATOSPHERE. OZONE DEPLETION ALLOWS MORE ULTRA-VIOLET RAYS FROM THE SUN TO REACH THE EARTH. THE CONSEQUENCES INCLUDE A POSSIBILITY OF A SIGNIFICANT INCREASE IN HUMAN CANCER AND CLIMATIC CHANGES. THE COMMISSION REQUIRES MFRS TO GIVE PERFORMANCE AND TECHNICAL DATA TO PURCHASERS WHICH INCLUDES AN IDENTIFICATION AND WARNING STATEMENT CONCERNING CHLOROFLUOROCARBONS.

REPORT 02 : EPA FYI OTS 0288-0602 FEB 88
ICI AMERICAS REPORTS ITS UK PARENT COMPANY STATEMENT TO ALL AEROSOL CUSTOMERS THAT ON BASIS OF CURRENTLY AVAILABLE DATA, AN INADEQUATE MARGIN OF SAFETY EXISTS TO RECOMMEND USE OF CFC 22 AS AEROSOL PROPELLANT FOR RETAIL MARKET, PARTICULARLY WITH REGARD TO EFFECTS SEEN AT 50,000 PPM IN RAT TERATOLOGY STUDY AND TWO-YEAR RAT BIOASSAY. THEREFORE A CONSERVATIVE APPROACH IS CONSIDERED APPROPRIATE IN REACHING DECISION NOT TO SELL THIS PRODUCT FOR USE AS AN AEROSOL PROPELLANT. IN PRACTICE, HEALTH EFFECTS NOTED SO FAR DUE TO ACUTE EXPOSURES WHERE BUILD UP TO VERY HIGH CONCENTRATIONS IN CONFINED SPACES OCCURRED AND WHERE OXYGEN DEFICIENCY LIKELY TO BE SEVERE.

REPORT 03 : EPA FEDERAL REGISTER NOTICE VOL.53, NO. 156 P.30566 AUG 88
FINAL RULE LIMITS PRODUCTION AND CONSUMPTION OF CFCS AND HALONS TO REDUCE RISKS OF STRATOSPHERIC OZONE DEPLETION. REQUIRES FREEZE AT 1986 LEVELS OF PRODUCTION AND CONSUMPTION OF CFC-11,-12,-113,-114, AND-115 BASED ON RELATIVE OZONE DEPLETION WEIGHTS, FOLLOWED BY REDUCTION TO 80% AND 50% OF 1986 LEVELS IN 1993 AND 1998. LIMITS PRODUCTION AND CONSUMPTION OF HALON 1211, 1301, AND 2402

TO 1986 LEVELS IN 1992. PROMULGATION OF RULE AUTHORIZED BY SECTION 157(B) OF CLEAN AIR ACT AND CONSTITUTES 45 IMPLEMENTATION OF "MONTREAL PROTOCOL ON SUBSTANCES THAT DEPLETE THE OZONE LAYER" RATIFIED ON APRIL 21, 1988. RULE ALLOCATES PRODUCTION AND CONSUMPTION ALLOWANCES TO THESE CHEMICALS IN 1986.

REPORT 06 : MULTI-AGENCY CROSS CHECK
R HHS MAGUIRE
R OSHA CARC
R EPA CAG
R EPA RCRA
Y EPA CLEAN AIR
R EPA CLEAN WATER

REPT'D USE : PROPELLANTS AND AEROSOLS

CAS NO : **593-70-4**

CHEM NAME : ***** CHLOROFLUOROMETHANE**

REPORT 01 : IARC VOL 41 1986
LIMITED EVIDENCE FOR CARCINOGENICITY OF CPD TO TEST ANIMALS BUT NO EVALUATION OF CARCINOGENICITY TO HUMANS COULD BE MADE. LIMITED EVIDENCE FOR GENETIC ACTIVITY IN SHORT-TERM TESTS. EFFECTS OF CPD ON REPRODUCTION OR PRENATAL TOXICITY IN TEST ANIMALS OR HUMANS COULD NOT BE EVALUATED. CPD MUTAGENIC TO S. TYPHIMURIUM AND CULTURED MAMMALIAN CELLS IN PRESENCE OF AN EXOGENOUS METABOLIC SYSTEM.

REPT'D USE : NO COMMERCIAL USE LISTED

CAS NO : **67-66-3**

CHEM NAME : ***** CHLOROFORM**

REPORT 01 : EPA AMBIENT WATER QUALITY CRITERIA OCT 80
CHLOROFORM IS SUSPECTED OF BEING A HUMAN CARCINOGEN. AMBIENT WATER CONCENTRATIONS OF 2.1 UG/L, .21 UG/L AND .021 UG/L ARE ESTIMATED TO RESULT IN AN INCREASED CANCER RISK OF 10 -5, 10 -6 AND 10 -8 RESPECTIVELY.

REPORT 02 : NIOSH CURRENT INTELLIGENCE BULLETIN 9 MAR 76
CHLOROFORM HAS BEEN SHOWN TO BE CARCINOGENIC BY INGESTION IN LABORATORY MICE AND RATS, ACCORDING TO A MARCH 1976 NCI BIOASSAY. THE CHEMICAL INDUCED MALIGNANT KIDNEY TUMORS IN RATS AND HEPATOCELLULAR CARCINOMA IN MICE. DATA ON TOXICITY AND OCCUPATIONAL EXPOSURE ARE GIVEN.

REPORT 03 : TOXICOLOGY INFORMATION RESPONSE CENTER DEC 77
THE REPORT PRESENTS A REVIEW OF THE HEALTH ASPECTS OF CHLOROFORM INCLUDING TOXICITY, TERATOGENICITY, AND CARCINOGENICITY. AN ABSTRACTED LITERATURE COLLECTION COVERING THE YEARS 1907-1977 IS ALSO INCLUDED.

REPORT 04 : NIOSH CRITERIA DOCUMENT JUN 76
NIOSH RECOMMENDS A CEILING OF 2 PPM BASED ON LIVER OR KIDNEY TUMORS AND CNS EFFECTS. THE PRESENT OSHA STANDARD IS 50 PPM CEILING.

REPORT 05 : REPORT ON CARCINOGENESIS BIOASSAY (NCI) MAR 76
USP GRADE CHLOROFORM WAS ADMINISTERED ORALLY IN CORN OIL TO OSBORNE-MENDEL RATS AND B6C3F1 MICE. KIDNEY EPITHELIAL TUMORS WERE OBSERVED IN MALE RATS AT A SIGNIFICANT LEVEL. THERE WAS A STATISTICALLY INSIGNIFICANT INCREASE IN THYROID TUMORS OBSERVED IN TREATED FEMALE RATS. HIGHLY SIGNIFICANT INCREASES IN HEPATOCELLULAR CARCINOMA WERE OBSERVED IN BOTH SEXES OF MICE. ALSO, NODULAR HYPERPLASIA OF THE LIVER WAS SEEN IN MANY LOW DOSE MALE MICE THAT HAD NOT DEVELOPED HEPATOCELLULAR CARCINOMA.

REPORT 06 : MULTI-AGENCY CROSS CHECK
Y HHS MAGUIRE
N OSHA CARC
Y EPA CAG
Y EPA RCRA
Y EPA CLEAN AIR
Y EPA CLEAN WATER

REPORT 07 : IARC MONOGRAPH 20 1979
CHLOROFORM WAS TESTED IN THREE EXPERIMENTS IN MICE AND ONE IN RATS BY ORAL ADMINISTRATION. IT PRODUCED HEPATOMAS AND HEPATOCELLULAR CARCINOMAS IN MICE, MALIGNANT KIDNEY TUMORS IN MALE RATS AND TUMORS OF THE THYROID IN FEMALE RATS. IN A 1979 SUPPLEMENT, IARC DETERMINED THAT THERE IS SUFFICIENT EVIDENCE FOR THE CARCINOGENICITY OF CHLOROFORM IN ANIMALS.

REPORT 08 : EPA RCRA HEALTH AND ENV. EFFECTS PROFILE OCT 80
CHLOROFORM HAS BEEN FOUND TO INDUCE HEPATOCELLULAR CARCINOMAS IN MICE AND KIDNEY EPITHELIAL TUMORS IN RATS. HEPATOMAS HAVE ALSO BEEN INDUCED IN MICE, BUT NECROSIS MAY BE A PREREQUISITE TO TUMOR FORMATION. BACTERIAL ASSAYS INVOLVING CHLOROFORM HAVE YIELDED NO MUTAGENIC EFFECTS. CHLOROFORM HAS PRODUCED TERATOGENIC EFFECTS WHEN ADMINISTERED TO PREGNANT RATS.

REPORT 09 : IARC MONOGRAPH 20 1979
CHLOROFORM IS FETOTOXIC. IN ASSAYS WITH SALMONELLA

TA1535 AND TA1538 AND WITH E. COLI K12 CHLOROFORM WAS NOT MUTAGENIC IN THE PRESENCE OF METABOLIC ACTIVATION.

REPT'D USE : SOLVENT, CLEANING AGENT, CHEMICAL INTERMEDIATE, PRODUCTION OF FLUOROCARBON-22 AND PESTICIDE

CAS NO : **3327-22-8**

CHEM NAME : ***** 3-CHLORO-2-HYDROXYPROPYLTRIMETHYL AMMONIUM CHLORIDE**

REPORT 01 : EPA TSCA 8(E) 8EHQ-1283-0503 DEC 83
DOW PROVIDED RESULTS FROM SHORT-TERM IN-VITRO GENOTOXICITY TESTS SHOWING POSITIVE RESPONSES IN AMES S. TYMPHIMURIUM MUTAGENICITY ASSAY, IN UNSCHEDULED DNA SYNTHESIS ASSAY IN CULTURED RAT HEPATOCYTES, AND IN HYPOXANTHINE GUANINE PHOSPHORIBOSYL TRANSFERASE MUTAGENICITY ASSAY USING CHO CELLS FOR TECHNICAL PURIFIED GRADES OF 3-CHLORO-2-HYDROXY-PROPYLTRIMETHYL AMMONIUM CHLORIDE.

REPORT 02 : EPA TSCA 8(E) 8EHQ-1283-0503 FEB 84
CHEMICAL PRODUCED POSITIVE RESULTS IN AMES S. TYPHIMURIUM MUTAGENICITY ASSAY, IN UNSCHEDULED DNA SYNTHESIS, IN CULTURED RAT HEPATOCYTES AND IN HYPOXANTHINE GUANINE PHOSPHORIBISYL TRANSFERASE MUTAGENICITY ASSAY.

REPT'D USE : PAPER, STRENGTH ADDITIVE, SHAMPOOS AND COSMETICS

CAS NO : **74-87-3**

CHEM NAME : ***** CHLOROMETHANE**

REPORT 01 : EPA RESPONSE TO ITC RECOMMS IN 45FR48524 JUL 80
IN RESPONSE TO THE ITC RECOMMENDATIONS IN 42FR55026 THE EPA PROPOSES THAT MANUFACTURERS AND PROCESSORS OF CHLOROMETHANE CONDUCT ONCOGENICITY AND TERATOGENICITY TESTS. THE EPA WILL CONTINUE TO ASSESS THE NEED FOR TESTING FOR OTHER CHRONIC EFFECTS AND FOR MUTAGENICITY. THE EPA ACKNOWLEDGES THAT THE CIIT IS CURRENTLY SPONSORING A CARCINOGENICITY ASSAY. HOWEVER, FLAWS IN THE STUDY MAY LIMIT ITS USEFULNESS.

REPORT 02 : EPA RCRA HEALTH AND ENV. EFFECTS PROFILE OCT 80
CHLOROMETHANE IS TOXIC TO HUMANS BY ITS ACTION ON THE CNS. INFORMATION IS NOT AVAILABLE ON CHRONIC TOXICITY, TERATOGENICITY OR CARCINOGENICITY. CHLOROMETHANE IS HIGHLY MUTAGENIC TO S. TYPHIMURIUM.

REPORT 03 : CIIT STRUCTURAL TERATOGENICITY STUDY APR 81
CIIT CONCLUDES THAT METHYL CHLORIDE INHALATION IN PREG-

NANT RATS IS NOT TERATOGENIC AT DOSES WHICH ELICIT MATERNAL AND FETAL TOXICITY. METHYL CHLORIDE AT 500, BUT NOT 100 PPM, WAS TERATOGENIC IN MICE LEADING TO A MALFORMATION OF THE HEART. NO EMBRYOTOXICITY OR TERATOGENICITY WAS ASSOCIATED WITH EXPOSURE OF MICE TO 100 PPM.

REPORT 04 : TSCA INTER AGENCY TESTING COMMITTEE (42FR55026) OCT 77
THE ITC RECOMMENDED THAT CHLOROMETHANE BE GIVEN PRIORITY CONSIDERATION FOR TESTING FOR CARCINOGENICITY, MUTAGENICITY, TERATOGENICITY, AND OTHER CHRONIC EFFECTS. THE ITC EXPRESSED PARTICULAR CONCERN FOR CHLOROMETHANE'S EFFECTS ON THE CNS, LIVER, KIDNEY, BONE MARROW, AND CV SYSTEM.

REPORT 05 : CIIT STRUCTURAL TERATOGENICITY STUDY DEC 81
CIIT CONCLUDED THAT METHYL CHLORIDE INHALATION IN PREGNANT C57BL/6 MICE RESULTED IN INDICATIONS OF MATERNAL TOXICITY ONLY AT THE 750 PPM DOSE, AND WAS TERATOGENIC TO B6C3F1 CONCEPTUSES AT 750 AND 500 PPM, LEADING TO HEART MALFORMATIONS. NO MATERNAL EMBRYO OR FETOTOXICITY OR TERATOGENICITY WAS ASSOCIATED WITH EXPOSURES TO 250 PPM.

REPORT 06 : MULTI-AGENCY CROSS CHECK
N HHS MAGUIRE
N OSHA CARC
N EPA CAG
Y EPA RCRA
N EPA CLEAN AIR
Y EPA CLEAN WATER

REPORT 07 : EPA TSCA 8(E) NOTICE 8EHQ-1182-0464 NOV 82
VULCAN CHEMIALS PROVIDED SUMMARIZED RESULTS FROM TWO IN VIVO STUDIES OF CHLOROMETHANE. MALE FISCHER F344 RAT REPRODUCTION WAS TOTALLY IMPAIRED FOLLOWING INHALATION EXPOSURE TO 1500 PPM CHLOROMETHANE FOR 6H/D, 5D/W FOR 10 WKS. 100% INCIDENCE OF TESTICULAR ATROPHY AND A 20% INCIDENCE OF SPERM GRANULOMA FORMATION. B6C3F1 MOUSE INHALATION STUDY AT 250, 500, OR 750 PPM REVEALED A DOSE-DEPENDENT AND STATISTICALLY SIGNIFICANT INCREASE IN THE INCIDENCE OF HEART DEFECTS AT MID AND HIGH DOSE EXPOSURES.

REPORT 09 : EPA TSCA8(E) 8EHQ-1083-0497 APR 82
IN A STUDY BY CIIT FOR VULCAN CHEMICAL CO., CHLOROMETHANE PRODUCED A STRONG BUT REVERSIBLE DOMINANT LETHAL EFFECT IN MALE F-344 RATS AND LOWERED SPERM NUMBER AND MOBILITY IN ANIMALS EXPOSED AT 3000 PPM.

REPORT 10 : EPA TSCA 8(E) OTS-1281-0149 APR 82
DOW CHEMICAL REPORT ON INTERIM RESULTS FROM A CHRONIC INHALATION STUDY OF WISTAR RATS, EXPOSED TO 0, 30, 100, OR

300 PPM CHLOROMETHANE FOR 6 HRS/DAY, 5 DAYS/WK, FOR 28 MONTHS, SHOWED DEGENERATIVE CHANGES IN OLFACTORY PARTS OF THE NASAL MUCOSA AND INDICATIONS OF INCREASED INCIDENCE AND NUMBERS OF BENIGN MAMMARY TUMORS AT THE TWO HIGHER DOSE LEVELS.

REPORT 11 : NIOSH CURRENT INTELLIGENCE BULLETIN 43 SEP 84
THE MONOHALOMETHANES (METHYL CHLORIDE, METHYL BROMIDE, AND METHYL IODIDE) ARE DIRECT-ACTING MUTAGENS IN AMES ASSAY. STUDIES IN RATS AND MICE BY VARIOUS ROUTES OF ADMINISTRATION, HAVE DEMONSTRATED ABIILITY TO PRODUCE CANCER. METHYL CHLORIDE PRODUCED TERATOGENIC EFFECT (HEART DEFECTS) IN OFFSPRING OF PREGNANT MICE EXPOSED BY INHALATION AT 500 AND 750 PPM. NIOSH RECOMMENDS THEY BE CONSIDERED POTENTIAL OCCUPATIONAL CARCINOGENS AND THAT METHYL CHLORIDE BE CONSIDERED A POTENTIAL OCCUPATIONAL TERATOGEN.

REPORT 12 : IARC VOL 41 1986
INADEQUATE EVIDENCE FOR CARCINOGENICITY OF MECL TO TEST ANIMALS AND TO HUMANS. INHALATION EXPOSURE TO MECL CAUSES FETAL GROWTH RETARDATION AND IMPAIRED MALE REPRODUCTIVE CAPACITY IN RATS AND MALFORMATION OF HEART IN FETAL MICE. MECL MUTAGENIC TO BACTERIA AND INDUCES CHROMOSOMAL ABERRATIONS IN PLANTS. INDUCES DNA DAMAGE IN MAMMALIAN CELLS IN VITRO BUT NOT IN VIVO. INDUCES SCE'S AND MUTATIONS AND ENHANCES VIRAL CELL TRANSFORMATION IN CULTURED MAMMALIAN CELLS. IT INDUCES DOMINANT LETHAL MUTATIONS IN RATS. IARC CONSIDERS EVIDENCE SUFFICIENT FOR GENETIC ACTIVITY IN SHORT-TERM TESTS.

REPT'D USE : CHEMICAL INTERMEDIATE FOR PROPELLANTS, AEROSOLS, BLOWING AGENT FOR POLYSTYRENE FOAMS AND HERBICIDE

CAS NO : **UNKNOWN-02**

CHEM NAME : ***** 2-CHLORO-1-METHYLETHYL(2-CHLOROPROPYL) ETHER**

REPORT 01 : NTP BOARD OF SCIENTIFIC COUNSELORS DEC 81
UNDER THE CONDITIONS OF THIS BIOASSAY (TEMPERATURE EXCURSIONS) TECHNICAL GRADE BIS (2-CHLORO-l-METHYLETHYL) ETHER CONTAINING 26%-29% OF THIS COMPOUND WAS CARCINOGENIC FOR B6C3F1 MICE, CAUSING AN INCREASED INCIDENCE OF ALVEOLAR/BRONCHEOLAR ADENOMAS IN BOTH SEXES AND HEPATOCELLULAR CARCINOMAS AND ADENOMAS OR CARCINOMAS OF THE STOMACH OR FORESTOMACH IN BOTH SEXES. A RARE TUMOR IN B6C3F1 MICE MAY BE RELATED TO ADMINISTRATION BY GAVAGE (SEE BIOASSAY FOR CAVEATS).

CAS NO : **107-30-2**
CHEM NAME : ***** CHLOROMETHYL METHYL ETHER (CMME)**

REPORT 01 : IARC MONOGRAPH 4 1974
CMME IS ALMOST INVARIABLY CONTAMINATED BY BCME, AND THE LATTER MAY BE RESPONSIBLE FOR AT LEAST PART OF THE OBSERVED CARCINOGENIC ACTIVITY. SUCH CONTAMINATED CMME HAS BEEN FOUND TO BE CARCINOGENIC ON SUBCUTANEOUS INJECTION IN THE MOUSE AND POSSIBLY TO BE AN INITIATOR FOR SKIN TUMORS. INHALATION IN MICE AND SUBCUTANEOUS INJECTION PRODUCED EQUIVOCAL EVIDENCE OF CARCINOGENIC ACTIVITY. ONE STUDY BASED ON 4 CASES OF OAT CELL LUNG CANCER AMONG 111 WORKERS EXPOSED TO CMME, FOLLOWED FOR 5 YEARS, SUGGESTS AN INCREASED RISK OF LUNG CANCER.

REPORT 02 : IARC SUPPLEMENT 1 SEP 79
IN A 1979 SUPPLEMENT, THE IARC CONCLUDED THAT CMME IS CARCINOGENIC FOR HUMANS.

REPORT 06 : MULTI-AGENCY CROSS CHECK
Y HHS MAGUIRE
R OSHA CARC
Y EPA CAG
Y EPA RCRA
Y EPA CLEAN AIR
N EPA CLEAN WATER

RPT'D USE : CHEMICAL INTERMEDIATE AND RESEARCH CHEMICAL

CAS NO : **563-47-3**
CHEM NAME : ***** 3-CHLORO-2-METHYLPROPENE**

REPORT 01 : NTP DRAFT TECHNICAL REPORT 300 NIH PUB 85-2556 OCT 84
UNDER CONDITIONS OF THESE 2-YEAR GAVAGE STUDIES, WITH DOSES OF 0, 75 OR 150 MG/KG FOR RATS AND 0, 100 OR 200 MG/KG FOR MICE THERE WAS CLEAR EVIDENCE OF CARCINOGENICITY AS SHOWN BY THE INCREASED INCIDENCE OF SQUAMOUS CELL NEOPLASMS IN THE FORESTOMACH OF MALE AND FEMALE F-344/N RATS AND MALE AND FEMALE B6C3F1 MICE.

REPORT 02 : NTP BIOASSAY REPORT 300 NIH PUB. 86-2556 1986
TECH. GRADE OF CPD ADMINISTERED BY GAVAGE TO MALE AND FEMALE RATS 5 DAYS/WEEK FOR 103 WEEKS. DOSE-RELATED BASAL CELL OR EPITHELIAL HYPERPLASIA OF FORESTOMACH AND SQUAMOUS CELL PAPILLOMAS OF FORESTOMACH SEEN IN RATS AND MICE. SQUAMOUS CELL CARCINOMAS OF THE FORESTOMACH SEEN IN HIGH DOSE GROUP MICE.

REPORT 06 : MULTI-AGENCY CROSS CHECK
Y HHS MAGUIRE

N OSHA CARC
N EPA CAG
N EPA RCRA
N EPA CLEAN AIR
N EPA CLEAN WATER

REPT'D USE : INTERMEDIATE IN PRODUCTION OF CARBOFURAN, TEXTILE AND PERFUME ADDITIVE, PHARMACEUTICAL INTERMEDIATE, INSECTICIDE, FUMIGANT AND CHEMICAL INTERMEDIATE
FEB 85 SCR/DISCONTINUE
SEP 85 EPA CHIP/DISCONTINUE
APR 86 CSC/DISCONTINUE
APR 87 CRC/DECISION/DISCONTINUE/CONSUMER EXPOSURE UNLIKELY

CAS NO : **6959-47-3**
CHEM NAME : ***** 2-(CHLOROMETHYL)PYRIDINE HYDROCHLORIDE**

REPORT 01 : NCI CARC TRS 178 DHEW PUB NO 79-1734
UNDER THE CONDITIONS OF THIS BIOASSAY, ADMINISTRATION OF 2-(CHLOROMETHYL)PYRIDINE HYDROCHLORIDE WAS NOT CARCINOGENIC TO FISCHER 344 RATS OR B6C3F1 MICE (GIVEN BY STOMACH INTUBATION).

REPORT 02 : NTP TECHNICAL BULLETIN 7/8 APR 82, JUL 82
2-(CHLOROMETHYL) PYRIDINE HCL WAS POSITIVE IN THE SALMONELLA ASSAY AND IN THE L5178Y MOUSE LYMPHOMA ASSAY.
REPT'D USE : INTERMEDIATE IN MFR OF UV ABSORBERS, SUNBURN CREAMS, ANALGESICS AND HERBICIDES

CAS NO : **6959-48-4**
CHEM NAME : ***** 3-(CHLOROMETHYL)PYRIDINE HYDROCHLORIDE**

REPORT 01 : NCI CARC TRS 95 DHEW PUB NO 78-1345
UNDER THE CONDITIONS OF THIS BIOASSAY 3-(CHLOROMETHYL) PYRIDINE HYDROCHLORIDE WAS CARCINOGENIC IN MALE FISCHER 344 RATS AND IN B6C3F1 MICE OF BOTH SEXES, PRODUCING PAPILLOMAS AND CARCINOMAS AT THE SITE OF TOPICAL APPLICATION, THE STOMACH.

REPORT 02 : NTP TECHNICAL BULLETIN 9 APR 83
3-(CHLOROMETHYL)PYRIDINE HD WAS POSITIVE IN ONE SALMONELLA TYPHIMURIUM MUTAGENICITY ASSAY.
REPT'D USE : INTERMEDIATE IN MFR OF UV ABSORBERS, SUNTAN CREAMS, PHARMACEUTICALS AND NEMATOCIDE

CAS NO : **89-63-4**
CHEM NAME : ***** 4-CHLORO-2-NITROANILINE**

REPORT 01 : NTP TECHNICAL BULLETIN NO. 7 APR 82
THE CHEMICAL WAS POSITIVE IN THE SALMONELLA TYPHIMURIUM MUTAGENICITY ASSAY.

REPORT 02 : 49(1) FR 108 ANPR JAN 84
EPA IS INITIATING RULEMAKING TO REQUIRE TESTING OF ANILINES FOR HEALTH AND ENVIRONMENTAL EFFECTS. THIS IS IN RESPONSE TO ITC'S DESIGNATION OF ANILINES FOR TESTING.
REPT'D USE : DYE INTERMEDIATE AND CHEMICAL INTERMEDIATE

CAS NO : **100-00-5**
CHEM NAME : ***** 1-CHLORO-4-NITROBENZENE**

REPORT 01 : NTP TECHNICAL BULLETIN 2/8 APR 80, JUL 82
1-CHLORO-4-NITROBENZENE WAS POSITIVE IN THE SALMONELLA ASSAY, BUT NEGATIVE IN DROSOPHILA WHEN TESTED FOR SEX-LINKED RECESSIVE LETHALS.
REPT'D USE : DYE INTERMEDIATE, INTERMEDIATE IN SYNTHESIS OF ETHYL METHYL PARATHION AND AGRICULTURAL CHEMICAL

CAS NO : **88-73-3**
CHEM NAME : ***** 2-CHLORONITROBENZENE**

REPORT 01 : NTP TECHNICAL BULLETIN NO. 3 DEC 80
THE CHEMICAL WAS POSITIVE IN THE SALMONELLA TYPHIMURIUM MUTAGENICITY ASSAY IN TWO LABORATORIES.
REPT'D USE : DYE INTERMEDIATE AND CHEMICAL INTERMEDIATE

CAS NO : **CLASS-29**
CHEM NAME : ***** CHLOROPHENOLS**

REPORT 01 : IARC VOL 41 1986
LIMITED EVIDENCE FOR CARCINOGENICITY OF OCCUPATIONAL EXPOSURES TO CHLOROPHENOLS TO HUMANS. ANIMAL AND SHORT-TERM TESTS HAVE BEEN EVALUATED IN IARC (1982), SUPPL 4, PP 205-206 AND 249-250.
REPT'D USE : PESTICIDES, WOOD PRESERVATIVES AND CHEMICAL INTERMEDIATE

CAS NO : **CLASS-30**
CHEM NAME : ***** CHLOROPHENOXY HERBICIDES**

REPORT 01 : IARC VOL 41 1986
LIMITED EVIDENCE THAT OCCUPATIONAL EXPOSURES TO CHLOROPHENOXY HERBICIDES ARE CARCINOGENIC TO HUMANS. EVALUATION OF EVIDENCE FOR CARCINOGENICITY IN ANIMALS AND SHORT-TERM TEST ACTIVITY ARE DISCUSSED IN IARC 1982 SUPPLEMENT 4, PP 101-103, 211-212 AND 235-238 AND IN IARC 1983 VOL 30, PP 255-269.
REPT'D USE : HERBICIDES

CAS NO : **5131-60-2**
CHEM NAME : ***** 4-CHLORO-M-PHENYLENEDIAMINE**

REPORT 01 : NCI CARC TRS 85 DHEW PUB NO 78-1335
UNDER THE CONDITIONS OF THIS BIOASSAY, DIETARY ADMINISTRATION OF 4-CHLORO-M-PHENYLENEDIAMINE WAS CARCINOGENIC TO THE EXPERIMENTAL ANIMALS, CAUSING AN INCREASED INCIDENCE OF HEPATOCELLULAR TUMORS IN MALE B6C3F1 MICE AND AN INCREASED INCIDENCE OF ADRENAL PHEOCHROMOCYTOMAS IN MALE FISCHER 344 RATS.

REPORT 02 : IARC MONOGRAPH 27 1982
NO MUTAGENICITY DATA OR DATA CONCERNING EFFECTS ON REPRODUCTION OR PRENATAL TOXICITY WERE AVAILABLE FOR 4-CHLORO-M-PDA.

REPORT 03 : EPA RESPONSE TO ITC RECOMMS (47FR973) JAN 81
IN RESPONSE TO THE ITC RECOMMENDATIONS (45FR35897), THE EPA PROPOSES AN ANPR REQUESTING COMMENTS CONCERNING A PROPOSAL TO CONDUCT MUTAGENICITY AND CARCINOGENICITY TESTING FOR 11 SELECTED PHENYLENEDIAMINES, AND TERATOGENICITY, REPRODUCTIVE EFFECTS AND OTHER CHRONIC EFFECTS TESTING, AS WELL AS AN EPIDEMIOLOGIC STUDY, FOR 12 PHENYLENEDIAMINES. THE EPA WILL CONSIDER FURTHER TESTING FOR AN ADDITIONAL 37 PHENYLENEDIAMINES AS APPROPRIATE.

REPORT 04 : IARC MONOGRAPH 27 1982
4-CHLORO-M-PHENYLENEDIAMINE WAS TESTED FOR CARCINOGENICITY IN MICE AND RATS BY DIETARY ADMINISTRATION. THE RESULTS OF THE STUDY IN MICE WERE INCONCLUSIVE; THOSE IN RATS WERE NOT INDICATIVE OF A CARCINOGENIC EFFECT. NO DATA ARE AVAILABLE TO EVALUATE THE CARCINOGENICITY OF 4-CHLORO-M-PHENYLENEDIAMINE. NO CASE REPORTS OR EPA STUDY WERE AVAILABLE TO THE WORKING GROUP FOR EVALUATION OF THE RISK TO HUMANS.

REPORT 06 : MULTI-AGENCY CROSS CHECK
N HHS MAGUIRE
Y OSHA CARC
N EPA CAG
N EPA RCRA
N EPA CLEAN AIR
N EPA CLEAN WATER

REPT'D USE : DYE INTERMEDIATE, COMPONENT OF HAIR DYES AND PHOTOGRAPHIC CHEMICAL POLYMER

CAS NO : **95-83-0**

CHEM NAME : ***** 4-CHLORO-O-PHENYLENEDIAMINE**

REPORT 01 : NCI CARC TRS 63 DHEW PUB NO 78-1313
UNDER THE CONDITIONS OF THIS BIOASSAY 4-CHLORO-O-PHENYLENEDIAMINE WAS CARCINOGENIC IN FISCHER 344 RATS AND B6C3F1 MICE, INDUCING TUMORS OF THE URINARY BLADDER AND FORESTOMACH IN BOTH SEXES OF RATS AND HEPATOCELLULAR CARCINOMAS IN BOTH SEXES OF MICE (FEEDING STUDY).

REPORT 02 : IARC MONOGRAPH 27 1982
4-CHLORO-O-PHENYLENEDIAMINE WAS TESTED FOR CARCINOGENICITY IN MICE AND RATS BY DIETARY ADMINISTRATION 4-CHLORO-O-PDA WAS CARCINOGENIC IN MICE, PRODUCING HEPATOCELLULAR CARCINOMAS, AND IN RATS, PRODUCING BENIGN AND MALIGNANT TUMORS OF THE URINARY BLADDER. THERE IS SUFFICIENT EVIDENCE FOR THE CARCINOGENICITY OF 4-CHLORO-O-PDA IN ANIMALS.

REPORT 03 : IARC MONOGRAPH 27 1982
4-CHLORO-O-PDA INDUCED REVERSE MUTATIONS IN SALMONELLA STRAIN TA98 WHEN ASSAYED IN THE PRESENCE OF A RAT LIVER ACTIVIATION SYSTEM. NO DATA CONCERNING EFFECTS ON REPRODUCTION OR PRENATAL TOXICITY WERE AVAILABLE.

REPORT 04 : NTP TECHNICAL BULLETIN 5 AUG 81
4-CHLORO-O-PDA WAS POSITIVE IN THE SALMONELLA TYPHIMURIUM MUTAGENICITY ASSAY IN ONE LABORATORY.

REPORT 06 : MULTI-AGENCY CROSS CHECK
N HHS MAGUIRE
Y OSHA CARC
N EPA CAG
N EPA RCRA
N EPA CLEAN AIR
N EPA CLEAN WATER

REPT'D USE : DYE INTERMEDIATE AND INTERMEDIATE IN MFR OF EXPERIMENTAL DRUGS

CAS NO : **61702-44-1**
CHEM NAME : ***** 2-CHLORO-P-PHENYLENEDIAMINE SULFATE**

REPORT 01 : NCI CARC TRS 113 DHEW PUB NO 78-1368
UNDER THE CONDITIONS OF THIS BIOASSAY THERE WAS INSUFFICIENT EVIDENCE THAT DIETARY ADMINISTRATION OF 2-CHLORO-P-PHENYLENEDIAMINE SULFATE WAS CARCINOGENIC TO FISCHER 344 RATS OR B6C3F1 MICE (FEEDING STUDY).

REPORT 02 : NTP TECHNICAL BULLETIN 5 AUG 81
2-CHLORO-P-PDA SULFATE WAS POSITIVE IN THE SALMONELLA TYPHIMURIUM MUTAGENICITY ASSAY IN ONE LABORATORY.

REPORT 03 : EPA RESPONSE TO ITC RECOMMS (47FR973) JAN 81
IN RESPONSE TO THE ITC RECOMMENDATIONS (45FR35897), THE EPA PROPOSES AN ANPR REQUESTING COMMENTS CONCERNING A PROPOSAL TO CONDUCT MUTAGENICITY AND CARCINOGENICITY TESTING FOR 11 SELECTED PHENYENEDIAMINES, AND TERATOGENICITY, REPRODUCTIVE EFFECTS AND OTHER CHRONIC EFFECTS TESTING, AS WELL AS AN EIPDEMIOLOGIC STUDY, FOR 12 PHENYLENEDIAMINES. THE EPA WILL CONSIDER ADDITIONAL TESTING FOR AN ADDITIONAL 37 PHENYLENEDIAMINES, AS APPROPRIATE.
REPT'D USE : COMPONENT OF HAIR DYE FORMULATIONS. NO LISTING ON TSCA INVENTORY

CAS NO : **76-06-2**
CHEM NAME : ***** CHLOROPICRIN**

REPORT 01 : NCI CARC TRS 65 DHEW PUB NO 78-1315
THE BIOASSAY OF CHLOROPICRIN USING OSBORNE-MENDEL RATS DID NOT PERMIT AN EVALUATION OF CARCINOGENICITY BECAUSE OF THE SHORT SURVIVAL TIME OF DOSED ANIMALS. THE BIOASSAY OF CHLOROPICRIN USING B6C3F1 MICE DID NOT PROVIDE CONCLUSIVE STATISTICAL EVIDENCE FOR THE CARCINOGENICITY OF THIS COMPOUND (ADMINISTERED BY GAVAGE).
REPT'D USE : GRAIN FUMIGANT, NEMATOCIDE AND SYNTHESIS OF METHYL VIOLET CHEMICAL WARFARE AGENT

CAS NO : **126-99-8**
CHEM NAME : ***** CHLOROPRENE**

REPORT 01 : NIOSH CURRENT INTELLIGENCE BULLETIN 1 JAN 75
E.I. DU PONT INFORMS NIOSH OF THE POTENTIAL CARCINOGENICITY OF CHLOROPRENE. TWO RUSSIAN STUDIES SUGGEST AN INCREASED INCIDENCE OF SKIN AND LUNG CANCER IN WORKERS EXPOSED TO CHLOROPRENE. TWO OTHER RUSSIAN STUDIES DESCRIBE THE ADVERSE EFFECT OF CHLOROPRENE ON EMBRYO

DEVELOPMENT IN RATS AND MICE. DATA ON TOXICITY IS ALSO GIVEN.

REPORT 02 : NIOSH CRITERIA DOCUMENT AUG 77
NIOSH RECOMMENDS A CEILING OF 1 PPM BASED ON REPRODUCTIVE EFFECTS AND POTENTIAL CARCINOGENICITY. THE PRESENT OSHA STANDARD IS 25 PPM (TWA).

REPORT 03 : IARC MONOGRAPH 19 1979
CHLOROPRENE WAS TESTED IN RATS BY ORAL, SUBCUTANEOUS AND INTRATRACHEAL ADMINISTRATION AND IN MICE BY SKIN APPLICATION. ALTHOUGH NO CARCINOGENIC EFFECTS WERE FOUND, THE WORKING GROUP CONSIDERED THESE STUDIES TO BE INADEQUATE TO ALLOW AN EVALUATION OF THE CARCINOGENICITY OF CHLOROPRENE IN EXPERIMENTAL ANIMALS. REPORTS OF INCREASED FREQUENCIES OF LUNG AND SKIN CANCER IN WORKERS EXPOSED TO CHLOROPRENE RAISE THE POSSIBILITY THAT CHLOROPRENE IS CARCINOGENIC TO HUMANS. THESE STUDIES, HOWEVER, HAVE METHODOLOGICAL DEFICIENCIES.

REPORT 04 : NTP TECHNICAL BULLETIN 6 JAN 82
THE COMPOUND WAS NEGATIVE IN THE SALMONELLA TYPHIMURIUM MUTAGENICITY ASSAY IN ONE LABORATORY.

REPORT 05 : IARC MONOGRAPH 19 1979
CHLOROPRENE IS EMBRYOTOXIC AND TERATOGENIC. THE CHEMICAL INDUCED REVERSE MUTATIONS IN SALMONELLA TA100 AND TA1530. IN DROSOPHILA, RECESSIVE LETHAL MUTATIONS WERE INDUCED. CHLOROPRENE INDUCED DOMINANT LETHAL MUTATIONS AND CHROMOSOMAL ABBERATIONS IN BONE MARROW CELLS OF RATS.

REPORT 06 : MULTI-AGENCY CROSS CHECK
N HHS MAGUIRE
N OSHA CARC
N EPA CAG
N EPA RCRA
Y EPA CLEAN AIR
N EPA CLEAN WATER

REPORT 08 : EPA RESPONSE TO ITC RECOMMENDATIONS 50 (FR34547) AUG 85
IN RESPONSE TO ITC RECOMMENDATION (49 FR46931), EPA DOES NOT PROPOSE TO TEST CHLOROPRENE BECAUSE OF AVAILABILITY OF ADEQUATE WATER SOLUBILITY DATA. ALSO FATE DATA PREDICTS LIMITED PERSISTENCE IN THE ENVIRONMENT. FURTHER, IT WILL NOT ENTER THE AQUATIC ENVIRONMENT IN SUFFICIENT QUANTITIES TO WARRANT FURTHER TOXICITY TESTING.

REPORT 09 : EPA STATUS REPORT 8EHQ-0887-0689 SEP 87
MALE B6C3F1 MICE EXPOSED TO 12, 32 AND 80 PPM FOR 6 HRS/DAY FOR 12 DAYS BY INHALATION DID NOT SHOW SIGNIFICANT

CHANGES IN CA, SCE OR MICR INUCLEATED ERYTHROCYTES. THERE WAS COMPLETE MORTALITY AMONG MICE EXPOSED AT 200 PPM.

REPT'D USE : CHEMICAL INTERMEDIATE, ELASTOMER MONOMERS, ADHESIVES AND MOLDED RUBBER PRODUCTS

CAS NO : **101-21-3**

CHEM NAME : ***** CHLOROPROPHAM**

REPORT 01 : IARC MONOGRAPH 12 1976
CHLOROPROPHAM PRODUCED NO CARCINOGENIC EFFECTS WHEN ADMINISTERED ORALLY TO MICE, RATS AND HAMSTERS OR SUBCUTANEOUSLY TO MICE. DOSES USED IN THE STUDIES IN MICE AND HAMSTERS DID NOT NECESSARILY CORRESPOND TO THE MAXIMUM TOLERATED DOSES.

REPORT 02 : IARC MONOGRAPH 12 1976
CHLOROPROPHAM INDUCED PETITE MUTATIONS IN SACCHAROMYCES, BUT NO REVERSE MUTATIONS WERE DETECTED IN B. SUBTILIS OR IN SALMONELLA. METABOLIC ACTIVATION SYSTEMS WERE NOT USED.

REPT'D USE : HERBICIDE, PAINTS AND VARNISHES

CAS NO : **54-05-7**

CHEM NAME : ***** CHLOROQUINE**

REPORT 01 : IARC MONOGRAPH 13 1977
THE NEGATIVE RESULTS OBTAINED IN THE ONLY AVAILABLE STUDY IN RATS GIVEN CHLOROQUINE BY ORAL ADMINISTRATION DID NOT FORM AN ADEQUATE BASIS ON WHICH TO MAKE AN EVALUATION OF THE CARCINOGENICITY OF THIS COMPOUND.

REPORT 02 : IARC MONOGRAPH 13 1977
CHLOROQUINE INTERACTS NON-COVALENTLY WITH DNA IN VITRO AND INHIBITS THE ACTIVITY OF DNA POLYMERASE, DNA SYNTHESIS IN MAMMALIAN CELLS AND DNA REPAIR BOTH IN BACTERIA AND IN MAMMALIAN CELLS.

REPT'D USE : ANTIMALARIAL AGENT AND ANTIAMOEBIC AGENT

CAS NO : **1897-45-6**

CHEM NAME : ***** CHLOROTHALONIL**

REPORT 01 : NCI CARC TRS 41 DHEW PUB NO 78-841
UNDER THE CONDITIONS OF THIS BIOASSAY, TECHNICAL-GRADE CHLOROTHALONIL WAS CARCINOGENIC TO OSBORNE-MENDEL RATS, PRODUCING TUMORS OF THE KIDNEY. CHLOROTHALONIL WAS NOT CARCINOGENIC FOR B6C3F1 MICE IN A FEEDING STUDY.

REPORT 02 : NTP TECHNICAL BULLETIN 5/8
CHLOROTHALONIL WAS NEGATIVE WHEN TESTED IN THE SALMONELLA AND DROSOPHILIA MUTAGENICITY ASSAYS.

REPORT 03 : IARC MONOGRAPH 30 JAN 83
CHLOROTHALONIL WAS TESTED FOR CARCINOGENICITY IN MICE AND RATS BY ADMINISTRATION IN DIET; IT PRODUCED ADENOMAS AND ADENOCARCINOMAS OF KIDNEY IN RATS. NO EVIDENCE OF CARCINOGENICITY WAS FOUND IN MICE.

REPORT 04 : IARC MONOGRAPH 30 JAN 83
CHLOROTHALONIL WAS NOT MUTAGENIC IN TWO FUNGAL SYSTEMS AND WAS REPORTED TO BE NONMUTAGENIC IN SALMONELLA TYPHIMURIUM. AVAILABLE DATA ON OTHER SYSTEMS ARE INADEQUATE, AND NO OVERALL EVALUATION OF MUTAGENICITY OF CHLOROTHALONIL COULD BE MADE.

REPORT 06 : MULTI-AGENCY CROSS CHECK
N HHS MAGUIRE
Y OSHA CARC
N EPA CAG
N EPA RCRA
N EPA CLEAN AIR
N EPA CLEAN WATER
REPT'D USE : FUNGICIDE AND MILDEWICIDE IN PAINTS/VARNISHES

CAS NO : **95-49-8**
CHEM NAME : ***** 2-CHLOROTOLUENE**

REPORT 01 : 50FR40445 OCT 85
EPA IS ISSUING A DECISION NOT TO REQUIRE FURTHER TESTING OF 2-CT FOR HEALTH AND ENVIRONMENTAL EFFECTS AND CHEMICAL FATE UNDER TSCA 4(A). EPA HAS RECEIVED DATA THAT ADEQUATELY CHARACTERIZE 2-CT FOR THESE EFFECTS.

REPORT 02 : 48FR34119 JUL 83
EPA ANNOUNCES RECEIPT OF DATA FROM TESTING OF 2-CT FOR TERATOGENICITY METABOLISM, CHRONIC TOXICITY AND ENVIRONMENTAL TOXICITY. BASED ON THIS, EPA DECIDED NOT TO PROPOSE A TEST RULE FOR 2-CT UNDER TSCA 4.

CAS NO : **95-74-9**
CHEM NAME : ***** 3-CHLORO-P-TOLUIDINE**

REPORT 01 : NCI CARC TRS 145 DHEW PUB NO. 78-1701
UNDER THE CONDITIONS OF THIS BIOASSAY THERE WAS NO CONVINCING EVIDENCE FOR THE CARCINOGENICITY OF 3-CHLORO-

P-TOLUIDINE IN A FEEDING STUDY IN FISCHER F-344 RATS OR B6C3F1 MICE.

REPORT 02 : NTP TECHNICAL BULLETIN 7 APR 82
3-CHLORO-P-TOLUIDINE WAS NEGATIVE IN THE SALMONELLA TYPHIMURIUM MUTAGENICITY ASSAY.
REPT'D USE : DYE INTERMEDIATE, PESTICIDE AND AVICIDE

CAS NO : **95-79-4**
CHEM NAME : ***** 5-CHLORO-O-TOLUIDINE**

REPORT 01 : NCI CARC TRS 187 DHEW PUB NO 79-1743
UNDER THE CONDITIONS OF THIS BIOASSAY, 5-CHLORO-O-TOLUIDINE WAS CARCINOGENIC TO B6C3F1 MICE, INDUCING HEMANGIOSARCOMAS AND HEPATOCELLULAR CARCINOMAS IN BOTH MALES AND FEMALES. THERE WAS NO CONCLUSIVE EVIDENCE OF THE CARCINOGENICITY OF THE COMPOUND IN FISCHER 344 RATS (FEEDING STUDY).

REPORT 02 : NTP TECHNICAL BULLETIN 7 APR 82
5-CHLORO-O-TOLUIDINE WAS NEGATIVE IN THE SALMONELLA TYPHIMURIUM MUTAGENICITY ASSAY.

REPORT 05 : EPA RESPONSE TO ITC RECOMMS (47FR973) JAN 81
IN RESPONSE TO THE ITC RECOMMENDATIONS (45FR35897), EPA PROPOSES AN ANPR REQUESTING COMMENTS CONCERNING A PROPOSAL TO CONDUCT MUTAGENICITY AND CARCINOGENICITY TESTING FOR 11 SELECTED PHENYLENEDIAMINES, AND TERATOGENICITY, REPRODUCTIVE EFFECTS AND OTHER CHRONIC EFFECTS TESTING, AND AN EPIDEMIOLOGIC STUDY FOR 12 PHENYLENEDIAMINES. THE EPA WILL CONSIDER ADDITIONAL TESTING FOR AN ADDITIONAL 37 PHENYLENEDIAMINES, AS APPROPRIATE.

REPORT 06 : MULTI-AGENCY CROSS CHECK
N HHS MAGUIRE
Y OSHA CARC
N EPA CAG
N EPA RCRA
N EPA CLEAN AIR
N EPA CLEAN WATER
REPT'D USE : PIGMENT INTERMEDIATE AND AZOIC DIAZO COMPONENT FOR TEXTILE DYEING

CAS NO : **3165-93-3**
CHEM NAME : ***** 4-CHLORO-O-TOLUIDINE HYDROCHLORIDE**

REPORT 01 : NCI CARC TRS 165 1979 DHEW PUB NO 79-1721
UNDER THE CONDITIONS OF THIS BIOASSAY, 4-CHLORO-O-TOLUIDINE HCL WAS NOT CARCINOGENIC FOR F344 RATS BUT WAS CARCINOGENIC FOR B6C3F1 MICE. EVIDENCE INCLUDED INDUCTION OF HEMANGIOSARCOMAS AND HEMANGIOMAS IN BOTH MALE AND FEMALE MICE IN A FEEDING STUDY.

REPORT 02 : NTP TECHNICAL BULLETIN 5 AUG 81
4-CHLORO-O-TOLUIDINE HCL WAS NEGATIVE IN THE SALMONELLA TYPHIMURIUM MUTAGENICITY ASSAY IN ONE LABORATORY.

REPORT 03 : IARC MONOGRAPH 16 1978
THE ONLY DATA ON P-CHLORO-O-TOLUIDINE HYDROCHLORIDE AVAILABLE TO THE WORKING GROUP WERE REPORTED IN AN ABSTRACT. ALTHOUGH THE RESULTS OF THE EXPERIMENT INDICATED A CARCINOGENIC EFFECT, THE INADEQUACY OF REPORTING DOES NOT ALLOW AN EVALUATION OF THE CARCINOGENICITY OF P-CHLORO-O-TOLUIDINE OR ITS HYDROCHLORIDE.

REPORT 06 : MULTI-AGENCY CROSS CHECK
N HHS MAGUIRE
Y OSHA CARC
N EPA CAG
Y EPA RCRA
N EPA CLEAN AIR
N EPA CLEAN WATER
REPT'D USE : DYE AND DYE INTERMEDIATE

CAS NO : **569-57-3**
CHEM NAME : ***** CHLOROTRIANISENE**

REPORT 01 : IARC MONOGRAPH 21 1979
CHLOROTRIANISENE WAS TESTED IN ONLY ONE EXPERIMENT IN RATS BY ORAL ADMINISTRATION, BUT THE DATA WERE INSUFFICIENT TO EVALUATE THE CARCINOGENICITY OF THIS CPD. STUDIES IN HUMANS STRONGLY SUGGEST THAT ADMINISTRATION OF ESTROGENS IS CAUSALLY RELATED TO AN INCREASED INCIDENCE OF ENDOMETRIAL CARCINOMA. THERE IS NO EVIDENCE THAT CPD IS DIFFERENT FROM OTHER ESTROGENS IN THIS RESPECT.

REPORT 02 : IARC MONOGRAPH 21 1979
NO DATA CONCERNING THE EMBRYOTOXICITY OR MUTAGENICITY OF CHLOROTRIANISENE WERE AVAILABLE.
REPT'D USE : ESTROGEN AND ANTINEOPLASTIC AGENT

CAS NO : **75-88-7**
CHEM NAME : ***** 2-CHLORO-1,1,1-TRIFLUOROETHANE**

REPORT 01 : IARC VOL 41 1986
LIMITED EVIDENCE FOR CARCINOGENICITY OF CPD TO TEST ANIMALS, BUT NO EVALUATION COULD BE MADE OF ITS CARCINOGENICITY TO HUMANS. DEGREE OF EVIDENCE IN SHORT-TERM TESTS FOR GENETIC ACTIVITY IS INADEQUATE. NO DATA AVAILABLE TO EVALUATE REPRODUCTIVE EFFECTS OR PRENATAL TOXICITY OF CPD TO HUMANS.
REPT'D USE : CHEMICAL INTERMEDIATE

CAS NO : **113-92-8**
CHEM NAME : ***** CHLORPHENIRAMINE MALEATE**

REPORT 01 : NTP TR 317 NIH PUB 86-2573 DEC 85
NO EVIDENCE OF CARCINOGENICITY IN F344/N RATS OR B6C3F1 MICE OF EITHER SEX ADMINISTERED CHLORPHENIRAMINE MALEATE IN DISTILLED WATER BY GAVAGE, 5 DAYS PER WEEK FOR 2 YEARS. DUE TO HIGH MORTALITY RATES IN HIGH DOSE FEMALE RATS AND HIGH DOSE MALE MICE, SENSITIVITY OF THESE GROUPS TO DETECT CARCINOGENIC RESPONSE WAS REDUCED. CHLORPHENIRAMINE MALEATE HAD PROLIFERATIVE EFFECT ON THYROID GLAND OF FEMALE MICE, AS SHOWN BY INCREASED INCIDENCES OF FOLLICULAR CELL CYSTS AND HYPERPLASIA IN BOTH LOW DOSE AND HIGH DOSE GROUPS.
REPT'D USE : ANTIHISTAMINE DRUG

CAS NO : **94-20-2**
CHEM NAME : ***** CHLORPROPAMIDE**

REPORT 01 : NCI CARC TRS 45 DHEW PUB NO 78-845
IT IS CONCLUDED THAT UNDER THE CONDITIONS OF THIS BIOASSAY, CHLORPROPAMIDE WAS NOT CARCINOGENIC FOR FISCHER 344 RATS OR B6C3F1 MICE IN A FEEDING STUDY.

REPORT 02 : NTP TECHNICAL BULLETIN 5 AUG 81
CHLORPROPAMIDE WAS NEGATIVE IN THE S. TYPHIMURIUM MUTAGENICITY ASSAY IN ONE LABORATORY.
REPT'D USE : ORAL HYPOGLYCEMIC AGENT

CAS NO : **57-88-5**
CHEM NAME : ***** CHOLESTEROL**

REPORT 01 : IARC MONOGRAPH 10 1976
ON THE BASIS OF THE EXPERIMENTAL EVIDENCE AVAILABLE NO

ASSESSMENT OF THE CARCINOGENICITY OF CHOLESTEROL IS POSSIBLE.

REPORT 02 : NAS REPORT ON DIET, NUTRITION AND CANCER 1982
THE RELATIONSHIP BETWEEN DIETARY CHOLESTEROL AND CANCER IN HUMANS IS NOT YET CLEAR. MANY STUDIES OF SERUM CHOLESTEROL LEVELS AND CANCER MORTALITY HAVE INDICATED THAT THERE IS AN INVERSE ASSOCIATION WITH COLON CANCER IN MALES, BUT THE EVIDENCE IS INCONSISTENT AND IS NOT SUFFICIENT TO ESTABLISH A CAUSAL RELATIONSHIP.

REPORT 03 : IARC MONOGRAPH 31 JUL 83
SERUM CHOLESTEROL DOES NOT HAVE STRONG OR DIRECT RELATIONSHIP WITH HUMAN CARCINOGENESIS WITH CORONARY HEART DISEASE. INVERSE RELATIONSHIP WITH TOTAL CANCER RISK, WHEN PRESENT, IS NOT GRADED AND APPLIES PREDOMINANTLY TO MEN. FOR CANCER OF COLON, STUDIES SHOW AN INVERSE ASSOCIATION WITH SERUM CHOLESTEROL. CARCINOGENICITY STUDIES IN MICE IN DIET, BY SUBCUTANEOUS ADMINISTRATION AND BY BLADDER IMPLANTATION, ARE INADEQUATE FOR EVALUATION. CPD TESTED IN COMBINATION WITH OTHER CARCINOGENS, BUT RESULTS WERE INSUFFICENT TO ASSESS CO-CARCINOGENICITY.

REPORT 04 : IARC MONOGRAPH 31 JUL 83
CHOLESTEROL, WHEN FREE OF OXIDATION PRODUCTS, NOT MUTAGENIC IN BACTERIA AND DID NOT CAUSE CELL TRANSFORMATION. CHOLESTEROL INDUCED CLEFT PLATE IN RATS.

REPT'D USE : COSMETICS, PHARMACEUTICALS AND MANUFACTURE OF VITAMINS, STEROIDS, ESTERS

CAS NO : **1066-30-4**

CHEM NAME : ***** CHROMIC ACETATE**

REPORT 01 : IARC MONOGRAPH 23 1980
EXPERIMENTS IN MICE AND RATS EXPOSED TO CHROMIC (III) ACETATE BY ORAL ADMINISTRATION, INTRAMUSCULAR, AND INTRAPLEURAL IMPLANTATION WERE INADEQUATE TO EVALUATE CARCINOGENICITY.

REPORT 02 : IARC MONOGRAPH 23
THERE IS GOOD EVIDENCE THAT CHROMIUM (III) COMPOUNDS ARE NOT MUTAGENIC IN A WIDE RANGE OF PROKARYOTIC AND EUKARYOTIC SYSTEMS BOT IN VITRO AND IN VIVO. THESE COMPOUNDS APPARENTLY CROSS THE PLACENTA TO ONLY A BEEN REPORTED WITH EXTREMELY ANIMALS. SOME TERATOGENIC EFFECTS HAVE BEEN REPORTED WITH EXTREMELY HIGH DOSES.

REPT'D USE : DYE MORDANT, TANNING, PHOTOGRAPHIC CHEMICAL AND CHEMICAL CATALYST

CAS NO : **1308-31-2**
CHEM NAME : ***** CHROMITE ORE**

REPORT 01 : IARC MONOGRAPH 23 1980
EXPERIMENTS IN MICE, RATS, AND RABBITS EXPOSED TO CHROMITE (III) ORE BY INTRAPLEURAL, INTRAMEDULLARY, AND I.V. INJECTIONS WERE INADEQUATE TO EVALUATE CARCINOGENICITY.

REPORT 06 : MULTI-AGENCY CROSS CHECK
Y HHS MAGUIRE
N OSHA CARC
N EPA CAG
N EPA RCRA
N EPA CLEAN AIR
N EPA CLEAN WATER
REPT'D USE : CORROSION INHIBITORS, PIGMENTS, LEATHER TANNING, TEXTILES, WOOD PRESERVATIVES, DYES, PAPER, PIGMENTS, STEEL, ALLOYS AND RESEARCH

CAS NO : **7440-47-3**
CHEM NAME : ***** CHROMIUM (ELEMENTAL)**

REPORT 04 : IARC MONOGRAPH 23 1980
EXPERIMENTS IN MICE, RATS, AND RABBITS EXPOSED TO CHROMIUM POWDER BY IP IV IM, INTRAPLEURAL, INTRATRACHEAL, AND INTRAMEDULLARY INJECTION WERE INADEQUATE TO EVALUATE CARCINOGENICITY.

REPORT 06 : MULTI-AGENCY CROSS CHECK
N HHS MAGUIRE
N OSHA CARC
Y EPA CAG
N EPA RCRA
N EPA CLEAN AIR
Y EPA CLEAN WATER
REPT'D USE : ALLOYS

CAS NO : **1333-82-0**
CHEM NAME : ***** CHROMIUM TRIOXIDE**

REPORT 01 : IARC MONOGRAPH 23 1980
AN EXPERIMENT IN RATS EXPOSED TO CHROMIUM (VI) TRIOXIDE BY BRONCHIAL IMPLANTATION WAS INADEQUATE TO EVALUATE CARCINOGENICITY. SINTERED CHROMIUM (VI) TRIOXIDE PRODUCED SARCOMAS AT THE SITE OF INTRAMUSCULAR ADMINISTRATION IN RATS. IN A 1979 SUPPLEMENT, IARC CONCLUDED THAT

THERE IS SUFFICIENT EVIDENCE FOR THE CARCINOGENICITY OF SINTERED CHROMINUM TRIOXIDE IN RATS.

REPORT 02 : IARC MONOGRAPH 23 1980
CHROMIUM (VI) COMPOUNDS CAUSE MUTATIONS AND ALLIED EFFECTS IN A VERY WIDE RANGE OF PROKARYOTIC AND EUKARYOTIC SYSTEMS BOTH IN VITRO AND IN VIVO. EMBRYOTOXICITY AND TERATOGENIC EFFECTS OCCURRED IN HAMSTERS FOLLOWING EXPOSURE TO CHROMIUM TRIOXIDE.

REPORT 03 : NIOSH CRITERIA DOCUMENT ON CHROMIC ACID JUL 73
NIOSH RECOMMENDS A STANDARD OF 0.05 MG/CU M AND CHROMATES BASED ON NASAL ULCERATION. THE PRESENT OSHA STANDARD IS 1 MG/CU M CEILING.

REPORT 06 : MULTI-AGENCY CROSS CHECK
Y HHS MAGUIRE
N OSHA CARC
N EPA CAG
N EPA RCRA
N EPA CLEAN AIR
Y EPA CLEAN WATER

REPT'D USE : PHOTOGRAPHY, CORROSION INHIBITOR, CHROME PLATING, ALUMINUM ANODIZING, TOPICAL ANTISEPTIC/ASTRINGENT, WART REMOVAL AND WOOD TREATMENT/PRESERVATIVES: PIGMENT

CAS NO : **218-01-9**
CHEM NAME : ***** CHRYSENE**

REPORT 01 : IARC MONOGRAPH 3 1973
CHRYSENE HAS PRODUCED SKIN TUMORS IN MICE FOLLOWING REPEATED PAINTING IN THE ONLY STUDY FOUND. IN THIS STUDY, A CONCENTRATION AS HIGH AS 1% ACETONE WAS USED. HIGH DOSES (2-20 MG) GIVEN SC PRODUCED LOW INCIDENCE OF TUMORS WITH A LONG INDUCTION TIME.

REPORT 02 : NIOSH SPECIAL HAZARD REVIEW JUN 78
EVIDENCE OF CHRYSENE'S CARCINOGENIC POTENTIAL HAS COME PRIMARILY FROM EXPERIMENTS IN WHICH THE ISOLATED CHEMICAL (AND CERTAIN DERIVATIVES) WAS ADMINISTERED TO ANIMALS BY SKIN PAINTING OR BY S.C. INJECTION. CHRYSENE IS ALSO ACTIVE IN SOME MAMMALIAN CELL CULTURE TESTS AND IN MICROBIAL ASSAYS FOR MUTAGENICITY. NIOSH RECOMMENDS THAT CHRYSENE BE CONTROLLED AS AN OCCUPATIONAL CARCINOGEN. THERE IS NO CURRENT OSHA STANDARD FOR CHRYSENE.

REPORT 03 : EPA AMBIENT WATER QUALITY CRITERIA OCT 80
PRESENTLY AVAILABLE DATA CANNOT SUPPORT INDIVIDUAL CRITERIA FOR EACH POLYNUCLEAR AROMATIC HYDROCARBON

(PAH), THE AWQC FOR CARCINOGENIC PAH CPDS IS BASED ON THE ASSUMPTION THAT EACH COMPOUND IS AS POTENT AS B(A)P AND THAT THE CARCINOGENIC EFFECT OF THE COMPOUNDS IS PROPORTIONAL TO THE SUM OF THEIR CONCENTRATIONS. THIS SUM SHOULD NOT EXCEED 28 NG/L, 2.8 NG/L, AND .28 NG/L TO KEEP THE LIFETIME CANCER RISK BELOW 10 -5, 10 -6, AND 10 -7 RESPECTIVELY.

REPORT 04 : EPA RCRA HEALTH AND ENV. EFFECTS PROFILE OCT 80
CHRYSENE IS GENERALLY REGARDED AS ONLY A WEAK CARCINOGEN TO ANIMALS.

REPORT 05 : IARC MONOGRAPH 32 DEC 83
PRODUCED SKIN TUMORS WHEN TESTED IN SKIN STUDIES BY SKIN APPLICATION TO MICE. IN ONE STUDY, CPD ENCHANCED THE TUMORIGENIC EFFECT WHEN TESTED SIMULTANEOUSLY WITH N-DODECANE. ACTIVE AS AN INITIATOR WHEN TESTED IN MOUSE-SKIN INITIATION-PROMOTION ASSAY. LOCAL TUMORS SEEN AFTER SC INJECTION IN MICE. PERINATAL ADMINISTRATION TO MICE SC OR IP INCREASED INCIDENCE OF LIVER TUMORS.

REPORT 06 : MULTI-AGENCY CROSS CHECK
N HHS MAGUIRE
N OSHA CARC
Y EPA CAG
Y EPA RCRA
N EPA CLEAN AIR
Y EPA CLEAN WATER

REPORT 07 : IARC MONOGRAPH 32 DEC 83
MUTAGENIC TO S. TYPHIMURIUM IN PRESENCE OF EXOGENOUS METABOLIC SYSTEM. DID NOT INDUCE MITOTIC RECOMBINATION IN YEAST, UNSCHEDULED DNA SYNTHESIS IN PRIMARY RAT HEPATOCYTES, OR MUTATIONS IN CHINESE HAMSTER V79 CELLS. HOWEVER, IN ONE STUDY EACH IN MICE AND HAMSTERS, SISTER CHROMATID EXCHANGES AND CHROMOSOMAL ABERRATIONS WERE INDUCED RESPECTIVELY. CPD POSITIVE IN ONE OF TWO REPORTED STUDIES OF MORPHOLOGICAL TRANSFORMATION IN MAMMALIAN CELLS.
REPT'D USE : COMBUSTION POLLUTANT AND CHEMICAL INTERMEDIATE

CAS NO : **12001-29-5**
CHEM NAME : ***** CHRYSOTILE ASBESTOS**

REPORT 01 : NTP TR 246 NIH PUB. 90-2502 JUL 90
SHORT RANGE NOR INTERMEDIATE RANGE CHRYSOTILE ASBESTOS CARCINOGENIC INGESTED AT 1% LEVELS IN DIET BY MALE AND FEMALE SYRIAN GOLDEN HAMSTERS. INCREASES IN RATES OF ADRENAL CORTICAL ADENOMAS EXPOSED TO IR CHRYSOTILE NOT

DIFFERENT WHEN COMPARED WITH CONTROL GROUPS. STUDIES USING IR CHRYSOTILE AND 1,2-DIMETHYLHYDRAZINE DIHYDROCHLORIDE CONSIDERED INADEQUATE BECAUSE NO INCREASE IN INTESTINAL NEOPLASIA IN DMH GROUP.

REPT'D USE : CEMENT, FLAME RETARDANT, TEXTILE, BRAKE LININGS, BUILDING MATERIALS AND HEAT RESISTANT INSULATOR

CAS NO : **129-17-9**

CHEM NAME : ***** C.I. ACID BLUE 1 (BLUE VRS)**

REPORT 01 : IARC MONOGRAPH 17 1978
BLUE VRS IS CARCINOGENIC IN RATS FOLLOWING SC OR IM INJECTION, PRODUCING SARCOMAS AT THE SITE OF REPEATED INJECIONS. NO DATA AVAILABLE ON THE EMBRYOTOXICITY, TERATOGENICITY OR MUTAGENICITY OF BLUE VRS (C.I. ACID BLUE 1) WERE AVAILABLE.

REPORT 02 : NTP TECHNICAL BULLETIN
THE COMPOUND WAS POSITIVE IN THE SALMONELLA TYPHIMURIUM MUTAGENICITY ASSAY IN ONE LABORATORY.

REPORT 03 : NTP TECHNICAL BULLETIN 9 APR 83
SUFAN BLUE WAS POSITIVE IN ONE SALMONELLA TYPHIMURIUM MUTAGENICITY ASSAY.

REPT'D USE : PAPER/TEXTILE/RESIN DYE, INKS, BIOLOGICAL STAIN AND PHARMACEUTICAL COLOR

CAS NO : **2650-18-2**

CHEM NAME : ***** C.I. ACID BLUE 9 (BRILLIANT BLUE FCF)**

REPORT 01 : IARC MONOGRAPH NO 16 1978
BRILLIANT BLUE FCF, DISODIUM SALT IS CARCINOGENIC IN RATS AFTER ITS SUBCUTANEOUS INJECTION: IT PRODUCED FIBROSARCOMAS FOLLOWING REPEATED INJECTIONS. IT ALSO PRODUCED AN INCREASED INCIDENCE OF KIDNEY TUMORS IN MICE AFTER ITS ORAL ADMINISTRATION.

REPORT 02 : IARC MONOGRAPH 16 1978
NO ADEQUATE DATA ON THE EMBRYOTOXICITY, TERATOGENICITY OR MUTAGENICITY OF BRILLIANT BLUE FCF WERE AVAILABLE.

REPT'D USE : DYE, DRUGS, COSMETICS, WOOD STAINS, CLEANING AGENTS, BIOLOGICAL STAIN AND TOILET BOWL CLEANERS
RESULTS FROM A SALMONELLA ASSAY WERE REPORTED TO BE EQUIVOCAL.

CAS NO : **2150-60-9**
CHEM NAME : ***** C.I. ACID BLUE 43**

REPORT 01 : CPSC ANTHRAQUINONE DYES MONOGRAPH OCT 81
C.I. ACID BLUE 43 INDUCED FRAMESHIFT MUTATIONS IN SALMONELLA AFTER S-9 ACTIVATION.
REPT'D USE : NO AVAILABLE PRODUCTION OR USE DATA

CAS NO : **4680-78-8**
CHEM NAME : ***** C.I. ACID GREEN 3 (GUINEA GREEN B)**

REPORT 02 : IARC MONOGRAPH 16 1978
GUINEA GREEN TESTED IN MICE AND RATS BY ORAL ADMINISTRATION AND IN RATS BY AMBIENT EXPOSURE AND SC INJECTION. IT WAS CARCINOGENIC IN RATS, PRODUCING HEPATIC TUMORS AFTER ORAL ADMINISTRATION, LOCAL SARCOMAS AFTER SC INJECTION AND BENIGN MAMMARY TUMORS FOLLOWING AMBIENT EXPOSURE.

REPORT 03 : IARC MONOGRAPH 16 1978
NO DATA ON THE EMBRYOTOXICITY, TERATOGENICITY OR MUTAGENICITY OF GUINEA GREEN B WERE AVAILABLE.

REPORT 04 : EPA TSCA 8(E) NOTICE 8EHQ-0780-0350
PROCTOR AND GAMBLE PROVIDED RESULTS FROM A BATTERY OF MUTAGENICITY TESTS ON ACID GREEN 3. SUBMITTER REPORTED THAT RESULTS FROM IN-VIVO CYTOGENETICS TESTS IN RATS WERE NEGATIVE; HOWEVER, ACID GREEN 3 SHOWN TO BE MUTAGENIC IN AN IN-VITRO MOUSE LYMPHOMA CELL ASSAY.

CAS NO : **5141-20-8**
CHEM NAME : ***** C.I. ACID GREEN 5 (LIGHT GREEN SF)**

REPORT 01 : IARC MONOGRAPH 16 1978
LIGHT GREEN SF TESTED IN MICE AND RATS BY ORAL ADMINISTRATION AND IN RATS BY SC AND IP INJECTION. IT WAS CARCINOGENIC IN RATS AFTER SC INJECTION, PRODUCING LOCAL SARCOMAS.

REPORT 02 : IARC MONOGRAPH 16 1978
NO DATA ON THE EMBRYOTOXICITY, TERATOGENICITY OR MUTAGENICITY OF LIGHT GREEN SF WERE AVAILABLE.
REPT'D USE : TEXTILE DYE, BIOLOGICAL STAIN AND PRINTING INKS

CAS NO : **6373-74-6**
CHEM NAME : ***** C.I. ACID ORANGE 3**

REPORT 01 : NTP DRAFT TR 335 NIH PUB 88-2591 JUL 87
UNDER CONDITIONS OF 2-YR GAVAGE STUDIES, NO EVIDENCE OF CARCINOGENIC ACTIVITY FOR MALE F344/N RATS ADMINISTERED 375 MG/KG OF CPD. BECAUSE OF REDUCTION IN SURVIVAL RATE, THE 750 MG/KG GROUP WAS INADEQUATE T0 ASSESS CARCINOGENCICITY. CLEAR EVIDENCE FOR CARCINOGENICITY FOR FEMALE F344/N RATS IN 750 MG/KG GROUP. NO EVIDENCE OF CARCINOGENIC ACTIVITY FOR MALE B6C3F1 MICE ADMINISTERED 125 OR 250 MG/KG OF CPD OR FOR FEMALE B6C3F1 MICE ADMINISTERED 250 OR 500 MG/KG. NONNEOPLASTIC LESIONS OF KIDNEY SEEN IN BOTH GROUPS OF BOTH SEXES OF MICE AND RATS.
REPT'D USE : HAIR DYE

CAS NO : **1936-15-8**
CHEM NAME : ***** C.I. ACID ORANGE 10 (ORANGE G)**

REPORT 01 : NTP BOARD OF SCIENTIFIC COUNSELORS FEB 81
THE BIOASSAY REPORT CONCLUSIONS FOR ORANGE G WERE REVIEWED BY THE PEER REVIEW PANEL AND APPROVED WITH MAJOR MODIFICATIONS. UNDER CONDITIONS OF THIS BIOASSAY, ORANGE G ADMINISTERED IN THE FEED WAS NOT CARCINOGENIC FOR F344 RATS OR B6C3F1 MICE OF EITHER SEX.

REPORT 02 : IARC MONOGRAPH 8 1975
ORANGE G WAS TESTED IN MICE AND RATS BY THE ORAL ROUTE ONLY. THE AVAILABLE STUDIES IN MICE DO NOT ALLOW AN EVALUATION, SINCE IN ONE STUDY THE ADEQUACY OF THE DOSE COULD NOT BE ASSESSED AND, IN THE OTHER STUDY, INFORMATION ON PATHOLOGY AND SURVIVAL RATES WAS INSUFFICIENT. THE STUDY IN RATS CANNOT BE EVALUATED DUE TO LIMITED REPORTING.
REPT'D USE : DRUG/COSMETIC, COLORANT, BIOLOGICAL STAINS, TEXTILE, LEATHER AND PAPER DYE

CAS NO : **1934-20-9**
CHEM NAME : ***** C.I. ACID ORANGE 12**

REPORT 01 : CPSC MONOGRAPH ON MONOAZO DYES OCT 81
C.I. ACID ORANGE 12 WAS SHOWN TO SERIOUSLY EFFECT THE LIVER AND RED BLOOD CELLS IN PIGS CHRONICALLY EXPOSED BY THE ORAL ROUTE.

CAS NO : **523-44-4**
CHEM NAME : ***** C.I. ACID ORANGE 20 (ORANGE 1)**

REPORT 01 : IARC MONOGRAPH 8 1975
ORANGE 1 HAS BEEN TESTED BY THE ORAL ROUTE IN MICE, RATS AND DOGS, BUT THE STUDIES COULD NOT BE EVALUATED DUE TO INADEQUATE REPORTING IT WAS ALSO TESTED BY THE SUBCUTANEOUS ROUTE IN RATS, PRODUCING INJECTION SITE TUMORS.

REPORT 02 : CPSC MONOGRAPH ON MONOAZO DYES OCT 81
A FEEDING STUDY WITH RATS RESULTED IN ACID ORANGE 20 CAUSING ANEMIA, ENLARGED SPLEENS, DIARRHEA, AND LEUCOCYTOSIS: THE ANIMLS FAILED TO THRIVE AND DIED EARLY. MILD CATHARSIS WAS SEEN IN EXPOSED DOGS AND HUMANS. THE DYE WAS NON-MUTAGENIC IN TWO BACTERIAL TESTS.
REPT'D USE : TEXTILE/LEATHER DYE

CAS NO : **3734-67-6**
CHEM NAME : ***** C.I. ACID RED 1**

REPORT 01 : CPSC MONOGRAPH ON MONOAZO DYES AUG 81
C.I. ACID RED 1 WAS MUTAGENIC TO E. COLI AND S. TYPHIMURIUM. THE CHEMICAL WAS ALSO POSITIVE IN A REC ASSAY WITH MICROSOMES.
REPT'D USE : TEXTILE DYE

CAS NO : **493-52-7**
CHEM NAME : ***** C.I. ACID RED 2 (METHYL RED)**

REPORT 01 : IARC MONOGRAPH 8 1975
METHYL RED WAS TESTED ONLY BY THE ORAL ROUTE IN RATS. THE INADEQUACY OF THE EXPERIMENT DOES NOT ALLOW AN EVALUATION OF THE DATA WITH REGARD TO THE CARCINOGENICITY OF THIS COMPOUND.
REPT'D USE : CHEMICAL INDICATOR

CAS NO : **3567-69-9**
CHEM NAME : ***** C.I. ACID RED 14 (CARMOISINE)**

REPORT 01 : NTP CARC TEST PROGRAM JUL 82
IN A CURRENT BIOASSAY, CARMOISINE WAS ADMINISTERED IN FEED TO F344 RATS AND B6C3F1 MICE AND WAS NOT CARCINOGENIC. THE BIOASSAY HAS BEEN COMPLETED AND THE REPORT IS UNDER REVIEW.

REPORT 02 : IARC MONOGRAPH 8 1975
CARMOISINE WAS TESTED IN MICE AND RATS BY ORAL AND SC ROUTES. THE TEST BY THE ORAL ROUTE IN MICE WAS NEGATIVE. THE OTHER STUDIES COULD NOT BE EVALUATED BECAUSE OF THE SMALL NUMBERS OF ANIMALS USED.

REPT'D USE : ANODIZED ALUMINUM, TEXTILE/LEATHER, DYE AND WOOD STAINS

CAS NO : **915-67-3**
CHEM NAME : ***** C.I. ACID RED 27 (AMARANTH)**

REPORT 01 : IARC MONOGRAPH 8 1975
TWO ORAL STUDIES IN RATS INDICATING A CARCINOGENIC EFFECT DO NOT LEAD TO A DEFINITE EVALUATION. OTHER ORAL STUDIES IN MICE, RATS AND DOGS GAVE NEGATIVE RESULTS, BUT WERE INADEQUATELY REPORTED. SC INJECTION EXPERIMENTS IN RATS COULD NOT BE EVALUATED, BECAUSE OF THE SMALL NUMBERS OF ANIMALS USED OR INSUFFICIENT REPORTING. THE CARCINOGENICITY OF THIS COMPOUND COULD NOT BE EVALUATED.

REPORT 02 : CPSC MONOGRAPH ON MONOAZO DYES AUG 81
AMARANTH WAS FOUND TO BE CARCINOGENIC IN TWO RUSSIAN REPORTS, BUT DATA WERE INSUFFICIENT TO ESTABLISH THIS CONCLUSIVELY. TERATOGENICITY STUDIES HAVE BEEN NEGATIVE IN SEVERAL SPECIES OF ANIMALS. ALLEGATIONS OF MUTAGENICITY HAVE BEEN PERSISTENT, BUT THE STUDIES SHOW NO STATISTICAL SIGNIFICANCE.

REPORT 03 : IARC MONOGRAPH 8 1975
IN A THREE-GENERATION STUDY WITH FOUR DIFFERENT DOSE LEVELS, NO INDICATION WAS FOUND OF ADVERSE EFFECTS OF AMARANTH ON THE REPRODUCTIVE SYSTEM OR OF TERATOGENICITY.

REPT'D USE : CHEMICAL INDICATOR, TEXTILE, LEATHER, PAPER DYE AND PHOTOGRAPHIC DYE

CAS NO : **1694-09-3**
CHEM NAME : ***** C.I. ACID VIOLET 49 (BENZYL VIOLET 4B)**

REPORT 01 : IARC MONOGRAPH 16 1978
BENZYL VIOLET 4B IS CARCINOGENIC IN RATS FOLLOWING ITS ORAL OR SC ADMINISTRATION. IT PRODUCED MAMMARY CARCINOMAS AND SQUAMOUS CELL CARCINOMAS OF THE SKIN AFTER ITS ORAL ADMINISTRAITON TO FEMALE RATS AND LOCAL FIBROSARCOMAS FOLLOWING SC INJECTION IN MALE AND FEMALE RATS. IN A 1979 SUPPLEMENT, IARC DETERMINED THAT THERE IS SUFFICIENT EVIDENCE FOR THE CARCINOGENICITY OF VIOLET 4B IN ANIMALS.

REPORT 02 : IARC MONOGRAPH 16 1978
NO DATA ON THE EMBRYOTOXICITY, TERATOGENICITY OR MUTAGENICITY OF BENZYL VIOLET 4B WERE AVAILABLE.
REPT'D USE : TEXTILE/PAPER, DYE, INKS, WOOD STAINS AND BIOLOGICAL STAINS

CAS NO : **1934-21-0**
CHEM NAME : ***** C.I. ACID YELLOW 23 (TARTRAZINE)**

REPORT 01 : CPSC/EVIRON CONTROL DYES MONOGRAPH AUG 81
ALLERGENIC REACTIONS IN HUMANS REPORTED IN SOME EXPOSURES.
REPT'D USE : TEXTILES LEATHER, PAPER AND WOOD STAINS

CAS NO : **532-82-1**
CHEM NAME : ***** C.I. BASIC ORANGE 2 (CHRYSOIDINE)**

REPORT 02 : IARC MONOGRAPH 8 1975
CHRYSOIDINE IS CARCINOGENIC IN MICE FOLLOWING ITS ORAL ADMINISTRATION PRODUCING LIVER CELL TUMORS, LEUKEMIA, AND RETICULUM CELL SARCOMAS.
REPT'D USE : INKS, WOOD/BIOLOGICAL STAINS, ANTISEPTIC/DISINFECTANT AND TEXTILE/PAPER DYE

CAS NO : **989-38-8**
CHEM NAME : ***** C.I. BASIC RED 1 (RHODAMINE 6G)**

REPORT 01 : NTP TR364 NIH 89-2819 SEP 89
2 YR. FEED STUDIES SHOWED EQUIVOCAL EVIDENCE OF CARCINOGENIC ACTIVITY FOR MALE F344/N RATS ADMINISTERED RHODAMINE 6G, INDICATED BY INCREASED INCIDENCE OF INTEGUMENTARY KERATOACANTHOMAS. EQUIVOCAL EVIDENCE OF CARCINOGENIC ACTIVITY FOR FEMALE F344/N RATS INDICATED BY MARGINAL INCREASE IN PHEOCHROMOCYTOMAS OR MALIGNANT PHEOCHROMOCYTOMAS (COMBINED) OF ADRENAL GLAND. NO EVIDENCE OF CARCINOGENIC ACTIVITY FOR MALE B6C3F1 MICE ADMINISTERED 1,000 OR 2,000 PPM. NO EVIDENCE OF CARCINOGENIC ACTIVITY FOR FEMALE B6C3F1 MICE ADMINISTERED 500 OR 1,000 PPM. NO SIGNIFICANT NONNEOPLASTIC LESIONS ATTRIBUTED TO RHODAMINE 6G ADMINISTRATION TO MALE OR FEMALE RATS OR MALE OR FEMALE MICE.

REPORT 02 : IARC MONOGRAPH 16 1978
RHODAMINE 6G IS CARCINOGENIC IN RATS AFTER ITS SUBCUTANEOUS INJECTION; PRODUCING SARACOMAS. IT HAS BEEN INADEQUATELY TESTED IN MICE BY ORAL ADMINISTRATION. NO ADEQUATE DATA ON THE EMBRYOTOXICITY, TERATOGENICITY OR MUTAGENICITY OR RHODAMINE 6G WERE AVAILABLE.
REPT'D USE : CHEMICAL ABSORPTION INDICATOR, PRINTING INKS, TRACING

AGENT IN WATER POLLUTION STUDIES, TEXTILE/PAPER, DYE AND PHOTOGRAPHIC CHEMICAL

CAS NO : **569-61-9**
CHEM NAME : ***** C.I. BASIC RED 9 MONOHYDROCHLORIDE**

REPORT 01 : NTP TECHNICAL REPORT 285 NIH PUB 86-2541 JAN 86
2-YEAR FEEDING STUDIES SHOW CLEAR EVIDENCE OF CARCINOGENICITY OF C.I. BASIC RED 9 MONOHYDROCHLORIDE FOR MALE AND FEMALE F344/N RATS AND FOR MALE AND FEMALE B6C3F1 MICE. CPD CAUSED SQUAMOUS CELL, ZYMBAL GLAND, HEPATOCELLULAR AND OTHER FORMS OF CARCINOMAS, FIBROMAS AND ADENOMAS. EXPOSURE MAY HAVE BEEN RELATED TO INCREASED INCIDENCES OF MAMMARY GLAND TUMORS IN FEMALE RATS AND HEMATPOIETIC SYSTEM TUMORS IN FEMALE MICE.

REPORT 06 : MULTI-AGENCY CROSS CHECK
Y HHS MAGUIRE
N OSHA CARC
N EPA CAG
N EPA RCRA
N EPA CLEAN AIR
N EPA CLEAN WATER

REPT'D USE : DYE FOR TEXTILES, LEATHER, PAPER AND BIOLOGIC STAIN; AGAR NUTRIENT FOR BACTERIUM I.D. AND PHOTOGRAPHY FILTER DYE

CAS NO : **81-88-9**
CHEM NAME : ***** C.I. BASIC VIOLET 10 (RHODAMINE B)**

REPORT 01 : IARC MONOGRAPH 16 1978
RHODAMINE B WAS TESTED IN MICE AND RATS BY SC INJECTIONS AND BY ORAL ADMINISTRATION. THE ORAL STUDIES ARE INADEQUATE FOR RISK EVALUATION; SC INJECTION PRODUCED LOCAL SARCOMAS. DATA ON THE EMBRYOTOXICITY, TERATOGENICITY OR MUTAGENICITY OF CPD WERE NOT AVAILABLE.

REPORT 02 : EPA TSCA 8(E) NOTICE 8EHQ-0282-0434 FEB 82
SUN CHEMICAL PROVIDED PRELIMINARY RESULTS FROM A CTFA-SPONSORED CHRONIC FEEDING STUDY IN MICE AND RATS. THE SUBMITTING COMPANY STATED THAT IT HAD LEARNED THAT THERE WAS AN INCREASED INCIDENCE OF MICE WITH HEPATIC NEOPLASMS AND AN INCREASED INCIDENCE OF RATS WITH THYROID TUMORS. THE STUDY WILL BE COMPLETED IN 1982.

REPORT 03 : FDA LISTING FOR EXTERNAL USE AUG 86
IN THE FR VOL. 51 P. 28346, FDA STATES THAT D+C RED NO. 19 INDUCES CANCER IN LABORATORY ANIMALS, HOWEVER, THE CTKA PRESENTED RISK ASSESSMENT ARGUMENTS SHOWING IT CAN BE

REGULATED FOR SAFE USE IN EXTERNALLY APPLIED DRUGS AND COSMETICS.

REPT'D USE : DYE, DRUG/COSMETIC COLOR, LAB REAGENT AND PAINTS

CAS NO : **632-99-5**

CHEM NAME : ***** C.I. BASIC VIOLET 14 (MAGENTA)**

REPORT 02 : IARC MONOGRAPH 4 1974

P-MAGENTA, ONE OF THE COMPONENTS OF COMMERCIAL MAGENTA, INDUCED LOCAL SARCOMAS IN RATS FOLLOWING SUBCUTANEOUS ADMINISTRATION. PHYSICAL PROPERTIES OF P-MAGENTA MAY BE RESPONSIBLE FOR THE SARCOMAS. ORALLY ADMINISTERED COMMERCIAL MAGENTA PRODUCED ADVERSE EFFECTS IN MICE IN ONE STUDY, BUT IT WAS BASED ON AN INSUFFICIENT NUMBER OF SURVIVING ANIMALS. ONE EPIDEMIOLOGICAL STUDY APPEARS TO ESTABLISH THE RISK TO MAGENTA MANUFACTURE WORKERS. IT IS NOT POSSIBLE TO ATTRIBUTE INDUSTRIAL BLADDER CANCER TO MAGENTA OR ITS INTERMEDIATES.

REPT'D USE : DYE INTERMEDIATE FOR CHINA/CLAY PRODUCTS, PRINTING INKS, PHARMACEUTICAL, TEXTILE DYE AND PHOTOGRAPHIC CHEMICAL

CAS NO : **492-80-8**

CHEM NAME : *****C.I. BASIC YELLOW 2 (AURAMINE)**

REPORT 02 : IARC MONOGRAPH 1 1972

AURAMINE IS CARCINOGENIC IN THE MOUSE AND RAT. GIVEN ORALLY, IT HAS PRODUCED LIVER TUMORS IN BOTH SPECIES. NO TUMORS WERE SEEN IN THE ONLY EXPERIMENT IN DOGS AND RABBITS. IN A 1979 SUPPLEMENT, IARC CONCLUDED THAT AURAMINE IS PROBABLY CARCINOGENETIC FOR HUMANS.

REPORT 06 : MULTI-AGENCY CROSS CHECK

Y HHS MAGUIRE
Y OSHA CARC
Y EPA CAG
Y EPA RCRA
N EPA CLEAN AIR
N EPA CLEAN WATER

REPT'D USE : DYE INTERMEDIATE, PAPER/TEXTILE, DYE, INKS, CARBON PAPER AND TYPEWRITER RIBBON

CAS NO : **95-69-2**

CHEM NAME : ***** C.I. 37085 (P-CHLORO-O-TOLUIDINE)**

REPORT 01 : IARC MONOGRAPH 16 JAN 78

THE ONLY DATA ON P-CHLORO-O-TOLUIDINE HCL THAT WERE AVAILABLE WERE REPORTED IN AN ABSTRACT. ALTHOUGH THE

RESULTS OF THIS EXPERIMENT INDICATE A CARCINOGENIC EFFECT, THE INADEQUACY OF REPORTING DOES NOT ALLOW AN EVALUATION OF THE CARCINOGENICITY OF THIS CHEMICAL.

REPORT 02 : IARC MONOGRAPH 16 JAN 78
NO DATA ON THE METABOLISM, EMBRYOTOXICITY, TERATOGENICITY OR MUTAGENICITY OF P-CHLORO-O-TOLUININE WERE AVAILABLE.

REPORT 03 : STATUS REPORT 8EHQ-0986-0634 OCT 86
EHQ-1086-0634-SUPPLEMENT
HOECHST AG DID AN EPA STUDY ON RETROSPECTIVE MORTALITY OF POTENTIAL CAUSATION OF BLADDER CANCER BY CPD. WITHIN FRAMEWORK OF STUDY, NO INDICATION OF NEOPLASTIC ACTIVITY. AFTER CLOSING DATE, HOECHST AG LEARNED OF FIVE CASES OF BLADDER CANCER WITHIN THE EPA STUDY COHORT. ONE OF THE PERSONS DIED. BECAUSE OF THIS, HOECHST AG CANNOT EXCLUDE POSSIBILITY THAT CPD CONTRIBUTED TO THE ABOVE FINDINGS.

CAS NO : **10101-53-8**
CHEM NAME : ***** C.I. 77305 (CHROMIUM SULFATE)**

REPORT 01 : IARC MONOGRAPH 23 1980
AN EXPERIMENT IN MICE EXPOSED TO CHROMIUM (III) SULFATE BY IP INJECTIONS WAS INADEQUATE TO EVALUATE CARCINOGENICITY.
REPT'D USE : DYE MORDANT, LEATHER TANNING, INKS, PAINTS/VARNISHES AND CERAMIC GLAZES

CAS NO : **1937-37-7**
CHEM NAME : ***** C.I. DIRECT BLACK 38**

REPORT 01 : NCI CARC TRS NO 108 DHEW PUB NO NIH 78-1358
UNDER THE CONDITIONS OF THESE 13-WEEK SUBCHRONIC TOXICITY STUDIES, DIRECT BLACK 38 WAS CARCINOGENIC IN MALE AND FEMALE FISCHER 344 RATS, THIS DYE INDUCING HEPATOCELLULAR CARCINOMAS AND NEOPLASTIC NODULES IN THE LIVER. THE TEST DYE WAS NOT CARCINOGENIC FOR B6C3F1 MICE IN THE 13-WEEK SUBCHRONIC FEEDING STUDY.

REPORT 02 : IARC MONOGRAPH 29 1982
THERE IS SUFFICIENT EVIDENCE THAT COMMERICAL DIRECT BLACK 38 IS CARCINOGENIC TO RATS. ALTHOUGH THE EPA DATA WERE INADEQUATE TO EVALUATE THE CARCINOGENICITY IN MAN OF DIRECT BLACK 38 ALONE, THE COMBINATION OF THE CPD AND BENZIDINE IN THE URINE OF EXPOSED WORKERS PROVIDES SUFFICIENT EVIDENCE THAT OCCUPATIONAL EXPOSURE TO

BENZIDINE-BASED DYES REPRESENTS A CARCINOGENIC RISK TO HUMANS.

REPORT 04 : NIOSH SPECIAL HAZARD REVIEW NOV 79
NIOSH CONCLUDES THAT ALL BENZIDINE-BASED DYES, REGARDLESS OF THEIR PHYSICAL STATE OR PROPORTION IN A MIXTURE, SHOULD BE RECOGNIZED AS POTENTIAL HUMAN CARCINOGENS.

REPORT 05 : EPA TSCA 8(E) NOTICE 8EHQ-0779-0295 JUL 79
IBM CORP SUBMITTED SUMMARIZED RESULTS OF A SKIN ABSORPTION STUDY IN RABBITS USING DIRECT BLACK 38 DYE WHICH WAS RADIOACTIVELY LABELLED WITH C-14 AT THE BIPHENYL MOIETY. SUBMITTER STATES THAT PRELIMINARY DATA SUGGEST THE DYE OR A PORTION OF ITS MOLECULE PENETRATES THE UNBROKEN SKIN OF EXPERIMENTAL ANIMALS AND APPEARS IN THE URINE AND FECES.

REPORT 06 : MULTI-AGENCY CROSS CHECK
Y HHS MAGUIRE
Y OSHA CARC
N EPA CAG
N EPA RCRA
N EPA CLEAN AIR
N EPA CLEAN WATER

REPORT 07 : EPA RESPONSE TO ITC RECOMS (46FR55004) NOV 81
IN RESPONSE TO THE ITC RECOMMENDATIONS (44FR70664), EPA HAS DECIDED NOT TO PURSUE TESTING PROPOSAL FOR BENZIDINE DYES AT THIS TIME BECAUSE TESTING PROGRAMS EXPECTED TO SUPPLY DATA SUFFICIENT FOR EPA'S NEEDS, ARE BEING PLANNED AND CONDUCTED, AND REGULATORY EFFORTS ARE ARE BEING PURSUED BY OSHA AND CPSC.

REPORT 08 : NTP TECHNICAL BULLETIN NO 3/7 DEC 80, APR 82
TERATOLOGY ASSAYS SHOW THAT SUBCUTANEOUS ADMINISTRATION OF DIRECT BLACK 38 CAUSED PRENATAL DEATH AND INCREASED MALFORMATIONS AT DOSES NONTOXIC TO THE MOTHER. THE CHEMICAL IS POSITIVE IN THE SALMONELLA TYPHIMURIUM MUTAGENICITY ASSAY IN TWO LABORATORIES.

REPORT 09 : NIOSH CURRENT INTELLIGENCE BULLETIN 24 APR 78
THE NIOSH RECOMMENDS THAT DIRECT BLACK 38 BE HANDLED IN THE WORKPLACE AS IF IT WERE A HUMAN CARCINOGEN. RECOMMENDATION IS BASED PRIMARILY ON A PRELIMINARY ANALYSIS OF NCI DATA FROM SHORT-TERM FEEDING STUDIES WHICH INDICATE THAT CANCEROUS AND PRECANCEROUS LIVER CONDITION WERE FOUND IN RATS, SIMILAR TO DAMAGE PRODUCED BY KNOWN LIVER CARCINOGENS.

REPT'D USE : TEXTILE, LEATHER AND PAPER DYE

CAS NO : **2602-46-2**
CHEM NAME : ***** C.I. DIRECT BLUE 6**

REPORT 01 : NCI CARC TRS NO 108 DHEW PUB NO NIH 78-1358
UNDER THE CONDITIONS OF THESE 13-WEEK SUBCHRONIC TOXICITY STUDIES, DIRECT BLUE 6 WAS CARCINOGENIC IN MALE AND FEMALE FISCHER 344 RATS; THIS DYE INDUCED HEPATOCELLULAR CARCINOMAS AND NEOPLASTIC NODULES IN THE LIVER. THE TEST DYES WERE NOT CARCINOGENIC FOR B6C3F1 MICE IN THE 13-WEEK SUBCHRONIC TOXICITY STUDIES (FEEDING STUDY).

REPORT 02 : NIOSH SPECIAL HAZARD- REVIEW NOV 79
NIOSH CONCLUDES THAT ALL BENZIDINE-BASED DYES, REGARDLESS OF THEIR PHYSICAL STATE OR PROPORTION IN A MIXTURE, SHOULD BE RECOGNIZED AS POTENTIAL HUMAN CARCINOGENS.

REPORT 03 : IARC MONOGRAPH 29 1982
THERE IS SUFFICIENT EVIDENCE THAT COMMERCIAL DIRECT BLUE 6 IS CARCINOGENIC TO RATS. ALTHOUGH THE EPA DATA WERE INADEQUATE TO EVALUATE THE CARCINOGENICITY IN MAN OF DIRECT BLUE 6 ALONE, THEY, TOGETHER WITH THE PRESENCE OF BENZIDINE IN THE URINE OF EXPOSED WORKERS, PROVIDE SUFFICIENT EVIDENCE THAT OCCUPATIONAL EXPOSURE TO BENZIDINE-BASED DYES REPRESENTS A CARCINOGENIC RISK TO HUMANS.

REPORT 04 : IARC MONOGRAPH 29 1982
DIRECT BLUE 6 IS TERATOGENIC IN RATS ONLY WHEN INJECTED DURING THE FIRST HALF OF PREGNANCY. NO DATA WERE AVAILABLE TO ASSESS THE MUTAGENICITY OF THIS CHEMICAL.

REPORT 05 : NTP TECHNICAL BULLETIN NO. 3/7 DEC 80, APR 82
TERATOLOGY ASSAYS SHOW THAT SUBCUTANEOUS ADMINISTRATION OF DIRECT BLUE 6 CAUSED INCREASED MALFORMATIONS AT DOSES NON-TOXIC TO THE MOTHER. THE CHEMICAL WAS NEGATIVE IN THE SALMONELLA TYPHIMURIUM MUTAGENICITY ASSAY IN TWO LABORATIORIES.

REPORT 06 : MULTI-AGENCY CROSS CHECK
Y HHS MAGUIRE
Y OSHA CARC
N EPA CAG
N EPA RCRA
N EPA CLEAN AIR
N FPA CLEAN WATER

REPORT 07 : NIOSH CURRENT INTELLIGENCE BULLETIN 24 APR 78
THE NIOSH RECOMMENDS THAT DIRECT BLUE 6 BE HANDLED IN THE WORKPLACE AS IF IT WERE A HUMAN CARCINOGEN. RECOMMENDATION IS BASED PRIMARILY ON A PRELIMINARY

ANALYSIS OF NCI DATA FROM SHORT-TERM FEEDING STUDIES WHICH INDICATE THAT CANCEROUS AND PRECANCEROUS LIVER CONDITIONS WERE FOUND IN RATS, SIMILAR TO DAMAGE PRODUCED BY KNOWN LIVER CARCINOGENS.

REPORT 08 : EPA RESPONSE TO ITC RECOMMS (46FR55004) NOV 81
IN RESPONSE TO THE ITC RECOMMENDATIONS (44FR70664) THE EPA HAS DECIDED NOT TO PURSUE TESTING PROPOSAL FOR BENZIDINE DYES AT THIS TIME BECAUSE TESTING PROGRAMS, WHICH ARE EXPECTED TO SUPPLY DATA SUFFICIENT FOR EPA NEEDS, ARE BEING PLANNED AND CONDUCTED, AND REGULATORY EFFORTS ARE BEING PURSUED BY THE OSHA AND CPSC.

REPT'D USE : TEXTILE, LEATHER AND PAPER DYE

CAS NO : **72-57-1**
CHEM NAME : ***** C. I. DIRECT BLUE 14 (TRYPAN BLUE)**

REPORT 01 : EPA RESPONSE TO ITC RECOMMS (46FR55004) NOV 81
IN RESPONSE TO THE ITC RECOMMENDATIONS (44FR70664) THE EPA HAS DECIDED NOT TO PURSUE TESTING PROPOSAL FOR O-TOLIDINE DYES AT THIS TIME BECAUSE TESTING PROGRAMS, WHICH ARE EXPECTED TO SUPPLY DATA SUFFICIENT FOR EPA NEEDS ARE BEING PLANNED AND CONDUCTED, AND REGULATORY EFFORTS ARE BEING PURSUED BY THE OSHA AND THE CPSC.

REPORT 02 : IARC MONOGRAPH 8 1975
TRYPAN BLUE IS CARCINOGENIC IN RATS FOLLOWING ITS SUBCUTANEOUS OR INTRAPERITONEAL INJECTION, PRODUCING RETICULUM-CELL SARCOMAS, MAINLY OF THE LIVER, AS WELL AS FIBROSARCOMAS AT THE SITE OF INJECTION IN A 1979 SUPPLEMENT, IARC DETERMINED THAT THERE IS SUFFICIENT EVIDENCE FOR THE CARCINOGENICITY OF TRYPAN BLUE IN ANIMALS.

REPORT 03 : IARC MONOGRAPH 8 1975
TRYPAN BLUE OF UNDEFINED PURITY EXERTS DIRECT MUTAGENIC ACTIVITY IN ANIDULANS. NO MUTAGENICITY DATA INVOLVING MAMMALIAN METABOLIC SYSTEMS WERE AVAILABLE.

REPORT 06 : MULTI-AGENCY CROSS CHECK
N HHS MAGUIRE
N OSHA CARC
Y EPA CAG
Y EPA RCRA
N EPA CLEAN AIR
N EPA CLEAN WATER

REPT'D USE : TEXTILE, LEATHER AND PAPER DYE

CAS NO : **2429-74-5**
CHEM NAME : ***** C.I. DIRECT BLUE 15**

REPORT 01 : NTP TR 397 NIH PUB. 90-2852 NOV 90
UNDER CONDITIONS OF 22-MONTH DRINKING WATER STUDIES. CLEAR EVIDENCE OF CARCINOGENIC ACTIVITY OF C.I. DIRECT BLUE 15 (DESALTED INDUSTRIAL GRADE) FOR MALE F344/N RATS INDICATED BY BENIGN AND MALIGNANT NEOPLAS MS OF SKIN, ZYMBAL'S GLAND, PREPUTIAL GLAND, LIVERI ORAL CAVITY, AND SMALL AND LARGE INTESTINE. INCREASED INCIDENCES OF MONONUCLEAR CELL LEUKEMIA AND NEOPLASMS OF BRAIN AND SMALL INSTESTINE RELATED TO CHEMICAL ADMINISTRATION. CLEAR EVIDENCE OF CARCINOGENIC ACTIVITY FOR F344 /N RATS INDICATED BY BENIGN AND MALIGNANT NEOPLASMS OF SKIN, ZYMBAL'S GLAND, CLITORAL GLAND, LIVER, ORAL CAVITY, SMALL AND LARGE INTESTINE, AND UTERUS, AND MONONUCLEAR CELL LEUKEMIA.

REPT'D USE : DYE IN TEXTILES, PAPER, PLASTIC, RUBBER AND LEATHER

CAS NO : **314-13-6**
CHEM NAME : ***** C.I. DIRECT BLUE 53 (EVANS BLUE)**

REPORT 01 : IARC MONOGRAPH 8
C.I. DIRECT BLUE 53 IS TERATOGENIC IN RATS FOLLOWING I.P. INJECTION.

REPORT 02 : IARC MONOGRAPH 8
EVANS BLUE WAS CARCINOGENIC IN ONE STUDY IN RATS WHEN ADMINISTERED INTRAPERITONEALLY, THE ONLY SPECIES AND ROUTE TESTED. PRODUCED SARCOMAS OF THE RETICULOENDOTHELIAL SYSTEM IN THE LIVER.

REPORT 03 : EPA RESPONSE TO ITC RECOMS (46FR55004) NOV 81
IN RESPONSE TO THE ITC RECOMMENDATIONS (44FR70664) THE EPA HAS DECIDED NOT TO PURSUE TESTING PROPOSAL FOR O-TOLIDINE DYES AT THIS TIME BECAUSE TESTING PROGRAMS, WHICH ARE EXPECTED TO SUPPLY DATA SUFFICIENT FOR EPA NEEDS, ARE BEING PLANNED AND CONDUCTED, AND REGULATORY EFFORTS ARE BEING PURSUED BY THE OSHA AND THE CPSC.

REPT'D USE : TEXTILE, LEATHER AND PAPER DYE

CAS NO : **16071-86-6**
CHEM NAME : ***** C.I. DIRECT BROWN 95**

REPORT 01 : NCI CARC TRS 108 DHEW PUB NO 78-1358
UNDER THE CONDITIONS OF THESE 13-WEEK SUBCHRONIC TOXICITY STUDIES, DIRECT BROWN 95 WAS CARCINOGENIC IN FEMALE FISCHER RATS; THIS DYE INDUCED HEPATOCELLULAR CARCINO-

MAS AND NEOPLASTIC NODULES IN THE LIVER. TEST DYES WERE NOT CARCINOGENIC FOR B6C3F1 MICE IN THE 13-WEEK SUBCHRONIC TOXICITY STUDIES (FEEDING STUDY).

REPORT 02 : NTP TECHNICAL BULLETIN 3/7/8 DEC 80, JUL 82
DIRECT BROWN 95 WAS NEGATIVE IN SALMONELLA AND IN DROSOPHILA ASSAYS. TERATOLOGY ASSAYS SHOW THAT SUBCUTANEOUS ADMINISTRATION TO MICE WAS NOT ASSOCIATED WITH PRENATAL DEATH OR INCREASED MALFORMATIONS AT DOSES TOXIC TO THE MOTHER.

REPORT 03 : NIOSH SPECIAL HAZARD REVIEW NOV 79
NIOSH CONCLUDES THAT ALL BENZIDINE-BASED DYES, REGARDLESS OF THEIR PHYSICAL STATE OR PROPORTION IN A MIXTURE, SHOULD BE RECOGNIZED AS POTENTIAL HUMAN CARCINOGENS.

REPORT 04 : NIOSH CURRENT INTELLIGENCE BULLETIN 24 APR 78
THE NIOSH RECOMMENDS THAT DIRECT BROWN 95 BE HANDLED IN THE WORKPLACE AS IF IT WERE A HUMAN CARCINOGEN. RECOMMENDATION IS BASED PRIMARILY ON A PRELIMINARY ANALYSIS OF NCI DATA FROM SHORT-TERM FEEDING STUDIES WHICH INDICATE CANCEROUS AND PRECANCEROUS LIVER CONDITIONS IN RATS, SIMILAR TO DAMAGE PRODUCED BY KNOWN LIVER CARCINOGENS.

REPORT 05 : IARC MONOGRAPH 29 1982
THE PRESENT DATA PROVIDE LIMITED EVIDENCE THAT COMMERCIAL DIRECT BROWN 95 IS CARCINOGENIC TO RATS. ALTHOUGH THE EPA DATA WERE INADEQUATE TO EVALUATE THE CARCINOGENCICITY TO MAN, TOGETHER WITH THE PRESENCE OF BENZIDINE IN THE URINE OF WORKERS, PROVIDE SUFFICIENT EVIDENCE THAT OCCUPATIONAL EXPOSURE TO BENZIDINE-BASED DYES PRESENTS A CARCINOGENIC RISK TO MAN.

REPORT 06 : MULTI-AGENCY CROSS CHECK
N HHS MAGUIRE
Y OSHA CARC
N EPA CAG
N EPA RCRA
N EPA CLEAN AIR
N EPA CLEAN WATER

REPORT 07 : IARC MONOGRAPH 29 1982
NO DATA WERE AVAILABLE TO ASSESS THE MUTAGENICITY OR TERATOGENICITY OF THIS CHEMICAL.

REPORT 08 : EPA. RESPONSE TO ITC RECOMMS (46FR55004) NOV 81
IN RESPONSE TO THE ITC RECOMMENDATIONS (44FR70664), THE EPA HAS DECIDED NOT TO PURSUE TESTING PROPOSAL FOR BENZIDINE DYES AT THIS TIME BECAUSE TESTING PROGRAMS, WHICH ARE EXPECTED TO SUPPLY DATA SUFFICIENT FOR EPA NEEDS, ARE BEING PLANNED AND CONDUCTED, AND REGULATORY EFFORTS ARE BEING PURSUED BY THE OSHA AND THE CPSC.

REPT'D USE : TEXTILE, LEATHER AND PAPER DYE

CAS NO : **10190-66-6**
CHEM NAME : ***** C.I. DIRECT BROWN 191**

REPORT 01 : CPSC MONOGRAPH ON MONOAZO DYES AUG 81
C.I. DIRECT BROWN 191 WAS HIGHLY MUTAGENIC TO SOME STRAINS OF SALMONELLA FOLLOWING METABOLIC ACTIVATION.
REPT'D USE : NO AVAILABLE PRODUCTION OR USE DATA. NO LISTING ON TSCA INVENTORY.

CAS NO : **8005-72-9**
CHEM NAME : *****C I. DIRECT YELLOW 28**

REPORT 01 : STATUS REPORT 8EHQ-1286-0645 DEC 86
CPD INDUCED A "WEAK" MUTAGENIC EFFECT WHEN TESTED WITH AND WITHOUT EXOGENOUS METABOLIC ACTIVATION IN STRAINS TA98 AND TA1538 OF S. TYPHIMURIUM.

REPORT 02 : TSCA SECT 8(E) STATUS REPORT 8EHQ-0687-0677 JUN 87
PYRAZOL YELLOW BG 250% (C.I. DIRECT YELLOW 28) FOUND TO BE WEAK MUTAGEN IN MODIFIED AMES S. TYPHIMURIUM ASSAY. CPD FOUND TO BE POSITIVE BOTH WITH AND WITHOUT ACTIVATION IN STRAINS TA98, TA1537, TA1538; STRAINS TA100 AND TA1535 REPORTED TO BE NEGATIVE.
REPT'D USE : DYEING AND STAINING NATURAL AND SYNTHETIC MATERIALS

CAS NO : **2475-45-8**
CHEM NAME : ***** C.I. DISPERSE BLUE 1**

REPORT 01 : NTP DRAFT REPORT TR299 MAR 85
FEEDING STUDIES OF C.I. DISPERSE BLUE 1 SHOW CLEAR EVIDENCE OF CARCINOGENICITY FOR MALE AND FEMALE F344/N RATS AS SHOWN BY INCREASED TRANSITIONAL CELL PAPILLOMAS AND CARCINOMAS, OF LEIOMYOMAS, LEIOMYOSARCOMAS AND OF SQUAMOUS CELL PAPILLOMAS OR CARCINOMAS OF THE URINARY BLADDER. URINARY BLADDER CALCULI SEEN IN RATS WITH INCREASED BLADDER NEOPLASMS. ALSO EQUIVOCAL EVIDENCE OF CARCINOGENICITY OF C.I. DISPERSE BLUE 1 IN MALE AND FEMALE B6C3F1 MICE SEEN.

REPORT 02 : NTP DRAFT REPORT TR299 MAR 85
C.I. DISPERSE BLUE 1 WAS MUTAGENIC IN TA1537 OF SALMONELLA TYPHIMURIUM AND IN TA97 AND TA98 IN PRESENCE OR ABSENCE OF 59, AND IN TA1535, IN THE PRESENCE OF 59; IT WAS NOT MUTAGENIC IN TA100.
REPT'D USE : HAIR DYE

CAS NO : **3179-90-6**
CHEM NAME : ***** C.I. DISPERSE BLUE 7**

REPORT 01 : EPA TSCA 8(E) NOTICE 8EHQ-1179-0321 NOV 79
TOMS RIVER CHEMICAL SUBMITTED SUMMARIZED FINAL RESULTS FROM A NUMBER OF IN-VITRO AND IN-VIVO MUTAGENICITY ASSAYS OF C.I. DISPERSE BLUE 7. SUBMITTER REPORTS THAT RESULTS FROM SEVERAL IN-VITRO TESTS INDICATE THAT THIS DYE CAN INDUCE POINT MUTATIONS. RESULTS FROM IN-VIVO TESTING INDICATE THAT IT DOES NOT PRODUCE CHROMOSOMAL ABERRATIONS.
REPT'D USE : TEXTILE DYE AND SURFACE DYEING OF THERMOPLASTICS

CAS NO : **3956-55-6**
CHEM NAME : ***** C.I. DISPERSE BLUE 79**

REPORT 01 : TSCA ITC 19TH REPORT TO EPA OCT 86
RECOMMENDED FOR TESTING WITHOUT BEING DESIGNATED FOR RESPONSE WITHIN 12 MONTHS. RECOMMENDED IT BE TESTED FOR CHEMICAL FATE, HEALTH EFFECTS AND ECOLOGICAL EFFECTS. ADDED TO THE 4(E) PRIORITY LIST, NOV 86.
REPT'D USE : TEXTILE DYE AND COMPONENT OF OTHER DYES

CAS NO : **3618-72-2**
CHEM NAME : ***** C.I. DISPERSE BLUE 79:1**

REPORT 01 : EPA FR 54 #223 NOTICE NOV 89
ITC TEST REFERRAL. SUBCHRONIC ORAL TOXICITY, METABOLISM, DEVELOPMENTAL TESTING INDICATE LOW TOXICITY FOR NON-CANCER END POINTS AND LOW CONCERN FOR ENVIRONMENTAL RISK.
REPT'D USE : DYEING POLYESTER AND TRIACETATE FIBERS

CAS NO : **UNKNOWN-10**
CHEM NAME : *****C.I. DISPERSE BLUE 291**

REPORT 01 : EPA TSCA 8 (E) NOTICE 8EHQ -0481-0396 APR 81
MARTIN MARIETTA REPORTED THAT C.I. DISPERSE BLUE 291 HAS BEEN FOUND POSITIVE IN THE SALMONELLA ASSAY. IN ADDITION, A DICHLOROMETHANE EXTRACT OF YARN DYED WITH C.I. DISPERSE BLUE 291 WAS MUTAGENIC WITH SALMONELLA.
REPT'D USE : DYE

CAS NO : **82-28-0**
CHEM NAME : ***** C.I. DISPERSE ORANGE 11**

REPORT 01 : NCI CARC TRS NO 111 DHEW PUB NO NIH 78-1366
UNDER THE CONDITIONS OF THIS BIOASSAY 1-AMINO-2-METHYLANTHRAQUINONE WAS CARCINOGENIC IN FISCHER 344 RATS, INDUCING HEPATOCELLULAR CARCINOMAS IN RATS OF BOTH SEXES, AND KIDNEY TUMORS IN MALES. THE CPD WAS CARCINOGENIC IN FEMALE B6C3F1 MICE, PRODUCING AN INCREASED COMBINED INCIDENCE OF HEPATOCELLULAR CARCINOMAS AND NEOPLASTIC NODULES. THERE WAS NO EVIDENCE OF CARCINOGENICITY IN MALE MICE.

REPORT 02 : IARC MONOGRAPH 27 1982
TECH GRADE 1-AMINO-2-METHYLANTHRAQUINONE WAS TESTED IN RATS AND MICE BY DIETARY ADMINISTRATION. THIS MATERIAL INCREASED THE INCIDENCE OF HEPATOCELLULAR CARCINOMAS IN RATS AND OF KIDNEY ADENOMAS AND ADENOCARCINOMAS IN MALE RATS. THE EXPERIMENT IN MICE WAS INADEQUATE FOR EVALUATION. NO EVALUATION OF THE CARCINOGENICITY OF THE PURE COMPOUND CAN BE MADE.

REPORT 03 : IARC MONOGRAPH 27 1982
NO DATA ON THE MUTAGENICITY OF THE COMPOUND WERE AVAILABLE. NO DATA CONCERNING EFFECTS ON REPRODUCTION OR PRENATAL TOXICITY WERE AVAILABLE.

REPORT 06 : MULTI-AGENCY CROSS CHECK
Y HHS MAGUIRE
Y OSHA CARC
N EPA CAG
N EPA RCRA
N EPA CLEAN AIR
N EPA CLEAN WATER

REPT'D USE : DYE INTERMEDIATE, TEXTILE DYE AND SURFACE DYEING OF THERMOPLASTICS. NO LISTING IN 1981 AATCC OR USITC.

CAS NO : **12226-87-8**
CHEM NAME : ***** C.I. DISPERSE RED 9**

REPORT 01 : CPSC ANTHRAQUINONE DYES MONOGRAPH OCT 81
C.I. DISPERSE RED 9 EXHIBITED MUTAGENIC ACTIVITY WHEN TESTED FOR SPECIFIC LOCUS FORWARD MUTATIONS IN THE MOUSE LYMPHOMA CELL ASSAY AND UNSCHEDULED DNA SYNTHESIS. DOMINANT LETHAL ASSAY RESULTS WERE INCONCLUSIVE. MAA WAS NOT MUTAGENIC IN SALMONELLA OR SACCHAROMYCES WITH OR WITHOUT ACTIVATION.

REPT'D USE : TEXTILE DYE

CAS NO : **116-85-8**
CHEM NAME : ***** C.I. DISPERSE RED 15**

REPORT 01 : CPSC ANTHRAQUINONE DYES MONOGRAPH OCT 81
C.I. DISPERSE RED 15 WAS MUTAGENIC IN SALMONELLA, AS WELL AS IN B. SUBTILIS.
REPT'D USE : TEXTILE DYE

CAS NO : **2832-40-8**
CHEM NAME : ***** C.I. DISPERSE YELLOW NO. 3**

REPORT 01 : NTP CARC TRS 222 NIH PUB NO 82-1778
UNDER THE CONDITIONS OF THIS BIOASSAY, DISPERSE YELLOW 3 ADMINISTERED IN THE FEED WAS CARCINOGENIC FOR MALE F344 RATS, BUT NOT FOR FEMALE F344 RATS. THE TEST CHEMICAL WAS CARCINOGENIC FOR FEMALE B6C3F1 MICE, AS EVIDENCED BY THE INCREASED INCIDENCE OF HEPATOCELLULAR ADENOMAS, BUT WAS NOT CARCINOGENIC FOR MALE B6C3F1 MICE.

REPORT 02 : IARC MONOGRAPH 8 1975
C.I. DISPERSE YELLOW 3 WAS TESTED IN MICE BY BLADDER IMPLANTATION IN CHOLESTEROL. THE RESULTS WERE OF BORDERLINE SIGNIFICANCE. AN EVALUATION OF THE CARCINOGENICITY OF THIS COMPOUND COULD NOT BE MADE IN THE ABSENCE OF OTHER INFORMATION.

CAS NO : **2353-45-9**
CHEM NAME : ***** C.I. FOOD GREEN 3 (FAST GREEN FCF)**

REPORT 01 : IARC MONOGRAPH 16 1978
FAST GREEN FCF HAS BEEN TESTED IN MICE AND RATS BY ORAL ADMINISTRATION AND IN RATS BY SC INJECTION. IT WAS CARCINOGENIC IN RATS, PRODUCING SARCOMAS AT THE SITE OF REPEATED SUBCUTANEOUS INJECTIONS NO ADEQUATE DATA ON THE EMBRYOTOXICITY, TERATOGENICITY OR MUTAGENICITY OF FAST GREEN FCF WERE AVAILABLE.
REPT'D USE : FOOD, COSMETIC COLOR, HISTOLOGICAL STAIN AND PHARMACEUTICAL COLOR

CAS NO : **4548-53-2**
CHEM NO : ***** C. I. FOOD RED 1 (PONCEAU SX)**

REPORT 01 : IARC MONOGRAPH 8 1975
PONCEAU SX WAS TESTED BY THE ORAL ROUTE IN MICE, RATS AND DOGS AND BY SC INJECTION IN MICE AND RATS. THE EXPERIMENTS DID NOT INDICATE A CARCINOGENIC EFFECT.

REPT'D USE : COLORING FOR MARASCHINO CHERRIES, EXTERNALLY APPLIED DRUGS AND COSMETICS

CAS NO : **3761-53-3**
CHEM NAME : ***** C.I. FOOD RED 5 (PONCEAU MX)**

REPORT 01 : IARC MONOGRAPH 8 1975
PONCEAU MX IS CARCINOGENIC, PRODUCING LIVER-CELL TUMORS IN MICE AND RATS AND POSSIBLY INTESTINAL TUMORS IN MICE, FOLLOWING ITS ADMINISTRATION BY THE ORAL ROUTE. IN A 1979 SUPPLEMENT, IARC DETERMINED THAT THERE WAS SUFFICIENT EVIDENCE FOR THE CARCINOGENICITY OF PONCEAU MX IN ANIMALS.

REPORT 02 : CPSC MONOGRAPH ON MONOAZO DYES OCT 81
THE EVIDENCE FOR THE CARCINOGENICITY OF FOOD RED 5 IS EQUIVOCAL. EARLY FEEDING AND BLADDER IMPLANTATION EXPERIMENTS IN RATS AND MICE WERE NEGATIVE. A LATER EXTENSIVELY CONTROLLED STUDY IN BOTH MICE AND RATS RESULTED IN A DOSE-RELATED INCIDENCE OF LIVER TUMORS, BUT THESE FINDINGS ARE DISPUTED. THE DYE WAS NONMUTAGENIC IN B. SUBTILIS AND S. TYPHIMURIUM ASSAYS.
REPT'D USE : INKS, TEXTILE/LEATHER DYE, WOOD STAINS AND FOOD COLORING

CAS NO : **3564-09-8**
CHEM NAME : ***** C.I. FOOD RED 6 (PONCEAU 3R)**

REPORT 01 : IARC MONOGRAPH 8 1975
PONCEAU 3R IS CARCINOGENIC IN RATS FOLLOWING ITS ORAL ADMINISTRATION, PRODUCING LIVER-CELL TUMORS. IT ALSO PRODUCED BLADDER TUMORS IN MICE FOLLOWING ITS IMPLANTATION IN THE URINARY BLADDER. IN A 1979 SUPPLEMENT, IARC DETERMINED THAT THERE IS SUFFICIENT EVIDENCE FOR CARCINOGENICITY OF PONCEAU 3R IN ANIMALS.
REPT'D USE : DYE. NO 1980 LISTING IN AATCC OR USITC OR TSCA INVENTORY.

CAS NO : **25956-17-6**
CHEM NAME : ***** C.I. FOOD RED 17 (FD&C RED 40)**

REPORT 01 : REPORT OF INTERAGENCY WORKING GROUP JUN 81
THE WORKING GROUP EVALUATED THE DESIGNN AND RESULTS OF TWO LIFETIME STUDIES IN MICE AND IN RATS INVOLVING FEEDING OF FD&C RED 40. N0 PATTERN OF STATISTICALLY SIGNIFICANT INCREASES IN THE INCIDENCE OF CANCER IN RELATION TO FD&C RED 40 WERE FOUND IN ANY OF THE THREE STUDIES WHEN ANALYZED AT THEIR COMPLETION OR SUITABLY ADJUSTED FOR REPEATED INTERIM TESTING. SOME FACTORS COMPLICATED BOTH

EXPERIMENTS IN MICE BUT ARE NOT DEEMED SUFFICIENT TO INVALIDATE THE DATA.

CAS NO : **2783-94-0**
CHEM NAME : ***** C.I. FOOD YELLOW 3 (SUNSET YELLOW FCF)**

REPORT 01 : NTP CARC TRS 208 NIH PUB NO. 81-1764
UNDER THE CONDITIONS OF THIS BIOASSAY THERE WAS NO CLEAR EVIDENCE OF THE CARCINOGENICITY OF C.I. FOOD YELLOW 1 IN F344 RATS OR B6C3F1 MICE OF EITHER SEX (FEED).

REPORT 02 : IARC MONOGRAPH 8 1975
IN THE ORAL EXPERIMENTS IN MICE THERE WAS NO EVIDENCE OF CARCINOGENICITY AS COMPARED WITH CONTROLS. TESTS IN RATS BY THE ORAL ROUTE SHOWED NEGATIVE RESULTS, BUT THE EXPERIMENTS WERE INADEQUATELY REPOTED. REPEATED SUBCUTANEOUS INJECTIONS IN RATS LED TO NEITHER LOCAL NOR DISTANT TUMORS.

REPORT 03 : IARC MONOGRAPH 8 1975
SUNSET YELLOW FCF OF UNDEFINED PURITY HAS BEEN SHOWN TO PRODUCE DOMINANT LETHALS IN LABORATORY RATS.
REPT'D USE : FOOD AND COSMETIC COLOR. NO LISTING IN 1980 USITC

CAS NO : **2538-85-4**
CHEM NAME : ***** C.I. MORDANT BLACK**

REPORT 01 : CPSC MONOGRAPH ON MONOAZO DYES AUG 81
C.I. MORDANT BLACK 7 WAS NEGATIVE IN A LIMITED AMES TEST. THE CHEMICAL WAS WEAKLY POSITIVE WHEN RIBOFLAVIN WAS ADDED.
REPT'D USE : NO AVAILABLE PRODUCTION OR USE DATA

CAS NO : **2243-76-7**
CHEM NAME : ***** C.I. MORDANT ORANGE 1**

REPORT 01 : CPSC MONOGRAPH ON MONOAZO DYES AUG 81
C.I. MORDANT ORANGE 1 WAS FOUND TO BE MUTAGENIC IN THE AMES TEST UNDER SEVERAL SETS OF CONDITIONS.

CAS NO : **1836-22-2**
CHEM NAME : ***** C.I. MORDANT RED 9**

REPORT 01 : CPSC MONOGRAPH ON MONAZO DYES AUG 81
SIGNIFICANTLY POSITIVE RESULTS WERE OBTAINED IN A MUTAGE-

NICITY ASSAY USING SEVERAL STRAINS OF SALMONELLA WITH AND WITHOUT REDUCTION BY DITHIONITE, BUT ONLY AFTER S-9 ACTIVATION.

CAS NO : **72-48-0**
CHEM NAME : ***** C.I. MORDANT RED 11**

REPORT 01 : CPSC ANTHRAQUINONE DYES MONOGRAPH OCT 81
C.I. MORDANT RED 11 BINDS TO DNA IN THE B. SUBTILIS REC ASSAY AND PRODUCES CHROMOSOMAL FRAGMENTATION IN ONION PLANTS. THE DYE PRODUCED MUTATIONS IN SALMONELLA WITH OR WITHOUT ACTIVATION.
REPT'D USE : PAINTS/VARNISHES AND AEROSOLS

CAS NO : **81-54-9**
CHEM NAME : ***** C.I. NATURAL RED 8 (PURPURIN)**

REPORT 01 : CPSC ANTHRAQUINONE DYES MONOGRAPH OCT 81
PURPURIN WAS MUTAGENIC IN SALMONELLA AND WAS TOXIC AT HIGHER CONCENTRATIONS. MUTAGENICITY OF THIS CHEMICAL DECREASED AFTER TREATMENT WITH LIVER MICROSOMAL FRACTION. PURPURIN BINDS TO, BUT DOES NOT DAMAGE, THE DNA OF B. SUBTILIS AND E. COLI.
REPT'D USE : NO AVAILABLE PRODUCTION OR USE DATA. NO LISTING ON TSCA INVENTORY

CAS NO : **1308-38-9**
CHEM NAME : ***** C.I. PIGMENT GREEN 17 (CHROMIC OXIDE)**

REPORT 01 : IARC MONOGRAPH 23 1980
EXPERIMENTS IN RATS EXPOSED TO CHROMIC (III) OXIDE BY ORAL ADMINISTRATION, BRONCHIAL IMPLANTATION, IP, INTRATRACHEAL, AND INTRAPLEURRAL INJECTION WERE INADEQUATE TO EVALUATE CARCINOGENICITY.
REPT'D USE : PIGMENT, ABRASIVES, REFRACTORY MATERIAL, PAINT/VARNISH, AND AEROSOL

CAS NO : **3468-63-1**
CHEM NAME : ***** C.I. PIGMENT ORANGE 5**

REPORT 01 : CPSC MONOGRAPH ON MONOAZO DYES AUG 81
C.I. PIGMENT ORANGE 5 WAS NEGATIVE IN DERMAL AND ORAL CARCINOGENICITY STUDIES IN DOGS AND MICE. FEEDING STUDIES EXAMINING REPRODUCTIVE AND TERATOGENIC EFFECTS WERE NEGATIVE IN RATS AND RABBITS. THERE WERE POSITIVE RESULTS IN THREE SEPARATE AMES TESTS FOR MUTAGENICITY.

REPORT 02 : FDA LISTING FOR EXTERNAL USE AUG 86
IN THE FR VOL. 51 P-28331, FDA STATES IT CONCLUDES D+C ORANGE NO. 17 INDUCES CANCER IN LABORATORY ANIMALS, HOWEVER THE CTRA PRESENTED RISK ASSESSMENT ARGUMENTS SHOWING IT CAN BE REGULATED FOR SAFE USE IN EXTERNALLY APPLIED DRUGS AND COSMETICS.

REPT'D USE : COSMETICS, DRUGS, AND PRINTING/GRAPHIC ARTS

CAS NO : **6410-10-2**

CHEM NAME : ***** C.I. PIGMENT RED 1**

REPORT 01 : CPSC MONOGRAPH ON MONOAZO DYES AUG 81
C.I. PIGMENT RED 1 WAS MUTAGENIC TO SOME STRAINS OF SALMONELLA, FOLLOWING METABOLIC ACTIVATION.

REPT'D USE : PAINTS/VARNISHES AND AEROSOLS

CAS NO : **5160-02-1**

CHEM NAME : ***** C.I. PIGMENT RED 53, BARIUM SALT**

REPORT 01 : NTP BOARD OF SCIENTIFIC COUNSELORS FEB 81
THE BIOASSAY REPORT CONCLUSIONS FOR D&C RED #9 WERE REVIEWED BY THE PEER REVIEW PANEL AND APPROVED WITH MINOR MODIFICTIONS. UNDER CONDITIONS OF THIS BIOASSAY, D&C RED #9 ADMINISTERED IN THE FEED WAS CARCINOGENIC IN MALE F344 RATS INDUCING SARCOMAS OF THE SPLEEN AND INCREASED INCIDENCE OF NEOPLASTIC NODULES IN THE LIVER. THE CHEMICAL WAS NOT CARCINOGENIC FOR FEMALE F344 RATS OR B6C3F1 MICE OF EITHER SEX.

REPORT 02 : IARC MONOGRAPH 8 1975
D AND C RED #9 WAS TESTED IN RATS BY THE ORAL ROUTE. NO EVIDENCE OF CARCINOGENICITY WAS OBSERVED IN THIS TEST.

REPORT 03 : CPSC MONOGRAPH ON MONOAZO DYES OCT 81
NO TERATOGENIC EFFECTS WERE NOTED IN A GAVAGE STUDY ON RATS AND RABBITS GIVEN THIS DYE DURING ORGANOGENESIS. METAGENICITY TESTS WERE NEGATIVE USING S. TYPHIMURIUM WITH AND WITHOUT 2-9 MIXTURE, BUT THERE WAS CONFUSION OVER THE IDENTITY OF THE COMPOUND TESTED.

REPT'D USE : PIGMENT, PRINTING INKS, PLASTICS/RUBBERS AND LIPSTICK

CAS NO : **6358-85-6**
CHEM NAME : ***** C.I. PIGMENT YELLOW 12 (DIARYLANILIDE YELLOW)**

REPORT 01 : NCI CARC TRS 30 DHEW PUB NO 78-830
THE RESULTS OF THIS STUDY DID NOT PROVIDE EVIDENCE FOR THE CARCINOGENICITY OF DIARYLANILIDE YELLOW IN FISHER 344 RATS OR B6C3F1 MICE IN A FEEDING STUDY.
REPT'D USE : PIGMENT, PAINTS/VARNISHES, PRINTING INKS, TEXTILES, RUBBERS, AEROSOLS, LINOLEUM/FLOOR TILES AND POLYMERIC RESINS

CAS NO : **10294-40-3**
CHEM NAME : ***** C.I. PIGMENT YELLOW 31 (BARIUM CHROMATE)**

REPORT 01 : IARC MONOGRAPH 23 1980
EXPERIMENTS IN RATS EXPOSED TO BARIUM CHROMATE (VI) BY IM AND INTRAPLUEURAL IMPLANTATION WERE INADEQUATE TO EVALUATE CARCINOGENICITY. CHROMIUM (VI) COMPOUNDS CAUSE MUTATIONS AND ALLIED EFFECTS IN A VERY WIDE RANGE OF PROKARYOTIC AND EUKARYOTIC SYSTEMS BOTH IN-VITRO AND IN-VIVO.
REPT'D USE : PIGMENT, PAINTS, CERAMIC OVERGLAZES, GLASS, METAL PRIMERS, AND SAFETY MATCHES

CAS NO : **7789-06-2**
CHEM NAME : ***** C.I. PIGMENT YELLOW 32 (STRONTIUM CHROMATE)**

REPORT 01 : IARC MONOGRAPH 23 1980
STRONTIUM CHROMATE (VI) PRODUCED SARCOMAS AT THE SITE OF IM INJECTION IN RATS. THERE IS SUFFICIENT EVIDENCE FOR THE CARCINOGENICITY OF STRONTIUM CHROMATE IN RATS.

REPORT 02 : IARC MONOGRAPH 23 1980
CHROMIUM (VI) COMPOUNDS CAUSE MUTATIONS AND ALLIED EFFECTS IN A VERY WIDE RANTGE OF PROKARYOTIC AND EUKARYOTIC SYSTEMS BOTH IN VITRO AND IN VIVO.

REPORT 03 : EPA TSCA 8(E) NOTICE 8EHQ-1179-0318
CIBA-GEIGY PROVIDED INTERIM RESULTS FROM AN IMPLANTATION STUDY OF STRONTIUM CHROMATE IN RATS WHICH INDICATE THAT LESIONS CONSISTENT WITH THE APPEARANCE OF SQUAMOUS CELL CARCINOMAS WERE OBSERVED GROSSLY IN THE LEFT LUNG OF 21/39 DEAD ANIMALS AFTER ONE YEAR. SUBMITER NOTES THAT THE DIAGNOSIS OF SQUAMOUS CELL CARCINOMAS HAS NOT YET BEEN CONFIRMED HISTOLOGICALLY.

REPORT 06 : MULTI-AGENCY CROSS CHECK
Y HHS MAGUIRE
N OSHA CARC
N EPA CAG
N EPA RCRA
N EPA CLEAN AIR
N EPA CLEAN WATER

REPT'D USE : ELECTROCHEMICAL PROCESSES, ARTISTS COLORS, WASH PRIMER, CORROSION INHIBITOR AND PYROTECHNICS

CAS NO : **13765-19-0**

CHEM NAME : ***** C.I. PIGMENT YELLOW 33 (CALCIUM CHROMATE)**

REPORT 01 : IARC MONOGRAPH 23 1980
CALCIUM CHROMATE (VI) IS CARCINOGENIC IN RATS WHEN GIVEN BY INHALATION, IMPLANTATION, OR INJECTION, PRODUCING TUMORS AT THE INJECTION SITES. SINTERED CALCIUM CHROMATE (VI) PRODUCED SARCOMAS AT SITES OF SC, IM, OR INTRAPLEURAL INJECTION IN RATS. THERE IS SUFFICIENT EVIDENCE FOR THE CARCINOGENICITY OF CALCIUM CHROMATE IN RATS.

REPORT 02 : IARC MONOGRAPH 23 1980
CHROMIUM (VI) COMPOUNDS CAUSE MUTATIONS AND ALLIED EFFECTS IN A VERY WIDE RANGE OF PROKARYOTIC AND EUKARYOTIC SYSTEMS BOTH IN-VITRO AND IN-VIVO.

REPORT 03 : NTP TECHNICAL BULLETIN 3/8 DEC 80, JUL 82
CALCIUM CHROMATE WAS POSITIVE IN THE SALMONELLA ASSAY, IN DROSOPHILA, AND IN THE L5178Y MOUSE LYMPHOMA ASSAY.

REPORT 06 : MULTI-AGENCY CROSS CHECK
Y HHS MAGUIRE
N OSHA CARC
N EPA CAG
Y EPA RCRA
N EPA CLEAN AIR
N EPA CLEAN WATER

REPT'D USE : CORROSION INHIBITOR, BATTERY DEPOLARIZER AND PIGMENTS

CAS NO. : **7758-97-6**

CHEM NAME : ***** C.I. PIGMENT YELLOW 34 (LEAD CHROMATE)**

REPORT 01 : NIOSH CURRENT INTELLIGENCE BULLETIN 4 JUN 75
A STUDY OF LEAD CHROMATE WORKERS CONDUCTED FOR THE DRY COLOR MANUFACTURERS ASSN CONCLUDES THAT THE APPARENT EXCESS OF RESPIRATORY CANCER IN THIS POPULATION IS CONSISTENT WITH THE HYPOTHESIS THAT LEAD CHROMATE IS A RESPIRATORY CARCINOGEN.

REPORT 02 : IARC MONOGRAPH 23 1980
LEAD CHROMATE (VI) PRODUCED SARCOMAS AT THE SITES OF SC, IN, AND INTRAPLEURAL ADMINISTRATION IN RATS. LEAD CHROMATE (VI) ALSO PRODUCED RENAL CARCINOMAS FOLLOWING IM INJECTION IN RATS. THERE IS SUFFICIENT EVIDENCE FOR THE CARCINOGENICITY OF LEAD CHROMATE (VI) IN RATS.

REPORT 03 : IARC MONOGRAPH 23 1980
CHROMIUM (VI) COMPOUNDS CAUSE MUTATIONS AND ALLIED EFFECTS IN A VERY WIDE RANGE OF PROKARYOTIC AND EUKARYOTIC SYSTEMS BOTH IN VITRO AND IN VIVO.

REPT'D USE : PIGMENTS, PAINTS, FABRIC PRINTING, CHINA PAINTS, PRINTING INKS, AND WATER COLORS

CAS NO : **13530-65-9**
CHEM NAME : ***** C.I. PIGMENT YELLOW 36 (ZINC CHROMATE)**

REPORT 01 : IARC MONOGRAPH 23 1980
ZINC CHROMATE (VI) PRODUCED SARCOMAS AT THE SITES OF SC, IM AND INTRAPLEURAL INJECTION IN RATS. THERE IS SUFFICIENT EVIDENCE FOR CARCINOGENICITY OF ZINC CHROMATE IN RATS.

REPORT 02 : IARC MONOGRAPH 23 1980
CHROMIUM (VI) COMPOUNDS CAUSE MUTATIONS AND ALLIED EFFECTS IN A VERY WIDE RANGE OF PROKARYOTIC AND EUKARYOTIC SYSTEMS BOTH IN VITRO AND IN VIVO.

REPORT 06 : MULTI-AGENCY CROSS CHECK
N HHS MAGUIRE
N OSHA CARC
N EPA CAG
N EPA RCRA
N EPA CLEAN AIR
N EPA CLEAN WATER

REPT'D USE : PAINTS/VARNISHES, CORROSION INHIBITOR, ARTISTS PAINTS, CONDITIONERS (WASH PRIMERS) APPLIED BEFORE PRIMING (INDUSTRIAL APPLICATIONS ONLY)

CAS NO : **81-64-1**
CHEM NAME : ***** C.I. 58050 (QUINIZARIN)**

REPORT 01 : CPSC ANTHRAQUINONE DYES MONOGRAPH OCT 81
QUINIZARIN IS MUTAGENIC IN SEVERAL STRAINS OF SALMONELLA.

REPT'D USE : SMOKE DYE, LUBRICANT, ANTIOXIDANT, MILDEWICIDE AND SYNTHESIS OF CHEMOTHERAPEUTIC AGENTS

CAS NO : **12225-25-1**
CHEM NAME : ***** C.I. REACTIVE BLACK 5**

REPORT 01 : CPSC MONOGRAPH ON FIBER REACTIVE DYES DEC 81
C.I. REACTIVE BLACK 5 WAS ADMINISTERED BY SC INJECTION AND IN THE DIET TO RATS AND MICE. THE CHEMICAL WAS AN EQUIVOCAL TUMORIGENIC AGENT IN RATS, BUT NOT IN MICE. DATA ON CONTROL ANIMALS WERE NOT REPORTED.

CAS NO : **2580-78-1**
CHEM NAME : ***** C.I. REACTIVE BLUE 19**

REPORT 01 : CPSC MONOGRAPH ON FIBER REACTIVE DYES DEC 81
C.I. REACTIVE BLUE 19 WAS ADMINISTERED BY SC INJECTION AND IN THE DIET TO RATS AND MICE. TUMORIGENIC RESPONSE WAS STATISTICALLY SIGNIFICANT ONLY FOR RATS FED THE DYE, COMPARED TO RATS INJECTED C.I. REACTIVE BLUE 19. THE DYE WAS NOT MUTAGENIC, WITH AND WITHOUT METABOLIC ACTIVATION, IN FIVE STRAINS OF SALMONELLA.

CAS NO : **12236-89-4**
CHEM NAME : ***** C.I. REACTIVE BLUE 36**

REPORT 01 : CPSC MONOGRAPH ON FIBER REACTIVE DYES DEC 81
C.I. REACTIVE BLUE 36 EXHIBITED MUTAGENIC ACTIVITY WITH AND WITHOUT METABOLIC ACTIVATIONS IN SALMONELLA STRAIN TA100, AND WAS WEAKLY MUTAGENIC IN TA98 WITHOUT METABOLIC ACTIVATION. A DOSE-RESPONSE RELATIONSHIP OBSERVED WITH METABOLIC ACTIVATION. THE TWO HIGHEST DOSES IN TA100 DEMONSTRATED REVERTANTS AT SLIGHTLY MORE THAN TWICE BACKGROUND.

CAS NO : **12225-61-5**
CHEM NAME : ***** C.I. REACTIVE BLUE 50**

REPORT 01 : CPSC MONOGRAPH ON FIBER REACTIVE DYES DEC 81
C.I. REACTIVE BLUE 50 WAS FOUND TO BE MUTAGENIC, WITH AND WITHOUT METABOLIC ACTIVATION, IN SALMONELLA STRAINS TA98, TA100, AND TA1537. A DOSE-RESPONSE RELATIONSHIP WAS OBSERVED WITH METABOLIC ACTIVATION.

CAS NO : **19526-81-9**
CHEM NAME : ***** C.I. REACTIVE RED 22**

REPORT 01 : CPSC MONOGRAPH ON FIBER REACTIVE DYES
FOUR OF 42 SURVIVING RATS INJECTED SC WITH C.I. REACTIVE RED 22 DEVELOPED LUNG AND LIVER TUMORS. AMONG RATS FED THE DYE, ONE DEVELOPED A BONE MARROW CARCINOMA, ONE A SUBCUTANEOUS TUMOR, AND ONE A LUNG TUMOR. RESULTS OF BOTH STUDIES ARE NOT SIGNIFICANT UNDER CONDITIONS OF THE TEST. DATA ON CONTROL ANIMALS WERE NOT REPORTED.

CAS NO : **12270-86-9**
CHEM NAME : ***** C.I. REACTIVE RED 78**

REPORT 01 : CPSC MONOGRAPH ON FIBER REACTIVE DYES DEC 81
C.I. REACTIVE RED 78 EXHIBITED MUTAGENIC ACTIVITY, WITH AND WITHOUT METABOLIC ACTIVATION, IN SALMONELLA STRAINS TA98, TA100 AND TA1537. A DOSE-RESPONSE RELATIONSHIP WAS OBSERVED WITH METABOLIC ACTIVATION. THE NUMBER OF REVERTANTS AT ANY DOSE WAS NOT TWICE BACKGROUND.

CAS NO : **61969-26-4**
CHEM NAME : ***** C.I. REACTIVE RED 83**

REPORT 01 : CPSC MONOGRAPH ON FIBER REACTIVE DYES DEC 81
C.I. REACTIVE RED 83 EXHIBITED MUTAGENIC ACTIVITY, WITH AND WITHOUT METABOLIC ACTIVATION, IN SALMONELLA STRAINS TA98, TA100 AND TA1535. A DOSE-RESPONSE RELATIONSHIP WAS OBSERVED WITH METABOLIC ACTIVATION. ONLY ONE VALUE AT THE MAXIMUM DOSE WAS TWICE BACKGROUND.

CAS NO : **61969-27-5**
CHEM NAME : ***** C.I. REACTIVE RED 84**

REPORT 01 : CPSC MONOGRAPH ON FIBER REACTIVE DYES DEC 81
C.I. REACTIVE RED 84 EXHIBITED MUTAGENIC ACTIVITY WITH AND WITHOUT METABOLIC ACTIVATION, IN SALMONELLA STRAINS TA98, TA100 AND TA1537. A DOSE-RESPONSE RELATIONSHIP WAS OBSERVED WITH METABOLIC ACTIVATION. ONLY ONE VALUE AT THE MAXIMUM DOSE WAS TWICE BACKGROUND.

CAS NO : **18976-74-4**
CHEM NAME : ***** C.I. REACTIVE YELLOW 14**

REPORT 01 : CPSC MONOGRAPH ON FIBER REACTIVE DYES DEC 81
C.I. REACTIVE YELLOW 14 WAS ADMINISTERED BY SC INJECTION AND IN THE DIET TO RATS AND MICE. THE CHEMICAL WAS AN EQUIVOCAL TUMORIGENIC AGENT BY BOTH ROUTES OF ADMINISTRATION IN BOTH SPECIES. DATA ON CONTROL ANIMALS WERE NOT REPORTED.

CAS NO : **12237-13-7**
CHEM NAME : ***** C.I. REACTIVE YELLOW 31**

REPORT 01 : CPSC MONOGRAPH ON FIBER REACTIVE DYES DEC 81
C.I. REACTIVE YELLOW 31 WAS INACTIVE OR AT MOST WEAKLY ACTIVE AS A MUTAGEN, WITH AND WITHOUT METABOLIC ACTIVATION IN SALMONELLA STRAINS TA98, TA100, AND TA1537.

CAS NO : **12226-61-8**
CHEM NAME : ***** C.I. REACTIVE YELLOW 39**

REPORT 01 : CPSC MONOGRAPH ON FIBER REACTIVE DYES DEC 81
C.I. REACTIVE YELLOW 39 EXHIBITED MUTAGENIC ACTIVITY WITH AND WITHOUT METABOLIC ACTIVATION IN SALMONELLA STRAIN TA100. A DOSE-RESPONSE RELATIONSHIP WAS OBSERVED WITH METABOLIC ACTIVATION.

CAS NO : **6416-57-5**
CHEM NAME : ***** C.I. SOLVENT BROWN 1 (SUDAN BROWN RR)**

REPORT 01 : IARC MONOGRAPH 8 1975
SUDAN BROWN RR WAS TESTED IN RATS BY THE ORAL ROUTE. IT PRODUCED NO EVIDENCE OF A CARCINOGENIC EFFECT, BUT THE ADEQUACY OF THE DOSE LEVEL USED WAS NOT KNOWN, AND NO EVALUATION COULD BE MADE.
REPT'D USE : DYE FOR ALCOHOLIC SOLVENTS, OILS, FATS, WAXES, PRINTING INKS AND SPIRIT LACQUERS. NO LISTING IN USITC OR AATCC

CAS NO : **2051-85-1**
CHEM NAME : ***** C.I. SOLVENT ORANGE 1**

REPT'D USE : TEXTILES AND LEATHER

CAS NO : **2646-17-5**
CHEM NAME : ***** C.I. SOLVENT ORANGE 2 (OIL ORANGE SS)**

REPORT 01 : IARC MONOGRAPH 8 1975
OIL ORANGE SS IS CARCINOGENIC IN MICE FOLLOWING ITS ORAL AND SC-INJECTION, PRODUCING INTESTINAL AND LOCAL TUMORS. IT ALSO PRODUCED CARCINOMAS OF THE BLADDER IN MICE FOLLOWING BLADDER IMPLANTATION. IN 1979 SUPPLEMENT, IARC STATES THAT THERE IS SUFFICIENT EVIDENCE FOR CARCINOGENICITY OF OIL ORANGE SS IN ANIMALS.

REPORT 02 : CPSC MONOGRAPH ON MONOAZO DYES OCT 81
C.I. SOLVENT ORANGE 2 WAS FOUND TO BE CARCINOGENIC IN TWO MOUSE SC INJECTION AND INGESTION STUDIES. THESE RESULTS WOULD NOT BE CONSIDERED STATISTICALLY SIGNIFICANT BY CURRENT STANDARDS. A BLADDER IMPLANTATION STUDY IN MICE SHOWED AN INCREASE IN CARCINOMAS IN THE TREATED ANIMALS, BUT THE CONTROLS ALSO HAD A HIGH INCIDENCE OF CARCINOMAS. THE DYE WAS NON-MUTAGENIC IN REC AND AMES ASSAYS.

REPORT 06 : MULTI-AGENCY CROSS CHECK
N HHS MAGUIRE
Y OSHA CARC
N EPA CAG
N EPA RCRA
N EPA CLEAN AIR
N EPA CLEAN WATER

REPT'D USE : DYE AND WAXES/VARNISHES

CAS NO : **3118-97-6**
CHEM NAME : ***** C.I. SOLVENT ORANGE 7 (SUDAN II)**

REPORT 01 : IARC MONOGRAPH 8 1975
SUDAN II WAS TESTED IN MICE AND RATS BY ORAL AND SUBCUTANEOUS ROUTES. RESULTS OF THESE STUDIES CANNOT BE EVALUATED BECAUSE OF VARIOUS INADEQUACIES. SUDAN II WAS ALSO TESTED IN MICE BY BLADDER IMPLANTATION, RESULTING IN A HIGH INCIDENCE OF BLADDER CARCINOMAS.

REPORT 02 : CPSC MONOGRAPH ON MONOAZO DYES OCT 81
C.I. SOLVENT ORANGE 7 IS TOXIC TO THE RED BLOOD SYSTEM AND CAN ACT AS A PRIMARY SKIN IRRITANT. IT WAS A WEAKLY POSITIVE MUTAGEN IN SOME STRAINS OF S. TYPHIMURIUM FOLLOWING MICROSOMAL ACTIVATION, BUT NOT IN OTHERS.

REPT'D USE : DYE FOR SOLVENTS, POLISHES, CANDLES AND POLYSTYRENE

CAS NO : **494-38-2**
CHEM NAME : ***** C.I. SOLVENT ORANGE 15 (ACRIDINE ORANGE)**

REPORT 01 : IARC MONOGRAPH 16 1978
ACRIDINE ORANGE HAS BEEN TESTED IN MICE BY SKIN APPLICATION, IN MICE AND RATS BY REPEATED SUBCUTANEOUS INJECTION AND IN MICE BY SINGLE INTRAPERITONEAL INJECTION FOLLOWED BY SKIN APPLICATION OF CROTON OIL. THE AVAILABLE DATA ARE INSUFFICIENT FOR AN EVALUATION OF THE CARCINOGENICITY OF THIS COMPOUND TO BE MADE.

REPORT 02 : IARC MONOGRAPH 16 1978
ACRIDINE ORANGE INDUCED REVERSE MUTATIONS IN SALMONELLA TA100, TA98 AND TA1537, WITH METABOLIC ACTIVATION, AND INDUCED MITOTIC GENE CONVERSION IN SACCHAROMYCES. NO DATA ON ITS EMBRYOTOXICITY OR TERATOGENICITY WERE AVAILABLE.
REPT'D USE : DYE. NO LISTING ON TSCA INVENTORY

CAS NO : **6368-72-5**
CHEM NAME : ***** C.I. SOLVENT RED 19 (SUDAN RED 7B)**

REPORT 01 : IARC MONOGRAPH 8 1975
SUDAN RED 7B WAS TESTED BY THE ORAL ROUTE IN RATS. ALTHOUGH THE AVAILABLE STUDIES DID NOT DEMONSTRATE AN INCREASE IN TUMOR INCIDENCE, THEY WERE CONSIDERED INADEQUATE FOR EVALUATION.

REPORT 02 : IARC MONOGRAPH 8 1975
4-HYDROXYAZOBENZENE HAS BEEN REPORTED TO BE NONMUTAGENIC WHEN TESTED IN DROSOPHILA FOR THE PRODUCTION OF X-LINKED RECESSIVE LETHALS.

REPORT 03 : EPA TSCA 8(E) NOTICE 8EHQ-0982-0455
CELANESE SUBMITTED COPIES OF FINAL REPORTS FROM SEVERAL IN-VITRO SCREENING ASSAYS OF SOLVENT RED 19. CPD WAS REPORTED TO BE ACTIVE IN SALMONELLAS AND MOUSE LYMPHOMA CELL TESTS WITH METABOLIC ACTIVATION. DYE WAS REPORTED TO BE NEGATIVE IN A CULTURED MAMMALIAN CELL TRANSFORMATION TEST.
REPT'D USE : COLORING AGENT FOR ALCOHOLIC ESTER AND HYDROCARBON SOLVENTS, VEGETABLE OILS, FATS, WAXES, PRINTING INKS, CANDLES AND CELLULOSE. NO LISTING IN 1981 USITC OR AATCC

CAS NO : **85-86-9**
CHEM NAME : ***** C.I. SOLVENT RED 23 (SUDAN III)**

REPORT 01 : IARC MONOGRAPH 8 1975
SUDAN III WAS ADMINISTERED TO MICE AND RATS BY THE ORAL ROUTE AND TO MICE BY THE SUBCUTANEOUS ROUTE. ALL EXPERIMENTS WERE INADEQUATE WITH REGARD TO EITHER THE DOSE ADMINISTERED, THE DEGREE OF REPORTING OR THE NUMBER OF ANIMALS USED.
REPT'D USE : DYE FOR SOAPS, SUNTAN OILS AND OIL-BASED DRUG SOLUTIONS, OINTMENTS, HAIR OILS AND POMANDER

CAS NO : **85-83-6**
CHEM NAME : ***** C.I. SOLVENT RED 24 (SCARLET RED)**

REPORT 01 : IARC MONOGRAPH 8 1975
SCARLET RED INDUCED LOCAL SARCOMAS IN RATS FOLLOWING ITS SUBCUTANEOUS INJECTION. HOWEVER NO DATA ARE REPORTED FOR CONTROLS GIVEN SOLVENT ONLY. AN EXPERIMENT IN MICE TREATED ORALLY GAVE NEGATIVE RESULTS, BUT THE ADEQUACY OF THE DOSE COULD NOT BE ASSESSED. OTHER EXPERIMENTS WERE CONSIDERED INADEQUATE FOR EVALUATION.
REPT'D USE : DYE, PAINTS/VARNISHES, SHOE/FLOOR POLISH, VETERINARY/MEDICINAL APPLICATIONS, CHEMICAL INDICATOR, SOAPS AND PLASTIC RESINS.

CAS NO : **6358-53-8**
CHEM NAME : ***** C.I. SOLVENT RED 80 (CITRUS RED NO. 2)**

REPORT 01 : IARC MONOGRAPH 8 1975
CITRUS RED NO. 2 IS CARCINOGENIC IN MICE AND RATS. FOLLOWING ITS ORAL ADMINISTRATION IT PRODUCED HYPERPLASIA AND TUMORS OF THE BLADDER. GIVEN SUBCUTANEOUSLY, IT PRODUCED ADENOCARCINOMAS OF THE LUNG AND LYMPHOSARCOMAS IN FEMALE MICE. IN A 1979 SUPPLEMENT, IARC DETERMINED THAT THERE IS SUFFICIENT EVIDENCE FOR THE CARCINOGENICITY OF CITRUS RED NO. 2 IN ANIMALS.

REPORT 06 : MULTI-AGENCY CROSS CHECK
N HHS MAGUIRE
N OSHA CARC
Y EPA CAG
N EPA RCRA
N EPA CLEAN AIR
N EPA CLEAN WATER
REPT'D USE : DYE FOR CITRUS FRUIT

CAS NO : **60-09-3**
CHEM NAME : ***** C.I. SOLVENT YELLOW 1 (P-AMINOAZOBENZENE)**

REPORT 01 : IARC MONOGRAPH 8 1975
AAB HAS BEEN REPORTED TO BE NON-MUTAGENIC WHEN TESTED IN DROSOPHILA FOR THE PRODUCTION OF X-LINKED RECESSIVE LETHALS.

REPORT 02 : IARC MONOGRAPH 8 1975
P-AMINOAZOBENZENE IS CARCINOGENIC IN RATS FOLLOWING ITS ORAL ADMINISTRATION, PRODUCING LIVER TUMORS, AND BY APPLICATION TO THE SKIN, PRODUCING EPIDERMAL TUMORS.

REPORT 03 : CPSC MONOGRAPH ON MONOAZO DYES OCT 81
THE RESULTS OF THREE MUTAGENICITY ASSAYS DEMONSTRATED SOLVENT YELLOW 1 TO BE MUTAGENIC FOLLOWING METABOLIC ACTIVATION.

REPORT 06 : MULTI-AGENCY CROSS CHECK
N HHS MAGUIRE
Y OSHA CARC
N EPA CAG
N EPA RCRA
N EPA CLEAN AIR
N EPA CLEAN WATER
REPT'D USE : DYE, DYE INTERMEDIATE, VARNISHES AND STYRENE RESINS

CAS NO : **60-11-7**
CHEM NAME : *****C.I. SOLVENT YELLOW 2 (P-DIMETHYLAMINO-AZOBENZENE)**

REPORT 03 : IARC MONOGRAPH 8 1975
PARA-DIMETHYLAMINOAZOBENZENE (DAB) IS CARCINOGENIC IN RATS, PRODUCING LIVER TUMORS AFTER ITS ADMINISTRATION BY SEVERAL ROUTES, AND IN DOGS, PRODUCING BLADDER TUMORS FOLLOWING ITS ADMINISTRATION BY THE ORAL ROUTE. IN A 1979 SUPPLEMENT, IARC DETERMINED THAT THERE WAS SUFFICIENT EVIDENCE FOR THE CARCINOGENICITY OF DAB IN ANIMALS.

REPORT 06 : MULTI-AGENCY CROSS CHECK
Y HHS MAGUIRE
R OSHA CARC
Y EPA CAG
Y EPA RCRA
N EPA CLEAN AIR
N EPA CLEAN WATER
REPT'D USE : DYE FOR WAXES, POLISHES, POLYSTYRENE, GASOLINE, RESEARCH CHEMICAL AND CHEMICAL INDICATOR. NO LISTING ON TSCA INVENTORY OR 1989 USITC

CAS NO : **97-56-3**
CHEM NAME : ***** C.I. SOLVENT YELLOW 3 (O-AMINOAZOTOLUENE)**

REPORT 01 : IARC MONOGRAPH 8 1975
CPD OF UNDEFINED PURITY WAS FOUND TO INDUCE MUTANTS RESISTANCE TO T1 BACTERIOPHAGE IN E. COLI, AND AUXOTROPHIC MUTANTS IN NEUROSPORA.

REPORT 02 : IARC MONOGRAPH 8 1975
ORTHO-AMINOAZOTOLUENE IS CARCINOGENIC IN MICE, RATS, HAMSTERS AND DOGS FOLLOWING ITS ORAL ADMINISTRATION, PRODUCING MAINLY TUMORS OF THE LIVER, GALL-BLADDER, LUNG AND URINARY BLADDER. IT ALSO PRODUCED A CARCINOGENIC EFFECT FOLLOWING ITS ADMINISTRATION BY OTHER ROUTES IN MICE AND RATS. IT IS EFFECTIVE IN SINGLE DOSES IN NEWBORN MICE. IN A 1979 SUPPLEMENT, IARC DETERMINED THAT THERE IS SUFFICIENT EVIDENCE FOR THE CARCINOGENICITY OF AMINOAZOTOLUENE IN ANIMALS.

REPORT 03 : CPSC MONOGRAPH ON MONOAZO DYES OCT 81
SOLVENT YELLOW 3 SHOWN TO BE MUTAGENIC IN S. TYPHIMURIUM, E. COLI, T1-BACTERIOPHAGE, AND NEUROSPORA TEST SYSTEMS, AND TO BE TERATOGENIC TO MICE. THE DYE ALSO PRODUCED ALLERGIC REACTIONS IN HUMANS.

REPORT 06 : MULTI-AGENCY CROS CHECK
Y HHS MAGUIRE
Y OSHA CARC
N EPA CAG
N EPA RCRA
N EPA CLEAN AIR
N EPA CLEAN WATER

REPT'D USE : DYE INTERMEDIATE, DYE FOR OILS, FATS, WAXES, AND PHARMACEUTICALS, LABORATORY EXPERIMENTAL USE ONLY.

CAS NO : **85-84-7**
CHEM NAME : ***** C.I. SOLVENT YELLOW 5 (YELLOW AB)**

REPORT 01 : IARC MONOGRAPH 8 1975
YELLOW AB WAS TESTED BY THE ORAL ROUTE IN RATS AND BY SUBCUTANEOUS INJECTION IN MICE AND RATS. THE SUBSTANCE WAS NOT FOUND TO BE CARCINOGENIC IN THESE STUDIES.

REPORT 02 : CPSC MONOGRAPH ON MONOZAO DYES OCT 81
SEVERE DAMAGE TO RED BLOOD CELLS, TESTES AND THE HEART WERE SEEN IN RATS EXPOSED TO YELLOW AB. THE CHEMICAL WAS NOT CARCINOGENIC, TERATOGENIC, OR NEUROTOXIC. A SMALL, BUT STATISTICALLY SIGNIFICANT INCREASE IN MUTATION

RATES WAS SEEN IN AMES TESTS USING LIVER HEMOGENATES FROM PCB-TREATED MICE.
REPT'D USE : LABORATORY CHEMICAL

CAS NO : **131-79-3**
CHEM NAME : ***** C.I. SOLVENT YELLOW 6 (YELLOW OB)**

REPORT 01 : IARC MONOGRAPH 8 1975
ALTHOUGH THE FEEDING EXPERIMENTS IN RATS AND THE SUBCUTANEOUS INJECTION STUDIES IN MICE WERE NEGATIVE, YELLOW OB PRODUCED LOCAL TUMORS IN RATS FOLLOWING ITS SUBCUTANEOUS ADMINISTRATION.
REPT'D USE : COSMETIC COLORANT. APPROVAL WITHDRAWN BY FDA IN 1960. NO LISTING ON TSCA INVENTORY

CAS NO : **1689-82-3**
CHEM NAME : ***** C.I. SOLVENT YELLOW 7 (4-HYDROXYAZOBENZENE)**

REPORT 01 : IARC MONOGRAPH 8 1974
4-HYDROXYAZOBENZENE WAS TESTED IN RATS BY THE ORAL ROUTE. AN EVALUATION OF ITS CARCINOGENICITY WAS NOT POSSIBLE DUE TO THE INADEQUATE DURATION OR INADEQUACY OF REPORTING OF THE STUDIES.

REPORT 02 : CPSC MONOGRAPH ON MONOAZO DYES OCT 81
C.I. SOLVENT YELLOW 7 WAS TESTED FOR CARCINOGENICITY AND MUTAGENICITY, BUT THE RESULTS WERE INCONSISTENT. THE DYE WAS MUTAGENIC TO FRUITFLIES, BUT NOT TO BACTERIA. THE CARCINOGENICITY STUDIES WERE INADEQUATE FOR EVALUATION.
REPT'D USE : COLOR FOR VARNISHES, WAXES, AND RESINS. NO LISTING IN 1980 USITC OR AATCC

CAS NO : **842-07-9**
CHEM NAME : ***** C.I. SOLVENT YELLOW 14 (SUDAN I)**

REPORT 01 : NTP BOARD OF SCIENTIFIC COUNSELORS FEB 81
THE BIOASSAY REPORT CONCLUSIONS FOR SUDAN I WERE REVIEWED BY THE PEER REVIEW PANEL AND APPROVED WITH MAJOR MODIFICATIONS. UNDER CONDITIONS OF THIS BIOASSAY, SUDAN I ADMINISTERED IN THE FEED RESULTED IN STATISTICALLY SIGNIFICANT INCREASED INCIDENCE OF ENOPLASTIC NODULES OF THE LIVER OF F344 RATS OF EITHER SEX BUT WAS NOT CARCINOGENIC FOR B6C3F1 MICE OF EITHER SEX.

REPORT 02 : IARC MONOGRAPH 8 1975
SUDAN I IS CARCINOGENIC IN MICE FOLLOWING SC ADMINISTRATION, PRODUCING TUMORS OF THE LIVER. IT ALSO PRODUCED BLADDER TUMORS IN MICE FOLLOWING ITS IMPLANTATION INTO THE URINARY BLADDER.

REPORT 03 : CPSC MONOGRAPH ON MONOAZO DYES OCT 81
SUDAN 1 WAS NOT MUTAGENIC IN THREE SEPARATE AMES TESTS. IT WAS SHOWN TO BE A STRONG ALLERGEN IN HUMANS.
REPT'D USE : DYE, SHOE/FLOOR POLISH, WAXES/VARNISHES, POLYMER RESINS, SOAPS AND FOOD COLOR

CAS NO : **81-77-6**
CHEM NAME : ***** C.I. VAT BLUE 4**

REPORT 01 : CPSC ANTHRAQUINONE DYES MONOGRAPH OCT 81
INDANTHRENE BLUE WAS ADMINISTERED IN THE FEED TO WISTAR RATS AT 1 PERCENT OF 10 G/KG FEED ONCE A DAY FOR 24 MONTHS. ALTHOUGH A VARIETY OF TUMORS WERE EXHIBITED, THE LEVEL OF INCIDENCE WAS NOT SIGNIFICANT OVER CONTROLS.
REPT'D USE : TEXTILE DYE

CAS NO : **128-60-9**
CHEM NAME : ***** C.I. VAT GREEN 9**

REPORT 01 : EPA TSCA 8(E) NOTICE 8EEHQ-479-0278 APR 79
TOMS RIVER CHEMICAL REPORTED THAT TWO PROPRIETARY DYE FORMULATIONS WERE FOUND TO INDUCE POINT MUTATIONS IN ONE OR MORE SALMONELLA STRAINS. A COMPOSITE SAMPLE OF THESE TWO PRODUCTS WAS ACTIVE IN THE MOUSE LYMPHOMA FORWARD MUTATION ASSAY, BUT WAS NEGATIVE IN A BHK CELL TRANSFORMATION TEST. C.I. VAT GREEN 9 REPRESENTS A MAJOR PORTION OF THE PROPRIETARY PRODUCTS.
REPT'D USE : TEXTILE DYE

CAS NO : **128-66-5**
CHEM NAME : ***** C.I. VAT YELLOW 4**

REPORT 02 : NCI CARC TRS 134 DHEW PUB NO 79-1389
UNDER THE CONDITIONS OF THIS BIOASSAY, THE FORMULATED PRODUCT CONTAINING C.I. VAT YELLOW 4 WAS NOT CARCINOGENIC FOR MALE OR FEMALE FISCHER 344 RATS OR FOR FEMALE B6C3F1 MICE, BUT WAS CARCINOGENIC FOR MALE B6C3F1 MICE, CAUSING AN INCREASED INCIDENCE OF LYMPHOMAS IN A FEEDING STUDY.
REPT'D USE : MILITARY SIGNAL AGENT, TEXTILE DYE AND PIGMENTS

CAS NO : **87-29-6**
CHEM NAME : ***** CINNAMYL ANTHRANILATE**

REPORT 01 : NCI CARC TRS 196 NIH PUB NO 80-1752
UNDER THE CONDITIONS OF THIS BIOASSAY, CINNAMYL ANTHRANILATE WAS CARCINOGENIC FOR MALE AND REMALE B6C3F1 MICE, INDUCING AN INCREASED OF HEPATOCELLULAR CARCINOMAS OR ADENOMAS. THE TEST CHEMICAL WAS ALSO CARCINOGENIC FOR MALE F344 RATS, INDUCING A LOW INCIDENCE OF ACINAR-CELL CARCINOMAS OR ADENOMAS OF THE PANCREAS AND ADENOCARCINOMAS OR ADENOMAS OF THE RENAL CORTEX. CINNAMYL ANTHRANILATE WAS NOT CARCINOGENIC FOR REMALE F344 RATS.

REPORT 02 : IARC MONOGRAH 16 1978
IN THE ONLY AVAILABLE STUDY, CINNAMYL ANTHRANILATE INCREASED THE INCIDENCE OF LUNG TUMORS IN MICE FOLLOWING INTRAPERITONEAL INJECTION. THESE DATA ARE INSUFFICIENT FOR EVALUATION OF THE CARCINOGENICITY OF THIS COMPOUND.

REPORT 03 : IARC MONOGRAPH 16 1978
NO DATA ON THE METABOLISM, EMBRYOTOXICITY, TERATOGENICITY OR MUTAGENICITY OF CINNAMYL ANTHRANILATE WERE AVAILABLE.

REPORT 04 : IARC MONOGRAPH 31 JUL 83
TESTED FOR CARCINOGENICITY IN MICE AND RATS IN DIET AND IN MICE BY INTRAPERITONEAL INJECTION. IN MICE, DOSE-RELATED INCREASED INCIDENCE OF HEPATOCELLULAR TUMORS FOUND FOLLOWING ORAL ADMINISTRATION AND AN INCREASED INCIDENCE OF LUNG TUMORS FOLLOWING INTRAPERITONEAL INJECTION. STUDY IN RATS COULD NOT BE EVALUATED.

REPORT 05 : IARC MONOGRAPH 31 JUL 83
CINNAMYL ANTHRANILATE WAS NOT MUTAGENIC TO SALMONELLA TYPHIMURIUM.
REPT'D USE : DETERGENTS, COSMETICS/PERFUMES, SYNTHETIC GRAPE/CHERRY FLAVORING AGENT AND FRAGRANCE IN SOAPS/DETERGENTS

CAS NO : **15663-27-1**
CHEM NAME : ***** CISPLATIN**

REPORT 01 : IARC MONOGRAPH 26 MAY 81
CISPLATIN WAS TESTED BY IP ADMINISTRATION IN MICE, INCREASING THE INCIDENCE OF LUNG TUMORS. WHEN CISPLATIN WAS ADMINISTERED IP, ALTERNATELY WITH CROTON OIL TO THE SKIN, PAPILLOMAS AND CARCINOMAS OF THE SKIN WERE PRODUCED ALONG WITH A SMALL NUMBER OF INTERNAL NEOPLASMS. THERE IS SUFFICIENT EVIDENCE THAT CISPLATIN IS CARCINOGENIC IN MICE.

REPORT 02 : IARC MONOGRAPH 26 MAY 81
HIGH DOSES OF CISPLATIN CAN INDUCE EMBRYOLETHALITY IN MICE. THE LIMITED DATA AVAILABLE DO NOT ALLOW AN EVALUATION OF THE TERATOGENICITY OF THIS DRUG. IT IS A MUTAGEN IN BACTERIAL AND MAMMALIAN CELLS. IT ALSO INDUCES CHROMOSOMAL ABERRATIONS IN VARIOUS CELLS IN CULTURE AND IN MICE IN-VIVO.

REPORT 06 : MULTI-AGENCY CROSS CHECK
Y HHS MAGUIRE
N OSHA CARC
N EPA CAG
N EPA RCRA
N EPA CLEAN AIR
N EPA CLEAN WATER
REPT'D USE : ANTINEOPLASTIC AGENT

CAS NO : **518-75-2**
CHEM NAME : ***** CITRININ**

REPORT 01 : IARC VOL 40 1986
LIMITED EVIDENCE OF CARCINOGENICITY TO TEST ANIMALS, BUT CARCINOGENICITY TO HUMANS COULD NOT BE EVALUATED. INADEQUATE EVIDENCE FOR GENETIC ACTIVITY IN SHORT-TERM TESTS. EMBROYOTOXICITY OCCURRED IN RODENTS AFTER INJECTION OF MATERNALLY TOXIC DOSES OF CPD. RESULTS OF ASSAYS IN B. SUBTILLIS WERE EQUIVOCAL. CPD DID NOT INDUCE RECOMBINATION IN S. CEREVISIAE NOR UNSCHEDULED DNA SYNTHESIS IN MAMMALIAN CELLS IN VITRO. CPD WAS NOT MUTAGENIC IN S. TYPHIMURIUM WITH OR WITHOUT AN EXOGENOUS METABOLIC SYSTEM. CPD INDUCED SCE'S BUT NO CHROMOSOMAL ABERRATIONS IN CHINESE HAMSTER V79 CELLS WITH METABOLIC ACTIVATION.
REPT'D USE : RESEARCH

CAS NO : **637-07-0**
CHEM NAME : ***** CLOFIBRATE**

REPORT 01 : IARC MONOGRAPH 24 1980
CLOFIBRATE HAS BEEN TESTED IN TWO STUDIES BY ORAL ADMINISTRATION TO MALE RATS; IT PRODUCED HEPATOCELLULAR CARCINOMAS. IN ADDITION, AN INCREASED INCIDENCE OF TUMORS AT SITES OTHER THAN THE LIVER WAS OBSERVED IN TREATED RATS COMPARED WITH CONTROLS.

REPORT 02 : IARC MONOGRAPH 24 1980
CLIFIBRATE WAS NOT MUTAGENIC TO SALMONELLA STRAINS TA1535, TA1537, TA1538, TA98, OR TA100, WITH OR WITHOUT

METABOLIC ACTIVATION. THERE IS NO EVIDENCE THAT IT IS TERATOGENIC IN RATS AND RABBITS; HOWEVER, SOME EVIDENCE OF PERINATAL TOXICITY IS PROVIDED BY THE FINDING OF ABNORMAL POSTNATAL THROMBOSIS IN RATS.

REPT'D USE : PHARMACEUTICAL AND INHIBITOR FOR CHOLESTEROL SYNTHESIS

CAS NO : **50-41-9**

CHEM NAME : ***** CLOMIPHENE CITRATE**

REPORT 01 : IARC MONOGRAPH 21 1979
CLOMIPHENE CITRATE WAS INADEQUATELY TESTED IN ONE EXPERIMENT IN NEW-BORN RATS BY SUBCUTANEOUS INJECTION. UTERINE AND OVARIAN TUMORS WERE REPORTED. THE AVAILABLE EXPERIEMNTAL AND HUMAN DATA ARE NOT SUFFICIENT TO EVALUATE THE CARCINOGENICTIY OF CLOMIPHENE CITRATE.

REPORT 03 : IARC MONOGRAPH 21 1979
NO DATA ON THE MUTAGENICITY OF CLOMIPHENE AND CLOMIPHENE CITRATE WERE AVAILABLE. CLOMIPHENE CITRATE IS EMBRYOLETHAL FOR PRE- AND POSTIMPLANTATION EMBRYOS IN SEVERAL SPECIES AND HAS VARIOUS TERATOGENIC EFFECTS IN RATS.

REPT'D USE : PHARMACEUTICAL OVULATION STIMULANT

CAS NO : **1420-04-8**

CHEM NAME : ***** CLONITRALID**

REPORT 01 : NCI CARC TRS 91 DHEW PUB NO 78-1341
UNDER THE CONDITIONS OF THIS BIOASSAY, THERE WAS NO CONVINCING EVIDENCE THAT CLONITRALID WAS CARCINOGENIC TO OSBORNE-MENDEL RATS OR TO FEMALE B6C3F1 MICE. POOR SURVIVAL OF MALE MICE DID NOT PERMIT AN EVALUATION OF CARCINOGENICITY IN THESE ANIMALS (FEEDING STUDY).

REPT'D USE : MOLLUSCIDE LAMPRICIDE

CAS NO : **8007-45-2**

CHEM NAME : ***** COAL-TARS AND DERIVED PRODUCTS**

REPORT 01 : NIOSH CRITERIA DOCUMENT SEP 77
NIOSH RECOMMENDS A STANDARD OF 0.1 MG/CU M (TWA) BASED ON LUNG AND SKIN CANCER. THERE IS NO CURRENT OSHA STANDARD FOR COAL-TAR PRODUCTS.

REPORT 02 : IARC MONOGRAPH 35 JAN 85
THERE IS SUFFICIENT EVIDENCE FOR CARCINOGENICITY IN EXPERIMENTAL ANIMALS OF COAL-TARS, CREOSOTES, CREOSOTE OILS, ANTHRACENE OILS AND COAL-TAR PITCHES. ALSO THE DATA

INDICATE COAL-TARS AND COAL-TAR PITCHES CAUSALLY ASSOCIATED WITH CANCER IN HUMANS AND CREOSOTES DERIVED FROM COAL-TARS ARE PROBABLY CARCINOGENIC IN HUMANS.

REPORT 03 : IARC MONOGRAPH 35 JAN 85
SAMPLES OF THERAPEUTIC COAL-TAR AND EXTRACTS OF COAL-TAR SHAMPOOS MUTAGENIC IN S. TYPHIMURIUM IN PRESENCE OF EXOGENOUS METABOLIC SYSTEM. EXTRACTS OF URINE FROM PATIENTS UNDERGOING COMBINED TREATMENT WITH COAL-TAR PREPARATIONS AND ULTRAVIOLET LIGHT MUTAGENIC IN S. TYPHIMURIUM. EXTRACTS OF COAL-TAR PITCH MUTAGENIC IN S.TYPHIMURIUM IN PRESENCE OF EXOGENOUS METABOLIC SYSTEM.

REPORT 04 : IARC MONOGRAPH 35 JAN 85
EXTRACTS OF ROOFING-TAR EMISSIONS MUTAGENIC IN S.TYPHIMURIUM IN PRESENCE OF EXOGENOUS METABOLIC SYSTEM AND MUTAGENIC IN TWO MUTATION ASSAYS AND INDUCED SISTER CHROMATID EXCHANGES IN CULTURED MAMMALIAN CELLS, BOTH IN THE PRESENCE AND ABSENCE EXOGENOUS METABOLIC SYSTEM. VIRAL TRANSFORMATION ENHANCED IN SYRIAN HAMSTER EMBRYO CELLS. CREOSOTE AND COAL-TAR-CREOSOTE MIXTURE MUTAGENIC IN S. TYPHIMURIUM AND POSITIVE IN MOUSE LYMPHOMA L5178Y SYSTEM AND URINE FROM RATS ADMINISTERED CREOSOTE MUTAGENICIN S. TYPHIMURIUM WITH ACTIVATION.

REPORT 05 : IARC SUPPLEMENT SEP 79
SOOTS, COAL-TARS, CREOSOTE OILS, SHALE OILS, AND CUTTING OILS ARE CARCINOGENIC IN EXPERIMENTAL ANIMALS AFTER SKIN PAINTING OR SC INJECTIONS. OCCUPATIONAL EXPOSURE TO THESE CHEMICALS CAUSES CANCER AT SEVERAL SITES, INCLUDING SKIN, LUNG, BLADDER AND GASTROINTESTINAL TRACT. EFFECT MAY BE DUE TO POLYCYCLIC AROMATIC HYDROCARBONS IN THESE MATERIALS.

REPORT 06 : MULTI-AGENCY CROSS CHECK
Y HHS MAGUIRE
N OSHA CARC
N EPA CAG
N EPA RCRA
N EPA CLEAN AIR
N EPA CLEAN WATER

REPT'D USE : MEDICINAL ALCOHOL, SURFACE-COATINGS, REFINING, OILS, FUELS, CREOSOTE, ROOFING AND BINDER

CAS NO : **11114-92-4**
CHEM NAME : ***** COBALT-CHROMIUM ALLOY**

REPORT 01 : IARC MONOGRAPH 23 1980
COBALT-CHROMIUM ALLOY PRODUCED SARCOMAS AT THE SITE OF INTRAMUSCULAR INJECTION IN RATS. THERE IS LIMITED EVIDENCE FOR THE CARCINOGENICITY OF COBALT-CHROMIUM ALLOY IN RATS.
REPT'D USE : INDUSTRIAL MACHINERY AND EQUIPMENT

CAS NO : **61789-51-3**
CHEM NAME : ***** COBALT NAPHTHENATE**

REPORT 01 : 49FR21411 MAY 84
EPA'S RESPONSE TO ITC'S RECOMMENDATION THAT EPA CONSIDER HEALTH AND ENVIRONMENTAL EFFECTS AND CHEMICAL FATE OF CALCIUM NAPHTHENATE UNDER TSCA 4(A). EPA IS NOT INITIATING RULEMAKING BECAUSE AGENCY HAS DETERMINED THAT DATA BEING DEVELOPED IN CURRENT TESTS SHOULD PROVIDE ADEQUATE INFORMATION TO PREDICT OR DETERMINE HEALTH RISKS. EPA ALSO BELIEVES THERE IS INSUFFICIENT BASIS TO FIND THAT IT PRESENTS AN UNREASONABLE RISK TO THE ENVIRONMENT AND DOES NOT PLAN TO SEEK FURTHER TESTING.
REPT'D USE : MARINE DIESEL OILS, PAINT AND INK DRIERS, FATTY-ACID LUBRICANTS FOR EXTREME PRESSURE SERVICE, ANTI-CORROSION AGENT AND CATALYST

CAS NO : **UNKNOWN-15**
CHEM NAME : ***** COFFEE**

REPORT 01 : NAS REPORT ON DIET, NUTRITION AND CANCER 1982
SPRAGUE-DAWLEY RATS WERE FED A DIET CONTAINING 5% INSTANT COFFEE FOR TWO YEARS. NO BLADDER TUMORS WERE NOTED IN RATS FED DIETS CONTAINING THE EQUIVALENT OF UP TO 85 CUPS OF COFFEE PER DAY. MAXIMUM TOLERATED DOSES OF REGULAR AND DECAFFEINATED INSTANT COFFEE (6% OF DIET) FED TO S-D RATS FOR TWO YEARS PRODUCED NO EVIDENCE OF CARCINOGENESIS.

CAS NO : **CLASS-07**
CHEM NAME : ***** COKE OVEN EMISSIONS (POLYCYCLIC ORGANIC MATTER)**

REPORT 01 : NIOSH CRITERIA DOCUMENT FEB 73
NIOSH RECOMMENDS WORK PRACTICES TO MINIMIZE EXPOSURE TO

EMISSIONS BASED ON LUNG CANCER. THE PRESENT OSHA STANDARD IS 150 MICROGRAMS/CU M (TWA) - SEE CFR 1910.1029.

REPORT 06 : MULTI-AGENCY CROSS CHECK
Y HHS MAGUIRE
R OSHA CARC
Y EPA CAG
N EPA RCRA
N EPA CLEAN AIR
N EPA CLEAN WATER

CAS NO : **64-86-8**
CHEM NAME : ***** COLCHICINE**

REPORT 01 : CPSC SCHOOL LAB CHEMICAL SURVEY JAN 82
THE CHEMICAL IS REPORTED IN ONE MAJOR TEXT TO BE AN EXPERIMENTAL CARCINOGEN. SUBSEQUENT CPSC STAFF REVIEW SHOWS THAT THERE IS NO EVIDENCE FOR THE CARCINOGENICITY OF COLCHICINE IN EXPERIMENTAL ANIMALS OR HUMANS. HOWEVER, NONE OF THE PUBLISHED STUDIES HAVE BEEN SUFFICIENTLY LONG FOR SUBJECTS TO BE AT RISK FROM LATE-DEVELOPING TUMORS. COLCHICINE HAS BEEN REPORTED TO CAUSE GENTOXIC EFFECTS SUCH AS MUTATIONS, IN HUMAN, ANIMAL AND BACTERIAL CELLS.
REPT'D USE : GOUT SUPPRESSANT AND REAGENT IN GENETIC RESEARCH

CAS NO : **CLASS-10**
CHEM NAME : ***** CONJUGATED ESTROGENS**

REPORT 01 : IARC MONOGRAPH 21 1979
CONJUGATED ESTROGENS WERE TESTED IN ONE EXPERIMENT IN RATS BY ORAL ADMINISTRATION. THE DATA WERE INSUFFICIENT TO EVALUATE THE CARCINOGENICITY OF THIS COMPOUND. STUDIES IN HUMANS STRONGLY SUGGEST THAT THE ADMINISTRATION OF CONJUGATED ESTROGENS IS CAUSALLY RELATED TO AN INCREASED INCIDENCE OF ENDOMETRIAL CARCINOMA.

REPORT 02 : IARC MONOGRAPH 21 1979
NO DATA ON THE TOXICITY, EMBRYOTOXICITY, OR MUTAGENICITY OF CONJUGATED ESTROGENS WERE AVAILABLE.

REPORT 06 : MULTI-AGENCY CROSS CHECK NOV 86
Y HHS MAGUIRE
N OSHA CARC
N EPA CAG
N EPA RCRA
N EPA CLEAN AIR
N EPA CLEAN WATER

CAS NO : **10380-28-6**
CHEM NAME : ***** COPPER 8-HYDROXYQUINOLINE**

REPORT 01 : IARC MONOGRAPH 15
COPPER 8-HYDROXYQUINOLINE HAS BEEN TESTED IN TWO STRAINS OF MICE BY ORAL AND BY SINGLE SUBCUTANEOUS ADMINISTRATION. ALTHOUGH A SIGNIFICANTLY INCREASED INCIDENCE OF RETICULUM-CELL SARCOMAS WAS OBSERVED ONLY IN MALES OR ONE STRAIN OF MICE FOLLOWING SINGLE SUBCUTANEOUS INJECTION, NO EVALUATION OF THE CARCINOGENICITY OF THIS COMPOUND CAN BE MADE ON THE BASIS OF THE AVAILABLE DATA.
REPT'D USE : WOOD, PAPER PRESERVATION, FUNGICIDE IN PAINTS AND TREATMENTS OF TEXTILES

CAS NO : **191-07-1**
CHEM NAME : ***** CORONENE**

REPORT 01 : IARC MONOGRAPH 32 DEC 83
CORONENE TESTED FOR CARCINOGENICITY IN ONE EXPERIMENT IN MICE BY SKIN APPLICATION. NO SIGNIFICANT INCREASE IN INCIDENCE OF SKIN TUMORS OBSERVED. TESTED IN MOUSE-SKIN INITIATION-PROMOTION ASSAY AND WAS ACTIVE AS INITIATOR. MUTAGENIC TO SALMONELLA TYPHIMURIUM IN PRESENCE OF EXOGENOUS METABOLIC SYSTEM.
REPT'D USE : COMBUSTION BY-PRODUCT

CAS NO : **56-72-4**
CHEM NAME : ***** COUMAPHOS**

REPORT 01 : NCI CARC TRS 96 DHEW PUB NO 79-1346
UNDER THE CONDITIONS OF THIS BIOASSAY COUMAPHOS WAS NOT CARCINOGENIC FOR EITHER F344 RATS OR B6C3F1 MICE (FEEDING STUDY).
REPT'D USE : PESTICIDE AND VETERINARY ECTOPARASITICIDE

CAS NO : **91-64-5**
CHEM NAME : ***** COUMARIN**

REPORT 01 : NTP TECHNICAL BULLETIN 3/8 DEC 80, JUL 82
COUMARIN WAS POSITIVE IN THE SALMONELLA ASSAY, BUT NEGATIVE FOR SEX-LINKED RECESSIVE LETHALS IN DROSOPHILA.

REPORT 02 : IARC MONOGRAPH 10 OCT 75
COUMARIN IS CARCINOGENIC IN RATS FOLLOWING ITS ORAL

ADMINISTRATION, THE ONLY SPECIES AND ROUTE OF ADMINISTRATION TESTED; IT PRODUCED BILE-DUCT CARCINOMAS.

REPORT 03 : NTP CARC TEST PROGRAM JUL 82
IN A CURRENT NTP BIOASSAY, COUMARIN IS BEING ADMINISTERED BY GAVAGE TO F344 RATS AND B6C3F1 MICE. THE PRECHRONIC TEST PHASE IS IN PROGRESS.

REPORT 04 : NAS REPORT ON DIET, NUTRITION AND CANCER 1982
COUMARIN HAS INDUCED BILE DUCT CARCINOMAS IN RATS FED 0.35% TO 0.5% OF THE CHEMICAL IN THE DIET FOR 18 MONTHS. COUMARIN WAS NEGATIVE IN THE E. COLI POL A ASSAY.
REPT'D USE : DEODORIZING AGENTS, FRAGRANCE ENHANCING AGENT FOR SOAPS, DETERGENTS AND PAPER PRODUCTS

CAS NO : **8001-58-9**
CHEM NAME : ***** CREOSOTE**

REPORT 01 : EPA RCRA HEALTH AND ENV. EFFECTS PROFILE OCT 80
THE EPA CARCINOGEN ASSESSMENT GROUP HAS EVALUATED CREOSOTE AND HAS FOUND SUFFICIENT EVIDENCE TO INDICATE THAT THIS COMPOUND IS CARCINOGENIC. CREOSOTE HAS BEEN ASSOCIATED WITH SEVERAL OCCUPATIONAL CASES OF SKIN CANCER OVER A 50-YEAR PERIOD.

REPORT 02 : IARC SUPPLEMENT 1 SEP 79
SOOTS, COAL-TARS, CREOSOTE OILS, SHALE OILS, AND CUTTING OILS ARE CARCINOGENIC IN EXPERIMENTAL ANIMALS AFTER SKIN PAINTING OR SC INJECTION. OCCUPATIONAL EXPOSURE TO THESE CHEMICALS CAUSES CANCER AT SEVERAL SITES, INCLUDING SKIN, LUNG, BLADDER AND GASTROINTESTINAL TRACT. EFFECT MAY BE DUE TO POLYCYCLIC AROMATIC HYDROCARBONS IN THESE MATERIALS.

REPORT 06 : MULTI-AGENCY CROSS CHECK
Y HHS MAGUIRE
N OSHA CARC
Y EPA CAG
Y EPA RCRA
N EPA CLEAN AIR
Y EPA CLEAN WATER
REPT'D USE : WOOD PRESERVATIVE AND HERBICIDE

CAS NO : **102-50-1**
CHEM NAME : ***** M-CRESIDINE**

REPORT 01 : NCI CARC TRS NO 105 DHEW PUB NO NIH 78-1355
UNDER THE CONDITIONS OF THIS BIOASSAY, M-CRESIDINE WAS

CARCINOGENIC TO FISCHER 344 RATS, CAUSING PAPILLARY TRANSITIONAL CELL CARCINOMAS OF THE URINARY BLADDER IN BOTH SEXES. NO CONVINCING EVIDENCE WAS PROVIDED FOR CARCINOGENICITY IN FEMALE B6C3F1 MICE. POOR SURVIVAL OF MALE B6C3F1 MICE RECEIVING M-CRESIDINE PRECLUDED EVALUATION OF POSSIBLE CARCINOGENICITY OF THE COMPOUND IN THESE ANIMALS. (THE COMPOUND WAS ADMINISTERED BY STOMACH INTUBATION).

REPORT 02 : IARC MONOGRAPH 27 APR 82
IN A DIETARY ADMINISTRATION EXPERIMENT IN RATS, TECHNICAL GRADE META-CRESIDINE PRODUCED AN INCREASED INCIDENCE OF TRANSITIONAL-CELL CARCINOMAS OF THE URINARY BLADDER IN MALES. THE RESULTS OF A STUDY IN MICE WERE INCONCLUSIVE. THE AVAILABLE DATA ARE INADEQUATE TO EVALUATE CARCINOGENICITY OF M-CRESIDINE.

REPORT 03 : IARC MONOGRAPH 27 APR 82
M-CRESIDINE WAS MUTAGENIC IN NEITHER E COLI NOR SALMONELLA STRAINS TA1537, TA1538, TA98 OR TA100 WITH OR WITHOUT METABOLIC ACTIVATION. NO DATA WERE AVAILABLE CONCERNING EFFECTS ON REPRODUCTION OR PRENATAL TOXICITY.

REPT'D USE : DYE INTERMEDIATE

CAS NO : **120-71-8**
CHEM NAME : ***** P-CRESIDINE**

REPORT 01 : NCI CARC TRS 142 DHEW PUB NO 79-1397
UNDER THE CONDITIONS OF THIS BIOASSAY, P-CRESIDINE WAS CARCINOGENIC TO FISCHER 344 RATS CAUSING INCREASED INCIDENCES OF CARCINOMAS AND OF PAPILLOMAS OF THE URINARY BLADDER IN BOTH SEXES, INCREASED INCIDENCES OF OLFACTORY NEUROBLASTOMAS IN BOTH SEXES, AND OF LIVER TUMORS IN MALES. P-CRESIDINE WAS ALSO CARCINOGENIC IN B6C3F1 MICE, CAUSING CARCINOMAS OF THE URINARY BLADDERS IN BOTH SEXES AND HEPATOCELLULAR CARCINOMAS IN FEMALES (FEEDING STUDY).

REPORT 02 : IARC MONOGRAPH 27 1982
P-CRESIDINE GIVEN BY DIETARY ADMINISTRATION, PRODUCED MALIGNANT TUMORS OF THE URINARY BLADDER IN BOTH MICE AND RATS, OLFACTORY NFUROBLASTOMAS IN RATS. IT ALSO PRODUCED NASAL CAVITY TUMORS IN MALE MICE AND LIVER-CELL TUMORS IN FEMALE MICE. CAVITY TUMORS IN MALE MICE AND LIVER-CELL TUMORS IN FEMALE MICE. THERE IS SUFFICIENT EVIDENCE FOR THE CARCINOGENICITY OF P-CRESIDENE IN ANIMALS.

REPORT 03 : IARC MONOGRAPH 27 1982
P-CRESIDINE INDUCED REVERSE MUTATIONS IN SALMONELLA STRAINS TA1538, TA98, AND TA100 IN THE PRESENCE OF ACTIVATED RAT LIVER MICROSOMES. NO DATA WERE AVAILABLE CONCERNING EFFECTS ON REPRODUCTION OR PRENATAL TOXICITY.

REPORT 06 : MULTI-AGENCY CROSS CHECK
Y HHS MAGUIRE
Y OSHA CARC
N EPA CAG
N EPA RCRA
N EPA CLEAN AIR
N EPA CLEAN WATER
REPT'D USE : DYE INTERMEDIATE

CAS NO : **108-39-4**
CHEM NAME : ***** M-CRESOL**

REPORT 01 : EPA PROPOSES TE ST RULE, 48FR52 2 JUL 83
UNDER TSCA 4 EPA PROPOSES THAT MFRS AND PROCESSORS OF M-CRESOL TEST FOR CARCINOGENICITY, MUTAGENICITY, TERATOGENICITY, REPRODUCTIVE EFFECTS, NEUROTOXICITY, SKIN SENSITIZATION AND SUBCHRONIC TOXICITY. THIS IS BASED ON LARGE VOLUME OF M-CRESOL PRODUCED IN US.
REPT'D USE : WIRE ENAMEL SOLVENTS, INTERMEDIATE IN MFR OF PHENOLIC RESINS, PHOSPHATE ESTERS AND HERBICIDES, CLEANING CPDS DISINFECTANTS, AND ORE FLOTATION

CAS NO : **1319-77-3**
CHEM NAME : ***** MIXED CRESOLS**

REPORT 01 : NIOSH CRITERIA DOCUMENT FEB 78
NIOSH RECOMMENDS A STANDARD OF 10 MG/CU M (TWA) BASED ON SKIN, LIVER, KIDNEY, AND PANCREAS EFFECTS. THE PRESENT OSHA STANDARD IS 22 MG/CU M TWA.

REPORT 02 : EPA RCRA HEALTH AND ENV. EFFECTS PROFILE OCT 80
NO CARCINOGENICITY TESTS CONDUCTED WITH CRESOLS HAVE BEEN FOUND. O-, M-, AND P-CRESOLS HAVE BEEN REPORTED TO PROMOTE THE CARCINOGENICITY OF DIMETHYLBENZANTHRACENE IN SKIN TESTS WITH MICE.

REPORT 03 : ITC FIRST REPORT TO EPA (42FR55026) OCT 77
THE INTERAGENCY TESTING COMMITTEE RECOMMENDS THAT THE EPA ADMINISTRATOR CONSIDER THE DEVELOPMENT OF TEST RULES TO TEST CRESOLS FOR CARCINOGENICITY, MUTAGENICITY, TERATOGENICITY AND OTHER CHRONIC EFFECTS.

REPORT 06 : MULTI-AGENCY CROSS CHECK
N HHS MAGUIRE
N OSHA CARC
N EPA CAG
N EPA RCRA
Y EPA CLEAN AIR
Y EPA CLEAN WATER

REPORT 07 : EPA RESPONSE TO ITC (84FR31812) JUL 83
EPA IS PROPOSING THAT MANUFACTURERS AND PROCESSORS OF CRESOLS TEST THESE CHEMICALS FOR CARCINOGENICITY, MUTAGENICITY, TERATOGENICITY, REPRODUCTIVE EFFECTS, NEUROTOXICITY, SKIN SENSITIZATION AND SUBCHRONIC TOXICITY. EPA IS NOT PROPOSING TO REQUIRE ANY ADDITIONAL ENVIRONMENTAL EFFECTS TESTING AT THIS TIME.
REPT'D USE : DISINFECTANT/ANTISEPTIC, HERBICIDE, FLAVORING AGENT, SOLVENT, INDUSTRIAL CLEANING AGENTS AND PHOTOGRAPHIC DEVELOPER

CAS NO : **95-48-7**
CHEM NAME : ***** O-CRESOL**

REPORT 01 : EPA PROPOSES TEST RULE 48FR522 JUL 83
UNDER TSCA 4 EPA PROPOSES THAT MFRS AND PROCESSORS OF CREOLS TEST FOR CARCINOGENICITY, METAGENICITY, TERATOGENICITY, REPRODUCTIVE EFFECTS, NEUROTOXICITY, SKIN SENSITIZATION AND SUBCHRONIC TOXICITY.
REPT'D USE : WIRE ENAMEL SOLVENTS, INTERMEDIATES IN MFR OF PHENOLIC RESINS, PHOSPHATE ESTERS AND HERBICIDES, CLEANING CPDS AND DISINFECTANTS AND ORE FLOTATION

CAS NO : **106-44-5**
CHEM NAME : ***** P-CRESOL**

REPORT 01 : EPA PROPOSES TEST RULE 48FR522 JUL 83
EPA PROPOSES THAT MFRS AND PROCESSORS TEST P-CRESOL FOR CARCINOGENICITY,TERATOGENICITY,METAGENICITY,REPRODUCTIVE EFFECTS, NEUROTOXICITY, SKIN SENSITIZATION AND SUBCHRONIC TOXICITY. THIS IS BASED ON LARGE VOLUME OF P-CRESOL PRODUCED IN US.

CAS NO : **14464-46-1**
CHEM NAME : ***** CRISTOBALITE**

REPORT 01 : NIOSH CRITERIA DOCUMENT NOV 74
NIOSH RECOMMENDS A STANDARD OF 50 UG/CU M TWA BASED ON SILICOSIS. THE PRESENT OSHA STANDARD IS 10 MG/CU M TWA.

REPT'D USE : SILICULOSE REFRACTORY MATERIALS, MFR OF GLASS, CERAMICS, AND ADHESIVES

CAS NO : **12001-28-4**
CHEM NAME : ***** CROCIDOLITE ASBESTOS**

REPORT 01 : NTP TR 280 NIH PUB 89-2536 DEC 88
UNDER THE CONDITIONS OF THESE FEED STUDIES, CROCIDOLITE ASBESTOS WAS NOT OVERTLY TOXIC AND DID NOT CAUSE A CARCINOGENIC RESPONSE WHEN INGESTED AT A CONCENTRATION OF 1% IN THE DIET BY MALE AND FEMALE F344/N RATS FOR THEIR LIFETIME.
REPT'D USE : FOUND IN DRINKING WATER/FOODS

CAS NO : **4170-30-3**
CHEM NAME : ***** CROTONALDEHYDE**

REPORT 01 : TECHNICAL BULLETIN 3 DEC 80
THE COMPOUND WAS FOUND TO BE POSITIVE IN THE SALMONELLA TYPHIMURIUM MUTAGENICITY ASSAY IN ONE LABORATORY.

REPORT 02 : EPA RCRA HEALTH AND ENV. EFFECTS PROFILE OCT 80
AN INCREASED INCIDENCE OF MALIGNANT NEOPLASMS HAS BEEN OBSERVED IN WORKERS IN AN ALDEHYDE FACTORY WHO WERE EXPOSED TO CROTONALDEHYDE, AMONG OTHER SUBSTANCES. THERE IS, HOWEVER, NO INDICATION THAT CROTONALDEHYDE WAS THE CAUSATIVE FACTOR IN THE EXCESS INCIDENCE OF CANCER.

REPORT 06 : MULTI-AGENCY CROSS CHECK
N HHS MAGUIRE
N OSHA CARC
N EPA CAG
Y EPA RCRA
N EPA CLEAN AIR
N EPA CLEAN WATER
REPT'D USE : CHEMICAL INTERMEDIATE, SOLVENT AND RUBBER ACCELERATOR

CAS NO : **98-82-8**
CHEM NAME : ***** CUMENE (1-METHYLETHYL)BENZENE)**

REPORT 01 : EPA TSCA FINAL TEST RULE (53FR28195) JUL 88
EPA REQUIRES MANUFACTURERS AND PROCESSORS OF CUMENE TO PERFORM ORAL AND INHALATION COMPARATIVE PHARMACOKINETICS, SUBCHRONIC INHALATION TOXICITY, DEVELOPMENTAL TOXICITY, NEUROTOXICITY, AND, IF TRIGGERED, TWO GENERATION REPRODUCTIVE EFFECTS TESTS. AND TO TEST FOR ACUTE

TOXICITY TO FISH AND INVERTEBRATES, BIODEGRADATION IN AN AQUATIC SYSTEM, AND, IF TRIGGERED, CHRONIC TOXICITY TO FISH AND INVERTEBRATES.

REPORT 02 : 50FR46106, EPA PROPOSED TEST RULE NOV 85
EPA PROPOSES THAT MFRS AND PROCESSORS OF CUMENE BE REQUIRED TO TEST FOR PHARMOKINETICS, SUBCHRONIC TOXICITY, ONGOGENICITY, MUTAGENICITY, DEVELOPMENTAL TOXICITY AND REPRODUCTIVE TOXICITY.

REPORT 03 : EPA TSCA 8(E) 8EHQ-1184-0536 JUL 85
BALB/353 MOUSE EMBRYO CELL TRANSFORMATION TEST WAS POSITIVE AT 60 MCG/ML. THE HEPATOCYTE PRIMARY CULTURE/DNA REPAIR TEST WITH HEPATOCYTES FROM FISHER 344 RATS WAS POSITIVE.

REPT'D USE : MFR OF PHENOL, ACETONE AND A-METHYL STYRENE, CHAIN INHIBITOR IN POLYMER INDUSTRY, LIQUID DETERGENTS AND SURFACTANTS, HIGH OCTANE AVIATION FUEL AND SOLVENT

CAS NO : **135-20-6**
CHEM NAME : ***** CUPFERRON**

REPORT 01 : NCI CARC TRS 100 DHEW PUB NO 78-1350
UNDER THE CONDITIONS OF THIS BIOASSAY CUPFERRON WAS CARCINOGENIC IN FISCHER 344 RATS, CAUSING HEMANGIOSARCOMAS, HEPATOCELLULAR CARCINOMAS AND SQUAMOUS-CELL CARCINOMAS OF THE FORESTOMACH IN MALES AND FEMALES, AS WELL AS CARCINOMAS OF THE AUDITORY SEBACEOUS GLAND IN FEMALES. THE CHEMICAL WAS ALSO CARCINOGENIC IN B6C3F1 MICE, CAUSING HEMANGIOSARCOMAS IN MALES AND HEPATOCELLULAR CARCINOMAS, CARCINOMAS OF THE AUDITORY SEBACEOUS GLAND, A COMBINATION OF HEMANGIOSARCOMAS AND HEMANGIOMAS AND ADENOMAS OF THE HARDERIAN GLAND IN FEMALES IN A FEEDING STUDY.

REPORT 06 : MULTI-AGENCY CROSS CHECK
Y HHS MAGUIRE
Y OSHA CARC
N EPA CAG
N EPA RCRA
N EPA CLEAN AIR
N EPA CLEAN WATER

REPT'D USE : CHEMICAL INDICATOR AND INDUSTRIAL CHELATING AGENT

CAS NO : **108-80-5**
CHEM NAME : ***** CYANURIC ACID**

REPT'D USE : STABILIZER, CONDITIONER, CHLORINE CARRIER IN SWIMMING POOLS AND TOX. DATA

CAS NO : **14901-08-7**
CHEM NAME : ***** CYCASIN**

REPORT 01 : IARC MONOGRAPH 1 1972
CYCASIN IS CARCINOGENIC IN 5 ANIMAL SPECIES, INDUCING TUMORS IN VARIOUS ORGANS. FOLLOWING ORAL EXPOSURE, IT IS CARCINOGENIC IN THE RAT, HAMSTER, GUINEA-PIG AND FISH. IT IS ACTIVE IN SIGNLE-DOSE EXPERIMENTS AND FOLLOWING PRENATAL EXPOSURE. THE NEGATIVE RESULT CONCERNING CANCER IN THE ONLY EPIDEMIOLOGIC STUDY PERFORMED TO DATE IS INSUFFICIENT TO EXCLUDE A POSSIBLE CARCINOGENIC EFFECT OF CYCASIN ON MAN. IN A 1979 SUPPLEMENT, IARC DETERMINED THAT THERE IS SUFFICIENT EVIDENCE FOR THE CARCINOGENICITY OF CYCASIN IN ANIMALS.

REPORT 02 : NAS REPORT ON DIET, NUTRITION AND CANCER 1982
WHEN ADMINISTERED ORALLY CYCASIN IS HIGHLY CARCINOGENIC TO THE LIVER, KIDNEY AND COLON OF RATS AND ALSO INDUCES TUMORS IN OTHER SPECIES. IN A DESCRIPTIVE STUDY IN OKINAWA, NO CORRELATION WAS FOUND BETWEEN MORTALITY FROM HEPATOMA AND THE INGESTION OF CYCAD NUTS, THEREFORE, THERE IS NO EVIDENCE FOR THE CARCINOGENICITY OF CYCASIN IN HUMANS.

REPORT 03 : 4TH ANNUAL REPORT ON CARCINOGENS (MAGUIRE REPT) 1985
THERE IS SUFFICIENT EVIDENCE FOR CARCINOGENICITY IN EXPERIMENTAL ANIMALS. HOWEVER, DROPPED FROM THE ANNUAL REPORT, BECAUSE NO U.S. RESIDENTS EXPOSED.

REPORT 06 : MULTI-AGENCY CROSS CHECK
Y HHS MAGUIRE
N OSHA CARC
Y EPA CAG
N EPA RCRA
N EPA CLEAN AIR
N EPA CLEAN WATER
REPT'D USE : NO AVAILABLE PRODUCTION OR USE DATA

CAS NO : **12663-46-6**
CHEM NAME : ***** CYCLOCHLOROTINE**

REPORT 01 : IARC MONOGRAPH 10 1976
CYCLOCHLOROTINE IS CARCINOGENIC IN MALE MICE FOLLOWING ITS ORAL ADMINISTRATION; IT PRODUCED LIVER TUMORS AND RETICULOENDOTHELIOMAS.
REPT'D USE : NATURALLY OCCURRING MYCOTOXIN

CAS NO : **110-82-7**
CHEM NAME : ***** CYCLOHEXANE**

REPORT 01 : 52 FR 19096 MAY 87
IN RESPONSE TO ITC'S DESIGNATION OF CPD FOR HEALTH EFFECTS TESTING, EPA PROPOSING THAT MFRS AND PROCESSORS OF CPD PERFORM HEALTH EFFECTS TESTING FOR SUBCHRONIC TOXICITY, ONGOGENICITY, REPRODUCTIVE TOXICITY, DEVELOPMENTAL TOXICITY, NEUROTOXICITY, DERMAL ABSORPTION AND DERMAL SENSITIZATION.
REPT'D USE : SOLVENT

CAS NO : **108-94-1**
CHEM NAME : ***** CYCLOHEXANONE**

REPORT 01 : EPA RESPONSE TO ITC, 49FR136 JAN 74
EPA'S RESPONSE TO ITC'S RECOMMENDATION THAT CY BE CONSIDERED FOR HEALTH AND ENVIRONMENTAL EFFECTS TESTING. EPA STUDIES INDUSTRY PROPOSAL FOR TESTING AND CONCLUDED THAT IT WAS ADEQUATE TO ADDRESS ITC'S CONCERNS AND THAT TESTING RULE IS NOT NECESSARY AT THIS TIME.

REPORT 02 : 52FR1512 - RECEIPT OF TEST DATA JAN 87
EPA ANNOUNCES RECEIPT OF CHEMICAL TEST DATA FROM INDUSTRY VOLUNTARY TESTING PROGRAM.
REPT'D USE : CHEMICAL INTERMEDIATE AND SOLVENT FOR RESINS, LACQUERS, DYES AND INSECTICIDES

CAS NO : **108-91-8**
CHEM NAME : ***** CYCLOHEXYLAMINE (CHA)**

REPORT 01 : IARC MONOGRAPH 22 1980
CYCLOHEXYLAMINE HAS BEEN TESTED BY ORAL ADMINISTRATION IN TWO EXPERIMENTS IN MICE, ONE OF WHICH WAS A MULTIGENERATION STUDY; AND IN FOUR EXPERIMENTS IN RATS, THERE WERE NO DIFFERENCES IN TUMOR INCIDENCE BETWEEN TREATED AND CONTROL ANIMALS. THE AVAILABLE EXPERIMEN-

TAL DATA PROVIDE NO EVIDENCE FOR THE CARCINOGENICITY OF CYCLOHEXYLAMINE.

REPORT 02 : IARC MONOGRAPH 22 1980
CYCLOHEXYLAMINE CAN CAUSE CHROMOSOME DAMAGE. THERE IS NO EVIDENCE THAT CYCLOHEXYLAMINE IS TERATOGENIC.

REPORT 03 : NTP TECHNICAL BULLETIN 9 APR 83
CYCLOHEXYLAMINE WAS NEGATIVE IN ONE SALMONELLA TYPHIMURIUM MUTAGENICITY ASSAY.

REPT'D USE : MFR OF RUBBER PROCESSING CHEMICALS, INDUSTRIAL WATER TREATMENT (CORROSION INHIBITOR), DYE/PLASTICIZER INTERMEDIATE AND DRY CLEANING SOAPS

CAS NO : **17796-82-6**

CHEM NAME : ***** M-(CYCLOHEXYLTHIO) PHTHALIMIDE**

REPORT 01 : STATUS REPORT FYI-OTS-0786- 0500 DEC 86
A 2-YR CHRONIC TOXICITY STUDY IN SPRAGUE-DAWLEY RATS CONDUCTED AS A DIETARY STUDY AT LEVELS OF 0, 50, 150, OR 500 PPM/MG/KG BD. WT/DAY. RESULTS FROM THE MID- AND HIGH-DOSE LEVELS REPRESENT APPARENT ADVERSE EFFECT LEVELS. RATS FED AT LEVELS OF 50 MG/KG BD. WT/DAY AND LOWER DID NOT SHOW ANY ADVERSE EFFECTS.

REPORT 02 : TSCA SECT 8(E) STATUS REPORT 8EHQ-0786-0681 JUNE 87
8EHQ-0187-0681
IN A 2-YR CHRONIC TOXICITY STUDY IN SPRAGUE-DAWLEY RATS CONDUCTED AS A DIETARY STUDY AT LEVELS OF 0, 50, 150, OR 500 MG/KG BODY WT/DAY. THE SUBMITTOR CLAIMS THAT THE FINDINGS DO NOT INDICATE A SIGNIFICANT HEALTH RISK. MID- AND HIGH-DOSE LEVELS (150-500 MG/KG/DAY) REPRESENT ADVERSE EFFECTS LEVEL. 50 MG/KG/DAY REPRESENTS AN APPARENT NO DOSE-EFFECT LEVEL

REPT'D USE : PRE-VULCANIZATION INHIBITOR

CAS NO : **27208-37-3**

CHEM NAME : ***** CYCLOPENTA(CD)PYRENE**

REPORT 01 : IARC MONOGRAPH 32 DEC 83
LIMITED EVIDENCE THAT CYCLOPENTA(CD)PYRENE IS CARCINOGENIC TO EXPERIMENTAL ANIMALS. TESTED FOR CARCINOGENICITY IN TWO STUDIES IN FEMALE MICE BY SKIN APPLICATION; PRODUCED SKIN TUMORS IN ONE STUDY. TESTED IN MOUSE-SKIN INITIATION-PROMOTION ASSAY IN THREE STUDIES AND ACTIVE AS AN INITIATOR.

REPORT 02 : IARC MONOGRAPH 32 DEC 83
CYCLOPENTA(CD)PYRENE MUTAGENIC TO SALMONELLA TYPHIMURIUM AND TO MAMMALIAN CELLS IN VITRO IN PRESENCE OF EXOGENOUS METABOLIC SYSTEM. INDUCED MORPHOLOGICAL TRANSFORMATION IN MAMMALIAN CELLS.

REPT'D USE : AIR POLLUTANT, GASOLINE ENGINE, AND EXHAUST COMPONENT

CAS NO : **50-18-0**
CHEM NAME : ***** CYCLOPHOSPHAMIDE**

REPORT 01 : IARC SUPPLEMENT 1 1979
CYCLOPHOSPHAMIDE IS CARCINOGENIC IN MICE AND RATS FOLLOWING INTRAPERITONEAL INJECTION, IN RATS FOLLOWING INTRAVENOUS INJECTION, AND IN MICE FOLLOWING SUBCUTANEOUS INJECTION. THERE ARE CASE REPORTS OF BLADDER CANCER AND ACUTE MYELOID LEUKEMIA IN PERSONS TREATED WITH CYCLOPHOSPHAMIDE FOR MEDICAL REASONS. IN A 1979 SUPPLEMENT, IARC CONCLUDED THAT CYCLOPHOSPHAMIDE IS PROBABLY CARCINOGENIC FOR HUMANS.

REPORT 02 : NCI BIOASSAY PROGRAM, CANCER, VOL. 40 1977
UNDER THE CONDITIONS OF THIS BIOASSAY, CYCLOPHOSPHOSAMIDE WAS CARCINOGENIC TO MALE AND FEMALE SWISS-WEBSTER MICE BY IP INJECTION. CAUSING LUNG TUMORS IN BOTH SEXES AND URINARY BLADDER PAPILLOMAS IN MALE MICE. IT WAS ALSO CARCINOGENIC TO SPRAGUE-DAWLEY (CD) FEMALE RATS, CAUSING BREAST CARCINOMAS AND ADENOMAS. IT WAS NOT CARCINOGENIC TO MALE RATS.

REPORT 03 : TECHNICAL BULLETIN 3 DEC 80
THE COMPOUND WAS FOUND TO BE POSITIVE IN THE SALMONELLA TYPHIMURIUM MUTAGENICITY ASSAY IN TWO LABORATORIES.

REPORT 04 : NTP TECHNICAL BULLETIN 3/8 DEC 80, JUL 82
CYCLOPHOSPHAMIDE WAS POSITIVE IN THE SALMONELLA ASSAY AND IN THE L5178Y MOUSE LYMPHOMA ASSAY.

REPORT 05 : IARC MONOGRAPH 26 1981
THERE IS EPIDEMIOLOGICAL EVIDENCE THAT CYCLOPHOSPHAMIDE INCREASES THE INCIDENCE OF BLADDER CANCER, AND THERE IS A SUGGESTION THAT THE INCIDENCE OF OTHER CANCERS MAY ALSO BE INCREASED. THERE IS SUFFICIENT EVIDENCE THAT CYCLOPHOSPHAMIDE IS CARCINOGENIC IN HUMANS.

REPORT 06 : MULTI-AGENCY CROSS CHECK
Y HHS MAGUIRE
N OSHA CARC
Y EPA CAG
Y EPA RCRA

N EPA CLEAN AIR
N EPA CLEAN WATER

REPORT 07 : IARC SURVEY OF CARC TESTING #9 AUG 81
IN A CURRENT STUDY CONDUCTED AT THE GERMAN CANCER RESEARCH CENTER, CYCLOPHOSPHAMIDE WAS ADMINISTERED IN THE DRINKING WATER TO S-D RATS. THE STUDY HAS BEEN COMPLETED AND PUBLISHED.

REPORT 08 : IARC MONOGRAPH 26 1981
THE TERATOGENIC EFFECTS OF CYCLOPHOSPHAMIDE ARE WELL ESTABLISHED IN MANY ANIMAL SPECIES. THE DRUG IS ALSO EMBRYOLETHAL AT DOSES NONTOXIC TO THE MOTHER. CYCLOPHOSPHAMIDE DEMONSTRATED MUTAGENIC ACTIVITY IN SEVERAL DIFFERENT ASSAYS, AND INDUCED CHROMOSOMAL ABERRATIONS IN MAMMALIAN CELLS OF SEVERAL SPECIES IN VITRO AND IN VIVO.

REPORT 09 : IARC MONOGRAPH 9 1975
CYCLOPHOSPHAMIDE IS TERATOGENIC IN SEVERAL SPECIES INCLUDING MICE, RATS, RABBITS AND CHICKS. CYCLOPHOSPHAMIDE INDUCES POINT MUTATIONS, IN E. COLI, MITOTIC GENE COVERSIONS IN SACCHAROMYCES, DOMINANT LETHAL MUTATIONS IN MICE, RECESSIVE LETHAL MUTATIONS IN DROSOPHILA, AND CHROMOSOME ABERRATION IN CULTURED HUMAN LEUCOCYTES.
REPT'D USE : ANTINEOPLASTIC AGENT AND CHEMICAL SHEARING OF SHEEP

CAS NO : **21739-91-3**
CHEM NAME : ***** CYTEMBENA**

REPORT 01 : NTP BIOASSAY TRS 207 NIH PUB. NO. 81-1763
UNDER THE CONDITIONS OF THIS BIOASSAY CYTEMBENA WAS CARCINOGENIC FOR MALE AND FEMALE F344 RATS CAUSING INCREASED INCIDENCES OF MESOTHELIOMAS IN THE TUNICA VAGINALIS AND IN MULTIPLE ORGANS OF MALES AND FIBROADENOMAS IN THE MAMMARY GLAND OF FEMALES. CYTEMBENA WAS NOT CARCINOGENIC FOR MALE OR FEMALE B6C3F1 MICE BY IP INJECTION.
REPT'D USE : CYTOSTATIC AGENT

CAS NO : **4465-94-5**
CHEM NAME : ***** CYTOXAL ALCOHOL**

REPORT 01 : NCI BIOASSAY PROGRAM, CANCER VOL. 40 1977
UNDER THE CONDITIONS OF THE BIOASSAY, I.P. INJECTION OF CYTOXAL ALCOHOL WAS CARCINOGENIC TO MALE AND FEMALE SWISS-WEBSTER MICE, CAUSING LUNG TUMORS IN BOTH SEXES

AND LYMPHOSARCOMAS AND UTERINE TUMORS IN FEMALES. IT WAS NOT CARCINOGENIC TO MALE AND FEMALE SPRAGUE-DAWLEY (CD) RATS.

REPORT 02 : NTP TECHNICAL BULLETIN 6 JAN 82
THE COMPOUND WAS POSITIVE IN THE SALMONELLA TYPHIMURIUM MUTAGENICITY ASSAY IN ONE LABORATORY.

REPT'D USE : NO PRODUCTION OR USE DATA AVAILABLE

CAS NO : **94-75-7**

CHEM NAME : ***** 2,4-D**

REPORT 01 : IARC MONOGRAPH 15 1977
2,4-D AND SEVERAL OF ITS ESTERS WERE TESTED IN RATS AND MICE BY ORAL ADMINISTRATION AND IN MICE BY SUBCUTANEOUS ADMINISTRATION. ALL OF THESE STUDIES HAD LIMITATIONS, DUE EITHER TO INADEQUATE REPORTING OR TO THE SMALL NUMBER OF ANIMALS USED. NO EVALUATION OF THE CARCINOGENICITY OF THIS COMPOUND COULD BE MADE.

REPORT 02 : IARC MONOGRAPH 15 1977
RESULTS OF TERATOLOGIC STUDIES WITH 2,4-D WERE VARIABLE; TERATOGENIC EFFECTS WERE OBSERVED WITH DOSES CLOSE TO THOSE WHICH CAUSE MATERNAL TOXICITY. 2,4-D WAS NOT MUTAGENIC IN E. COLI OR IN SALMONELLA STRAINS TA1535, TA1536, TA1537, OR TA1538. NO INCREASE IN RECESSIVE LETHALS WAS OBSERVED IN DROSOPHILA. 2,4-D DID NOT INCREASE DOMINANT LETHAL MUTATIONS IN MICE.

REPORT 03 : EPA RCRA HEALTH AND ENV. EFFECTS PROFILE OCT 80
ORAL ADMINISTRATION OF 2,4-D FAILED TO PRODUCE CARCINOGENIC EFFECTS IN MICE OR DOGS; HOWEVER, FEEDING TECHNICAL GRADE 2,4-D DID PRODUCE TUMORS IN A STUDY WITH RATS. A SINGLE STUDY HAS INDICATED THAT 2,4-D PRODUCED MUTAGENIC EFFECTS IN SACCHAROMYCES. OTHER INVESTIGATIONS HAVE FAILED TO SHOW MUTAGENIC EFFECTS OF THE COMPOUND IN SALMONELLA, DROSOPHILIA, SACCHAROMYCES, OR THE DOMINANT LETHAL ASSAY IN MICE.

REPORT 04 : NTP TECHNICAL BULLETIN 3/8 DEC 80, JUL 82
2,4-D WAS NEGATIVE IN THE SALMONELLA ASSAY AND IN DROSOPHILA.

REPT'D USE : HERBICIDE AND FUNGICIDE

CAS NO : **4342-03-4**
CHEM NAME : ***** DACARBAZINE**

REPORT 01 : NCI BIOASSAY PROGRAM, CANCER VOL. 40 1977
UNDER THE CONDITIONS OF THE BIOASSAY, IP INJECTION OF DACARBAZINE WAS CARCINOGENIC TO SWISS-WEBSTER MALE AND FEMALE MICE, CAUSING LUNG TUMORS, LYMPHOMAS, AND HEMANGIOMAS IN MALE MICE, AND LUNG AND UTERINE TUMORS IN FEMALE MICE. IT WAS ALSO CARCINOGENIC TO MALE AND FEMALE SPRAGUE-DAWLEY (CD) RATS, CAUSING LYMPHOMAS AND HEART TUMORS. IN ADDITION, MALE RATS HAD RENAL TUMORS AND FEMALE RATS HAD BREAST CARCINOMAS.

REPORT 02 : IARC MONOGRAPH 26 1981
DACARBAZINE IS CARCINOGENIC IN MICE AND RATS. FOLLOWING ITS ORAL OR I.P. ADMINISTRATION TO RATS, DACARBAZINE PRODUCED TUMORS AT VARIOUS SITES, INCLUDING BREAST, THYMUS, SPLEEN AND BRAIN. AFTER ITS I.P. ADMINISTRATION TO MICE, DACARBAZINE PRODUCED TUMORS AT VARIOUS SITES, INCLUDING LUNG, HEMATOPOIETIC TISSUE AND UTERUS. THERE IS SUFFICIENT EVIDENCE FOR THE CARCINOGENICITY OF DACARBAZINE IN RATS AND MICE.

REPORT 03 : IARC MONOGRAPH 26 1981
DACARBAZINE CAN INDUCE TERATOGENIC EFFECTS IN SEVERAL SPECIES. THE DRUG WAS NOT MUTAGENIC IN SALMONELLA STRAINS TA1535, TA1537, TA1538 AND TA98 WHEN TESTED IN THE PRESENCE OR ABSENCE OF METABOLIC ACTIVATION. IN THE MOUSE LYMPHOMA CELL ASSAY, DACARBAZINE WAS DOSE-DEPENDENTLY MUTAGENIC AFTER ACTIVATION.

REPORT 06 : MULTI-AGENCY CROSS CHECK NOV 86
Y HHS MAGUIRE
Y OSHA CARC
N EPA CAG
N EPA RCRA
N EPA CLEAN AIR
N EPA CLEAN WATER
REPT'D USE : ANTINEOPLASTIC AGENT

CAS NO : **1596-84-5**
CHEM NAME : ***** DAMINOZIDE**

REPORT 01 : NCI CARC TRS 83 DHEW PUB NO 78-1333
UNDER THE CONDITIONS OF THIS BIOASSAY, DAMINOZIDE WAS NOT CARCINOGENIC IN THE MALE FISCHER 344 RATS OR IN THE FEMALE B6C3F1 MALE B6C3F1 MICE. THE INDUCTION OF HEPATOCELLULAR CARCINOMAS MAY HAVE BEEN ASSOCIATED WITH THE ADMINISTRATION OF THE TEST CHEMICAL. DAZ WAS CARCINO-

GENIC IN FEMALE FISCHER 344 RATS, INDUCING ADENOCARCINOMAS OF THE ENDOMETRIUM OF THE UTERUS AND LEIOMYOSARCOMAS OF THE UTERUS IN A FEEDING STUDY.

REPORT 02 : NTP TECHNICAL BULLETIN 6 JAN 82
THE COMPOUND WAS NEGATIVE IN THE SALMONELLA TYPHIMURIUM METAGENICITY ASSAY IN ONE LABORATORY.

REPORT 06 : MULTI-AGENCY CROSS CHECK
N HHS MAGUIRE
Y OSHA CARC
N EPA CAG
N EPA RCRA
N EPA CLEAN AIR
N EPA CLEAN WATER
REPT'D USE : PLANT GROWTH REGULATOR AND HERBICIDE

CAS NO : **80-08-0**
CHEM NAME : ***** DAPSONE**

REPORT 01 : NCI CARC TRS 20 DHEW PUB NO 77-820
IT IS CONCLUDED THAT UNDER THE CONDITIONS OF THIS BIOASSAY, DAPSONE WAS NOT CARCINOGENIC FOR FEMALE FISHER 344 RATS OR B6C3F1 MICE OF EITHER SEX. DAPSONE WAS CARCINOGENIC (SARCOMAGENIC) FOR MALE FISCHER 344 RATS, CAUSING MESENCHYMAL TUMORS IN THE SPLEEN AND THE PERITONEUM (FEEDING STUDY).

REPORT 02 : IARC MONOGRAPH 24 1980
DAPSONE HAS BEEN TESTED BY ORAL ADMINISTRATION IN MICE AND RATS, BY INTRAPERITONEAL ADMINISTRATION IN MICE AND BY PRENATAL AND LIFETIME ORAL EXPOSURE IN MICE AND RATS. IN THREE DIFFERENT STUDIES IN RATS, HIGH DOSES OF DAPSONE INDUCED MESENCHYMAL TUMORS OF THE SPLEEN IN MALES (AND OF THE PERITONEUM IN TWO STUDIES). AN INCREASED INCIDENCE OF TUMORS OF THE THYROID WAS FOUND IN RATS OF BOTH SEXES IN ONE STUDY AND IN MALES IN ANOTHER STUDY.

REPORT 03 : IARC MONOGRAPH 24 1980
NEITHER DAPSONE NOR ITS N-ACETYL AND N,N-DIACETYL METABOLITES WERE MUTAGENIC IN SALMONELLA STRAINS TA1535, TA1537, TA1538, TA98, OR TA100 WITH OR WITHOUT METABOLIC ACTIVATION. ATTENTION IS DRAWN TO THE ABSENCE OF STUDIES ON THE TERATOGENICITY OF THIS COMPOUND.

REPORT 04 : NTP TECHNICAL BULLETIN 5 AUG 81
DAPSONE WAS NEGATIVE IN THE SALMONELLA TYPHIMURIUM METAGENICITY ASSAY IN ONE LABORATORY.

REPORT 06 : MULTI-AGENCY CROSS CHECK
N HHS MAGUIRE
Y OSHA CARC
N EPA CAG
N EPA RCRA
N EPA CLEAN AIR
N EPA CLEAN WATER

REPT'D USE : THERAPEUTIC AGENT FOR LEPROSY AND ANTIMALARIAL DRUG

CAS NO : **20830-81-3**
CHEM NAME : ***** DAUNOMYCIN**

REPORT 01 : IARC MONOGRAPH 10 1976
DAUMOMYCIN IS CARCINOGENIC IN RATS FOLLOWING IV INJECTION OF SINGLE DOSES AND IN MICE FOLLOWING ITS REPEATED SC INJECTION. IT PRODUCED MAMMARY AND KIDNEY TUMORS IN RATS AND LOCAL SARCOMAS IN MICE. IN A 1979 SUPPLEMENT, IARC DETERMINED THAT THERE IS SUFFICIENT EVIDENCE FOR THE CARCINOGENICITY OF DAUNOMYCIN IN ANIMALS.

REPORT 02 : NCI BIOASSAY PROGRAM- CANCER VOL. 40 1977
UNDER THE CONDITIONS OF THE BIOASSAY, I.P. INJECTION OF DAUNOMYCIN WAS CARCINOGENIC TO MALE SPRAGUE-DAWLEY (CD) RATS CAUSING PERITONEAL SARCOMAS. IT WAS NOT CARCINOGENIC TO FEMALE RATS AND MALE AND FEMALE SWISS-WEBSTER MICE.

REPORT 03 : IARC MONOGRAPH 10 1976
DAUNOMYCIN HYDROCHLORIDE INDUCES HIGH FREQUENCIES OF REVERSE MUTATIONS IN SALMONELLA. DAUNOMYCIN DID NOT INDUCE UNSCHEDULED DNA SYNTHESIS IN CULTURED HUMAN FIBROBLASTS. THE CHEMICAL-INDUCED HIGH FREQUENCIES OF CHROMOSOME AND CHROMATID BREAKS AND TRANSLOCATIONS IN CULTURES OF HUMAN PERIPHERAL LEUCOCYTES.

REPORT 04 : IARC MONOGRAPH 20 1979
1,2 DIBROMO-3-CHLOROPROPANE WAS MUTAGENIC IN SALMONELLA STRAINS TA100, TA1530, AND TA1535 BUT NOT IN TA1538, BOTH IN THE PRESENCE AND ABSENCE OF METABOLIC ACTIVATION. NO DATA ON THE EMBRYOTOXICITY OR TERATOGENICITY OF DBCP WERE AVAILABLE.

REPORT 06 : MULTI-AGENCY CROSS CHECK
N HHS MAGUIRE
H OSHA CARC
Y EPA CAG
Y EPA RCRA

N EPA CLEAN AIR
N EPA CLEAN WATER
REPT'D USE : CHEMOTHERAPEUTIC AGENT

CAS NO : **53-19-0**
CHEM NAME : ***** O,P'-DDD**

REPORT 01 : NCI BIOASSAY PROGRAM- CANCER VOL. 40 1977
UNDER THE CONDITIONS OF THE BIOASSAY, I.P. INJECTION OF 0,P'-DDD WAS CARCINOGENIC TO MALE AND FEMALE SPRAGUE-DAWLEY (CD) RATS, CAUSING PARATHYROID TUMORS IN BOTH SEXES AND BREAST TUMORS IN FEMALES. IT WAS ALSO CARCINOGENIC TO MALE MICE.

REPORT 02 : EPA RCRA HEALTH AND ENV. EFFECTS PROFILE OCT 80
THERE IS SOME EVIDENCE THAT DDD IS CARCINOGENIC IN MICE; HOWEVER, IN OTHER SPECIES, IT APPEARS TO BE NON-CARCINOGENIC. P,P-DDD HAS BEEN SHOWN TO BE MUTAGENIC IN DROSOPHILIA, BUT NOT IN YEAST OR BACTERIA. IN CELL CULTURE, P,P-DDD CAUSES CHROMOSOMAL BREAKS.
REPT'D USE : ANTINEOPLASTIC AGENT

CAS NO : **72-55-9**
CHEM NAME : ***** P,P'-DDE**

REPORT 01 : NCI CARC TRS 131 DHEW PUB NO 78-1386
UNDER THE CONDITIONS OF THESE BIOASSAYS THERE WAS NO EVIDENCE FOR THE CARCINOGENICITY OF P,P'-DDE IN OSBORNE-MENDEL RATS, ALTHOUGH P,P'-DDE WAS HEPATOTOXIC IN OSBORNE-MENDEL RATS. RESULTS OF A FEEDING STUDY USING DDE, TDE AND DDT GIVEN SEPARATELY IN FEED ARE ALSO GIVEN.

REPORT 02 : IARC MONOGRAPH 5 1974
TWO DDT METABOLITES, P,P'-DDE AND TDE (DDD), WERE TESTED BY ORAL ADMINISTRATION TO MICE. THE LATTER PRODUCED A SIGNIFICANT INCREASE IN LUNG TUMORS, WHILE DDE WAS FOUND TO PRODUCE A HIGH INCIDENCE OF LIVER CELL TUMORS.

REPORT 03 : EPA AMBIENT WATER QUALITY CRITERIA OCT 80
FOR METABOLITES OF DDT, THE CRITERION OF .001 UG/L IS ESTABLISHED TO PROTECT AQUATIC LIFE. A HUMAN HEALTH EFFECTS CRITERION, BASED ON CARCINOGENICITY, HAS BEEN DERIVED FOR DDT.

REPORT 04 : EPA RCRA HEALTH AND ENV. EFFECTS PROFILE OCT 80
P,P-DDE HAS BEEN SHOWN TO BE CARCINOGENIC IN MICE, BUT NOT IN RATS. IN CELL CULTURE IT CAUSES CHROMOSOMAL BREAKS.

REPORT 06 : MULTI-AGENCY CROSS CHECK
N HHS MAGUIRE
Y OSHA CARC
N EPA CAG
N EPA RCRA
N EPA CLEAN AIR
Y EPA CLEAN WATER
REPT'D USE : PESTICIDE

CAS NO : **50-29-3**
CHEM NAME : ***** DDT**

REPORT 01 : NCI CARC TRS 131 DHEW PUB NO 78-1386
UNDER THE CONDITIONS OF THESE BIOASSAYS THERE WAS NO EVIDENCE FOR THE CARCINOGENICITY OF DDT IN OSBORNE-MENDEL RATS OR B6C3F1 MICE. REPORT INCLUDES RESULTS ON TDE AND P,P'-DDE (FEEDING STUDY).

REPORT 02 : NIOSH SPECIAL HAZARD REVIEW SEP 78
NIOSH RECOMMENDS A STANDARD OF 0.5 MG/CU M (TWA) BASED ON POTENTIAL CARCINOGENICITY. THE PRESENT OSHA STANDARD IS 1 MG/CU M TWA.

REPORT 06 : MULTI-AGENCY CROSS CHECK
Y HHS MAGUIRE
N OSHA CARC
N EPA CAG
N EPA RCRA
N EPA CLEAN AIR
N EPA CLEAN WATER
REPT'D USE : INSECTICIDE AND PEDICULICIDE

CAS NO : **8017-34-3**
CHEM NAME : ***** DDT (TECHNICAL GRADE-MIXTURE OF ISOMERS)**

REPORT 01 : IARC MONOGRAPH 5 1974
DDT IS CARCINOGENIC IN MICE CAUSING LIVER TUMORS FOLLOWING ORAL ADMINISTRATION. NON-METASTASIZING LIVER TUMORS OCCUR IN RATS FED DDT. THE EPIDEMIOLOGIC STUDIES AVAILABLE WERE CONSIDERED INADEQUATE TO EVALUATE HUMAN CARCINOGENICITY.

REPORT 02 : EPA AMBIENT WATER QUALITY CRITERIA OCT 80
AMBIENT WATER CONCENTRATIONS, BASED ON THE POTENTIAL CARCINOGENIC EFFECTS OF DDT, OF .24 NG/L, .024 NG/L, AND .002 NG/L ARE ESTIMATED TO RESULT IN AN INCREMENTAL INCREASE OF CANCER RISK OVER THE LIFETIME OF 10-5, 10-6, AND 10-7 RESPECTIVELY.

REPORT 03 : EPA RCRA HEALTH AND ENV. EFFECTS PROFILE OCT 80
DDT HAS PRODUCED ADVERSE REPRODUCTIVE EFFECTS IN RODENTS AND BIRDS, BUT ADVERSE EFFECTS HAVE NOT BEEN NOTED IN MAN. THERE IS SUGGESTIVE EVIDENCE THT DDT MIGHT BE A WEAK CARCINOGEN, AND WEAK EVIDENCE THAT IT MIGHT BE A TERATOGEN. CHROMOSOMAL BREAKS HAVE BEEN OBSERVED WITH DDT EXPOSURE IN VITRO AND IN VIVO.

REPORT 06 : MULTI-AGENCY CROSS CHECK
Y HHS MAGUIRE
N OSHA CARC
Y EPA CAG
Y EPA RCRA
N EPA CLEAN AIR
Y EPA CLEAN WATER
REPT'D USE : PESTICIDE

CAS NO : **13654-09-6**
CHEM NAME : ***** DECABROMOBIPHENYL**

REPORT 01 : IARC MONOGRAPH 18 1978
ONLY ONE COMMERCIAL PREPARATION OF PBB (TECH GRADE HEXABROMOBIPHENYL) COMPOSED PRIMARILY OF PENTA-, HEXA-, AND HEPTABROMINATED CONGENERS, WAS TESTED IN ONE EXPERIMENT IN RATS BY ORAL ADMINISTRATION OF SINGLE DOSES. ALTHOUGH NEOPLASTIC CHANGES OF THE LIVER WERE REPORTED, THE STUDY WAS CONSIDERED TO BE INADEQUATE. NO EVALUATION OF THE CARCINOGENICITY OF PBB COULD BE MADE ON THE BASIS OF THE AVAILABLE DATA.
REPT'D USE : FR CHEMICAL. NO KNOWN COMMERCIAL PRODUCTION OR USE. NO LISTING ON TSCA INVENTORY.

CAS NO : **1163-19-5**
CHEM NAME : ***** DECABROMODIPHENYL OXIDE**

REPORT 01 : NTP TR 309, NIH PUB. # 86-2565 MAY 86
UNDER CONDITIONS OF 2-YR FEED STUDIES, SOME EVIDENCE OF CARCINOGENICITY IN MALE AND FEMALE F344/N RATS AS SHOWN BY INCREASED INCIDENCE OF NEOPLASTIC NODULES IN LIVER. EQUIVOCAL EVIDENCE OF CARCINOGENICITY FOR MALE B6C3F1 MICE AS SHOWN BY INCREASED INCIDENCE OF HEPATOCELLULAR ADENOMAS OR CARCINOMAS.
REPT'D USE : FLAME RETARDANT IN POLYSTYRENE, OTHER PLASTICS, ADHESIVES, AND EPOXY RESINS

CAS NO : **91-17-8**
CHEM NAME : ***** DECALIN**

REPORT 01 : STAFF REFERRAL OCT 84
STUDY AT PROCTOR AND GAMBLE BY ALDEN ET AL, DEMONSTRATED SAME ALPHA GLOBULIN WHICH PRODUCED KIDNEY DAMAGE FOLLOWING EXPOSURE TO NUMBER OF CARCINOGENIC FUELS ALSO PRODUCED FOLLOWING EXPOSURE TO DECALIN.

REPORT 03 : EPA TSCA 8(E) 0283-0471 AND 0383-0471 MAR 83
FERRO CORP REPORTED ITS INDUSTRIAL LUBRICANT ADDITIVE PRODUCT (MAJOR COMPONENT BIS(2-CHLOROOCTYL) DISULFIDE) EXHIBITED MODERATE TO EXTREME DERMAL SENSITIZATION IN GUINEA PIG. CONTACT DERMATITIS REPORTED AMONG METALWORKERS USING PRODUCT WITH 1,1,1-TRICHLOROETHANE, STODDARD SOLVENT, AND IN MOST CASES KEROSENE. PRODUCT MUTAGENIC IN SACCHAROMYCES CEREVISIAE WITHOUT ACTIVATION, CAUSED SOME MUTANTS WITH ACTIVATION.

REPT'D USE : SOLVENT IN LACQUERS, SHOE POLISHES, FLOOR WAXES, RUBBER RESINS, STAIN-REMOVER AND STOVE FUEL

CAS NO : **123-42-2**
CHEM NAME : ***** DIACETONE ALCOHOL (DA)**

REPORT 01 : HS STAFF REPORT SEP 89
DA EXHIBITS ACUTE TOXICITY AT HIGH DOSE LEVELS CAUSED EYE, NOSE AND THROAT IRRITATION IN HUMANS WHEN EXPOSED FOR 15 MINUTES AT 100 PPM. GAVE WEAK POSITIVE RESPONSE IN RAT LIVER CHROMOSOME ASSAY. NO DATA TO PERMIT JUDGEMENT CONCERNING CHRONIC (CARCINOGENIC) EFFECTS.

REPT'D USE : SOLVENT FOR NITROCELLULOSE, CELLULOSE ACETATE, CELLULOSE ESTERS EPOXY RESINS, HYDROCARBONS, OILS, FATS, RESINS, DYES; HYDRAULIC FLUIDS, METAL CLEANING COMPOUNDS, ANTIFREEZE SOLUTION, PHARMACEUTICAL PRESERVATIVES; MANUFACTURE OF PHOTOGRAPHIC FILM, ANTIFICIAL SILK, LEATHER, COATINGS, TEXTILES; CHEMICAL INTERMEDIATE AND SUBSTITUTE FOR METHYLENE

CAS NO : **83-63-6**
CHEM NAME : ***** DIACETYLAMINOAZOTOLUENE**

REPORT 01 : IARC MONOGRAPH 8 1975
DIACETYLAMINOAZOTOLUENE WAS ONLY TESTED BY THE ORAL ROUTE IN RATS. ALTHOUGH THE AUTHORS CLAIMED HEPATOCARCINOGENICITY, THE STUDIES COULD NOT BE EVALUATED DUE TO INADEQUATE REPORTING.

REPT'D USE : TREATMENT FOR WOUND EPITHELIZATION AND VETERINARY DRUGS

CAS NO : **613-35-4**
CHEM NAME : ***** N,N-DIACETYLBENZIDINE**

REPORT 01 : IARC MONOGRAPH 16 1978
N,N'-DIACETYLBENZIDINE IS CARCINOGENIC IN RATS AFTER ITS ORAL, SC, OR IP ADMINISTRATION. IT PRODUCED CARCINOMAS OF THE LIVER, EAR DUCT AND MAMMARY GLAND. IN A 1979 SUPPLEMENT, IARC DETERMINED THAT THERE IS SUFFICIENT EVIDENCE FOR THE CARCINOGENICITY N,N'-DIACETYLBENZIDINE IN ANIMALS.

REPORT 02 : IARC MONOGRAPH 16 1978
NO DATA ON THE EMBRYOTOXICITY, TERATOGENICITY OR MUTAGENICITY OF N,N-DIACETYLBENZIDINE WERE AVAILABLE.

REPORT 06 : MULTI-AGENCY CROSS CHECK
N HHS MAGUIRE
Y OSHA CARC
N EPA CAG
N EPA RCRA
N EPA CLEAN AIR
N EPA CLEAN WATER
REPT'D USE : ORGANIC SYNTHESIS OF CHEMICAL/DYE INTERMEDIATES

CAS NO : **2303-16-4**
CHEM NAME : ***** DIALLATE**

REPORT 01 : IARC MONOGRAPH 12 1976
DIALLATE IS CARCINOGENIC IN MICE AFTER ITS ORAL ADMINISTRATION: IT PRODUCED AN INCREASED INCIDENCE OF LIVER CELL TUMORS IN MALES OF TWO STRAINS AND A SLIGHT INCREASE IN THE NUMBER OF LIVER CELL TUMORS IN FEMALES OF ONE STRAIN. A SINGLE SUBCUTANEOUS INJECTION IN MICE PRODUCED A SLIGHT INCREASE IN THE INCIDENCE OF SYSTEMIC RETICULUM-CELL SARCOMAS IN MALES OF ONE STRAIN.

REPORT 02 : IARC MONOGRAPH 12 1976
TESTS OF DIALLATE FOR REVERSE MUTATIONS IN SALMONELLA WERE NEGATIVE FOR BASE SUBSTITUTION AND FRAME-SHIFT MUTATIONS, METABOLIC ACTIVATION SYSTEMS WERE NOT USED IN THESE TESTS.

REPORT 03 : IARC MONOGRAPH 30 JAN 83
DIALLATE TESTED FOR CARCINOGENICITY IN TWO STRAINS OF MICE BY ORAL AND SUBCUTANEOUS ADMINISTRATION AND IN ONE STRAIN OF RATS BY ORAL ADMINISTRATION. FOLLOWING ADMINISTRATION IN DIET, INCIDENCE OF HEPATOMAS WAS INCREASED IN MALE MICE OF BOTH STRAINS, AND INCIDENCE OF LUNG

ADENOMAS INCREASED IN MALE MICE OF ONE STRAIN. STUDY IN RATS CONSIDERED INADEQUATE FOR EVALUATION.

REPORT 04 : IARC MONOGRAPH 30 JAN 83
MUTAGENIC IN BACTERIA AND YEASTS. INDUCED MUTATIONS IN DROSOPHILA MELANOGASTER, AND UNSCHEDULED DNA SYNTHESIS, SISTER CHROMATID EXCHANGE, GENE MUTATIONS AND CHROMOSOMAL ABERRATIONS IN CULTURED MAMMALIAN CELLS. EXOGENOUS METABOLIC ACTIVATION USUALLY REQUIRED TO PRODUCE OR ENHANCE MUTAGENIC ACTIVITY. INEFFECTIVE IN DOMINANT LETHAL TEST IN MICE. SUFFICIENT EVIDENCE THAT DIALLATE MUTAGENIC IN CELLULAR SYSTEMS, BUT INSUFFICIENT EVIDENCE FOR ITS MUTAGENICITY IN MAMMALS.

REPORT 06 : MULTI-AGENCY CROSS CHECK
N HHS MAGUIRE
N OSHA CARC
Y EPA CAG
Y EPA RCRA
N EPA CLEAN AIR
N EPA CLEAN WATER
REPT'D USE : HERBICIDE

CAS NO : **142-22-1**
CHEM NAME : ***** DIALLYL DIGLYCOL CARBONATE (CR39 MONOMER)**

REPORT 01 : TSCA SECT 8(E) STATUS REPORT 8EHQ-0487-0666 S MAY 87
A TERATOLOGY STUDY USING SKIN EXPOSURE OF PREGNANT RABBITS PRODUCED SIGNIFICANT TOXICITIES IN UNBORN AT DOSE LEVELS WHICH ALSO CAUSED SIGNIFICANT MATERNAL TOXICITIES. NON-MATERNALLY TOXIC DOSES TO THE SKIN OF PREGNANT RABBITS DID NOT CAUSE HARMFUL EFFECTS ON THE UNBORN. SKIN IRRITATION WAS PRESENT IN ALL MONOMER TREATED GROUPS. MATERNAL TOXICITIES CAUSED BY BOTH 0.5 AND 1.0 ML MONOMER/KG DOSE LEVELS. CR39 MONOMER CAN BE MODERATELY TO SEVERELY IRRITATING TO HUMAN EYES AND SEVERELY IRRITATING TO HUMAN SKIN. 90% OF AMOUNT ABSORBED THRU SKIN OF RHESUS MONKEYS ELIMINATED WITHIN 4 DAYS.
REPT'D USE : MANUFACTURE OF OPTICAL QUALITY CLEAR PLASTIC SHEET

CAS NO : **131-17-9**
CHEM NAME : ***** DIALLYL PHTHALATE**

REPORT 01 : NTP TECHNICAL REPORT NO. 284 NIH PUB 85-2540 AUG 85
UNDER CONDITIONS OF THIS STUDY, ADMINISTRATION BY GAVAGE OF DIALLYL-PHTHALATE TO MALE AND FEMALE F344/N RATS FOR 2-YEARS CAUSED CHRONIC LIVER DISEASE CHARACTERIZED BY PERIPORTAL FIBROSIS AND PIGMENT ACCUMULATION, AND

INCREASED SEVERITY OF BILE DUCT HYPERPLASIA. INCIDENCE OF MONONUCLEAR CELL LEUKEMIA WAS SIGNIFICANTLY INCREASED IN FEMALE RATS RECEIVING 100 MG/KG, PROVIDING EQUIVOCAL EVIDENCE OF CARCINOGENICITY IN FEMALE RATS. THERE WAS NO EVIDENCE OF CARCINOGENICITY IN MALE RATS.

REPORT 02 : NTP TR 242 NIH PUB 83-1798 APR 83
DEVELOPMENT OF CHRONIC INFLAMMATION AND HYPERPLASIA OF FORESTOMACH IN MALE AND FEMALE B6C3F1 MICE CONSIDERED RELATED TO ADMINISTRATION OF DIALLYL PHTHALATE. DEVELOPMENT OF SQUAMOUS CELL PIPILLOMAS OF FORESTOMACH ALSO RELATED BUT AVAILABLE DATA INSUFFICIENT TO INDICATE CLEAR CAUSE AND EFFECT RELATIONSHIP. INCREASED INCIDENCE OF MALE MICE WITH LYMPHOMAS OBSERVED CONSIDERED EQUIVOCALLY RELATED. ALTHOUGH MAXIMALLY TOLERATED DOSE MAY NOT HAVE BEEN ACHIEVED, CPD NOT CONSIDERED CARCINOGENIC TO MICE.

CAS NO : **39156-41-7**
CHEM NAME : ***** 2,4-DIAMINOANISULE SULFATE**

REPORT 01 : NCI CARC TRS 84 DHEW PUB NO 78-1334
UNDER THE CONDITIONS OF THIS BIOASSAY, TECHNICAL-GRADE 2,4-DIAMINOANISOLE SULFATE WAS CARCINOGENIC TO BOTH SEXES OF BOTH SPECIES. IN FISCHER 344 RATS DIETARY ADMINISTRATION OF THE CHEMICAL INDUCED INCREASED INCIDENCE OF MALIGNANT TUMORS OF THE SKIN AND ITS ASSOCIATED GLANDS AND MALIGNANT THYROID TUMORS IN EACH SEX. IN B6C3F1 MICE, DIETARY ADMINISTRATION OF 2,4-DIAMINOANISOLE SULFATE INDUCED THYROID TUMORS IN EACH SEX.

REPORT 02 : IARC MONOGRAPH 27 APR 82
2,4-DIAMINOANISOLE SULFATE WAS TESTED BY DIETARY ADMINISTRATION IN ONE EXPERIMENT IN MICE AND IN TWO EXPERIMENTS IN ONE STRAIN OF RATS. BENIGN OR MALIGNANT TUMORS OF THE THYROID GLAND WERE INDUCED IN RATS AND MICE WITH THE HIGHEST DOSE TESTED; TUMORS OF THE SKIN AND OF THE PREPUTIAL, CLITORAL AND AYMBAL GLANDS WERE ALSO INDUCED IN RATS. THERE IS SUFFICIENT EVIDENCE FOR THE CARCINOGENICITY OF 2,4-DIAMINOANISOL SULFATE IN ANIMALS.

REPORT 06 : MULTI-AGENCY CROSS CHECK
Y HHS MAGUIRE
Y OSHA CARC
N EPA CAG
N EPA RCRA
N EPA CLEAN AIR
N EPA CLEAN WATER

REPT'D USE : COUPLER IN PERMANENT HAIR DYE FORMULATIONS, TEXTILE DYE INTERMEDIATE AND FUR DYE

CAS NO : **137-09-7**
CHEM NAME : ***** 2,4-DIAMINOPHENOL:2HCL**

REPORT 01 : NTP CARC TEST PROGRAM JUL 82
IN A CURRENT NTP BIOASSAY, 2,4-DIAMINOPHENOL HCL IS BEING ADMINISTERED BY GAVAGE TO F344 RATS AND B6C3F1 MICE. THE PRECHRONIC TEST PHASE IS IN PROGRESS.

REPORT 02 : EPA RESPONSE TO ITC RECOMMS (47FR973) JAN 81
IN RESPONSE TO THE ITC RECOMMENDATIONS (45FR35897) THE EPA PROPOSES AN ADVANCED NOTICE REQUESTING COMMENTS CONCERNING A PROPOSAL TO CONDUCT MUTAGENICITY AND CARCINOGENICITY TESTING FOR 11 SELECTED PHENYLENEDIAMINES, AND TERATOGENICITY/REPRODUCTIVE EFFECTS AND OTHER CHRONIC EFFECTS TESTING, AS WELL AS AN EPIDEMIOLOGIC STUDY, FOR 12 PHENYLENEDIAMINES. THE EPA WILL CONSIDER ADDITIONAL TESTING FOR AN ADDITIONAL 37 PHENYLENEDIAMINES, AS APPROPRIATE.

REPORT 03 : NTP TECHNICAL BULLETIN 2/8 APR 80, JUL 82
2,4-DIAMINOPHENOL WAS POSITIVE IN THE SALMONELLA ASSAY, BUT NEGATIVE IN DROSOPHILA WHEN TESTED FOR SEX-LINKED RECESSIVE LETHALS.
REPT'D USE : PHOTOGRAPHIC DEVELOPER, DYE AND CHEMICAL REAGENT

CAS NO : **94-78-0**
CHEM NAME : ***** 2,6-DIAMINO-3-(PHENYLAZO)PYRIDINE**

REPORT 01 : IARC MONOGRAPH 8 1975
2,6-DIAMINO-3-(PHENYLAZO)PYRIDINE WAS TESTED IN MICE BY IMPLANTATION OF CHOLESTEROL PELLETS CONTAINING THIS SUBSTANCE INTO THE URINARY BLADDERS, IN WHICH IT PRODUCED BENIGN AND MALIGNANT TUMORS. HOWEVER, THE STATISTICAL SIGNIFICANCE OF THIS EXPERIMENT IS DOUBTFUL. NO EVALUATION CAN BE MADE.

REPORT 02 : IARC MONOGRAPH 27 1982
2,4-DIAMINOANISOLE SULPHATE WAS MUTAGENIC FOR CULTURED L5178Y MOUSE LYMPHOMA CELLS AND DROSOPHILA. THE SULPHATE PRODUCED MITOTIC RECOMBINATION IN YEAST. IT DID NOT INDUCE MICRONUCLEI, SPERM ABNORMALITIES OR DOMINANT LETHAL MUTATIONS IN RODENTS. NO DATA WERE AVAILABLE CONCERNING EFFECTS ON REPRODUCTION OR PRENATAL TOXICITY.

REPORT 03 : NIOSH CURRENT INTELLIGENCE BULLETIN 19 JAN 78
THE NIOSH RECOMMENDS THAT 2,4-DIAMINOANISOLE AND ITS SALTS BE HANDLED IN THE WORKPLACE AS IF THEY WERE HUMAN CARCINOGENS. RECOMMENDATION IS BASED ON A PRELIMINARY ANALYSIS OF NCI DATA INDICATING LABORATORY RATS AND MICE FED THE 2,4-DIAMINOANISOLE SULFATE EXPERIENCED A STATISTICALLY SIGNIFICANT EXCESS OF SITE-SPECIFIC TUMORS COMPARED TO CONTROLS.

REPT'D USE : VETERINARY MEDICINE, TREATMENT OF CYSTITIS, PROSTATIS AND LOCAL ANTISEPTIC

CAS NO : **95-80-7**
CHEM NAME : ***** 2,4-DIAMINOTOLUENE**

REPORT 01 : NCI CARC TRS 162 DHEW PUB NO 79-1718
UNDER THE CONDITIONS OF THIS BIOASSAY, 2,4-DIAMINOTOLUENE WAS CARCINOGENIC FOR F344 RATS, INDUCING HEPATOCELLULAR CARCINOMAS OR NEO-PLASTIC NODULES IN BOTH MALES AND FEMALES AND CARCINOMAS OR ADENOMAS IN THE MAMMARY GLAND IN FEMALES. THE TEST CHEMICAL WAS ALSO CARCINOGENIC FOR FEMALE B6C3F1 MICE, INDUCING HEPATOCELLULAR CARCINOMAS. THE INCIDENCE OF LYMPHOMAS IN THE FEMALE MICE SUGGESTED THAT THESE TUMORS ALSO MAY HAVE BEEN RELATED TO ADMINISTRATION OF THE TEST CHEMICAL. THE COMPOUND WAS ADMINISTERED IN FEED IN THIS STUDY.

REPORT 02 : EPA RCRA HEALTH AND ENV. EFFECTS PROFILE OCT 80
2,4-TOLUENEDIAMINE PRODUCED CARCINOGENIC EFFECTS IN RATS AND MICE IN AN NCI BIOASSAY. 2,4-TDA WAS FOUND TO BE MUTAGENIC, USING MUTANTS OF SALMONELLA, HAMSTER EMBRYO CELL SYSTEMS, AND DROSPHILLA.

REPORT 03 : IARC MONOGRAPH 16 JUN 77
2,4-DIAMINOTOLUENE IS CARCINOGENIC IN RATS AFTER ITS ORAL ADMINISTRATION, PRODUCING HEPATOCELLULAR CARCINOMAS, AND AFTER ITS SUBCUTANEOUS INJECTION, INDUCING LOCAL SARCOMAS. IN A 1979 SUPPLEMENT, IARC DETERMINED THAT THERE IS SUFFICIENT EVIDENCE FOR THE CARCINOGENICITY OF 2,4-DIAMINOTOLUENE IN ANIMALS.

REPORT 04 : EPA TSCA 8(E) NOTICE 0367 OCT 80
OLIN CHEMICALS GROUP SUBMITTED A HEALTH HAZARD EVALUATION CONDUCTED BY NIOSH AT THE REQUEST OF WORKERS. MEDICAL FINDINGS INCLUDED A SIGNIFICANT REDUCTION IN SPERM COUNT IN THE EXPOSED GROUP. NIOSH DETERMINED THAT WHILE THIS STUDY CANNOT BE INTERPRETED AS CONCLUSIVE FOR TOXIC EFFECTS OF DNA OR DNT ON THE MALE REPRODUCTIVE SYSTEM, IT IS STRONGLY SUGGESTIVE OF A PROBLEM. EFFECTS WERE OBSERVED IN THE PRESENCE OF DECLINING DNT AND TDA

EXPOSURES, NONE OF WHICH EXCEEDED CURRENT OSHA EXPOSURE LIMITS DURING THE SURVEY.

REPORT 05 : EPA RESPONSE TO ITC (47FR973) JAN 82
IN RESPONSE TO ITC TESTING RECOMMENDATION EPA PROPOSES TO REQUIRE TESTING OF 13 PDAS INCULDING 2,4-DIAMINOTOLUENE FOR ONCOGENICITY, MUTAGENICITY, TERATOGENICITY, NEUROTOXICITY, REPRODUCTIVE EFFECTS, AND ENVIRONMENTAL EFFECTS AND FATE.

REPORT 06 : MULTI-AGENCY CROSS CHECK
Y HHS MAGUIRE
Y OSHA CARC
N EPA CAG
Y EPA RCRA
N EPA CLEAN AIR
N EPA CLEAN WATER

REPORT 07 : NTP TECHNICAL BULLETIN 5 AUG 81
2,4-DIAMINOTOLUENE WAS POSITIVE IN THE SALMONELLA TYPHIMURIUM MUTAGENICITY ASSAY IN ONE LABORATORY.

REPORT 08 : IARC MONOGRAPH 16 1978
NO DATA ON THE EMBRYOTOXICITY OR TERATOGENICITY OF 2, 4-DIAMINOTOLUENE WERE AVAILABLE. THE CHEMICAL WAS MUTAGENIC IN SALMONELLA TA1538 AND TA98, WITH METABOLIC ACTIVATION, AND IS A WEAK MUTAGEN IN DROSOPHILA. IT INDUCED MORPHOLOGICAL TRANSFORMATION IN SYRIAN GOLDEN HAMSTER EMBRYO CELLS.

REPORT 09 : EPA FRYI-OTS 0288-0600 FEB 88
CHEVRON REPORTS 2,4-DIAMINOTOLUENE UNINTENTIONALLY FORMED IN THE MANUFACTURE OF CHEVRON AUTOMOTIVE GREASE.

RFPT'D USE : INTERMEDIATE IN TOLUENE DIISOCYANATE PRODUCTION AND DYES

CAS NO : **95-70-5**

CHEM NAME : ***** 2,5-DIAMINOTOLUENE**

REPORT 01 : IARC MONOGRAPH 16 JAN 78
NO DATA ON 2,5-DIAMINOTOLUENE OR 2,5-DIAMINOTOLUENE SULFATE ALONE WERE AVAILABLE. NO EVALUATION OF THEIR CARCINOGENICITY CAN BE MADE. THE CHEMICAL INDUCES REVERSE MUTATIONS IN SALMONELLA TA1538 WITH METABOLIC ACTIVATION; NO DOMINANT LETHALS WERE INDUCED IN RATS INJECTED IP.

CAS NO : **119-90-4**
CHEM NAME : ***** O-DIANISIDINE (3,3'-DIMETHOXYBENZIDINE)**

REPORT 01 : EPA RESPONSE TO ITC RECOMMS (46FR5004) NOV 81
IN RESPONSE TO THE ITC RECOMMENDATIONS (44FR70664) THE EPA HAS DECIDED NOT TO PURSUE TESTING PROPOSAL FOR O-DIANISIDINE DYES AT THIS TIME BECAUSE TESTING PROGRAMS WHICH ARE EXPECTED TO SUPPLY DATA SUFFICIENT FOR EPA NEEDS ARE BEING PLANNED AND CONDUCTED, AND REGULATORY EFFORTS ARE BEING PURSUED BY THE OSHA AND THE CPSC .

REPORT 02 : IARC MONOGRAPH 4 1974
3,3'-DIMETHOXYBENZIDINE (O-DIANISIDINE) WAS SHOWN TO HAVE A CARCINOGENIC EFFECT IN RATS FOLLOWING ORAL ADMINISTRATION. THE FINDINGS OBTAINED IN THE HAMSTER BY THE SAME ROUTE SUGGEST A SIMILAR EFFECT. IN A 1979 SUPPLEMENT, IARC DETERMINED THAT THERE IS SUFFICIENT EVIDENCE FOR THE CARCINOGENICITY OF 3,3'-DIMETHOXYBENZIDINE IN ANIMALS.

REPORT 03 : NTP CARC TEST PROGRAM JUL 82
IN A CURRENT NTP BIOASSAY, O-DIANISIDINE IS BEING ADMINISTERED IN THE WATER TO F344 RATS. THE PRECHRONIC TEST PHASE IS IN PROGRESS.

REPORT 04 : NTP TECHNICAL BULLETIN 2/8 APR 80, JUL 82
O-DIANISIDINE WAS POSITIVE IN THE SALMONELLA ASSAY BUT NEGATIVE IN DROSOPHILA WHEN TESTED FOR SEX-LINKED RECESSIVE LETHALS.

REPORT 06 : MULTI-AGENCY CROSS CHECK
Y HHS MAGUIRE
N OSHA CARC
Y EPA CAG
Y EPA RCRA
N EPA CLEAN AIR
N EPA CLEAN WATER

REPORT 07 : IARC SUPPLEMENT 4 OCT 82
THE EVIDENCE FOR CARCINOGENICITY TO HUMANS IS INADEQUATE. A RUSSIAN STUDY DID NOT FIND ANY CASES OF OCCUPATIONAL URINARY BLADDER NEOPLASM WHICH COULD BE TRACED DIRECTLY TO THE CPD. THE CPD PRODUCED TUMORS IN RATS AT VARIOUS SITES, INCLUDING THE BLADDER, INTESTINE, SKIN AND ZYMBAL GLAND WHEN ADMINISTERED ORALLY,

REPT'D USE : DYE INTERMEDIATE, DYE, CHEMICAL INTERMEDIATE IN PROD. OF DISOCYANATE AND METALLURGICAL ANALYSIS

CAS NO : **439-14-5**
CHEM NAME : ***** DIAZEPAM**

REPORT 01 : IARC MONOGRAPH 13 1977
IN THE ONLY REPORT AVAILABLE, OXAZEPAM, A MAJOR METABOLITE OF DIAZEPAM, WAS CARCINOGENIC IN MICE AFTER ITS ORAL ADMINISTRATION: IT PRODUCED LIVER-CELL TUMORS.
REPT'D USE : TRANQUILIZER AND ANTIEPILEPTIC AGENT

CAS NO : **333-41-5**
CHEM NAME : ***** DIAZINON**

REPORT 01 : NCI CARC TRS 137 DHEW PUB NO 79-1392
UNDER THE CONDITIONS OF THIS BIOASSAY, DIAZINON WAS NOT CARCINOGENIC FOR F344 RATS OR B6C3F1 MICE OF EITHER SEX (FEEDING STUDY).
REPT'D USE : PESTICIDE, ACARICIDE AND NEMATOCIDE

CAS NO : **334-88-3**
CHEM NAME : ***** DIAZOMETHANE**

REPORT 01 : IARC MONOGRAPH 7 1974
LIMITED STUDIES INDICATE THAT DIAZOMETHANE IS CARCINOGENIC IN MICE AND RATS, THE ONLY SPECIES TESTED. IN MICE IT INCREASED THE INCIDENCE OF LUNG TUMORS FOLLOWING SKIN APPLICATION. EXPOSURE TO THE GAS INDUCED LUNG TUMORS IN RATS.
REPT'D USE : LABORATORY METHYLATING AGENT

CAS NO : **226-36-8**
CHEM NAME : ***** DIBENZ(A,H)ACRIDINE**

REPORT 01 : IARC MONOGRAPH 3 1973
IN MICE, PAINTING WITH DB(A,H)AC INDUCES SKIN TUMORS. SUBCUTANEOUS ADMINISTRATION TO MICE INDUCED LOCAL SARCOMAS AND INCREASED THE INCIDENCE OF LUNG TUMORS. IN A 1979 SUPPLEMENT, IARC DETERMINED THAT THERE IS SUFFICIENT EVIDENCE FOR THE CARCINOGENICITY OF DB(A,H)AC IN ANIMALS.

REPORT 02 : EPA AMBIENT WATER QUALITY CRITERIA OCT 80
AVAILABLE DATA CANNOT SUPPORT INDIVIDUAL CRITERIA FOR EACH POLYNUCLEAR AROMATIC HYDROCARBON. THE AWQC FOR CARCINOGENIC PAH COMPOUNDS IS BASED ON THE ASSUMPTION THAT EACH COMPOUND IS AS POTENT AS B(A)P AND THAT THE CARCINOGENIC EFFECT OF THE COMPOUNDS IS PROPORTIONAL TO

THE SUM OF THEIR CONCENTRATIONS. THIS SUM SHOULD NOT EXCEED 28 NG/L, 2.8 NG/L, AND .28 NG/L TO KEEP THE LIFETIME CANCER RISK BELOW 10 -5, 10 -6, AND 10 -7 RESPECTIVELY.

REPORT 03 : IARC MONOGRAPH 32 DEC 83
DIBENZ(A,H)ACRIDINE HAS BEEN SHOWN TO BE CARCINOGENIC TO EXPERIMENTAL ANIMALS (SEE IARC, 1973). BOTH POSITIVE AND NEGATIVE RESULTS OBTAINED IN TESTS OF MUTAGENICITY IN SALMONELLA TYPHIMURIUM.

REPORT 06 : MULTI-AGENCY CROSS CHECK
Y HHS MAGUIRE
N OSHA CARC
Y EPA CAG
N EPA RCRA
N EPA CLEAN AIR
N EPA CLEAN WATER
REPT'D USE : TOBACCO SMOKE

CAS NO : **224-42-0**
CHEM NAME : *****DIBENZ(A,J)ACRIDINE**

REPORT 01 : IARC MONOGRAPH 3 1973
DB(A,J)AC INDUCED SKIN TUMORS IN MICE FOLLOWING TOPICAL APPLICATION AND PRODUCED LOCAL SARCOMAS AT THE HIGHEST DOSE TESTED FOLLOWING SUBCUTANEOUS ADMINISTRATION. IT INCREASED THE INCIDENCE OF LUNG TUMORS AFTER S.C. ADMINISTRATION. IN A 1979 SUPPLEMENT, IARC DETERMINED THAT THERE IS SUFFICIENT EVIDENCE FOR THE CARCINOGENICITY OF DB(A,J)AC IN ANIMALS.

REPORT 02 : IARC MONOGRAPH 32 DEC 83
DIBENZ(A,J)ACRIDINE SHOWN TO BE CARCINOGENIC TO EXPERIMENTAL ANIMALS (SEE IARC 3, 1973). IN ABSENCE OF ADEQUATE DATA ON HUMANS IT IS REASONABLE, FOR PRACTICAL PURPOSES, TO REGARD THERE IS CARCINOGENIC RISK TO HUMANS. MUTAGENIC TO SALMONELLA TYPHIMURIUM IN PRESENCE OF EXOGENOUS METABOLIC SYSTEM. DID NOT INDUCE UNSCHEDULED DNA SYNTHESIS IN RAT'S HEPATOCYTES.

REPORT 06 : MULTI-AGENCY CROSS CHECK
Y HHS MAGUIRE
N OSHA CARC
Y EPA CAG
N EPA RCRA
N EPA CLEAN AIR
N EPA CLEAN WATER
REPT'D USE : COMBUSTION POLLUTANT AND RESEARCH CHEMICAL

CAS NO : **215-58-7**
CHEM NAME : ***** DIBENZ(A,C)ANTHRACENE**

REPORT 01 : IARC MONOGRAPH 32 DEC 83
TESTED FOR CARCINOGENICITY BY SKIN APPLICATION TO FEMALE MICE AND PRODUCED SKIN TUMORS. TESTED IN THREE EXPERIMENTS USING MOUSE SKIN INTIATION-PROMOTION ASSAY AND SHOWED INITIATING ACTIVITY.

REPORT 02 : IARC MONOGRAPH 32 DEC 83
IN PRESENCE OF AN EXOGENOUS METABOLIC SYSTEM, DIBENZ(A, C)ANTHRACENE WAS MUTAGENIC TO S. TYPHIMURIUM AND MAMMALIAN CELLS IN CULTURE, POSITIVE IN ASSAY FOR DIFFERENTIAL SURVIVAL, USING DNA REPAIR--PROFICIENT/DEFICIENT STRAINS OF B. SUBTILIS, AND INDUCED UNSCHEDULED DNA REPAIR SYNTHESIS IN CULTURED MAMMALIAN CELLS. MORPHOLOGICAL TRANSFORMATION INDUCED IN MAMMALIAN CELLS IN TWO STUDIES.
REPT'D USE : POLLUTANT AND TOBACCO SMOKE

CAS NO : **53-70-3**
CHEM NAME : ***** DIBENZ(A,H)ANTHRACENE (DBA)**

REPORT 01 : IARC MONOGRAPH 3 1973
DB(A,H)A HAS PRODUCED TUMORS BY DIFFERENT ROUTES OF ADMINISTRATION IN MICE, RATS, GUINEA PIGS, FROGS, PIGEONS AND CHICKENS. IT HAS BOTH LOCAL AND SYSTEMIC CARCINOGENIC EFFECTS. IN A 1979 SUPPLEMENT, IARC DETERMINED THAT THERE IS SUFFICIENT EVIDENCE FOR THE CARCINOGENICITY OF DB(A,H)A IN ANIMALS.

REPORT 02 : EPA RCRA HEALTH AND ENV. EFFECTS PROFILE OCT 80
DB(A,H)A WAS THE FIRST PURE CHEMICAL SHOWN TO PRODUCE TUMORS IN ANIMALS. IT IS CARCINOGENIC BY SKIN APPLICATION, BY INJECTION, AND BY ORAL ADMINISTRATION TO RODENTS.

REPORT 03 : EPA AMBIENT WATER QUALITY CRITERIA OCT 80
AVAILABLE DATA CANNOT SUPPORT INDIVIDUAL CRITERIA FOR EACH PAH. THE AWQC FOR CARCINOGENIC PAH COMPOUNDS IS BASED ON THE ASSUMPTION THAT EACH COMPOUND IS AS POTENT AS B(A)P AND THAT THE CARCINOGENIC EFFECT OF THE COMPOUNDS IS PROPORTIONAL TO THE SUM OF THEIR CONCENTRATIONS. THIS SUM SHOULD NOT EXCEED 28 NG/L, 2.8 NG/L AND 0.28 NG/L TO KEEP THE LIFETIME CANCER RISK BELOW 10 -5, 10 -6 AND 10 -7 RESPECTIVELY.

REPORT 04 : NAS REPORT ON DIET, NUTRITION AND CANCER 1982
DIETARY ADMINISTRATION OF DBA FOR 5 TO 7 MONTHS RESULTED IN THE APPEARANCE OF FORESTOMACH TUMORS IN MICE AFTER

ONE YEAR. ADMINISTRATION OF DBA BY STOMACH TUBE TO SEVERAL STRAINS OF MICE RESULTED IN A SIGNIFICANT INCREASE IN MAMMARY CARCINOMAS. DBA WAS POSITIVE IN THE SALMONELLA ASSAY AND INDUCED TRANSFORMATION IN SYRIAN GOLDEN HAMSTER EMBRYO CELLS.

REPORT 05 : IARC MONOGRAPH 32 DEC 83
DIBENZ(A,H)ANTHRACENE WAS SHOWN TO BE CARCINOGENIC TO EXPERIMENTAL ANIMALS (SEE IARC 3, 1973). EMBRYOTOXIC TO RATS WHEN GIVEN AT HIGH DOSES, BUT AVAILABLE DATA ON TERATOGENICITY WERE INADEQUATE FOR EVALUATION. POSITIVE IN DIFFERENTIAL SURVIVAL ASSAYS USING DNA-REPAIR-PROFICIENT/DEFICIENT STRAINS OF BACTERIA AND MUTAGENIC TO SALMONELLA TYPHIMURUIUM IN PRESENCE OF EXOGENOUS METABOLIC SYSTEM. IN CULTURED MAMMALIAN CELLS, DIBENZ(A,H)ANTHRACENE WAS MUTAGENIC AND INDUCED DNA SYNTHESIS IN PRESENCE OF EXOGENOUS METABOLIC SYSTEM. IT WAS POSITIVE IN ASSAYS FOR MORPHOLOGICAL TRANSFORMATION AND IT INDUCED SISTER CHROMATID EXCHANGE BUT NOT CHROMOSOMAL ABERRATIONS IN VIVO.

REPORT 06 : MULTI-AGENCY CROSS CHECK
Y HHS MAGUIRE
N OSHA CARC
Y EPA CAG
Y EPA RCRA
N EPA CLEAN AIR
Y EPA CLEAN WATER

REPT'D USE : COMBUSTION POLLUTANT

CAS NO : **224-41-9**
CHEM NAME : ***** DIBENZ(A,J)ANTHRACENE**

REPORT 01 : IARC MONOGRAPH 32 DEC 83
DIBENZ(A,J)ANTHRACENE TESTED FOR CARCINOGENICITY IN FEMALE MICE BY SKIN APPLICATION AND PRODUCED SKIN TUMORS. WHEN TESTED IN MICE BY SC INJECTION, IT PRODUCED A FEW TUMORS AT THE SITE OF INJECTION. IT WAS MUTAGENIC TO S. TYPHIMURIUM IN PRESENCE OF EXOGENOUS METABOLIC SYSTEM.

REPT'D USE : AIR POLLUTANT

CAS NO : **194-59-2**
CHEM NAME : ***** 7H-DIBENZO(C,G)CARBAZOLE**

REPORT 01 : IARC MONOGRAPH 3 1973
7H-DB(C,G)C IS CARCINOGENIC IN THE MOUSE, RAT, HAMSTER AND POSSIBLY IN THE DOG. IT HAS BOTH A LOCAL AND SYSTEMIC

CARCINOGENIC EFFECT. IN A 1979 SUPPLEMENT, IARC DETERMINED THAT THERE IS SUFFICIENT EVIDENCE FOR THE CARCINOGENICITY OF 7H-DB(C,G)C IN ANIMALS.

REPORT 02 : EPA AMBIENT WATER QUALITY CRITERIA OCT 80
AVAILABLE DATA CANNOT SUPPORT INDIVIDUAL CRITERIA FOR EACH POLYNUCLEAR AROMATIC HYDROCARBON. THE AWQC FOR CARCINOGENIC PAH COMPOUNDS IS BASED ON THE ASSUMPTION THAT EACH COMPOUND IS AS POTENT AS B(A)P AND THAT THE CARCINOGENIC EFFECT OF THE COMPOUNDS IS PROPORTIONAL TO THE SUM OF THEIR CONCENTRATIONS. THIS SUM SHOULD NOT EXCEED 28 NG/L, 2.8 NG/L, AND .28 NG/L TO KEEP THE LIFETIME CANCER RISK BELOW 10 -5, 10 -6, AND 10 -7 RESPECTIVELY.

REPORT 03 : IARC MONOGRAPH 32 DEC 83
7H-DIBENZO(C,G)CARBAZOLE HAS BEEN SHOWN TO BE CARCINOGENIC TO EXPERIMENTAL ANIMALS (SEE IARC 3, 1973). THERE WERE INSUFFICIENT DATA AVAILABLE TO EVALUATE THE MUTAGENICITY OF 7H-DIBENZO-(C,G)CARBAZOLE TO SALMONELLA TYPHIMURIUM.

REPORT 06 : MULTI-AGENCY CROSS CHECK
Y HHS MAGUIRE
N OSHA CARC
Y EPA CAG
N EPA RCRA
N EPA CLEAN AIR
N EPA CLEAN WATER
REPT'D USE : CIGARETTE TAR/SMOKE

CAS NO : **262-12-4**
CHEM NAME : ***** DIBENZO-P-DIOXIN**

REPORT 01 : NCI CARC TRS 122 DHEW PUB NO 79-1377
UNDER THE CONDITIONS OF THIS BIOASSAY, UDD WAS NOT CARCINOGENIC FOR OSBORNE-MENDEL RATS OR B6C3F1 MICE (FEEDING STUDY).
REPT'D USE : BY PRODUCT DURING MFR OF CHLOROPHENOLS

CAS NO : **5385-75-1**
CHEM NAME : ***** DIBENZO(A,E)FLUORANTHENE**

REPORT 01 : IARC MONOGRAPH 32 DEC 83
DIBENZO(A,E)FLUORANTHENE PRODUCED SKIN TUMORS WHEN TESTED BY SKIN APPLICATION TO MICE. ACTIVE AS AN INITIATOR IN MOUSE SKIN INITIATION-PROMOTION ASSAY. SC ADMINISTRATION IN MICE PRODUCED SARCOMAS AT THE INJECTION SITE. IT WAS MUTAGENIC TO S. TYPHIMURIUM TA100 (HIS-/HIS+) WHEN

ADDED AT A CONCENTRATION OF 100 UG/PLATE IN PRESENCE OF AN EXOGENOUS METABOLIC SYSTEM (POST-MITOCHONDRIAL SUPERNATANT FROM AROCLOR-INDUCED RAT LIVER).

REPT'D USE : TOBACCO SMOKE

CAS NO : **194-47-2**

CHEM NAME : ***** DIBENZO(H,RST)PENTAPHENE**

REPORT 01 : IARC MONOGRAPH 3 1973
IN THE ONLY EXPERIMENT REPORTED, SUBCUTANEOUS INJECTION OF DIBENZO(H,RST)PENTAPHENE IN MICE INDUCED SARCOMAS.

REPORT 02 : EPA AMBIENT WATER QUALITY CRITERIA OCT80
AVAILABLE DATA CANNOT SUPPORT INDIVIDUAL CRITERIA FOR EACH PAH. THE AW4C FOR CARCINOGENIC PAH COMPOUNDS IS BASED ON THE ASSUMPTIONS THAT EACH COMPOUND IS AS POTENT AS B(A)P AND THAT THE CARCINOGENIC EFFECT OF THE COMPOUNDS IS PROPORTIONAL TO THE SUM OF THEIR CONCENTRATIONS THIS SUM SHOULD NOT EXCEED 28 NG/L, 2.8 NG/L AND 0.28 NG/L TO KEEP LIFETIME CANCER RISK BELOW 10 -5, 10 -6 AND 10 -7 RESPECTIVELY.

REPT'D USE : AIR POLLUTANT

CAS NO : **192-65-4**

CHEM NAME : ***** DIBENZO(A,E)PYRENE**

REPORT 01 : IARC MONOGRAPH 3 1973
DB(A,E)P HAS PRODUCED TUMORS IN MICE FOLLOWING SKIN PAINTINGS AND S.C. INJECTION. IN SKIN PAINTING EXPERIMENTS, DB(A,E)P WAS LESS ACTIVE THAN B(A)P. IN A 1979 SUPPLEMENT, IARC DETERMINED THAT THERE IS SUFFICIENT EVIDENCE FOR THE CARCINOGENICITY OF DB(A,E)P IN ANIMALS.

REPORT 02 : EPA AMBIENT WATER QUALITY CRITERIA OCT 80
AVAILABLE DATA CANNOT SUPPORT INDIVIDUAL CRITERIA FOR EACH PAH. THE AWQC FOR CARCINOGENIC PAH COMPOUNDS IS BASED ONT HE ASSUMPTIONS THAT EACH COMPOUND IS AS POTENT AS B(A)P AND THAT THE CARCINOGENIC EFFECT OF THE COMPOUNDS IS PROPORTIONAL TO THE SUM OF THEIR CONCENTRATIONS. THIS SUM SHOULD NOT EXCEED 28 NG/L, 2.8 NG/L AND 0.28 NG/L TO KEEP THE LIFETIME CANCER RISK BELOW 10 -5, 10 -6 AND 10 -7 RESPECTIVELY.

REPORT 03 : IARC MONOGRAPH 32 DEC 83
DIBENZO(A,E)PYRENE HAS BEEN SHOWN TO BE CARCINOGENIC TO EXPERIMENTAL ANIMALS (SEE IARC, 3, 1973). MUTAGENIC TO SALMONELLA TYPHIMURIUM IN THE PRESENCE OF EXOGENOUS METABOLIC SYSTEM.

REPORT 06 : MULTI-AGENCY CROSS CHECK
Y HHS MAGUIRE
N OSHA CARC
Y EPA CAG
N EPA RCRA
N EPA CLEAN AIR
N EPA CLEAN WATER
RFPT'D USE : TOBACCO SMOKE

CAS NO : **189-64-0**
CHEM NAME : ***** DIBENZO(A,H)PYRENE**

REPORT 01 : IARC MONOGRAPH 3 1973
CARCINOGENIC EFFECTS OF DB(A,H)P WERE DEMONSTRATED FOLLOWING REPEATED SKIN PAINTING IN MICE AND INJECTIONS IN MICE AND RATS. IN A 1979 SUPPLEMENT, IARC DETERMINED THAT THERE IS SUFFICIENT EVIDENCE FOR THE CARCINOGENICITY OF DB(A,H)P IN ANIMALS.

REPORT 02 : EPA AMBIENT WATER QUALITY CRITERIA
AVAILABLE DATA CANNOT SUPPORT INDIVIDUAL CRITERIA FOR EACH PAH. THE AWQC FOR CARCINOGENIC PAH COMPOUNDS IS BASED ON THE ASSUMPTIONS THAT EACH COMPOUND IS AS POTENT AS B(A)P AND THAT THE CARCINOGENIC EFFECT OF THE COMPOUNDS IS PROPORTIONAL TO THE SUM OF THEIR CONCENTRATIONS. THIS SUM SHOULD NOT EXCEED 28 NG/L, 2.8 NG/L AND 0.28 NG/L TO KEEP THE LIFETIME CANCER RISK BELOW 10 -5, 10 -6 AND 10 -7 RESPECTIVELY.

REPORT 03 : IARC MONOGRAPH 32 DEC 83
DIBENZO(A,H)PYRENE SHOWN TO BE CARCINOGENIC TO EXPERIMENTAL ANIMALS (SEE IARC 3, 1973). MUTAGENIC TO SALMONELLA TYPHIMURIUM IN PRESENCE EXOGENOUS METABOLIC SYSTEM.

REPORT 06 : MULTI-AGENCY CROSS CHECK
Y HHS MAGUIRE
N OSHA CARC
Y EPA CAG
N EPA RCRA
N EPA CLEAN AIR
N EPA CLEAN WATER
REPT'D USE : COMBUSTION POLLUTANT

CAS NO : **189-55-9**
CHEM NAME : ***** DIBENZO(A,I)PYRENE**

REPORT 01 : IARC MONOGRAPH 3 1973
S.C. INJECTIONS OF DB(A,I)P RESULTED IN THE RAPID APPEARANCE

OF LOCAL SARCOMAS IN THE HAMSTER AND THE MOUSE. REPEATED SKIN APPLICATION IN MICE WAS ALSO EFFECTIVE, BUT DB(A,I)P WAS LESS ACTIVE THAN B(A)P. IN A 1979 SUPPLEMENT, IARC DETERMINED THAT THERE IS SUFFICIENT EVIDENCE FOR THE CARCINOGENICITY OF DB(A,I)P IN ANIMALS.

REPORT 02 : EPA AMBIENT WATER QUALITY CRITERIA OCT 80
AVAILABLE DATA CANNOT SUPPORT INDIVIDUAL CRITERIA FOR EACH PAH. THE AWQC FOR CARCINOGENIC PAH COMPOUNDS IS BASED ON THE ASSUMPTIONS THAT EACH COMPOUND IS AS POTENT AS B(A)P AND THAT THE CARCINOGENIC EFFECT OF THE COMPOUNDS IS PROPORTIONAL TO THE SUM OF THEIR CONCENTRATIONS. THIS SUM SHOULD NOT EXCEED 28 NG/L, 2.8 NG/L AND 0.28 NG/L TO KEEP THE LIFETIME CANCER RISK BELOW 10 -5, 10 -6 AND 10 -7 RESPECTIVELY.

REPORT 03 : IARC MONOGRAPH VOLUME 32 DEC 83
DIBENZO(A,I)PYRENE HAS BEEN SHOWN TO BE CARCINOGENIC TO EXPERIMENTAL ANIMALS (SEE IARC, 1973.) MUTAGENIC TO SALMONELLA TYPHIMURIUM IN PRESENCE OF EXOGENOUS METABOLIC SYSTEM. POSITIVE IN ASSAYS FOR DIFFERENTIAL KILLING IN STRAINS OF DNA-REPAIR-PROFICIENT/DEFICIENT BACTERIA. DID NOT INDUCE UNSCHEDULED DNA SYNTHESIS IN RAT HEPATOCYTES.

REPORT 06 : MULTI-AGENCY CROSS CHECK
Y HHS MAGUIRE
N OSHA CARC
Y EPA CAG
Y EPA RCRA
N EPA CLEAN AIR
Y EPA CLEAN WATER

REPT'D USE : CHEMICAL INTERMEDIATE, PAINT REMOVERS, COMBUSTION POLLUTANT AND EXPERIMENTAL CARCINOGEN. NO LISTING ON TSCA INVENTORY

CAS NO : **191-30-0**
CHEM NAME : ***** DIBENZO(A,L)PYRENE**

REPORT 01 : IARC MONOGRAPH 3 1973
IN THE ONLY STUDY CARRIED OUT WITH DB(A,L)P, SARCOMAS WERE INDUCED IN ALL ANIMALS FOLLOWING SUBCUTANEOUS ADMINISTRATION TO MICE.

REPORT 02 : EPA AMBIENT WATER QUALITY CRITERIA OCT 80
AVAILABLE DATA CANNOT SUPPORT INDIVIDUAL CRITERIA FOR EACH PAH. THE AWQC FOR CARCINOGENIC PAH COMPOUNDS IS BASED ON THE ASSUMPTIONS THAT EACH COMPOUND IS AS POTENT AS B(A)P AND THAT THE CARCINOGENIC EFFECT OF THE

COMPOUNDS IS PROPORTIONAL TO THE SUM OF THEIR CONCENTRATIONS. THIS SUM SHOULD NOT EXCEED 28 NG/L, 2.8 NG/L AND 0.28 NG/L TO KEEP THE LIFETIME CANCER RISK BELOW 10 -5, 10 -6 AND 10 -7 RESPECTIVELY.

REPORT 03 : IARC MONOGRAPH 32 DEC 83
DIBENZO(A,L)PYRENE TESTED FOR CARCINOGENICITY IN MICE IN ONE STUDY BY SKIN APPLICATION AND IN ONE STUDY BY SUBCUTANEOUS ADMINISTRATION. INDUCED TUMORS AT SITES OF APPLICATION.

REPORT 06 : MULTI-AGENCY CROSS CHECK
Y HHS MAGUIRE
N OSHA CARC
N EPA CAG
N EPA RCRA
N EPA CLEAN AIR
N EPA CLEAN WATER
REPT'D USE : COMBUSTION PRODUCT AND CIGARETTE SMOKE

CAS NO : **96-12-8**
CHEM NAME : ***** 1, 2-DIBROMO-3-CHLOROPROPANE, DBCP**

REPORT 01 : NCI CARC TRS 28 DHEW PUB NO 78-828
UNDER THE CONDITIONS OF THIS STUDY, DBCP IS A STOMACH CARCINOGEN IN RATS AND MICE OF BOTH SEXES AND IS CARCINOGENIC TO THE MAMMARY GLAND IN FEMALE RATS. DBCP WAS ADMINISTERED BY GAVAGE IN THIS STUDY.

REPORT 02 : NIOSH CRITERIA DOCUMENT SEP 77
NIOSH RECOMMENDS A 10 PPB CEILING BASED ON RENAL AND LIVER EFFECTS AND MALE STERILITY. THE PRESENT OSHA STANDARD IS 1 PPB TWA.

REPORT 03 : EPA RCRA HEALTH AND ENV. EFFECTS PROFILE OCT 80
THE EPA CARCINOGEN ASSESSMENT GROUP HAS EVALUATED DBCP AND HAS FOUND SUFFICIENT EVIDENCE THAT THIS COMPOUND IS CARCINOGENIC.

REPORT 05 : IARC MONOGRAPH 20 1979
1,2-DIBROMO-3-CHLOROPROPANE TESTED IN ONE EXPERIMENT IN MICE AND ONE IN RATS BY ORAL ADMINISTRATION PRODUCED SQUAMOUS-CELL CARCINOMAS OF THE FORESTOMACH IN ANIMALS OF BOTH SPECIES AND ADENOCARCINOMAS OF THE MAMMARY GLAND IN FEMALE RATS. IN A 1979 SUPPLEMENT, IARC DETERMINED THAT THERE IS SUFFICIENT EVIDENCE FOR THE CARCINOGENICITY OF DBCP IN ANIMALS.

REPORT 06 : MULTI-AGENCY CROSS CHECK
Y HHS MAGUIRE
R OSHA CARC
Y EPA CAG
Y EPA RCRA
N EPA CLEAN AIR
N EPA CLEAN WATER

REPORT 07 : EPA TSCA8(E) NOTICE 8EHQ-0478-0128 MAY 78
RESULTS OF STERILITY RETESTING OF EMPLOYEES SHOWS THAT NONE OF FOUR WORKERS EXPOSED TO DBCP HAD NORMAL SPERM COUNTS ON RETEST, ALTHOUGH THE RETEST SHOWED SOME IMPROVEMENT, POSSIBLY SIGNIFICANT, IN ONE CASE.

REPORT 08 : NTP BIOASSAY TRS 206 NIH PUB. NO. 82-1762
UNDER THE CONDITIONS OF THIS BIOASSAY, DBCP ADMINISTERED BY INHALATION WAS CARCINOGENIC FOR MALE AND FEMALE F344 RATS, INDUCING INCREASED INCIDENCES OF NASAL CAVITY TUMORS AND TUMORS OF THE TONGUE IN BOTH SEXES, AND CORTICAL ADENOMAS IN THE ADRENAL GLANDS OF FEMALES. DBCP WAS CARCINOGENIC IN MALE AND FEMALE B6C3F1 MICE, INDUCING INCREASED INCIDENCES OF NASAL CAVITY TUMORS AND LUNG TUMORS.

REPORT 09 : EPA TSCA 8(E) NOTICE 8EH4-0278-0056
VELSICOL PROVIDED RESULTS OF STERILITY TESTS CONDUCTED ON EMPLOYEES WHO WERE OCCUPATIONALLY EXPOSED TO DBCP AND TDBP. TWELVE WORKERS HAD OBVIOUSLY INADEQUATE REPRODUCTIVE ABILITY AND 11/12 WORKERS ALSO WERE DEFICIENT IN SPERM MOTILITY. IN ADDITION, EIGHT OF THE MEN FROM THIS GROUP HAD INADEQUATE SPERM MORPHOLOGY.

REPT'D USE : SOIL FUMIGANT AND FR CHEMICAL

CAS NO : **3322-93-8**
CHEM NAME : ***** 1,2-DIBROMO-4-(1,2-DIBROMOETHYL)CYCLOHEXANE**

REPORT 01 : 50FR19460 MAY85
THIS IS EPA'S RESPONSE TO ITC'S DESIGNATION OF TBEC FOR PRIORITY TESTING FOR HEALTH EFFECTS, CHEMICAL FATE AND ECOLOGICAL EFFECTS. EPA IS NOT INITIATING RULEMAKING UNDER TSCA 4(A) BECAUSE FEW PEOPLE ARE EXPOSED AND THEN ONLY AT LOW CONCENTRATIONS.

CAS NO : **10318-26-0**
CHEM NAME : ***** DIBROMODULCITOL**

REPORT 01 : NCI BIOASSAY PROGRAM- CANCER VOL. 40 1977
UNDER THE CONDITIONS OF THE BIOASSAY, IP INJECTION OF

DIBROMODULCITOL WAS CARCINOGENIC TO SWISS-WEBSTER MALE AND FEMALE MICE, CAUSING LYMPHOMAS AND LUNG TUMORS. IT WAS ALSO CARCINOGENIC TO MALE SPRAGUE-DAWLEY (CD) RATS, CAUSING SUBCUTANEOUS TUMORS. IT WAS NOT CARCINOGENIC TO FEMALE RATS.

REPT'D USE : NO AVAILABLE PRODUCTION OR USE DATA

CAS NO : **106-93-4**

CHEM NAME : ***** 1,2-DIBROMOETHANE (EDB)**

REPORT 01 : NCI CARC TRS 86 DHEW PUB NO 78-1336
UNDER THE CONDITIONS OF THIS BIOASSAY, 1,2-DIBROMOETHANE WAS CARCINOGENIC TO OSBORNE-MENDEL RATS AND B6C3F1 MICE. THE COMPOUND INDUCED SQUAMOUS-CELL CARCINOMAS OF THE FORESTOMACH IN RATS OF BOTH SEXES, HEPATOCELLULAR CARCINOMAS IN FEMALE RATS, AND HEMANGIOSARCOMAS IN MALE RATS. IN MICE OF BOTH SEXES THE COMPOUND INDUCED SQUAMOUS-CELL CARCINOMAS OF THE FORESTOMACH AND ALVEOLAR/BRONCHIOLAR ADENOMAS (FEEDING STUDY).

REPORT 02 : NIOSH CURRENT INTELLIGENCE BULLETINS 3 AND 23 — JUL 75,APR 78
INITIALLY, NIOSH ALERTED THE PUBLIC TO A PRELIMINARY REPORT BY NCI INDICATING THAT EDB IS CARCINOGENIC IN LABORATORY RODENTS. LATER, NIOSH RECOMMENDS THAT NO WORKER BE EXPOSED TO BOTH EDB AND DISULFURAM. THE RECOMMENDATION IS BASED ON PRELIMINARY RESULTS WHICH SUGGEST A VERY SERIOUS TOXIC INTERACTION BETWEEN INHALED ETHYLENE DIBROMIDE AND INGESTED DISULFURAM, RESULTING IN AN EXCEDINGLY HIGH MORTALITY OF LABORATORY RATS.

REPORT 03 : CARC TEST PROGRAM — JUL 82
IN A CURRENT BIOASSAY, EDB WAS ADMINISTERED BY INHALATION TO F344 RATS AND B6C3F1 MICE. THE BIOASSAY HAS BEEN COMPLETED AND THE REPORT IS UNDER REVIEW.

REPORT 04 : NIOSH CRITERIA DOCUMENT — AUG 77
NIOSH RECOMMENDS A CEILING OF 1 MG/CU M TO PROTECT AGAINST CARCINOGENICITY, TERATOGENICITY, STERILITY, AND ADVERSE EFFECTS TO THE LIVER OR KIDNEYS OR HEART. THE PRESENT OSHA STANDARD IS 20 PPM (TWA) WITH A 30 PPM CEILING.

REPORT 05 : IARC MONOGRAPH 15 — 1977
ETHYLENE DIBROMIDE IS CARCINOGENIC IN MICE AND RATS AFTER ITS ORAL ADMINISTRATION, THE ONLY ROUTE TESTED; IT PRODUCED SQUAMOUS-CELL CARCINOMAS OF THE FORESTOMACH. IN A 1979 SUPPLEMENT, IARC DETERMINED THAT THERE IS

SUFFICIENT EVIDENCE FOR THE CARCINOGENICITY OF ETHYLENE DIBROMIDE IN ANIMALS.

REPORT 06 : MULTI-AGENCY CROSS CHECK
Y HHS MAGUIRE
N OSHA CARC
Y EPA CAG
Y EPA RCRA
Y EPA CLEAN AIR
N EPA CLEAN WATER

REPORT 07 : NIOSH CURRENT INTELLIGENCE BULLETIN 37 OCT 81
NIOSH REAFFIRMS ITS 1977 RECOMMENDATION THAT EDB BE TREATED AS A POTENTIAL OCCUPATIONAL CARCINOGEN IN THE WORKPLACE. RECENT ANIMAL STUDIES INVOLVING EXPOSURE TO EDB BY SKIN APPLICATION, ORAL ADMINISTRATION AND INHALATION, SHOWED STATISTICALLY SIGNIFICANT INCREASES IN TUMORS OF THE RESPIRATORY TRACT, MAMMARY GLAND, SPLEEN, AND NASAL CAVITY. INHALATION STUDIES WITH RATS AND MICE AT EDB CONCENTRATIONS BELOW THE OSHA PERMISSIBLE EXPOSURE LIMIT OF 20 PPM DEMONSTRATED A CARCINOGENIC RISK.

REPORT 08 : IARC MONOGRAPH 15 1977
NO DATA ON THE EMBRYOTOXICITY OR TERATOGENICITY OF EDB WERE AVAILABLE. EDB WAS MUTAGENIC TO SALMONELLA TA1530, TA1535, AND TA100 WITHOUT METABOLIC ACTIVATION. SIGNIFICANT INCREASES IN RECESSIVE LETHALS WERE OBSERVED IN DROSOPHILA. EDB CAUSED NO DOMINANT LETHAL MUTATIONS IN MICE.

REPORT 09 : NTP TECHNICAL BULLETIN 8 JUL 82
1,2-DIBROMOETHANE WAS POSITIVE IN THE L5178Y MOUSE LYMPHOMA ASSAY.

REPT'D USE : SOLVENT, PESTICIDE, CHEMICAL INTERMEDIATE AND LEAD SCAVENGER IN FUEL

CAS NO : **488-41-5**

CHEM NAME : ***** DIBROMOMANNITOL**

REPORT 01 : NCI BIOASSAY PROGRAM - CANCER VOL. 40 1977
UNDER THE CONDITIONS OF THE BIOASSAY, I.P. INJECTION OF DIBROMOMANNITOL WAS CARCINOGENIC TO SWISS-WEBSTER MALE AND FEMALE MICE, CAUSING LUNG TUMORS IN BOTH SEXES AND LYMPHOMAS IN FEMALES. IT WAS ALSO CARCINOGENIC TO MALE AND FEMALE SPRAGUE-DAWLEY (CD) RATS, CAUSING PERITONEAL SARCOMAS. IN ADDITION, MALE RATS ALSO HAD SUBCUTANEOUS TUMORS.

REPT'D USE : NO AVAILABLE PRODUCTION OR USE DATA

CAS NO : **96-13-9**
CHEM NAME : ***** 2,3-DIBROMO-1-PROPANOL**

REPORT 01 : NTP TECHNICAL BULLETIN 8 JUL 8
2,3-DIBROMO-1-PROPANOL CAUSED AN INCREASED NUMBER OF SEX-LINKED RECESSIVE LETHALS IN DROSOPHILIA.

REPORT 02 : CPSC FLAME RETARDANTS MONOGRAPH JUL 82
THE COMPOUND WAS MUTAGENIC IN SALMONELLA TA100 AND TA1535. ACTIVITY WAS INCREASED FURTHER IN THE PRESENCE OF METABOLIC ACTIVATION. 2,3-DIBROMOPROPANOL WAS NEGATIVE IN SALMONELLA TA1538, WITH AND WITHOUT ACTIVATION, AND SHOWED WEAK ACTIVITY IN TA98 WITH ACTIVATION.
REPT'D USE : INTERMEDIATE IN PREPARATION OF FLAME RETARDANTS

CAS NO : **19660-16-3**
CHEM NAME : ***** 2,3-DIBROMOPROPYL ACRYLATE**

REPORT 01 : NTP TECHNICAL BULLETIN 9 APR 83
2,3-DIBROMOPROPYL ACRYLATE WAS POSITIVE IN ONE SALMONELLA TYPHIMURIUM MUTAGENICITY ASSAY.

CAS NO : **3066-70-4**
CHEM NAME : ***** 2,3-DIBROMOPROPYL METHACRYLATE**

REPORT 01 : NTP TECHNICAL BULLETIN 9 APR 83
2,3-DIBROMOPROPYL METHACRYLATE WAS POSITIVE IN ONE SALMONELLA TYPHIMURIUM MUTAGENICITY ASSAY.

CAS NO : **84-74-2**
CHEM NAME : ***** DI-N-BUTYL PHTHALATE (PHTHALIC ACID, DIBUTYL ESTER) (DBP)**

REPORT 01 : EPA RCRA HEALTH AND ENV. EFFECTS PROFILE OCT 80
TERATOGENIC EFFECTS IN RATS HAVE BEEN REPORTED IN TESTING OF DI-N-BUTYL PHTHALATE FOLLOWING I.P. ADMINISTRATION, BUT NOT AFTER ORAL ADMINISTRATION AT HIGH DOSES. RATS FED DI-N-BUTYL PHTHALATE OR ITS MONOESTER METABOLITE HAVE DEVELOPED TESTICULAR DAMAGE AND ATROPHY.

REPORT 02 : EPA AMBIENT WATER QUALITY CRITERIA OCT 80
DI-N-BUTYL PHTHALATE HAS NOT BEEN FOUND TO BE CARCINOGENIC IN ANIMALS OR MAN. A CRITERION, BASED ON CHRONIC TOXICOLOGIC DATA IN RATS, AND AN UNCERTAINTY FACTOR OF 100, IS 34.1 MG/L.

REPORT 03 : NTP TECHNICAL BULLETIN 6 JAN 82
THE COMPOUND WAS NEGATIVE IN THE SALMONELLA TYPHIMURIUM MUTAGENICITY ASSAY IN ONE LABORATORY.

REPORT 04 : EPA RESPONSE TO ITC RECOMMS (46FR53775) OCT 81
IN RESPONSE TO THE ITC RECOMMENDATIONS (42FR55026), EPA HAS RECEIVED A PROPOSAL FROM THE CHEMICAL MANUFACTURERS ASSN (CMA) IN WHICH THE PHTHALATE ESTER INDUSTRY HAS AGREED TO TEST VOLUNTARILY AND EXPEDITIOUSLY, ALKYL PHTHALATES FOR THE ENVIRONMENTAL HEALTH EFFECTS OF CONCERN TO ITC AND EPA. CONSEQUENTLY, THE EPA HAS DECIDED NOT TO PROPOSE ADDITIONAL TESTING AT THIS TIME.

REPORT 05 : 49FR44142 NOV 84
ANNOUNCES SUBMISSIONS RECEIVED BY EPA DURING FY 84 FROM NEGOTIATED AND OTHER TESTING PROGRAMS. THESE SUBMISSIONS INCLUDE BIODEGRADATION, VAPOR PRESSURE, OCTANOL-WATER PARTITION COEFF. AND ACUTE AOUATIC TOXICITY TESTING.

REPORT 06 : MULTI-AGENCY CROSS CHECK
N HHS MAGUIRE
N OSHA CARC
N EPA CAG
Y EPA RCRA
N EPA CLEAN AIR
Y EPA CLEAN WATER

REPORT 07 : EPA 8(E) STATUS REPORT 8EHQ-0886-0620 AUG86
CELL TRANSFORMATION AND AMES ASSAYS WERE NEGATIVE; MOUSE LYMPHOMA FOR WARD MUTATION ASSAYS IN L5178Y CELLS WITH ACTIVATION GAVE WEAK DOSE-RELATED POSITIVE RESPONSES.

REPORT 08 : 52FR2152 - RECEIPT OF TEST DATA JAN 87
EPA ANNOUNCES RECEIPT OF TEST DATA FROM INDUSTRY'S VOLUNTARY TESTING PROGRAM.

REPT'D USE : PLASTICIZER, PAINTS/VARNISHES, CARPET BACKINGS ADHESIVES, FLOOR WAXES, INSECT REPELLANTS, FLAME RETARDANTS, CLEANING AGENTS AND AEROSOLS

CAS NO : **1067-33-0**
CHEM NAME : ***** DIBUTYLTIN DIACETATE**

REPORT 01 : NCI CARC TRS 183 DHEW PUB NO 79-1739
UNDER THE CONDITIONS OF THIS BIOASSAY, THERE WAS NO CONCLUSIVE EVIDENCE FOR THE CARCINOGENICITY OF DIBUTYLTIN DIACETATE IN MALE 344 RATS OR B6C3F1 MICE OF EITHER SEX. THE LOSS OF TISSUES TAKEN FROM HIGH DOSE

FEMALE RATS IN THIS BIOASSAY PRECLUDED AN EVALUATION OF THE CARCINOGENICITY OF DIBUTYLTIN DIACETATE TO FEMALE FISCHER 344 RATS IN A FEEDING STUDY.

REPORT 02 : NTP TECHNICAL BULLETIN 9 APR 83
DIBUTYLTIN DIACETOTE WAS NEGATIVE IN ONE SALMONELLA TYPHIMURIUM MUTAGENICITY ASSAY.

REPT'D USE : CHEMICAL INTERMEDIATE IN PRODUCTION OF URETHANE

CAS NO : **1649-08-7**

CHEM NAME : ***** 1,2-DICHLORO-1,1-DIFLUOROETHANE**

REPORT 01 : TSCA SEC. 8 (E) JUN.87
INHALATION EXPOSURE TO THIS CHLOROFLUOROCARBON CAN ADVERSELY AFFECT THE REPRODUCTIVE ORGANS/FUNCTION IN MALE RATS; FURTHER EVALUATION OF THE FINDINGS SHOULD BE POSSIBLE UPON EPS'S RECEIPT OF A FULL COPY OF THE FINAL REPORT OF THE ONGOING 90-DAY INHALATION STUDY.

CAS NO : **7572-29-4**

CHEM NAME : ***** DICHLOROACETYLENE**

REPORT 01 : IARC MONOGRAPH 39, 1986 JUN 85
INHALATION TESTS IN MICE INDUCED DOSE-RELATED INCREASES IN ADENOCARCINOMAS IN MALES AND BENIGN TUMORS OF LIVER AND KIDNEY AND AN INCREASED NUMBER OF LYMPHOMAS IN RATS. NO DATA AVAILABLE TO EVALUATE REPRODUCTIVE EFFECTS OR PRENATAL TOXICITY IN HUMANS OR ANIMALS. DCA MUTAGENIC TO S. TYPHIMURIUM WITH OR WITHOUT ACTIVATION. DCA IS A NEUROTOXIN IN HUMANS WITH AN AFFINITY FOR CRANIAL NERVES. THERE ARE NO CASE REPORTS OR EPA STUDIES FOR EVALUATION OF CARCINOGENICITY TO HUMANS. THERE IS LIMITED EVIDENCE FOR CARCINOGENICITY OF DCA TO EXPERIMENTAL ANIMALS.

REPT'D USE : NOT KNOWN TO BE USED COMMERCIALLY

CAS NO : **95-50-1**

CHEM NAME : ***** 1,2-DICHLOROBENZENE (ODCB)**

REPORT 01 : EPA RESPONSE TO ITC RECQMMS (45FR48524) JUL80
IN RESPONSE TO THE ITC RECOMMENDATIONS (42FR55026) THE EPA PROPOSES THAT MANUFACTURERS AND PROCESSORS OF CHLORINATED BENZENES, EXCEPT HEXACHLOROBENZENE CONDUCT TESTING FOR TERATOGENICITY/REPODUCTIVE EFFECTS AND

OTHER CHRONIC EFFECTS. THE EPA WILL CONDUCT MUTAGENICITY TESTS AND DECIDE ON FURTHER TIER TESTING AS APPROPRIATE. THE EPA BELIEVES IT NOT REASONABLE TO REQUIRE AN EPIDEMIOLOGIC STUDY FOR THESE CHEMICALS.

REPORT 02 : IARC SUPPLEMENT 1 1979
NO ADEQUATE DATA ON THE CARCINOGENICITY OF 1,2- OR 1,4-DICHLOROBENZENE IN EXPERIMENTAL ANIMALS WERE AVAILABLE. THERE IS ONE REPORT OF FOUR CASES OF LEUKEMIA IN SUBJECTS EXPOSED TO 1,2- OR 1,4-DICHLOROBENZENE AS A SOLVENT FOR OTHER CHEMICALS OR IN CHLORINATED BENZENE MIXTURES.

REPORT 03 : CMA INHALATION TERATOLOGY STUDY JUN 82
IN A STUDY SPONSORED BY THE US CHLOROBENZENE PRODUCERS, F344 RATS AND INSEMINATED NEW ZEALAND WHITE RABBITS WERE EXPOSED TO 0, 100, 200 OR 400 PPM OF ODCB BY INHALATION ON DAYS 6-15 (RATS) OR 6-18 (RABBITS) OF GESTATION. LOW INCIDENCES OF MALFORMATIONS WERE OBSERVED AMONG FETUSES FROM ALL TEST GROUPS, INCLUDING CONTROLS. EXPOSURE TO ODCB WAS NOT EMBRYOTOXIC OR TERATOGENIC IN EITHER SPECIES AT THE CONCENTRATIONS TESTED.

REPORT 04 : EPA RCRA HEALTH AND ENV. EFFECTS PROFILE OCT 80
THERE HAVE NOT BEEN STUDIES AVAILABLE TO DETERMINE THE CARCINOGENIC OR TERATOGENIC POTENTIAL OF 1,2-DICHLOROBENZENE. 1,2-DCB WAS MUTAGENIC WHEN TESTED WITH THE MOLD ASPERGILLIS AND NEGATIVE WHEN TESTED WITH THE AMES SALMONELLA ASSAY.

REPORT 05 : IARC MONOGRAPH 29 APR 80
NO ADEQUATE STUDY ON ODCB WAS AVAILABLE. ODCB SHOWED NO MUTAGENIC ACTIVITY IN A NUMBER OF SALMONELLA STRAINS TESTED WITH AND WITHOUT METABOLIC ACTIVATION.

REPORT 06 : MULTI-AGENCY CROSS CHECK
N HHS MAGUIRE
N OSHA CARC
N EPA CAG
Y EPA RCRA
N EPA CLEAN AIR
Y EPA CLEAN WATER

REPORT 07 : NTP TECHNICAL BULLETIN 2 APR80
ODCB WAS NEGATIVE IN THE SALMONELLA TYPHIMURIUM MUTAGENICITY ASSAY IN ONE LABORATORY.

REPORT 08 : NTP TR 255 NIH PUB 86-2511 OCT85
UNDER CONDITIONS OF THESE TWO-YEAR STUDIES, NO EVIDNENCE OF CARCINOGENICITY OF 1,2-DICHLOROBENZENE FOR MALE OR FEMALE F344/N RATS OR B6C3F1 MICE RECEIVING 60 OR 120 MG/KG PER DAY BY GAVAGE.

REPORT 09 : EPA TSCA 4(E) TEST RULE (51 FR11728) APR 86
IN RESPONSE TO ITC TESTING RECOMMENDATIONS EPA HEREBY REQUIRES MANUFACTURERS AND PROCESSORS OF 1,2-DICHLOROBENZENE TO CONDUCT CHEMICAL FATE TESTING.

REPORT 10 : EPA TSCA 4(E) TEST RULE-51FR2 4657 JUL 86
IN COMPLIANCE WITH A COURT ORDER EPA HEREBY REQUIRES MFRS AND PROCESSORS OF 1,2-DCB TO CONDUCT REPRODUCTIVE TOX EFFECTS TESTS ACCORDING TO TSCA HEALTH EFFECTS TESTING GUIDELINES IN 40CFR798. THESE RULES ARE TO BE PROMULGATED FOR JUDICIAL REVIEW ON JULY 22, 1986 AND BECOME EFFECTIVE ON AUG 21, 1986.

REPT'D USE : SOLVENT, MFR OF ISOCYANATES, PAINTS, ENGINE CLEANERS, CLEANING/POLISHING AGENTS, DEGREASING AGENTS, WOOD PRESERVATIVES, INSECTICIDE AND DYE INTERMEDIATE

CAS NO : **106-46-7**
CHEM NAME : ***** 1,4-DICHLOROBENZENE (PDCB)**

REPORT 01 : EPA RESPONSE TO ITC RECOMMS (45FR48524) JUL 80
IN RESPONSE TO THE ITC RECOMMENDATIONS (42FR55026) THE EPA PROPOSES THAT MANUFACTURERS AND PROCESSORS OF CHLORINATED BENZENES, EXCEPT HEXACHLOROBENZENE, CONDUCT TESTING FOR TERATOGENICITY/REPRODUCTIVE EFFECTS AND OTHER CHRONIC EFFECTS. THE EPA WILL CONDUCT MUTAGENICITY TESTS AND DECIDE ON FURTHER TESTING AS APPROPRIATE. THE EPA BELIEVES IT IS NOT REASONABLE TO REQUIRE AN EPIDEMIOLOGIC STUDY FOR THESE CHEMICALS.

REPORT 02 : IARC SUPPLEMENT 1 1979
NO ADEQUATE DATA WERE AVAILABLE ON THE CARCINOGENICITY OF 1,2- OR 1,4-DICHLOROBENZENE IN EXPERIMENTAL ANIMALS. THERE IS ONE REPORT OF FOUR CASES OF LEUKEMIA IN SUBJECTS EXPOSED TO 1,2- OR 1,4-DICHLOROBENZENE AS A SOLVENT FOR OTHER CHEMICALS OR IN CHLORINATED BENZENE MIXTURES.

REPORT 03 : NTP TECHNICAL BULLETIN 2 APR 80
PDCB IS NEGATIVE IN THE SALMONELLA TYPHIMURIUM MUTAGENICITY ASSAY IN ONE LABORATORY.

REPORT 04 : RCRA HEALTH AND ENV. EFFECTS PROFILE OCT 80
STUDIES TO DETERMINE THE CARCINOGENIC OR TERATOGENIC POTENTIAL OF 1,4-DICHLOROBENZENE COULD NOT BE LOCATED IN

THE AVAILABLE LITERATURE. 1,4-DCB PRODUCES CHROMOSOMAL ABERRATIONS IN ROOT TIPS AND HAS BEEN SHOWN TO INCREASE THE MUTATION RATE IN THE MOLD ASPERGILLIS.

REPORT 05 : INDUSTRY TEST PROGRAM DOW CHEMICAL /CMA DEC 80
IN A CURRENT STUDY, DOW CHEMICAL IS EVALUATING THE TERATOGENIC POTENTIAL OF PDCB IN FISCHER 344 RATS EXPOSED TO THREE DOSE LEVELS OF THE COMPOUND. THE STUDY IS SCHEDULED FOR COMPLETION IN APRIL 1982.

REPORT 06 : MULTI-AGENCY CROSS CHECK
Y HHS MAGUIRE
N OSHA CARC
N EPA CAG
Y EPA RCRA
Y EPA CLEAN AIR
Y EPA CLEAN WATER

REPORT 07 : NTP CARC TEST PROGRAM JUL 82
IN A CURRENT NTP BIOASSAY, PDCB WAS ADMINISTERED BY GAVAGE TO F344 RATS AND B6C3F1 MICE. THE CHRONIC EXPOSURE PHASE HAS BEEN COMPLETED AND HISTOPATHOLOGY IS IN PROGRESS.

REPORT 08 : IARC MONOGRAPH 29 1982 NO
DATA WERE AVAILABLE FOR REVIEW OF THE CARCINOGENICITY OF PDCB. PDCB SHOWED NO MUTAGENIC ACTIVITY FOR SALMONELLA STRAINS TA1535, TA1537, TA1538, TA98, OR TA100, WITH OR WITHOUT METABOLIC ACTIVATION. NO DATA WERE AVAILABLE TO ASSESS THE TERATOGENICITY OF THIS CHEMICAL.

REPORT 09 : DOW CHEMICAL - CMA SEP 82
INHALATION OF 100, 300 OR 800 PPM OF PDCB VAPORS FOR 6 HRS/DAY ON DAY 6 THROUGH 18 OF GESTATION WAS NOT TERATOGENIC IN NEW ZEALAND WHITE RABBITS.

REPORT 10 : EPA TSCA 4(A) TEST RULE (51FR 11728) APR 86
IN RESPONSE TO ITC TESTING RECOMMENDATIONS EPA HEREBY REQUIRES MANUFACTURERS AND PROCESSORS OF 1,4-DICHLOROBENZENE TO CONDUCT CHEMICAL FATE TESTING.

REPORT 11 : 51 CFR 130 JUL 86
EPA IS ISSUING A FINAL RULE UNDER TSCA 4, REQUIRING PROCESSORS AND MFRS OF PDCB TO CONDUCT REPRODUCTIVE TOXICITY TESTS ACCORDING TO TSCA HEALTH EFFECTS TESTING GUIDELINES IN 40 CFR 798. THIS RULE IS PROMULGATED FOR JUDICIAL REVIEW AT 1300 HRS, JULY 22, 1986 AND BECOMES EFFECTIVE ON AUGUST 21, 1986.

REPORT 12 : EPA NAT'L DW REG'S - VSOC'S, 49 FR 24330 JUN 84
PROPOSED RULE ESTABLISHES RECOMMENDED MAXIMUM CON-

TAMINANT LEVELS (RMC LS) IN DRINKING WATER. RMCLS ARE NON-ENFORCEABLE HEALTH GOALS SET AT "NO KNOWN EFFECT" OR "NO ANTICIPATED EFFECT" LEVEL.

REPORT 13 : PROPOSED RULEMAKING - NAT'L DRINKING WATER REGS, 49FR24330 — JUN 84
EPA PROPOSES TO SET LIMIT OF 0.75 MG/L IN DRINKING WATER FOR PDCB.

REPORT 14 : 51CFR24657 - CHLORINATED BENZENES-FINAL TEST RULE-EPA — JUL 86
EPA IS PROMULGATING A FINAL TEST RULE FOR CHLORINATED BENZENES TO BE DONE IN ACCORDANCE WITH TSCA HEALTH EFFECTS TESTING GUIDELINES IN 40 CFR PART 798.

REPORT 15 : NTP TR 319 NIH PUB # 87-2575 — JAN 87
IN 2-YEAR GAVAGE STUDIES, CPD PRODUCED CLEAR EVIDENCE OF CARCINOGENICITY FOR MALE F344/N RATS AS SHOWN BY INCREASED INCIDENCE OF RENAL TUBULAR CELL ADENOCARCINOMAS. NO EVIDENCE OF CARCINOGENICITY FOR FEMALE F344/N RATS RECEIVING DOSES OF 300 OR 600 MG/KG. CLEAR EVIDENCE OF CARCINOGENICITY FOR BOTH MALE AND FEMALE B6C3F1 MICE, AS SHOWN BY INCREASED INCIDENCES OF HEPATOCELLULAR CARCINOMAS AND HEPATOCELLULAR ADENOMAS. ONLY MARGINAL INCREASES SEEN FOR OTHER EFFECTS.

REPT'D USE : CHEMICAL INTERMEDIATE, FUMIGANT, SYNTHESIS OF DYE INTERMEDIATES, INSECTICIDES, CLEANING AGENTS, MFR OF PHARMACEUTICALS AND SPACE DEODORANT

CAS NO : **91-94-1**
CHEM NAME : ***** 3,3'-DICHLOROBENZIDINE (DCB)**

REPORT 01 : EPA RCRA HEALTH AND ENV. EFFECTS PROFILE — OCT 80
THE ADVERSE HEALTH EFFECTS ASSOCIATED WITH 3,3'-DICHLOROBENZIDINE INCLUDE THE ELEVATED RISK OF CARCINOGENICITY BASED UPON DATA FROM SEVERAL EXPERIMENTAL BIOASSAYS.

REPORT 02 : IARC MONOGRAPH 29 — 1982
3,3-DICHLOROBENZIDINE WAS TESTED IN MICE, RATS, HAMSTERS AND DOGS BY ORAL ADMINISTRATION. IT PRODUCED LIVER-CELL TUMORS IN MICE, HEPATOCELLULAR TUMORS IN DOGS, MAMMARY AND ZYMBAL-GLAND TUMORS IN RATS, AND CARCINOMAS OF THE BLADDER IN HAMSTERS AND DOGS. THERE IS SUFFICIENT EVIDENCE THAT 3,3-DCB IS CARCINOGENIC IN ANIMALS.

REPORT 03 : EPA AMBIENT WATER QUALITY CRITERIA — OCT 80
DICHLOROBENZIDINE IS CONSIDERED TO BE A HUMAN CARCINOGEN. AMBIENT WATER CONCENTRATIONS OF .103 UG/L, .01 UG/L,

AND .001 UG/L ARE ESTIMATED TO RESULT IN AN INCREMENTAL INCREASE OF CANCER RISK OVER THE LIFETIME OF 10 -5, 10 -6, AND 10 -7, RESPECTIVELY.

REPORT 04 : IARC MONOGRAPH 29 1982
3,3-DICHLOROBENZIDINE IS MUTAGENIC TO SALMONELLA STRAINS TA1538, TA98, & TA100, IN THE PRESENCE OR ABSENCE OF METABOLIC ACTIVATION. IT INDUCES UNSCHEDULED DNA SYNTHESIS IN HELA CELLS. NO DATA WERE AVAILABLE TO ASSESS THE TERATOGENICITY OF THIS CHEMICAL.

REPORT 06 : MULTI-AGENCY CROSS CHECK
Y HHS MAGUIRE
R OSHA CARC
Y EPA CAG
Y EPA RCRA
N EPA CLEAN AIR
Y EPA CLEAN WATER
REPT'D USE : DYE INTERMEDIATE, CHEMICAL INDICATOR AND CURING AGENT

CAS NO : **328-84-7**
CHEM NAME : ***** 3,4-DICHLOROBENZOTRIFLUORIDE**

REPORT 01 : 49FR44218 NOV 84
EPA PROPOSES TO ADD THIS CHEMICAL, RECOMMENDED FOR TESTING CONSIDERATION IN THE 14TH REPORT OF ITC, TO PRELIMINARY ASSESSMENT INFORMATION RULE. EPA WILL USE INFORMATION SUBMITTED BY MFRS TO EVALUATE RISKS ASSOCIATED WITH THIS CHEMICAL AND DETERMINE IF IT SHOULD BE INCLUDED IN TESTING RULES ISSUED UNDER TSCA 4(A).

RFPORT 02 : 52 FR 23547 JUN 87
EPA HAS SIGNED AN ENFORCEABLE TESTING CONSENT ORDER WITH OCCIDENTAL CHEM. CO. THE COMPANY AGREES TO DO SPECFIED ENVIRONMENTAL EFFECTS TESTS OF CPD. THIS DOCUMENT IS IN RESPONSE TO ITC'S RECOMMENDATION THAT EPA CONSIDER HEALTH EFFECTS CHEMICAL FATE AND ENVIRONMENTAL EFFECTS OF CPD.

CAS NO : **28434-86-8**
CHEM NAME : ***** 3,3'-DICHLORO-4,4'-DIAMINODIPHENYL ETHER**

REPORT 01 : IARC MONOGRAPH 16 1978
3,3'-DICHLORO-4,4'-DIAMINODIPHENYLETHER IS CARCINOGENIC IN RATS; IT PRODUCED CARCINOMAS OF THE EAR DUCT AFTER ITS SUBCUTANEOUS INJECTION. IN A 1979 SUPPLEMENT, IARC DETERMINED THAT THERE IS SUFFICIENT EVIDENCE FOR THE CARCINOGENICITY OF THIS COMPOUND IN ANIMALS.

REPORT 02 : IARC MONOGRAPH 16 1978
NO DATA ON THE METABOLISM, EMBRYOTOXICITY, TERATOGENICITY OR MUTAGENICITY OF 3, 3'-DICHLORO-4, 4'-DIAMINODIPHENYL WERE AVAILABLE.

REPT'D USE : NO AVAILABLE PRODUCTION OR USE DATA

CAS NO : **33857-26-0**

CHEM NAME : ***** 2,7-DICHLORODIBENZO-P-DIOXIN (DCDD)**

REPORT 02 : NCI CARC TRS 123 DHEW PUB NO 79-1378 1977
UNDER THE CONDITIONS OF THIS BIOASSAY, DCDD WAS NOT CARCINOGENIC FOR OSBORNE-MENDEL RATS OF EITHER SEX OR FOR FEMALE B6C3F1 MICE. THE MARGINALLY INCREASED INCIDENCES OF COMBINATIONS OF LEUKEMIAS AND LYMPHOMAS, OF HEMANGIOMAS AND HEMANGIOSARCOMAS, AND OF HEPATOCELLULAR CARCINOMAS AND ADENOMAS IN MALE B6C3F1 MICE ARE CONSIDERED SUGGESTIVE OF THE CARCINOGENIC EFFECT OF 2,7-DICHLORODIBENZO-P-DIOXIN.

REPORT 03 : NTP TECHNICAL BULLETIN 9 APR 83
2,7-DICHLORODIBENZO-P-DIOXIN WAS NEGATIVE IN ONE SALMONELLA TYPHIMURIUM MUTAGENICITY ASSAY.

REPT'D USE : HERBICIDE WOOD PRESERVATIVES AND HEAT TRANSFER AGENT

CAS NO : **75-71-8**

CHEM NAME : ***** DICHLORODIFLUOROMETHANE (CFC-12)**

REPORT 01 : EPA AMBIENT WATER QUALITY CRITERIA OCT 80
A CRITERION BASED ON A NO OBSERVED ADVERSE EFFECT LEVEL IN DOGS, AND CALCULATED ON AN ACCEPTABLE DAILY HUMAN INTAKE OF 5.6 MG AND AN UNCERTAINTY FACTOR OF 1000, IS 2.8 MG/L. HOWEVER, FOR THE MAXIMUM PROTECTION OF HUMAN HEALTH FROM THE POTENTIAL CARCINOGENIC EFFECTS OF DCDFM, CONCENTRATIONS OF 1.9 UG/L, .19 UG/L AND .019 UG/L ARE ESTIMATED TO RESULT IN AN INCREMENTAL INCREASE OF CANCER RISK OVER A LIFETIME OF 10 -5, 10 -6 AND 10 -7 RESPECTIVELY.

REPORT 02 : EPA RCRA HEALTH AND ENV. EFFECTS PROFILE OCT 80
FULLY HALOGENATED CHLOROFLUORALKANES, INCLUDING TRICHLOROTRIFLUOROETHANE (F-113) ARE ALLEGED TO DEPLETE THE STRATOSPHERIC OZONE, LEADING TO VARIOUS ADVERSE EFFECTS. F-113 HAS TESTED NEGATIVE IN SALMONELLA ASSAYS.

REPORT 06 : MULTI-AGENCY CROSS CHECK
N HHS MAGUIRE
N OSHA CARC
N EPA CAG

Y EPA RCRA
N EPA CLEAN AIR
Y EPA CLEAN WATER

REPT'D USE : AEROSOL PROPELLANT, REFRIGERANT, FOAMING AGENT, SOLVENT/DEGREASER, PLASTIC MONOMER, PAINTS/VARNISHES, WATER SOFTENERS, SPACE DEODORANTS, TARNISH REMOVERS, CLEANERS/SANITIZERS AND DRAIN CLEANERS

CAS NO : **75-34-3**
CHEM NAME : ***** 1,1-DICHLOROETHANE**

REPORT 01 : NCI CARC TRS 66 DHEW PUB NO 78-1316
UNDER THE CONDITIONS OF THIS BIOASSAY, THERE WAS NO CONCLUSIVE EVIDENCE FOR THE CARCINOGENICITY OF 1,1-DICHLOROETHANE IN OSBORNE-MENDEL RATS OR B6C3F1 MICE (THE COMPOUND WAS ADMINISTERED BY GAVAGE).

REPORT 02 : EPA RCRA HEALTH AND ENV. EFFECTS PROFILE OCT 80
THERE IS NO AVAILABLE EVIDENCE TO INDICATE THAT 1,1-DICHLOROETHANE PRODUCES CARCINOGENIC OR MUTAGENIC EFFECTS. A SINGLE STUDY IN RATS FAILED TO SHOW TERATOGENIC EFFECTS FOLLOWING INHALATION EXPOSURE.

REPORT 03 : NIOSH CURRENT INTELLIGENCE BULLETIN 27 AUG 78
NCI BIOASSAY RETESTING RECOMMENDED BECAUSE INITIAL RESULTS WERE INCONCLUSIVE. OSHA EXPOSURE STANDARD IS 100 PPM FOR WORKERS. ALL CHLOROETHANE COMPOUNDS ARE KNOWN TO CAUSE CNS DEPRESSION IN LABORATORY ANIMALS. THIS COMPOUND CAUSES KIDNEY DAMAGE IN CATS; LIVER INJURY IN DOGS AND RATS; AND RETARDED FETAL DEVELOPMENT IN RATS. DICHLOROETHANE SHOULD BE TREATED IN THE WORKPLACE WITH CAUTION BECAUSE OF ITS STRUCTURAL SIMILARITY TO FOUR CARCINOGENIC CHLOROMETHANES.

REPORT 06 : MULTI-AGENCY CROSS CHECK
N HHS MAGUIRE
N OSHA CARC
N EPA CAG
Y EPA RCRA
N EPA CLEAN AIR
Y EPA CLEAN WATER

REPT'D USE : DEGREASING AGENT, EXTRACTION SOLVENT, RUBBER CEMENTS, FUMIGANT/INSECTICIDE SPRAYS, CHEMICAL INTERMEDIATE AND FIRE EXTINGUISHERS

CAS NO : **107-06-2**
CHEM NAME : *****1,2-DICHLOROETHANE (EDC)**

REPORT 01 : NCI CARC TRS 55 DHEW PUB NO 78-1361
UNDER CONDITIONS OF THIS STUDY, 1,2-DICHLOROETHANE WAS CARCINOGENIC TO OSBORNE-MENDEL RATS, CAUSING SQUAMOUS-CELL CARCINOMAS OF THE FORESTOMACH, HEMANGIOSARCOMAS AND SUBCUTANEOUS FIBROMAS IN MALE RATS AND MAMMARY ADENOCARCINOMAS IN FEMALE RATS. THIS COMPOUND WAS WAS ALSO FOUND TO BE CARCINOGENIC TO B6C3F1 MICE, CAUSING MAMMARY ADENOCARCINOMAS AND ENDOMETRIAL TUMORS IN FEMALE MICE, AND ALVEOLAR AND BRONCHIOLAR ADENOMAS IN MICE OF BOTH SEXES. EDC WAS ADMINISTERED BY GAVAGE.

REPORT 02 : NIOSH CRITERIA DOCUMENT 25 SEP 78
NIOSH RECOMMENDS, ON THE BASIS OF AN NCI BIOASSAY DEMONSTRATING THAT ETHYLENE DICHLORIDE IS CARCINOGENIC IN RATS AND MICE AND STUDIES DEMONSTRATING MUTAGENICITY, THAT ETHYLENE DICHLORIDE BE CONTROLLED AS AN OCCUPATIONAL CARCINOGEN AND THAT THE PREVIOUS RECOMMENDED STANDARD BE REVISED DOWNWARD FROM 5 PPM TO 1 PPM DETERMINED AS A TWA EXPOSURE FOR UP TO A 10-HOUR WORKSHIFT. NIOSH FURTHER RECOMMENDS A CEILING CONCENTRATION OF 2 PPM AS DETERMINED OVER A 15-MINUTE SAMPLING PERIOD. THE PRESENT OSHA STANDARD IS 50 PPM TWA.

REPORT 03 : EPA AMBIENT WATER QUALITY CRITERIA OCT 80
EDC IS CONSIDERED TO BE CARCINOGENIC IN AT LEAST ONE RODENT SPECIES. AMBIENT WATER CONCENTRATIONS OF 9.4 UG/L, .94 UG/L, AND .094 UG/L ARE ESTIMATED TO KEEP THE LIFETIME CANCER RISK BELOW 10 -5, 10 -6, AND 10 -7 RESPECTIVELY.

REPORT 04 : NIOSH CURRENT INTELLIGENCE BULLETIN 25 AUG 78
NCI BIOASSAY RESULTS INDICATE THAT THE COMPOUND IS CARCINOGENIC IN MICE INDUCING LIVER CANCER IN BOTH SEXES AS WELL AS CAUSING CANCER IN MALE AND FEMALE RATS. OSHA EXPOSURE STANDARD IS 50 PPM BUT NIOSH RECOMMENDS A 5 PPM EXPOSURE STANDARD. ALL CHLOROETHANE COMPOUNDS CAUSE CNS DEPRESSION IN LABORATORY ANIMALS. 1,2-DICHLOROETHANE IS A BACTERIAL MUTAGEN; IS EMBRYOTOXIC IN RATS; AND INDUCES LIVER DYSFUNCTION AND CHANGES IN TEST ANIMALS. NIOSH RECOMMENDS THAT OCCUPATIONAL EXPOSURE TO 1,2-DICHLOROETHANE BE MINIMIZED.

REPORT 05 : COLD SPRING HARBOR BANBURY REPORT
ALTHOUGH STUDIES OF THE HEALTH EFFECTS OF EDC ARE CONFLICTING, THEY DO SUPPORT THE GENERAL CONCERN ABOUT POTENTIAL HEALTH RISKS. PRESENT STUDIES DO NOT PROVIDE ANY EVIDENCE THAT EDC CAUSES BIRTH DEFECTS OR REPRODUC-

		TIVE EFFECTS. HOWEVER OTHER STUDIES SUGGEST THAT EDC MAY POSE A CARCINOGENIC AND A MUTAGENIC HAZARD.
REPORT 06	:	MULTI-AGENCY CROSS CHECK Y HHS MAGUIRE N OSHA CARC Y EPA CAG Y EPA RCRA Y EPA CLEAN AIR Y EPA CLEAN WATER
REPORT 07	:	IARC MONOGRAPH 20 1979 IN MICE, EDC PRODUCED BENIGN AND MALIGNANT TUMORS OF THE LUNG AND MALIGNANT LYMPHOMAS IN ANIMALS OF BOTH SEXES, HEPATOCELLULAR CARCINOMAS IN MALES AND MAMMARY AND UTERINE ADENOCARCINOMAS IN FEMALES. IN MALE RATS, IT PRODUCED CARCINOMAS OF THE FORESTOMACH; IN FEMALES, BENIGN AND MALIGNANT MAMMARY TUMORS; AND HEMANGIOSARCOMAS IN ANIMALS OF BOTH SEXES. IN A 1979 SUPPLEMENT, IARC DETERMINED THAT THERE IS SUFFICIENT EVIDENCE FOR THE CARCINOGENICITY OF EDC IN ANIMALS.
REPORT 08	:	EPA RCRA HEALTH AND ENV. EFFECTS PROFILE OCT 80 RESULTS OF AN NCI CARCINOGENESIS BIOASSAY IN RAT AND MICE HAVE SHOWN THAT 1,2-DICHLOROETHANE MAY PRODUCE A WIDE VARIETY OF TUMORS, INCLUDING SQUAMOUS CELL CARCINOMAS, HEMANGIOSARCOMAS, MAMMARY ADENOCARCINOMAS, AND HEPATOCELLULAR CARCINOMAS. MUTAGENIC EFFECTS HAVE BEEN SHOWN IN THE AMES SALMONELLA SYSTEM AND IN E. COLI. ONE STUDY HAS FAILED TO INDICATE TERATOGENIC EFFECTS FOLLOWING INHALATION EXPOSURE TO 1,2-DICHLOROETHANE, ALTHOUGH REPRODUCTIVE TOXICITY WAS DEMONSTRATED.
REPORT 09	:	IARC MONOGRAPH 20 1979 EDC IS MUTAGENIC IN SALMONELLA TA1530, TA1535 AND TA100, PRESUMABLY CAUSING BASE-PAIR SUBSTITUTION MUTATIONS. THE CHEMICAL-INDUCED SEX-LINKED RECESSIVE LETHALS IN DROSOPHILA. NO DATA ON THE EMBRYOTOXICITY OR TERATOGENICITY OF EDC WERE AVAILABLE.
REPORT 10	:	CMA MALTON REPORT NOV 82 EDC GIVEN BY INHALATION AT 5,10,50,150, AND 250 PPM TO MICE FOR 18 MONTHS AND RATS FOR 24 MONTHS FAILED TO SHOW SPECIFIC CARCINOGENIC EFFECTS.
REPORT 11	:	NAT'L DRINKING WATER STD RULEMAKING 49FR24330 JUN 84 EPA PROPOSING TO SET WATER LEVELS OF EDC AT 0 MG/L UNDER SAFE WATER ACT.
REPT'D USE	:	CHEMICAL INTERMEDIATE, METAL DEGREASER, ADHESIVE, FUEL CLEANERS, PAINT REMOVER, ANTIKNOCK FUEL ADDITIVE, SOLVENT, DEGREASING AGENT AND ACRYLIC PLASTIC CEMENT

CAS NO : **540-59-0**
CHEM NAME : ***** 1,2-DICHLOROETHYLENE**

REPORT 01 : NTP TECHNICAL BULLETIN 3 DEC 80
THE COMPOUND WAS FOUND TO BE NEGATIVE IN THE SALMONELLA TYPHIMURIUM MUTAGENICITY ASSAY IN ONE LABORATORY.

REPORT 02 : EPA RCRA HEALTH & ENV. EFFECTS PROFILE OCT 80
TRANS-1,2-DCE HAS BEEN SHOWN TO BE A MUTAGEN IN BACTERIAL SYSTEMS. CIS-1,2-DCE WAS MUTAGENIC IN SALMONELLA TESTER STRAINS, AND PROMOTED CHROMOSOMAL ABERRATIONS IN CYTOGENIC ANALYSIS OF BONE MARROW CELLS.

REPORT 06 : MULTI-AGENCY CROSS CHECK
N HHS MAGUIRE
N OSHA CARC
N EPA CAG
Y EPA RCRA
N EPA CLEAN AIR
Y EPA CLEAN WATER
REPT'D USE : SOLVENT, REFRIGERATION COOLANT AND FERMENTATION RETARDANT

CAS NO : **528-74-5**
CHEM NAME : ***** DICHLOROMETHOTREXATE**

REPORT 01 : NCI BIOASSAY PROGRAM CANCER VOL. 40 1977
UNDER THE CONDITIONS OF THE BIOASSOY, I.P. INJECTION OF DICHLOROMETHOTREXATE WAS CARCINOGENIC TO MALE SPRAGUE-DAWLEY (CD) RATS, CAUSING PITUITARY AND ADRENAL TUMORS. IT WAS NOT CARCINOGENIC TO FEMALE RATS AND MALE AND FEMALE SWISS-WEBSTER MICE.
REPT'D USE : NO AVAILABLE PRODUCTION OR USE DATA

CAS NO : **611-06-3**
CHEM NAME : ***** 2,4-DICHLORONITROBENZENE**

REPORT 01 : NTP TECHNICAL BULLETIN 2/8 APR 80, JUL 82
2,4-DICHLORONITROBENZENE WAS POSITIVE IN THE SALMONELLA ASSAY BUT NEGATIVE IN DROSOPHIAL WHEN TESTED FOR SEX-LINKED RECESSIVE LETHALS.
REPT'D USE : CHEMICAL INTERMEDIATE FOR HERBICIDES, DYES AND PHARMACEUTICALS - IMPORTS ONLY

CAS NO : **99-54-7**
CHEM NAME : ***** 3,4-DICHLORONITROBENZENE**

REPORT 01 : NTP TECHNICAL BULLETIN 2/8 APR 80, JUL 82
3,4-DICHLORONITROBENZENE WAS POSITIVE IN SALMONELLA, BUT NEGATIVE IN DROSOPHILA WHEN TESTED FOR SEX LINKED RECESSIVE LETHALS.
REPT'D USE : NO AVAILABLE USE DATA

CAS NO : **120-83-2**
CHEM NAME : ***** 2,4-DICHLOROPHENOL**

REPORT 01 : NTP 353 NIH PUB 89-2808 JUN 89
UNDER THE CONDITION OF THESE 2-YEAR FEED STUDIES, THERE WAS NO EVIDENCE OF CARCINOGENIC ACTIVITY FOR MALE F344/N RATS FED DIETS CONTAINING 5,000 OR 10,000 PPM 2,4-DICHLOROPHENOL OR FOR FEMALE F344/N RATS FED DIETS CONTAINING 2,500 OR 5,000 PPM 2,4-DICHLOROPHENOL. THERE WAS NO EVIDENCE OF CARCINOGENIC ACTIVITY FOR MALE OR FEMALE B6C3F1 MICE FED DIETS CONTAINING 5,000 OR 10,000 PM 2,4-DICHLORPHENOL.
REPT'D USE : CHEMICAL INTERMEDIATE IN MANUFACTURE OF HERBICIDE 2,4-DICHLOROPHENOXYACETIC ACID, AND METHYLATED CHLOROPHENOLS, MOTHPROOFING, ANTISEPTICS AND SEED DISINFECTANTS

CAS NO : **609-20-1**
CHEM NAME : ***** 2,6-DICHLORO-P-PHENYLENEDIAMINE**

REPORT 01 : NTP TECHNICAL BULLETIN 7 APR 82
2,6-DICHLORO-PHENYLENEDIAMINE WAS POSITIVE IN THE S. TYPHIMURIUM MUTAGENICITY ASSAY.

REPORT 02 : NTP BIOASSAY TRS 219 NIH PUB NO 82-1775
UNDER THE CONDITIONS OF THIS BIOASSAY, 2,6-DICHLORO-P-PHENYLENEDIAMINE WAS CARCINOGENIC FOR MALE AND FEMALE B6C3F1 MICE, CAUSING INCREASED INCIDENCES OF COMBINED HEPATOCELLULAR ADENOMAS AND CARCINOMAS AND FOR MALE B6C3F1 MICE, CAUSING AN INCREASED INCIDENCE OF HEPATOCELLULAR ADENOMAS ALONE. COMPOUND WAS NOT CARCINOGENIC FOR F344 RATS OF EITHER SEX (FEED).

REPORT 03 : IARC MONOGRAPH 39, 1986 JUN 85
MICE TESTED WITH CPD IN DIET SHOWED INCREASED NUMBERS OF LIVER-CELL TUMORS; NO TREATMENT-RELATED TUMORS FOUND IN RATS TESTED SIMILARLY. CPD MUTAGENIC TO S. TYPHIMURIUM WITH AND WITHOUT ACTIVATION. NO DATA TO EVALUATE

REPRODUCTIVE EFFECTS OR PRENATAL TOXICITY IN HUMANS OR ANIMALS. NO CASE REPORTS OR EPA STUDIES FOUND TO EVALUATE CARCINOGENICITY OF CPD IN HUMANS. IARC SAYS THERE IS LIMITED EVIDENCE FOR CARCINOGENICITY OF CPD TO EXPERIMENTAL ANIMALS, BUT NO DATA ON HUMANS.

REPT'D USE : LIMITED USE AS INTERMEDIATE IN DYE AND RESIN MFR. POTENTIAL FOR EXPOSURE LIES IN ITS FORMATION AS METABOLITE OF PESTICIDE 2,4-DICHLORO-4-NITROANILINE

CAS NO : **78-87-5**
CHEM NAME : ***** 1,2-DICHLOROPROPANE**

REPORT 02 : EPA RCRA HEALTH AND ENV. EFFECTS PROFILE OCT 80
DICHLOROPROPANES HAVE PRODUCED MUTATIONS IN BACTERIA AND CAUSED CHROMOSOMAL ABERRATIONS IN RATS.

REPORT 04 : EPA AMBIENT WATER QUALITY CRITERIA OCT 80
DICHLOROPROPANE HAS NOT BEEN ADEQUATELY TESTED FOR CARCINOGENICITY. A CRITERION, BASED ON CHANGES IN SERUM ENZYME LEVELS IN RATS AND A SAFETY FACTOR OF 1000, IS 483 UG/L.

REPORT 05 : ITC THIRD REPORT TO EPA (43FR50631) OCT 78
THE INTERAGENCY TESTING COMMITTEE RECOMMENDS THAT THE EPA ADMINISTRATOR CONSIDER THE DEVELOPMENT OF TEST RULES TO TEST 1,2-DICHLOROPROPANE FOR CARCINOGENICITY, MUTAGENICITY, TERATOGENICITY/REPRODUCTIVE EFFECTS; AND THAT AN EPIDEMIOLOGIC STUDY BE CONDUCTED.

REPORT 06 : MULTI-AGENCY CROSS CHECK
N HHS MAGUIRE
N OSHA CARC
N EPA CAG
Y EPA RCRA
N EPA CLEAN AIR
Y EPA CLEAN WATER

REPORT 07 : 48FR899 JAN 84
UNDER TSCA 4 EPA PROPOSES THAT MFRS AND PROCESSORS CONDUCT TESTS FOR ENVIRONMENTAL AND HEALTH EFFECTS AND CHEMICAL FATE.

REPORT 08 : 49FR2948 JAN 84
GIVES NOTICE OF A TWO-DAY MEETING OF THE ENVIRONMENTAL ENGINEERING COMMITTEE OF THE SCIENCE ADVISORY BOARD ON FEB 8-9, 1984.

REPORT 09 : NTP-TR263, NIH PUB.#86-2519, FINAL REPORT AUG 86
NO EVIDENCE OF CARCINOGENICITY FOR MALE F344/N RATS

RECEIVING 65 OR 125 MG/KG; EQUIVOCAL EVIDENCE OF CARCINOGENICITY AT 250 MG/KG LEVEL IN FEMALE RATS; SOME EVIDENCE OF CARCINOGENICITY FOR MALE AND FEMALE B6C3F1 MICE. CPD ADMINISTERED BY GAVAGE.

REPORT 10 : 51 CFR 32107 - PROPOSED TEST RULE AND TESTING STDS — SEP 86
EPA PROPOSES THAT PHARMACOKINETICS TESTING BE CONDUCTED USING TSCA GUIDELINES.

REPORT 11 : IARC VOL 41 — 1986
THERE IS LIMITED EVIDENCE OF CARCINOGENICITY TO TEST ANIMALS. NO EVALUATION COULD BE MADE OF THE CARCINOGENICITY TO HUMANS. ORAL ADMINISTRATION BY GAVAGE TO MICE AND RATS CAUSED DOSE-RELATED INCREASE IN INCIDENCE OF HEPATOCELLULER TUMORS IN MICE. RESULTS FROM TESTS OF FEMALE RATS WERE INCONCLUSIVE. NO EFFECT SEEN IN MALE RATS. NO DATA AVAILABLE TO EVALUATE REPRODUCTIVE EFFECTS OR PRENATAL TOXICITY TO EXPERIMENTAL ANIMALS.

REPT'D USE : INSECTICIDE, INDUSTRIAL SOLVENT, SOIL FUMIGANT, DRY CLEANING FLUIDS, SPOTTING AGENTS, INTERMEDIATE IN RUBBER PROCESSING AND PESTICDE

CAS NO : **542-75-6**
CHEM NAME : ***** 1,3-DICHLOROPROPENE**

REPORT 01 : EPA AMBIENT WATER QUALITY CRITERIA — OCT 80
DICHLOROPENE HAS NOT BEEN ADEQUATELY TESTED FOR CARCINOGENICITY. A CRITERION, BASED ON CHANGES IN BLOOD SERUM ENZYMES IN THE LIVER FUNCTION OF RATS AND A SAFETY FACTOR OF 1000, IS 87 UG/L.

REPORT 02 : EPA RCRA HEALTH AND ENV. EFFECTS PROFILE — OCT 80
CHRONIC EXPOSURE OF RATS TO DICHLOROPROPENE RESULTED IN MILD KIDNEY DAMAGE. DICHLOROPROPENE HAS PRODUCED SUBCUTANEOUS TUMORS AT THE SITE OF INJECTION, AND HAS BEEN SHOWN TO BE MUTAGENIC IN BACTERIA. HOWEVER, NOT ENOUGH INFORMATION IS AVAILABLE TO CLASSIFY THIS COMPOUND AS A CARCINOGEN.

REPORT 03 : NTP TECHNICAL BULLETIN 6 — JAN 82
THE COMPOUND WAS POSITIVE IN THE SALMONELLA TYPHIMURIUM MUTAGENICITY ASSAY IN ONE LABORATORY.

REPORT 04 : NTP TECHNICAL REPORT 269 NIH PUB 85-2525 — MAY 85
UNDER CONDITIONS OF GAVAGE STUDIES, THERE WAS CLEAR EVIDENCE OF CARCINOGENICITY FOR MALE F344/N RATS. IN FEMALE F344XN RATS, THERE WAS SOME EVIDENCE OF CARCINO-

GENICITY. EXPERIMENT IN MALE B6C3F1 MICE WAS CONSIDERED AN INADEQUATE STUDY OF CARCINOGENICITY BECAUSE OF REDUCED SURVIVAL IN THE VEHICLE CONTROL GROUP. THERE WAS CLEAR EVIDENCE OF CARCINOGENICITY FOR FEMALE B6C3F1 MICE.

REPORT 05 : IARC VOL 41 1986
SUFFICIENT EVIDENCE FOR CARCINOGENICITY OF TECH GRADE OF CPD TO TEST ANIMALS, BUT INADEQUATE EVIDENCE FOR THE CARCINOGENICITY OF CPD TO HUMANS. ORAL ADMINISTRATION CAUSED DOSE-RELATED INCREASES IN INCIDENCE OF BENIGN AND/OR MALIGNANT TUMORS OF URINARY BLADDER, LUNG AND FORESTOMACH IN MICE AND RATS. ONE STUDY USING THE CIS ISOMER BY SC INJECTION PRODUCED TUMORS AT INJECTION SITE. CIS ISOMER NOT ACTIVE AS AN INITIATOR IN 2-STAGE APPLICATION STUDY. NO DATA AVAILABLE TO ASSESS REPRODUCTIVE EFFECTS OR PRENATAL TOXICITY TO TEST ANIMALS.

REPORT 06 : MULTI-AGENCY CROSS CHECK
N HHS MAGUIRE
N OSHA CARC
N EPA CAG
Y EPA RCRA
N EPA CLEAN AIR
Y EPA CLEAN WATER

REPT'D USE : FUMIGANT AND PESTICIDE INTERMEDIATE

CAS NO : **62-73-7**
CHEM NAME : ***** DICHLORVOS**

REPORT 01 : NCI CARC TRS 10 DHEW PUB NO 77-810
DICHLORVOS ADMINISTERED IN THE DIET SHOWS NO STATISTICALLY SIGNIFICANT INCREASE IN INCIDENCE OF TUMORS IN OSBORNE-MENDEL RATS OR B6C3F1 MICE.

REPORT 03 : IARC MONOGRAPH 20 1979
DICHLORVOS WAS TESTED IN ONE EXPERIMENT IN MICE AND IN ONE IN RATS BY ORAL ADMINISTRATION NO STATISTICALLY SIGNIFICANT EXCESS OF TUMORS WAS OBSERVED. THE AVAILABLE DATA DO NOT ALLOW AN EVALUATION OF THE CARCINOGENICITY OF DICHLORVOS TO BE MADE.

REPORT 04 : IARC MONOGRAPH 20 1979
DICHLORVOS IS AN ALKYLATING AGENT AND BINDS TO BACTERIAL AND MAMMALIAN NUCLEIC ACIDS. THE CHEMICAL IS A MUTAGEN IN A NUMBER OF MICROBIAL SYSTEMS, BUT THERE IS NO EVIDENCE OF ITS MUTAGENICITY IN ANIMALS, IN WHICH IT IS RAPIDLY DEGRADED.

REPORT 05 : NTP TR 342 NIH PUB 89-2598 SEP 89
UNDER CONDITIONS OF 2-YR GAVAGE STUDIES SOME EVIDENCE OF CARCINOGENIC ACTIVITY OF DICHLORVOS FOR MALE F344/N RATS (ADENOMAS OF THE EXOCRINE PANCREAS AND MONONUCLEAR CELL LEUKEMIA). EQUIVOCAL EVIDENCE OF CARCINOGENIC ACTIVITY FOR FEMALE F344XN RATS (ADENOMAS OF THE EXOCRINE PANCREAS AND MAMMARY GLAND FIBROADENOMAS). SOME EVIDENCE OF CARCINOGENIC ACTIVITY FOR MALE B6C3F1 MICE (FORESTOMACH SQUAMOUS CELL PAPILLOMAS). CLEAR EVIDENCE OF CARCINOGENIC ACTIVITY FOR FEMALE B6C3F1 MICE (FORESTOMACH SQUAMOUS CELL PAPILLOMAS).

REPT'D USE : VETERINARY ANTIELMINTHIC, INSECTICIDE, FLEA/PEST COLLARS, AEROSOLS, SPACE DEODORIZERS AND PESTICIDES

CAS NO : **115-32-2**

CHEM NAME : ***** DICOFOL**

REPORT 01 : NCI CARC TRS 90 DHEW PUB NO 78-1340
UNDER THE CONDITIONS OF THIS BIOASSAY, TECHNICAL-GRADE DICOFOL WAS CARCINOGENIC IN MALE B6C3F1 MICE, CAUSING HEPATOCELLULAR CARCINOMAS. NO EVIDENCE FOR CARCINOGENICITY WAS OBTAINED FOR THIS COMPOUND IN OSBORNE-MENDEL RATS OF EITHER SEX OR IN FEMALE B6C3F1 MICE (FEEDING STUDY).

REPORT 02 : NTP TECHNICAL BULLETIN 6 JAN 82
THE COMPOUND WAS NEGATIVE IN THE SALMONELLA TYPHIMURIUM MUTAGENICITY ASSAY IN ONE LABORATORY.

REPORT 03 : IARC MONOGRAPH 30 JAN 83
DICOFOL (TECHNICAL-GRADE) TESTED FOR CARCINOGENICITY IN MICE AND RATS BY ADMINISTRATION IN DIET. IT INDUCED HEPATOCELLULAR CARCINOMAS IN MALE MICE. THE STUDY IN RATS WAS CONSIDERED INADEQUATE FOR EVALUATION.

REPORT 04 : IARC MONOGRAPH 30 JAN 83
DICOFOL (TECHNICAL-GRADE), WHEN GIVEN AT HIGH DOSES, HAD NO EFFECT ON REPRODUCTION OR FOETAL DEVELOPMENT IN MICE; HOWEVER, HIGH DOSES IN RATS APPEARED TO HAVE ADVERSE EFFECT ON PREIMPLANTATION STAGES.

REPORT 05 : IARC MONOGRAPH 30 JAN 83
DICOFOL NEGATIVE IN BACTERIAL TESTS FOR MUTAGENICITY AND FOR DNA DAMAGE WITH OR WITHOUT EXOGENOUS METABOLIC ACTIVATION. EXPERIMENTAL PROTOCOL AND RESULTS OF STUDIES WITH EUKARYOTES NOT PRESENTED IN ADEQUATE DETAIL FOR EVALUATION. NO OVERALL EVALUATION OF MUTAGENICITY OF DICOFOL COULD BE MADE.

REPT'D USE : ACARACIDE AND MITICIDE

CAS NO : **1212-29-9**
CHEM NAME : ***** N,N'-DICYCLOHEXYLTHIOUREA**

REPORT 01 : NCI CARC TRS 56 DHEW PUB NO 78-1362
UNDER THE CONDITIONS OF THIS BIOASSAY N,N'-DICYCLOHEXYLTHIOUREA WAS NOT DEMONSTRATED TO BE CARCINOGENIC IN EITHER SPECIES (F344 RATS AND B6C3F1 MICE) (FEEDING STUDY).

REPORT 02 : NTP TECHNICAL BULLETIN 8 JUL 82
N,N-DICYCLOHEXYLTHIOUREA WAS NEGATIVE IN THE L5178Y MOUSE LYMPHOMA ASSAY.

REPT'D USE : CHEMICAL INTERMEDIATE, CHEMICAL REAGENT AND BIOCIDAL ADDITIVE TO CUTTING FLUIDS

CAS NO : **60-57-1**
CHEM NAME : ***** DIELDRIN**

REPORT 01 : EPA AMBIENT WATER QUALITY CRITERIA OCT 80
DIELDRIN IS SUSPECTED OF BEING A HUMAN CARCINOGEN. FOR DIELDRIN, AMBIENT WATER CONCENTRATIONS OF .71 NG/L, .071 NG/L AND .007 NG/L ARE ESTIMATED TO RESULT IN A RESPECTIVE 10 -5, 10 -6, AND 10 -7 INCREASE IN CANCER RISK OVER A LIFETIME.

REPORT 02 : NCI CARC TRS 21 AND 22 DHEW PUB NO 78-822
TWO STUDIES WERE COMPLETED. IN ONE STUDY, TECHNICAL GRADE DIELDRIN WAS GIVEN IN FEED TO FISCHER 344 RATS AND FOUND NOT TO BE CARCINOGENIC. IN THE SECOND STUDY, DIELDRIN WAS GIVEN IN FEED TO OSBORNE-MENDEL RATS AND B6C3F1 MICE. DIELDRIN WAS FOUND TO CAUSE CANCER IN MALE MICE, BUT IN FEMALE MICE INCREASES IN LIVER CANCER COULD NOT BE RELATED TO TREATMENT. IN O-M RATS, TUMORS COULD NOT BE CLEARLY ASSOCIATED WITH TREATMENT.

REPORT 03 : EPA RCRA HEALTH AND ENV. EFFECTS PROFILE OCT 80
THE INDUCTION OF HEPATOCELLULAR CARCINOMAS IN MICE BY DIELDRIN LEADS TO THE CONCLUSION THAT IT IS LIKELY TO BE A HUMAN CARCINOGEN. DIELDRIN HAS BEEN FOUND TO BE NON-MUTAGENIC IN SEVERAL TEST SYSTEMS.

REPORT 04 : IARC SUPPLEMENT 1 1979
DIELDRIN IS CARCINOGENIC IN MICE, CAUSING LIVER TUMORS FOLLOWING ORAL ADMINISTRATION. FEEDING STUDIES IN RATS HAVE SHOWN NO CARCINOGENIC EFFECT. A STUDY OF WORKERS EXPOSED TO DIELDRIN WAS INSUFFICIENT TO ALLOW AN EVALUATION OF HUMAN CARCINOGENICITY.

REPORT 05 : NIOSH SPECIAL HAZARD REVIEW
NIOSH RECOMMENDS THAT DIELDRIN BE CONTROLLED AND HAN-

DLED IN THE WORKPLACE AS A SUSPECTED OCCUPATIONAL CARCINOGEN. NIOSH RECOMMENDS A STANDARD OF 0.15 MG/CU M. THE PRESENT OSHA STANDARD IS 0.25 MG/CU M.

REPORT 06 : MULTI-AGENCY CROSS CHECK
N HHS MAGUIRE
N OSHA CARC
Y EPA CAG
N EPA RCRA
N EPA CLEAN AIR
Y EPA CLEAN WATER

REPORT 07 : NTP TECHNICAL BULLETIN 5 AUG 81
DIELDRIN WAS NEGATIVE IN THE SALMONELLA TYPHIMURIUM MUTAGENICITY ASSAY IN ONE LABORATORY.
REPT'D USE : INSECTICIDE

CAS NO : **84-17-3**
CHEM NAME : ***** DIENOESTROL**

REPORT 01 : IARC MONOGRAPH 21 1979
DIENOESTROL WAS TESTED IN FEMALE GUINEA PIGS BY SUBCUTANEOUS INJECTION AND IN FEMALE MICE BY INTRAVAGINAL ADMINISTRATION. ALTHOUGH POINTING TO THE INDUCTION OF 'TERINE TUMORS' IN GUINEA PIGS AND OVARIAN TUMORS IN MICE, THESE EXPERIMENTS WERE INSUFFICIENT TO EVALUATE THE CARCINOGENICITY OF THIS COMPOUND. STUDIES IN HUMANS STRONGLY SUGGEST THAT THE ADMINISTRATION OF ESTROGENS IS CAUSALLY RELATED TO AN INCREASED INCIDENCE OF ENDOMETRIAL CARCINOMA; NO EVIDENCE THIS IS DIFFERENT FROM OTHER ESTROGENS IN THIS RESPECT.

REPORT 02 : IARC MONOGRAPH 21 1979
DIENOESTROL INDUCED SISTER CHROMATID EXCHANGES IN CULTURED HUMAN FIBROPLASTS; HOWEVER IT WAS NOT MUTAGENIC IN 4 SALMONELLA STRAINS EITHER IN THE PRESENCE OR ABSENCE OF METABOLIC ACTIVATION. NO DATA ON THE EMBRYOTOXICITY OR TERATOGENICITY OF DIENOESTROL WERE AVAILABLE.
REPT'D USE : ESTROGEN AND VETERINARY MEDICINE

CAS NO : **1464-53-5**
CHEM NAME : ***** DIEPOXYBUTANE**

REPORT 01 : IARC MONOGRAPH 11 1976
D,L- AND MESO-1,2:3,4-DIEPOXYBUTANE ARE CARCINOGENIC IN MICE BY SKIN APPLICATION: BOTH COMPOUNDS PRODUCED SQUAMOUS-CELL SKIN CARCINOMAS. THE D,L-RACEMATE ALSO PRODUCED LOCAL SARCOMAS IN MICE AND RATS BY SUBCUTA-

NEOUS INJECTION. L-1,2:3,4-DIEPOXYBUTANE IS CARCINOGENIC IN MICE BY INTRAPERITONEAL INJECTION. NO HUMAN DATA IN THE FORM OF CASE REPORTS OR EPIDEMIOLOGICAL STUDIES WERE AVAILABLE. IN A 1979 SUPPLEMENT IARC DETERMINED THAT THERE WAS SUFFICIENT EVIDENCE FOR THE CARCINOGENICITY OF DIEPOXYBUTANE IN ANIMALS.

REPORT 02 : IARC MONOGRAPH 11 1976
DIEPOXYBUTANE INDUCED REVERSE MUTATIONS IN SALMONELLA TA1535, IN S. POMBE, AND E. COLI. THE CHEMICAL PRODUCED MITOTIC GENE CONVERSION IN SACCHAROMYCES. IN DROSOPHILIA IT HAS PRODUCED SEX-LINKED RECESSIVE LETHALS AND TRANSLOCATIONS. DIEPOXYBUTANE PRODUCED CHROMOSOMAL ABERRATIONS IN HUMAN CELLS TREATED IN VIVO.

REPORT 03 : 50FR36476 SEP 85
EPA FOUND NO EVIDENCE OF TOXICITY OR INFORMATION ON CHEMICAL FATE OF 1,2-EOB. EPA DECIDED NOT TO PROCEED WITH RULE MAKING TO REQUIRE HEALTH EFFECTS OR CHEMICAL FATE TESTING FOR 1,2-EOB.

REPORT 04 : 49FR449 JAN 84
THIS ANPR IS EPA'S RESPONSE TO IRC'S DESIGNATION OF ALKYL EPOXIDES, INCLUDING 2,3-EOB, FOR HEALTH EFFECTS AND CHEMICAL FATE TESTING. EPA HAS NOT FOUND ANY TOXICITY DATA FOR 2,3-EOB. EPA DOES NOT PLAN TO INITIATE TESTING PROGRAM FOR 2,3-EOB BECAUSE OF THE EXTREMELY LOW PRODUCTION AND USE RATES.

REPORT 06 : MULTI-AGENCY CROSS CHECK
Y HHS MAGUIRE
N OSHA CARC
Y EPA CAG
Y EPA RCRA
N EPA CLEAN AIR
N EPA CLEAN WATER

REPT'D USE : TEXTILE CROSSLINKING AGENT, CHEMICAL INTERMEDIATE, AND POLYMER CURING AGENT AND ANTIMICROBIAL AGENT

CAS NO : **564-00-1**
CHEM NAME : ***** 1,2,3,4-DIEPOXYBUTANE, DL**

REPORT 01 : NTP TECHNICAL BULLETIN 8 JUL 82
1,2,3,4-DIEPOXYBUTANE WAS POSITIVE IN THE L5178Y MOUSE LYMPHOMA ASSAY.

CAS NO : **2238-07-5**
CHEM NAME : ***** DI(2,3-EPOXYPROPYL) ETHER (DGE)**

REPORT 01 : NIOSH CRITERIA DOCUMENT JUN 78
DGE PRODUCED A 40% INCREASE IN SKIN PAPILLOMAS IN MICE SURVIVING A DOSE OF .75 MILLIMOLE. IT HAS ALSO SHOWN MUTAGENIC ACTIVITY IN SALMONELLA. BECAUSE DGE HAS SHOWN TUMORIGENIC ACTIVITY IN MICE AND PRODUCED MUTATIONS IN BACTERIA, IT SHOULD BE REGARDED AS A POTENTIAL OCCUPATIONAL CARCINOGEN. NIOSH RECOMMENDS THAT EXPOSURE TO AIRBORNE DGE NOT EXCEED .2 PPM AS A 15 MIN CEILING CONCENTRATION.

CAS NO : **68334-30-5**
CHEM NAME : ***** DIESEL FUELS**

REPORT 01 : IARC MONOGRAPHS VOLUME 45 1989
INADEQUATE EVIDENCE FOR CARCINOGENICITY IN HUMANS OF DIESEL FUELS. MARINE DIESEL FUEL IS POSSIBLY CARCINOGENIC TO HUMANS. DISTILLATE (LIGHT) DIESEL FUELS ARE NOT CLASSIFIABLE AS TO CARCINOGENICITY TO HUMANS. EVIDENCE OF INCREASED RISK FOR SQUAMOUS-CELL CARCINOMA OF LUNG IN MEN ESTIMATED TO HAVE SUBSTANTIAL EXPOSURE TO DIESEL FUEL; ALSO INCREASED RISK FOR CANCER OF PROSTATE. INHALATION OR INGESTION OF DIESEL FUEL RESULTED IN ACUTE AND PERSISTENT LUNG DAMAGE IN HUMANS.

REPORT 02 : IARC MONOGRAPH 45 1989
LIMITED EVIDENCE FOR CARCINOGENICITY IN ANIMALS OF MARINE DIESEL FUEL. ONE SAMPLE TESTED IN ONE STRAIN OF MICE BY SKIN APPLICATION, PRODUCING FEW SQUAMOUS-CELL CARCINOMAS AND PAPILLOMAS AT APPLICATION SITE OF BOTH SEXES AND FEW CARCINOMAS AT ADJACENT INGUINAL REGION IN MALES. TWO SAMPLES OF STRAIGHT-RUN KEROSENE, ONE SAMPLE OF LIGHT VACUUM DISTILLATE, AND THREE SAMPLES OF LIGHT CATALYTICALLY CRACKED DISTILLATE PRODUCED SKIN TUMORS IN MICE. RESIDUES FROM THERMAL CRACKING PRODUCED BENIGN AND MALIGNANT SKIN TUMORS IN MICE. APPLICATION OF MARINE DIESEL FUEL TO SKIN OF MICE RESULTED IN ULCERATION.

REPORT 03 : IARC MONOGRAPH 45 1989
DIESEL FUEL INDUCED CHROMOSOMAL ABERRATIONS IN BONE-MARROW CELLS OF RATS; NO MUTATION IN CULTURED MAMMALIAN CELLS BUT WEAKLY MUTAGENIC TO BACTERIA. ANOTHER SAMPLE DID NOT INDUCE MUTATION IN BACTERIA OR ALGAE; SAMPLE OF MARINE DIESEL FUEL AND ALIPHATIC AND AROMATIC FRACTIONS OF UNSPECIFIED DIESEL FUEL NOT MUTAGENIC TO BACTERIA.
REPT'D USE : TRANSPORTATION FUEL AND GAS TURBINES

CAS NO : **6175-45-7**
CHEM NAME : ***** 2,2'-DIETHOXYACETOPHENONE (DEAP)**

REPORT 01 : EPA TSCA 8(E) NOTICE 8EHQ-0581-0400 JUN 81
UPJOHN CO. REPORTED THAT A SINGLE SQUAMOUS CELL CARCINOMA HAS BEEN VERIFIED AT THE SITE OF REPEATED APPLICATION DURING A LIFETIME SKIN-PAINTING STUDY OF DEAP IN MICE. DEAP WAS REPORTED TO BE NEGATIVE, BOTH WITH AND WITHOUT METABOLIC ACTIVATION, IN SALMONELLA.
REPT'D USE : PHOTOSENSITIVE INITIATOR IN UV-CURED INKS, COATINGS AND ADHESIVES

CAS NO : **88-10-8**
CHEM NAME : ***** DIETHYLCARBAMOYL CHLORIDE (DECC)**

REPORT 01 : NIOSH CURRENT INTELLIGENCE BULLETIN 12 JUL 76
A STUDY BY A RESEARCHER AT THE NEW YORK UNIVERSITY MEDICAL CENTER HAS SHOWN DECC TO BE MUTAGENIC IN TWO E. COLI STRAINS.
REPT'D USE : ANTIHELINTIC

CAS NO : **112-34-5**
CHEM NAME : ***** DIETHYLENE GLYCOL BUTYL ETHER (DGBE)**

REPORT 01 : EPA PROPOSED TEST RULE, (51FR27880) AUG 86
EPA IS PROPOSING THAT MFRS AND PROCESSORS OF DGBE PERFORM HEALTH EFFECTS TESTING FOR SUBCHRONIC TOXICITY WITH EMPHASIS ON REPRODUCTIVE; HEMATOLOGICAL; LIVER AND KIDNEY EFFECTS; NEUROLOGICAL/BEHAVIORAL EFFECTS; DEVELOPMENTAL NEUROTOXICITY; PHARMACOKINETIC MUTAGENICITY; AND ONCOGENICITY. EPA IS USING EXPOSURE TO DGBE IN CLEANING AND OTHER CONSUMER PRODUCTS AS THE BASIS FOR SECT 4(A)(1) FINDING.

REPORT 02 : CMA GLYCOL ETHERS PROGRAM PANEL AUG 87
A CHROMOSOMAL TEST OF CPD IN MOUSE BONE MARROW NUCLEUS WAS NEGATIVE. THE CHINESE HAMSTER OVARY CELL/HYPOXANTHINE-GUANINE-PHOSPHORIBOSYL TRANSFERASE (CHO/HGPRT) FORWARD MUTATION ASSAY WAS ALSO NEGATIVE.

REPORT 04 : EPA RESPONSE TO ITC 49 FR 45606 NOV 84
IN RESPONSE TO ITC RECOMMENDATIONS (48 FR 55674) EPA EXPANDS SCOPE OF RULEMAKING TO INCLUDE DGBE AND PROPOSES A 4 (A) TSCA TEST RULE.
REPT'D USE : LATEX PAINTS, CLEANERS, PAINT REMOVERS, BRAKE FLUID, CUTTING OIL AND PENETRATING OILS

CAS NO : **111-40-0**
CHEM NAME : ***** DIETHYLENETRIAMINE (DETA)**

REPORT 01 : ITC 8TH REPORT TO EPA (46FR281 38) APR 81
THE INTERAGENCY TESTING COMMITTEE RECOMMENDS THAT THE EPA ADMINISTRATOR CONSIDER THE DEVELOPMENT OF RULES TO TEST FOR SUBCHRONIC EFFECTS, MUTAGENICITY AND CHEMICAL FATE (AEROBIC AND ANAEROBIC).

REPORT 02 : EPA RESPONSE TO ITC RECOMMS (50 FR 21398) MAY 85
UNDER SECTION 4 (A) TSCA EPA ESTABLISHES TESTING REQUIREMENTS FOR MANUFACTURERS AND PROCESSORS AS FOLLOWS; (1) ORAL SUBCHRONIC, (2) DERMAL ABSORPTION, (3) AEROBIC CHEMICAL FATE, (4) MUTAGENICITY (IF POSITIVE THEN ONCOGENICITY).

REPORT 03 : 50FR21398 - ESTABLISHING TESTING REQUIREMENTS MAY 85
RULE ESTABLISHES TESTING REQUIREMENTS UNDER TSCA 4(A) CONSISTING OF ORAL SUBCHRONIC 90 DAY TOXICITY IN AT LEAST ONE MAMMALIAN SPECIES, DERMAL ABSORPTION IN SAME SPP, CHEMICAL FATE UNDER AEROBIC CONDITIONS AND GENE MUTATIONS AND CHROSOMAL ABERRATIONS.

REPORT 04 : FINAL TEST STDS AND REPORTING JAN 87
REQUIREMENTS - 52FR3230
EPA HAS REVIEWED COMMENTS ON PREVIOUS PROPOSAL THAT PROCESSORS AND MFRS TEST SUBSTANCE IN ORAL SUBCHRONIC, DERMAL ABSORPTION, CHEMICAL FATE UNDER AEROBIC CONDITIONS AND MUTAGENICITY. EPA IS PROMULGATING THESE REVISED RULES TO GO INTO EFFECT ON FEB 17, 1987.

REPORT 05 : 52FR2152 - RECEIPT OF TEST DATA JAN 87
EPA ANNOUNCES RECEIPT OF TEST DATA FROM INDUSTRY'S VOLUNTARY TESTING PROGRAM.
REPT'D USE : WET-STRENGH RESINS FOR PAPER/PAPERBOARD, EPOXY-CURING AGENTS, CHELATING AGENTS, LUBRICATING OIL, FABRIC SOFTENERS, FUEL ADDITIVES, SURFACTANTS, CORROSION INHIBITORS AND EMULSIFIERS

CAS NO : **103-23-1**
CHEM NAME : ***** DI(2-ETHYLHEXYL)ADIPATE**

REPORT O1 : NTP BIOASSAY TRS 212 NIH PUB NO 81-1768
UNDER THE CONDITIONS OF THIS BIOASSAY, D1(2-ETHYLHEXYL)ADIPATE ADMINISTERED IN THE FEED WAS NOT CARCINOGENIC FOR F344 RATS. THE COMPOUND WAS CARCINOGENIC FOR FEMALE B6C3F1 MICE CAUSING AN INCREASED INCIDENCE OF

HEPATOCELLULAR CARCINOMA AND WAS PROBABLY CARCINOGENIC FOR MALE B6C3F1 MICE CAUSING HEPATOCELLULAR ADENOMAS.

REPORT 02 : NTP TECHNICAL BULLETIN 6 JAN 82
THE COMPOUND WAS NEGATIVE IN THE SALMONELLA TYPHIMURIUM MUTAGENICITY ASSAY IN TWO LABORATORIES.

REPORT 03 : IARC MONOGRAPH 29 1982
DI(2-ETHYLHEXYL)ADIPATE WAS TESTED IN MICE AND RATS BY ORAL ADMINISTRATION. IT INDUCED SIGNIFICANT INCREASES IN THE INCIDENCE OF LIVER-CELL TUMORS IN MICE. NO INCREASE IN TUMOR INCIDENCE WAS OBSERVED IN MICE. THERE IS LIMITED EVIDENCE THAT DEHA IS CARCINOGENIC IN ANIMALS.

REPORT 04 : IARC MONOGRAPH 29 1982
DEHA INDUCES DOMINANT LETHAL MUTATIONS AND REDUCES FERTILITY IN MICE. IT DID NOT INDUCE REVERSE MUTATIONS IN SALMONELLA STRAINS TA1535, TA1537, TA1538, TA98 AND TA100, WITH OR WITHOUT METABOLIC ACTIVATION. NO DATA WERE AVAILABLE TO ASSESS THE TERATOGENICITY OF THIS CHEMICAL.
REPT'D USE : RESIN PLASTICIZER FOR VINYL FOOD PACKAGING

CAS NO : **117-81-7**
CHEM NAME : ***** DI(2-ETHYLHEXYL)PHTHALATE (DEHP)**

REPORT 01 : NTP TECHNICAL BULLETIN 4/8 APR 81, JUL 82
DEHP WAS NEGATIVE IN SALMONELLA AND DROSOPHILA MUTAGENICITY ASSAY.

REPORT 02 : NTP BIOASSAY TRS 217 NIH PUB NO 82-1773
UNDER THE CONDITIONS OF THIS BIOASSAY DEHP WAS CARCINOGENIC FOR F344 RATS AND B6C3F1 MICE CAUSING INCREASED INCIDENCES OF HEPATOCELLULAR CARCINOMAS IN FEMALE RATS AND MALE AND FEMALE MICE AND INDUCING AN INCREASED INCIDENCE OF MALE RATS WITH EITHER HEPATOCELLULAR CARCINOMA OR NEOPLASTIC NODULES (FEED).

REPORT 03 : EPA RCRA HEALTH AND ENV. EFFECTS PROFILE OCT 80
DEHP HAS BEEN SHOWN TO PRODUCE MUTAGENIC EFFECTS IN THE AMES SALMONELLA ASSAY AND IN THE DOMINANT LETHAL ASSAY. TERATOGENIC EFFECTS IN RATS WERE REPORTED FOLLOWING I.P. ADMINISTRATION AND ORAL ADMINISTRATION OF DEHP. TESTICULAR DAMAGE AND DECREASED SPERMATOGENESIS HAVE BEEN REPORTED IN RATS FOLLOWING I.P. OR ORAL ADMINISTRATION, AND IN MICE GIVEN DEHP BY INTUBATION.

REPORT 04 : EPA AMBIENT WATER QUALITY CRITERIA OCT 80
DEHP HAS NOT BEEN FOUND TO BE CARCINOGENIC IN ANIMALS OR

MAN. A CRITERION, BASED ON CHRONIC TOXICOLOGIC DATA IN RATS AND AN UNCERTAINTY FACTOR OF 100, IS 14.8 MG/L.

REPORT 05 : EPA RESPONSE TO ITC RECOMMS (46FR53775) OCT 81
IN RESPONSE TO THE ITC RECOMMENDATIONS (42FR55026) THE EPA HAS RECEIVED A PROPOSAL FROM THE CHEMICAL MANUFACTURERS ASSN (CMA) THAT THE PHTHALATE ESTER INDUSTRY HAS AGREED TO VOLUNTARILY AND EXPEDITIOUSLY TEST ALKYL PHTHALATES FOR THE ENVIRONMENTAL AND HEALTH EFFECTS OF CONCERN TO THE ITC AND EPA. CONSEQUENTLY, THE EPA HAS DECIDED NOT TO PROPOSE ADDITIONAL TESTING AT THIS TIME.

REPORT 06 : MULTI-AGENCY CROSS CHECK
Y HHS MAGUIRE
N OSHA CARC
N EPA CAG
Y EPA RCRA
N EPA CLEAN AIR
Y EPA CLEAN WATER

REPORT 07 : IARC MONOGRAPH 29 1982
DEHP WAS TESTED IN RATS AND MICE BY ORAL ADMINISTRATION. IT SIGNIFICANTLY INCREASED THE INCIDENCE OF BENIGN AND MALIGNANT LIVER-CELL TUMORS IN ANIMALS OF BOTH SPECIES, AND A DOSE-REPONSE RELATIONSHIP WAS OBSERVED. THERE IS SUFFICIENT EVIDENCE FOR THE CARCINOGENICITY OF DEHP IN EXPERIMENTAL ANIMALS.

REPORT 08 : IARC MONOGRAPH 29 1982
DEHP CAN CAUSE TESTICULAR DAMAGE IN RATS. THERE IS EVIDENCE THAT THIS CHEMICAL AND ITS METABOLITE ARE TERATOGENIC AND EMBRYOLETHAL TO RODENTS. DEHP WAS NOT MUTAGENIC TO SALMONELLA, BUT DID CAUSE DOMINANT LETHAL MUTATIONS IN MICE AFTER SYSTEMIC BUT NOT ORAL ADMINISTRATION.

REPORT 09 : TSCA SECT 8(E) STATUS REPORT 8EH4-0587-0672 JUN 87
DEHP ACTS AS A TERATOGEN BY IN VIVO HYDROLYSIS TO 2-ETHYL-HEXANOL (2-EHXO), WHICH IN TURN IS METABOLIZED TO 2-ETHYLHEXANOIC ACID (2-EHXA), THE PROXIMATE TERATOGEN. ON AN EQUIMOLAR BASIS, DEHP WAS CONSIDERED THE LEAST POTENT, 2-EXHO WAS INTERMEDIATE AND 2-EHXA WAS THE MOST POTENT TERATOGEN. ALL 3 TESTS WERE POTENTIATED BY CAFFIENE. VALPROIC ACID, AN ISOMER OF 2-EHXA PRODUCED SIMILAR DEFECTS AND WAS APPROXIMATELY 2X AS POTENT AS 2-EHXA.

REPT'D USE : POLYMER PLASTICIZER AND OIL DIFFUSION PUMP FLUID

CAS NO : **1615-80-1**
CHEM NAME : ***** 1,2-DIETHYLHYDRAZINE (SDEH)**

REPORT 01 : IARC MONOGRAPH 4 1974
1,2-DIETHYLHYDRAZINE (SDEH) IS CARCINOGENIC IN RATS BY SUBCUTANEOUS ADMINISTRATION AND TRANSPLACENTAL EXPOSURE. IN A 1979 SUPPLEMENT, IARC DETERMINED THAT THERE IS SUFFICIENT EVIDENCE FOR THE CARCINOGENICITY OF SDEH IN ANIMALS.

REPORT 06 : MULTI-AGENCY CROSS CHECK
N HHS MAGUIRE
N OSHA CARC
Y EPA CAG
Y EPA RCRA
N EPA CLEAN AIR
N EPA CLEAN WATER
REPT'D USE : LABORATORY AGENT AND EXPERIMENTAL ROCKET FUEL

CAS NO : **84-66-2**
CHEM NAME : ***** DIETHYL PHTHALATE**

REPORT 02 : EPA RCRA HEALTH AND ENV. EFFECTS PROFILE OCT 80
DIETHYL PHTHALATE HAS BEEN SHOWN TO PRODUCE MUTAGENIC EFFECTS IN THE AMES SALMONELLA ASSAY. TERATOGENIC EFFECTS WERE REPORTED FOLLOWING I.P. ADMINISTRATION OF DIETHYL PHTHALATE TO PREGNANT RATS. THIS SAME STUDY HAS ALSO INDICATED FETAL TOXICITY AND INCREASED RESORPTION AFTER I.P. ADMINISTRATION OF DEP.

REPORT 03 : EPA AMBIENT WATER QUALITY CRITERIA OCT 80
DIETHYL PHTHATE HAS NOT BEEN FOUND TO BE CARCINOGENIC IN ANIMALS OR MAN. A CRITERION, BASED ON CHRONIC TOXICOLOGIC DATA IN RATS AND AN UNCERTAINTY FACTOR OF 100, IS 350 MG/L.

REPORT 04 : EPA RESPONSE TO ITC RECOMMS (46FR53775) OCT 81
IN RESPONSE TO THE ITC RECOMMENDATIONS (42FR55026) THE EPA HAS RECEIVED A PROPOSAL FROM THE CHEMICAL MANUFACTURERS ASSN (CMA) WHEREBY THE PHTHALATE ESTER INDUSTRY HAS AGREED TO TEST, VOLUNTARILY AND EXPEDITIOUSLY, ALKYL PHTHALATES FOR THE ENVIRONMENTAL AND HEALTH EFFECTS OF CONCERN TO THE ITC AND EPA. CONSEQUENTLY, THE EPA HAS DECIDED NOT TO PROPOSE ADDITIONAL TESTING AT THIS TIME.

REPORT 05 : NTP TECHNICAL BULLETIN 5 AUG 81
DIETHYL PHTHALATE WAS NEGATIVE IN THE SALMONELLA TYPHIMURIUM MUTAGENICITY ASSAY IN TWO LABORATORIES.

REPORT 06 : MULTI-AGENCY CROSS CHECK
N HHS MAGUIRE
N OSHA CARC
N EPA CAG
Y EPA RCRA
N EPA CLEAN AIR
Y EPA CLEAN WATER

REPORT 07 : 49FR44142 NOV 84
ANNOUNCES SUBMISSIONS RECEIVED BY EPA DURING 3RD QUARTER FY 84 FROM NEGOTIATED TEST PROGRAMS AND OTHER TESTING PROGRAMS. THESE INCLUDE BIODEGRADATION, VAPOR PRESSURE, OCTANOL-WATER PARTITION COEFF. AND ACUTE AQUATIC TOXICITY TESTING.

REPT'D USE : PLASTICIZER FOR CELLULOSE FILM/SHEET, PHOTOGRAPHIC FILM MOLDED PLASTIC ARTICLES, AEROSOLS, CHEMICAL DEODORIZERS AND ADHESIVES

CAS NO : **56-53-1**
CHEM NAME : ***** DIETHYL STILBESTROL**

REPORT 01 : IARC MONOGRAPH 21 1979
DES DID NOT INDUCE REVERSE MUTATIONS IN 5 SALMONELLA STRAINS. IT DOES NOT SIGNIFICANTLY INCREASE SISTER CHROMATID EXCHANGES IN CHINESE HAMSTER CELL LINE. DES INDUCED MUTATIONS IN A MOUSE LYPHOMA SYSTEM. DES IS EMBRYOLETHAL FOR PRE- AND POSTIMPLANTATION EMBRYOS IN SOME SPECIES AND USES TERATOGENIC EFFECTS ON THE GENITAL TRACT.

REPORT 02 : IARC MONOGRAPH 21 1979
DES IS CARCINOGENIC IN SEVERAL SPECIES OF ANIMALS, PRODUCING TUMORS IN ESTROGEN-RESPONSIVE TISSUES. THE ADMINISTRATION OF DES TO WOMEN DURING PREGNANCY IS ASSOCIATED WITH AN INCREASED RISK OF VAGINAL OR CERVICAL ADENOCARCINOMA IN THEIR EXPOSED FEMALE OFFSPRING. IN A 1979 SUPPLEMENT. THE IARC CONCLUDED THAT DES IS CARCINOGENIC FOR HUMANS.

REPORT 03 : NAS REPORT ON DIET, NUTRITION, AND CANCER 1982
THERE IS SUFFICIENT EVIDENCE THAT DES USED IN THERAPERUTIC DOSES PRODUCES VAGINAL AND CERVICAL CANCER IN THE FEMALE OFFSPRING OF TREATED WOMEN. IN ANIMALS, IT PRODUCES MAINLY MAMMARY TUMORS IN VARIOUS SPECIES.

REPORT 04 : NTP TECHNICAL BULLETIN 8 JUL 82
DES WAS POSITIVE IN THE L5178Y MOUSE LYMPHOMA ASSAY.

REPORT 06 : MULTI-AGENCY CROSS CHECK
Y HHS MAGUIRE

N OSHA CARC
Y EPA CAG
Y EPA RCRA
N EPA CLEAN AIR
N EPA CLEAN WATER

REPT'D USE : DRUG

CAS NO : **130-80-3**
CHEM NAME : ***** DIETHYLSTILBESTROL DIPROPIONATE**

REPORT 06 : MULTI-AGENCY CROSS CHECK
N HHS MAGUIRE
Y OSHA CARC
N EPA CAG
N EPA RCRA
N EPA CLEAN AIR
N EPA CLEAN WATER

REPT'D USE : PHARMACEUTICAL AND VETERINARY ESTROGEN

CAS NO : **64-67-5**
CHEM NAME : ***** DIETHYL SULFATE**

REPORT 01 : EPA TSCA 8(E) 8EHQ-1083-0509 MAR 83
A SKIN PAINTNG STUDY INVOLVING MALE CEH/HEJ MICE FOR THEIR LIFESPAN SHOWED DERMAL CARCINOGENICITY. IN A BATTERY OF IN-VITRO TESTS: CHO MUTATION ASSAY, THE SISTER CHROMATID EXCHANGE TEST AND UNSCHEDULED DNA SYNTHESIS IN RAT LIVER CELLS, THE PATTERN OF POSITIVE RESPONSES INDICATED THAT THE CPD POSSESSED A SIGNIFICANT MUTAGENIC POTENTIAL.

REPORT 02 : IARC MONOGRAPH 4 1974
DIETHYL SULFATE IS CARCINOGENIC IN THE RAT, THE ONLY SPECIES TESTED, FOLLOWING SUBCUTANEOUS ADMINISTRATION AND PRE-NATAL EXPOSURE. IN A 1979 SUPPLEMENT, IARC DETERMINED THAT THERE IS SUFFICIENT EVIDENCE FOR THE CARCINOGENICITY OF DIETHYL SULFATE IN ANIMALS.

REPORT 03 : EPA TSCA 8(E) 8EHQ-0585-0554 MAY 85
MUTAGENIC IN VARIETY OF IN VITRO AND IN VIVO TESTS. STUDIES IN LAB ANIMALS INDICATES DES CAN CAUSE CANCER BY INJECTION (RATS) & REPEATED SKIN APPLICATION (MICE).

REPORT 04 : 8EHQ-1983-0509 EPA STATUS RPT MAR 84
DIETHYL SULFATE SHOWED DERMAL CARCINOGENIC ACTIVITY IN A LIFETIME SKIN PAINTING STUDY IN C3H/HEJ MALE MICE. TUMOR AND CANCER INDICES FOR DES GROUP WERE 87.5, COMPARED TO

100 AND 94.4 RESPECTIVELY FOR A GROUP OF CEH/HEJ MALE MICE PAINTED WITH METHYLCHOLANTHRENE.

REPORT 06 : MULTI-AGENCY CROSS CHECK
Y HHS MAGUIRE
Y OSHA CARC
N EPA CAG
N EPA RCRA
N EPA CLEAN AIR
N EPA CLEAN WATER
REPT'D USE : ALKYLATING AGENT AND CHEMICAL INTERMEDIATE

CAS NO : **105-55-5**
CHEM NAME : ***** N,N'-DIETHYLTHIOUREA**

REPORT 01 : NCI CARC TRS 149 DHEW PUB NO 79-1705
UNDER THE CONDITIONS OF THIS BIOASSAY, N,N'-DIETHYLTHIOUREA WAS CARCINOGENIC TO FISCHER 344 RATS, CAUSING FOLLICULAR-CELL CARCINOMAS OF THE THYROID IN MALES AND FOLLICULAR-CELL NEOPLASMS OF THE THYROID IN FEMALES. THERE WAS NO EVIDENCE FOR THE CARCINOGENICITY OF THE COMPOUND IN B6C3F1 MICE (FEEDING STUDY).

REPORT 02 : NTP TECHNICAL BULLETIN 7 APR 82
N,N-DIETHYLTHIOUREA WAS NEGATIVE IN THE SALMONELLA TYPHIMURIUM MUTAGENICITY ASSAY.

REPORT 06 : MULTI-AGENCY CROSS CHECK
N HHS MAGUIRE
Y OSHA CARC
N EPA CAG
N EPA RCRA
N EPA CLEAN AIR
N EPA CLEAN WATER
REPT'D USE : CORROSION INHIBITOR AND RUBBER ACCELERATOR

CAS NO : **68479-98-1**
CHEM NAME : ***** DIETHYL TOLUENE DIAMINE**

REPORT 01 : EPA TSCA 8(E) 83HQ-0883-0489 AUG 83
SHORT-TERM TESTS WITH AND WITHOUT METABOLIC ACTIVATION SHOW CPD CAUSES DOSE-DEPENDENT MUTAGENIC RESPONSE WITH ACTIVATION.
REPT'D USE : EXTENDER AND CURING AGENT FOR POLYURETHANE, CURING AGENT FOR EPOXY RESINS AND CHEMICAL INTERMEDIATE

CAS NO : **101-90-6**
CHEM NAME : ***** DIGLYCIDYL RESORCINOL ETHER (DGRE)**

RERORT 01 : IARC MONOGRAPH 11 1976
DIGLYCIDYL RESORCINOL ETHER WAS REPORTED IN A PRELIMINARY COMMUNICATION TO BE CARCINOGENIC IN MICE BY SKIN APPLICATION AND IN RATS BY SUBCUTANEOUS INJECTION. NO CARCINOGENIC EFFECT WAS OBSERVED IN A FURTHER LONG-TERM SKIN APPLICATION STUDY IN MICE. NO EVALUATION CAN BE MADE.

REPORT 02 : IARC MONOGRAPH 36 DEC 87
CPD WAS MUTAGENIC TO S. TYPHIMURIUM TA1535 AND TA100 BUT NOT TA1537 OR TA98 IN A PREINCUBATION ASSAY, BOTH IN PRESENCE AND ABSENCE OF AROCLOR-INDUCED RAT OR HAMSTER LIVER METABOLIC SYSTEM. CPD PRODUCES SEVERE BURNS ON CONTACT WITH SKIN AND SKIN SENSITIZATION HAS OCCURRED IN LIMITED NUMBER OF CASES.

REPORT 03 : IARC MONOGRAPH 36 FEB 85
DIGLYCIDYL RESORCINOL ETHER (OF TECHNICAL GRADE) TESTED BY INTRAGASTRIC INTUBATION IN MICE AND IN RATS INDUCED SQUAMOUS-CELL CARCINOMAS AND PAPILLOMAS OF FORESTOMACH IN ANIMALS OF BOTH SPECIES. IN FEMALE MICE, INCREASED INCIDENCE OF HEPATOCELLULAR TUMORS OBSERVED. IN ONE EXPERIMENT IN MICE, NO SKIN TUMOR OBSERVED AFTER SKIN APPLICATION.

REPORT 04 : NTP TR 257 NIH PUB # 87-2513 OCT 86
IN 2 YR GAVAGE STUDIES, TECH GRADE DGRE CAUSED HYPERKERATOSIS AND HYPERPLASIA OF FORESTOMACH IN RATS AND MICE. DGRE CARCINOGENIC FOR MALE AND FEMALE F344/N RATS AND B6C3F1 MICE.

REPORT 06 : MULTI-AGENCY CROSS CHECK
Y HHS MAGUIRE
N OSHA CARC
N EPA CAG
N EPA RCRA
N EPA CLEAN AIR
N EPA CLEAN WATER
REPT'D USE : EPOXY RESINS, RUBBER CURING AGENT AND REACTIVE DILUENT

CAS NO : **94-58-6**
CHEM NAME : ***** DIHYDROSAFROLE**

REPORT 01 : IARC MONOGRAPH 10 1976
DIHYDROSAFROLE GIVEN ORALLY IS CARCINOGENIC IN RATS, IN WHICH IT PRODUCES TUMORS OF THE ESOPHAGUS, AND IN MICE,

IN WHICH IT PRODUCES LIVER TUMORS IN MALES AND AN INCREASED INCIDENCE OF LUNG TUMORS IN BOTH MALES AND FEMALES. IN A 1979 SUPPLEMENT, IARC DETERMINED THAT THERE IS SUFFICIENT EVIDENCE FOR THE CARCINOGENICITY OF DIHYDROSAFROLE IN ANIMALS.

REPORT 06 : MULTI-AGENCY CROSS CHECK 1985
Y HHS MAGUIRE
N OSHA CARC
Y EPA CAG
Y EPA RCRA
N EPA CLEAN AIR
N EPA CLEAN WATER
REPT'D USE : FLAVORING AGENT-ROOT BEER (BANNED 1960 BY FDA), FRAGRANCE, CHEMICAL INTERMEDIATE AND PESTICIDE INTERMEDIATE

CAS NO : **117-12-4**
CHEM NAME : ***** 1,5-DIHYDROXYANTHRAQUINONE**

REPORT 01 : CPSC ANTHRAQUINONE DYES MONOGRAPH OCT 81
ANTHRARUFIN IS A HIGHLY POSITIVE MUTAGEN IN ACTIVATED OR NON-ACTIVATED SYSTEMS USING SALMONELLA.
REPT'D USE : DYE INTERMEDIATE

CAS NO : **117-10-2**
CHEM NAME : ***** 1,8-DIHYDROXYANTTHRAQUINONE (CHRYSAZIN)**

REPORT Ol : CPSC ANTHRAQUINONE DYES MONOGRAPH OCT 81
DANTHRON CAUSED FRAMESHIFT MUTATIONS IN SALMONELLA AND HAS BEEN SHOWN TO BIND AND CAUSE DAMAGE TO DNA.

REPORT 02 : CRC MEETING MINUTES 10/7/87 OCT 87
REPORT IN JAPANESE J. CANCER RESEARCH THAT CPD CARCINOGENIC IN LARGE INTESTINE AND LIVER OF MICE. TOXLINE SEARCH SHOWS IT TO BE CARCINOGENIC, MUTAGENIC AND EYE IRRITANT.
REPT'D USE : DYES AND MEDICINES

CAS NO : **131-53-3**
CHEM NAME : ***** 2,2,-DIHYDROXY-4-METHOXYBENZOPHENONE**

REPORT 01 : NTP TECHNICAL BULLETIN 9 APR 83
2,2-DIHYDROXY-4-METHOXYBENZOPHENONE WAS POSITIVE IN ONE SALMONELLA TYPHIMURIUM MUTAGENICITY ASSAY.

CAS NO : **794-93-4**
CHEM NAME : ***** DIHYDROXYMETHYLFURATRIZINE**

REPORT 01 : IARC MONOGRAPH 24 1980
DIHYDROXYMETHYLFURATRIZINE WAS TESTED ALONE IN FEMALE RATS BY ORAL ADMINISTRATION. ALTHOUGH A FEW TUMORS OF THE SMALL INTESTINE WERE OBSERVED IN TREATED ANIMALS, THERE WAS NO DIFFERENCE IN OVERALL TUMOR INCIDENCE AS COMPARED WITH CONTROLS. THERE ARE INSUFFICIENT DATA TO EVALUATE THE CARCINOGENICITY OF DIHYDROXYMETHYL-FURATRIZINE ALONE IN EXPERIMENTAL ANIMALS.

REPORT 02 : IARC MONOGRAPH 24 1980
DIHYDROXYMETHYLFURATRIZINE INHIBITED THE GROWTH OF S. AUREUS AND OF E. COLI. THE MUCH HIGHER SENSITIVITY OF DIFFERENT REPAIR-DEFICIENT MUTANTS OF E. COLI AS COMPARED WITH REPAIR-PROFICIENT ONES INDICATES THAT THIS COMPOUND INDUCES DNA DAMAGE. ATTENTION IS DRAWN TO THE ABSENCE OF STUDIES ON THE TERATOGENICITY OF THIS COMPOUND.
REPT'D USE : ANTIBACTERIAL AGENT

CAS NO : **2973-10-6**
CHEM NAME : ***** DIISOPROPYL SULFATE (DIPS)**

REPORT 01 : EPA DRAFT CHIP REPORT SEP 86
STUDIES PROVIDE CONCLUSIVE EVIDENCE OF CARCINOGENICITY OF DIPS TO MICE AND RATS BY SUBCUTANEOUS INJECTION AND TO MICE BY SKIN APPLICATION. STATISTICALLY SIGNIFICANT INCREASE ($P<0.01$) IN THE INCIDENCE OF TUMORS IN MICE RECEIVING 13 INJECTIONS OF 0.025 ML. 14 OF 15 RATS DEVELOPED SARCOMAS AFTER RECEIVING 15 INJECTIONS OF 100 MG/KG. IN SKIN-PAINTING STUDY, 83% OF SURVIVING MICE DEVELOPED PAPILLOMAS. STUDIES WITH BYPRODUCTS (ISOPROPYL OIL) FROM BOTH THE STRONG-ACID AND WEAK-ACID PROCESSES PRESENT INADEQUATE EVIDENCE FOR CARCINOGENICITY.

REPORT 02 : EPA DRAFT CHIP REPORT SEP 86
OTHER HEALTH EFFECTS STUDIES INDICATE 0.01 ML OF UNDILUTED ISOPROPYL SULFATE PRODUCED NECROSIS OF TISSUE TO RABBIT SKIN WITHIN 24-HOURS. INSTILLATION OF 0.005 ML OF ISOPROPYL SULFATE PRODUCED SEVERE CORNEAL BURNS IN RABBIT EYES.
REPT'D USE : MANUFACTURE OF ISOPROPYL ALCOHOL AND ALKYLATING AGENT

CAS NO : **108-18-9**
CHEM NAME : ***** DIISOPROPYLAMINE**

REPORT 01 : EPA TSCA 8(E) 1287-0705 JAN 88
MONSANTO PROVIDED REPORT OF 1-MONTH INHALATION STUDY OF DIISOPROPYLAMINE IN FOUR GROUPS OF 15 MALES AND FEMALE SPRAGUE-DAWLEY RATS EXPOSED FOR 6 HRS/DAY AT CONCENTRATIONS OF 0.0, 0.10, 0.60, AND 2.00 MG/LITER OF AIR FOR 23 EXPOSURES IN ONE MONTH. THREE HIGH EXPOSURE LEVEL ANIMALS DIED. GROSS SIGNS OR IRRITATION TO EYES AND RESPIRATORY TRACT AND DECREASED BODY WTS. OCCURRED AT MID AND HIGH EXPOSURES. OPHTHALMIC EXAMINATIONS SHOWED DOSE RELATED CORNEAL KERATOPATHY. INCREASED RED BLOOD CELL COUNTS, HEMOGLOBIN AND HEMATOCRIT MEASUREMENTS IN HIGH LEVEL MALES AND AMONG MID AND HIGH LEVEL FEMALES. LOWER WHITE BLOOD CELL COUNTS PRESENT IN MALES AT ALL EXPOSURES.

REPORT 02 : EPA TSCA 8(E) 287-0705 JAN 88
"OCCURRENCES" OF DECREASED CORPUSCULAR VOLUME, SERUM CHLORIDE, TOTAL PROTEIN, ALBUMIN AND INCREASED SERUM CHOLESTEROL DIRECTLY ASSOCIATED TO COMPOUND EXPOSURE. THYMIC ATROPHY AND SEMINAL VESICLE ATROPHY AT HIGH EXPOSURES. INFLAMMATORY LESIONS IN NASAL PASSAGES OF ALL EXPOSURE GROUPS AND IN TRACHEA, LUNG AND CORNEAS AT HIGH EXPOSURE. NO-EFFECT LEVEL COULD NOT BE DETERMINED. MONSANTO NOTED PRIMARY HAZARD FROM DIISOPROPYLAMINE DUE TO ITS MARKED IRRITANCY TO SKIN, EYES AND MOCOSAE.
REPT'D USE : CATALYST AND CHEMICAL INTERMEDIATE. NO LONGER MANUFACTURED

CAS NO : **79-64-1**
CHEM NAME : ***** DIMETHISTERONE**

REPORT 01 : IARC MONOGRAPH 21 1979
DIMETHISTERONE WAS TESTED IN DOGS IN COMBINATION WITH ETHINYL-OESTRADIOL BY ORAL ADMINISTRATION. NO INCREASE IN INCIDENCE OF MAMMARY TUMORS WAS FOUND. OWING TO THE LACK OF EXPERIMENTAL AND HUMAN DATA ON DIMETHISTERONE ALONE, NO EVALUATION OF THE CARCINOGENICITY OF DIMETHISTERONE COULD BE MADE.

REPORT 02 : IARC MONOGRAPH 21 1979
NO CHROMOSOMAL EFFECTS WERE OBSERVED WHEN DIMETHISTERONE WAS ADDED TO CULTURES OF LYMPHOCYTES GROWN FROM THE BLOOD OF HEALTHY WOMAN. NO DATA ON THE EMBRYOTOXICITY OR TERATOGENICITY OF THIS COMPOUND WERE AVAILABLE.
REPT'D USE : PROGESTRIN AND ORAL CONTRACEPTIVES

CAS NO : **60-51-5**
CHEM NAME : ***** DIMETHOATE**

REPORT 01 : NCI CARC TRS 4 DHEW PUB NO 77-804
DIMETHOATE GIVEN IN FEED TO OSBORNE-MENDEL RATS AND B6C3Fl MICE SHOWED NO STATISTICALLY SIGNIFICANT INCREASE IN TUMORS.

REPORT 02 : NTP TECHNICAL BULLETIN 6 JAN 82
THE COMPOUND WAS POSITIVE IN THE SALMONELLA TYPHIMURIUM MUTAGENICITY ASSAY IN TWO LABORATORIES.
REPT'D USE : INSECTICIDE, ACARICIDE AND MITICIDE

CAS NO : **828-00-2**
CHEM NAME : ***** DIMETHOXANE**

REPORT 01 : NTP TECHNICAL BULLETIN 3 DEC 80
THE COMPOUND WAS FOUND TO BE POSITIVE IN THE SALMONELLA TYPHIMURIUM MUTAGENICITY ASSAY IN ONE LABORATORY.

REPORT 02 : IARC MONOGRAPH 15 1977
DIMETHOXANE IS CARCINOGENIC IN MALE RATS AFTER ORAL ADMINISTRATION, THE ONLY SPECIES, SEX AND ROUTE TESTED; IT PRODUCED MALIGNANT TUMORS, PREDOMINANTLY IN THE LIVER.

REPORT 03 : NTP CARC TEST PROGRAM JUL 82
IN A CURRENT NTP BIOASSAY, DIMETHOXANE IS BEING ADMINISTERED BY GAVAGE TO F344 RATS AND B6C3Fl MICE. THE PRECHRONIC TEST PHASE IS IN PROGRESS.

REPORT 04 : NTP TR 354 NIH PUB.89-2809 SEP 89
UNDER THE CONDITIONS OF THESE 2-YEAR CORN OIL GAVAGE STUDIES, THERE WAS NO EVIDENCE OF CARCINOGENIC ACTIVITY OF DIMETHOXANE FOR MALE F344/N RATS RECEIVING 62.5 OR FEMALE 125 MG/KG 4344/N RATS RECEIVING 125 OR 250 MG/KG. THERE WAS EQUIVOCAL EVIDENCE OF CARCINOGENIC ACTIVITY OF DIMETHOXANE FOR MALE B6C3F1, MICE (FORESTOMACH NEOPLASMS) AND NO EVIDENCE OF CARCINOGENIC ACTIVITY FOR FEMALE B6C3Fl MICE RECEIVING 250 OR 500 MG/KG.

REPORT 05 : NTP TR 354 NIH 89-2809 SEP 89
DIMETHOXANE WAS MUTAGENIC IN SALMONELLA TYPHIMURIUM STRAIN TA100 WHEN EXPOSURE OCCURRED IN THE PRESENCE OF EXOGENOUS METABOLIC ACTIVATION; IT WAS NOT MUTAGENIC IN STRAIN TA100 WITHOUT METABOLIC ACTIVATION OR IN STRAINS TA98, TA1535, OR TA 1537 WITH OR WITHOUT ACTIVATION. DIMETHOXANE INDUCED SEX-LINKED RECESSIVE LETHAL MUTATIONS IN DROSOPHILA WHEN ADMINISTERED BY ABDOMINAL

INJECTION TO ADULT MALE CANTONS FLIES, NO INDUCTION OF RECIPROCAL TRANSLOCATIONS WAS OBSERVED.

REPORT 06 : MULTI-AGENCY CROSS CHECK
N HHS MAGUIRE
Y OSHA CARC
N EPA CAG
N EPA RCRA
N EPA CLEAN AIR
N EPA CLEAN WATER

REPT'D USE : INDUSTRIAL CUTTING OILS, TEXTILE RESIN EMULSIONS, ANTIMICROBIAL AGENT, LATEX PAINTS, FABRIC SOFTENERS, ADHESIVE SINK AND PRESERVATIVE

CAS NO : **54150-69-5**
CHEM NAME : ***** 2,4-DIMETHOXYANILINE HYDROCHLORIDE**

REPORT 01 : NCI CARC TRS 171 DHEW PUB NO 79-1727
UNDER THE CONDITIONS OF THIS BIOASSAY, THERE WAS NO CONVINCING EVIDENCE FOR THE CARCINOGENICITY OF 2,4-DIMETHOXYANILINE HCL IN FISCHER 344 RATS OR B6C3F1 MICE (FEEDING STUDY). (UNABLE TO CONFIRM REGISTRY NUMBER.)

REPT'D USE : DYE INTERMEDIATE

CAS NO : **20325-40-0**
CHEM NAME : ***** 3,3'-DIMETHOXYBENZIDINE DIHYDROCHLORIDE**

REPORT 01 : NTP TR 372 NIH PUB 89-2827 JAN 90
UNDER 21-MONTH DRINKING WATER STUDIES, CLEAR EVIDENCE OF CARCINOGENIC ACTIVITY OF 3,3'-DIMETHOXYBENZIDINE DIHYDROCHLORIDE FOR MALE F344/N RATS INDICATED BY BENIGN AND MALIGNANT NEOPLASMS OF SKIN, ZYMBAL GLAND, PREPUTIAL GLAND, ORAL CAVITY, INTESTINE, LIVER, AND MESOTHELIUM. INCREASED INCIDENCES OF ASTROCYTOMAS OF BRAIN MAY HAVE BEEN RELATED TO CHEMICAL ADMINISTRATION. CLEAR EVIDENCE OF CARCINOGENIC ACTIVITY FOR FEMALE F344/N RATS INDICATED BY BENIGN AND MALIGNANT NEOPLASMS OF ZYMBAL GLAND, CLITORAL GLAND, AND MAMMARY GLAND. INCREASES IN NEOPLASMS OF SKIN, ORAL CAVITY, INTESTINE, LIVER, AND UTERUS/CERVIX RELATED TO CHEMICAL ADMINISTRATION.

REPORT 06 : MULTI-AGENCY CROSS CHECK
Y HHS MAGUIRE
Y OSHA CARC
N EPA CAG
N EPA RCRA
N EPA CLEAN AIR
N EPA CLEAN WATER

REPT'D USE : INTERMEDIATE IN PRODUCTION OF BISAZOBIPHENYL DYES AND OF O-DIANISIDINE DIISOCYANATE

CAS NO : **91-93-0**
CHEM NAME : ***** 3,3'-DIMETHOXYBENZIDINE-4,4'-DIISOCYANATE**

REPORT 01 : NCI CARC TRS 128 DHEW PUB NO 79-1383
GIVEN BY GAVAGE IN THIS BIOASSAY, ADMINISTRATION OF THE COMPOUND WAS CARCINOGENIC TO FISCHER 344 RATS, CAUSING NEOPLASMS OF THE SKIN (EXCLUDING SKIN OF THE EAR) IN MALES, ENDOMETRIAL STROMAL POLYPS IN FEMALES AND LEUKEMIA AND MALIGNANT LYMPHOMA IN BOTH SEXES. THE COMPOUND WAS ALSO ASSOCIATED WITH THE DEVELOPMENT OF A COMBINATION OF SQUAMOUS-CELL CARCINOMAS AND SEBACEOUS ADENOCARCINOMAS OF THE ZYMBAL GLAND AND SKIN OF THE EAR IN RATS OF BOTH SEXES. THERE WAS NO EVIDENCE FOR THE CARCINOGENICITY OF THE COMPOUND IN B6C3F1 MICE.

REPORT 02 : NTP TECHNICAL BULLETIN 5 AUG 81
THE CHEMICAL WAS POSITIVE IN THE SALMONELLA TYPHIMURIUM MUTAGENICITY ASSAY IN ONE LABORATORY.

REPORT 03 : IARC MONOGRAPH 39, 1986 JUN 85
CPD ADMINISTERED VIA GAVAGE AND DIET TO MICE AND RATS; NO DOSE-RELATED TUMOR SEEN IN MICE. STATISTICALLY SIGNIFICANT INCREASE IN INCIDENCE OF LEUKEMIA AND MALIGNANT LYMPHOMAS SEEN IN BOTH MALE AND FEMALE RATS PLUS INCREASED INCIDENCE OF SKIN AND ZIMBAL GLAND TUMORS. INCREASED INCIDENCE OF ENDOMETRIAL STROMAL POLYPS IN FEMALE RATS. CPD MUTAGENIC TO S. TYPHIMURIUM WITH ACTIVATION. NO DATA AVAILABLE TO EVALUATE REPRODUCTIVE EFFECTS OR PRENATAL TOXICITY IN HUMANS OR ANIMALS. NO CASE REPORTS OR EPI STUDIES FOUND TO EVALUATE CARCINOGENICITY OF CPD TO HUMANS. IARC FOUND LIMITED EVIDENCE FOR CARCINOGENICITY OF CPD TO ANIMALS.
REPT'D USE : POLYMER SYNTHESIS

CAS NO : **117-82-8**
CHEM NAME : ***** DIMETHOXYETHYL PHTHALATE**

REPORT 01 : EPA DRAFT CHIP/ OAK RIDGE LAB AUG 85
CARCINOGENICITY DATA FOR DMEP NOT FOUND. FETOTOXIC AND TERATOGENIC EFFECTS OBSERVED IN RATS GIVEN DMEP VIA SINGLE IP INJECTION AT CONCENTRATION OF 0.374 ML/KG; W/RESORPTIONS AT 27.6%. TERATOGENIC EFFECTS INCLUDED PRIMARILY SKELETAL ABNORMALITIES. HIGHER CONCENTRATIONS DRAMATICALLY INCREASED THE INCIDENCE OF RESORPTIONS. ADVERSE DEVELOPMENTAL EFFECTS REPORTED IN A SEPARATE

STUDY IN PREGNANT RATS ADMINISTERED SINGLE IP INJECTION OF 0.6 ML/KG. SINGLE IP DMEP INJECTION OF 2.38 ML/KG TO MICE RESULTED IN MARKED REDUCTION IN PREGNANCIES AND LITTER SIZE. REDUCED TESTES WEIGHTS AND INCREASES IN ABNORMAL SPERM HEADS IN RATS EXPOSED TO SINGLE ORAL DOSE OF 1000 MG DMEP/KG.

REPORT 02 : EPA DRAFT CHIP/ OAK RIDGE LAB AUG 85
HEMATOLOGICAL EFFECTS REPORTED IN LABORATORY ANIMALS (BASED ON SIGNIFICANT DECREASE IN HEMOGLOBIN CONCENTRATION AND HEMATOCRIT AND WHITE BLOOD CELL COUNTS FOLLOWING ADMINISTRATION OF 100 OR 1000 MG/KG/DAY FOR TWO WEEKS), AND INTRADERMAL IRRITATION. DMEP WAS NOT IRRITATING TO EYES OF RABBITS AND DID NOT CAUSE SKIN SENSITIZATION IN GUINEA PIGS.
REPT'D USE : PLASTICIZER, SOLVENT

CAS NO : **627-93-0**
CHEM NAME : ***** DIMETHYL ADIPATE (DMA)**

REPORT 01 : HS STAFF REPORT AUG 89
DIMETHYL ADIPATE, A DIBASIC ESTER, PRIMARILY USED AS PLASTICIZER FOR CELLULOSE TYPE RESINS SLIGHTLY TOXIC TO RATS AND CYTOTOXIC TO HELA CELLS. A POTENTIAL ANIMAL TERATOGEN. MILD TO MODERATE SQUAMOUS METAPLASIA (PRE-NEOPLASTIC LESION) IN NASAL TISSUE OF RATS AT ALL DOSE LEVELS BY INHALATION SHOWS POTENTIAL FOR CARCINOGENIC ACTIVITY EXPOSURE FROM DIBASIC ESTER SOLVENT MIXTURE (DIMETHY GLUTARATE (66.6%), DIMETHYL ADIPATE (16.5%), DIMETHYL SUCCINATE (16.9%)) AT 0.16, 0.40, 1.0 MG/L; 0.02, 0.08, 0.40 MG/L. FURTHER TESTING NEEDED TO EVALUATE ADVERSE EFFECTS OF DMA.
REPT'D USE : RESINS AND SUBSTITUTE FOR METHYLENE CHLORIDE

CAS NO : **124-40-3**
CHEM NAME : ***** DIMETHYL AMINE (DMA)**

REPORT 01 : MAY 84
REPT'D USE : CHEMICAL INTERMEDIATE, RUBBER VULCANIZING ACCELERATOR, GH, ASOLINE STABILIZER, FLOTATION AGENT, ACID GAS ABSORBENT AND PESTICIDE

CAS NO : **121-69-7**
CHEM NAME : ***** N,N-DIMETHYLANILINE**

REPORT 01 : NTP TR 360 NIH PUB. 90-2815 0CT 89
IN 2-YEAR GAVAGE STUDIES SOME EVIDENCE OF CARCINOGENIC

		ACTIVITY OF N,N-DIMETHYLANILINE FOR MALE F344/N RATS INDICATED BY INCREASED INCIDENCES OF SARCOMAS OR OSTEO-SARCOMAS (COMBINED) OF SPLEEN. NO EVIDENCE OF CARCINO-GENIC ACTIVITY FOR FEMALE F344/N RATS GIVEN 3 OR 30 MG/KG BODY WEIGHT BY GAVAGE. NO EVIDENCE OF CARCINOGENIC ACTIVITY FOR MALE B6C3F1 MICE GIVEN 15 OR 30 MG/KG BODY WEIGHT BY GAVAGE. EQUIVOCAL EVIDENCE OF CARCINOGENIC ACTIVITY FOR FEMALE B6C3F1 MICE INDICATED BY INCREASED INCIDENCE OF SQUAMOUS CELL PAPILLOMAS OF FORESTOMACH. BOTH RATS AND MICE COULD HAVE TOLERATED DOSES HIGHER THAN USED.
REPT'D USE	:	CHEMICAL INTERMEDIATE IN PRODUCTION OF DYES, VANILLIN, MICHLER'S KETONE; SOLVENT; METHYLATION AID; ALKYLATING AGENT; VULCANIZING AGENT AND POLYESTER ACTIVATOR

CAS NO	:	**75-60-5**
CHEM NAME	:	***** DIMETHYLARSINIC ACID**

REPORT 01	:	IARC MONOGRAPH 23 1980 THERE IS INADEQUATE EVIDENCE FOR THE CARCINOGENICITY OF ARSENIC COMPOUNDS IN ANIMALS. THERE IS SUFFICIENT EVI-DENCE THAT INORGANIC ARSENIC COMPOUNDS ARE SKIN AND LUNG CARCINOGENS IN HUMANS. THE EVIDENCE THAT ARSENIC COMPOUNDS CAUSE MUTATIONS AND ALLIED EFFECTS IN BACTE-RIA IS INCONCLUSIVE. HOWEVER, ARSENIC COMPOUNDS INDUCE CHROMOSOMAL ABERRATIONS AND MORPHOLOGICAL TRANSFOR-MATION IN MAMMALIAN CELLS.
REPORT 02	:	NIOSH CRITERIA DOCUMENT JUN 75 NIOSH RECOMMENDS A CEILING OF 2 MG/CU M FOR INORGANIC ARSENIC CPDS, BASED ON DERMATITIS, LUNG, AND LYMPHATIC CANCER. THE CURRENT OSHA STANDARD FOR INORGANIC ARSENIC IS 10 MICROGRAMS/CU M TWA. (SEE CFR 1910.1018.)
REPORT 06	:	MULTI-AGENCY CROSS CHECK Y HHS MAGUIRE N OSHA CARC Y EPA CAG N EPA RCRA N EPA CLEAN AIR N EPA CLEAN WATER
REPT'D USE	:	HERBICIDE, DEFOLIANT(AGENT BLUE) AND CAUSTICS/DRAIN CLEANERS

CAS NO : **57-97-6**
CHEM NAME : ***** 7,12-DIMETHYL-BENZO(A)ANTHRACENE**

REPORT 01 : EPA AMBIENT WATER QUALITY CRITERIA OCT 80
AVAILABLE DATA CANNOT SUPPORT INDIVIDUAL CRITERIA FOR EACH PAH. THE AWQC FOR CARCINOGENIC PAH COMPOUNDS IS BASED ON ASSUMPTIONS THAT EACH COMPOUND IS AS POTENT AS B(A)P AND THAT THE CARCINOGENIC EFFECT IS PROPORTIONAL TO THE SUM OF THEIR CONCENTRATIONS. THIS SUM SHOULD NOT EXCEED 28 NG/L, 2.8 NG/L AND .28 NG/L TO KEEP THE LIFETIME CANCER RISK BELOW 10 -5, 10 -6 AND 10 -7, RESPECTIVELY.

REPORT 02 : NTP CARC TEST PROGRAM JUL 82
IN A CURRENT BIOASSAY, DMBA/TPA WAS ADMINISTERED BY SKIN PAINTING TO F344 RATS. THE STUDY HAS BEEN COMPLETED AND THE REPORT IS BEING DRAFTED.

REPORT 03 : NTP TECHNICAL BULLETIN 8 JUL 82
DMBA/TMA WAS POSITIVE IN THE L5178Y MOUSE LYMPHOMA ASSAY.

REPORT 06 : MULTI-AGENCY CROSS CHECK
N HHS MAGUIRE
N OSHA CARC
Y EPA CAG
Y EPA RCRA
N EPA CLEAN AIR
N EPA CLEAN WATER
REPT'D USE : VETERINARY MEDICINE AND EXPERIMENTAL CARCINOGEN

CAS NO : **79-44-7**
CHEM NAME : ***** DIMETHYLCARBAMOYL CHLORIDE (DMCC)**

REPORT 01 : NIOSH CURRENT INTELLIGENCE BULLETIN 11 JUL 76
REFERS TO ACUTE TOXIC EFFECTS OF DMCC, AS WELL AS ITS CARCINOGENIC POTENTIAL IN ICR/HA SWISS MICE. HIGH INCIDENCE OF NASAL CANCER AND SHORT LATENCY PERIOD IN RATS SUGGESTS A POTENTIALLY SERIOUS OCCUPATIONAL HAZARD.

REPORT 02 : IARC MONOGRAPH 12 1976
DIMETHYLCARBAMOYL CHLORIDE IS CARCINOGENIC IN MICE BY SKIN APPLICATION OR BY SUBCUTANEOUS OR INTRAPERITONEAL INJECTION: IT PRODUCED LOCAL TUMORS BY EACH OF THE THREE ROUTES OF ADMINISTRATION. IN A 1979 SUPPLEMENT, IARC DETERMINED THAT THERE IS SUFFICIENT EVIDENCE FOR THE CARCINOGENICITY OF DIMETHYLCARBAMOYL CHLORIDE IN ANIMALS AND THAT DIMETHYLCARBAMOYL IS PROBABLY CARCINOGENIC FOR HUMANS.

REPORT 03 : IARC MONOGRAPH 12 1976
DIMETHYLCARBAMOYL CHLORIDE INDUCED REVERSE MUTATIONS IN SALMONELLA TA1535 AND TA100. NO METABOLIC ACTIVATION WAS REQUIRED.

REPORT 04 : NTP TECHNICAL BULLETIN 3/8 DEC 80, JUL 82
DMCC WAS POSITIVE IN SALMONELLA AND IN DROSOPHILA WHEN TESTED FOR SEX-LINKED RECESSIVE LETHALS.

REPORT 06 : MULTI-AGENCY CROSS CHECK
Y HHS MAGUIRE
N OSHA CARC
Y EPA CAG
Y EPA RCRA
N EPA CLEAN AIR
N EPA CLEAN WATER

REPT'D USE : INTERMEDIATE IN MFR OF PHARMACEUTICALS, PESTICIDES AND DYES

CAS NO : **1119-40-0**

CHEM NAME : ***** DIMETHYL GLUTARATE (DMG)**

REPORT 01 : HS STAFF REPORT AUG 89
DIMETHYL GLUTARATE (DMG), A DIBASIC ESTER, USED AS CHEMICAL INTERMEDIATE FOR POLYAMIDE AVAILABLE ON HUMANS. SUBCHRONIC INHALATION EXPOSURE TO DMG AT 0.02, 0.08, 0.04 MG/L INDUCED MILD TO MODERATE SQUAMOUS METAPLASIA (A PRE-NEOPLASTIC LESION) IN NASAL TISSUE OF RATS AT ALL DOSE LEVELS (SHOWS POTENTIAL CARCINOGENIC ACTIVITY). EXPOSURE FROM DIBASIC ESTER SOLVENT MIXTURE (DIMETHYL GLUTARATE 66.6%), DIMETHYL ADIPATE (16.5%), DIMETHYL SUCCINATE (16.9%) AT 0.16, 0.40, 1.0 AND 0.02, 0.08, 0.40 MG/L. REPRODUCTIVE EFFECTS IN RATS SEEN ONLY AT HIGHEST DOSE (1.0 MG/L) IN PRESENCE OF MATERNAL TOXICITY. FURTHER TESTING NECESSARY TO DETERMINE TOXICITY OF DMG.

REPT'D USE : CHEMICAL INTERMEDIATE IN POLYAMIDE, EPICHLBRHYDRIN, POLYESTER RESINS AND SUBSTITUTE FOR METHYLENE CHLORIDE

CAS NO : **57-14-7**

CHEM NAME : ***** 1,1-DIMETHYLHYDRAZINE (UDMH)**

REPORT 02 : IARC MONOGRAPH 4 1974
UDMH IS CARCINOGENIC IN MICE AFTER ORAL ADMINISTRATION. THE OBSERVATION OF A FEW LIVER TUMORS AFTER HIGH ORAL DOSES OF UDMH OCCURRING IN RATS AFTER A LONG LATENT PERIOD DOES NOT ALLOW A PROPER EVALUATION OF THE CARCINOGENIC EFFECT IN THIS SPECIES. IN A 1979 SUPPLEMENT,

IARC DETERMINED THAT THERE IS SUFFICIENT EVIDENCE FOR THE CARCINOGENICITY OF UDMH IN ANIMALS.

REPORT 03 : NIOSH CRITERIA DOCUMENT JUN 78
NIOSH RECOMMENDS A CEILING OF 0.15 MG/CU M BASED ON LIVER, BLOOD, EYE, SKIN EFFECTS AND POTENTIAL CARCINOGENICITY. THE PRESENT OSHA STANDARD IS 1.0 MG/CU M TWA.

REPORT 06 : MULTI-AGENCY CROSS CHECK
Y HHS MAGUIRE
N OSHA CARC
Y EPA CAG
Y EPA RCRA
N EPA CLEAN AIR
N EPA CLEAN WATER
REPT'D USE : PROPELLANT FOR LIQUID-FUELED ROCKETS, CHEMICAL INTERMEDIATE AND PLANT GROWTH REGULATOR

CAS NO : **540-73-8**
CHEM NAME : ***** 1,2-DIMETHYLHYDRAZINE (SDMH)**

REPORT 01 : IARC MONOGRAPH 4 1974
SDMH IS CARCINOGENIC IN MICE, RATS AND HAMSTERS FOLLOWING ORAL AND SUBCUTANEOUS OR INTRAMUSCULAR ADMINISTRATION. IN A 1979 SUPPLEMENT, IARC DETERMINED THAT THERE IS SUFFICIENT EVIDENCE FOR THE CARCINOGENICITY OF SDMH IN ANIMALS.

REPORT 06 : MULTI-AGENCY CROSS CHECK
N HHS MAGUIRE
N OSHA CARC
Y EPA CAG
Y EPA RCRA
N EPA CLEAN AIR
N EPA CLEAN WATER
REPT'D USE : EXPERIMENTAL ROCKET FUEL

CAS NO : **306-37-6**
CHEM NAME : ***** 1,2-DIMETHYLHYDRAZINE DIHYDROCHLORIDE**

REPORT 06 : MULTI-AGENCY CROSS CHECK
N HHS MAGUIRE
Y OSHA CARC
N EPA CAG
N EPA RCRA
N EPA CLEAN AIR
N EPA CLEAN WATER
REPT'D USE : NO AVAILABLE USE DATA

CAS NO : **868-85-9**
CHEM NAME : ***** DIMETHYL HYDROGEN PHOSPHITE**

REPORT 01 : EPA TSCA 8(E) NOTICE 8EHQ-1080-0366 OCT 80
SUBMITTED RESULTS FROM LAB STUDIES SHOWING DIMETHYL HYDROGEN PHOSPHITE TO BE POSITIVE IN DROSOPHILIA AND IN AN IN-VITRO MOUSE LYMPHOMA ASSAY (ACTIVATED). THE CHEMICAL WAS NEGATIVE IN BOTH THE MOUSE LYMPHOMA ASSAY WITHOUT ACTIVATION, AND IN THE AMES TEST (W/WO ACTIVATION). RESULTS OF A DNA REPAIR TEST WERE INCONCLUSIVE.

REPORT 02 : NTP TR 287 NIH PUB 86-2543 NOV 85
UNDER CONDITIONS OF GAVAGE STUDIES, THERE WAS CLEAR EVIDENCE OF CARCINOGENICITY IN MALE F344/N RATS, AS SHOWN BY INCREASED INCIDENCES OF ALVEOLAR/BRONCHIOLAR ADENOMAS, ALVEOLAR/BRONCHIOLAR CARCINOMAS, SQUAMOUS CELL CARCINOMAS OF LUNG, AND NEOPLASMS OF FORESTOMACH. THERE WAS EQUIVOCAL EVIDENCE OF CARCINOGENICITY IN FEMALE F344N/N RATS, SHOWN BY MARGINALLY INCREASED INCIDENCES OF ALVEOLAR/BRONCHIOLAR CARCINOMAS OF LUNG AND NEOPLASMS OF FORESTOMACH. NO EVIDENCE OF CARCINOGENICITY FOR MALE OR FEMALE B6C3F1 MICE NOTED.
REPT'D USE : FR CHEMICAL, LUBRICANT ADDITIVE AND ADHESIVES

CAS NO : **20120-33-6**
CHEM NAME : ***** DIMETHYL N-METHYLOLPHOSPHONOPIONAMIDE (DMMP)**

REPORT 01 : CPS FLAME RETARDANTS MONOGRAPH JUL 82
IN THE SALMONELLA MICROSOME ASSAY, DMMPP WAS MUTAGENIC IN TA100, BUT NOT IN TA98, OR SILAWORMS DMMPP WAS MUTAGENIC IN THE REC ASSAY WITH B. SUBTILIS WITH METABOLIC ACTIVATION, AND PRODUCED CHROMOSOMAL ABERRATIONS IN HAMSTER LUNG FIBROBLAST CELLS W/O METABOLIC ACTIVATION. RESULTS IN HUMAN EMBRYO FIBROBLASTS WERE INCONCLUSIVE.

CAS NO : **756-79-6**
CHEM NAME : ***** DIMETHYL METHYLPHOSPHONATE**

REPORT 01 : NTP BIOASSAY TR REPORT 323 NIH PUB. 88-2579 NOV 87
IN 2-YEAR GAVAGE STUDIES THERE WAS SOME EVIDENCE OF CARCINOGENICITY FOR MALE F344/N RATS, NO EVIDENCE OF CARCINOGENICITY FOR FEMALE F344/N RATS GIVEN DOSES OF 500 OR 1000 MG/KG. INADEQUATE EVIDENCE OF CARCINOGENICITY IN MALE B6C3F1 MICE BECAUSE OF DECREASED SURVIVAL IN DOSED GROUPS. NO EVIDENCE OF CARCINOGENICITY FOR FEMALE B6C3F1 MICE RECEIVING 1000 MG/KG. DECREASED SURVIVAL OF THIS

GROUP AT 2000 MG/KG MADE THEM INADEQUATE FOR DETERMINATION OF CARCINOGENIC ACTIVITY.

REPT'D USE : FLAME RETARDANT, ANTIFOAM AGENT, GASOLINE ADDITIVE, TEXTILE CONDITIONER, SOLVENT ADDITIVE AND PLASTICIZER STABILIZER

CAS NO : **141-91-3**

CHEM NAME : ***** 2,6-DIMETHYL MORPHOLINE**

REPORT 01 : NTP TECHNICAL BULLETIN 3/8 DEC 80, JUL 82
2,6-DIMETHYL MORPHOLINE WAS POSITIVE IN SALMONELLA BUT NEGATIVE IN DROSOPHILA WHEN TESTED FOR SEX-LINKED RECESSIVE LETHALS.

REPT'D USE : CHEMICAL INTERMEDIATE, STABILIZER FOR CHLORINATED SOLVENTS, RUBBER ACCELERATOR, CORROSION INHIBITOR AND GERMICIDE

CAS NO : **138-89-6**

CHEM NAME : ***** N,N-DIMETHYL-P-NITROSOANILINE**

REPORT 01 : NTP TECHNICAL BULLETIN 6/8 JAN 82, JUL 82
N,N-DIMETHYL-P-NITROSOANILINE WAS POSITIVE IN THE SALMONELLA ASSAY, BUT NEGATIVE IN DROSOPHILA WHEN TESTED FOR SEX-LINKED RECESSIVE LETHALS.

REPORT 06 : MULTI-AGENCY CROSS CHECK
N HHS MAGUIRE
Y OSHA CARC
N EPA CAG
N EPA RCRA
N EPA CLEAN AIR
N EPA CLEAN WATER

REPT'D USE : ORGANIC SYNTHESIS, RUBBER VULCANIZING AGENT, TEXTILE PRINTING AND DYE INTERMEDIATE

CAS NO : **22349-59-3**

CHEM NAME : ***** 1,4-DIMETHYLPHENANTHRENE**

REPORT 01 : IARC MONOGRAPH 32 DEC 83
1,4-DIMETHYLPHENANTHRENE TESTED ONLY FOR TUMOUR-INITIATING ACTIVITY IN MOUSE-SKIN INITIATION-PROMOTION ASSAY. ACTIVE AS AN INITIATOR, MUTAGENIC TO SALMONELLA TYPHIMURIUM IN PRESENCE OF EXOGENOUS METABOLIC SYSTEM AND INDUCED UNSCHEDULED DNA SYNTHESIS IN CULTURED PRIMARY RAT HEPATOCYTES.

REPT'D USE : TOBACCO SMOKE, USED ENGINE OIL COMBUSTION PRODUCT

CAS NO : **131-11-3**
CHEM NAME : ***** DIMETHYL PHTHALATE**

REPORT 01 : EPA RCRA HEALTH AND ENV. EFFECTS PROFILE OCT 80
DIMETHYL PHTHALATE HAS BEEN SHOWN TO PRODUCE MUTAGENIC EFFECTS IN THE AMES SALMONELLA ASSAY. ADMINISTRATION OF DIMETHYL PHTHALATE TO PREGNANT RATS BY I.P. INJECTION HAS BEEN REPORTED TO PRODUCE TERATOGENIC EFFECTS IN A SINGLE STUDY.

REPORT 02 : EPA AMBIENT WATER QUALITY CRITERIA OCT 80
DIMETHYL PHTHALATE HAS NOT BEEN FOUND TO BE CARCINOGENIC IN ANIMALS OR MAN. A CRITERION, BASED ON CHRONIC TOXICOLOGIC DATA IN RATS AND AN UNCERTAINTY FACTOR OF 100, IS 313 MG/L.

REPORT 03 : NTP TECHNICAL BULLETIN 6 JAN 82
THE COMPOUND WAS NEGATIVE IN THE SALMONELLA TYPHIMURIUM MUTAGENICITY ASSAY IN ONE LABORATORY.

REPORT 04 : EPA RESPONSE TO ITC RECOMM'NS 46FR53775 OCT 81
IN RESPONSE TO THE ITC RECOMMENDATIONS (42FR55026), EPA HAS RECEIVED A PROPOSAL FROM THE CHEMICAL MANUFACTURERS ASSN (CMA) IN WHICH THE PHTHALATE ESTER INDUSTRY HAS AGREED TO VOLUNTARILY AND EXPEDITIOUSLY TEST ALKYL PHTHALATES FOR THE ENVIRONMENTAL AND HEALTH EFFECTS OF CONCERN TO THE ITC AND EPA. CONSEQUENTLY, THE EPA HAS DECIDED NOT TO PROPOSE ADDITIONAL TESTING AT THIS TIME.

REPORT 05 : 49FR44142 NOV 84
ANNOUNCES DATA SUBMISSIONS RECEIVED BY EPA DURING 3RD QUARTER 1984 FROM NEGOTIATED OTHER TESTING PROGRAMS, INCLUDING BIODEGRADATION, VAPOR PRESSURE, OCTANOL-WATER PARTITION COEFF. AND ACUTE AQUATIC TOXICITY TESTING.

REPORT 06 : MULTI-AGENCY CROSS CHECK
N HHS MAGUIRE
N OSHA CARC
N EPA CAG
Y EPA RCRA
N EPA CLEAN AIR
Y EPA CLEAN WATER

REPORT 07 : EPA 8(E) STATUS 8EHQ-0886-0620 AUG 86
CELL TRANSFORMATIONS AND AMES ASSAYS FOR ALL TEST SUBSTANCES COMPLETED AND WERE NEGATIVE. DMP GAVE WEAK DOSE-RELATED INCREASE IN MUTANT FREQUENCY IN PRESENCE OF AROCLOR-INDUCED RAT LIVER S-9.

REPORT 08 : 52FR2152 - RECEIPT OF TEST DATA JAN 87
EPA ANNOUNCES RECEIPT OF TEST DATA FROM INDUSTRY'S VOLUNTARY TESTING PROGRAM.

REPT'D USE : PLASTICIZER FOR CELLULOSE FILM/SHEET, PHOTOGRAPHIC FILM, MOLDED PLASTIC ARTICLES, PAINTS/VARNISHES, AEROSOLS AND INSECT REPELLANTS

CAS NO : **77-78-1**

CHEM NAME : ***** DIMETHYL SULFATE**

REPORT 02 : IARC MONOGRAPH 4 1974
DIMETHYL SULFATE HAS BEEN SHOWN TO BE CARCINOGENIC IN THE RAT, THE ONLY SPECIES TESTED, BY INHALATION, SUBCUTANEOUS INJECTION AND FOLLOWING PRE-NATAL EXPOSURE. IT IS CARCINOGENIC TO THE RAT IN A SINGLE-DOSE EXPOSURE. IN A 1979 SUPPLEMENT, IARC DETERMINED THAT THERE IS SUFFICIENT EVIDENCE FOR THE CARCINOGENICITY OF DIMETHYL SULFATE IN ANIMALS AND THAT DIMETHYL SULFATE IS PROBABLY CARCINOGENIC FOR HUMANS.

REPORT 06 : MULTI-AGENCY CROSS CHECK
Y HHS MAGUIRE
N OSHA CARC
Y EPA CAG
Y EPA RCRA
N EPA CLEAN AIR
N EPA CLEAN WATER

REPT'D USE : ALKYLATING AGENT FOR DYES, STABILIZER FOR SO3, SOLVENT IN SEPARATION OF MINERAL OILS, INDUSTRIAL METHYLATING AGENT, PHARMACEUTICAL INTERMEDIATES AND POLYURETHANE-ADHESIVES

CAS NO : **67-68-5**

CHEM NAME : ***** DIMETHYL SULFOXIDE**

REPORT 01 : NIH RESEARCH RESOURCES REPORTER SEP 82
(DMSO) CAUSED SIGNIFICANT CHROMOSOMAL DAMAGE IN CHINESE HAMSTER OVARY (CHO) CELL CULTURES IN A STUDY CONDUCTED BY RESEARCHERS AT ATLANTA UNIVERSITY IN GEORGIA. DMSO CONCENTRATIONS OF 1 TO 4 PERCENT INCREASED CHROMOSOMAL ABERRATIONS UP TO 27 PERCENT AND THE EXTENT OF CHROMOSOMAL ABERRATION WAS DOSE-DEPENDENT.

REPT'D USE : SOLVENT, ANTIFREEZE, PAINT/VARNISH REMOVER

CAS NO : **120-61-6**
CHEM NAME : ***** DIMETHYL TEREPHTHALATE**

REPORT 01 : NCI CARC TRS 121 DHEW PUB 79-1376
ALTHOUGH IT IS RECOGNIZED THAT BOTH RATS AND MICE MAY NOT HAVE RECEIVED A DOSE OF THE TEST CHEMICAL SUFFICIENTLY HIGH TO PROVIDE MAXIMUM TEST SENSITIVITY, IT IS CONCLUDED THAT UNDER THE CONDITIONS OF THIS BIOASSAY, DIMETHYL TEREPHTHALATE WAS NOT CARCINOGENIC FOR F344 RATS OR B6C3F1 MICE (FEEDING STUDY).

REPORT 02 : NTP BOARD OF SCIENTIFIC COUNSELORS JUN 81
THE NTP PROVIDED THE PEER REVIEW COMMITTEE WITH A RE-ANALYSIS OF THE DATA OF TRS NO. 121. AFTER A RE-ANALYSIS OF THE ORIGINAL DATA AND A RE-EXAMINATION OF THE SLIDES OF LUNG TUMORS FROM THE MALE MICE, THE NTP COUCLUDED THAT THE INCIDENCE OF LUNG TUMORS IN THE HIGH DOSE GROUP OF MICE WAS SIGNIFICANTLY INCREASED BUT THE BIOLOGICAL MEANING WAS EQUIVOCAL.

REPORT 03 : NTP TECHNICAL BULLETIN 7 APR 82
DIMETHYL TEREPHTHALATE WAS NEGATIVE IN THE SALMONELLA TYPHIMURIUM MUTAGENICITY ASSAY.
REPT'D USE : INTERMEDIATE IN PRODUCTION OF POLYESTER

CAS NO : **55-80-1**
CHEM NAME : ***** N,N-DIMETHYL-P-(M-TOLYLAZO)-ANILINE**

REPORT 06 : MULTI-AGENCY CROSS CHECK
N HHS MAGUIRE
Y OSHA CARC
N EPA CAG
N EPA RCRA
N EPA CLEAN AIR
N EPA CLEAN WATER
REPT'D USE : NO AVAILABLE USE DATA

CAS NO : **513-37-1**
CHEM NAME : ***** DIMETHYLVINYL CHLORIDE**

REPORT 01 : NTP-TR 316, NIH PUB.# 86-2572 FINAL REPORT AUG 86
CLEAR EVIDENCE OF CARCINOGENICITY SEEN IN 2 YEAR STUDIES IN BOTH SEXES OF F344/N RATS AND B6C3F1 MICE, BASED ON INCREASED INCIDENCES OF NEOPLASMS OF NASAL CAVITY, ORAL CAVITY, ESOPHAGUS AND FORESTOMACH IN F344/N RATS AND INCREASED INCIDENCE OF SQUAMOUS CELL NEOPLASMS OF FORESTOMACH IN MALES AND FEMALES AND SQUAMOUS CELL CARCINOMAS OF PREPUTIAL GLAND IN B6C3F1 MALE MICE.

REPORT 06 : MULTI-AGENCY CROSS CHECK
Y HHS MAGUIRE
N OSHA CARC
N EPA CAG
N EPA RCRA
N EPA CLEAN AIR
N EPA CLEAN WATER
REPT'D USE : LABORATORY CHEMICAL, BYPRODUCT IN PRODUCTION OF 3-CHLORO-2-METHYLPROPENE AND ORGANIC SYNTHESES

CAS NO : **97-02-9**
CHEM NAME : ***** 2,4-DINITROANILINE**

REPORT 01 : NTP TECHNICAL BULLETIN NO. 7 APR 82
THE CHEMICAL WAS POSITIVE IN THE SALMONELLA TYPHIMURIUM MUTAGENICITY ASSAY.

REPORT 02 : 49(1) FR 108 ANPR JAN 84
EPA INITIATING RULEMAKING TO REQUIRE TESTING OF ANILINES FOR HEALTH AND ENVIRONMENTAL EFFECTS. THIS IS IN RESPONSE TO ITC'S DESIGNATION OF ANILINES FOR TESTING.
REPT'D USE : DYE/PIGMENT INTERMEDIATE

CAS NO : **97-00-7**
CHEM NAME : ***** 2,4-DINITRO-1-CHLOROBENZENE**

REPORT 01 : CHEMICAL HAZARD INFORMATION PROFILE MAR 81
STRONG MUTAGEN WITHOUT ACTIVATION IN AMES S. TYPHIMURIUM ASSAY AND A POTENTIAL CARCINOGEN.
REPT'D USE : INTERMEDIATE IN PRODUCTION OF AZO AND SULFUR DYES; AND RUBBER CHEMICALS

CAS NO : **534-52-1**
CHEM NAME : ***** 4,6-DINITRO-O-CRESOL**

REPORT 01 : NIOSH CRITERIA DOCUMENT FEB 78
NIOSH RECOMMENDS A STANDARD OF 0.2 MG/CU M TWA BASED ON CNS AND METABOLIC EFFECTS. THE PRESENT OSHA STANDARD IS 0.1 MG/CU M TWA.

REPORT 02 : EPA RCRA HEALTH AND ENV. EFFECTS PROFILE OCT 80
THIS COMPOUND HAS PRODUCED SOME DNA DAMAGE IN PROTEUS BUT FAILED TO SHOW MUTAGENIC EFFECTS IN THE AMES ASSAY OR IN E. COLI. AVAILABLE INFORMATION DOES NOT INDICATE THAT DNOC PRODUCES TERATOGENIC OR ADVERSE REPRODUCTIVE EFFECTS.

REPORT 06 : MULTI-AGENCY CROSS CHECK
N HHS MAGUIRE
N OSHA CARC
N EPA CAG
N EPA RCRA
N EPA CLEAN AIR
Y EPA CLEAN WATER

REPT'D USE : BLOSSOM-THINNING AGENT, FUNGICIDE, INSECTICIDE AND MITICIDE

CAS NO : **51-28-5**
CHEM NAME : ***** 2,4-DINITROPHENOL**

REPORT 01 : EPA RCRA HEALTH AND ENV. EFFECTS PROFILE OCT 80
GENETIC TOXICITY TESTING HAS SHOWN POSITIVE EFFECTS IN MOUSE BONE MARROW CELLS AND IN E. COLI. IN-VITRO CELL CULTURE ASSAYS FAILED TO SHOW THE POTENTIAL FOR MUTAGENIC ACTIVITY OF 2,4-DINITROPHENOL AS MEASURED BY UNSCHEDULED DNA SYNTHESIS. TERATOGENIC EFFECTS HAVE BEEN OBSERVED IN THE CHICK EMBRYO ASSAY FOLLOWING ADMINISTRATION OF 2,4-DINITROPHENOL. STUDIES IN MAMMALS FAILED TO SHOW THAT THE CPD PRODUCED ANY TERATOGENIC EFFECTS.

REPORT 06 : MULTI-AGENCY CROSS CHECK
N HHS MAGUIRE
N OSHA CARC
N EPA CAG
Y EPA RCRA
N EPA CLEAN AIR
Y EPA CLEAN WATER

REPT'D USE : DYE INTERMEDIATE, CHEMICAL INDICATOR/REAGENT INTERMEDIATE FOR MFR OF PHOTOCHEMICALS AND EXPLOSIVES

CAS NO : **42397-65-9**
CHEM NAME : ***** 1,8-DINITROPYRENE**

REPORT 01 : IARC MONOGRAPH 33 APR 84
NO DATA AVAILABLE TO EVALUATE CARCINOGENICITY OF 1,8-DINITROPYRENE TO HUMANS OR TO EXPERIMENTAL ANIMALS. CPD MUTAGENIC TO S.TYPHIMURIUM IN ABSENCE OF EXOGENOUS METABOLIC SYSTEM UNDER REDUCED OXYGEN TENSION. CPD INDUCED SISTER CHROMATID EXCHANGE AND MUTAGENIC AND CLASTOGENIC CHANGES TO MAMMALIAN CELLS IN CULTURE WITHOUT EXOGENOUS METABOLIC SYSTEM. INDUCED MORPHOLOGICAL TRANSFORMATION IN SYRIAN HAMSTER EMBRYO CELLS.

REPT'D USE : FOUND IN CARBON BLACKS AND DIESEL ENGINE EXHAUST. NOT PRODUCED COMMERCIALLY

CAS NO : **101-25-7**
CHEM NAME : ***** DINITROSOPENTAMETHYLENETETRAMINE (DNPMT)**

REPORT 01 : IARC MONOGRAPH 11 1976
DINITROSOPENTAMETHYLENETETRAMINE WAS TESTED IN RATS BY ORAL ADMINISTRATION AND INTRAPERITONEAL INJECTION. NO CARCINOGENIC EFFECTS WERE OBSERVED.
REPT'D USE : BLOWING AGENT FOR RUBBER AND PLASTIC

CAS NO : **140-79-4**
CHEM NAME : ***** 1,4-DINITROSOPIPERAZINE**

REPORT 06 : MULTI-AGENCY CROSS CHECK
N HHS MAGUIRE
Y OSHA CARC
N EPA CAG
N EPA RCRA
N EPA CLEAN AIR
N EPA CLEAN WATER
REPT'D USE : NO AVAILABLE USE DATA

CAS NO : **121-14-2**
CHEM NAME : ***** 2,4-DINITROTOLUENE (DNT)**

REPORT 01 : NCI CARC TRS 54 DHEW PUB NO 78-1360
UNDER THE CONDITIONS OF THIS BIOASSAY, DIETARY ADMINISTRATION OF 2,4-DINITROTOLUENE TO FISCHER 344 RATS INDUCED BENIGN TUMORS SUCH AS FIBROMA OF THE SKIN AND SUBCUTANEOUS TISSUE IN MALES AND OF THE MAMMARY GLAND IN FEMALES. NO EVIDENCE WAS PROVIDED FOR THE CARCINOGENICITY OF THE COMPOUND IN B6C3F1 MICE OF EITHER SEX.

REPORT 02 : EPA AMBIENT WATER QUALITY CRITERIA OCT 80
2,4-DNT IS SUSPECTED OF BEING A HUMAN CARCINOGEN. AMBIENT WATER CONCENTRATIONS OF 1.1 UG/L, .11 UG/L, AND .011 UG/L ARE ESTIMATED TO RESULT IN A RESPECTIVE 10 -5, 10 -6, AND 10 -7 INCREASE IN CANCER RISK OVER A LIFETIME.

REPORT 03 : CIIT CURRENT STATUS REPORT 2 MAY 77
LONG-TERM EXPOSURE TO DINITROTOLUENES IN HUMANS HAS RESULTED IN NORMOCYTIC-NORMOCHROMIC AND, RARELY, APLASTIC ANEMIA AND JAUNDICE. BUT THESE MANIFESTATIONS ARE INFREQUENTLY ENCOUNTERED NOW SINCE EFFECTIVE MEANS ARE AVAILABLE FOR MONITORING AND LIMITING EXPOSURE. SINCE TDA, A METABOLITE OF DNT, HAS BEEN SHOWN TO BE CARCINOGENIC IN RATS, THE EFFECTS OF LONG-TERM LOW LEVEL EXPOSURE TO DNT SHOULD BE PROBED.

REPORT 04 : NTP TECHNICAL BULLETIN 2/8 APR 80, JUL 82
2,4-DINITROTOLUENE WAS POSITIVE IN THE SALMONELLA ASSAY, BUT NEGATIVE IN DROSOPHILA WHEN TESTED FOR SEX-LINKED RECESSIVE LETHALS.

REPORT 05 : EPA RCRA HEALTH AND ENV. EFFECTS PROFILE OCT 80
2,4-DINITROTOLUENE INDUCES LIVER CANCER AND MAMMARY TUMORS IN MICE AND IS MUTAGENIC IN SOME ASSAY SYSTEMS. CHRONIC EXPOSURE TO 2,4-DINITROTOLUENE INDUCES LIVER DAMAGE, JAUNDICE, METHEMOGLOBINEMIA AND ANEMIA IN HUMANS AND ANIMALS.

REPORT 06 : MULTI-AGENCY CROSS CHECK
N HHS MAGUIRE
N OSHA CARC
Y EPA CAG
Y EPA RCRA
N EPA CLEAN AIR
Y EPA CLEAN WATER

REPORT 07 : EPA TSCA 8(E) NOTICE 8EHQ-1080-0367
OLIN CHEMICALS GROUP SUBMITTED A HEALTH HAZARD EVALUATION CONDUCTED BY NIOSH AT THE REQUEST OF WORKERS. FINDINGS INCLUDED A SIGNIFICANT REDUCTION IN SPERM COUNT IN THE EXPOSED GROUP. NIOSH DETERMINED THAT WHILE THIS STUDY CANNOT BE INTERPRETED AS CONCLUSIVE PROOF OF TOXIC EFFECTS OF TDA OR DNT ON THE MALE REPRODUCTIVE SYSTEM, IT IS STRONGLY SUGGESTIVE OF A PROBLEM. EFFECTS WERE OBSERVED IN THE PRESENCE OF DECLINING DNT AND TDA EXPOSURES, NONE OF WHICH EXCEEDED CURRENT OSHA EXPOSURE LIMITS DURING THE SURVEY.

REPORT 08 : CIIT 24 MONTH TOXICOLOGY STUDY
DINITROTOLUENE IS BEING EVALUATED FOR CARCINOGENIC AND TOXICOLOGIC EFFECTS IN CD-F RATS FOLLOWING DIETARY ADMINISTRATION AT DOSAGE LEVELS OF 3.5 AND 14 MG/KG/DAY. NECROPSY OF ANIMALS SACRIFICED AT WEEKS 26, 52 AND 78 REVEALED GROSS ALTERATIONS OF THE LIVER AND NODULES ON THE LUNGS OF SEVERAL MALES AND ONE FEMALE.

REPT'D USE : INTERMEDIATE IN MFR OF TD1 GELATINIZING/WATERPROOFING AGENT IN EXPLOSIVES AND DYE INTERMEDIATE

CAS NO : **606-20-2**

CHEM NAME : ***** 2,6-DINITROTOLUENE**

REPORT 01 : EPA RCRA HEALTH AND ENV. EFFECTS PROFILE OCT 80
2,6-DNT IS KNOWN TO CAUSE METHEMOGLOBINEMIA IN CATS, DOGS, RATS, AND MICE. 13-WEEK FEEDING STUDIES SHOW THE MAJOR EFFECTS, IN ADDITION TO BLOOD EFFECTS, TO BE DE-

PRESSED SPERMATOGENESIS, LIVER AND KIDNEY DEGENERATION, BILE DUCT HYPERPLASIA, INCOORDINATION AND PARALYSIS OF THE HIND LIMBS. POSITIVE RESULTS WERE OBTAINED WITH MUTAGENICITY TESTING IN SALMONELLA.

REPORT 06 : MULTI-AGENCY CROSS CHECK
N HHS MAGUIRE
N OSHA CARC
N EPA CAG
Y EPA RCRA
N EPA CLEAN AIR
Y EPA CLEAN WATER
REPT'D USE : EXPLOSIVES INTERMEDIATE IN MFR OF TDI AND DYE INTERMEDIATES

CAS NO : **84-76-4**
CHEM NAME : ***** DINONYL PHTHALATE ESTER (1,2-BENZENEDICARBOXYLIC ACID, DINONYL ESTER)**

REPORT 01 : TSCA 'FYI' DRAFT OCT 84
S-D RATS FED DIET CONTAINING CPD AT 500, 5,000 AND 10,000 PPM FOR 2 YEARS SHOWED FOLLOWING: HIGH DOSE MALES HAD NON-NEOPLASTIC LESIONS, HEPATOCELLULAR NECROSIS, INTERSTITIAL CELL HYPERPLASIA IN TESTES AND MINERAL DEPOSITS IN KIDNEY; MID AND HIGH DOSE GROUPS SHOWED INCREASE IN HEPATOCELLULAR CARCINOMAS; INCIDENCE OF LIVER TUMORS EXCEEDED THAT OF CONTROLS. CPD ALSO KNOWN AS SANITICIZER 900 (DOW CHEM. CO.)
REPT'D USE : PLASTICIZER

CAS NO : **117-84-0**
CHEM NAME : ***** DI-N-OCTYL PHTHALATE (DOP)**

REPORT 01 : EPA RCRA HEALTH & ENV. EFFECTS PROFILE
DI-N-OCTYL-PHTHALATE HAS PRODUCED TERATOGENIC EFFECTS FOLLOWING I.P. INJECTION OF PREGNANT RATS. THIS SAME STUDY HAS ALSO INDICATED SOME INCREASED RESORPTIONS AND FETAL TOXICITY.

REPORT 02 : EPA RESPONSE TO ITC RECOMMS (46FR53775) OCT 81
IN RESPONSE TO THE ITC RECOMMENDATIONS (42FR55026) THE EPA HAS RECEIVED A PROPOSAL FROM THE CHEMICAL MANUFACTURERS ASSN (CMA) WHEREBY THE PHTHALATE ESTER INDUSTRY HAS AGREED TO TEST, VOLUNTARILY AND EXPEDITIOUSLY, ALKYL PHTHALATES FOR THE ENVIRONMENTAL AND HEALTH EFFECTS OF CONCERN TO THE ITC AND EPA. COSEQUENTLY, THE EPA HAS DECIDED NOT TO PROPOSE ADDITIONAL TESTING AT THIS TIME.

REPORT 03 : NTP TECHNICAL BULLETIN 7 APR 82
DI-N-OCTYL-PHTHALATE WAS NEGATIVE IN THE SALMONELLA TYPHIMURIUM MUTAGENICITY ASSAY.

REPORT 04 : EPA FR NOTICE VOL 58 NO 8 JAN 93
1/13/93 PAGE 4133 EPA PROPOSES TO REMOVE FROM SECTION 313 EPCRA RT. TO KNOW LIST, BECAUSE CHEMICAL NOT ANTICIPATED TO CAUSE SIGNIFICANT ADVERSE ACUTE OR CHRONIC EFFECTS TO HUMANS OR AQUATIC LIFE AT LIKELY LEVELS OF EXPOSURE AND DOES NOT BIOCONCENTRATE.

REPORT 06 : MULTI-AGENCY CROSS CHECK
N HHS MAGUIRE
N OSHA CARC
N EPA CAG
Y EPA RCRA
N EPA CLEAN AIR
Y EPA CLEAN WATER

REPT'D USE : DYE CARRIER, PRODUCTION OF ADHESIVES, PLASTISOLS AND NITROCELLULOSE

CAS NO : **505-22-6**
CHEM NAME : ***** 1,3-DIOXANE**

REPORT 01 : NTP TECHNICAL BULLETIN 3/8 DEC 80, JUL 82
1,3-DIOXANE WAS POSITIVE IN THE SALMONELLA ASSAY, BUT NEGATIVE IN DROSOPHILA WHEN TESTED FOR SEX-LINKED RECESSIVE LETHALS.

REPT'D USE : NO AVAILABLE PRODUCTION OR USE DATA. NO LISTING ON TSCA INVENTORY

CAS NO : **123-91-1**
CHEM NAME : ***** 1,4-DIOXANE**

REPORT 01 : NCI CARC TRS 80 DHEW PUB NO 78-1330
UNDER THE CONDITIONS OF THIS BIOASSAY, 1,4-DIOXANE INDUCED HEPATOCELLULAR ADENOMAS IN FEMALE OSBORNE-MENDEL RATS. 1,4-DIOXANE WAS CARCINOGENIC IN BOTH SEXES OF RATS, PRODUCING SQUAMOUS-CELL CARCINOMAS OF THE NASAL TURBINATES, AND IN BOTH SEXES OF B6C3F1 MICE, PRODUCING HEPATOCELLULAR CARCINOMAS. 1,4-DIOXANE WAS GIVEN IN DRINKING WATER.

REPORT 02 : NTP TECHNICAL BULLETIN 3/8 DEC 80, JUL 82
1,4-DIOXANE WAS NEGATIVE IN THE SALMONELLA ASSAY AND IN DROSOPHILA WHEN TESTED FOR SEX-LINKED RECESSIVE LETHALS.

REPORT 03 : NIOSH CRITERIA DOCUMENT SEP 77
NIOSH RECOMMENDS A 1 PPM CEILING BASED ON LIVER AND KIDNEY EFFECTS AND POTENTIAL CARCINOGENICITY. THE PRESENT OSHA STANDARD IS 100 PPM TWA.

REPORT 04 : IARC MONOGRAPH 11 1976
1,4-DIOXANE IS CARCINOGENIC IN RATS AND GUINEA-PIGS BY ORAL ADMINISTRATION, PRODUCING MALIGNANT TUMORS OF THE NASAL CAVITY AND LIVER IN RATS AND TUMORS OF THE LIVER AND GALL BLADDER IN GUINEA-PIGS. IT WAS ALSO ACTIVE AS A PROMOTER IN A TWO-STAGE SKIN CARCINOGENESIS STUDY IN MICE. NO CARCINOGENIC EFFECT WAS OBSERVED IN ONE INHALATION STUDY IN RATS. IN A 1979 SUPPLEMENT, IARC DETERMINED THAT THERE IS SUFFICIENT EVIDENCE FOR THE CARCINOGENICITY OF 1,4-DIOXANE IN ANIMALS.

REPORT 05 : NTP TECHNICAL BULLETIN 3/8 DEC 80, JUL 82
1,4-DIOXANE WAS NEGATIVE IN THE SALMONELLA ASSAY AND IN DROSOPHILA WHEN TESTED FOR SEX-LINKED RECESSIVE LETHALS.

REPORT 06 : MULTI-AGENCY CROSS CHECK
Y HHS MAGUIRE
N OSHA CARC
Y EPA CAG
Y EPA RCRA
Y EPA CLEAN AIR
N EPA CLEAN WATER

REPORT 07 : EPA TSCA 8(E) 1088-0761 NOV 88
GELMAN SCIENCES REPORTED PRESENCE OF 1,4-DIOXANEIN OF NEW AND USED FOREIGN AND DOMESTIC CAR AND TRUCK COOLANTS OF VARIOUS AGES WITH CONCENTRATIONS BETWEEN 10 AND 22,000 PARTS PER BILLION (PPB) AND 100 TO 3400 PPB IN "OFF-THE-SHELF" AUTOMOTIVE ENGINE COOLANTS, AND IN "RADIATOR BOIL-OVER POOLS" AT REST AREAS IN MICHIGAN FROM <10 PPB TO OVER 2,000 PPB.

REPT'D USE : SOLVENT STABILIZER FOR CHLORINATED SOLVENTS; BY-PRODUCT IN DETERGENTS, SHAMPOOS, AUTOMOTIVE COOLANTS; SOLVENT FOR CELLULOSE ACETATE, ETHYL RESINS, OILS, WOKS, GREASES, POLYVINYL POLYMERS; TEXTILE WETTING AND DISPERSING AGENT; PAINTS AND VARNISH STRIPPERS; AND PRINTING COMPOSITIONS

CAS NO : **78-34-2**
CHEM NAME : ***** DIOXATHION**

REPORT 01 : NCI CARC TRS 125 DHEW PUB NO 78-1380
UNDER THE CONDITIONS OF THIS BIOASSAY, DIETARY ADMINISTRATION OF DIOXATHION WAS NOT CARCINOGENIC IN

OSBORNE-MENDEL RATS OR B6C3F1 MICE.
REPT'D USE : INSECTICIDE, ACARICIDE AND MITICIDE

CAS NO : **646-06-0**
CHEM NAME : ***** 1,3-DIOXOLANE**

REPORT 01 : EPA TSCA SECTION 8(D) FR NOTICE 47FR54624 DEC 82
1, 3-DIOXOLANE ADDED TO THE LIST OF CHEMICALS FOR WHICH COPIES OF UNPUBLISHED HEALTH AND SAFETY STUDIES MUST BE SUBMITTED TO EPA, BECAUSE CHEMICAL RECOMMENDED BY ITC FOR PRIORITY CONSIDERATION.

REPORT 02 : EPA TSCA TEST PROGRA FR NOTICE 49FR32113 AUG 84
EPA ANNOUNCED NOT TO INITIATE TESTING RULE SINCE FERO CORP. AND PPG INDUSTRIES PROPOSED TO TEST 1, 3-DIOXOLANE FOR MUTAGENICITY AND WILL AUDIT AND REVIEW 2-YEAR DRINKING WATER CHRONIC TOXICITY WITH RATS BEING COMPLETED. ADDITIONAL TESTING WILL BE DETERMINED AFTER THESE TESTS RESULTS INTERPRETED. CEASING PRODUCTION IS ALSO POSSIBLE.
REPT'D USE : STABILIZER FOR 1,1,1-TRICHLOROETHANE PRODUCTION

CAS NO : **147-24-0**
CHEM NAME : ***** DIPHENHYDRAMINE HYDROCHLORIDE**

REPORT 01 : NTP TECHNICAL BULLETIN 8 JUL 82
RESULTS FROM A SALMONELLA TYPHIMURIUM ASSAY FOR DIPHENHYDRAMINE HCL IN ONE LABORATORY WERE NEGATIVE.

REPORT 02 : NTP DRAFT TR355 NIH 89-2810 OCT 88
2-YEAR FEED STUDIES SHOWED EQUIVOCAL EVIDENCE OF CARCINOGENIC ACTIVITY OF DIPHENHYDRAMINE HYDROCHLORIDE FOR MALE F344/N RATS, SHOWN BY INCREASED INCIDENCES OF UNCOMMON BRAIN NEOPLASMS (ASTROCYTOMAS OR GLIOMAS) AND OF ALVEOLAR/BRONCHIOLAR NEOPLASMS. THERE WAS EVIDENCE OF CARCINOGENIC ACTIVITY FOR FEMALE F344/N RATS, SHOWN BY AN INCREASE IN THE INCIDENCE OR PITUITARY GLAND ADENOMAS. NO EVIDENCE OF CARCINOGENIC ACTIVITY FOR MALE OR FEMALE B6C3F1 MICE FED DIETS CONTAINING 156 OR 313 PPM DIPHENHYDRAMINE HYDROCHLORIDE.

REPORT 03 : NTP TR 355 NIH 89-2810 SEP 89
UNDER THE CONDITIONS OF THESE 2-YEAR FEED STUDIES, THERE WAS EQUIVOCAL EVIDENCE OF CARCINOGENIC ACTIVITY OF DIPHENHYDRAMINE HYDROCHLORIDE FOR MALE F344/N RATS, BASED ON MARGINALLY INCREASED INCIDENCES OF UNCOMMON BRAIN NEOPLASMS (ASTROCYTOMAS OR GLIOMAS) AND OF ALVELOLAR/BRONCHIOLAR NEOPLASMS. THERE WAS EQUIVOCAL

EVIDENCE OF CARCINOGENIC ACTIVITY FOR FEMALE F344/N RATS, BASED ON A MARGINAL INCREASE IN THE INCIDENCE OF PITUITARY GLAND ADENOMAS. THERE WAS NO EVIDENCE OF CARCINOGENIC ACTIVITY FOR MALE OR FEMALE B6C3F1 MICE FED DIETS CONTAINING 156 OR 313 PPM DIPHENHYDRAMINE HYDROCHLORIDE.

REPT'D USE : DRUG

CAS NO : **492-17-1**

CHEM NAME : ***** 2,4'-DIPHENYLDIAMINE**

REPORT 01 : IARC MONOGRAPH 16 1978

2,4'-DIPHENYLDIAMINE HAS BEEN INADEQUATELY TESTED IN RATS AND DOGS BY ORAL ADMINISTRATION. NO EVALUATION OF THE CARCINOGENICITY OF THIS COMPOUND CAN BE MADE. NO DATA ON THE METABOLISM, EMBRYOTOXICITY, TERATOGENICITY OR MUTAGENICITY OF 2,4'-DIPHENYLDIAMINE WERE AVAILABLE.

REPT'D USE : DYE INTERMEDIATE AND TUNGSTEN DETECTION

CAS NO : **102-06-7**

CHEM NAME : ***** 1,3-DIPHENYLGUANIDINE (DPC)**

REPORT 01 : HS STAFF INFORMATION OCT 87

1,3-DIPHENYLGUANIDINE RANKED BY NTP OF SCIENTIFIC COUNSELORS AS MODERATE PRIORITY FOR CARCINOGENESIS TESTING BECAUSE OF POTENTIAL FOR HUMAN EXPOSURE IN ITS USE AS ACCELERATOR IN RUBBER PROCESSING AND BIOLOGICAL ACTIVITY OF ARYLAMIDINE GROUP. TOXLINE SHOWS REPORTS ON MUTAGENICITY AND REPRODUCTIVE EFFECTS.

REPT'D USE : ACCELERATOR IN RUBBER VULCANIZING

CAS NO : **57-41-0**

CHEM NAME : ***** 5,5'-DIPHENYLHYDANTOIN (PHENYTOIN)**

REPORT 01 : NTP TECHNICAL BULLETIN 3 DEC 80

THE COMPOUND WAS FOUND TO BE NEGATIVE IN THE SALMONELLA TYPHIMURIUM MUTAGENICITY ASSAY BY TWO LABORATORIES.

REPORT 02 : IARC MONOGRAPH 13 1977

PHENYTOIN PRODUCED LYMPHOMAS AND LEUKEMIAS FOLLOWING INTRAPERITONEAL INJECTION IN MICE. STUDIES IN WHICH RATS WERE ADMINISTERED PHENYTOIN ORALLY WERE OF TOO SHORT DURATION TO ALLOW AN EVALUATION TO BE MADE. AN ASSOCIATION HAS BEEN OBSERVED IN EPILEPTIC PATIENTS BETWEEN THE OCCURRENCE OF LYMPHOMAS AND LONG-TERM ANTICONVULSANT THERAPY IN WHICH PHENYTOIN ALONE OR IN COMBINATION WITH OTHER ANTICONVULSANTS SUCH AS PHENOBARBITAL WAS GIVEN.

REPORT 06 : MULTI-AGENCY CROSS CHECK
Y HHS MAGUIRE
N OSHA CARC
N EPA CAG
N EPA RCRA
N EPA CLEAN AIR
N EPA CLEAN WATER
REPT'D USE : ANTICONVULSANT AGENT

CAS NO : **122-66-7**
CHEM NAME : ***** 1,2-DIPHENYLHYDRAZINE (HYDRAZOBENZENE)**

REPORT 01 : NCI CARC TRS 92 DHEW PUB NO 78-1342
UNDER THE CONDITIONS OF THIS BIOASSAY, HYDRAZOBENZENE WAS CARCINOGENIC TO FISCHER 344 RATS OF BOTH SEXES CAUSING INCREASED INCIDENCES OF HEPATOCELLULAR CARCINOMA AND ZYMBAL'S GLAND SQUAMOUS-CELL NEOPLASMS IN MALE RATS, NEOPLASTIC NODULES OF THE LIVER IN FEMALE RATS, AND MAMMARY ADENOCARCINOMAS IN FEMALE RATS. HYDRAZOBENZENE WAS ALSO CARCINOGENIC TO FEMALE B6C3F1 MICE, CAUSING AN INCREASED INCIDENCE OF HEPATOCELLULAR CARCINOMAS. THE COMPOUND WAS NOT CARCINOGENIC TO MALE B6C3F1 MICE (FEEDING STUDY).

REPORT 02 : NTP TECHNICAL BULLETIN 3 DEC 80
THE COMPOUND WAS FOUND TO BE POSITIVE IN THE SALMONELLA TYPHIMURIUM MUTAGENICITY ASSAY.

REPORT 03 : EPA AMBIENT WATER QUALITY CRITERIA OCT 80
DIPHENYLHYDRAZINE IS SUSPECTED OF BEING A HUMAN CARCINOGEN. AMBIENT WATER CONCENTRATIONS OF 422 NG/L, 42 NG/L, AND 4 NG/L ARE ESTIMATED TO RESULT IN A RESPECTIVE 10 -5, 10 -6, AND 10 -7 INCREASE IN CANCER RISK OVER A LIFETIME.

REPORT 06 : MULTI-AGENCY CROSS CHECK
Y HHS MAGUIRE
N OSHA CARC
Y EPA CAG
Y EPA RCRA
N EPA CLEAN AIR
Y EPA CLEAN WATER
REPT'D USE : INTERMEDIATE IN MFR OF BENZIDINE, PHENYLBUTAZONE, MOTOR OIL ADDITIVE, PHOTOCHROMIC RESINS AND POLYMERIZATION REACTIONS

CAS NO : **530-47-2**
CHEM NAME : ***** 1,1-DIPHENYLHYDRAZINE HYDROCHLORIDE**

REPORT 01 : NTP TECHNICAL BULLETIN 3 DEC 80
TERATOLOGY ASSAYS SHOW THAT ORAL ADMINISTRATION OF 1,1-DIPHENYLHYDRAZINE HYDROCHLORIDE CAUSES INCREASED MALFORMATIONS AT DOSES NOT TOXIC TO THE MOTHER.
REPT'D USE : CHEMICAL REAGENT

CAS NO : **101-68-8**
CHEM NAME : ***** 4,4'-DIPHENYLMETHANE DIISOCYANATE (MDI)**

REPORT 01 : IARC MONOGRAPH 19 1979
NO DATA ON THE CARCINOGENICITY OR MUTAGENICITY OF 2,4- AND 2,6-TOLUENE DIISOCYANATES, 1,5-NAPHTHALENE DIISOCYANATE, 4,4'-METHYLENE-DIPHENYL DIISOCYANATE OR POLYMETHYLENE POLYPHENYL ISOCYANATE WERE AVAILABLE.

REPORT 02 : NIOSH CRITERIA DOCUMENT SEP 78
NIOSH RECOMMENDS A STANDARD OF 50 UG/CU M TWA BASED ON RESPIRATORY EFFECTS AND SENSITIZATION. THE PRESENT OSHA STANDARD IS 0.2 NG/CU M, WHICH IS ALSO THE CEILING.

REPORT 03 : NTP CARC TEST PROGRAM JUL 82
IN A CURRENT NTP BIOASSAY, 4,4'-DIPHENYLMETHANE DIISOCYANATE IS BEING ADMINISTERED TO F344 RATS AND B6C3F1 MICE IN THE DIET. THE CHRONIC TESTING PHASE BEGAN JUNE 80.

REPORT 04 : NTP TECHNICAL BULLETIN 6 JAN 82
THE COMPOUND WAS NEGATIVE IN THE SALMONELLA TYPHIMURIUM MUTAGENICITY ASSAY IN ONE LABORATORY.
REPT'D USE : INTERMEDIATE FOR POLYURETHANE RESINS

CAS NO : **57472-68-1**
CHEM NAME : ***** DIPROPYLENEGLYCOL DIACRYLATE**

REPORT 01 : EPA TSCA 8(E) NOTICE 8EHQ-1082-0459
ATLANTIC RICHFIELD PROVIDED SUMMARY RESULTS FROM ACUTE ANIMAL EYE AND SKIN IRRITATION STUDIES AND A BATTERY OF SHORT-TERM IN-VITRO TESTS. IT WAS SHOWN TO BE "EXTREMELY TOXIC" TO CULTURED MOUSE LYMPHOMA AND CHO CELLS WITHOUT METABOLIC ACTIVATION. AN EXCESS NUMBER OF CHROMOSOMAL BREAKS WERE NOTED IN THE CULTURED CHO CELLS TREATED WITH DPGDA. THE CHEMICAL WAS NEGATIVE IN THE S. TYPHIMURIUM ASSAY.

CAS NO : **128-39-2**
CHEM NAME : ***** 2,6-DI-TERT-BUTYLPHENOL**

REPORT 01 : 52 FR 23862 JUN 87
EPA PROPOSING THAT MFRS AND PROCESSORS OF CPD BE REQUIRED UNDER TSCA 4(A) TO PERFORM TESTING FOR CHEMICAL FATE AND ENVIRONMENTAL EFFECTS. THIS RULE PROPOSED IN RESPONSE TO ITC'S DESIGNATION OF DTBP FOR PRIORITY CONSIDERATION FOR CHEMICAL FATE, HEALTH EFFECTS AND ENVIRONMENTAL EFFECTS TESTING.

CAS NO : **142-46-1**
CHEM NAME : ***** 2,5-DITHIOBIUREA**

REPORT 01 : NCI CARC TRS 132 DHEW PUB NO 79-1387
UNDER THE CONDITIONS OF THIS BIOASSAY, THE EVIDENCE SUGGESTED, BUT WAS INSUFFICIENT TO ESTABLISH THE CARCINOGENICITY OF 2,5-DITHIOBIUREA FOR FEMALE B6C3F1 MICE. THE COMPOUND WAS NOT CARCINOGENIC TO MALE B6C3F1 MICE OR TO MALE OR FEMALE FISCHER 344 RATS (FEEDING STUDY).

REPORT 02 : NTP TECHNICAL BULLETIN 6 JAN 82
THE COMPOUND WAS NEGATIVE IN THE SALMONELLA TYPHIMURIUM MUTAGENICITY ASSAY IN ONE LABORATORY.
REPT'D USE : PHOTOGRAPHIC CHEMICAL, MILITARY ORDINANCE AND ELECTROPLATING

CAS NO : **480-22-8**
CHEM NAME : ***** DITHRANOL**

REPORT 01 : IARC MONOGRAPH 13 1977
DITHRANOL IS A TUMOR-PROMOTING AGENT IN MOUSE SKIN CARCINOGENESIS EXPERIMENTS FOLLOWING INITIATION WITH EITHER 7,12-DIMETHYLBENZ(A)-ANTHRACENE OR URETHANE. AN INCREASED INCIDENCE OF LYMPHOMAS WAS ALSO OBSERVED IN MICE PAINTED WITH DITHRANOL AFTER URETHANE INITIATION.

REPORT 02 : IARC MONOGRAPH 13 1977
DITHRANOL INDUCES PETITE MUTANTS, BUT NO GENE CONVERSION, IN SACCHAROMYCES. NO POINT MUTATIONS WERE FOUND IN SALMONELLA TA98, TA100, TA15 35 OR TA1537 IN THE PRESENCE OR ABSENCE OF METABOLIC ACTIVATION.
REPT'D USE : TREATMENT OF PSORIASIS

CAS NO : **60-13-9**
CHEM NAME : ***** DL-AMPHETAMINE SULFATE**

REPORT 01 : NTP DRAFT TR387 NIH-90-2842 NOV 89
UNDER 2-YEAR FEED STUDIES NO EVIDENCE OF CARCINOGENIC ACTIVITY OF DL-AMPHETAMINE SULFATE FOR MALE OR FEMALE F344/N RATS OR MALE OR FEMALE B6C3F1 MICE FED DIETS CONTAINING 20 OR 100 PPM. ADMINISTRATION ASSOCIATED WITH DECREASED BODY WEIGHT. DECREASED INCIDENCES OF TOTAL NEOPLASMS IN DOSED RATS AND MICE, OF ADRENAL PHEOCHROMOCYTOMAS IN MALE RATS, OF MAMMARY GLAND FIBROADENOMAS AND UTERINE POLYPS IN FEMALE RATS. OF PITUITARY GLAND ADENOMAS IN MALE AND FEMALE RATS AND FEMALE MICE, AND OF HARDERIAN GLAND ADENOMAS, LIVER NEOPLASMS, AND LUNG NEOPLASMS IN MALE AND FEMALE MICE OBSERVED.
REPT'D USE : DRUG

CAS NO : **27193-86-8**
CHEM NAME : ***** DODECYLPHENOL**

REPORT 01 : TSCA SECT 8(E) STATUS REPORT 8EHQ-0687-0682 JUL 87
MATERNAL TOXICITY, FETAL TOXICITY, EMBRYOTOXICITY AND SKELETAL ABNORMALITIES OBSERVED AT HIGHEST DOSE (300 MG/KG/DAY) OF CPD ADMINISTERED VIA GAVAGE TO PREGNANT RATS. NO MATERNAL OR DEVELOPMENTAL TOXICITY SEEN AT LOWER DOSES TESTED (100 MG/KG/DAY AND 20 MG/KG/DAY).
REPT'D USE : DETERGENTS

CAS NO : **483-18-1**
CHEM NAME : ***** EMETINE**

REPORT 01 : NCI CARC TRS 43 DHEW PUB NO 78-843
RESULTS OF THE STUDY DO NOT PERMIT EVALUATION OF EMETINE GIVEN BY ABDOMINAL INJECTION TO RATS AND MICE BECAUSE OF EARLY TERMINATION OF THE STUDY AND POOR SURVIVAL OF TREATED MICE.
REPT'D USE : PHARMACEUTICAL AMEBICIDE

CAS NO : **115-29-7**
CHEM NAME : ***** ENDOSULFAN**

REPORT 01 : NCI CARC TRS 62 DHEW PUB NO 78-1312
IT IS CONCLUDED THAT UNDER THE CONDITIONS OF THIS BIOASSAY, TECHNICAL GRADE ENDOSULFAN WAS NOT CARCINOGENIC IN FEMALE OSBORNE-MENDEL RATS OR IN FEMALE B6C3F1 MICE (FEEDING STUDY).

REPORT 02 : EPA RCRA HEALTH AND ENV. EFFECTS PROFILE OCT 80
CHRONIC FEEDING OF ENDOSULFAN TO RATS AND MICE PRODUCED KIDNEY DAMAGE, PARATHYROID HYPERPLASIA, TESTICULAR ATROPHY, HYDROPIC CHANGE OF LIVER AND LOWERED SURVIVAL. ORAL ADMINISTRATION TO PREGNANT RATS INCREASED FETAL MORTALITY AND RESORPTIONS.

REPORT 03 : NTP TECHNICAL BULLETIN 9 APR 83
ENDOSULFAN WAS NEGATIVE IN ONE SALMONELLA TYPHIMURIUM MUTAGENICITY ASSAY.
REPT'D USE : INSECTICIDE

CAS NO : **72-20-8**
CHEM NAME : ***** ENDRIN**

REPORT 01 : NCI CARC TRS 12 DHEW PUB NO 79-812
UNDER THE CONDITIONS OF THIS BIOASSAY, ENDRIN WAS NOT CARCINOGENIC FOR OSBORNE-MENDEL RATS OR FOR B6C3F1 MICE (FEEDING STUDY).

REPORT 02 : IARC MONOGRAPH 5 1974
IN ORAL EXPERIMENTS IN MICE, INSUFFICIENT DETAILS WERE AVAILABLE TO ALLOW AN EVALUATION TO BE MADE. TWO FEEDING STUDIES IN RATS WERE REPORTED. ONE WAS NEGATIVE, ALTHOUGH THE TIME WHEN THE TUMORS OCCURRED IN BOTH TREATED AND CONTROL GROUPS WAS NOT GIVEN; THE OTHER PROVIDED NO DETAILS OF PATHOLOGICAL STUDIES AND THEREFORE CANNOT BE EVALUATED.

REPORT 03 : EPA RCRA HEALTH AND ENV. EFFECTS PROFILE OCT 80
CHRONIC ADMINISTRATION OF ENDRIN CAUSES DAMAGE TO THE LIVER, LUNG, KIDNEY, AND HEART OF EXPERIMENTAL ANIMALS. ENDRIN IS TERATOGENIC AND EMBYOTOXIC IN HIGH DOSES AND PRODUCES GROSS CHROMOSOMAL ABNORMALITIES WHEN ADMINISTERED INTRATESTICULARLY.

REPORT 06 : MULTI-AGENCY CROSS CHECK
N HHS MAGUIRE
N OSHA CARC
N EPA CAG
N EPA RCRA
N EPA CLEAN AIR
Y EPA CLEAN WATER
REPT'D USE : INSECTICIDE, AVICIDE, RODENTICIDE AND CHIPMUNK KILLER

CAS NO : **15086-94-9**
CHEM NAME : ***** EOSIN**

REPORT 01 : IARC MONOGRAPH 15 1977
EOSIN HAS BEEN TESTED ONLY IN RATS BY ORAL AND SUBCUTANEOUS ADMINISTRATION; THE DATA ARE INSUFFICIENT FOR AN EVALUATION OF CARCINOGENICITY OF THIS COMPOUND TO BE MADE.

REPORT 02 : IARC MONOGRAPH 15 1977
NO DATA ON EMBRYOTOXICITY OR TERATOGENICITY OF EOSIN WERE AVAILABLE.
REPT'D USE : INKS, COSMETICS, LAB TITRATION INDICATOR AND DYE

CAS NO : **17372-87-1**
CHEM NAME : ***** EOSIN DISODIUM SALT**

REPORT 01 : IARC MONOGRAPH 15 1977
EOSIN DISODIUM SALT HAS BEEN ADMINISTERED TO RATS BY SUBCUTANEOUS INJECTION; THE DATA ARE INSUFFICIENT FOR AN EVALUATION OF CARCINOGENICITY OF THIS COMPOUND TO BE MADE.

REPORT 02 : IARC MONOGRAPH 15 1977
NO DATA ON EMBRYOTOXICITY OR TERATOGENICITY OF EOSIN DISODIUM SALT WERE AVAILABLE.
REPT'D USE : DYE FOR PAPER, INKS, WOODSTAINS, SOAPS, COSMETICS AND BIOLOGICAL STAIN

CAS NO : **134-72-5**
CHEM NAME : ***** EPHEDRINE SULFATE**

REPORT 01 : NTP TR 307 NIH PUB 86-2563 MAY 86
UNDER CONDITIONS OF THESE STUDIES, THERE WAS NO EVIDENCE OF CARCINOGENICITY FOR F344/N RATS OR B6C3F1 MICE OF EITHER SEX RECEIVING 125 OR 250 PPM EPHEDRINE SULFATE IN DIET FOR 2 YEARS.
REPT'D USE : PHARMACEUTICAL

CAS NO : **3132-64-7**
CHEM NAME : ***** EPIBROMOHYDRIN (1,2-EPOXY-3-BROMOPROPANE)**

REPORT 01 : CPSC FLAME RETARDANTS MONOGRAPH JUL 82
EPIBROMOHYDRIN SHOWED MUTAGENIC ACTIVITY IN SALMONELLA TA1535 AND TA100, BUT NOT IN TA1537 AND TA98. THE CHEMICAL WAS MUTAGENIC IN E. COLI.

REPORT 02 : NTP TECHNICAL BULLETIN 9 APR 83
1,2-EPOXY-3-BROMOPROPANE WAS POSITIVE IN ONE SALMONELLA TYPHIMURIUM MUTAGENICITY ASSAY.

CAS NO : **106-89-8**
CHEM NAME : ***** EPICHLOROHYDRIN**

REPORT 01 : EPA RCRA HEALTH AND ENV. EFFECTS PROFILE OCT 80
STUDIES HAVE DEMONSTRATED EPICHLOROHYDRIN TO BE A POTENT CARCINOGEN TO NASAL TISSUE IN EXPERIMENTAL ANIMALS. CYTOGENIC STUDIES BOTH IN VITRO AND IN VIVO IN HUMANS AND EXPERIMENTAL ANIMALS HAVE INDICATED EPICHLOROHYDRIN TO BE AN ACTIVE CLASTOGENIC AGENT.

REPORT 02 : NIOSH CRITERIA DOCUMENT SEP 76
NIOSH RECOMMENDS A STANDARD OF 2 MG/CU M TWA TO PROTECT AGAINST CARCINOGENESIS, MUTAGENESIS, STERILITY, AND DAMAGE TO KIDNEYS, LIVER, RESPIRATORY TRACT AND SKIN. THE PRESENT OSHA STANDARD IS 5 PPM TWA.

REPORT 03 : NIOSH CURRENT INTELLIGENCE BULLETIN OCT 78
NIOSH RECOMMENDS THAT EPICHLOROHYDRIN BE HANDLED IN THE WORKPLACE AS IF IT WERE A HUMAN CARCINOGEN. A RECENT LONG TERM EPIDEMIOLOGIC STUDY SHOWING A SIGNIFICANT INCREASE IN RESPIRATORY CANCER DEATHS OF EXPOSED WORKERS, AND AN INHALATION STUDY SHOWING AN INCREASE IN NASAL CARCINOMAS IN RATS. IN ADDITION, CYTOGENIC STUDIES OF HUMAN PERIPHERAL LYMPHOCYTES HAVE SHOWN A HIGHLY SIGNIFICANT INCREASE IN CHROMOSOME ABNORMALITIES IN EXPOSED WORKERS.

REPORT 04 : IARC MONOGRAPH 11 1976
RECESSIVE LETHAL MUTATIONS HAVE BEEN OBTAINED IN DROSOPHILA AFTER TREATMENT WITH EPICHLOROHYDRIN. THE CHEMICAL ADMINISTERED BY I.P. INJECTION DID NOT INDUCE DOMINANT LETHAL MUTATIONS IN SWISS MICE.

REPORT 05 : IARC MONOGRAPH 11 1976
EPICHLOROHYDRIN IS CARCINOGENIC IN MICE BY SUBCUTANEOUS INJECTION: IT PRODUCED LOCAL SARCOMAS. IT WAS ACTIVE AS AN INITIATOR IN A TWO-STAGE SKIN CARCINOGENESIS STUDY IN MICE.

REPORT 06 : MULTI-AGENCY CROSS CHECK
Y HHS MAGUIRE
N OSHA CARC
Y EPA CAG
Y EPA RCRA
Y EPA CLEAN AIR
N EPA CLEAN WATER

REPORT 07 : ITC SECOND REPORT TO EPA (43FR16684) APR78
THE INTERAGENCY TESTING COMMITTEE RECOMMENDS THAT THE EPA ADMINISTRATOR CONSIDER THE DEVELOPMENT OF RULES TO TEST HALOGENATED ALKYL EPOXIDES FOR CARCINOGENICITY, MUTAGENICITY, TERATOGENICITY AND OTHER CHRONIC EFFECTS; AND THAT AN EPIDEMIOLOGIC STUDY BE CONDUCTED.

REPORT 08 : NTP TECHNICAL BULLETIN 9 APR 83
EPICHLOROHYDRIN WAS POSITIVE IN ONE SALMONELLA THYPHIMURIUM MUTAGENICITY ASSAY.

REPORT 09 : RESPONSE TO ITC 48FR57695 DEC 83
EPA'S RESPONSE TO ITC'S RECOMMENDATION REQUIRING HEALTH EFFECTS TESTING OF HALOGENATED ALKYL EPOXIDES UNDER TSCA 4(A). EPA IS NOT INITIATING RULEMAKING FOR ECH NOW BECAUSE RECENT AND CURRENT DATA ARE CONSIDERED ADEQUATE TO DEFINE RISK, ASSESS EXPOSURE AND DEVELOP CONTROLS TO PROTECT AGAINST EXPOSURE.

REPORT 11 : TSCA SECT 8(E) STATUS REPORT 8EHQ-0687-0679 JUL 87
REACTION PRODUCT OF D-GLUCITOL AND EPICHLORIHYDRIN (CAS NO. 68412-01-1) GAVE POSITIVE RESULTS IN GENOTOXICOLOGIC STUDIES IN AMES S. TYPHIMURIUM BACTERIAL ASSAY, A POINT MUTATION ASSAY IN CULTURED V79 CHINESE HAMSTER CELLS AND CHROMOSOMAL ABERRATION ASSAY IN HUMAN LYMPHOCYTES. NEGATIVE RESULTS REPORTED IN DNA REPAIR ASSAYS IN CULTURED RAT HEPATOCYTES AND HUMAN FIBROBLASTS AND A CELL TRANSFORMATION ASSAY IN CULTURED BALB/3T3 MOUSE CELLS.

REPT'D USE : CHEMICAL INTERMEDIATE, EPOXY RESINS, ADHESIVES, MARINE SURFACE COATING, PAINTS/VARNISHES AND STABILIZER

CAS NO : **106-88-7**
CHEM NAME : ***** 1,2-EPOXYBUTANE**

REPORT 01 : NTP TECHNICAL BULLETIN 6/7/8 JAN 82, JUL 82
1,2-EPOXYBUTANE WAS POSITIVE IN THE SALMONELLA ASSAY AND THE L5178Y MOUSE LYMPHOMA ASSAY. THE CHEMICAL WAS NEGATIVE IN THE RAT DOMINANT LETHAL ASSAY, THE DROSOPHILA SEX-LINKED RECESSIVE LETHAL ASSAY, AND IN VITRO UNSCHEDULED DNA SYNTHESIS IN HUMAN FIBROBLASTS. RESULTS IN THE RAT BONE MARROW CYTOLOGY ASSAY WERE SUGGESTIVE BUT NOT CLEARLY POSITIVE.

REPORT 02 : 49FR504 JAN 84
EPA'S RESPONSE TO THE ITC'S RECOMMENDATION DESIGNATING 1,2-BUTYLENE OXIDE FOR CONSIDERATION FOR HEALTH AND ENVIRONMENTAL EFFECTS TESTING. AS A MEMBER OF THE ALKYL EPOXIDES FAMILY, EPA HAS SUFFICIENT DATA TO REASONABLY

PREDICT THE ENVIRONMENTAL FATE OF CPD AND IS NOT INITIATING RULEMAKING UNDER TSCA 4(A).

REPORT 03 : NTP DRAFT BIOASSAY TR REPORT AUG 86
THIS 2-YEAR INHALATION STUDY SHOWED CLEAR EVIDENCE OF CARCINOGENICITY OF CPD FOR MALE RATS AS SHOWN BY INCREASED INCIDENCE OF PAPILLARY ADENOMAS OF NASAL CAVITY AND ALVEOLAR AND BRIONCHIOLAR ADENOMAS OR COMBINED CARCINOMAS. SOME EVIDENCE OF CARCINOGENICITY FOR FEMALE F344/N RATS, AS SHOWN BY PRESENCE OF PAPILLARY ADENOMAS OF NASAL CAVITY OF RATS IN 400 PPM GROUP. NO EVIDENCE OF CARCINOGENICITY FOR MALE OR FEMALE B6C3F1 MICE EXPOSED AT 50 OR 100 PPM. EXPOSURE TO CPD ASSOCIATED WITH NONNEOPLASTIC LESIONS OF NASAL CAVITY IN RATS AND MICE.

REPORT 04 : NTP TR 329 NIH PUB. 88-2585 MAR 88
AFTER 2 YEAR INHALATION STUDIES, EVIDENCE OF CARCINOGENIC ACTIVITY OF 1,2-EPOXYBUTANE FOR MALE F344/N RATS PROVEN BY INCREASED INCIDENCE OF PAPILLARY ADENOMAS OF THE NASAL CAVITY, ALVEOLAR/BRONCHIOLAR ADENOMAS OR CARCINOMAS (COMBINED). EQUIVOCAL EVIDENCE OF CARCINOGENIC ACTIVITY FOR FEMALE F344/N RATS PROVEN BY PRESENCE OF PAPILLARY ADENOMAS OF NASAL CAVITY. NO EVIDENCE OF CARCINOGENIC ACTIVITY FOR MALE OR FEMALE B6C3F1 MICE EXPOSED AT 50 OR 100 PPM. 1,2-EPOXYBUTANE EXPOSURE WAS ASSOCIATED WITH ADENOMATOUS HYPERPLASIA AND INFLAMMATORY LESIONS OF THE NASAL CAVITY IN RATS AND NASAL CAVITY IN MICE.

REPT'D USE : CHEMICAL INTERMEDIATE, PLASTIC MONOMER, CLEANING AGENTS AND STABILIZER IN CHLORINATED SOLVENTS

CAS NO : **3388-04-3**

CHEM NAME : ***** BETA-(3,4-EPOXYCYCLOHEXYL)-ETHYLTRIMETHOXY SILANE**

REPORT 01 : EPA TSCA 8(E) NOTICE 8EHQ-0681-0402
UNION CARBIDE PROVIDED PRELIMINARY RESULTS OF A NEARLY COMPLETECHRONIC SKIN-PAINTINGSTUDYOFB-(3,4-EPOXYCYCLOHEXYL)ETHYLTRIMETHOXYSILANE IN MICE. OF THE THREE SURVIVING ANIMALS AT 25 MONTHS, TWO HAD SKIN CARCINOMAS AND THE THIRD HAD A SKIN PAPILLOMA. THE DERMAL STUDY WAS UNDERTAKEN AFTER THE CHEMICAL WAS SHOWN ACTIVE IN SALMONELLA, SISTER CHROMATID EXCHANGE, AND MOUSE LYMPHOMA ASSAY, AND WEAKLY ACTIVE IN AN UNSCHEDULED DNA REPAIR TEST.

CAS NO : **141-37-7**
CHEM NAME : ***** 3,4-EPOXY-6-METHYLCYCLOHEXYLMETHYL -3,4-EPOXY-6-METHYL CARBOXYLATE**

REPORT 01 : IARC MONOGRAPH 11 1976
3,4-EPOXY-6-METHYLCYCLOHEXYLMETHYL-3,4-EPOXY-6-METHYL CARBOXYLATE IS CARCINOGENIC IN MICE BY SKIN APPLICATION: IT PRODUCED SKIN CARCINOMAS.
REPT'D USE : CURING AGENT FOR RESIN SYSTEMS. NO KNOWN CURRENT PRODUCTION OR USE. NO LISTING ON TSCA INVENTORY.

CAS NO : **3033-77-0**
CHEM NAME : ***** 2,3-EPOXYPROPYLTRIMETHYL-AMMONIUMCHLORIDE**

REPORT 01 : EPA TSCA 8(E) EHQ-1283-0503 DEC 83
DOW CHEMICAL PROVIDED COPIES OF SHELL INTERNATIONAL CHEMICAL COMPANY LETTERS REPORTING MUTAGENIC ACTIVITY IN BACTERIA AND IN MAMMALIAN CELLS AND IN LIFE-TIME MOUSE SKIN APPLICATION STUDY. AT TWO DOSES IT PRODUCED SKIN TUMORS. AT LOWEST DOSES THERE WERE NO SKIN TUMORS, WHILE AT INTERMEDIATE DOSES THE INCIDENCE WAS EXTREMELY LOW.

REPORT 02 : EPA TSCA 8(E) 8E HQ-1283-0503 FEB 84
FOUND IN 3-CHLORO-2-HYDROXYPROPYLTRIMETHYLAMMONIUM CHLORIDE AS A CONTAMINANT AND REPORTED BY PRODUCER TO DISPLAY MUTAGENIC ACTIVITY BACTERIA AND MAMMALIAN CELLS. IN EXPOSURE TESTS, FOUND TO INDUCE TUMORS IN LIFE-TIME MOUSE STUDY.
REPT'D USE : PAPER STRENGTH ADDITIVE AND SHAMPOOS COSMETICS

CAS NO : **25068-38-6**
CHEM NAME : ***** EPOXY RESIN**

REPORT 03 : EPA TSCA 8(E) 0584-0518S MAY 84
A COMPANY PROVIDED RESULTS FROM SEVERAL IN-VITRO GENOTOXICITY STUDIES OF A PROPRIETARY MIXTURE OF EPOXY RESINS THAT SHOWED THAT IT PRODUCED A POSITIVE REPONSE WITH AND WITHOUT METABOLIC ACTIVATION IN THE AMES TEST, THE MOUSE LYMPHOMA TEST AND THE SISTER CHROMATID EXCHANGE TEST.
REPT'D USE : AMINES, ANHYDRIDES, ACIDS, FILLERS AND REINFORCEMENTS

CAS NO : **2443-39-2**
CHEM NAME : ***** CIS-9,10-EPOXYSTEARIC ACID**

REPORT 01 : IARC MONOGRAPH 11 1976
CIS-9,10-EPOXYSTEARIC ACID HAS BEEN TESTED IN MICE BY SKIN APPLICATION AND IN MICE AND RATS BY SUBCUTANEOUS INJECTION. THE LIMITED AMOUNT OF DATA DO NOT INDICATE THAT THE COMPOUND IS CARCINOGENIC.

REPORT 02 : NTP CARC TEST PROGRAM JUL 82
IN A CURRENT NTP BIOASSAY, 1,2-EPOXYBUTANE IS BEING ADMINISTERED BY INHALATION TO F344 RATS AND B6C3F1 MICE. THE CHRONIC EXPOSURE PHASE IS IN PROGRESS.
REPT'D USE : LABORATORY CHEMICAL

CAS NO : **3083-23-6**
CHEM NAME : ***** 1-2-EPOXY-3,3,3-TRICHLOROPROPANE**

REPORT 01 : NTP TECHNICAL BULLETIN NO. 7 APR 82
THE CHEMICAL WAS POSITIVE IN THE SALMONELLA TYPHIMURIUM MUTAGENICITY ASSAY.
REPT'D USE : NO AVAILABLE PRODUCTION OR USE DATA. NO LISTING ON TSCA INVENTORY

CAS NO : **66733-21-9**
CHEM NAME : ***** ERIONITE**

REPORT 01 : IARC MONOGRAPH 42 1987
SUFFICIENT EVIDENCE FOR CARCINOGENICITY OF ERIONITE TO EXPERIMENTAL ANIMALS AND TO HUMANS. WHEN CPD FROM NATURAL SOURCES TESTED FOR CARCINOGENICITY IN RATS VIA INHALATION AND INTRAPLUERAL IMPLANTATION AND IN MICE BY INTRAPERITONEAL INJECTION, HIGH INCIDENCES OF MESOTHELIOMA INDUCED. DESCRIPTIVE STUDIES HAVE SHOWN HIGH MORTALITY FROM MALIGNANT MESOTHELIOMA IN 3 TURKISH VILLAGES WHERE CONTAMINATION FROM CPD FROM BIRTH. ERIONITE INDUCED UNSCHEDULED DNA SYNTHESIS AND MORPHOLOGICAL TRANSFORMATION IN CULTURED MAMMALIAN CELLS.

REPORT 06 : MULTI-AGENCY CROSS CHECK
Y HHS MAGUIRE
N OSHA CARC
N EPA CAG
N EPA RCRA
N EPA CLEAN AIR
N EPA CLEAN WATER

REPT'D USE : METAL-IMPREGNATED CATALYST, COMMERCIAL ABSORBENT REF. CHECK ON HS BUILDING MATERIALS; NOT MARKETED FOR COMMERCIAL PURPOSES

CAS NO : **643-22-1**
CHEM NAME : ***** ERYTHROMYCIN STEARATE**

REPORT 01 : NTP TR 338 NIH PUB 89-2594 DEC 88
UNDER 2-YEAR STUDIES, NO EVIDENCE OF CARCINOGENIC ACTIVITY FOR MALE OR FEMALE F344/N RATS ADMINISTERED ERYTHROMYCIN STEARATE IN DIET AT 5,000 OR 10,000 PPM. NO EVIDENCE OF CARCINOGENIC ACTIVITY FOR MALE OR FEMALE B6C3F1 MICE ADMINISTERED ERYTHROMYCIN STEARATE IN DIET AT 2,500 OR 5,000 PPM. DOSE-RELATED INCREASES IN INCIDENCES OF GRANULOMAS OF LIVER OBSERVED IN MALE AND FEMALE RATS. ABSENCE OF ANY BIOLOGICALLY IMPORTANT CHEMICAL-ASSOCIATED EFFECTS IN MICE SUGGESTS HIGHER DOSES COULD HAVE BEEN GIVEN TO MALE AND FEMALE MICE.
REPT'D USE : DRUG

CAS NO : **50-28-2**
CHEM NAME : ***** ESTRADIOL-17B**

REPORT 01 : IARC MONOGRAPH 21 1979
ESTRADIOL-17B AND ITS ESTERS WERE TESTED IN MICE, RATS, HAMSTERS, GUINEA-PIGS AND MONKEYS BY SUBCUTANEOUS INJECTION OR IMPLANTATION AND IN MICE BY ORAL ADMINISTRATION. SUBCUTANEOUS ADMINISTRATION RESULTED IN INCREASED INCIDENCES OF VARIOUS TUMORS IN MICE AND RATS. MALIGNANT TUMORS OCCURRED IN MALE AND OVARIECTOMIZED FEMALE HAMSTERS, BUT NOT IN INTACT FEMALES. IT IS REASONABLE TO REGARD ESTRADIOL-17B AS IF IT PRESENTED A CARCINOGENIC RISK TO HUMANS. THE IARC 1979 SUPPLEMENT SAYS THERE IS SUFFICIENT EVIDENCE FOR CARCINOGENICITY IN ANIMALS.

REPORT 02 : IARC MONOGRAPH 21 1979
A TWOFOLD INCREASE IN THE NUMBER OF CHROMOSOMAL ABERRATIONS WAS OBSERVED IN CULTURES OF HUMAN FIBROBLASTS AND RENAL EPITHELIAL CELLS. ESTRADIOL 17B HAS TERATOGENIC ACTIONS ON THE GENITAL TRACT AND POSSIBLE OTHER ORGANS AND IMPAIRS FERTILITY.

REPORT 06 : MULTI-AGENCY CROSS CHECK
Y HHS MAGUIRE
N OSHA CARC
N EPA CAG
N EPA RCRA
N EPA CLEAN AIR

N EPA CLEAN WATER
REPT'D USE : TREATMENT OF CLIMACTERIC AND DRUG

CAS NO : **22966-79-6**
CHEM NAME : ***** ESTRADIOL MUSTARD**

REPORT 01 : NCI CARC TRS NO 59 DHEW PUB NO NIH 78-1309
UNDER THE CONDITIONS OF THIS BIOASSAY ESTRADIOL MUSTARD ADMINISTERED IN A BUFFERED SALINE VEHICLE WAS NOT CARCINOGENIC IN SPRAGUE-DAWLEY RATS. ESTRADIOL MUSTARD WAS CARCINOGENIC IN BOTH MALE AND FEMALE B6C3F1 MICE, INDUCING LYMPHOMA, SARCOMA OF THE MYOCARDIUM, ALVEOLAR ADENOMA OR CARCINOMA, AND SQUAMOUS-CELL CARCINOMA OF THE STOMACH. ESTRADIOL MUSTARD WAS GIVEN BY GAVAGE.

REPORT 02 : IARC MONOGRAPH 9 1975
ESTRADIOL MUSTARD IS CARCINOGENIC IN MICE FOLLOWING ITS INTRAPERITONEAL INJECTION, THE ONLY SPECIES AND ROUTE TESTED, PRODUCING A DOSE RELATED INCREASE IN THE INCIDENCE OF LUNG TUMORS.
REPT'D USE : EXPERIMENTAL ANTINEOPLASTIC AGENT

CAS NO : **53-16-7**
CHEM NAME : ***** ESTRONE**

REPORT 01 : IARC MONOGRAPH 21 1979
ESTRONE WAS TESTED IN MICE BY ORAL ADMINISTRATION; IN MICE, RATS AND HAMSTERS BY SUBCUTANEOUS INJECTION AND IMPLANTATION; AND IN MICE BY SKIN PAINTING. THERE WAS AN INCREASED INCIDENCE OF MAMMARY TUMORS IN MICE; AN INCREASED INCIDENCE IN PITUITARY, ADRENAL AND MAMMARY TUMORS AS WELL AS BLADDER TUMORS ASSOCIATED WITH STONES IN RATS; AND AN INCREASED INCIDENCE IN RENAL TUMORS IN BOTH CASTRATED AND INTACT MALE HAMSTERS. IT IS REASONABLE TO REGARD ESTRONE AS IF IT PRESENTED A CANCER RISK TO HUMANS. 1979 SUP./SUF. EVIDENCE ANIMAL CARCINOGEN

REPORT 02 : IARC MONOGRAPH 21 1979
NO DATA ON THE MUTAGENICITY OF ESTRONE OR ESTRONE BENZOATE WERE AVAILABLE. ESTRONE IS EMBRYOLETHAL FOR PREIMPLANTATION EMBRYOS IN SOME ANIMAL SPECIES.

REPORT 06 : MULTI-AGENCY CROSS CHECK
Y HHS MAGUIRE
N OSHA CARC
N EPA CAG
N EPA RCRA

N EPA CLEAN AIR
N EPA CLEAN WATER
REPT'D USE : TREATMENT OF CLIMACTERIC AND DRUG

CAS NO : **26914-40-9**
CHEM NAME : ***** 1,2-ETHANEDITHIOL**

REPORT 01 : EPA TSCA 8(E) 83HQ-0683-0481 JUL 83
CPD NEGATIVE IN AMES S. TYPHIMURIUM BACTERIA ASSAY BOTH WITH AND WITHOUT METABOLIC ACTIVATION; POSITIVE WITHOUT METABOLIC ACTIVATION AND NEGATIVE WITH METABOLIC ACTIVATION IN A MOUSE LYMPHOMA CELL FORWARD MUTATION ASSAY. CPD FOUND TO CAUSE INCREASED SCES IN CULTURED CHO CELLS.

CAS NO : **36443-68-2**
CHEM NAME : *****1,2-ETHANEDIYL-BIS (OXY-2, 1-ETHANEDIYL) ESTER**

REPORT 01 : EPA TSCA 8(E) 0388-0725 APR 88
CIBA-GEIGY LTD REPORTED ON 24-MONTH FEEDING STUDY IN RATS GIVEN DOSES OF 0, 5, 15, 50 AND 100 MG/KG/DAY. RESULTS SHOWED DISTURBANCE OF LIPID METABOLISM AND LIVER ENZYME ACTIVITIES, INDICATIVE OF LIVER TOXICITY, AND INCREASED INCIDENCE OF CYSTIC DILATATION OR HYPERPLASIA OF THYROID FOLLICLES IN 15 (MALES ONLY), 50 AND 100 MG/KG GROUPS AND INCREASED INCIDENCE OF THYROID GLAND FOLLICULAR ADENOMAS AND CARCINOMAS OBSERVED IN 50 (MALES ONLY) AND 100 MG/KG/DAY GROUPS.
REPT'D USE : POLYMERS STABILIZER

CAS NO : **141-43-5**
CHEM NAME : ***** ETHANOLAMINE**

REPORT 01 : STAFF REFERRAL AUG 89
INHALATION STUDIES OF MONOETHANOLAMINE SHOWED DOGS AND CATS SURVIVED EXPOSURES TO 2.47 MG/LITER FOR 7 HOURS/4 DAYS. GUINEA PIGS DIED AT 0.58 MG/LITER FOR 1 HOUR. RATS, RABBITS, AND MICE SURVIVED EXPOSURE TO 0.26 TO 0.27 MG/LITER FOR 7 HOURS/5 DAYS AND 25, 7-HOURS EXPOSURES OVER 5 WEEKS) TO CONCENTRATIONS OF 0.26 MG/LITER (104 P.P.M.). RESPIRATORY TRACT IRRITATION OBSERVED EYE IRRITATION NEGLIGIBLE. PATHOLOGICAL CHANGES AT HIGHER CONCENTRATIONS WERE CHIEFLY PULMONARY IRRITATION, WITH SOME DEGENERATIVE CHANGES IN LIVER AND KIDNEYS (TREON ET AL IN TUSTRIAL HYGIENE AND TOX., PATTY, 1963).

REPORT 02 : STAFF REFERRAL AUG 89
DOGS, RATS, AND GUINEA PIGS SURVIVED INHALATION OF 12 TO 25 P.P.M. FOR 90 DAYS, FRACTIONAL MORTALITY OCCURRED IN 24 TO 30 DAYS AT 100 P.P.M. (DOGS) AND 66 TO 75 P.P.M. (RODENTS). SKIN IRRITATION AND LETHARGY OCCURRED AT 5 AND 12 P.P.M. THE MEDIAN DETECTIBLE (ODOR) CONCENTRATION FOR HUMANS 3 TO 4 P.P.M. INSTILLED IN RABBIT EYE PRODUCES SEVERE INJURY. UNDILUTED APPLICATION TO HUMAN SKIN SHOWED REDNESS AND INFILTRATION OF THE SKIN (WEEKS INDUSTRIAL HYGIENE AND TOX., PATTY, 1963). OSHA STANDARD TWA 3 PPM; ACG1H TLV-TWA 3 P.P.M.

REPT'D USE : MILDLY AMMONIACAL CO2 EXTRACTOR, ALKALINE CONDITIONING AGENT, INTERMEDIATE FOR SOAPS, DETERGENTS AND TEXTILE AGENTS

CAS NO : **57-63-6**
CHEM NAME : ***** ETHINYLESTRADIOL**

REPORT 01 : IARC MONOGRAPH 21 1979
ETHINYLESTRADIOL, WHEN ADMINISTERED ALONE TO MICE INCREASED THE INCIDENCE OF PITUITARY TUMORS AND MALIGNANT MAMMARY TUMORS IN BOTH MALES AND FEMALES AND PRODUCED MALIGNANT TUMORS OF THE UTERUS AND THE CERVIX IN FEMALES. IN RATS, AN INCREASED INCIDENCE OF BENIGN LIVER CELL TUMORS IN BOTH MALES AND FEMALES AND MALIGNANT LIVER CELL TUMORS IN FEMALES WAS OBSERVED. IT IS REASONABLE TO REGARD ETHINYLESTRADIOL AS IF IT PRESENTED A CARCINOGENIC RISK TO HUMANS.

REPORT 02 : IARC MONOGRAPH 21 1979
NO MUTAGENIC ACTIVITY WAS REPORTED IN ONE TEST IN SALMONELLA WITH METABOLIC ACTIVATION. NO DOMINANT LETHAL MUTATIONS WERE OBSERVED IN FEMALE MICE. ETHINYLESTRADIOL IS EMBRYOLETHAL FOR PREIMPLANTATION EMBRYOS IN SOME SPECIES.

REPORT 06 : MULTI-AGENCY CROSS CHECK
Y HHS MAGUIRE
N OSHA CARC
N EPA CAG
N EPA RCRA
N EPA CLEAN AIR
N EPA CLEAN WATER

REPT'D USE : ORAL CONTRACEPTIVES AND TREATMENT OF CLIMACTERIC

CAS NO : **536-33-4**
CHEM NAME : ***** ETHIONAMIDE**

REPORT 01 : NCI CARC TRS 46 DHEW PUB NO 78-846
UNDER THE CONDITIONS OF THIS BIOASSAY ETHIONAMIDE WAS NOT CARCINOGENIC IN EITHER FISCHER 344 RATS OR B6C3F1 MICE (FEEDING STUDY).

REPORT 02 : IARC MONOGRAPH 13 1977
ETHIONAMIDE IS CARCINOGENIC IN MICE AFTER ITS ORAL ADMINISTRATION, THE ONLY SPECIES AND ROUTE OF ADMINISTRATION TESTED: IT PRODUCED THYROID CARCINOMAS.
REPT'D USE : ANTI-TUBERCULIN AGENT

CAS NO : **13073-35-3**
CHEM NAME : ***** ETHIONINE**

REPORT 06 : MULTI-AGENCY CROSS CHECK
N HHS MACUIRE
Y OSHA CARC
N EPA CAG
N EPA RCRA
N EPA CLEAN AIR
N EPA CLEAN WATER
REPT'D USE : NO AVAILABLE USE DATA

CAS NO : **67-21-0**
CHEM NAME : ***** DL-ETHIONINE**

REPORT 06 : MULTI-AGENCY CROSS CHECK
N HHS MAGUIRE
Y OSHA CARC
N EPA CAG
N EPA RCRA
N EPA CLEAN AIR
N EPA CLEAN WATER
REPT'D USE : RESEARCH CHEMICAL

CAS NO : **111-15-9**
CHEM NAME : ***** 2-ETHOXYETHANOL ACETATE**

REPORT 01 : EPA TSCA 8(E) NOTICE 8EHQ-0682-0450 JUL 82
TENNESSEE EASTMAN SUBMITTED SUMMARIZED RESULTS OF A NIOSH STUDY OF INHALATION TERATOGENICITY OF THREE GLYCOL ETHERS AT 600, 300, AND 130 PPM, IN RATS. THE DATA INDICATE THAT 2-EEA IS EMBRYOLETHAL FOR THE FETUS AT THE 600 AND

300 PPM DOSES, AND INDUCED CARDIAC MALFORMATIONS IN THE 300 AND 130 PPM GROUPS. NO MATERNAL TOXICITY WAS APPARENT AT ANY DOSE LEVEL. THE CURRENT OSHA STANDARD FOR 2-EEA IS 100 PPM AS 8 HOUR TWA.

REPORT 02 : EPA TSCA 8(E) NOTICE 8EHQ-0381-0386 MAR 81
SHELL OIL SUBMITTED A COPY OF A NIOSH PREPUBLICATION REPORT CONTAINING THE RESULTS FROM A TESTING OF SELECTED WORKPLACE CHEMICALS FOR TERATOGENIC POTENTIAL. SUBMITTER REPORTED THAT "AT A 2-ETHOXYETHANOL INHALATION LEVEL OF 160 PPM FOR RABBITS AND 200 PPM FOR RATS, TERATOGENIC EFFECTS WERE OBSERVED." THESE EFFECTS WERE CHARACTERIZED BY SKELETAL AND/OR SOFT TISSUE ABNORMALITIES.

REPORT 03 : EPA FEDERAL REGISTOR NOTICE 51FR18488 MAY 86
UNDER TSCA 9(A) EPA CONCLUDES RISK OF INJURY TO WORKER HEALTH FROM EXPLOSIVE TO THIS GLYCOL ETHER IS UNREASONABLE AND MAY BE PREVENTED OR REDUCED BY OSHA.

REPT'D USE : CLEANING/AGENT, PAINTS/VARNISHES, SPACE DEODORIZERS, PHOTOGRAPHIC CHEMICALS, LAB REAGENT, WOOD STAINS AND TEXTILE/LEATHER TREATMENTS

CAS NO : **150-69-6**
CHEM NAME : ***** P-ETHOXYPHENYLUREA**

REPORT 01 : IARC MONOGRAPH 12 1976
DULCIN HAS BEEN TESTED ONLY IN RATS BY ORAL ADMINISTRATION. INCREASED INCIDENCES OF LIVER CELL ADENOMAS WERE OBSERVED IN ONE STUDY, AND PAPILLOMAS OF THE RENAL PELVIS AND BLADDER, ASSOCIATED WITH STONE FORMATION, WERE SEEN IN ANOTHER STUDY, BUT THESE RESULTS WERE NOT CONFIRMED IN TWO OTHER TRIALS. NO EVALUATION OF THE CARCINOGENICITY OF DULCIN CAN BE MADE.

REPORT 02 : NTP TECHNICAL BULLETIN NO. 6 JAN 82
THE COMPOUND WAS NEGATIVE IN THE SALMONELLA TYPHIMURIUM MUTAGENICITY ASSAY IN ONE LABORATORY.

REPT'D USE : NON-NUTRITIVE SWEETENER. NOT APPROVED FOR USE BY FDA.

CAS NO : **140-88-5**
CHEM NAME : ***** ETHYL ACRYLATE (ACRYLIC ACID, ETHYL ESTER)**

REPORT 01 : IARC MONOGRAPH 19 FEB 79
NO DATA ON THE EMBRYOTOXICITY, TERATOGENICITY, OR MUTAGENICITY OF ETHYL ACRYLATE WERE AVAILABLE.

REPORT 02 : IARC MONOGRAPH 19 1979
ETHYL ACRYLATE WAS TESTED IN RATS BY ORAL ADMINISTRATION; THE REPORTING OF THE EXPERIMENT WAS CONSIDERED INADEQUATE.

REPORT 03 : NTP CARC TEST PROGRAM JUL 82
IN A CURRENT NTP BIOASSAY, ETHYL ACRYLATE WAS ADMINISTERED BY GAVAGE TO F344 RATS AND B6C3F1 MICE. THE STUDY HAS BEEN COMPLETED AND THE REPORT IS BEING DRAFTED.

REPORT 04 : NTP TECHNICAL BULLETIN 6 JAN 82
THE COMPOUND WAS NEGATIVE IN THE SALMONELLA TYPHIMURIUM MUTAGENICITY ASSAY IN TWO LABORATORIES.

REPORT 06 : MULTI-AGENCY CROSS CHECK
Y HHS MAGUIRE
N OSHA CARC
N EPA CAG
Y EPA RCRA
N EPA CLEAN AIR
N EPA CLEAN WATER

REPORT 07 : EPA TSCA 8(E) NOTICE 8EHQ-1282-0467 DEC 82
UNION CARBIDE REPORTED ON NTP DRAFT REPORT ON ETHYL ACRYLATE ADMINISTERED CHRONICALLY BY GAVAGE TO MALE AND FEMALE MICE AND RATS SHOWS COMPOUND RELATED INCREASES IN INFLAMMATION, HYPERPLASIA, AND HYPERKERATOSIS OF FORESTOMACH IN TREATED ANIMALS; FOR ANIMALS OF EACH SEX, INCIDENCES OF SQUAMOUS CELL PAPILLOMAS, SQUAMOUS CELL CARCINOMAS, OR PAPILLOMAS OR CARCINOMAS WITH STATISTICALLY SIGNIFICANT POSITIVE TRENDS, AND INCIDENCES IN TREATED GROUPS SIGNIFICANTLY HIGHER THAN IN VEHICLE (CORN OIL) CONTROLS.

REPORT 08 : IARC MONOGRAPH 39, 1986 JUN 85
MICE AND RATS TESTED BY GAVAGE SHOWED INCREASED PAPILLOMAS AND CARCINOMAS. INHALATION IN SAME STRAINS OF MICE AND RATS SHOWED NO LESIONS. LONG-TERM SKIN APPLICATION TEST WAS NEGATIVE ON MALE MICE. ORAL DOSAGE IN RATS GAVE SIGNS OF EMBRYO- AND FETOTOXICITY AT "MILDLY TOXIC DOSES". CPD NOT MUTAGENIC TO S.TYPHIMURIUM OR D. MELANOGASTER WITH OR WITHOUT ACTIVATION. CHROMOSOMAL ABERRATIONS IDUCED IN CH LUNG CELLS IN-VITRO AND IN MOUSE BONE MAROW MICRONUCLEI IN-VIVO. NO DATA TO EVALUATE HUMAN OR ANIMAL REPRODUCTIVE EFFECTS OR PRENATAL TOXICITY. IARC: SUFFICIENT EVIDENCE FOR CARCINOGENICITY OF CPD; NO DATA ON HUMANS AVAILABLE.

REPT'D USE : PLASTIC MONOMER, LATEX PAINTS, FOOD ADJUVANT, ADHESIVES/SEALANTS, DENTAL/MEDICAL APPLICATIONS AND CHEMICAL INTERMEDIATE

CAS NO : **43130-12-7**
CHEM NAME : ***** ETHYL AURAMINE NITRATE**

REPORT 01 : EPA TSCA 8(E)-0730
GROUPS OF RATS WERE ADMINISTERED 0, 1000 OR 2000 PPM OF ETHYL AURAMINE, NITRATE SALT IN THE FEED FOR NINE MONTHS. ANIMALS WERE SACRIFICED AT 3, 6 AND 9 MONTHS FOR HISTOPATHOLOGICAL EXAMINATION. MEAN BODY WEIGHTS AND FOOD INTAKE WERE LOWER IN THE TREATED ANIMALS AND INCREASED ACTIVITY OF GAMMA-GLUTAMYL TRANSPEPTIDASE WAS OBSERVED IN LIVER HOMOGENATE. IN ADDITION, HIGHER INCIDENCES OF FOCI AND ALTERED HEPATOCYTES WERE NOTED IN THE TREATED ANIMALS AFTER 6 MONTHS AND 9 MONTHS. AT NINE MONTHS, HYPERPLASTIC NODULES WERE NOTED IN TWO OF 5 HIGH DOSE MALES.
REPT'D USE : YELLOW DYE FOR PAPER, TEXTILES, LEATHER; ALSO AN ANTISEPTIC AND FUNGICIDE

CAS NO : **100-41-4**
CHEM NAME : ***** ETHYLBENZENE**

REPORT 01 : 20TH REPORT OF THE TSCA ITC MAY 87
CPD IS RECOMMENDED WITH INTENT TO DESIGNATE FOR TESTING FOR ECOLOGICAL EFFECTS, ACUTE TOXICITY TO FRESHWATER ALGAE AND AQUATIC INVERTEBRATES; AND ACUTE TOXICITY TO SALTWATER ALGAE, AQUATIC INVERTEBRATES AND FISH.
REPT'D USE : CHEMICAL INTERMEDIATE

CAS NO : **51-79-6**
CHEM NAME : ***** ETHYL CARBAMATE (URETHANE)**

REPORT 01 : EPA RCRA HEALTH & ENV. EFFECTS PROFILE OCT 80
THE EPA CARCINOGEN ASSESSMENT GROUP HAS EVALUATED URETHANE AND HAS FOUND SUFFICIENT EVIDENCE TO INDICATE THAT THIS COMPOUND IS CARCINOGENIC IN EXPERIMENTAL ANIMALS.

REPORT 02 : IARC MONOGRAPH 7 1974
URETHANE HAS BEEN SHOWN TO BE CARCINOGENIC IN MICE, RATS, AND HAMSTERS FOLLOWING ADMINISTRATION BY THE ORAL, INHALATION, SUBCUTANEOUS, OR INTRAPERITONEAL ROUTES, PRODUCING, AMONG OTHERS, LUNG TUMORS, LYMPHOMAS, HEPATOMAS, MELANOMAS, AND VASCULAR TUMORS. IT IS AN INITIATOR FOR SKIN CARCINOGENESIS IN MICE BOTH WHEN GIVEN ORALLY AND TOPICALLY. IT IS CARCINOGENIC IN SINGLE DOSE EXPERIMENTS AND FOLLOWING PRENATAL EXPOSURE.

REPORT 03 : NAS REPORT ON DIET, NUTRUTION AND CANCER 1982
THE ABILITY OF URETHANE TO INDUCE TUMORS HAS BEEN DEMONSTRATED BY ADMINISTERING THE CHEMICAL DURING THE PRENATAL AND PREWEANLING PERIODS AS WELL AS TO ADULT ANIMALS. THE SIGNIFICANCE OF NATURALLY OCCURRING ETHYL CARBAMATE IN FOODS IN THE DEVELOPMENT OF HUMAN CANCER IS UNKNOWN, BUT THE LEVELS ARE VERY LOW IN COMPARISON TO THOSE USED TO INDUCE TUMORS IN EXPERIMENTAL ANIMALS.

REPORT 06 : MULTI-AGENCY CROSS CHECK
Y HHS MAGUIRE
N OSHA CARC
Y EPA CAG
Y EPA RCRA
N EPA CLEAN AIR
N EPA CLEAN WATER
REPT'D USE : INTERMEDIATE IN PRODUCTION OF PHARMACEUTICALS, PESTICIDES AND FUNGICIDES; RESEARCH CHEMICAL AND FABRIC FINISH INTERMEDIATE

CAS NO : **72-56-0**
CHEM NAME : ***** P,P'-ETHYL-DDD**

REPORT 01 : NCI CARC TRS 156 DHEW PUB NO 79-1712
UNDER THE CONDITIONS OF THIS BIOASSAY, P,P'-ETHYL-DDD WAS NOT CARCINOGENIC FOR MALE OR FEMALE F344 RATS OR MALE B6C3F1 MICE. HOWEVER, THE OCCURRENCE OF HEPATOCELLULAR CARCINOMAS AND ADENOMAS IN FEMALE MICE WAS SUGGESTIVE OF A CARCINOGENIC EFFECT (FEEDING STUDY).

REPORT 03 : NTP TECHNICAL BULLETIN 7 APR 82
PERTHANE WAS POSITIVE IN THE SALMONELLA TYPHIMURIUM MUTAGENICITY ASSAY.

REPORT 06 : MULTI-AGENCY CROSS CHECK
N HHS MAGUIRE
Y OSHA CARC
N EPA CAG
N EPA RCRA
N EPA CLEAN AIR
N EPA CLEAN WATER
REPT'D USE : CONTACT INSECTICIDE

CAS NO : **74-85-1**
CHEM NAME : ***** ETHYLENE**

REPORT 01 : IARC MONOGRAPH 19 1979
NO DATA ON THE CARCINOGENICITY OR MUTAGENICITY OF ETHYLENE WERE AVAILABLE TO THE WORKING GROUP.

REPORT 02 : CITT 24-MONTH EXECUTIVE SUMMARY OCT 80
UNDER THE CONDITIONS OF THE BIOASSAY, CHRONIC INHALATION EXPOSURE OF ETHYLENE GAS AT CONCENTRATIONS OF 3000, 1000, AND 300 PPM TO F344 RATS, RESULTED IN NO EXPOSURE-RELATED LESIONS. IT IS CONCLUDED THAT ETHYLENE GAS IS NON-TOXIC AT CONCENTRATIONS AS HIGH AS 3000 PPM FOR UP TO TWO YEARS OF EXPOSURE IN RATS.
REPT'D USE : CHEMICAL INTERMEDIATE, REFRIGERANT, PLASTIC MONOMER, FRUIT RIPENER AND ANESTHETIC

CAS NO : **111- 54-6**
CHEM NAME : ***** ETHYLENE BIS DITHIOCARBAMATE**

REPORT 06 : MULTI-AGENCY CROSS CHECK
N HHS MAGUIRE
N OSHA CARC
N EPA CAG
Y EPA RCRA
N EPA CLEAN AIR
N EPA CLEAN WATER
REPT'D USE : NO AVAILABLE PRODUCTION OR USE DATA. NO LISTING ON TSCA INVENTORY.

CAS NO : **111-21-7**
CHEM NAME : ***** ETHYLENE BIS(OXYETHYLENE) DIACETATE**

REPORT 01 : EPA'S RESPONSE TO ITC FOR PRIORITY TEST'G 49FR45651 NOV 84
EPA'S RESPONSE TO ITC'S DESIGNATING EBOD (TGD) FOR PRIORITY TESTING UNDER TSCA 4(A). EPA FINDS THAT AVAILABLE DATA ARE SUFFICIENT TO REASONABLY DETERMINE OR PREDICT THE EFFECTS ON HEALTH OF HUMANS UNDER CURRENT MFG CONDITIONS AND USE. EPA IS NOT ISSUING A TEST RULE AT THIS TIME.
REPT'D USE : PLASTICIZER IN PRODUCTION OF CIGARETTE FILTER TIPS

CAS NO : **10310-38-0**
CHEM NAME : ***** ETHYLENEBIS TRIS (2-CYANOETHYL) PHOSPHONIUM BROMIDE**

REPORT 01 : EPA TSCA 8(E) NOTICE 8EHQ-0780-0369
AMERICAN CYANAMID PROVIDED RESULTS FROM SEVERAL SHORT TERM IN VIVO STUDIES ON ETHYLENBIS TRIS(2-CYANOETHYL) PHOSPHONIUM BROMIDE. SUBMISSION INDICATED THAT NEUROTOXIC EFFECTS WERE OBSERVED IN RATS FOLLOWING ORAL ADMINISTRATION OF THE CHEMICAL, EITHER AS PART OF THE DAILY DIET OR BY GAVAGE. HOWEVER, THE COMPOUND DID NOT APPEAR TO CAUSE NEUROTOXIC EFFECTS IN RATS FOLLOWING DERMAL APPLICATION.

CAS NO : **107-15-3**
CHEM NAME : ***** ETHYLENEDIAMINE**

REPORT 01 : NTP TECHNICAL BULLETIN 3/8 DEC 80, JUL 82
ETHYLENEDIAMINE WAS POSITIVE IN THE SALMONELLA ASSAY, BUT NEGATIVE IN DROSOPHILA WHEN TESTED FOR SEX LINKED RECESSIVE LETHALS.
REPT'D USE : PHOTOGRAPHIC, CHEMICAL INTERMEDIATE, SOLVENT ANTIFREEZE, CHEMICAL STABILIZER, FOOD ADDITIVE, EMULSIFIER AND FUNGICIDES

CAS NO : **1429-50-1**
CHEM NAME : ***** ETHYLENEDIAMINETETRA(METHYL PHOSPHONIC ACID)**

REPORT 01 : EPA TSCA 8(E) 8EHQ-0683-0483S AUG 83
OSTEOGENIC SARCOMAS WERE OBSERVED AFTER ONE YEAR INTO TWO CHRONIC GAVAGE STUDIES IN RATS ADMINISTERED 15, 50, 75, AND 150 MG/KG OF CPD. FEMALE RATS AT HIGHEST DOSE SHOWED WEAK TUMORIGENIC EFFECT TOWARD BONE IN ONE STUDY. IN 1-MONTH RANGE-FINDING STUDY, MALE RATS DOSED AT 500 MG/KG SHOWED A STATISTICALLY SIGNIFICANT INCREASE IN ALKALINE PHOSPHATASE LEVELS. FEMALE RATS IN SAME STUDY ALSO SHOWED AN INCREASE, BUT IT WAS NOT STATISTICALLY SIGNIFICANT.
REPT'D USE : DETERGENTS AND UNSPECIFIED HUMAN ORAL USE (EXPERIMENTAL)

CAS NO : **15142-96-8**
CHEM NAME : ***** ETHYLENEDIAMINETETRA(METHYL PHOSPHONIC ACID) HEXASODIUM SALT**

REPT'D USE : DETERGENTS AND UNSPECIFIED NEW HUMAN ORAL USE

CAS NO : **68188-96-5**
CHEM NAME : ***** ETHYLENEDIAMINETETRA(METHYL PHOSPHONIC ACID) TETRAPOTASS. SALT**

REPORT 01 : CSC CHEMICAL PRODUCT REVIEW RECORD MAY 84
OSTEOSARCOMAS SEEN IN MALE CHARLES RIVER RATS IN CHRONIC ORAL TOX STUDY. DOSE-DEPENDENCE AND SEX SPECIFICITY SEEN IN THIS STUDY. NO TUMORS SEEN IN ONGOING 18-MONTH STUDY IN B6C3F1 MICE OF BOTH SEXES. NON-CANCEROUS BONE DISTURBANCES OBSERVED IN RATS AND MICE IN THESE STUDIES. IN HUMANS CPD RAPIDLY CLEARED FROM BLOOD AND COULD BE VISUALIZED IN BONE, BLADDER AND KIDNEY WHEN INJECTED.
REPT'D USE : DETERGENTS AND UNSPECIFIED NEW HUMAN ORAL USE

CAS NO : **110-80-5**
CHEM NAME : ***** ETHYLENE GLYCOL MONOMETHYL ETHER (2-ETHOXYETHANOL) (2-EE)**

REPORT 01 : NTP TECHNICAL BULLETIN 3/7 DEC 80, APR 82
ETHYL CELLOSOLVE WAS NEGATIVE IN THE SALMONELLA ASSAY, THE RAT DOMINANT LETHAL ASSAY, THE MOUSE SPERM HEAD MORPHOLOGY ASSAY, AND THE DROSOPHILA SEX-LINKED RECESSIVE LETHAL ASSAY.

REPORT 02 : EPA TSCA 8(E) NOTICE 8EHQ-0682-0450 JUL 82
TENNESSEE EASTMAN SUBMITTED SUMMARIZED RESULTS OF A NIOSH STUDY OF INHALATION TERATOGENICITY OF THREE GLYCOL ETHERS AT 600, 300, AND 130 PPM, IN RATS. THE DATA INDICATE THAT 2-EEA IS EMBRYOLETHAL FOR THE FETUS AT THE 600 AND 300 PPM DOSES, AND INDUCED CARDIAC MALFORMATIONS IN THE 300 AND 130 PPM GROUPS. NO MATERNAL TOXICITY WAS APPARENT AT ANY DOSE LEVEL. THE CURRENT OSHA STANDARD FOR 2-EEA IS 100 PPM AS 8 HOUR TWA.

REPORT 03 : NTP CARC TEST PROGRAM JUL 82
IN A CURRENT NTP BIOASSAY, 2-ETHOXYETHANOL IS BEING ADMINISTERED BY GAVAGE TO F344 RATS AND B6C3F1 MICE. THE CHRONIC EXPOSURE PHASE HAS BEEN COMPLETED AND HISTOPATHOLOGY IS IN PROGRESS.

REPORT 05 : EPA TSCA 8(E) NOTICE 8EHQ-0381-0386 MAR 81
SHELL OIL SUBMITTED A COPY OF A NIOSH PREPUBLICATION REPORT CONTAINING THE RESULTS FROM A TESTING OF SELECTED WORKPLACE CHEMICALS FOR TERATOGENIC POTENTIAL. SUBMITTER REPORTED THAT "AT A 2-ETHOXYETHANOL INHALATION LEVEL OF 160 PPM FOR RABBITS, AND 200 PPM FOR RATS, TERATOGENIC EFFECTS WERE OBSERVED. THESE EFFECTS WERE CHARACTERIZED BY SKELETAL AND/OR SOFT TISSUE ABNORMALITIES."

REPORT 06 : MULTI-AGENCY CROSS CHECK MAY 83
N HHS MAGUIRE
N OSHA CARC
N EPA CAG
N EPA RCRA
N EPA CLEAN AIR
N EPA CLEAN WATER

REPORT 07 : NIOSH CURRENT INTELLIGENCE BULLETIN 39 MAY 83
TERATOGENIC AND EMBRYOTOXIC TO PREGNANT RATS AND RABBITS. FERTILITY OF NON-PREGNANT FEMALE RATS NOT AFFECTED BY EXPOSURE TO CPD, BUT IT DID PRODUCE VARYING DEGREES OF TOXICITY IN PREGNANT RATS AND RABBITS. IN MALES PRODUCED TESTICULAR ATROPHY IN MICE AND TESTICULAR CHANGES IN MICE, RATS AND DOGS. BASED ON THESE FINDINGS AND A CONTINUED CONCERN FOR ADVERSE EFFECTS AFTER PC ABSORPTION, NIOSH RECOMMENDS IT BE REGARDED AS HAVING POTENTIAL TO CAUSE ADVERSE REPRODUCTIVE EFFECTS IN BOTH SEXES AND EMBRYOTOXIC EFFECTS, INCLUDING TERATOGENICITY, IN OFFSPRING OF EXPOSED, PREGNANT FEMALES.

REPT'D USE : SOLVENT, VARNISH REMOVERS, CLEANING AGENTS, CHEMICAL INTERMEDIATE, LEATHER FINISHING, DYE BATHS, LACQUERS, WOOD STAINS, ADHESIVES, PAINTS/VARNISHES, CAULKING/SPACKLING CMPDS, AEROSOLS CHEMICALS, SPACE DEODORIZERS AND PHOTOGRAPHIC CHEMICAL

CAS NO : **75-21-8**
CHEM NAME : ***** ETHYLENE OXIDE (ETO)**

REPORT 01 : NIOSH CURRENT INTELLIGENCE BULLETIN 35 MAY 81
NIOSH RECOMMENDS THAT ETO BE REGARDED IN THE WORKPLACE AS A POTENTIAL OCCUPATIONAL CARCINOGEN. CONCLUSIONS ARE BASED PRIMARILY UPON AN INDUSTRY-SPONSORED STUDY DEMONSTRATING THAT ETO IS CARCINOGENIC IN EXPERIMENTAL ANIMALS. THERE HAS BEEN WIDESPREAD RECOGNITION OF THE MUTAGENIC POTENTIAL OF ETO, AND RECENT EVIDENCE DEMONSTRATES ADVERSE REPRODUCTIVE EFFECTS IN MAMMALS. IN ADDITION, LIMITED EPIDEMIOLOGIC INVESTIGATIONS AT TWO

WORKSITES PROVIDE EVIDENCE THAT EXCESS RISK OF CANCER MORTALITY MAY EXIST FOR THE ETO WORKERS STUDIED.

REPORT 02 : TSCA SECT 8(E) STATUS REPORT 8EHQ-1086-0641 DEC 86
8EHQ-1186-0641
14 PERSONS CHRONICALLY EXPOSED TO ETO VAPOR OVER PERIOD OF 18 MONTHS SHOWED SENSORY AND MOTOR DISTURBANCES AND DISTURBANCES IN COGNITIVE FUNCTIONS SUCH AS MEMORY, MOODS, LANGUAGE AND SPEECH. ALSO THEY REPORTED HEADACHES AND SEIZURES. FOLLOW-UP AFTER 12 TO 18 MONTHS SHOWED MARKED IMPROVEMENT IN MOOD AND MEMORY.

REPORT 03 : EPA TSCA 8(E) NOTICE 8EHQ-0979-0305 FEB 81
IN A STUDY CONDUCTED BY UNION CARBIDE, FISCHER 344 RATS WERE EXPOSED TO 100, 33 OR 10 PPM OF ETO VAPOR BY INHALATION FOR 24 MONTHS. AT FINAL SACRIFICE, HISTOLOGIC FINDINGS CONFIRMED HEMATOLOGIC EVIDENCE OF INCREASED PREVALENCE OF MONO-NUCLEAR LEUKEMIA. THE PREVALENCE OF THIS NEOPLASM AND OTHER NEOPLASMS WAS SIGNIFICANT FOR ALL THREE CONCENTRATIONS.

REPORT 04 : IARC MONOGRAPH 11 1976
TESTED IN MICE BY SKIN APPLICATION AND IN RATS BY SC INJECTION. RESULTS INADEQUATE TO EVALUATE CARCINOGENICITY. IN 1979 SUPPLEMENT, IARC SAID CPD PROBABLY CARCINOGENIC FOR HUMANS. INDUCED RECESSIVE LETHALS, TRANS-LOCATIONS AND MINUTE MUTATIONS IN D. MELANOGASTER. CHROMOSOMAL ABERRATIONS SEEN IN BONE-MARROW CELLS OF TREATED MALE MICE.

REPORT 05 : ITC FIRST REPORT TO EPA (42FR55026) OCT 77
THE ITC RECOMMENDS THAT THE EPA ADMINISTRATOR CONSIDER DEVELOPING RULES TO TEST ALKYL EPOXIDES FOR CARCINOGENICITY, MUTAGENICITY, TERATOGENICITY, AND OTHER CHRONIC EFFECTS AND THAT AN EPIDEMIOLOGIC STUDY BE CONDUCTED.

REPORT 06 : MULTI-AGENCY CROSS CHECK
Y HHS MAGUIRE
N OSHA CARC
Y EPA CAG
Y EPA RCRA
Y EPA CLEAN AIR
N EPA CLEAN WATER

REPORT 07 : NTP TECHNICAL BULLETIN 7 APR 82
ETO WAS NEGATIVE IN THE MOUSE SPERM HEAD MORPHOLOGY ASSAY.

REPORT 08 : EPA TSCA NOTICE FYI-0382-0123 SUPP MAR 82
JOHNSON & JOHNSON REPORTED PRELIMINARY RESULTS OF A PILOT RESEARCH CHROMOSOME STUDY OF WORKERS AT SITES

WHERE ETO IS USED AS A STERILANT. THE STUDY SUGGEST A DOSE-RESPONSE ASSOCIATION BETWEEN EXPOSURE TO ETC AND SCE'S IN HUMANS. THERE DOES NOT APPEAR TO BE A DOWNWARD TREND IN GROUP SCE LEVELS FOR ALL EMPLOYEE GROUPS TESTED AT ONE PLANT FOLLOWING 6 MONTHS WITHOUT EXPOSURE.

REPORT 09 : NTP TR 326 NIH 88-2582 DEC 87
2 YEAR INHALATION STUDIES EVIDENCE OF CARCINOGENIC ACTIVITY FOR B6C3F1 MICE INDICATED BY DOSE-RELATED INCREASED INCIDENCES OF BENIGN OR MALIGNANT NEOPLASMS OF LUNG AND BENIGN NEOPLASMS OF HARDERIAN GLAND IN BOTH MALE AND FEMALE B6C3F1 MICE FOLLOWING EXPOSURE TO ETHYLENE OXIDE VAPORS AT 50 AND 100 PPM. IN FEMALE MICE, ETHYLENE OXIDE CAUSED ADDITIONAL MALIGNANT NEOPLASMS OF UTERUS, MAMMARY GLAND, AND HEMATOPOIETIC SYSTEM (LYMPHOMA).

REPORT 10 : EPA TSCA 8(E) NOTICE 8EHQ-1080-0305S
UNION CARBIDE SUBMITTED FINAL RESULTS OF A 24 MONTH INHALATION STUDY OF ETHYLENE OXIDE IN RATS. SUBMITTER STATED THAT HISTOLOGICAL EXAMINATION SHOWED AN INCREASE IN MONONUCLEAR CELL LEUKEMIA AND MALIGNANT NEOPLASMS IN FEMALE RATS AND AN INCREASED INCIDENCE OF PERITONEAL MESOTHELIMA IN MALE RATS COMPARED TO CONTROL GROUPS.

REPORT 11 : EPA RESPONSE TO ITC RECOMMENDATION (49FR200) JAN 84
IN RESPONSE TO ITC 1977 DESIGNATION OF ETHYLENE OXIDE (OF ALKYL EPOXIDE CATEGORY) FOR HEALTH AND ENVIRONMENTAL FATE TESTS, EPA WILL PURSUE ADDITIONAL TESTING ONLY IF OSHA AND OFFICE OF PESTICIDE PROGRAMS CONCLUDE IT IS NECESSARY, SINCE BOTH HAVE REGULATORY ACTIVITIES UNDERWAY AND EXISTING DATA ARE ADEQUATE TO PRODUCT ENVIRONMENTAL FATE.

REPORT 12 : IARC MONOGRAPH 36 FEB 85
INTRAGASTRIC INTUBATION IN RATS INDUCED DOSE-RELATED SQUAMOUS CELL CARCINOMAS OF FORESTOMACH IN RATS. RATS FED DIET FUMIGATED WITH ETO SHOWED NO INCREASED INCIDENCE OF TUMORS. INHALATION INCREASED INCIDENCE OF MONONUCLEAR CELL LEUKEMIA IN BOTH SEXES AND PERITONEAL MESOTHELIOMA IN MALES. ALSO GLIOMAS OF BRAIN AND HIGH INCIDENCE OF LESIONS OF ADRENAL CORTEX WERE INDUCED. SC INJECTION IN MICE PRODUCED LOCAL TUMORS. CPD HAD TERATOGENIC EFFECTS IN MICE AND RATS. A SIGNIFICANT EXCESS OF SPONTANEOUS ABORTIONS WAS SEEN IN A STUDY OF WOMEN ON THE STERILIZING STAFF OF A HOSPITAL WHO WERE EXPOSED TO CPD DURING PREGNANCY.

REPORT 13 : TSCA SECT 8(E) STATUS REPORT 8EHQ-0487-0665 S MAY 87
BSC-125 IS AN ETHYLENE OXIDE/PROPYLENE OXIDE BLOCKED POLYMER USED AS A SURFACTANT. UNDER CONDITIONS OF THE TEST, BSC-125 PRODUCED CORNEAL OPACITY AND CONJUNCTIVAL IRRITATION TO RABBITS' EYES WITHOUT IRRIGATION. THESE EFFECTS PERSISTED TO DAY 21 IN ALL ANIMALS. CONJUNCTIVAL IRRITATION WAS PRODUCED BY BSC-125 IN 3 RABBITS' EYES IRRIGATED 30 SECONDS AFTER ADMINISTRATION. CORNEAL OPACITY AND CONJUNCTIVAL IRRITATION WERE SEEN TO BE REVERSIBLE BY DAYS 7 AND 19, RESPECTIVELY.

REPORT 14 : IARC MONOGRAPH 36 FEB 85
CAUSED DNA DAMAGE IN BACTERIA AND WAS MUTAGENIC TO BACTERIA, PLANTS, FUNGI AND INSECTS. CAUSED CHROMOSOMAL ABERRATIONS IN PLANTS AND HERITABLE TRANSLOCATIONS IN INSECTS. MUTAGENIC IN CULTURED MAMMALIAN CELLS AND INDUCED DNA DAMAGE, CHROMOSOMAL ABERRATIONS AND SCE'S. CPD ALKYLATED DNA IN MICE AND RATS IN VIVO. INDUCED SCE'S, CHROMOSOMAL ABERRATIONS, MICRONUCLEI, DOMINANT LETHAL MUTATIONS AND HERITABLE TRANSLOCATION IN MAMMALS IN-VIVO. INDUCED CHROMOSOMAL ABERRATIONS SCE'S IN PERIPHERAL LYMPHOCYTES.

REPORT 15 : IARC MONOGRAPH 36 FEB 85
SIGNIFICANT INCREASES IN FREQUENCIES OF SISTER CHROMATID EXCHANGES IN PERIPHERAL LYMPHOCYTES ASSOCIATED WITH OCCUPATIONAL EXPOSURE TO ETHYLENE OXIDE. ALSO INDICATIONS OF INCREASES IN FREQUENCIES OF CHROMOSOMAL ABERRATIONS IN PERIPHERAL LYMPHOCYTES.

REPT'D USE : CHEMICAL INTERMEDIATE IN MANUFACTURE OF ANTIFREEZE, POLYESTER RESINS, NON-IONIC SURFACTANTS SPECIALTY SOLVENTS, MEDICAL PRODUCTS, HOSPITAL GASEOUS STERILANT, AND INSECTICIDAL FUMIGANT.

CAS NO : **420-12-2**

CHEM NAME : ***** ETHYLENE SULFIDE**

REPORT 01 : IARC MONOGRAPH 11 1976
ETHYLENE SULFIDE WAS CARCINOGENIC IN A LIMITED STUDY IN RATS BY SC INJECTION: IT PRODUCED LOCAL SARCOMAS.

REPORT 02 : IARC MONOGRAPH 11
ETHYLENE SULFIDE INDUCE SEX-LINKED RECESSIVE LETHALS IN DROSOPHILA.

REPT'D USE : CHEMICAL INTERMEDIATE, ANTIMICROBIAL AGENT AND PLASTIC MONOMER

CAS NO : **96-45-7**
CHEM NAME : ***** ETHYLENE THIOUREA (ETU)**

REPORT 01 : NTP CARC TEST PROGRAM JUL 82
IN A CURRENT NTP BIOASSAY, ETHYLENE THIOUREA IS BEING ADMINISTERED IN THE DIET TO F344 RATS AND B6C3F1 MICE. THE PRECHRONIC TEST PHASE IS IN PROGRESS.

REPORT 02 : NIOSH CURRENT INTELLIGENCE BULLETIN 224 NOV 78
NIOSH LITERATURE REVIEW CLEARLY SHOWS ETHYLENE THIOUREA TO BE A TERATOGEN AND A CARCINOGEN IN LABORATORY RATS WITH SUPPORTIVE STUDIES IN OTHER SPECIES. NIOSH RECOMMENDS THAT ETHYLENE THIOUREA BE HANDLED IN THE WORKPLACE AS IF IT WERE A HUMAN CARCINOGEN AND TERATOGEN. THERE IS NO CURRENT OSHA EXPOSURE STANDARD FOR ETU.

REPORT 03 : IARC MONOGRAPH 7 JUN 74
ETHYLENETHIOUREA (ETU) HAS BEEN TESTED ONLY BY THE ORAL ROUTE IN RATS PRODUCING THYROID CARCINOMAS. THE REPORTED INCREASED INCIDENCE OF LIVER CELL TUMORS IN 2 STRAINS OF MICE FOLLOWING ORAL ADMINISTRATION AWAITS CONFIRMATION. IN A 1979 SUPPLEMENT, IARC DETERMINED THAT THERE IS SUFFICIENT EVIDENCE FOR THE CARCINOGENICITY OF ETU IN ANIMALS.

REPORT 04 : NIOSH SPECIAL HAZARD REVIEW OCT 78
NIOSH RECOMMENDS THAT THE SYNTHETIC RUBBER INDUSTRY CONVERT FROM THE USE OF THE POWDERED FORM OF ETU TO THE ENCAPSULATED FORMULATION TO MINIMIZE ESCAPE OF ETU INTO THE ENVIRONMENT. NIOSH ALSO RECOMMENDS THAT NON-ENCAPSULATED ETU BE LABELLED AS CARCINOGENIC AND TERATOGENIC. EMPLOYEES SHOULD BE INFORMED OF THE POSSIBLE ADDITIVE EFFECTS FROM TAKING ANTITHYROIDAL MEDICATION (PRESENTLY NO OSHA STANDARD).

REPORT 05 : NTP TECHNICAL BULLETIN 2 APR 80
THE COMPOUND WAS FOUND TO BE NEGATIVE IN THE SALMONELLA TYPHIMURIUM MUTAGENICITY ASSAY IN ONE LABORATORY.

REPORT 06 : MULTI-AGENCY CROSS CHECK
Y HHS MAGUIRE
N OSHA CARC
Y EPA CAG
Y EPA RCRA
N EPA CLEAN AIR
N EPA CLEAN WATER

REPORT 07 : IARC MONOGRAPH 7 JUN 74
ETU WAS TERATOGENIC IN RATS AT DOSES THAT PRODUCED NO APPARENT MATERNAL TOXICITY OR FETAL DEATHS, SUGGESTING

THAT TRANSPLACENTAL TRANSFER OF THE SUBSTANCE MAY TAKE PLACE.

REPORT 09 : IARC SUPP 4 OCT 82
NO CASES OF THYROID CANCER NOTED IN STUDY OF 1,929 WORKERS KNOWN TO HAVE WORKED WITH ETU AT SOME TIME IN SEVERAL RUBBER MFG PLANTS. ETU PRODUCED CARCINOMAS IN MICE AND RATS AFTER ORAL ADMINISTRATION. LOWER DOSES PRODUCED FOLLICULAR HYPOPLASIA IN MICE. A FEEDING STUDY IN HAMSTERS SHOWED NO EFFECT. THE EVIDENCE FOR ACTIVITY OF ETU IN SHORT-TERM TESTS IS LIMITED DUE TO CONFLICTING REPORTS FROM DIFFERENT INVESTIGATORS.

REPT'D USE : NEOPRENE RUBBER ACCELERATOR, SHOE SOLES, CONTAINER CLOSURES, AEROSOL, PESTICIDES, ELECTROPLATING AND CHEMICAL INTERMEDIATE

CAS NO : **149-57-5**
CHEM NAME : ***** 2-ETHYLHEXANOIC ACID**

REPORT 01 : PROPOSED TEST RULE, 50FR20678 MAY 85
UNDER TSCA 4(A) EPA IS PROPOSING THAT MFRS AND PROCESSORS OF CPD CONDUCT HEALTH EFFECTS TESTS, INCLUDING PHARMACOKINETIC STUDIES AND 90-DAY SUBCHRONIC TOXICITY AND DEVELOPMENTAL TOXICITY TESTS. THIS IS IN RESPONSE TO ITC'S DESIGNATION OF CPD FOR PRIORITY TESTING

REPORT 02 : EPA FINAL TEST RULE 51FR40318 NOV 86
EPA IS ISSUING A FINAL TEST RULE UNDER TSCA SECT 4 REQUIRING MFRS AND PROCESSORS OF 2-EHA TO CONDUCT 90-DAY SUBCHRONIC TOXICITY, DEVELOPMENTAL TOXICITY AND PHARMACOKINETICS STUDIES.

REPORT 03 : TSCA SECT 8(E) STATUS REPORT 8EHQ-0587-067 S JUN 87
DEHP ACTS AS A TERATOGEN BY IN VIVO HYDROLYSIS TO 2-ETHYLHEXANOL (2-EHXO) WHICH IN TURN IS METABOLIZED TO 2-ETHYLHEXANOIC ACID (2-EHXA), THE PROXIMATE TERATOGEN. ON AN EQUIMOLAR BASIS, DEHP WAS LEAST POTENT, 2-EHXO WAS INTERMEDIATE AND 2-EXHA WAS MOST POTENT AS A TERATOGEN. ALL 3 WERE POTENTIATED BY CAFFEINE. VALPROIC ACID, AN ISOMER OF 2-EXHA, PRODUCED SIMILAR DEFECTS AND WAS APPROXIMATELY 2X AS POTENT AS 2-EXHA.

REPT'D USE : VINYL STABILIZER, PAINT AND INK DRYERS, PEROXIDE CATALYSTS AND CATALYST

CAS NO : **104-76-7**
CHEM NAME : ***** 2-ETHYLHEXANOL**

REPORT 01 : 52 FR 28609 AUG 87
EPA ISSUING FINAL TEST RULE UNDER SECT 4 TSCA REQUIRING MANUFACTURERS AND PROCESSORS TO CONDUCT 2-YR ONCOGENICITY BIOASSAY.

CAS NO : **97-63-2**
CHEM NAME : ***** ETHYL METHACRYLATE**

REPORT 01 : EPA RCRA HEALTH AND ENV. EFFECTS PROFILE OCT 80
INFORMATION ON THE CARCINOGENIC AND MUTAGENIC EFFECTS OF ETHYL METHACRYLATE WAS NOT FOUND IN THE AVAILABLE LITERATURE. ETHYL METHACRYLATE HAS, HOWEVER, BEEN SHOWN TO CAUSE TERATOGENIC EFFECTS IN RATS.

REPORT 06 : MULTI-AGENCY CROSS CHECK
Y HHS MAGUIRE
N OSHA CARC
N EPA CAG
Y EPA RCRA
N EPA CLEAN AIR
N EPA CLEAN WATER
REPT'D USE : PLASTIC MONOMER, DENTAL APPLIANCES AND GLASS SUBSTITUTES

CAS NO : **62-50-0**
CHEM NAME : ***** ETHYL METHANESULFONATE (EMS)**

REPORT 01 : IARC MONOGRAPH 7 1974
ETHYL METHANESULFONATE (EMS) IS CARCINOGENIC IN MICE AND RATS FOLLOWING SUBCUTANEOUS OR INTRAPERITONEAL INJECTION, THE ONLY SPECIES AND ROUTES TESTED. IT PRODUCED MAINLY LUNG AND KIDNEY TUMORS IN BOTH SPECIES. IT IS CARCINOGENIC FOLLOWING ADMINISTRATION OF A SINGLE DOSE. IN A 1979 SUPPLEMENT, IARC DETERMINED THAT THERE IS SUFFICIENT EVIDENCE FOR THE CARCINOGENICITY OF EMS IN ANIMALS.

REPORT 06 : MULTI-AGENCY CROSS CHECK
Y HHS MAGUIRE
N OSHA CARC
Y EPA CAG
Y EPA RCRA
N EPA CLEAN AIR
N EPA CLEAN WATER

REPT'D USE : LABORATORY AGENT, EXPERIMENTAL INSECT CHEMOSTERILANT, EXPERIMENTAL DRUG AND CHEMOSTERILANT

CAS NO : **15336-82-0**
CHEM NAME : ***** 5-ETHYL-5-METHYL-1,3-BIS-(OXIRANYLMETHYL) -2,4-IMIDAZOLIDINEDIONE**

REPORT 01 : EPA TSCA 8(E) NOTICE 8EHQ-1081-0418 NOV 81
CIBA-GEIGY PROVIDED RESULTS FROM IN VITRO BALB/3T3 MOUSE CELL TRANSFORMATION ASSAY AND IN VIVO CHINESE HAMSTER BONE MARROW CYTOGENETICS ASSAY. SIGNIFICANT DOSE-DEPENDENT TREND IN TRANSFORMATION FREQUENCY OBSERVED IN MOUSE CELL TRANSFORMATION ASSAY. IN HAMSTER BONE MARROW CYTOGENETICS ASSAY, A SIGNIFICANT DIFFERENCE IN CHROMOSOMAL ABERRATION FREQUENCY OBSERVED BETWEEN CONTROL ANIMALS AND ANIMALS RECEIVING HIGHEST ORAL DOSE. OTHER PREVIOUSLY PERFORMED MUTAGENICITY TESTS, WITH EXCEPTION OF IN VITRO MOUSE LYMPHOMA ASSAY, HAD GIVEN RESULTS WHICH WERE EITHER NEGATIVE OR EQUIVOCAL.
REPT'D USE : INDUSTRIAL ADHESIVE SYSTEMS

CAS NO : **993-43-1**
CHEM NAME : ***** ETHYLPHOSPHONOTHIOIC DICHLORIDE (EPTD)**

REPORT 01 : EPA TSCA 8(E) NOTICE 8EHQ-0981-0412
THE ETHYL CORP SUBMITTED REPORTS FROM A BATTERY OF IN VITRO AND IN VIVO GENETIC TOXICOLOGY ASSAYS OF EPTD. EPTD WAS POSITIVE WITH METABOLIC ACTIVATION, IN BOTH A MAMMALIAN CELL SPECIFIC LOCUS MUTATION ASSAY AND A MAMMALIAN CELL TRANSFORMATION ASSAY. EPTD WAS CLEARLY NEGATIVE IN SALMONELLA, WITH AND WITHOUT ACTIVATION, AND IN AN IN VIVO MOUSE MICRONUCLEUS ASSAY. RESULTS FROM A RAT DOMINANT LETHAL ASSAY WERE INCONCLUSIVE.
REPT'D USE : INTERMEDIATE IN MFR OF INSECTICIDES

CAS NO : **12367-47-4**
CHEM NAME : ***** ETHYL SELENAC**

REPORT 01 : IARC MONOGRAPH 12 1976
ETHYL SELENAC HAS BEEN TESTED BY ORAL ADMINISTRATION AND BY SINGLE SUBCUTANEOUS INJECTION IN TWO STRAINS OF MICE. IT PRODUCED AN INCREASED INCIDENCE OF LIVER-CELL TUMORS IN MALES OF ONE STRAIN BUT DID NOT SIGNIFICANTLY INCREASE THE INCIDENCE OF TUMORS IN THE OTHER. THE AVAILABLE DATA ARE INSUFFICIENT TO ALLOW AN EVALUATION

		OF THE CARCINOGENICITY OF THIS COMPOUND TO BE MADE.
REPT'D USE	:	RUBBER ACCELERATOR. NO KNOWN CURRENT PRODUCTION. NO LISTING ON TSCA INVENTORY.

CAS NO : **20941-65-5**

CHEM NAME : ***** ETHYL TELLURAC**

REPORT 01 : NCI CARC TRS 152 DHEW PUB NO 79-1708
UNDER THE CONDITIONS OF THIS BIOASSAY, ETHYL TELLURAC WAS NOT CARCINOGENIC FOR F344 RATS OR B6C3F1 MICE OF EITHER SEX. THE INCIDENCE OF MESOTHELIOMAS IN DOSED MALE RATS AND THE INCIDENCE OF ADENOMAS OF THE LACRIMAL (HARDERIAN) GLAND OF THE EYE IN DOSED MICE OF EITHER SEX PROVIDED EVIDENCE WHICH WAS SUGGESTIVE BUT UNDER THE CONDITIONS OF THE BIOASSAY INSUFFICIENT TO ESTABLISH THE CARCINOGENICITY OF ETHYL TELLURAC IN THESE ANIMALS (FEEDING STUDY).

REPORT 02 : IARC MONOGRAPH 12 1976
ETHYL TELLURAC HAS BEEN TESTED BY ORAL ADMINISTRATION AND BY SINGLE SUBCUTANEOUS INJECTION IN TWO STRAINS OF MICE. THERE WAS A SMALL BUT SIGNIFICANT INCREASE IN THE NUMBER OF LUNG TUMORS IN MALES AND FEMALES OF ONE STRAIN COMBINED AFTER ITS ORAL ADMINISTRATION. THE DATA ARE INSUFFICIENT TO ALLOW AN EVALUATION OF THE CARCINOGENICITY OF THIS COMPOUND.

REPORT 03 : NTP TECHNICAL BULLETIN 5 AUG 81
ETHYL TELLURAC WAS NEGATIVE IN THE SALMONELLA TYPHIMURIUM MUTAGENICITY ASSAY IN TWO LABORATORIES.

REPORT 06 : MULTI-AGENCY CROSS CHECK
N HHS MAGUIRE
Y OSHA CARC
N EPA CAG
N EPA RCRA
N EPA CLEAN AIR
N EPA CLEAN WATER

REPT'D USE : RUBBER CURING AGENT

CAS NO : **611-14-3**

CHEM NAME : ***** O-ETHYL TOLUENE**

REPORT 01 : 52FR2522 FINAL TEST RULE JAN 87
EPA IS ISSUING A FINAL TEST RULE UNDER SECT 4(A) TSCA THAT SPECIFIES TEST STANDARDS AND REPORTING REQUIREMENTS FOR C9 FRACTION OF AROMATIC HYDROCARBONS.

CAS NO : **25550-14-5**
CHEM NAME : ***** ETHYLTOLUENES**

REPORT 01 : EPA PROPOSED TEST RULE, 48FR23088 MAY 83
EPA'S RESPONSE TO ITC'S DESIGNATING ET FOR PRIORITY CONSIDERATION FOR ENVIRONMENTAL AND HEALTH EFFECTS TESTING UNDER TSCA 4(A). EPA IS PROPOSING THAT MFRS AND PROCESSORS OF THE AROMATIC HYDROCARBON FRACTIONS CONTAINING THE O-,M- AND P- ISOMERS AS PRIMARY COMPONENTS, TEST FOR HEALTH EFFECTS INCLUDING MUTAGENICITY, TERATOGENICITY, NEUROTOXICITY, REPRODUCTIVE EFFECTS AND CARCINOGENICITY.

REPORT 02 : 52FR2522 FINAL TEST STANDARDS AND JAN 87
REPORTING REQUIREMENTS
EPA IS ISSUING A FINAL TEST RULE UNDER TSCA 4(A) FOR TESTING OF THE C9 AROMATIC HYDROCARBON FRACTION.
REPT'D USE : GASOLINE BLENDING, SOLVENT, PAINT AND VARNISH FORMULATIONS, HYDROCARBON LUBRICANTS FOR REFRIGERATORS, PAINT THINNERS, PRINTING INKS AND PESTICIDES

CAS NO : **297-76-7**
CHEM NAME : ***** ETHYNODIOL DIACETATE**

REPORT 01 : IARC MONOGRAPH 21 1979
ETHYNODIOL DIACETATE WAS TESTED IN MICE, RATS, MONKEYS ALONE OR IN COMBINATION WITH ESTROGENS BY ORAL ADMINISTRATION. IN CASTRATED MALE MICE. IT INCREASED THE INCIDENCE OF MAMMARY TUMORS, AND IN MALE RATS IT PRODUCED BENIGN MAMMARY TUMORS. IN COMBINATION WITH ESTROGENS, IT INCREASED THE INCIDENCE OF PITUITARY TUMORS IN MICE AND OF MALIGNANT MAMMARY TUMORS IN MALE AND FEMALE RATS.

REPORT 02 : IARC MONOGRAPH 21 1979
CYCLIC INTAKE OF A MIXTURE OF ETHYNDIOL DIACETATE AND MESTRANOL DAILY FOR AT LEAST SIX MONTHS DID NOT SIGNIFICANTLY INCREASE THE FREQUENCY OF CHROMOSOME ABERRATIONS IN CULTURED LYMPHOCTES FROM 35 WOMEN. THE COMPOUND WAS REPORTED TO BE EMBRYOLETHAL FOR PRE- AND POSTIMPLANTATION EMBRYOS AND TO HAVE TERATOGENIC EFFECTS IN SOME SPECIES.
REPT'D USE : ORAL CONTRACEPTIVES

CAS NO : **97-53-0**
CHEM NAME : ***** EUGENOL**

REPORT 01 : NTP TECHNICAL REPORT 223 NIH PUB 84-1779 DEC 83
UNDER CONDITIONS OF THIS BIOASSAY, EUGENOL ADMINISTERED IN THE FEED WAS NOT CARCINOGENIC FOR F344 RATS OF EITHER

SEX. AN INCREASE IN LIVER TUMORS IN B6C3F1 MICE OF BOTH SEXES WAS CONSIDERED EQUIVOCAL.

REPORT 02 : NTP TECHNICAL BULLETIN 3 DEC 80
THE COMPOUND WAS FOUND TO BE NEGATIVE IN THE SALMONELLA TYPHIMURIUM MUTAGENICITY ASSAY IN ONE LABORATORY.

REPORT 04 : IARC MONOGRAPH 36 FEB 85
TESTED IN MICE AND RATS BY ORAL ADMINISTRATION OF DIET CONTAINING HIGH DOES OF EUGENOL. IN MICE, SIGNIFICANT INCREASE IN INCIDENCE OF LIVER TUMORS IN FEMALES; IN MALES, INCREASE SIGNIFICANT ONLY FOR THOSE RECEIVING LOWER DOSE. IN RATS, NO INCREASED INCIDENCE OF TUMORS OBSERVED. OTHER STUDIES IN MICE BY ORAL ADMINISTRATION, SKIN APPLICATION AND INTRAPERITONEAL INJECTION INADEQUATE FOR EVALUATION OF CARCINOGENICITY.

REPORT 05 : IARC MONOGRAPH 36 FEB 85
GAVE BOTH POSITIVE AND NEGATIVE RESULTS IN TESTS FOR DNA DAMAGE IN BACTERIA. WAS NOT MUTAGENIC IN SEVERAL STUDIES IN BACTERIA. WAS NOT ACTIVE IN HOST-MEDIATED ASSAY IN MICE, NOR URINE OF RATS TREATED WITH EUGENOL MUTAGENIC. INDUCED CHROMOSOMAL ABERRATIONS AND SMALL INCREASE IN SISTER CHROMATID EXCHANGES IN MAMMALIAN CELLS IN VITRO.

REPORT 07 : IARC MONOGRAPH 36 FEB 85
IN TWO-STAGE MOUSE-SKIN ASSAY, 2',3'-EPOXYEUGENOL, IN-VITRO METABOLITE OF EUGENOL, SHOWED INITIATING ACTIVITY. 2',3'-EPOXYEUGENOL MUTAGENIC TO BACTERIA. TWO URINARY METABOLITES OF EUGENOL, 3-PIPERIDYL-1-(3'-METHOXY-4'-HYDROXYPENYL)-1-PROPANONE AND 3-PYRROLIDINYL-1-(3'-METHOXY-4'-HYDROXYPHENYL)-1-PROPANONE NOT MUTAGENIC TO BACTERIA OR IN HOSTMEDIATED ASSAY.

REPT'D USE : FLAVORING AGENT-CLOVE, PERFUMES, DENTAL ANALGESIC, PAINTS/VARNISHES AND MFR OF VANILLAN

CAS NO : **55-38-9**
CHEM NAME : ***** FENTHION**

REPORT 01 : NCI CARC TRS 103 DHEW PUB NO 79-1353
UNDER THE CONDITIONS OF THIS BIOASSAY, FENTHION WAS NOT CARCINOGENIC FOR MALE OR FEMALE F344 RATS OR FOR FEMALE B6C3F1 MICE. THE INCREASED INCIDENCE OF SARCOMAS, FIBROSARCOMAS, AND ESPECIALLY RHABDOMYOSARCOMAS OF THE INTEGUMENTARY SYSTEM IN THE MALE B6C3F1 MICE SUGGESTED THAT THE TEST CHEMICAL WAS CARCINOGENIC IN THESE ANIMALS (FEEDING STUDY).

REPT'D USE : INSECTICIDE AND PARASITICIDE

CAS NO : **14484-64-1**
CHEM NAME : ***** FERBAM**

REPORT 01 : IARC MONOGRAPH 12 1976
FERBAM HAS BEEN TESTED BY ORAL ADMINISTRATION IN MICE AND RATS AND BY SINGLE SUBCUTANEOUS INJECTION IN MICE. ALTHOUGH NO CARCINOGENIC EFFECT WAS OBSERVED IN THESE TESTS, THE AVAILABLE DATA ARE INSUFFICIENT FOR AN EVALUATION OF THE CARCINOGENICITY OF THIS COMPOUND TO BE MADE.

REPORT 02 : IARC MONOGRAPH 12 1976
COLOR MUTANTS AND REVERSE MUTATIONS WERE IDENTIFIED IN A. NIGER TREATED WITH FERBAM; METABOLIC ACTIVATION SYSTEMS WERE NOT USED.
REPT'D USE : FUNGICIDE

CAS NO : **1309-37-1**
CHEM NAME : ***** FERRIC OXIDE**

REPORT 01 : IARC MONOGRAPH 1 1972
FERRIC OXIDE GIVEN BY INHALATION OR BY INTRATRACHEAL ROUTE HAS NOT BEEN FOUND TO BE CARCINOGENIC IN THE HAMSTER, THE MOUSE OR THE GUINEA PIG. SOME STUDIES OF METAL WORKERS EXPOSED TO FERRIC OXIDE DUSTS HAVE SHOWN AN INCREASED INCIDENCE OF LUNG CANCER, WHILE OTHERS HAVE NOT. THE INFLUENCE OF FACTORS IN THE WORKPLACE OTHER THAN FERRIC OXIDE CANNOT BE ELIMINATED.
REPT'D USE : PAINTS/VARNISHES, FR CHEMICALS, AEROSOLS, FEED ADDITIVE, PIGMENTS, POLISHING AGENT FOR GLASS/METALS, MAGNETS/MAGNETIC TAPE AND CHEMICAL REAGENT/CATALYST

CAS NO : **102-54-5**
CHEM NAME : ***** FERROCENE**

REPORT 01 : NTP TECHNICAL BULLETIN 8 JUL 82
FERROCENE CAUSED AN INCREASED NUMBER OF SEX-LINKED RECESSIVE LETHALS IN DROSOPHILA.

CAS NO : **443-48-1**
CHEM NAME : ***** FLAGYL**

REPORT 02 : IARC MONOGRAPH 13 1977
METRONIDAZOLE IS CARCINOGENIC IN MICE AFTER ITS ORAL ADMINISTRATION: IT SIGNIFICANTLY INCREASED THE INCIDENCE OF LUNG TUMORS IN BOTH SEXES AND THE INCIDENCE OF LYMPHOMAS IN FEMALES. ITS ORAL ADMINISTRATION TO RATS

INCREASED THE INCIDENCE AND MULTIPLICITY OF MAMMARY FIBROADENOMAS.

REPORT 06 : MULTI-AGENCY CROSS CHECK
Y HHS MAGUIRE
N OSHA CARC
N EPA CAG
N EPA RCRA
N EPA CLEAN AIR
N EPA CLEAN WATER

REPT'D USE : ANTIPROTOZOAL AGENT

CAS NO : **2164-17-2**

CHEM NAME : ***** FLUOMETURON**

REPORT 01 : IARC MONOGRAPH 30 JAN 83
TESTED FOR CARCINOGEICITY IN MICE AND RATS BY ADMINISTRATION IN DIET. AN INCREASED INCIDENCE OF LIVER-CELL TUMOURS WERE FOUND IN MALE MICE. RESULTS WERE NOT INDICATIVE OF A CARCINOGENIC EFFECT IN FEMALE MICE AND BOTH SEXES OF RATS.

REPORT 02 : IARC MONOGRAPH 30 JAN 83
WEAKLY OR NONMUTAGENIC TO SALMONELLA TYPHIMURIUM. NOT ACTIVE IN INDUCING MITOTIC GENE CONVERSION IN YEAST. MITOTIC CHROMOSOMAL ABERRATIONS OBSERVED IN TREATED VICIA FABA ROOT TIPS AND COTTON SEEDS. IN ALL POSITIVE STUDIES COMMERCIAL FORMULATIONS KNOWN OR SUSPECTED TO HAVE BEEN USED, SUGGESTING THAT THE OBSERVED ACTIVITY MAY BE ATTRIBUTABLE TO OTHER INGREDIENTS. INSUFFICIENT DATA TO EVALUATE MUTAGENICITY OF PURE FLUOMETURON, BUT LIMITED EVIDENCE FOR COMMERCIAL FORMULATIONS OF THIS COMPOUND MUTAGENIC IN ANIMALS.

REPT'D USE : HERBICIDE

CAS NO : **206-44-0**

CHEM NAME : ***** FLUORANTHENE**

REPORT 01 : IARC MONOGRAPH 32 DEC 83
TESTED FOR CARCINOGENICITY BY SKIN APPLICATION IN MICE IN TWO STUDIES AND NO TUMORIGENIC EFFECT OBSERVED. TESTED IN MOUSE-SKIN INITIATION PROMOTION ASSAY AND INACTIVE AS INITIATOR. A STUDY IN MICE BY SUBCUTANEOUS ADMINISTRATION CONSIDERED INADEQUATE FOR EVALUATION. WHEN FLUORANTHENE ADMINISTERED TO MICE BY SKIN APPLICATION TOGETHER WITH BENZO(A)PYRENE, AN EXCESS OF SKIN TUMORS PRODUCED OVER THAT INDUCED BY THE SAME DOSE OF BENZO(A)PYRENE ALONE. MUATAGENIC TO SALMONELLA

TYPHIMURIUM AND TO CULTURED HUMAN LYMPHOBIASTOID CELLS IN PRESENCE OF AN EXOGENOUS METABOLIC SYSTEM.
REPT'D USE : INCOMPLETE COMBUSTION PRODUCT, CIGARETTE SMOKE, RAIN WATER, SLUDGE, AND ENGINE EXHAUST

CAS NO : **86-73-7**
CHEM NAME : ***** FLUORENE**

REPORT 01 : IARC MONOGRAPH 32 DEC 83
TESTED FOR CARCINOGENICITY IN MICE BY SKIN APPLICATION AND BY SUBCUTANEOUS ADMINISTRATION AND IN FEMALE RATS BY ORAL ADMINISTRATION IN DIET. STUDIES CONSIDERED INADEQUATE FOR EVALUATION. NOT MUTAGENIC TO SALMONELLA TYPHIMURIUM. IN THE ONE STUDY, IT DID NOT INDUCE UNSCHEDULED DNA SYNTHESIS IN PRIMARY RAT HEPATOCYTE CULTURES
REPT'D USE : INCOMPLETE COMBUSTION PRODUCT, CIGARETTE SMOKE AND SLUDGE

CAS NO : **CLASS-21**
CHEM NAME : ***** FLUOROALKANES**

REPORT 01 : PROPOSED TEST RULE - EXT'N OF COMMENT PERIOD; DEC 85
50FR53348
EPA EXTENDS COMMENT PERIOD FOR PROPOSED TEST RULE IN RESPONSE TO REQUEST FROM AN INDUSTRY GROUP. PROPOSED TESTING IS MORE EXTENSIVE THAN WAS PROPOSED IN PREVIOUS NEGOTIATED TESTING AGREEMENT, AND EXTRA TIME IS NEEDED TO REVIEW APPROPRIATENESS OF REQUIRED TESTS.
REPT'D USE : FLUORINATED PLASTICS

CAS NO : **393-52-2**
CHEM NAME : ***** 2-FLUOROBENZOYL CHLORIDE**

REPORT 01 : NTP TECHINICAL BULLETIN 3 DEC 80
THE COMPOUND WAS FOUND TO BE POSITIVE IN THE SALMONELLA TYPHIMURIUM MUTAGENICITY ASSAY IN ONE LABORATORY.
REPT'D USE : ORGANIC SYNTHESIS

CAS NO : **70-34-8**
CHEM NAME : ***** 1-FLUORO-2,4-DINITROBENZENE**

REPORT 01 : NTP TECHNICAL BULLETIN 3/8DEC 80, JUL 82
1-FLUORO-2,4-DINITROBENZENE WAS POSITIVE IN THE SALMONELLA ASSAY, BUT NEGATIVE IN DROSOPHILA WHEN TESTED FOR

SEX-LINKED RECESSIVE LETHALS.
REPT'D USE : LABORATORY CHEMICAL

CAS NO : **CLASS-27**
CHEM NAME : ***** FLUOROSILICON FLUIDS AND GREASES**

REPORT 01 : 8EHQ-1186- 0644, EPA STATUS REPORT DEC 86
WHEN THESE CPDS WERE HELD AT HIGH TEMPERATURES (400 -500 DEG. F) IN AIR FOR SEVERAL HOURS, TRIFLUOROPROPOPION-ALDEHYDE WAS DETECTED BY IR SPECTROSCOPY. ACCORDING TO DOW CHEM. CO., AN INITIAL RANGE FINDING ACUTE ORAL TOXICITY STUDY IN RATS SHOWED THE CPD TO BE HIGHLY TOXIC. THE ALD50 APPEARS TO BE SOMEWHAT LESS THAN 15 MG/KG BODY WEIGHT.
REPT'D USE : SYNTHETIC MOTOR OILS AND GREASES

CAS NO : **51-21-8**
CHEM NAME : ***** 5-FLUOROURACIL**

REPORT Ol : IARC MONOGRAPH 26 MAY 81
5-FLUOROURACIL WAS TESTED BY IV ADMINISTRATION IN MICE AND RATS AND BY ORAL ADMINISTRATION IN RATS. NO EVIDENCE OF CARCINOGENICITY WAS FOUND, BUT THE STUDIES SUFFERED FROM LIMITATIONS WITH REGARD TO DURATION OR DOSE. THERE WAS NO EVIDENCE FOR THE CARCINOGENICITY OF 5-FLUOROURACIL IN THE LIMITED STUDIES IN EXPERIMENTAL ANIMALS. THE DATA FROM CASE REPORTS IN HUMANS ARE INSUFFICIENT TO ARRIVE AT A CONCLUSION.

REPORT 02 : IARC MONOGRAPH 26 MAY 81
5-FLUOROURACIL CAN INDUCE EMBRYOTOXIC AND TERATOGENIC EFFECTS IN SEVERAL ANIMAL SPECIES AND MAY BE EMBRYOLETHAL IN MONKEYS AT DOSES NOT TOXIC TO THE MOTHER. THE AVAILABLE EXPERIMENTAL DATA ON THE MUTAGE-NICITY OF 5-FLUOROURACIL ARE INCONCLUSIVE; THAT DRUG DID NOT INDUCE TRANSFORMATION IN A MOUSE CELL LINE.
REPT'D USE : ANTINEOPLASTIC AGENT

CAS NO : **50-00-0**
CHEM NAME : ***** FORMALDEHYDE**

REPORT 01 : CIIT 24 MONTH TOXICOLOGY STUDY DEC 81
F344 RATS AND B6C3F1 MICE WERE EXPOSED TO 15, 6, 2, AND 0 PPM. SQUAMOUS CELL CARCINOMAS OF THE NASAL CAVITY WERE OBSERVED IN 2 MICE IN THE 15 PPM GROUP, 2 RATS IN THE 6 PPM,

AND IN 103 RATS FROM THE 15 PPM GROUP. TWO NASAL CARCINOMAS AND 2 UNDIFFERENTIATED CANCERS WERE ALSO OBSERVED IN THE 15 PPM GROUP.

REPORT 02 : NIOSH CURRENT INTELLIGENCE BULLETIN 34 APR 81
THE NIOSH RECOMMENDS THAT FORMALDEHYDE BE HANDLED AS A POTENTIAL OCCUPATIONAL CARCINOGEN AND THAT APPROPRIATE CONTROLS BE USED TO REDUCE WORKER EXPOSURE. THESE RECOMMENDATIONS ARE BASED PRIMARILY ON A CIIT STUDY IN WHICH LAB RATS AND MICE EXPOSED TO HCHO VAPOR DEVELOPED NASAL CANCER, AND ARE SUPPORTED BY AN NYU STUDY WHERE RATS EXPOSED TO A MIXTURE OF HCHO AND HCL ACID VAPORS DEVELOPED NASAL CANCER. HCHO HAS ALSO BEEN SHOWN TO BE A MUTAGEN IN SEVERAL SHORT TERM LAB STUDIES.

REPORT 03 : ENV SCI LAB REPORT TO AMER CANCER SOC FEB 81
FORMALDEHYDE IS A CARCINOGEN IN RATS AND, DATA SUGGEST, IN MICE, AT EXPOSURE LEVELS COMPARABLE TO THOSE FOUND IN SOME HOME AND WORK ENVIRONMENTS. THESE FINDINGS INDICATE THAT EFFECTIVE CONTROLS SHOULD BE INITIATED TO REDUCE OR ELIMINATE HUMAN EXPOSURES TO HCHO. IN RECOMMENDING CONTROLS, WE EMPHASIZE THAT CANCER OCCURRED IN ANIMALS AT LEVELS NOT UNLIKE THOSE WHICH CAN BE FOUND IN SOME HUMAN EXPOSURES.

REPORT 04 : NIOSH CRITERIA DOCUMENT DEC 76
NIOSH RECOMMENDS THAT EXPOSURE TO FORMALDEHYDE BE CONTROLLED SO THAT NO EMPLOYEE IS EXPOSED TO FORMALDEHYDE AT A CONCENTRATION GREATER THAN 1.2 MILLIGRAMS PER CUBIC METER OF AIR (1 PPM) FOR ANY 30-MINUTE SAMPLING PERIOD. OCCUPATIONAL EXPOSURE TO FORMALDEHYDE IS DEFINED AS EXPOSURE TO FORMALDEHYDE IN AIR AT A CONCENTRATION IN EXCESS OF 0.6 MG/CUBIC METER (0.5 PPM), BASED ON A 30-MINUTE SAMPLING PERIOD, OR BY CONTACT WITH FORMALDEHYDE IN LIQUID OR SOLID FORM BASED ON EYE AND THROAT IRRITATION AND RESPIRATORY DISORDERS. OSHA STANDARD-3 PPM TWA.

REPORT 05 : REPORT OF THE FEDERAL PANEL ON FORMALDEHYDE NOV 80
THE PANEL CONCLUDED THAT DEFINITIVE EXPERIMENTS EXIST WHICH DEMONSTRATE THE MUTAGENICITY AND CARCINOGENICITY OF FORMALDEHYDE UNDER LABORATORY CONDITIONS. FORMALDEHYDE INDUCED BOTH GENE MUTATIONS AND CHROMOSOMAL ABERRATIONS IN A VARIETY OF TEST SYSTEMS. BY INHALATION, FORMALDEHYDE CAUSED CANCER OF THE NOSE IN RATS. EPIDEMIOLOGIC STUDIES ON EXPOSED HUMAN POPULATIONS ARE IN PROGRESS AS ARE TERATOGENICITY AND REPRODUCTIVE DISORDER STUDIES. THE PANEL CONCLUDES THAT IT IS PRUDENT TO REGARD FORMALDEHYDE AS POSING A CARCINOGENIC RISK TO MAN.

REPORT 06 : MULTI-AGENCY CROSS CHECK
Y HHS MAGUIRE
N OSHA CARC
Y EPA CAG
Y EPA RCRA
Y EPA CLEAN AIR
N EPA CLEAN WATER

REPORT 07 : EPA RCRA HEALTH AND ENV. EFFECTS PROFILE OCT 80
THE EPA CAG RECENTLY CONCLUDED THAT THERE IS SUBSTANTIAL EVIDENCE THAT FORMALDEHYDE IS LIKELY TO BE A HUMAN CARCINOGEN. THERE IS AN EXTENSIVE DATA BASE SHOWING THAT HCHO IS MUTAGENIC IN MICROORGANISMS, PLANTS, INSECTS, CULTURED MAMMALIAN CELLS, AND MICE.

REPORT 08 : NTP TECHNICAL BULLETIN 3 DEC 80
FORMALDEHYDE WAS FOUND TO BE POSITIVE IN THE SALMONELLA TYPHIMURIUM MUTAGENICITY ASSAY IN THREE LABORATORIES. TERATOLOGY ASSAYS SHOWED THAT ORAL ADMINISTRATION OF FORMALDEHYDE TO MICE WAS NOT ASSOCIATED WITH PRENATAL DEATH OR INCREASED MALFORMATIONS AT DOSES TOXIC TO THE MOTHER.

REPORT 09 : IARC MONOGRAPH 29 1982
FORMALDEHYDE WAS TESTED BY INHALATION EXPOSURE IN TWO STRAINS OF RATS AND ONE STRAIN OF MICE. A SIGNIFICANT INCIDENCE OF SQUAMOUS-CELL CARCINOMAS OF THE NASAL CAVITY WAS INDUCED IN BOTH STRAINS OF RATS BUT NOT IN MICE. THERE IS SUFFICIENT EVIDENCE THAT FORMALDEHYDE GAS IS CARCINOGENIC TO RATS.

REPORT 10 : IARC MONOGRAPH 29 1982
DNA DAMAGE HAS BEEN OBSERVED IN BACTERIA, YEAST, AND MAMMALIAN CELLS EXPOSED TO FORMALDEHYDE. THE CHEMICAL IS MUTAGENIC TO BACTERIA YEASTS, AND DROSOPHILA, BUT NOT TO THE SILKWORM NOR TO MAMMALIAN CELLS IN CULTURE. ONE POSITIVE AND ONE NEGATIVE STUDY ON THE INDUCTION OF DOMINANT LETHAL MUTATIONS IN MICE HAVE BEEN REPORTED. THE AVAILABLE DATA ARE INSUFFICIENT TO EVALUATE THE EMBRYOTOXICITY OR TERATOGENICITY OF FORMALDEHYDE.

REPORT 11 : EPA TSCA 8(E) NOTICE 8EHQ-1079-0314
CHEMICAL INDUSTRY INSTITUTE OF TOXICOLOGY (CIIT) PROVIDED A SUMMARY REPORT OF INTERIM SACRIFICE PATHOLOGY RESULTS FROM A 24-MONTH INHALATION STUDY OF FORMALDEHYDE IN MICE AND RATS. IT IS REPORTED THAT NASAL SQUAMOUS-CELL CARCINOMAS WERE DETECTED IN THREE RATS SACRIFICED AT THE HIGHEST DOSE TESTED - 15 PPM. NO SQUAMOUS CELL CARCINOMAS OF THE NASAL CAVITY HAVE BEEN OBSERVED IN THE EXPOSED OR CONTROL MICE, NOR IN CONTROL RATS IN THIS STUDY.

REPORT 12 : EPA TSCA 8(E) NOTICE 8EHQ-1179-0314
CIBA-GEIGY PROVIDED INTERIM RESULTS FROM AN IMPLANTATION STUDY OF STRONTIUM CHROMATE IN RATS WHICH INDICATE THAT LESIONS CONSISTENT WITH THE APPEARANCE OF SQUAMOUS CELL CARCINOMAS WERE OBSERVED GROSSLY IN THE LEFT LUNG OF 21/39 ANIMALS WHO DIED AFTER ONE YEAR. SUBMITTER NOTES THAT THE DIAGNOSIS OF SQUAMOUS CELL CARCINOMAS HAS NOT YET BEEN CONFIRMED HISTOLOGICALLY.

REPT'D USE : PLASTIC MONOMER, ADHESIVES, LEATHER TANNING, PHOTOGRAPHIC, TEXTILE FINISHES, AGRICULTURAL, DEODORIZING AGENT, PAINTS, SHAMPOO, FOAM INSULATION, PLYWOOD/PARTICLEBOARD, LAB REAGENT AND BIOLOGICAL SPECIMENS

CAS NO : **75-12-7**
CHEM NAME : ***** FORMAMIDE**

REPORT 01 : EPA DRAFT TECHNICAL REPORT 68-01-6530 OCT 82
EMBRYOTOXICITY AND WEAK TERATOGENETIC EFFECTS SEEN WHEN ORAL DOSES OF 80 MG/KG GIVEN TO RABBITS ON GESTATION DAYS 6 TO 18. FOLLOWING A 2 MG/KG DOSE ON GESTATION DAY 7, ONE-HALF OF IMPLANTED FETUSES RESORBED AND GROWTH WAS STUNTED IN 26% OF SURVIVORS. IN MICE, TWO DERMAL APPLICATIONS OF 86 MG/KG ON DAYS 10 AND 11 YIELDED 36% FETAL MALFORMATIONS AND 61% OF THE ANIMALS. IP INJECTIONS OF 215 MG/KG ON GESTATION DAYS 6 - 15 RESULTED IN 25% MALFORMED FETUSES AND 27% FETAL LOSS. MUTAGENICITY ASSAYS WERE NEGATIVE.

REPORT 02 : RECEIPT OF TEST DATA - EPA 49FR5187 FEB 84
EPA ANNOUNCES RECEIVING RESULTS OF RANGE-FINDING STUDY TO ASSESS DERMAL TOXICITY OF FORMAMIDE FROM BASF. BASF PLANS TO DO A 90 DAY STUDY.

REPORT 03 : EPA RESPONSE TO ITC 48FR23098 MAY 83
BASF'S PLANS FOR SUBCHRONIC TESTING AND NTP'S TESTING PROGRAM TO DEFINE GENOTOXICITY OF CPD ARE CONSIDERED ADEQUATE TO EVALUATE CHRONIC EFFECTS OF CPD. EPA BELIEVES THAT ONCOGENICITY TESTING NOT WARRANTED, BASED ON AVAILABLE DATA. THEREFORE EPA NOT INITIATING RULE MAKING UNDER TSCA 4(A).

REPORT 04 : EPA DECISION NOT TO TEST FORMAMIDE 51FR6929 FEB 86
EPA HAS DETERMINED THAT SUFFICIENT DATA NOW AVALABLE TO EVALUATE CHRONIC TOXICITY. THEREFORE NO FURTHER TESTING IS REQUIRED AT THIS TIME.

REPT'D USE : CHEMICAL INTERMEDIATE, ADDITIVES FOR OIL WELL DRILLING, INK SOLVENTS, SOLVENT FOR GLUE, SOFTENER IN PAPER, ANOMAL GLUES, WATER-SOLUBLE GUMS AND SOLUBILIZER IN LATEX PAINTS

CAS NO : **140-56-7**
CHEM NAME : ***** FORMULATED FENAMINOSULF**

REPORT 01 : NCI CARC TRS 101 DHEW PUB NO 78-1351
UNDER THE CONDITIONS OF THIS BIOASSAY, DIETARY ADMINISTRATION OF FORMULATED FENAMINOSULF WAS NOT CARCINOGENIC IN EITHER FISCHER 344 RATS OR B6C3F1 MICE.
REPT'D USE : AGRICULTURAL SEED FUNGICIDE

CAS NO : **3570-75-0**
CHEM NAME : ***** 2-(2-FORMYLHYDRAZINO)-4-(5-NITRO-2-FURYL)THIAZOLE**

REPORT 01 : IARC MONOGRAPH 7 1974
2-(2-FORMYLHYDRAZINO)-4-(5-NITRO-2-FURYL)THIAZOLE IS CARCINOGENIC IN MICE, RATS AND HAMSTERS FOLLOWING ORAL ADMINISTRATION, THE ONLY SPECIES AND ROUTE TESTED. IN MICE, IT PRODUCED MAINLY STOMACH AND PULMONARY TUMORS AND LYMPHOCYTIC LEUKEMIAS. IN RATS, IT PRODUCED MAINLY MAMMARY, RENAL AND HEPATIC GASTROINTESTINAL TRACT TUMORS AND LYMPHOCYTIC LEUKEMIAS; IN HAMSTERS, IT PRODUCED MAINLY FORESTOMACH AND URINARY BLADDER TUMORS. IN A 1979 SUPPLEMENT, IARC DETERMINED THAT THERE IS SUFFICIENT EVIDENCE FOR THE CARCINOGENICITY OF IT IN ANIMALS.
REPT'D USE : ANTIBACTERIAL AGENT

CAS NO : **68476-30-2**
GHEM NAME : ***** NO. 2 FUEL OIL**

REPORT 01 : EPA TSCA 8(E) NOTICE 8EHQ-0778-0215
EXXON PROVIDED RESULTS OF A SKIN-PAINTING STUDY, OF NO. 2 FUEL OIL IN MICE, WHICH SHOWS A DEFINITE INCIDENCE OF MALIGNANT NEOPLASMS OF THE SKIN FOLLOWING LIFETIME APPLICATION TO MICE.

REPORT 02 : EPA FYI AX-0385-0389 MAR 85
API REPORTS MIDDLE DISTILLATE FUELS ARE WEAKLY TO MODERATELY CARCINOGENIC IN ANIMALS, PRODUCING TUMOR INCIDENCES IN 5 TO 30% RANGE (SKIN-PAINTINGS). TWO YEAR

USAF/NAVY INHALATION STUDIES ON VOLATILE FRACTIONS IN PROGRESS.

REPORT 03 : EPA FYI AX-0385-0389 MAR 85
API REPORTS SOME MIDDLE DISTILLATE FUEL MATERIALS MUTAGENIC IN VIVO RAT BONE MARROW CYTOGENETICS AND MOUSE LYMPHOMA ASSAYS; BUT RESULTS NOT CONSISTENT. MODIFIED AMES ASSAY PRODUCED POSITIVE RESULTS FOR SOME MATERIALS.

CAS NO : **CLASS-33**
CHEM NAME : ***** FUEL OILS (HEATING OILS)**

REPORT 01 : IARC MONOGRAPH 45 1989
DISTILLATE FUEL OILS (LIGHT FUEL OILS) ARE: CHEMICALLY-NEUTRALIZED LIGHT DISTILLATE (CAS NO. 64742-31-0); DOMESTIC FUEL OILS; DOMESTIC HEATING OIL; FUEL OIL NO. 1; FUEL OIL NO. 2 (CAS NO. 68476-30-2); FURNACE OIL NO. 1: FURNACE OIL NO. 2; HEATING OIL; HOME HEATING OIL; HYDROTREATED LIGHT DISTILLATE (CAS NO. 64742-47-8); KEROSENE; KEROSENE, STRAIGHT RUN(5) (CAS NO. 8008-20-6); KEROSINE; LAMP OIL; LIGHT HEATING OIL; PARAFFIN OIL; PETROLEUM DISTILLATE; ROTARY BURNER FUEL; STOVE OIL.

REPORT 02 : IARC MONOGRAPH 45 1989
RESIDUAL (HEAVY) FUEL OILS POSSIBLY CARCINOGENIC TO HUMANS. THERE IS INADEQUATE EVIDENCE FOR CARCINOGENICITY IN HUMANS OF FUEL OILLS. THERE IS SUFFICIENT EVIDENCE FOR CARCINOGENICITY IN EXPERIMENTAL ANIMALS OF FUEL OIL NO. 2. IN FORMULATING EVALUATION FOLLOWING EVIDENCE IN PETROLEUM CONSITERED. THERE IS SUFFICIENT EVIDENCE FOR CARCINOGENICITY IN EXPERIMENTAL ANIMALS OF LIGHT AND HEAVY CATALYTICALLY CRACKED DISTILLATES, OF LIGHT AND HEAVY VACUUM DISTILLATES AND OF CRACKED RESIDUES DERIVED FROM REFINING OF CRUDE OIL. THERE IS LIMITED EVIDENCE CARCINOGENICITY IN ANIMALS OF ST.-RUN KEROSENE.

REPORT 03 : IARC MONOGRAPH 45 1989
TWO LARGE HISTORICAL COHORT STUDIES IN JAPAN SHOWED EXCESS OF LUNG CANCER AMONG MEN EXPOSED TO KEROSENE; DIESEL OIL, CRUDE PETROLEUM AND MINERAL OIL CONSIDERED AND EXCESS OF STOMACH CANCER AMONG WORKERS POSSIBLY EXPOSED TO KEROSENE, MACHINE OIL OR GREASE. EXCESS LEUKEMIA IN INDUSTRIES WHERE KEROSENE, PARAFFIN OIL OR PETROLEUM COMBUSTIBLES USED OR PRODUCED, BUT EXPOSURES NOT DEFINED CLEARLY. IN CASE-CONTROL STUDIES EXCESS OF COLORECTAL CANCER ASSOCIATED WITH EXPOSURE TO SOLVENTS AND FUEL OIL; AND EXCESS OF STOMACH CANCER WITH EXPOSURE TO KEROSENE, AND EXCESSES OF RECTAL CANCER AND OAT-CELL LUNG CANCER WITH EXPOSURE TO HEATING OIL.

REPORT 04 : IARC MONOGRAPH 45 1989
CRACKED BUNKER FUEL TESTED ALONE AND BLENDED WITH RESIDUE FROM THERMAL CRACKING OF CATALYTICALLY CRACKED CLARIFIED OIL (31) BY SKIN APPLICATION TO MICE, INDUCED BENIGN AND MALIGNANT SKIN TUMORS; FURTHER INCREASE WHEN CRACKED RESIDUE ADDED. UNCRACKED RESIDUE (8 OR 21) TESTED ALONE OR IN COMBINATION WITH RESIDUE (31) ABOVE PRODUCED ONE SKIN PAPILLOMA, AND HIGH INCIDENCE OF SKIN PAPILLOMAS AND CARCINOMAS WHEN CRACKED RESIDUE ADDED. SAMPLE OF FUEL OIL NO. 2 TESTED BY SKIN APPLICATION TO MICE PRODUCED SKIN CARCINOMAS AND PAPILLOMAS.

REPORT 05 : IARC MONOGRAPH 45 1989
KEROSENE INGESTION COMMON CAUSE OF CHILDHOOD POISONING AND MAY RESULT IN LUNG DAMAGE. KEROSENE DID NOT INDUCE CHROMOSOMAL ABERRATIONS IN RAT BONE MARROW, NOR INDUCE MUTATION IN CULTURED MAMMALIAN CELLS OR IN BACTERIA. FUEL OIL NO. 2 INDUCED CHROMOSOMAL ABERRATIONS IN RAT BONE MARROW AND MUTATION IN CULTURED MAMMALIAN CELLS AND IN BACTERIA. AROMATIC FRACTIONS OF FUEL OIL NO. 2 INDUCED SISTER CHROMATID EXCHANGE, BUT NOT CHROMOSOMAL ABERRATIONS, IN CULTURED MAMMALIAN CELLS. ONE FOUR-TO-SEVEN RING PAH FRACTION OF FUEL OIL NO. 2 INDUCED MUTATION IN BACTERIA. A HEAVY FUEL OIL INDUCED CHROMOSOMAL ABERRATIONS IN CULTURED MAMMALIAN CELLS; BUNKER FUEL DID NOT INDUCE.

REPORT 07 : IARC MONOGRAPH 45 1989
RESIDUAL FUEL OILS (HEAVY FUEL OILS) ARE: BUNKER C; BUNKER C FUEL OIL; BUNKER FUEL OIL; BUNKER OIL; FUEL OIL LOURD; FUEL OIL NO. 4 (CAS NO. 68476-31-3); FUEL OIL NO. 5; FUEL OIL NO. 6. TWO SAMPLES OF STRAIGHT-RUN KEROSENE (5), ONE SAMPLE OF LIGHT VACUUM DISTILLATE (19), SEVERAL SAMPLES OF HEAVY VACUUM DISTILLATES (20) THREE SAMPLES OF LIGHT CATALYTICALLY CRACKED DISTILLATES PRODUCED SKIN TUMORS IN MICE. SEVERAL HEAVY CATALYTICALLY CRACKED DISTILLATES, RESIDUES OF CATALYTICALLY CRACKED CLARIFIED OILS, THERMALLY CRACKED RESIDUES AND STEAM-CRACKED RESIDUES PRODUCED HIGH INCIDENCES OF BENIGN AND MALIGNANT SKIN TUMORS IN MICE.
REPT'D USE : HEATING OILS, ROAD OILS, ASPHALT CEMENT AND LIGHTING OIL

CAS NO : **67-45-8**
CHEM NAME : *****FURAZOLIDONE**

REPORT 01 : IARC MONOGRAPH 31 JUL 83
DATA ON CARCINOGENICITY OF FURAZOLIDONE REPORTED ONLY IN SECONDARY SOURCES AND THEREFORE COULD NOT BE EVALUATED. FURAZOLIDONE HAS BEEN SHOWN TO INDUCE DNA

DAMAGE AND MUTATIONS IN BACTERIA AND MUTATIONS IN FUNGI AND INSECTS. CAUSED UNSCHEDULED DNA SYNTHESIS IN MAMMALIAN CELLS IN VITRO AND YIELDED CONFLICTING RESULTS WHEN TESTED FOR CHROMOSOMAL ANOMALIES IN HUMAN CELLS IN VITRO. FURAZOLIDONE INDUCED ABORTION IN MICE.

REPT'D USE : ANTIBACTERIAL AND ANTIPROTOZOAL AGENT

CAS NO : **98-01-1**

CHEM NAME : ***** FURFURAL**

REPORT 01 : NTP TECHNICAL REPORT 382. NIH PUB. #90-2837 MAR 90
UNDER 2-YR. GAVAGE STUDIES, SOME EVIDENCE OF CARCINOGENIC ACTIVITY OF FURFURAL FOR MALE F344/N RATS BASED ON UNCOMMON CHOLANGIOCARCINOMAS AND BILE DUCT DYSPLASIA AND FIBROSIS. NO EVIDENCE OF CARCINOGENIC ACTIVITY FOR FEMALE F344/N RATS AT DOSES OF 0, 30, OR 60 MG/KG. CLEAR EVIDENCE OF CARCINOGENIC ACTIVITY FOR MALE B6C3F1 MICE BASES ON INCREASED HEPATOCELLULAR ADENOMAS AND HEPATOCELLULAR CARCINOMAS. SOME EVIDENCE OF CARCINOGENIC ACTIVITY IN FEMALE B6C3F1 MICE BASED ON INCREASED HEPATOCELLULAR ADENOMAS. RENAL CORTICAL ADENOMAS OR CARCINOMAS IN MALE MICE AND SQUAMOUS CELL PAPILLOMAS OF FORESTOMACH IN FEMALE MICE MAY HAVE BEEN RELATED TO EXPOSURE.

REPT'D USE : PRODUCTION OF FURAN, FURFURYL ALCOHOL, TETRAHYDROFURAN; SYNTHETIC RUBBER MANUFACTURE, WETTING AGENT AND INDUSTRIAL SOLVENT

CAS NO : **623-17-6**

CHEM NAME : ***** FURFURYL ACETATE**

REPORT 01 : NTP TECHNICAL BULLETIN 9 APR 83
FURFURYL ACETATE WAS POSITIVE IN ONE SALMONELLA TYPHIMURIUM MUTAGENICITY ASSAY.

CAS NO : **7440-44-0**

CHEM NAME : ***** FURNACE CARBON BLACK**

REPORT 01 : TSCA SECT 8(E) STATUS REPORT 8EHQ-0487-0668 JUN 87
XEROX SUBMITTED CHRONIC STUDY FINDINGS OF XEROX 9000-TYPE PHOTOCOPYING TONER ADMINISTERED AT 1.0, 4.0 AND 16.0 MG/M3 INHALATION TO FISCHER 344 RATS. NO UNUSUAL HISTOPATHOLOGICAL FINDINGS EXCEPT THE 16 MG/M3 GROUP SHOWED INCREASE IN COLLAGEN, POSSIBLY RESULT OF ARTIFACT ASSOCIATED WITH THICKLY CUT HISTOLOGIC SECTIONS, AND INCREASED LUNG WEIGHT, RETENTION OF TEST MATERIAL, AND

RETARDATION OF ALVEOLAR MACROPHAGE MEDIATED CLEARANCE. TONER MATERIAL USED ENRICHED TEN-FOLD IN RESPIRABLE SIZE PARTICLES WITH RESPECT TO COMMERCIALLY AVAILABLE TONER MATERIAL.

REPORT 02 : EPA TSCA 8(E) 0487-0668 JUN 87
XEROX PROVIDED REPORT FROM SUBCHRONIC INHALATION STUDY OF XEROX 9000- TYPE PHOTOCOPYING TONER IN FISCHER 344 RATS. ACGIH RESPIRABLE FRACTION OF THE TONER ENRICHED TO 35%. EXPOSURE CONCENTRATIONS USED WERE 0, 1.0, 4.0, 16.0 AND 64.0 MG/M3 INCREASE IN LUNG WEIGHT OBSERVED AT 16 AND 64 MG/M3 DOSE LEVELS. AT 64 MG/M3, FOOD COMSUMPTION IN BOTH MALE AND FEMALE RATS SLIGHTLY DECREASED, BUT BODY WEIGHT NOT AFFECTED. DOSE-RELATED INCREASE IN PARTICLELADEN ALVEOLAR MACROPHAGES. THICKENING OF ALVEOLAR WALLS OBSERVED IN HIGH EXPOSURE GROUPS. MFTD MAXIMUM LUNG BURDEN FOR WHICH MACROPHAGE MEDIATED LUNG CLEARANCE NOT SIGNIFICANTLY IMPAIRED EXCEEDED 64 MG/M3 EXPOSURE LEVEL.

REPORT 03 : EPA TSCA 8(E) 1289-0668
ABSENCE OF SYSTEMIC OR UPPER RESPIRATORY SYSTEM TOXICITY, COUPLED WITH CHANGES IN LUNGS OR LYMPH NODES IN CHRONIC TONER INHALATION STUDY IN RATS SHOWED RESPIRATORY SYSTEM TARGET ORGAN AND PULMONARY CHANGES ATTRIBUTED TO "LUNG OVERLOADING", A GENERIC RESPONSE OF RESPIRATORY SYSTEM TO EXCESSIVE AMOUNTS OF DUST RETAINED IN LUNGS FOR PROLONGED INTERVAL. EXPOSURE AT HIGH LEVEL RESULTED IN EXCEEDING LUNG OVERLOADING WITH CHRONIC AND PERSISTENT INFLAMMATORY RESPONSE IN RAT LUNG. EXPOSURE AT LOW LEVEL RESULTED IN BELOW LUNG OVERLOADING TRANSIENT AND REVERSIBLE INFLAMMATORY RESPONSE IN RAT LUNG.

REPT'D USE : XEROGRAPHIC TONER WITH 1-BUTYLMETHACRYLATE/STYRENE RANDOM COPOLYMER

CAS NO : **UNKNOWN-25**
CHEM NAME : ***** FURNITURE AND CABINET-MAKING INDUSTRY**

REPORT 01 : IARC MONOGRAPH 25 FEB 81
THERE IS SUFFICIENT EVIDENCE THAT NASAL ADENOCARCINOMAS HAVE BEEN CAUSED BY EMPLOYMENT IN THE FURNITURE-MAKING INDUSTRY. THE EXCESS RISK OCCURS MAINLY AMONG THOSE EXPOSED TO WOOD DUST. ALTHOUGH ADENOCARCINOMAS PREDOMINATE, AN INCREASED RISK OF OTHER NASAL CANCERS AMONG FURNITURE WORKERS IS ALSO SUGGESTED. NO EVALUATION OF THE RISK OF LUNG CANCER IS POSSIBLE.

CAS NO : **54-31-9**
CHEM NAME : ***** FUROSEMIDE**

REPORT 01 : DRAFT TR 356 NIH PUB. 88-2811 APR 88
IN FEED STUDIES EQUIVOCAL EVIDENCE OR CARCINOGENIC ACTIVITY FOR MALE F344/N RATS SHOWN BY MARGINAL INCREASES IN UNCOMMON TUBULAR CELL NEOPLASMS Of THE KIDNEY AND MENINGIOMAS OF THE BRAIN. NO EVIDENCE OF CARCINOGENIC ACTIVITY FOR FEMALE F344/N RATS FED 350 ON 700 PPM. NO EVIDENCE OF CARCINOGENIC ACTIVITY FOR MALE B6C3F1 MICE FED DIETS CONTAINING 700 OR 1,400 PPM. SOME EVIDENCE OF CARCINOGENIC ACTIVITY OF FUROSEMIDE FOR FEMALE MICE, SHOWN BY AN INCREASE IN MALIGNANT TUMORS OF THE MAMMARY GLAND.
REPT'D USE : HUMAN AND VETERINARY MEDICINE

CAS NO : **23255-69-8**
CHEM NAME : ***** FURSARENON-X**

REPORT 01 : IARC MONOGRAPH 11 1976
FURSARENON-X WAS INADEQUATELY TESTED IN MICE AND RATS. NO ASSESSMENT OF ITS CARCINOGENICITY CAN BE MADE.

REPORT 02 : IARC MONOGRAPH 31 JUL 83
FURSARENON-X TESTED FOR CARCINOGENICITY IN MALE RATS BY ADMINISTRATION IN DIET AND IN MALE MICE AND MALE RATS BY SUBCUTANEOUS INJECTION. STUDIES INADEQUATE FOR EVALUATION.

REPORT 03 : IARC MONOGRAPH 31 JUL 83
FURSARENON-X FOUND TO BE MUTAGENIC TO SALMONELLA TYPHIMURIUM AND CAUSED PETITE MUTATIONS IN YEAST. NO MUTAGENIC ACTIVITY REPORTED IN MAMMALIAN CELLS IN VITRO. INDUCED ABORTION IN MICE AND WAS FETOTOXIC BUT NOT TERATOGENIC TO THE SURVIVING FETUSES.
REPT'D USE : NO PRODUCTION OR USE DATA AVAILABLE. A TOXIN PRODUCED BY FUNGI

CAS NO : **8006-61-9**
CHEM NAME : ***** GASOLINE**

REPORT 01 : EPA TSCA 8(E) STATUS REPORT 8EHQ-1281-0422 JAN 82
ATLANTIC RICHFIELD CO. REPORTED PRELIMINARY HISTOPATHALOGIC FINDINGS FROM AN AMERICAN PETROLEUM INSTITUTE CHRONIC INHALATION STUDY IN WHICH F344 RATS AND B6C3F1 MICE OF BOTH SEXES WERE EXPOSED TO LEADED GASOLINE AT 0, 67, 292, OR 2056 PPM 6 HOURS PER DAY, 5 DAYS PER WEEK FOR 2 YEARS. THERE WAS AN INCREASED INCIDENCE OF

RENAL TUBULAR EPITHELIAL NEOPLASMS IN MALE RATS. NO TUMORIGENIC CHANGES WERE NOTED IN FEMALE RATS OR MICE OF EITHER SEX.

REPORT 02 : EPA FYI AX-0487 -0488 NOV 87
FINAL REPORT RECEIVED FROM LOVELACE BIOMEDICAL AND ENVIRONMENTAL RESEARCH INSTITUTE ON STUDIES ON INHALATION TOXICITY OF HYDROCARBPON VAPORS.

REPORT 03 : EPA FYI-AX- 0688-0618 JUN 88
AMERICAN PETROLEUM INSTITUTE REPORTED ON CASE-CONTROL STUDY OF KIDNEY CANCER AMONG PETROLEUM WORKERS A 30% INCREASE IN KIDNEY CANCER ASSOCIATED WITH LONG-TERM EXPOSURE TO GASOLINE DISTILLATES, PARTICULARLY FOR WORKERS EXPOSED FOR 15-24 YEARS AND WHO HAD REACHED THEIR 5TH YEAR OF EXPOSURE MORE THAN 15 YEARS PRIOR TO DIAGNOSIS. WHEN EXPOSURE TO AROMATICS WAS CONTROLLED, INCIDENCE DOUBLED FOR WORKERS WITH LONG EXPOSURES.

REPORT 04 : FYI-1288-0576 OCT 88
THE AMERICAN PETROLEUM INSTITUTE REPORTED THAT RESULTS OF UNSCHEDULED DNA SYNTHESIS IN RAT PRIMARY HEPATOCYTES WITH P5-6 UNLEADED GASOLINE, ITS EVAPORATION RESIDUE AND DMSO EXTRACT SHOWED PS-6 AND PS-6 EXTRACT DID NOT CAUSE SIGNIFICANT INCREASES IN MEAN NUMBER OF NET NUCLEAR GRAIN COUNTS AT ANY DOSE LEVEL (WAS CONSIDERED NEGATIVE). HOWEVER, PS-6 RESIDUE CAUSED SIGNIFICANT INCREASE IN MEAN NUMBER OF NET NUCLEAR GRAIN COUNTS AT THREE DOSE LEVELS AND, THEREFORE, WAS CONSIDERED POSITIVE.

REPORT 05 : EPA TSCA FYI-576 SEP 88
THE RESULTS OF THE UDS ASSAY INDICATE THAT UNDER THE TEST CONDITIONS, PS-6 DID NOT CAUSE A SIGNIFICANT INCREASE IN THE UNSCHEDULED DNA SYNTHESIS AS MEASURED BY THE MEAN NUMBER OF NET NUCLEAR GRAIN COUNTS (I.E., AN INCREASE OF AT LEAST 5 COUNTS OVER THE CONTROL), AT ANY DOSE LEVEL. IN THIS STUDY THE POSITIVE CONTROL, 7,12-DIMETHYLBENZATHRACENE (DMBA), INDUCED SIGNIFICANT INCREASES IN THE MEAN NUMBER OF NET NUCLEAR GRAIN COUNTS OVER THAT IN THE SOLVENT CONTROL.

REPORT 07 : IARC MONOGRAPH 45 1989
GASOLINE POSSIBLY CARCINOGENIC TO HUMANS (EVIDENCE INADEQUATE). LIMITED EVIDENCE FOR CARCINOGENICITY IN EXPERIMENTAL ANIMALS OF UNLEADED AUTOMOTIVE GASOLINE. UNLEADED GASOLINE INDUCES UNSCHEDULED DNA SYNTHESIS IN HEPATOCYTES FROM MALE AND FEMALE MICE TREATED IN VIVO AND IN CULTURED MOUSE, RAT AND HUMAN HEPATOCYTES. IN UK COHORT STUDY ON OIL DISTRIBUTION WORKERS, SLIGHTLY ELEVATED DEATHS FROM NEOPLASMS OF LYMPHATIC AND HAEMATOPOIETIC TISSUES. SWEDISH COHORT STUDY ON PANCRE-

ATIC CANCER SHOWED MODERATE INCREASE AMONG SERVICE STATION WORKERS. IN U.S. STUDY ON KIDNEY CANCER EVIDENCE OF A POSITIVE TREND IN RISK WITH DURATION OF EMPLOYMENT AS SERVICE STATION WORKERS.

REPORT 08 : IARC MONOGRAPH 45 1989
DANISH STUDY ON BLADDER CANCER ELEVATED RISK ASSOCIATED WITH 'OIL OR GASOLINE WORK'. U.S. STUDY ON CANCER OF RENAL PELVIS SUGGESTED RISK FOR WORKERS EXPOSED TO UNSPECIFIED PETROLEUM, TAR OR PITCH PRODUCTS. SWEDISH STUDY INDICATED RISK FOR ACUTE NONLYMPHOCYTIC LEUKAEMIA IN MEN WITH EXPOSURE TO PETROLEUM PRODUCTS. U.S. STUDY REVEALED RISK FOR TESTICULAR CANCER IN SERVICE STATION ATTENDANTS AND GARAGE WORKERS; ANOTHER SHOWED INCREASED RISK FOR PANCREATIC CANCER IN MEN WITH EXPOSURE TO DRY CLEANING AGENTS OR GASOLINE. ANOTHER U.S. STUDY DEMONSTRATED RISK FOR LIVER CANCER IN SERVICE STATION ATTENDANTS, PARTICULARLY FOR HEPATOCELLULAR CARCINOMA.

REPORT 09 : IARC MONOGRAPH 45 1989
VOLATILIZED UNLEADED GASOLINE TESTED FOR CARCINOGENICITY IN MICE AND RATS BY INHALATION PRODUCED INCREASE OF HEPATOCELLULAR ADENOMAS AND CARCINOMAS IN FEMALE MICE; NO INCREASE OBSERVED IN MALES. EXPOSURE OF MALE RATS RESULTED IN INCREASED INCIDENCE OF ADENOMAS AND CARCINOMAS OF KIDNEY; NO TUMOR FOUND IN FEMALES. ONE SAMPLE OF LIGHT STRAIGHT-RUN NAPHTHA (3) AND ONE SAMPLE OF LIGHT CATALYTICALLY CRACKED NAPHTHA (22) PRODUCED SKIN TUMORS IN MICE (SEE MONOGRAPH ON OCCUPATIONAL EXPOSURES IN PETROLEUM REFINING). URINARY THIOETHER EXCRETION INCREASED IN SAMPLES FROM SERVICE STATION ATTENDANTS.

REPT'D USE : AUTOMOTIVE FUEL

CAS NO : **67730-11-4**

CHEM NAME : ***** GLU-P-1 (2-AMINO-6-METHYLPYRIDO) 1,2-A:3',2'-D(IMIDAZOLE)**

REPORT 01 : IARC VOL 40 1986
SUFFICIENT EVIDENCE FOR CARCINOGENICITY OF GLU-P-1 TO TEST ANIMALS, BUT NO DATA AVAILABLE ON THE CARCINOGENICITY OF CPD TO HUMANS. THE DEGREE OF EVIDENCE IN SHORT-TERM TESTS FOR GENETIC ACTIVITY IS SUFFICIENT. NO DATA AVAILABLE TO EVALUATE REPRODUCTIVE EFFECTS OR PRENATAL TOXICITY TO TEST ANIMALS. NO CASE REPORT OR EPIDEMIOLOGICAL STUDIES OF CARCINOGENICITY TO HUMANS AVAILABLE TO WORKING GROUP.

REPT'D USE : NO COMMERICAL USE

CAS NO : **67730-10-3**
CHEM NAME : ***** GLU-P-2(2-AMINOPYRIDO)1,2-A:3'2-D(IMIDAZOLE)**

REPORT 01 : IARC VOL 40 1986
SUFFICIENT EVIDENCE FOR THE CARCINOGENICITY OF CPD TO TEST ANIMALS, BUT THERE WERE NO DATA ON CARCINOGENICITY TO HUMANS. THERE WERE NO DATA TO EVALUATE THE REPRODUCTIVE EFFECTS OR PRENATAL TOXICITY OF CPD TO TEST ANIMALS. CPD INDUCED PROPHAGE IN BACTERIA, WAS MUTAGENIC IN S. TYPHIMURIUM WITH ACTIVATION AND IN CULTURED MAMMALIAN CELLS. IT GAVE POSITIVE RESULTS IN THE D. MELANOGASTER WING SPOT TEST. RESULTS OF EXPERIMENTS FOR THE INDUCTION OF SCE'S AND CHROMOSOMAL ABERRATIONS IN CULTURED MAMMALIAN CELLS WERE INADEQUATE.
REPT'D USE : NO COMMERCIAL USE

CAS NO : **111-30-8**
CHEM NAME : ***** GLUTERALDEHYDE**

REPORT 01 : NTP TECHNICAL BULLETIN NO. 3 DEC 80
THE CHEMICAL WAS WEAKLY POSITIVE IN THE SALMONELLA TYPHIMURIUM MUTAGENICITY ASSAY IN ONE LABORATORY.

REPORT 02 : BIBRA TOX PROFILE
GA IS A SKIN AND MUCOUS MEMBRANE IRRITANT, SKIN SENSITIZER AND CAN CAUSE EYE DAMAGE. MODERATE ACUTE ORAL AND DERMAL ACUTE TOXICITIES IN ANIMALS. TERATOGENIC EFFECTS IN MICE ONLY AT MATERNALLY TOXIC DOSES. CAN INDUCE POINT MUTATIONS IN BACTERIA AND CULTURED MAMMALIAN CELLS.

REPORT 03 : BIBRA TOX PROFILE 1987
NINE OF 11 EXPOSED PEOPLE TO GA AS A DISINFECTANT HAD SYMPTOMS OF IRRITATION OF SKIN, THROAT, EYES, PULMONARY DISCOMFORT, COUGH, HEADACHE. NIOSH HAS OTHER REPORTS OF SIMILAR SYMPTOMS.

REPORT 04 : NTP TECH BULLETIN #3 DEC 80
ONE LABORATOTY SHOWED GA TO BE WEAKLY POSITIVE IN S. TYPHIMURIUM ASSAY.

REPORT 05 : CRC MINUTES 11/4/87 NOV 87
CAN CAUSE SKIN AND MUCOUS MEMBRANE IRRITATION, SKIN SENSITIZATION, EYE DAMAGE. REPORTED TO PRODUCE MODERATE ORAL AND DERMAL ACUTE TOXICITY. IS TERATOGENIC IN MICE AT MATERNALLY TOXIC DOSES AND CAN INDUCE POINT MUTATIONS IN BACTERIA AND IN CULTURED MAMMALIAN CELLS.

HOSPITAL WORKERS COMPLAINED IT CAUSED IRRITATIONS.
REPT'D USE : INTERMEDIATE IN MFR OF ADHESIVES, SEALANTS, DISINFECTANT, EMBALMING FLUID, BIOLOGICAL TISSUE FIXATIVE AND LEATHER TANNING AGENT

CAS NO : **765-34-4**
CHEM NAME : ***** GLYCIDALDEHYDE**

REPORT 01 : IARC MONOGRAPH 11 1976
GLYCIDALDEHYDE IS CARCINOGENIC IN MICE BY SKIN APPLICATION AND BY SUBCUTANEOUS INJECTION AND IN RATS BY SUBCUTANEOUS INJECTION. IT PRODUCED MALIGNANT TUMORS AT THE SITE OF APPLICATION IN BOTH SPECIES. IN A 1979 SUPPLEMENT, IARC DETERMINED THAT THERE IS SUFFICIENT EVIDENCE FOR THE CARCINOGENICITY OF GLYCIDALDEHYDE IN ANIMALS.

REPORT 02 : IARC MONOGRAPH 11 1976
GLYCIDALDEHYDE HAS PRODUCED BASE-PAIR MUTATIONS IN SALMONELLA TA1535 AND TA100.

REPORT 06 : MULTI-AGENCY CROSS CHECK
N HHS MAGUIRE
N OSHA CARC
Y EPA CAG
Y EPA RCRA
N EPA CLEAN AIR
N EPA CLEAN WATER
REPT'D USE : ANTIVIRAL AGENT

CAS NO : **556-52-5**
CHEM NAME : ***** GLYCIDOL**

REPORT 01 : NTP TR 374 NIH 90-2829 MAR 90
UNDER THE CONDITIONS OF THESE 2-YEAR GAVAGE STUDIES, THERE WAS CLEAR EVIDENCE OF CARCINOGENIC ACTIVITY OF GLYCIDOL FOR MALE AND FEMALE F344/N RATS AND MALE AND FEMALE B6C3F1 MICE, BASED ON INCREASED INCIDENCES OF A VARIETY OF NEOPLASMS AT NUMEROUS TISSUE SITES.
REPT'D USE : STABILIZER IN MANUFACTURE OF VINYL POLYMERS, OIL AND SYNTHETIC HYDRAULIC FLUIDS ADDITIVE AND EPOXY RESINS DILUENT

CAS NO : **CLASS-12**
CHEM NAME : ***** GLYCIDOL AND DERIVATIVES (ETHERS, ESTERS)**

REPORT 01 : EPA RESPONSE TO ITC RECOMS (48 FR 57562) DEC 83
IN REPONSE TO ITC RECOMMENDATIONS (OCT 30, 1978, 43 FR 50630)

EPA AGREES THAT MEMBERS OF GLYCIDOL CATEGORY BE CONSIDERED FOR HEALTH EFFECTS TESTING AND INITIATES RULEMAKING BY AN ANPR TO DEFINE THE TESTING IT MAY PROPOSE AND TO SEEK COMMENTS ON ITS PLAN TO PROPOSE A TEST RULE FOR CHEMICALS REPRESENTATIVE OF SEVEN SUBCATEGORIES BASED ON PHYSICAL STRUCTURE. THE ITC RECOMMENDED CARCINOGENIC, MUTAGENIC TERATOGENIC, AND OTHER ADVERSE HEALTH EFFECTS TESTING.

REPORT 02 : EPA TSCA 8(E) 0584-0518S MAY 84
A COMPANY PROVIDED RESULTS FROM SEVERAL IN VITRO GENOTOXICITY STUDIES OF A PRODUCT THAT IS A PROPRIETARY MIXTURE OF EPOXY RESINS THAT SHOWED THAT IT PRODUCED A POSITIVE RESPONSE WITH AND WITHOUT (METABOLIC) ACTIVATION IN THE AMES TEST, THE MOUSE LYMPHOMA TEST AND THE SISTER CHROMATID EXCHANGE TEST.
REPT'D USE : EXPOXY RESINS, DENTAL FILLINGS, STABILIZER, DEMULSIFIER AND DYE-LEVELING AGENTS

CAS NO : **106-90-1**
CHEM NAME : ***** GLYCIDYL ACRYLATE (GA)**

REPORT 01 : EPA TSCA 8(E) NOTICE 8EH4-0179-0270 MAY 79
THIOKOL CORP SUBMITTED RESULTS OF A BATTERY OF SALMONELLA ASSAYS CONDUCTED ON GLYCIDYL ACRYLATE AND GLYCIDYL METHACRYLATE. STUDY INDICATES THAT BOTH GA AND GMA ARE DIRECT ACTING MUTAGENS AND THAT THEY CAN ALSO BE METABOLIZED TO AN ACTIVE MUTAGEN IN BACTERIA.
REPT'D USE : POLYMERIZING AGENT, ACRYLIC AND RESINS

CAS NO : **106-91-2**
CHEM NAME : ***** GLYCIDYL METHACRYLATE (GMA)**

REPORT 01 : EPA TSCA 8(E) NOTICE 8EHQ-0179-0270 MAY 79
THIOKOL CORP SUBMITTED RESULTS OF A BATTERY OF SALMONELLA ASSAYS CONDUCTED ON GLYCIDYL ACRYLATE AND GLYCIDYL METHACRYLATE. STUDY INDICATES THAT BOTH GA AND GMA ARE DIRECT ACTING MUTAGEN AND THAT THEY CAN ALSO BE METABOLIZED TO AN ACTIVE MUTAGEN IN BACTERIA.
REPT'D USE : POLYMERIZING AGENT, ACRYLIC AND RESINS

CAS NO : **5431-33-4**
CHEM NAME : ***** GLYCIDYL OLEATE**

REPORT 01 : IARC MONOGRAPH 11 1976
GLYCIDYL OLEATE HAS BEEN TESTED IN A LIMITED STUDY IN MICE IN WHICH SUBCUTANEOUS INJECTION PRODUCED A LOW INCI-

DENCE OF LOCAL SARCOMAS. THIS RESULT IS INSUFFICIENT FOR EVALUATION.
REPT'D USE : NO PRODUCTION OR USE DATA AVAILABLE

CAS NO : **7460-84-6**
CHEM NAME : ***** GLYCIDYL STEARATE**

REPORT 01 : IARC MONOGRAPH 11 1976
GLYCIDYL STEARATE WAS TESTED IN MICE BY SUBCUTANEOUS INJECTION, IT PRODUCED NO SIGNIFICANT INCREASE IN THE INCIDENCE OF LOCAL TUMORS.
REPT'D USE : CHEMICAL INTERMEDIATE. NO KNOWN CURRENT PRODUCTION. NO LISTING ON TSCA INVENTORY.

CAS NO : **303-45-7**
CHEM NAME : ***** GOSSYPOL**

REPORT 01 : GAO SUVERY ON TOXICANTS AUG 90
CHEMICAL PAST OF GAO'S SURVEY OF AGENCIES CONTROL OF REPRODUCTIVE AND DEVELOPMENTAL TOXICANTS.
REPT'D USE : STABILIZER FOR VINYL, POLYMERS, INSECTICIDE AND CONTRACEPTIVE

CAS NO : **126-07-8**
CHEM NAME : ***** GRISEOFULVIN**

REPORT 01 : IARC MONOGRAPH 10 1976
GRISEOFULVIN IS HEPATOCARCINOGENIC FOLLOWING ITS ORAL ADMINISTRATION TO ADULT MICE OR ITS SUBCUTANEOUS ADMINISTRATION TO MALE INFANT MICE.
REPT'D USE : DRUG

CAS NO : **1455-77-2**
CHEM NAME : ***** GUANAZOLE**

REPORT 01 : NTP TECHNICAL BULLETIN 5 AUG 81
GUANAZOLE WAS NEGATIVE IN THE SALMONELLA TYPHIMURIUM ASSAY IN ONE LABORATORY.

REPORT 02 : NCI BIOASSAY PROGRAM-CANCER VOL. 40 1977
UNDER THE CONDITIONS OF THE BIOASSAY, I.P. INJECTION OF GUANAZOLE WAS CARCINOGENIC TO FEMALE SPRAGUE-DAWLEY (CD) RATS, CAUSING BREAST, PITUITARY, AND THYROID TUMORS. IT WAS NOT CARCINOGENIC TO MALE RATS AND MALE AND

FEMALE SWISS-WEBSTER MICE.
REPT'D USE : CHEMOTHERAPEUTIC AGENT

CAS NO : **9000-30-0**
CHEM NAME : ***** GUAR GUM**

REPORT 01 : NTP BOARD OF SCIENTIFIC COUNSELORS DRAFT FEB 81
THE BIOASSAY REPORT CONCLUSIONS FOR GUAR GUM WERE REVIEWED BY THE PEER REVIEW PANEL AND APPROVED WITH MINOR MODIFICATIONS. UNDER CONDITIONS OF THIS BIOASSAY, GUAR GUM ADMINISTERED IN THE FEED WAS NOT CARCINOGENIC FOR F344 RATS OR B6C3F1 MICE OF EITHER SEX.

REPORT 02 : FDA GRAS LISTING 21CFR184.1339
GUAR GUM IS GENERALLY RECOGNIZED AS SAFE FOR DIRECT ADDITION TO HUMAN FOODS AT CONCENTRATIONS UP TO 2 PERCENT IN SPECIFIC FOOD PRODUCTS.
REPT'D USE : FOOD STABILIZER, PAPER SIZING, TOOTHPASTES, LAXATIVES, PHARMACEUTICAL PREPS AND THICKENING AGENT

CAS NO : **9000-01-5**
CHEM NAME : ***** GUM ARABIC**

REPORT 01 : NTP CARC TRS 227 NIH PUB NO. 82-1783
UNDER THE CONDITIONS OF THIS BIOASSAY, GUM ARABIC WAS NOT CARCINOGENIC FOR F344 RATS OR B6C3F1 MICE OF EITHER SEX.
REPT'D USE : FOOD STABILIZER/FLAVOR FIXATIVE, EXCIPIENT FOR PILLS/TABLETS, ADHESIVES/GLUES, INKS, PAINTS AND FOAM STABILIZER FOR SOFT DRINKS

CAS NO : **16568-02-8**
CHEM NAME : ***** GYROMITRIN (ACETALDEHYDE FORMYLMETHYLHYDRAZONE)**

REPORT 01 : IARC MONOGRAPH 31 JUL 83
TESTED IN MICE BY GAVAGE, PRODUCING INCREASED INCIDENCES OF LUNG, FORESTOMACH AND CLITORAL GLAND TUMORS IN FEMALES AND OF PREPUTIAL TUMORS IN MALES. N-METHYL-N-FORMYLHYDRAZINE, A METABOLITE, TESTED IN MICE IN DRINKING-WATER, INCREASING INCIDENCE OF TUMORS OF LIVER, LUNG, GALL BLADDER AND BILE DUCT. N-METHYLHYDRAZINE, ANOTHER METABOLITE, TESTED IN MICE AND HAMSTERS IN DRINKING-WATER; IT PRODUCED INCREASED INCIDENCES OF HISTIOCYTOMAS AND CAECAL TUMORS IN HAMSTERS.

REPORT 02 : IARC MONOGRAPH 31 JUL 83
GYROMITRIN NOT MUTAGENIC IN BACTERIA, BUT METABOLITE N-METHYLHYDRAZINE GAVE POSITIVE RESULTS.
REPT'D USE : NOT PRODUCED OR USED COMMERCIALLY. FOUND IN MUSHROOM GYROMITRA ESCULENTA

CAS NO : **UNKNOWN-21**
CHEM NAME : ***** HAIR SERVICE OCCUPATION AND USERS OF HAIR DYES**

REPORT 01 : IARC MONOGRAPH 27 APR 82
THE EVIDENCE RELATING BLADDER CANCER OR ANY OTHER CANCER TO THE OCCUPATION OF HAIRDRESSING OR TO PERSONAL USE OF HAIR DYES IS INCONCLUSIVE. IN VIEW OF THE POSSIBLE HAZARD SUGGESTED BY EXPERIMENTAL STUDIES OF SOME COMPONENTS OF HAIR DYES, THE SUBJECT MERITS FURTHER STUDY.

CAS NO : **80-13-7**
CHEM NAME : ***** HALAZONE**

REPORT 01 : NTP TECHNICAL BULLETIN 9 APR 83
HALAZONE WAS POSITIVE IN ONE SALMONELLA TYPHIMURIUM MUTAGENICITY ASSAY.

CAS NO : **68916-39-2**
CHEM NAME : ***** HAMAMELIS WATER**

REPORT 01 : NTP BOARD OF SCIENTIFIC COUNSELORS MAR 84
WITH HAZEL TESTED IN DERMAL STUDIES.

REPORT 02 : CLINICAL TOXICOLOGY OF COMMERCIAL PRODUCTS 1984
FLUID-EXTRACT OF HAMAMELIS VIRGINIANA LEAVES (CONTAINING 2 TO 9 PERCENT TANNIN) HAS ETHYL ALCOHOL CONTENT OF 70 TO 78 PERCENT AND 2 TO 4 ML. TANNIN. WITCH HAZEL IS FROM DRIED TWIGS; CONTAINS 15 PERCENT ETHANOL AND SMALL AMOUNTS OF SESQUITERPENES.
REPT'D USE : ASTRINGENT AND DRUGS

CAS NO : **2784-94-3**
CHEM NAME : ***** HC BLUE NO. 1**

REPORT 01 : NTP TECHNICAL REPORT 271 NIH PUB 85-2527 AUG 85
THERE WAS EQUIVOCAL EVIDENCE OF CARCINOGENICITY IN MALE F344/N RATS; THERE WAS SOME EVIDENCE OF CARCINOGENICITY

FOR FEMALE RATS; CLEAR EVIDENCE OF CARCINOGENICITY IN MALE B6F3C1 MICE; AND CLEAR EVIDENCE OF CARCINOGENICITY IN FEMALE MICE.

REPORT 02 : NTP TECHNICAL REPORT 271 NIH PUB 85-2527 AUG 85
FEEDING STUDIES GAVE EQUIVOCAL EVIDENCE OF CARCINOGENICITY IN MALE RATS, CAUSING A MARGINAL INCREASE IN INCIDENCE OF HEPATOCELLULAR NEOPLASTIC NODULES AND CARCINOMAS. IN FEMALE F344/N RATS, THERE WAS SOME EVIDENCE OF CARCINOGENICITY BECAUSE OF INCREASED INCIDENCES OF ALVEOLAR AND BRONCHIOLAR NEOPLASMS. THERE WAS CLEAR EVIDENCE OF CARCINOGENICITY OF CPD FOR MALE AND FEMALE B6C3F1 MICE, AS SHOWN BY INCREASED INCIDENCES OF HEPATOCELLULAR CARCINOMAS. INCIDENCES OF FOLLICULAR CELL ADENOMAS OF THYROID GLAND ALSO INCREASED IN MALE MICE.
REPT'D USE : HAIR DYE

CAS NO : **33229-34-4**
CHEM NAME : ***** HC BLUE NO. 2**

REPORT 01 : NTP TECHNICAL REPORT 293 NIH PUB 85-2549 JUL 84
UNDER THE CONDITIONS OF THESE STUDIES, THERE WAS NO EVIDENCE OF CARCINOGENICITY IN MALE OR FEMALE F344/N RATS AND IN MALE AND FEMALE B6C4F1 MICE RECEIVING HC BLUE NO. 2 IN THE DIET FOR TWO YEARS IN DOSES OF .5 AND 1% FOR MALES AND 1 AND 2% FOR FEMALES. THERE WAS RELATED INCREASE IN INCIDENCE OF HYPEROSTOSIS OF SKULL IN MALE AND FEMALE RATS.
REPT'D USE : HAIR DYE

CAS NO : **2871-01-4**
CHEM NAME : ***** HC RED NO. 3**

REPORT 01 : NTP TECHNICAL REPORT 281, NIH PUBLICAT'N 86-2537
IN 2-YEAR GAVAGE STUDIES THERE WAS NO EVIDENCE OF CARCINOGENICITY FOR MALE OR FEMALE F344/N RATS GIVEN 250 OR 500 MG/KG PER DAY. EVIDENCE OF CARCINOGENICITY FOR MALE B6C3F1 MICE WAS EQUIVOCAL, AS INDICATED BY HEPATOCELLULAR ADENOMAS OR CARCINOMAS IN THE 250 MG/KG GROUP. POOR SURVIVAL COUPLED WITH LACK OF SIGNIFICANT FINDINGS RENDERED STUDY IN FEMALE B6C3F1 MICE INADEQUATE. BOTH SEXES OF EACH SPECIES MAY HAVE BEEN ABLE TO TOLERATE HIGHER DOSES. THEREFORE, SENSITIVITY OF STUDIES FOR DETECTING CARCINOGENESIS MAY HAVE BEEN LIMITED.
REPT'D USE HAIR DYE

CAS NO : **52551-67-4**
CHEM NAME : ***** HC YELLOW 4**

REPORT 01 : NTP TECHNICAL BULLETIN NO. 7 APR 82
THE CHEMICAL WAS POSITIVE IN THE SALMONELLA TYPHIMURIUM MUTAGENICITY ASSAY.
REPT'D USE : COMPONENT OF PERMANENT HAIR DYES

CAS NO : **UNKNOWN-05**
CHEM NAME : ***** HEMATITE MINING**

REPORT 01 : IARC SUPPLEMENT 1
UNDERGROUND HEMATITE MAINERS HAVE A HIGH INCIDENCE OF LUNG CANCER. WHEREAS SURFACE HEMATITE MINERS DO NOT. IT IS NOT KNOWN WHETHER THIS EXCESS RISK MAY BE DUE TO HEMATITE; TO RADON (A KNOWN LUNG CARCINOGEN); TO INHALATION OF FERRIC OXIDE OR SILICA; OR TO A COMBINATION OF THESE OR OTHER FACTORS. THE IARC CONCLUDED THAT THE UNDERGROUND MINING OF HEMATITE IS CARCINOGENIC FOR HUMANS.

REPORT 06 : MULTI-AGENCY CROSS CHECK
Y HHS MAGUIRE
N OSHA CARC
N EPA CAG
N EPA RCRA
N EPA CLEAN AIR
N EPA CLEAN WATER

CAS NO : **76-44-8**
CHEM NAME : ***** HEPTACHLOR**

REPORT 01 : NCI CARC TRS 9 DHEW PUB NO 77-809
UNDER CONDITIONS OF THIS STUDY, DIETARY ADMINISTRATION OF HEPTACHLOR TO OSBORNE-MENDEL RATS AND B6C3F1 MICE RESULTED IN HEPATOCELLULAR CARCINOMAS IN MICE OF BOTH SEXES. NO HEPATIC TUMORS WERE OBSERVED IN RATS.

REPORT 02 : IARC SUPPLEMENT 1 1979
CHLORDANE AND HEPTACHLOR ARE CONSIDERED TOGETHER BECAUSE THEY ARE STRUCTURALLY SIMILAR AND OFTEN CONTAMINATED WITH EACH OTHER. CHLORDANE AND HEPTACHLOR ARE CARCINOGENIC IN MICE, PRODUCING LIVER TUMORS FOLLOWING ORAL ADMINISTRATION. CASE REPORTS OF NEUROBLASTOMA AND ACUTE LEUKEMIA HAVE BEEN ASSOCIATED WITH EXPOSURE TO CHLORDANE.

REPORT 03 : EPA AMBIENT WATER QUALITY CRITERIA OCT 80
HEPTACHLOR IS SUSPECTED OF BEING A HUMAN CARCINOGEN. AMBIENT WATER CONCENTRATIONS OF .23 NG/L, .023 NG/L, AND .002 NG/L ARE ESTIMATED TO RESULT IN AN INCREASED CANCER RISK OVER A LIFETIME OF 10 -5, 10 -6, AND 10 -7 RESPECTIVELY.

REPORT 04 : EPA RCRA HEALTH AND ENV. EFFECTS PROFILE OCT 80
IT HAS BEEN FOUND THAT HEPTACHLOR AND ITS METABOLITE, HEPTACHLOR EPOXIDE, INDUCED LIVER CANCER IN MICE AND RATS. HEPTACHLOR WAS MUTAGENIC IN TWO MAMMALIAN ASSAYS BUT NOT IN THE AMES TEST.

REPORT 05 : NAS REPORT ON DIET NUTRITION AND CANCER 1982
ON THE BASIS OF STUDIES IN ANIMALS AND IN THE ABSENCE OF ADEQUATE DATA FROM EPA STUDIES, IT APPEARS THAT HEPTACHLOR PERHAPS PRESENTS A CARCINOGENIC RISK TO HUMANS. HOWEVER IT IS REASONABLE TO ASSUME THAT AMOUNTS PRESENT IN THE AVERAGE US DIET DO NOT MAKE A MAJOR CONTRIBUTION TO THE OVERALL RISK OF CANCER FOR HUMANS.

REPORT 06 : MULTI-AGENCY CROSS CHECK
N HHS MAGUIRE
N OSHA CARC
Y EPA CAG
N EPA RCRA
N EPA CLEAN AIR
Y EPA CLEAN WATER

REPORT 07 : IARC MONOGRAPH 20 1979
HEPTACHLOR SHOWED NO ACTIVITY IN THE REC ASSAY WITH B. SUBTILIS, AND WAS NONMUTAGENIC TO SALMONELLA STRAINS TA1535, TA1536, TA1537, AND TA1538, WITH OR WITHOUT METABOLIC ACTIVATION. THERE WAS NO EVIDENCE OF INDUCTION OF X-LINKED RECESSIVE LETHALS IN DROSOPHILA. HEPTACHLOR WAS NEGATIVE IN DOMINANT LETHAL TESTS IN MICE.

REPORT 08 : NTP TECHNICAL BULLETIN 9 APR 83
HEPTACHLOR WAS NEGATIVE IN ONE SALMONELLA TYPHIMURIUM MUTAGENICITY ASSAY.

REPT'D USE : INSECTICIDE

CAS NO : **1024-57-3**
CHEM NAME : ***** HEPTACHLOR EPOXIDE**

REPORT 01 : IARC MONOGRAPH 20
HEPTACHLOR EPOXIDE WAS NOT MUTAGENIC TO SALMONELLA STRAINS TA1535, TA1536, TA1537, AND TA1538, WITH OR WITHOUT METABOLIC ACTIVATION. THERE WAS NO EVIDENCE OF INDUCTION

OF X-LINKED RECESSIVE LETHALS IN DROSOPHILA. HEPTACHLOR EPOXIDE WAS NEGATIVE IN DOMINANT LETHAL TESTS IN MICE.

REPORT 02 : IARC MONOGRAPH 20 1979
A RE-EVALUATION OF UNPUBLISHED STUDIES SUGGESTED THAT HEPTACHLOR EPOXIDE PRODUCED LIVER CARCINOMAS IN MICE OF BOTH SEXES AND HEPATOMAS IN RATS OF BOTH SEXES.

REPORT 03 : EPA RCRA HEALTH RCRA EFFECTS PROFILE OCT 80
HEPTACHLOR EPOXIDE (MOSTLY OF UNSPECIFIED PURITY) HAS INDUCED LIVER CANCER IN MICE AND RATS AND WAS MUTAGENIC IN A MAMMALIAN ASSAY SYSTEM, BUT NOT IN A BACTERIAL SYSTEM.

REPORT 06 : MULTI-AGENCY CROSS CHECK
N HHS MAGUIRE
N OSHA CARC
N EPA CAG
N EPA RCRA
N EPA CLEAN AIR
Y EPA CLEAN WATER
REPT'D USE : METABOLITE OF HEPTACHLOR

CAS NO : **59080-40-9**
CHEM NAME : ***** HEXABROMOBIPHENYL (PURE)**

REPORT 01 : IARC MONOGRAPH 18 1978
ONLY ONE COMMERCIAL PREPARATION OF PBB (TECH GRADE HEXABROMOBIPHENYL) COMPOSED PRIMARILY OF PENTA-, HEXA-, AND HEPTABROMINATED CONGENERS, WAS TESTED IN ONE EXPERIMENT IN RATS BY ORAL ADMINISTRATION OF SINGLE DOSES. ALTHOUGH NEOPLASTIC CHANGES OF THE LIVER WERE REPORTED, THE STUDY WAS CONSIDERED TO BE INADEQUATE. NO EVALUATION OF THE CARCINOGENICITY OF PBB COULD BE MADE ON THE BASIS OF THE AVAILABLE DATA.
REPT'D USE : FR CHEMICAL FOR PLASTICS. NO KNOWN CURRENT PRODUCTION.

CAS NO : **36355-01-8**
CHEM NAME : ***** HEXABROMOBIPHENYL (TECH)**

REPORT 01 : IARC MONOGRAPH 18 1978
ONLY ONE COMMERCIAL PREPARATION OF PBB (TECH GRADE HEXABROMOBIPHENYL) COMPOSED PRIMARILY OF PENTA-, HEXA-, AND HEPTABROMINATED CONGENERS, WAS TESTED IN ONE EXPERIMENT IN RATS BY ORAL ADMINISTRATION OF SINGLE DOSES. ALTHOUGH NEOPLASTIC CHANGES OF THE LIVER WERE REPORTED, THE STUDY WAS CONSIDERED TO BE INADEQUATE. NO

EVALUATION OF THE CARCINOGENICITY OF PBB COULD BE MADE ON THE BASIS OF THE AVAILABLE DATA.

REPORT 02 : NTP BOARD OF SCIENTIFIC COUNSELORS JUN 81
UNDER THE CONDITIONS OF THIS BIOASSAY, A POLYBROMINATED BIPHENY MIXTURE (FIREMASTER FF-l) WAS CARCINOGENIC FOR FISCHER 344 RATS - AND F6C3F1 MICE OF BOTH SEXES INDUCING NEOPLASTIC NODULES. HEPATOCELLULAR CARCINOMAS, AND CHOLANGIOMAS (BASED ON MORPHOTOGICAL APPEARANCE) IN RATS AND HEPATOCELLULAR CARCINOMAS IN MICE (INGESTION IN FEED).

REPORT 03 : IARC MONOGRAPH 18 1978
FIREMASTER FF-1 IS EMBRYOTOXIC AND TERATOGENIC. CYTOGENIC TESTS IN MICE WERE NEGATIVE.

REPORT 04 : NTP TECHNICAL BULLETIN 3 DEC 80
HEXABROMOBIPHENYL WAS NEGATIVE IN THE SALMONELLA TYPHIMURIUM MUTAGENICITY ASSAY IN ONE LABORATORY.

REPORT 05 : NTP TECHNICAL BULLETIN 3 DEC 80
HEXABROMOBIPHENYL WAS NEGATIVE IN THE SALMONELLA TYPHIMURIUM MUTAGENICITY ASSAY IN ONE LABORATORY.
REPT'D USE : FR CHEMICAL. NO KNOWN CURRENT PRODUCTION. NO LISTiNG ON TSCA INVENTORY.

CAS NO : **118-74-1**
CHEM NAME : ***** HEXACHLOROBENZENE**

REPORT 01 : IARC MONOGRAPH 20 1979
HEXACHLOROBENZENE WAS TESTED BY ORAL ADMINISTRATION IN ONE EXPERIMENT IN MICE AND ONE IN HAMSTERS. IN MICE, IT PRODUCED LIVER CELL TUMORS IN ANIMALS OF BOTH SEXES. IN HAMSTERS OF BOTH SEXES, IT PRODUCED HEPATOMAS, LIVER HEMANGIOTHELIOMAS, AND THYROID TUMORS. IN A 1979 SUPPLEMENT, IARC DETERMINED THAT THERE IS SUFFICIENT EVIDENCE FOR THE CARCINOGENICITY OF HEXACHLOROBENZENE IN ANIMALS.

REPORT 03 : EPA AMBIENT WATER QUALITY CRITERIA DEC 80
HCB IS SUSPECTED OF BEING A HUMAN CARCINOGEN. AMBIENT WATER CONCENTRATIONS OF 7.2 NGZL .72 NG/L, AND .072 NG/L ARE ESTIMATED TO RESULT IN A RESPECTIVE 10 -5, 10 -6, AND 10 -7 INCREASE IN CANCER OVER A LIFETIME.

REPORT 04 : EPA RCRA HEALTH AND ENV. EFFECTS PROFILE OCT 80
HEXACHLOROBENZENE APPEARS TO BE NEITHER A TERATOGEN NOR A MUTAGEN. HOWEVER, THIS COMPOUND HAS PRODUCED TUMORS IN BOTH RATS AND MICE.

REPORT 05 : IARC MONOGRAPH 20 1979
HEXACHLOROBENZENE IS FETOTOXIC AND PRODUCES SOME TERATOGENIC EFFECTS. IT WAS NOT MUTAGENIC IN DROSOPHILA AND DID NOT INDUCE DOMINANT LETHAL MUTATIONS IN MALE RATS.

REPORT 06 : MULTI-AGENCY CROSS CHECK
Y HHS MAGUIRE
N OSHA CARC
Y EPA CAG
N EPA RCRA
N EPA CLEAN AIR
Y EPA CLEAN WATER

REPORT 07 : NTP TECHNICAL BULLETIN 3/7 DEC 80, APR 82
HCB WAS NEGATIVE IN THE SALMONELLA ASSAY, THE RAT DOMINANT LETHAL ASSAY, THE MOUSE SPERM HEAD MORPHOLOGY ASSAY, THE DROSOPHILA SEX-LINKED RECESSIVE LETHAL ASSAY, AND THE RAT BONE MARROW CYTOLOGY ASSAY. RESULTS IN THE IN VITRO UNSCHEDULED DNA SYNTHESIS IN HUMAN FIBROBLASTS WERE SUGGESTIVE BUT NOT CLEARLY POSITIVE.

REPORT 08 : NAS REPORT ON DIET NUTRITION AND CANCER 1982
ON THE BASIS OF STUDIES IN ANIMALS AND IN THE ABSENCE OF ADEQUATE DATA FROM EPA STUDIES, IT APPEARS THAT HEXACHLOROBENZENE PRESENTS A CARCINOGENIC RISK TO HUMANS. HOWEVER, IT IS REASONABLE TO ASSUME THAT AMOUNTS PRESENT IN THE AVERAGE U.S. DIET DO NOT MAKE A SUBSTANTIAL CONTRIBUTION TO THE OVERALL RISK OF CANCER FOR HUMANS.

REPT'D USE : CHEMICAL INTERMEDIATE, DYE INTERMEDIATE, FUNGICIDE, FR CHEMICAL AND WOOD PRESERVATIVE

CAS NO : **35694-04-3**

CHEM NAME : ***** 2,2',3,3',5,5'-HEXACHLOROBIPHENYL**

REPORT 01 : NTP TECHNICAL BULLETIN 3 DEC 80
TERATOLOGY ASSAYS SHOW THAT ORAL ADMINISTRATION OF 2,2',3,3',5,5'-HEXACHLOROBIPHENYL CAUSED PRENATAL DEATH IN MICE.

REPT'D USE : NO LISTING ON TSCA INVENTORY

CAS NO : **87-68-3**

CHEM NAME : ***** HEXACHLORO-1,3-BUTADIENE**

REPORT 01 : IARC MONOGRAPH 20 1979
HEXACHLOROBUTADIENE WAS TESTED IN ONE EXPERIMENT IN RATS BY ORAL ADMINISTRATION: IT PRODUCED BENIGN AND

MALIGNANT TUMORS IN THE KIDNEYS OF ANIMALS OF BOTH SEXES.

REPORT 02 : ITC FIRST REPORT TO EPA (42FR55026) OCT 77
THE INTERAGENCY TESTING COMMITTEE RECOMMENDS THAT THE EPA ADMINISTRATOR CONSIDER THE DEVELOPMENT OF TEST RULES TO TEST HEXACHLOROBUTADIENE FOR ENVIRONMENTAL EFFECTS. HUMAN HEALTH EFFECTS OF HEXACHLOROBUTADIENE ARE BEING STUDIED IN DEPTH.

REPORT 03 : EPA AMBIENT WATER QUALITY CRITERIA OCT 80
THE EVIDENCE OF CARCINOGENICITY IS SUFFICIENT TO CONCLUDE THAT HCBD IS A SUSPECT HUMAN CARCINOGEN. AMBIENT WATER CONCENTRATIONS OF 4.46 UG/L, .45 UG/L, AND .045 UG/L ARE ESTIMATED TO RESULT IN AN INCREASED CANCER RISK OVER A LIFETIME OF 10 -5, 10 -6, AND 10 -7 RESPECTIVELY.

REPORT 04 : EPA RCRA HEALTH AND ENV. EFFECTS PROFILE OCT 80
HCBD HAS BEEN FOUND TO INDUCE RENAL NEOPLASMS IN RATS. THE MUTAGENICITY OF HCBD HAS NOT BEEN PROVEN CONCLUSIVELY, BUT A BACTERIAL ASSAY SUGGESTS A POSITIVE RESULT. TWO STUDIES ON THE POSSIBLE TERATOGENIC EFFECTS OF HCBD PRODUCED CONFLICTING RESULTS.

REPORT 05 : IARC MONOGRAPH 20 1979
HEXACHLOROBUTADIENE WAS REPORTED TO BE MUTAGENIC IN SPOT TESTS WITH SALMONELLA STRAIN TA100, BUT NO DATA WERE GIVEN.

REPORT 06 : MULTI-AGENCY CROSS CHECK
N HHS MAGUIRE
N OSHA CARC
Y EPA CAG
N EPA RCRA
N EPA CLEAN AIR
Y EPA CLEAN WATER

REPORT 07 : NTP TECHNICAL BULLETIN 3/7 DEC 80, APR 82
HEXACHLOROBUTADIENE WAS NEGATIVE IN THE SALMONELLA, RAT DOMINANT LETHAL, THE MOUSE SPERM HEAD MORPHOLOGY, THE DROSOPHILA SEX-LINKED RECESSIVE LETHAL AND RAT BONE MARROW CYTOLOGY ASSAYS. RESULTS IN THE UNSCHEDULED DNA SYNTHESIS IN HUMAN FIBROBLASTS WERE SUGGESTIVE BUT NOT CLEARLY POSITIVE.

REPORT 08 : EPA RESPONSE TO ITC 47FR58029 12/29/82 DEC 82
IN RESPONSE TO ITC'S 1977 RECOMMENDATION THAT HCBD BE TESTED FOR ENVIRONMENTAL EFFECTS UNDER TSCA 4(A), EPA CONCLUDES THAT CURRENT, MANUFACTURE, USE AND DISPOSAL DOES NOT PRESENT AN UNREASONABLE RISK TO THE ENVIRONMENT. ALSO HCBD KNOWN TO BE A HAZARD AT LOW LEVELS AND

REGULATORY AND VOLUNTARY MEASURES ADEQUATE TO CONTROL RELEASE TO ENVIRONMENT HAVE BEEN TAKEN

REPT'D USE : ELASTOMER SOLVENT, TRANSFORMER/HYDRAULIC FLUID, GYROSCOPE FLUID AND GAS RECOVERY

CAS NO : **319-84-6**

CHEM NAME : ***** ALPHA-HEXACHLOROCYCLOHEXANE**

REPORT 01 : EPA AMBIENT WATER QUALITY CRITERIA OCT 80
A-HCH IS SUSPECTED OF BEING A HUMAN CARCINOGEN. AMBIENT WATER CONCENTRATIONS OF 92 NG/L, 9.2 NG/L, AND .92 NG/L ARE ESTIMATED TO RESULT IN AN INCREASED CANCER RISK OVER A LIFETIME OF 10 -5, 10 -6, AND 10 -7 RESPECTIVELY.

REPORT 02 : IARC MONOGRAPH 20 1979
THERE IS SUFFICIENT EVIDENCE THAT ALPHA-HCH, AND TECHNICAL HCH ARE CARCINOGENIC IN MICE.

REPORT03 : EPA RCRA HEALTH AND ENV. EFFECTS PROFILE OCT 80
TECHNICAL HCH, ALPHA-HCH, BETA-HCH, AND LINDANE (GAMMA-HCH) HAVE ALL BEEN SHOWN TO INDUCE LIVER TUMORS IN MICE.

REPORT 04 : IARC MONOGRAPH 20 1979
REVERSION STUDIES USING SACCHAROMYCES WERE NEGATIVE WITH A-HCH, BOTH IN THE PRESENCE AND ABSENCE OF METABOLIC ACTIVATION. A-HCH WAS NOT MUTAGENIC IN BACTERIA OR DROSOPHILA.

REPORT 06 : MULTI-AGENCY CROSS CHECK
Y HHS MAGUIRE
N OSHA CARC
N EPA CAG
N EPA RCRA
N EPA CLEAN AIR
Y EPA CLEAN WATER

REPT'D USE : PESTICIDE CONTAMINANT

CAS NO : **319-85-7**

CHEM NAME : ***** BETA-HEXACHLOROCYCLOHEXANE**

REPORT 01 : EPA AMBIENT WATER QUALITY CRITERIA OCT 80
B-HCH IS SUSPECTED OF BEING A HUMAN CARCINOGEN. AMBIENT WATER CONCENTRATIONS OF 163 NG/L, 16.3 NG/L, AND 1.63 NG/L ARE ESTIMATED TO RESULT IN AN INCREASED CANCER RISK OVER A LIFETIME OF 10 -5, 10 -6, AND 10 -7 RESPECTIVELY.

REPORT 02 : IARC MONOGRAPH 20 1979
THERE IS SUFFICIENT EVIDENCE THAT ALPHA-HCH, LINDANE, AND TECHNICAL HCH ARE CARCINOGENIC IN MICE; THERE IS LIMITED EVIDENCE THAT BETA-HCH IS CARCINOGENIC IN MICE. B-HCH WAS NOT MUTAGENIC IN BACTERIA, YEAST, OR DROSOPHILA.

REPORT 03 : EPA RCRA HEALTH AND ENV. EFFECTS PROFILE OCT 80
TECHNICAL-HCH, ALPHA-HCH, BETA-HCH, AND LINDANE (GAMMA-HCH) HAVE ALL BEEN SHOWN TO INDUCE LIVER TUMORS IN MICE.

REPORT 06 : MULTI-AGENCY CROSS CHECK
Y HHS MAGUIRE
N OSHA CARC
N EPA CAG
N EPA RCRA
N EPA CLEAN AIR
Y EPA CLEAN WATER
REPT'D USE : PESTICIDE CONTAMINANT

CAS NO : **608-73-1**
CHEM NAME : ***** 1,2,3,4,5,6-HEXACHLOROCYCLOHEXANE (HCH)**

REPORT 01 : IARC MONOGRAPH 20 1979
THERE IS SUFFICIENT EVIDENCE THAT ALPHA-HCH, LINDANE, AND TECHNICAL HCH ARE CARCINOGENIC IN MICE; THERE IS LIMITED EVIDENCE THAT BETA-HCH IS CARCINOGENIC IN MICE.

REPORT 02 : EPA AMBIENT WATER QUALITY CRITERIA OCT 80
TECHNICAL HCH IS SUSPECTED OF BEING A HUMAN CARCINOGEN. AMBIENT WATER CONCENTRATIONS OF 123 NG/L, 12.3 NG/L, AND 1.23 NG/L ARE ESTIMATED TO RESULT IN AN INCREASED CANCER RISK OVER A LIFETIME OF 10 -5, 10 -6, AND 10 -7 RESPECTIVELY.

REPORT 03 : EPA RCRA HEALTH AND ENV. EFFECTS PROFILE OCT 80
TECHNICAL HCH, ALPHA-HCH, BETA-HCH, AND LINDANE (GAMMA-HCH) HAVE ALL BEEN SHOWN TO INDUCE LIVER TUMORS IN MICE.

REPORT 06 : MULTI-AGENCY CROSS CHECK
Y HHS MAGUIRE
Y OSHA CARC
Y EPA CAG
Y EPA RCRA
N EPA CLEAN AIR
Y EPA CLEAN WATER
REPT'D USE : INSECTICIDE

CAS NO : **77-47-4**
CHEM NAME : ***** HEXACHIOROCYCLOPENTADIENE**

REPORT 01 : 47FR58023: EPA RESPONSE TO ITC DEC 82
IN RESPONSE TO ITC JUNE 1979 DESIGNATING HCCP FOR HEALTH & ENVIRONMENTAL EFFECTS TESTING UNDER TSCA 4(A), EPA CONCLUDES THERE IS INSUFFICENT EVIDENCE THAT CURRENT USAGE AND EXPOSURE PRESENT AN UNREASONABLE RISK TO THE ENVIRONMENT OR PEOPLE OR THAT THERE ARE MUTAGENIC OR TERATOGENIC HEALTH EFFECTS. ALSO ADEQUATE DATA EXIST PREDICT CHRONIC HEALTH EFFECTS AND AN ONCOGENICITY BIOASSAY IS UNDER WAY.

CAS NO : **57653-85-7**
CHEM NAME : *****1,2,3,6,7,8-HEXACHLORODIBENZO-P-DIOXIN (HCDD)**

REPORT 01 : NCI CARC TRS 202 NIH PUB NO 80-1758
A BIOASSAY OF A MIXTURE OF 1,2,3,6,7,8- AND 1,2,3,7,8,9-HEXACHLORODIBENZO-P-DIOXINS (HCDD) FOR POSSIBLE CARCINOGENICITY WAS CONDUCTED BY DERMAL APPLICATION OF A SUSPENSION OF THIS SUBSTANCE TO SWISS-WEBSTER MICE. UNDER THE CONDITIONS OF THIS BIOASSAY, HCDD WAS NOT CARCINOGENIC FOR MALE OR FEMALE SWISS-WEBSTER MICE.

REPORT 02 : NCI CARC TRS 198 NIH PUB NO 80-1754
UNDER THE CONDITIONS OF THIS BIOASSAY, HCDD ADMINISTERED BY GAVAGE WAS CARCINOGENIC, CAUSING INCREASED INCIDENCES OF HEPATOCELLULAR CARCINOMAS OR NEOPLASTIC NODULES IN FEMALE OSBORNE-MENDEL RATS AND INDUCING HEPATOCELLULAR CARCINOMAS AND ADENOMAS IN MALE AND FEMALE B6C3F1 MICE. HCDD WAS NOT DEMONSTRATED TO BE CARCINOGENIC FOR MALE RATS.

REPT'D USE : IN SITU CONTAMINANT IN CHLOROPHENOL WOOD PRESERVATIVES. NO KNOWN COMMERCIAL PRODUCTION. NO LISTING ON TSCA INVENTORY.

CAS NO : **19408-74-3**
CHEM NAME : *****1,2,3,7,8,9-HEXACHLORODIBENZO-P-DIOXIN (HCDD)**

REPORT 01 : NCI CARC TRS 202 NIH PUB NO 80-1758
A BIOASSAY OF A MIXTURE OF 1,2,3,6,7,8- AND 1,2,3,7,8,9-HEXACHLORODIBENZO-P-DIOXINS (HCDD) FOR POSSIBLE CARCINOGENICITY WAS CONDUCTED BY DERMAL APPLICATION OF A SUSPENSION OF THIS SUBSTANCE TO SWISS-WEBSTER MICE. UNDER THE CONDITIONS OF THIS BIOASSAY, HCDD WAS NOT CARCINOGENIC FOR MALE OR FEMALE SWISS-WEBSTER MICE.

REPORT 02 : NCI CARC TRS 198 NIH PUB NO 80-1754
UNDER THE CONDITIONS OF THIS BIOASSAY, HCDD ADMINISTERED BY GAVAGE WAS CARCINOGENIC, CAUSING INCREASED INCIDENCES OF HEPATOCELLULAR CARCINOMAS OR NEOPLASTIC NODULES IN FEMALE OSBORNE-MENDEL RATS AND INDUCING HEPATOCELLULAR CARCINOMAS AND ADENOMAS IN MALE AND FEMALE B6C3F1 MICE. HCDD WAS NOT DEMONSTRATED TO BE CARCINOGENIC FOR MALE RATS.

REPT'D USE : IN SITU CONTAMINANT IN CHLOROPHENOL WOOD PRESERVATIVES. NO KNOWN COMMERCIAL PRODUCTION. NO LISTING ON TSCA INVENTORY.

CAS NO : **67-72-1**
CHEM NAME : ***** HEXACHLOROETHANE**

REPORT 01 : NCI CARC TRS 68 DHEW PUB NO 78-1318
NO EVIDENCE WAS PROVIDED FOR THE CARCINOGENICITY OF THE COMPOUND IN OSBORNE-MENDEL RATS. IT IS CONCLUDED THAT UNDER THE CONDITIONS OF THIS BIOASSAY, HEXACHLOROETHANE WAS CARCINOGENIC IN B6C3F1 MICE, INDUCING HEPATOCELLULAR CARCINOMAS IN BOTH SEXES. HEXACHLOROETHANE WAS ADMINISTERED BY GAVAGE.

REPORT 02 : NIOSH CURRENT INTELLIGENCE BULLETIN 27 AUG 78
NCI BIOASSAY RESULTS INDICATE THAT HEXACHLOROETHANE IS CARCINOGENIC IN MICE, INDUCING LIVER CANCER IN BOTH SEXES. NO CONCLUSIVE EVIDENCE OF CANCER IN RATS WAS FOUND. HEXACHLOROETHANE ACCELERATED THE MORTALITY OF TESTED RATS WITH INCREASED DOSAGE AND MAY HAVE OBSCURED A CARCINOGENIC EFFECT. THE COMPOUND INDUCES LIVER AND KIDNEY DAMAGE IN TEST ANIMALS. OSHA EXPOSURE STANDARD IS 1 PPM BUT NIOSH HAS NO CURRENT RECOMMENDED EXPOSURE STANDARD. NIOSH RECOMMENDS THAT OCCUPATIONAL EXPOSURE TO HEXACHLOROETHANE BE MINIMIZED.

REPORT 03 : NTP TECHNICAL BULLETIN 2 APR 82
THE COMPOUND WAS FOUND TO BE NEGATIVE IN THE SALMONELLA TYPHIMURIUM MUTAGENICITY ASSAY IN ONE LABORATORY.

REPORT 04 : IARC MONOGRAPH 20 1979
HEXACHLOROETHANE WAS TESTED IN ONE EXPERIMENT IN MICE AND IN ONE IN RATS BY ORAL ADMINISTRATION. IN MICE, IT PRODUCED MALIGNANT LIVER TUMORS IN MALES AND FEMALES. IN RATS, NO STATISTICALLY SIGNIFICANT EXCESS OF TUMORS WAS OBSERVED; HOWEVER, A FEW RENAL TUMORS, RARELY SEEN IN UNTREATED ANIMALS, WERE FOUND. NO DATA ON THE EMBRYOTOXICITY, TERATOGENICITY OR MUTAGENICITY OF HEXACHLOROETHANE WERE AVAILABLE.

REPORT 05 : EPA AMBIENT WATER QUALITY CRITERIA OCT 80
HCE IS CONSIDERED TO BE CARCINOGENIC IN AT LEAST ONE RODENT SPECIES. AMBIENT WATER CONCENTRATIONS OF 19 UG/L, 1.9 UG/L, AND .19 UG/L ARE ESTIMATED TO KEEP THE LIFETIME CANCER RISK BELOW 10 -5, 10 -6, AND 10 -7 RESPECTIVELY.

REPORT 06 : MULTI-AGENCY CROSS CHECK
N HHS MAGUIRE
N OSHA CARC
Y EPA CAG
Y EPA RCRA
N EPA CLEAN AIR
Y EPA CLEAN WATER

REPORT 07 : EPA RCRA HEALTH AND ENV. EFFECTS PROFILE OCT 80
RESULTS OF AN NTP BIOASSAY SHOWED THAT HEXA-CHLOROROETHANE PRODUCED AN INCREASE IN HEPATOCELLULAR CARCINOMA INCIDENCE IN MICE. TESTING OF HEXACHLORO-ETHANE IN THE AMES SALMONELLA ASSAY SHOWED NO MUTA-GENIC EFFECTS. NO TERATOGENIC EFFECTS WERE OBSERVED FOLLOWING ORAL OR INHALATION EXPOSURE OF RATS TO HEXACHLOROETHANE, BUT SOME TOXIC EFFECTS ON FETAL DEVELOPMENT WERE OBSERVED.

REPORT 08 : EPA RESPONSE TO ITC RECOMMS (47FR18175) APR 82
IN RESPONSE TO THE ITC RECOMMENDATIONS (46FR28138) THE EPA IS NOT INITIATING RULEMAKING TO REQUIRE FURTHER HEALTH EFFECTS TESTING OF HCE BECAUSE, GIVEN THE SMALL AMOUNTS OF THE SUBSTANCE WHICH WILL BE RELEASED TO THE ENVIRON-MENT, THERE IS NO BASIS FOR BELIEVING THAT THE CHEMICAL MAY PRESENT AN UNREASONABLE RISK.

REPORT 11 : NTP TR361 OCT 89
UNDER THE CONDITIONS OF THESE 2-YEAR GAVAGE STUDIES, THERE WAS CLEAR EVIDENCE OF CARCINOGENIC ACTIVITY OF HEXACHLOROETHANE FOR MALE F344/N RATS, BASED ON THE INCREASED INCIDENCES OF RENAL NEOPLASMS. THE MARGINALLY INCREASED INCIDENCES OF PHEOCHROMOCYTOMAS MAY HAVE BEEN RELATED TO HEXACHLOROETHANE ADMINISTRATION TO MALE RATS. THERE WAS NO EVIDENCE OF CARCINOGENIC ACTIVITY OF HEXACHLOROETHANE FOR FEMALE F344/N RATS ADMINISTERED 80 OR 160 MG/KG BY GAVAGE FOR 103 WEEKS.

REPT'D USE : CHEMICAL INTERMEDIATE, INSECTICIDE, PLASTICIZER, DEGASSING AGENT FOR MAGNESIUM, IGNITION SUPPRESSANT, FIRE-EXTINGUISHING FLUIDS, RUBBER VULCANIZING, ACCEL-ERATOR, VETERINARY MEDICINE, SUBMARINE PAINT AND PYRO-TECHNICS

CAS NO : **3389-71-7**
CHEM NAME : ***** 1,2,3,4,7,7-HEXACHLORONORBORNADIENE**

REPORT 01 : EPA RESPONSE TO ITC, 49FR45654 NOV 84
THIS IS EPA'S RESPONSE TO ITC'S RECOMMENDATION THAT EPA CONSIDER HEALTH EFFECTS TESTING OF CPD. EPA IS NOT CONSIDERING TESTING UNDER TSCA 4(A) AT THIS TIME BECAUSE THE LIMITED MFR AND CONTROLLED DISPOSAL OF CHEMICAL IS NOT EXPECTED TO CAUSE SIGNIFICANT RISK OF INJURY TO HUMAN HEALTH OR THE ENVIRONMENT. EPA PLANNING TO INITIATE RULE MAKING UNDER TSCA 8(A) OR 5(A)(2) TO MONITOR ANY SIGNIFICANT CHANGES IN THESE CONDITIONS.

REPT'D USE : INTERMEDIATE IN MFR PESTICIDE ENDRIN

CAS NO : **70-30-4**
CHEM NAME : ***** HEXACHLOROPHENE**

REPORT 01 : NCI CARC TRS 40 DHEW PUB NO 78-840
UNDER THE CONDITIONS OF THIS BIOASSAY, HEXACHLOROPHENE DID NOT INDUCE MALIGNANT OR BENIGN TUMORS IN FISCHER 344 RATS (FEEDING STUDY).

REPORT 02 : IARC MONOGRAPH 20 1979
HEXACHLOROPHENE WAS TESTED IN ONE EXPERIMENT IN RATS BY ORAL ADMINISTRATION; IT HAD NO CARCINOGENIC EFFECT. IT WAS INADEQUATELY TESTED IN ONE EXPERIMENT IN MICE BY SKIN APPLICATION. HEXACHLOROPHENE IS EMBRYOTOXIC AND PRODUCES SOME TERATOGENIC EFFECTS. IT WAS NOT MUTAGENIC IN SALMONELLA TYPHIMURIUM AND WAS NEGATIVE IN A DOMINANT LETHAL ASSAY IN MALE MICE.

REPORT 03 : NTP TECHNICAL BULLETIN 3 DEC 80
THE COMPOUND WAS FOUND TO BE NEGATIVE IN THE SALMONELLA TYPHIMURIUM MUTAGENICITY ASSAY IN ONE LABORATORY.

REPORT 04 : EPA RCRA HEALTH AND ENV. EFFECTS PROFILE OCT 80
MUTAGENIC EFFECTS OF HEXACHLOROPHENE HAVE BEEN REPORTED IN ONE STUDY WHICH INDICATED INCREASED CHROMOSOME ABERRATIONS IN RATS. SEVERAL REPORTS HAVE INDICATED HEXACHLOROPHENE MAY PRODUCE SOME TERATOGENIC AND EMBRYOTOXIC EFFECTS. HEXACHLOROPHENE HAS SHOWN SOME ADVERSE EFFECTS ON MALE REPRODUCTIVE PERFORMANCE.

REPORT 05 : IARC MONOGRAPH 20 1979
HEXACHLOROPHENE IS EMBRYOTOXIC AND PRODUCES SOME TERATOGENIC EFFECTS. IT WAS NOT MUTAGENIC IN SALMONELLA AND WAS NEGATIVE IN A DOMINANT LETHAL ASSAY IN MALE MICE. CYTOGENIC TESTS WITH CULTURED HUMAN LYMPHOCYTES WERE ALSO NEGATIVE.

REPORT 06 : MULTI-AGENCY CROSS CHECK
N HHS MAGUIRE
N OSHA CARC
N EPA CAG
Y EPA RCRA
N EPA CLEAN AIR
N EPA CLEAN WATER

REPT'D USE : LUBRICANTS, INGNITION SUPRESSANT, VETERINARY ANTIHELMINTHIC, FERMENTATION RETARDANT, TOPICAL ANTISEPTICS, ACARICIDE, SOAPS/SHAMPOOS AND COSMETICS

CAS NO : **116-15-4**
CHEM NAME : ***** HEXAFLUOROPROPENE**

REPORT 01 : PROPOSED TEST RULE, 50FR78432 NOV 85
DOCUMENT PROPOSES REQUIRING SUBCHRONIC TOXICITY TESTING AND TIERED MUTAGENICITY TESTING FOR HFP. THIS IS IN RESPONSE TO ITC'S DESIGNATING FLUOROALKANES FOR PRIORITY TESTING CONSIDERATION IN ITS 7TH ANNUAL REPORT, PUBLISHED IN 45FR78432.

REPORT 04 : PROPOSED TEST RULE, 50F546133 NOV 85
DOCUMENT PROPOSES TIERED MUTAGENICITY TESTING AND CHRONIC ONCOGENICITY BIOASSAYS, DEPENDING ON OUTCOME OF MUTAGENICITY BIOASSAYS. THIS IS EPA'S RESPONSE TO ITC'S DESIGNATING FLUORALKANES FOR PRIORITY TESTING CONSIDERATION IN ITS 7TH ANNUAL REPORT, PUBLISHED IN 45FR78432.

REPORT 05 : EXTENSION OF COMMENT PERIOD, 50FR53348 DEC 85
IN RESPONSE TO A REQUEST FROM THE FLUOROALKANES INDUSTRY GROUP, EPA IS EXTENDING THE COMMENT PERIOD ON THE PROPOSED TEST RULE ON FLUOROALKANES, ORIGINALLY PUBLISHED IN 50FR78432.

REPORT 07 : FLUOROALKENES; FINAL TEST RULE 52 FR 21516 JUN 87
EPA IS ISSUING A FINAL TEST RULE UNDER TSCA TO REQUIRE HEALTH EFFECTS TESTING FOR CERTAIN FLUOROALKENES. THIS TEST RULE BECOMES EFFECTIVE ON JULY 22, 1987.

REPT'D USE : MFR OF POLYMERS AND ELASTOMERS

CAS NO : **428-59-1**
CHEM NAME : ***** HEXAFLUOROPROPYLENE OXIDE**

REPORT 01 : PROPOSED TEST RULE, 48FR1422 DEC 83
IN RESPONSE TO ITC'S RECOMMENDATION TO REQUIRE HEALTH EFFECTS TESTING OF HALOGENATED ALKYL EPOXIDES UNDER TSCA 4(A), EPA IS NOT PROPOSING TESTING BECAUSE EPA HAS NOT FOUND EVIDENCE OF TERATOGENICITY. AN EPA STUDY IS NOT

BEING PROPOSED BECAUSE NO SUITABLE COHORT IS AVAILABLE.
REPT'D USE : INTERMEDIATE IN MFR OF FLUORINATED CPDS.

CAS NO : **680-31-9**
CHEM NAME : ***** HEXAMETHYLPHOSPHORAMIDE (HMPA)**

REPORT 01 : NIOSH CURRENT INTELLIGENCE BULLETIN 6 OCT 75
DU PONT REPORTED TO NIOSH IN 1975 THAT NASAL TUMORS HAVE BEEN OBSERVED IN RATS EXPOSED TO HEXAMETHYLPHOSPHORIC TRIAMIDE. NO DATA AVAILABLE ON HUMAN TOXICITY. PRESENTLY, OSHA HAS NO STANDARD.

REPORT 02 : IARC MONOGRAPH 15 1977
HEXAMETHYLPHOSPHORAMIDE IS CARCINOGENIC IN RATS, THE ONLY SPECIES TESTED, FOLLOWING ITS ADMINISTRATION BY INHALATION; IT PRODUCED SQUAMOUS CELL CARCINOMAS OF THE NASAL CAVITY. IN A 1979 SUPPLEMENT, IARC DETERMINED THAT THERE IS SUFFICIENT EVIDENCE FOR THE CARCINOGENICITY OF HMPA IN ANIMALS.

REPORT 03 : IARC MONOGRAPH 15 1977
NO CHROMOSOME ABERRATIONS WERE OBSERVED IN CHINESE HAMSTER LUNG CELLS TREATED WITH HEXAMETHYLPHOSPHORAMIDE. HMPA HAS BEEN SHOWN TO INDUCE DOMINANT LETHALS IN MALE MICE INJECTED I.P.

REPORT 04 : EPA TSCA FYI NOTICE OTS-0382-0400 SUPP MAR 82
DUPONT REPORTED FINAL RESULTS FROM A TWO-YEAR INHALATION STUDY IN C-D RATS. NASAL TUMORS WERE FIRST DETECTED AFTER 7 MONTHS EXPOSURE AT 4000 AND 400 PPB, 9 MONTHS AT 100 PPB, AND 12 MONTHS AT 50 PPB. A TOTAL OF 473 NASAL CARCINOMAS AND PAPILLOMAS WERE REPORTED. NO HMPA-RELATED LESIONS WERE FOUND AT 10 PPB.

REPORT 05 : EPA TSCA 8(E) NOTICE 8EHQ-0679-0040
DUPONT PROVIDED PATHOLOGY RESULTES FROM A 2 YEAR INHALATION TEST OF HEXAMETHYLPHOSPHORAMIDE IN C-D RATS. NASAL TUMORS WERE FIRST OBSERVED AFTER APPROXIMATELY 7 MONTHS EXPOSURE AT 4000 AND 400 PPB, 9 MONTHS AT PPB, AND 12 MONTHS AT 50 PPB LEVEL. NO HMPA-RELATED LESIONS WERE FOUND AT 10 PPB.

REPORT 06 : MULTI-AGENCY CROSS CHECK
Y HHS MAGUIRE
Y OSHA CARC
N EPA CAG
N EPA RCRA
N EPA CLEAN AIR
N EPA CLEAN WATER

REPT'D USE : JET FUEL DE-ICER, POLYMERIZATION CATALYST/STABILIZER, INSECT CHEMOSTERILANT RESEARCH CHEMICAL AND POLYMER PROCESSING SOLVENT

CAS NO : **110-54-3**
CHEM NAME : ***** HEXANE**

REPORT 01 : NIOSH CRITERIA DOCUMENT ON ALKANES MAR 77
NIOSH RECOMMENDS A STANDARD OF 100 PPM HEXANE TWA BASED ON SKIN AND CNS EFFECTS. THE PRESENT OSHA STANDARD IS 500 PPM TWA.

REPORT 02 : PETITION RECEIVED
AUG 1979 PETITION REQUESTING A BAN OF CONSUMER PRODUCTS CONTAINING N-HEXANE BECAUSE OF THE POTENTIAL FOR SEVERE NEUROLOGICAL DAMAGE WAS DENIED BY THE COMMISSION BECAUSE AVAILABLE INFORMATION WAS NOT SUFFICIENT TO INDICATE THAT EXPOSURE PRESENTED AN UNREASONABLE RISK OF INJURY. STAFF WAS DIRECTED TO CONTINUE RESEARCH ON TOXICITY AND EXPOSURE.

REPORT 03 : CIIT MONTHLY ACTIVITIES REPORT DOCKET APR 77
IN A PRECHRONIC EXPOSURE STUDY, CONCENTRATIONS OF 3,000 PPM, 6,500 PPM, AND 10,000 PPM OF N-HEXANE WERE ADMINISTERED TO F344 RATS BY INHALATION. FINDINGS FROM THIS 90-DAY SUBCHRONIC INHALATION TOXICITY STUDY HAVE BEEN CAREFULLY REVIEWED BY CIIT. NO SIGNIFICANT TOXIC EFFECTS WERE SEEN. IN LIGHT OF THIS, AND OTHER DATA, DECISION WAS MADE TO CANCEL THE PROPOSED LIFETIME INHALATION TOXICITY STUDY PLANNED FOR N-HEXANE.

REPORT 04 : NTP TEST REGIMEN
IN A CURRENT NTP BIOASSAY, N-HEXANE IS BEING ADMINISTERED TO F344 RATS BY INHALATION. THE PRECHRONIC TESTING PHASE BEGAN AUGUST 80.

REPORT 05 : NTP TECHNICAL BULLETIN 6 JAN 82
THE COMPOUND WAS NEGATIVE IN THE SALMONELLA TYPHIMURIUM MUTAGENICITY ASSAY IN ONE LABORATORY.

REPORT 07 : 53 FR 3382 FEB 88
UNDER TSCA 4(A) EPA ISSUING FINAL TEST RULE REQUIRING MANUFACTURERS TO PERFORM TESTING FOR SUBCHRONIC TOXICITY, ONCOGENICITY, REPRODUCTIVE TOXICITY, MUTAGENICITY, DEVELOPMENTAL TOXICITY, NEUROTOXICITY, AND INHALATION AND DERMAL PHARMACOKINETICS.

REPORT 09 : EPA FYI-OTS- 0888-0633 AUG 88
INHALATION DEVELOPMENT STUDIES CARRIED OUT FOR NIEHS

SHOWED RATS AT EXPOSURE GIVING MATERIAL TOXICITY PRODUCED NO INCREASE IN INCIDENCE OF INTRAUTERINE DEATHS ON FETAL MALFORMATIONS. THERE WAS A REDUCTION IN FETAL WEIGHT AT 1000 AND 5000 PPM EXPOSIVES, AND SLIGHT GROWTH RETARDATION AT THE 5000 PPPM. IN MICE THERE WAS INCREASE IN RESONBED FETUSES AT 200, 1000 AND 5000 PPM. THERE WAS NO TERATOGENIC EFFECT.

REPT'D USE : SOLVENT, ADHESIVES, CLEANING AGENTS, PAINTS/VARNISHES AND FUELS

CAS NO : **136-77-6**

CHEM NAME : ***** 4-HEXYLRESORCINOL**

REPORT 01 : NTP TR 330 NIH PUB. 88-2586 MAY 88

IN 2 YEAR GAVAGE STUDIES, NO EVIDENCE OF CARCINOGENIC ACTIVITY OF 4-HEXYLRESORCINOL FOR MALE OR FEMALE F344/N RATS GIVEN DOSES OF 62.5 OR 125 MG/KG. EQUIVOCAL EVIDENCE OF CARCINOGENIC ACTIVITY FOR MALE B6C3F1 MICE, SHOWN BY MARGINALLY INCREASED INCIDENCES OF PHEOCHROMOCYTOMAS (& HYPERPLASIA) OF THE ADRENAL MEDULLA & OF HARDERIAN GLAND NEOPLASMS. NO EVIDENCE OF CARCINOGENIC ACTIVITY FOR FEMALE B6C3F1 MICE GIVEN DOSES OF 62.5 OR 125 MG/KG. DECREASED INCIDENCES OF THREE TUMOR TYPES CONSIDERED RELATED TO 4-HEXYLRESORCINOL ADMINISTRATION: MONONUCLEAR CELL LEUKEMIA IN MALE & FEMALE RATS, HEPATOCELLULAR PEOPLASMS IN MALE MICE, AND CIRCULATORY SYSTEM TUMORS IN MALE & FEMALE MICE.

REPT'D USE : ANTISEPTIC; ANTHELMINTIC DRUG

CAS NO : **23255-93-8**

CHEM NAME : ***** HYCANTHONE MESYLATE**

REPORT 01 : IARC MONOGRAPH 13 1977

HYCANTHONE MESYLATE IS CARCINOGENIC IN MICE PREVIOUSLY INFECTED WITH SCHISTOSOMA MANSONI: A SIGNIFICANT INCREASE IN THE INCIDENCE OF HEPATOCELLULAR CARCINOMAS WAS OBSERVED IN THESE MICE FOLLOWING REPEATED INTRAMUSCULAR INJECTIONS OF THIS COMPOUND. THE LIMITED DATA AVAILABLE ARE NOT SUFFICIENT TO EVALUATE WHETHER HYCANTHONE MESYLATE IS CARCINOGENIC IN NON-INFECTED ANIMALS.

REPORT 02 : IARC MONOGRAPH 13 1977

HYCANTHONE MESYLATE INDUCES POINT MUTATIONS IN SALMONELLA TA1538 WITH OUT METABOLIC ACTIVATION, AND SEX-LINKED RECESSIVE LETHALS IN DROSOPHILA. DOMINANT LETHALS HAVE BEEN REPORTED IN RATS TREATED BY I.P. INJECTION. CYTOGENIC EFFECTS HAVE ALSO BEEN REPORTED IN VIVO IN RAT BONE-MARROW CELLS.

REPT'D USE : ANTI-SCHISTOSOMAL AGENT

CAS NO : **86-54-4**
CHEM NAME : ***** HYDRALAZINE**

REPORT 01 : IARC MONOGRAPH 24 1980
HYDRALAZINE IS MUTAGENIC FOR SALMONELLA TYPHIMURIUM. ATTENTION IS DRAWN TO THE ABSENCE OF PUBLISHED STUDIES ON THE TERATOGENICITY OF THIS COMPOUND. TWO STUDIES HAVE SUGGESTED AN ASSOCIATION BETWEEN HYDRALAZINE AND HUMAN CANCER. THE EPIDEMIOLOGICAL DATA WERE INSUFFICIENT.

REPORT 02 : IARC MONOGRAPH 24 1980
HYDRALAZINE IS MUTAGENIC IN SALMONELLA TA1530, AND TA1537, AND TA100 WITH OR WITHOUT METABOLIC ACTIVATION. THE COMPOUND IS ALSO ACTIVE IN E. COLI INDICATING AN INTERACTION WITH DNA. ATTENTION IS DRAWN TO THE ABSENCE OF PUBLISHED STUDIES ON THE TERATOGENICITY OF THIS COMPOUND.
REPT'D USE : CHEMICAL INTERMEDIATE AND ANTI-HYPERTENSIVE AGENT

CAS NO : **304-20-1**
CHEM NAME : ***** HYDRALAZINE HYDROCHLORIDE**

REPORT 01 : IARC MONOGRAPH 24 1980
HYDRALAZINE HYDROCHLORIDE WAS TESTED IN ONE EXPERIMENT IN MICE BY ORAL ADMINISTRATION. A SIGNIFICANT INCREASE IN THE INCIDENCE OF LUNG TUMORS WAS REPORTED. THE EXPERIMENTAL DATA, WHILE PROVIDING LIMITED EVIDENCE FOR THE CARCINOGENICITY OF HYDRALAZINE HYDROCHLORIDE, WERE DIFFICULT TO INTERPRET DUE TO CERTAIN ASPECTS OF EXPERIMENTAL DESIGN AND ANALYSIS.
REPT'D USE : ANTIHYPERTENSIVE AGENT

CAS NO : **302-01-2**
CHEM NAME : ***** HYDRAZINE**

REPORT 02 : IARC MONOGRAPH 4 1974
HYDRAZINE OR HYDRAZINE SALTS HAVE BEEN SHOWN TO BE CARCINOGENIC IN MICE AFTER ORAL AND INTRAPERITONEAL ADMINISTRATION, AND IN RATS FOLLOWING ORAL ADMINISTRATION. IN A 1979 SUPPLEMENT, IARC DETERMINED THAT THERE IS SUFFICIENT EVIDENCE FOR THE CARCINOGENICITY OF HYDRAZINE IN ANIMALS.

REPORT 03 : NIOSH EXPOSURE STANDARD JUN 78
NIOSH RECOMMENDS THAT OCCUPATIONAL EXPOSURE TO

HYDRAZINES BE CONTROLLED SO THAT EMPLOYEES ARE NOT EXPOSED TO CONCENTRATIONS GREATER THAN THE FOLLOWING LIMITS, EXPRESSED AS MG/CU M, DETERMINED AS CEILING CONCENTRATIONS IN ANY 2-HOUR PERIOD: HYDRAZINE-0.04 MG/CU M (0.03 PPM); METHYLHYDRAZINE-0.08 MG/CU M (0.04 PPM); 1,1-DIMETHYLHYDRAZINE-0.15 MG/CU M (0.06 PPM); PHENYLHYDRAZINE-0.6 MG/CU M (0.14 PPM). NO LIMIT IS RECOMMENDED FOR 1,2-DIMETHYLHYDRAZINE, SINCE AN ACCEPTABLE METHOD OF SAMPLING AND ANALYSIS IS PRESENTLY UNAVAILABLE.

REPORT 06 : MULTI-AGENCY CROSS CHECK
Y HHS MAGUIRE
N OSHA CARC
Y EPA CAG
Y EPA RCRA
N EPA CLEAN AIR
N EPA CLEAN WATER
REPT'D USE : CHEMICAL INTERMEDIATE, ROCKET FUEL, OXYGEN SCAVINGER IN BOILER SYSTEMS, METAL PLATING, POLYMERIZATION CATALYST AND BLOWING AGENT

CAS NO : **7803-57-8**
CHEM NAME : ***** HYDRAZINE MONOHYDRATE**

REPORT 06 : MULTI-AGENCY CROSS CHECK
N HHS MAGUIRE
Y OSHA CARC
N EPA CAG
N EPA RCRA
N EPA CLEAN AIR
N EPA CLEAN WATER
REPT'D USE : LABORATORY SOLVENT

CAS NO : **10034-93-2**
CHEM NAME : ***** HYDRAZINE SULFATE (1:1)**

REPORT O1 : NTP TECHNICAL BULLETIN 8 JUL 82
HYDRAZINE SULFATE WAS NEGATIVE IN DROSOPHILA WHEN TESTED FOR SEX-LINKED RECESSIVE LETHALS.

REPORT 06 : MULTI-AGENCY CROSS CHECK
Y HHS MAGUIRE
Y OSHA CARC
N EPA CAG
N EPA RCRA
N EPA CLEAN AIR
N EPA CLEAN WATER

REPT'D USE : CATALYST, FUNGICIDE, GERMICIDE, METAL REFINING, ANTIOXIDANT AND HYDRAZINE HYDRATE PREPARATIONS

CAS NO : **109-84-2**
CHEM NAME : ***** 2-HYDRAZINOETHANOL**

REPORT 06 : MULTI-AGENCY CROSS CHECK
N HHS MAGUIRE
Y OSHA CARC
N EPA CAG
N EPA RCRA
N EPA CLEAN AIR
N EPA CLEAN WATER
REPT'D USE : ORGANIC SYNTHESIS

CAS NO : **58-93-5**
CHEM NAME : ***** HYDROCHLOROTHIAZIDE**

REPORT 01 : NTP 357 NIH 357 PUB. 89-2812 JUL 89
UNDER 2-YEAR FEED STUDIES, NO EVIDENCE OF CARCINOGENIC ACTIVITY OF HYDROCHLOROTHIAZIDE FOR MALE OR FEMALE F344/N RATS FED 250, 500, OR 2,000 PPM. EQUIVOCAL EVIDENCE OF CARCINOGENIC ACTIVITY OF CARCINOGENIC ACTIVITY OF HYDROCHLOROTHIAZIDE FOR MALE B6C3F1 MICE. BASED ON HEPATOCELLULAR NEOPLASMS. NO EVIDENCE OF CARCINOGENIC ACTIVITY FOR FEMALE B6C3F1 MICE GIVEN DIETS CONTAINING 2,500 OR 5,000 PPM. CHRONIC RENAL DISEASE AND INCREASED SECONDARY LESIONS (PARATHYROID HYPERPLASIA, FIBROUS OSTEODYSTROPHY, AND MINERALIZATION IN MULTIPLE ORGANS) IN DOSED RATS.
REPT'D USE : PRESCRIPTION DRUG

CAS NO : **64742-86-5**
CHEM NAME : ***** HYDRODESULFURIZED VACUUM GAS OIL**

REPORT 01 : EPA TSCA REPORT 8EHQ-887-0687 SEP 87
TEXACO PROVIDED FINDINGS FROM CHRONIC MOUSE SKIN-PAINTING STUDY THAT SHOWED CPD CAUSED SKIN CANCER ON LAB MICE WHEN 50 UL WAS APPLIED UNDILUTED 2 TIMES/WEEK FOR SIX MONTHS ON SHAVED BACKS OF MICE. TEXACO STATED THAT THERE WAS A SIGNIFICANT INCREASE IN INCIDENCE OF PAPILLOMAS AND ADVANCED TUMORS IN THE BENZO(A)PYRENE AND HYDRODESULFURIZED VACUUM OIL GROUPS COMPARED TO NEGATIVE CONTROL GROUPS.
REPT'D USE : SITE LIMITED REFINERY PROCESS STREAM AND USED AS FEED TO CATALYTIC CRACKING OR RECYCLING IN H-OIL PROCESS

CAS NO : **7664-39-3**
CHEM NAME : ***** HYDROGEN FLUORIDE**

REPORT 01 : NIOSH CRITERIA DOCUMENT MAR 76
NIOSH RECOMMENDS A STANDARD OF 2.5 MG/CU M TWA BASED ON SKIN, EYE, AND AIRWAY IRRITATION AND BONE EFFECTS. THE PRESENT OSHA STANDARD 1S 3 PPM TWA.

REPORT 02 : EPA RCRA HEALTH AND ENV. EFFECTS PROFILE OCT 80
HYDROFLUORIC ACID HAS PRODUCED MUTAGENIC EFFECTS IN PLANTS AND DROSPHILIA AND LYMPHOCYTE CHROMOSOME ABERRATIONS IN RATS. CHRONIC EXPOSURE TO THE COMPOUND HAS PRODUCED SKELETAL FLUOROSIS, DENTAL MOTTLING, AND PULMONARY IMPAIRMENT.

REPORT 06 : MULTI-AGENCY CROSS CHECK
N HHS MAGUIRE
N OSHA CARC
N EPA CAG
Y EPA RCRA
N EPA CLEAN AIR
N EPA CLEAN WATER

REPT'D USE : INTERMEDIATE IN MFR OF FLUOROCARBONS, URANIUM PROCESSING, PETROLEUM ALKYLATION, PICKLING AGENT FOR STAINLESS STEEL, CLEANING AGENTS, RUST REMOVERS, GLASS ETCHING AND CHEMICAL INDICATOR

CAS NO : **7722-84-1**
CHEM NAME : ***** HYDROGEN PEROXIDE**

REPORT O1 : IRPTC BULLETIN 3/2 JUL 80
THE MINISTRY OF HEALTH AND WELFARE IN JAPAN HAS PROMULGATED A REGULATION TO RESTRICT USE OF H2O2 AS A FOOD ADDITIVE. THIS ACTION WAS TAKEN AFTER EVALUATING THE RESULTS OF A TWO YEAR FEEDING STUDY WHICH SHOWED THAT H2O2 WAS CARCINOGENIC FOR C57BL/6J MICE BY ORAL ADMINISTRATION.

REPORT 02 : IARC MONOGRAPH 36 FEB 85
TESTED ONLY IN MICE, BY ORAL ADMINISTRATION IN DRINKING-WATER, BY SKIN APPLICATION AND BY SUBCUTANEOUS ADMINISTRATION. ADENOMAS AND CARCINOMAS OF DUODENUM REPORTED FOLLOWING ITS ORAL ADMINISTRATION. OTHER STUDIES INADEQUATE FOR EVALUATION OF CARCINOGENICITY. ONE STUDY BY SKIN APPLICATION INDICATED NO PROMOTING ACTIVITY.

REPORT 03 : IARC MONOGRAPH 36 FEB 85
HYDROGEN PEROXIDE INDUCED DNA DAMAGE IN BACTERIA;

MUTAGENIC TO BACTERIA AND FUNGI. NOT MUTAGENIC TO INSECTS OR TO MAMMALIAN CELLS IN VITRO. INDUCED DNA DAMAGE, SISTER CHROMATID EXCHANGES AND CHROMOSOMAL ABERRATIONS IN MAMMALIAN CELLS IN VITRO.

REPT'D USE : BLEACHING AGENT, ANTI-INFECTIVE TOPICAL ANTISEPTIC, CLEANING AGENTS/CMPDS, PHOTOGRAPHIC CHEMICAL, WINE AGING, MOUTHWASHES/PHARMACEUTICALS, DENTIFRICES, CHEMICAL INTERMEDIATE AND WASTE TREATMENT

CAS NO : **123-31-9**
CHEM NAME : ***** HYDROQUINONE**

REPORT 01 : NTP TECHNICAL BULLETIN 2 APR 80
HYDROQUINONE WAS NEGATIVE IN THE SALMONELLA TYPHIMURIUM MUTAGENICITY ASSAY IN ONE LABORATORY.

REPORT 02 : IARC MONOGRAPH 15 1977
IN BLADDER IMPLANTATION STUDIES, HYDROQUINONE IN CHOLESTEROL PELLETS INCREASED THE INCIDENCE OF BLADDER CARCINOMAS IN MICE. THE AVAILABLE DATA DO NOT ALLOW AN EVALUATION OF THE CARCINOGENICITY OF THE COMPOUND.

REPORT 04 : IARC MONOGRAPH 15 1977
NO DATA ON EMBRYOTOXICITY OR TERATOGENICITY OF HYDROQUINONE WERE AVAILABLE. HYDROQUINONE INDUCED CHROMOSOME ABERRATIONS IN A. CELA, V. FABA, AND C. FRAGRANS.

REPORT 05 : NIOSH CRITERIA DOCUMENT APR 78
NIOSH RECOMMENDS A STANDARD OF 2 MG/CU M CEILING BASED ON EYE AND SKIN EFFECTS. THE PRESENT OSHA STANDARD IS 2 MG/CU M TWA.

REPORT 07 : ITC FIFTH REPORT TO EPA (44FR70664) DEC 79
THE INTERAGENCY TESTING COMMITTEE RECOMMENDS THAT THE EPA ADMINISTRATOR CONSIDER THE DEVELOPMENT OF TEST RULES TO TEST THE CPD FOR CARCINOGENICITY AND TERATOGENICITY; AND THAT AN EPIDEMIOLOGIC STUDY BE CONDUCTED.

REPORT 08 : PROPOSED TEST RULE, 49FR438 JAN 84
EPA IS PROPOSING THAT MFRS AND PROCESSORS OF HQ EVALUATE ITS TOXICOKINETICS; ITS POTENTIAL TO HARM THE NERVOUS SYSTEM AND REPRODUCTIVE, TERATOGENETIC AND MUTAGENETIC EFFECTS. EPA ALSO PROPOSES THAT EPA STUDIES BE DONE TO EVALAUATE EXPLICIT TYPES AND RISKS OF ADVERSE EFFECTS FROM HUMAN EXPOSURE AND THAT CHEMICAL FATE AND ENVIRONMENTAL EFFECTS TESTING BE DONE. THIS IS PROPOSED UNDER TSCA 4(A).

REPORT 09 : TESTING REQUIREMENTS 50FR53145 DEC 85
BASED ON COMMENTS RECEIVED FOLLOWING PROPOSAL TO ISSUE TEST RULE (49 FR438), EPA IS PROMULGATING FINAL TEST RULE THAT REQUIRES MFRS AND PROCESSORS TO EVALUATE THE TOXICOKINETICS OF HQ AND DETERMINE ITS POTENTIAL FOR CNS, TERATOGENETIC AND REPRODUCTIVE EFFECTS.

REPORT 11 : NTP TR 366 NIH 90-2821 OCT 89
2-YEAR GAVAGE STUDIES SHOWED SOME EVIDENCE OF CARCINOGENIC ACTIVITY OF HYDROQUINONE FOR MALE F344/N RATS EXPOSED TO HYDROQUINONE, AS SHOWN BY MARKED INCREASES IN TUBULAR CELL ADENOMAS OF THE KIDNEY. THERE WAS SOME EVIDENCE OF CARCINOGENICACTIVITY OF HYDROQUINONE FOR FEMALE F344/N RATS, AS SHOWN BY INCREASES IN MONONUCLEAR CELL LEUKEMIA. THERE WAS NO EVIDENCE OF CARCINOGENIC ACTIVITY OF HYDROQUINONE FOR MALE B6C3F1 MICE ADMINISTERED 50 OR 100 MG/KG IN WATER BY GAVAGE. THERE WAS SOME EVIDENCE OF CARCINOGENIC ACTIVITY OF HYDROQUINONE FOR FEMALE B6C3F1 MICE, AS SHOWN BY INCREASES IN HEPATOCELLULAR NEOPLASMS, MAINLY ADENOMAS.

REPORT 12 : EPA FYI-AX- 1188-0468 NOV 88
CMA REPORTED THAT IN A SUBCHRONIC STUDY HYDROQUINONE ADMINISTERED IN WATER TO MALE AND FEMALE RATS AT 0, 20, 64, OR 200 MG/KG/DAY BY GAVAGE, AT THE 64 AND 200 DOSE LEVELS SHOWED TREMORS AND REDUCED ACTIVITY. ALSO BODY WEIGHT GROWTH REDUCED 7% IN THE 200 DOSED MALE RATS.

REPT'D USE : PAINTS/VARNISHES AND PHOTOGRAPHIC DEVELOPER, AEROSOLS, SPACE DEODORIZERS, LABORATORY AGENT, DERMATOLOGICAL PREPARATIONS, MOTOR FUELS AND OILS, CHEM. INTERMEDIATE, SKIN DEPIGMENTING AGENT, ANTIOXIDANTS AND POLYMERIZATION INHIBITOR

CAS NO : **3445-11-2**
CHEM NAME : ***** 1-(2-HYDROXYETHYL)-2-PYRROLIDONE**

REPORT 01 : EPA TSCA 8(E) NOTICE 8EHQ-0682-0448S AUG 82
GAF CORP SUBMITTED RESULTS FROM SEVERAL SHORT-TERM IN-VITRO ASSAYS. 1-(2-HYDROXYETHYL)-2-PYRROLIDONE DEMONSTRATED A POSITIVE MUTAGENIC RESPONSE IN SALMONELLA, A SUGGESTIVE MUTAGENIC RESPONSE IN CHINESE HAMSTER OVERY/HGPRT GENE MUTATION ASSAY, AND NEGATIVE RESPONSES IN CHO/SISTER CHROMATID EXCHANGE, CHO/CHROMOSOMAL ABERRATION, AND BALB/C3T3 MOUSE CELL TRANSFORMATION ASSAYS.

CAS NO : **7803-49-8**
CHEM NAME : ***** HYDROXYLAMINE**

REPORT 01 : CHEM HAZARD INORMATION PROFILE - 9/11/84 JAN 85
CAN CAUSE CHROMOSOMAL ABNORMALITIES IN MAMMALIAN CELL CULTURES AND INHIBIT DNA SYNTHESIS IN HUMAN HELA S3 CELLS. WEAKLY MUTAGENIC IN CULTURED MAMMALIAN CELLS. CARCINOGENICITY NOT DEMONSTRATED. NO TERATOGENICITY SEEN WHEN PREGNANT RATS EXPOSED BY IP INJECTION.
REPT'D USE : REDUCING AGENT, PRODUCTION OF SYNTHETIC RUBBERS, ORGANIC ANALYSES AND CHEMICAL SYNTHESES AND PHOTOGRAPHIC DEVELOPERS

CAS NO : **5470-11-1**
CHEM NAME : ***** HYDROXYLAMINE HYDROCHLORIDE**

REPORT 01 : CHEMICAL HAZARD INFORMATION PROFILE 9/11/84 JAN 85
CAN CAUSE CHROMOSOMAL ABNORMALITIES IN MAMMALIAN CELL CULTURES AND INHIBIT DNA SYNTHESIS IN HUMAN HELA 2E CELLS. CPD WEAKLY MUTAGENIC IN CULTURED MAMMALIAN CELLS; CARCINOGENICITY NOT DEMONSTRATED YET. INCREASED INCIDENCE OF MAMMARY TUMORS SEEN IN OLDER MICE, DECREASED INCIDENCE SEEN IN YOUNG MICE. TERATOGENICITY SEEN IN RABBIT AND CHICKEN EMBRYOS EXPOSED VIA INTRACOELOMIC INJECTION AND INTO AIR SAC, RESPECTIVELY. HIGH CONCENTRATIONS CAUSE CYANOSIS, CONVULSIONS AND DEATH.
REPT'D USE : REDUCING AGENTS, PRODUCTION OF SYNTHETIC RUBBERS, ORGANIC ANALYSES AND SYNTHESES AND PHOTOGRAPHIC DEVELOPERS

CAS NO : **10046-00-1**
CHEM NAME : ***** HYDROXYLAMINE SULFATE (1:1) (SALT)**

REPORT 01 : CHEMICAL HAZARD INFORMATION PROFILE - 9/11/84 JAN 85
CAN CAUSE CHROMOSOMAL ABNORMALITIES IN MAMMALIAN CELL CULTURES AND INHIBIT DNA SYNTHESIS IN HUMAN HELA S3 CELLS. CARCINOGENICITY NOT DEMONSTRATED. INCREASED INCIDENCE OF MAMMARY TUMORS IN POST-PUBESCENT MICE, DECREASE IN YOUNGER MICE. HIGH CONCENTRATIONS CAUSE CYANOSIS, CONVULSIONS AND DEATH.
REPT'D USE : CHEMICAL INTERMEDIATE

CAS NO : **10039-54-0**
CHEM NAME : ***** HYDROXYLAMINE SULFATE (2:1) SALT**

REPORT 01 : CHEMICAL HAZARD INFORMATION PROFILE 9/11/84 JAN 85
CAN CAUSE CHROMOSOMAL ABNORMALITIES IN MAMMALIAN CELL CULTURES AND INHIBIT DNA SYNTHESIS IN HUMAN HELA S3 CELLS. WEAKLY MUTAGENIC IN CULTURED MAMMALIAN CELLS. CARCINOGENICITY NOT DEMONSTRATED. INCREASED INCIDENCE OF MAMMARY TUMORS SEEN IN OLDER MICE, DECREASED INCIDENCE SEEN IN YOUNGER ANIMALS. HIGH CONCENTRATIONS CAN CAUSE CYANOSIS, CONVULSIONS AND DEATH.
REPT'D USE : INTERMEDIATE IN PRODUCTION OF CAPROLACTAM AND BASIC INGREDIENT IN NYLON-6 FIBERS.

CAS NO : **630-56-8**
CHEM NAME : ***** 17-ALPHA-HYDROXYPROGESTERONE CAPROATE**

REPORT 01 : IARC MONOGRAPH 21 1979
THIS COMPOUND WAS TESTED IN RABBITS BY REPEATED SUBCUTANEOUS INJECTION. THE DATA WERE INSUFFICIENT TO EVALUATE THE CARCINOGENICITY OF THIS COMPOUND. THE AVAILABLE EXPERIMENTAL AND HUMAN DATA ARE INADEQUATE TO EVALUATE THE CARCINOGENICITY OF 17-ALPAHYDROXYPROGESTERONE CAPROATE.

REPORT 02 : IARC MONOGRAPH 21 1979
NO DATA ON THE MUTAGENICITY OF 17A-HYDROXPROGESTERONE CAPROATE WERE AVAILABLE.
REPT'D USE : DRUG

CAS NO : **148-24-3**
CHEM NAME : ***** 8-HYDROXYQUINOLINE**

REPORT 01 : IARC MONOGRAPH 13 1977
8-HYDROXYQUINOLINE HAS BEEN TESTED IN MICE AND RATS BY ORAL, SUBCUTANEOUS AND INTRAVAGINAL ADMINISTRATION AND IN MICE BY SKIN APPLICATION AND BLADDER IMPLANTATION. MOST OF THESE EXPERIMENTS WERE OF LIMITED VALUE. NO EVALUATION OF THE CARCINOGENICITY OF 8-HYDROXYQUINOLINE CAN BE MADE ON THE BASIS OF THE AVAILABLE DATA.

REPORT 02 : NTP TECHNICAL REPORT 276 NIH PUB 5-2532 APR 85
8-HYDROXYQUINOLINE WAS ADMINISTERED TO MALE AND F344 RATS AND B6C3F1 MICE IN THE FEED AT CONCENTRATIONS OF 1,500 AND 3,000 PPM. THERE WAS NO EVIDENCE OF CARCINOGENICITY.

REPORT 03 : IARC MONOGRAPH 13 1977
8-HYDROXYQUINOLINE CAUSED POINT MUTATIONS IN SALMONEL-

LA TA100, WITH METABOLIC ACTIVATION. IN A BRIEF NOTE, A VARIETY OF CHROMOSOME ABNORMALITIES WERE REPORTED IN MOUSE BONE-MARROW CELLS AFTER TREATMENT BY I.P. INJECTION.

REPT'D USE : FUNGICIDE, ANTISEPTIC, DYE MANUFACTURING (IARC), COSMETICS, COLORIMETRIC REAGENT, AEROSOLS, AND ANTIMICROBIAL AGENT

CAS NO : **26782-43-4**
CHEM NAME : ***** HYDROXYSENKIRKINE**

REPORT 01 : IARC MONOGRAPH 10 1976
HYDROXYSENKIRKINE WAS TESTED ONLY BY THE INTRAPERITONEAL ROUTE IN FIVE MALE RATS, PRODUCING TUMORS OF THE BRAIN IN ONE ANIMAL.

REPT'D USE : TANZANIAN MEDICINAL HERB. NO KNOWN U.S. COMMERCIAL PRODUCTION

CAS NO : **127-07-1**
CHEM NAME : ***** HYDROXYUREA**

REPORT 01 : NCI BIOASSAY PROGRAM, CANCER VOL. 40 1977
UNDER THE CONDITIONS OF THE BIOASSAY, I.P. INJECTION OF HYDROXYUREA WAS CARCINOGENIC TO FEMALE SPRAGUE-DAWLEY (DE) RATS, CAUSING BREAST AND ADRENAL TUMORS. IT WAS NOT CARCINOGENIC TO MALE RATS AND MALE AND FEMALE SWISS-WEBSTER MICE.

REPT'D USE : NO AVAILABLE PRODUCTION OR USE DATA

CAS NO : **21416-87-5**
CHEM NAME : ***** ICRF - 159**

REPORT 01 : NCI CARC TRS 78 DHEW PUB NO 78-1328
UNDER THE CONDITIONS OF THIS BIOASSAY, ICRF-159 WAS CARCINOGENIC FOR FEMALE SPRAGUE-DAWLEY RATS, PRODUCING UTERINE ADENOCARCINOMAS, AND WAS ALSO CARCINOGENIC FOR FEMALE B6C3F1 MICE, PRODUCING LYMPHOMAS. ICRF-159 WAS GIVEN BY ABDOMINAL INJECTION.

REPT'D USE : EXPERIMENTAL ANTINEOPLASTIC AGENT

CAS NO : **5034-77-5**
CHEM NAME : ***** IMIDAZOLE MUSTARD**

REPORT 01 : NCI BIOASSAY PROGRAM CANCER VOL. 40 1977
UNDER THE CONDITIONS OF THE BIOASSAY, I.P. INJECTION OF

IMIDAZOLE MUSTARD WAS CARCINOGENIC TO MALE AND FEMALE SWISS-WEBSTER MICE, CAUSING LYMPHOSARCOMAS IN BOTH SEXES AND LUNG TUMORS IN FEMALES. IT WAS ALSO CARCINOGENIC TO SPRAGUE-DAWLEY (CD) MALE RATS, CAUSING PERITONEAL SARCOMAS, GENERAL SARCOMAS AND LUNG TUMORS. NO DATA WERE GIVEN REGARDING THE TUMOR INCIDENCE IN FEMALE RATS.

REPT'D USE : NO AVAILABE PRODUCTION OR USE DATA.

CAS NO : **193-39-5**

CHEM NAME : ***** INDENO(1,2,3-CD)PYRENE (IP)**

REPORT 01 : IARC MONOGRAPHS 3 AND 32 1973, DEC 83
IP IS A COMPLETE CARCINOGEN AND AN INITIATOR FOR SKIN CARCINOGENESIS IN THE MOUSE. IT PRODUCES LOCAL SARCOMAS IN THE SAME SPECIES AFTER SUBCUTANEOUS INJECTION. IN A 1979 SUPPLEMENT, IARC DETERMINED THAT THERE IS SUFFICIENT EVIDENCE FOR THE CARCINOGENICITY OF IP IN ANIMALS.

REPORT 02 : EPA AMBIENT WATER QUALITY CRITERIA OCT 80
AVAILABLE DATA CANNOT SUPPORT INDIVIDUAL CRITERIA FOR EACH PAH. THE AWQC FOR CARCINOGENIC PAH COMPOUNDS IS BASED ON THE ASSUMPTIONS THAT THE COMPOUND IS AS POTENT AS B(A)P AND THE CARCINOGENIC EFFECT OF THE COMPOUNDS IS PROPORTIONAL TO THE SUM OF THEIR CONCENTRATIONS. THIS SUM SHOULD NOT EXCEED 28.0 NG/L, 2.8 NG/L AND 0.28 NG/L TO KEEP THE LIFETIME CANCER RISK BELOW 10 -5, 10 -6 AND 10 -7, RESPECTIVELY.

REPORT 03 : EPA RCRA HEALTH AND ENV. EFFECTS PROFILE OCT 80
IP IS GENERALLY REGARDED AS ONLY A WEAK CARCINOGEN TO ANIMALS OR MAN.

REPORT 04 : IARC MONOGRAPH 32 DEC 83
INDENO(1,2,3-CD)PYRENE MUTAGENIC TO SALMONELLA TYPHIMURIUM IN PRESENCE OF EXOGENOUS METABOLIC SYSTEM.

REPORT 06 : MULTI-AGENCY CROSS CHECK
Y HHS MAGUIRE
N OSHA CARC
Y EPA CAG
Y EPA RCRA
N EPA CLEAN AIR
Y EPA CLEAN WATER

REPT'D USE : COMBUSTION POLLUTANT, FOOD CONTAMINANT AND USED MOTOR OIL

CAS NO : **75432-59-6**
CHEM NAME : ***** INTRACRON BLUE 3G**

REPORT 01 : CPSC MONOGRAPH ON FIBER REACTIVE DYES DEC 81
INTRACRON BLUE 3G EXHIBITED MUTAGENIC ACTIVITY, WITH AND WITHOUT METABOLIC ACTIVATION, IN SALMONELLA STRAINS TA98, TA100, AND TA1537. A DOSE-RESPONSE RELATIONSHIP WAS OBSERVED WITH METABOLIC ACTIVATION.
REPT'D USE : TEXTILE DYE

CAS NO : **UNKNOWN-20**
CHEM NAME : ***** INTRACRON ORANGE G**

REPORT 01 : CPSC MONOGRAPH ON FIBER REACTIVE DYES DEC 81
INTRACRON ORANGE G WAS INACTIVE OR AT MOST WEAKLY ACTIVE AS A MUTAGEN IN SALMONELLA STRAINS TA98, TA100 AND TA1537, WITH AND WITHOUT METABOLIC ACTIVATION.
REPT'D USE : TEXTILE DYE. NO KNOWN CURRENT PRODUCTION.

CAS NO : **5634-39-9**
CHEM NAME : ***** IODINATED GLYCEROL**

REPORT 01 : NTP TR 340 NIH 90-2596 MAR 90
2-YEAR GAVAGE STUDIES SHOWED SOME EVIDENCE OF CARCINOGENICITY FOR MALE F344/RATS ADMINISTERED IODINATED GLYCEROL. INDICATED BY MONONUCLEAR CELL LEUKEMIA & FOLLICULAR CELL CARCINOMAS OF THYROID GLAND; ADENOMAS OF NASAL CAVITY IN TWO HIGH DOSE MALE RATS. NO EVIDENCE OF CARCINOGENICITY FOR FEMALE F344/N RATS ADMINISTERED 62 OR 125 MG/KG. NO EVIDENCE OF CARCINCGENICITY FOR MALE B6C3F1 MICE ADMINISTERED 125 OR 250 MG/KG. SOME EVIDENCE OF CARCINOGENICITY FOR FEMALE B6C3F1 MICE INDICATED BY ADENOMAS OF ANTERIOR PITUITARY GLAND AND NEOPLASMS OF HARDERIAN GLAND. ALSO SQUAMOUS CELL PAPILLOMAS OF FORESTOMACH POSSIBLY RELATED TO IODINATED GLYCEROL.
REPT'D USE : DRUG

CAS NO : **75-47-8**
CHEM NAME : ***** IODOFORM**

REPORT 01 : NCI CARC TRS 110 DHEW PUB NO 78-1365
UNDER THE CONDITIONS OF THIS BIOASSAY, NO CONVINCING EVIDENCE WAS PROVIDED FOR THE CARCINOGENICITY OF IODOFORM IN OSBORNE-MENDEL RATS OR B6C3F1 MICE. IODOFORM WAS GIVEN BY GAVAGE.

REPORT 02 : NTP TECHNICAL BULLETIN 9 APR 83
IODOFORM WAS POSITIVE IN ONE SALMONELLA TYPHIMURIUM MUTAGENICITY ASSAY.
REPT'D USE : TOPICAL ANTISEPTIC

CAS NO : **3458-22-8**
CHEM NAME : ***** IPD**

REPORT 01 : NCI CARC REPT TRS 18- DHEW PUB NO 78-818
CONCLUSIONS FROM THIS STUDY ARE LIMITED BY EARLY DEATHS AND TOXICITY, BUT THE APPEARANCE OF TUMORS IN THE PERITONEUM NEAR THE INJECTION SITES IN BOTH RATS AND MICE INDICATE THE CARCINOGENIC POTENTIAL OF IPD. IPD WAS GIVEN INTRAPERITONEALLY.
REPT'D USE : CHEMOTHERAPEUTIC AGENT

CAS NO : **76180-96-6**
CHEM NAME : ***** IQ(2-AMINO-3-METHYLIMIDAZO) (4,5-F QUINOLINE)**

REPORT 01 : IARC VOL 40 1986
SUFFICIENT EVIDENCE FOR CARCINOGENICITY OF IQ TO TEST ANIMALS, BUT NO DATA AVAILABLE ON THE CARCINOGENICITY TO HUMANS. IQ CROSSES THE PLACENTA OF MICE. INDUCED PROPHAGE IN BACTERIA AND MUTAGENIC TO S. TYPHIMURIUM WITH ACTIVATION. POSITIVE IN WING SPOT TEST IN D. MELANOGASTER AND INDUCED SEX-LINKED RECESSIVE LETHAL MUTATIONS. INDUCED DNA DAMAGE, MUTATIONS AND SCE'S IN MAMMALIAN CELLS IN VITRO. INDUCE MUTATIONS OR CHROMOSOMAL EFFECTS IN VIVO.
REPT'D USE : NOT USED COMMERCIALLY

CAS NO : **9004-66-4**
CHEM NAME : ***** IRON DEXTRAN**

REPORT 01 : IARC SUPPLEMENT 1 1979
IRON DEXTRAN IS CARCINOGENIC IN MICE AND RATS AFTER SUBCUTANEOUS OR INTRAMUSCULAR INJECTION, PRODUCING LOCAL TUMORS. THERE HAVE BEEN CASE REPORTS OF SARCOMAS ASSOCIATED WITH INJECTIONS OF IRON DEXTRAN. THE SIMILARITY OF THE LOCAL EFFECT IN HUMANS AND ANIMALS WAS NOTED. IN A 1979 SUPPLEMENT IARC CONCLUDED THAT IRON DEXTRAN IS PROBABLY CARCINOGENIC FOR HUMANS.

REPORT 06 : MULTI-AGENCY CROSS CHECK
Y HHS MAGUIRE
N OSHA CARC
Y EPA CAG

Y EPA RCRA
N EPA CLEAN AIR
N EPA CLEAN WATER
REPT'D USE : TREATMENT OF IRON-DEFICIENCY ANEMIA AND HEMATINIC AGENT

CAS NO : **8050-93-9**
CHEM NAME : ***** IRON DEXTRIN**

REPORT 01 : IARC MONOGRAPH 2 1973
BOTH IRON-DEXTRIN COMPLEX AND SACCHARATED IRON OXIDE PRODUCE LOCAL SARCOMAS IN MICE AFTER REPEATED SUBCUTANEOUS INJECTIONS.
REPT'D USE : TREATMENT OF IRON-DEFICIENCY ANEMIA

CAS NO : **15503-86-3**
CHEM NAME : ***** ISATIDINE**

REPORT 01 : IARC MONOGRAPH 10 OCT 76
ISATIDINE IS CARCINOGENIC IN RATS AS SHOWN BY A LIMITED STUDY IN WHICH IT PRODUCED LIVER TUMORS FOLLOWING ITS ORAL ADMINISTRATION.
REPT'D USE : AFRICAN MEDICINAL HERB

CAS NO : **71-41-0**
CHEM NAME : ***** ISOAMYL ALCOHOL**

REPORT 01 : CPSC SCHOOL LAB CHEMICAL SURVEY JAN 82
ISOAMYL ALCOHOL IS REPORTED IN ONE MAJOR TEXT TO BE AN EXPERIMENTAL CARCINOGEN. SUBSEQUENT CPSC STAFF REVIEW SHOWS THAT THE RESULTS OF ONE STUDY SUGGEST THAT THE CHEMICAL MAY BE CARCINOGENIC WHEN ADMINISTERED ORALLY AND SUBCUTANEOUSLY TO WISTAR RATS. NO DATA ON MUTAGENICITY TESTS WERE FOUND IN THE LITERATURE.
REPT'D USE : SOLVENT FOR FATS, RESINS INTERMEDIATE FOR ARTIFICIAL SILK, LACQUERS, SYNTHETIC FLAVORING AND PHARMACEUTICALS

CAS NO : **78-83-1**
CHEM NAME : ***** ISOBUTYL ALCOHOL**

REPORT 01 : CPSC SCHOOL LAB CHEMICAL SURVEY JAN 82
ISOBUTYL ALCOHOL IS REPORTED IN ONE MAJOR TEXT TO BE AN EXPERIMENTAL CARCINOGEN. SUBSEQUENT CPSC STAFF REVIEW SHOWS THAT ONE STUDY SUGGESTS THAT THE CHEMICAL IS CARCINOGENIC IN RATS AFTER EITHER ORAL OR S.C. ADMINISTRATION. A SINGLE TEST FOR MUTAGENICITY IN AN E. COLI

REVERSION ASSAY WAS POSITIVE. THE AVAILABLE DATA ARE INADEQUATE FOR A DETERMINATION OF THE CARCINOGENICITY OF ISOBUTYL ALCOHOL.

CAS NO : **542-56-3**
CHEM NAME : ***** ISOBUTYL NITRITE**

REPORT 01 : NTP TECHNICAL BULLETIN NO. 7 APR 82
THE CHEMICAL WAS FOUND TO BE POSITIVE IN THE SALMONELLA TYPHIMURIUM MUTAGENICITY ASSAY.
REPT'D USE : NO KNOWN CURRENT PRODUCTION

CAS NO : **54-85-3**
CHEM NAME : ***** ISONICOTINIC ACID HYDRAZIDE (INH)**

REPORT 01 : IARC MONOGRAPH 4 1974
INH IS CARCINOGENIC IN MICE AFTER ORAL, SUBCUTANEOUS, AND INTRAPERITONEAL ADMINISTRATION. THE OBSERVATION OF TUMORS IN RATS IN ONLY ONE OF SEVERAL ORAL STUDIES IS INCONCLUSIVE.
REPT'D USE : ANTI-TUBERCULAR AGENT AND ANTIBACTERIAL AGENT

CAS NO : **78-59-1**
CHEM NAME : ***** ISOPHORONE (3,5,5-TRIMETHYL-2-CYCLOHEXEN-1-ONE)**

REPORT 01 : ITC REPORT TO EPA (44FR31867) JUN 79
THE INTERAGENCY TESTING COMMITTEE RECOMMENDS THAT ISOPHORONE BE GIVEN PRIORITY CONSIDERATION FOR TESTING FOR CARCINOGENICITY, MUTAGENICITY, TERATOGENICITY, AND OTHER CHRONIC EFFECTS, AND THAT AN EPIDEMIOLOGY STUDY BE PERFORMED.

REPORT 02 : EPA RESPONSE TO ITC RECOMMS JAN 83, JAN 84
(48FR727) (49FR2012)
IN RESPONSE TO THE ITC RECOMMENDATIONS (JUNE 1, 1979; 44FR31867) THE EPA HAS DECIDED NOT TO INITIATE RULEMAKING, UNDER TSCA SECTION 4 (TO REQUIRE TESTING OF ISOPHORONE SINCE THE CMA TESTING PROGRAM AND NTP BIOASSAY RESULTS WILL PROVIDE SUFFICIENT DATA TO DETERMINE HEALTH EFFECTS OF ISOPHORONE.

REPORT 03 : NTP TECHNICAL REPORT 291 NIH PUB 86-2547 JAN 86
FROM 2-YEAR GAVAGE STUDIES SOME EVIDENCE OF CARCINOGENICITY IN MALE F344/N RATS GIVEN 250 OR 500 MG/KG PER DAY. SHOWN BY RENAL TUBULAR CELL ADENOMAS AND ADENOCARCI-

NOMAS. ALSO CARCINOMAS OF PREPUTIAL GLAND OBSERVED AT 500 MG/KG. FOR MALE B6C3F1 MICE EQUIVOCAL EVIDENCE OF CARCINOGENICITY SHOWN BY HEPATOCELLULAR ADENOMAS OR CARCINOMAS AND TUMORS IN INTEGUMENTARY SYSTEM AT 500 MG/KG AND BY MALIGNANT LYMPHOMAS AT 250 MG/KG PER DAY. NO EVIDENCE OF CARCINOGENICITY IN FEMALE F344/N RATS OR B6C3F1 MICE GIVEN THE SAME DOSES.

REPORT 04 : CMA/TSCA VOLUNTARY TESTING PROGRAM OCT 84
ISOPHORONE FAILED TO GIVE ANY EVIDENCE OF MUTAGENIC ACTIVITY IN THESE TEST SYSTEMS: FORWARD GENE MUTATION ACTIVITY IN MOUSE LYMPHOMA ASSAY IN PRESENCE AND ABSENCE OF METABOLIC ACTIVATION SYSTEM; CLASTOGENIC ACTIVITY IN VIVO MICRONUCLEUS TEST IN MOUSE; CHANGES IN HERITABLE MATERIAL AT MOLECULAR LEVEL INVESTIGATED IN UNSCHEDULED DNA SYNTHESIS TEST IN PRIMARY RAT LIVER HEPATOCYTE CELL CULTURES.

REPORT 05 : CMA/EXXON PROJECT 323772 NOV 84
AT INHALATION VAPOR DOSE LEVELS 0, 25, 50, 115 PPM ISOPHORONE WAS NOT TERATOGENIC OR FETOTOXIC FOR CD-1 MICE OR F344 RATS.

REPT'D USE : SOLVENT FOR POLYVINYL AND NITROCELLULOSE RESINS, LACQUERS, FINISHES, PESTICIDES, HERBICIDES, FATS, OILS, AND GUMS, CHEMICAL INTERMEDIATE IN MANUFACTURE OF 3,5-XYLENOL 3,3,5-TRIMETHYLCYCLOHEXANOL AND PLANT GROWTH RETARDANTS

CAS NO : **3778-73-2**
CHEM NAME : ***** ISOPHOSPHAMIDE**

REPORT 01 : NCI CARC TRS 32 DHEW PUB NO 77-832
UNDER THE CONDITIONS OF THIS BIOASSAY, ISOPHOSPHAMIDE WAS NOT CARCINOGENIC IN MALE SPRAGUE-DAWLEY RATS OR IN MALE B6C3F1 MICE. HOWEVER, THE INCIDENCE OF LEIOMYOSARCOMAS OF THE UTERUS INDICATES THAT ISOPHOSPHAMIDE WAS CARCINOGENIC IN FEMALE SPRAGUE-DAWLEY RATS, AND THE INCIDENCE OF FIBROADENOMA OF THE MAMMARY GLAND IN FEMALE RATS WAS ASSOCIATED WITH TREATMENT WITH ISOPHOSPHAMIDE. ISOPHOSPHAMIDE WAS CARCINOGENIC IN FEMALE B6C3F1 MICE, PRODUCING MALIGNANT LYMPHOMAS OF THE HEMATOPOIETIC SYSTEM. ISOPHOSPHAMIDE WAS GIVEN BY INJECTION.

REPORT 02 : IARC MONOGRAPH 26 1981
CPD CAN INDUCE TERATOGENIC EFFECTS IN MICE AND EMBRYOLETHALITY AT DOSES NONTOXIC TO THE MOTHER. CPD WAS MUTAGENIC IN SALMONELLA STRAIN TA1535 WHEN TESTED WITH METABOLIC ACTIVATION. CPD ALSO PRODUCED A

DOSE-DEPENDENT INCREASE IN CHROMOSOMAL ABERRATIONS IN CH BONE MARROW CELLS.

REPORT 03 : IARC MONOGRAPH 26 1981
ISOPHOSPHAMIDE WAS TESTED IN FOUR STUDIES IN MICE AND ONE STUDY IN RATS BY SC OR IP ADMINISTRATION. IN ONE STUDY IN MICE WITH IP INJECTION IT PRODUCED AN INCREASED INCIDENCE OF LUNG ADENOMAS. THE OTHER FOUR STUDIES, ALTHOUGH INDICATING A CARCINOGENIC EFFECT, COULD NOT BE EVALUATED. THERE IS LIMITED EVIDENCE FOR THE CARCINOGENICITY OF ISOPHOSPHAMIDE IN MICE AND RATS.
REPT'D USE : CHEMOTHERAPEUTIC AGENT

CAS NO : **78-79-5**
CHEM NAME : ***** ISOPRENE**

REPORT 01 : EPA STATUS REPORT 8EHQ-0887-0689 SEP 87
B6C3F1 MICE EXPOSED BY INHALATION TO 438, 1750 AND 7,000 PPM 6 HRS PER DAY FOR 12 DAYS SHOWED SIGNIFICANT INCREASE IN SISTER CHROMATID EXCHANGE FREQUENCY AND LEVEL OF MICRONUCLEATED POLYCHROMATIC ERYTHROCYTES AND OF MICRONUCLEATED NONCHROMIC ERYTHROCYTES IN PERIPHERAL BLOOD.
REPT'D USE : PRODUCTION OF SYNTHETIC ELASTOMERS

CAS NO : **67-63-0**
CHEM NAME : ***** ISOPROPYL ALCOHOL (ISOPROPANOL)**

REPORT 01 : IARC MONOGRAPH 15 1977
NO DATA ON THE MUTAGENICITY OF ISOPROPYL ALCOHOL WERE AVAILABLE.

REPORT 02 : IARC MONOGRAPH 15 1977
ISOPROPYL ALCOHOL HAS BEEN INADEQUATELY TESTED IN MICE BY SKIN APPLICATION, INHALATION EXPOSURE AND SUBCUTANEOUS INJECTION. DUE TO VARIOUS LIMITATIONS OF THESE STUDIES, NO EVALUATION OF CARCINOGENICITY OF THIS COMPOUND CAN BE MADE.

REPORT 03 : IARC SUPPLEMENT NO. 1 1977
AN INCREASED INCIDENCE OF CANCER OF THE PARANASAL SINUSES HAS BEEN FOUND IN WORKERS IN FACTORIES MANUFACTURING ISOPROPYL ALCOHOL BY THE STRONG-ACID PROCESS IN WHICH ISOPROPYL OILS WERE FORMED AS BY-PRODUCTS.

REPORT 04 : EPA TSCA FINAL TEST RULE (SECTION 4 (A)) OCT 89
IN FR NOTICE (54 FR 43252) EPA ISSUED FINAL TEST RULE, UNDER SECTION 4 OF TSCA REQUIRING MANUFACTURERS AND PROCES-

SORS OF ISOPROPANOL TO TEST FOR HEALTH EFFECTS. TEST REQUIREMENTS INCLUDE SUBCHRONIC TOXICITY, REPRODUCTIVE TOXICITY, DEVELOPMENTAL TOXICITY, NEUROTOXICITY, DEVELOPMENTAL NEUROTOXICITY, MUTAGENICITY, ONCOGENICITY, AND PHARMACOKINETICS. THIS IS IN RESPONSE TO ITC DESIGNATION OF ISOPROPANOL FOR PRIORITY TESTING CONSIDERATION.

REPT'D USE : CHEMICAL INTERMEDIATE CLEANING AGENTS, PAINTS/VARNISHES, ADHESIVES, DEODORIZERS, SOLVENT, ELECTRONICS, DEICING COMPOUND, MEDICINE, DISINFECTANT, ANTIFREEZE, FR CHEMICALS, POLISHES, CAULKING AND FLAVORING AGENT

CAS NO : **4016-14-2**

CHEM NAME : ***** ISOPROPYL GLYCIDYL ETHER**

REPORT 01 : NTP TECHNICAL BULLETIN 9 APR 83
ISOPROPYL GLYCIDYL ETHER WAS POSITIVE IN ONE SALMONELLA TYPHIMURIUM MUTAGENICITY ASSAY.

CAS NO : **51-30-9**

CHEM NAME : ***** ISOPROTERENOL HYDROCHLORIDE**

REPORT 01 : NTP TECHNICAL BULLETIN 3 DEC 80
THE COMPOUND WAS FOUND TO BE WEAKLY POSTIVE IN THE SALMONELLA TYPHIMURIUM MUTAGENICITY ASSAY IN ONE LABORATORY.

REPORT 02 : NTP CARC TEST PROGRAM JUL 82
IN A CURRENT NTP BIOASSAY, ISOPROTERENOL HC1 IS BEING ADMINISTERED BY INHALATION TO F344 RATS AND B6C3F1 MICE. THE PRECHRONIC TEST PHASE IS IN PROGRESS.

REPT'D USE : NO AVAILABLE PRODUCTION OR USE DATA. NO LISTING ON TSCA INVENTORY.

CAS NO : **120-58-1**

CHEM NAME : ***** ISOSAFROLE**

REPORT 01 : IARC MONOGRAPH 1 1972
SAFROLE AND ISOSAFROLE ARE LIVER CARCINOGENS FOR THE MOUSE AND THE RAT WHEN ADMINISTERED ORALLY OR SUBCUTANEOUSLY. IN A 1979 SUPPLEMENT, IARC DETERMINED THAT THERE IS SUFFICIENT EVIDENCE FOR THE CARCINOGENICITY OF ISOSAFROLE IN ANIMALS.

REPORT 06 : MULTI-AGENCY CROSS CHECK
N HHS MAGUIRE
N OSHA CARC
Y EPA CAG

Y EPA RCRA
N EPA CLEAN AIR
N EPA CLEAN WATER

REPT'D USE : FLAVORING AGENT IN DRUGS, INTERMEDIATE IN MFR OF HELLOTROPIN AND FRAGRANCE AGENT

CAS NO : **471-14-7**
CHEM NAME : ***** JACOBINE**

REPORT 01 : IARC MONOGRAPH 10 1976
NO DATA WERE AVAILABLE CONCERNING THE CARCINOGENICITY OF PURE JACOBINE .

REPORT 02 : IARC MONOGRAPH 10 1976
JACOBINE OF UNSPECIFIED PURITY IN SODIUM CHLORIDE WAS INJECTED INTO THE ABDOMEN OF DROSOPHILA MALES. WEAK MUTAGENIC EFFECTS ON THE INDUCTION OF SEX-LINKED RECESSIVE LETHALS WERE SCORED.

REPT'D USE : EUROPEAN HERBAL MEDICATION

CAS NO : **CLASS-32**
CHEM NAME : ***** JET FUEL**

REPORT 01 : IARC MONOGRAPH 45 1989
JET FUELS PRODUCED MAINLY FROM STRAIGHT-RUN AND HYDROTREATED KEROSENE OR KEROSENE BLENDED WITH HEAVY NAPHTHA STREAMS FROM ATMOSPHERIC DISTILLATION OF CRUDE OIL. JET FUELS COMPOSED MAINLY OF ALIPHATIC AND AROMATIC HYDROCARBONS WITH BOILING RANGES OF 150-300 DEGREE C (KEROSENE TYPE) AND 45-280 DEGREE C (WIDE-CUT TYPE). FORMULATED PRODUCTS USED IN TURBINE ENGINES OF CIVIL AND MILITARY AIRCRAFT. LIMITED EVIDENCE FOR CARCINOGENICITY IN EXPERIMENTAL ANIMALS OF STRAIGHT-RUN KEROSENE AND OF HYDROTREATED KEROSENE. TWO SAMPLES OF STRAIGHT-RUN KEROSENE AND ONE SAMPLE OF HYDROTREATED KEROSENE PRODUCED SKIN TUMORS IN MICE.

REPORT 02 : IARC MONOGRAPH 45 1989
INADEQUATE EVIDENCE FOR CARCINOGENICITY IN HUMANS AND IN EXPERIMENTAL ANIMALS OF JET FUEL. JET FUEL TESTED BY SKIN APPLICATION IN ONE EXPERIMENT IN MALE AND FEMALE MICE SHOWED NO SKIN TUMOR AT APPLICATION SITE. MEN EXPOSED TO JET FUEL, AVIATION KEROSENE, AND OTHER FUELS IN SWEDISH AIR FORCE HAD NO INCREASED CANCER RISK DURING 10-YRS. FOLLOW-UP. STUDY OF CANCER AT SITES IN CANADA REVEALED ELEVATED RISK FOR KIDNEY CANCER, WITH POSITIVE DOSE-RESPONSE RELATIONSHIP, IN MEN EXPOSED TO JET FUEL. INDUCED CHROMOSOMAL ABERRATIONS IN BONE-MARROW CELLS

OF RATS AND MUTATIONS IN CULTURED MAMMALIAN CELLS IN PRESENCE OF EXPGENOUS METABOLIC SYSTEM BUT DID NOT INDUCE MUTATION IN BACTERIA.

REPT'D USE : AIRCRAFT ENGINE FUEL

CAS NO : **520-18-3**

CHEM NAME : ***** KAEMPFEROL**

REPORT 01 : IARC MONOGRAPH 31 JUL 83
ONE EXPERIMENT IN RATS BY ADMINISTRATION IN DIET INADEQUATE TO MAKE EVAULATION. MUTAGENIC IN BACTERIA AND INSECTS AND IN MAMMALIAN CELLS IN VITRO; IT INDUCED MICRONUCLEI IN MICE. NO DATA WERE AVAILABLE TO EVAULATE THE TERATOGENICITY OF KAEMPFEROL TO EXPERIMENTAL ANIMALS.

REPT'D USE : NATURAL OCCURRENCE IN FRUIT AND OTHER EDIBLE PLANTS AND TEXTILE DYES

CAS NO : **8008-20-6**

CHEM NAME : ***** KEROSENE**

REPORT 01 : CPSC SCHOOL LAB CHEMICAL SURVEY JAN 82
THE CHEMICAL IS REPORTED IN ONE MAJOR TEXT TO BE AN EXPERIMENTAL CARCINOGEN. SUBSEQUENT CPSC STAFF REVIEW SHOWS THAT THERE IS NO EVIDENCE THAT KEROSENE IS CARCINOGENIC, MUTAGENIC OR TERATOGENIC. NIOSH RECOMMENDED STANDARD TWA 100 MG/CUBIC METER FOR PETROLEUM SOLVENTS (7/77); SEE 68477-31-6.

REPORT 02 : EPA TSCA 8(E) 8EHQ-0283-0472S FEB 83
SUBMITTING COMPANY (NAME CLAIMED CBI) REPORTED ON LIFETIME MOUSE SKIN-PAINTING STUDY OF LOW-BOILING CATAYTIC REFORMER FRACTIONATOR RESIDUE PETROLEUM DISTILLATE WITH KEROSENE. FIVE MICE TREATED WITH 3:1 MIXTURE DEVELOPED ONE MASS AT SITE OF APPLICATION, WITH A MEAN LATENCY PERIOD OF 95.8 WEEKS. HISTOPATHOLOGIC EXAMINATIONS HAVE NOT BEEN DONE. THE LONG LATENCY PERIOD AND THE SMALL NUMBER OF TUMORS INDICATED THAT THE TESTED MATERIALS ARE WEAK ANIMAL CARCINOGENS.

REPORT 03 : EPA FYI-AX -0385 -0389 MAR 85
API REPORTS MIDDLE DISTILLATE FUELS ARE WEAKLY TO MODERATELY CARCINOGENIC IN ANIMALS, PRODUCING TUMOR INCIDENCES IN 5 TO 30% RANGE (SKIN-PAINTINGS). TWO-YEAR USAF/NAVY INHALATION STUDIES ON VOLATILE FRACTIONS IN PROGRESS.

REPORT 04 : EPA FYI-AX -0385 -0389 MAR 85
API REPORTS SOME MIDDLE DISTILLATE FUEL MATERIALS MUTAGENIC IN-VIVO RAT BONE MARROW CYTOGENETICS AND MOUSE LYMPHOMA ASSAYS, BUT RESULTS NOT CONSISTENT. MODIFIED AMES ASSAY PRODUCED POSITIVE RESULTS FOR SOME MATERIALS.

REPORT 05 : NTP TR 310 NIH PUB. 86-2566 SEP 86
UNDER 2-YEAR DERMAL STUDIES, MARINE DIESEL FUEL AT DOSES OF 250 AND 500 MG/KG RESULTED IN SQUAMOUS CELL NEOPLASMS OF SKIN (PRIMARILY CARCINOMAS), PROVIDING EQUIVOCAL EVIDENCE OF CARCINOGENICITY FOR MALE AND FEMALE B6C3F1 MICE. JP-5 NAVY FUEL AT DOSES OF 250 AND 500 MG/KG PROVIDED NO EVIDENCE OF CARCINOGENICITY FOR MALE AND FEMALE B6C3F1 MICE.

REPT'D USE : LAB REAGENT CLEANING AGENTS, PAINTS/VARNISHES AND FUEL FOR PLANES, VEHICLES, SHIPS AND POWER PLANTS

CAS NO : **55-31-2**

CHEM NAME : ***** L-EPINEPHRINE HYDROCHLORIDE**

REPORT 01 : NTP DRAFT TR380 NIH-90-2835 MAR 90
UNDER 2 YEAR STUDIES NO EVIDENCE OF CARCINOGENIC ACTIVITY OF L-EPINEPHRINE HYDROCHLORIDE FOR MALE OR FEMALE F344/N RATS EXPOSED TO AEROSOLS CONTAINING 1.5 OR 5 MG/M3 FOR 2 YEARS OR FOR MALE OR FEMALE B6C3F1 MICE EXPOSED TO AEROSOLS CONTAINING 1.5 OR 3 MG/M3 FOR 2 YEARS. INADEQUATE STUDIES OF CARCINOGENIC ACTIVITY BECAUSE CONCENTRATIONS CONSIDERED TOO LOW FOR ANIMALS TO HAVE RECEIVED ADEQUATE SYSTEMIC CHALLENGE.

REPT'D USE : DRUG

CAS NO : **303-34-4**

CHEM NAME : ***** LASIOCARPINE**

REPORT 01 : NCI CARC TRS 39, 1978 DHEW PUB NO 78-839
UNDER THE CONDITIONS OF THIS BIOASSAY, LASIOCARPINE WAS CARCINOGENIC IN FISCHER 344 RATS PRODUCING HEPATOCELLULAR TUMORS AND ANGIOSARCOMAS OF THE LIVER IN BOTH SEXES AND HEMATOPOIETIC TUMORS IN FEMALE ANIMALS (FEEDING STUDY).

REPORT 02 : IARC MONOGRAPH 10 1976
LASIOCARPINE IS CARCINOGENIC IN RATS FOLLOWING ITS INTRAPERITONEAL INJECTION; IT PRODUCED MALIGNANT TUMORS OF THE LIVER, SKIN, AND INTESTINE. IN A 1979 SUPPLEMENT, IARC DETERMINED THAT THERE WAS SUFFICIENT EVIDENCE FOR THE CARCINOGENICITY OF LASIOCARPINE IN ANIMALS.

REPORT 03 : NTP TECHNICAL BULLETIN 3 DEC 80
THE COMPOUND WAS FOUND TO BE POSITIVE IN THE SALMONELLA TYPHIMURIUM MUTAGENICITY ASSAY IN ONE LABORATORY.

REPORT 04 : IARC MONOGRAPH 10 1976
LASIOCARPINE OF UNSPECIFIED PURITY IN SODIUM CHLORIDE WAS INJECTED IN TO THE ABDOMEN OF DROSOPHILA MALES. STRONG MUTAGENIC EFFECTS ON THE INDUCTION OF RECESSIVE SEX-LINKED LETHALS WERE SCORED.

REPORT 06 : MULTI-AGENCY CROSS CHECK
N HHS MAGUIRE
N OSHA CARC
Y EPA CAG
Y EPA RCRA
N EPA CLEAN AIR
N EPA CLEAN WATER

REPT'D USE : NO AVAILABLE PRODUCTION ON OR USE DATA.

CAS NO : **105-74-8**

CHEM NAME : ***** LAUROYL PEROXIDE**

REPORT 01 : IARC MONOGRAPH 36 FEB 85
TESTED BY SUBCUTANEOUS ADMINISTRATION IN MICE AND RATS AND BY SKIN APPLICATION IN MICE. IN ONE STUDY IN MICE BY SUBCUTANEOUS ADMINISTRATION, EVIDENCE ON CARCINOGENIC EFFECT INCONCLUSIVE; IN TWO OTHER STUDIES, NO INCREASE IN TUMOR INCIDENCE OBSERVED. TWO STUDIES IN MICE BY SKIN APPLICATION INADEQUATE FOR EVALUATION OF CARCINOGENICITY. ONE STUDY INDICATED THAT LAUROYL PEROXIDE HAS PROMOTING ACTIVITY IN MOUSE SKIN.

REPT'D USE : POLYMER PRODUCTION AND FOOD PACKAGING

CAS NO : **7439-92-1**

CHEM NAME : ***** LEAD**

REPORT 01 : EPA AMBIENT WATER QUALITY CRITERIA OCT 80
ALTHOUGH CERTAIN INORGANIC LEAD COMPOUNDS ARE CARCINOGENIC TO SOME SPECIES OF EXPERIMENTAL ANIMALS, A CLEAR ASSOCIATION BETWEEN LEAD EXPOSURE AND CANCER DEVELOPMENT HAS NOT BEEN SHOWN IN HUMAN POPULATIONS. THE CRITERION FOR LEAD IS RECOMMENDED TO BE IDENTICAL TO THE EXISTING WATER STANDARD WHICH IS 50 UG/L.

REPORT 02 : IARC MONOGRAPH 23 1980
NO EVALUATION COULD BE MADE OF THE CARCINOGENICITY OF METALLIC LEAD POWDER. LEAD SALTS HAVE BEEN REPORTED TO CROSS THE PLACENTA AND TO INDUCE EMBRYO- AND

FETOMORTALITY. THEY ALSO HAVE A TERATOGENIC EFFECT IN SOME ANIMAL SPECIES. NO TERATOGENIC EFFECTS HAVE BEEN REPORTED WITH EXPOSURE TO ORGANOMETALLIC LEAD COMPOUNDS.

REPORT 03 : NIOSH CRITERIA DOCUMENT MAR 77
NIOSH RECOMMENDS A STANDARD LESS THAN 100 UG/CU M TWA BASED ON KIDNEY, BLOOD AND CNS EFFECTS. THE PRESENT OSHA STANDARD IS 50 UG/CU M TWA.

REPORT 04 : EPA RCRA HEALTH AND ENV. EFFECTS PROFILE OCT 80
THE HEMATOPOIETIC SYSTEM IS THE MOST SENSITIVE TARGET ORGAN FOR LEAD IN HUMANS; SUBTLE NEUROBEHAVIORAL EFFECTS ARE SUSPECTED IN CHILDREN AT SIMILAR EXPOSURE LEVELS. THE MORE SERIOUS HEALTH EFFECTS OF CHRONIC LEAD EXPOSURE INVOLVE NEUROLOGICAL DAMAGE, IRREVERSIBLE RENAL DAMAGE AND ADVERSE REPRO. EFFECTS OBSERVED ONLY AT HIGHER EXP. LEVELS. ALTHOUGH CERTAIN INORGANIC LEAD COM. ARE CARCINOGENIC TO SOME SPECIES OF EXP. ANIMALS, A CLEAR ASSOCIATION BETWEEN LEAD EXP. AND CANCER DEVELOPMENT HAS NOT BEEN SHOWN IN HUMAN POPULATIONS.

REPORT 05 : 16 CFR PART 1303 LEAD IN PAINT RULE (42FR44199) SEP 77
CPSC DECLARED AS BANNED HAZARDOUS PRODUCTS: 1) PAINT AND OTHER SIMILAR SURFACE COATING MATERIALS CONTAINING MORE THAN 0.06% LEAD, 2) TOYS AND OTHER CHILDREN'S PRODUCTS WHICH CONTAIN SUCH PAINT, AND 3) FURNITURE ARTICLES WHICH CONTAIN SUCH PAINT.

REPORT 06 : MULTI-AGENCY CROSS CHECK 1985
Y HHS MAGUIRE
N OSHA CARC
N EPA CAG
N EPA RCRA
N EPA CLEAN AIR
Y EPA CLEAN WATER

REPORT 07 : NAS REPORT ON DIET NUTRITION AND CANCER 1982
THE INGESTION OF HIGH LEVELS OF LEAD COMPOUNDS INDUCES RENAL TUMORS IN MICE AND RATS. THE ONLY STUDY THAT CORRELATED LEAD LEVELS IN DRINKING WATER AND CANCER MORTALITY SUGGESTED THAT LEAD INCREASED THE RISK OF CANCER. EXPOSURE TO LEAD IN INDUSTRIAL SETTINGS HAS NOT BEEN CLEARLY ASSOCIATED WITH AN INCREASED RISK FOR ANY FROM OF CANCER.

REPORT 08 : EPA TSCA 8(E) STATUS REPORT 8EHQ-0285- 0546 FEB 85
INTERNATIONAL LEAD ZINC RESEARCH ORGANIZATION REPORTED UPDATED RESULTS OF STUDY OF MORTALITY OF BATTERY AND SMELTER WORKERS. EXCESS DEATHS FROM STOMACH CANCER IN BATTERY WORKERS WAS SIGNIFICANT. THERE WAS INCREASED

MORTALITY FROM ALL NEOPLASMS, FROM CHRONIC NEPHRITIS, AND FROM HYPERTENSIVE DISEASE. SIGNIFICANT TREND BETWEEN MORTALITY AND CUMULATIVE LEAD EXPOSURE.

REPT'D USE : PAINTS/VARNISHES

CAS NO : **301-04-2**

CHEM NAME : ***** LEAD ACETATE**

REPORT 01 : IARC MONOGRAPH 23 MAR 89
LEAD ACETATE INDUCED MORPHOLOGICAL TRANSFORMATIONS IN SYRIAN HAMSTER CELLS. THERE IS NO EVIDENCE THAT LEAD ACETATE INDUCED MUTATIONS OR ALLIED EFFECTS IN BACTERIA, SOME CHROMOSOMAL ABERRATION TESTS IN MAMMALIAN SYSTEMS HAVE GIVEN POSITIVE RESULTS.

REPORT 02 : IARC MONOGRAPH 23 1980
LEAD ACETATE PRODUCED RENAL TUMORS AND GLIOMAS IN RATS FOLLOWING ORAL ADMINISTRATION. LEAD ACETATE INDUCED MORPHOLOGICAL TRANSFORMATIONS IN SYRIAN HAMSTER CELLS. THERE IS SUFFICIENT EVIDENCE THAT LEAD ACETATE IS CARCINOGENIC TO RATS. IN A 1979 SUPPLEMENT, IARC DETERMINED THAT THERE IS SUFFICIENT EVIDENCE FOR THE CARCINOGENICITY OF LEAD ACETATE IN ANIMALS.

REPORT 03 : FDA APPROVAL ANNOUNCEMENT OCT 80
FDA ANNOUNCED APPROVAL OF LEAD ACETATE AS A COLOR ADDITIVE IN HAIR DYES ON THE BASIS THAT THE AMOUNTS EXPECTED TO BE ABSORBED THROUGH THE SKIN IN NORMAL USE WILL NOT INCREASE A PERSON'S RISK OF LEAD POISONING OR CANCER. A STUDY BY COMBE INC. SHOWED THAT THE EXPECTED ABSORPTION OF LEAD FROM HAIR DYE IS 0.5 MICROGRAMS PER APPLICATION. THIS COMPARES TO AN EXPECTED INTAKE OF AS MUCH AS 100 TO 500 MICROGRAMS PER DAY FROM FOOD ALONE AND AS MUCH AS 20 TO 400 MICROGRAMS PER DAY FROM AIR.

REPORT 04 : NTP TECHNICAL BULLETIN 8 JUL 82
LEAD ACETATE WAS NEGATIVE IN THE L5178 MOUSE LYMPHOMA ASSAY.

REPORT 06 : MULTI-AGENCY CROSS CHECK
Y HHS MAGUIRE
N OSHA CARC
N EPA CAG
Y EPA RCRA
N EPA CLEAN AIR
N EPA CLEAN WATER

REPT'D USE : HAIR COLOR ADDITIVE, CHEMICAL INTERMEDIATE, SWIMMING POOL CHEMICALS, PAINTS, DYES, METAL COATINGS, WATERPROOFING, PRINTING INKS, EXPLOSIVES AND MEDICINE

CAS NO : **7784-40-9**
CHEM NAME : ***** LEAD ARSENATE**

REPORT 01 : IARC MONOGRAPH 23 MAR 89
THERE IS INADEQUATE EVIDENCE FOR THE CARCINOGENICITY OF ARSENIC COMPOUNDS IN ANIMALS. THERE IS SUFFICIENT EVIDENCE THAT INORGANIC ARSENIC COMPOUNDS ARE SKIN AND LUNG CARCINOGENS IN HUMANS. THE EVIDENCE THAT ARSENIC COMPOUNDS CAUSE MUTATIONS AND ALLIED EFFECTS IN BACTERIA IS INCONCLUSIVE. HOWEVER, ARSENIC COMPOUNDS INDUCE CHROMOSOMAL ABERRATIONS AND MORPHOLOGICAL TRANSFORMATION IN MAMMALIAN CELLS.

REPORT 02 : NIOSH CRITERIA DOCUMENT SEP 75
NIOSH RECOMMENDS A CEILING OF 2 MICROGRAMS/CU M FOR INORGANIC ARSENIC COMPOUNDS BASED ON DERMATITIS, LUNG, AND LYMPHATIC CANCER. THE CURRENT OSHA STANDARD FOR INORGANIC ARSENIC IS 10 MICROGRAMS/CU M TWA. SEE CFR 1910.1018.

REPORT 06 : MULTI-AGENCY CROSS CHECK
Y HHS MAGUIRE
N OSHA CARC
Y EPA CAG
N EPA RCRA
N EPA CLEAN AIR
N EPA CLEAN WATER
REPT'D USE : HERBICIDE, INSECTICIDE, VETERINARY AND HELMINTIC

CAS NO : **598-63-0**
CHEM NAME : ***** LEAD CARBONATE**

REPORT 01 : IARC MONOGRAPH 23 MAR 89
NO EVALUATION COULD BE MADE OF THE CARCINOGENICITY OF LEAD CARBONATE. THERE WAS INSUFFICIENT EVIDENCE TO EVALUATE THE MUTAGENICITY OF ORGANOMETALLIC LEAD COMPOUNDS. NO TERATOGENIC EFFECTS HAVE BEEN REPORTED WITH EXPOSURE TO ORGANOMETALLIC LEAD COMPOUNDS.
REPT'D USE : PIGMENT, CEMENTS AND PAINTS

CAS NO : **7758-95-4**
CHEM NAME : ***** LEAD CHLORIDE**

REPORT 01 : IARC MONOGRAPH 23 MAR 89
LEAD CHLORIDE GAVE POSITIVE RESULTS IN DNA MISINCORPORATION TESTS. THERE IS NO EVIDENCE THAT LEAD CHLORIDE INDUCES MUTATIONS OR ALLIED EFFECTS IN BACTERIA. SOME CHROMOSOMAL ABERRATION TESTS IN MAMMALIAN

SYSTEMS HAVE GIVEN POSITIVE RESULTS.
REPT'D USE : PIGMENT, SOLDER FLUXES AND DRYING AGENT IN PANTS

CAS NO : **1344-38-3**
CHEM NAME : ***** LEAD CHROMATE (VI) OXIDE**

REPORT 01 : IARC MONOGRAPH 23 1980
LEAD CHROMATE (VI) OXIDE PRODUCED SARCOMAS AT THE SITE OF SUBCUTANEOUS ADMINISTRATION IN RATS. THERE IS LIMITED EVIDENCE FOR THE CARCINOGENICITY OF LEAD CHROMATE OXIDE IN RATS. CHROMIUM(VI) COMPOUNDS CAUSE MUTATIONS AND ALLIED EFFECTS IN A VERY WIDE RANGE PROKARYOTIC AND EUKARYOTIC SYSTEMS BOTH IN VITRO AND IN VIVO.

REPORT 06 : MULTI-AGENCY CROSS CHECK
Y HHS MAGUIRE
N OSHA CARC
N EPA CAG
N EPA RCRA
N EPA CLEAN AIR
N EPA CLEAN WATER
REPT'D USE : PAINTS, CORROSION INHIBITORS AND PIGMENT

CAS NO : **19010-66-3**
CHEM NAME : ***** LEAD DIMETHYLDITHIOCARBAMATE**

REPORT 01 : NCI CARC TRS 151 DHEW PUB NO 79-1707 MAR 89
UNDER THE CONDITIONS OF THIS BIOASSAY, LEAD DIMETHYLDITHIOCARBAMATE WAS NOT CARCINOGENIC FOR F344 RATS OR B6C3F1 MICE OF EITHER SEX (FEEDING STUDY).

REPORT 02 : IARC MONOGRAPH 12 1976
LEAD HAS BEEN TESTED BY ORAL ADMINISTRATION AND BY SINGLE SUBCUTANEOUS INJECTION IN TWO STRAINS OF MICE. ALTHOUGH AN INCREASE IN THE INCIDENCE OF RETICULUM-CELL SARCOMAS WAS OBSERVED IN MALES OF ONE STRAIN FOLLOWING ORAL ADMINISTRATION OF THE COMPOUND, THE AVAILABLE DATA ARE INSUFFICIENT FOR EVALUATION OF ITS CARCINOGENICITY.

REPORT 03 : NTP TECHNICAL BULLETIN 6 JAN 82
THE COMPOUND WAS POSITIVE IN THE SALMONELLA TYPHIMURIUM MUTAGENICITY ASSAY IN ONE LABORATORY.
REPT'D USE : RUBBER VULCANIZATION RETARDER

CAS NO : **50825-29-1**
CHEM NAME : ***** LEAD NAPHTHENATE**

REPORT 01 : IARC MONOGRAPH 23 MAR 89
NO EVALUATION COULD BE MADE OF THE CARCINOGENICITY OF LEAD NAPHTHENATE. THERE WAS INSUFFICIENT EVIDENCE TO EVALUATE THE MUTAGENICITY OF ORGANOMETALLIC LEAD COMPOUNDS. NO TERATOGENIC EFFECTS HAVE BEEN REPORTED WITH EXPOSURE TO ORGANOMETALLIC LEAD COMPOUNDS.

REPORT 02 : EPA'S RESPONSE TO ITC, 49FR214 11 MAY 84
EPA'S RESPONSE TO ITC'S RECOMMENDATION THAT EPA CONSIDER CHEMICAL FATE, HEALTH EFFECTS AND ENVIRONMENTAL EFFECTS TESTING OF CPD UNDER TSCA 4(A). EPA NOT INITIATING RULEMAKING NOW BECAUSE SUFFICIENT DATA AVAILLABLE IN ONGOING TESTS TO DETERMINE IF IT PRESENTS AN UNREASONABLE RISK TO THE ENVIRONMENT.
REPT'D USE : PAINT DRIER, LUBRICANTS AND WOOD PRESERVATTVES

CAS NO : **10099-74-8**
CHEM NAME : ***** LEAD NITRATE**

REPORT 01 : IARC MONOGRAPH 23 MAR 89
NO EVALUATION COULD BE MADE OF THE CARCINOGENICITY OF LEAD NITRATE.
REPT'D USE : MATCHES, PYROTECHNICS, DYE MORDANT, ENGRAVING, PHOTOGRAPHIC CHEMICAL AND VETERINARY APPLICATIONS

CAS NO : **1317-36-8**
CHEM NAME : ***** LEAD OXIDE**

REPORT 01 : IARC MONOGRAPH 23 MAR 89
NO EVALUATION COULD BE MADE OF THE CARCINOGENICITY OF LEAD OXIDE.
REPT'D USE : PAINT/VARNISHES, CHEMICAL INTERMEDIATE, GLASS MAKING, LINSEED OIL AND BATTERIES

CAS NO : **7446-27-7**
CHEM NAME : ***** LEAD PHOSPHATE**

REPORT 01 : IARC MONOGRAPH 23 MAR 89
LEAD PHOSPHATE PRODUCED RENAL TUMORS IN RATS FOLLOWING SUBCUTANEOUS INJECTIONS. THERE IS SUFFICIENT EVIDENCE THAT LEAD PHOSPHATE IS CARCINOGENIC TO RATS. IN A 1979 SUPPLEMENT, IARC DETERMINED THAT THERE IS SUFFICIENT

EVIDENCE FOR THE CARCINOGENICITY OF LEAD PHOSPHATE IN ANIMALS.

REPORT 06 : MULTI-AGENCY CROSS CHECK
Y HHS MAGUIRE
N OSHA CARC
N EPA CAG
Y EPA RCRA
N EPA CLEAN AIR
N EPA CLEAN WATER

REPT'D USE : STABILIZER IN STYRENE AND CASEIN PLASTICS

CAS NO : **1335-32-6**
CHEM NAME : ***** LEAD SUBACETATE**

REPORT 01 : IARC MONOGRAPH 23 MAR 89
LEAD SUBACETATE PRODUCED RENAL TUMORS AND GLIOMAS IN RATS FOLLOWING ORAL ADMINISTRATION; IT ALSO PRODUCED RENAL TUMORS IN MICE FOLLOWING ORAL ADMINISTRATION. THERE IS SUFFICIENT EVIDENCE THAT LEAD SUBACETATE IS CARCINOGENIC TO MICE AND RATS. IN A 1979 SUPPLEMENT, IARC DETERMINED THAT THERE WAS SUFFICIENT EVIDENCE FOR THE CARCINOGENICITY OF LEAD SUBACETATE IN ANIMALS.

REPORT 02 : IARC MONOGRAPH 23 1980
THERE WAS INSUFFICIENT EVIDENCE TO EVALUATE THE MUTAGENICITY OF ORGANOMETALLIC LEAD COMPOUNDS. NO TERATOGENIC EFFECTS HAVE BEEN REPORTED WITH EXPOSURE TO ORGANOMETALLIC LEAD COMPOUNDS.

REPORT 06 : MULTI-AGENCY CROSS CHECK
N HHS MAGUIRE
N OSHA CARC
N EPA CAG
Y EPA RCRA
N EPA CLEAN AIR
N EPA CLEAN WATER

REPT'D USE : LABORATORY AGENT

CAS NO : **UNKNOWN-16**
CHEM NAME : ***** LEATHER GOODS MANUFACTURE**

REPORT 01 : IARC MONOGRAPH 25 SEP 81
THERE ARE A FEW REPORTS OF CASES OF LEUKEMIA FOLLOWING EXPOSURE TO BENZENE IN THE MANUFATURE OF LEATHER GOODS OTHER THAN BOOTS AND SHOES OR IN TANNING. A POSITIVE ASSOCIATION BETWEEN EXPLOYMENT IN THE LEATHER INDUSTRY WITH BLADDER CANCER IS SUPPORTED BY A NUMBER OF STUDIES;

BUT THE SPECIFIC ROLE OF THE PRODUCTION OF LEATHER GOODS CANNOT BE EVALUATED.

CAS NO : **UNKNOWN-27**
CHEM NAME : ***** LEATHER TANNING AND PROCESSING INDUSTRIES**

REPORT 01 : IARC MONOGRAPH 25 SEP 81
VERY FEW EPIDEMIOLOGIC STUDIES OR CASE REPORTS DEAL SPECIFICALLY WITH WORKERS ENGAGED IN LEATHER TANNING OR PROCESSING. THERE IS NO EVIDENCE TO SUGGEST AN ASSOCIATION BETWEEN LEATHER TANNING & NASAL CANCER, A POSITIVE ASSOCIATION BETWEEN EMPLOYMENT IN THE LEATHER INDUSTRY & BLADDER CANCER IS SUPPORTED BY A NUMBER OF STUDIES. THE ONLY STUDY THAT DEALS SPECIFICALLY WITH LEATHER TANNERS, HOWEVER, REVEALED A RELATIVE RISK OF 1.5, WHICH IS NOT STATISTICALLY SIGNIFICANT.

CAS NO : **5989-27-5**
CHEM NAME : ***** D-LIMONENE**

REPORT 01 : NTP TR 347 NIH PUB #90- 2802 JAN 90
UNDER CONDITIONS OF 2-YR GAVAGE STUDY, CLEAR EVIDENCE OF CARCINOGENIC ACTIVITY FOR MALE F344/N RATS AT 75 OR 150 MG/KG. SHOWN BY INCREASED INCIDENCES OF TUBULAR CELL HYPERPLOSIA, ADEMONAS, AND ADENOCARCINOMAS OF THE KIDNEY. NO EVIDENCE OF CARCINOGENIC ACTIVITY FOR FEMALE F344/N RATS AT 300 OR 600 MG/KG. NO EVIDENCE OF CARCINOGENIC ACTIVITY FOR MALE B6C3F1 MICE AT 250 OR 500 MG/KG. NO EVIDENCE OF CARCINOGENIC ACTIVITY OF D-LIMONENE FOR FEMALE B6C3F1 MICE AT 500 OR 1,000 MG/KG.
REPT'D USE : FLAVOR AND FRAGRANCE ADDITIVE FOR FOOD, BEVERAGES AND HOUSEHOLD CLEANERS AND INDUSTRIAL SOLVENT

CAS NO : **58-89-9**
CHEM NAME : ***** LINDANE (GAMMA-HEXACHLOROCYCLOHEXANE)**

REPORT 01 : IARC MONOGRAPH 20 1979
LINDANE WAS TESTED IN SIX EXPERIMENTS IN MICE BY ORAL ADMINISTRATION. IT PRODUCED BENIGN AND MALIGNANT LIVER TUMORS IN ANIMALS OF BOTH SEXES IN TWO EXPERIMENTS, ONE OF WHICH INVOLVED ONLY SMALL GROUPS OF ANIMALS. THE OTHER EXPERIMENTS WERE CONSIDERED INADEQUATE FOR AN EVALUATION OF CARCINOGENICITY. THERE IS SUFFICIENT EVIDENCE THAT ALPHA-HCH, LINDANE, AND TECHNICAL HCH ARE CARCINOGENIC IN MICE.

REPORT 02 : NCI CARC TRS TRS14 DHEW PUB NO 77-814
UNDER THE CONDITIONS OF THIS BIOASSAY, LINDANE WAS NOT CARCINOGENIC FOR OSBORNE-MENDEL RATS OR B6C3F1 MICE.

REPORT 03 : EPA AMBIENT WATER QUALITY CRITERIA OCT 80
GAMMA-HCH IS SUSPECTED OF BEING A HUMAN CARCINOGEN. AMBIENT WATER CONCENTRATIONS OF 186 NG/L, 18.6 NG/L, AND 1.86 NG/L ARE ESTIMATED TO RESULT IN AN INCREASED CANCER RISK OVER A LIFETIME OF 10 -5, 10 -6, AND 10 -7 RESPECTIVELY.

REPORT 04 : EPA RCRA HEALTH & ENV. EFFECTS PROFILE OCT 80
GAMMA-HCH CAN INDUCE LIVER TUMORS IN MICE. EVIDENCE FOR MUTAGENICITY OF LINDANE IS EQUIVOCAL. LINDANE WAS NOT TERATOGENIC FOR RATS, ALTHOUGH IT REDUCED REPRODUCTIVE CAPACITY OVER FOUR GENERATIONS.

REPORT 05 : IARC MONOGRAPH 20 1979
LINDANE IS EMBRYOTOXIC. LINDANE WAS NOT MUTAGENIC IN SALMONELLA, SAC CHAROMYCES, OR DROSOPHILA. LINDANE INDUCES CHROMOSOMAL ABERRATIONS, POLYPLOIDY AND MITOTIC ARREST IN A NUMBER OF PLANT SYSTEMS. THE CHEMICAL ALSO INDUCED CHROMATID BREAKS IN HUMAN LYMPHOCYTES INVITRO.

REPORT 06 : MULTI-AGENCY CROSS CHECK
Y HHS MAGUIRE
N OSHA CARC
Y EPA CAG
N EPA RCRA
N EPA CLEAN AIR
Y EPA CLEAN WATER

REPORT 07 : NTP TECHINCAL BULLETIN 5 AUG 81
LINDANE WAS NEGATIVE IN THE SALMONELLA TYPHIMURIUM MUTAGENICITY ASSAY IN ONE LABORATORY.

REPORT 08 : NAS REPORT ON DIET NUTRITION AND CANCER 1982
ON THE BASIS OF STUDIES IN ANIMALS AND IN THE ABSENCE OF ADEQUATE DATA FROM EPA STUDIES, IT APPEARS THAT LINDANE PERHAPS PRESENTS A CARCINOGENIC RISK TO HUMANS. HOWEVER IT IS REASONABLE TO ASSUME THAT AMOUNTS PRESENT IN THE AVERAGE U.S. DIET DO NOT MAKE A MAJOR CONTRIBUTION TO THE OVERALL RISK OF CANCER FOR HUMANS.

REPORT 09 : IARC SUPPL. 4 OCT 82
CPD IS CARCINOGENIC TO MICE WHEN GIVEN ORALLY, PRODUCING LIVER TUMORS STUDIES IN RATS AND DOGS ARE INADEQUATE TO DETERMINE CARCINOGENICITY. CASES OF LEUKEMIA AND APLASTIC ANEMIA RECORDED AS ASSOCIATED WITH CPD IN HUMANS. EXPOSURE TO CPD NOTED IN CASE-CONTROL STUDIES OF SOFT-TISSUE SARCOMAS AND LYMPHOMAS, BUT INSUFFICIENTLY FRE-

QUENT FOR CONCLUSIONS TO BE DRAWN.
REPT'D USE : INSECTICIDE AND SWIMMING POOL CHEMICAL

CAS NO : **7439-93-2**
CHEM NAME : ***** LITHIUM**

REPORT 01 : GAO SURVEY ON TOXICANTS AUG 90
CHEMICAL PART OF GAO SURVEY OF AGENCIES CONTROL OF REPRODUCTIVE AND DEVELOPMENTAL TOXICANTS. CONDENSED CHEMICAL DICTIONARY IGNITES IN AIR NEAR ITS MELTING POINT; DANGEROUS FIRE AND EXPLOSION RISK WHEN EXPOSED TO WATER, ACIDS, OR OXICIZING AGENTS. LITHIUM IN SOLUTION IS TOXIC TO CENTRAL NERVOUS SYSTEM.
REPT'D USE : PRODUCTION OF TRITIUM, REDUCING AND HYDROGENATING AGENTS, ALLOY HARDENERS, PHARMACEUTICALS; GRIGNARD REAGENTS, SCAVENGER AND DEGASIFIER FOR STEELS, MODULAR IRON, SOAPS AND GREASES, DEOXIDIZER IN COPPER, CATALYST, HEAT-TRANSFER LIQUID, STORAGE BATTERIES, ROCKET PROPELLANTS, VITAMIN A SYNTHESIS AND SILVER SOLDERS

CAS NO : **434-13-9**
CHEM NAME : ***** LITHOCHOLIC ACID**

REPORT 01 : NCI CARC TRS 175 DHEW PUB NO 79-1731
UNDER THE CONDITIONS OF THIS BIOASSAY, LITHOCHOLIC ACID WAS NOT CARCINOGENIC WHEN ADMINISTERED BY GAVAGE TO FISCHER 344 RATS OR B6C3F1 MICE.

REPORT 02 : NTP TECHNICAL BULLETIN 8 JUL 82
LITHOCHOLIC ACID WAS NEGATIVE IN THE L5178Y MOUSE LYMRHOMA ASSAY.
REPT'D USE : ENDOGENOUS CHEMICAL AND EXPERIMENTAL CARCINOGEN

CAS NO : **9000-40-2**
CHEM NAME : ***** LOCUST BEAN GUM**

REPORT 01 : NTP BIOASSAY TRS 221 DHHS PUB. NO. 82-1777
UNDER THE CONDITIONS OF THIS BIOASSAY, LOCUST BEAN GUM WAS NOT CARCINOGENIC FOR MALE OR FEMALE F344 RATS OR B6C3F1 MICE (FEEDING STUDY).
REPT'D USE : FOOD STABILIZER, MFR OF PHARMACEUTICALS, COSMETICS, PAINTS AND TOOTHPASTE

CAS NO : **13010-47-4**
CHEM NAME : ***** LOMUSTINE1-(2-CHLOROETHYL)-3-CYCLOHEXYL-1-NITROSOUREA**

REPORT 01 : NCI BIOASSAY PROGRAM, CANCER VOL. 40 1977
UNDER THE CONDITIONS OF THE BIOASSAY, I.P. INJECTION OF LOMUSTINE WAS CARCINOGENIC TO MALE AND FEMALE SWISS-WEBSTER MICE, CAUSING LYMPHO-SARCOMAS, LEUKEMIAS AND LUNG TUMORS. IT WAS NOT CARCINOGENIC TO MALE AND FEMALE SPRAGUE-DAWLEY (CD) RATS.

REPORT 02 : IARC MONOGRAPH 26 1981
CCNU IS CARCINOGENIC IN RATS FOLLOWING ITS IP OR IV INJECTION. IT WAS ALSO TESTED IN MICE BY IP INJECTION; A SLIGHT INCREASE IN THE INCIDENCE OF LYMPHOMAS WAS OBSERVED. THERE IS SUFFICIENT EVIDENCE FOR THE CARCINOGENICITY OF CCNU IN RATS. THE DATA FROM STUDIES IN HUMANS ARE INADEQUATE.

REPORT 03 : IARC MONOGRAPH 26 1981
CCNU CAN INDUCE EMBRYO- AND FETOLETHALITY IN RATS AND RABBITS AT DOSES NONTOXIC TO THE MOTHER AND A VARIETY OF TERATOGENIC EFFECTS IN RATS. CCNU INDUCED MUTATIONS IN SALMONELLA STRAINS TA1530, TA1535 AND TA100 WITHOUT METABOLIC ACTIVATION, AND IN THE MOUSE HOST-MEDIATED ASSAY. CCNU ALSO INDUCED MUTATIONS IN CHINESE HAMSTER CELLS IN-VITRO.
REPT'D USE : NO AVAILABLE PRODUCTION OR USE DATA; ANTINEOPLASTIC AGENT

CAS NO : **UNKNOWN-23**
CHEM NAME : ***** LUMBER AND SAWMILL INDUSTRIES**

REPORT 01 : IARC MONOGRAPH 25 FEB 81
THE EPIDEMIOLOGIC DATA ARE NOT SUFFICIENT TO MAKE A DEFINITE ASSESSMENT OF THE CARCINOGENIC RISKS OF EMPLOYMENT IN THE LUMBER AND SAWMILL INDUSTRIES. SOME STUDIES SUGGEST THAT THE INCIDENCES OF NASAL CANCER AND HODGKINS DISEASE MAY BE INCREASED. THE HYPOTHESIZED LINK TO HODGKINS DISEASE IS NOT ADEQUATELY SUPPORTED. SOFT-TISSUE SARCOMAS AND HISTOCYTIC LYMPHOMAS HAVE BEEN REPORTED FOLLOWING EXPOSURE TO CHLOROPHENOLS. THE USE PATTERNS OF CHLOROPHENOLS SUGGESTS THAT SAWMILL WORKERS IN THIS STUDY WERE AT INCREASED RISK FOR BOTH OF THESE MALIGNANCIES.

CAS NO : **21884-44-6**
CHEM NAME : ***** LUTEOSKYRIN**

REPORT 01 : IARC MONOGRAPH 10 OCT 75
LUTEOSKYRIN IS CARCINOGENIC IN MICE FOLLOWING ITS ORAL ADMINISTRATION, THE ONLY SPECIES AND ROUTE OF ADMINISTRATION TESTED; IT PRODUCED BENIGN AND MALIGNANT TUMORS OF THE LIVER.

REPORT 02 : IARC MONOGRAPH 10 1976
LUTEOSKYRIN IN THE PRESENCE OF ABSENCE OF A RAT LIVER-MICROSOME SYSTEM DID NOT INDUCE UNSCHEDULED DNA-REPAIR SYNTHESIS IN CULTURED HUMAN FIBROBLASTS.
REPT'D USE : MYCOTOXIC CONTAMINANT OF FOODS

CAS NO : **52-76-6**
CHEM NAME : ***** LYNOESTRENOL**

REPORT 01 : IARC MONOGRAPH 21 1979
LYNOESTRENOL WAS TESTED BY ORAL ADMINISTRATION IN RATS AND MICE, ALONE OR IN COMBINATION WITH MESTRANOL. IT DID NOT INCREASE THE INCIDENCE OF TUMORS. NO EVALUATION OF THE CARCINOGENICITY OF LYNOESTRENOL COULD MADE.

REPORT 02 : IARC MONOGRAPH 21 1979
DOMINANT LETHAL MUTATIONS WERE OBSERVED IN FEMALE MICE TREATED WITH LYNOESTRENOL. NO DATA ON THE TERATOGENICITY AND EMBRYOTOXICITY OF LYNOESTRENOL WERE AVAILABLE.
REPT'D USE : ESTROGEN AND ORAL CONTRACEPTIVES

CAS NO : **1634-78-2**
CHEM NAME : ***** MALAOXON**

REPORT 01 : NCI CARC TRS 135 DHEW PUB NO NIH 79-1390
IT WAS CONCLUDED THAT UNDER THE CONDITIONS OF THIS BIOASSAY MALAOXON WAS NOT CARCINOGENIC B6C3F1 MICE (FEEDING STUDY). THERE WAS EQUIVOCAL EVIDENCE OF CARCINOGENICITY IN MALE AND FEMALE RATS BASED ON THE C-CELL LESIONS OF THYROID.
REPT'D USE : METABOLITE OF MALATHION

CAS NO : **121-75-5**
CHEM NAME : ***** MALATHION**

REPORT 01 : NTP TECHNICAL BULLETIN 7 APR 82
MALATHION WAS NEGATIVE IN THE SALMONELLA TYPHIMURIUM MUTAGENICITY ASSAY.

REPORT 02 : NCI CARC TRS 24 DHEW PUB NO 78-824
IT IS CONCLUDED THAT UNDER THE CONDITIONS OF THIS BIOASSAY, THERE WAS NO CLEAR EVIDENCE OF THE ASSOCIATION OF THE TUMOR INCIDENCE WITH THE ADMINISTRATION OF MALATHION TO OSBORNE-MENDEL RATS OR B6C3F1 MICE (FEEDING STUDY).

REPORT 03 : NCI CARC TRS 192 DHEW PUB NO 79-1748
IT WAS CONCLUDED THAT UNDER THE CONDITIONS OF THIS BIOASSAY, MALATHION WAS NOT CARCINOGENIC IN MALE OR FEMALE RATS, BUT THE FEMALES MAY NOT HAVE RECEIVED A MAXIMUM TOLERATED DOSE (FEEDING STUDY IN F344 RATS ONLY).

REPORT 04 : IARC MONOGRAPH 30 JAN 83
MALATHION AND ITS METABOLITE MALAOXON TESTED FOR CARCINOGENICITY IN MICE AND RATS BY ADMINISTRATION IN DIET. NO EVIDENCE OF CARCINOGENICITY FOUND.

REPORT 05 : IARC MONOGRAPH 30 JAN 83
NO EVIDENCE OF TERATOGENICITY OR EMBRYOTOXICITY IN RATS GIVEN MATERNALLY TOLERATED DOSES OF MALATHION DURING PREGNANCY. IN ANIMALS FED VERY HIGH DOSES FOR TWO GENERATIONS, MALATHION HAD NO EFFECT ON REPRODUCTION OTHER THAN TO DECREASE SURVIVAL IN FIRST GENERATION AND TO REDUCE GROWTH IN SECOND GENERATION OF RATS.

REPORT 07 : IARC MONOGRAPH 30 JAN 83
NO MUTAGENIC ACTION OBSERVED IN DROSOPHILA MELANOGASTER. MALATHION INCREASED SISTER CHROMATID EXCHANGE FREQUENCY IN CULTURED MAMMALIAN CELLS BUT DID NOT INDUCE UNSCHEDULED DNA SYNTHESIS. CHROMOSOMAL ABERRATIONS INCREASED IN CERTAIN TYPES OF CULTURED MAMMALIAN CELLS AND IN MICE TREATED IN VIVO WITH MALATHION. DOMINANT LETHAL TESTS IN MICE WERE REPORTED NEGATIVE. THUS LIMITED EVIDENCE FOR THE MUTAGENICITY OF MALATHION.
REPT'D USE : INSECTICIDE

CAS NO : **108-31-6**
CHEM NAME : ***** MALEIC ANHYDRIDE**

REPORT 01 : NTP TECHNICAL BULLETIN 3 DEC 80
THE COMPOUND WAS FOUND TO BE NEGATIVE IN THE SALMONELLA TYPHIMURIUM MUTAGENICITY ASSAY IN ONE LABORATORY.

REPORT 02 : CIIT 24 MONTH TOXICOLOGY STUDY APR 79
MALEIC ANHYDRIDE IS BEING EVALUATED FOR TOXICOLOGIC AND/OR CARCINOGENIC EFFECTS IN F344 RATS FOLLOWING DIETARY ADMINISTRATION AT LEVELS OF 100, 32 AND 10 MG/KG/DAY. GROSS AND MICROSCOPIC PATHOLOGICAL EVALUATION AT WEEK 52 REVEALED NO DISTINCT TREATMENT-RELATED FINDINGS ALTHOUGH A VARIETY OF SPONTANEOUS LESIONS WERE OBSERVED.

REPORT 06 : MULTI-AGENCY CROSS CHECK
N HHS MAGUIRE
N OSHA CARC
N EPA CAG
Y EPA RCRA
N EPA CLEAN AIR
N EPA CLEAN WATER
REPT'D USE : CHEMICAL INTERMEDIATE, MFR OF DYE INTERMEDIATES/PHARMACEUTICALS, AGRICULTURAL CHEMICALS, PAINTS/VARNISHES AND CLEANING AGENTS

CAS NO : **123-33-1**
CHEM NAME : ***** MALEIC HYDRAZIDE**

REPORT 01 : IARC MONOGRAPH 4 1974
NO CARCINOGENIC EFFECT WAS OBSERVED IN ADULT MICE AND RATS FOLLOWING ORAL OR SUBCUTANEOUS ADMINISTRATION OF MALEIC HYDRAZIDE. THE SIGNIFICANCE OF HEPATOMAS OBTAINED IN NEWBORN MICE CANNOT BE ASSESSED BECAUSE OF THE CONTAMINATION OF MALEIC HYDRAZIDE WITH HYDRAZINE.

REPORT 02 : NTP TECHNICAL BULLETIN 8 JUL 82
MALEIC HYDRAZIDE WAS NEGATIVE IN DROSOPHILA WHEN TESTED FOR SEX-LINKED RECESSIVE LETHALS.

REPORT 06 : MULTI-AGENCY CROSS CHECK
N HHS MAGUIRE
N OSHA CARC
N EPA CAG
Y EPA RCRA
N EPA CLEAN AIR
N EPA CLEAN WATER
REPT'D USE : HERBICIDE AND PLANT GROWTH REGULATOR

CAS NO : **542-78-9**
CHEM NAME : ***** MALONALDEHYDE**

REPORT 01 : IARC MONOGRAPH 36 FEB 85
MALONALDEHYDE AND ITS BIS(DIMETHYLACETAL) AND SODIUM SALTS WERE TESTED IN MICE BY SKIN APPLICATION. BIS(DIMETHYLACETAL) AND SODIUM SALTS WERE TESTED IN MICE BY ORAL ADMINISTRATION IN DRINKING WATER. STUDIES BY ORAL ADMINISTRATION WERE INADEQUATE FOR EVALUATION. AFTER TOPICAL APPLICATION, AN INCREASE IN INCIDENCE OF SKIN TUMORS OBSERVED IN ONE STUDY. IN ONE 2-STAGE MOUSE-SKIN ASSAY, HIGH DOESE OF MALONALDEHYDE, POSSIBLY CONTAINING IMPURITIES, SHOWED INITIATING ACTIVITY. IN TWO OTHER ASSAYS USING LOW DOSES, NO INITIATING OR PROMOTING ACTIVITY WAS OBSERVED.

REPORT 02 : IARC MONOGRAPH 36 FEB 85
MALONALDEHYDE MUTAGENIC TO BACTERIA AND INDUCED SOMATIC MUTATIONS IN INSECTS; MUTAGENIC TO MAMMALIAN CELLS IN VITRO AND INDUCED MICRONUCLEI, CHROMOSOMAL ABERRATIONS AND ANEUPLOIDY IN THESE CELLS.
REPT'D USE : NOT PRODUCED COMMERCIAL

CAS NO : **4341-85-9**
CHEM NAME : ***** MALONONITRILE**

REPORT 01 : NIOSH CRITERIA DOCUMENT ON NITRILES SEP 78
NIOSH RECOMMENDS A STANDARD OF 3 PPM TWA BASED ON LIVER, KIDNEY, RESPIRATORY, CARDIOVASCULAR, GI, AND CNS EFFECTS. THERE IS NO PRESENT OSHA STANDARD.

REPORT 02 : EPA RCRA HEALTH AND ENV. EFFECTS PROFILE OCT 80
THE ONLY AVAILABLE HUMAN TOXICITY DATA ON MALONONITRILE ARE THOSE REPORTED DURING THE CLINICAL USE OF THIS COMPOUND IN THE TREATMENT OF MENTAL ILLNESS. A THOROUGH DOCUMENTATION OF THE SIDE EFFECTS OF THIS COMPOUND EXISTS.

REPORT 06 : MULTI-AGENCY CROSS CHECK
N HHS MAGUIRE
N OSHA CARC
N EPA CAG
Y EPA RCRA
N EPA CLEAN AIR
N EPA CLEAN WATER
REPT'D USE : CHEMICAL INTERMEDIATE AND LEACHING AGENT FOR GOLD. NO LISTING ON TSCA INVENTORY

CAS NO : **12427-38-2**
CHEM NAME : ***** MANEB**

REPORT 01 : IARC MONOGRAPH 12 1976
MANEB HAS BEEN TESTED IN MICE AND RATS BY ORAL ADMINISTRATION AND BY SINGLE SUBCUTANEOUS INJECTION. ITS ORAL ADMINISTRATION PRODUCED AN INCREASED INCIDENCE OF LUNG TUMORS IN MICE OF ONE STRAIN, BUT NO INCREASE WAS OBSERVED IN THREE OTHER STRAINS. THE STUDIES IN RATS CANNOT BE EVALUATED DUE TO THE SMALL NUMBERS OF SURVIVING ANIMALS. NO EVALUATION OF THE CARCINOGENICITY OF THIS COMPOUND CAN BE MADE.

REPORT 02 : IARC MONOGRAPH 12 1976
A SLIGHT INCREASE IN THE NUMBER OF CHROMOSOME ABERRATIONS IN METAPHASES OF BONE-MARROW CELLS WAS OBSERVED IN MICE TREATED WITH MANEB.
REPT'D USE : FUNGICIDE AND INSECTICIDE

CAS NO : **69-65-8**
CHEM NAME : ***** D-MANNITOL**

REPORT 01 : NTP BOARD OF SCIENTIFIC COUNSELORS DEC 81
UNDER THE CONDITIONS OF THIS BIOASSAY, D-MANNITOL ADMINISTERED IN THE FEED WAS NOT CARCINOGENIC FOR F344 RATS OR B6C3F1 MICE OF EITHER SEX.

REPORT 02 : NTP TECHNICAL BULLETIN 3 DEC 80
D-MANNITOL WAS NEGATIVE IN THE SALMONELLA TYPHIMURIUM MUTAGENICITY ASSAY IN ONE LABORATORY.
REPT'D USE : SUGAR SUBSTITUTE, FOOD HUMECTANT, DIURETIC AND MFR OF ELECTROLYTIC CONDENSORS/RESIN PLASTICIZERS

CAS NO : **551-74-6**
CHEM NAME : ***** MANNOMUSTINE (DIHYDROCHLORIDE)**

REPORT 01 : IARC MONOGRAPH 9 1975
MANNOMUSTINE ADMINISTERED AS THE DIHYDROCHLORIDE IS CARCINOGENIC IN MICE FOLLOWING ITS INTRAPERITONEAL INJECTION, PRODUCING AN INCREASED INCIDENCE OF LEUKEMIA AND A DOSE-RELATED INCREASE IN MANNOMUSTINE 2HC1 INDUCES CHROMOSOME ABERRATIONS IN CULTURED PERIPHERAL HUMAN LYMPHOCYTES AND IN EMBRYONIC FIBROBLASTS.
REPT'D USE : EXPERIMENTAL ANTINEOPLASTIC AGENT

CAS NO : **UNKNOWN-03**
CHEM NAME : ***** MANUFACTURE OF AURAMINE (C.I. BASIC YELLOW 2)**

REPORT 06 : MULTI-AGENCY CROSS CHECK
Y HHS MAGUIRE
N OSHA CARC
N EPA CAG
N EPA RCRA
N EPA CLEAN AIR
N EPA CLEAN WATER

CAS NO : **UNKNOWN-01**
CHEM NAME : ***** MANUFACTURE OF CHROMATE**

REPORT 01 : IARC SUPPLEMEMT 1 1979
THERE IS AN INCREASED INCIDENCE OF LUNG CANCER AMONG WORKERS IN THE CHROMATE-PRODUCING INDUSTRY, AND POSSIBLY ALSO AMONG CHROMIUM PLATERS AND CHROMIUM ALLOY WORKERS. THERE IS ALSO A SUGGESTION OF INCREASED INDICENCE OF CANCERS AT OTHER SITES. THE CHROMIUM COMPOUND(S) RESPONSIBLE HAS NOT BEEN SPECIFIED PRECISELY.

REPORT 06 : MULTI-AGENCY CROSS CHECK
Y HHS MAGUIRE
N OSHA CARC
N EPA CAG
N EPA RCRA
N EPA CLEAN AIR
N EPA CLEAN WATER

CAS NO : **CLASS-38**
CHEM NAME : **MATE**

REPORT 01 : IARC MONOGRAPHS 51 1991
4.5 EVALUATION 1, LIMITED EVIDENCE FOR CARCINOGENICITY OF HOT MATE DRINKING IN HUMANS. NO DATA AVAILABLE ON DRINKING OF COLD MATE. NO DATA ON CARCINOGENICITY OF MATE IN EXPERIMENTAL ANIMALS. THREE CASE-CONTROL STUDIES IN SOUTH AMERICA INVESTIGATED ASSOCIATION BETWEEN MATE DRINKING AND OESOPHAGEAL CANCER. TWO FROM URUGUAY REPORTED INCREASED RISK AMONG DRINKERS AND DOSE-RESPONSE RELATIONSHIPS, AFTER ADJUSTMENT FOR ALCOHOL CONSUMPTION AND SMOKING. HEAVY DRINKERS OF MATE TEN TIMES MORE LIKELY TO DEVELOP CANCER THAN PEOPLE WHO DID NOT DRINK, INTAKE LEVELS LOWER THAN IN PREVIOUS STUDIES.

REPORT 02 : IARC MONOGRAPHS 51 1991
MATE IN ORAL CANCER SUBJECT OF CASE-CONTROL INVESTIGATION IN BRAZIL. ANALYSIS SHOWED DOSE-RESPONSE EFFECT WITH FREQUENCY OF MATE DRINKING, EFFECT NO LONGER PRESENT AFTER ADJUSTMENT FOR SMOKING AND ALCOHOL CONSUMPTION. AFTER ADJUSTMENT, MATE DRINKERS 1.6 TIMES MORE LIKELY TO HAVE ORAL CANCER THAN NONDRINKERS. CASE-CONTROL STUDY FROM URUGUAY REPORTED DOSE-RESPONSE ASSOCIATION BETWEEN MATE DRINKING AND OROPHARYNGEAL CANCER, WHICH REMAINED AFTER ADJUSTMENT FOR AGE, ALCOHOL & SMOKING.

REPORT 03 : IARC MONOGRAPHS 51 1991
ONE STUDY FROM URUGUAY REPORTED THREE-FOLD INCREASED RISK FOR LARYNGEAL CANCER AMONG MATE DRINKERS, WITH SIGNIFICANT DOSE-RESPONSE RELATIONSHIP AFTER ADJUSTMENT FOR AGE, TOBACCO AND ALCOHOL. RESULTS OF CASE-CONTROL STUDY OF BLADDER CANCER IN ARGENTINA SHOWED NO EVIDENCE OF TREND IN RISK WITH INCREASING CONSUMPTION OF MATE. ENDOSCOPIC SURVEY FROM BRAZIL SHOWED DAILY DRINKERS OF HOT MATE HAD PREVALENCE OF HISTOLOGICALLY CONFIRMED OESOPHAGITIS THREE TIMES HIGHER THAN THAT OF NONDRINKERS.
REPT'D USE : BEVERAGE

CAS NO : **94-74-6**
CHEM NAME : ***** MCPA**

REPORT 01 : IARC MONOGRAPH 30 JAN 83
NO ADEQUATE STUDY OF THE CARCINOGENICITY OF MCPA TO ANIMALS WAS AVAILABLE.

REPORT 02 : IARC MONOGRAPH 30 JAN 83
MCPA AND MATERNALLY TOXIC DOSES OF ITS ETHYL ESTER WERE TERATOGENIC AND EMBRYOTOXIC TO RATS.

REPORT 03 : IARC MONOGRAPH 30 JAN 83
MCPA NOT MUTAGENIC IN BACTERIAL SYSTEMS IN PLATE OR HOST-MEDICATED ASSAYS. QUESTIONABLE RESULT OBTAINED WITH MCPA IN YEAST, BUT ITS METHYL ESTER WAS MUTAGENIC. RECESSIVE LETHAL MUTATIONS, BUT NO CHROMOSOME LOSS, NON-DISJUNCTION OR X-Y CHROMOSOME REARRANGEMENT, INDUCED IN DROSOPHILA MELANOGASTER. NO OVERALL EVALUATION OF MUTAGENICITY OF MCPA COULD BE MADE.
REPT'D USE : HERBICIDE

CAS NO : **26148-68-5**
CHEM NAME : ***** MEA-A-C(2-AMINO-3-METHYL-9H-PYRIDO(2,3-B)INDOLE**

REPORT 01 : IARC VOL 40 1986
SUFFICIENT EVIDENCE FOR CARCINOGENICITY OF CPD TO TEST ANIMALS, BUT NO DATA AVAILABLE ON CARCINOGENICITY TO HUMANS. NO DATA AVAILABLE TO EVALUATE REPRODUCTIVE EFFECTS OR PRENATAL TOXICITY OF CPD TO TEST ANIMALS.
REPT'D USE : NO COMMERCIAL USE

CAS NO : **71-58-9**
CHEM NAME : ***** MEDROXYPROGESTERONE ACETATE**

REPORT 01 : IARC MONOGRAPH 21 1979
MEDROXYPROGESTERONE ACETATE WAS TESTED BY THE INTRA-MUSCULAR ROUTE IN DOGS AND PRODUCED MALIGNANT MAMMARY TUMORS.

REPORT 02 : IARC MONOGRAPH 21 1979
NO DATA ON THE MUTAGENICITY OF MEDROXYPROGESTERONE ACETATE WERE AVAILABLE. THE COMPOUND WAS REPORTED TO HAVE TERATOGENIC EFFECTS IN SOME ANIMAL SPECIES.
REPT'D USE : PHARMACEUTICAL AND VETERINARY MEDICINE

CAS NO : **595-33-5**
CHEM NAME : ***** MEGESTROL ACETATE**

REPORT 01 : IARC MONOGRAPH 21
MEGESTROL ACETATE ALONE OR WITH ETHINYLOESTRADIOL WAS TESTED IN MICE, RATS AND DOGS BY ORAL ADMINISTRATION AND IN RATS BY SUBCUTANEOUS ADMINISTRATION. IT PRODUCED MAMMARY TUMORS IN DOGS WHEN TESTED ALONE AND IN MICE WHEN TESTED IN COMBINATION WITH ETHINYLOESTRADIOL. EXPERIMENTS IN WHICH IT WAS TESTED IN RATS IN COMBINATION WITH ETHINYLOESTRADIOL WERE NEGATIVE OR INADEQUATE.

REPORT 02 : IARC MONOGRAPH 21 1979
NO CHROMOSOMAL EFFECTS WERE OBSERVED WHEN MEGESTROL ACETATE WAS ADDED TO CULTURES OF LYMPHOCYTES GROWN FROM THE BLOOD OF HEALTHY WOMEN. NO DATA ON THE EMBRYOTOXICITY OR TERATOGENICITY OF MEGESTROL ACETATE WERE AVAILABLE.
REPT'D USE : ANTINEOPLASTIC, ORAL CONTRACEPTIVES AND DERMATOLOGICAL AGENT

CAS NO : **77094-11-2**
CHEM NAME : ***** MEIQ (2-AMINO-3,4-DIMETHYLIMIDAZO(4,5-F) QUINOLINE (MEIQ)**

REPORT 01 : IARC VOL 40 1896
INADEQUATE EVIDENCE FOR CARCINOGENICITY OF MEIQ IN TEST ANIMALS, BUT NO EVALUATION OF THE CARCINOGENICITY TO HUMANS COULD BE MADE. THE DEGREE OF EVIDENCE IN SHORT-TERM TESTS FOR GENETIC ACTIVITY IS SUFFICIENT. MEIQ CROSSES THE PLACENTA OF MICE. CPD INDUCED PROPHAGE IN BACTERIA, WAS MUTAGENIC TO S. TYPHIMURIUM WITH ACTIVATION AND WAS POSITIVE IN D. MELANOGASTER WING SPOT TESTS. INDUCED DNA DAMAGE AND WITH ACTIVATION, MUTATIONS IN CULTURED MAMMALIAN CELLS.
REPT'D USE : NO COMMERCIAL USE

CAS NO : **77500-04-0**
CHEM NAME : ***** MEIQX (2-AMINO-3,4-DIMETHYLIMIDAZO (4,5-F)QUINOXALINE)**

REPORT 01 : IARC VOL 40 1896
INADEQUATE EVIDENCE FOR CARCINOGENICITY OF CPD IN TEST ANIMALS, BUT NO EVALUATION OF CARCINOGENICITY TO HUMANS COULD BE MADE. NO DATA AVAILABLE TO EVALUATE REPRODUCTIVE EFFECTS OR PRENATAL TOXICITY TO EXPERIMENTAL ANIMALS. CPD MUTAGENIC TO S. TYPHIMURIUM AND GAVE POSITIVE RESULTS IN D. MELANOGASTER WING SPOT TESTS. WAS MUTAGENIC IN CULTURED CHINESE HAMSTER LUNG CELLS WITH ACTIVATION.
REPT'D USE : NO COMMERCIAL USE

CAS NO : **108-78-1**
CHEM NAME : ***** MELAMINE**

REPORT 01 : NTP TR 245 NIH PUB 83-2501
UNDER THE CONDITIONS OF THIS BIOASSAY, MELAMINE WAS CARCINOGENIC FOR MALE F344/N RATS, CAUSING TRANSITIONAL-CELL CARCINOMAS IN THE URINARY BLADDER. WITH ONE EXCEPTION, URINARY BLADDER STONES WERE OBSERVED IN MALE RATS THAT HAD TRANSITIONAL-CELL CARCINOMAS. MELAMINE WAS NOT CARCINOGENIC FOR FEMALE F344/N RATS OR FOR B6C3F1 MICE OF EITHER SEX.

REPORT 02 : IARC MONOGRAPH 39, 1986 JUN 85
MICE FED CPD SHOWED NO TREATMENT RELATED NEOPLASMS. RATS FED CPD DEVELOPED URINARY BLADDER TUMORS AND BLADDER STONES. NO INITIATING ACTIVITY SEEN TO RESULT

FROM SKIN PAINTING TEST IN MICE. DATA WERE INSUFFICIENT TO EVALUATE REPRODUCTIVE EFFECTS OR PRENATAL TOXICITY IN HUMANS OR ANIMALS. CPD WAS NOT MUTAGENIC TO S. TYPHIMURIUM WITH OR WITHOUT ACTIVATION. NO CASE REPORTS OR EPA STUDIES WERE AVAILABLE TO EVALUATE HUMAN CARCINOGENICITY. IARC HAS DETERMINED THAT THERE IS INADEQUATE EVIDENCE FOR CARCINOGENICITY OF CPD IN EXPERIMENTAL ANIMALS.

REPT'D USE : ANTI-CREASE AGENTS FOR PERMANENT-PRESS FABRICS, PLASTIC MOLDING RESIN, SILVER TARNISH CLEANERS, TEXTILE FINISHES, SURFACE COATING RESINS, OR THOPEDIC CASTS, PAPER PRODUCTS, LAMINATES, LEATHER TANNING, ADHESIVES AND PAINTS

CAS NO : **148-82-3**
CHEM NAME : ***** MELPHALAN**

REPORT 01 : IARC MONOGRAPH 9 1975
MELPHALAN IS CARCINOGENIC IN MICE AND RATS FOLLOWING INTRAPERITONEAL INJECTION, PRODUCING LYMPHOSARCOMAS, LUNG TUMORS IN MICE, AND PERITONEAL SARCOMAS IN RATS. EPIDEMIOLOGIC STUDIES SHOWED SUBSTANTIALLY INCREASED RATES OF LEUKEMIA IN PATIENTS TREATED WITH MELPHALAN. IN A 1979 SUPPLEMENT, IARC DETERMINED THAT THERE IS SUFFICIENT EVIDENCE FOR THE CARCINOGENICITY OF MELPHALAN, AND THAT MELPHALAN IS CARCINOGENIC FOR HUMANS.

REPORT 02 : NCI BIOASSAY PROGRAM CANCER VOL. 40 1977
UNDER THE CONDITIONS OF THIS BIOASSAY, I.P. INJECTION OF MELPHALAN WAS CARCINOGENIC TO MALE AND FEMALE SWISS-WEBSTER MICE, CAUSING LUNG TUMORS IN BOTH SEXES AND LYMPHOSARCOMAS IN MALES. IT WAS ALSO CARCINOGENIC TO MALE AND FEMALE SPRAGUE-DAWLEY (CD) RATS CAUSING PERITONEAL SARCOMAS.

REPORT 03 : IARC MONOGRAPH 9 1975
MELPHALAN PRODUCES STRUCTURAL ABERRATIONS OF THE CHROMATID AND CHROMOSOME TYPES IN BONE-MARROW CELL OF TREATED WISTAR RATS AND CHROMATID-TYPE ABERRATIONS IN PERIPHERAL BLOOD LYMPHOCYTES TAKEN FROM CANCER PATIENTS.

REPORT 06 : MULTI-AGENCY CROSS CHECK
Y HHS MAGUIRE
N OSHA CARC
Y EPA CAG
Y EPA RCRA
N EPA CLEAN AIR
N EPA CLEAN WATER

REPORT 07 : IARC SUPPL. 4 OCT 82
EVIDENCE IS SUFFICIENT FOR CARCINOGENICITY OF CPD TO HUMANS. EPA AND OTHER STUDIES HAVE SHOWN INCREASED INCIDENCE OF LEUKEMIA IN PATIENTS TREATED WITH CPD. CPD IS CARCINOGENIC TO MICE AND RATS FOLLOWING IP INJECTION, CAUSING LYMPHOSARCOMAS AND A DOSE-RELATED INCREASE IN LUNG TUMORS IN MICE AND PERITONEAL SARCOMAS IN RATS. CPD REACTED WITH DNA TO PRODUCE MUTATIONS IN BACTERIA BOTH WITH AND WITHOUT ACTIVATION. INDUCED CHROMOSOMAL ABERRATIONS AND SCE'S. ALSO PRODUCED CHROMOSOME DAMAGE IN LYMPHOCYTES OF PATIENTS TREATED THERAPEUTICALLY.
REPT'D USE : ANTINEOPLASTIC AGENT AND INSECT CHEMOSTERILANT

CAS NO : **89-78-1**
CHEM NAME : ***** DL-MENTHOL**

REPORT 01 : NCI CARC TRS 98 DHEW PUB NO NIH 79-1348
UNDER THE CONDITIONS OF THIS BIOASSAY, DL-MENTHOL WAS NOT CARCINOGENIC FOR EITHER FISCHER 344 RATS OR B6C3F1 MICE (FEEDING STUDY).

REPORT 02 : CHEMICAL DOSSIER STAGE 2 DEC 73
REPT'D USE : CIGARETTE MENTHOL, LINIMENTS, NASAL SPRAYS, MOUTHWASHES, COUGH DROPS, TOOTHPASTES, SPACE DEODORIZERS AND LOCAL ANESTHETIC/ANTISEPTIC

CAS NO : **149-30-4**
CHEM NAME : ***** 2-MERCAPTOBENZOTHIAZOLE**

REPORT 01 : 50FR46121 NOV 85
EPA PROPOSES THAT MBT BE TESTED UNDER TOSCA 4 FOR ENVIRONMENTAL EFFECTS, HEALTH EFFECTS AND CHEMICAL FATE AS SPECIFIED IN 40 CFR (AS GIVEN IN 50FR39252).

REPORT 02 : NTP TR 332 NIH PUB.88-2588 MAY 88
SOME EVIDENCE OF CARCINOGENIC ACTIVITY OF 2-MERCAPTOBENZOTHIAZOLE FOR MALE F344/N RATS, INDICATED BY INCREASED INCIDENCES OF MONONUCLEAR CELL LEUKEMIA, PANCREATIC ACINAR CELL ADENOMAS, ADRENAL GLAND PHEOCHROMOCYTOMAS AND PREPUTIAL GLAND ADENOMAS OR CANCINOMAS (COMBINED) SOME EVIDENCE OF CARCINOGENIC ACTIVITY FOR FEMALE F344/N RATS, INDICATED BY INCREASED INCIDENCES OF ADRENAL GLAND PHEOCHROMOCYTOMAS AND PITUITARY GLAND ADENOMAS. NO EVIDENCE OF CARCINOGENIC ACTIVITY FOR MALE B6C3F1 MICE DOSED WITH 375 OR 750 MG/KG. EQUIVOCAL EVIDENCE OF CARCINOGENIC ACTIVITY FOR FEMALE B6C3F1 MICE INDICATED BY INCIDENCES OF HEPATOCELLULAR

ADENOMAS OR CANCINOMAS (COMBINED).

REPT'D USE : VULCANIZATION ACCELERATOR, CHEMICAL INTERMEDIATE, CORROSION INHIBITOR IN CUTTING OILS AND PETROLEUM PRODUCTS AND FUNGICIDE FOR USE IN TROPICAL CLOTHING

CAS NO : **50-44-2**

CHEM NAME : ***** 6-MERCAPTOPURINE**

REPORT 01 : NCI BIOASSAY PROGRAM CANCER VOL. 40 1977
UNDER THE CONDITIONS OF THE BIOASSAY, I.P. INJECTION OF 6-MERCAPTOPURINE WAS CARCINOGENIC TO MALE AND FEMALE SWISS-WEBSTER MICE, CAUSING LYMPHOSARCOMAS, LEUKEMIA, AND LUNG TUMORS. IT WAS ALSO CARCINOGENIC TO FEMALE SPRAGUE-DAWLEY (CD) RATS CAUSING PITUITARY AND BREAST TUMORS. IT WAS NOT CARCINOGENIC TO MALE RATS.

REPORT 02 : IARC MONOGRAPH 26 1981
6-MERCAPTOPURINE WAS TESTED BY IP ADMINISTRATION AND BY SKIN PAINTING IN MICE AND BY IP, SC AND IV ROUTES OF ADMINISTRATION IN RATS. LIMITATIONS TO THE DATA IN ALL REPROTS PRECLUDED EVALUTAION OF THE POSSIBLE CARCINOGENICITY OF 6-MERCAPTOPURINE.

REPORT 03 : IARC MONOGRAPH 26 1981
6-MERCAPTOPURINE AND 6-MERCAPTOPURINE RIBOSIDE WERE PROVEN TO CAUSE EMBRYOLETHALITY AT DOSES NONTOXIC TO THE MOTHER AND TO INDUCE SEVERE TERATOGENIC EFFECTS IN SEVERAL ANIMAL SPECIES. 6-MERCAPTOPURINE IS MUTAGENIC WITHOUT ACTIVATION IN SALMONELLA STRAIN TA1535. IT PRODUCED DOMINANT LETHAL EFFECTS IN VIVO IN MALE MICE.

REPT'D USE : ANTINEOPLASTIC AGENT

CAS NO : **574-25-4**

CHEM NAME : ***** 6-MERCAPTOPURINE-RIBOSIDE**

REPORT 01 : NCI BIOASSAY PROGRAM- CANCER VOL. 40 1977
UNDER THE CONDITIONS OF THE BIOASSAY, I.P. INJECTION OF 6-MERCAPTOPURINE-RIBOSIDE WAS CARCINOGENIC TO MALE AND FEMALE SWISS-WEBSTER MICE, CAUSING LUNG AND ADRENAL TUMORS IN MALES, AND LUNG AND OVARIAN TUMORS IN FEMALES. IT WAS NOT CARCINOGENIC TO SPRAGUE-DAWLEY (CD) RATS OF EITHER SEX.

REPORT 02 : IARC MONOGRAPH 26 1981
6-MERCAPTOPURINE AND 6-MERCAPTOPURINE RIBOSIDE WERE PROVEN TO CAUSE EMBRYOLETHALITY AT DOSES NONTOXIC TO THE MOTHER AND TO INDUCE SEVERE TERATOGENIC EFFECTS IN SEVERAL ANIMAL SPECIES. 6-MERCAPTOPURINE IS MUTAGENIC

WITHOUT ACTIVATION IN SALMONELLA STRAIN TA1535. IT PRODUCED DOMINANT LETHAL EFFECTS IN VIVO IN MALE MICE.

REPT'D USE : NO AVAILABLE PRODUCTION OR USE DATA

CAS NO : **7439-97-6**

CHEM NAME : ***** MERCURY**

REPORT 01 : EPA AMBIENT WATER QUALITY CRITERIA OCT 80
A CRITERION, CALCULATED ON THE LOWEST OBSERVED EFFECT LEVEL OF 200 UG/DAY AND AN UNCERTAINTY FACTOR OF 10, IS 144 NG/L. IN HUMANS, MERCURIALS HAVE BEEN ASSOCIATED WITH NEUROLOGICAL DISORDERS, SENSORY IMPAIRMENT, AND TREMORS. MERCURIALS HAVE INDUCED CHROMOSOME BREAKAGE IN PLANT CELLS AND POINT MUTATIONS IN DROSOPHILIA.

REPORT 02 : NIOSH CRITERIA DOCUMENT JAN 73
NIOSH RECOMMENDS A STANDARD OF 0.05 MG/CU M TWA BASED ON CNS EFFECTS. THE PRESENT OSHA STANDARD IS 0.1 MG/CU M CEILING. (NO TWA)

REPORT 03 : EPA RCRA HEALTH AND ENV. EFFECTS PROFILE OCT 80
MERCURIALS HAVE INDUCED CHROMOSOME BREAKAGE IN PLANT CELLS AND POINT MUTATIONS IN DROSOPHILA. MERCURIALS HAVE NOT BEEN SHOWN TO PRODUCE CARCINOGENIC EFFECTS OTHER THAN NON-SPECIFIC INJECTION SITE SARCOMAS.

REPORT 06 : MULTI-AGENCY CROSS CHECK
N HHS MAGUIRE
N OSHA CARC
N EPA CAG
Y EPA RCRA
N EPA CLEAN AIR
Y EPA CLEAN WATER

REPT'D USE : PREPARATION OF CHLORINE/CAUSTIC SODA, MFR OF ELECTRICAL APPARATUS, BAROMETERS/THERMOMETERS, MERCURY-ARC LAMPS,MIRRORS,CATALYST,AMALGAMS,FUNGICIDES/ALGACIDES AND SOLDERS

CAS NO : **531-76-0**

CHEM NAME : ***** MERPHALAN**

REPORT 01 : IARC MONOGRAPH 9 1975
MERPHALAN (A MIXTURE OF MEDPHALAN AND MELPHALAN) PRODUCED AN INCREASED INCIDENCE OF MAMMORY FIBROADENOMAS IN RATS FOLLOWING ITS INTRAPERITONEAL INJECTION IN SINGLE DOSES. IN A 1979 SUPPLEMENT, IARC DETERMINED THAT THERE IS SUFFICIENT EVIDENCE FOR THE CARCINOGENICITY OF MERPHALEN IN ANIMALS.

REPORT 02 : NCI BIOASSAY PROGRAM, CANCER VOL. 40 1977
UNDER THE CONDITIONS OF THE BIOASSAY, I.P. INJECTION OF MERPHALAN WAS CARCINOGENIC TO FEMALE SWISS-WEBSTER MICE, CAUSING LYMPHOSARCOMAS. IT WAS NOT CARCINOGENIC TO MALE MICE AND FEMALE SPRAGUE-DAWLEY (CD) RATS.

REPORT 03 : IARC MONOGRAPH 9 1975
MERPHALAN WAS TERATOGENIC IN RATS WHEN GIVEN DURING THE FIRST TEN DAYS OF PREGNANCY. MERPHALAN IS MUTAGENIC IN E. COLI.

REPT'D USE : ANTINEOPLASTIC AGENT

CAS NO : **108-67-8**
CHEM NAME : ***** MESITYLENE**

REPORT 01 : 52FR2522 FINAL TEST STDS AND REPORTING REQUIREMENTS JAN 87
EPA IS ISSUING A FINAL TEST RULE UNDER SECT 4(A) OF TSCA THAT SPECIFIES TEST STANDARDS AND REPORTING REQUIREMENTS FOR TESTING OF AROMATIC HYDROCARBON FRACTION.

CAS NO : **141-97-7**
CHEM NAME : ***** MESITYL OXIDE**

REPORT 01 : PROPOSED TEST RULE, 48FR30699 JUL 85
EPA'S RESPONSE TO ITC
EPA'S RESPONSE TO ITC'S DESIGNATION OF MO FOR PRIORITY CONSIDERATION FOR HEALTH EFFECTS TESTING (44FR31866), TO INCLUDE TESTING FOR CARCINOGENICITY, MUTAGENICITY, TERATOGENICITY AND OTHER CHRONIC EFFECTS WITH EMPHASIS ON BLOOD DISORDERS. AN EPA STUDY WAS ALSO RECOMMENDED. ALL OF THIS FALLS UNDER TSCA 4(A).

REPORT 02 : FINAL TEST RULE - 50FR51857 DEC 85
EPA IS ESTABLISHING FINAL TEST REQUIREMENTS UNDER TSCA 4(A) FOR MFRS AND PROCESSORS OF MO. TESTING REQUIREMENTS INCLUDE INHALATION SUBCHRONIC (90-DAY) TOXICITY IN AT LEAST 1 MAMMALIAN SPECIES, MUTAGENICITY AND ONCOGENICITY, IF CERTAIN MUTAGENICITY TESTS ARE POSITIVE.

REPT'D USE : CHEMICAL INTERMEDIATE AND SOLVENT (USE BEING PHASED OUT)

CAS NO : **72-33-3**
CHEM NAME : ***** MESTRANOL**

REPORT 01 : IARC MONOGRAPH 21 1979
MESTRANOL WAS TESTED IN MICE, RATS, DOGS AND MONKEYS BY ORAL ADMINISTRATION; IN MOST STUDIES IT WAS ADMINISTERED IN COMBINATION WITH PROGESTINS. WHEN ADMINISTERED ALONE, IT INCREASED THE INCIDENCES OF PITUITARY TUMORS IN ONE

STRAIN OF MICE AND INCREASED THE INCIDENCE OF MALIGNANT MAMMARY TUMORS OF OTHER MICE STRAINS. AN INCREASED INCIDENCE OF MALIGNANT MAMMARY TUMORS WAS OBSERVED IN FEMALE RATS. IT IS REASONABLE TO REGARD MESTRANOL AS IF IT PRESENTED A CARCINOGENIC RISK TO HUMANS. 1979 SUPPLEMENT-SUFFICIENT EVIDENCE ANIMAL CARCINOGENICITY.

REPORT 02 : IARC MONOGRAPH 21 1979
DOMINANT LETHAL MUTATIONS WERE OBSERVED IN FEMALE MICE TREATED WITH MESTRANOL. IT IS EMBRYOLETHAL FOR PRE- AND POSTIMPLANTATION EMBRYOS IN SOME ANIMAL SPECIES.

REPORT 03 : IARC SUPP 4, OCT 1982 1985
THERE IS SUFFICIENT EVIDENCE FOR CARCINOGENICITY OF MESTRANOL IN EXPERIMENTAL ANIMALS.

REPORT 06 : MULTI-AGENCY CROSS CHECK
Y HHS MAGUIRE
N OSHA CARC
N EPA CAG
N EPA RCRA
N EPA CLEAN AIR
N EPA CLEAN WATER

REPT'D USE : ORAL CONTRACEPTIVES, TREATMENT OF AMENORRHEA AND ENDOMETRIOSIS

CAS NO : **CLASS-06**
CHEM NAME : ***** METAL-WORKING FLUIDS**

REPORT 01 : NIOSH BULLETIN 15 OCT 76
DIETHANOLNITROSAMINE, AN ANIMAL CARCINOGEN AND A SUSPECTED HUMAN CARCINOGEN, WAS FOUND IN COMMERCIAL CUTTING FLUIDS. INDUSTRIAL HYGIENE PRACTICES CAN HELP MINIMIZE DERMAL AND RESPIRATORY EXPOSURES TO THE FLUIDS.

REPORT 02 : EPA TSCA 8(E) 283-0469S FEB 83
A COMPANY REPORTED A NITRITE/AMINE SYNTHETIC UNDILUTED METALWORKING FLUID CONTAINING A HIGH LEVEL OF NITROSODIETHANOLAMINE AND A NONNITRITE SYNTHETIC FLUID WITHOUT DETECTABLE NITROSAMINES WERE FOUND TO BE ONCOGENIC TO MOUSE LIVER IN LIFETIME SKIN-PAINTING STUDY CONDUCTED BY NIOSH.

REPORT 03 : EPA TSCA 8(E) 0283-0471 AND 0383-0471 MAR 83
FERRO CORP REPORTED ITS INDUSTRIAL LUBRICANT ADDITIVE PRODUCT (MAJOR COMPONENT BIS (2-CHLOROOCTYL) DISULFIDE) EXHIBITED MODERATE TO EXTREME DERMAL SENSITIZATION IN GUINEA PIG. CONTACT DERMATITIS REPORTED ON METALWORKERS USING PRODUCT WITH 1,1,1-TRICHLOROETHANE STODDARD

SOLVENT, AND IN MOST CASES KEROSENE. PRODUCT MUTAGENIC IN SACCHAROMYCES CEREVISIA WITHOUT ACTIVATION, AND CAUSED SOME MUTANTS WITH ACTIVATION.

REPORT 04 : EPA TSCA 8(E) 0185-0542S JAN 85
A COMPANY REPORTS THAT NITROSODIETHANOLAMINE CAN FORM IN NITRITE-CONTAINING METAL-WORKING FLUIDS.

REPORT 05 : IARC SUPPLEMENT 1 SEP 79
SOOTS, COAL-TARS, CREOSOTE OILS, SHALE OILS, AND CUTTING OILS ARE CARCINOGENIC IN EXPERIMENTAL ANIMALS AFTER SKIN PAINTING OR SUBCUTANEOUS INJECTIONS. OCCUPATIONAL EXPOSURE TO THESE CHEMICALS CAUSES CANCER AT SEVERAL SITES, INCLUDING SKIN, LUNG, BLADDER AND GASTROINTESTINAL TRACT. EFFECT MAY BE DUE TO POLYCYCLIC AROMATIC HYDROCARBONS IN THESE MATERIALS.

REPORT 07 : EPA TSCA 8(E) 0788-0742 JUL 88
ARCO REPORTS RABBITS WITH 0.5 ML OF METAL WORKING OIL F-82 APPLIED UNDILUTED TO THEIR SHAVED BACKS AT FOUR SITES SHOWED IT TO BE CORROSIVE.

REPT'D USE : COMMERCIAL CUTTING/GRINDING FLUIDS

CAS NO : **13414-54-5**

CHEM NAME : ***** METHALLY-2-NITROPHENYL ETHER**

REPORT 01 : PROPOSED TEST RULE; JUL 85
50FR29761 EPA'S RESPONSE TO ITC
THIS IS EPA'S RESPONSE TO ITC'S RECOMMENDATION FOR PRIORITY CONSIDERATION FOR CHEMICAL FATE AND ENVIRONMENTAL EFFECTS TESTING UNDER TSCA 4(A). EPA DOES NOT BELIEVE BASIS FOR FINDING THAT THIS CHEMICAL PRESENTS AN UNREASONABLE RISK OF INJURY TO THE ENVIRONMENT NOR THAT THERE WILL BE A SUBSTANTIAL RELEASE OF THE CPD TO THE ENVIRONMENT.

REPT'D USE : INTERMEDIATE IN PRODUCTION OF INSECTICIDE-NEMATICIDE

CAS NO : **2163-80-6**

CHEM NAME : ***** METHANEARSONIC ACID.NA**

REPORT 01 : IARC MONOGRAPH 23 OCT 79
THERE IS INADEQUATE EVIDENCE FOR THE CARCINOGENICITY OF ARSENIC COMPOUNDS IN ANIMALS. THERE IS SUFFICIENT EVIDENCE THAT INORGANIC ARSENIC COMPOUNDS ARE SKIN AND LUNG CARCINOGENS IN HUMANS.

REPORT 02 : NIOSH CRITERIA DOCUMENT JUN 75
NIOSH RECOMMENDS A CEILING OF 2 MICROGRAMS/CU M FOR INORGANIC ARSENIC COMPOUNDS BASED ON DERMATITIS, LUNG, AND LYMPHATIC CANCER. THE CURRENT OSHA STANDARD FOR

INORGANIC ARSENIC IS 10 MICROGRAMS/CU M TWA. SEE CFR 1910.1018.

REPORT 06 : MULTI-AGENCY CROSS CHECK
Y HHS MAGUIRE
N OSHA CARC
Y EPA CAG
N EPA RCRA
N EPA CLEAN AIR
N EPA CLEAN WATER

CAS NO : **144-21-8**
CHEM NAME : ***** METHANEARSONIC ACID.2NA**

REPORT 01 : IARC MONOGRAPH 23 1980
THERE IS INADEQUATE EVIDENCE FOR THE CARCINOGENICITY OF ARSENIC COMPOUNDS IN ANIMALS. THERE IS SUFFICIENT EVIDENCE THAT INORGANIC ARSENIC COMPOUNDS ARE SKIN AND LUNG CARCINOGENS IN HUMANS.THE EVIDENCE THAT ARSENIC COMPOUNDS CAUSE MUTATIONS AND ALLIED EFFECTS IN BACTERIA IS INCONCLUSIVE. HOWEVER, ARSENIC COMPOUNDS INDUCE CHROMOSOMAL ABERRATIONS AND MORPHOLOGICAL TRANSFORMATION IN MAMMALIAN CELLS.

REPORT 02 : NIOSH CRITERIA DOCUMENT JUN 75
NIOSH RECOMMENDS A CEILING OF 2 MICROGRAMS/CU M FOR INORGANIC ARSENIC COMPOUNDS BASED ON DERMATITIS, LUNG, AND LYMPHATIC CANCER. THE CURRENT OSHA STANDARD FOR INORGANIC ARSENIC IS 10 MICROGRAMS/CU M TWA. SEE CFR 1910.1018.

REPORT 06 : MULTI-AGENCY CROSS CHECK
Y HHS MAGUIRE
N OSHA CARC
Y EPA CAG
N EPA RCRA
N EPA CLEAN AIR
N EPA CLEAN WATER
REPT'D USE : HERBICIDE, PESTICIDE AND CAUSTICS/DRAIN CLEANERS

CAS NO : **67-56-1**
CHEM NAME : ***** METHANOL**

REPORT 01 : EPA TSCA FYI- 1288-0658 NOV 88
METHANOL, IN CONTRAST TO GASOLINE, POSES HIGH RISK OF SYSTEMIC TOXICITY, WITH MISLEADING DELAY IN APPEARANCE OF TOXIC MANIFESTATIONS, AND CAN CAUSE DEATH, PERMANENT BLINDNESS, AND NEUROLOGIC IMPAIRMENT FOLLOWING FAILURE

TO OBTAIN TREATMENT PROMPTLY. INGESTED DOSE WHICH REQUIRES HEMODIALYSIS IS 0.4 ML/KG, A MERE TEASPOONFUL (5 ML) OF 85% METHANOL FUEL IN 10 KG ONE-YEAR-OLD CHILD, 7 ML IN 15 KG THREE-YEAR-OLD, AND 28 ML IN A 70 KG ADULT. POTENTIALLY LETHAL DOSE IS 1 ML/KG. METHANOL FATALITIES CAN INCREASE FROM 24 TO 195 A YEAR.

REPT'D USE : SOLVENT, WINDSHIELD WASHER SOLUTIONS AND AUTOMOTIVE FUEL

CAS NO : **91-80-5**

CHEM NAME : ***** METHAPYRILENE**

REPORT 01 : NTP CARC TEST PROGRAM JUL 82

IN A CURRENT NTP BIOASSAY, METHAPYRILINE WAS ADMINISTERED IN THE DIET TO F344 RATS AND B6C3F1 MICE. THE CHRONIC EXPOSURE PHASE HAS BEEN COMPLETED AND HISTOPATHOLOGY IS IN PROGRESS. IN A SECOND BIOASSAY THE CHEMICAL IS BEING ADMINISTERED IN THE DIET TO F344 RATS AND B6C3F1 MICE. THE PRECHRONIC TEST PHASE IS IN PROGRESS.

REPORT 06 : MULTI-AGENCY CROSS CHECK

N HHS MAGUIRE
N OSHA CARC
Y EPA CAG
N EPA RCRA
N EPA CLEAN AIR
N EPA CLEAN WATER

REPT'D USE : VETERINARY MEDICINE

CAS NO : **59-05-2**

CHEM NAME : ***** METHOTREXATE**

REPORT 01 : IARC MONOGRAPH 26 1981

METHOTREXATE WAS TESTED BY ORAL ADMINISTRATION IN MICE AND HAMSTERS, BY IP INJECTION IN MICE AND RATS, AND BY IV INJECTION IN RATS. ONE STUDY IN MICE BY ORAL ADMINISTRATION REPORTED A HIGH INCIDENCE OF LUNG CARCINOMAS, BUT IT DID NOT INCLUDE MATCHED CONTROLS. ALL OTHER STUDIES FAILED TO REVEAL A CARCINOGENIC EFFECT, BUT THE SIGNIFICANCE OF SEVERAL WAS LIMITED BECAUSE OF DEFICIENCIES IN EXPERIMENTAL DESIGN OR REPORTING OF DATA.

REPORT 02 : NCI BIOASSAY PROGRAM - CANCER VOL. 40 1977

UNDER THE CONDITIONS OF THE BIOASSAY, I.P. INJECTION OF METHOTREXATE WAS NOT CARCINOGENIC TO SWISS-WEBSTER MICE OR SPRAGUE-DAWLEY (CD) RATS OF EITHER SEX.

REPORT 03 : IARC MONOGRAPH 26 1981
METHOTREXATE CAN INDUCE TERATOGENIC EFFECTS IN SEVERAL SPECIES AND EMBRYOLETHALITY AT DOSES NONTOXIC TO THE MOTHER. METHOTREXATE IS MUTAGENIC IN MICE IN VIVO. IN VARIOUS MAMMALIAN CELLS IN CULTURE THE DRUG INDUCES CHROMOSOMAL ABERRATIONS AND INCREASES SISTER CHROMATID EXCHANGES. IT ALSO INDUCES MORPHOLOGICAL TRANSFORMATION IN MOUSE CELLS.
REPT'D USE : ANTINEOPLASTIC AGENT AND INSECT CHEMOSTERILANT

CAS NO : **72-43-5**
CHEM NAME : ***** METHOXYCHLOR**

REPORT 01 : NCI CARC TRS NO 35 DHEW PUB NO NIH 78-835
UNDER THE CONDITIONS OF THIS STUDY, METHOXYCHLOR WAS NOT FOUND TO BE CARCINOGENIC IN OSBORNE-MENDEL RATS OR B6C3F1 MICE OF EITHER SEX (FEEDING STUDY).

REPORT 02 : IARC MONOGRAPH 20 OCT 79
METHOXYCHLOR WAS TESTED IN ONE EXPERIMENT IN MICE AND IN SEVERAL EXPERIMENTS IN RATS BY ORAL ADMINISTRATION. THE STUDY IN MICE GAVE NEGATIVE RESULTS. A SUGGESTION THAT IT WAS HEPATOCARCINOGENIC, MADE IN AN EARLIER STUDY THAT WAS INADEQUATELY REPORTED, WAS NOT CONFIRMED IN THREE MORE RECENT EXPERIMENTS.

REPORT 03 : IARC MONOGRAPH 20 1979
METHOXYCHLOR WAS NOT MUTAGENIC IN SALMONELLA STRAIN TA1535, TA1537, TA1538, TA98, AND TA100, WITH AND WITHOUT METABOLIC ACTIVATION. THE CHEMICAL WAS ALSO NEGATIVE IN YEAST AND DROSOPHILA, CYTOGENIC AND DOMINANT LETHAL TESTS IN MICE WERE ALSO NEGATIVE.

REPORT 04 : NTP TECHNICAL BULLETIN 8 JUL 82
METHOXYCHLOR WAS POSITIVE IN THE L5178Y MOUSE LYMPHOMA ASSAY.

REPORT 05 : NTP TECHNICAL BULLETIN 9 APR 83
METHOXYCHLOR WAS NEGATIVE IN ONE SALMONELLA TYPHIMURIUM MUTAGENICITY ASSAY.
REPT'D USE : ECTOPARASTITICIDE, PET FLEA/TICK SPRAYS AND AEROSOLS

CAS NO : **109-86-4**
CHEM NAME : ***** 2-METHOXYETHANOL (2-ME)**

REPORT 01 : EPA TSCA 8(E) NOTICE 0281-0384 FEB 81
DOW CHEMICAL REPORTED INTERIM RESULTS FROM AN INHALATION TERATOLOGY STUDY OF EGME IN RATS, MICE, AND RABBITS.

THE DATA INDICATE THAT EXPOSURE TO 50 PPM, OF EGME IS TERATOGENIC IN RABBITS, AND SUGGEST SOME EVIDENCE OF FETOTOXICITY IN MICE. AN EXPOSURE LEVEL OF 10 PPM WAS NOT CONSIDERED TERATOGENIC OR FETOTOXIC IN EITHER RABBITS OR MICE, THE INTERIM DATA ALSO INDICATE THAT EXPOSURE OF RATS TO LEVELS UP TO 50 PPM PRODUCE NO EVIDENCE OF TERATOGENICITY OR FETOTOXICITY.

REPORT 02 : NTP TECHNICAL BULLETIN 7 APR 82
EGME INDUCED ANTIFERTILITY EFFECTS, WITHOUT EVIDENCE OF DOMINANT LETHAL MUTATIONS, IN RATS. THE CHEMICAL WAS POSITIVE IN THE MOUSE SPERM HEAD MORPHOLOGY ASSAY, BUT WAS NEGATIVE IN VITRO UNSCHEDULED DNA SYNTHESIS IN HUMAN FIBROBLASTS AND THE RAT BONE MAROW CYTOLOGY ASSAY. RESULTS IN THE DROSOPHILA SEX-LINKED RECESSIVE LETHAL ASSAY WERE SUGGESTIVE BUT NOT CLEARLY POSITIVE.

REPORT 03 : CMA INHALATION TERATOLOGY STUDY
EXPOSURE OF PREGNANT RABBITS TO 50 PPM OF EGME PRODUCED CONCLUSIVE EVIDENCE OF TERATOGENICITY AND FETOTOXICITY. THERE WAS NO EVIDENCE OF MATERNAL OR FETAL EFFECTS AMONG RATS EXPOSED TO 10 OR 3 PPM OF EGME. THE INCIDENCES OF MAJOR MALFORMATIONS AMONG FETAL RATS AT EACH EXPOSURE LEVEL, INCLUDING 50 PPM, WERE NOT SIGNIFICANTLY INCREASED IN ANY EXPOSURE GROUP.

REPORT 04 : EPA TSCA 8(E) 8EHQ-1184-0534 NOV 84
DOSE-RELATED ADVERSE TESTICULAR EFFECTS NOTED IN RATS AFTER DOSING WITH 2-METHOXYETHYL PALMITATE PROBABLY CAUSED BY 2-METHOXYETHANOL, A SUSPECTED IN VIVO METABOLITE OF 2-METHOXYETHYL PALMITATE.

REPORT 05 : NIOSH CURRENT INTELLIGENCE BULLETIN 39 MAY 83
TERATOGENIC AND EMBRYOTOXIC TO PREGNANT RATS AND RABBITS. EXPOSURE OF NON-PREGNANT FEMALES CAUSED NO DISCERNIBLE CHANGES IN REPRODUCTIVE ORGANS. EXPOSURE OF MALES RESULTED IN TESTICULAR ATROPHY, HISTOPATHOLOGICAL TESTICULAR CHANGES, INFERTILITY AND ABNORMAL SPERMHEAD MORPHOLOGY. BASED ON THESE FINDINGS AND CONTINUED CONCERN FOR ADVERSE EFFECTS AFTER PERCUTANEOUS ABSORPTION, NIOSH RECOMMENDS THAT CPD BE REGARDED AS HAVING POTENTIAL TO CAUSE ADVERSE REPRODUCTIVE EFFECTS IN BOTH SEXES AND EMBRYOTOXIC EFFECTS, INCLUDING TERATOGENESIS OF OFFSPRING OF EXPOSED, PREGNANT WORKERS.

REPORT 07 : EPA FEDERAL REGISTOR NOTICE 51FR18488 MAY 86
UNDER TSCA 9(A) EPA CONCLUDES RISK OF INJURY TO WORKER HEALTH FROM EXPOSURE TO THIS GLYCOL ETHER IS UNREASONABLE AND MAY BE PREVENTED OR REDUCED BY OSHA.

REPT'D USE : SOLVENT FOR RESINS/DYES, PAINTS/VARNISHES, NAIL POLISHES, WOOD STAINS, AEROSOLS, CHEMICAL REAGENT AND DE-ICING ADDITIVE FOR JET FUELS

CAS NO : **110-49-6**
CHEM NAME : ***** 2-METHOXYETHANOL ACETATE**

REPORT 01 : EPA TSCA. 8(E) NOTICE 0281-0384 FEB 81
DOW CHEMICAL REPORTED INTERIM RESULTS FROM AN INHALATION TERATOLOGY STUDY OF EGME IN RATS, MICE, AND RABBITS. THE DATA INDICATE THAT EXPOSURE TO 50 PPM, OF EGME IS TERATOGENIC IN RABBITS, AND SUGGEST SOME EVIDENCE OF FETOTOXICITY IN MICE. AN EXPOSURE LEVEL OF 10 PPM WAS NOT CONSIDERED TERATOGENIC OR FETOTOXIC IN EITHER RABBITS OR MICE. THE INTERIM DATA ALSO INDICATE THAT EXPOSURE OF RATS TO LEVELS UP TO 50 PPM PRODUCE NO EVIDENCE OF TERATOGENICITY OR FETOTOXICITY.

REPORT 02 : NTP TECHNICAL BULLETIN 7 APR 82
EGME INDUCED ANTIFERTILITY EFFECTS, WITHOUT EVIDENCE OF DOMINANT LETHAL MUTATIONS, IN RATS. THE CHEMICAL WAS POSITIVE IN THE MOUSE SPERM HEAD MORPHOLOGY ASSAY, BUT WAS NEGATIVE IN VITRO UNSCHEDULED DNA SYNTHESIS IN HUMAN FIBROBLASTS AND THE RAT BONE MARROW CYTOLOGY ASSAY. RESULTS IN THE DROSOPHILA SEX-LINKED RECESSIVE LETHAL ASSAY WERE SUGGESTIVE BUT NOT CLEARLY POSITIVE.

REPORT 03 : CMA INHALATION TERATOLOGY STUDY SEP 82
EXPOSURE OF PREGNANT RABBITS TO 50 PPM OF EGME PRODUCED CONCLUSIVE EVIDENCE OF TERATOGENICITY AND FETOXICITY. THERE WAS NO EVIDENCE OF MATERNAL OR FETAL EFFECTS AMONG RATS EXPOSED TO 10 OR 3 PPM OF EGME. THE INCIDENCES OF MAJOR MALFORMATIONS AMONG FETAL RATS AT EACH EXPOSURE LEVEL, INCLUDING 50 PPM, WERE NOT SIGNIFICANTLY INCREASED IN ANY EXPOSURE GROUP.

REPORT 04 : EPA FEDERAL REGISTORNOTICE 51FR18488 MAY 86
UNDER TSCA 9(A) EPA CONCLUDES RISK OF INJURY TO WORKER HEALTH FROM EXPOSIVE TO THIS GLYCOL ETHER IS UNREASONABLE AND MAY BE PREVENTED OR REDUCED BY OSHA.

REPT'D USE : TEXTILE PRINTING, PHOTO FILMS, ADHESIVES, DOPES, WAXES, OILS AND LACQUERS

CAS NO : **111-07-9**
CHEM NAME : ***** 2-METHOXYETHYL PALMITATE**

REPORT 01 : EPA TSCA 8(E) 8EHQ-1084-0534 NOV 84
CPD SHOWN TO CAUSE DOSE-RELATED ADVERSE TESTICULAR EFFECTS IN RATS AFTER 11 ORAL EXPOSURES AT 100 MG/KG/DAY AND 1000 MG/KG/DAY OVER 15 DAYS.

REPT'D USE : PLASTICIZER AND LUBRICANT COMPONENT

CAS NO : **298-81-7**
CHEM NAME : ***** METHOXYPSORALEN**

REPORT 01 : IARC MONOGRAPH Z4 1980
THE AVAILABLE DATA ARE INSUFFICIENT TO ALLOW A CONCLUSION AS TO THE CARCINOGENICITY OF METHOXYPSORALEN.

REPORT 02 : IARC MONOGRAPH 24 1980
METHOXYPSORALEN, MAINLY IN COMBINATION WITH LONG-WAVE ULTRA-VIOLET LIGHT, BUT ALSO IN THE DARK, WAS MUTAGENIC IN A VARIETY OF PROKARYOTIC AND EUKARYOTIC CELLS. ATTENTION IS DRAWN TO THE ABSENCE OF STUDIES ON THE TERATOGENICITY OF THIS COMPOUND.

REPORT 04 : IARC SUPPL. 2 OCT 82
METHOXYPSORALEN IN CONJUNCTION WITH LONG-WAVE ULTRA-VIOLET LIGHT (A) ASSOCIATED WITH DEVELOPMENT OF VARIOUS CARCINOMAS, NEOPLASMS, BASAL CELL SKIN CANCERS AND SQUAMOUS-CELL SKIN CANCER. IN A STUDY LACKING ADEQUATE STATISTICAL POWER, A 3.6 YEAR FOLLOW-UP OF 525 PSORIASIS PATIENTS TREATED WITH LONG-WAVE LENGTH UV LIGHT SHOWED ONLY ONE SKIN CANCER, BUT IT PROBABLY WAS PRESENT BEFORE TREATMENT STARTED. CPD TESTED IN CONJUNCTION WITH UV IN MICE BY ORAL AND IP INJECTION, INCREASED INCIDENCES OF EPIDERMAL AND DERMAL TUMORS. THERE WAS SUFFICIENT EVIDENCE FOR ACTIVITY IN SHORT-TERM TESTS FOR METHOXYPSORALEN AND METHOXYPSORALEN IN CONJUNCTION WITH UV LIGHT.

REPORT 06 : MULTI-AGENCY CROSS CHECK 1985
Y HHS MAGUIRE
N OSHA CARC
N EPA CAG
N EPA RCRA
N EPA CLEAN AIR
N EPA CLEAN WATER

REPORT 07 : NTP TR 359 NIH PUB. 89-2814 JUL 89
UNDER 2-YEAR GAVAGE STUDIES CLEAR EVIDENCE OF CARCINOGENIC ACTIVITY OF 8-METHOXYPSORALEN (WITHOUT ULTRAVIOLET RADIATION) FOR MALE F344/N RATS, SHOWN BY INCREASED INCIDENCES OF TUBULAR CELL HYPERPLASIA, ADENOMAS, AND ADENOCARCINOMAS OF KIDNEY AND CARCINOMAS OF ZYMBAL GLAND. SUBCUTANEOUS TISSUE FIBROMAS AND ALVEOLAR/BRONCHIOLAR ADENOMAS OF LUNG IN MALE F344/N RATS MAY BE RELATED TO CHEMICAL ADMINISTRATION. DOSE-RELATED NONNEOPLASTIC LESIONS IN MALE F344/N RATS INCLUDED INCREASED SEVERITY OF NEPHROPATHY AND MINERALIZATION OF KIDNEY AND FORESTOMACH LESIONS. NO EVIDENCE OF CARCINOGENIC ACTIVITY FOR FEMALE F344/N RATS GIVEN

CHEMICAL AT 37.5 OR 75 MG/KG PER DAY FOR 2 YEARS.
REPT'D USE : SKIN REPIGMENTATION, PSORIASIS, MYCOSIS, T-CELL LYMPHOMAS DRUG AND EUROPEAN SUNSCREEN LOTIONS

CAS NO : **484-20-8**
CHEM NAME : ***** 5-METHOXYPSORALEN**

REPORT 01 : IARC VOL 40 1986
THERE IS SUFFICIENT EVIDENCE FOR CARCINOGENICITY OF CPD TO MOUSE SKIN IN COMBINATION WITH ULTRAVIOLET A (UVA) RADIATION. THERE IS INADEQUATE EVIDENCE FOR CARCINOGENICTY OF CPD TO MOUSE SKIN IN ABSENCE OF UVA RADIATION. THERE IS INADEQUATE EVIDENCE FOR CARCINOGENICITY OF CPD TO HUMANS. THERE IS SUFFICIENT EVIDENCE TO DETERMINE GENETIC ACTIVITY IN SHORT-TERM TESTS IN COMBINATION WITH UVA. MATERNALLY TOXIC DOSES OF CPD DID NOT INCREASE NUMBER OF ANOMALIES IN SURVIVING FETUSES OF TREATED RATS. ONE SMALL SURVEY SHOWED NO EXCESS OF SKIN TUMORS IN WORKERS IN THE BERGAMOT OIL PRODUCTION INDUSTRY, BUT STUDY HAD METHODOLOGICAL WEAKNESSES.
REPT'D USE : THERAPEUTIC AGENT

CAS NO : **96-37-7**
CHEM NAME : ***** METHYCYCLOPENTANE (MCP)**

REPORT 01 : EPA FED. REG. NOTICE 51 FR 17854 MAY 86
IN RESPONSE TO ITC DESIGNATION OF MCP FOR HEALTH EFFECTS TESTING EPA UNDER TSCA 4(A) AUTHORITY PROPOSES THAT MANUFACTURERS AND PROCESSORS OF MCP AND COMMERCIAL HEXANE PERFORM NEUROTOXICITY, SUBCHRONIC TOXICITY, INHALATION AND DERMAL PHARMACOKINETICS TESTS ON MCP BECAUSE OF HUMAN EXPOSURE.
REPT'D USE : CHEMICAL INTERMEDIATE

CAS NO : **96-33-3**
CHEM NAME : ***** METHYL ACRYLATE**

REPORT 01 : IARC MONOGRAPH 19 FEB 79
NO CARCINOGENICITY OR MUTAGENICITY DATA ON METHYL ACRYLATE WERE AVAILABLE.

REPORT 02 : IARC MONOGRAPH 39, 1986 JUN 85
RATS EXPOSED TO MA FOR 2 YRS BY INHALATION SHOWED NO NEOPLASTIC EFFECTS. NO DATA TO EVALUATE REPRODUCTIVE EFFECTS OR PRENATAL TOXICITY. MA NOT MUTAGENIC TO BACTERIA WITH OR WITHOUT ACTIVATION. MA INDUCED CHROMOSOMAL ABERRATIONS IN CHINESE HAMSTER LUNG CELLS IN-VITRO AND IN

MICE BONE MARROW IN VIVO. NO DATA TO EVALUATE HUMAN REPRODUCTIVE EFFECTS OR PRENATAL TOXICITY. NO EPA OR CASE STUDIES AVAILABLE TO EVALUATE CARCINOGENICITY OF MA TO HUMANS. INADEQUATE EVIDENCE FOR EVALUATION OF CARCINOGENICITY OF MA TO EXPERIMENTAL ANIMALS.

REPT'D USE : COMONOMER WITH ACRYLONITRILE IN PREPARATION OF FIBERS FOR CLOTHING, CARPETING, YARN, HOSIERY, FIRE RETARDANT SLEEPWEAR, DRAPES AND CURTAINS, SHAMPOOS, ADHESIVES AND COATINGS

CAS NO : **55-55-0**

CHEM NAME : ***** 4-METHYL-P-AMINOPHENOL SULFATE**

REPORT 01 : NTP TECHNICAL BULLETIN 6 JAN 82
THE COMPOUND WAS WEAKLY POSITIVE IN THE SALMONELLA TYPHIMURIUM MUTAGENICITY ASSAY IN ONE LABORATORY.

REPT'D USE : NO AVAILABLE PRODUCTION OR USE DATA.

CAS NO : **592-62-1**

CHEM NAME : ***** METHYLAZOXYMETHANOL ACETATE**

REPORT 01 : IARC MONOGRAPH 10 1976
METHYLAZOXYMETHANOL ACETATE, A SYNTHETIC SUBSTANCE CLOSELY RELATED TO METHYLAZOXYMETHANOL, THE AGLYCONE OF CYCASIN, IS CARCINOGENIC IN RATS BY VARIOUS ADMINISTRATION ROUTES. IN A 1979 SUPPLEMENT, IARC DETERMINED THAT THERE IS SUFFICIENT EVIDENCE FOR THE CARCINOGENICITY OF METHYLAZOXYMETHANOL ACETATE IN ANIMALS.

REPT'D USE : NO AVAILABLE PRODUCTION OR USE DATA.

CAS NO : **98-85-1**

CHEM NAME : ***** A-METHYLBENZYL ALCOHOL**

REPORT 01 : NTP TR 369 NIH PUB NO. 89-2824 JAN 90
UNDER 2-YEAR GAVAGE STUDIES EVIDENCE OF CARCINOGENIC ACTIVITY OF A-METHYLBENZYL ALCOHOL FOR MALE F344/N RATS SHOWN BY RENAL TUBULAR CELL ADENOMAS AND ADENOMAS OR ADENOCARCINOMAS (COMBINED). NO EVIDENCE OF CARCINOGENIC ACTIVITY FOR FEMALE F344/N RATS ADMINISTERED 375 OR 750 MG/KG. RENAL TOXICITY CHARACTERIZED BY SEVERE NEPHROPATHY AND RELATED SECONDARY LESIONS. EXCESSIVE MORTALITY DURING LAST QUARTER OF STUDIES. POOR SURVIVAL REDUCED DETECTING PRESENCE OF CARCINOGENIC RESPONSE IN MALE RATS AND IN HIGH DOSE FEMALE RATS. NO EVIDENCE OF CARCINOGENIC ACTIVITY FOR MALE OR FEMALE B6C3F1 MICE AT 375 OR 750 MG/KG DOSE.

REPT'D USE : FRAGRANCE ADDITIVE IN PERFUMES, CREAMS, SOAPS, STYRENE PRODUCTION INTERMEDIATE AND FOOD FLAVORING AGENT

CAS NO : **1634-04-4**
CHEM NAME : ***** METHYL TERT-BUTYL ETHER**

REPORT 01 : TSCA ITC 19TH REPORT TO EPA OCT 86
RECOMMENDED FOR ADDITION TO 4(E) PRIORITY LIST, NOV 86, WITH INTENT TO DESIGNATE. TESTS REQUIRED: CHRONIC TOXICITY SUCH AS ONCOGENIC, REPRODUCTIVE AND DEVELOPMENTAL EFFECTS; CHEMICAL FATE; CHRONIC ECOLOGICAL EFFECTS ON AQUATIC AND TERRESTRIAL PLANTS AND AQUATIC INVERTEBRATES; ACUTE AND CHRONIC EFFECTS ON BENTHIC ORGANISMS AND SOIL INVERTEBRATES IF FOUND PERSISTENT UNDER ANAEROBIC CONDITIONS.

REPORT 02 : 20TH REPORT OF THE TSCA ITC MAY 87
CPD IS DESIGNATED FOR RESPONSE WITHIN 12 MONTHS. RECOMMENDED TESTS INCLUDE CHRONIC INHALATION TOXICITY INCLUDING NEUROTOXIC, HEMATOLOGIC AND ONCOGENIC EFFECTS, FOR HEALTH EFFECTS. CHEMICAL EFFECTS TESTING INCLUDES MONITORING AT REPRESENTATIVE GASOLINE TERMINALS AND SERVICE STATIONS.
REPT'D USE : GASOLINE ADDITIVE

CAS NO : **591-78-6**
CHEM NAME : ***** METHYL-N-BUTYL KETONE (MBK)**

REPORT 01 : NIOSH CRITERIA DOCUMENT ON KETONES JUN 78
NIOSH RECOMMENDS A STANDARD OF 4 MG/CU M TWA BASED ON LIVER, KIDNEY, AND CNS EFFECTS. THE PRESENT OSHA STANDARD IS 410 MG/CU M TWA.

REPORT 02 : HS STAFF REPORT SEP 89
MBK NOT EXPECTED TO CAUSE ACUTE TOXICITY AND ADVERSE REPRODUCTIVE EFFECTS FOLLOWING LOW LEVEL INFREQUENT EXPOSURES. MAY CAUSE NEUROTOXICITY (MUSCLE WEAKNESS IN LIMBS, SENSORY DEFICITS) FOLLOWING DAILY EXPOSURES AT RELATIVELY HIGH DOSE LEVELS (100, 200, 400, 600 PPM). MAY INCREASE TOXICITY OF CARBON TETRACHLORIDE FOLLOWING SIMULTANEOUS EXPOSURES. NO DATA AVAILABLE TO PERMIT ANY JUDGEMENT CONCERNING CHRONIC EFFECTS (CARCINOGENICITY).
REPT'D USE : SOLVENT, PAINT/VARNISH REMOVERS AND SUBSTITUTE FOR METHYLENE CHLORIDE. SEP 81 - EPA CHEMICAL HAZARD INFORMATION PROFIIE

CAS NO : **598-55-0**
CHEM NAME : ***** METHYL CARBAMATE**

REPORT 01 : IARC MONOGRAPH 12 1976
METHYL CARBAMATE HAS BEEN TESTED FOR SKIN AND LUNG TUMOR INDUCTION IN MICE BY SUBCUTANEOUS AND INTRAPERITONEAL ADMINISTRATION AND BY SKIN APPLICATION. ALTHOUGH NO SIGNIFICANT CARCINOGENIC EFFECT WAS OBSERVED, NO EVALUATION OF CARCINOGENICITY CAN BE MADE DUE TO THE SHORT DURATION OF THESE EXPERIMENTS.

REPORT 02 : IARC MONOGRAPH 12 1976
METHYL CARBAMATE DID NOT INDUCE AN INCREASE IN THE NUMBER OF STREPTOMYCIN-INDEPENDENT MUTATIONS IN E. COLI AND NO REVERSE MUTATIONS WERE RECORDED IN B. SUBTILIS. NO REVERSE MUTATIONS WERE OBSERVED IN SALMONELLA TA100, TA98, TA1535, OR TA1537. METABOLIC ACTIVATION SYSTEMS WERE NOT USED IN THESE TESTS.

REPORT 03 : NTP DRAFT BIOASSAY TR REPORT AUG 86
THESE 6-, 12-, 18-, AND 24-MO GAVAGE STUDIES GAVE CLEAR EVIDENCE OF CARCINOGENICITY FOR MALE AND FEMALE F344/N RATS AS SHOWN BY INCREASED INCIDENCES OF NEOPLASTIC NODULES AND CARCINOMAS OF LIVER. NO EVIDENCE OF CARCINOGENICITY FOR MALE AND FEMALE B6C3F1 MICE DOSED AT 500 OR 1000 MG/KG. MEC ALSO INDUCED RETINAL ATROPHY, CATARACTS AND INFLAMMATION OF THE HARDERIAN GLAND IN MALE AND FEMALE RATS AND ADENOMATOUS HYPERPLASIA AND HISTIOCYSTOSIS OF LUNG IN MALE AND FEMALE SWISS-WEBSTER MICE.
REPT'D USE : INTERMEDIATE FOR FABRIC FINISHES

CAS NO : **56-49-5**
CHEM NAME : ***** 3-METHYLCHOLANTHRENE**

REPORT 01 : EPA AMBIENT WATER QUALITY CRITERIA OCT 80
AVAILABLE DATA CANNOT SUPPORT INDIVIDUAL CRITERIA FOR EACH POLYNUCLEAR AROMATIC HYDROCARBON. THE AWQC FOR CARCINOGENIC PAH COMPOUNDS IS BASED ON THE ASSUMPTION THAT EACH COMPOUND IS AS POTENT AS B(A)P AND THAT THE CARCINOGENIC EFFECT OF THE COMPOUNDS IS PROPORTIONAL TO THE SUM OF THEIR CONCENTRATIONS. THIS SUM SHOULD NOT EXCEED 28 NG/L, 2.8 NG/L, AND .28 NG/L TO KEEP THE LIFETIME CANCER RISK BELOW 10 -5, 10 -6, AND 10 -7 RESPECTIVELY.

REPORT 02 : NTP TECHNICAL BULLETIN 3
THE COMPOUND WAS FOUND TO BE POSITIVE IN THE SALMONELLA TYPHIMURIUM MUTAGENICITY ASSAY IN TWO LABORATORIES.

REPORT 06 : MULTI-AGENCY CROSS CHECK
N HHS MAGUIRE
N OSHA CARC
Y EPA CAG
N EPA RCRA
N EPA CLEAN AIR
N EPA CLEAN WATER
REPT'D USE : LABORAT0RY REAGENT AND ENZYME INDUCING AGENT

CAS NO : **CLASS-13**
CHEM NAME : ***** METHYLCHRYSENES**

REPORT 01 : IARC MONOGRAPH 32 DEC 83
1-,2-,3-,4-,5-, AND 6-METHLCHRYSENES TESTED FOR CARCINOGENICITY BY SKIN APPLICATION TO FEMALE MICE AND IN MOUSE SKIN INITIATION-PROMOTION ASSAY. 5-METHYLCHRYSENE INDUCED HIGHEST INCIDENCE OF MALIGNANT SKIN TUMORS, WHEN TESTED ALONE OR TOGETHER WITH PROMOTER. INTERMEDIATE RESPONSE OBSERVED WITH 2-,3-,4- AND 6-METHYLCHRYSENES WHEN TESTED AS CARCINOGENS 1-METHYLCHRYSENE WAS INACTIVE. ALL DERIVATIVES SHOWED VARYING DEGREES OF INITIATING ACTIVITY. REASONABLE FOR PRACTICAL PURPOSES TO REGARD CARCINOGENIC RISK TO HUMANS. 5-METHYLCHRYSENE, WHEN TESTED BY SUBCUTANEOUS INJECTION IN MICE, PRODUCED HIGH INCIDENCE OF SARCOMAS AT SITE OF INJECTION.

REPORT 02 : IARC MONOGRAPH 32 DEC 83
1-,2-,3-,4,5- AND 6-METHYLCHRYSENES MUTAGENIC TO SALMONELLA TYPHIMURIUM IN PRESENCE OF EXOGENOUS METABOLIC SYSTEM. 5-METHLCHRYSENE INDUCED DNA DAMAGE IN PRIMARY RAT HEPATOCYTES.

REPORT 06 MULTI-AGENCY CROSS CHECK
Y HHS MAGUIRE
N OSHA CARC
N EPA CAG
N EPA RCRA
N EPA CLEAN AIR
N EPA CLEAN WATER
REPT'D USE : TOBACCO SMOKE, ENGINE EXHAUST, LUBRICATING OILS AND USED MOTOR OIL

CAS NO : **137-05-3**
CHEM NAME : ***** METHYL-2-CYANOACRYLATE**

REPORT 01 : NTP TECHNICAL BULLETIN 9 APR 83
METHL-2-CYANOACRYLATE WAS POSITIVE IN ONE SALMONELLA TYPHIMURIUM MUTAGENICITY ASSAY.

CAS NO : **41372-08-1**
CHEM NAME : ***** ALPHA-METHYLDOPA SESQUIHYDRATE**

REPORT 01 : NTP TR 348 NIH PUB # 89-2803 MAR 89
UNDER 2-YEAR FEED STUDIES NO EVIDENCE OF CARCINOGENIC ACTIVITY OF A-METHYLDOPA SESQUIHYDRATE FOR MALE OR FEMALE F344/N RATS FED DIETS CONTAINING 3,100 OR 6,300 PPM. EQUIVOCAL EVIDENCE OF CARCINOGENIC ACTIVITY FOR MALE B6C3F1 MICE, SHOWN BY THREE DOSED MICE HAVING TUBULAR CELL TUMORS OF KIDNEY. EVIDENCE OF CARCINOGENIC ACTIVITY FOR FEMALE B6C3F1 MICE FED DIETS CONTAINING 6,300 OR 12,500 PPM. NONNEOPLASTIC LESIONS OF KIDNEY (KARYOMEGALY) OBSERVED IN DOSED FEMALE MICE. DECREASED INCIDENCES TUMORS IN ADRENAL GLAND IN MALE RATS, UTERUS IN FEMALE RATS, LIVER IN MALE AND FEMALE MICE, AND ANTERIOR PITUITARY GLAND IN FEMALE MICE CONSIDERED RELATED TO CHEMICAL.
REPT'D USE : DRUG

CAS NO : **101-14-4**
CHEM NAME : ***** 4,4'-METHYLENEBIS(2-CHLOROANILINE) (MBOCA)**

REPORT 01 : IARC MONOGRAPH 4 1974
4,4'-METHYLENE BIS(2-CHLOROANILINE) IS CARCINOGENIC IN THE MOUSE AND RAT AFTER ORAL ADMINISTRATION AND PRODUCES DISTANT TUMORS IN THE RAT AFTER SUBCUTANEOUS ADMINISTRATION. IN A 1979 SUPPLEMENT, IARC DETERMINED THAT THERE IS SUFFICIENT EVIDENCE FOR THE CARCINOGENICITY OF THIS COMPOUND IN ANIMALS.

REPORT 02 : NIOSH SPECIAL HAZARD REVIEW SEP 78
NIOSH RECOMMENDS THAT EXPOSURE TO CPD IN THE WORKPLACE BE CONTROLLED SO THAT NO WORKER WILL BE EXPOSED TO CONCENTRATIONS IN EXCESS OF 3 MG/M3 OF AIR, DETERMINED AS A TWA CONCENTRATION FOR UP TO A 10-HOUR WORKSHIFT, 40-HOUR WORK WEEK, OVER A WORKING LIFETIME. THE EXPOSURE LEVEL FOR MBOCA IS THE LOWEST LEVEL WHICH CAN BE RELIABLY MEASURED AT THIS TIME. THIS RECOMMENDATION IS BASED ON CPD'S CARCINOGENICITY. AT PRESENT, THERE IS NO OSHA STANDARD.

REPORT 03 : NTP TECHNICAL BULLETIN 3/8 DEC 80, JUL 82
MBOCA WAS POSITIVE IN THE SALMONELLA ASSAY AND IN THE L5178Y MOUSE LYMPHOMA ASSAY.

REPORT 04 : IARC SUPPL. 4 1985
MBOCA CONSIDERED TO BE CARCINOGENIC IN EXPERIMENTAL ANIMALS.

REPORT 05 : FED REG. MAY 88
SECTION 8 (A) OF THE TOXIC SUBSTANCES CONTROL ACT (TSCA), REQUIRES PERSONS WHO INTEND TO MANUFACTURE MBOCA IN THE UNITED STATES TO NOTIFY EPA AND SUBMIT INFORMATION TO THE AGENCY BEFORE BEGINNING PRODUCTION. THE RULE ALSO REQUIRES PERSONS TO REPORT IF THEY ARE MANUFACTURING MBOCA IN THE UNITED STATES AS OF THE INEFFECTIVE DATE OF THE RULE. IN ADDITION, THE RULES REQUIRES PERSONS WHO ARE MANUFACTURING MBOCA TO NOTIFY THE AGENCY OF AN INTENT ALTER THEIR MANNER OR METHOD OF MANUFACTURING THE SUBSTANCE. THE AGENCY CLASSIFIES MBOCA AS A PROBABLE HUMAN CARCINOGEN. MBOCA HAS NOT BEEN MANUFACTURED IN THE UNITED STATES SINCE 1979.

REPORT 06 : MULTI-AGENCY CROSS CHECK
Y HHS MAGUIRE
N OSHA CARC
Y EPA CAG
Y EPA RCRA
N EPA CLEAN AIR
N EPA CLEAN WATER
REPT'D USE : CURING AGENT FOR URETHANE/EPOXY RESINS

CAS NO : **101-61-1**
CHEM NAME : ***** 4,4'-METHYLENEBIS(N,N-DIMETHYL)ANILINE**

REPORT 01 : NCI CARC TRS 186 DHEW PUB NO NTH 79-1742
UNDER THE CONDITIONS OF THIS BIOASSAY, 4.4'-METHYLENEBIS(N,N-DIMETHYL)BENZENAMINE WAS CARCINOGENIC IN FISCHER 344 RATS. INDUCING THYROID FOLLICULAR-CELL CARCINOMAS IN BOTH MALES AND FEMALES. ADMINISTRATION OF THE COMPOUND WAS CARCINOGENIC IN FEMALE B6C3F1, INDUCING LIVER NEOPLASMS. THERE WAS NO CONCLUSIVE EVIDENCE THAT 4,4'-METHYLENEBIS(N,N-DIMETHYL)BENZENAMINE WAS CARCINOGENIC IN MALE B6C3F1 MICE (FEEDING STUDY).

REPORT 02 : NTP TECHNICAL BULLETIN 9 APR 83
4,4-METHYLENEBIS-(N,N-DIMETHYL)ANILINE WAS POSITIVE IN THE L5178Y MOUSE LYMPHOMA ASSAY, RAUSCHER LEUKEMIA VIRUS, UNSCHEDULED DNA SYNTHESIS, BALB/C 3TC TRANSFORMATION ASSAYS.

REPORT 03 : IARC MONOGRAPH 27 1982
4,4'-METHYLENEBIS(N,N-DIMETHYL)BENZENAMINE WAS TESTED IN ONE EXPERIMENT IN MICE AND ONE EXPERIMENT IN RATS BY DIETARY ADMINISTRATION. IT PRODUCED LIVER CELL ADENOMAS IN FEMALE MICE AND THYROID FOLLICULAR CELL CARCINOMAS IN MALE AND FEMALE RATS. THERE IS LIMITED EVIDENCE FOR THE CARCINOGENICITY OF THIS CHEMICAL IN ANIMALS.

REPORT 04 : IARC MONOGRAPH 27 1982
4,4'-METHYLENEBIS(N,N-DIMETHYL)BENZENAMINE WAS POSITIVE IN E. COLI REPAIR TESTS. IT WAS MUTAGENIC TO SALMONELLA, BOTH IN VITRO AND IN THE MURINE HOST MEDIATED ASSAY. IT DID NOT INDUCE MITOTIC RECOMBINATION IN YEAST, BUT CAUSED MORPHOLOGICAL TRANSFORMATION IN HAMSTER EMBRYO CELLS. NO DATA CONCERNING EFFECTS ON REPRODUCTION OR PRENATAL TOXICITY WERE AVAILABLE.

REPORT 06 : MULTI-AGENCY CROSS CHECK
Y HHS MAGUIRE
Y OSHA CARC
N EPA CAG
N EPA RCRA
N EPA CLEAN AIR
N EPA CLEAN WATER

REPT'D USE : ANALYTICAL REAGENT FOR LEAD AND DYE INTERMEDIATE

CAS NO : **1807-55-2**

CHEM NAME : ***** 4,4'-METHYLENEBIS(2-METHYLANILINE)**

REPORT 01 : IARC MONOGRAPH 4 1974
4,4'-METHYLENEBIS(2-METHYLANILINE) IS CARCINOGENIC IN THE RAT AFTER ORAL ADMINISTRATION, THE ONLY SPECIES AND ROUTE TESTED. IN A 1979 SUPPLEMENT, IARC DETERMINED THAT THERE IS SUFFICIENT EVIDENCE FOR THE CARCINOGENICITY OF THIS COMPOUND IN ANIMALS.

REPT'D USE : CHEMICAL INTERMEDIATE, DYE INTERMEDIATE AND EPOXY CURING AGENT. NO KNOWN CURRENT U.S. PRODUCTION.

CAS NO : **75-09-2**

CHEM NAME : ***** METHYLENE CHLORIDE**

REPORT 01 : EPA AMBIENT WATER QUALITY CRITERIA OCT 80
A CRITERION, BASED ON THE TLV FOR EXCESSIVE COHB FORMATION OF 12.4 MG/L, IS RECOMMENDED TO PROTECT HUMANS FROM THE POTENTIAL CARCINOGENIC EFFECTS OF EXPOSURE TO HALOMETHANES, INCLUDING DICHLOROMETHANE. AMBIENT WATER CONCENTRATIONS OF 1.9 UG/L, 0.19 UG/L,AND 0.019 UG/L ARE ESTIMATED TO RESULT IN A RESPECTIVE 10 -5, 10 -6 AND 10 -7 INCREASE IN CANCER RISK OVER THE LIFETIME.

REPORT 02 : NIOSH CRITERIA DOCUMENT MAR 76
NIOSH RECOMMENDS A STANDARD OF 75 PPM TWA TO PROTECT AGAINST OXYGEN TRANSPORT INTERFERENCE AND ABNORMALITIES IN CNS FUNCTION. THE PRESENT OSHA STANDARD IS 500 PPM TWA.

REPORT 04 : EPA TSCA FY1 NOTICE OTS-1180-0097
IN A 2-YEAR INHALATION STUDY CONDUCTED BY DOW CHEMICAL, RATS AND HAMSTERS WERE EXPOSED TO 0, 500, 1500, AND 3500 PPM MECL. PRELIMINARY FINDINGS INCLUDE SLIGHT LIVER TOXICITY IN MALE/FEMALE RATS AT ALL EXPOSURE LEVELS, INCREASED NUMBERS OF BENIGN MAMMARY TUMORS IN TUMOR-BEARING FEMALE RATS AT ALL EXPOSURES, INCREASED INCIDENCE OF BENIGN MAMMARY TUMORS IN MALE RATS AT 3500 PPM, AND INCREASED INCIDENCE OF SARCOMAS IN THE SALIVARY GLANDS OF MALE RATS.

REPORT 05 : IARC MONOGRAPH 20 OCT 79
DICHLOROMETHANE WAS TESTED IN MALE MICE BY INTRAPERITONEAL INJECTION. THIS EXPERIMENT WAS CONSIDERED TO BE INADEQUATE, ALTHOUGH THE RESULTS SUGGESTED AN INCREASED INCIDENCE OF LUNG TUMORS. THE AVAILABLE DATA DO NOT PERMIT AN EVALUATION OF THE CARCINOGENICITY OF DICHLOROMETHANE TO BE MADE.

REPORT 06 : MULTI-AGENCY CROSS CHECK
N HHS MAGUIRE
N OSHA CARC
N EPA CAG
Y EPA RCRA
Y EPA CLEAN AIR
Y EPA CLEAN WATER

REPORT 07 : NTP CARC TEST PROGRAM JUL 82
IN A NTP BIOASSAY METHYLENE CHLORIDE WAS ADMINISTERED BY GAVAGE TO F344 RATS AND B6C3F1 MICE. THE STUDY HAS BEEN COMPLETED AND THE FINAL TECHNICAL REPORT IS IN PREPARATION.

REPORT 08 : EPA RCRA HEALTH AND ENV. EFFECTS PROFILE OCT 80
ON THE BASIS OF PRESENT EVIDENCE, DCM CANNOT BE FIRMLY IDENTIFIED AS AN ANIMAL OR HUMAN CARCINOGEN. DCM HAS BEEN SHOWN TO BE MUTAGENIC TO SALMONELLA, BUT NOT TO SACCHAROMYCES AND DROSOPHILA, AND CAUSES CELL TRANSFORMATION.

REPORT 09 : IARC MONOGRAPH 20 1979
DICHLOROMETHANE WAS MUTAGENIC IN SALMONELLA TA100 AND TA98, BOTH IN THE PRESENCE AND ABSENCE OF METABOLIC ACTIVATION.

REPORT 10 : EPA RESPONSE TO ITC RECOMMS (46FR30300) JUN 81
IN RESPONSE TO THE ITC RECOMMENDATIONS (43FR16684), THE EPA PROPOSES THAT MANUFACTURERS AND PROCESSORS OF DICHLOROMETHANE CONDUCT TESTING FOR ACUTE DERMAL SENSITIZATION AND REPRODUCTIVE EFFECTS. THE EPA WILL CONDUCT MUTAGENICITY TESTING AND DECIDE ON FURTHER TIER

TESTING AS APPROPRIATE. THE NTP IS CONDUCTING AN ONCOGENICITY STUDY. INDUSTRY IS SPONSORING AN EPIDEMIOLOGIC STUDY.

REPORT 11 : EPA RESPONSE TO ITC (49FR25009) JUNE 84
IN RESPONSE TO APRIL 1978 ITC RECOMMENDATIONS (43FR16884) EPA DECIDED NOT TO PROMULGATE A RULE TO REQUIRE TESTING BECAUSE THERE IS SUFFICIENT BASIS TO DETERMINE MANUFACTURE, USE; OR DISPOSAL OF DICHLOROMETHANE DOES NOT PRESENT UNREASONABLE RISK OF ADVERSE EFFECTS TO ENVIRONMENT; INFORMATION ON CARDIOVASCULOR EFFECTS DO NOT SUPPORT UNREASONABLE RISK FINDING; AND INITIATED TESTING IS EXPECTED TO PROVIDE DATA TO DETERMINE EFFECTS ON HUMAN REPRODUCTION.

REPORT 12 : HALOGENATED SOLVENTS INDUSTRY ALLIANCE SUBMITTED PRELIMINARY, RESULTS FROM NATIONAL TOXICOLOGY PROGRAM BIOASSAY CURRENTLY IN HISTOPATHOLOGY PHASE OF METHYLENE CHLORIDE ADMINISTERED VIA INHALATION TO MALE AND FEMALE F344 RATS AT DOSES OF 0, 1000, 2000, OR 4000 PPM AND TO MALE AND FEMALE B6C3F1 MICE AT DOSES OF 0, 2000, OR 4000 PPM. WHICH SHOWED THE CHEMICAL INDUCED A VERY HIGH INCIDENCE OF LUNG AND LIVER TUMORS IN BOTH SEXES OF MICE.

REPORT 13 : NTP TECHNICAL REPORT 306 NIH PUB 86-2562 JAN 86
FROM INHALATION STUDIES, SOME EVIDENCE OF CARCINOGENICITY OF DICHLOROMETHANE SEEN FOR MALE F344/N RATS. CLEAR EVIDENCE OF CARCINOGENICTY FOR FEMALE RATS SHOWN BY NEOPLASMS OF MAMMARY GLAND CLEAR EVIDENCE OF CARCINOGENICITY FOR MALE AND FEMALE B6C3F1 MICE SHOWN BY ALVEOLAR/BRONCHIOLAR NEOPLASMS AND OF HEPATOCELLULAR NEOPLASMS.

REPORT 14 : IARC VOL 41 1986
STUDY OF SPONTANEOUS ABORTIONS AMONG WOMEN WORKING IN PHARMACEUTICAL INDUSTRY SHOWED AN INCREASED RISK FROM EXPOSURE TO SEVERAL CHEMICALS, INCLUDING DCM. NO INCREASED RISK OF DEATH SEEN IN 3 STUDIES OF WORKERS EXPOSED TO DCM. STUDIES OF EXPOSURE ORALLY IN MICE AND RATS NEGATIVE OR INCONCLUSIVE. INHALATION EXPOSURES INCREASED INCIDENCE OF BENIGN AND MALIGNANT LUNG AND LIVER TUMORS IN MICE OF EACH SEX AND AN INCREASED INCIDENCE OF SARCOMAS OF NECK. IN RATS, POSTNATAL BEHAVIORAL DEVELOPMENT IS AFFECTED BY PRENATAL EXPOSURE TO DCM. IARC SAYS THERE IS SUFFICIENT EVIDENCE FOR GENETIC ACTIVITY AND CARCINOGENICITY TO TEST ANIMALS BUT INADEQUATE EVIDENCE FOR HUMANS.

REPT'D USE : SOLVENT, INDUSTRIAL EXTRACTION AGENT, VAPOR-PRESSURE DEPRESSANT IN AEROSOLS, PAINT STRIPPERS, CLEANING AGENTS, PAINT/VARNISH REMOVERS, DRAIN CLEANERS, ADHESIVES,

REFRIGERANT, AEROSOLS, BLOWING BLOWING AGENT FOR PLASTIC FOAMS AND IN FOOD AND DRUG PROCESSING

CAS NO : **101-77-9**
CHEM NAME : ***** 4,4'-METHYLENEDIANILINE (MDA)**

REPORT 01 : NIOSH INTELLIGENCE BULLETIN 8 JAN 76
IN 1965 HEPATOTOXIC EFFECTS IN HUMANS OBSERVED (HEPATOCELLULAR DAMAGE). PRODUCED LIVER LESIONS IN INTRAGASTRICALLY FED RATS AND LIVER DEGENERATION AND SPLEEN LESIONS IN ANOTHER GROUP OF FED RATS. OCCUPATIONALLY IMPLICATED IN TOXIC HEPATITIS.

REPORT 02 : ITC FOURTH REPORT TO EPA (44FR31866) JUN 79
THE INTERAGENCY TESTING COMMITTEE RECOMMENDS THAT THE EPA ADMINISTRATOR CONSIDER THE DEVELOPMENT OF TEST RULES TO TEST 4,4'-DIAMINODIPHENYLMETHANE FOR CARCINOGENICITY, MUTAGENICITY, TERATOGENICITY AND REPRODUCTIVE EFFECTS; AND THAT AN EPIDEMIOLOGIC STUDY BE CONDUCTED.

REPORT 03 : NTP TECHNICAL REPORT TR 248 JUN 83
NIH PUBLICATION 83-2504
UNDER CONDITIONS OF STUDIES 4,4'-METHYLENEDIANILINE DIHYDROCHORIDE CARCINOGENIC FOR 344/N RATS AND B6C3F1 MICE OF EACH SEX, CAUSING SIGNIFICANTLY INCREASED INCIDENCES OF THYROID FOLLICULAR CELL CARCINOMAS IN MALE RATS, THYROID FOLLICULAR CELL ADENOMAS IN FEMALE RATS AND IN MICE OF EACH SEX, C-CELLADENO-MAS OF THYROID GLAND IN FEMALE RATS, NEOPLASTIC NODULES IN LIVER OF MALE RATS, HEPATOCELLULAR CARCINOMAS IN MICE OF EACH SEX, ADENOMAS OF LIVER AND MALIGNANT LYMPHOMAS IN FEMALE MICE, AND ADRENAL PHEOCHROMOCYTOMAS IN MALE MICE (SEE 13552-44-8).

REPORT 04 : NTP TSCA SECTION 9 FR NOTICE 50FR 27674 JUL 85
THE EPA REPORTED ITS CONCLUSION THAT MANUFACTURE AND USE OF 4',4'-METHYLENEDIANILINE (4',4'-MDA) PRESENT UNREASONABLE SICK OF INJURY TO HEALTH OF EXPOSED WORKERS, AND THIS RISK MAY BE PREVENTED OR REDUCED BY OSHA UNDER THE OSHACT. EPA SUBMITTING TO OSHA REPORT UNDER SEC. 9(A) OF TSCA OF RISKS OF 4,4'-MDA AND REQUESTS OSHA RESPOND TO EPA WITHIN 180 DAYS.

REPORT 05 : EPA REPONSE TO ITC (48FR31806) JUL 83
EPA IS NOT INITIATING RULEMAKING UNDER TSCA SECTION 4(A) TO REQUIRE TESTING OF MDA FOR HEALTH EFFECTS BECAUSE A COMPLETED 2 YEAR CARCINOGENICITY STUDY SHOWED MDA CARCINOGENIC IN RATS AND MICE; AN EXPOSURE EVALUATION HAS BEEN INITIATED; MDA NOT ANTICIPATED TO ENTER ENVIRON-

MENT IN SUBSTANTIAL QUANTITIES OR IF SO TO PERSIST; AND SUITABLE POPULATION FOR EPIDEMIOLOGICAL STUDIES NOT IDENTIFIED.

REPORT 06 : MULTI-AGENCY CROSS CHECK
Y HHS MAGUIRE
N OSHA CARC
N EPA CAG
N EPA RCRA
N EPA CLEAN AIR
N EPA CLEAN WATER

REPORT 07 : OSHA REPONSE TO EPA TSCA SEC 9 (50FR27674) OCT 85
OSHA ESTABLISHES NEGOTIATED RULEMAKING ADVISORY COMMITTEE ON MDA TO ADVISE SECRETARY OF LABOR ON PROPOSED STANDARD ON MDA BECAUSE IT WAS PRELIMINARILY DETERMINED THAT SIGNIFICANT RISK IS ASSOCIATED WITH WORKER EXPOSURE TO MDA BECAUSE IT IS A POTENTIAL HUMAN CARCINOGEN (SEE 50FR42787).

REPORT 08 : IARC MONOGRAPH 39, 1986 JUN 85
ORAL ADMINISTRATION IN MICE CAUSED DOSE-RELATED INCREASES IN INCIDENCE OF ADENOMAS AND NEOPLASMS IN BOTH MALES AND FEMALES AND DOSE-RELATED INCREASES IN CARCINOMAS AND HEPATIC NODULES IN MALE RATS AND THYROID CARCINOMAS IN FEMALE RATS. DATA INADEQUATE TO EVALUATE REPRODUCTIVE EFFECTS AND PRENATAL TOXICITY IN HUMANS AND ANIMALS. MDA MUTAGENIC TO S. TYPHIMURIUM WITH ACTIVATION AND HEPATOTOXIC IN HUMANS. NO CASE REPORTS OR EPA STUDIES AVAILABLE TO EVALUATE CARCINOGENICITY OF MDA TO HUMANS. CPD INDUCED DNA DAMAGE IN CHV79 CELLS. IARC SAYS THAT THERE IS SUFFICIENT DATA FOR CARCINOGENICITY OF CPD AND ITS HCL SALT TO EXPERIMENTAL ANIMALS.

REPORT 09 : OSHA FR NOTICE 54FR 20672 MAY 89
OSHA PROPOSES NEW STANDARD REGULATING EXPOSURE TO MDA BASED ON ANIMAL AND HUMAN DATA. EXPOSURE TO MDA AT CURRENT OCCUPATIONAL EXPOSURE LEVEL CAUSES ADVERSE EFFECTS ON EMPLOYEE HEALTH INCLUDING RISK OF CANCER. THEREFORE OCCUPATIONAL EXPOSURE IS LIMITED TO EIGHT-HR. TIME WEIGHTED AVERAGE (TWA) OF 10 PPB SHORT-TERM EXPOSURE AND IS LIMITED TO ESTABLISHING PERMISSIBLE EXPOSURE LIMITS (PELS) INCLUDES PROPER MEDICAL SURVEILLANCE, EXPOSURE MONITORING, HYGIENE FACILITIES, ENGINEERING CONTROLS AND WORK PRACTICES, RESPIRATOR USE, AND RECORDKEEPING, AN ACTION LEVEL OF 5 PPB IS INCLUDED FOR EXEMPTING EMPLOYERS FROM EXPOSURE MONITORING WHERE THERE IS EXPOSURE AT OR BELOW THIS LEVEL.

REPT'D USE : SOIL FUMIGANT, CONTAMINANT IN TDCP, DYE INTERMEDIATE, EPOXY RESINS AND COATINGS, FR CONTAMINANT, ANTI-OXIDANT AND NEOPRENE CURATIVE

CAS NO : **78-93-3**
CHEM NAME : ***** METHYL ETHYL KETONE**

REPORT 01 : NIOSH CRITERIA DOCUMENT ON KETONES JUN 78
NIOSH RECOMMENDS A STANDARD OF 590 MC/CU M TWA BASED ON LIVER, KIDNEY, AND CNS EFFECTS. THE PRESENT OSHA STANDARD IS 590 MG/CU M.

REPORT 02 : CIIT 90-DAY TOXICOLOGY STUDY NOV 81
IN A PRECHRONIC EXPOSURE STUDY, CONCENTRATIONS OF 1.250 PPH, 2.500 PPM AND 5.000 PPH OF MEK WERE ADMINISTERED TO 1344 RATS BY INHALATION. FINDINGS FROM THIS 90-DAY SUBCHRONIC INHALATION TOXICITY STUDY HAVE BEEN CAREFULLY RIVIEWED BY CIIT. NO SIGNIFICANT TOXIC EFFECTS WERE SEEN. IN LIGHT OF THIS, AND OTHER DATA, DECISION WAS MADE TO CANCEL THE PROPOSED LIFETIME INHALATION TOXICITY STUDY PLANNED FOR N-HEXANE.

REPORT 03 : ITC FOURTH REPORT TO EPA (44FR31866) JUN 79
THE INTERAGENCY TESTING COMMITTEE RECOMMENDS THAT THE EPA ADMINISTRATOR CONSIDER THE DEVELOPMENT OF TEST RULES TO TEST METHYL ETHYL KETONE FOR NEUROLOGICAL EFFECTS; AND THAT AN EPIDEMIOLOGIC STUDY BE CONDUCTED.

REPORT 04 : 47FR58025: EPA RESPONSE TO ITC DEC 82
IN RESPONSE TO EPA'S 1979 DESIGNATION OF MEK FOR PRIORITY TESTING FOR HEALTH EFFECTS, EPA CONCLUDED THAT EXISTING DATA SUFFICIENT TO EVALUATE EFFECTS ITC WANTS TESTED IN SOME CASES AND THAT IN OTHERS, TESTING IS NOT WARRANTED. EPA RESPONDED UNDER TSCA 4(E) BY AGREEING TO ACCEPT INDUSTRY'S TESTING PROPOSAL IN LIEU OF RULEMAKING TO FILL GAPS IN DATA.

REPORT 05 : EPA DECISION TO ADOPT NEGOTIATE TESTING SEP 83
PROGRAM - 48FR44905
EPA WILL NOT INITIATE RULEMAKING UNDER TSCA 4(A) TO REQUIRE HEALTH EFFECTS TESTING OF MEK. EPA HAS ACCEPTED A TESTING PROGRAM SUBMITTED BY THE KETONES PROGRAM PANEL OF THE CMA.

REPORT 06 : MULTI-AGENCY CROSS CHECK
N HHS MAGUIRE
N OSHA CARC
N EPA CAG
Y EPA RCRA
N EPA CLEAN AIR
N EPA CLEAN WATER

REPORT 07 : CMA KETONES PROGRAM PANEL REPORT II, OCT 84
MEK MUTAGENICITY
RESULTS OF FIVE MUTAGENICITY ASSAYS WERE NEGATIVE. ASSAYS INCLUDED WERE THE AMES SALMONELLA PLATE, MOUSE LYMPHOMA, UNSCHEDULED DNA SYNTHESIS, MOUSE MICRONUCLEUS, AND CELL TRANSFORMATION. REPORT CONCLUDES THAT RESULTS OF SHORT-TERM TESTING DO NOT PROVIDE A BASIS FOR CONCERN ABOUT MUTAGENIC OR ONCOGENIC EFFECTS AND THAT THERE IS NO NEED FOR FURTHER TESTING.

REPORT 08 : PESTICIDE AND TOXIC CHEMICAL NEWS, NOV 14, 1984 NOV 84
CMA KETONES PANEL REPORT INDICATES THAT MEK, MIBK OR ISOPHORONE DO NOT POSE UNREASONABLE RISK TO HUMAN HEALTH DURING CURRENT OR ANTICIPATED CONDITIONS OF USE AND EXPOSURE. NONE OF THE DATA PROVIDE A BASIS FOR INITIATING FURTHER TESTING WHEN EVALUATED BY CRITERIA SPECIFIED IN NEGOTIATED TESTING AGREEMENTS.

REPORT 09 : EPA NOT TO PROPOSE TSCA 4(A) RULE; 48FR44905 SEP 83
EPA HAS REVIEWED INITIAL DECISION NOT TO PROPOSE A TSCA 4(A) RULE TO REQUIRE HEALTH EFFECTS TESTING OF MEK AND FINDS NO REASON TO CHANGE ITS DECISION.

REPORT 10 : HS STAFF REPORT SEP 89
MEK EXPOSURE ALONE FOUND TO HAVE RELATIVELY LOW LEVEL OF ACUTE AND SUBCHRONIC TOXICITY. ONE CASE OF TOXIC NEUROPATHY AND ONE CASE OF SEVERE METABOLIC ACIDOSIS AFTER INGESTION WHERE MEK ALONE HAS BEEN IMPLICATED CAUSE INCREASED LEVEL OF TOXICITY WHEN SIMULTANEOUS EXPOSURES TO OTHER CHEMICALS SUCH AS N-HEXANE OR METHYL N-BUTYL KETONE, CHLOROFORM, ETHYL N-BUTYL KETONE, AND CARBON TETRACHLORIDE. NO DATA CONCERNING CARCINOGENIC EFFECTS OF MEK. ALTHOUGH POSSIBILITY OF FETOTOXICITY DATA ARE INCONCLUSIVE.

REPT'D USE : SOLVENT, ADHESIVES, ANTIQUING AGENTS, LACQUER, CHINA CEMENTS, AEROSOLS, CAUSTICS/DRAIN CLEANERS, CLEANING AGENTS, PAINTS/VARNISHES, PAINT/VARNISH REMOVERS AND SUBSTITUTE FOR METHYLENE CHLORIDE

CAS NO : **96-29-7**
CHEM NAME : ***** METHYL ETHYL KETOXIME**

REPORT 01 : ITC'S 19TH REPORT TO EPA OCT 86
RECOMMENDED FOR ADDITION TO THE 4(E) PRIORITY LIST, NOV 86, WITHOUT BEING DESIGNATED FOR RESPONSE WITHIN 12 MONTHS. TESTS RECOMMENDED: HEALTH EFFECTS - CHRONIC TOXICITY.

REPORT 02 : EPA FEDERAL REGISTER NOT~CE 53FR35838 SEP 88
EPA IS PROPOSING THAT MANUFACTURERS AND PROCESSORS BE

REQUIRED UNDER TSCA SECTION 4 TO PERFORM TESTING FOR ONCOGENICITY, MUTAGENICITY, DEVELOPMENTAL TOXICITY, REPRODUCTIVE EFFECTS, NEUROTOXICITY AND PHARMACOKINETICS, IN RESPONSE TO ITC'S RECOMMENDATION FOR TESTING.

REPT'D USE : ALKYD PAINTS, CAULK, PASTE FILTERS, PENTANE GEL AND ANTISKINNING PRODUCTS

CAS NO : **33543-31-6**

CHEM NAME : ***** 2-METHYLFLUORANTHENE**

REPORT 01 : IARC MONOGRAPH 32 DEC 83

2-METHYLFLUORANTHENE TESTED FOR CARCINOGENICITY IN ONE EXPERIMENT IN MICE BY SKIN APPLICATION AND PRODUCED BENIGN AND MALIGNANT SKIN TUMORS. ALSO TESTED IN MOUSESKIN INITIATION-PROMOTION ASSAY IN ONE EXPERIMENT AND ACTIVE AS AN INITIATOR. THUS, LIMITED EVIDENCE THAT 2-METHYLFLUORANTHENE CARCINOGENIC TO EXPERIMENTAL ANIMALS. IN THE ONE STUDY, 2-AND 3-METHYLFLUORANTHENES MUTAGENIC TO SALMONELLA TYPHIMURIUM IN PRESENCE OF EXOGENOUS METABOLIC SYSTEM.

REPT'D USE : CIGARETTE SMOKE

CAS NO : **1706-10-0**

CHEM NAME : ***** 3-METHYLFLUORANTHENE**

REPORT 01 : IARC MONOGRAPH 32 DEC 83

3-METHYLFLUORANTHENE TESTED IN ONLY ONE EXPERIMENT IN MOUSESKIN INITIATION-PROMOTION ASSAY, AND BENIGN SKIN TUMORS OCCURRED. 2-AND 3-METHYLFLUORANTHENES WERE MUTAGENIC TO SALMONELLA TYPHIMURIUM WITH METABOLIC ACTIVATION.

REPT'D USE : CIGARETTE SMOKE

CAS NO : **60-34-4**

CHEM NAME : ***** METHYLHYDRAZINE**

REPORT 01 : NIOSH CRITERIA DOCUMENT JUN 78

NIOSH RECOMMENDS THAT OCCUPATIONAL EXPOSURE TO METHYLHYDRAZINE BE CONTROLLED SO THAT NO EMPLOYEE BE EXPOSED TO A CEILING CONCENTRATION GREATER THAN 0.08 MG/CU M (0.04 PPM). BEING A REDUCING AGENT, IT CAUSES OXIDATIVE DAMAGE TO THE ERYTHROCYTES, SUCH AS THE FORMATION OF HEINZ BODIES, PRODUCTION OF METHEMOGLOBIN, AND A DECREASE OF REDUCED GLUTATHIONE. THE PRESENT OSHA STANDARD IS 0.35 MG/CU M CEILING.

REPORT 02 : NTP TECHNICAL BULLETIN 3 DEC 80
THE COMPOUND WAS FOUND TO BE NEGATIVE IN THE SALMONELLA TYPHIMURIUM MUTAGENICITY ASSAY IN ONE LABORATORY.

REPORT 06 : MULTIAGENCY CROSS CHECK
N HHS MAGUIRE
Y OSHA CARC
N EPA CAG
N EPA RCRA
N EPA CLEAN AIR
N EPA CLEAN WATER
REPT'D USE : CHEMICAL INTERMEDIATE AND ROCKET FUEL

CAS NO : **302-15-8**
CHEM NAME : ***** METHYLHYDRAZINE SULFATE (1:1)**

REPORT 06 : MULTI-AGENCY CROSS CHECK
N HHS MAGUIRE
Y OSHA CARC
N EPA CAG
N EPA RCRA
N EPA CLEAN AIR
N EPA CLEAN WATER
REPT'D USE : NO AVAILABLE PRODUCTION OR USE DATA

CAS NO : **74-88-4**
CHEM NAME : ***** METHYL IODIDE**

REPORT 01 : IARC MONOGRAPH 15 1977
METHYL IODIDE IS CARCINOGENIC IN RATS, INDUCING LOCAL SARCOMAS AFTER SINGLE OR REPEATED SUBCUTANEOUS INJECTIONS. IN A STUDY IN WHICH ONLY A FEW MICE WERE USED, IT CAUSED AN INCREASED INCIDENCE OF LUNG TUMORS AFTER ITS INTRAPERITONEAL INJECTION. IN A 1979 SUPPLEMENT, IARC DETERMINED THAT THERE IS SUFFICIENT EVIDENCE FOR THE CARCINOGENICITY OF METHYL IODIDE IN ANIMALS.

REPORT 02 : IARC MONOGRAPH 15 1977
DATA PM EMBRYOTOXICITY OR TERATOGENICITY OF METHYL IODIDE WERE NOT AVAILABLE. METHYL IODIDE IS MUTAGENIC TO SALMONELLA TA100.

REPORT 03 : NIOSH CURRENT INTELLIGENCE BULLETIN 43 SEP 84
THE MONOHALOMETHANES (METHYL CHLORIDE, METHYL BROMIDE, AND METHYL IODIDE) ARE DIRECT-ACTING MUTAGENS IN AMES ASSAY. IN STUDIES IN RATS OR MICE BY VARIOUS ROUTES OF ADMINISTRATION, HAVE DEMONSTRATED ABILITY TO PRODUCE CANCER. METHYL CHLORIDE PRODUCED TERATOGENIC EFFECT

(HEART DEFECTS) IN OFFSPRING OF PREGNANT MICE EXPOSED BY INHALATION AT 500 AND 750 PPM. NIOSH RECOMMENDS THEY BE CONSIDERED POTENTIAL OCCUPATIONAL CARCINOGENS AND THAT METHYL CHLORIDE BE CONSIDERED A POTENTIAL OCCUPATIONAL TERATOGEN.

REPORT 04 : 4TH ANNUAL REPORT ON CARCINOGENS 1985
(MAGUIRE REPT)
SUFFICIENT EVIDENCE FOR CARCINOGENICITY IN EXPERIMENTAL ANIMALS. WHEN RATS AND MICE WERE GIVEN MI SUBCUTANEOUSLY, ONLY THE RATS DEVELOPED LOCAL CANCERS. MICE GIVEN MI INTRAPERITONEALLY, DEVELOPED LUNG TUMORS. HOWEVER, DROPPED FROM THE ANNUAL REPORT, BECAUSE EVALUATED BY IARC AND NOW CONSIDERED "EQUIVOCAL" CARCINOGENICITY DATA.

REPORT 05 : IARC VOL 41 1986
LIMITED EVIDENCE FOR CARCINOGENICITY OF MI TO TEST ANIMALS, BUT NO EVALUATION COULD BE MADE OF THE CARCINOGENICITY OF MI TO HUMANS. SUFFICIENT EVIDENCE FOR GENETIC ACTIVITY IN SHORT-TERM TESTS. NO DATA TO EVALUATE REPRODUCTIVE OR FETAL TOXICITY, MUTAGENICITY OR CHROMOSOMAL EFFECTS OF MI TO TEST ANIMALS OR HUMANS. CASE STUDIES OF HUMAN EXPOSURES INDICATE LONG-LASTING NEUROLOGICAL EFFECTS IN WORKERS EXPOSED TO MI. ONE DEATH HAS BEEN REPORTED FROM OCCUPATIONAL EXPOSURE.

REPORT 06 : MULTI-AGENCY CROSS CHECK 1985
Y HHS MAGUIRE
N OSHA CARC
Y EPA CAG
Y EPA RCRA
Y EPA CLEAN AIR
N EPA CLEAN WATER

REPT'D USE : METHYLATING AGENT IN MFR OF PHARMACEUTICAL INTERMEDIATES, CHEMICAL INDICATOR, MICROSCOPY IMBEDDING AGENT AND FUMIGANT AND SOIL DISINFECTANT

CAS NO : **108-10-1**
CHEM NAME : ***** METHYL ISOBUTYL KETONE**

REPORT 01 : NIOSH CRITERIA DOCUMENT ON KETONES JUN 78
NIOSH RECOMMENDS A STANDARD OF 200 MG/CU M TWA BASED ON LIVER, KIDNEY, AND CNS EFFECTS. THE PRESENT OSHA STANDARD IS 410 MG/CU M TWA.

REPORT 02 : ITC FOURTH REPORT TO EPA (44FR31866) JUN 79
THE INTERAGENCY TESTING COMMITTEE RECOMMENDS THAT THE EPA ADMINISTRATOR CONSIDER THE DEVELOPMENT OF TEST

RULES TO TEST METHYL ISOBUTYL KETONE FOR MUTAGENICITY, TERATOGENICITY AND OTHER CHRONIC EFFECTS; AND THAT AN EPIDEMIOLOGIC STUDY BE CONDUCTED.

REPORT 03 : 47FR58025: EPA RESPONSE TO ITC DEC 82
IN RESPONSE TO ITC'S 1979 DESIGNATION OF MIBK UNDER PRIORITY TESTING FOR HEALTH EFFECTS, EPA'S RESPONSE UNDER TSCA 4(E), CONCLUDES THAT EXISTING DATA SUFFICIENT TO EVALUATE SOME EFFECTS ITC WANTS TESTED AND THAT FURTHER TESTING IN NOT WARRANTED IN OTHERS. EPA WILL ACCEPT INDUSTRY'S TESTING PROPOSAL IN LIEU OF RULEMAKING TO FILL GAPS IN DATA.

REPORT 04 : NOTICE OF TEST PROGRAM BY CMA PANEL (48FR53161) NOV 83
KETONES PROGRAM PANEL OF CMA CONDUCTING TESTING PROGRAM TO DESCRIBE ITS POTENTIAL HEALTH EFFECTS. PROGRAM CONSISTS OF 5 SHORT-TERM TESTS, INHALATION TERATOLOGY TESTS AND 90-DAY SUBCHRONIC STUDY ON RATS AND MICE. THE KETONE PROGRAM PANEL SUBMITTED RESULTS OF PROBE TERATOLOGY STUDY WHICH SHOWED SOME MATERNAL TOXICITY AT HIGH LEVELS OF MIBK. NO ADVERSE MATERNAL EFFECTS SEEN AT 500 PPM IN MICE AND NO SIGNIFICANT EMBRYO OR FETAL TOXICITY OBSERVED IN RATS OR MICE. THE PANEL WILL USE DOSES OF 3,000, 1000 AND 500 PPM IN MAIN TERATOLOGY STUDY.

REPORT 05 : 48 FR 53161 NOV 83
KETONES PROGRAM PANEL OF CMA CONDUCTING TESTING PROGRAM ON MIBK TO CHARACTERIZE POTENTIAL HEALTH EFFECTS. THE TESTING CONSISTS OF 5 SHORT-TERM TESTS ON BOTH CPDS, PLUS AN INHALATION TERATOLOGY TEST AND A 90-DAY SUBCHRONIC STUDY. SHORT-TERM TESTS INCLUDE AMES TEST, MOUSE LYMPHOMA ASSAY, BALB/3T3 MOUSE EMBRYO CELL TRANSFORMATION, CYTOGENIC STUDY W/ MOUSE MICRO-NUCLEUS IN VIVO AND UNSCHEDULED DNA SYNTHESIS.

REPORT 06 : MULTI-AGENCY CROSS CHECK
N HHS MAGUIRE
N OSHA CARC
N EPA CAG
Y EPA RCRA
N EPA CLEAN AIR
N EPA CLEAN WATER

REPORT 07 : 49FR5190 FEB 84
KETONES PROGRAM PANEL OF CMA CONDUCTING A TESTING PROGRAM ON MIBK TO CHARACTERIZE THEIR POTENTIAL HEALTH EFFECTS.

REPORT 08 : 49FR5190 FEB 84
KETONES PROGRAM PANEL REPORTED THAT MALE MICE AND RATS

SHOWED INCREASED LIVER WEIGHTS FOLLOWING 90-DAY INHALATION EXPOSURE STUDY. FEMALES SHOWED NO WEIGHT INCREASE IN THIS STUDY. NO GROSS OR MICROSCOPIC LESIONS RELATED TO MIBK EXPOSURE.

REPORT 09 : CMA KETONES PROGRAM PANEL TEST PROGRAM OCT 84
IN VIEW OF ABSENCE OF CLEARCUT POSITIVE RESULTS IN SHORT-TERM ASSAYS AND TAKING INTO ACCOUNT THE SUBSTANTIAL EVIDENCE THAT MIBK LACKS GENOTOXIC ACTIVITY, THE PANEL BELIEVES THAT FURTHER TESTING FOR MUTAGENICITY AND CARCINOGENICITY IS NOT NEEDED.

REPORT 10 : 48FR44905 EPA TO ADOPT NEGOTIATE TEST PROGRAM SEP 83
EPA TO ADOPT NEGOTIATED TEST PROGRAM AND WILL NOT PROCEDE WITH A TSCA 4(A) RULE TO REQUIRE HEALTH EFFECTS TESTING OF MIBK.

REPT'D USE : SOLVENT, ADHESIVES, MODEL AIRCRAFT, DOPES, CHINA CEMENTS, LACQUERS AND RUBBER CEMENTS

CAS NO : **80-62-6**

CHEM NAME : ***** METHYL METHACRYLATE**

REPORT 01 : EPA RCRA HEALTH AND ENV. EFFECTS PROFILE OCT 80
TERATOGENIC EFFECTS (HEMANGIOMAS) HAVE BEEN REPORTED FOLLOWING IP ADMINISTRATION OF METHYL METHACRYLATE TO PREGNANT RATS. INHALATION EXPOSURE OF PREGNANT RATS TO ACRYLIC CEMENT CONTAINING MM FAILED TO PRODUCE SIGNIFICANT TERATOGENIC EFFECTS.

REPORT 02 : IARC MONOGRAPH 19 1979
METHYL METHACRYLATE WAS TESTED BY ORAL ADMINISTRATION AND BY SKIN APPLICATION IN RATS. THESE STUDIES WERE CONSIDERED INADEQUATE BY THE WORKING GROUP FOR AN EVALUATION OF THE CARCINOGENICITY OF THIS COMPOUND.

REPORT 03 : CARC TEST PROGRAM JUL 82
IN A CURRENT NTP BIOASSAY, METHYL METHACRYLATE IS BEING ADMINISTERED TO F344 AND B6C3F1 MICE BY INHALATION. THE CHRONIC EXPOSURE TEST PHASE BEGAN IN JAN 81.

REPORT 04 : IARC MONOGRAPH 19 1979
NO DATA ON THE MUTAGENICITY OF METHYL METHACRYLATE ALONE WERE AVAILABLE.

REPORT 05 : NTP TR 314 NIH PUB # 87-2570 OCT 86
NO EVIDENCE OF CARCINOGENICITY OF MMA FOR MALE OR FEMALE RATS OR MICE. INHALATION ASSOCIATED WITH INFLAMMATION OF NASAL CAVITY AND DEGENERATION OF OLFACTORY

SENSORY EPITHELIUM IN RATS AND MICE. EPITHELIAL HYPERPLASIA OF THE NASAL CAVITY ALSO SEEN IN EXPOSED MICE.

REPORT 06 : MULTI-AGENCY CROSS CHECK
N HHS MAGUIRE
N OSHA CARC
N EPA CAG
Y EPA RCRA
N EPA CLEAN AIR
N EPA CLEAN WATER

REPT'D USE : PLASTIC MONOMER, COSMETICS, PAINTS, DENTURE MATERIALS AND PROSTHETICS

CAS NO : **66-27-3**

CHEM NAME : ***** METHYL METHANESULFONATE (MMS)**

REPORT 01 : IARC MONOGRAPH 7 1974
METHYL METHANESULFONATE (MMS) IS CARCINOGENIC IN RATS FOLLOWING SUBCUTANEOUS AND INTRAPERITONEAL INJECTION, PRODUCING LOCAL TUMORS AND TUMORS OF THE NERVOUS SYSTEM. FOLLOWING ORAL ADMINISTRATION IN MICE IT INCREASED THE INCIDENCE OF LUNG TUMORS AND OF LYMPHOMAS. IN RATS, IT IS CARCINOGENIC UPON ADMINISTRATION OF A SINGLE DOSE AS WELL AS FOLLOWING PRENATAL EXPOSURE. IN A 1979 SUPPLEMENT, IARC DETERMINED THAT THERE IS SUFFICIENT EVIDENCE FOR THE CARCINOGENICIY OF MMS IN ANIMALS.

REPORT 02 : NTP TECHNICAL BULLETIN 9 APR 83
METHYL METHANE SULFONATE WAS POSITIVE IN ONE MUTAGENICITY ASSAY.

REPORT 06 : MULTI-AGENCY CROSS CHECK
Y HHS MAGUIRE
N OSHA CARC
Y EPA CAG
N EPA RCRA
N EPA CLEAN AIR

REPT'D USE : INSECT CHEMOSTERILANT, CHEMOTHERAPEUTIC AGENT, CATALYST AND RESEARCH CHEMICAL. NO AVAILABLE PRODUCTION DATA

CAS NO : **129-15-7**

CHEM NAME : ***** 2-METHYL-1-NITROANTHRAQUINONE (MNA)**

REPORT 01 : NCI CARC TRS 29 DHEW PUB NO NIH 78-829
UNDER THE CONDITIONS OF THIS BIOASSAY, THE RESULTS INDICATE THAT ORALLY ADMINISTERED 2-METHYL-1-NITROANTHRAQUINONE IS CARCINOGENIC IN MALE FISCHER 344

RATS, PRODUCING HEPATOCELLULAR CARCINOMAS. INCREASED INCIDENCES OF SUBCUTANEOUS FIBROMAS IN BOTH MALE AND FEMALE FISCHER 344 RATS WERE ALSO ASSOCIATED WITH THE ADMINISTRATION OF THE COMPOUND TUMORS OF THE FORESTOMACH AND BLADDER IN THESE ANIMALS MAY ALSO HAVE BEEN RELATED TO THE ADMINISTRATION OF THE TEST CHEMICAL.

REPORT 02 : IARC MONOGRAPH 27 1982
2-METHYL-1-NITROANTHRAQUINONE OF UNSPECIFIED PURITY INDUCED REVERSE MUTATIONS IN SALMONELLA STRAINS TA1537, TA1538, TA98 AND TA100 WITH AND WITHOUT METABOLIC ACTIVATION. NO DATA CONCERNING EFFECTS ON REPRODUCTION OR PRENATAL TOXICITY WERE AVAILABLE.

REPORT 03 : CPSC ANTHRAQUINONE DYES MONOGRAPH OCT 81
MNA IS MUTAGENIC TO SALMONELLA WITH AND WITHOUT METABOLIC ACTIVATION. MICROSOMAL ACTIVATION WAS SHOWN TO INCREASE MNA MUTAGENICITY.

REPORT 04 : NTP TECHNICAL BULLETIN 6/8 JAN 82, JUL 82
2-METHYL-1-NITROANTHRAQUINONE WAS POSITIVE IN THE SALMONELLA ASSAY, BUT NEGATIVE IN DROSOPHILA WHEN TESTED FOR SEX-LINKED RECESSIVE LETHALS.

REPORT 05 : IARC MONOGRAPH 27 1982
2-METHYL-1-NITROANTHRAQUINONE OF UNCERTAIN PURITY WAS TESTED IN MICE AND RATS BY DIETARY ADMINISTRATION. IT WAS CARCINOGENIC IN BOTH SPECIES TESTED. IN VIEW OF THE UNCERTAIN PURITY OF THE COMPOUND TESTED, NO EVALUATION OF THE CARCINOGENICITY OF THIS CHEMICAL CAN BE MADE.
REPT'D USE : DYE INTERMEDIATE. NO KNOWN CURRENT US PRODUCTION; NO LISTING ON TSCA INVENTORY

CAS NO : **70-25-7**
CHEM NAME : ***** N-METHYL-N'-NITRO-N-NITROSOGUANIDINE (MNNG)**

REPORT 01 : IARC MONOGRAPH 4 1974
MNNG IS CARCINOGENIC IN ALL SPECIES TESTED: MOUSE, RAT, HAMSTER, RABBIT, AND DOG. IT HAS A PREDOMINANTLY LOCAL CARCINOGENIC EFFECT FOLLOWING ADMINISTRATION BY ORAL AND OTHER ROUTES. IT IS CARCINOGENIC IN SINGLE DOSE EXPERIMENTS. IN A 1979 SUPPLEMENT, IARC DETERMINED THAT THERE IS SUFFICIENT EVIDENCE FOR THE CARCINOGENICITY OF MNNG IN ANIMALS.

REPORT 06 : MULTI-AGENCY CROSS CHECK
Y HHS MAGUIRE
N OSHA CARC
Y EPA CAG
Y EPA RCRA

N EPA CLEAN AIR
N EPA CLEAN WATER
REPT'D USE : RESEARCH CHEMICAL

CAS NO : **64091-90-3**
CHEM NAME : ***** 4-(METHYLNITROSAMINO)-4-(3-PYRIDYL) BUTANAL (NNA)**

REPORT 01 : IARC MONOGRAPH 37 SEP 85
TESTED BY INTRAPERITONEAL INJECTION IN FEMALE MICE OF ONE STRAIN AT ONE DOSE LEVEL ONLY. ALTHOUGH INCIDENCE OF LUNG ADENOMAS IN TREATED MICE EXCEEDED THOSE IN CONTROLS, EXPERIMENT DOES NOT ALLOW EVALUATION OF CARCINOGENICITY OF NNA.
REPT'D USE : LABORATORY CHEMICAL

CAS NO : **64091-91-4**
CHEM NAME : ***** 4-(METHYLNITROSAMINO)-1-(3-PYRIDYL)-1-BUTANONE (NNK)**

REPORT 01 : IARC MONOGRAPH 37 SEP 85
TESTED FOR CARCINOGENICITY IN SEVERAL STUDIES BY SUBCUTANEOUS INJECTION IN RATS AND HAMSTERS AND BY INTRAPERITONEAL INJECTION IN MICE. IN RATS, INDUCED CARCINOMAS OF NASAL CAVITY, LUNG AND LIVER, WITH A CLEAR DOSE-RESPONSE RELATIONSHIP. IN HAMSTERS, INDUCED BENIGN AND MALIGNANT TUMORS OF NASAL CAVITY, TRACHEA AND LUNG, EVEN AFTER A SINGLE ADMINISTRATION. IN MICE, NNK AND ITS METABOLITES 4-(METHYLNITROSAMINO)-1-(3-PYRIDYL-N-OXIDE)-1-BUTANONE AND 4-(METHYL-NITROSAMINO)-1-(3-PYRIDYL)BUAN-1-OL INDUCED BENIGN AND MALIGNANT TUMORS OF LUNG. NNK AND ITS METABOLITES CAN CROSS THE PLACENTAL BARRIER IN MICE. NNK CAN BE METABOLICALLY ACTIVATED BY MOUSE FETAL TISSUES.

REPORT 02 : IARC MONOGRAPH 37 SEP 85
ADMINISTRATION OF NNK TO RATS RESULTS IN ABNORMAL DNA METHYLATION IN LIVER AND LUNG. NNK MUTAGENIC TO SALMONELLA TYPHIMURIUM IN PRESENCE OF EXOGENOUS METABOLIC SYSTEM. IT INDUCES UNSCHEDULED DNA SYNTHESIS IN PRIMARY CULTURES OF RAT HEPATOCYTES.

REPORT 06 : MULTI-AGENCY CROSS CHECK
Y HHS MAGUIRE
N OSHA CARC
N EPA CAG
N EPA RCRA
N EPA CLEAN AIR
N EPA CLEAN WATER

REPT'D USE : FOUND IN TOBACCO PRODUCTS AND SMOKE AND LABORATORY CHEMICAL

CAS NO : **614-00-6**
CHEM NAME : ***** N-METHYL-N-NITROSOANILINE**

REPORT 06 : MULTI-AGENCY CROSS CHECK
N HHS MAGUIRE
Y OSHA CARC
N EPA CAG
N EPA RCRA
N EPA CLEAN AIR
N EPA CLEAN WATER
REPT'D USE : NO AVAILABLE PRODUCTION OR USE DATA

CAS NO : **298-00-0**
CHEM NAME : ***** METHYL PARATHION**

REPORT 01 : IARC MONOGRAPH 30 JAN 83
METHYL PARATHION TESTED FOR CARCINOGENICITY IN MICE AND RATS BY ADMINISTRATION IN THE DIET. NO EVIDENCE OF CARCINOGENICITY FOUND.

REPORT 02 : IARC MONOGRAPH 30 JAN 83
METHYL PARATHION, ADMINISTERED INTRAPERITONEALLY AT MATERNALLY LETHAL DOSES, TERATOGENIC TO MICE. IN RATS, DECREASED VIABILITY AND BODY WEIGHT SEEN IN PROGENY OF TREATED ANIMALS, BUT NO TERATOGENICITY OBSERVED.

REPORT 03 : IARC MONOGRAPH 30 JAN 83
METHYL PARATHION WEAKLY OR NOT MUTAGENIC IN BACTERIAL SYSTEMS AND IN DROSOPHILA MELANOGASTER, BUT IT MUTAGENIC IN YEASTS. IN MAMMALIAN CELLS, SISTER CHROMATID EXCHANGE AND PRESUMED GENE MUTATIONS INDUCED, BUT NEITHER CHROMOSOMAL ABERRATION NOR UNSCHEDULED DNA SYNTHESIS ELICITED. CHROMOSOMAL ABERRATIONS AND DOMINANT LETHAL MUTATIONS NOT INCREASED IN MICE TREATED WITH METHYL PARATHION. SUFFICIENT EVIDENCE THAT METHYL PARATHION IS MUTAGENIC IN VARIETY OF CELLULAR SYSTEMS, BUT INSUFFICIENT EVIDENCE THAT IT IS MUTAGENIC IN MAMMALS.
REPT'D USE : INSECTICIDE

CAS NO : **832-69-9**
CHEM NAME : ***** 1-METHYLPHENANTHRENE**

REPORT 01 : IARC MONOGRAPH 32 DEC 83
TESTED IN MOUSE SKIN INITIATION-PROMOTION ASSAY IN ONE STUDY AND WAS INACTIVE AS AN INITIATOR MUTAGENIC TO SALMONELLA TYPHIMURIUM IN PRESENCE OF EXOGENOUS METABOLIC SYSTEM. INDUCED UNSCHEDULED DNA SYNTHESIS IN CULTURED PRIMARY HEPATOCYTES FROM RATS AND MUTAGENIC TO HUMAN LYMPHOBLASTS IN VITRO IN PRESENCE OF EXOGENOUS METABOLIC SYSTEM.
REPT'D USE : INCOMPLETE COMBUSTION PRODUCT AND CIGARETTE SMOKE

CAS NO : **144-34-3**
CHEM NAME : ***** METHYL SELENAC**

REPORT 01 : IARC MONOGRAPH 12 1976
METHYL SELENAC HAS BEEN TESTED BY ORAL ADMINISTRATION AND BY SINGLE S.C. INJECTION IN TWO STRAINS OF MICE. ALTHOUGH NO CARCINOGENIC EFFECT WAS OBSERVED IN THESE TESTS, THE AVAILABLE DATA ARE INSUFFICIENT TO ALLOW AN EVALUATION OF THE CARCINOGENICITY OF THIS COMPOUND.
REPT'D USE : RUBBER VULCANIZING AGENT

CAS NO : **56-04-2**
CHEM NAME : ***** METHYLTHIOURACIL**

REPORT 01 : IARC MONOGRAPH 7 1974
MTU ADMINISTERED TO MICE, RATS AND HAMSTERS BY THE ORAL ROUTE PRODUCED THYROID TUMORS IN ALL THREE SPECIES. IT WAS SIMILARLY EFFECTIVE IN RATS FOLLOWING S.C. IMPLANTATION. KIDNEY TUMORS WERE INDUCED IN FEMALE RATS FOLLOWING ORAL ADMINISTRATION. IN A 1979 SUPPLEMENT, IARC DETERMINED THAT THERE IS SUFFICIENT EVIDENCE FOR THE CARCINOGENICITY OF METHYLTHIOURACIL IN ANIMALS.

REPORT 06 : MULTI-AGENCY CROSS CHECK
N HHS MAGUIRE
N OSHA CARC
Y EPA CAG
Y EPA RCRA
N EPA CLEAN AIR
N EPA CLEAN WATER
REPT'D USE : ANTI-THYROID AGENT

CAS NO : **CLASS-26**
CHEM NAME : ***** MICA DUST**

REPORT 01 : STATUS REPORT 8EHQ-0986-0632 OCT 86
TWO PHYSICIANS HAVE INDICATED THAT A PERSON'S SUSCEPTIBILITY TO EAR INFECTIONS WAS RELATED TO INHALATION OF MICA DUST.
REPT'D USE : ELECTRICAL AND BUILDING APPLICATIONS, COSMETICS, DRILLING MUDS AND RUBBER PRODUCTS

CAS NO : **90-94-8**
CHEM NAME : ***** MICHLER'S KETONE**

REPORT 01 : NCI CARC TRS 181 DHEW PUB NO NIH 79-1737
UNDER THE CONDITIONS OF THIS BIOASSAY, DIETARY ADMINISTRATION OF MICHLER'S KETONE WAS CARCINOGENIC TO MALE AND FEMALE FISCHER 344 RATS AND FEMALE B6C3F1 MICE, CAUSING HEPATOCELLULAR CARCINOMAS, AND TO MALE B6C3F1 MICE, CAUSING HEMANGIOSARCOMAS.

REPORT 02 : NTP TECHNICAL REPORT 8 JUL 82
MICHLER'S KETONE WAS POSITIVE IN THE L5178Y MOUSE LYMPHOMA ASSAY.

REPORT 06 : MULTI-AGENCY CROSS CHECK
Y HHS MAGUIRE
Y OSHA CARC
N EPA CAG
N EPA RCRA
N EPA CLEAN AIR
N EPA CLEAN WATER
REPT'D USE : DYE INTERMEDIATE

CAS NO : **2385-85-5**
CHEM NAME : ***** MIREX**

REPORT 01 : IARC MONOGRAPH 20 1979
MIREX HAS BEEN TESTED IN ONE EXPERIMENT IN TWO STRAINS OF MICE AND IN ONE EXPERIMENT IN RATS BY ORAL ADMINISTRATION. TESTS IN TWO STRAINS OF MICE BY SC INJECTION OF SINGLE DOSES WERE ALSO PERFORMED. VIA ORAL ADMINISTRATION, THE COMPOUND PRODUCED BENIGN AND MALIGNANT LIVER TUMORS IN MICE AND RATS OF BOTH SEXES. SC INJECTION RESULTED IN AN EXCESS OF LIVER TUMORS IN MALE MICE OF ONE STRAIN. THE EXPERIMENT ALSO SUGGESTED THAT MIREX PRODUCED RETICULUM-CELL SARCOMAS IN BOTH STRAINS OF MALE MICE. IARC 1979 SUPPLEMENT SAYS THERE IS SUFFICIENT EVIDENCE FOR CARCINOGENICITY IN EXPERIMENTAL ANIMALS.

REPORT 03 : NTP TECHNICAL BULLETIN 7 APR 82
MIREX WAS NEGATIVE IN THE SALMONELLA TYPHIMURIUM MUTAGENICITY ASSAY.

REPORT 04 : IARC MONOGRAPH 20 1979
MIREX IS FETOTOXIC AND PRODUCES TERATOGENIC EFFECTS IT WAS NEGATIVE IN A DOMINANT LETHAL ASSAY IN MICE.

REPORT 06 : MULTI-AGENCY CROSS CHECK
Y HHS MAGUIRE
Y OSHA CARC
N EPA CAG
N EPA RCRA
N EPA CLEAN AIR
N EPA CLEAN WATER

REPORT 07 : NTP TR 313 NIH 90-2569 FEB 90
UNDER 2-YEAR FEED STUDIES OF MIREX CLEAR EVIDENCE OF CARCINOGENIC ACTIVITY FOR MALE AND FEMALE F344/N RATS, INDICATED BY MARKED BENIGN NEOPLASTIC NODULES OF LIVER, AND PHEOCHROMOCYTOMAS OF ADRENAL GLAND AND TRANSITIONAL CELL PAPILLOMAS OF KIDNEY IN MALES AND BY MONONUCLEAR CELL LEUKEMIA IN FEMALES. NONNEOPLASTIC EFFECTS INCLUDE CYTOMEGALY, FATTY METAMORPHOSIS, ANGIECTASIS (MALES ONLY), AND CELLULAR NECROSIS IN LIVER.
REPT'D USE : INSECTICIDE AND FIRE RETARDANT

CAS NO : **50-07-7**
CHEM NAME : ***** MITOMYCIN C**

REPORT 01 : IARC MONOGRAPH 10 1976
MITOMYCIN C IS CARCINOGENIC IN MICE FOLLOWING ITS SUBCUTANEOUS INJECTION AND IN RATS FOLLOWING ITS INTRAPERITONEAL OR INTRAVENOUS INJECTION. IN RATS, IT PRODUCED BOTH LOCAL AND DISTANT TUMORS. IN A 1979 SUPPLEMENT, IARC DETERMINED THAT THERE IS SUFFICIENT EVIDENCE OF CARCINOGENICITY FOR MITOMYCIN C IN ANIMALS.

REPORT 02 : NCI BIOASSAY PROGRAM - CANCER VOL. 40 1977
UNDER THE CONDITIONS OF THIS BIOASSAY, I.P. INJECTION OF MITOMYCIN C WAS CARCINOGENIC TO MALE AND FEMALE SPRAGUE-DAWLEY (CD) RATS, CAUSING PERITONEAL SARCOMAS. IT WAS NOT CARCINOGENIC TO SWISS-WEBSTER MICE OF EITHER SEX.

REPORT 03 : IARC MONOGRAPH 10 1976
MITOMYCIN C INDUCED REVERSE MUTATIONS IN SALMONELLA; IT INDUCED MITOTIC CROSSING OVER IN SACCHAROMYCES. THE CHEMICAL INDUCED CHROMOSOMAL ABERRATIONS IN DROSOPHI-

LA AND DOMINANT AND RECESSIVE MUTATIONS IN THE WASP. MITOMYCIN C INDUCED CHROMOSOMAL BREAKS AND REARRANGEMENTS IN CULTURES OF HUMAN PERIPHERAL LEUCOCYTES.

REPORT 06 : MULTI-AGENCY CROSS CHECK
N HHS MAGUIRE
N OSHA CARC
Y EPA CAG
Y EPA RCRA
N EPA CLEAN AIR
N EPA CLEAN WATER

REPT'D USE : ANTIBIOTIC AND ANTINEOPLASTIC AGENT

CAS NO : **79-11-8**

CHEM NAME : ***** MONOCHLOROACETIC ACID**

REPORT 01 : NTP DRAFT TR 396 NIH PUB NO 90-2851 NOV 90
UNDER CONDITIONS OF 2 YEAR STUDIES NO EVIDENCE OF CARCINOGENIC ACTIVITY OF MONOCHLOROACETIC ACID FOR MALE OR FEMALE F344/N RATS GIVEN 15 OR 30 MG/KG IN WATER BY GAVAGE FOR 104 WEEKS. NO EVIDENCE OF CARCINOGENIC ACTIVITY OF MONOCHLOROACETIC ACID FOR MALE OR FEMALE B6C3F1 MICE GIVEN 50 OR 100 MG/KG IN WATER BY GAVAGE FOR 104 WEEKS. MONOCHLOROACETIC ACID ADMINISTRATION WAS ASSOCIATED WITH INFLAMMATORY LESIONS OF NASAL MUCOSA, METAPLASIA OF OLFACTORY EPITHELIUM, AND SQUAMOUS CELL HYPERPLASIA OF FORESTOMACH IN MALE AND FEMALE MICE.

REPT'D USE : HERBICIDE AND INTERMEDIATE

CAS NO : **315-22-0**

CHEM NAME : ***** MONOCROTALINE**

REPORT 01 : IARC MONOGRAPH 10 1976
MONOCROTALINE PRODUCED CARCINONOMAS OF THE LIVER IN RATS FOLLOWING ORAL ADMINISTRATION, THE ONLY SPECIES AND ROUTE OF ADMINISTRATION TESTED. IN A 1979 SUPPLEMENT, IARC DETERMINED THAT THERE IS SUFFICIENT EVIDENCE FOR THE CARCINOGENICITY OF MONOCROTALINE IN EXPERIMENTAL ANIMALS.

REPORT 02 : IARC MONOGRAPH 10 1976
MONOCROTALINE OF UNSPECIFIED PURITY IN SODIUM CHLORIDE WAS INJECTED INTO THE ABDOMEN OF DROSOPHILA MALES. STRONG EFFECTS ON THE INDUCTION OF SEX-LINKED RECESSIVE LETHALS WERE SCORED.

REPT'D USE : EXPERIMENTAL CYTOSTATIC AGENT

CAS NO : **1746-81-2**
CHEM NAME : ***** MONOLINURON**

REPORT 01 : EPA TSCA 8(E) 1088-0755 OCT 88
ELI LILLY REPORTED THREE MALE EMPLOYEES WITH SYMPTOMS COMPATIBLE WITH HYPOXIA; TWO DIAGNOSED WITH METHEMOGLOBINEMIA; OBTAINED WHILE PREPARING PESTICIDE CONTAINING MONOLINURON AND ETHALFLURALIN. COPY OF HOECHST A.G. DATA SHEET FOR MONOLINURON PROVIDED: ORAL LD50: 1800 MG/KG (RAT), DERMAL LD50: 1500 MG/KG (FEMALE RAT); NONIRRITANT IN DERMAL AND EYE TESTS (RABBIT).
REPT'D USE : PESTICIDE

CAS NO : **150-68-5**
CHEM NAME : ***** MONURON**

REPORT 01 : IARC MONOGRAPH 12 1976
IN ONE STUDY BY ORAL ADMINISTRATION IN MICE, AN INCREASED INCIDENCE OF LUNG TUMORS WAS OBSERVED IN MALES OF ONE OF TWO STRAINS. IN ONE STUDY BY ORAL ADMINISTRATION IN MALE RATS, TUMORS WERE OBSERVED AT VARIOUS SITES; NONE WERE OBSERVED IN CONCURRENT CONTROLS. THE AVAILABLE DATA SUGGEST THAT MONURON IS CARCINOGENIC.

REPORT 03 : NTP TECHNICAL BULLETIN 7 APR 82
MONURON WAS NEGATIVE IN THE SALMONELLA TYPHIMURIUM MUTAGENICITY ASSAY.

REPORT 04 : NTP TR 266 NIH PUBLICATION NO. 88-2522 AUG 88
UNDER CONDITIONS OF 2-YEAR FEED STUDIES CLEAR EVIDENCE OF CARCINOGENICITY FOR MALE F344/N RATS IN THAT MONURON CAUSED INCREASED INCIDENCES OF TUBULAR CELL ADENOCARCINOMAS OF THE KIDNEY, TUBULAR CELL ADENOMAS OF KIDNEY, AND NEOPLASTIC NODULES OR CARCINOMAS (COMBINED) OF LIVER. MONURON INDUCED CYTOMEGALY OF RENAL TUBULAR EPITHELIAL CELLS IN BOTH MALE AND FEMALE F344/N RATS. NO EVIDENCE OF CARCINOGENICITY FOR FEMALE F344/N RATS OR FOR MALE OR FEMALE B6C3F1 MICE.

REPORT 06 : MULTI-AGENCY CROSS CHECK
N HHS MAGUIRE
Y OSHA CARC
N EPA CAG
N EPA RCRA
N EPA CLEAN AIR
N EPA CLEAN WATER
REPT'D USE : HERBICIDE

CAS NO : **139-91-3**
CHEM NAME : ***** 5-(MORPHOLINOMETHYL)-3-((5-NITROFURFURYL-IDENE)-AMINOL)-2-OXA**

REPORT 01 : IARC MONOGRAPH 7 1974
THIS COMPOUND IS CARCINOGENIC IN RATS FOLLOWING ORAL ADMINISTRATION OF ITS HYDROCHLORIDE. IT PRODUCED MAINLY MAMMARY CARCINOMAS AND LYMPHOBLASTIC LYMPHOMAS. IN A 1979 SUPPLEMENT, IARC DETERMINED THAT THERE IS SUFFICIENT EVIDENCE FOR THE CARCINOGENICITY OF THIS COMPOUND IN ANIMALS.
REPT'D USE : VETERINARY ANTIBACTERIAL DRUG

CAS NO : **597-25-1**
CHEM NAME : ***** MORPHOLINOPHOSPHORAMIDATE**

REPORT 01 : NTP TECHICAL REPORT 298 NIH PUB 86-2554 JAN 86
FROM 2-YEAR GAVAGE STUDIES, SOME EVIDENCE OF CARCINOGENICITY FOR MALE AND FEMALE F344/N RATS GIVEN DIMETHYL MORPHOLINOPHOSPHORAMIDATE, AS INDICATED BY INCREASED INCIDENCES OF MONONUCLEAR CELL LEUKEMIA. NO EVIDENCE OF CARCINOGENICITY FOR MALE AND FEMALE B6C3F1 MICE GIVEN DIMETHYL MORPHOLINOPHOSPHORAMIDATE AT DOSES OF 150 (MALE), 300, OR 600 (FEMALE) MG/KG FOR 2 YEARS.
REPT'D USE : MILITARY CHEMICAL DEFENSE NERVE AGENT, STIMULANT

CAS NO : **505-60-2**
CHEM NAME : ***** MUSTARD GAS**

REPORT 01 : IARC MONOGRAPH 9 1975
MUSTARD GAS IS CARCINOGENIC IN MICE, THE ONLY SPECIES TESTED, FOLLOWING THEIR EXPOSURE TO ITS VAPORS OR FOLLOWING ITS INTRAVENOUS INJECTION, INDUCING AN INCREASE IN THE INCIDENCE OF LUNG TUMORS. FOLLOWING ITS SUBCUTANEOUS INJECTION IN MICE, IT PRODUCED A LOW INCIDENCE OF SARCOMAS AT THE INJECTION SITE. THERE IS EVIDENCE OF AN INCREASED INCIDENCE OF CANCERS OF THE RESPIRATORY TRACT IN MEN EXPOSED TO MUSTARD GAS. IN A 1979 SUPPLEMENT, THE IARC CONCLUDED THAT MUSTARD GAS IS CARCINOGENIC FOR HUMANS.

REPORT 02 : IARC MONOGRAPH 9 1975
MUSTARD GAS WAS THE FIRST CHEMICAL REPORTED TO INDUCE MUTATIONS IN DROSOPHILA. IT INDUCES CHROMOSOME ABERRATIONS AND REVERSE MUTATIONS IN HOST-MEDIATE ASSAY IN MALE MICE AND DOMINANT LETHAL MUTATIONS IN RATS.

REPORT 06 : MULTI-AGENCY CROSS CHECK
Y HHS MAGUIRE
N OSHA CARC
Y EPA CAG
N EPA RCRA
N EPA CLEAN AIR
N EPA CLEAN WATER

REPT'D USE : CHEMICAL WARFARE AGENT AND LABORATORY ALKYLATING AGENT

CAS NO : **924-42-5**
CHEM NAME : ***** N-METHYLOLACRYLAMIDE**

REPORT 02 : NTP TR 352 NIH 89-2807 SEP 89
UNDER THE CONDITIONS OF THESE 2-YEAR AQUEOUS GAVAGE STUDIES, THERE WAS NO EVIDENCE OF CARCINOGENIC ACTIVITY OF N-METHYLOLACRYLAMIDE FOR MALE OR FEMALE F344/N RATS RECEIVING DOSES OF 6 OR 12 MG/KG PER DAY. THERE WAS CLEAR EVIDENCE OF CARCINOGENIC ACTIVITY OF N-METHYLOLACRYLAMIDE FOR MALE B6C3F1 MICE, BASED ON INCREASED INCIDENCES OF NEOPLASMS OF THE HARDERIAN GLAND, LIVER, AND LUNG. THERE WAS CLEAR EVIDENCE OF CARCINOGENIC ACTIVITY OF N-METHYLOLACRYLAMIDE FOR FEMALE B6C3F1 MICE, BASED ON INCREASED INCIDENCES OF NEOPLASMS OF THE HARDERIAN GLAND, LIVER, LUNG, AND OVARY.

REPT'D USE : PAPER, TEXTILES/ADHESIVES, BINDERS, SURFACE COATINGS, VARNISH RESINS, LATEX FILMS, SIZING AGENTS AND FINISHED COTTON

CAS NO : **872-50-4**
CHEM NAME : ***** N-METHYLPYRROLIDONE (NMP)**

REPORT 01 : EPA 8EHQ-1087- 0695 NOV 87
UNDER CONDITIONS OF TERATOLOGY STUDY, CPD ADMINISTERED IP TO PREGNANT NMRI MICE AT 610 AND 1525 MM3/KG OR ORALLY AT 1026 AND 2565 MM3/KG ON THE 11TH THRU 15TH DAY OF GESTATION, INCREASED RESORPTION RATE, # OF RUNTS, REDUCTIONS IN WEIGHTS AND LENGTHS OF FETUSES. ALSO INCREASE IN RATE OF BIRTH DEFECTS. CPD ADMINISTERED ORALLY AT 320 OR 970 MM3/KG TO PREGNANT SPRAGUE-DAWLEY RATS ON DAYS 6 THRU 15 OF GESTATION. HIGH DOSE PRODUCED 95% EMBRYOLETHALITY AND DEFORMITIES IN 8/15 OF SURVIVING FETUSES.

REPORT 02 : HS STAFF REPORT AUG 89
ALTHOUGH NMP STUDIED FOR CARCINOGENICITY, REPRODUCTIVE EFFECTS AND METABOLISM DATA GAPS EXIST. NEED FOR CANCER BIOASSAY IN AT LEAST ONE MORE SPECIES AT TWO DOSE LEVELS. MULTIGENERATIONAL STUDIES NEEDED TO UNDER-

STAND REPRODUCTIVE EFFECTS. METABOLISM STUDIES IN RATS AND MICE FOLLOWING SEVERAL ROUTES OF ADMINISTRATION, A 90-DAY INHALATION STUDY IN RODENTS, NEUROTOXICITY, MUTAGNICITY, MUTAGENICITY TESTING IN MAMMALIAN CELL CULTURES, DROSOPHILA SEX-LINKED RECESSIVE LETHAL TEST, AND BIOCHEMICAL SPECIFIC LOCUS TEST ARE NEED EPA DEVELOPING TEST RULE FOR HEALTH EFFECTS.

REPORT 03 : HS STAFF REPORT AUG 89
NMP NOT EXPECTED TO CAUSE ACUTE TOXIE EFFECTS FOLLOWING INFREQUENT LOW EXPOSURES OF SHORT DURATION. MAY CAUSE ADVERSE EFFECTS ON BLOOD FORMING ORGANS FOLLOWING MULTIPLE EXPOSURES AT HIGH DOSES. MAY CAUSE ADVERSE REPRODUCTIVE EFFECT AT HIGHER DOSES.
REPT'D USE : SOLVENT, CHEMICAL INTERMEDIATE, SPINNING AGENT FOR PVC, DISPERSANT FOR PIGMENTS AND SUBSTITUTE FOR METHYLENE CHLORITE

CAS NO : **3771-19-5**
CHEM NAME : ***** NAFENOPIN**

REPORT 01 : IARC MONOGRAPH 24 1980
NAFENOPIN WAS TESTED IN ACATALASEMIC MICE (A STRAIN WITH AN UNSTABLE CATALASE GENE) AND IN MALE RATS BY ORAL ADMINISTRATION: IT PRODUCED HEPATOCELLULAR CARCINOMAS IN BOTH SPECIES. A LOW INCIDENCE OF PANCREATIC TUMORS WAS ALSO OBSERVED IN RATS.

REPORT 02 : IARC MONOGRAPH 24 1980
NAFENOPIN WAS NOT MUTAGENIC TO SALMONELLA STRAINS TA98, TA100, TA1535, TA1537, OR TA1538, WITH OR WITHOUT METABOLIC ACTIVATION BY RAT LIVER MICROSOMES. ATTENTION IS DRAWN TO THE ABSENCE OF STUDIES ON THE TERATOGENICITY OF THIS COMPOUND.

REPORT 03 : IARC MONOGRAPHS SUPPL 4, 1982 1982
THERE IS SUFFICIENT EVIDENCE FOR CARCINOGENICITY OF NAFENOPIN IN EXPERIMENTAL ANIMALS.
REPT'D USE : ANTIHYPERCHLOESTEROLAEMIA AGENT

CAS NO : **389-08-2**
CHEM NAME : ***** NALDIXIC**

REPORT 01 : NTP TR 368 NIH PUB. 90-2823 OCT 89
UNDER 2-YEAR FEED STUDIES CLEAR EVIDENCE OF CARCINOGENIC ACTIVITY OF NALIDIXIC ACID FOR F344/N RATS INDICATED BY INCREASED INCIDENCES OF PREPUTIAL GLAND NEOPLASMS IN MALES AND CLITORAL GLAND NEOPLASMS IN FEMALES. EQUIVO-

CAL EVIDENCE OF CARCINOGENICITY FOR MALE B6C3F1 MICE FED DIETS CONTAINING NALIDIXIC ACID INDICATED BY MARGINALLY INCREASED INCIDENCES OF SUBCUTANEOUS TISSUE NEOPLASMS. NO EVIDENCE OF CARCINOGENICITY FOR FEMALE B6C3F1 MICE FED DIETS CONTAINING 2,000 OR 4,000 PPM NALIDIXIC ACID FOR 2 YEARS.

REPT'D USE : ANTIMICROBIAL DRUG

CAS NO : **91-20-3**

CHEM NAME : ***** NAPHTHALENE**

REPORT 01 : EPA TSCA FYI 8EHQ-1287-0704
NATIONAL TOXICOLOGY PROGRAM BIOASSAY OF NAPHTHALENE IN MALE AND FEMALE B6C3F1 MICE SHOWED THAT NAPHTHALENE CAUSED LUNG TUMOR IN MALE MICE.

REPT'D USE : INSECTICIDES, DYES, MOTH REPELLENTS, SURFACTANTS/LEATHER TANNING AND CHEMICALS

CAS NO : **2243-62-1**

CHEM NAME : ***** 1,5-NAPHTHALENEDIAMINE**

REPORT 01 : NCI CARC TRS 143 DHEW PUB NO NIH 78-1398
UNDER THE CONDITIONS OF THIS BIOASSAY, 1,5-NAPHTHALENE-DIAMINE WAS CARCINOGENIC IN FEMALE FISCHER 344 RATS, CAUSING CLITORAL AND UTERINE NEOPLASMS. 1,5-NAPHTHA-LENE-DIAMINE WAS ALSO CARCINOGENIC FOR B6C3F1 MICE, PRODUCING THYROID NEOPLASMS IN MALES AND NEOPLASMS OF THE THYROID, LIVER, AND LUNG IN FEMALES. INSUFFICIENT EVIDENCE WAS PROVIDED FOR THE CARCINOGENICITY OF THE COMPOUND IN MALE FISCHER 344 RATS (FEEDING STUDY).

REPORT 02 : IARC MONOGRAPH 27 1982
1,5-NAPHTHALENEDIAMINE WAS TESTED IN ONE EXPERIMENT IN MICE AND IN ONE EXPERIMENT IN RATS BY DIETARY ADMINISTRATION. IT PRODUCED ADENOMAS OF THE THYROID IN MALE MICE AND CARCINOMAS AND ADENOMAS OF THE THYROID AND LUNGS AND CARCINOMAS OF THE LIVER IN FEMALE MICE. THE EXPERIMENT IN RATS WAS INADEQUATE FOR EVALUATION. THERE IS LIMITED EVIDENCE FOR THE CARCINOGENICITY OF 1,5-NAPHTHALENEDIAMINE IN ANIMALS.

REPORT 03 : IARC MONOGRAPH 27 1982
1,5-NAPHTHALENEDIAMINE WAS MUTAGENIC TO SALMONELLA STRAIN TA100 WITHOUT METABOLIC ACTIVATION. NO DATA WERE AVAILABLE CONCERNING EFFECTS ON REPRODUCTION OR PRENATAL TOXICITY.

REPT'D USE : DYE INTERMEDIATE AND ORGANIC SYNTHESIS. NO LISTING ON TSCA INVENTORY

CAS NO : **3173-72-6**
CHEM NAME : ***** 1,5-NAPHTHALENE DIISOCYANATE**

REPORT 01 : IARC MONOGRAPH 19 1979
NO DATA ON THE CARCINOGENICITY OR MUTAGENICITY OF 1,5-NAPHTHALENE DIISOCYANATE, WAS AVAILABLE.
REPT'D USE : NO EVIDENCE OF CURRENT PRODUCTION. RESEARCH CHEMICAL

CAS NO : **64741-52-2**
CHEM NAME : ***** NAPHTHENIC DISTILLATE**

REPORT 01 : EPA TSCA 8(E) NOTICE 8EHQ-0682-0447
SUN REFINING CO PROVIDED SUMMARIZED FINAL RESULTS, FROM A CHRONIC MOUSE SKIN-PAINTING STUDY OF LIGHT NAPHTHENIC DISTILLATE WHICH INDICATE THAT THE TESTED DISTILLATE INDUCED A STATISTICALLY SIGNIFICANT INCREASE IN BENIGN EPIDERMAL TUMORS WHEN COMPARED TO USP WHITE MINERAL OIL (CONTROL).

CAS NO : **130-15-4**
CHEM NAME : ***** 1,4-NAPHTHOQUINONE**

REPORT 01 : EPA RCRA HEALTH AND ENV. EFFECTS PROFILE OCT 80
THE MOST CONSISTENT FINDINGS REPORTED IN THE LITERATURE FOR HEALTH EFFECTS OF 1,4-NAPTHOQUINONE INVOLVE HEMATOLOGICAL CHANGES, IRRITANT AND ALLERGENIC ACTIVITY, AND INHIBITION OF BIOCHEMICAL OXIDATION PROCESSES. ONE STUDY FOUND 1.4-NQ TO BE ONCOGENIC.

REPORT 06 : MULTI-AGENCY CROSS CHECK
N HHS MAGUIRE
N OSHA CARC
N EPA CAG
Y EPA RCRA
N EPA CLEAN AIR
N EPA CLEAN WATER
REPT'D USE : CHEMICAL/DYE INTERMEDIATE AND POLYMERIZATION INHIBITOR

CAS NO : **134-32-7**
CHEM NAME : ***** 1-NAPHTHYLAMINE**

REPORT 02 : IARC MONOGRAPH 4 1974
NO CARCINOGENIC EFFECT WAS OBSERVED IN HAMSTERS FOLLOWING ORAL ADMINISTRATION. THE RESULTS OBTAINED AFTER ORAL AND SUBCUTANEOUS ADMINISTRATION TO MICE ARE INCONCLUSIVE. EXPERIMENTS IN DOGS DEMONSTRATE THAT 1-NAPHTHYL-

AMINE, IF CARCINOGENIC AT ALL, IS LESS SO TO THE BLADDER IN THIS SPECIES THAN IS THE 2-ISOMER. OCCUPATIONAL EXPOSURE TO COMMERCIAL 1-NAPHTHYLAMINE CONTAINING 4-10% 2-NAPTHYLAMINE IS STRONGLY ASSOCIATED WITH BLADDER CANCER. IT IS NOT POSSIBLE TO DECIDE IF 1-NAPHTHYLAMINE, FREE FROM THE 2-ISOMER, IS CARCINOGENIC TO MAN.

REPORT 06 : MULTI-AGENCY CROSS CHECK
N HHS MAGUIRE
R OSHA CARC
Y EPA CAG
Y EPA RCRA
N EPA CLEAN AIR
N EPA CLEAN WATER

REPT'D USE : CHEMICAL/DYE INTERMEDIATE, MFR OF HERBICIDES/ANTI-OXIDANTS AND PRINT TONERS

CAS NO : **91-59-8**
CHEM NAME : ***** 2-NAPHTHYLAMINE**

REPORT 01 : NIOSH CURRENT INTELLIGENCE BULLETIN 16 DEC 76
BOTH N-PHENYL-B-NAPHTHYLAMINE (A WIDELY USED RUBBER ANTIOXIDANT) AND 2-NITRONAPHTHALENE (A BY-PRODUCT OF A-NAPHTHYLAMINE) ARE METABOLIZED TO THE KNOWN HUMAN CARCINOGEN, B-NAPHTHYLAMINE.

REPORT 02 : IARC MONOGRAPH 4 1974
2-NAPHTHYLAMINE IS CARCINOGENIC IN THE MOUSE, HAMSTER, DOG, AND MONKEY. GIVEN ORALLY IT PRODUCED BLADDER CARCINOMAS IN THE DOG AND MONKEY, AND, AT HIGH DOSAGE LEVELS, IN THE HAMSTER. BY THIS ROUTE, IT HAS INCREASED THE INCIDENCE OF HEPATOMAS IN THE MOUSE. EPIDEMIOLOGICAL STUDIES HAVE SHOWN THAT OCCUPATIONAL EXPOSURE TO 2-NAPHTHYLAMINE, EITHER ALONE OR WHEN PRESENT AS AN IMPURITY IN OTHER COMPOUNDS, IS STRONGLY ASSOCIATED WITH THE OCCURRENCE OF BLADDER CANCER. THERE IS NO DOUBT THAT 2-NAPHTHYLAMINE IS A HUMAN BLADDER CARCINOGEN.

REPORT 03 : REMOVAL FROM TSCA INVENTORY (46FR30563) JUN 81
EPA, IN REVIEWING THE CHEMICAL SUBSTANCES INCLUDED ON THE TSCA INVENTORY, HAS CONCLUDED THAT 2-NAPHTHYLAMINE WAS INCORRECTLY REPORTED. EPA HAS DETERMINED THAT THERE IS NO ON GOING MANUFACTURE, IMPORTATION, OR PROCESSING OF THIS SUBSTANCE, AND THE CHEMICAL IS CONSIDERED TO BE NOT ON THE INVENTORY.

REPORT 06 : MULTI-AGENCY CROSS CHECK
Y HHS MAGUIRE
R OSHA CARC

Y EPA CAG
Y EPA RCRA
N EPA CLEAN AIR
N EPA CLEAN WATER

REPT'D USE : ANTIOXIDANT CHEMICAL, DYE INTERMEDIATE AND RESEARCH CHEMICAL. NO KNOWN CURRENT US PRODUCTION. REMOVED FROM TSCA INVENTORY

CAS NO : **1465-25-4**
CHEM NAME : ***** N-(-1-NAPHTHYL)ETHYLENEDIAMINE DIHYDROCHLORIDE**

REPORT 01 : NCI CARC TRS 168 DHEW PUB NO NIH 79-1724
UNDER THE CONDITIONS OF THIS BIOASSAY, DIETARY ADMINISTRATION OF N(-1-NAPHTHYL)ETHYLENEDIAMINE DIHYDROCHLORIDE WAS NOT CARCINOGENIC IN FISCHER 344 RATS OR B6C3F1 MICE.

REPORT 02 : NTP TECHNICAL BULLETIN 5 AUG 81
THE CHEMICAL WAS POSITIVE IN THE SALMONELLA TYPHIMURIUM MUTAGENICITY ASSAY IN ONE LABORATORY.

REPT'D USE : CHEMICAL INDICATOR

CAS NO : **86-88-4**
CHEM NAME : ***** 1-NAPHTHYLTHIOUREA**

REPORT 01 : IARC MONOGRAPH 30 JAN 83
1-NAPHTHYLTHIOUREA TESTED FOR CARCINOGENICITY IN MICE AND RATS BY ADMINISTRATION IN DIET. THE STUDIES WERE CONSIDERED TO BE INADEQUATE FOR EVALUATION.

REPORT 02 : IARC MONOGRAPH 30 JAN 83
1-NAPHTHYLTHIOUREA MUTAGENIC TO SALMONELLA TYPHIMURIUM IN PRESENCE OF EXOGENOUS METABOLIC ACTIVATION SYSTEM. INDUCED TRANSFORMED PHENOTYPE IN SYRIAN HAMSTER EMBRYO CELLS IN VITRO. LIMITED EVIDENCE THAT 1-NAPHTHYLTHIOUREA IS MUTAGENIC IN CELLULAR SYSTEMS. NO DATA AVAILABLE TO EVALUATE ITS MUTAGENICITY TO MAMMALS.

REPORT 04 : EPA WATER QUALITY CRITERIA - 51FR43665 DEC 86
EPA ANNOUNCES AVAILIBILITY OF "QUALTIY CRITERIA FOR WATER 1986".

REPORT 06 : MULTI-AGENCY CROSS CHECK
N HHS MAGUIRE
N OSHA CARC
N EPA CAG

N EPA RCRA
N EPA CLEANAIR
Y EPA CLEAN WATER
REPT'D USE : RODENTICIDE

CAS NO : **7440-02-0**
CHEM NAME : ***** NICKEL (INCL CERTAIN NI CPDS)**

REPORT 01 : EPA RCRA HEALTH AND ENV. EFFECTS PROFILE OCT 80
ALTHOUGH NICKEL IS TOXIC AND APPEARS TO BE A CARCINOGEN TO MAN, THERE IS AN INCREASINGLY STRONG INDICATION THAT NICKEL IS AN ESSENTIAL ELEMENT. THE ROUTE OF EXPOSURE TO NICKEL IS VERY IMPORTANT, SINCE ORAL INTAKE OF NICKEL METAL IS COMPARATIVELY NONTOXIC. HOWEVER, EXPOSURE TO NICKEL BY INHALATION OR PARENTERAL ADMINISTRATION AS WELL AS CUTANEOUS CONTACT CAN PRODUCE TOXIC EFFECTS.

REPORT 02 : IARC MONOGRAPH 11
INJECTION I.M. OF NICKEL POWDER, NICKEL SUBSULFIDE, NICKEL OXIDE OR NICKELOCENE INTO MICE OR RATS MAY RESULT IN THE APPEARANCE OF FIBROSARCOMAS AND/OR RHABDOMYOSARCOMAS. SINCE A VARIETY OF NICKEL COMPOUNDS PRODUCE LOCAL TUMORS, THE SUGGESTION IS THAT NICKEL IN SOME FORM IS THE ACTIVE AGENT. AN EXCESSIVE RISK OF CANCERS OF THE NASAL SINUS AND LUNG AMONG NICKEL REFINERY WORKERS, WHO INHALE NICKELCONTAINING DUSTS FROM CRUDE ORES, HAS BEEN FOUND. IT IS PROBABLE THAT NICKEL IN SOME FORM IS CARCINOGENIC.

REPORT 03 : NIOSH CRITERIA DOCUMENT MAY 77
NIOSH RECOMMENDS A STANDARD OF 15 UG/CU M TWA BASED ON THE CONCLUSION THAT INORGANIC NICKEL SUBSTANCES ARE CARCINOGENIC. THE PRESENT OSHA STANDARD IS 1 MG/CU M TWA.

REPORT 05 : EPA HEALTH ASSESSMENT DOC FOR SEP 86
NICKEL AND NI C PDS
NICKEL HAS POTENTIAL TO CAUSE REPRODUCTIVE AND DEVELOPMENTAL TOXICITY IN VARIOUS ANIMALS. NICKEL CARBONYL MOST ACUTELY TOXIC FORM. NICKEL CPDS HAVE CAUSED DERMATOLOGICAL PROBLEMS. CAUSES DIRECT AND INDIRECT RESPIRATORY EFFECTS IN HUMANS, REPRODUCTIVE AND DEVELOPMENTAL EFFECTS IN ANIMALS. NI MAY INDUCE CARCINOGENIC EFFECTS IN HUMANS AND MUTAGENIC EFFECTS IN ANIMALS.

REPORT 06 : MULTI-AGENCY CROSS CHECK
Y HHS MAGUIRE
N OSHA CARC
Y EPA CAG

N EPA RCRA
Y EPA CLEAN AIR
Y EPA CLEAN WATER

REPT'D USE : CATALYST, CERAMICS, DRAIN CLEANERS, SPARK PLUGS, COINS, BATTERIES, MAGNETS, ELECTRICAL GOODS, FABRICATED, METAL PRODUCTS AND SURGICAL/DENTAL INSTRUMENTS

CAS NO : **30947-30-9**
CHEM NAME : ***** NICKEL-BIS(ETHYL(3,5-DI-T-BUTYL-4-HYDROXY-BENZYL)PHOSPHONATE**

REPORT 01 : EPA TSCA 8(E) NOTICE 8EHQ-0482-0443
CIBA-GEIGY REPORTED RESULTS OF A TWO-YEAR INHALATION STUDY OF NICKEL BIS(ETHYL(3,5-DI-T-BUTYL-4-HYDROXYBENZYL)PHOSPHONATE IN RATS. TWO OF THE THREE MALES SURVIVING TO FINAL SACRIFICE EACH DEVELOPED A SQUAMOUS CELL CARCINOMA OF THE LUNG. NO MENTION OF NEOPLASTIC CHANGES WERE EVIDENT IN ANY FEMALE UP TO FINAL SACRIFICE.

CAS NO : **3333-67-3**
CHEM NAME : ***** NICKEL CARBONATE**

REPORT 01 : IARC MONOGRAPH 11 FEB 76
INTRAMUSCULAR INJECTION AND/OR IMPLANTATION OF NICKEL CARBONATE PRODUCED LOCAL RHABDOMYOSARCOMAS AND/OR FIBROSARCOMAS IN RATS.

REPORT 06 : MULTI-AGENCY CROSS CHECK
Y HHS MAGUIRE
N OSHA CARC
N EPA CAG
N EPA RCRA
N EPA CLEAN AIR
N EPA CLEAN WATER

REPT'D USE : ELECTRONIC FERRITES/THERMISTORS, MFR OF NICKEL CATALYSTS, CHEMICAL INTERMEDIATE AND LAB REAGENT

CAS NO : **13463-39-3**
CHEM NAME : ***** NICKEL CARBONYL**

REPORT O1 : IARC MONOGRAPH 11 1976
ADMINISTRATION OF NICKEL CARBONYL TO RATS BY REPEATED INTRAVENOUS INJECTION WAS ASSOCIATED WITH AN INCREASED INCIDENCE OF MALIGNANT TUMORS IN VARIOUS TISSUES. INHALATION EXPOSURE OF RATS TO NICKEL CARBONYL WAS ASSOCIATED WITH A FEW PULMONARY MALIGNANCIES.

REPORT 03 : NIOSH SPECIAL HAZARD REVIEW MAY 77
NIOSH RECOMMENDS A STANDARD OF 7 UG/CU M TWA BASED ON CARCINOGENICITY. THE PRESENT OSHA STANDARD IS 7 UG/CU M TWA.

REPORT 06 : MULTI-AGENCY CROSS CHECK
Y HHS MAGUIRE
N OSHA CARC
N EPA CAG
N EPA RCRA
N EPA CLEAN AIR
N EPA CLEAN WATER
REPT'D USE : INTERMEDIATE PRODUCT IN NICKEL REFINING AND CATALYST IN CHEMICAL REACTIONS

CAS NO : **1271-28-9**
CHEM NAME : ***** NICKELOCENE**

REPORT 01 : IARC MONOGRAPH 11 1976
INJECTION I.M. OF NICKEL POWDER, NICKEL SUBSULFIDE, NICKEL OXIDE OR NICKELOCENE INTO MICE OR RATS MAY RESULT IN THE APPEARANCE OF FIBROSARCOMAS AND/OR RHABDOMYOSARCOMAS. THE FACT THAT A VARIETY OF NICKEL COMPOUNDS PRODUCE LOCAL TUMORS SUGGESTS THAT NICKEL, IN SOME FORM, IS THE ACTIVE AGENT.

REPORT 02 : TECHNICAL BULLETIN 3 DEC 80
THE COMPOUND WAS FOUND TO BE NEGATIVE IN THE SALMONELLA TYPHIMURIUM MUTAGENICITY ASSAY IN ONE LABORATORY.

REPORT 06 : MULTI-AGENCY CROSS CHECK
Y HHS MAGUIRE
N OSHA CARC
N EPA CAG
Y EPA RCRA
N EPA CLEAN AIR
Y EPA CLEAN WATER
REPT'D USE : LABORATORY REAGENT AND CATALYST

CAS NO : **1313-99-1**
CHEM NAME : ***** NICKEL OXIDE**

REPORT 02 : IARC MONOGRAPH 11 FEB 76
INJECTION I.M. OF NICKEL POWDER, NICKEL SUBSULFIDE, NICKEL OXIDE OR NICKELOCENE INTO MICE OR RATS MAY RESULT IN THE

APPEARANCE OF FIBROSARCOMAS AND/OR RHABDOMYOSARCOMAS. THE FACT THAT A VARIETY OF NICKEL COMPOUNDS PRODUCE LOCAL TUMORS SUGGESTS THAT NICKEL IN SOME FORM IS THE ACTIVE AGENT.

REPORT 06 : MULTI-AGENCY CROSS CHECK
Y HHS MAGUIRE
N OSHA CARC
N EPA CAG
N EPA RCRA
N EPA CLEANAIR
N EPA CLEAN WATER
REPT'D USE : ELECTRONICS COMPONENTS, CERAMIC GLAZE AND GLASS

CAS NO : **UNKNOWN-06**
CHEM NAME : ***** NICKEL REFINING**

REPORT 01 : IARC SUPPLEMENT 1 1979
EPIDEMIOLOGICAL STUDIES HAVE DEMONSTRATED INCREASED INCIDENCES OF CANCER OF THE NASAL CAVITY LUNG, AND POSSIBLY LARYNX IN WORKERS IN NICKEL REFINERIES. IT IS NOT POSSIBLE, HOWEVER, TO STATE WITH CERTAINTY WHICH SPECIFIC NICKEL COMPOUND(S) IS CARCINOGENIC FOR HUMANS. THE IARC CONCLUDED THAT NICKEL REFINING IS CARCINOGENIC FOR HUMANS.

REPORT 06 : MULTI-AGENCY CROSS CHECK
Y HHS MAGUIRE
N OSHA CARC
N EPA CAG
N EPA RCRA
N EPA CLEAN AIR
N EPA CLEAN WATER

CAS NO : **12035-72-2**
CHEM NAME : ***** NICKEL SUBSULFIDE**

REPORT 01 : IARC MONOGRAPH 11 1976
NICKEL SUBSULFIDE IS CARCINOGENIC IN RATS FOLLOWING ITS INHALATION: IT PRODUCED MALIGNANT LUNG TUMORS. INTRAMUSCULAR INJECTION AND/OR IMPLANTATION OF NICKEL SUBSULFIDE, NICKEL POWDER, NICKELOCENE, OR NICKEL OXIDE, CARBONATE OR HYDROXIDE PRODUCED LOCAL RHABDOMYOSARCOMAS AND/OR FIBROSARCOMAS IN RATS; NICKEL SUBSULFIDE AND NICKEL OXIDE PRODUCED SIMILAR TUMORS IN MICE AND NICKEL POWDER AND NICKELOCENE, IN HAMSTERS. IN A 1979 SUPPLEMENT, IARC DETERMINED THAT THERE IS SUFFICIENT

EVIDENCE FOR THE CARCINOGENICITY OF THIS COMPOUND IN ANIMALS.

REPORT 06 : MULTI-AGENCY CROSS CHECK
Y HHS MAGUIRE
N OSHA CARC
Y EPA CAG
Y EPA RCRA
N EPA CLEAN AIR
Y EPA CLEAN WATER
REPT'D USE : NO AVAILABLE PRODUCTION OR USE DATA

CAS NO : **61-57-4**
CHEM NAME : ***** NIRIDAZOLE**

REPORT 01 : IARC MONOGRAPH 13 1977
NIRIDAZOLE IS CARCINOGENIC IN MICE AND HAMSTERS AFTER ITS ORAL ADMINISTRATION: IN MICE IT INDUCED LYMPHOMAS AND TUMORS OF THE LUNG, STOMACH, MAMMARY GLAND, OVARY AND BLADDER; IN HAMSTERS IT PRODUCED TUMORS OF THE FORE-STOMACH AND PAPILLOMAS OF THE URINARY BLADDER. IN A 1979 SUPPLEMENT, IARC DETERMINED THAT THERE IS SUFFICIENT EVIDENCE FOR THE THE CARCINOGENICITY OF NIRIDAZOLE IN ANIMALS.

REPORT 02 : IARC MONOGRAPH 13 1977
MUTAGENICITY WAS REPORTED FOR NIRIDAZOLE IN SALMONELLA TAl00 AND TA1538, WITHOUT METABOLIC ACTIVATION AND THE INDUCTION OF DUPLICATIONS IN E. COLI. THERE IS NO INDUCATION OF AN ABILITY TO CAUSE CHROMOSOME ABERRATIONS AFTER I.P. INJECTION IN BONE MARROW CELLS OF RATS. IN THE DOMINANT LETHAL TEST IN MICE A SIGNIFICANT DECREASE IN THE NUMBER OF PREGNANT ANIMALS AND TOTAL IMPLANTS, BUT NO DIFFERENCE IN POST-IMPLANTATION LOSSES WAS OBSERVED.
REPT'D USE : AMOEBACIDE

CAS NO : **139-94-6**
CHEM NAME : ***** NITHIAZIDE**

REPORT 01 : NCI CARC TRS 146 DHEW PUB NO NIH 79-1702
UNDER THE CONDITIONS OF THIS BIOASSAY, NITHIAZIDE WAS CARCINOGENIC IN MALE AND PROBABLY FEMALE B6C3F1 MICE, CAUSING A COMBINATION OF HEPATOCELLULAR CARCINOMAS AND HEPATOCELLULAR ADENOMAS. NITHIAZIDE WAS ALSO CARCINOGENIC IN FEMALE FISCHER 344 RATS, CAUSING AN INCREASE IN THE INCIDENCE OF MAMMARY NEOPLASMS. THE

COMPOUND WAS NOT CARCINOGENIC IN MALE FISCHER 344 RATS (FEEDING STUDY)

REPORT 02 : NTP TECHNICAL BULLETIN 6 JAN 82
THE COMPOUND WAS POSITIVE IN THE SALMONELLA TYPHIMURIUM MUTAGENICITY ASSAY IN ONE LABORATORY.

REPORT 03 : IARC MONOGRAPH 31 JUL 83
TESTED IN MICE AND RATS BY IN DIET. INCREASED INCIDENCE OF HEPATOCELLULAR CARCINOMAS AND ADENOMAS IN MALE MICE. IN FEMALE RATS, INCREASED INCIDENCES OF FIBROADENOMAS AND CYSTADENOMAS OF SKIN, SUBCUTANEOUS TISSUE AND MAMMARY GLAND (SIGNIFICANT ONLY IF INDIVIDUAL INCIDENCES FOR EACH SITE COMBINED) AND INCIDENCE OF ENDOMETRIAL STROMAL POLYPS OF UTERUS.
REPT'D USE : ANTIPROTOZOAL

CAS NO : **139-13-9**
CHEM NAME : ***** NITRILOTRIACETIC ACID (NTA)**

REPORT 01 : NCI CARC TRS 6 DHEW PUB NO NIH 77-806
NTA WAS SHOWN TO BE CARCINOGENIC TO THE URINARY TRACTS OF BOTH RATS AND MICE AT THE HIGHER DOSES TESTED. LOWER DOSES, AS DELINEATED IN THIS REPORT, DID NOT INDUCE SIGNIFICANT NUMBERS OF SUCH LESIONS (FEEDING STUDY).

REPORT 02 : NTP COMMITTEE REPORT OF CARCINOGENESIS JAN 80
THE SUBCOMMITTEE CONCLUDES FROM ITS REVIEW THAT THERE IS NO EVIDENCE THAT NTA IS A MUTAGEN AND THAT NTA IS A RENAL URINARY TRACT CARCINOGEN IN RODENTS. THE MECHANISM OF NTA CARCINOGENESIS IS UNKNOWN AND THE SUBCOMMITTEE CANNOT ESTABLISH WHAT LEVEL OF EXPOSURE CONSTITUTES A THRESHOLD, IF THERE IS A THRESHOLD. USING A LINEAR MODEL, THE MAXIMUM URINARY TRACT TUMOR RISK IN RODENTS AT THE AVERAGE DRINKING WATER EXPOSURE LEVEL (0.07 UG/KG/DAY) IS ESTIMATED TO BE LESS THAN 2X10 -7.

REPORT 03 : INTERNATIONAL JOINT COMM. FINAL REPORT ON NTA DEC 78
THE TASK FORCE CONCLUDES THAT THE EXTENSIVE LITERATURE AND CANADIAN EXPERIENCE INDICATES THAT THE USE OF NTA DOES NOT CONSTITUTE AN OBVIOUS ENVIRONMENTAL HAZARD AND RECOMMENDS THAT NTA SHOULD NOT BE PROHIBITED FROM USE AS A DETERGENT BUILDER. THE TASK FORCE FURTHER RECOMMENDS THAT ITS USE BE ACCOMPANIED BY STUDIES TO ADDRESS SEVERAL REMAINING SCIENTIFIC QUESTIONS.

REPORT 04 : EPA DENIAL OF PETITION TO REGULATE (45FR72778) NOV 80
EPA FINDS THAT PROJECTED LEVELS OF EXPOSURE FROM USE OF NTA IN LAUNDRY DETERGENTS ARE GENERALLY LOW AND THERE-

FORE THAT ASSOCIATED HUMAN RISKS WOULD BE LOW. BASED ON THESE FINDINGS AND THE NEED TO DEAL WITH HIGHER PRIORITY PROBLEMS, EPA DENIES THE PETITION. EPA RECOMMENDS THAT NTA NOT BE USED IN CONSUMER PRODUCTS TO WHICH THERE IS DIRECT DERMAL OR ORAL EXPOSURE.

REPORT 06 : MULTI-AGENCY CROSS CHECK
Y HHS MAGUIRE
N OSHA CARC
N EPA CAG
N EPA RCRA
N EPA CLEAN AIR
N EPA CLEAN WATER
REPT'D USE : CHELATING AGENT, DETERGENTS, CLEANING AGENTS AND WATER TREATMENT CHEMICAL

CAS NO : **18662-53-8**
CHEM NAME : ***** NITRILOTRIACETIC ACID, TRISODIUM SALT, MONOHYDRATE**

REPORT 01 : NCI CARC TRS 6 DHEW PUB NO NIH 77-806
NA3NTA.H20 WAS SHOWN TO BE CARCINOGENIC TO THE URINARY TRACT OF BOTH RATS AND MICE AT THE HIGHER DOSES TESTED. LOWER DOSES, AS DELINEATED IN THIS REPORT, DID NOT INDUCE SIGNIFICANT NUMBERS OF SUCH LESIONS (FEEDING STUDY).

REPORT 02 : NTP COMMITTEE REPORT OF CARCINOGENESIS JAN 80
THE SUBCOMMITTEE CONCLUDES FROM ITS REVIEW THAT THERE IS NO EVIDENCE THAT NTA IS A MUTAGEN AND THAT NTA IS A RENAL URINARY TRACT CARCINOGEN IN RODENTS. THE MECHANISM OF NTA CARCINOGENESIS IS UNKNOWN AND THE SUBCOMMITTEE CANNOT ESTABLISH WHAT LEVEL OF EXPOSURE CONSTITUTES A THRESHOLD, IF THERE IS A THRESHOLD. USING A LINEAR MODEL, THE MAXIMUM URINARY TRACT TUMOR RISK IN RODENTS AT THE AVERAGE DRINKING WATER EXPOSURE LEVEL (0.07 UG/KG/DAY) IS ESTIMATED TO BE LESS THAN 2X10 -7.

REPORT 03 : INTERNATIONAL JOINT COMM. FINAL REPORT ON NTA DEC 78
THE TASK FORCE CONCLUDES THAT THE EXTENSIVE LITERATURE AND CANADIAN EXPERIENCE INDICATES THAT THE USE OF NTA DOES NOT CONSTITUTE AN OBVIOUS ENVIRONMENTAL HAZARD AND RECOMMENDS THAT NTA SHOULD NOT BE PROHIBITED FROM USE AS A DETERGENT BUILDER. THE TASK FORCE FURTHER RECOMMENDS THAT ITS USE BE ACCOMPANIED BY STUDIES TO ADDRESS SEVERAL REMAINING SCIENTIFIC QUESTIONS.

REPORT 04 : EPA DENIAL OF PETITION TO REGULATE (45FR2778) NOV 80
EPA FINDS THAT PROJECTED LEVELS OF EXPOSURE FROM USE OF NTA IN LAUNDRY DETERGENTS ARE GENERALLY LOW AND THERE-

FORE THAT ASSOCIATED RISKS WOULD BE LOW. BASED ON THESE FINDINGS AND THE NEED TO DEAL WITH HIGHER PRIORITY PROBLEMS, EPA DENIES THE PETITION. EPA RECOMMENDS NTA NOT BE USED IN CONSUMER PRODUCTS TO WHICH THERE IS DIRECT DERMAL OR ORAL EXPOSURE.

REPT'D USE : DETERGENTS

CAS NO : **602-87-9**

CHEM NAME : ***** 5-NITROACENAPHTHENE**

REPORT 01 : NCI CARC TRS 118 DHEW PUB NO NIH 78-1373
UNDER THE CONDITIONS OF THIS BIOASSAY, 5-NITROACENAPHTHENE WAS CARCINOGENIC TO FISCHER 344 RATS, CAUSING INCREASED INCIDENCES OF MALIGNANT TUMORS OF THE EAR CANAL AND LUNG IN BOTH SEXES, AND OF THE CLITORAL GLAND AND MAMMARY GLAND IN FEMALES. 5-NITROACENAPHTHENE WAS ALSO CARCINOGENIC TO FEMALE BUT NOT MALE B6C3F1 MICE, CAUSING CARCINOMAS OF THE LIVER AND OVARIAN TUMORS (FEEDING STUDY).

REPORT 02 : IARC MONOGRAPH 16 1978
5-NITROACENAPHTHENE IS CARCINOGENIC IN FEMALE RATS AND FEMALE HAMSTERS FOLLOWING ITS ORAL ADMINISTRATION; IT PRODUCED MAINLY ADENOCARCINOMAS OF THE SMALL INTESTINE AND MAMMARY CARCINOMAS IN FEMALE BUT NOT IN MALE RATS AND CHOLANGIOMAS IN FEMALE BUT NOT IN MALE HAMSTERS. IN ADDITION, IT PRODUCED LEUKEMIA AND RETICULUM CELL SARCOMAS IN MICE FOLLOWING ITS INTRAPERITONEAL INJECTION. IN A 1979 SUPPLEMENT, IARC DETERMINED THAT THERE IS SUFFICIENT EVIDENCE FOR THE CARCINOGENICITY OF 5-NITROACENAPHTHENE IN ANIMALS.

REPORT 03 : NTP TECHNICAL BULLETIN 6 JAN 82
THE COMPOUND WAS POSITIVE IN THE SALMONELLA TYPHIMURIUM MUTAGENICITY ASSAY IN ONE LABORATORY.

EPORT 04 : IARC MONOGRAPH 16 1978
REVERSE MUTATIONS WERE NOT INCREASED IN E. COLI BY 5-NITROACENAPHTHENE. THE CHEMICAL INDUCED REVERSE MUTATIONS IN SALMONELLA TA98 AND TA100. NO DATA ON THE EMBRYOTOXICITY OR TERATOGENICITY WERE AVAILABLE.

REPORT 06 : MULTI-AGENCY CROSS CHECK
N HHS MAGUIRE
Y OSHA CARC
N EPA CAG
N EPA RCRA
N EPA CLEAN AIR
N EPA CLEAN WATER

REPT'D USE : CHEMICAL INTERMEDIATE FOR FLUORESCENT WHITENING AGENTS. NO PRODUCTION OR USE DATA AVAILABLE

CAS NO : **1777-84-0**
CHEM NAME : ***** 5-NITRO-P-ACETOPHENETIDE**

REPORT 01 : NCI CARC TRS 133 DHEW PUB NO 79-1388
UNDER THE CONDITIONS OF THIS BIOASSAY, DIETARY ADMINISTRATION OF 5-NITRO-P-ACETOPHENETIDE WAS NOT CARCINOGENIC IN FISCHER 344 RATS OF EITHER SEX OR IN FEMALE B6C3F1 MICE. THE COMPOUND, HOWEVER, WAS CONSIDERED CARCINOGENIC IN MALE B6C3F1 MICE, BASED ON A SIGNIFICANT INCREASE IN THE COMBINED INCIDENCE OF HEPATOCELLULAR CARCINOMAS AND HEPATOCELLULAR ADENOMAS IN THESE ANIMALS.

REPORT 02 : NTP TECHNICAL BULLETIN 8 JUL 82
5-NITRO-P-ACETOPHENETIDE WAS NEGATIVE IN DROSOPHILA WHEN TESTED FOR SEX-LINKED RECESSIVE LETHALS.

REPORT 06 : MULTI-AGENCY CROSS CHECK
N HHS MAGUIRE
Y OSHA CARC
N EPA CAG
N EPA RCRA
N EPA CLEAN AIR
N EPA CLEAN WATER
REPT'D USE : ANALGESIC

CAS NO : **100-01-6**
CHEM NAME : ***** P-NITROANILINE**

REPORT 01 : NTP TECHNICAL BULLETIN NO. 7 APR 82
THE CHEMICAL WAS FOUND TO BE POSITIVE IN THE SALMONELLA TYPHIUMRIUM MUTAGENICITY ASSAY.
REPT'D USE : CHEMICAL INTERMEDIATE AND DYE INTERMEDIATE

CAS NO : **99-59-2**
CHEM NAME : ***** 5-NITRO-O-ANISIDINE**

REPORT 01 : NCI CARC TRS 127 DHEW PUB NO NIH 78-1382
UNDER THE CONDITIONS OF THIS BIOASSAY, DIETARY ADMINISTRATION OF 5-NITRO-O-ANISIDINE WAS CARCINOGENIC IN F344 RATS, CAUSING TUMORS OF THE INTEGUMENTARY SYSTEM IN MALES AND FEMALES AND OF THE CLITORAL GLAND IN FEMALES. THE COMPOUND WAS ALSO CARCINOGENIC TO FEMALE B6C3F1 MICE, CAUSING HEPATOCELLULAR CARCINOMAS.

REPORT 02 : IARC MONOGRAPH 27 1982
5-NITRO-O-ANISIDINE WAS TESTED FOR CARCINOGENICITY IN RATS AND MICE BY DIETARY ADMINISTRATION. IT PRODUCED SKIN CARCINOMAS IN MALE RATS, CARCINOMAS OF THE ZYMBAL GLAND IN MALE AND FEMALE RATS, MAMMARY ADENOCARCINOMAS IN FEMALE RATS AND ADENOMAS AND CARCINOMAS OF THE PREPUTIAL AND CLITORAL GLANDS IN MALE AND FEMALE RATS RESPECTIVELY. THE EXPERIMENT IN MICE WAS INADEQUATE FOR EVALUATION.

REPORT 03 : IARC MONOGRAPH 27 1982
5-NITRO-O-ANISIDINE INDUCED REVERSE MUTATIONS IN SALMONELLA STRAIN TA98 IN THE ABSENCE OF METABOLIC ACTIVATION. NO DATA WERE AVAILABLE CONCERNING EFFECTS ON REPRODUCTION OR PRENATAL TOXICITY.

REPORT 04 : NTP TECHNICAL BULLETIN 8 JUL 82
5-NITRO-O-ANISIDINE WAS NEGATIVE IN DROSOPHILA WHEN TESTED FOR SEXLINKED RECESSIVE LETHALS.

REPORT 05 : 5TH ANNUAL REPORT ON CARCINOGENS 1989
(MAGUIRE REPT)
5-NA CARCINOGENIC IN RATS WHEN GIVEN IN DIET, CAUSING SKIN TUMORS AND IN THE FEMALES, TUMORS OF THE CLITORAL GLAND. IN FEMALE MICE, IT CAUSED LIVER CANCER. THIS WAS CONSIDERED LIMITED EVIDENCE OF CARCINOGENICITY BY IARC. IN VIEW OF AN NCI/OTA INTERPRETATION, THIS MAY BE CONSIDERED AS SUFFICIENT EVIDENCE FOR CARCINOGENICITY IN EXPERIMENTAL ANIMALS. HOWEVER, DROPPED FROM ANNUAL REPORT, BECAUSE OF INSUFFICIENT EVIDENCE OF CARCINOGENICITY.

REPORT 06 : MULTI-AGENCY CROSS CHECK
Y HHS MAGUIRE
Y OSHA CARC
N EPA CAG
N EPA RCRA
N EPA CLEAN AIR
N EPA CLEAN WATER

REPT'D USE : INKS, PAINTS/VARNISHES, CHEMICAL INTERMEDIATE FOR PIGMENTS/AZOIC DYE COMPONENTS, PAPER/TEXTILES AND RUBBERS/PLASTICS

CAS NO : **91-23-6**
CHEM NAME : ***** O-NITROANISOLE**

REPORT 01 : NTP TECHNICAL BULLETIN NO. 7 APR 82
THE CHEMICAL WAS POSITIVE IN THE SALMONELLA TYPHIMURIUM MUTAGENICITY ASSAY.

REPT'D USE : MFR OF INTERMEDIATES FOR DYES AND PHARMACEUTICALS

CAS NO : **602-60-8**
CHEM NAME : **9-NITROANTHRACENE**

REPORT 01 : IARC MONOGRAPH 33 APR 84
9-NITROANTHRACENE MUTAGENIC TO S. TYPHIMURIUM STRAINS TA98 AND TA10 BOTH IN THE PRESENCE AND ABSENCE OF AN ARCLOR-INDUCED RAT-LIVER SUPERNATANT. AN INCREASE IN MUTAGENIC ACTIVITY WAS OBSERVED WHEN INCUBATION WAS CARRIED OUT UNDER ANAEROBIC CONDITIONS. NO DATA WERE AVAILABLE TO EVALUATE THE CARCINOGENICITY OF 9-NITROANTHRACENE TO HUMANS OR EXPERIMENTAL ANIMALS.
REPT'D USE : FOUND IN DIESEL EXHAUST, URBAN AIR AND ONE CARBON BLOCK (NOT PRODUCED) RESEARCH CHEMICAL

CAS NO : **619-17-0**
CHEM NAME : ***** 4-NITROANTHRANILIC ACID**

REPORT 01 : NCI CARC TRS 109 DHEW PUB NO NIH 78-1364
UNDER THE CONDITIONS OF THIS BIOASSAY EVIDENCE WAS NOT PROVIDED FOR THE CARCINOGENICITY OF 4-NITROANTHRANILIC ACID IN FISCHER 344 RATS OR B6C3F1 MICE (FEEDING STUDY).

REPORT 02 : NTP TECHNICAL BULLETIN 7 APR 82
4-NITROANTHRANILIC ACID WAS FOUND TO BE POSITIVE IN THE SALMONELLA TYPHIMURIUM MUTAGENICITY ASSAY.
REPT'D USE : DYE INTERMEDIATE. LABORATORY AGENT IS ONLY KNOWN CURRENT USE

CAS NO : **98-95-3**
CHEM NAME : ***** NITROBENZENE**

REPORT 01 : NTP TEST REGIMEN
THE COMPOUND WAS FOUND TO BE NEGATIVE IN THE SALMONELLA ASSAY.

REPORT 02 : EPA RCRA HEALTH AND ENV. EFFECTS PROFILE OCT 80
THERE IS LITTLE OR NO INFORMATION AVAILABLE ON THE TERATOGENIC MUTAGENIC OR CARCINOGENIC EFFECTS OF NITROBENZENE. IT HAS YIELDED NEGATIVE RESULTS IN THE AMES ASSAY FOR MUTAGENICITY. GROSS ABNORMALITIES WERE OBSERVED IN 4 FETUSES OF 30 RATS ADMINISTERED NITROBENZENE.

REPORT 03 : NTP CARC TEST PROGRAM JUL 82
IN A CURRENT NTP BIOASSAY, NITROBENZENE IS BEING ADMINISTERED BY GAVAGE TO F344 RATS AND B6C3F1 MICE. IN A SECOND BIOASSAY, NITROBENZENE IS BEING APPLIED BY SKIN PAINTING TO

F344 RATS AND B6C3F1 MICE. BOTH STUDIES ARE IN THE PRECHRONIC TEST PHASE.

REPORT 04 : EPA RESPONSE TO ITC RECOMMS (46FR30300) JUN 81
IN RESPONSE TO THE ITC RECOMMENDATIONS (42FR55026) THE EPA PROPOSES THAT MANUFACTURERS AND PROCESSORS OF NITROBENZENE CONDUCT TESTING FOR TERATOGENICITY/REPRODUCTIVE EFFECTS. THE NTP IS CONDUCTING ONCOGENICITY TESTING. THE EPA WILL CONDUCT MUTAGENICITY TESTS AND DECIDE ON FURTHER TIER TESTING AS APPROPRIATE. NO STUDY POPULATION FOR AN EPIDEMIOLOGIC STUDY COULD BE IDENTIFIED.

REPORT 05 : EPA RESPONSE TO ITC (49FR25013) JUN 84
IN RESPONSE TO OCTOBER 1977 ITC RECOMMENDATION EPA DECIDED NOT TO PROMULGATE A RULE TO REQUIRE TESTING BECAUSE ONGOING TESTS AND EXISTING INFORMATION ARE EXPECTED TO PROVIDE DATA ON HUMAN HEALTH EFFECTS AND DETERMINE ANY NEED FOR FURTHER TESTING. EPA ALSO CONCLUDED THE RELEASE OF NITROBENZENE TO THE ENVIRONMENT ARE NOT SUBSTANTIAL AND CURRENT DATA DOES NOT SUPPORT A NEED TO TEST FOR ENVIRONMENTAL EFFECTS.

REPORT 06 : MULTI-AGENCY CROSS CHECK
N HHS MAGUIRE
N OSHA CARC
N EPA CAG
Y EPA RCRA
Y EPA CLEAN AIR
Y EPA CLEAN WATER

REPORT 07 : TSCA SECT 8(E) STATUS REPORT 8EHQ-0487-0670 S MAY 87
SUBMITTER STATED THERE WAS A SIGNIFICANT INCREASE IN CORNEAL OPACITY AT ALL SUBSTITUTED NITROBENZENE DOSE LEVELS IN A FEEDING STUDY USING SPRAGUE-DAWLEY RATS. SUBMITTER STATED THAT "WHILE CORNEAL OPACITY IS UNUSUAL, IT IS NOT UNIQUE TO THIS (SUBSTITUTED NITROBENZENE) SUBSTANCE, BUT HAS BEEN REPORTED TO OCCUR FROM THE ADMINISTRATION OF SEVERAL DRUGS, INCLUDING INDOMETHACIN, CHLOROQUINE, AMIODARONE AND OTHERS."

REPT'D USE : SOLVENT, CHEMICAL INTERMEDIATE AND AEROSOL

CAS NO : **94-52-0**

CHEM NAME : ***** 6-NITROBENZIMIDAZOLE**

REPORT 01 : NCI CARC TRS 117 DHEW PUB NO 79-1372
UNDER THE CONDITIONS OF THIS BIOASSAY, DIETARY ADMINISTRATION OF 6-NITROBENZIMIDAZOLE WAS NOT CARCINOGENIC TO

FISCHER 344 RATS; HOWEVER, THE COMPOUND WAS CARCINOGENIC TO B6C3F1 MICE, CAUSING HEPATOCELLULAR CARCINOMAS IN BOTH SEXES.

REPORT 02 : NTP TECHNICAL BULLETIN 6 JAN 82
THE COMPOUND WAS POSITIVE IN THE SALMONELLA TYPHIMURIUM MUTAGENICITY ASSAY IN ONE LABORATORY.

REPORT 06 : MULTI-AGENCY CROSS CHECK
N HHS MAGUIRE
Y OSHA CARC
N EPA CAG
N EPA RCRA
N EPA CLEAN AIR
N EPA CLEAN WATER

REPT'D USE : ANTIFOGGING AGENT IN COMMERCIAL PHOTOGRAPHY

CAS NO : **63041-90-7**
CHEM NAME : ***** 6-NITROBENZO(A)PYRENE**

REPORT 01 : IARC MONOGRAPH 33 APR 84
AVAILABLE DATA INADEQUATE TO PERMIT EVALUATION OF CARCINOGENICITY OF 6-NITROBENZO(A)PYRENE TO EXPERIMENTAL ANIMALS. NO DATA ON HUMANS AVAILABLE. TESTED FOR INITIATING ACTIVITY AT ONLY ONE DOSE LEVEL IN MOUSE-SKIN INITIATION-PROMOTION EXPERIMENT. INCREASED INCIDENCE OF SKIN TUMORS NOT STATISTICALLY SIGNIFICANT. MUTAGENIC TO SALMONELLA TYPHIMURIUM IN PRESENCE OF EXOGENOUS METABOLIC SYSTEM. INDUCED MORPHOLOGICAL TRANSFORMATION OF SYRIAN HAMSTER EMBRYO CELLS; TRANSFORMED HUMAN FORESKIN FIBROBLASTS WHEN TESTED ANAEROBICALLY.

REPT'D USE : FOUND IN DIESEL ENGINE EXHAUST AND URBAN AIR. NOT PRODUCED COMMERCIALLY

CAS NO : **122-04-3**
CHEM NAME : ***** P-NITROBENZOYL CHLORIDE**

REPORT 01 : NTP TECHNICAL BULLETIN 9 APR 83
P-NITROBENZOYL CHLORIDE WAS POSITIVE IN ONE SALMONELLA TYPHIMURIUM MUTAGENICITY ASSAY.

CAS NO : **100-14-1**
CHEM NAME : ***** P-NITROBENZYL CHLORIDE**

REPORT O1 : NTP TECHNICAL BULLETIN NO. 7 APR 82
THE CHEMICAL WAS FOUND TO BE POSITIVE IN THE SALMONELLA TYPHIMURIUM MUTAGENICITY ASSAY.

REPT'D USE : NO KNOWN CURRENT PRODUCTION

CAS NO : **92-93-3**
CHEM NAME : ***** 4-NITROBIPHENYL**

REPORT 01 : IARC MONOGRAPH 4 1974
4-NITROBIPHENYL INDUCED CARCINOMAS OF THE BLADDER WHEN GIVEN ORALLY TO DOGS, THE ONLY SPECIES AND ROUTE KNOWN TO HAVE BEEN TESTED. THERE ARE NO DATA ON THE CARCINOGENICITY OF 4-NITROBIPHENYL IN MAN. HOWEVER, IT HAS BEEN USED IN THE PRODUCTION OF 4-AMINOBIPHENYL, WHICH IS A RECOGNIZED HUMAN BLADDER CARCINOGEN.
REPT'D USE : INTERMEDIATE FOR RUBBER ANTIOXIDANT. NO LISTING ON TSCA INVENTORY

CAS NO : **13414-55-6**
CHEM NAME : ***** 7-NITRO-2,2-DIMETHYL-2,3-DIHYDROBENZOFURAN**

REPORT 01 : EPA'S RESPONSE TO ITC - 50FR29 761
EPA'S RESPONSE TO ITC'S RECOMMENDATION FOR PRIORITY CONSIDERATION FOR CHEMICAL FATE AND ENVIRONMENTAL IMPACT TESTING UNDER TSCA 4(A). EPA DOES NOT BELIEVE SUBSTANCE PRESENTS AN UNREASOABLE RISK OF INJURY TO THE ENVIRONMENT OR FOR SUBSTANTIAL RELEASE INTO THE ENVIRONMENT.
REPT'D USE : INTERMEDIATE IN PRODUCTION OF INSECTICIDE-NEMATICIDE

CAS NO : **1836-75-5**
CHEM NAME : ***** NITROFEN**

REPORT 01 : NCI CARC TRS 184 DHEW PUB NO NIH 79-1740
UNDER THE CONDITIONS OF THIS BIOASSAY, DIETARY ADMINISTRATION OF NITROFEN WAS CARCINOGENIC TO B6C3F1 MICE, CAUSING HEPATOCELLULAR CARCINOMAS IN BOTH SEXES. THERE WAS NO EVIDENCE FOR CARCINOGENICITY IN FISCHER 344 RATS.

REPORT 02 : NCI CARC TRS 26 DHEW PUB NO NIH 78-826
UNDER THE CONDITIONS OF THIS BIOASSAY, ORALLY ADMINISTERED TECHNICAL GRADE NITROFEN IS A LIVER CARCINOGEN IN B6C3F1 MICE OF BOTH SEXES. TOK IS ALSO CARCINOGENIC TO FEMALE OSBORNE-MENDEL RATS.

REPORT 03 : IARC MONOGRAPH 30 JAN 83
NITROFEN (TECHNICAL GRADE) TESTED FOR CARCINOGENICITY IN MICE AND RATS BY ADMINISTRATION IN DIET. IN MICE STUDIES IT SIGNIFICANTLY INCREASED INCIDENCE OF HEPATOCELLULAR CARCINOMAS; IN ONE, HAEMANGIOSARCOMAS AT VARIOUS

ANATOMICAL SITES OCCURRED IN SIGNIFICANTLY HIGHER INCIDENCE AMONG MALE MICE GIVEN HIGHER DOSE. IN RATS OF ONE STRAIN, ADENOCARCINOMAS OF PANCREAS PRODUCED IN FEMALES. NITROFEN SHOUD BE REGARDED, FOR PRACTICAL PURPOSES, AS IF IT PRESENTED A CARCINOGENIC RISK TO HUMANS.

REPORT 04 : IARC MONOGRAPH 30 JAN 83
NITROFEN TERATOGENIC TO MICE AND RATS WHEN GIVEN AT VERY HIGH DOSES; DECREASED NEONATAL AND POSTNATAL VIABILITY IN THREE-GENERATION REPRODUCTION STUDIES IN RATS.

REPORT 05 : IARC MONOGRAPH 30 JAN 83
NITROFEN MUTAGENIC TO SALMONELLA TYPHIMURIUM. IN TWO STUDIES, THERE WAS NO EVIDENCE FOR CHROMOSOMAL DAMAGE IN MAMMALS. THE DATA ARE INSUFFICIENT TO EVALUATE MUTAGENICITY OF NITROFEN IN MAMMALS.

REPORT 06 : MULTI-AGENCY CROSS CHECK
Y HHS MAGUIRE
Y OSHA CARC
N EPA CAG
N EPA RCRA
N EPA CLEAN AIR
N EPA CLEAN WATER
REPT'D USE : PESTICIDE AND HERBICIDE

CAS NO : **892-21-7**
CHEM NAME : ***** 3-NITROFLUORANTHENE**

REPORT 01 : IARC MONOGRAPH 33 APR 84
MALE RATS TESTED BY SUBCUTANEOUS INJECTION OF 3-NITROFLUORANTHENE RESULTED IN SARCOMAS AT INJECTION SITE. 3-NITROFLUORANTHERE WAS MUTAGENIC TO SALMONELLA TYPHIMURIUM IN PRESENCE AND ABSENCE OF EXOGENOUS METABOLIC SYSTEM. ALSO INDUCED MORPHOLOGICAL TRANSFORMATION IN SYRIAN HAMSTER EMBRYO CELLS.
REPT'D USE : DIESEL-EXHAUST AND AIRBORNE PARTICULATE. NOT PRODUCED OR USED COMMERCIALLY

CAS NO : **607-57-8**
CHEM NAME : ***** 2-NITROFLUORENE**

REPORT 06 : MULTI-AGENCY CROSS CHECK
N HHS MAGUIRE
Y OSHA CARC
N EPA CAG

N EPA RCRA
N EPA CLEAN AIR
N EPA CLEAN WATER
REPT'D USE : NO AVAILABLE PRODUCTION OR USE DATA

CAS NO : **67-20-9**
CHEM NAME : ***** NITROFURANTOIN**

REPORT 01 : NTP TECHNICAL BULLETIN 3/8 DEC 80, JUL 82
NITROFURANTOIN WAS POSITIVE IN THE SALMONELLA ASSAY, BUT NEGATIVE IN DROSOPHILA WHEN TESTED FOR SEX-LINKED RECESSIVE LETHALS.

REPORT 03 : NTP TR 341 NIH PUB 89-2597 SEP 89
UNDER 2-YR CONDITIONS 2-YEAR FEED STUDIES SOME EVIDENCE OF CARCINOGENIC ACTIVITY OF NITROFURANTOIN FOR MALE F344/N RATS (KIDNEY TUBULAR CELL NEOPLASMS). NO EVIDENCE OF CARCINOGENIC ACTIVITY FOR FEMALE F344/N RATS AND B6C3F1 MICE. THERE WAS CLEAR EVIDENCE OF CARCINGENIC ACTIVITY OF NITROFURANTOIN FOR FEMALE B6C3F1 MICE (TUBLAR ADENOMAS, BENIGN MIXED TUMORS, AND GRANULOSA CELL TUMORS OF THE OVARY).
REPT'D USE : BACTERIOCIDE, MEDICINE, FOOD PRESERVATIVE AND FEED ADDITIVE

CAS NO : **59-87-0**
CHEM NAME : ***** NITROFURAZONE**

REPORT 01 : IARC MONOGRAPH 7 1974
NITROFURAZONE WAS TESTED ONLY BY THE ORAL ROUTE IN RATS WHERE IT INCREASED THE INCIDENCE OF BENIGN MAMMARY TUMORS. THE AVAILABLE EVIDENCE IS INSUFFICIENT TO EVALUATE THE CARCINOGENICITY OF THIS COMPOUND.

REPORT 02 : NTP TECHNICAL BULLETIN 6 JAN 82
THE COMPOUND WAS POSITIVE IN THE SALMONELLA TYPHIMURIUM MUTAGENICITY ASSAY IN ONE LABORATORY.

REPORT 03 : NTP TR 337 NIH 88-2593 JUN 88
TWO YEAR FEED STUDIES SHOWED EQUIVOCAL EVIDENCE OF CARCINOGENIC ACTIVITY FOR MALE F344/N RATS SHOWN BY OCCURENCE OF SEBACEOUS GLAND ADENOMAS AND TRICHOEPITHELIOMAS OF THE SKIN, MESOTHELIOMAS OF TUNICA VAGINALIS, AND PREPUTIAL GLAND TUMORS. CLEAR EVIDENCE OF CARCINOGENIC ACTIVITY FOR FEMAL F344/N RATS SHOWN BY MARKEDLY INCREASED INCIDENCE OF FIBROADENOMAS OF MAMMARY GLAND. NO EVIDENCE OF CARCINOGENIC ACTIVITY FOR MALE B6C3F1 MICE FED DIETS CONTAINING NITROFURAZONE AT 150 OR

310 PPM. CLEAR EVIDENCE OF CARCINOGENIC ACTIVITY FOR FEMALE B6C3F1 MICE SHOWN BY INCREASED INCIDENCES OF BENIGN MIXED TUMORS AND GRANULOSA CELL TUMORS OF OVARY.

REPORT 06 : MULTI-AGENCY CROSS CHECK
N HHS MAGUIRE
Y OSHA CARC
N EPA CAG
N EPA RCRA
N EPA CLEAN AIR
N EPA CLEAN WATER
REPT'D USE : TOPICAL ANTI-INFECTIVE, VETERINARY AND HUMAN MEDICINE AND ANTIMICROBIAL AGENT

CAS NO : **555-84-0**
CHEM NAME : ***** 1-((5-NITROFURFURYLIDENE)AMINO)-2-IMIDAZOLIDINONE**

REPORT 01 : IARC MONOGRAPH 7 1974
THIS COMPOUND IS CARCINOGENIC IN RATS FOLLOWING ORAL ADMINISTRATION, THE ONLY SPECIES AND ROUTE TESTED. IT PRODUCED MAMMARY CARCINOMAS AND LYMPHOBLASTIC LYMPHOMAS. IN A 1979 SUPPLEMENT, IARC DETERMINED THAT THERE WAS SUFFICIENT EVIDENCE FOR THE CARCINOGENICITY OF THIS COMPOUND IN ANIMALS.
REPT'D USE : ANTIMICROBIAL AGENT

CAS NO : **531-82-8**
CHEM NAME : ***** N-(4-(5-NITRO-2-FURYL)-2-THIAZOLYL) ACETAMIDE**

REPORT 01 : IARC MONOGRAPH 7 1974
NFTA IS CARCINOGENIC IN MICE, RATS (INDEPENDENT OF THE CRYSTALLINE FORM) AND HAMSTERS FOLLOWING ORAL ADMINISTRATION, THE ONLY ROUTE TESTED. IT PRODUCED GENERALIZED LYMPHOSARCOMAS AND FORESTOMACH TUMORS IN MICE; MAINLY MAMMARY CARCINOMAS, SALIVARY GLAND CARCINOMAS, LUNG CARCINOMAS AND TRANSITIONAL-CELL CARCINOMAS OF THE RENAL PELVIS IN RATS; AND, IN HAMSTERS, URINARY BLADDER CARCINOMAS AND FORESTOMACH TUMORS. TWO DOGS GIVEN NFTA ORALLY DEVELOPED GALL-BLADDER ADENOMAS. 1979 IARC SUPPL-SUFFICIENT EVIDENCE OF CARCINOGENICITY IN ANIMALS.
REPT'D USE : VETERINARY MEDICINE

CAS NO : **10102-44-0**
CHEM NAME : ***** NITROGEN DIOXIDE**

REPORT 01 : CPSC REPORT TO EPA CLEAN AIR COMMITTEE SEP 85
ALTHOUGH HUMAN DATA (PULMONARY FUNCTION AND RESPONSE AND NONPULMONARY METABOLIC PROCESSES) QUALITATIVELY DEMONSTRATES HEALTH EFFECTS IN SENSITIVE POPULATIONS FOLLOWING EXPOSURE TO NO2 FROM 0.1 PPM TO 1 PPM NO UNEQUIVOCAL EPIDEMIOLOGICAL DATA TO SUBSTANTIATE EFFECTS AT THESE CONCENTRATIONS. HOWEVER, ANIMAL DATA BASE DEMONSTRATES POTENTIALLY SERIOUS EFFECTS FROM EXPOSURE TO SUCH CONCENTRATIONS OF NO2. RESPONSES IN ANIMALS SUPPORT NEED FOR FURTHER INVESTIGATION ON RELATIONSHIP BETWEEN NO2 AND LUNG DISEASE AND IMPAIRED LUNG FUNCTION.
REPT'D USE : INDOOR COMBUSTION POLLUTANT

CAS NO : **51-75-2**
CHEM NAME : ***** NITROGEN MUSTARD**

REPORT 01 : IARC SUPPL. 4 OCT 82
PATIENTS WITH HODGKINS DISEASE OR OTHER SOLID TUMORS TREATED WITH NITROGEN MUSTARD + OTHER CYTOTOXIC DRUGS AND/OR RADIATION HAVE DEVELOPED ACUTE NONLYMPHOCYTIC LEUKEMIA. SQUAMOUS-CELL CARCINOMAS HAVE BEEN NOTED TO OCCUR AFTER LONG-TERM TOPICAL TREATMENT WITH NITROGEN MUSTARD AS THE HYDROCHLORIDE. IT ALSO PRODUCES MAINLY LUNG TUMORS IN MICE FOLLOWING SC, IV OR IP INJECTION. IV INJECTION IN RATS INDUCED TUMORS IN DIFFERENT ORGANS. SKIN-PAINTING PRODUCED LOCAL TUMORS. THERE WAS SUFFICIENT EVIDENCE FOR ACTIVITY IN SHORT-TERM TESTS.

REPORT 06 : MULTI-AGENCY CROSS CHECK
Y HHS MAGUIRE
N OSHA CARC
Y EPA CAG
N EPA RCRA
N EPA CLEAN AIR
N EPA CLEAN WATER
REPT'D USE : GAS WARFARE AGENT AND CHEMCIAL INTERMEDIATE

CAS NO : **55-86-7**
CHEM NAME : ***** NITROGEN MUSTARD HYDROCHLORIDE**

REPORT 01 : IARC MONOGRAPH 9 1975
NITROGEN MUSTARD, ADMINISTERED MAINLY AS THE HYDROCHLORIDE, IS CARCINOGENIC IN MICE AND RATS. FOLLOWING ITS SUBCUTANEOUS, INTRAPERITONEAL OR INTRAVENOUS INJECTION,

IT PRODUCED AN INCREASED INCIDENCE OF LUNG TUMORS AND THYMIC LYMPHOMAS IN MICE. IT PRODUCED A VARIETY OF MALIGNANT TUMORS IN RATS FOLLOWING ITS INTRAVENOUS INJECTION. IN A 1979 SUPPLEMENT, IARC DETERMINED THAT THERE WAS SUFFICIENT EVIDENCE FOR THE CARCINOGENICITY OF NITROGEN MUSTARD HYDROCHLORIDE IN ANIMALS.

REPORT 02 : IARC MONOGRAPH 9 1975
HN2 INDUCES POINT MUTATIONS IN E. COLI AND NEUROSPORA, AND CAUSES CHROMOSOME ABERRATIONS IN A VARITY OF PLANTS AND MAMMALS. INDUCTIONS OF DOMINANT LETHAL MUTATIONS IN MICE HAVE BEEN REPORTED.

REPORT 06 : MULTI-AGENCY CROSS CHECK
Y HHS MAGUIRE
Y OSHA CARC
Y EPA CAG
N EPA RCRA
N EPA CLEAN AIR
N EPA CLEAN WATER
REPT'D USE : ANTINEOPLASTIC AGENT AND EXPERIMENTAL IMMUNOSUPPRESSANT

CAS NO : **126-85-2**
CHEM NAME : ***** NITROGEN MUSTARD N-OXIDE**

REPORT 06 : MULTI-AGENCY CROSS CHECK
N HHS MAGUIRE
N OSHA CARC
Y EPA CAG
N EPA RCRA
N EPA CLEAN AIR
N EPA CLEAN WATER
REPT'D USE : INTERMEDIATE IN MFR OF NEOPLASTIC AGENTS

CAS NO : **302-70-5**
CHEM NAME : ***** NITROGEN MUSTARD N-OXIDE HYDROCHLORIDE**

REPORT 01 : IARC MONOGRAPH 9 1975
NITROGEN MUSTARD N-OXIDE HYDROCHLORIDE IS CARCINOGENIC IN MICE AND RATS. FOLLOWING ITS SUBCUTANEOUS INJECTION IN MICE, IT PRODUCED LUNG TUMORS, THYMIC LYMPHOMAS AND HARDERIAN GLAND ADENOMAS; FOLLOWING ITS INTRAVENOUS INJECTION IN RATS IT PRODUCED MAINLY LYMPHORETICULAR TUMORS AND SARCOMAS. IN A 1979 SUPPLEMENT, IARC DETERMINED THAT THERE IS SUFFICIENT EVIDENCE FOR THE CARCINOGENICITY OF NITROGEN MUSTARD N-OXIDE HYDROCHLORIDE IN ANIMALS.

REPORT 02 : IARC MONOGRAPH 9 1976
HN2 N-OXIDE HYDROCHLORIDE INDUCES CHROMATID ABERRATIONS IN TUMOR CELLS GROWN IN MOUSE PERITONEUM POINT MUTATIONS IN BACTERIA, AND DOMINANT LETHAL MUTATIONS IN MICE.

REPORT 06 : MULTI-AGENCY CROSS CHECK
N HHS MAGUIRE
N OSHA CARC
Y EPA CAG
N EPA RCRA
N EPA CLEAN AIR
N EPA CLEAN WATER

REPT'D USE : INTERMEDIATE IN MFR OF ANTINEOPLASTIC AGENT

CAS NO : **86-57-7**
CHEM NAME : ***** 1-NITRONAPHTHALENE**

REPORT 01 : NCI CARC TRS 64 DHEW PUB NO NIH 78-1314
UNDER THE CONDITIONS OF THIS BIOASSAY 1-NITRONAPHTHALENE WAS NOT DEMONSTRATED TO BE CARCINOGENIC IN FISCHER 344 RATS OR B6C3F1 MICE (FEEDING STUDY).

REPORT 02 : NTP TECHNICAL BULLETIN 3 DEC 80
THE COMPOUND WAS FOUND TO BE POSITIVE IN THE SALMONELLA TYPHIMURIUM MUTAGENICITY ASSAY IN ONE LABORATORY.

REPT'D USE : DYE INTERMEDIATE AND WOOD PRESERVATIVES

CAS NO : **581-89-5**
CHEM NAME : ***** 2-NITRONAPHTHYLENE**

REPORT 01 : NIOSH CURRENT INTELLIGENCE BULLETIN 16 DEC 76
THE NIOSH LEARNED THAT 2-NITRONAPHTHYLENE IS METABOLIZED TO THE KNOWN HUMAN CARCINOGEN, 2-NAPHTHYLAMINE. BULLETIN EMPHASIZES THE POTENTIAL PROBLEM OF METABOLIC CONVERSION OF MATERIALS BELIEVED TO BE RELATIVELY INNOCUOUS INTO KNOWN CARCINOGENS.

REPT'D USE : NO AVAILABLE PRODUCTION OR USE DATA. NO LISTING ON TSCA INVENTORY

CAS NO : **100-02-7**
CHEM NAME : ***** P-NITROPHENOL**

REPORT 01 : EPA RCRA HEALTH AND ENV. EFFECTS PROFILE OCT 80
WEAK MUTAGENIC EFFECTS IN SACCHAROMYCES AND IN PROTEUS HAVE BEEN OBSERVED. RESULTS FROM THE AMES ASSAY, E. COLI,

AND DOMINANT LETHAL ASSAY FAILED TO SHOW MUTAGENIC EFFECTS FROM P-NITROPHENOL.

REPORT 02 : NTP CARC TEST PROGRAM JUL 82
IN A CURRENT NTP BIOASSAY, P-NITROPHENOL IS BEING APPLIED BY SKIN-PAINTING TO B6C3F1 MICE. THE PRECHRONIC TEST PHASE IS IN PROGRESS.

REPORT 03 : NTP TECHNICAL BULLETIN 3/8 DEC 80, JUL 82
P-NITROPHENOL WAS NEGATIVE IN THE SALMONELLA ASSAY AND IN DROSOPHILA WHEN TESTED FOR SEX-LINKED RECESSIVE LETHALS.

REPORT 06 : MULTI-AGENCY CROSS CHECK
N HHS MAGUIRE
N OSHA CARC
N EPA CAG
Y EPA RCRA
N EPA CLEAN AIR
Y EPA CLEAN WATER

REPT'D USE : DYE INTERMEDIATE, INTERMEDIATE IN MFR OF PHARMACEUTICALS, WOOD PRESERVATIVES, PHOTOGRAPHIC CHEMICALS, LEATHER PRESERVATIVES, PESTICIDAL/FUNGICIDAL AGENTS AND CHEMICAL INDICATOR

CAS NO : **5307-14-2**
CHEM NAME : ***** 2-NITRO-P-PHENYLENEDIAMINE**

REPORT 01 : NCI CARC TRS 169 DHEW PUB NO NIH 79-1725
UNDER THE CONDITIONS OF THIS BIOASSAY, DIETARY ADMINISTRATION OF 2-NITRO-P-PHENYLENEDIAMINE WAS CARCINOGENIC TO FEMALE B6C3F1 MICE, CAUSING AN INCREASED INCIDENCE OF HEPATOCELLULAR NEOPLASMS, PRIMARILY HEPATOCELLULAR ADENOMAS. THERE WAS NO CONVINCING EVIDENCE FOR THE CARCINOGENICITY OF THE COMPOUND IN FISCHER 344 RATS OR IN MALE B6C3F1.

REPORT 02 : IARC MONOGRAPH 16 1978
NO DATA ON 1,4-DIAMINO-2-NITROBENZENE ALONE WERE AVAILABLE. NO EVALUATION OF THE CARCINOGENICITY OF THIS COMPOUND CAN BE MADE. 1,4-DIAMINO-2-NITROBENZENE IS MUTAGENIC IN SALMONELLA TA1538 WITHOUT METABOLIC ACTIVATION. NO DOMINANT LETHALS WERE INDUCED IN RATS TREATED I.P. CHROMATID BREAKS AND GAPS HAVE BEEN FOUND IN CULTURED HUMAN PERIPHERAL LYMPHOCYTES AFTER EXPOSURE TO THIS CHEMICAL.

REPORT 03 : EPA RESPONSE TO ITC RECOMMS (47FR973) JAN 81
IN RESPONSE TO THE ITC RECOMMENDATIONS (45FR35897) THE EPA

PROPOSES AN ADVANCED NOTICE REQUESTING COMMENTS CONCERNING A PROPOSAL TO CONDUCT MUTAGENICITY AND CARCINOGENICITY TESTING FOR 11 SELECTED PHENYLENEDIAMINES, AND TERATOGENICITY/REPRODUCTIVE EFFECTS AND OTHER CHRONIC EFFECTS TESTING, AS WELL AS EPIDEMIOLOGIC STUDY, FOR 12 PHENYLENEDIAMINES. THE EPA WILL CONSIDER ADDITIONAL TESTING FOR AN ADDITIONAL 37 PHENYLENEDIAMINES, AS APPROPRIATE.

REPORT 04 : NTP TECHNICAL BULLETIN 8 JUL 82
2-NITRO-P-PDA WAS POSITIVE IN THE L5178Y MOUSE LYMPHOMA ASSAY.

REPORT 06 : MULTI-AGENCY CROSS CHECK
N HHS MAGUIRE
Y OSHA CARC
N EPA CAG
N EPA RCRA
N EPA CLEAN AIR
N EPA CLEAN WATER
REPT'D USE : HAIR DYE AND DYE

CAS NO : **99-56-9**
CHEM NAME : ***** 4-NITRO-O-PHENYLENEDIAMINE**

REPORT 01 : NCI CARC TRS 180 DHEW PUB NO NIH 79-1736
UNDER THE CONDITIONS OF THIS BIOASSAY, DIETARY ADMINISTRATION OF 4-NITRO-O-PHENYLENEDIAMINE WAS NOT CARCINOGENIC IN F344 RATS OR B6C3F1 MICE.

REPORT 02 : IARC MONOGRAPH 16 1978
NO DATA ON 1,2-DIAMINO-4-NITROBENZENE ALONE WERE AVAILABLE. NO EVALUATION OF THE CARCINOGENICITY OF THIS COMPOUND CAN BE MADE. 1,2-DIAMINO-4-NITROBENZENE IS MUTAGENIC IN SALMONELLA TA1538 AND INDUCED SEX-LINKED RECESSIVE LETHALS IN DROSOPHILA. NO DOMINANT LETHALS WERE INDUCED IN RATS TREATED I.P. NO CHROMOSOMAL ABERRATIONS WERE OBSERVED WHEN THIS COMPOUND WAS INCUBATED WITH CULTURED HUMAN PERIPHERAL LYMPHOCYTES.

REPORT 03 : EPA RESPONSE TO ITC RECOMMS (47FR973) JAN 81
IN RESPONSE TO THE ITC RECOMMENDATIONS (45FR35897), THE EPA PROPOSES AN ADVANCED NOTICE REQUESTING COMMENTS CONCERNING A PROPOSAL TO CONDUCT MUTAGENICITY AND CARCINOGENICITY TESTING FOR 11 SELECTED PHENYLENEDIAMINES, AND TERATOGENICITY/REPRODUCTIVE EFFECTS AND OTHER CHRONIC EFFECTS TESTING AS WELL AS AN EPIDEMIOLOGIC STUDY, FOR 12 PHENYLENEDIAMINES. THE EPA WILL CONSIDER ADDITIONAL

TESTING FOR AN ADDITIONAL 37 PHENYLENEDIAMINES, AS APPROPRIATE.

REPORT 04 : NTP TECHNICAL BULLETIN 8 JUL 82
4-NITRO-O-PDA WAS POSITIVE IN THE L5178Y MOUSE LYMPHOMA ASSAY.

REPORT 05 : NTP TECHNICAL BULLETIN 9 APR 83
4-NITRO-O-PHENYLENEDIAMINE WAS POSITIVE IN ONE SALMONELLA TYPHIMURIUM MUTAGENICITY ASSAY.

REPT'D USE : CHEMICAL INDICATOR, COMPONENT IN HAIR DYE FORMULATIONS AND FUR DYE

CAS NO : **79-46-9**

CHEM NAME : ***** 2-NITROPROPANE**

REPORT 01 : NIOSH CURRENT INTELLIGENCE BULLETIN 17 APR 77
IN A NIOSH SPONSORED STUDY, ALL LABORATORY RATS EXPOSED TO 200 PPM 2-NITROPROPANE OVER A SIX MONTH PERIOD DEVELOPED HEPATOCELLULAR CARCINOMA OR HEPATIC ADENOMA. NIOSH BELIEVES IT WOULD BE PRUDENT TO HANDLE 2-NITROPROPANE IN THE WORKPLACE AS IF IT WAS A HUMAN CARCINOGEN.

REPORT 02 : NTP TECHNICAL BULLETIN 2/7 APR 80, APR 82
2-NITROPROPANE WAS POSITIVE IN THE SALMONELLA ASSAY, AND NEGATIVE IN THE RAT DOMINANT LETHAL ASSAY, THE MOUSE SPERM HEAD MORPHOLOGY ASSAY THE DROSOPHILA SEX-LINKED RECESSIVE LETHAL ASSAY, AND IN VITRO UNSCHEDULED DNA SYNTHESIS IN HUMAN FIBROBLASTS, AND THE RAT BONE MARROW CYTOLOGY ASSAY.

REPORT 03 : OSHA/NIOSH HEALTH HAZARD ALERT OCT 80
2-NITROPROPANE SHOULD BE HANDLED IN THE WORKPLACE AS A POTENTIAL HUMAN CARCINOGEN. OSHA'S CURRENT PERMISSIBLE EXPOSURE LIMIT FOR 2-NP IS 25 PPM OR 90 MG/M3 (8-HOUR, TIME-WEIGHTED AVERAGES). EVIDENCE OF CARCINOGENICITY WAS NOT CONSIDERED IN SETTING THIS LIMIT. BECAUSE 2-NP HAS NOW BEEN SHOWN TO BE CARCINOGENIC IN RATS, OCCUPATIONAL EXPOSURE TO IT SHOULD BE REDUCED TO THE LOWEST FEASABLE LEVELS.

REPORT 04 : IARC MONOGRAPH 29 1982
2-NITROPROPANE WAS TESTED IN TWO EXPERIMENTS IN RATS BY INHALATION. HEPATOCELLULAR CARCINOMAS WERE PRODUCED IN ONE EXPERIMENT, AND AN INCREASED INCIDENCE OF HEPATOCELLULAR NODULES IN THE OTHER. THERE IS SUFFICIENT EVIDENCE FOR THE CARCINOGENICITY OF 2-NITROPROPANE IN RATS.

REPORT 05 : IARC MONOGRAPH 29 1982
2-NITROPROPANE INDUCED REVERSE MUTATIONS IN SALMONELLA STRAINS TA1537, TA92, TA98, AND TA100, WITH OR WITHOUT METABOLIC ACTIVATION. IT WAS NEGATIVE IN THE MOUSE MICRONUCLEUA ASSAY. NO ADEQUATE DATA WERE AVAILABLE TO ASSESS THE TERATOGENICITY OF THIS CHEMICAL.

REPORT 06 : MULTI-AGENCY CROSS CHECK
Y HHS MAGUIRE
Y OSHA CARC
N EPA CAG
N EPA RCRA
N EPA CLEAN AIR
N EPA CLEAN WATER

REPORT 07 : EPA TSCA 8(E) NOTICE 8EHQ-0578-0170
INTERNATIONAL MINERALS AND CHEMICALS PROVIDED RESULTS OF A 6-MONTH INHALATION STUDY OF 2-NITROPROPANE IN RATS. AN EARLIER OBSERVATION THAT 2-NP CAN PRODUCE MALIGNANT CHANGES IN LIVER CELLS, AND THAT THESE METASTASIZE TO THE LUNG, HAS BEEN CONFIRMED IN THE PRESENT STUDY.

REPORT 09 : EPA NOTICE OF POTENTIAL RISK DEC 86
SEVERAL DEATHS REPORTED FROM SHORT EXPOSURES TO "HIGH LEVELS" OF 2-NP ATTRIBUTED TO SHORT-TERM EXPOSURES IN ENCLOSED SPACES WHEN WORKERS NOT WEARING PROTECTIVE BREATHING APPARATUS. ADVISORY SUGGESTS WAYS TO REDUCE EXPOSURES TO 2-NP AND REDUCE CHANCES FOR HARMFUL EFFECTS.

REPT'D USE : PAINTS/VARNISHES, SOLVENT AND CHEMICAL INTERMEDIATE

CAS NO : **504-88-1**

CHEM NAME : ***** 3-NITROPROPIONIC ACID**

REPORT 01 : NCI CARC TRS 52 DHEW PUB NO NIH 78-1302
UNDER THE CONDITIONS OF THIS BIOASSAY, THERE WAS AN ELEVATED INCIDENCE OF HEPATOCELLULAR NEOPLASMS, PRIMARILY BENIGN, AND OF ISLET-CELL ADENOMAS OF THE PANCREAS IN MALE FISCHER 344 RATS RECEIVING 3-NITROPROPIONIC ACID AS COMPARED WITH CONTROLS; HOWEVER, THERE WAS NO CONCLUSIVE EVIDENCE THAT 3-NITROPROPIONIC ACID WAS CARCINOGENIC IN THESE ANIMALS. THE CHEMICAL WAS NOT CARCINOGENIC IN FEMALE RATS OR IN MALE OR FEMALE B6C3F1 MICE. 3-NITROPROPIONIC ACID WAS GIVEN BY GAVAGE.

REPORT 02 : NTP TECHNICAL BULLETIN 8 JUL 82
3-NITROPROPRIONIC ACID WAS POSITIVE IN THE L5178Y MOUSE LYMPHOMA ASSAY.

REPORT 06 : MULTI-AGENCY CROSS CHECK 1985
Y HHS MAGUIRE
N OSHA CARC
N EPA CAG
N EPA RCRA
N EPA CLEANAIR
N EPA CLEAN WATER

REPORT 07 : IARC SUPPL. 4 OCT 82
3-NITROPROPIONIC ACID CONSIDERED TO BE CARCINOGENIC IN ANIMALS.
REPT'D USE : FOOD CONTAMINANT

CAS NO : **5522-43-0**
CHEM NAME : ***** 1-NITROPYRENE**

REPORT 01 : IARC MONOGRAPH 33 APR 84
1-NITROPYRENE TESTED IN MOUSE SKIN INITIATION-PROMOTION ASSAY; NO SIGNIFICANT INCREASE IN SKIN TUMOR INCIDENCE OBSERVED. IN ONE EXPERIMENT TESTED IN RATS BY SUBCUTANEOUS INJECTION, TUMORS OBSERVED AT THE INJECTION SITE.

REPORT 02 : IARC MONOGRAPH 33 APR 84
1-NITORPYRENE BOUND COVALENTLY TO DNA IN VITRO AND TO CELLULAR DNA IN PRESENCE OF APPROPRIATE ACTIVATION ENZYMES. MUTAGENIC TO SALMONELLA TYPHIMURIUM IN PRESENCE AND ABSENCE OF EXOGENOUS METABOLIC SYSTEM. BOTH POSITIVE AND NEGATIVE MUTAGENCITY RESULTS OBSERVED IN MAMMALIAN CELLS.
REPT'D USE : DIESEL ENGINE EXHAUST PARTICULATE, FURNACE BLACK AND CHEMICAL INTERMEDIATE

CAS NO : **607-35-2**
CHEM NAME : ***** 8-NITROQUINOLINE**

REPORT 06 : MULTI-AGENCY CROSS CHECK
N HHS MAGUIRE
Y OSHA CARC
N EPA CAG
N EPA RCRA
N EPA CLEAN AIR
N EPA CLEAN WATER
REPT'D USE : ORGANIC SYNTHESIS

CAS NO : **56-57-5**
CHEM NAME : ***** 4-NITROQUINOLINE-1-OXIDE**

REPORT 01 : NTP TECHNICAL BULLETIN 8 JUL 82
4-NITROQUINOLINE-1-OXIDE WAS POSITIVE IN THE L5178Y MOUSE LYMPHOMA ASSAY.

REPORT 06 : MULTI-AGENCY CROSS CHECK
N HHS MAGUIRE
N OSHA CARC
Y EPA CAG
N EPA RCRA
N EPA CLEAN AIR
N EPA CLEAN WATER
REPT'D USE : NO AVAILABLE PRODUCTION OR USE DATA. NO LISTING ON TSCA INVENTORY

CAS NO : **CLASS-18**
CHEM NAME : ***** N-NITROSAMINES DERIVED FROM ARECA-NUT ALKALOIDS**

REPORT 01 : IARC MONOGRAPH 37 SEP 85
N-NITROSOGUVACOLINE TESTED IN ONE EXPERIMENT IN RATS BY ADMINISTRATION IN DRINKING-WATER. NO INCREASE IN INCIDENCE OF TUMORS DETECTED, BUT EXPERIMENT NOT ADEQUATE TO ALLOW EVALUATION OF CARCINOGENICTY. 3-METHYLNITROSAMINOPROPIONITRILE TESTED IN ONE 24-WEEK EXPERIMENT IN RATS BY SUBCUTANEOUS INJECTION. PRODUCED MULTIPLE PAPILLOMAS AND CARCINOMAS OF OESOPHAGUS AND TONGUE, AND PAPILLOMAS OF NASAL CAVITY. SUFFICIENT EVIDENCE FOR CARCINOGENICITY OF 3-METHYLNITROSAMINOPROPIONITRILE, AND INADEQUATE EVIDENCE FOR SUCH OF N-NITROSOGUVACOLINE TO EXPERIMENTAL ANIMALS.

REPORT 02 : IARC MONOGRAPH 37 SEP 85
ONE STUDY INADEQUATE TO EVALUATE THE MUTAGENICITY OF N-NITROSOGUVACINE TO SALMONELLA TYPHIMURIUM. N-NITROSOGUVACOLINE MUTAGENIC TO SALMONELLA TYPHIMURIUM IN PRESENCE OF METABOLIC SYSTEM. IT DID NOT INDUCE SEX-LINKED RECESSIVE LETHAL MUTATINS IN DROSOPHILA MELANOGASTER.
REPT'D USE : FOUND IN SALIVA OF TOBACCO CHEWERS

CAS NO : **37620-20-5**
CHEM NAME : ***** N'-NITROSOANABASINE (NAB)**

REPORT 01 : IARC MONOGRAPH 37 SEP 85
TESTED FOR CARCINOGENICITY BY ORAL ADMINISTRATION IN TWO STRAINS OF RATS AND BY SUBCUTANEOUS INJECTION IN HAMSTERS. IN RATS, INDUCED OESOPHAGEAL CARCINOMAS AND/OR PAPILLOMAS. STUDY IN HAMSTERS INADEQUATE FOR EVALUATION. THUS LIMITED EVIDENCE FOR CARCINOGENICITY OF N-NITROSOANABASINE TO EXPERIMENTAL ANIMALS.
REPT'D USE : LABORATORY CHEMICAL AND FOUND IN TOBACCO PRODUCTS AND SMOKE

CAS NO : **71267-22-6**
CHEM NAME : ***** N'-NITROSOANATABINE**

REPORT 01 : IARC MONOGRAPH 37 SEP 85
NAT TESTED FOR CARCINOGENICITY BY SUBCUTANEOUS INJECTION IN RATS OF ONE STRAIN AT THREE DOSE LEVELS. THERE WAS NO INCREASE IN TUMOR INCIDENCE. THUS, DATA ARE INADEQUATE TO EVALUATE CARCINOGENICITY OF N'-NITROSOANATABINE TO EXPERIMENTAL ANIMALS.
REPT'D USE : LABORATORY CHEMICAL

CAS NO : **924-16-3**
CHEM NAME : ***** N-NITROSODI-N-BUTYLAMINE (DBNA)**

REPORT 01 : IARC MONOGRAPH 17 1978
DBNA IS CARCINOGENIC IN ALL ANIMAL SPECIES TESTED: MICE, RATS, SYRIAN GOLDEN, CHINESE AND EUROPEAN HAMSTERS, RABBITS AND GUINEA-PIGS, AFTER ITS ORAL, SUBCUTANEOUS, INTRAPERITONEAL OR INTRAVENOUS ADMINISTRATION. IT PRODUCES BENIGN AND MALIGNANT TUMORS IN THE URINARY BLADDER, ESOPHAGUS, LIVER, RESPIRATORY TRACT, STOMACH AND INTESTINE, AND ALSO LEUKEMIA; IT IS PARTICULARLY EFFECTIVE AS A BLADDER CARCINOGEN. IT IS CARCINOGENIC FOLLOWING ITS ADMINISTRATION PRENATALLY AND IN SINGLE DOSES. 1979 SUPPLEMENT/SUFFICIENT EVIDENCE FOR ANIMAL CARCINOGENICITY.

REPORT 02 : EPA AMBIENT WATER QUALITY CRITERIA 10/80 OCT 80
NITROSAMINES ARE SUSPECTED OF BEING HUMAN CARCINOGENS, AMBIENT WATER CONCENTRATIONS OF 64 NG/L, 6.4 NG/L, AND .64 NG/L ARE ESTIMATED TO RESULT IN AN INCREASED CANCER RISK OVER A LIFETIME OF 10 -5, 10 -6, AND 10 -7 RESPECTIVELY.

REPORT 03 : NAS REPORT ON DIET, NUTRITION AND CANCER 1982
AS A GROUP, THE N-NITROSO COMPOUNDS ARE CLEARLY CARCINO-

GENIC IN NUMEROUS SPECIES OF ANIMALS IN WHICH THEY HAVE BEEN TESTED. POSITIVE RESULTS HAVE BEEN OBTAINED FOR NEARLY ALL OF THE APPROXIMATELY 300 N-NITROSO COMPOUNDS TESTED FOR CARCINOGENICITY IN ONE OR MORE SPECIES. MANY OF THESE COMPOUNDS ARE ALSO MUTAGENIC.

REPORT 04 : IARC MONOGRAPH 17 1978
NO TERATOGENIC EFFECTS WERE OBSERVED FOR NDBA IN RATS WHEN A SINGLE MTD OF 100 MG/KG WAS GIVEN I.P. OR WHEN 1200 MG/KG WERE GIVEN ORALLY DURING PREGNANCY. NDBA IS MUTAGENIC IN SALMONELLA TA100, TA153 AND TA1535, WITH METABOLIC ACTIVATION, AND IN E. COLI.

REPORT 05 : STATUS REPORT 8EHQ-1186-0643 DEC 86
DOW CORNING REPORTED THAT THEY RECENTLY FOUND THAT LOW MOLECULAR WEIGHT NITROSAMINS ARE PRODUCED ND LIBERATED DURING FOAMING OF 2 OF DC'S SILICONE ELASTOMERS FOAMS.

REPORT 06 : MULTI-AGENCY CROSS CHECK
Y HHS MAGUIRE
N OSHA CARC
Y EPA CAG
Y EPA RCRA
N EPA CLEAN AIR
N EPA CLEAN WATER

REPT'D USE : COMBUSTION POLLUTANT, FOOD CONTAMINANT, RESEARCH CHEMICAL AND ORGANIC SYNTHESIS

CAS NO : **1116-54-7**
CHEM NAME : ***** N-NITROSODIETHANOLAMINE**

REPORT 01 : IARC MONOGRAPH 17 1978
N-NITROSODIETHANOLAMINE IS CARCINOGENIC IN RATS AFTER ITS ORAL ADMINISTRATION AND IN HAMSTERS AFTER ITS SUBCUTANEOUS INJECTION. IT PRODUCES HEPATOCELLULAR CARCINOMAS AND RENAL ADENOMAS IN RATS, AND ADENOCARCINOMAS OF THE NASAL CAVITY, PAPILLARY TUMORS OF THE TRACHEA, HEPATOCELLULAR ADENOMAS AND LOCAL FIBROSARCOMAS IN HAMSTERS. IN A 1979 SUPPLEMENT, IARC DETERMINED THAT THERE IS SUFFICIENT EVIDENCE FOR THE CARCINOGENCIITY OF N-NITROSODIETHANOLAMINE IN ANIMALS.

REPORT 02 : IARC MONOGRAPH 17 1978
NO DATA ON THE MUTAGENICITY, EMBRYOTOXICITY OR TERATOGENICITY OF NDEL A WERE AVAILABLE.

REPORT 03 : EPA AMBIENT WATER QUALITY CRITERIA OCT 80
NITROSAMINES ARE SUSPECTED OF BEING HUMAN CARCINOGENS.

AMBIENT WATER CONCENTRATIONS OF 8 NG/L, .8 NG/L AND .08 NG/L ARE ESTIMATED TO RESULT IN AN INCREASED CANCER RISK OVER A LIFETIME OF 10 -5, 10 -6 AND 10 -7 RESPECTIVELY.

REPORT 04 : NAS REPORT ON DIET, NUTRITION AND CANCER 1982
AS A GROUP, THE N-NITROSO COMPOUNDS ARE CLEARLY CARCINOGENIC IN NUMEROUS SPECIES OF ANIMALS IN WHICH THEY HAVE BEEN TESTED. POSITIVE RESULTS HAVE BEEN OBTAINED FOR NEARLY ALL OF THE APPROXIMATELY 300 N-NITROSO COMPOUNDS TESTED FOR CARCINOGENICITY IN ONE OR MORE SPECIES. MANY OF THESE COMPOUNDS ARE ALSO MUTAGENIC.

REPORT 05 : NTP CARCINOGEN TEST PROGRAM JUL 82
IN A CURRENT NTP BIOASSAY, N-NITROSODIETHANOLAMINE WAS ADMINISTERED IN WATER TO F344 RATS AND B6C3F1 MICE. THE EXPOSURE COMPLETED AND HISTOPATHOLOGY IS IN PROGRESS.

REPORT 06 : MULTI-AGENCY CROSS CHECK
Y HHS MAGUIRE
N OSHA CARC
Y EPA CAG
Y EPA RCRA
N EPA CLEAN AIR
N EPA CLEAN WATER

REPORT 07 : EPA TSCA 8(E) 283-0469S FEB 83
A COMPANY REPORTED A NITRITE/AMINE SYNTHETIC UNDILUTED METAL-WORKING FLUID CONTAINING A HIGH LEVEL OF NITROSODIETHANOLAMINE AND A NON-NITRITE SYNTHETIC FLUID WITHOUT DETECTABLE NITROSAMINES WERE FOUND TO BE ONCOGENIC TO MOUSE LIVER IN LIFETIME SKIN-PAINTING STUDY CONDUCTED BY NIOSH.

REPT'D USE : COMBUSTION POLLUTANT, RESEARCH CHEMICAL, CONTAMINANT IN CUTTING FLUIDS, PESTICIDES, COSMETICS AND TOBACCO. NO EVIDENCE OF COMMERCIAL PRODUCTION OR USE

CAS NO : **55-18-5**
CHEM NAME : ***** N-NITROSODIETHYLAMINE (NDEA)**

REPORT 01 : IARC MONOGRAPH 17 1978
N-NITROSODIETHYLAMINE (DEN) IS CARCINOGENIC IN ALL TEN ANIMAL SPECIES TESTED, INCLUDING SUB-HUMAN PRIMATES. THE MAIN TARGET ORGANS ARE THE NASAL CAVITY, TRACHEA, LUNG, ESOPHAGUS, AND LIVER. IT INDUCES TUMORS FOLLOWING DIFFERENT ROUTES OF ADMINISTRATION, INCLUDING INGESTION, INHALATION AND SKIN PAINTING. IT IS CARCINOGENIC IN SINGLE-DOSE EXPERIMENTS AND FOLLOWING PRENATAL EXPOSURE. IN A 1979 SUPPLEMENT, IARC DETERMINED THAT THERE

IS SUFFICIENT EVIDENCE FOR THE CARCINOGENICITY OF DEN IN ANIMALS.

REPORT 02 : EPA AMBIENT WATER QUALITY CRITERIA OCT 80
NITROSAMINES ARE SUSPECTED OF BEING HUMAN CARCINOGENS. AMBIENT WATER CONCENTRATIONS OF 8 NG/L, .8 NG/L, AND .08 NG/L ARE ESTIMATED TO RESULT IN AN INCREASED CANCER RISK OVER A LIFETIME OF 10 -5, 10 -6, AND 10 -7 RESPECTIVELY.

REPORT 03 : STATUS REPORT 8EHQ-1186-0643 DEC 86
DOW CORNING REPORTED THAT THEY RECENTLY DISCOVERED THAT LOW MOLECULAR WEIGHT NITROSAMINES PRODUCED AND LIBERATED DURING CURING AND FOAMING OF 2 OF DC'S SILICONE ELASTOMER FOAMS.

REPORT 06 : MULTI-AGENCY CROSS CHECK
Y HHS MAGUIRE
N OSHA CARC
Y EPA CAG
Y EPA RCRA
Y EPA CLEAN AIR
N EPA CLEAN WATER

REPORT 07 : NAS REPORT ON DIET, NUTRITION AND CANCER 1982
AS A GROUP, THE N-NITROSO COMPOUNDS ARE CLEARLY CARCINOGENIC IN NUMEROUS SPECIES OF ANIMALS IN WHICH THEY HAVE BEEN TESTED. POSITIVE RESULTS HAVE BEEN OBTAINED FOR NEARLY ALL OF THE APPROXIMATELY 300 N-NITROSO COMPOUNDS TESTED FOR CARCINOGENICITY IN ONE OR MORE SPECIES. MANY OF THESE COMPOUNDS ARE ALSO MUTAGENIC.

REPORT 08 : IARC MONOGRAPH 17 1978
TERATOGENIC EFFECTS WERE NOT NOTED FOR NDEA IN RATS GIVEN SINGLE DOSES OF 180 MG/KG I.P. OR 200 MG/KG ORALLY DURING PREGNANCY. NDEA CAUSED FORWARD MUTATIONSA IN SACCHAROMYCES ABD NEUROSPORA, AND REVERSE MUTATIONS IN SALMONELLA TAIO0 AND TA1530, WITH METABOLIC ACTIVATION. NDEA WAS MUTAGENIC IN RECESSIVE LETHAL TESTS IN DROSOPHILA. DOMINANT LETHAL MUTATIONS WERE NOT OBSERVED IN MICE. NDEA DID NOT INDUCE TRANSFORMATION IN HAMSTER EMBRYO CELLS.

REPT'D USE : COMBUSTION POLLUTANT, PLASTIC LUBRICANT ADDITIVE/ANTIOXIDANT AND RESEARCH CHEMICAL. NO EVIDENCE OF COMMERICAL PRODUCTION OR USE

CAS NO : **62-75-9**
CHEM NAME : ***** N-NITROSODIMETHYLAMINE (NDMA)**

REPORT 01 : EPA AMBIENT WATER QUALITY CRITERIA OCT 80
NITROSAMINES ARE SUSPECTED OF BEING HUMAN CARCINOGENS. AMBIENT WATER CONCENTRATIONS OF 14 NG/L, 1.4 NG/L, AND .14 NG/L ARE ESTIMATED TO RESULT IN AN INCREASED CANCER RISK OVER A LIFETIME OF 10 -5, 10 -6, 10 -7 RESPECTIVELY.

REPORT 02 : IARC MONOGRAPH 17 1978
DMN IS CARCINOGENIC IN ALL ANIMAL SPECIES TESTED: MICE, RATS, SYRIAN GOLDEN, CHINESE AND EUROPEAN HAMSTERS, GUINEA PIGS, RABBITS, DUCKS, MASTOMYS, VARIOUS FISH, NEWTS AND FROGS. IT INDUCES BENIGN AND MALIGNANT TUMORS FOLLOWING ITS ADMINISTRATION IN VARIOUS ORGANS IN VARIOUS SPECIES. IT PRODUCES TUMORS, MAINLY OF THE LIVER, KIDNEY AND RESPIRATORY TRACT. IT IS CARCINOGENIC FOLLOWING ITS ADMINISTRATION PRENATALLY AND IN SINGLE DOSES. IN SEVERAL STUDIES, DOSE-RESPONSE RELATIONSHIPS WERE ESTABLISHED. 1979 SUP./SUFF. EVIDENCE ANIMAL CARCINOGEN.

REPORT 03 : NAS REPORT ON DIET, NUTRITION AND CANCER 1982
AS A GROUP, THE N-NITROSO COMPOUNDS ARE CLEARLY CARCINOGENIC IN NUMEROUS SPECIES OF ANIMALS IN WHICH THEY HAVE BEEN TESTED. POSITIVE RESULTS HAVE BEEN OBTAINED FOR NEARLY ALL OF THE APPROXIMATELY 300 N-NITROSO COMPOUNDS TESTED FOR CARCINOGENICITY IN ONE OR MORE SPECIES. MANY OF THESE COMPOUNDS ARE ALSO MUTAGENIC.

REPORT 04 : EPA RCRA HEALTH AND ENV. EFFECTS PROFILE OCT 80
DIMETHYLNITROSAMINE PRODUCES LIVER AND KIDNEY TUMORS IN RATS. IT IS MUTAGENIC IN SEVERAL ASSAY SYSTEMS.

REPORT 05 : IARC MONOGRAPH 17 1978
NO TERATOGENIC EFFECTS FOR NDMA WERE OBSERVED IN RATS. NDMA IS MUTAGENIC, WITH METABOLIC ACTIVATION IN SALMONELLA, E. COLI, B. SUBTILIS, SACCHAROMYCES AND NEUROSPORA. CHROMOSOME ABERRATIONS WERE FOUND IN LIVER CELLS OF CHINESE HAMSTER INJECTED I.P. WITH NDMA.

REPORT 06 : MULTI-AGENCY CROSS CHECK
Y HHS MAGUIRE
Y OSHA CARC
Y EPA CAG
Y EPA RCRA
Y EPA CLEAN AIR
Y EPA CLEAN WATER

REPORT 07 : NTP TECHNICAL BULLETIN 3/8 DEC 80, JUL 82
NDMA WAS POSITIVE IN THE SALMONELLA ASSAY AND IN DROSOPHILA WHEN TESTED FOR SEX-LINKED RECESSIVE LETHALS.

REPORT 09 : EPA TSCA REPORT 8EHQ-1186-0643 DEC 86
DOW CORNING RECENTLY FOUND OUT THAT LOW MOLECULAR WEIGHT NITROSAMINES PRODUCED AND LIBERATED DURING FOAMING OF 2 OF DC'S SILICONE ELASTOMER FOAMS.

REPT'D USE : INTERMEDIATE IN MFR OF ROCKET FUEL, RESEARCH CHEMICAL, AIR POLLUTANT, TOBACCO SMOKE CONDENSATE AND RESEARCH CHEMICAL. NO EVIDENCE OF COMMERICAL PRODUCTION OR USE.

CAS NO : **86-30-6**
CHEM NAME : ***** N-NITROSODIPHENYLAMINE**

REPORT 01 : NCI CARC TRS 164 DHEW PUB NO NIH 79-1720
UNDER THE CONDITIONS OF THIS BIOASSAY, N-NITROSODIPHENYLAMINE WAS CARCINOGENIC FOR BOTH SEXES OF F344 RATS, INDUCING TRANSITIONAL-CELL CARCINOMAS OF THE URINARY BLADDER, BUT WAS NOT CARCINOGENIC FOR B6C3F1 MICE OF EITHER SEX (FEEDING STUDY).

REPORT 02 : EPA AMBIENT WATER QUALITY CRITERIA OCT 80
NITROSAMINES ARE SUSPECTED OF BEING HUMAN CARCINOGENS. AMBIENT WATER CONCENTRATIONS OF 49 MG/L, 4.9 MG/L, AND .49 MG/L ARE ESTIMATED TO RESULT IN AN INCREASED CANCER RISK OVER A LIFETIME OF 10 -5, 10 -6, AND 10 -7 RESPECTIVELY.

REPORT 03 : EPA RCRA HEALTH AND ENV. EFFECTS PROFILE OCT 80
IN AN NCI BIOASSAY, NDPHA WAS FOUND TO INDUCE A SIGNIFICANT INCIDENCE OF URINARY BLADDER TUMORS IN BOTH MALES AND FEMALES. FEW URINARY BLADDER TUMORS WERE OBSERVED IN MICE IN A SIMILAR EXPERIMENT. ALTHOUGH THERE WAS A HIGH INCIDENCE OF NON-NEOPLASTIC BLADDER LESIONS. N-NITROSODIPHENYLAMINE HAS CONSISTENTLY BEEN FOUND NEGATIVE IN A VARIETY OF MUTAGENICITY ASSAY.

REPORT 04 : NAS REPORT ON DIET, NUTRITION AND CANCER 1982
AS A GROUP, THE N-NITROSO COMPOUNDS ARE CLEARLY CARCINOGENIC IN NUMEROUS SPECIES OF ANIMALS IN WHICH THEY HAVE BEEN TESTED. POSITIVE RESULTS HAVE BEEN OBTAINED FOR NEARLY ALL OF THE APPROXIMATELY 300 N-NITROSO COMPOUNDS TESTED FOR CARCINOGENICITY IN ONE OR MORE SPECIES. MANY OF THESE COMPOUNDS ARE ALSO MUTAGENIC.

REPORT 05 : IARC MONOGRAPH 27 1982
N-NITROSODIPHENYLAMINE WAS ADEQUATELY TESTED IN MICE AND RATS BY DIETARY ADMINISTRATION. NO CARCINOGENIC EFFECTS WERE OBSERVED IN MICE. IT WAS CARCINOGENIC FOR

RATS, PRODUCING TRANSITIONAL-CELL CARCINOMAS OF THE URINARY BLADDER. THERE IS LIMITED EVIDENCE FOR THE CARCINOGENICITY OF N-NITROSODIPHENYLAMINE IN ANIMALS.

REPORT 06 : MULTI-AGENCY CROSS CHECK
N HHS MAGUIRE
Y OSHA CARC
N EPA CAG
N EPA RCRA
N EPA CLEAN AIR
Y EPA CLEAN WATER

REPORT 07 : IARC MONOGRAPH 27 1982
N-NITROSODIPHENYLAMINE HAS BEEN EXTENSIVELY TESTED IN SHORT-TERM ASSAYS IN PROKARYOTES AND EUKARYOTES FOR MUTAGENICITY AND OTHER CHROMOSOME EFFECTS. IT WAS NOT MUTAGENIC AND DID NOT INDUCE DNA SYNTHESIS OR CHROMOSOMAL DAMAGE UNDER ANY TEST CONDITION. NO DATA CONCERNING EFFECTS ON REPRODUCTION OR PRENATAL TOXICITY WERE AVAILABLE.

REPORT 08 : EPA AMBIENT WATER QUALITY CRITERIA OCT 80
NITROSAMINES ARE SUSPECTED OF BEING HUMAN CARCINOGENS. AMBIENT WATER CONCENTRATIONS OF 8 NG/L, .8 NG/L, AND .08 NG/L ARE ESTIMATED TO RESULT IN AN INCREASED CANCER RISK OVER A LIFETIME OF 10 -5, 10 -6, AND 10-7 RESPECTIVELY.

REPT'D USE : CHEMICAL INTERMEDIATE AND RUBBER VULCANIZATION RETARDANT

CAS NO : **156-10-5**

CHEM NAME : ***** P-NITROSODIPHENYLAMINE**

REPORT 01 : NCI CARC TRS 190 DHEW PUB NO NIH 79-1746
UNDER THE CONDITIONS OF THIS BIOASSAY, P-NITROSODIPHENYL AMINE WAS CARCINOGENIC WHEN ADMINISTERED IN THE DIET TO MALE B6C3F1 MICE, CAUSING HEPATOCELLULAR CARCINOMAS. THE CHEMICAL WAS ALSO CARCINOGENIC IN MALE FISCHER 344 RATS, CAUSING LIVER NEOPLASMS. NO EVIDENCE WAS PROVIDED FOR THE CARCINOGENICITY OF P-NITROSODIPHENYLAMINE IN FEMALE B6C3F1 MICE OR IN FEMALE FISCHER 344 RATS.

REPORT 02 : IARC MONOGRAPH 27 1982
TECH GRADE P-NITROSODIPHENYLAMINE WAS TESTED IN RATS AND MICE BY DIETARY ADMINISTRATION. IT PRODUCED LESIONS DESCRIBED AS NEOPLASTIC NODULES IN THE LIVER OF MALE RATS. THE RESULTS OF THE STUDY IN MICE WERE INCONCLUSIVE. THE AVAILABLE DATA ARE INSUFFICIENT FOR EVALUATION.

REPORT 03 : IARC MONOGRAPH 27 1982
P-NITROSODIPHENYLAMINE DID NOT TRANSFORM CELLS OF EMBRYOS EXCISED FROM PREGNANT HAMSTERS TREATED IP WITH UP TO 20 MG/BW. NO DATA CONCERNING REPRODUCTIVE EFFECTS OR PRENATAL TOXICITY WERE AVAILABLE.

REPORT 04 : 5TH ANNUAL REPORT ON CARCINOGENS (MAGUIRE REPT)1989
CPD CAUSED LIVER CANCERS IN MALE MICE AND LIVER NEOPLASMS IN MALE RATS. NO EVIDENCE OF CARCINOGENICITY IN FEMALE MICE OR FEMALE RATS. IARC CONSIDERED THIS LIMITED EVIDENCE OF CARCINOGENICITY, BUT NCI/OTA INTERPRETED IT AS BEING SUFFICIENT EVIDENCE. HOWEVER, DROPPED FROM ANNUAL REPORT BECAUSE OF INSUFFIENT EVIDENCE OF CARCINOGENICITY.

REPORT 06 : MULTI-AGENCY CROSS CHECK
Y HHS MAGUIRE
Y OSHA CARC
N EPA CAG
N EPA RCRA
N EPA CLEAN AIR
Y EPA CLEAN WATER

REPT'D USE : RUBBER VULCANIZATION ACCELERATOR(N-NITROSO-N-PHENYLBENZENEAMINE)

CAS NO : **621-64-7**
CHEM NAME : ***** N-NITROSODI-N-PROPYLAMINE**

REPORT 01 : IARC MONOGRAPH 17 1978
N-NITROSODI-N-PROPYLAMINE IS CARCINOGENIC IN RATS AFTER ITS ORAL ADMINISTRATION AND IN RATS AND HAMSTERS AFTER ITS SUBCUTANEOUS INJECTION. IT PRODUCES BENIGN AND MALIGNANT TUMORS OF THE LIVER, KIDNEY, ESOPHAGUS AND RESPIRATORY TRACT. IN A 1979 SUPPLEMENT, IARC DETERMINED THAT THERE IS SUFFICIENT EVIDENCE FOR THE CARCINOGENICITY OF THIS COMPOUND IN ANIMALS.

REPORT 02 : EPA RCRA HEALTH AND ENV. EFFECTS PROFILE OCT 80
THE IARC HAS CONCLUDED THAT "N-NITROSODI-N-PROPYLANINE SHOULD BE REGARDED FOR PRACTICAL PURPOSES AS IF IT WERE CARCINOGENIC IN HUMANS." THE CONCLUSION IS BASED ON POSITIVE FINDINGS IN SEVERAL LONG-TERM ANIMAL STUDIES WITH THE COMPOUND. IT HAS ALSO BEEN FOUND MUTAGENIC IN SEVERAL TEST SYSTEMS WITH ACTIVATION.

REPORT 03 : NAS REPORT ON DIET, NUTRITION AND CANCER 1982
AS A GROUP, THE N-NITROSO COMPOUNDS ARE CLEARLY CARCINOGENIC IN NUMEROUS SPECIES OF ANIMALS IN WHICH THEY HAVE BEEN TESTED. POSITIVE RESULTS HAVE BEEN OBTAINED FOR NEARLY ALL OF THE APPROXIMATELY 300 N-NITROSO COMPOUNDS

TESTED FOR CARCINOGENICITY IN ONE OR MORE SPECIES. MANY OF THESE COMPOUNDS ARE ALSO MUTAGENIC.

REPORT 04 : EPA AMBIENT WATER QUALITY CRITERIA OCT 80
NITROSAMINES ARE SUSPECTED OF BEING HUMAN CARCINOGENS. AMBIENT WATER CONCENTRATIONS OF 8 NG/L, .8 NG/L, AND .08 NG/L ARE ESTIMATED TO RESULT IN AN INCREASED CANCER RISK OVER A LIFETIME OF 10 -5, 10 -6, AND 10 -7 RESPECTIVELY.

REPORT 06 : MULTI-AGENCY CROSS CHECK
Y HHS MAGUIRE
N OSHA CARC
Y EPA CAG
Y EPA RCRA
N EPA CLEAN AIR
Y EPA CLEAN WATER

REPT'D USE : NO LISTING ON TSCA INVENTORY; NO KNOWN COMMERICAL USE.

CAS NO : **29291-35-8**
CHEM NAME : *****N-NITROSOFOLIC ACID**

REPORT 01 : IARC MONOGRAPH 17 1978
N-NITROSOFOLIC ACID HAS BEEN TESTED ONLY IN NEWBORN MICE BY INTRAPERITONEAL INJECTION, WITH INCONCLUSIVE RESULTS. NO EVALUATION OF THE CARCINOGENICITY OF N-NITROSOFOLIC ACID COULD BE MADE ON THE BASIS OF AVAILABLE DATA.

REPORT 02 : IARC MONOGRAPH 17 1978
NO DATA ON THE EMBRYOTOXICITY OR TERATOGENICITY OF N-NITROSOFOLIC ACID WERE AVAILABLE. THE CHEMICAL WAS MUTAGENIC IN SALMONELLA TA1538, TA98 AND TA100 WITH METABOLIC ACTIVATION.

REPT'D USE : NO PRODUCTION OR USE DATA AVAILABLE; NO LISTING ON TSCA INVENTORY

CAS NO : **30310-80-6**
CHEM NAME : ***** N-NITROSOHYDROXYPROLINE (NHPRO)**

REPORT 01 : IARC MONOGRAPH 17 1978
N-NITROSOHYDROXYPROLINE HAS BEEN TESTED IN RATS BY ORAL ADMINISTRATION. ALTHOUGH THIS STUDY DID NOT INDICATE A CARCINOGENIC EFFECT, IT WAS INADEQUATE WITH REGARD TO DOSE AND/OR DURATION. NO EVALUATION OF THE CARCINOGENICITY OF THIS COMPOUND COULD BE MADE ON THE BASIS OF THE AVAILABLE DATA.

REPT'D USE : LAB CHEMICAL; NO LISTING ON TSCA INVENTORY

CAS NO : **10595-95-6**
CHEM NAME : ***** N-NITROSOMETHYLETHYLAMINE**

REPORT 01 : IARC MONOGRAPH 17 1978
N-NITROSOMETHYLETHYLAMINE IS CARCINOGENIC IN RATS AFTER ITS ORAL ADMINISTRATION, THE ONLY SPECIES AND ROUTE TESTED: IT PRODUCES HEPATOCELLULAR CARCINOMAS. IN A 1979 SUPPLEMENT, IARC DETERMINED THAT THERE IS SUFFICIENT EVIDENCE FOR THE CARCINOGENICITY OF THIS COMPOUND IN ANIMALS.

REPORT 02 : NAS REPORT ON DIET, NUTRITION AND CANCER 1982
AS A GROUP, THE N-NITROSO COMPOUNDS ARE CLEARLY CARCINOGENIC IN NUMEROUS SPECIES OF ANIMALS IN WHICH THEY HAVE BEEN TESTED. POSITIVE RESULTS HAVE BEEN OBTAINED FOR NEARLY ALL OF THE APPROXIMATELY 300 N-NITROSO COMPOUNDS TESTED FOR CARCINOGENICITY IN ONE OR MORE SPECIES. MANY OF THESE COMPOUNDS ARE ALSO MUTAGENIC.

REPORT 03 : IARC MONOGRAPH 17 1978
NO DATA ON THE EMBRYOTOXICITY, TERTOGENICITY METABOLISM OR MUTAGENICITY OF NMEA WERE AVAILABLE.

REPORT 04 : EPA AMBIENT WATER QUALITY CRITERIA OCT 80
NITROSAMINES ARE SUSPECTED OF BEING HUMAN CARCINOGENS. AMBIENT WATER CONCENTRATIONS OF 8 NG/L, .8 NG/L, AND .08 NG/L ARE ESTIMATED TO RESULT IN AN INCREASED CANCER RISK OVER A LIFETIME OF 10 -5, 10 -6, AND 10 -7 RESPECTIVELY.

REPORT 06 : MULTI-AGENCY CROSS CHECK
Y HHS MAGUIRE
N OSHA CARC
Y EPA CAG
N EPA RCRA
N EPA CLEAN AIR
N EPA CLEAN WATER

REPT'D USE : NO AVAILABLE PRODUCTION OR USE DATA; NO LISTING ON TSCA INVENTORY

CAS NO : **684-93-5**
CHEM NAME : ***** N-NITROSO-N-METHYLUREA (NMU)**

REPORT 01 : IARC MONOGRAPH 1
NMU IS CARCINOGENIC IN ALL SIX ANIMAL SPECIES TESTED. IT HAS A LOCAL AS WELL AS A SYSTEMIC CARCINOGENIC EFFECT, PRODUCING TUMORS AT DIFFERENT SITES, INCLUDING THE NERVOUS TISSUE. IT INDUCES TUMORS FOLLOWING DIFFERENT ROUTES OF ADMINISTRATION, INCLUDING INGESTION. IT IS CARCINOGENIC IN SINGLE-DOSE EXPERIMENTS AND FOLLOWING

PRENATAL EXPOSURE. IN A 1979 SUPPLEMENT, IARC DETERMINED THAT THERE IS SUFFICIENT EVIDENCE FOR THE CARCINOGENICITY OF NMU IN ANIMALS.

REPORT 02 : IARC MONOGRAPH 17 MAY 78
THE GENETIC ACTIVITY OF NMU HAS BEEN DEMONSTRATED IN BACTERIAL PHAGE, E. COLI, SALMONELLA, SACCHAROMYCES, S. MARCESCENS, CHINESE HAMSTER CELLS AND DROSOPHILA, INDUCING FORWARD AND REVERSE MUTATIONS AND GENE CONVERSIONS. SISTER CHROMATID EXCHANGES WERE FOUND IN CHINESE HAMSTER CELLS EXPOSED TO NMU.

REPORT 03 : NAS REPORT ON DIET, NUTRITION AND CANCER 1982
AS A GROUP, THE N-NITROSO COMPOUNDS ARE CLEARLY CARCINOGENIC IN NUMEROUS SPECIES OF ANIMALS IN WHICH THEY HAVE BEEN TESTED. POSITIVE RESULTS HAVE BEEN OBTAINED FOR NEARLY ALL OF THE APPROXIMATELY 300 N-NITROSO COMPOUNDS TESTED FOR CARCINOGENICITY IN ONE OR MORE SPECIES. MANY OF THESE COMPOUNDS ARE ALSO MUTAGENIC.

REPORT 06 : MULTI-AGENCY CROSS CHECK
Y HHS MAGUIRE
N OSHA CARC
Y EPA CAG
Y EPA RCRA
Y EPA CLEAN AIR
N EPA CLEAN WATER

REPT'D USE : POLYMERIZATION INITIATOR, ORGANIC SYNTHESIS AND CHEMOTHERAPEUTIC; NO COMMERICAL USE

CAS NO : **615-53-2**
CHEM NAME : ***** N-NITROSO-N-METHYLURETHANE (NMUT)**

REPORT 01 : IARC MONOGRAPH 17 1978
N-NITROSO-N-METHYLUREA IS CARCINOGENIC IN ALL ANIMAL SPECIES TESTED: MICE, RATS, SYRIAN GOLDEN, CHINESE AND EUROPEAN HAMSTERS, GUINEA-PIG, RABBITS, GERBILS, PIGS, DOGS AND MONKEYS. IT INDUCES BENIGN AND MALIGNANT TUMORS FOLLOWING ITS ADMINISTRATION BY DIFFERENT ROUTES, INCLUDING INGESTION. IT PRODUCES TUMORS AT DIFFERENT SITES, INCLUDING THE NERVOUS TISSUE, STOMACH, ESOPHAGUS, PANCREAS, RESPIRATORY TRACT, INTESTINE, LYMPHORETICULAR TISSUES, SKIN AND KIDNEY. IT IS CARCINOGENIC FOLLOWING ITS ADMINISTRATION PRENATALLY AND IN SINGLE DOSES.

REPORT 02 : NAS REPORT ON DIET, NUTRITION AND CANCER 1982
AS A GROUP, THE N-NITROSO COMPOUNDS ARE CLEARLY CARCINOGENIC IN NUMEROUS SPECIES OF ANIMALS IN WHICH THEY HAVE BEEN TESTED. POSITIVE RESULTS HAVE BEEN OBTAINED FOR

NEARLY ALL OF THE APPROXIMATELY 300 N-NITROSO COMPOUNDS TESTED FOR CARCINOGENICITY IN ONE OR MORE SPECIES. MANY OF THESE COMPOUNDS ARE ALSO MUTAGENIC.

REPORT 06 : MULTI-AGENCY CROSS CHECK
N HHS MAGUIRE
N OSHA CARC
Y EPA CAG
Y EPA RCRA
N EPA CLEAN AIR
N EPA CLEAN WATER
REPT'D USE : ORGANIC SYNTHESIS

CAS NO : **4549-40-0**
CHEM NAME : ***** N-NITROSOMETHYLVINYLAMINE**

REPORT 01 : IARC MONOGRAPH 17 1978
N-NITROSOMETHYLVINYLAMINE IS CARCINOGENIC IN RATS, THE ONLY SPECIES TESTED. IT PRODUCES CARCINOMAS OF THE ESOPHAGUS, TONGUE AND PHARYNX AFTER ITS CONTINUOUS ORAL ADMINISTRATION AND CARCINOMAS OF THE NASAL CAVITIES AFTER INHALATION EXPOSURE. IN A 1979 SUPPLEMENT, IARC DETERMINED THAT THERE IS SUFFICIENT EVIDENCE FOR THE CARCINOGENICITY OF N-NITROSOMETHYLVINYLAMINE IN ANIMALS.

REPORT 02 : NAS REPORT ON DIET, NUTRITION AND CANCER 1982
AS A GROUP, THE N-NITROSO COMPOUNDS ARE CLEARLY CARCINOGENIC IN NUMEROUS SPECIES OF ANIMALS IN WHICH THEY HAVE BEEN TESTED. POSITIVE RESULTS HAVE BEEN OBTAINED FOR NEARLY ALL OF THE APPROXIMATELY 300 N-NITROSO COMPOUNDS TESTED FOR CARCINOGENICITY IN ONE OR MORE SPECIES. MANY OF THESE COMPOUNDS ARE ALSO MUTAGENIC.

REPORT 03 : EPA AMBIENT WATER QUALITY CRITERIA OCT 80
NITROSAMINES ARE SUSPECTED OF BEING HUMAN CARCINOGENS. AMBIENT WATER CONCENTRATIONS OF 8 NG/L, .8 NG/L, AND .08 NG/L ARE ESTIMATED TO RESULT IN AN INCREASED CANCER RISK OVER A LIFETIME OF 10 -5, 10 -6, AND 10 -7 RESPECTIVELY.

REPORT 04 : IARC MONOGRAPH 17 1978
NO DATA ON THE EMBRYOTOXICITY OR TERATOGENICITY OF NMVA WERE AVAILABLE. NMVA CAUSES SEX-LINKED RECESSIVE LETHAL MUTATIONS IN DROSOPHIIA.

REPORT 06 : MULTI-AGENCY CROSS CHECK
Y HHS MAGUIRE
N OSHA CARC
Y EPA CAG

N EPA RCRA
N EPA CLEAN AIR
N EPA CLEAN WATER

REPORT 07 : 4TH ANNUAL REPORT ON CARCINOGENS (MAGUIRE REPT)
NMVA CARCINOGENIC IN RATS, ONLY SPECIES TESTED, PRODUCING CANCER OF THE NASAL CAVITIES AFTER INHALATION AND CANCERS OF ESOPHAGUS, PHARYNX AND TONGUE AFTER CONTINUOUS ADMINISTRATION IN DRINKING WATER.

REPT'D USE : NO AVAILABLE PRODUCTION OR USE DATA; NO LISTING ON TSCA INVENTORY

CAS NO : **59-89-2**
CHEM NAME : ***** N-NITROSOMORPHOLINE**

REPORT 01 : IARC MONOGRAPH 17 1978
N-NITROSOMORPHOLINE IS CARCINOGENIC IN MICE, RATS, SYRIAN GOLDEN, CHINESE AND EUROPEAN HAMSTERS AND VARIOUS FISH. FOLLOWING ITS ORAL ADMINISTRATION, IT PRODUCES BENIGN AND MALIGNANT TUMORS OF THE LIVER AND LUNG IN MICE, OF THE LIVER, KIDNEY AND BLOOD VESSELS IN RATS AND OF THE LIVER IN HAMSTERS. AFTER ITS SUBCUTANEOUS INJECTION, IT PRODUCES TUMORS OF THE UPPER DIGESTIVE AND RESPIRATORY TRACTS IN HAMSTERS; IT IS CARCINOGENIC AFTER ITS ADMINISTRATION IN SINGLE DOSES. 1979 SUPPLEMENT/SUFFICIENT EVIDENCE FOR CARCINOGENICITY IN ANIMALS.

REPORT 02 : NAS REPORT ON DIET, NUTRITION AND CANCER 1982
AS A GROUP, THE N-NITROSO COMPOUNDS ARE CLEARLY CARCINOGENIC IN NUMEROUS SPECIES OF ANIMALS IN WHICH THEY HAVE BEEN TESTED. POSITIVE RESULTS HAVE BEEN OBTAINED FOR NEARLY ALL OF THE APPROXIMATELY 300 N-NITROSO COMPOUNDS TESTED FOR CARCINOGENICITY IN ONE OR MORE SPECIES. MANY OF THESE COMPOUNDS ARE ALSO MUTAGENIC.

REPORT 03 : IARC MONOGRAPH 17 1978
NO DATA ON THE EMBRYOTOXICITY OR TERATOGENICITY OF NMOR WERE AVAILABLE. NMOR WAS MUTAGENIC IN SALMONELLA TA1530, TA1535, TA1537 AND TA1538 WITH METABOLIC ACTIVATION. THE CHEMICAL INDUCED REVERSE MUTATIONS IN E. COLI. INDUCTION OF SEX-LINKED RECESSIVE LETHAL MUTATIONS AND TRANSLOCATIONS WERE NOTED IN DROSOPHILA EXPOSED TO NMOR.

REPORT 04 : EPA AMBIENT WATER QUALITY CRITERIA OCT 80
NITROSAMINES ARE SUSPECTED OF BEING HUMAN CARCINOGENS. AMBIENT WATER CONCENTRATIONS OF 8 NG/L, .8 NG/L, AND .08 NG/L ARE ESTIMATED TO RESULT IN AN INCREASED CANCER RISK OVER A LIFETIME OF 10 -5, 10 -6, AND 10 -7 RESPECTIVELY.

REPORT 06 : MULTI-AGENCY CROSS CHECK
Y HHS MAGUIRE
N OSHA CARC
Y EPA CAG
N EPA RCRA
Y EPA CLEAN AIR
N EPA CLEAN WATER

REPORT 07 : STATUS REPORT 8EHQ-1186-643 DEC 86
DOW CORNING REPORTED THAT THEY RECENTLY FOUND THAT LOW MOLECULAR WEIGHT NITROSAMINES ARE PRODUCED AND LIBERATED DURING FOAMING OF 2 OF DC'S SILICONE ELASTOMER FOAMS.

REPT'D USE : SOLVENT FOR PAN; NO CURRENT PRODUCTION DATA.

CAS NO : **759-73-9**
CHEM NAME : ***** N-NITROSO-N-ETHYLUREA**

REPORT 01 : ICRDB CANCER GRAM SERIES CK 01 N0 3, ITEM 41 MAR 79
STUDIES WERE DONE OF ORGAN CULTURES OF EMBRYONIC BRAIN TISSUE FROM THE HIPPOCAMPUS REGION OF MONGREL RATS TREATED TRANSPLACENTALLY WITH A SINGLE IV DOSE DURING THE LAST WEEK OF PREGNANCY. ORGAN CULTURES PREPARED ON THE 20TH DAY WERE HIGHLY SENSITIVE TO CPD AS SHOWN BY THE INCREASED SURVIVAL RATE OF THE EXPERIMENTAL CULTURES COMPARED TO THE CONTROLS AND THE APPEARANCE OF EPITHELIAL PROLIFERATIONS OF THE VASCULAR PLEXUS WHICH RESEMBLED ADENOMAS.

REPORT 02 : IARC MONOGRAPH 17 OCT 77
N-NITROSO-N-ETHYLUREA IS CARCINOGENIC IN ALL ANIMAL SPECIES TESTED: MICE, RATS, SYRIAN GOLDEN HAMSTERS, RABBITS, OPOSSUMS, PIGS AND MONKEYS. IT PRODUCES BENIGN AND MALIGNANT TUMORS FOLLOWING ITS ADMINISTRATION BY DIFFERENT ROUTES, INCLUDING SINGLE ORAL DOSES, THE MAIN TARGET ORGAN APPEARS TO VARY WITH THE ROUTE OF ADMINISTRATION. PRENATAL EXPOSURE TO THE SUBSTANCE HAS BEEN SHOWN TO BE PARTICULARLY EFFECTIVE IN PRODUCING TUMORS OF THE NERVOUS SYSTEM. IARC 1979 SUPPLEMENT CLAIMS THERE IS SUFFICIENT EVIDENCE OF CARCINOGENICITY IN ANIMALS.

REPORT 03 : IARC MONOGRAPH 17 1978
NEU INDUCES MUTATIONS IN SALMONELLA, E. COLI, LOWER AND HIGHER PLANTS. IN DROSOPHILA AND IN OTHER ORGANISMS. TREATMENT OF CULTURED HUMAN FIBROBLASTS WITH NEU INDUCED BOTH CHROMATID AND CHROMOSOME ABERRATIONS.

REPORT 06 : MULTI-AGENCY CROSS CHECK
Y HHS MAGUIRE

N OSHA CARC
Y EPA CAG
Y EPA RCRA
Y EPA CLEAN AIR
N EPA CLEAN WATER

REPT'D USE : RESEARCH CHEMICAL AND INTERMEDIATE IN MFR OF DIAZOMETHANE

CAS NO : **16543-55-8**
CHEM NAME : ***** N-NITROSONORNICOTINE**

REPORT 01 : IARC MONOGRAPH 17 1978
N-NITROSONORNICOTINE IS CARCINOGENIC IN RATS, MICE, AND SYRIAN GOLDEN HAMSTERS. FOLLOWING ITS ORAL ADMINISTRATION TO RATS, IT PRODUCES CARCINOMAS OF THE UPPER DIGESTIVE TRACT, MAINLY THE ESOPHAGUS, AND OF THE NASAL CAVITIES. IN HAMSTERS, ITS SUBCUTANEOUS INJECTION PRODUCES MAINLY TRACHEAL TUMORS. IN MICE, ITS INTRAPERITONEAL INJECTION PRODUCES LUNG TUMORS. IN A 1979 SUPPLEMENT, IARC DETERMINED THAT THERE IS SUFFICIENT EVIDENCE FOR THE CARCINOGENICITY OF N-NITROSONORNICOTINE IN ANIMALS.

REPORT 02 : NAS REPORT ON DIET, NUTRITION AND CANCER 1982
AS A GROUP, THE N-NITROSO COMPOUNDS ARE CLEARLY CARCINOGENIC IN NUMEROUS SPECIES OF ANIMALS IN WHICH THEY HAVE BEEN TESTED. POSITIVE RESULTS HAVE BEEN OBTAINED FOR NEARLY ALL OF THE APPROXIMATELY 300 N-NITROSO COMPOUNDS TESTED FOR CARCINOGENICITY IN ONE OR MORE SPECIES. MANY OF THESE COMPOUNDS ARE ALSO MUTAGENIC.

REPORT 03 : IARC MONOGRAPH 17 1978
NO DATA ON THE EMBRYOTOXICITY, TERATOGENICITY, METABOLISM OR MUTAGENICITY OF NNN WERE AVAILABLE.

REPORT 04 : EPA AMBIENT WATER QUALITY CRITERIA OCT 80
NITROSAMINES ARE SUSPECTED OF BEING HUMAN CARCINOGENS. AMBIENT WATER CONCENTRATIONS OF 8 NG/L, .8 NG/L, AND .08 NG/L ARE ESTIMATED TO RESULT IN AN INCREASED CANCER RISK OVER A LIFETIME OF 10 -5, 10 -6, AND 10 -7 RESPECTIVELY.

REPORT 05 : IARC MONOGRAPH 37 SEP 85
FOLLOWING ORAL ADMINISTRATION, PRODUCED CARCINOMAS OF UPPER DIGESTIVE TRACT, MAINLY OESOPHAGUS, AND OF NASAL CAVITY IN RATS AND NASAL-CAVITY TUMORS IN HAMSTERS. FOLLOWING SUBCUTANEOUS ADMINISTRATION, PRODUCED PRIMARILY TUMORS OF NASAL CAVITY IN RATS AND TUMORS OF TRACHEA IN HAMSTERS. INTRAPERITONEAL INJECTION PRODUCED LUNG TUMORS IN MICE AND TUMORS OF NASAL CAVITY AND TRACHEA IN HAMSTERS. EVIDENCE OF DOSE-RESPONSE RELATION-

SHIP AFTER SUBCUTANEOUS ADMINISTRATION TO RATS. SEVERAL METABOLITES TESTED IN MICE BY INTRAPERITONEAL INJECTION, PRODUCING LUNG TUMORS. NNN-1-N-OXIDE TESTED IN RATS AND HAMSTERS BY ORAL ADMINISTRATION; PRODUCED NASAL-CAVITY AND OESOPHAGEAL TUMORS IN RATS.

REPORT 06 : MULTI-AGENCY CROSS CHECK
Y HHS MAGUIRE
N OSHA CARC
Y EPA CAG
N EPA RCRA
N EPA CLEAN AIR
N EPA CLEAN WATFR

REPORT 07 : IARC MONOGRAPH 37 SEP 85
MUTAGENIC TO SALMONELLA TYPHIMURIUM IN PRESENCE OF EXOGENOUS METABOLIC SYSTEM. INDUCES UNSCHEDULED DNA SYNTHESIS IN PRIMARY CULTURES OF RAT HEPATOCYTES.
REPT'D USE : NATURAL CONSTITUENT OF TOBACCO, COMBUSTION POLLUTANT AND RESEARCH CHEMICAL; NO LISTING ON TSCA INVENTORY

CAS NO : **100-75-4**
CHEM NAME : ***** N-NITROSOPIPERIDINE**

REPORT 01 : IARC MONOGRAPH 17 1978
N-NITROSOPIPERIDINE IS CARCINOGENIC IN MICE, RATS, SYRIAN GOLDEN, EUROPEAN AND CHINESE HAMSTERS AND MONKEYS AFTER ITS ADMINISTRATION BY ORAL AND OTHER ROUTES. IT PRODUCES BENIGN AND MALIGNANT TUMORS OF THE LIVER, LUNG, FORESTOMACH AND ESOPHAGUS IN MICE, OF THE LIVER,ESOPHAGUS AND RESPIRATORY SYSTEM IN RATS, AND OF THE UPPER DIGESTIVE TRACT, RESPIRATORY SYSTEM AND LIVER IN HAMSTERS; IT PRODUCES HEPATOCELLULAR CARCINOMAS IN MONKEYS. IT IS CARCINOGENIC IN MICE AND HAMSTERS AFTER SINGLE DOSE ADM. 1979 SUP.-SUFFICIENT EVIDENCE ANIMAL CARCINOGENICITY.

REPORT 02 : NTP TECHNICAL BULLETIN 3
THE COMPOUND WAS FOUND TO BE POSITIVE IN THE SALMONELLA TYPHIMURIUM MUTAGENICITY ASSAY IN ONE LABORATORY.

REPORT 03 : IARC MONOGRAPH 17 1978
NO DATA ON THE EMBRYOTOXICITY OR TERATOGENICITY OF NPIP WERE AVAILABLE. NPIP INDUCED REVERSE MUTATIONS IN E. COLI AND IN SALMONELLA TA100, TA1530 AND TA1535, WITH METABOLIC ACTIVATION.

REPORT 04 : EPA AMBIENT WATER QUALITY CRITERIA OCT 80
NITROSAMINES ARE SUSPECTED OF BEING HUMAN CARCINOGENS.

AMBIENT WATER CONCENTRATIONS OF 8 NG/L, .8 NG/L, AND .08 NG/L ARE ESTIMATED TO RESULT IN AN INCREASED CANCER RISK OVER A LIFETIME OF 10 -5, 10 -6, AND 10 -7 RESPECTIVELY.

REPORT 05 : NAS REPORT ON DIET, NUTRITION AND CANCER 1982
AS A GROUP, THE N-NITROSO COMPOUNDS ARE CLEARLY CARCINOGENIC IN NUMEROUS SPECIES OF ANIMALS IN WHICH THEY HAVE BEEN TESTED. POSITIVE RESULTS HAVE BEEN OBTAINED FOR NEARLY ALL OF THE APPROXIMATELY 300 N-NITROSO COMPOUNDS TESTED FOR CARCINOGENICITY IN ONE OR MORE SPECIES. MANY OF THESE COMPOUNDS ARE ALSO MUTAGENIC.

REPORT 06 : MULTI-AGENCY CROSS CHECK
Y HHS MAGUIRE
N OSHA CARC
Y EPA CAG
Y EPA RCRA
N EPA CLEAN AIR
N EPA CLEAN WATER

REPT'D USE : CIGARETTE SMOKE CONDENSATE, FOOD CONTAMINANT, EPOXY CURING AGENT FOR ELECTRICAL, LEADS AND CARDIOVASCULAR IMPLANTS

CAS NO : **7519-36-0**
CHEM NAME : ***** N-NITROSOPROLINE (NPRO)**

REPORT 01 : IARC MONOGRAPH 17 1978
N-NITROSOPROLINE HAS BEEN TESTED IN MICE AND RATS BY ORAL ADMINISTRATION. ALTHOUGH THESE STUDIES DID NOT INDICATE A CARCINOGENIC EFFECT, THEY WERE INADEQUATE WITH REGARD TO DOSE AND/OR DURATION. NO EVALUATION OF THE CARCINOGENICITY OF THIS COMPOUND COULD BE MADE ON THE BASIS OF THE AVAILABLE DATA.

REPORT 02 : IARC MONOGRAPH 17 1978
NO DATA ON THE EMBRYOTOXICITY OR TERATOGENICITY OF NPRO WERE AVAILABLE. NPRO WAS NOT MUTAGENIC IN SALMONELLA TA1535 EITHER WITH OR WITHOUT METABOLIC ACTIVATION.

REPORT 03 : EPA AMBIENT WATER QUALITY CRITERIA OCT 80
NITROSAMINES ARE SUSPECTED OF BEING HUMAN CARCINOGENS. AMBIENT WATER CONCENTRATIONS OF 8 NG/L, .8 NG/L, AND .08 NG/L ARE ESTIMATED TO RESULT IN AN INCREASED CANCER RISK OVER A LIFETIME OF 10 -5, 10 -6, AND 10 -7 RESPECTIVELY.

REPT'D USE : LABORATORY CHEMICAL; NO LISTING ON TSCA INVENTORY

CAS NO : **930-55-2**
CHEM NAME : ***** N-NITROSOPYRROLIDINE (NPYR)**

REPORT 01 : IARC MONOGRAPH 17 1978
N-NITROSOPYRROLIDINE IS CARCINOGENIC IN RATS AFTER ITS ORAL ADMINISTRATION: IT PRODUCES HEPATOCELLULAR CARCINOMAS. IT ALSO INCREASES THE INCIDENCE OF LUNG ADENOMAS IN MICE FOLLOWING ITS ORAL ADMINISTRATION. A STUDY IN RATS HAS BEEN REPORTED IN WHICH A DOSE-RESPONSE RELATIONSHIP WAS ESTABLISHED. IN A 1979 SUPPLEMENT, IARC DETERMINED THAT THERE WAS SUFFICIENT EVIDENCE FOR THE CARCINOGENICITY OF N-NITROSOPYRROLIDINE IN ANIMALS.

REPORT 02 : IARC MONOGRAPH 17 1978
NO DATA ON THE EMBRYOTOXICITY OR TERATOGENICITY OF NPYR WERE AVAILABLE. NPYR PRODUCES REVERSE MUTATIONS IN E. COLI AND SALMONELLA TA1530, TA1535, TA1537 AND TA1538, WITH METABOLIC ACTIVATION.

REPORT 03 : NAS REPORT ON DIET, NUTRITION AND CANCER 1982
AS A GROUP, THE N-NITROSO COMPOUNDS ARE CLEARLY CARCINOGENIC IN NUMEROUS SPECIES OF ANIMALS IN WHICH THEY HAVE BEEN TESTED. POSITIVE RESULTS HAVE BEEN OBTAINED FOR NEARLY ALL OF THE APPROXIMATELY 300 N.-NITROSO COMPOUNDS TESTED FOR CARCINOGENICITY IN ONE OR MORE SPECIES. MANY OF THESE COMPOUNDS ARE ALSO MUTAGENIC.

REPORT 04 : EPA AMBIENT WATER QUALITY CRITERIA OCT 80
NITROSAMINES ARE SUSPECTED OF BEING HUMAN CARCINOGENS. AMBIENT WATER CONCENTRATIONS OF 160 NG/L, 16 NG/L, AND 1.6 NG/L ARE ESTIMATED TO RESULT IN AN INCREASED CANCER RISK OVER A LIFETIME OF 10 -5, 10 -6, AND 10 -7 RESPECTIVELY.

REPORT 06 : MULTI-AGENCY CROSS CHECK
Y HHS MAGUIRE
N OSHA CARC
Y EPA CAG
Y EPA RCRA
N EPA CLEAN AIR
N EPA CLEAN WATER

REPT'D USE : POLLUTION, CIGARETTE SMOKE CONDENSATE AND FOOD CONTAMINANT; NO EVIDENCE OF COMMERCIAL PRODUCTION OR USE

CAS NO : **13256-22-9**
CHEM NAME : ***** N-NITROSOSARCOSINE**

REPORT 01 : IARC MONOGRAPH 17 1978
N-NITROSOSARCOSINE IS CARCINOGENIC IN MICE AND RATS. IT PRODUCES CARCINOMAS OF THE NASAL CAVITIES IN MICE AND

ESOPHAGEAL CARCINOMAS IN RATS AFTER ITS ORAL ADMINISTRATION, AND LIVERCELL CARCINOMAS IN NEWBORN MICE AFTER ITS IP ADMINISTRATION. IN A 1979 SUPPLEMENT, IARC DETERMINED THAT THERE IS SUFFICIENT EVIDENCE FOR THE CARCINOGENICITY N-NITROSOSARCOSINE IN ANIMALS.

REPORT 02 : EPA AMBIENT WATER QUALITY CRITERIA OCT 80
NITROSAMINES ARE SUSPECTED OF BEING HUMAN CARCINOGENS. AMBIENT WATER CONCENTRATIONS OF 8 NG/L, .8 NG/L, AND .08 NG/L ARE ESTIMATED TO RESULT IN AN INCREASED CANCER RISK OVER A LIFETIME OF 10 -5, 10 -6, AND 10 -7 RESPECTIVELY.

REPORT 03 : IARC MONOGRAPH 17 1978
NO DATA ON THE EMBRYOTOXICITY OR TERATOGENICITY OF NSAR WERE AVAILABLE. NSAR ADMINISTERED TO MICE BY GAVAGE WAS NOT MUTAGENIC IN THE HOST-MEDIATED ASSAY USING SALMONELLA.

REPORT 06 : MULTI-AGENCY CROSS CHECK
Y HHS MAGUIRE
N OSHA CARC
Y EPA CAG
N EPA RCRA
N EPA CLEAN AIR
N EPA CLEAN WATER
REPT'D USE : NO PRODUCTION OR USE DATA AVAILABLE. NO LISTING ON TSCA INVENTORY

CAS NO : **99-55-8**
CHEM NAME : ***** 5-NITRO-O-TOLUIDINE**

REPORT 01 : NCI CARC TRS 107 DHEW PUB NO NIH 78-1357
UNDER THE CONDITIONS OF THIS BIOASSAY 5-NITRO-O-TOLUIDINE WAS CARCINOGENIC IN B6C3F1 MICE, CAUSING HEPATOCELLULAR CARCINOMAS IN BOTH SEXES, AN INCREASE IN THE COMBINED INCIDENCE OF HEMANGIOMAS AND HEMANGIOSARCOMAS IN MALE MICE, AND AN INCREASED INCIDENCE OF HEMANGIOSARCOMAS IN FEMALE MICE. THE COMPOUND WAS NOT CARCINOGENIC IN FISCHER 344 RATS (FEEDING STUDY).

REPORT 06 : MULTI-AGENCY CROSS CHECK
N HHS MAGUIRE
N OSHA CARC
Y EPA CAG
Y EPA RCRA
N EPA CLEAN AIR
N EPA CLEAN WATER
REPT'D USE : DYE INTERMEDIATE

CAS NO : **10024-97-2**
CHEM NAME : ***** NITROUS OXIDE (NITROGEN OXIDE)**

REPORT 01 : EPA 8EHQ-0886- 0628 STATUS REPORT OCT 86
PRENATAL EXPOSURE OF RATS TO 75% NO FOR 8 HRS CAN PERMANENTLY ALTER THE NEUROBEHAVIORAL OUTPUT OF THE CNS.
REPT'D USE : ANESTHETIC

CAS NO : **804-36-4**
CHEM NAME : ***** NITROVIN**

REPORT 01 : IARC MONOGRAPH 31 JUL 83
TEST IN FEMALE RATS BY ADMINISTRATION IN DIET INADEQUATE FOR EVALUATION. INDUCES DNA DAMAGE AND MUTATIONS IN BACTERIA AND MUTATIONS IN INSECTS.
REPT'D USE : VETERINARY BACTERIOSTATIC AND BACTERICIDAL AND CHICKEN GROWTH PROMOTER

CAS NO : **68-22-4**
CHEM NAME : ***** NORETHISTERONE**

REPORT 01 : IARC MONOGRAPH 21 1979
WHEN ADMINISTERED ALONE TO MICE, NORETHISTERONE INCREASED THE INCIDENCE OF BENIGN LIVER CELL TUMORS IN MALES AND OF PITUITARY TUMORS IN FEMALES AND PRODUCED GRANULOSA-CELL TUMORS OF THE OVARY IN FEMALES.

REPORT 02 : IARC MONOGRAPH 21 1979
DOMINANT LETHAL MUTATIONS WERE OBSERVED IN FEMALE MICE TREATED WITH NORETHISTERONE ACETATE DAILY FOR 4 WEEKS. NORETHISTERONE IS EMBRYO LETHAL IN SOME SPECIES AND PRODUCES VIRILIZATION IN FEMALE FETUSES.

REPORT 04 : IARC SUPPL. 4 OCT 83
EVIDENCE FOR ACTIVITY IN SHORT-TERM TESTS IS INADEQUATE TO EVALUATE. NO DATA AVAILABLE TO EVALUATE CARCINOGENICITY IN HUMANS.

REPORT 06 : MULTI-AGENCY CROSS CHECK
Y HHS MAGUIRE
N OSHA CARC
N EPA CAG
N EPA RCRA
N EPA CLEAN AIR
N EPA CLEAN WATER
REPT'D USE : PROGESTIN IN ORAL CONTRACEPTIVES

CAS NO : **51-98-9**
CHEM NAME : ***** NORETHISTERONE ACETATE**

REPORT 01 : IARC MONOGRAPH 21 1979
ADMINISTRATION OF NORETHISTERONE ACETATE INCREASED THE INCIDENCE OF BENIGN LIVER CELL TUMORS IN MALE MICE. IN MALE RATS, ADMINISTRATION OF NORETHISTERONE ALONE INCREASED THE INCIDENCE OF BENIGN LIVER CELL TUMORS.

REPORT 02 : 4TH ANNUAL REPORT ON CARCINOGENS 1985
(MAGUIRE REPT)
NORETHISTERONE ACETATE IN COMBINATION WITH ETHINYLESTRADIOL INCREASED INCIDENCE OF BENIGN TUMORS IN MALE AND FEMALE RATS IN TWO STUDIES.

REPORT 06 : MULTI-AGENCY CROSS CHECK
Y HHS MAGUIRE
N OSHA CARC
N EPA CAG
N EPA RCRA
N EPA CLEAN AIR
N EPA CLEAN WATER
REPT'D USE : TREATMENT OF AMENORRHEA AND ENDOMETRIOSIS

CAS NO : **68-23-5**
CHEM NAME : ***** NORETHYNODREL**

REPORT 01 : IARC MONOGRAPH 21 1979
NORETHYNODREL WAS TESTED BY THE ORAL ROUTE AND BY SUBCUTANEOUS IMPLANTATION IN MICE AND BY THE ORAL ROUTE IN RATS. NORETHYNODREL INCREASED THE INCIDENCE OF PITUITARY TUMORS IN MICE OF BOTH SEXES AND OF MAMMARY TUMORS IN CASTRATED MALES OF ONE STRAIN; IT ALSO INCREASED THE INCIDENCE OF LIVER CELL, PITUITARY AND MAMMARY TUMORS IN MALE RATS.

REPORT 02 : IARC MONOGRAPH 21 1979
NORETHYNODREL DID NOT INDUCE MUTATIONS IN SALMONELLA STRAINS TA98, TA100, TA1535, TA1537 AND TA1538, WITH OR WITHOUT METABOLIC ACTIVATION. NORETHYNODREL WAS REPORT TO BE EMBRYOLETHAL IN SOME SPECIES AND TO HAVE TERATOGENIC EFFECTS IN MICE.
REPT'D USE : PROGESTIN IN ORAL CONTRACEPTIVES

CAS NO : **797-63-7**
CHEM NAME : ***** NORGESTREL**

REPORT 01 : IARC MONOGRAPH 21 1979
NORGESTREL WAS TESTED IN MICE AND RATS ALONE OR IN COMBINATION WITH ETHINYLESTRADIOL BY ORAL ADMINISTRATION. THERE WAS NO INCREASE IN THE INCIDENCE OF TUMORS IN EITHER SPECIES. THE AVAILABLE DATA IN EXPERIMENTAL ANIMALS AND HUMANS ARE INSUFFICIENT TO EVALUATE THE CARCINOGENICITY OF NORGESTREL.

REPORT 02 : IARC MONOGRAPH 21 1979
NO DATA ON THE MUTAGENICITY OF NORGESTREL WERE AVAILABLE. NORGESTREL IS EMBRYOLETHAL FOR PRE-AND POST-IMPLANTATION EMBRYOS IN RATS.
REPT'D USE : ORAL CONTRACEPTIVE - PROGESTIN

CAS NO : **25038-54-4**
CHEM NAME : ***** NYLON 6**

REPORT 01 : IARC MONOGRAPH 19 1979
INTRAPERITONEAL IMPLANTATION OF NYLON 6 IN RATS INDUCED LOCAL SARCOMAS. DATA FROM ONLY ONE EXPERIMENTAL STUDY WITH NYLON 6, AND THE ABSENCE OF BOTH ANIMAL AND HUMAN DATA ON CAPROLACTAM, PRECLUDE A DEFINITE ASSESSMENT OF THE CARCINOGENICITY OF CAPROLACTAM AND OF ITS POLYMER.
REPT'D USE : PLASTICS, HOME FURNISHINGS, CLOTHING, FOOD PACKAGING, ROPES, HOSES AND TIRE CORDING

CAS NO : **2698-41-1**
CHEM NAME : ***** O-CHLOROBENZALMALONONITRILE (CS2)**

REPORT 01 : NTP TR377 NIH- PUB. 90-2832 NOV 89
UNDER THESE INHALATION STUDIES NO EVIDENCE OF CARCINOGENIC ACTIVITY OF CS2 FOR MALE OR FEMALE F344/N RATS EXPOSED TO 0.075, 0.25, OR 0.75 MG/M3 IN AIR FOR UP TO 2 YEARS. NO EVIDENCE OF CARCINOGENIC ACTIVITY FOR MALE OR FEMALE B6C3F1 MICE EXPOSED TO 0.75 OR 1.5 MG/M3 IN AIR FOR UP TO 2 YEARS. CONCENTRATION-RELATED DECREASES IN INCIDENCES OF PITUITARY GLAND ADENOMAS AND LYMPHOMAS OBSERVED IN FEMALE MICE. EXPOSURE CAUSED DEGENERATION AND SQUAMOUS METAPLASIA OF OLFACTORY EPITHELIUM, HYPERPLASIA AND METAPLASIA OF RESPIRATORY EPITHELIUM, AND PROLIFERATION OF PERIOSTEUM OF NASAL PASSAGE OF RATS. IN MICE, EXPOSURE CAUSED SUPPURATIVE INFLAMMATION AND HYPERPLASIA.
REPT'D USE : RIOT CONTROL AGENT

CAS NO : **303-47-9**
CHEM NAME : ***** OCHRATOXIN A**

REPORT 01 : IARC MONOGRAPH 10 1976
OCHRATOXIN A HAS BEEN TESTED ORALLY IN RATS AND TROUT AND BY SUBCUTANEOUS INJECTION IN MICE AND RATS. ALL STUDIES WERE INADEQUATE IN TERMS OF THE NUMBERS OF ANIMALS USED AND SURVIVAL RATES; NO EVALUATION CAN BE MADE.

REPORT 02 : NTP TR 358 NIH PUB. 89-2813 MAY 89
UNDER THE CONDITIONS OF THESE 2-YEAR GAVAGE STUDIES, THERE WAS CLEAR EVIDENCE OF CARCINOGENIC ACTIVITY OF OCHRATIOXIN A FOR MALE F344/N RATS (TUBULAR CELL ADENOMAS AND TUBLAR CELL CARCINOMAS OF THE KIDNEY). FOR FEMALE F344/N RATS AND (TUBULAR CELL ADENOMAS AND TUBLAR CELL CARCINOMAS OF THE KIDNEY AND FIBROADENOMAS OF THE MAMMARY GLAND).

REPORT 03 : IARC MONOGRAPH 31 JUL 83
TESTED IN MICE AND RATS BY ORAL ADMINISTRATION AND SUBCUTANEOUS INJECTION. IN LIMITED FEEDING STUDY IN MICE, HEPATIC-CELL ADENOMAS, SOLID RENAL-CELL TUMORS AND CYSTIC ADENOMAS OF THE KIDNEY OBSERVED. THE OTHER EXPERIMENTS CONSIDERED INADEQUATE FOR EVALUATION.

REPORT 04 : IARC MONOGRAPH 31 JUL 83
NOT MUTAGENIC IN BACTERIA OR YEAST OR IN MAMMALIAN CELLS IN VITRO; NEGATIVE IN A BACTERIAL REPAIR TEST. EMBRYOLETHAL AND TERATOGENIC TO MICE, RATS AND HAMSTERS.

REPORT 06 : MULTI-AGENCY CROSS CHECK
Y HHS MAGUIRE
N OSHA CARC
N EPA CAG
N EPA RCRA
N EPA CLEAN AIR
N EPA CLEAN WATER

REPT'D USE : FUNGAL TOXIN FOUND IN CORN, PEANUTS, STORED GRAINS, AND OTHER DECAYING VEGETATION; NOT PRODUCED OR USED COMMERCIALLY

CAS NO : **27858-07-7**
CHEM NAME : ***** OCTABROMOBIPHENYL**

REPORT 01 : IARC MONOGRAPH 18 1978
ONLY ONE COMMERCIAL PREPARATION OF PBB (TECH GRADE HEXABROMOBIPHENYL) COMPOSED PRIMARILY OF PENTA-, HEXA-,

AND HEPTABROMINATED CONGENERS, WAS TESTED IN ONE EXPERIMENT IN RATS BY ORAL ADMINISTRATION OF SINGLE DOSES. ALTHOUGH NEOPLASTIC CHANGES OF THE LIVER WERE REPORTED, THE STUDY WAS CONSIDERED TO BE INADEQUATE. NO EVALUATION OF THE CARCINOGENICITY OF PBB COULD BE MADE ON THE BASIS OF THE AVAILABLE DATA.

REPT'D USE : FR CHEMICAL; NO KNOWN CURRENT PRODUCTION

CAS NO : **556-67-2**

CHEM NAME : ***** OCTAMETHYLCYCLOTETRASILOXANE**

REPORT 01 : EPA TSCA 4(A) PROPOSED RULE OCT 85
NOTICE 50FR45123
EPA PROPOSES THAT MFRS AND PROCESSORS CONDUCT CHEMICAL FATE AND ENVIRONMENTAL EFFECTS TESTING FOR OMCTS. REQUESTED TESTS INCLUDE BIODEGRADATION STUDIES; ENVIRONMENTAL EFFECTS STUDIES IN FISH AND INVERTEBRATES; AND A MALLARD REPRODUCTION TEST. THIS IS EPA'S RESPONSE TO ITC'S DESIGNATION OF OMCTS FOR PRIORITY TESTING UNDER TSCA 4(A).

REPORT 02 : EPA TSCA 8(E) 0288-0718 FEB 88
DOW CORNING REPORTED RESULTS FROM CHRONIC DAPHNIA MAGNA REPRODUCTIVE LIMIT TEST OF MIXTURE OF POLYETHYLENE GLYCOL SORBITAN MONOLAURATE (CAS NO. 9005-64-5) AND OCTAMETHYLCYCLOTETRASILOXANE (CAS NO. 556-67-2) SURVIVAL OF DAPHNIA EXPOSED TO 10 PPM PGSM/100 PPB OMCTS SIGNIFICANTLY LESS THAN CONTROLS. A NO OBSERVABLE EFFECT LEVEL CAN NOT BE DETERMINED FROM TEST, NOR CAN SYNERGISTIC EFFECT BETWEEN PGSM AND OMCTS BE ELIMINATED FINAL REPORT BEING SUBMITTED TO EPA UNDER SECTION 8(D) OF TSCA.

REPT'D USE : CHEM INTERMEDIATE (80-85%) IN PRODUCTION OF SILICONE RUBBERS, SURFACE ACTIVE AGENTS, LUBRICANTS, HIGH TEMPERATURE AUTOMOBILE HOSES, SEALS, WIRE COATINGS, COSMETICS, CLEANERS AND MEDICINES

CAS NO : **111-87-5**

CHEM NAME : ***** 1-OCTANOL**

REPORT 01 : EPA TSCA 8(E)-1088-0762 NOV 88
AMOCO REPORTS ACUTE INHALATION STUDY INDICATES VERY HIGH CONCENTRATIONS OF 1-OCTANOL VAPOR FOR AN EXTENDED PERIOD OF TIMS IS CAPABLE OF PRODUCING TEMPORARY LUNG DAMAGE. HOWEVER, TO PRODUCE THE VAPORS USED IN THIS STUDY, IT WAS NECESSARY TO HEAT THE SAMPLE TO APPROXIMATELY 300 DEGREES C (THE BOILING POINT OF 1-OCTANOL IS 196 DEGREES C). GIVEN THAT THE BOILING POINT OF 1-OCTANOL IS SO HIGH, THE POTENTIAL FOR HUMAN EXPOSURE TO THE MASSIVE CONCENTRATIONS USED IN THIS STUDY IS UNLIKELY. ALSO, IN

RATS AN EXPOSURE OF GREATER THAN ONE HOUR WAS REQUIRED TO PRODUCE THE EFFECTS SEEN; IT IS ALSO UNLIKELY THAT HUMANS WOULD BE EXPOSED FOR SUCH A LONG DURATION.

REPT'D USE : PERFUMERY, COSMETICS, ORGANIC SYNTHESIS, SOLVENT, MANUFACTURE OF HIGH-BOILING ESTERS, ANTIFOAMING AGENT AND FLAVORING AGENT

CAS NO : **50-27-1**

CHEM NAME : ***** OESTRIOL**

REPORT O1 : IARC MONOGRAPH 21 1979

OESTRIOL WAS TESTED BY SUBCUTANEOUS IMPLANTATION IN CASTRATED MICE AND IN RATS AND HAMSTERS. IT INCREASED THE INCIDENCE AND ACCELERATED THE APPEARANCE OF MAMMARY TUMORS IN BOTH MALE AND FEMALE MICE AND PRODUCED KIDNEY TUMORS IN HAMSTERS. STUDIES IN HUMANS STRONGLY SUGGEST THAT THE ADMINISTRATION OF OESTROGENS IS CAUSALLY RELATED TO AN INCREASED INCIDENCE OF ENDOMETRIAL CARCINOMA; THERE IS NO EVIDENCE TIIAT OESTRIOL IS DIFFERENT FROM OTHER OESTROGENS IN THIS RESPECT.

REPT'D USE : CLIMACTERIC MEDICATION

CAS NO : **112-90-3**

CHEM NAME : ***** OLEYLAMINE (9-OCTADECENYLAMINE, ODA)**

REPORT 01 : EPA TSCA 4(A) PROPOSED RULE NOTICE 49FR45610 NOV 84

IN RESPONSE TO ITC'S DESIGNATION OF ODA FOR PRIORITY CONSIDERATION OF HEALTH EFFECTS TESTING, EPA IS PROPOSING THAT MFRS AND PROCESSORS OF ODA BE REQUIRED UNDER TSCA 4(A) TO PERFORM TESTS FOR DEVELOPMENTAL TOXICITY; A 90-DAY DERMAL SUBCHRONIC TOXICITY WHICH WILL INCLUDE OBSERVATIONS ON NEUROBEHAVIORAL EFFECTS AND REPRODUCTIVE SYSTEM HISTPATHOLOGY; DETERMINATION OF DERMAL ABSORPTION AND A TIERED MUTAGENICITY SCHEME WITH TRIGGERS TO ONCOGENICITY TESTING.

REPORT 02 : CMA 14-DAY DERMAL RANGE FINDING STUDY APR 85

TEST CONCENTRATIONS OF ODA OF 1.5 AND 3.0% (V/V) IN MINERAL OIL PRODUCED MODERATE TO SEVERE SKIN IRRITATION IN MALE AND FEMALE SPRAGUE-DAWLEY RATS. BODY WEIGHT WAS ALSO SIGNIFICANTLY REDUCED. A CONCENTRATION OF 0.3% (V/V) ODA IN MINERAL OIL PRODUCED MILD TO MODERATE IRRITATION AND FLAKING OF THE OUTER LAYERS OF THE EPIDERMIS. THIS LATTER CONCENTRATION IS RECOMMENDED FOR FUTURE TESTS.

REPORT 03 : EPA TSCA 4(A) NOTICE 52FR31962 AUG 87

EPA ISSUING FINAL RULE UNDER TSCA SECT 4(A) REQUIRING MANUFACTURERS AND PROCESSORS OF CPD TO TEST CHEMICAL

FOR DEVELOPMENTAL TOXICITY AND FOR MUTAGENICITY WITH A 2-TIERED TEST SCHEME. EPA TERMINATING RULEMAKING FOR PROPOSED 90-DAY DERMAL SUBCHRONIC TESTING.

REPT'D USE : ADDITIVE FOR LUBRICANTS, CHEMICAL INTERMEDIATE, GLUES AND IN ASPHALT PREPARATION

CAS NO : **UNKNOWN-22**

CHEM NAME : ***** ORANGE HISALONIC**

REPORT 01 : CPSC MONOGRAPH ON FIBER REACTIVE DYES DEC 81
RATS TREATED WITH ORANGE HISALONIC EXHIBITED DECREASED WEIGHT GAIN AND A STATISTICALLY SIGNIFICANT DECREASE IN SERUM PROTEINS. MORPHOLOGICAL ALTERATIONS OF THE LIVER AND KIDNEY WERE ALSO OBSERVED.

REPT'D USE : NO KNOWN CURRENT PRODUCTION

CAS NO : **CLASS-16**

CHEM NAME : ***** ORGANOTIN COMPOUNDS**

REPORT 01 : NIOSH CRITERIA DOCUMENT NOV 76
NIOSH RECOMMENDS A STANDARD OF 0.1 MG/CU M TWA BASED ON EYE, SKIN, LIVER, CNS, AND CARDIAC EFFECTS. THE PRESENT OSHA STANDARD IS 0.1 MG/CU M TWA.

REPT'D USE : CATALYST, STERILIZER PESTICIDE, PRESERVATIVE, HEAT STABILIZER AND RODENT REPELLANT

CAS NO : **2386-87-0**

CHEM NAME : ***** 7-OXABICYCLO(4.1.0)HEPTANE-3-CARBOXYLIC ACID**

REPORT 01 : EPA TSCA 8(E) NOTICE 8EHQ-0784-0522 JUL 84
WITCO CHEMICAL PROVIDED REPORTS FROM BATTERY OF IN VITRO GENOTOXICITY STUDIES WITH MIXTURE OF 7-OXABICYCLO-(4.1.0)HEPTANE-3-CARBOXYLIC ACID, 7-OXABICYCLO(4.1.0)-HEPT-3-YLMETHYL ESTER AND PHOSPHOROUS ACID, ISODECYL DIPHENYL ESTER SHOWING INCREASED REVERTANT FREQUENCY IN SALMONELLA TYPHIMURIUM STRAIN TA 1535 WITH ACTIVATION, INCREASED MUTATION FREQUENCY IN CULTURED MOUSE LYMPHOMA CELLS WITH AND WITHOUT ACTIVATION, AND INCREASED NUMBER OF SISTER CHROMATID EXCHANGES IN CULTURED CHO CELLS WITH AND WITHOUT ACTIVATION.

CAS NO : **604-75-1**
CHEM NAME : ***** OXAZEPAM**

REPORT 01 : IARC MONOGRAPH 13 1977
IN THE ONLY AVAILABLE REPORT, OXAZEPAM, A MAJOR METABOLITE OF DIAZEPAM, WAS CARCINOGENIC IN MICE AFTER ITS ORAL ADMINISTRATION, PRODUCING LIVER-CELL TUMORS. CPD DID NOT CAUSE NON-DISJUNCTION AND CROSSING-OVER IN A. MIDULANS. MAMMALIAN METABOLIC ACTIVATION SYSTEMS WERE NOT USED IN THIS TEST.

CAS NO : **5026-74-4**
CHEM NAME : ***** N-(4-(OXIRANYLMETHOXY)PHENYL)-N-(OXIRANYLMETHYL)-OXIRANEMETHANAMINE**

REPORT 01 : EPA TSCA 8(E) NOTICE 8EHQ-1281-0426 JAN 82
CIBA-GEIGY CORPORATION SUBMITTED REPORTS FROM A BATTERY OF SHORT-TERM IN VITRO AND IN VIVO MUTAGENICITY STUDIES. POSITIVES RESULTS WERE OBTAINED IN A CHINESE HAMSTER (BONE MARROW) SISTER CHROMATID EXCHANGE STUDY, CHINESE HAMSTER (BONE MARROW) NUCLEUS ANOMALY TEST (SOMATIC INTERPHASE NUCLEI), MOUSE LYMPHOMA CELL POINT MUTATION ASSAY, AND SALMONELLA AND SACCHAROMYCES TESTS WITH AND WITHOUT METABOLIC ACTIVATION.
REPT'D USE : EPOXY RESIN MONOMER FOR ADHESIVES/LAMINATES

CAS NO : **101-80-4**
CHEM NAME : ***** 4,4'-OXYDIANILINE**

REPORT 01 : IARC MONOGRAPH 29 1982
4,4-DIAMINODIPHENYL ETHER WAS TESTED IN MICE AND RATS BY ORAL ADMINISTRATION AND BY SC INJECTION. IT PRODUCED BENIGN AND MALIGNANT LEVEL-CELL TUMORS IN BOTH SPECIES AND ADENOMAS OF THE HARDERIAN GLAND IN MICE . THERE IS SUFFICIENT EVIDENCE FOR THE CARCINOGENICITY OF 4,4-DIAMINODIPHENYL ETHER IN ANIMALS.

REPORT 02 : NCI/NTP CARC TRS NO 205 DHHS PUB. NO. NIH 80-1761
THE COMPOUND WAS CARCINOGENIC FOR MALE AND FEMALE F344 RATS PRODUCING HEPATOCELLULAR CARCINOMAS OR NEOPLASTIC NODULES AND FOLLICULAR-CELL ADENOMAS OR CARCINOMAS OF THE THYROID. IT WAS ALSO CARCINOGENIC FOR MALE AND FEMALE B6C3F1 MICE PRODUCING ADENOMAS IN THE HARDERIAN GLANDS, HEPATOCELLULAR ADENOMAS OR CARCINOMAS, AND (IN FEMALES) FOLLICULAR-CELL ADENOMAS IN THE THYROID (FEEDING STUDY).

REPORT 03 : IARC MONOGRAPH 29 1982
4,4'-DIAMINODIPHENYL ETHER WAS MUTAGENIC TO SALMONELLA STRAINS TA98 AND TA100 WHEN TESTED IN THE PRESENCE OF A METABOLIC ACTIVATION SYSTEM. NO DATA WERE AVAILABLE TO ASSESS THE TERATOGENICITY OF THIS CHEMICAL.

REPORT 06 : MULTI-AGENCY CROSS CHECK
Y HHS MAGUIRE
N OSHA CARC
N EPA CAG
N EPA RCRA
N EPA CLEAN AIR
N EPA CLEAN WATER

REPT'D USE : INTERMEDIATE IN MFR OF POLYIMID/POLYESTERIOMIDE AND HI TEMP INDUSTRIAL RESINS

CAS NO : **102-77-2**

CHEM NAME : ***** N-OXYDIETHYLENE-2-BENZOTHIAZYL SULFENAMIDE**

REPORT 01 EPA TSCA 8(E) NOTICE 8EHQ-0280-0334
BF GOODRICH SUBMITTED A SUMMARY OF THE RESULTS FROM A BATTERY OF IN VITRO MUTAGENICITY TESTS ON N-OXYDI-ETHYLENE THIOCARBAMYL-N-OXYDIETHLENE AND N-OXYDIETHYLENE-2-BENZOTHIAZYL SULFENAMIDE. SUBMITTER REPORTS THAT EACH COMPOUND WAS SHOWN TO GIVE POSITIVE RESULTS IN SEVERAL OF THE PERFORMED SHORT-TERM ASSAYS.

CAS NO : **13752-51-7**

CHEM NAME : ***** N-OXYDIETHYLENE THIOCARBAMYL-N-OXYDIETHYLENE**

REPORT 01 : EPA TSCA 8(E) NOTICE 8EHQ-0280-0334
BF GOODRICH SUBMITTED A SUMMARY OF THE RESULTS FROM A BATTERY OF IN-VITRO MUTAGENICITY TESTS ON N-OXDIETHYLENE THIOCARBAMYL-N-OXYDIETHYLENE AND N-OXYDIETHYLENE-2-BENZOTHIAZYL SULFENAMIDE. BFG REPORTS THAT EACH COMPOUND WAS SHOWN TO GIVE POSITIVE RESULTS IN SEVERAL OF THE SHORT-TERM ASSAYS.

CAS NO : **434-07-1**

CHEM NAME : ***** OXYMETHOLONE**

REPORT 01 : IARC SUPPLEMENT 1 1979
NO DATA FROM EXPERIMENTAL ANIMAL STUDIES WERE AVAILABLE. TEN CASES OF LIVER CELL TUMORS HAVE BEEN REPORT-

ED IN PATIENTS TREATED WITH OXYMETHOLONE ALONE OR IN COMBINATION WITH OTHER DRUGS; HOWEVER, A CAUSAL RELATIONSHIP CANNOT BE ESTABLISHED. IN A 1979 SUPPLEMENT, IARC CONCLUDED THAT OXYMETHOLONE IS PROBABLY CARCINOGENIC FOR HUMANS.

REPORT 06 : MULTI-AGENCY CROSS CHECK
Y HHS MAGUIRE
N OSHA CARC
N EPA CAG
N EPA RCRA
N EPA CLEAN AIR
N EPA CLEAN WATER
REPT'D USE : ANDROGENIC-ANABOLIC STEROID HORMONE

CAS NO : **129-20-4**
CHEM NAME : ***** OXYPHENBUTAZONE**

REPORT 01 : IARC MONOGRAPH 13 1977
A NUMBER OF CASES OF LEUKEMIA WERE REPORTED BETWEEN 1960 AND 1966 IN SUBJECTS TREATED WITH OXYPHENBUTAZONE. THE AVAILABLE EVIDENCE IS, HOWEVER, INSUFFICIENT TO SUBSTANTIATE THE SUGGESTION THAT USE OF OXYPHENBUTAZONE IS RELATED TO SUBSEQUENT DEVELOPMENT OF LEUKEMIA.
REPT'D USE : ANTI-INFLAMMATORY AGENT

CAS NO : **2058-46-0**
CHEM NAME : ***** OXYTETRACYCLINE HYDROCHLORIDE**

REPORT 01 : NTP TR 315 NIH PUB NO. 87-2571 JAN 87
2-YR FEED STUDIES GAVE EQUIVOCAL EVIDENCE OF CARCINOGENICITY FOR MALE F344/N RATS. THIS IS INDICATED BY INCREASED INCIDENCE OF PHEOCHROMOCYTOMAS OF ADRENAL GLAND. ALSO EQUIVOCAL EVIDENCE OF CARCINOGENICITY FOR FEMALE F344/N RATS AS SHOWN BY INCREASED INCIDENCE OF ADENOMAS OF THE PITUITARY GLAND. NO EVIDENCE OF CARCINOGENICITY FOR MALE OR FEMALE B6C3F1 MICE FED DIETS CONTAINING 6300 OR 12500 PPM OF THE CPD FOR 2 YEARS.
REPT'D USE : ANTIBIOTIC

CAS NO : **20265-96-7**
CHEM NAME : ***** PARA-CHLOROANILINE HYDROCHLORIDE**

REPORT 01 : NTP TR 351 NIH PUB. 89-2806 JUL 89
UNDER 2-YEAR WATER GAVAGE STUDIES CLEAR EVIDENCE OF CARCINOGENIC ACTIVITY OF P-CHLOROANILINE HYDROCHLORIDE

FOR MALE F344/N RATS INDICATED BY INCREASED INCIDENCES OF UNCOMMON SARCOMAS OF SPLEEN. PHEOCHROMOCYTOMAS OF ADRENAL GLAND MAY HAVE BEEN ASSOCIATED WITH CHEMICAL ADMINISTRATION. EQUIVOCAL EVIDENCE OF CARCINOGENIC ACTIVITY FOR FEMALE F344/N RATS INDICATED BY PRESENCE OF UNCOMMON SARCOMAS OF SPLEEN IN ONE MID AND ONE HIGH DOSE ANIMAL AND PHEOCHROMOCYTOMAS OF ADRENAL GLAND, SOME EVIDENCE OF CARCINOGENIC ACTIVITY FOR MALE B6C3F1 MICE INDICATED BY INCREASED INCIDENCES OF HEPATOCELLULAR NEOPLASMS AND OF HEMANGIOSARCOMAS OF LIVER OR SPLEEN. NO EVIDENCE OF CARCINOGENIC.

REPT'D USE : DYE INTERMEDIATE AND DEGRADATION PRODUCT IN PHARMACEUTICALS

CAS NO : **10048-32-5**

CHEM NAME : ***** PARASORBIC ACID**

REPORT 01 : IARC MONOGRAPH 10 1976

PARASORBIC ACID ADMINISTERED BY SUBCUTANEOUS INJECTION TO RATS PRODUCED LOCAL SARCOMAS.

REPORT 02 : NAS REPORT ON DIET, NUTRITION AND CANCER 1982

SARCOMAS RESULTED WITHIN TWO YEARS IN RATS THAT HAD RECEIVED REPEATED SUBCUTANEOUS INJECTIONS OF PARASORBIC ACID. NO STUDIES CONCERNING THE MUTAGENICITY OF PARASORBIC ACID COULD BE IDENTIFIED.

REPT'D USE : FRUIT DRINKS AND MILK ACIDIFIER

CAS NO : **56-38-2**

CHEM NAME : ***** PARATHION**

REPORT 01 : IARC MONOGRAPH 30 JAN 83

TESTED FOR CARCINOGENICITY IN ONE STRAIN OF MICE AND IN THREE STRAINS OF RATS BY ADMINISTRATION IN DIET. ALTHOUGH DOSE-RELATED INCREASE IN INCIDENCE OF ADRENAL CORTICAL ADENOMAS OBSERVED IN MALE AND FEMALE RATS OF ONE STRAIN, SIGNIFICANCE OF OCCURRENCE OF LESIONS IN AGED RATS IS NOT WELL UNDERSTOOD. THE OTHER EXPERIMENTS IN RATS AND THE EXPERIMENT IN MICE WERE CONSIDERED TO BE INADEOUATE FOR EVALUATION.

REPORT 02 : NCI CARC TRS 70 DHEW PUB NO NIH 79-1320

UNDER THE CONDITIONS OF THIS BIOASSAY, PARATHION WAS NOT CARCINOGENIC TO B6C3F1 MICE. IN THE MALE AND FEMALE OSBORNE-MENDEL RATS RECEIVING PARATHION IN THEIR DIET, THERE WAS A HIGHER INCIDENCE OF CORTICAL TUMORS OF THE ADRENAL THAN IN POOLED OR HISTORICAL CONTROLS, SUGGESTING THAT PARATHION IS CARCINOGENIC TO THIS STRAIN OF RAT.

REPORT 03 : NIOSH CRITERIA DOCUMENT JUN 76
NIOSH RECOMMENDS A STANDARD OF 0.05 MG/CU M TWA BASED ON CNS EFFECTS. THE PRESENT OSHA STANDARD IS 0.11 MG/CU M TWA.

REPORT 04 : IARC MONOGRAPH 30 JAN 83
PARATHION PRODUCED EMBRYOCIDAL EFFECTS AND FOETAL GROWTH REDUCTION BUT NO MALFORMATION IN MICE AND RATS AT DOSES THAT WERE GENERALLY BELOW LEVEL THAT IS TOXIC FOR MATERNAL ORGANISM.

REPORT 05 : IARC MONOGRAPH 30 JAN 83
PARATHION NOT MUTAGENIC TO WIDE RANGE OF MICROORGANISMS IN PRESENCE OR ABSENCE OF RAT LIVER MICROSOMAL PREPARATIONS. IT DID NOT INDUCE UNSCHEDULED DNA SYNTHESIS IN CULTURED MAMMALIAN CELLS, RECESSIVE LETHAL MUTATIONS IN DROSOPHILA MELANOGASTER OR DOMINANT LETHAL MUTATIONS IN MICE. NO EVIDENCE HAS BEEN FOUND THAT PARATHION IS MUTAGENIC.

REPORT 06 : MULTI-AGENCY CROSS CHECK DEC 86
N HHS MAGUIRE
N OSHA CARC
N EPA CAG
N EPA RCRA
N EPA CLEAN AIR
Y EPA CLEAN WATER

REPT'D USE : INSECTICIDE AND ACARICIDE

CAS NO : **149-29-1**
CHEM NAME : ***** PATULIN**

REPORT 01 : IARC MONOGRAPH 10 1976
IN THE ONLY STUDY AVAILABLE, PATULIN WAS SHOWN TO PRODUCE SARCOMAS IN RATS AT THE SITE OF ITS SUBCUTANEOUS INJECTION. PATULIN OF UNSPECIFIED PURITY AND ORIGIN INDUCED SINGLE AND DOUBLE STRAND BREAKS IN DNA OF HELA CELLS.

REPORT 02 : IARC VOL 40 1986
INADEQUATE EVIDENCE FOR CARCINOGENICITY IN TEST ANIMALS, BUT NO EVALUATION COULD BE MADE OF CARCINOGENICITY TO HUMANS. DEGREE OF EVIDENCE INADEQUATE FOR GENETIC ACTIVITY IN SHORT-TERM TESTS. DATA WAS INADEQUATE TO EVALUATE REPRODUCTIVE EFFECTS OR PRENATAL TOXICITY OF THE CPD TO TEST ANIMALS. INDUCED DNA DAMAGE IN B. SUBTILIS BUT NOT IN E. COLI. CPD WAS NOT MUTAGENIC TO S. TYPHIMURIUM, BUT INDUCED DNA STRAND BREAKS IN HELA CELLS BUT NOT UNSCHEDULED DNA SYNTHESIS IN CULTURED

MAMMALIAN CELLS. INDUCED SCE'S IN CULTURED MAMMALIAN CELLS. INDUCED CHROMOSOMAL ABERRATIONS BUT NOT SCE'S IN BONE-MARROW CELLS OF CHIN. HAMSTERS IN VIVO.
REPT'D USE : EXPERIMENTAL ANTIBIOTIC

CAS NO : **90-65-3**
CHEM NAME : ***** PENICILLIC ACID**

REPORT 01 : IARC MONOGRAPH 10 1976
PENICILLIC ACID WAS TESTED BY SUBCUTANEOUS INJECTION IN MICE AND RATS. IT PRODUCED LOCAL SARCOMAS. PENICILLIC ACID INDUCED SINGLE AND DOUBLE STRAND BREAKS IN DNA OF HELA CELLS.
REPT'D USE : NO AVAILABLE PRODUCTION OR USE DATA

CAS NO : **132-98-9**
CHEM NAME : ***** PENICILLIN VK**

REPORT 01 : NTP TR 336 NIH PUB 88-2592 JUL 88
UNDER CONDITIONS OF 2-YR GAVAGE STUDY, NO EVIDENCE OF CARCINOGENICITY FOR F344/N RATS OR B6C3F1 MICE ADMINISTERED 500 OR 1,000 MG/KG CPD IN CORN OIL BY GAVAGE, 5 DAYS/WEEK FOR 103 WEEKS. DECREASED SURVIVAL RATE OF LOW AND HIGH DOSE FEMALE RATS REDUCED SENSITIVITY FOR DETERMINING PRESENCE OR ABSENCE OF CARCINOGENIC RESPONSE.
REPT'D USE : DRUG

CAS NO : **85-22-3**
CHEM NAME : ***** PENTABROMOETHYLBENZENE (PEB)**

REPORT 01 : PROPOSED TEST RULE, 50FR46785 NOV 85
IN RESPONSE TO ITC'S DESIGNATION OF PEB FOR PRIORITY TESTING FOR HEALTH AND ENVIRONMENTAL EFFECTS UNDER TSCA 4(A), EPA IS PROPOSING THAT MFRS AND PROCESSORS BE REQUIRED TO TEST PEB FOR CHEMICAL FATE AND ENVIRONMENTAL EFFECTS.
REPT'D USE : FLAME RETARDANT IN PLASTICS

CAS NO : **3555-11-1**
CHEM NAME : ***** PENTABROMOPHENYL ALLYL ETHER**

REPORT 01 : EPA TSCA 8(E) NOTICE 8EHQ-0179-0268S
CONFIDENTIAL SUBMITTER PROVIDED THE FINAL REPORTS FROM TWO SHORT-TERM IN VITRO TESTS ON PENTABROMOPHENYL ALLYL ETHER. RESULTS FROM A SALMONELLA ASSAY AND A BHK

21 CELL TRANSFORMATION TEST INDICATE THAT THE CHEMICAL IS A WEAK MUTAGEN.

CAS NO : **608-93-5**
CHEM NAME : ***** PENTACHLOROBENZENE**

REPORT 01 : NTP TEST REGIMEN
THE COMPOUND WAS FOUND TO BE NEGATIVE IN THE SALMONELLA TYPHIMURIUM MUTAGENICITY ASSAY.

REPORT 02 : EPA RCRA HEALTH AND ENV. EFFECTS PROFILE OCT 80
ORAL FEEDING OF QCB TO PREGNANT RATS HAS PRODUCED DEVELOPMENTAL EFFECTS AND DECREASED BODY WEIGHTS IN FETUSES. A SINGLE STUDY HAS ALLUDED TO CARCINOGENIC EFFECTS OF QCB IN MICE AND LACK OF CARCINOGENIC EFFECTS IN DOGS AND RATS. THE DETAILS OF THIS STUDY WERE NOT AVAILABLE FOR EVALUATION.

REPORT 03 : EPA RESPONSE TO ITC RECOMMS 45FR4852 JUL 80
IN RESPONSE TO THE ITC RECOMMENDATIONS (42FR55026), THE EPA PROPOSES THAT MANUFACTURES AND PROCESSORS OF CHLORINATED BENZENES, EXCEPT HEXACHCHLOROBENZENE, CONDUCT TESTING FOR TERATOGENICITY/REPRODUCTIVE EFFECTS AND OTHER CHRONIC EFFECTS. THE EPA WILL CONDUCT MUTAGENICITY TESTS AND DECIDE ON FURTHER TIER TESTING AS APPROPRIATE. THE EPA BELIEVES IT IS NOT REASONABLE TO REQUIRE AN EPIDEMIOLOGIC STUDY FOR THESE CHEMICALS.

REPORT 06 : MULTI-AGENCY CROSS CHECK
N HHS MAGUIRE
N OSHA CARC
N EPA CAG
Y EPA RCRA
N EPA CLEAN AIR
N EPA CLEAN WATER
REPT'D USE : INTERMEDIATE IN MFR OF AGRICULTURAL CHEMICALS

CAS NO : **76-01-7**
CHEM NAME : ***** PENTACHLOROETHANE**

REPORT 01 : NIOSH CURRENT INTELLIGENCE BULLETIN 27 AUG 78
NCI BIOASSAY TESTING IN PROGRESS AS OF JULY 1978 BUT NO REPORT IS AVAILABLE. NO OSHA EXPOSURE STANDARD EXISTS. ALL CHLOROETHANE COMPOUNDS ARE KNOWN TO CAUSE CNS DEPRESSION IN LABORATORY ANIMALS. THIS COMPOUND INDUCES LIVER, KIDNEY, AND LUNG CHANGES IN CATS; FATTY DEGENERATION OF THE LIVER AND KIDNEY AND LUNG INJURIES IN DOGS; AND LIVER DYSFUNCTION IN SHEEP. NIOSH RECOMMENDS THAT

THE COMPOUND SHOULD BE TREATED IN THE WORKPLACE WITH CAUTION BECAUSE OF ITS RELATION TO FOUR CARCINOGENIC CHLOROETHANE COMPOUNDS.

REPORT 02 : IARC VOL 41 1986
NO DATA AVAILABLE TO EVALUATE REPRODUCTIVE EFFECTS OR PRENATAL TOXICITY OF CPD TO TEST ANIMALS OR HUMANS. NO DATA AVAILABLE ON CARCINOGENICITY OF PURE CPD TO TEST ANIMALS OR TO HUMANS. CPD IS NOT MUTAGENIC TO S. TYPHIMURIUM.

REPORT 03 : NTP TECHNICAL BULLETIN 6 JAN 82
THE COMPOUND WAS NEGATIVE IN THE SALMONELLA TYPHIMURIUM MUTAGENICITY ASSAY IN ONE LABORATORY.

REPORT 05 : NTP TR 232 NIH PUB 83-1788
UNDER THE CONDITIONS OF THIS BIOASSAY, TECHNICAL GRADE PENTACHLOROETHANECONTAINING4.2%HEXACHLOROETHANE(A KNOWN CARCINOGEN IN MICE) WAS NOT CARCINOGENIC IN F344/N RATS. THE DECREASED SURVIVAL OF DOSED RATS MIGHT HAVE REDUCED THE SENSITIVITY FOR A CARCINOGENIC RESPONSE IN THIS SPECIES. PENTACHLOROETHANE WAS NEPHROTOXIC TO MALE RATS. TECH. GRADE PENTACHLOROETHANE WAS CARCINOGENIC FOR B6C3F1 MICE, CAUSING HEPATOCELLULAR CARCINOMAS IN MALES AND FEMALES, AND ADENOMAS IN FEMALES.

REPORT 06 : MULTI-AGENCY CROSS CHECK
N HHS MAGUIRE
N OSHA CARC
N EPA CAG
Y EPA RCRA
N EPA CLEAN AIR
Y FPA CLEAN WATER

REPORT 07 : 51 CFR 32077 - DETERMINATION OF SIGNIFICANT NEW USE SEP 86
EPA IS PROMULGATING A SIGNIFICANT NEW USE RULE UNDER TSCA 5(AA)(2) THAT REQUIRES MFR TO NOTIFY EPA AT LEAST 90 DAYS BEFORE BEGINNING TO MFR, IMPORT OR PROCESS PENTACHLOROETHANE. EPA BELIEVES PCE IS HAZARDOUS TO HUMAN HEALTH AND MAY LEAD TO SIGNIFICANT HUMAN EXPOSURE.

REPT'D USE : NO REPORTED COMMERCIAL USE.

CAS NO : **82-68-8**
CHEM NAME : ***** PENTACHLORONITROBENZENE (PCNB)**

REPORT 01 : NCI CARC TRS 61 DHEW PUB NO NIH 78-1311
UNDER THE CONDITIONS OF THIS BIOASSAY, PCNB WAS NOT CARCI-

NOGENIC IN EITHER OSBORNE-MENDEL RATS OR B6C3F1 MICE (FEEDING STUDY).

REPORT 02 : IARC MONOGRAPH 5 1975
PENTACHLORONITROBENZENE WAS TESTED IN A PRELIMINARY STUDY BY THE ORAL ROUTE IN TWO STRAINS OF MICE AND PRODUCED AN INCREASED INCIDENCE OF HEPATOMAS IN MALES OF ONE STRAIN.

REPORT 03 : NTP TECHNICAL BULLETIN 2 APR 80
THE COMPOUND WAS FOUND TO BE NEGATIVE IN THE SALMONELLA TYPHIMURIUM MUTAGENICITY ASSAY.

REPORT 04 : EPA RCRA HEALTH AND ENV. EFFECTS PROFILE OCT 80
INCREASED INCIDENCE OF HEPATOMA FORMATION WAS REPORTED IN HYBRID MICE TREATED WITH PCNB. PCNB WAS FOUND TO BE MUTAGENIC IN THE HCR STRAIN OF E. COLI OCHRE, BUT NOT IN ANOTHER E. COLI STRAIN. NO SIGNIFICANT TERATOGENIC EFFECTS WERE DETECTED IN STUDIES AT DOSAGES AS HIGH AS 1500 PPM.

REPORT 05 : NTP CARC TEST PROGRAM JUL 82
IN A CURRENT NTP BIOASSAY, PCNB IS BEING ADMINISTERED IN THE DIET TO B6C3F1 MICE. THE CHRONIC EXPOSURE PHASE IS IN PROGRESS.

REPORT 06 : MULTI-AGENCY CROSS CHECK
N HHS MAGUIRE
N OSHA CARC
Y EPA CAG
Y EPA RCRA
N EPA CLEAN AIR
N EPA CLEAN WATER

REPORT 07 : NTP TECH REPORT NTP TR 325, NIH PUB # 87-2581 JAN 87
IN A 2-YEAR GAVAGE STUDY, THERE WAS NO EVIDENCE OF CARCINOGENICITY FOR MALE OR FEMALE B6C3F1 MICE RECEIVING 2500 OR 5000 PPM OF CPD. INFECTION IS BELIEVED TO HAVE DECREASED SURVIVAL OF MICE AND REDUCED SENSITIVITY FOR DETERMINING CARCINOGENIC RESPONSE.

REPORT 08 : NTP TECH REPORT #325 NIH PUB #87-2581 JAN 87
UNDER CONDITIONS OF 2-YR FEED STUDIES, NO EVIDENCE OF CARCINOGENICITY FOR EITHER MALE OR FEMALE B6C3F1 MICE RECEIVING 2,500 OR 5,000 PPM PCNB. INFECTION REDUCED SURVIVAL OF FEMALE MICE, THUS REDUCING THE SENSITIVITY FOR DETERMING ABSENCE OR PRESENCE OF A CARCINOGENIC RESPONSE.

REPT'D USE : INSECTICIDE, FUNGICIDE, SLIMICIDE AND SOIL FUMIGANT

CAS NO : **87-86-5**
CHEM NAME : ***** PENTACHLOROPHENOL**

REPORT 01 : IARC MONOGRAPH 20 OCT 79
PENTACHLOROPHENOL WAS TESTED IN ONE EXPERIMENT IN TWO STRAINS OF MICE AND IN ONE EXPERIMENT IN RATS BY ORAL ADMINISTRATION. NO CARCINOGENIC EFFECT WAS SEEN IN EITHER SPECIES. PENTACHLOROPHENOL WAS ALSO TESTED IN ONE EXPERIMENT IN MICE OF TWO STRAINS BY SUBCUTANEOUS INJECTION OF SINGLE DOSES; IT PRODUCED HEPATOMAS IN MALES OF ONE STRAIN.

REPORT 02 : DRAFT NTP TR 349 NIH PUB. 89-2804 MAR 89
UNDER 2 YEAR FEED STUDIES CLEAR EVIDENCE OF CARCINOGENIC ACTIVITY FOR MALE B6C3F1 MICE FED DIETS CONTAINING TECHNICAL-GRADE PENTACHLOROPHENOL, SHOWN BY ADRENAL MEDULLARY AND HEPATOCELLULAR NEOPLASMS. SOME EVIDENCE OF CARCINOGENIC ACTIVITY FOR FEMALE B6C3F1 MICE EXPOSED TO TECHNICAL-GRADE SHOWN BY HEMANGIOSARCOMAS AND HEPATOCELLULAR NEOPLASMS. CLEAR EVIDENCE OF CARCINOGENIC ACTIVITY FOR MALE B6C3F1 MICE EXPOSED TO PENTACHLOROPHENOL, EC-7 SHOWN BY ADRENAL MEDULLARY AND HEPATOCELLULAR NEOPLASMS. CLEAR EVIDENCE OF CARCINOGENIC ACTIVITY FOR FEMALE B6C3F1 MICE EXPOSED TO PENTACHLOROPHENOL, EC-7, SHOWN BY ADRENAL MEDULLARY AND HEPATOCELLULAR NEOPLASMAS AND HEMANGIOSARCOMAS.

REPORT 03 : EPA RCRA HEALTH AND ENV. EFFECTS PROFILE OCT 80
PENTACHLOROPHENOL HAS SHOWN NO EVIDENCE OF CARCINOGENICITY. EVIDENCE FOR MUTAGENICITY IS EQUIVOCAL. PENTACHLOROPHENOL IS TERATOGENIC IN EXPERIMENTAL ANIMALS AT LEVELS WHICH PRODUCE MATERNAL OR FETAL TOXICITY.

REPORT 04 : EPA AMBIENT WATER QUALITY CRITERIA OCT 80
BASED ON AVAILABLE LITERATURE, PCP IS NOT CONSIDERED TO BE CARCINOGENIC. A CRITERION, BASED ON OTHER CHRONIC TOXICITY DATA, CORRESPONDING TO A CALCULATED DAILY INTAKE OF 0.3 MG/KG, IS 1.01 MG/LITER.

REPORT 05 : NTP TECHNICAL BULLETIN 3 DEC 80
PENTACHLOROPHENOL WAS NEGATIVE IN THE SALMONELLA TYPHIMURIUM MUTAGENICITY ASSAY IN ONE LABORATORY.

REPORT 06 : MULTI-AGENCY CROSS CHECK
N HHS MAGUIRE
N OSHA CARC
N EPA CAG
N EPA RCRA
N EPA CLEAN AIR
Y FPA CLEAN WATER

REPORT 07 : IARC MONOGRAPH 20 1979
PENTACHLOROPHENOL DID NOT INDUCE SEX-LINKED RECESSIVE LETHALS IN DROSOPHILA.
REPT'D USE : WOOD PRESERVATIVE

CAS NO : **UNKNOWN-09**
CHEM NAME : ***** PENTAERYTHRITOL MONOOLEATE TRIACRYLATE**

REPORT 01 : EPA TSCA 8(E) NOTICE 8EHQ-1082-0460
ATLANTIC RICHFIELD PROVIDED SUMMARY RESULTS FROM ACUTE ANIMAL EYE/SKIN IRRITATION STUDIES AN A BATTERY OF SHORT-TERM IN VITRO TESTS. IT WAS SHOWN TO BE "EXTREMELY TOXIC" TO CULTURED MOUSE LYMPHOMA AND CHINESE HAMSTER OVARY CELLS WITHOUT METABOLIC ACTIVATION. EXCESS NUMBERS OF CHROMOSOMAL BREAKS WERE NOTED IN THE CULTURED CHINESE HAMSTER OVARY CELLS TREATED WITH PEMOTA. THE CHEMICAL WAS NEGATIVE IN THE SALMONELLA ASSAY.

CAS NO : **78-11-5**
CHEM NAME : ***** PENTAERYTHRITOL TETRANNITRATE (PETN, NF)**

REPORT 01 : NTP DRAFT TR 365 NIH 89-2820 OCT 88
UNDER THE CONDITIONS OF THESE 2-YEAR FEED STUDIES, THERE WAS EQUIVOCAL EVIDENCE OF CARCINOGENIC ACTIVITY OF PETN, NF, FOR MALE AND FEMALE F344/N RATS (NEOPLASMS OF THE ZYMBAL GLAND). THERE WAS NO EVIDENCE OF CARCINOGENIC ACTIVITY OF PETN, NF, FOR MALE OR FEMALE B6C3F1 MICE FED DIETS CONTAINING 25,000 OR 50,00 PPM.

REPORT 02 : NTP DRAFT TR 365 NIH 89-2820 OCT 88
UNDER THE CONDITIONS OF THESE 2-YEAR FEED STUDIES, THERE WAS EQUIVOCAL EVIDENCE OF CARCINOGENIC ACTIVITY OF PENTAERYTHRITOL TETRANITRATE (PETN), NF, FOR MALE AND FEMALE F344/N RATS, BASED ON A MARGINAL INCREASE IN NEOPLASMS OF THE ZYMBAL GLAND. FEMALE RATS MIGHT HAVE TOLERATED A HIGHER DOSE. THERE WAS NO EVIDENCE OF CARCINOGENIC ACTIVITY OF PETN, NF, FOR MALE OR FEMALE B6C3F1 MICE FED DIETS CONTAINING 25,000 OR 50,000 PPM FOR 2 YEARS. NO NONNEOPLASTIC LESIONS WERE ATTRIBUTED TO PETN, NF, ADMINISTRATION.
REPT'D USE : EXPLOSIVE AND MEDICINE

CAS NO : **57116-45-7**
CHEM NAME : ***** PENTAERYTHRITOL-TRIS-(B-(N-AZIRIDINYL) PROPIONATE)**

REPORT 01 : EPA 8(E) STATUS REPORT 8EHQ-0986- 0630 SEP 86
IN-VITRO CHROMOSOMAL ABERRATION AND IN VITRO CELL TRANSFORMATION ASSAYS BOTH POSITIVE. CHROMOSOMAL ABERRATION TEST POSITIVE WITHOUT ACTIVATION.
REPT'D USE : CROSS-LINKER IN COATING SYSTEMS

CAS NO : **3524-68-3**
CHEM NAME : ***** PENTAERYTHRITOL TRIACRYLATE (PETA)**

REPORT 01 : EPA TSCA 8(E) REPORT 8EHQ-0882-0454 OCT 82
FINAL RESULTS OF A SKIN PAINTING STUDY ON MICE SHOWED THAT THE INCIDENCE OF LYMPHOMAS SEEN IN THE TREATED ANIMALS IS PROBABLY RELATED TO THE TREATMENT AND THE DATA SUGGEST THAT PETA IS ABSORBED THROUGH THE SKIN AND MAY ACT AS AN INTERNAL CARCINOGEN TO LYMPHOID SYSTEMS. 6/43 ANIMALS IN TEST GROUP DEVELOPED LYMPHOMAS WHILE 0/48 OF ANIMALS IN THE "NO TREATMENT " CONTROL GROUP AND 1/49 ANIMALS IN THE MINERAL OIL CONTROL GROUP DEVELOPED LYMPHOMAS.

REPORT 02 : EPA TSCA 8(E) REPORT 0488-0454
DUPONT PROVIDED PATHOLOGY PEER GROUP REVIEW OF INTERNAL ORGAN HISTOPATHOLOGY SLIDES ON CHRONIC DERMAL TOXICITY STUDY OF MICE REPORTED IN TSCA 8(E) REPORT 0882-0454 THAT SHOW LYMPHOMAS WERE MISDIAGNOSED AND THERE WAS NO EVIDENCE OF ANY COMPOUND-RELATED INCREASE IN TUMORS. THE TECHNICAL QUALITY OF THE HISTOLOGICAL MATERIAL WAS CONSIDERED INSUFFICIENT TO SUPPORT ANYDEFINITIVE CONCLUSION AS TO THE CARCINOGENICITY OF PETA.

CAS NO : **109-52-4**
CHEM NAME : ***** N-PENTANOIC ACID (N-VALERIC ACID)**

REPORT 01 : EPA TSCA 8(E) 8EHQ-0584-0513 JUL 84
CHRONIC DERMAL APPLICATION STUDY RESULTED IN A SIGNIFICANT INCREASE IN TUMOR INCIDENCE RATE IN MICE.
REPT'D USE : INTERMEDIATE IN MANUFACTURE OF DRUGS, FLAVORING AGENTS, PERFUMES, PLASTICIZERS, LUBRICANTS AND STABILIZERS

CAS NO : **127-18-4**
CHEM NAME : ***** PERCHLOROETHYLENE**

REPORT 01 : NCI CARC TRS 13 DHEW PUB NO NIH 77-813
THE RESULTS OF THE BIOASSAY OF TETRACHLOROETHYLENE IN OSBORNE-MENDEL RATS DO NOT ALLOW AN EVALUATION OF THE CARCINOGENICITY OF THIS COMPOUND DUE TO THE HIGH RATE OF EARLY DEATH AMONG THE TREATED ANIMALS. HOWEVER, UNDER THE CONDITION OF THIS STUDY, TETRACHLOROETHYLENE IS A LIVER CARCINOGEN IN B6C3F1 MICE OF BOTH SEXES. PCE WAS ADMINISTERED BY GAVAGE.

REPORT 02 : NIOSH CURRENT INTELLIGENCE BULLETIN 20 JAN 78
NIOSH RECOMMENDS THAT IT IS PRUDENT TO HANDLE TETRACHLOROETHYLENE (PERCHLOROETHYLENE) IN THE WORKPLACE AS IF IT WERE A HUMAN CARCINOGEN. THE RECOMMENDATION IS BASED ON A RECENT STUDY BY NCI INDICATING THAT TETRACHLOROETHYLENE CAUSES LIVER CANCER IN LABORATORY MICE. ANOTHER STUDY SUGGESTS THE TERATOGENIC POTENTIAL OF TETRACHLOROETHYLENE IN MICE AND RATS.

REPORT 03 : NIOSH CRITERIA DOCUMENT JUL 76
NIOSH RECOMMENDS A STANDARD OF 50 PPM TWA BASED ON CNS, CARDIAC, RESPIRATORY, AND LIVER EFFECTS. THE PRESENT OSHA STANDARD IS 100 PPM TWA.

REPORT 04 : EPA RCRA HEALTH AND ENV. EFFECTS PROFILE OCT 80
TETRACHLOROETHYLENE HAS NOT BEEN SHOWN TO BE TERATOGENIC, BUT IT HAS BEEN SHOWN TO BE MUTAGENIC IN BACTERIAL ASSAYS AND CARCINOGENIC IN MICE.

REPORT 05 : NTP TECHNICAL BULLETIN 2 APR 80
THE COMPOUND WAS FOUND TO BE NEGATIVE IN THE SALMONELLA TYPHIMURIUM MUTAGENICITY ASSAY.

REPORT 06 : MULTI-AGENCY CROSS CHECK
Y HHS MAGUIRE
N OSHA CARC
Y EPA CAG
Y EPA RCRA
Y EPA CLEAN AIR
Y EPA CLEAN WATER

REPORT 07 : EPA AMBIENT WATER QUALITY CRITERIA OCT 80
TETRACHLOROETHYLENE IS SUSPECTED OF BEING A HUMAN CARCINOGEN. AMBIENT WATER CONCENTRATIONS OF 8.0 UG/L, .8 UG/L, AND 0.08 UG/L ARE ESTIMATED TO RESULT IN A RESPECTIVE 10 -5, 10 -6, AND 10 -7 INCREASE IN CANCER RISK OVER A LIFETIME.

REPORT 08 : IARC MONOGRAPH 20 OCT 79
TETRACHLOROETHYLENE WAS TESTED IN ONE EXPERIMENT IN MICE AND IN ONE IN RATS BY ORAL ADMINISTRATION. IN MICE, IT PRODUCED HEPATOCELLULAR CARCINOMAS IN ANIMALS OF BOTH SEXES. THE EXPERIMENT IN RATS WAS CONSIDERED TO BE INADEQUATE.

REPORT 09 : NTP CARC TEST PROGRAM JUL 82
IN A CURRENT NTP BIOASSAY, PCE IS BEING ADMINISTERED BY INHALATION TO F344 RATS AND B6C3F1 MICE. THE CHRONIC EXPOSURE PHASE BEGAN JAN 81. IN ANOTHER STUDY, PCE WAS ADMINISTERED BY GAVAGE TO VARIOUS RAT STRAINS AND B6C3F1 MICE. THE STUDY HAS BEEN COMPLETED AND THE REPORT IS BEING DRAFTED.

REPORT 10 : IARC MONOGRAPH 20 1979
RESULTS REPORTED IN AN ABSTRACT SUGGEST THAT PCE IS MUTAGENIC IN SALMONELLA TA100. PCE WAS NOT MUTAGENIC IN E. COLI K12 AND THERE WAS NO INDUCTION OF CHROMOSOMAL ABERRATIONS IN BONE MARROW CELLS OF MICE TREATED WITH THE CHEMICAL.

REPORT 11 : INHALATION STUDY OF PCE IN RATS OCT 80
IN AN EVALUATION OF THE TOXICITY AND CARCINOGENICITY OF PCE, DOW CHEMICAL CO. EXPOSED SPRAGUE-DAWLEY RATS TO VAPORS OF THE CPD AT 600 OR 300 PPM FOR 12 MONTHS. THIS WAS FOLLOWED BY OBSERVATION OF THE RATS FOR THEIR LIFETIME. THERE WAS NO EVIDENCE OF A TUMORIGENIC RESPONSE IN MALE OR FEMALE RATS EXPOSED TO 600 OR 300 PPM PCE VAPORS.

REPORT 12 : NTP TR 311 NIH PUB NO. 86-2567 AUG 86
2-YEAR GAVAGE STUDIES GAVE CLEAR EVIDENCE OF CARCINOGENICITY OF PCE AS SHOWN BY INCREASED INCIDENCE OF MONONUCLEAR CELL LEUKEMIA AND UNCOMMON RENAL TUBULAR CELL NEOPLASMS. SOME EVIDENCE OF CARCINOGENICITY IN FEMALE F344/N RATS, SHOWN BY INCREASED INCIDENCE OF MONONUCLEAR CELL LEUKEMIA. CLEAR EVIDENCE OF CARCINOGENICITY IN B6C3F1 MICE, SHOWN BY INCREASED INCIDENCES OF BOTH HEPATOCELLULAR ADENOMAS AND CARCINOMAS IN MALES AND HEPATOCELLULAR CARCINOMAS IN FEMALES.

REPORT 13 : EPA PROPOSED NAT'L DRINKING REG'S - JUN 84
VSOC'S 49 FR24330
EPA PROPOSES TO ESTABLISH RECOMMENDED MAXIMUM CONTAMINANT LEVELS (RMCLS) FOR DINKING WATER FOR PERC AT 0 MG/L UNDER SAFE DRINKING WATER ACT.

REPORT 14 : EPA ANPR (CURRENT REPORT 6/8/84 P 305) OCT 83
RECOMMENDED MAXIMUM CONTAMINANT LEVELS OF ZERO ARE BEING PROPOSED UNDER THE SAFE WATER DRINKING ACT.

REPT'D USE : SOLVENT, CHEMICAL INTERMEDIATE, AEROSOLS, SPOT REMOVERS, FR CHEMICALS, RUST REMOVERS, DRY CLEANING, RUG/UPHOLSTERY CLEANERS, FIRE EXTINGUISHERS, ANTIHELMINTIC, METAL DEGREASING HEAT EXCHANGE FLUID AND PAINTS/VARNISHES

CAS NO : **68298-78-2**
CHEM NAME : ***** PERFLUOROALKYL RESINS**

REPORT 01 : EPA STATUS REPORT, 8EHQ 0986-0631 S SEP 86
A COMPANY PROVIDED PRELIMINARY FINDINGS FROM ACUTE INHALATION STUDIES IN RATS EXPOSED TO A RESPIRABLE AEROSOL OF MINERAL SPIRITS (CONTROL) OR A MIXTURE OF 0.5% PERFLUOROALKYL RESINS, 4.5% CHLOROFLUOROCARBON 113 AND 95% MINERAL SPIRITS. THESE STUDIES SHOWED THAT THE MIXTURE WAS MORE TOXIC THAN MINERAL SPIRITS ALONE. THE COMPANY ALSO REPORTED THERE WERE NO DEATHS FOR RATS EXPOSED VIA INHALATION TO 102 PPM OF THE MIXTURE, BUT A 4-HOUR, WHOLE BODY EXPOSURE GAVE 100% MORTALITY.

REPORT 02 : EPA STATUS REPORT, 8EHQ- 0000-0535
TESTS SHOW THAT PFA RESINS EXTREMELY TOXIC TO BIRDS WHEN INHALED IN A SOLVENT MEDIA. THE LC50 FOR QUAIL IS 0.25 MG/M3 AND THE LC50 FOR RATS IS 2 MG/M3.

REPT'D USE : FABRIC STAIN REPELLENTS

CAS NO : **12427-27-9**
CHEM NAME : ***** PERLITE**

REPORT 01 : OSHA AIR CONTAMINANTS FINAL RULE FR NOTICE JAN 89
OSHA REPORTED THERE ARE NO PUBLISHED REPORTS OF ADVERSE PHYSIOLOGIC EFFECTS FROM EXPOSURE TO PERLITE DUST. ITS TWA FOR PERLITE IN TOTAL DUST 15 MG/M3 CONTAINING LESS THAN ONE PERCENT QUARTZ. AS RESPIRABLE FRACTION OF PERLITE DUST AT 5 MG/M3. OSHA CONCLUDES THESE LIMITS PROTECT WORKERS FROM RISK OF EYE, SKIN AND OTHER PHYSICAL IRRITATION. (JANUARY 19, 1989, P. 2948, 2597-2598). ACGIH HAS TLV-TWA OF 10 MG/M3, THE LEVEL SET FOR NUISANCE DUSTS. (ACGIH, THRESHOLD LIMIT, VOLUMES, 5TH ED., P. 467).

REPORT 02 : VARIOUS REPORTS (1972 TO) 1986
CHEST ROENTGENOGRAMS OF PERLITE WORKERS EMPLOYED IN INDUSTRY SHOWED NO EVIDENCE OF PNEUMOCONIOSIS ASSOCIATED WITH PERLITE EXPOSURES (COOPER, 1975, 1976, 1986). GROUPS OF 21 MALE AND 21 FEMALE MICE GIVEN DIETS CONTAINING 10 OR 20 PERCENT PERLITE FOR 28 WEEKS NOT EFFECTED, EXCEPT SLIGHT REDUCE GROWTH AMONG MALES (SAKAI AND NAGAO, 1985). GUINEA PIGS EXPOSED TO FIR BARK OR PERLITE DUST SHOWED PATHOLOGIC CHANGES IN LUNGS (LYMPHOID AGGREGAT-

ED AND PERIVASCULAR INFLAMMATORY RESPONSE) SUGGESTING MORE THAN NUISANCE DUSTS (MCMICHAEL, ET AL, 1983).

REPT'D USE : GROWING MEDIUM, FILTERS, DEODORANTS, DENTIFRICES, WATER SUPPLY SYSTEMS (MEDLARS) AND INSULATION AND PLASTER (ELMES)

CAS NO : **79-21-0**
CHEM NAME : ***** PEROXYACETIC ACID**

REPORT 06 : MULTI-AGENCY CROSS CHECK
N HHS MAGUIRE
Y OSHA CARC
N EPA CAG
N EPA RCRA
N EPA CLEAN AIR
N EPA CLEAN WATER

REPT'D USE : OXIDIZING AGENT, POLYMERIZATION CATALYST, INDUSTRIAL BLEACH AND EGG WASH

CAS NO : **198-55-0**
CHEM NAME : ***** PERYLENE**

REPORT 01 : IARC MONOGRAPH 32 DEC 83
TESTED FOR CARCINOGICITY BY SKIN APPLICATION TO MICE IN ONE EXPERIMENT AND NO CARCINOGENIC EFFECT OBSERVED. ALSO TESTED IN MOUSESKIN INITIATION-PROMOTION ASSAY IN ONE STUDY, WITH NEGATIVE RESULTS.

REPORT 02 : IARC MONOGRAPH 32 DEC 83
MUTAGENIC TO SALMONELLA TYPHIMURIUM IN PRESENCE OF EXOGENOUS METABOLIC SYSTEM. IN ONE STUDY, DID NOT INDUCE MUTATIONS IN CULTURED HUMAN LYMPHOBLASTOID CELLS. DATA FROM ONE STUDY ON CHROMOSOMAL EFFECTS INADEQUATE TO MAKE EVALUATION.

REPT'D USE : INCOMPLETE COMBUSTION PRODUCT, CIGARETTE SMOKE, ENGINE EXHAUSTS, SLUDGE AND RAIN WATER

CAS NO : **60102-37-6**
CHEM NAME : ***** PETASITENINE**

REPORT 01 : IARC MONOGRAPH 31 JUL 83
PETASITENINE ISOLATED FROM YOUNG FLOWER STALKS OF WILD PETASITES JAPONICUS MAXIM. TESTED IN RATS BY ADMINISTRATION IN DRINKING-WATER; HAEMANGIOENDOTHELIAL SARCOMAS OF LIVER AND LIVER-CELL ADENOMAS OBSERVED. IN MICE, RATS AND HAMSTERS IN DIET INCREASED INCIDENCE OF LUNG ADENOMAS AND ADENOCARCINOMAS IN MICE OF ONE STRAIN AND OF

HAEMANGIOENDOTHELIAL SARCOMAS OF LIVER AND OF HEPATOCELLULAR ADENOMAS AND HEPATOCELLULAR CARCINOMAS IN RATS. NO INCREASE IN TUMOR INCIDENCE OBSERVED IN HAMSTERS.

REPORT 02 : IARC MONOGRAPH 31 JUL 83
PETASITENINE IS MUTAGENIC IN BACTERIAL AND IN MAMMALIAN CELLS IN VITRO. INDUCED CHROMOSOMAL ABERRATIONS, UNSCHEDULED DNA SYNTHESIS AND TRANSFORMATION IN MAMMALIAN CELLS IN VITRO.

REPT'D USE : FOOD AND HERBAL REMEDY IN JAPAN. NOT PRODUCED COMMERCIALLY

CAS NO : **68477-31-6**

CHEM NAME : ***** PETROLEUM DISTILLATES**

REPORT 01 : EPA TSCA 8(E) 8EHQ-0283-0472S FEB 83
PRELIMINARY RESULTS FROM LIFETIME MOUSE SKIN PAINTING STUDY OF LOW-BOILING CATALYTIC REFORMER FRACTIONATOR RESIDUE PETROLEUM DISTILLATE. TWO TREATED MICE DEVELOPED ONE TISSUE MASS AT APPLICATION SITE WITH MEAN LATENCY PERIOD OF 76.5 WKS. HISTOPATH EXAMS DONE. RESULTS INDICATE CPD IS WEAKLY CARCINOGENIC IN ANIMALS. NIOSH RECOMMENDED STANDARD TWA 350 MG/CUBIC METER FOR PETROLEUM SOLVENTS (7/77); SEE 68477-31-6.

REPORT 02 : EPA TSCA 8(E) 83HQ-0283-0472S FEB 83
HEAVY CATALYTIC REFORMED PETROLEUM NAPHTHA, (CAS NO 64741-68-0), USED IN LIFETIME MOUSE SKIN-PAINTING STUDY. TREATED MICE DEVELOPED ONE TISSUE MASS EACH AT APPLICATION SITE, WITH MEAN LATENCY OF 67.5 WKS. HISTOPATH EXAMS DONE. RESULTS INDICATE CPD IS WEAKLY CARCINOGENIC IN ANIMALS.

REPORT 03 : EPA TSCA 8(E) 8EHQ-0283-0472S FEB 83
A LOW BOILING CATALYTIC REFORMER FRACTIONATOR RESIDUE PETROLEUM DISTILLATE INDUCED ONE TISSUE MASS EACH AT THE APPLICATION SITE IN A LIFETIME MOUSE SKIN PAINTING STUDY. MEAN LATENCY PERIOD WAS 67.5 WEEKS. HISTOPATH EXAMS WERE DONE. LONG LATENCY PERIOD AND SMALL NUMBER OF TUMORS INDICATE THAT MATERIAL IS WEAK ANIMAL CARCINOGEN.

REPORT 04 : IARC MONOGRAPH 33 APR 84
UNTREATED VACUUM DISTILLATES, ACID-TREATED OILS AND AROMATIC OILS, INCLUDING EXTRACTS FROM SOLVENT TREATMENT OF DISTILLATES AND HIGH-BOILING FRACTION OF CATALYTICALLY CRACKED OILS (CLASSES 1, 2 AND 6); MILDLY SOLVENT-REFINED OILS (CLASS 3); MILDLY HYDROTREATED OILS (CLASS 4)-

AND ONE SAMPLE USED GASOLINE ENGINE OIL (CLASS 7.2) - SUFFICIENT EVIDENCE OF CARCINOGENICITY TO HUMANS BECAUSE OF CARCINOGENICITY TO ANIMALS.

REPORT 05 : IARC MONOGRAPH 33 APR 84
VACUUM DISTILLATION (CLASS 1) AND SOLVENT-REFINED SAMPLES FROM CORRESPONDING DISTILLATES (CLASS 3) WERE MUTAGENIC IN S. TYPHIMURIUM STRAIN TA98 IN PRESENCE OF AROCLOR-INDUCED RAT LIVER METABOLIC SYSTEM (S9).

REPORT 07 : IARC MONOGRAPH 33 APR 84
DMSO EXTRACTS OF POOLED SAMPLE OF 15 COMMERCIAL, UNUSED 10W-40 GASOLINE ENGINE OILS (CLASS 7.1) SHOWED NO MUTAGENIC ACTIVITY IN S. TYPHIMURIUM BUT A POOLED SAMPLE OF USED OIL (CLASS 7.2) WAS MUTAGENIC IN STRAINS TA1537, TA1538 AND TA100 IN PRESENCE AND ABSENCE OF S9. SIMILAR RESULTS WERE OBTAINED IN 3 OTHER STUDIES.

REPORT 08 : EPA FYI-AX-0385 -0389 MAR 85
API REPORTS THAT MIDDLE DISTILLATE FUELS ARE WEAKLY TO MODERATELY CARCINOGENETIC IN ANIMALS, INDUCING TUMORS IN 5 - 30% OF EXPERIMENTAL ANIMALS IN SKIN PAINTING STUDY. 2-YEAR USAF/NAVY INHALATION STUDIES ON VOLATILE FRACTIONS IN PROGRESS. SOME OF THESE FUELS MUTAGENIC IN INVIVO RAT BONE MARROW CYTOGENETICS AND MOUSE LYMPHOMA ASSAYS. MODIFIED AMES ASSAY PRODUCED POSITIVE RESULTS FOR SOME MATERIALS.

REPORT 09 : EPA TSCA 8(E) 8EHQ-0683-0480 JUL 83
A CHEMICALLY NEUTRALIZED MIDDLE PETROLEUM DISTILLATE, TRADE-NAMED MEN TOR 28 WAS USED IN A SKIN PAINTING STUDY OF 50 C3H MALE MICE. BY THE 756TH DAY OF THE STUDY, 8 OF THE 50 HAD BEEN SEEN TO HAVE A TOTAL OF 3 PAPILLOMAS AND 5 CARCINOMAS.

REPORT 10 : EPA TSCA 8(E) 8EHQ-0384-0507 APR 84
GULF PROCESS 65 OIL (CAS NO 64742-11-6), GULF 100 TEXAS DISTILLATE (CAS NO 64741-53-3) AND GULF TEXAS 100 OIL (CAS NO 64742-52-5) USED IN SKIN PAINTING STUDY ON C3H MICE. PROCESS 65 OIL INDUCED ONE SQUAMOUS CELL CARCINOMA AND ONE BENIGN PAPILLOMA. TEXAS 100 OIL DISTILLATE PRODUCED ONE SQUAMOUS CELLCARCINOMA AND 2 BENIGN PAPILLOMAS. TEXAS 100 OIL DID NOT INDUCE ANY NEOPLASMS. TEXAS 100 OIL DID PRODUCE MILD TO MODERATE EPIDERMAL ACANTHOSIS AND HYPERKERATOSIS IN BOTH SEXES. 50 ANIMALS EACH SEX USED PER STUDY.

CAS NO : **8002-05-9**
CHEM NAME : ***** PETROLEUM MINERAL AND CRUDE OILS**

REPORT 02 : IARC MONOGRAPH 33 APR 84
THERE IS SUFFICIENT EVIDENCE FROM STUDIES IN HUMANS THAT MINERAL OILS (CONTAINING VARIOUS ADDITIVES AND IMPURITIES) USED IN OCCUPATIONS SUCH AS MULESPINNING, METAL MACHINING AND JUTE PROCESSING ARE CARCINOGENIC TO HUMANS. NIOSH RECOMMENDED STANDARD TWA 350 MG/CUBIC METER FOR PETROLEUM SOLVENTS (7/77)/ SEE 68477-31-6.

REPORT 03 : IARC MONOGRAPH 33 APR 84
THERE IS LIMITED EVIDENCE FOR CARCINOGENICITY OF SOME CUTTING OILS (CLASSES 7.1 AND 7.2) TO EXPERIMENTAL ANIMALS. DATA INADEQUATE TO EVALUATE CARCINOGENICITY TO EXPERIMENTAL ANIMALS OF FORMULATED PRODUCTS (CLASS 7.1) AS A CLASS, SINCE POSSIBLE CARCINOGENIC ACTIVITY OF INDIVIDUAL PRODUCTS DEPENDENT UPON SEVERITY OF PROCESSING OF BASE OILS AND NATURE AND CONCENTRATION OF ADDITIVES.

REPORT 04 : IARC MONOGRAPH 33 APR 84
THE DATA ARE INADEQUATE TO EVALUATE THE CARCINOGENICITY TO EXPERIMENTAL ANIMALS OF USED FORMULATED PRODUCTS (CLASS 7.2) AS A CLASS, SINCE THE POSSIBLE CARCINOGENIC ACTIVITY OF INDIVIDUAL PRODUCTS DEPENDENT UPON QUALITY OF BASE OILS USED, THE NATURE AND CONCENTRATION OF ADDITIVES AND CONTAMINANTS, AND THE CONDITIONS OF USE.

REPORT 05 : IARC MONOGRAPH 33 APR 84
AVAILABLE DATA ON SEVERELY HYDROTREATED OILS (CLASS 4) ARE INADEQUATE FOR EVALUATION OF CARCINOGENICITY TO EXPERIMENTAL ANIMALS. THERE IS NO EVIDENCE FOR THE CARCINOGENICITY TO EXPERIMENTAL ANIMALS OF SEVERELY SOLVENT-REFINED OILS (CLASS 3) OR OF WHITE OILS (CLASS 5) WHEN ADMINISTERED BY ROUTES OTHER THAN INTRAPERITONEAL INJECTION.

REPORT 06 : MULTI-AGENCY CROSS CHECK
Y HHS MAGUIRE
N OSHA CARC
N EPA CAG
N EPA RCRA
N EPA CLEANAIR
N EPA CLEAN WATER

REPORT 07 : IARC MONOGRAPH 33 APR 84
URINE OF 17 MEN EXPOSED OCCUPATIONALLY TO MINERAL OILS (CLASS 8) AND IRON OXIDE PARTICLES AND OF 16 MEN EXPOSED ONLY TO MINERAL OILS EXHIBITED MUTAGENICITY IN SALMONELLA TYPHIMURIUM STRAINS TA98 AND TA100 IN PRESENCE OF AN

AROCLOR-INDUCED RAT-LIVER METOBOLIC SYSTEM AND B-GLUCURONIDASE; BUT SUITABLE CONTROLS LACKING.

REPORT 08 : EPA TSCA 8(E) 8EHQ-0585-0557 JUL 85
A RETROSPECTIVE COHORT MORTALITY STUDY OF MORE THAN 10,000 REFINERY WORKERS "REVEALED STATISTICALLY SIGNIFICANT ELEVATIONS IN MORTALITY IN THREE DISEASE CATEGORIES"; SKIN CANCER, BENIGN NEOPLASMS AND DIGESTIVE SYSTEM CANCERS.

REPORT 09 : IARC MONOGRAPH 33 APR 84
A HYDROTREATED PETROLEUM RESIDUE (CLASS 4) WAS MUTAGENIC TO S. TYPHIMURIUM STRAIN TA98 IN THE PRESENCE OR S9; A MEDICINAL QUALITY (CLASS 5) WHITE OIL WAS NOT MUTAGENIC IN S. TYPHIMURIUM STRAIN TA98 IN PRESENCE OF S9.

REPORT 10 : IARC MONOGRAPH 33 APR 84
A HIGHLY-REFINED STEEL-HARDENING OIL (CLASS 7.1) UNUSED WAS NOT MUTAGENIC IN S. TYPHIMURIUM STRAIN TA98 IN THE PRESENCE OF S9, WHILE THE USED OIL WAS MUTAGNEIC BOTH IN THE PRESENCE AND ABSENCE OF THE METABOLIC SYSTEM.

REPORT 12 : EPA FYI AX-0385-0389 MAR 85
API REPORTS MIDDLE DISTILLATE FUELS ARE WEAKLY TO MODERATLY ÇONGINOGEIC IN ANIMALS, PRODUCING TUMOR INCIDENCES IN 5 TO 30% RANGE (SKIN-PAINTINGS). TWO-YEAR USAF/NAVY INHALATION STUDIES ON VOLATILE FRACTIONS IN PROGRESS.

REPORT 13 : EPA FYI AX-0385-0389 MAR 85
API REPORTS SOME MIDDLE DISTILLATE FUEL MATERIALS MUTAGENIC IN VIVO RAT BONE MARROW CYTOGENETICS AND MOUSE LYMPHOMA ASSAYS; BUT RESULTS NOT CONSISTENT. MODIFIED AMES ASSAY PRODUCED POSITIVE RESULTS FOR SOME MATERIALS.

REPORT 14 : IARC MONOGRAPH 45 1989
CRUDE OIL IS NOT CLASSIFIABLE AS TO ITS CARCINOGENICITY TO HUMANS. INADEQUATE EVIDENCE FOR CARCINOGENICITY IN HUMANS FROM CRUDE OIL. RETROSPECTIVE COHORT STUDY IN PETROLEUM PRODUCING AND PIPELINE OPERATIONS. MORTALITY FROM ALL TYPES OF CANCER LOW, EXCEPT THYROID CANCER. POPULATION-BASED CASE-CONTROL STUDY ELEVATED LUNG CANCER RISK OBSERVED AMONG OLDER MEN IN PETROLEUM EXPLORATION AND PRODUCTION. IN PETROLEUM MINING AND REFINING IDUSTRY SHOWED ELEVATED RISK FOR LUNG CANCER AMONG WELDERS, OPERATORS, BOILER MAKERS, PAINTERS, OIL-FIELD WORKERS. IN CASE-CONTROL STUDY EXCESS RISK FOR TESTICULAR CANCER OBSERVED AMONG PETROLEUM AND NATURAL GAS EXTRACTION WORKERS.

REPT'D USE : LUBRICANT AND BINDER

CAS NO : **64741-68-0**
CHEM NAME : ***** PETROLEUM NAPHTHA, HEAVY CATALYTIC REFORMED**

REPORT 01 : EPA TSCA 8(E) 8EHQ-0283-0472S FEB 83
SUBMITTING COMPANY (NAME CLAIMED CBI) PROVIDED PRELIMINARY RESULTS FROM A LIFETIME MOUSE SKIN-PAINTING STUDY OF HEAVY CATALYTIC REFORMED PETROLEUM NAPHTHA. TWO TREATED MICE DEVELOPED ONE TISSUE MASS EACH AT THE APPLICATION SITE, WITH A MEAN LATENCY PERIOD OF 67.5 WEEKS. HISTOPATHOLOGIC EXAMINATIONS HAVE BEEN DONE. THE LONG LATENCY PERIOD AND THE SMALL NUMBER OF TUMORS INDICATED THAT THE TESTED MATERIALS ARE WEAK ANIMAL CARCINOGENS. NIOSH RECOMMENDED STANDARD TWA 350 MG/CUBIC METER FOR PETROLEUM SOLVENTS (7/77); SEE 68477-31-6.

CAS NO : **64742-95-6**
CHEM NAME : ***** PETROLEUM NAPHTHA SOLVENT**

REPORT 01 : TSCA SECT 8(E) STATUS REPORT 8EHQ-0587-0673 JUN 87
N-1386 HAN (A MIXTURE OF BIS(TRICHLOROMETHYL) SULFONE AND PETROLEUM NAPHTHA SOLVENT) PRODUCED SIGNIFICANT PULMONARY EFFECTS LEADING TO DEATHS OF ALL EXPOSED ANIMALS AFTER 1-HOUR INHALATION EXPOSURES AT 6.96 MG/L OR 1.91 MG/L. THEREFORE, THE 1-HOUR LC50 IS LESS THAN 1.9 MG/L. NIOSH RECOMMENDED STANDARD TWO 350 MG/CUBIC METER FOR PETROLEUM SOLVENTS (7/77); SEE 68477-31-6.
REPT'D USE : CHEMICAL INTERMEDIATE FOR PESTICIDE PRODUCTION

CAS NO : **CLASS-23**
CHEM NAME : ***** PETROLEUM REFINING OCCUPATIONAL EXPOSURES**

REPORT 01 : IARC MONOGRAPH 45 MAR 89
OCCUPATIONAL EXPOSURES IN PETROLEUM REFINING ARE PROBABLY CARCINOGENIC TO HUMANS. THERE IS SUFFICIENT EVIDENCE FOR CARCINOGENICITY IN EXPERIMENTAL AIMALS OF SEVERAL POLYCYCLIC AROMATIC HYDROCARBONS. LIMITED EVIDENCE WORKING IN PETROLEUM REFINERIES ENTAILS CARCINOGENIC RISK; APPLIES TO SKIN CANCER AND LEUKEMIA; FOR ALL OTHER CANCER SITES EVIDENCE IS INADEQUATE. SUFFICIENT EVIDENCE FOR CARCINOGENICITY IN EXPERIMENTAL ANIMALS OF LIGHT AND HEAVY VACUUM DISTILLATES, OF LIGHT AND HEAVY CATALYTICALLY CRACKED DISTILLATES AND OF CRACKED RESIDUES DERIVED FROM THE REFINING OF CRUDE OIL. LIMITED EVIDENCE FOR CARCINOGENICITY IN EXPERIMENTAL ANIMALS OF

LIGHT STRAIGHT-RUN NAPHTHA
REPT'D USE : PETROCHEMICAL; FEED STOCK

CAS NO : **62-44-2**
CHEM NAME : ***** PHENACETIN**

REPORT 01 : IARC SUPPLEMENT 1 19/9
RATS FED A DIET CONTAINING PHENACETIN HAD AN EXCESS OF NASAL AND URINARY TRACT TUMORS. SEVERAL STUDIES INDICATE THAT CHRONIC ABUSE OF ANALGESIC MIXTURES CONTAINING PHENACETIN IS ASSOCIATED WITH PAPILLARY NECROSIS OF THE KIDNEY. IN A 1979 SUPPLEMENT, IARC CONCLUDED THAT PHENACETIN IS PROBABLY CARCINOGENIC FOR HUMANS.

REPORT 02 : IARC MONOGRAPH 24 1980
PHENACETIN IS MUTAGENIC TO SALMONELLA IN THE PRESENCE OF A HAMSTER LIVER MICROSOME PREPARATION, AND IT PRODUCES CHROMOSOME ABERRATIONS IN CHINESE HAMSTER FIBROPLASTS IN VITRO. NO MUTAGENIC EFFECTS WERE DETECTED IN DROSOPHILA OR IN MICE IN VIVO, OR IN BACTERIAL TEST SYSTEMS WHEN RAT OR MOUSE LIVER MICROSOME PREPARATIONS WERE USED. NO TERATOGENIC EFFECTS WERE FOUND IN RATS, ALTHOUGH EMBRYOTOXICITY WAS OBSERVED.

REPORT 05 : IARC SUPPL. 4 OCT 82
EVIDENCE FOR CARCINOGENICITY TO HUMANS LIMITED FOR PHENACETIN, BUT SUFFICIENT FOR MIXTURES CONTAINING PHENACETIN. RENAL PELVIC AND URINARY TRACT TUMORS HAVE BEEN REPORTED IN ASSOCIATION WITH ANALGESIC ABUSE, BUT ANALYTICAL STUDIES ARE NOT CONCLUSIVE. A LARGE-SCALE RX DRUG SCREENING PROGRAM FOUND A NON-STATISTICALLY SIGNIFICANT DEFICIT OF ALL CANCERS, AN EXCESS OF PF CANCERS OF MOUTH FLOOR, A DEFICIT OF BREAST CANCERS AND NO REPORTED DEVIATION FROM EXPECTATION OF URINARY TRACT CANCERS FOR AN ASPIRIN-PHENACETIN-CAFFEINE-BUTALBITAL MIXTURE. EVIDENCE FOR CARCINOGENICITY TO ANIMALS LIMITED FOR ANALGESIC MIXTURES, BUT SUFFICIENT FOR PHENACETIN.

REPORT 06 : MULTI-AGENCY CROSS CHECK
Y HHS MAGUIRE
N OSHA CARC
Y EPA CAG
Y EPA RCRA
N EPA CLEAN AIR
N EPA CLEAN WATER
REPT'D USE : ANALGESIC/ANTIPYRETIC AGENT AND PEROXIDE STABILIZER

CAS NO : **85-01-8**
CHEM NAME : ***** PHENANTHRENE**

REPORT 01 : IARC MONOGRAPH 32 DEC 83
TESTED FOR CARCINOGENICITY IN TWO FRAGMENTARY STUDIES IN MICE BY SKIN PAINTING AND NO SKIN TUMOR REPORTED. IN SIX STUDIES IN WHICH PHENANTHRENE TESTED IN MOUSE-SKIN INITIATION-PROMOTION ASSAY, IT WAS ACTIVE AS INITIATOR IN ONE STUDY, INACTIVE AS AN INITIATOR IN FOUR OTHERS, AND INACTIVE AS PROMOTER IN ONE STUDY. PHENANTHRENE ADMINISTERED BY INTRAPERITONEAL OR SUBCUTANEOUS INJECTION TO NEONATAL MICE DID NOT INCREASE INCIDENCE OF TUMORS OVER THAT IN CONTROLS. EXPERIMENTS INVOLVING SINGLE ORAL ADMINISTRATION TO RATS AND SINGLE SUBCUTANEOUS INJECTION TO MICE INADEQUATE FOR EVALUATION.

REPORT 02 : IARC MONOGRAPH 32
GENERALLY REPORTED TO BE NON-MUTAGENIC TO SALMONELLA TYPHIMURIUM; HOWEVER, IN ONE STUDY REPORTED TO BE MUTAGENIC TO SALMONELLA TYPHIMURIUM IN PRESENCE OF HIGH CONCENTRATION OF AN EXOGENOUS METABOLIC SYSTEM. GAVE NEGATIVE RESULTS FOR DIFFERENTIAL SURVIVAL USING DNA-REPAIR-PROFICIENT/-DEFICIENT BACILLUS SUBTILIS STRAINS. DID NOT INDUCE DNA REPAIR, CHROMOSOMAL ABERRATIONS OR SISTER CHROMATID EXCHANGE IN CULTURED MAMMALIAN CELLS. DID INDUCE MUTATION IN EXPERIMENT IN HUMAN CELLS IN CULTURE IN PRESENCE OF EXOGENOUS METABOLIC SYSTEM, AND INDUCED SISTER CHROMATID EXCHANGE IN CHINESEHAMSTER BONE-MARROW CELLS IN VIVO. FAILED TO INDUCE MORPHOLOGICAL TRANSFORMATION.

REPT'D USE : INCOMPLETE COMBUSTION PRODUCT, CIGARETTE SMOKE, USED MOTOR OIL AND TAP WATER

CAS NO : **136-40-3**
CHEM NAME : ***** PHENAZOPYRIDINE HYDROCHLORIDE**

REPORT 01 : NTP TECHNICAL BULLETIN 3 DEC 80
THE COMPOUND WAS FOUND TO BE NEGATIVE IN THE SALMONELLA TYPHIMURIUM MUTAGENICITY ASSAY IN ONE LABORATORY.

REPORT 02 : NCI CARC TRS 99 DHEW PUB NO NIH 78-1349
UNDER THE CONDITIONS OF THIS BIOASSAY, PHENAZOPYRIDINE HYDROCHLORIDE WAS CARCINOGENIC IN FISCHER 344 RATS, INDUCING ADENOCARCINOMAS OF THE COLON IN BOTH MALES AND FEMALES. ALTHOUGH PHENAZOPYRIDINE HYDROCHLORIDE WAS NOT CARCINOGENIC IN MALE B6C3Fl MICE, THE CHEMICAL WAS CARCINOGENIC IN FEMALES, INDUCING HEPATOCELLULAR ADENOMAS AND CARCINOMAS (FEEDING STUDY).

REPORT 03 : IARC MONOGRAPH 24 1980
PHENAZOPYRIDINE HYDROCHLORIDE WAS TESTED IN MICE AND RATS BY ORAL ADMINISTRATION AND IN MICE BY INTRAPERITONEAL ADMINISTRATION. AFTER ITS ORAL ADMINISTRATION IN FEMALE MICE, IT SIGNIFICANTLY INCREASED THE INCIDENCE OF HEPATOCELLULAR ADENOMAS AND CARCINOMAS. IN MALE AND FEMALE RATS. IT INDUCED TUMORS OF THE COLON AND RECTUM.

REPORT 06 : MULTI-AGENCY CROSS CHECK
Y HHS MAGUIRE
Y OSHA CARC
N EPA CAG
N EPA RCRA
N EPA CLEAN AIR
N EPA CLEAN WATER
REPT'D USE : URINARY TRACT ANALGESIC/ANTISEPTIC

CAS NO : **156-51-4**
CHEM NAME : ***** PHENELZINE SULFATE**

REPORT 01 : IARC MONOGRAPH 24 1980
PHENELZINE SULFATE WAS TESTED IN MICE BY ORAL ADMINISTRATION: IT SIGNIFICANTLY INCREASED THE INCIDENCE OF LUNG AND BLOOD VESSEL TUMORS IN FEMALES. THERE IS LIMITED EVIDENCE THAT IT IS CARCINOGENIC IN EXPERIMENTAL ANIMALS.

REPORT 02 : IARC MONOGRAPH 24 1980
PHENELZINE SULFATE WAS MUTAGENIC FOR SALMONELLA TA100 IN THE ABSENCE OF METABOLIC ACTIVATION. PHENELZINE SULFATE EXHIBITS REACTIVITY TOWARDS DNA, AS EVIDENCED BY POSITIVE RESULTS IN E.COLI. AND BY THE INACTIVATION OF B. SUBTILIS TRANSFORMING DNA.
REPT'D USE : INHIBITOR OXIDASE AND ANTIDEPRESSANT

CAS NO : **3546-10-9**
CHEM NAME : ***** PHENESTERIN**

REPORT 01 : NCI CARC TRS 60 DHEW PUB NO NIH 78-1310
UNDER THE CONDITIONS OF THIS BIOASSAY, PHENESTERIN WAS CARCINOGENIC IN FEMALE SPRAGUE-DAWLEY RATS, PRODUCING ADENOCARCINOMAS OF THE MAMMARY GLAND, AND IN BOTH SEXES OF B6C3F1 MICE, PRODUCING ALVEOLAR/ BRONCHIOLAR CARCINOMAS, HEMATOPOIETIC TUMORS, AND MYOCARDIAL SARCOMAS. PHENESTERIN WAS GIVEN BY GAVAGE.
REPT'D USE : EXPERIMENTAL ANTINEOPLASTIC AGENT

CAS NO : **114-86-3**
CHEM NAME : ***** PHENFORMIN HYDROCHLORIDE**

REPORT 01 : NCI CARC TRS 7 DHEW PUB NO NIH 77-807
TUMORS APPEARING IN TREATED RATS AND MICE WERE SIMILAR IN TYPE AND NUMBER TO THOSE IN CONTROLS, AND NO PATHOLOGIC OR STATISTICAL EVIDENCE OF INDUCTION OF TUMORS IN THESE SPECIES BY PHENFORMIN WAS FOUND (FEEDING STUDY).
REPT'D USE : HYPOGLYCEMIC DRUGS

CAS NO : **103-03-7**
CHEM NAME : ***** PHENICARBAZIDE**

REPORT 01 : IARC MONOGRAPH 12 1976
PHENICARBAZIDE IS CARCINOGENIC IN MICE WHEN ADMINISTERED ORALLY, THE ONLY SPECIES AND ROUTE INVESTIGATED. IT PRODUCED LUNG TUMORS AND ANGIOMAS AND/OR ANGIOSARCOMAS IN THE LIVER.
REPORT 06 : MULTI-AGENCY CROSS CHECK
N HHS MAGUIRE
Y OSHA CARC
N EPA CAG
N EPA RCRA
N EPA CLEAN AIR
N EPA CLEAN WATER
REPT'D USE : EXPERIMENTAL ANTIPYRETIC/NEUROLEPTIC AGENT

CAS NO : **57-30-7**
CHEM NAME : ***** PHENOBARBITAL SODIUM**

REPORT 01 : IARC MONOGRAPH 13 1977
PHENOBARBITAL SODIUM IS CARCINOGENIC IN MICE AND RATS AFTER ITS ORAL ADMINISTRATION FOR LIFETIME. IN MICE, IT PRODUCED BENIGN AND MALIGNANT LIVER-CELL TUMORS; IN RATS, BENIGN LIVER-CELL TUMORS WERE OBSERVED VERY LATE IN LIFE.

REPORT 02 : IARC MONOGRAPH 13 1977
PHENOBARBITOL BINDS TO RAT CHROMATIN IN-VIVO. IT HAS BEEN REPORTED TO BE WEAKLY MUTAGENIC IN DROSOPHILA AFTER ADDITION OF 2-3 MG TO THE NUTRIENT MEDIUM. IT DID NOT INDUCE POINT MUTATIONS IN SALMONELLA TA100, TA1535, TA1537 OR TA98.
REPT'D USE : BARBITURATE AND ANTICONVULSANT DRUG

CAS NO : **50-06-6**
CHEM NAME : ***** PHENOBARBITONE**

REPORT 01 : IARC MONOGRAPH 13 1977
THE AVAILABLE EVIDENCE IS INSUFFICIENT TO ALLOW AN EVALUATION OF THE CARCINOGENICITY OF PHENOBARBITAL TO BE MADE.
REPT'D USE : BARBITURATE

CAS NO : **108-95-2**
CHEM NAME : ***** PHENOL**

REPORT 01 : EPA AMBIENT WATER QUALITY CRITERIA OCT 80
A CRITERION, BASED ON CHRONIC TOXICITY DATA SHOWING KIDNEY DAMAGE IN RATS, AND AN UNCERTAINTY FACTOR OF 500, CORRESPONDING TO THE CALCULATED ACCEPTABLE DAILY INTAKE OF 0.1 MG/KG IS 3.5 MG/LITER.

REPORT 02 : NIOSH CRITERIA DOCUMENT JUN 76
NIOSH RECOMMENDS A STANDARD OF 20 MG/CU M TWA BASED ON SKIN, EYE, CNS, LIVER, AND KIDNEY EFFECTS. THE PRESENT OSHA STANDARD IS 5 PPM TWA.

REPORT 03 : NCI/NTP CARC TRS 203 DHEW PUB NO 80-15
UNDER THE CONDITIONS OF THIS BIOASSAY, PHENOL WAS NOT CARCINOGENIC FOR EITHER MALE OR FEMALE F344 RATS OR MALE AND FEMALE B6C3F1 MICE. PHENOL WAS GIVEN IN DRINKING WATER.

REPORT 04 : EPA RCRA HEALTH AND ENV. EFFECTS PROFILE OCT 80
INSUFFICIENT DATA EXIST TO INDICATE THAT PHENOL IS A CARCINOGENIC AGENT. INFORMATION ON THE MUTAGENICITY OF PHENOL IS EQUIVOCAL. PHENOL DOES NOT APPEAR TO BE TERATOGENIC.

REPORT 05 : NTP TECHNICAL BULLETIN 6 JAN 82
THE COMPOUND WAS NEGATIVE IN THE SALMONELLA TYPHIMURIUM MUTAGENICITY ASSAY IN TWO LABORATORIES.

REPORT 06 : MULTI-AGENCY CROSS CHECK
N HHS MAGUIRE
N OSHA CARC
N EPA CAG
Y EPA RCRA
Y EPA CLEAN AIR
Y EPA CLEAN WATER

REPT'D USE : MFR OF PHENOLIC RESINS, CAPROLACTAM, BISPHENOL A DISINFECTANT, AEROSOLS, FR CHEMICALS, ADHESIVES/GLUES, PHOTOGRAPHIC CHEMICAL, WOOD PRESERVATIVES AND PAINT/VARNISH REMOVERS

CAS NO : **59-96-1**
CHEM NAME : ***** PHENOXYBENZAMINE**

REPORT 01 : IARC MONOGRAPH 24 1980
PHENOXYBENZAMINE WAS TESTED INTRAPERITONEALLY IN MICE; IT PRODUCED AN INCREASED INCIDENCE OF LUNG TUMORS.

REPORT 02 : IARC MONOGRAPHS SUPP 4, 1982 1982
THERE IS SUFFICIENT EVIDENCE THAT PHENOXYBENZAMINE IS CARCINOGENIC IN EXPERIMENTAL ANIMALS.
REPT'D USE : ANTIHYPERTENSIVE AND TREATMENT OF PERIPHERAL VASCULAR DISORDERS

CAS NO : **63-92-3**
CHEM NAME : ***** PHENOXYBENZAMINE HCL**

REPORT 01 : NCI CARC TRS 72 DHEW PUB NO 78-1322
UNDER THE CONDITIONS OF THIS BIOASSAY, PHENOXYBENZAMINE HYDROCHLORIDE WAS CARCINOGENIC (SARCOMAGENIC) FOR THE PERITONEUM OF BOTH SEXES OF SPRAGUE-DAWLEY RATS AND B6C3F1 MICE. THE COMPOUND WAS ADMINISTERED BY ABDOMINAL INJECTION.

REPORT 02 : IARC MONOGRAPH 24 1980
PHENOXYBENZAMINE HYDROCHLORIDE WAS TESTED INTRAPERITONEALLY IN MICE AND RATS; IT PRODUCED LOCAL PERITONEAL SARCOMAS IN ANIMALS OF BOTH SEXES.

REPORT 06 : MULTI-AGENCY CROSS CHECK
Y HHS MAGUIRE
N OSHA CARC
N EPA CAG
N EPA RCRA
N EPA CLEAN AIR
N EPA CLEAN WATER
REPT'D USE : ANTIHYPERTENSIVE AGENT, TREATMENT OF PERIPHERAL VASCULAR DISORDERS

CAS NO : **122-99-6**
CHEM NAME : ***** 2-PHENOXYETHANOL**

REPORT 01 : EPA TSCA 8(E) NOTICE 8EHQ-0884-0528 AUG 84
DOW CHEMICAL COMPANY REPORTED SUMMARY RESULTS ON 2-WEEK TERATOLOGY STUDY IN WHICH NON-PREGNANT FEMALE RABBITS GIVEN 2-PHENOXYETHANOL DERMALLY AT DOSE LEVEL OF 1000 MG/KG/DAY FOR 14 DAYS SHOWED SENSITIVITY TO HEMOLYTIC EFFECTS OF 2-PHENOXYETHANOL. EACH EXHIBITED SIGNS SIMILAR TO AFFECTED PREGNANT RABBITS IN TERATOLOGY STUDY. THE PROMINENT SIGNS WERE HEMATOLOGIC CHANGES INDICATIVE OF RED BLOOD CELL BREAKDOWN.

REPORT 02 : EPA RESPONSE TO ITC (FR NOTICE 49FR21407) MAY 84
12TH REPORT OF ITC DESIGNATED 2-PE FOR PRIORITY TESTING. MFRS BEGAN TESTING PROGRAM AND EPA CONCLUDES THAT DATA FROM THIS PROGRAM PLUS EXISTING DATA IS SUFFICIENT FOR DETERMINATION OR PREDICTION OF POTENTIAL HEALTH EFFECTS. THEREFORE, EPA IS NOT INITIATING RULEMAKING.
REPT'D USE : LATEX PAINTS, CLEANERS, POLISHES, COSMETICS, PHOTOGRAPHIC DEVELOPERS, SOLVENT, PERFUMES, BACTERICIDE, ANTISEPTIC, INKS, DYES AND ORGANIC SYNTHESIS.

CAS NO : **4075-79-0**
CHEM NAME : ***** 4'-PHENYLACETANILIDE**

REPORT 06 : MULTI-AGENCY CROSS CHECK
N HHS MAGUIRE
Y OSHA CARC
N EPA CAG
N EPA RCRA
N EPA CLEAN AIR
N EPA CLEAN WATER
REPT'D USE : NO AVAILABLE PRODUCTION OR USE DATA

CAS NO : **50-33-9**
CHEM NAME : ***** PHENYLBUTAZONE**

REPORT 01 : IARC MONOGRAPH 13 1977
A NUMBER OF CASES OF LEUKEMIA WERE REPORTED BETWEEN 1960 AND 1966 IN SUBJECTS TREATED WITH PHENYLBUTAZONE. THE AVAILABLE EVIDENCE IS, HOWEVER, INSUFFICIENT TO SUBSTANTIATE THE SUGGESTION THAT USE OF PHENYLBUTAZONE IS RELATED TO SUBSEQUENT DEVELOPMENT OF LEUKEMIA.

REPORT 02 : NTP TR 367 NIH 90-2822 MAR 90
UNDER 2-YEAR GAVAGE STUDIES EQUIVOCAL EVIDENCE OF CARCINOGENIC ACTIVITY OF PHENYLBUTAZONE FOR MALE F344/N

RATS SHOWN BY SMALL NUMBERS OF RENAL TUBULAR CELL ADENOMAS AND CARCINOMAS. SOME EVIDENCE OF CARCINOGENIC ACTIVITY FOR FEMALE F344/N RATS SHOWN BY TWO TRANSITIONAL CELL CARCINOMAS AT 100 MG/KG DOSE. TUBULAR CELL ADENOMAS AND CARCINOMAS MAY BE ASSOCIATED WITH ADMINISTRATION OF PHENYLBUTAZONE TO FEMALE RATS. SOME EVIDENCE OF CARCINOGENIC ACTIVITY FOR MALE B6C3F1 MICE SHOWN BY INCREASED INCIDENCE OF HEPATOCELLULAR ADENOMAS OR CARCINOMAS (COMBINED). NO EVIDENCE OF CARCINOGENICITY FOR FEMALE B6C3F1 MICE GIVEN CHEMICAL IN CORN OIL.

REPORT 03 : IARC MONOGRAPH 13 1977
NEGATIVE RESULTS HAVE BEEN REPORTED WITH PHENYLBUTAZONE IN SALMONELLA TA100, TA98, TA1535 AND TA1537, WITH AND WITHOUT METABOLIC ACTIVATION. NEGATIVE RESULTS WERE ALSO OBTAINED IN THE DOMINANT LETHAL TEST FOLLOWING ADMINISTRATION EITHER TO MALE MICE BY I.P. INJECTION OR TO FEMALE MICE AS A SINGLE ORAL DOSE.

REPORT 04 : NTP TECHNICAL BULLETIN 5 AUG 81
PHENYL BUTAZONE WAS NEGATIVE IN THE S. TYPHIMURIUM MUTAGENICITY ASSAY IN ONE LABORATORY.
REPT'D USE : VETERINARY MEDICINE AND PHARMACEUTICAL

CAS NO : **4994-16-5**
CHEM NAME : ***** 4-PHENYLCYCLOHEXENE (4-PCH)**

REPORT 01 : EPA FYI-0689-0596
4-PHENYLCYCLOHEXENE (4-PCH) EVALUATED FOR DERMAL SENSITIZATION USING MODIFIED BUEHLER METHOD IN MALE HARTLEY ALBINO GUINEA PIGS BY THREE DERMAL APPLICATIONS OF 10% 4-PCH IN DIPROPYLENE GLYCOL MONOMETHYL ETHER ONCE PER WEEK FOR THREE WEEKS. POSITIVE CONTROL GROUP OF ANIMALS RECEIVED THREE DERMAL APPLICATIONS OF 10% PER 331 EPOXY RESIN OF THE DOW CHEMICAL COMPANY IN DPGME. CHALLENGE APPLICATION WITH POSITIVE CONTROL CAUSED SLIGHT ERYTHEMA AT DERMAL TEST SITE OF NINE OF TEN ANIMALS. CHALLENGE APPLICATION WITH 10% 4-PCH CAUSED NO RESPONSE. CONCLUDED 4-PCH DID NOT CAUSE DELAYED CONTACT HYPERSENSITIVITY IN GUINEA PIGS.
REPT'D USE : SOLVENT, PENETRATING AGENT, INTERMEDIATE AND LABORATORY REAGENT

CAS NO : **CLASS-22**
CHEM NAME : ***** PHENYLENEDIAMINES CATEGORY**

REPORT 01 : EPA TSCA DECISION NOT TO TEST FR NOTICE 50FR4267 JAN 85
THIS NOTICE IS EPA'S FINAL DISPOSITION OF 34 OF 47 PDA'S

RECOMMENDED FOR PRIORITY TESTING CONSIDERATION UNDER TSCA. IT REFLECTS COMMENTS RECEIVED IN RESPONSE TO ANPR (47FR973). THESE PDA'S NOT RECOMMENDED FOR TESTING BECAUSE OF LOW OR NO PRODUCTION OR LACK OF TSCA-RELATED EXPOSURE. 13 OTHER PDA'S BEING CONSIDERED FOR TESTING WILL BE ANNOUNCED IN FUTURE FR NOTICE.

REPT'D USE : HAIR DYES

CAS NO : **108-45-2**

CHEM NAME : ***** M-PHENYLENEDIAMINE**

REPORT 01 : IARC MONOGRAPH 16
META-PHENYLENEDIAMINE WAS INADEQUATELY TESTED IN RATS BY SUBCUTANEOUS INJECTION. NO EVALUATION OF THE CARCINOGENICITY OF THIS COMPOUND CAN BE MADE. M-PHENYLENEDIAMINE INDUCED REVERSE MUTATIONS IN SALMONELLA TA1538 WITH METABOLIC ACTIVATION. NO DOMINANT LETHALS WERE INDUCED IN CHARLES RIVER RATS INJECTED I.P. WITH M-PDA BEFORE MATING.

REPORT 02 : EPA RESPONSE TO ITC RECOMMS (47FR973) JAN 81
IN RESPONSE TO THE ITC RECOMMENDATIONS (45FR35897) THE EPA PROPOSES AN ADVANCED NOTICE REQUESTING COMMENTS CONCERNING A PROPOSAL TO CONDUCT MUTAGENICITY AND CARCINOGENICITY TESTING FOR 11 SELECTED PHENYLENEDIAMINES, AND TERATOGENICITY/REPRODUCTIVE EFFECTS AND OTHER CHRONIC EFFECTS TESTING, AS WELL AS AN EPIDEMIOLOGIC STUDY, FOR 12 PHENYLENEDIAMINES. THE EPA WILL CONSIDER ADDITIONAL TESTING FOR AN ADDITIONAL 37 PHENYLENEDIAMINES, AS APPROPRIATE.

REPORT 03 : EPA TSCA PROPOSED TEST RULE FR NOTICE 51FR472 JAN 86
EPA PROPOSING A TEST RULE UNDER TSCA 4(A) IN RESPONSE TO ITC'S DESIGNATION FOR PRIORITY TESTING IN ITS 6TH REPORT (45FR35897). EPA PROPOSING TESTING FOR AQUATIC OXIDATION AND AQUATIC TOXICITY AS WELL AS DROSOPHILA SEX-LINKED RECESSIVE LETHALITY.

REPORT 04 : EPA 8(E) STATUS REPORT - 8EH4-1285- 0577 JAN 86
MIXTURE CONTAINING 56.7 WT.% CPD, 30.3 WT.% 4,4'-METHYLENEDIANILINE AND 13.0 WT.% DIGLYCIDYL ETHER OF BISPHENOL A (ARALDITE 6010) OR CPD ALONE NOT CONSIDERED TO BE TERATOGENIC FOLLOWING ORAL ADMINISTRATION TO PREGNANT RATS AT DOSE LEVELS SUFFICIENT TO CAUSE MILD MATERNAL TOXICITY.

REPT'D USE : HEAT RESISTANT FIBERS, DYE INTERMEDIATE AND EPOXY CURING AGENT

CAS NO : **541-69-5**
CHEM NAME : ***** M-PHENYLENEDIAMINE HYDROCHLORIDE**

REPORT 01 : IARC MONOGRAPH 16 1978
META-PHENYLENEDIAMINE HYDROCHLORIDE WAS INADEQUATELY TESTED IN RATS BY SUBCUTANEOUS INJECTION. NO EVALUATION OF THE CARCINOGENICITY OF THIS COMPOUND CAN BE MADE.

REPORT 02 : EPA RESPONSE TO ITC RECOMMS 47FR973 JAN 81
IN RESPONSE TO THE ITC RECOMMENDATIONS (45FR35897) THE EPA PROPOSES AN ADVANCED NOTICE REQUESTING COMMENTS CONCERNING A PROPOSAL TO CONDUCT MUTAGENICITY AND CARCINOGENICITY TESTING FOR 11 SELECTED PHENYLENEDIAMINES, AND TERATOGENICITY/REPRODUCTIVE EFFECTS AND OTHER CHRONIC EFFECTS TESTING, AS WELL AS AN EPIDEMIOLOGIC STUDY, FOR 12 PHENYLENEDIAMINES. THE EPA WILL CONSIDER ADDITIONAL TESTING FOR AN ADDITIONAL 37 PHENYLENEDIAMINES, AS APPROPRIATE.
REPT'D USE : RUBBER CURING AGENT, PHOTOGRAPHIC, CORROSION INHIBITOR, DYE DEVELOPING AGENT, ANALYTICAL REAGENT AND HAIR DYE

CAS NO : **54-17-8**
CHEM NAME : ***** M-PHENYLENEDIAMINE.H2SO4**

REPORT 01 : EPA TSCA PROPOSED TEST RULE FR NOTICE 51FR472 JAN 86
EPA PROPOSING TEST RULE UNDER TSCA 4(A) IN RESPONSE TO ITC'S DESIGNATION OF M-PD.H2SO4 FOR PRIORITY TESTING (45FR35897). TESTING PROPOSED FOR AQUATIC OXIDATION, AQUATIC TOXICITY AND DROSOPHILA SEX-LINKED RECESSIVE LETHALITY.
REPT'D USE : INTERMEDIATE IN PRODUCTION OF ARAMID FIBERS AND DYE INTERMEDIATE

CAS NO : **95-54-5**
CHEM NAME : ***** O-PHENYLENEDIAMINE**

REPORT 01 : EPA TSCA PROPOSED TEST RULE FR NOTICE 51FR472 JAN 86
EPA PROPOSING TEST RULE UNDER TSCA 4(A) IN RESPONSE TO ITC'S DESIGNATION OF OPD FOR PRIORITY TESTING. IT REFLECTS COMMENTS RECEIVED BY EPA IN ANSWER TO ITS ANPR (47FR973). TESTING IS PROPOSED FOR AQUATIC OXIDATION RATE AND TOXICITY TO AQUATIC ORGANISMS.
REPT'D USE : INTERMEDIATE IN PRODUCTION OF FUNGICIDE, VAT DYES AND CORROSION INHIBITOR.

CAS NO : **615-28-1**
CHEM NAME : ***** O-PHENYLENEDIAMINE DIHYDROCHLORIDE**

REPORT 01 : EPA RESPONSE TO ITC RECOMMS (47FR973) JAN 81
IN RESPONSE TO THE ITC RECOMMENDATIONS (45FR35897) THE EPA PROPOSES AN ADVANCED NOTICE REQUESTING COMMENTS CONCERNING A PROPOSAL TO CONDUCT MUTAGENICITY AND CARCINOGENICITY TESTING FOR 11 SELECTED PHENYLENEDIAMINES AND TERATOGENICITY AND REPRODUCTIVE EFFECTS AND OTHER CHRONIC EFFECTS TESTING, AS WELL AS AN EPIDEMIOLOGIC STUDY, FOR 12 PHENYLENEDIAMINES. THE EPA WILL CONSIDER FURTHER TESTING FOR AN ADDITIONAL 37 PHENYLENEDIAMINES, AS APPROPRIATE.

REPORT 06 : MULTI-AGENCY CROSS CHECK
N HHS MAGUIRE
Y OSHA CARC
N EPA CAG
N EPA RCRA
N EPA CLEAN AIR
N EPA CLEAN WATER

REPT'D USE : NO KNOWN USE FOR THE DIHYDROCHLORIDE FORM

CAS NO : **106-50-3**
CHEM NAME : ***** P-PHENYLENEDIAMINE**

REPORT 01 : IARC MONOGRAPH NO. 16 JUN 77
PARA-PHENYLENEDIAMINE HAS BEEN INADEQUATELY TESTED IN MICE BY SKIN APPLICATION AND IN RATS BY ORAL AND SUBCUTANEOUS ADMINISTRATION. NO EVALUATION OF THE CARCINOGENICITY OF THIS COMPOUND CAN BE MADE.

REPORT 02 : EPA RESPONSE TO ITC RECOMMS (47FR973) JAN 81
IN RESPONSE TO THE ITC RECOMMENDATIONS (45FR35897) THE EPA PROPOSES AN ADVANCED NOTICE REQUESTING COMMENTS CONCERNING A PROPOSAL TO CONDUCT MUTAGENICITY AND CARCINOGENICITY TESTING FOR 11 SELECTED PHENYLENEDIAMINES, AND TERATOGENICITY AND REPRODUCTIVE EFFECTS AND OTHER CHRONICS EFFECTS TESTING, AS WELL AS AN EPIDEMIOLOGIC STUDY, FOR 12 PHENYLENEDIAMINES. THE EPA WILL CONSIDER FURTHER TESTING FOR AN ADDITIONAL 37 PHENYLENEDIAMINES, AS APPROPRIATE.

REPORT 03 : IARC MONOGRAPH 16 1978
P-PHENYLENEDIAMINE HAS BEEN REPORTED TO INDUCE REVERSE MUTATIONS IN SALMONELLA TA1538 WITH METABOLIC ACTIVATION. THE CHEMICAL DID NOT INDUCE MICRONUCLEATED CELLS IN BONE MARROW WHEN ADMINISTERED ORALLY TO RATS.

REPORT 04 : EPA TSCA PROPOSED TEST RULE FR NOTICE 51FR472 JAN 86
EPA PROPOSING TEST RULE UNDER TSCA 4(A) IN RESPONSE TO ITC'S DESIGNATION FOR PRIORITY TESTING IN ITS 6TH REPORT (45FR35897). EPA IS PROPOSING TESTING FOR AQUATIC OXIDATION AND AQUATIC TOXICITY.

REPT'D USE : PHOTOGRAPHIC DEVELOPERS, CHEMICAL INTERMEDIATE, FLAME RETARDANTS, HAIR DYES AND ANTIOXIDANT

CAS NO : **624-18-0**

CHEM NAME : ***** P-PHENYLENEDIAMINE DIHYDROCHLORIDE**

REPORT 01 : NCI CARC TRS NO 174 1979 DHEW PUB NO NIH 79-1730
UNDER THE CONDITIONS OF THIS BIOASSAY, THERE WAS NO CONVINCING EVIDENCE THAT DIETARY ADMINISTRATION OF P-PHENYLENEDIAMINE DIHYDROCHLORIDE WAS CARCINOGENIC IN FISCHER 344 RATS OR B6C3F1 MICE.

REPORT 02 : EPA RESPONSE TO ITC RECOMMS (47FR973)1 JAN 81
IN RESPONSE TO THE ITC RECOMMENDATIONS (45FR35897), THE EPA PROPOSES AN ADVANCED NOTICE REQUESTING COMMENTS CONCERNING A PROPOSAL TO CONDUCT MUTAGENICITY AND CARCINOGENICITY TESTING FOR 11 SELECTED PHENYLENEDIAMINES, AND TERATOGENICITY/REPRODUCTIVE EFFECTS AND OTHER CHRONIC EFFECTS TESTING, AS WELL AS AN EPIDEMIOLOGIC STUDY, FOR 12 PHENYLENEDIAMINES. THE EPA WILL CONSIDER ADDITIONAL TESTING FOR AN ADDITIONAL 37 PHENYLENEDIAMINES, AS APPROPRIATE.

REPORT 03 : NTP TECHNICAL BULLETIN 8 JUL 82
P-PDA 2HCL WAS POSITIVE IN THE L5178Y MOUSE LYMPHOMA ASSAY.

REPT'D USE : HAIR/FUR DYE, DYE INTERMEDIATE, CHEMICAL INDICATOR, PHOTOGRAPHIC CHEMICAL AND RUBBER ANTIOXIDANT

CAS NO : **1624-57-75**

CHEM NAME : ***** P-PHENYLENEDIAMINE.H2SO4 SALT**

REPORT 01 : EPA TSCA PROPOSED TEST RULE FR NOTICE 51FR472 JAN 86
EPA IS PROPOSING A TEST RULE UNDER TSCA 4(A) IN RESPONSE TO ITC'S DESIGNATION FOR PRIORITY TESTING IN ITS 6TH REPORT (45FR35897). ACTION REFLECTS COMMENTS RECEIVED AS RESULT OF ANPR (46FR973). TESTING IS PROPOSED FOR AQUATIC OXIDATION RATE AND AQUATIC TOXICITY.

REPT'D USE : PRODUCTION OF ARAMID FIBERS AND DYE INTERMEDIATE

CAS NO : **61-76-7**
CHEM NAME : ***** PHENYLEPHRINE HYDROCHLORIDE**

REPORT 01 : NTP TECH REPORT TR 322 JAN 87
UNDER CONDITIONS OF THESE 2-YEAR STUDIES, NO EVIDENCE OF CARCINOGENICITY FOR MALE OR FEMALE F344/N RATS GIVEN 620 OR 1250 PPM IN FEED, OR FOR MALE OR FEMALE B6C3F1 MICE GIVEN 1250 OR 2500 PPM IN FEED. SURVIVAL OF HIGH DOSE RATE MALE RATS GREATER THAN IN CONTROLS AND INCIDENCES OF MONONUCLEAR CELL LEUKEMIA AND PHEOCHROMOCYTOMAS LOWER IN DOSED THAN IN CONTROL MALE RATS. INFLAMMATION MORE FREQUENTLY SEEN IN LIVER AND PROSTATE GLAND OF DOSED MALE RATS THAN IN CONTROLS.
REPT'D USE : NASAL DECONGESTANT AND OPHTHALMIC APPLICATIONS

CAS NO : **60-12-8**
CHEM NAME : ***** 2-PHENYLETHANOL (2-PE)**

REPORT 01 : EPA CHIP MAY 85
PREGNANT RATS GIVEN 4.2 MG/KG 2-PHENYLETHANOL BY GAVAGE ON DAYS 6-15 OF GESTATION, SHOWED TERATOGENIC EFFECTS. 2-PE GAVE NEGATIVE RESULTS IN DNA REPAIR ASSAY IN E. COLI; GENE CONVERSION ASSAY IN S. TYPHIMURIUM TA98, TA1535 AND TA1537; MITOCHONDRIAL DNA ASSAY IN S. CERVISIAE AND SISTER CHROMATID ASSAY IN HUMAN LYMPHOCYTES.
REPT'D USE : FOOD FLAVORING AGENT AND COSMETIC AND DETERGENT FRAGRANCE

CAS NO : **122-60-1**
CHEM NAME : ***** PHENYL GLYCIDYL ETHER**

REPORT 01 : NIOSH CRITERIA DOCUMENT JUN 78
PGE HAS SHOWN MUTAGENIC ACTIVITY IN THE SALMONELLA ASSAY, BUT IT PRODUCED NO DOMINANT LETHAL OR TERATOGENIC EFFECTS IN MICE EXPOSED TO 11.5 PPM FOR 12-19 DAYS. INCONCLUSIVE EVIDENCE TO TESTICULAR DEGENERATION WAS REPORTED IN SOME RATS EXPOSED TO 1.75-11.2 PPM PGE, NIOSH RECOMMENDS A LIMIT OF 1 PPM FOR PGE, MEASURED AS A 15 MIN CEILING CONCENTRATION, TO PREVENT SKIN AND RESPIRATORY TRACT IRRITATION.

CAS NO : **59-88-1**
CHEM NAME : ***** PHENYLHYDRAZINE MONOHYDROCHLORIDE**

REPORT 06 : MULTI-AGENCY CROSS CHECK
N HHS MAGUIRE

Y OSHA CARC
N EPA CAG
N EPA RCRA
N EPA CLEAN AIR
N EPA CLEAN WATER
REPT'D USE : REDUCING AGENT AND HEMOLYTIC AGENT

CAS NO : **100-65-2**
CHEM NAME : ***** N-PHENYLHYDROXYLAMINE**

REPORT 01 : NTP TECHNICAL BULLETIN 6 JAN 82
THE COMPOUND WAS POSITIVE IN THE SALMONELLA TYPHIMURIUM MUTAGENICITY ASSAY IN ONE LABORATORY.
REPT'D USE : INTERMEDIATE IN MFR OF CUPFERRON, 4-AMINOPHENOL

CAS NO : **89-25-8**
CHEM NAME : ***** 1-PHENYL-3-METHYL-5-PYRAZOLONE**

REPORT 01 : NCI CARC TRS 141 DHEW PUB NO 78-1396
UNDER THE CONDITIONS OF THIS BIOASSAY, THERE WAS NO EVIDENCE FOR THE CARCINOGENICITY OF 1-PHENYL-3-METHYL-5-PYRAZOLONE TO FISCHER 344 RATS OR B6C3F1 MICE (FEEDING STUDY).

REPORT 02 : NTP TECHNICAL BULLETIN 7 APR 82
1-PHENYL-3-METHYL-5-PYRAZOLONE WAS NEGATIVE IN THE SALMONELLA TYPHIMURIUM MUTAGENICITY ASSAY.

REPORT 03 : NTP TECHNICAL BULLETIN NO. 7 APR 82
1-PHENLY-3-METHYL-5-PYRAZOLONE WAS NEGATIVE IN THE SALMONELLA TYPHIMURIUM MUTAGENICITY ASSAY.
REPT'D USE : DYE CHEMICAL, INTERMEDIATE AND CHEMICAL INDICATOR

CAS NO : **135-88-6**
CHEM NAME : ***** N-PHENYL-2-NAPHTHYLAMINE (PBNA)**

REPORT 01 : IARC MONOGRAPH 16 1978
N-PHENYL-2-NAPHTHYLAMINE HAS BEEN TESTED IN MICE BY ORAL AND SINGLE SC ADMINISTRATION AND ORALLY IN A SMALL NUMBER OF DOGS. RESULTS OF ORAL ADMINISTRATION IN MICE INDICATE A STATISTICALLY SIGNIFICANT INCREASE IN THE INCIDENCE OF ALL TUMORS AND OF HEPATOMAS IN MALE MICE OF ONE OF TWO STRAINS TESTED. THESE FINDINGS SUGGEST THAT THE CPD IS CARCINOGENIC, BUT THEY DO NOT ALLOW A CONCLUSIVE EVALUATION OF THE CARCINOGENICITY OF THIS CPD.

REPORT 02 : NIOSH CURRENT INTELLIGENCE BULLETIN 16 DEC 76
THE NIOSH LEARNED THAT N-PHENYL-2-NAPHYLAMINE IS METABOLIZED TO THE KNOWN CARCINOGEN, 2-NAPHTHLYAMIN. BULLETIN EMPHASIZES THE POTENTIAL PROBLEM OF METABOLIC CONVERSION OF MATERIALS BELIEVED TO BE RELATIVELY INNOCUOUS INTO KNOWN CARCINOGENS.

REPORT 03 : NTP TR 333 NIH PUB. 88-2589 JAN 88
FEED STUDIES SHOWS NO EVIDENCE OF CARCINOGENIC ACTIVITY FOR MALE OR FEMALE F 344/N RATS FED DIETS CONTAINING 2,500 OR 5,000 PPM N-PHENYL2-NAPHTHYLAMINE. DECREASED INCIDENCES OF SEVERAL NEOPLASMS WERE OBSERVED IN DOSED RATS: THYROID GLAND C-CELL NEOPLASMS IN MALES AND FEMALES AND MONONUCLEAR CELL LEUKEMIA, PITUITARY GLAND ADENOMAS, AND MAMMARY GLAND FIBROADENOMAS IN FEMALES. NO EVIDENCE OF CARCINOGENIC ACTIVITY FOR MALE B6C3F1 MICE FED DIETS CONTAINING 2,500 OR 5,000 PPM N-PHENYL-2-NAPHTHYLAMINE FOR FEMALE B6C3F1 MICE INDICATED BY THE OCCURRENCE OF TWO RARE KIDNEY NEOPLASMS. CHEMICAL-RELATED NONNEOPLASTIC LESIONS OCCURRED IN THE KIDNEY OF RATS AND MICE.

REPORT 04 : IARC MONOGRAPH 16 1978
NO DATA ON THE EMBRYOTOXICITY TERATOGENICITY OR MUTAGENICITY OF N-PHENYL-2-NAPHTHYLAMINE WERE AVAILABLE.

REPT'D USE : RUBBER ANTIOXIDANT/ACCELERATOR, ADHESIVES, STABILIZER IN SILICONE ENAMELS, ROCKET FUELS AND POLYMERIZATION INHIBITOR

CAS NO : **90-43-7**

CHEM NAME : ***** O-PHENYLPHENOL**

REPORT 01 : NTP TECHNICAL BULLETIN 6 JAN 82
THE COMPOUND WAS WEAKLY POSITIVE IN THE SALMONELLA TYPHIMURIUM MUTAGENICITY ASSAY IN ONE LABORATORY.

REPORT 02 : NTP TR 301 NIH PUB 86-2557 MAR 86
UNDER CONDITIONS OF 2 YR DERMAL APPLICATION STUDIES THERE WAS NO EVIDENCE OF CARCINOGENICITY IN MALE OR FEMALE SWISS CD-1 MICE ADMINISTERED O-PHENYLPHENOL ALONE OR AS PROMOTER FOLLOWING INITIATION WITH DMBA. O-PHENYLPHENOL CAUSED NONNEOPLASTIC LESIONS AT SITE OF APPLICATION WHICH INCLUDED HYPERKERATOSIS, ULCERATIONS, AND INFLAMMATION.

REPORT 03 : IARC MONOGRAPH 30 JAN 83
ORTHO-PHENYLPHENOL TESTED FOR CARCINOGENICITY IN MICE AND RATS BY ADMINISTRATION IN DIET. SODIUM ORTHO-PHENYLPHENATE TESTED IN RATS BY ADMINISTRATION IN DIET. NO

EVIDENCE OF CARCINOGENICITY OF ORTHO-PHENYLPHENOL BUT BOTH STUDIES HAD SOME LIMITATIONS. IN RATS, SODIUM ORTHO-PHENYPHENATE IS CARCINOGENIC TO URINARY TRACT, PRODUCING BOTH BENIGN AND MALIGNANT TUMORS.

REPORT 04 : IARC MONOGRAPH 30 JAN 83
THE AVAILABLE DATA WERE INSUFFICIENT TO EVALUATE THE TERATOGENICITY OF SODIUM ORTHO-PHENYLPHENATE TO EXPERIMENTAL ANIMALS.

REPORT 05 : IARC MONOGRAPH 30 JAN 83
ORTHO-PHENYLPHENOL AND SODIUM ORTHO-PHENYLPHENATE SUBJECTED TO VARIETY OF MUTAGENICITY TESTS, INCLUDING THOSE USING RODENTS, WITHOUT DEMONSTRATING ANY UNAMBIGUOUS MUTAGENIC OR DNA DAMAGING EFFECT. COMMERCIAL PREPARATIONS REPORTED TO CONTAIN UNCHARACTERIZED IMPURITIES CAPABLE OF INDUCING CHROMOSOMAL ABERRATIONS IN CULTURED MAMMALIAN CELLS.

REPT'D USE : FUNGICIDE, GERMICIDE, DYE INTERMEDIATE, DISHWASHING DETERGENTS, DISINFECTANT, PRESERVATIVE, COATINGS AND INTERMEDIATES

CAS NO : **101-54-2**

CHEM NAME : ***** N-PHENYL-P-PHENYLENEDIAMINE**

REPORT 01 : NCI CARC TRS 82 DHEW PUB NO 78-1332
UNDER THE CONDITIONS OF THIS BIOASSAY, N-PHENYL-P-PHENYLENEDIAMINE WAS NOT CARCINOGENIC FOR FISCHER 344 RATS OR FOR B6C3F1 MICE (FEEDING STUDY).

REPORT 02 : TSCA INTERAGENCY TESTING COMM. MAY 80
IN A JAN 1981 RULING, THE SOUTHERN DISTRICT COURT OF NEW YORK ORDERED EPA TO PROPOSE TEST RULES, OR DECISIONS NOT TO REQUIRE TESTING, BY THE END OF 1983 FOR ALL CHEMICALS LISTED IN THE FIRST FIVE ITC REPORTS. THE SCHEDULE SUBMITTED BY THE EPA, AND ACCEPTED BY THE COURT WITHOUT CHANGE, SHOWS THAT A DECISION FOR TESTING OF CARCINOGENICITY, MUTAGENICITY, TERATOGENICITY AND EPA STUDIES OF PHENYLENEDIAMINES WILL BE MADE IN 1982.

REPORT 03 : NTP TECHNICAL BULLETIN 6 JAN 82
THE COMPOUND WAS NEGATIVE IN THE SALMONELLA TYPHIMURIUM MUTAGENICITY ASSAY IN ONE LABORATORY.

REPORT 04 : EPA TSCA 8(E)-0888-0746 SEP 88
MONSANTO PROVIDED PRELIMINARY RESULTS OF TERATOLOGY STUDY OF 4-AMINODIPHENYLAMINE IN SPRAGUE-DAWLEY RATS FOR SINGLE DOSE ADMINISTERED BY GAVAGE AT 150 MG/KG/DAY IN CORN OIL. SIGNIFICANT MATERNAL TOXICITY, EMBRYO-

TOXICITY (I.E., INCREASED RESORPTIONS) AND DEVELOPMENTAL EFFECTS IN FETAL POPULATION, INCLUDING GROSS EXTERNAL AND INTERNAL MALFORMATIONS. OBSERVATIONS DIFFER SUBSTANTIALLY FROM RESULTS OF TERATOGENICITY TESTING OF 4-ADPA USED AS COMPONENT OF HAIR DYES (PICCIANO, ET AL., DRUG AND CHEM. TOXICOL. 7:167, 1984).

REPORT 05 : EPA TSCA 8(E)-0888-0746 SEP 88
N-PHENYL-P-PHENYLENEDIAMINE EVALUATED FOR TERATOGENIC POTENTIAL BY GAVAGE TO PREGNANT SPRAGUE-DAWLEY RATS AT DOSE LEVELS OF 50, 100, AND 200 MG/KG SHOWED NO SIGNS OF TOXICITY. SIGNIFICANT REDUCTION IN MEAN MATERNAL BODY WEIGHT GAIN NOTED AT HIGH DOSE OF 200 MG/KG NO EMBRYOTOXIC NOR FETAL OBSERVED TOXIC EFFECTS. EVALUATION OF FETAL EXTERNAL, VISCERAL, AND SKELETAL ANOMALIES REVEALED NO STATISTICALLY SIGNIFICANT DIFFERENCES BETWEEN DYE TREATED AND CONTROL GROUPS.

REPT'D USE : ORGANIC SYNTHESIS, DYE INTERMEDIATE AND HAIR DYE FORMULATIONS

CAS NO : **92-43-3**
CHEM NAME : ***** 1-PHENYL-3-PYRAZOLIDINONE (PHENIDONE)**

REPORT 01 : EPA TSCA 8(E) REPORT 8EHO-0984-0529 SEP 84
EASTMAN KODAK COMPANY SUBMITTED RESULTS FROM SUBCHRONIC STUDY IN MALE AND FEMALE RATS EXPOSED TO PHENIDONE IN DIET FOR 42 OR 90 DAYS AT DOSES FROM 12 TO 270 MG/KG/DAY. HEMATOLOGIC EFFECTS OBSERVED IN DOSE-RELATED MANNER, ALSO ENLARGEMENT OF SPLEENS AND TESTICULAR ATROPHY IN SOME MALES.

CAS NO : **103-85-5**
CHEM NAME : ***** 1-PHENYL-2-THIOUREA**

REPORT 01 : NCI CARC TRS 148 DHEW PUB NO 78-1704
UNDER THE CONDITIONS OF THIS BIOASSAY, 1-PHENYL-2-THIOUREA WAS NOT CARCINOGENIC TO FISCHER 344 RATS OR B6C3F1 MICE (FEEDING STUDY).

REPORT 02 : NTP TECHNICAL BULLETIN 6 JAN 82
THE COMPOUND WAS NEGATIVE IN THE SALMONELLA TYPHIMURIUM MUTAGENICITY ASSAY IN ONE LABORATORY.

REPT'D USE : RODENTICIDE AND GENETIC RESEARCH

CAS NO : **630-93-3**
CHEM NAME : ***** PHENYTOIN SODIUM**

REPORT 01 : IARC MONOGRAPH 13 1977
PHENYTOIN AND ITS SODIUM SALT ARE CARCINOGENIC IN MICE FOLLOWING THEIR INTRAPERITONEAL INJECTION AND ORAL ADMINISTRATION, RESPECTIVELY. THEY PRODUCED LYMPHOMAS AND LEUKEMIAS. AN ASSOCIATION HAS BEEN OBSERVED IN EPILEPTIC PATIENTS BETWEEN THE OCCURRENCE OF LYMPHOMAS AND LONG-TERM ANTICONVULSANT THERAPY IN WHICH PHENYTOIN ALONE OR IN COMBINATION WITH OTHER ANTICONVULSANTS SUCH AS PHENOBARBITAL WERE GIVEN.

REPORT 02 : NTP CARC TEST PROGRAM JUL 82
IN A CURRENT BIOASSAY, DIPHENYL HYDANTOIN IS BEING ADMINISTERED IN THE DIET TO F344 RATS AND B6C3F1 MICE. THE CHRONIC EXPOSURE PHASE IS IN PROGRESS.

REPORT 03 : IARC MONOGRAPH 13 1977
NEGATIVE RESULTS HAVE BEEN REPORTED IN THE DOMINANT LETHAL TEST IN MICE AFTER PHENYTOIN SODIUM WAS ADMINISTERED BY I.P. INJECTION.
REPT'D USE : ANTICONVULSANT AND ANTIARRHYTHMIC DRUG

CAS NO : **13171-21-6**
CHEM NAME : ***** PHOSPHAMIDON**

REPORT 01 : NCI CARC TRS 16 DHEW PUB NO 79-816
UNDER THE CONDITIONS OF THIS BIOASSAY, TECHNICAL-GRADE PHOSPHAMIDON WAS NOT CARCINOGENIC FOR B6C3F1 MICE. THE DATA OBTAINED IN THIS BIOASSAY WITH OSBORNE-MENDEL RATS ARE INSUFFICIENT TO ALLOW THE INTERPRETATION THAT TECHNICAL-GRADE PHOSPHAMIDON IS CARCINOGENIC IN THIS SPECIES (FEEDING STUDY).

REPORT 02 : NTP TECHNICAL BULLETIN 5 AUG 81
PHOSPHAMIDON WAS POSITIVE IN THE SALMONELLA TYPHIMURIUM MUTAGENICITY ASSAY IN ONE LABORATORY.
REPT'D USE : INSECTICIDE

CAS NO : **7723-14-0**
CHEM NAME : ***** PHOSPHORUS**

REPORT 01 : EPA 8EHQ-0889- 0820 NOV 89
ELEMENTAL PHOSPHORUS WAS ADMINISTERED BY GAVAGE (IN CORN OIL OR TRICAPRYLIN AT DOSE LEVLS 0, 0.005, 0.015, OR 0.075 MG/KG/DAY TO 100 RATS PRIOR TO MATING, DURING MATING AND EITHER THROUGH DAY 15 OF GESTATION OR THROUGOUT GESTA-

TION. AT THE HIGHEST DOSE, IT ADVERSELY AFFECTED PARTURITION FOR BOTH LITTERS WITH SLIGHT REDUCTION IN MEAN NUMBER OF LIVE PUPS. NO OTHER TOXIC EFFECTS WERE OBSERVED.

REPT'D USE : RODENTICIDES, SMOKE SCREEN AND GAS ANALYSIS

CAS NO : **26544-23-0**

CHEM NAME : ***** PHOSPHOROUS ACID, ISODECYL DIPHENYL ESTER**

REPORT 01 : EPA TSCA 8(E) NOTICE 8EHO-0784-0522 JUL 84
WITCO CHEMICAL PROVIDED REPORTS FROM BATTERY OF IN VITRO GENOTOXICITY STUDIES WITH MIXTURE OF 7-OXABICYCLO-(4.1.0)HEPTANE-3-CARBOXYLIC ACID, 7-OXABICYCLO(4.1.0)-HEPT-3-YLMETHYL ESTER AND PHOSPHOROUS ACID, ISODECYL DIPHENYL ESTER SHOWING INCREASED REVERTANT FREQUENCY IN SALMONELLA TYPHIMURIUM STRAIN TA 1535 WITH ACTIVATION, INCREASED MUTATION FREQUENCY IN CULTURED MOUSE LYMPHOMA CELLS WITH AND WITHOUT ACTIVATION, AND INCREASED NUMBER OF SISTER CHROMATID EXCHANGES IN CULTURED CHO CELLS WITH AND WITHOUT ACTIVATION.

CAS NO : **13366-73-9**

CHEM NAME : ***** PHOTODIELDRIN**

REPORT 01 : NCI CARC TRS 17 DHEW PUB NO 77-817
UNDER THE CONDITIONS OF THIS BIOASSAY, PHOTODIELDRIN WAS NOT CARCINOGENIC FOR OSBORNE-MENDEL RATS OR B6C3F1 MICE (FEEDING STUDY).

REPT'D USE : PHOTOCONVERSION PRODUCT OF DIELDRIN

CAS NO : **88-96-0**

CHEM NAME : ***** PHTHALAMIDE**

REPORT 01 : NCI CARC TRS 161 DHEW PUB NO 79-1717
UNDER THE CONDITIONS OF THIS BIOASSAY, PHTHALAMIDE WAS NOT CARCINOGENIC FOR F344 RATS OR B6C3F1 MICE OF EITHER SEX (FEEDING STUDY).

REPT'D USE : NO AVAILABLE PRODUCTION OR USE DATA

CAS NO : **CLASS-24**

CHEM NAME : ***** PHTHALATE ESTERS**

REPORT 01 : COMP. REVIEW OF ENVIRONMENTAL APR 81
MONITORING DATA SRC/CMA
COLLECTION OF ALL KNOWN PUBLISHED AND UNPUBLISHED

ENVIRONMENTAL DATA ON PHTHALATE ESTERS (PAE) PROVIDES A CRITICAL REVIEW AND EVALUATION OF ALL DATA PRESENTED. FOR MOST PART, DATA GROUPED BY SOURCE: AIR, WATER, SEDIMENTS, BIOTA. AMBIENT MONITORING DATA FROM STORET IS ALSO INCLUDED, BUT RELIABILITY OF DATA NOT KNOWN.

REPORT 02 : COMP. REV. OF ACUTE AQUATIC APR 81
TOX DATA ON PAE'S, SRC/CMA
ACUTE TOXICITY INFORMATION REPORTED FOR AT LEAST 1 AQUATIC SPP FOR PAE'S: DMP, DEP, DPP, DBP, DIBP, BBP, DOP, DEHP, DNP, DINP, DDP, DIDP. RESULTS SHOW THAT 7 PAE'S: DAP, DPP, DBP, DOP, DEHP AND DINP AND THE DBP/D-EHP MIXTURE CAUSED ACUTE EFFECTS IN 1 OR MORE SPP AT 1 MG/L OR LESS. LOWEST ACUTELY TOXIC CONCENTRATION FOR DMP, DEP AND DIBP = 1.7, 3.3 AND 2.5 MG/L RESPECTIVELY. 3 OTHER PAE'S NOT ACUTELY TOXIC IN TEST. EC50 AND LC50 VALUES REPORTED FOR 3 PAE'S: DMP, DEP AND BBP, FOR FRESH WATER AND MARINE SPP IN EACH GROUP.

REPORT 03 : SEL. CRIT REV. HUMAN HLTH EFFE CTS FEB 81
ASSOC W/PAE 'S - SRC/CMA
DISCUSSES PHARMACOKINETC, BIOCHEMICAL AND PHYSIOLOGIC EFFECTS OF CERTAIN PAE'S, THEIR METABOLITES AND SOME OTHER DIETHYLHEXYL ESTERS. SHORT-CHAIN ALKYL ESTERS ABSORBED MORE READILY BY ORAL ROUTE THAN LONGER-CHAIN ONES. TISSUE SAMPLES OBTAINED AT AUTOPSY MAY OCCASIONALLY C/PARTS PER MILLION LEVELS OF PAE'S. METABOLISM GOES VIA HYDROLYSIS TO MONOESTER, FOLLOWED BY GAMMA AND GAMMA-1 OXIDATION OF ALKYL CHAIN. SHORT-CHAIN ESTERS TEND TO BE EXCRETED IN URINE, LONG-CHAIN ESTERS IN THE FECES. ORAL ADMINISTRATION MAY LEAD TO INCREASED LIVER KIDNEY WEIGHTS, TESTICULAR ATROPHY WITH IMPAIRED SPERMATOGENESIS, AND ACUTE LUNG TOXICITY.

REPORT 04 : COMP. REV. ENV. MONITORING DATA APR 81
ON PAE'S SRC
SUMMARIZES CONCENTRATION RANGES THAT ARE APT TO BE FOUND IN VARIOUS ENVIRONMENTAL SAMPLES.

CAS NO : **85-44-9**
CHEM NAME : ***** PHTHALIC ANHYDRIDE**

REPORT 01 : NCI CARC TRS 159 DHEW PUB NO 79-1715
UNDER THE CONDITIONS OF THIS BIOASSAY, PHTHALIC ANHYDRIDE WAS NOT CARCINOGENIC FOR F344 RATS OR B6C3F1 MICE OF EITHER SEX (FEEDING STUDY).

REPORT 02 : NTP TECHNICAL BULLETIN 3 DEC 80
THE COMPOUND WAS FOUND TO BE NEGATIVE IN THE SALMONELLA TYPHIMURIUM MUTAGENICITY ASSAY IN ONE LABORATORY.

REPORT 03 : EPA RCRA HEALTH AND ENV. EFFECTS PROFILE OCT 80
CHRONIC OCCUPATIONAL EXPOSURE TO PHTHALIC ANHYDRIDE HAS BEEN REPORTED TO PRODUCE PROGRESSIVE RESPIRATORY DAMAGE IN WORKERS, INCLUDING MARKED FIBROSIS OF THE LUNGS.

REPORT 06 : MULTI-AGENCY CROSS CHECK
N HHS MAGUIRE
N OSHA CARC
N EPA CAG
Y EPA RCRA
N EPA CLEAN AIR
N EPA CLEAN WATER

REPT'D USE : PLASTIC, RESIN, DYE INTERMEDIATE, PAINTS/VARNISHES

CAS NO : **1918-02-1**

CHEM NAME : ***** PICLORAM**

REPORT 01 : NCI CARC TRS 23 DHEW PUB NO 78-823
UNDER THE CONDITIONS OF THE BIOASSAY, THE FINDINGS ARE SUGGESTIVE OF THE ABILITY OF THE COMPOUND TO INDUCE BENIGN TUMORS IN THE LIVERS OF FEMALE OSBORNE-MENDEL RATS (FEEDING STUDY).

REPORT 06 : MULTI-AGENCY CROSS CHECK
N HHS MAGUIRE
Y OSHA CARC
N EPA CAG
N EPA RCRA
N EPA CLEAN AIR
N EPA CLEAN WATER

REPT'D USE : HERBICIDE

CAS NO : **88-89-1**

CHEM NAME : ***** PICRIC ACID**

REPORT 01 : NTP TECHNICAL BULLETIN 2/8 APR 80, JUL 82
PICRIC ACID WAS POSITIVE IN THE SALMONELLA ASSAY, BUT NEGATIVE IN DROSOPHILA WHEN TESTED FOR SEX-LINKED RECESSIVE LETHALS.

REPT'D USE : TEXTILE MORDANT, BATTERIES, PHOTOGRAPHIC EMULSIONS, LAB AGENT, CHEMICAL INTERMEDIATE, ETCHING AGENT, ASTRINGENTS, EXPLOSIVES, ANTISEPTICS AND MATCHES

CAS NO : **51-03-6**
CHEM NAME : ***** PIPERONYL BUTOXIDE**

REPORT 01 : NCI CARC TRS 120 DHEW PUB NO 79-1375
UNDER THE CONDITIONS OF THIS BIOASSAY, PIPERONYL BUTOXIDE WAS NOT CARCINOGENIC FOR F344 RATS OR B6C3F1 MICE (FEEDING STUDY).

REPORT 02 : NTP TECHNICAL BULLETIN 5 AUG 81
PIPERONYL BUTOXIDE WAS NEGATIVE IN THE SALMONELLA TYPHIMURIUM MUTAGENICITY ASSAY IN ONE LABORATORY.

REPORT 03 : IARC MONOGRAPH 30 JAN 83
TESTED FOR CARCINOGENICITY IN TWO STUDIES IN RATS AND IN THREE STUDIES IN MICE BY ORAL ADMINISTRATION. ALSO TESTED IN MICE OF THREE STRAINS BY SUBCUTANEOUS ADMINISTRATION (IN ONE, NEONATALLY). NO STATISTICALLY SIGNIFICANT INCIDENCE OF TUMORS OBSERVED IN MICE OR RATS.

REPORT 04 : IARC MONOGRAPH JAN 83
TREATMENT OF RATS WITH HIGH DOSES OF PIPERONYL BUTOXIDE DURING ORGANOGENESIS DID NOT AFFECT FETAL DEVELOPMENT; HOWEVER, ITS CONTINUED ADMINISTRATION AT HIGH DOSE LEVELS FOR THREE GENERATIONS CAUSED REDUCTION IN NUMBER OF PREGNANCIES AND OFFSPRING.

REPORT 05 : IARC MONOGRAPH JAN 83
PIPERONYL BUTOXIDE REPORTED NONMUTAGENIC TO BACTERIA, SILKWORMS AND CULTURED MAMMALIAN CELLS. A DOMINANT LETHAL STUDY IN MICE WAS INCONCLUSIVE. LACK OF AVAILABILITY OF MUCH OF DETAILED DATA PREVENTED OVERALL EVALUATION OF MUTAGENICITY OF PIPERONYL BUTOXIDE, BUT ALL CITED RESULTS NEGATIVE.
REPT'D USE : INSECTICIDE, SYNERGIST AND AEROSOLS

CAS NO : **120-62-7**
CHEM NAME : ***** PIPERONYL SULFOXIDE**

REPORT 01 : NCI CARC TRS 124 DHEW PUB NO 79-1379
UNDER THE CONDITIONS OF THIS BIOASSAY, TECHNICAL-GRADE PIPERONYL SULFOXIDE WAS NOT CARCINOGENIC FOR MALE OR FEMALE FISCHER 344 RATS OR FOR FEMALE B6C3F1 MICE, BUT WAS CARCINOGENIC FOR MALE B6C3F1 MICE, PRODUCING AN INCREASED INCIDENCE OF HEPATOCELLULAR CARCINOMAS (FEEDING STUDY).

REPORT 02 : NTP TECHNICAL BULLETIN 5 AUG 81
PIPERONYL SULFOXIDE WAS NEGATIVE IN THE SALMONELLA TYPHIMURIUM MUTAGENICITY ASSAY IN ONE LABORATORY.

REPT'D USE : SYNERGISTIC INSECTICIDE

CAS NO : **1955-45-9**
CHEM NAME : ***** PIVALOLACTONE**

REPORT 01 : NCI CARC TRS 140 DHEW PUB NO 78-1395
UNDER THE CONDITIONS OF THIS BIOASSAY, PIVALOLACTONE WAS FOUND TO BE CARCINOGENIC TO BOTH MALE AND FEMALE FISCHER 344 RATS, PRODUCING SQUAMOUS-CELL CARCINOMAS AND SQUAMOUS-CELL PAPILLOMAS OF THE FORESTOMACH. THIS STUDY PROVIDED NO EVIDENCE FOR THE CARCINOGENICITY OF PIVALOLACTONE IN B6C3F1 MICE OF EITHER SEX. PIVALOLACTONE WAS GIVEN BY GAVAGE.
REPT'D USE : LABORATORY CHEMICAL

CAS NO : **9003-01-4**
CHEM NAME : ***** POLYACRYLIC ACID**

REPORT 01 : IARC MONOGRAPH 19 FEB 79
NO CARCINOGENICITY OR MUTAGENICITY DATA ON POLYACRYLIC ACID WERE AVAILABLE.

CAS NO : **67774-32-7**
CHEM NAME : ***** POLYBROMINATED BIPHENYL MIXTURE (PBB)**

REPORT 01 : NTP TECHNICAL REPORT TR 244 NIH PUB. 83-1800 JUN 83
UNDER THE CONDITIONS OF THESE STUDIES, POLYBROMINATED BIPHENYL MIXTURE (FIRE-MASTER FF-1) WAS CARCINOGENIC FOR FISCHER 344/N RATS AND B6C3F1 MICE OF EACH SEX, INDUCING NEOPLASTIC NODULES, HEPATOCELLULAR CARCINOMAS, AND CHOLANGIOCARCINOMAS IN RATS, AND HEPATOCELLULAR CARCINOMAS IN MICE. OTHER TOXICITIES INCLUDED PORPHYROGENIC EFFECTS AND HEPATO-TOXICITY.

REPORT 02 : IARC VOL 41 1986
NO RELEVANT DATA AVAILABLE TO EVALUATE REPRODUCTIVE EFFECTS OR PRENATAL TOXICITY OF PPB TO HUMANS. A SMALL COHORT STUDY OF CHEMICAL WORKERS POTENTIALLY EXPOSED TO PBB'S WAS UNINFORMATIVE WITH REGARD TO CANCER. THERE IS SUFFICIENT EVIDENCE FOR CARCINOGENICITY OF COMMERCIAL MIXTURES OF PPBS TO HUMANS AND NO EVIDENCE OF GENETIC ACTIVITY IN SHORT-TERM TESTS.

REPORT 06 : MULTI-AGENCY CROSS CHECK
Y HHS MAGUIRE
N OSHA CARC
N EPA CAG

N EPA RCRA
N EPA CLEAN AIR
N EPA CLEAN WATER

REPT'D USE : FIRE RETARDANTS FOR SYNTHETIC FIBERS AND MOULDED THERMOPLASTICS

CAS NO : **1336-36-3**
CHEM NAME : ***** POLYCHLORINATED BIPHENYLS (PCB'S)**

REPORT 01 : NIOSH CURRENT INTELLIGENCE BULLETIN 7 NOV 75
ACUTE TOXIC EFFECTS OF PCBS IN HUMANS INCLUDE AN ACNELIKE SKIN ERUPTION (CHLORACNE), PIGMENTATION OF THE SKIN AND NAILS, EXCESSIVE EYE DISCHARGE, SWELLING OF EYELIDS, AND DISTINCTIVE HAIR FOLLICLES. THE TOXIC EFFECTS OF THESE COMPOUNDS IN ANIMALS HAVE BEEN STUDIED EXTENSIVELY. THE LITERATURE REPORTS MALIGNANT, NONMALIGNANT, AND REPRODUCTIVE EFFECTS.

REPORT 02 : NIOSH CRITERIA DOCUMENT SEP 77
NIOSH RECOMMENDS A STANDARD OF 1 UG/CU M TWA BASED ON CARCINOGENICITY AND SKIN, LIVER, AND REPRODUCTIVE EFFECTS. THE PRESENT OSHA STANDARD IS 1 MG/CU M TWA.

REPORT 03 : TOX-TIPS DEC 80
PERSONNEL RECORDS GATHERED IN 1975 BY NIOSH AND THE NEW YORK STATE DEPT. OF HEALTH FOR A MORTALITY STUDY OF GENERAL ELECTRIC CAPACITOR WORKERS ARE BEING REVIEWED BY NIOSH FOR EVIDENCE OF LOW BIRTH WEIGHT INFANTS AND NEONATAL DEATHS. NIOSH ALSO PROPOSES TO GATHER, THROUGH INTERVIEWS, INFORMATION ON OTHER ADVERSE REPRODUCTIVE OUTCOMES INCLUDING MISCARRIAGES, STILLBORNS, AND CONGENITAL DEFECTS. THE STUDY IS SCHEDULED FOR COMPLETION IN JULY 1983.

REPORT 04 : NCI CARC TRS 38 DHEW PUB NO 78-838
UNDER THE CONDITIONS OF THIS BIOASSAY, AROCLOR 1254 WAS NOT CARCINOGENIC IN FISCHER 344 RATS; HOWEVER, A HIGH INCIDENCE OF HEPATOCELLULAR PROLIFERATIVE LESIONS IN BOTH MALE AND FEMALE RATS WAS RELATED TO ADMINISTRATION OF THE CHEMICAL. IN ADDITION, THE CARCINOMAS OF THE GASTROINTESTINAL TRACT MAY BE ASSOCIATED WITH ADMINISTRATION OF AROCLOR 1254 IN BOTH MALES AND FEMALES (FEEDING STUDY).

REPORT 05 : IARC MONOGRAPH 18 JUN 78
THERE IS EXPERIMENTAL EVIDENCE OF A CARCINOGENIC EFFECT OF SOME PCBS IN RODENTS. THE EPIDEMIOLOGICAL DATA PROVIDE SUGGESTIVE EVIDENCE OF A RELATIONSHIP BETWEEN EXPOSURE TO PCBS AND THE DEVELOPMENT OF MALIGNANT

MELANOMA. EFFORTS SHOULD BE MADE TO OBTAIN BOTH CONFIRMATORY EXPERIMENTAL AND EPIDEMIOLOGICAL EVIDENCE; IN PARTICULAR, CONTINUING FOLLOW-UP OF SURVIVORS OF THE YUSHO EPISODE IS NECESSARY. IN A 1979 SUPPLEMENT; IARC CONCLUDED THAT PCBS ARE PROBABLY CARCINOGENIC FOR HUMANS.

REPORT 06 : MULTI-AGENCY CROSS CHECK
Y HHS MAGUIRE
N OSHA CARC
Y EPA CAG
N EPA RCRA
Y EPA CLEAN AIR
Y EPA CLEAN WATER

REPORT 07 : EPA AMBIENT WATER QUALITY CRITERIA OCT 80
PCBS ARE SUSPECTED OF BEING HUMAN CARCINOGENS. AMBIENT WATER CONCENTRATIONS OF .79 NG/L, .08 NG/L, AND .01 NG/L ARE ESTIMATED TO RESULT IN A RESPECTIVE 10 -5, 10 -6, AND 10 -7 INCREASE IN CANCER RISK OVER A LIFETIME.

REPORT 08 : EPA TO AMEND FINAL RULE ON USE 48FR52402 NOV 83
EPA PROPOSING TO AMEND FINAL RULE PUBLISHED IN 44FR31514 BY AUTHORIZNG INDEFINITE USE OF SMALL QUANTITIES OF PCBS IN ART AND HISTORIC CONSERVATION AND IN RESEARCH AND DEVELOPMENT. THIS IS BASED ON THE FINDING THAT THESE USES DO NOT POSE UNREASONABLE RISKS TO PUBLIC HEALTH OR THE ENVIRONMENT.

REPORT 09 : DENIAL OF CITIZEN'S PETITION, 49FR1697 JAN 84
NOTICE ANNOUNCES EPA'S DECISION TO DENY PETITION FROM CANNELTON IND. UNDER TSCA 21 CONCERNING DISPOSAL OF NON-LIQUID PCB WASTE.

REPT'D USE : INDUSTRIAL HYDRAULIC FLUIDS, TRANSFORMER, CAPACITOR DIELECTRIC FLUID, PLASTICIZERS, CARBONLESS PAPER AND LUBRICANT. MANUFACTURE IN U.S. DISCONTINUED 1976; SINCE 1979 NO IMPORT/EXPORT ALLOWED

CAS NO : **CLASS-19**
CHEM NAME : ***** POLYCHLORINATED DIBENZOFURAN (PCDB)**

REPORT 01 : EPA HEALTH ASSESSMENT DOCUMENT, 600/8-86/018A JUN 86
PCDB'S ARE EXTREMELY TOXIC TO ANIMALS AND HUMANS. 2,3,7,8-TCDF IS TERATOGENIC IN MICE. HUMAN TOXICITY EFFECTS ARE SIMILAR TO THOSE CAUSED BY 2,3,7,8-TCDD. PREVALENCE AND STRUCTURAL SIMILARITIES TO PBB'S CAUSE HEATED PUBLIC CONCERN. FIRES INVOLVING TRANSFORMERS HAVE BEEN REPORTED TO PRODUCE PCDF AND PCDD; WHEN CAPACITORS WERE INVOLVED ONLY PCDF WAS FORMED. OTHER SOURCES OF

EXPOSURE ARE UNLEADED GASOLINE AND OFFICES USING CARBONLESS COPY PAPER CONTAINING PCB'S. ALSO ANALYSIS OF NIPPLES REVEALED PCDD'S AND PCDF'S TO BE PRESENT. WOOD PRESERVED BY CHLOROPHENOLS IS ANOTHER SOURCE OF PCDF'S.

REPT'D USE : LEADED GASOLINES AND CONTAMINATES IN OTHER CPDS.

CAS NO : **9010-98-4**

CHEM NAME : ***** POLYCHLOROPRENE**

REPORT 01 : IARC MONOGRAPH 19 1979

NO DATA ON THE CARCINOGENICITY OF POLYCHLOROPRENE WERE AVAILABLE TO THE WORKING GROUP.

REPT'D USE : FLAME RETARDANT, HOSE/TUBINGS, ADHESIVES, ELASTOMER AND WIRE/CABLES

CAS NO : **9002-88-4**

CHEM NAME : ***** POLYETHYLENE**

REPORT 01 : IARC MONOGRAPH 19 1979

POLYETHYLENE WAS TESTED BY SUBCUTANEOUS IMPLANTATION IN RATS AND MICE BY INTRAPERITONEAL IMPLANTATION IN RATS, USING DISCS, FILMS, RODS, FRAGMENTS AND POWDER, AND BY INTRAUTERINE INSERTION IN RATS OF POLYETHYLENE DEVICES. SUBCUTANEOUS IMPLANTATION OF ALL FORMS EXCEPT THE POWDER RESULTED IN LOCAL SARCOMAS. THE AVAILABLE DATA ON POLYETHYLENE DO NOT PERMIT AN EVALUATION OF ITS POTENTIAL CARCINOGENICITY.

REPORT 02 : IARC MONOGRAPH 19 1979

NO DATA ON THE EMBRYOTOXICITY, TERATOGENICITY OR MUTAGENICITY OF POLYETHYLENE WERE AVAILABLE.

REPT'D USE : PLASTIC, PACKAGING, BUILDING MATERIALS, CLEANING AGENTS, PAINTS/VARNISHES, FR CHEMICAL, AEROSOLS, FLOOR WAXES, ELECTRICAL WIRING, HOUSEWARES AND PLUMBING

CAS NO : **9005-64-5**

CHEM NAME : ***** POLYETHYLENE GLYCOL SORBITAN MONOLAURATE**

REPORT 01 : EPA TSCA 8(E) 0288-0718 FEB 88

DOW CORNING REPORTED RESULTS FROM CHRONIC DAPHNIA MAGNA REPRODUCTIVE LIMIT TEST OF MIXTURE OF POLYETHYLENE GLYCOL SORBITAN MONOLAURATE (CAS NO. 9005-64-5) AND OCTAMETHYLCYCLOTETRASILOXANE (CAS NO. 556-672). SURVIVAL OF DAPHNIA EXPOSED TO 10 PPM PGSM/100 PPB OMCTS SIGNIFICANTLY LESS THAN CONTROLS. A NO OBSERVABLE EFFECT LEVEL

		CAN'T BE DETERMINED FROM TEST, NOR CAN SYNERGISTIC EFFECT BETWEEN PGSM AND OMCTS BE ELIMINATED. FINAL REPORT BEING SUBMITTED TO EPA UNDER SECTION 8(D) OF TSCA.
REPT'D USE	:	CHEM. INTERMEDIATE, PROD. OF SILICONE RUBBERS; SURFACE AGENTS, LUBRICANTS, HOSES, SEALS AND WIRE COATINGS

CAS NO : **UNKNOWN-32**

CHEM NAME : ***** POLYHALOGENATED DIBENZO-P-DIOXINS (HDD)**

REPORT 01 : POLYHALOGENATED DIBENZO-P- DIOXINS; JUN 87
TEST AND REPORT RULES
EPA IS ISSUING TESTING AND REPORTING REQUIREMENTS FOR POLYHALOGENATED DIBENZO-P-DIOXINS. RULES REQUIRE MFRS AND IMPORTERS TO TEST 12 CURRENTLY TESTED ORGANIC CHEMICALS AND 20 CHEMICALS NOT NOW MFD OR IMPORTED INTO U.S. AND REPORT TO EPA IF HDDS FOUND IN EXISTING TEST RECORDS OR IN RESULTS FROM NEW TESTS.

REPT' USE : CONTAMINANT

CAS NO : **UNKNOWN-29**

CHEM NAME : ***** POLYHALOGENATED DIBENZOFURAN (HDF) (2,3,7,8-TCDF)**

REPORT 01 : CRC MEETING MINUTES, 8/18/86 AUG 86
2,3,7,8-TCDF IS CLAIMED TO BE TERATOGENIC IN MICE. HUMAN EFFECTS SIMIILAR TO THOSE OF 2,3,7,8-TCDD.

REPORT 02 : POLYHALOGENATED HDDS AND HDFS; JUN 87
TESTING AND REPORTING
UNDER TSCA 4 AND 8, EPA IS ESTABLISHING TESTING/REPORTING REQUIREMENTS FOR 12 ORGANIC CHEMICALS PLUS 20 ORGANIC CHEMICALS NOT NOW MFD OR IMPORTED INTO U.S. THESE CHEMICALS MUST BE TESTED FOR PRESENCE OF HDFS. IF NEW TESTS OR EXISTING TEST RESULTS SHOW PRESENCE OF HDFS ABOVE "LEVELS OF QUANTITATION" STATED IN THIS RULE, MFR OR IMPORTER MUST FORWARD CERTAIN INFORMATION TO EPA.

REPT'D USE : CONTAMINANT

CAS NO : **9016-87-9**

CHEM NAME : ***** POLYMETHYLENE POLYPHENYL ISOCYANATE**

REPORT 01 : IARC MONOGRAPH 19 1979
NO DATA ON THE CARCINOGENICITY OR MUTAGENICITY OF 2,4- AND 2,6-TOLUENE DIISOCYANATES, 1,5-NAPHTHALENE DIISOCYANATE, 4,4'-METHYLENE-DIPHENYL DIISOCYANATE OR POLYMETHYLENE POLYPHENYL ISOCYANATE WERE AVAILABLE.

REPT'D USE : PLASTIC MONOMER

CAS NO : **9011-14-7**
CHEM NAME : ***** POLYMETHYL METHACRYLATE**

REPORT 01 : IARC MONOGRAPH 19 1979
POLYMETHYL METHACRYLATE, WHEN IMPLANTED INTRAMUSCULARLY, SUBCUTANEOUSLY OR INTRAPERITONEALLY AS DISCS, PRODUCED LOCAL SARCOMAS IN RATS. IN MICE, SUBCUTANEOUS IMPLANTATION OF POLYMETHYL METHACRYLATE FILMS ALSO RESULTED IN LOCAL SARCOMAS. INTRAMUSCULAR IMPLANTATION OF POLYMETHYL METHCARYLATE IN GUINEA-PIGS AND IMPLANTATION INTO THE CHEEK POUCH IN HAMSTERS PRODUCED NO IMPLANTATION SITE TUMORS.

REPORT 02 : IARC MONOGRAPH 19 1979
NO DATA ON THE TOXICITY, TERATOGENICITY, EMBRYOTOXICITY, METABOLISM OR MUTAGENICITY OF POLYMETHYL METHACRYLATE WERE AVAILABLE.
REPT'D USE : ACRYLIC SHEETS, GLAZING AND ACRYLIC MOLDINGS

CAS NO : **9003-07-0**
CHEM NAME : ***** POLYPROPYLENE**

REPORT 01 : IARC MONOGRAPH 19 1979
POLYPROPYLENE WAS TESTED IN ONE EXPERIMENT IN RATS BY SUBCUTANEOUS IMPLANTATION OF DISCS OR POWDER; LOCAL SARCOMAS WERE INDUCED AT THE SITE OF IMPLANTATION. THE AVAILABLE DATA ON POLYPROPYLENE DO NOT PERMIT AN EVALUATION OF ITS CARCINOGENICITY.
REPT'D USE : PLASTIC, TEXTILES, LUGGAGE, PACKAGING AND WIRE/CABLE INSULATION

CAS NO : **9003-53-6**
CHEM NAME : ***** POLYSTYRENE**

REPORT 01 : IARC MONOGRAPH 19 1979
SUBCUTANEOUS IMPLANTATION OF POLYSTYRENE DISCS, RODS, SPHERES OR POWDER IN RATS INDUCED LOCAL SARCOMAS, THE INCIDENCES OF WHICH VARIED WITH THE SIZE AND FORM OF THE IMPLANT.
REPT'D USE : PACKAGING, HOUSEWARES AND APPLIANCES

CAS NO : **9002-84-0**
CHEM NAME : ***** POLYTETRAFLUOROETHYLENE (PTFE)**

REPORT 01 : IARC MONOGRAPH 19 1979
POLYTETRAFLUOROETHYLENE DISCS, SQUARES, FRAGMENTS OR POWDER IMPLANTED SUBCUTANEOUSLY OR INTRAPERITONEALLY IN MICE OR RATS INDUCED LOCAL SARCOMAS. THE INDUCTION OF SARCOMAS BY POLYTETRAFLUOROETHYLENE IMPLANTS IN MICE AND RATS, TOGETHER WITH A SINGLE CASE OF A FIBROSARCOMA IN A PATIENT TREATED WITH A POLYTETRAFLUOROLENE/DACRON IMPLANT, PROVIDE INSUFFICIENT EVIDENCE TO ASSESS THE CARCINOGENIC RISK OF EXPOSURE TO TETRAFLUOROETHYLENE AND POLYTETRAFLUOROETHYLENE.

REPORT 02 : IARC MONOGRAPH 19 1979
NO DATA ON THE EMBRYOTOXICITY, TERATOGENICITY, METABOLISM OR MUTAGENICITY OF POLYTETRAFLUOROETHYLENE WERE AVAILABLE.

REPORT 03 : EPA TSCA 8(E) STATUS REPORT AUG 86
INHALATION TESTS APPEARED TO SHOW THAT MIXTURE CONTAINING 0.5% PERFLUOROALKYL/METHYACRYLATE COPOLYMERS WAS MORE TOXIC THAN MINERAL SPIRITS ALONE.

REPT'D USE : COOKING UTENSILS, CHEMICAL/MECHANICAL/ELECTRICAL EQUIPMENT AND FIBERS

CAS NO : **9009-54-5**
CHEM NAME : ***** POLYURETHANE**

REPORT 01 : IARC MONOGRAPH 19 1979
POLYURETHANES OF VARIOUS CHEMICAL COMPOSITIONS WERE TESTED FOR CARCINOGENICITY IN A NUMBER OF PHYSICAL FORMS (DISCS, STRIPS, FOAMS, POWDER) BY S.C., INTRAPERITONEAL, INTRAVAGINAL, INTRATRACHEAL AND INTRABRONCHIAL ADMINISTRATION AND BY INHALATION OF THE DUST IN MICE AND RATS. FIBROSARCOMAS WERE INDUCED AT THE SITES OF S.C. AND INTRAPERITONEAL IMPLANTATION. THE INCIDENCE OF SARCOMAS VARIED WITH THE PHYSICAL AND CHEM. CHARACTERISTICS OF IMPLANTS. IN ONE EXPERIMENT WITH INTRAPERITONEAL IMPLANTATION, EPITHELIAL TUMORS WERE ALSO INDUCED.

REPORT 02 : IARC MONOGRAPH 19 1979
NO DATA ON THE TOXICITY, EMBRYOTOXICITY, TERATOGENICITY, METABOLISM OR MUTAGENICITY OF POLYURETHANE FOAMS WERE AVAILABLE.

REPT'D USE : BUILDING MATERIALS, POLYMER, FURNITURE, PACKAGING, INSULATION, BEDDING AND CARPET UNDERLAY

CAS NO : **9003-20-7**
CHEM NAME : ***** POLYVINYL ACETATE**

REPORT 01 : IARC MONOGRAPH 19 1979
ANIMAL STUDIES INVOLVING IMPLANTATION OF POLYVINYL ACETATE AND POLYVINYL ALCOHOL POWDER IN RATS DID NOT RESULT IN LOCAL SARCOMAS, WHEREAS IN SIMILAR EXPERIMENTS WITH POLYVINYL ALCOHOL SPONGES, LOCAL SARCOMAS WERE PRODUCED. FURTHER STUDIES ARE REQUIRED BEFORE AN EVALUATION CAN BE MADE OF THE CARCINOGENICITY OF THESE COMPOUNDS.

REPORT 02 : IARC MONOGRAPH 19 1979
NO DATA ON THE EMBRYOTOXICITY, TERATOGENICITY, METABOLISM OR MUTAGENICITY OF POLYVINYLACETATE WERE AVAILABLE.

REPT'D USE : ADHESIVES, PAINTS/VARNISHES, CAULKING/SPACKLING COMPOUNDS AND PAPER/TEXTILE FINISHES

CAS NO : **9002-89-5**
CHEM NAME : *****POLYVINYL ALCOHOL**

REPORT 01 : IARC MONOGRAPH 19 1979
ANIMAL STUDIES INVOLVING IMPLANTATION OF POLYVINYL ACETATE AND POLYVINYL ALCOHOL POWDER IN RATS DID NOT RESULT IN LOCAL SARCOMAS, WHEREAS IN SIMILAR EXPERIMENTS WITH POLYVINYL ALCOHOL SPONGES, LOCAL SARCOMAS WERE PRODUCED. FURTHER STUDIES ARE REQUIRED BEFORE AN EVALUATION CAN BE MADE OF THE CARCINOGENICITY OF THESE COMPOUNDS.

REPORT 02 : IARC MONOGRAPH 19 1979
NO DATA ON THE EMBRYOTOXICITY, TERATOGENICITY METABOLISM OR MUTAGENICITY OF POLYVINYLALCOHOL WERE AVAILABLE.

REPT'D USE : TEXTILE WARP, SIZING, ADHESIVES, PAPER SIZING, PESTICIDE SPRAYS, CLEANING AGENTS AND PAINTS/VARNISHES

CAS NO : **9002-86-2**
CHEM NAME : ***** POLYVINYL CHLORIDE**

REPORT 01 : IARC MONOGRAPH 19 1979
PVC WAS TESTED IN RATS BY S.C. AND INTRAPERITONEAL IMPLANTATION: LOCAL SARCOMAS WERE INDUCED, THE INCIDENCE OF WHICH VARIED WITH THE SIZE AND FORM OF THE IMPLANT. THE AVAILABLE STUDIES ON PVC, WHICH INDICATE AN ELEVATED PROPORTION OF DIGESTIVE SYSTEM CANCER IN MALE AND FEMALE WORKERS AND POSSIBLY OF CANCERS OF THE BREAST

AND URINARY ORGANS IN FEMALE WORKERS INVOLVED IN THE FABRICATION OF PLASTICS, INCLUDING PVC, ARE INSUFFICIENT TO EVALUATE THE CARCINOGENICITY OF THIS COMPOUND.

REPORT 02 : IARC MONOGRAPH 19 1979
NO DATA ON THE EMBRYOTOXICITY, TERATOGENICITY METABOLISM OR MUTAGENICITY OF POLYVINYLCHLORIDE WERE AVAILABLE.
REPT'D USE : BUILDING MATERIALS, PACKAGING, ELECTRIC WIRE INSULATION, SPACE DEODORIZERS, FR CHEMICALS, PAINTS/VARNISHES, SYNTHETIC LEATHER, RAINCOATS AND MYRIAD CONSUMER PRODUCTS

CAS NO : **9003-39-8**
CHEM NAME : ***** POLYVINYL PYRROLIDONE**

REPORT 01 : IARC MONOGRAPH 19 1979
PVP WAS TESTED IN MICE, RATS AND RABBITS BY SEVERAL ROUTES OF ADMINISTRATION, USING VARIOUS MOLECULAR WEIGHTS. REPEATED S.C. INJECTIONS OF AN AQUEOUS SOLUTION OF PVP TO RATS RESULTED IN LOCAL SARCOMAS. SINGLE OR SEVERAL S.C. OR INTRAPERITONEAL IMPLANTATIONS OF PVP POWDER RESULTED IN A LOW INCIDENCE OF LOCAL TUMORS. AFTER SEVERAL I.V. INJECTIONS OR AFTER I.P. IMPLANTATION OF PVP, TUMORS OCCURRED IN RATS AT DISTANT SITES; THE RESULTS OF THESE EXPERIMENTS DO NOT ALLOW AN EVALUATION OF A POSSIBLE ASSOCIATION OF DISTANT TUMORS W/ TREATMENT.
REPT'D USE : ADHESIVES, PLASTIC, INKS, COSMETICS, PHARMACEUTICAL, TABLET BINDER AND PLASMA EXTENDER

CAS NO : **7784-41-0**
CHEM NAME : ***** POTASSIUM ARSENATE**

REPORT 01 : IARC MONOGRAPH 23 1980
THERE IS INADEQUATE EVIDENCE FOR THE CARCINOGENICITY OF ARSENIC COMPOUNDS IN ANIMALS. THERE IS SUFFICIENT EVIDENCE THAT INORGANIC ARSENIC COMPOUNDS ARE SKIN AND LUNG CARCINOGENS IN HUMANS. THE EVIDENCE THAT ARSENIC COMPOUNDS CAUSE MUTATIONS AND ALLIED EFFECTS IN BACTERIA IS INCONCLUSIVE. HOWEVER, ARSENIC COMPOUNDS INDUCE CHROMOSOMAL ABERRATIONS AND MORPHOLOGICAL TRANSFORMATION IN MAMMALIAN CELLS.

REPORT 02 : NIOSH CRITERIA DOCUMENT JUN 75
NIOSH RECOMMENDS A CEILING OF 2 MICROGRAMS/CU M FOR INORGANIC ARSENIC COMPOUNDS BASED ON DERMATITIS, LUNG, AND LYMPHATIC CANCER. THE CURRENT OSHA STANDARD FOR

INORGANIC ARSENIC IS 10 MICROGRAMS/CU M TWA. SEE CFR 1910.1018.

REPORT 06 : MULTI-AGENCY CROSS CHECK
Y HHS MAGUIRE
N OSHA CARC
Y EPA CAG
N EPA RCRA
N EPA CLEAN AIR
N EPA CLEAN WATER

REPT'D USE : LAB REAGENT, LEATHER TANNING, TEXTILE PRINTING AND INSECTICIDES

CAS NO : **13464-35-2**
CHEM NAME : ***** POTASSIUM ARSENITE**

REPORT 01 : IARC MONOGRAPH 23 1980
THERE IS INADEQUATE EVIDENCE FOR THE CARCINOGENICITY OF ARSENIC COMPOUNDS IN ANIMALS. THERE IS SUFFICIENT EVIDENCE THAT INORGANIC ARSENIC COMPOUNDS ARE SKIN AND LUNG CARCINOGENS IN HUMANS. THE EVIDENCE THAT ARSENIC COMPOUNDS CAUSE MUTATIONS AND ALLIED EFFECTS IN BACTERIA IS INCONCLUSIVE. HOWEVER, ARSENIC COMPOUNDS INDUCE CHROMOSOMAL ABERRATIONS AND MORPHOLOGICAL TRANSFORMATION IN MAMMALIAN CELLS.

REPORT 02 : NIOSH CRITERIA DOCUMENT JUN 75
NIOSH RECOMMENDS A CEILING OF 2 MICROGRAMS/CU M FOR INORGANIC ARSENIC COMPOUNDS BASED ON DERMATITIS, LUNG, AND LYMPHATIC CANCER. THE CURRENT OSHA STANDARD FOR INORGANIC ARSENIC IS 10 MICROGRAMS/CU M TWA. SEE CFR 1910.1018.

REPORT 06 : MULTI-AGENCY CROSS CHECK
Y HHS MAGUIRE
N OSHA CARC
Y EPA CAG
N EPA RCRA
N EPA CLEAN AIR
N EPA CLEAN WATER

REPT'D USE : TREATMENT OF LEUKEMIA, VETERINARY MEDICINE AND SKIN LESIONS

CAS NO : **20762-60-1**
CHEM NAME : ***** POTASSIUM AZIDE**

REPORT 01 : EPA TSCA 8(E) 8EHQ-1083-0494 OCT 83
PPG REPORTED ON A 125-DAY SUBACUTE ORAL TOXICITY STUDY OF

		POTASSIUM AZIDE (KN) ON BEAGLE DOGS DOSED WITH 0.1, 1.0 AND 10.0 MG/KG/DAY. AT HIGH DOSE SYMPTOMS OF MOTOR ATAXIA OBSERVED IN FOUR OF EIGHT ANIMALS SACRIFICED IN EXTREMIS (A MALE AFTER 60 DAYS AND A FEMALE AFTER 111 DAYS). NO HISTOPATHOLOGICAL OR OTHER ABNORMALITIES NOTED.
REPT'D USE	:	UNKNOWN

CAS NO : **23746-34-1**
CHEM NAME : ***** POTASSIUM BIS(S-HYDROXYETHYL)-DITHIOCARBAMATE**

REPORT 01 : IARC MONOGRAPH 12 1976
POTASSIUM BIS(2-HYDROXYETHYL)-DITHIOCARBAMATE IS CARCINOGENIC IN MICE AFTER ITS ORAL ADMINISTRATION: IT INDUCED LIVER-CELL TUMORS IN MALES OF TWO STRAINS AND IN FEMALES OF ONE STRAIN.

REPORT 06 : MULTI-AGENCY CROSS CHECK
N HHS MAGUIRE
Y OSHA CARC
N EPA CAG
N EPA RCRA
N EPA CLEAN AIR
N EPA CLEAN WATER
REPT'D USE : METALLURGICAL INDICATOR

CAS NO : **7758-01-2**
CHEM NAME : ***** POTASSIUM BROMATE**

REPORT 01 : IARC 40 1986
NO DATA AVAILABLE TO WORKING GROUP ON THE EFFECTS ON REPRODUCTION AND PRENATAL TOXICITY; ABSORPTION, DISTRIBUTION EXCRETION AND METABOLISM; MUTAGENICITY AND CHROMOSOMAL EFFECTS. ALSO, NO CASE REPORTS OR EPIDEMIOLOGICAL STUDIES OF CARCINOGENICITY TO HUMANS. THERE IS SUFFICIENT EVIDENCE FOR GENETIC ACTIVITY IN SHORT-TERM TESTS AND FOR CARCINOGENICITY IN RATS.
REPT'D USE : FOOD ADDITIVE

CAS NO : **7447-40-7**
CHEM NAME : ***** POTASSIUM CHLORIDE**

REPORT 01 : NTP TECHNICAL BULLETIN NO. 7 APR 82
THE CHEMICAL WAS POSITIVE IN THE SALMONELLA TYPHIMURIUM MUTAGENICITY ASSAY.
REPT'D USE : CLEANING AGENTS/COMPOUNDS, SPACE DEODORIZERS, CAUS-

TICS/DRAIN CLEANERS, AGRICULTURAL CHEMICALS/FERTILIZERS, PHOTOGRAPHIC CHEMICAL, DIETARY SUPPLEMENT, CHEMICAL INTERMEDIATE AND SALT SUBSTITUTE

CAS NO : **7789-00-6**
CHEM NAME : ***** POTASSIUM CHROMATE**

REPORT 01 : IARC MONOGRAPH 23 1980
AN EXPERIMENT IN RATS EXPOSED TO POTASSIUM CHROMATE (VI) BY BRONCHIAL IMPLANTATION WAS INADEQUATE TO EVALUATE CARCINOGENICITY. CHROMIUM (VI) COMPOUNDS CAUSE MUTATIONS AND ALLIED EFFECTS IN A VERY WIDE RANGE OF PROKARYOTIC AND EUKARYOTIC SYSTEMS BOTH IN VITRO AND IN VIVO.
REPT'D USE : CORROSION INHIBITOR, TEXTILE PRINTING, CERAMIC ENAMELS AND LEATHER TANNING AGENT

CAS NO : **7778-50-9**
CHEM NAME : ***** POTASSIUM DICHROMATE**

REPORT 01 : IARC MONOGRAPH 23 1980
EXPERIMENTS IN RATS EXPOSED TO POTASSIUM DICHROMATE (VI) BY BRONCHIAL IMPLANTATION WERE INADEQUATE TO EVALUATE CARCINOGENICITY. CHROMIUM (VI) COMPOUNDS CAUSE MUTATIONS AND ALLIED EFFECTS IN A VERY WIDE RANGE OF PROKARYOTIC AND EUKARYOTIC SYSTEMS BOTH IN-VITRO AND IN-VIVO.
REPT'D USE : PHOTOLITHOGRAPHY, PIGMENTS/PRINTING, WOOD STAINS, SAFETY MATCHES AND LEATHER TANNING

CAS NO : **53-03-2**
CHEM NAME : ***** PREDNISONE**

REPORT 01 : NCI BIOASSAY PROGRAM- CANCER VOL. 40 1977
UNDER THE CONDITIONS OF THE BIOASSAY, I.P. INJECTION OF PREDNISONE WAS CARCINOGENIC TO FEMALE SPRAGUE-DAWLEY (CD) RATS, CAUSING BREAST, PITUITARY, AND ADRENAL TUMORS. IT WAS NOT CARCINOGENIC TO MALE RATS AND MALE AND FEMALE SWISS-WEBSTER MICE.

REPORT 02 : NTP TECHNICAL BULLETIN 6 JAN 82
THE COMPOUND WAS WEAKLY POSITIVE IN THE SALMONELLA TYPHIMURIUM MUTAGENICITY ASSAY IN ONE LABORATORY.

REPORT 03 : IARC MONOGRAPH 26 1981
PREDNISONE WAS TESTED IN MICE AND RATS BY IP ADMINISTRATION. LITTLE OR NO CARCINOGENIC EFFECT WAS OBSERVED, BUT THE STUDIES SUFFERED FROM LIMITATIONS IN DESIGN AND

REPORTING. THE AVAILABLE DATA WERE INADEOUATE TO EVALUATE THE CARCINOGENICITY OF PREDNISONE.

REPORT 04 : IARC MONOGRAPH 26 1981
PREDNISONE CAN INDUCE TERATOGENIC EFFECTS, PARTICULARLY CLEFT PALATE, IN RODENTS. PREDNISONE WAS NOT MUTAGENIC IN E COLI, AND IT CAUSED NO CHROMOSOMAL DAMAGE WHEN ADMINISTERED TO WISTAR RATS.

REPT'D USE : ADRENOCORTICAL STEROID AND ANTIINFLAMMATOR/AGENT. PROPOSED PPPA EXEMPTION FOR PREDNISONE (47FR10235)

CAS NO : **57-66-9**

CHEM NAME : ***** PROBENECID**

REPORT 01 : NTP DRAFT TR 395 NIH PUB NO. 90-2850 NOV 90
UNDER CONDITIONS OF 2-YEAR STUDIES, NO EVIDENCE OF CARCINOGENIC ACTIVITY OF PROBENECID FOR MALE OR FEMALE F344/N RATS RECEIVING 100 OR 400 MG/KG IN CORN OIL BY GAVAGE FOR 103 WEEKS - NO EVIDENCE OF CARCINOGENIC ACTIVITY FOR MALE B6C3F1 MICE GIVEN 100 OR 400 MG/KG PROBENECID IN CORN OIL BY GAVAGE FOR 103 WEEKS. SOME EVIDENCE OF CARCINOGENIC ACTIVITY OF PROBENECID FOR FEMALE B6C3F1 MICE BASED ON INCREASED INCIDENCE OF HEPATOCELLULAR NEOPLASMS.

REPT'D USE : DRUG

CAS NO : **366-70-1**

CHEM NAME : ***** PROCARBAZINE HYDROCHLORIDE**

REPORT 01 : NCI CARC TRS 19 DHEW PUB NO 79-819
IT IS CONCLUDED THAT UNDER THE CONDITIONS OF THIS BIOASSAY, PROCARBAZINE WAS CARCINOGENIC FOR BOTH SPRAGUE-DAWLEY RATS AND B6C3F1 MICE, PRODUCING SEVERAL TYPES OF TUMORS IN BOTH SEXES OF THESE TWO SPECIES. PROCARBAZINE WAS ADMINISTERED BY I.P. INJECTION.

REPORT 02 : NCI BIOASSAY PROGRAM, CANCER VOL. 40 1977
UNDER THE CONDITIONS OF THE BIOASSAY, I.P. INJECTION OF PROCARBAZINE WAS CARCINOGENIC TO SWISS-WEBSTER MALE AND FEMALE MICE, CAUSING LUNG TUMORS IN BOTH SEXES AND LYMPHOMAS, KIDNEY TUMORS, AND UTERINE TUMORS IN FEMALES. IT WAS ALSO CARCINOGENIC TO MALE AND FEMALE SPRAGUE-DAWLEY (CD) RATS, CAUSING LYMPHOMAS, LEUKEMIAS, AND MAMMARY TUMORS.

REPORT 03 : IARC MONOGRAPH 26 1981
PROCARBAZINE HCL IS CARCINOGENIC IN MICE AND RATS AFTER ITS IP ADMINISTRATION, PRODUCING MALIGNANT TUMORS OF THE

NERVOUS SYSTEM, HEMATOPOIETIC SYSTEM AND POSSIBLE OTHER ORGANS IN BOTH SPECIES, AND TUMORS OF THE MAMMARY GLAND IN RATS ONLY. EVIDENCE OF CARCINOGENICITY WAS ALSO FOUND IN RATS AND MICE FOLLOWING ITS ORAL ADMINISTRATION, IN RATS FOLLOWING IV ADMINISTRATION, AND IN ONE INSTANCE, FOLLOWING TRANSPLACENTAL EXPOSURE.

REPORT 04 : NTP TECHNICAL BULLETIN 8 JUL 82
PROCARBAZINE WAS POSITIVE IN THE L5178Y MOUSE LYMPHOMA ASSAY.

REPORT 05 : IARC MONOGRAPH 26 1981
PROCARBAZINE DID INDUCE TERATOGENIC EFFECTS IN RATS AND EMBRYOLETHALITY AT DOSES NONTOXIC TO THE MOTHER. THE COMPOUND IS MUTAGENIC IN SALMONELLA STRAINS TA1535, TA1537, TA98, AND TA100, AS WELL AS YEAST, CULTURED MAMMALIAN CELLS AND DROSOPHILA, AND IN MICE AND RATS IN VIVO.

REPORT 06 : MULTI-AGENCY CROSS CHECK
Y HHS MAGUIRE
Y OSHA CARC
N EPA CAG
N EPA RCRA
N EPA CLEAN AIR
N EPA CLEAN WATER
REPT'D USE : ANTINEOPLASTIC

CAS NO : **1811-28-5**
CHEM NAME : ***** PROFLAVINE HEMISULPHATE**

REPORT 01 : IARC MONOGRAPH 24 SEP 80
PROFLAVINE MONOHYDROCHLORIDE, AS THE HEMIHYDRATE, TESTED IN MICE AND RATS BY ORAL ADMINISTRATION. PROFLAVINE HEMISULPHATE TESTED IN MICE BY SKIN APPLICATION, INTRADERMAL ADMINISTRATION AND SUBCUTANEOUS IMPLANTATION. ALL TESTS INADEQUATE FOR EVALUATION.

REPORT 02 : IARC MONOGRAPH 24 SEP 80
PROFLAVINE IS MUTAGENIC IN VIRAL AND BACTERIAL SYSTEMS. IT INCREASED NUMBER OF CHROMATID BREAKS AND INDUCED SISTER CHROMATID EXCHANGES IN MAMMALIAN CELLS.
REPT'D USE : PHARACEUTICALS

CAS NO : **952-23-8**
CHEM NAME : ***** PROFLAVINE MONOHYDROCHLORIDE HEMIHYDRATE**

REPORT 01 : NCI CARC TRS 5 DHEW PUB NO 77-805
THE UNUSUALLY HIGH INCIDENCE OF HEPATOCELLULAR CARCINOMAS AND HEMANGIOSARCOMAS IN CONTROL MALE MICE AND THE UNUSUALLY HIGH INCIDENCE OF MALIGNANT LYMPHOMAS IN ALL GROUPS OF FEMALE MICE, IN CONJUNCTION WITH THE FACT THAT A POSITIVE-CONTROL CARCINOGEN WAS TESTED IN THE SAME ROOM WITH THESE ANIMALS, RAISES A QUESTION OF THE VALIDITY OF THESE BIOASSAY RESULTS (FEEDING STUDY).

REPORT 02 : IARC MONOGRAPH 24 1980
THE AVAILABLE EXPERIMENTAL RESULTS WERE INADEQUATE FOR AN EVALUATION OF THE CARCINOGENICITY OF PROFLAVINE IN EXPERIMENTAL ANIMALS, AND NO DATA WERE AVAILABLE FROM HUMAN STUDIES.
REPT'D USE : BACTERIOSTATIC/BACTERIOCIDAL AGENT

CAS NO : **57-83-0**
CHEM NAME : ***** PROGESTERONE**

REPORT 01 : IARC MONOGRAPH 21 1979
PROGESTERONE WAS TESTED BY SUBCUTANEOUS AND INTRAMUSCULAR INJECTION IN MICE AND DOGS. PROGESTERONE INCREASED THE INCIDENCE OF OVARIAN, UTERINE AND MAMMARY TUMORS IN MICE; THE DATA FROM DOGS WERE INSUFFICIENT TO EVALUATE CARCINOGENICITY.

REPORT 02 : IARC MONOGRAPH 21 1979
CHROMOSOMAL ABNORMALITIES, ANEUPLOIDY AND POLYPLOIDY WERE OBSERVED IN GERM CELLS OF CHINESE HAMSTERS AND DOGS TREATED WITH PROGESTERONE FOR SEVERAL WEEKS. TREATMENT OF CULTURES OF HUMAN EMBRYONIC FIBROPLASTS AND OF RENAL EPITHELIA RESULTED IN A TWO-FOLD INCREASE IN THE NUMBER OF CHROMOSOMAL ABERRATIONS.

REPORT 06 : MULTI-AGENCY CROSS CHECK
Y HHS MAGUIRE
N OSHA CARC
N EPA CAG
N EPA RCRA
N EPA CLEAN AIR
N EPA CLEAN WATER

REPORT 08 : IARC SUPPL. 4 1985
EVIDENCE FOR ACTIVITY IN SHORT-TERM TESTS IS INADEQUATE FOR EVALUATION.

REPT'D USE : ORAL CONTRACEPTIVES

CAS NO : **23950-58-5**
CHEM NAME : ***** PRONAMIDE**

REPORT 06 : MULTI-AGENCY CROSS CHECK
Y HHS MAGUIRE
N OSHA CARC
Y EPA CAG
Y EPA RCRA
N EPA CLEAN AIR
N EPA CLEAN WATER
REPT'D USE : PESTICIDE AND AGRICULTURAL CHEMICAL

CAS NO : **51-02-5**
CHEM NAME : ***** PRONETALOL HYDROCHLORIDE**

REPORT 01 : IARC MONOGRAPH 13 1977
PRONETALOL HYDROCHLORIDE IS CARCINOGENIC IN MICE FOLLOWING ITS ORAL ADMINISTRATION: IT PRODUCED THYMIC LYMPHOMAS.
REPT'D USE : ANDRENERGIC-BLOCKING AGENT

CAS NO : **63843-89-0**
CHEM NAME : ***** PROPANEDIOIC ACID COMPOUND**

REPORT 01 : EPA TSCA 8(E) 0186-0585 JAN 86
CIBA-GEIGY SUBMITTED RESULTS FROM A 28 DAY GAVAGE STUDY IN RATS THAT INDICATED A POTENTIAL IMMUNOSUPPRESSIVE ACTIVITY: SIX GROUPS OF 10 RATS EACH RECEIVED DAILY DOSES OF 0, 60, 100, 300, 600 AND 1000 MG/KG. ALL DOSE LEVELS PRODUCED ALTERATIONS IN WHITE BLOOD CELL PARAMETERS, THYMUS WEIGHT PLUS LESIONS IN THE SPLEEN, LYMPH NODES AND LIVER.
REPT'D USE : AUTOMATIVE COATINGS AND POLYPOPYLENE FILMS

CAS NO : **1120-71-4**
CHEM NAME : ***** 1,3-PROPANE SULTONE**

REPORT 01 : NCI SUMMARY DEC 75
ORAL ADMINISTRATION OF THE COMPOUND AS AN AQUEOUS SOLUTION TO RATS RESULTED IN MALIGNANT TUMORS. RATS AND MICE ADMINISTERED THE CHEMICAL BY SUBCUTANEOUS AND/OR INTRAMUSCULAR ROUTES DEVELOPED TUMORS AT THE INJECTION SITE. A SINGLE I.V. INJECTION OF THE CHEMICAL TO

PREGNANT RATS ON THE 15TH DAY OF GESTATION PRODUCED MALIGNANT TUMORS IN THE OFFSPRING. NO HUMAN EPIDEMIOLOGICAL DATA ARE AVAILABLE.

REPORT 02 : IARC MONOGRAPH 4 1974
1,3-PROPANE SULTONE HAS A CARCINOGENIC EFFECT IN THE RAT WHEN GIVEN ORALLY, INTRAVENOUSLY OR BY PRE-NATAL EXPOSURE, AND A LOCAL CARCINOGENIC EFFECT IN THE MOUSE AND THE RAT WHEN GIVEN SUBCUTANEOUSLY. IT IS CARCINOGENIC IN THE RAT AFTER SINGLE-DOSE EXPOSURES. IN A 1979 SUPPLEMENT, IARC DETERMINED THAT THERE IS SUFFICIENT EVIDENCE FOR THE CARCINOGENICITY OF 1,3-PROPANE SULTONE IN ANIMALS.

REPORT 06 : MULTI-AGENCY CROSS CHECK
Y HHS MAGUIRE
N OSHA CARC
Y EPA CAG
Y EPA RCRA
N EPA CLEAN AIR
N EPA CLEAN WATER

REPT'D USE : ANTISTATIC ADDITIVE FOR POLYAMIDE FIBERS, CATION-EXCHANGE RESINS, INSECTICIDES/FUNGICIDES, RUBBER VULCANIZATION AC-CELERATOR, INTERMEDIATE IN DETERGENTS/WETTING AGENTS AND SOLUBLE TEXTILE STARCHES

CAS NO : **122-42-9**
CHEM NAME : ***** PROPHAM**

REPORT 01 : IARC MONOGRAPH 12 1976
IN ONE STUDY IN WHICH PROPHAM WAS ADMINISTERED ORALLY TO TWO STRAINS OF MICE, NO EVIDENCE OF CARCINOGENICITY WAS OBSERVED.

REPORT 02 : IARC MONOGRAPH 12 1976
PROPHAM INHIBITS SPINDLE FIBER FORMATION AND INDUCED CHROMOSOME ABERRATIONS IN PLANTS. NO REVERSE MUTATIONS WERE DETECTED IN B. SUBTILIS OR IN SALMONELLA, METABOLIC ACTIVATION SYSTEMS WERE NOT USED.

REPT'D USE : HERBICIDE

CAS NO : **57-57-8**
CHEM NAME : ***** B-PROPIOLACTONE**

REPORT 01 : IARC MONOGRAPH 4 1974
B-PROPIOLACTONE IS CARCINOGENIC IN THE MOUSE BY SKIN APPLICATION, SUBCUTANEOUS OR INTRAPERITONEAL INJECTION, AND, IN THE RAT, BY SUBCUTANEOUS INJECTION. ORAL ADMINIS-

TRATION IN THE RAT GAVE SOME INDICATION OF CARCINOGENIC ACTIVITY. THE RESULTS OBTAINED IN THE HAMSTER AND GUINEA PIG ARE EQUIVOCAL. IT IS CARCINOGENIC TO MICE AFTER A SINGLE DOSE EXPOSURE. IN A 1979 SUPPLEMENT, IARC DETERMINED THAT THERE IS SUFFICIENT EVIDENCE FOR THE CARCINOGENICITY OF B-PROPIOLACTONE IN ANIMALS.

REPORT 02 : NTP TECHNICAL BULLETIN 3/8 DEC 80
B-PROPIOLACTONE WAS POSITIVE IN THE SALMONELLA ASSAY AND INDROSOPHILA WHEN TESTED FOR SEX-LINKED RECESSIVE LETHALS.

REPORT 06 : MULTI-AGENCY CROSS CHECK
Y HHS MAGUIRE
R OSHA CARC
Y EPA CAG
N EPA RCRA
N EPA CLEAN AIR
N EPA CLEAN WATER

REPT'D USE : INTERMEDIATE FOR ACRYLIC ACID AND ESTERS, CHEMOSTERILANT, DISINFECTANT, VACCINES, POLYMERS AND HALOQUINONE PRODUCTION

CAS NO : **79-09-4**
CHEM NAME : ***** PROPIONIC ACID**

REPORT 01 : STATUS REPORT 8EHQ-0486-0600 MAY 86
IN A FEEDING TEST NOW UNDERWAY WITH 3 GROUPS OF MALE WISTAR RATS EXPOSED TO PROPIONIC ACID, INDICATIONS ARE THAT THE CPD INDUCED CARCINOMATOUS CHANGES IN THE SQUAMOUS EPITHELIUM OF THE FORESTOMACH AFTER ONLY 20 WEEKS EXPOSURE.

REPT'D USE : MFR OF HERBICIDES AND PERFUMES, ARTIFICIAL FOOD FLAVORINGS, DRUGS AND A CHEMICAL INTERMEDIATE, GRAIN AND WOOD CHIP PRESERVATIVE, BREAD MOLD INHIBITORS AND ANTIFUNGAL AGENTS

CAS NO : **1569-01-3**
CHEM NAME : ***** 1-PROPOXY-2-PROPANOL**

REPORT 01 : EPA TSCA 8(E) 0686-0603 JUN 86
UNION CARBIDE CONDUCTED INHALATION STUDY OF 1-PROPOXY-2-PROPANOL VAPOR IN MALE AND FEMALE F344 RATS AT 0, 503, 983, AND 2000 PPM FOR NINE DAYS AT SIX HOURS A DAY. OPACIFICATION OF CORNEA DEVELOPED ONLY AT 2000 PPM. HOWEVER, HISTOLOGICAL EXAMINATION SHOWED LESIONS IN CORNEAS AT ALL EXPOSURE LEVELS, EXCEPT AIR ONLY CONTROLS. LESION CHARACTERIZED BY SEPARATION OF EPITHELIUM

FROM UNDERLYING SUBSTANTIA PROPRIA, LOSS OF EPITHELIM PRODUCING ULCERATION, INFLAMMATORY CELL INFILTRATION, AND VASCULARIZATION OF OTHER SUBSTANTIA PROPRIA. CHEMICAL IN COMMERCE FOR OVER 20 YEARS WITH NO REPORTS OF CHRONIC EYE INJURY ASCRIBED TO IT.

REPT'D USE : ENAMEL SOLVENT

CAS NO : **71-23-8**

CHEM NAME : ***** N-PROPYL ALCOHOL**

REPORT 01 : CPSC SCHOOL LAB CHEMICAL SURVEY JAN 82

PROPANOL IS REPORTED IN ONE MAJOR TEXT TO BE AN EXPERIMENTAL CARCINOGEN. SUBSEQUENT CPSC STAFF REVIEW SHOWS THAT THERE IS LIMITED EVIDENCE TO SUGGEST THAT N-PROPYL ALCOHOL, UPON ORAL ADMINISTRATION TO RATS, INDUCES BENIGN TUMORS OF THE ANTERIOR STOMACH OR MAMMARY GLAND; IT INDUCED LIVER SARCOMAS UPON SO INJECTION. PROPANOL WAS NEGATIVE, WITH AND WITHOUT ACTIVATION IN S POMBE AND V79 CHINESE HAMSTER CELLS.

REPT'D USE : LABORATORY REAGENT, SOLVENT, PAINTS/VARNISHES, AEROSOLS, CLEANING/POLISHING AGENTS, COSMETICS/PHARMACEUTICALS

CAS NO : **627-12-3**

CHEM NAME : ***** N-PROPYL CARBAMATE**

REPORT 01 : IARC MONOGRAPH 12 1976

NO REVERSE MUTATIONS WERE INDUCED BY N-PROPYL CARBAMATE IN B. SUBTILIS; METABOLIC ACTIVATIONS SYSTEMS WERE NOT USED IN THIS TEST.

REPORT 02 : IARC MONOGRAPH 12 1976

N-PROPYL CARBAMATE IS CARCINOGENIC IN MICE AFTER ITS INTRAPERITONEAL INJECTION: IT PRODUCED LUNG ADENOMAS IN THREE DIFFERENT STRAINS.

REPT'D USE : CHEMICAL INTERMEDIATE AND MFR OF DURABLE PRESS TEXTILE RESINS

CAS NO : **121-79-9**

CHEM NAME : ***** PROPYL GALLATE**

REPORT 01 : NTP TR 240 NIH PUB 83-1796 DEC 82

UNDER CONDITIONS OF BIOASSAY, PROPYL GALLATE WAS NOT CONSIDERED CARCINOGENIC FOR 344/N RATS, ALTHOUGH THERE WAS EVIDENCE OF INCREASED PROPORTION OF LOW-DOSE MALE RATS WITH PREPUTIAL GLAND TUMORS, ISLET-CELL TUMORS OF THE PANCREAS, AND PHEOCHROMOCYTOMAS OF ADERNAL GLANDS; RARE TUMORS OF THE BRAIN OCCURRED IN TWO

LOW-DOSE FEMALES. PROPYL GALLATE WAS NOT CONSIDERED TO BE CARCINOGENIC FOR B6C3F1 MICE OF EITHER SEX, ALTHOUGH THE INCREASED INCIDENCE OF MALIGNANT LYMPHOMAS IN MALE MICE MAY HAVE BEEN RELATED TO DIETARY ADMINISTRATION OF PROPYL GALLATE.

REPT'D USE : ANTIOXIDANT, HAIR GROOMING PRODUCTS, ADHESIVES, PRESERVATIVE, LUBRICATING OIL AND LABORATORY REAGENT

CAS NO : **115-07-1**
CHEM NAME : ***** PROPYLENE**

REPORT 01 : IARC MONOGRAPH 19 1979
NO INFORMATION WAS AVAILABLE TO THE WORKING GROUP FOR EVALUATING THE POSSIBLE CARCINOGENIC EFFECTS OF PROPYLENE.

REPORT 02 : NTP TECHICAL REPORT 272 NIH PUB #86- 2528 NOV 85
NO EVIDENCE OF CARCINOGENICITY IN MALE AND FEMALE F344/N RATS OR IN MALE AND FEMALE B6C3F1 MICE EXPOSED TO PROPYLENE BY INHALATION AT CONCENTRATIONS OF 5,000 OR 10,000 PPM FOR 103 WEEKS. IN NASAL CAVITY, PROPYLENE INDUCES SQUAMOUS METAPLASIA OF RESPIRATORY EPITHELIUM IN MALE AND FEMALE RATS AND EPITHELIAL HYPERPLASIA IN FEMALE RATS.

REPORT 03 : IARC MONOGRAPH 19 1979
NO DATA ON THE CARCINOGENICITY OR MUTAGENICITY OF PROPYLENE WERE AVAILABLE.

REPT'D USE : PLASTIC MONOMER AND CHEMICAL INTERMEDIATE

CAS NO : **CLASS-31**
CHEM NAME : ***** PROPYLENE BASED GLYCOL ETHERS**

REPORT 01 : HS STAFF REPORT SEP 89
PROPYLENE BASED GLYCOL ETHERS AND ACETATES PROPYLENE GLYCOL MONOMETHYL ETHER (PGME), DIPROPYLENE GLYCOL MONOMETHY ETHER (DPGME), TRIPROPYLENE GLYCOL MONOMETHYL ETHER (TPGME), PROPYLENE GLYCOL MONOMETHYL ETHER ACETATE (PGMEA), AND PROPYLENE GLYCOL MONOTERTIARY BUTYL ETHER (PGTBE) WIDELY USED AS SOLVENTS IN PAINTS, RESINS, CLEANERS, INKS, ELECTRONICS AND INTERMEDIATES. USED AS REPLACEMENT FOR ETHYLENE BASED GLYCOL ETHERS AND METHYLENE CHLORIDE (E.G. PGME AND PGMEA) NONE OF THESE ETHERS, TESTED FOR CARCINGENIC EFFECTS IN LONG-TERM BIOASSAYS.

REPORT 02 : HS STAFF REPORT SEP 89
LD50 OF PGME FROM 1.11 TO 10.78 G/KG, SUGGESTING LOW TOXICITY. SUBCHRONIC ORAL ADMINISTRATION PRODUCED MILD

INCREASES IN WEIGHTS OF LIVER AND KIDNEY OF RATS ONLY AT HIGH DOSE OF 2.77 G/KG. UNDILUTED PGME PRODUCED MILD TRANSITORY CONJUNCTIVAL IRRITATION IN EYES OF RABBITS. SINGLE DERMAL APPLICATION NOT IRRITATING TO RABBIT SKIN BUT CAUSED NARCOSIS AT ALL DOSES (4.61 TO 19.36 G/KG), AND DEATHS AT DOSES 11.06 G/KG AND ABOVE. SUBCHRONIC APPLICATION TO SKIN OF RABBITS (0.92 TO 9.68 G/KG) INDUCED NARCOSIS AND MORTALITY AT 6.45 G/KG DOSE AND ABOVE.

REPORT 03 : HS STAFF REPORT SEP 89
ACUTE/SUBCHRONIC INHALATION TOXICITY OF PGME LOW AND COMPARABLE TO THAT OF ORAL AND DERMAL TOXICITY. IN RATS LC50, 15,000 PPM. GUINEA PIGS AND RABBITS MORE RESISTANT TO LETHAL EFFECT OF PGME INHALATION; TRANSIENT NARCOSIS AND INCREASED LIVER WEIGHTS OBSERVED. KIDNEY WEIGHT INCREASED BUT ONLY AT 6000-10,000 PPM. INCREASES OF LIVER AND KIDNEY CONSIDERED PHYSIOLOGICAL ADAPTATIONS, NOT TOXIC EFFECTS.

REPORT 04 : HS STAFF REPORT SEP 89
REPRODUCTIVE/DEVELOPMENTAL TOXICITY OF PGME NEGLIGIBLE. ONLY EFFECT OBSERVED INCREASE (P = 0.05) IN DELAYED STERNEBRAL OSSIFICATION IN RATS AT 3000 PPM. AN ALTERATION, NOT MALFORMATION, AND NOT OBSERVED IN RABBITS AT SAME DOSE. PGME ACETATE ORAL LD50, 8.5-10.0 G/KG (MORE THAN 5.0 G/KG IN RABBITS) AND LOW ACUTE INHALATION TOXICITY CLC50 (MORE THAN 4350 PPM IN RATS). HOWEVER, PURE BETA ISOMER OF PGMEA EXHIBITED DEVELOPMENTAL TOXICITY AT 550 OR 2700 PPM IN RATS (INHALATION) AND 550 PPM (INHALATION) IN RABBITS.

REPORT 05 : HS STAFF REPORT SEP 89
LD50, OF DPGME 5.1 G/KG IN RATS. OCULAR ADMINISTRATION OF DPGME CAUSED MILD REVERSIBLE CONJUNCTIVAL IRRITATION IN RABBITS. ONLY AT VERY HIGH DOSE (19.36 G/KG) IN DERMAL APPLICATION NARCOSIS INDUCED IN RABBITS. SUBCHRONIC INHALATION TOXICITY OF MINIMAL, ONLY MILD NARCOSIS AND INCREASED LIVER WEIGHT OBSERVED IN RATS.

REPORT 07 : HS STAFF REPORT SEP 89
TPGME LD50 OF 3.2 G/KG IN RATS. OCULAR TOXICITY CAUSED MILD CONJUNCTIVAL IRRITATION IN RABBITS.

REPORT 08 : HS STAFF REPORT SEP 89
PGTBE FOUND PRACTICALLY NONTOXIC. ORAL LD50, 3.8 G/KG IN RATS AND DERMAL LD50 MORE THAN 2.0 G/KG IN RABBITS. 20% AQUEOUS SOLUTION ONLY SLIGHTLY IRRITATING TO RABBIT EYE. FOUND TO BE NONTOXIC IN SUBCHRONIC INHALATION ASSAY IN RATS.

REPORT 09 : HS STAFF REPORT SEP 89
ACUTE/SUBCHRONIC AND REPRODUCTIVE/DEVELOPMENTAL STUDIES OF COMMERCIALLY AVAILABLE PROPYLENE-BASED GLYCOL ETHERS (WHICH CONTAIN 95% ALPHA ISOMER; LESS THAN 5% BETA ISOMER) SHOW THEM PRACTICALLY NONTOXIC CHEMICALS. ONLY CONCERN LIES WITH DEVELOPMENTAL TOXICITY OF THE BETA ISOMER OF PGMEA AND THIS WOULD INCREASE IF BETA ISOMER OF ANY OF THESE GLYCOL ETHERS WAS ELEVATED ABOVE LESS THAN 5% LEVEL.

REPT'D USE : SOLVENTS, INTERMEDIATES AND REPLACEMENTS FOR ETHYLENE GLYCOL ETHERS AND METHYLENE CHLORIDE

CAS NO : **75-55-8**
CHEM NAME : ***** PROPYLENEIMINE (2-METHYLAZIRIDINE)**

REPORT 02 : IARC MONOGRAPH 9 1975
2-METHYLAZIRIDINE IS CARCINOGENIC IN RATS FOLLOWING ITS ORAL ADMINISTRATION, THE ONLY SPECIES AND ROUTE TESTED, PRODUCING A VARIETY OF MALIGNANT TUMORS. IN A 1979 SUPPLEMENT, IARC DETERMINED THAT THERE IS SUFFICIENT EVIDENCE FOR THE CARCINOGENICITY OF 2-METHYLAZIRIDINE IN ANIMALS.

REPORT 03 : IARC SUPPL. 4 OCT 82
SUFFICIENT EVIDENCE FOR CARCINOGENICITY OF PROPYLENEIMINE IN ANIMALS. NO INFORMATION ON CARCINOGENICITY IN HUMANS.

REPORT 06 : MULTI-AGENCY CROSS CHECK
Y HHS MAGUIRE
N OSHA CARC
N EPA CAG
N EPA RCRA
N EPA CLEAN AIR
N EPA CLEAN WATER

REPT'D USE : CHEMICAL INTERMEDIATE, PAPER BINDERS, RESIN CURING AGENT AND PHARACEUTICALS

CAS NO : **75-56-9**
CHEM NAME : ***** PROPYLENE OXIDE (1,2-EPOXYPROPANE)**

REPORT 01 : NTP REPORT TR 267 NIH PUB 85-2527 MAR 85
SOME EVIDENCE OF CARCINOGENICITY FOR F344/N RATS, AS INDICATED BY INCREASED PAPILLARY ADENOMAS OF NASAL TURBINATES IN MALE AND FEMALE RATS EXPOSED TO PROPYLENE OXIDE AT 400 PPM. FOR MALE AND FEMALE B6C3F1 MICE CLEAR EVIDENCE OF CARCINOGENICITY, AS INDICATED BY INCREASED HEMANGIOMAS OR HEMANGIOSARCOMAS OF NASAL TURBINATES

AT 400 PPM. IN THE RESPIRATORY EPITHELIUM OF NASAL TURBINATES, PROPYLENE OXIDE ALSO CAUSED SUPPURATIVE INFLAMMATION, HYPERPLASIA, AND SQUAMOUS METAPLASIA IN RATS AND INFLAMMATION IN MICE.

REPORT 02 : IARC MONOGRAPH 11 1976
PROPYLENE OXIDE WAS CARCINOGENIC IN A LIMITED STUDY IN RATS BY SUBCUTANEOUS INJECTION: IT PRODUCED LOCAL SARCOMAS.

REPORT 03 : IARC MONOGRAPH 11 1976
REVERSE MUTATIONS WERE INDUCED IN NEUROSPORA TREATED WITH PROPYLENE OXIDE. THE CHEMICAL HAS ALSO INDUCED RECESSIVE LETHAL MUTATIONS IN DROSOPHILA.

REPORT 04 : EPA TSCA 8(E) NOTICE 8EHQ-0282-0439 APR 82
DOW CHEMICAL SUBMITTED INTERIM RESULTS FROM CHRONIC INHALATION STUDY OF PROPYLENE OXIDE (PO) IN WISTAR RATS. INTERIM SACRIFICES CONDUCTED AT 12, 18, AND 24 MONTHS SHOW MICROSCOPICALLY DETECTABLE DEGENERATIVE CHANGES IN OLFACTORY PORTIONS OF THE NASAL MUCOSA OF RATS EXPOSED TO 100 AND 300 PPM PO. THERE ALSO APPEARS TO BE AN INCREASED NUMBER OF FEMALES WITH GROSSLY OBSERVED MAMMARY TUMOR IN BOTH 100 AND 300 PPM EXPOSURE GROUPS.

REPORT 05 : NTP TECHNICAL BULLETIN 3/7 DEC 80, APR 82
PROPYLENE OXIDE WAS POSITIVE IN THE SALMONELLA ASSAY IN TWO LABORATORIES AND SUGGESTIVE BUT NOT CLEARLY POSITIVE IN THE RAT LETHAL DOMINANT ASSAY. IT WAS NEGATIVE IN THE MOUSE SPERM HEAD MORPHOLOGY ASSAY.

REPORT 06 : MULTI-AGENCY CROSS CHECK
Y HHS MAGUIRE
N OSHA CARC
N EPA CAG
N EPA RCRA
Y EPA CLEAN AIR
N EPA CLEAN WATER

REPORT 07 : EPA TSCA 4(A) TEST RULE 49FR30) NOV 85
AS PART OF THE ITC OCTOBER 1977 DESIGNATION FOR TESTING ALKYL EPOXIDE S, EPA INDICATES THAT UNDER TSCA 4(A) PROPYLENE OXIDE WILL BE TESTED FOR INHALATION DEVELOPMENTAL TOXICITY IN THE RAT.

REPORT 08 : IARC MONOGRAPH 36 FEB 85
TESTS BY ORAL GAVAGE IN RATS PRODUCED LOCAL, MAINLY SQUAMOUS-CELL, CARCINOMAS AND PAPILLOMAS OF FORESTOMACH, IN DOSE-DEPENDENT MANNER. INHALATION IN MICE AND RATS PRODUCED HEMANGIOMAS AND HEMANGIOSARCOMAS OF NASAL TURBINATES IN MICE AND INCREASED INCIDENCE OF

PAPILLARY ADENOMAS OF NASAL TURBINATES IN RATS. SC ADMINISTRATION IN MICE INDUCED LOCAL SARCOMAS, MAINLY FIBROSARCOMAS, IN DOSE-DEPENDENT MANNER. EXPOSURE OF RATS AT MATERNALLY TOXIC DOSES BY INHALATION DECREASED NUMBER OF CORPORA LUTEA, IMPLANTATION SITES AND LIVE FETUSES. NO TERATOGENIC EFFECT SEEN IN RATS OR RABBITS.

REPORT 09 : IARC MONOGRAPH 36 FEB 85
MUTAGENIC TO BACTERIA, FUNGI AND INSECTS. INDUCED DNA DAMAGE AND CHROMOSOMAL ABERRATIONS IN CULTURED MAMMALIAN CELLS AND MICRONUCLEI IN BONE-MARROW CELLS OF MICE IN-VIVO. DOMINANT LETHAL MUTATIONS NOT INDUCED IN MICE OR RATS.

REPORT 10 : TSCA SECT 8(E) STATUS REPORT 8EHQ-0487-0665 S MAY 87
BSC-125 IS AN ETHYLENE OXIDE/PROPYLENE OXIDE BLOCKED POLYMER USED AS A SURFACTANT. UNDER THE CONDITIONS OF THIS STUDY, BSC-125 PRODUCED CORNEAL OPACITY AND CONJUNCTIVAL IRRITATION TO RABBITS' EYES WITHOUT IRRIGATION. THESE EFFECTS PERSISTED FOR 21 DAYS IN ALL ANIMALS. CONJUNCTIVAL IRRITATION WAS PRODUCED IN 3 RABBITS' EYES IRRIGATED 30 SECONDS AFTER ADMINISTRATION. CORNEAL OPACITY AND CONJUNCTIVAL IRRITATION (SIC) WERE REVERSIBLE BY DAYS 7 AND 19, RESPECTIVELY.

REPT'D USE : INTERMEDIATE IN PRODUCTION OF POLYOLS/PROPYLENE GLYCOL, HERBICIDES, MICROBICIDE IN COSMETICS, FR CHEMICAL, AEROSOLS CLEANING AGENTS, ADHESIVES/GLUES AND DICHLOROMETHANE STABILIZER

CAS NO : **51-52-5**
CHEM NAME : ***** PROPYLTHIOURACIL**

REPORT 01 : IARC M0NOGRAPH 7 1974
PROPYLTHIOURACIL PRODUCED THYROID TUMORS IN MICE, RATS, HAMSTERS, AND GUINEA-PIGS FOLLOWING ORAL ADMINISTRATION, THE ONLY ROUTE TESTED. IN MICE, PITUITARY ADENOMAS WERE ALSO OBSERVED. IN A 1979 SUPPLEMENT, IARC DETERMINED THAT THERE IS SUFFICIENT EVIDENCE FOR THE CARCINOGENICITY OF PROPYLTHIOURACIL IN ANIMALS.

REPORT 03 : IARC SUPPL. 4 OCT 82
EVIDENCE FOR ACTIVITY IN SHORT-TERM TESTS WAS INADEQUATE.

REPORT 06 : MULTI-AGENCY CROSS CHECK
Y HHS MAGUIRE
N OSHA CARC
Y EPA CAG
N EPA RCRA
N EPA CLEAN AIR

N EPA CLEAN WATER
REPT'D USE : ANTI-THYROID AGENT

CAS NO : **UNKNOWN-26**
CHEM NAME : ***** PULP AND PAPER INDUSTRY**

REPORT 01 : IARC MONOGRAPH 25 SEP 81
THE EPIDEMIOLOGIC DATA NOT SUFFICIENT TO MAKE DEFINITIVE ASSESSMENT OF CARCINOGENIC RISK OF EMPLOYMENT IN PAPER OR PULP MILL INDUSTRY. SEVERAL STUDIES SUGGEST INCREASED RISK OF LYMPHOPROLIFERATIVE NEOPLASMS, PARTICULARLY HODGKINS DISEASE AND PERHAPS LEUKEMIA, MAY BE LINKED TO EMPLOYMENT IN THESE INDUSTRIES. EXCESS OF ORAL AND PHARYNGEAL CANCERS REPORTED IN TWO STUDIES BUT HAVE NOT BEEN EVALUATED IN INDEPENDENT STUDIES. EXCESS RISK OF LUNG CANCER OBSERVED IN SOME SUBGROUPS OF WORKERS IN TWO OF THESE STUDIES CANNOT BE EVALUATED.

CAS NO : **98-96-4**
CHEM NAME : ***** PYRAZINAMIDE**

REPORT 01 : NCI CARC TRS 48 DHEW PUB NO 78-848
UNDER THE CONDITIONS OF THIS BIOASSAY, THE EARLY DEATHS AND SMALL SIZE OF THE CONTROL GROUP PRECLUDED A CONCLUSION REGARDING THE CARCINOGENICITY OF PYRAZINAMIDE IN FEMALE B6C3F1 MICE. PYRAZINAMIDE WAS NOT CARCINOGENIC FOR FISCHER 344 RATS OR FOR MALE MICE (FEEDING STUDY).
REPT'D USE : TUBERCULOSTATIC AGENT

CAS NO : **288-13-1**
CHEM NAME : ***** PYRAZOLE**

REPORT 01 : EPA AND EHQ-0789- 0807 AUG 89
PYRAZOLE CAUSED DEVELOPMENTAL AS WELL AS MATERNAL TOXICITY IN PRE-GRANT RATS AT DOSE LEVELS GREATER THAN OR EQUAL TO 10 MG/KG/DAY WHEN ADMINISTERED BY GAVAGE ON GESTATION DAYS 6-18 PREVIOUSLY, IT HAS BEEN REPORTED TO INDUCE DEVELOPMENTAL TOXICITY IN RATS AND MICE.
REPT'D USE : R & D CHEMICAL

CAS NO : **129-00-0**
CHEM NAME : ***** PYRENE**

REPORT 01 : IARC MONOGRAPH 32 DEC 83
TESTED FOR CARCINOGENICITY IN SEVERAL EXPERIMENTS BY SKIN APPLICATION TO MICE AND NO SKIN TUMOR OBSERVED. TESTED

IN SEVERAL STUDIES IN MOUSE-SKIN INITIATION-PROMOTION ASSAY, WITH INCONCLUSIVE RESULTS. WHEN TESTED ON MICE SKIN SIMULTANEOUSLY WITH BENZO(A)PYRENE ENHANCED CARCINOGENIC EFFECTS OF BENZO(A)PYRENE. A STUDY IN MICE BY SUBCUTANEOUS INJECTION INADEQUATE FOR EVALUATION OF CARCINOGENICITY. INTRATRACHEAL ADMINISTRATION TO HAMSTERS OF PYRENE ATTACHED TO HAEMATITE DID NOT PRODUCE TUMORS.

REPORT 02 : IARC MONOGRAPH 32 DEC 83
TESTED EXTENSIVELY IN BOTH IN-VITRO AND IN-VIVO SHORT-TERM TESTS. NEGATIVE IN ASSAYS FOR DIFFERENTIAL SURVIVAL IN DNA-REPAIR-PROFICIENT/DEFICIENTS STRAINS OF BACTERIA AND MUTAGENIC IN SOME ASSAYS IN SALMONELLA TYPHIMURIUM IN PRESENCE OF EXOGENOUS METABOLIC SYSTEM. TESTS FOR GENETIC ACTIVITY IN YEAST NEGATIVE. NOT MUTAGENIC TO DROSOPHILA MELANOGASTER. DID INDUCE MUTATIONS AND UN-SCHEDULED DNA SYNTHESIS IN SOME IN-VITRO ASSAYS IN MAMMALIAN CELLS. DID NOT INDUCE MORPHOLOGICAL TRANS-FORMATION. IN TESTS IN MAMMALS IN VIVO DID NOT INDUCE SISTER CHROMATID EXCHANGE OR MICRONUCLEI.

REPT'D USE : INCOMPLETE COMBUSTION PRODUCT, CIGARETTE SMOKE AND ENGINE EXHAUST

CAS NO : **110-86-1**
CHEM NAME : ***** PYRIDINE**

REPORT 01 : ITC SECOND REPORT TO EPA (43FR16684) APR 78
THE INTERAGENCY TESTING COMMITTEE RECOMMENDS THAT THE EPA ADMINISTRATOR CONSIDER THE DEVELOPMENT OF TEST RULES TO TEST PYRIDINE FOR CARCINOGENICITY MUTAGENICITY, TERATOGENICITY AND OTHER CHRONIC EFFECTS; AND THAT AN EMIOLOGIC STUDY BE CONDUCTED.

REPORT 02 : NTP TECHNICAL BULLETIN 3 DEC 80
PYRIDINE WAS NEGATIVE IN THE SALMONELLA TYPHIMURIUM MUTAGENICITY ASSAY IN ONE LABORATORY.

REPORT 03 : EPA RCRA HEALTH AND ENV. EFFECTS PROFILE OCT 80
A SINGLE STUDY HAS INDICATED THAT PYRIDENE PRODUCED DEVELOPMENTAL ABNORMALITIES WHEN ADMINISTERED TO CHICKEN EMBRYOS. CHRONIC EXPOSURE TO PYRIDENE PRODUCES CNS DISTURBANCES AND MAY PRODUCE ADVERSE HEPATIC AND RENAL EFFECTS.

REPORT 04 : NTP CARC TEST PROGRAM JUL 82
IN A CURRENT NTP BIOASSAY, PYRIDINE WAS ADMINISTERED BY GAVAGE TO F344 RATS AND B6C3F1 MICE. THE CHRONIC EXPOSURE

PHASE HAS BEEN COMPLETED AND HISTOPATHOLOGY IS IN PROGRESS.

REPORT 05 : 47FR580131: EPA RESPONSE TO ITC RECOMMEND'N TO TEST — DEC 82
IN RESPONSE TO ITC'S 1978 RECOMMENDATION THAT PYRIDINE BE TESTED FOR HEALTH AND ENVIRONMENTAL EFFECTS UNDER TSCA 4(A), EPA CONCLUDES THAT SUFFICIENT DATA HAVE BEEN OR ARE BEING DEVELOPED FOR CARCINOGENICITY AND CHRONIC EFFECTS BY NTP. ALSO, EPA HAS FOUND NO SUBSTANTIAL ENVIRONMENTAL RELEASE AND HAS NO REASON TO BELIEVE THAT PYRIDINE PRESENTS AN UNREASONABLE RISK TO THE ENVIRONMENT.

REPORT 06 : MULTI-AGENCY CROSS CHECK
N HHS MAGUIRE
N OSHA CARC
N EPA CAG
Y EPA RCRA
N EPA CLEAN AIR
N EPA CLEAN WATER

REPORT 07 : EPA RESPONSE TO ITC — DEC 82
IN RESPONSE TO APRIL 1978 RECOMMENDATION THAT PYRIDINE BE TESTED FOR HEALTH AND ENVIRONMENTAL EFFECTS UNDER TSCA 4(A), EPA HAS TENTATIVELY DECIDED THAT NO TESTS FOR CARCINOGENICITY OR CHRONIC EFFECTS ARE NEEDED BECAUSE SUFFICIENT TOXICITY DATA HAS BEEN OR ARE BEING DEVELOPED BY NTP. ALSO NO EVIDENCE OF SUBSTANTIAL ENVIRONMENTAL RELEASE OR PERSISTENCE OR BIOACCUMULATION; THEREFORE IT DOES NOT PRESENT UNREASONABLE ENVIRONMENTAL RISKS.

REPT'D USE : INDUSTRIAL/LABORATORY SOLVENT, LABORATORY AGENT, ALCOHOL DENATURANT, INTERMEDIATE IN MFR OF PHARMACEUTICALS, DYES, DISINFECTANTS, EXPLOSIVES AND WATER PROOFING AGENTS FOR TEXTILES

CAS NO : **85878-62-2**

CHEM NAME : ***** PYRIDO(3,4-C)PSORALEN AND 7-METHYLPYRIDO(3,4-C)PSORALEN**

REPORT 01 : IARC VOL 40 — 1986
INADEQUATE EVIDENCE THAT CPDS IN COMBINATION WITH ULTRAVIOLET A RADIATION (UVA) IS CARCINOGENIC TO TEST ANIMALS, BUT DATA INADEQUATE TO EVALUATE CARCINOGENICITY OF CPDS TO TEST ANIMALS WITHOUT EXPOSURE TO UVA. CARCINOGENICITY TO HUMANS FROM EXPOSURE TO CPD COULD NOT BE MADE. NO DATA AVAILABLE TO EVALUATE REPRODUCTIVE EFFECTS OR PRENATAL TOXICITY OF CPD TO EXPERIMENTAL ANIMALS.

REPT'D USE : RESEARCH

CAS NO : **58-14-0**
CHEM NAME : ***** PYRIMETHAMINE**

REPORT 01 : NCI CARC TRS 77 DHEW PUB NO 78-1327
UNDER THE CONDITIONS OF THIS BIOASSAY, PYRIMETHAMINE WAS NOT CARCINOGENIC FOR MALE OR FEMALE FISCHER 344 RATS OR FOR FEMALE B6C3F1 MICE. THE CARCINOGENIC POTENTIAL OF PYRIMETHAMINE FOR MALE B6C3F1 MICE CANNOT BE ASSESSED BY THIS BIOASSAY, BECAUSE OF THE MARKEDLY REDUCED LIFE SPAN (FEEDING STUDY).

REPORT 02 : IARC MONOGRAPH 13 1977
IN THE ONLY STUDY AVAILABLE, PYRIMETHAMINE PRODUCED A SIGNIFICANT INCREASE IN THE NUMBER OF LUNG TUMORS PER MOUSE WHEN IT WAS GIVEN INTRAPERITONEALLY AT HIGH DOSES. THIS LIMITED EVIDENCE OF CARCINOGENICITY AWAITS CONFIRMATION.

REPORT 03 : NTP TECHNICAL BULLETIN 6 JAN 82
THE COMPOUND WAS NEGATIVE IN THE SALMONELLA TYPHIMURIUM MUTAGENICITY ASSAY IN ONE LABORATORY.

REPORT 04 : IARC MONOGRAPH 13 1977
PYRIMETHAMINE CAUSED AN INCREASE IN THE NUMBER OF X-LINKED RECESSIVE LETHALS IN DROSOPHILA. CHROMOSOME ABERRATIONS HAVE BEEN OBSERVED IN MOUSE BONE MARROW CELLS AND IN ELASTOMERS FROM PREGNANT RATS FOLLOWING TREATMENT IN VIVO.
REPT'D USE : ANTI-MALARIAL AGENT

CAS NO : **87-66-1**
CHEM NAME : ***** PYROGALLOL**

REPORT 01 : EPA TSCA 8(E) 8EHQ-1284-0539 DEC 84
NOR-AM CHEMICAL CO. REPORTED IN BATTERY OF IN VITRO GENOTOXICITY ASSAYS PYROGALLOL FOUND POSITIVE IN PRESENCE AND ABSENCE OF METABOLIC ACTIVATION IN SALMONELLA TYPHIMURIUM BACTERIA STRAINS TA 100 AND TA 1537; CULTURED MOUSE LYMPHOMA CELL ASSAY; AND METAPHASE CHROMOSOME ANALYSIS OF CULTURED HUMAN LYMPHOCYTES.
REPT'D USE : INTERMEDIATE IN MANUFACTURE OF DYES, INKS, DRUGS, LUBRICANT OILS, ANTIOXIDANT, ANALYTICAL REAGENT, REDUCING AGENT, ANTIBACTERIAL AND PHOTOGRAPHIC CHEMICAL

CAS NO : **14808-60-7**
CHEM NAME : ***** QUARTZ**

REPORT 01 : NIOSH CRITERIA DOCUMENT NOV 74
NIOSH RECOMMENDS A STANDARD OF 50 UG/CU M TWA BASED ON SILICOSIS. THE PRESENT OSHA STANDARD IS 10 MG/CU M TWA.

CAS NO : **117-39-5**
CHEM NAME : ***** QUERCETIN**

REPORT 01 : NTP DRAFT TR409 NIH PUB. 91-3140 JUL 83
UNDER 2-YEAR STUDIES SOME EVIDENCE OF CARCINOGENIC ACTIVITY OF QUERCETIN FOR MALE F344/N RATS BASED ON RENAL TUBULAR CELL ADENOMAS IN 40,000 PPM GROUP. NO EVIDENCE OF CARCINOGENIC ACTIVITY FOR FEMALE F344/N RATS FED DIETS CONTAINING 1,000, 10,000, OR 40,000 PPM OF QUERCETIN. HYPERPLASIA OF RENAL TUBULAR EPITHELIUM AND SEVERITY OF NEPHROPATHY INCREASED IN DOSED MALE RATS.

REPORT 02 : IARC MONOGRAPH 31 JUL 83
QUERCETIN MUTAGENIC IN BACTERIA AND INSECTS AND CAUSED GENE CONVERSION IN YEAST. EQUIVOCAL RESULTS OBTAINED TO MUTATIONS IN MAMMALIAN CELLS IN VITRO INDUCED CHROMOSOMAL ANOMALIES IN CULTURED CELLS, BUT EQUIVOCAL RESULTS OBTAINED IN MICRONUCLEUS TEST IN VIVO. INDUCED MORPHOLOGICAL TRANSFORMATION OF HAMSTER EMBRYO CELLS IN VITRO. IN ONE EXPERIMENT WITH RATS, NO TERATOGENIC EFFECT SEEN.
REPT'D USE : NATURALLY OCCURS IN FRUIT AND OTHER EDIBLE PLANTS

CAS NO : **4637-56-3**
CHEM NAME : ***** 4-QUINOLINAMINE-N-HYDROXY-1-OXIDE**

REPORT 01 : NTP TECHNICAL BULLETIN 8 JUL 82
THE CHEMICAL WAS POSITIVE IN THE L5178Y MOUSE LYMPHOMA ASSAY.

CAS NO : **91-22-5**
CHEM NAME : ***** QUINOLINE**

REPORT 01 : NTP TECHNICAL BULLETIN 2/8 APR 80, JUL 82
QUINOLINE WAS POSITIVE IN THE SALMONELLA ASSAY BUT NEGATIVE IN DROSOPHILA WHEN TESTED FOR SEX-LINKED RECESSIVE LETHALS.

REPORT 06 : MULTI-AGENCY CROSS CHECK
N HHS MAGUIRE
Y OSHA CARC
N EPA CAG
N EPA RCRA
N EPA CLEAN AIR
N EPA CLEAN WATER

REPT'D USE : DYE INTERMEDIATE, SOLVENT FOR RESINS/TERPENES, BIOLOGICAL SPECIMEN, PRESERVATIVE, ANTIMALARIAL AGENT, LABORATORY AGENT, INTERMEDIATE IN MFR OF PESTICIDES AND PHARMACEUTICALS

CAS NO : **106-51-4**
CHEM NAME : ***** P-QUINONE**

REPORT 01 : NTP TEST REGIMEN
IN A CURRENT NTP BIOASSAY, P-QUINONE IS BEING ADMINISTERED BY GAVAGE TO F344 RATS AND B6C3F1 MICE. THE PRECHRONIC TESTING PHASE BEGAN MAY 80.

REPORT 02 : IARC MONOGRAPH 15
PARA-QUINONE HAS BEEN TESTED IN MICE BY SKIN APPLICATION AND INHALATION AND IN RATS BY SUBCUTANEOUS INJECTION. THE AVAILABLE DATA ARE INSUFFICIENT TO EVALUATE THE CARCINOGENICITY OF THIS COMPOUND.

REPORT 03 : IARC MONOGRAPH 15 1977
NO DATA ON THE EMBRYOTOXICITY OR TERATOGENICITY OF P-QUINONE WERE AVAILABLE. A SINGLE I.P. INJECTION OF P-QUINONE DID NOT INDUCE DOMINANT LETHAL MUTATIONS IN MALE MICE.

REPORT 04 : EPA RCRA HEALTH AND ENV. EFFECTS PROFILE OCT 80
QUINONE HAS BEEN REPORTED TO PRODUCE NEOPLASMS, BUT INSUFFICIENT DATA ARE AVAILABLE TO ACCESS ITS CARCINOGENIC POTENTIAL. QUINONE WAS NOT MUTAGENIC TO DROSOPHILIA, HUMAN LEUKOCYTES, OR NEUROSPORA.

REPORT 05 : NTP TECHNICAL BULLETIN 3 DEC 80
P-QUINONE WAS NEGATIVE IN THE SALMONELLA TYPHIMURIUM MUTAGENICITY ASSAY IN ONE LABORATORY.

REPORT 06 : MULTI-AGENCY CROSS CHECK
N HHS MAGUIRE
N OSHA CARC
N EPA CAG
Y EPA RCRA
N EPA CLEAN AIR
N EPA CLEAN WATER

REPORT 07 : ITC FIFTH REPORT TO EPA (44FR70664) DEC 79
THE INTERAGENCY TESTING COMMITTEE RECOMMENDS THAT THE EPA ADMINISTRATOR CONSIDER THE DEVELOPMENT OF TEST RULES TO TEST QUINONE FOR CARCINOGENICITY AND TERATOGENICITY/REPRODUCTIVE EFFECTS.

REPORT 08 : EPA TSCA PROPOSED TEST RULE FR NOTICE 49FR456 JAN 84
IN RESPONSE TO ITC'S DESIGNATION OF P-Q FOR TESTING UNDER TSCA 4(A), EPA PROPOSING THAT MFRS AND PROCESSORS PERFORM TESTING TO EVALUATE P-Q'S CARCINOGENIC POTENTIAL AND TO DETERMINE CHEMICAL FATE AND ENVIRONMENTAL EFFECTS.

REPORT 09 : EPA TSCA WITHDRAWAL OF TEST RULE DEC 85
FR NOTICE 49FR456
IN RESPONSE TO COMMENTS TO PROPOSED TEST RULE THAT INDICATED LITTLE OR NO HUMAN EXPOSURE, EPA IS WITHDRAWING ITS PROPOSED RULEMAKING UNDER TSCA 4(A).

REPT'D USE : INTERMEDIATE IN SYNTHESIS OF HYDROQUINONE, POLYMERIZATION INHIBITOR, OXIDIZING AGENT, PHOTOGRAPHIC CHEMICALS, TANNING AGENT, REAGENT, RUBBER ACCELERATOR AND DYE INTERMEDIATE

CAS NO : **105-11-3**
CHEM NAME : ***** P-QUINONE DIOXIME**

REPORT 01 : NCI CARC TRS NO 179 DHEW PUB NO NIH 79-1735
UNDER THE CONDITIONS OF THIS BIOASSAY, DIETARY ADMINISTRATION OF P-QUINONE DIOXIME WAS CARCINOGENIC TO FEMALE FISCHER 344 RATS, CAUSING NEOPLASMS OF THE URINARY BLADDER. THE COMPOUND WAS NOT CARCINOGENIC TO MALE FISCHER 344 RATS OR B6C3F1 MICE OF EITHER SEX.

REPORT 02 : NTP TECHNICAL BULLETIN NO. 3 DEC 80
THE COMPOUND WAS FOUND TO BE POSITIVE IN THE SALMONELLA TYPHIMURIUM MUTAGENICITY ASSAY IN ONE LABORATORY.

REPORT 03 : IARC MONOGRAPH 29 1982
P-BENZOQUINONE WAS TESTED FOR CARCINOGENICITY IN MICE AND RATS BY ORAL ADMINISTRATION. A SIGNIFICANT INCREASE IN THE NUMBER OF TRANSITIONAL-CELL PAPILLOMAS AND CARCINOMAS OF THE URINARY BLADDER OCCURRED IN FEMALE RATS. THERE IS LIMITED EVIDENCE THAT P-BENZOQUINONE IS CARCINOGENIC TO RATS.

REPORT 04 : IARC MONOGRAPH 29 1982
NO DATA WERE AVAILABLE TO ASSESS THE MUTAGENICITY OR TERATOGENICITY OF P-BENZOQUINONE..

REPORT 06 : MULTI-AGENCY CROSS CHECK
N HHS MAGUIRE
Y OSHA CARC
N EPA CAG
N EPA RCRA
N EPA CLEAN AIR
N EPA CLEAN WATER
REPT'D USE : RUBBER VULCANIZATION AGENT

CAS NO : **50-55-5**
CHEM NAME : ***** RESERPINE**

REPORT 01 : IARC MONOGRAPH 24 1980
RESERPINE WAS TESTED IN TWO EXPERIMENTS IN MICE BY ORAL ADMINISTRATION; IN ONE EXPERIMENT IT PRODUCED MALIGNANT MAMMARY TUMORS IN FEMALES AND CARCINOMAS OF THE SEMINAL VESICLES IN MALES. IT WAS TESTED IN THREE EXPERIMENTS IN RATS BY ORAL ADMINISTRATION; IN ONE EXPERIMENT IT INCREASED THE INCIDENCE OF PHEOCHROMOCYTOMAS IN MALES.

REPORT 02 : IARC MONOGRAPH 24 1980
RESERPINE HAS NO MUTAGENIC ACTIVITY IN SALMONELLA TA1950 OR TA1538 WITH OR WITHOUT METABOLIC ACTIVATION. IN MOUSE BONE-MARROW CELLS, RESERPINE WAS REPORTED TO STIMULATE MITOSIS AND INDUCE AN INCREASED NUMBER OF CHROMOSOME ABERRATIONS. RESERPINE IS EMBRYOTOXIC AND HAS EFFECTS ON REPRODUCTION. THERE IS LIMITED EVIDENCE THAT IT IS TERATOGENIC IN EXPERIMENTAL ANIMALS.

REPORT 04 : IARC SUPPL. 4 OCT 82
EVIDENCE FOR ACTIVITY IN SHORT-TERM TESTS IS INADEOUATE TO JUDGE IT.

REPORT 06 : MULTI-AGENCY CROSS CHECK
Y HHS MAGUIRE
N OSHA CARC
Y EPA CAG
Y EPA RCRA
N EPA CLEAN AIR
N EPA CLEAN WATER
REPT'D USE : TRANQUILIZER/SEDATIVE AND HYPOTENSIVE DRUG

CAS NO : **108-46-3**
CHEM NAME : ***** RESORCINOL**

REPORT 01 : NTP TECHNICAL BULLETIN 2 APR 80
THE COMPOUND WAS FOUND TO BE NEGATIVE IN THE SALMONELLA TYPHIMURIUM MUTAGENICITY ASSAY.

REPORT 02 : IARC MONOGRAPH 15 1977
WHEN TESTED BY REPEATED SKIN APPLICATION, RESORCINAL SHOWED NO CARCINOGENIC EFFECT IN MICE. THE AVAILABLE DATA DO NOT ALLOW AN EVALUATION OF THE CARCINOGENICITY OF THIS COMPOUND.

REPORT 03 : EPA RCRA HEALTH AND ENV. EFFECTS PROFILE OCT 80
RESORCINOL IS A STRONG MITOTIC SPINDLE POISON IN PLANTS. THIS EVIDENCE OF MUTAGENIC ACTIVITY AND THE STRONG ONCOGENIC ACTIVITY IN PLANTS HAVE NOT BEEN ADEQUATELY TESTED IN ANIMALS TO PROVIDE AN UNDERSTANDING OF THE PROCESSES.

REPORT 04 : IARC MONOGRAPH 15 1977
NO DATA ON THE EMBRYOTOXICITY OR TERATOGENICITY OF RESORCINOL WERE AVAILABLE. RESORCINOL WAS NOT MUTAGENIC TO SALMONELLA TAI535, TA1537, TA98 OR TA100, IN THE PRESENCE OR ABSENCE OF METABOLIC ACTIVATION. RESORCINOL INDUCED CHROMOSOME ABERRATIONS IN A. CEPA.

REPORT 05 : NTP CARC TEST PROGRAM JUL 82
IN A CURRENT NTP BIOASSAY, RESORCINOL IS BEING ADMINISTERED BY GAVAGE TO F344 RATS AND B6C3F1 MICE. THE PRECHRONIC TESTING PHASE BEGAN DEC 80.

REPORT 06 : MULTI-AGENCY CROSS CHECK
N HHS MAGUIRE
N OSHA CARC
N EPA CAG
Y EPA RCRA
N EPA CLEAN AIR
N EPA CLEAN WATER

REPT'D USE : RESIN MONOMER, CHEMICAL INTERMEDIATE AND TEXTILE PHOTOGRAPHICS

CAS NO : **480-54-6**
CHEM NAME : ***** RETROSINE**

REPORT 01 : IARC MONOGRAPH 10 1976
RETROSINE IS CARCINOGENIC IN RATS FOLLOWING ITS ORAL ADMINISTRATION; IT PRODUCED A VARIETY OF TUMORS.

REPT'D USE : MEDICINAL HERB, DIAPHORETIC AND DIURETIC

CAS NO : **64741-75-9**
CHEM NAME : *****RHU LIGHT VACUUM GAS OIL**

REPORT 01 : EPA TSCA 8(E) 1288-0774 DEC 88
AMOCO REPORTED THAT RESID HYDROPROCESSING UNIT (RHU) LIGHT VACUUM GAS OIL POSSESS INITIATION TUMORIGENIC POTENTIAL AS SHOWN IN INCREASED TUMORS AND TUMOR LATENCY IN MALE MICE TESTED DERMALLY WITH 50 UL OF THIS OIL WITH PHORBOL-12-MYRISTATE-13-ACETATE (PMA) IN ACETONE AS A PROMOTER. ALL INDUCED TUMORS WERE SQUAMOUS CELL PAPILLOMAS OR KERATOACANTHOMAS.
REPT'D USE : UNKNOWN

CAS NO : **64741-76-0**
CHEM NAME : ***** RHU MIDDLE DISTILLATE**

REPORT 01 : EPA TSCA 8(E) 1288-0773 DEC 88
AMOCO REPORTED THAT RESID HYDROPROCESSING UNIT (RHU) MIDDLE DISTILLATE POSSESS SKIN TUMOR PROMOTION POTENTIAL AS SHOWN BY INCREASED TUMORS DMBA INITIATED TEST IN COMPARED TO ACETONE-INITIATED CONTROLS AND DMBA-INITIATED/TEST MALE MICE COMPARED TO ACETONE-INITIATED CONTROLS AND DMBA-INITIATED SHAM CONTROLS. THE TUMORS WERE SQUAMOUS CELL PAPILLOMAS AND ONE KERATOACEANTHOMA. TUMOR LATENCIES WERE ALSO SIGNIFICANTLY SHORTENED IN THE DMBA-INITIATED/TEST GROUP.
REPT'D USE : UNKNOWN

CAS NO : **23246-96-0**
CHEM NAME : ***** RIDDELLIINE**

REPORT 01 : IARC MONOGRAPH 10 1976
THE AVAILABLE INFORMATION IS INSUFFICIENT TO EVALUATE THE CARCINOGENICITY OF RIDDELLIINE.
REPT'D USE : EAST AFRICAN HERBAL MEDICINE

CAS NO : **13292-46-1**
CHEM NAME : ***** RIFAMPICIN**

REPORT 01 : IARC MONOGRAPH 24 1980
RIFAMPICIN HAS BEEN TESTED IN TWO STRAINS OF MICE AND IN RATS BY ORAL ADMINISTRATION. IT WAS ALSO TESTED IN MICE BY SUBCUTANEOUS ADMINISTRATION. AFTER ORAL ADMINISTRATION, IT SIGNIFICANTLY INCREASED THE INCIDENCE OF BENIGN AND MALIGNANT LIVER-CELL TUMORS ONLY IN FEMALE

MICE OF ONE STRAIN; NO EVIDENCE OF CARCINOGENICITY WAS OBSERVED IN ANIMALS OF THE OTHER STRAIN. IN RATS, NO INCREASED TUMOR INCIDENCE WAS FOUND.

REPORT 02 : IARC MONOGRAPH 24 1980
THE AVAILABLE STUDIES OF RIFAMPICIN FOR MUTAGENESIS INDICATED THE ABSENCE OF A MUTAGENIC EFFECT. ORAL ADMINISTRATION OF RIFAMPICIN TO MICE, RATS, AND RABBITS DURING THE ACTIVE PERIOD OF ORGANOGENESIS 1/2 REDUCED SPINAL BIFIDA IN OFFSPRING OF RATS AND MICE AND CLEFT PALATES IN MOUSE FETUSES; NO EFFECTS WERE OBSERVED ON RABBITS FETUSES.
REPT'D USE : ANTIBIOTIC

CAS NO : **83-79-4**
CHEM NAME : ***** ROTENONE**

REPORT 01 : NTP TR 320 NIH PUB. 88-2576 JAN 88
TWO YEAR FEED STUDIES SHOW EQUIVOCAL EVIDENCE OF CARCINOGENIC ACTIVITY OF ROTENONE FOR MALE F 344/N RATS, INDICATED BY INCREASED INCIDENCE OF PARATHYROID GLAND ADENOMAS (UNCOMMON TUMORS). NO EVIDENCE OF CARCINOGENIC ACTIVITY FOR MALE OR FEMALE B6C3F1 MICE FED DIETS CONTAINING 600 OR 1,200 PPM ROTENONE FOR 2 YEARS. DECREASED INCIDENCE OF LIVER NEOPLASMS IN MALE MICE MAY HAVE BEEN RELATED TO THE ADMINISTRATION OF ROTENONE.
REPT'D USE : INSECTICIDE

CAS NO : **121-19-7**
CHEM NAME : ***** ROXARSONE**

REPORT 01 : NTP DRAFT TECH REPORT TR 345 NIH PUB #88-2800 MAR 89
UNDER 2-YEAR FEED STUDIES EQUIVOCAL EVIDENCE OF CARCINOGENIC ACTIVITY OF ROXARSONE FOR MALE F344/N RATS INDICATED BY MARGINALLY INCREASED OF ADENOMAS OF EXOCRINE PANCREAS. NO EVIDENCE OF CARCINOGENIC ACTIVITY FOR FEMALE F344/N RATS FED DIETS CONTAINING 50 OR 100 PPM ROXARSONE FOR 2 YEARS. NO EVIDENCE OF CARCINOGENIC ACTIVITY FOR MALE OR FEMALE B6C3F1 MICE FED DIETS CONTAINING 100 OR 200 PPM ROXARSONE FOR 2 YEARS.
REPT'D USE : GROWTH PROMOTER DRUG

CAS NO : **UNKNOWN-17**
CHEM NAME : ***** RUBBER INDUSTRY**

REPORT 01 : IARC MONOGRAPH 28 APR 82
AMONG RUBBER WORKERS, EXCESS MALIGNANCIES OF LYM-

PHATIC AND HAEMATOPOIETIC SYSTEMS, PARTICULARLY LYMPHATIC LEUKEMIA, ASSOCIATAED WITH JOBS ENTAILING EXPOSURE TO SOLVENTS. STOMACH CANCER ELEVATED IN U.S. AND UK RUBBER WORKERS, APPEARS ASSOCIATED WITH JOBS EARLY IN THE PRODUCTION LINE. LUNG CANCER POSITIVELY RELATED TO VARIETY OF JOBS WITHIN INDUSTRY. ATTRIBUTION TO SPECIFIC FACTORS IN WORKER'S ENVIRONMENT CANNOT BE MADE. MORTALITY FROM PROSTATE CANCER FOUND MODERATELY ELEVATED IN SEVERAL STUDIES, AND SOME ASSOCIATION FOUND WITH COMPOUNDING AND MIXING JOBS. NOISH RECOMMENDED STANDARD TWA 350 MG/CUBIC METER FOR PETROLEUM SOLVENTS (7/77); SEE 68477-31-6.

CAS NO : **23537-16-8**
CHEM NAME : ***** RUGULOSIN**

REPORT 01 : IARC VOL 40 1986
INADEQUATE EVIDENCE FOR CARCINOGENICITY IN TEST ANIMALS. CARCINOGENICITY OF CPD TO HUMANS COULD NOT BE EVALUATED. NO DATA AVAILABLE TO EVALUATE REPRODUCTIVE EFFECTS OR PRENATAL TOXICITY TO TEST ANIMALS OR HUMANS. INDUCED DNA DAMAGE IN B. SUBTILIS, DID NOT INDUCE REVERSE MUTATIONS IN S. TYPHIMURIUM BUT WAS REPORTED TO INDUCE FORWARD MUTATIONS. DID NOT INDUCE MUTATIONS IN E. COLI NOR UNSCHEDULED DNA SYSNTHESIS IN MAMMALIAN CELLS IN VITRO. NO CASE REPORT OR EPIDEMIOLOGIC STUDY AVAILABLE TO WORKING GROUP.
REPT'D USE : NO COMMERCIAL USE

CAS NO : **8047-67-4**
CHEM NAME : ***** SACCHARATED IRON**

REPORT 01 : IARC MONOGRAPH 2 1973
SACCHARATED IRON OXIDE PRODUCES LOCAL SARCOMAS IN MICE AFTER REPEATED SUBCUTANEOUS INJECTIONS.
REPT'D USE : HEMATINIC AGENT

CAS NO : **81-07-2**
CHEM NAME : ***** SACCHARIN**

REPORT 01 : IARC MONOGRAPH 22 1980
SACCHARIN HAS BEEN TESTED BY ORAL ADMINISTRATION IN MICE, RATS AND HAMSTERS. IN MICE, SACCHARIN PRODUCED NO DIFFERENCE IN TUMOR INCIDENCE BETWEEN TREATED AND CONTROL ANIMALS IN ONE SINGLE AND IN ONE MULTIGENERATION STUDY. A STUDY IN MICE BY BLADDER INSERTION PROVIDED EVIDENCE FOR THE INDUCTION OF BLADDER

CARCINOMAS. THERE IS SUFFICIENT EVIDENCE THAT SACCHARIN ALONE, GIVEN AT HIGH DOSES, PRODUCES TUMORS OF THE URINARY TRACT IN MALE RATS AND CAN PROMOTE THE ACTION OF KNOWN CARCINOGENS IN THE BLADDER OF RATS OF BOTH SEXES.

REPORT 02 : OTA REPORT: CANCER TESTING - SACCHARIN OCT 77
LABORATORY EVIDENCE DEMONSTRATES THAT SACCHARIN IS A CARCINOGEN. PROLONGED INGESTION OF SACCHARIN AT HIGH LEVELS CAUSED A SIGNIFICANT INCREASE IN THE INCIDENCE OF BLADDER CANCER IN RATS IN THREE INDEPENDENT EXPERIMENTS. THIS EVIDENCE LEADS TO THE CONCLUSION THAT SACCHARIN IS A POTENTIAL CAUSE OF CANCER IN HUMANS. THERE ARE NO RELIABLE QUANTITATIVE ESTIMATES OF THE RISK OF SACCHARIN TO HUMANS.

REPORT 03 : IARC MONOGRAPH 22 1980
THERE IS LITTLE EVIDENCE THAT SACCHARIN ITSELF INDUCES POINT MUTATIONS. DOMINANT LETHAL EFFECTS AND UNSCHEDULED DNA SYNTHESIS HAVE BEEN REPORTED. IT ALSO CAUSES SISTER CHROMATID EXCHANGES AND OTHER CHROMOSOMAL EFFECTS. IN THE MAJORITY OF STUDIES, NO INDICATION OF TERATOGENICITY WAS FOUND. IMPURITIES MAY BE RESPONSIBLE FOR THE OCCASIONAL EFFECTS REPORTED.

REPORT 04 : NAS REPORT ON DIET, NUTRITION AND CANCER 1982
EXPERIMENTAL STUDIES HAVE PROVIDED SUFFICIENT EVIDENCE THAT SACCHARIN ALONE, GIVEN AT HIGH DOSES, PRODUCES TUMORS OF THE URINARY TRACT IN MALE RATS AND CAN PROMOTE THE ACTION OF KNOWN CARCINOGENS OF THE BLADDER OF RATS. THERE IS LIMITED EVIDENCE FOR ITS CARCINOGENICITY IN MICE. THE EPA DATA DO NOT PROVIDE A CLEAR INDICATION OF AN ASSOCIATION BETWEEN THE USE OF NON-NUTRITIVE SWEETENERS AND CANCER. EXCEPTIONS ARE TWO STUDIES WHICH SHOW A DIRECT ASSOCIATION IN MEN, AND A POSSIBLE EFFECT IN CERTAIN SUBGROUPS.

REPORT 05 : NTP TECHNICAL BULLETIN 9 APR 83
SACCHARIN WAS NEGATIVE IN ONE SALMONELLA TYPHIMURIUM MUTAGENICITY ASSAY.

REPORT 06 : MULTI-AGENCY CROSS CHECK
Y HHS MAGUIRE
N OSHA CARC
Y EPA CAG
Y EPA RCRA
N EPA CLEAN AIR
N EPA CLEAN WATER

REPORT 08 : IARC SUPPL.4 OCT 82
EVIDENCE FOR ACTIVITY IN SHORT-TERM TESTS IS INADEQUATE.
RFPT'D USE : SYNTHETIC SWEETENER

CAS NO : **128-44-9**
CHEM NAME : ***** SACCHARIN, SODIUM**

REPORT 01 : IARC MONOGRAPH 22 1980
SODIUM SACCHARIN HAS BEEN TESTED BY ORAL ADMIN. IN MICE, RATS, AND MONKEYS. ONE SINGLE-GENERATION STUDY IN RATS SHOWED AN INCREASED INCIDENCE OF BLADDER TUMORS IN MALES; 2 FURTHER STUDIES SHOWED A FEW BLADDER TUMORS; 1 OTHER SHOWED NO DIFFERENCE IN TUMOR INCIDENCE BETWEEN TREATED AND CONTROL ANIMALS; 2 OTHERS: INADEQUATE FOR EVALUATION IN 3 TWO-GEN. STUDIES IN RATS, IT PRODUCED A STATISTICALLY SIGNIFICANT INCREASE IN BLADDER TUMORS IN MALES. IT HAS ALSO BEEN TESTED IN MICE BY BLADDER INSERTION: INCIDENCE OF BLADDER CARCINOMAS INCREASED.

REPORT 02 : OTA REPORT: CANCER TESTING - SACCHARIN OCT 77
LABORATORY EVIDENCE DEMONSTRATES THAT SACCHARIN IS A CARCINOGEN. PROLONGED INGESTION OF SACCHARIN AT HIGH LEVELS CAUSED A SIGNIFICANT INCREASE IN THE INCIDENCE OF BLADDER CANCER IN RATS IN THREE INDEPENDENT EXPERIMENTS. THIS EVIDENCE LEADS TO THE CONCLUSION THAT SACCHARIN IS A POTENTIAL CAUSE OF CANCER IN HUMANS. THERE ARE NO RELIABLE QUANTITATIVE ESTIMATES OF THE RISK OF SACCHARIN TO HUMANS.

REPORT 03 : IARC MONOGRAPH 22 1980
THERE IS LITTLE EVIDENCE THAT SACCHARIN ITSELF INDUCES POINT-MUTATIONS. DOMINANT LETHAL EFFECTS AND UNSCHEDULED DNA SYNTHESIS HAVE BEEN REPORTED; AND IT CAUSES SISTER CHROMATIDEX CHANGES AND OTHER CHROMOSOMAL EFFECTS. IN THE MAJORITY OF STUDIES, NO INDICATION OF TERATOGENICITY WAS FOUND; IMPURITIES MAY BE RESPONSIBLE FOR THE OCCASIONAL EFFECTS REPORT.
REPT'D USE : NON-NUTRITIVE SWEETENER, DENTURE CLEANING AGENTS AND PYROTECHNICS

CAS NO : **94-59-7**
CHEM NAME : ***** SAFROLE**

REPORT 01 : IARC MONOGRAPH 1 1972
SAFROLE AND ISOSAFROLE ARE LIVER CARCINOGENS FOR THE MOUSE AND THE RAT WHEN ADMINISTERED ORALLY OR SUBCUTANEOUSLY. IN A 1979 SUPPLEMENT, IARC DETERMINED THAT

THERE IS SUFFICIENT EVIDENCE FOR THE CARCINOGENICITY OF SAFROLE IN ANIMALS.

REPORT 06 : MULTI-AGENCY CROSS CHECK
Y HHS MAGUIRE
N OSHA CARC
Y EPA CAG
Y EPA RCRA
N EPA CLEAN AIR
N EPA CLEAN WATER
REPT'D USE : FLAVORING AGENT, FRAGRANCE, MFR OF HELIOTROPIN, PEDICULICIDE AND CARMINATIVE

CAS NO : **7782-49-2**
CHEM NAME : ***** SELENIUM**

REPORT O1 : IARC MONOGRAPH 9 1975
SELENIUM COMPOUNDS WERE TESTED IN MICE AND RATS BY THE ORAL ROUTE. ALTHOUGH IN ONE EXPERIMENT IN RATS SELENIUM PRODUCED AN INCREASE IN THE INCIDENCE OF LIVER TUMORS, THE AVAILABLE DATA ARE INSUFFICIENT TO ALLOW AN EVALUATION OF THE CARCINOGENICITY OF SELENIUM COMPOUNDS.

REPORT 02 : EPA RCRA HEALTH AND ENV. EFFECTS PROFILE OCT 80
ALTHOUGH IT HAS BEEN SHOWN TO PRODUCE TUMORS IN ANIMALS, THE FDA, IARC, AND NAS HAVE CONCLUDED THAT THE AVAILABLE ANIMAL DATA ARE INSUFFICIENT TO ALLOW AN EVALUATION OF THE CARCINOGENICITY OF SELENIUM COMPOUNDS.

REPORT 03 : NAS REPORT ON DIET, NUTRITION, AND CANCER 1982
NUMEROUS EXPERIMENTS IN ANIMALS HAVE DEMONSTRATED AN ANTITUMORIGENIC EFFECT OF SELENIUM. RECENT DATA DO NOT SUPPORT EARLIER REPORTS THAT SELENIUM FIX PERSE IS CARCINOGENIC. THE EPA EVIDENCE PERTAINING TO THE RELATIONSHIP BETWEEN SELENIUM AND CANCER IS DERIVED FROM A LIMITED NUMBER OF GEOGRAPHICAL CORRELATIONS STUDIES. ALTHOUGH THESE STUDIES GENERALLY DEMONSTRATE AN INVERSE RELATIONSHIP BETWEEN THE LEVEL OF SELENIUM AND CANCER, FIRM CONCLUSIONS CANNOT BE DRAWN ON THE BASIS OF AVAILABLE EVIDENCE.

REPORT 06 : MULTI-AGENCY CROSS CHECK
N HHS MAGUIRE
N OSHA CARC
N EPA CAG
N EPA RCRA
N EPA CLEAN AIR

Y EPA CLEAN WATER

REPT'D USE : PHOTOGRAPHY CHEMICAL, GLASS COLORING AGENT, RUBBER ADDITIVE AND ELECTRONIC COMPONENTS

CAS NO : **7446-34-6**

CHEM NAME : ***** SELENIUM SULFIDE**

REPORT 01 : NTP BIOASSAY TRS 194 NIH PUB NO. 80-1750
UNDER THE CONDITIONS OF THIS BIOASSAY SELENIUM SULFIDE ADMINISTERED BY GAVAGE WAS CARCINOGENIC FOR F344 RATS AND FEMALE B6C3F1 MICE, INDUCING HEPATOCELLULAR CARCINOMAS IN MALE AND FEMALE RATS AND FEMALE MICE AND ALVEOLAR/BRONCHIOLAR CARCINOMAS AND ADENOMAS IN FEMALE MICE. SELENIUM SULFIDE WAS NOT CARCINOGENIC FOR MALE MICE; BUT THEY MAY HAVE BEEN ABLE TO TOLERATE HIGHER DOSES.

REPORT 02 : NTP BIOASSAY TRS 197 NIH PUB NO. 80-1753
UNDER THE CONDITIONS OF THIS BIOASSAY, DERMAL APPLICATION OF SELENIUM SULFIDE DID NOT PRODUCE A CARCINOGENIC EFFECT IN ICR SWISS MICE, BUT THE STUDY WAS LIMITED BY THE RELATIVELY SHORT LIFESPAN OF THIS STRAIN OF MOUSE.

REPORT 06 : MULTI-AGENCY CROSS CHECK
Y HHS MAGUIRE
N OSHA CARC
Y EPA CAG
Y EPA RCRA
N EPA CLEAN AIR
N EPA CLEAN WATER

REPT'D USE : CYTOSTATIC AGENT FOR DANDRUFF TREATMENT, GLASS MANUFACTURING, PHOTOELECTRIC CELLS AND SHAMPOO

CAS NO : **57-56-7**

CHEM NAME : ***** SEMICARBAZIDE**

REPORT 01 : DANGEROUS PROPERTIES OF INDUSTRIAL MATERIALS REPORTED TO BE MUTAGENIC; AN EQUIVOCAL TUMORIGENIC AGENT IN EXPERIMENTAL TESTS; TO AFFECT HUMAN CENTRAL NERVOUS SYSTEM. CONSIDERED SEVERELY TOXIC BY SUBCUTANEOUS, INTRAPERITONEAL, ORAL, AND INTRAVENOUS ROUTES.

REPORT 06 : MULTI-AGENCY CROSS CHECK
N HHS MAGUIRE
Y OSHA CARC
N EPA CAG
N EPA RCRA
N EPA CLEAN AIR

N EPA CLEAN WATER
REPT'D USE : LABORATORY REAGENT

CAS NO : **563-41-7**
CHEM NAME : ***** SEMICARBAZIDE HYDROCHLORIDE**

REPORT 01 : IARC MONOGRAPH 12 1976
SEMICARBAZIDE HYDROCHLORIDE IS CARCINOGENIC IN MICE AFTER ITS ORAL ADMINISTRATION: IT PRODUCED ANGIOMAS, ANGIOSARCOMAS AND LUNG TUMOR.

REPORT 02 : NTP TECHNICAL BULLTIN 2 APR 80
THE COMPOUND WAS FOUND TO BE NEGATIVE IN THE SALMONELLA TYPHIMURIUM MUTAGENICITY ASSAY IN ONE LABORATORY.

REPORT 06 : MULTI-AGENCY CROSS CHECK
N HHS MAGUIRE
Y OSHA CARC
N EPA CAG
N EPA RCRA
N EPA CLEAN AIR
N EPA CLEAN WATER
REPT'D USE : CHEMICAL INDICATOR AND LABORATORY AGENT

CAS NO : **480-81-9**
CHEM NAME : ***** SENECIPHYLLINE**

REPORT 01 : IARC MONOGRAPH 10 1976
NO DATA ON THE CARCINOGENICITY OF PURE SENECIPHYLLINE WERE AVAILABLE.
REPT'D USE : MEDICINAL HERB EXTRACT

CAS NO : **2318-18-5**
CHEM NAME : ***** SENKIRKINE**

REPORT 01 : IARC MONOGRAPH 31 JUL 83
IN TESTS IN MALE RATS BY INTRAPERITONEAL ADMINISTRATION SIGNIFICANT INCREASED INCIDENCE OF LIVER-CELL ADENOMAS. IN RATS FED HERB TUSSILAGO FARFARA L., WHICH HAS BEEN SHOWN TO CONTAIN SENKIRKINE AS ONLY PYRROLIZIDINE ALKALOID, AN INCREASED INCIDENCE OF LIVER HAEMANGIOENDOTHELIAL SARCOMAS OBSERVED.

REPORT 02 : IARC MONOGRAPH 31 JUL 83
SENKIRKINE IS MUTAGENIC IN BACTERIA AND IN MAMMALIAN CELLS IN VITRO. INDUCED CHROMOSOMAL ABERRATIONS AND UNSCHEDULED DNA SYNTHESIS IN MAMMALIAN CELLS IN VITRO.

REPORT 03 : IARC MONOGRAPH 10 1976
NO CARCINOGENICITY STUDIES IN PURE SENKIRKINE WERE AVAILABLE.
REPT'D USE : FOODS, HERBAL REMEDIES AND TOILETRIES

CAS NO : **18307-23-8**
CHEM NAME : ***** SEPIOLITE**

REPORT 01 : IARC MONOGRAPH 42 1987
INADEQUATE EVIDENCE FOR CARCINOGENICITY OF SEPIOLITE TO EXPERIMENTAL ANIMALS. PULMONARY OPACITIES REPORTED IN ONE GROUP OF SEPIOLITE WORKERS. CLINICAL AND RADIOLOGICAL EVIDENCE OF PULMONARY FIBROSIS (SMALL IRREGULAR OPACITIES), SEEN IN 10 OF 63 SEPIOLITE TRIMMERS. RADIOLOGICAL EXAMINATON OF INHABITANTS OF 4 NEARBY VILLAGES WHERE CPD MINED >100 YEARS, SHOWED NO SIGNS OF PLEURAL DISEASE. CPD DID NOT INDUCE UNSCHEDULED DNA SYNTHESIS IN PRIMARY CULTURES OF RAT HEPATOCYTES.
REPT'D USE : PET LITTER ABSORBENT, ANIMAL FEEDS, PESTICIDE, OIL BLEACHING AGENT, DRILLING MUD, COSMETICS, DETERGENTS, PHARMACEUTICALS AND CIGARETTE FILTERS

CAS NO : **68308-34-9**
CHEM NAME : ***** SHALE-OILS**

REPORT 01 : IARC MONOGRAPH 35 JAN 85
THERE IS SUFFICIENT EVIDENCE FOR CARCINOGENICITY IN EXPERIMENTAL ANIMALS OF HIGH-TEMPERATURE AND LOW-TEMPERATURE CRUDE SHALE-OILS, FRACTIONS OF HIGH-TEMPERATURE SHALE-OIL, CRUDE SHALE-OIL DISTILLATION FRACTIONS, SHALE-OIL BITUMENS AND COMMERCIAL BLENDS OF SHALE-OILS; AND LIMITED EVIDENCE FOR CARCINOGENICITY IN ANIMALS OF RAW OIL SHALE, SPENT OIL SHALE AND RESIDUE OF SHALE-OIL DISTILLATION. THERE IS SUFFICIENT EVIDENCE THAT SHALE-OILS ARE CARCINOGENIC IN HUMANS.

REPORT 02 : IARC MONOGRAPH 35 JAN 85
PREPARATIONS OF SHALE-DERIVED CRUDE OILS FROM VARIOUS SOURCES AND RETORT PROCESSES MUTAGENIC IN BACTERIA, YEAST AND CULTURED MAMMALIAN CELLS FOLLOWING METABOLIC OR PHOTO-INDUCED ACTIVATION. THREE CRUDE SHALE-OIL PREPARATIONS DID NOT INDUCE MITOTIC GENE CONVERSION IN YEAST; TWO OTHERS INDUCED SISTER CHROMATID EXCHANGES IN CULTURED MAMMALIAN CELLS. POSITIVE AND NEGATIVE RESULTS OBTAINED IN MAMMALIAN IN-VIVO ASSAYS FOR CHROMOSOMAL EFFECTS.

REPORT 03 : IARC MONOGRAPH 35 JAN 85
OIL-SHALE RETORT PROCESS-WATERS ELICITED DNA DAMAGE AND MUTATIONS IN BACTERIA AND IN CULTURED MAMMALIAN CELLS FOLLOWING METABOLIC ACTIVATION OR PHOTOACTIVATION. THEY INDUCED CHROMOSOMAL ABERRATIONS IN CULTURED MAMMALIAN CELLS; AND INDUCED CHROMOSOMAL ABERRATIONS BUT NOT SISTER CHROMATID EXCHANGES IN MOUSE CELLS IN VIVO. EXTRACTS OF OIL-SHALE ASH MUTAGENIC IN BACTERIA BOTH IN REVERSION AND FOR FORWARD MUTATION ASSAYS IN ABSENCE OF METABOLIC SYSTEM. TESTS WITH EUKARYOTIC SYSTEMS GAVE NEGATIVE RESULTS.

REPORT 04 : IARC SUPPLMENT 1 SEP 79
SOOTS, COAL-TARS, CREOSOTE OILS, SHALE OILS, AND CUTTING OILS ARE CARCINOGENIC IN EXPERIMENTAL ANIMALS AFTER SKIN PAINTING OR SUBCUTANEOUS INJECITONS. OCCUPATIONAL EXPOSURE TO THESE CHEMICALS CAUSES CANCER AT SEVERAL SITES, INCLUDING SKIN, LUNG, BLADDER AND GASTROINTESITINAL TRACT. EFFECT MAY BE DUE TO POLYCYCLIC AROMATIC HYDROCARBONS IN THESE MATERIALS.

REPORT 05 : EPA TSCA 8(E) 8EHQ-0184-0504 FEB 84
LURGI LIGHT/MIDDLE SHALE OIL ADMINISTERED TO RATS VIA INHALATION: 57% MORTALITY AT AEROSOL CONCENTRATION OF 490 MG/M3, NO MORTALITY AT 56 OR 120 MG/M3; DOSE-RELATED DECREASES IN BODY WEIGHT GAIN, CHANGES IN VARIOUS ORGAN WEIGHTS, OCULAR AND NASAL DISCHARGES AND PERIANAL SOIL ING; DECREASES IN TOTAL ERYTHROCYTES, LEUKOCYTES AND PLATELET COUNTS AT HIGH DOSE LEVEL (490 MG/M3); ATROPHY OF LYMPH NODES AND SPLENIC PULP; EPITHELIAN NECROSIS OF KIDNEY TUBULES; ATROPHY OF THYROID FOLLICLES AND TESTICULAR EPITHELIUM; NECROSIS OF LIVER PARENCHYMA AND DERMATITIS.

REPT'D USE : LAMP OIL, GASOLINE, DIESEL, OILS, LUBRICANTS AND PARAFIN WAX PRODUCTION, WOOD PRESERVATIVES, NEROZIN, ELECTRODE COKE, MASTICS AND TANNING AGENTS

CASE NO : **7631-86-9**
CHEM NAME : ***** SILICA**

REPORT 01 : IARC MONOGRAPH 42 1987
SUFFICIENT EVIDENCE FOR CARCINOGENICITY OF CRYSTALLINE SILICA TO EXPERIMENTAL ANIMALS; INADEQUATE EVIDENCE FOR CARCINOGENICITY OF AMORPHOUS SILICA TO EXPERIMENTAL ANIMALS; LIMITED EVIDENCE FOR CARCINOGENICITY OF CRYSTALLINE SILICA TO HUMANS; INADEQUATE EVIDENCE FOR CARCINOGENICITY OF AMORPHOUS SILICA TO HUMANS. AMORPHOUS SILICA TESTED BY ORAL ADMINISTRATION TO RATS, INHALATION IN MICE, INTRATRACHEAL ADMINISTRATION TO

HAMSTERS, SUBCUTANEOUS AND INTRAPERITONEAL INJECTION TO MICE AND INTRAPLEURAL ADMINISTRATION TO RATS. MOST TESTS RESULTS GAVE NEGATIVE OR INADEQUATE RESULTS. SLIGHT INCREASE IN TWO TYPES TUMORS SEEN IN INTRAPERITONEAL AND INHALATION.

REPORT 02 : IARC MONOGRAPH 42 1987
DIFFERENT SAMPLES WITH PARTICLES IN RESPIRABLE RANGE TESTED BY INHALATION AND INTRATRACHEAL INSTILLATION IN RATS. SIGNIFICANT INCREASES IN INCIDENCE OF ADENOSARCOMAS AND SQUAMOUS CELL CA'S OF LUNGS NOTED. NO PULMONARY TUMOR SEEN IN HAMSTERS TREATED WITH INTRATRACHEAL INSTILLATION OF QUARTZ DUST. INTRAPLEURAL AND INTRAPERITONEAL INJECTIONS OF RATS WITH QUARTZ RESULTED IN MALIGNANT TUMORS OF THORAX AND ABDOMEN. INTRAPLEURAL INJECTION OF RESPIRABLE TRIDYMITE AND CRYSTOBALITE LEAD TO MALIGNANT LYMPHOMAS. NO TUMORIGENIC RESPONSE SEEN WITH QUARTZ IN STRAIN A MOUSE LUNG ADENOMA ASSAY.

REPORT 03 : IARC MONOGRAPH 42 1987
SILICA NOT MUTAGENIC TO S. TYPHIMURIUM OR E. COLI. QUARTZ INDUCED MICRONUCLEI BUT NOT SISTER CHROMATID EXCHANGES IN MAMMALIAN CELLS IN VITRO. TWO SAMPLES OF QUARTZ INDUCED TRANSFORMATION IN SYRIAN HAMSTER EMBRYO CELLS IN CULTURE. QUARTZ DID NOT INDUCE MICRONUCLEI IN MICE IN VIVO.

REPORT 06 : MULTI-AGENCY CROSS CHECK
Y HHS MAGUIRE
Y OSHA CARC
N EPA CAG
N EPA RCRA
N EPA CLEAN AIR
N EPA CLEAN WATER

REPT'D USE : ELECTRICAL AND OPTICAL COMPONENTS, FOUNDRY CASTINGS, GLASS, CERAMICS, SANDPAPER AND SAND BLASTING, GRINDING AND POLISHING, BRICKS AND TILES, FILTERS, JEWELRY, FILLER FOR PAINT, PAPER AND SCOURING POWDER, GRANITE INDUSTRY, DRY WALL AND PATCHING AND SPACKLING CPDS

CAS NO : **7631-89-2**
CHEM NAME : ***** SODIUM ARSENATE**

REPORT 01 : IARC MONOGRAPH 23 1980
THERE IS INADEQUATE EVIDENCE FOR THE CARCINOGENICITY OF ARSENIC COMPOUNDS IN ANIMALS. THERE IS SUFFICIENT EVIDENCE THAT INORGANIC ARSENIC COMPOUNDS ARE SKIN AND LUNG CARCINOGENS IN HUMANS.

REPORT 02 : NIOSH CRITERIA DOCUMENT JUN 75
NIOSH RECOMMENDS A CEILING OF 2 MICROGRAMS/CU M FOR INORGANIC ARSENIC COMPOUNDS BASED ON DERMATITIS, LUNG, AND LYMPHATIC CANCER. THE CURRENT OSHA STANDARD FOR INORGANIC ARSENIC IS 10 MICROGRAMS/CU M TWA. SEE CFR 1910.1018.

REPORT 03 : IARC MONOGRAPH 23 1980
SODIUM ARSENATE HAS EMBRYOLETHAL EFFECTS AND A TERATOGENIC POTENTIAL IN SEVERAL MAMMALIAN SPECIES. THE EVIDENCE THAT ARSENIC COMPOUNDS CAUSE MUTATIONS AND ALLIED EFFECTS IN BACTERIA IS INCONCLUSIVE. HOWEVER ARSENIC COMPOUNDS INDUCE CHROMOSOMAL ABERRATIONS AND MORPHOLOGICAL TRANSFORMATION IN MAMMALIAN CELLS.

REPORT 06 : MULTI-AGENCY CROSS CHECK
Y HHS MAGUIRE
N OSHA CARC
Y EPA CAG
N EPA RCRA
N EPA CLEAN AIR
N EPA CLEAN WATER

REPT'D USE : WOOD PRESERVATIVES, INSECTICIDE AND TEXTILE MORDANT

CAS NO : **7784-46-5**
CHEM NAME : ***** SODIUM ARSENITE**

REPORT 01 : IARC MONOGRAPH 23 1980
THERE IS INADEQUATE EVIDENCE FOR THE CARCINOGENICITY OF ARSENIC COMPOUNDS IN ANIMALS. THERE IS SUFFICIENT EVIDENCE THAT INORGANIC ARSENIC COMPOUNDS ARE SKIN AND LUNG CARCINOGENS IN HUMANS.

REPORT 02 : NIOSH CRITERIA DOCUMENT JUN 75
NIOSH RECOMMENDS A CEILING OF 2 MICROGRAMS/CU M FOR INORGANIC ARSENIC COMPOUNDS BASED ON DERMATITIS, LUNG, AND LYMPHATIC CANCER. THE CURRENT OSHA STANDARD FOR INORGANIC ARSENIC IS 10 MICROGRAMS/CU M TWA. SEE CFR 1910.1018.

REPORT 03 : IARC MONOGRAPH 23 1980
SODIUM ARSENITE HAS EMBRYOLETHAL EFFECTS AND A TERATOGENIC POTENTIAL IN SEVERAL MAMMALIAN SPECIES. THE EVIDENCE THAT ARSENIC COMPOUNDS CAUSE MUTATIONS AND ALLIED EFFECTS IN BACTERIA IS INCONCLUSIVE. HOWEVER ARSENIC COMPOUNDS INDUCE CHROMOSOMAL ABERRATIONS AND MORPHOLOGICAL TRANSFORMATION IN MAMMALIAN CELLS.

REPORT 06 : MULTI-AGENCY CROSS CHECK
Y HHS MAGUIRE
N OSHA CARC
Y EPA CAG
N EPA RCRA
N EPA CLEAN AIR
N EPA CLEAN WATER

REPT'D USE : HERBICIDE, PESTICIDE, CORROSION INHIBITOR, CHEMICAL INTERMEDIATE AND SOIL STERILANT

CAS NO : **26628-22-8**
CHEM NAME : ***** SODIUM AZIDE**

REPORT 01 : NTP TECHNICAL BULLETIN 3 DEC 80
THE COMPOUND WAS FOUND TO BE POSITIVE IN THE SALMONELLA TYPHIMURIUM MUTAGENICITY ASSAY IN ONE LABORATORY.

REPORT 02 : NTP TR 389 NIH PUB. 90-2844 MAR 90
UNDER 2-YEAR STUDIES NO EVIDENCE OF CARCINOGENIC ACTIVITY OF SODIUM AZIDE FOR MALE OR FEMALE F344/N RATS ADMINISTERED 5 OR 10 MG/KG SODIUM AZIDE IN DISTILLED WATER BY GAVAGE FOR UP TO 2 YEARS. SODIUM AZIDE INDUCED NECROSIS IN CEREBRUM AND THALAMUS OF BRAIN IN BOTH MALE AND FEMALE RATS.

REPORT 03 : EPA TSCA 8(E) 8EHQ-1083-0494 OCT 83
PPG REPORT ON SUBCHRONIC ORAL FEEDING STUDY IN DOGS USING SODIUM AZIDE 1, 3 AND 10 MG/KG/DAY. IN 27TH AND LAST WEEK OF STUDY, ATAXIA OBSERVED IN TWO HIGH-DOSE FEMALES AND ONE MID-DOSE MALE. COMPOUND-RELATED HISTOMORPHOLOGIC CHANGES FOUND IN THE TWO FEMALES. LESIONS IN ANTERIOR AND MIDDLE CEREBRAL SECTIONS ALSO REPORTED: DECREASED ALKALINE PHOSPHATASE, DECREASED LDH LEVELS, AND DECREASED SERUM GLUTAMIC OXALOCETIC TRANSAMINASE LEVELS.

REPT'D USE : CHEMICAL INTERMEDIATE, ANTIHYPERTENSIVE AGENT, HERBICIDE, HEMATOCIDE, AND INSECTICIDE

CAS NO : **124-65-2**
CHEM NAME : ***** SODIUM CACODYLATE**

REPORT 01 : IARC MONOGRAPH 23 1980
THERE IS INADEQUATE EVIDENCE FOR THE CARCINOGENICITY OF ARSENIC COMPOUNDS IN ANIMALS. THERE IS SUFFICIENT EVIDENCE THAT INORGANIC ARSENIC COMPOUNDS ARE SKIN AND LUNG CARCINOGENS IN HUMANS.

REPORT 02 : NIOSH CRITERIA DOCUMENT JUN 75
NIOSH RECOMMENDS A CEILING OF 2 MICROGRAMS/CU M FOR INORGANIC ARSENIC COMPOUNDS BASED ON DERMATITIS, LUNG, AND LYMPHATIC CANCER. THE CURRENT OSHA STANDARD FOR INORGANIC ARSENIC IS 10 MICROGRAMS/CU M TWA. SEE CFR 1910.1018.

REPORT 06 : MULTI-AGENCY CROSS CHECK
Y HHS MAGUIRE
N OSHA CARC
Y EPA CAG
N EPA RCRA
N EPA CLEAN AIR
N EPA CLEAN WATER

REPT'D USE : HERBICIDE, VETERINARY MEDICINE AND CAUSTICS/DRAIN CLEANERS

CAS NO : **7775-11-3**

CHEM NAME : ***** SODIUM CHROMATE**

REPORT 01 : IARC MONOGRAPH 23 1980
AN EXPERIMENT IN RATS EXPOSED TO SODIUM CHROMATE (VI) BY BRONCHIAL IMPLANTATION WAS INADEQUATE TO EVALUATE CARCINOGENICITY. CHROMIUM (VI) COMPOUNDS CAUSE MUTATIONS AND ALLIED EFFECTS IN A VERY WIDE RANGE OF PROKARYOTIC AND EUKARYOTIC SYSTEMS BOTH IN-VITRO AND IN-VIVO.

REPT'D USE : CHEMICAL INTERMEDIATE, INKS, PAINTS/VARNISH, LEATHER TANNING AGENT, TEXTILE PRINTING AND CORROSION INHIBITOR

CAS NO : **139-05-9**

CHEM NAME : ***** SODIUM CYCLAMATE**

REPORT 01 : IARC MONOGRAPH 22 1980
SODIUM CYCLAMATE HAS BEEN TESTED BY ORAL ROUTE IN 2 EXPERIMENTS IN MICE (ONE WAS MULTIGENERATIONAL) AND IN 3 EXPERIMENTS IN RATS. A FEW BENIGN AND MALIGNANT TUMORS WERE OBSERVED IN RATS: INCIDENCES WERE NOT STATISTICALLY GREATER THAN CONTROLS IN ANY SINGLE STUDY. AN INCREASED INCIDENCE OF LYMPHO-SARCOMAS WAS SEEN IN FEMALE MICE IN ONE EXPERIMENT. IT HAS ALSO BEEN TESTED IN MICE BY BLADDER INSERTION (IMPLANTATION) IN ONE EXPERIMENT: IT INCREASED THE INCIDENCE OF BLADDER CARCINOMAS.

REPORT 02 : IARC MONOGRAPH 22 1980
THE LIMITED NUMBER OF MUTAGENICITY STUDIES PUBLISHED GIVE NO EVIDENCE THAT CYCLAMATES CAUSE POINT MUTATIONS.

CYCLAMATES DO CAUSE CHROMOSOME DAMAGE. THERE IS NO EVIDENCE THAT CYCLAMATES ARE TERATOGENIC.

REPORT 03 : NAS REPORT ON DIET, NUTRITION AND CANCER 1982
THE EXPERIMENTAL DATA PROVIDE LIMITED EVIDENCE FOR THE CARCINOGENICITY OF CYCLAMATES IN MICE AND RATS. IN ADDITION THERE IS EVIDENCE THAT CYCLAMATES CAN PROMOTE THE ACTION OF KNOWN CARCINOGENS OF THE BLADDER. THERE ARE NO ADEQUATE EPA DATA ON THE EFFECT OF CYCLAMATE ALONE SINCE IT IS RARELY CONSUMED BY HUMANS IN THE ABSENCE OF SACCHARIN.

REPT'D USE : NON-NUTRITIVE SWEETENER

CAS NO : **10588-01-9**

CHEM NAME : ***** SODIUM DICHROMATE (VI)**

REPORT 01 : IARC MONOGRAPH 23 1980
EXPERIMENTS IN RATS EXPOSED TO SODIUM DICHROMATE (VI) BY I.M. AND INTRAPLEURAL INJECTION AND I.M., INTRAPLEURAL, AND BRONCHIAL IMPLANTATION WERE INADEQUATE TO EVALUATE CARCINOGENICITY. ONE EPIDEMIOLOGICAL STUDY THAT THE HIGHEST RISK OF LUNG CANCER OCCURS AMONG WORKERS IN AREAS WHERE DICHROMATE (VI) IS PROCESSED.

REPORT 02 : EPA TSCA 8(E) NOTICE 8EHQ-1182-0463
DAMOND SHAMROCK CORP PROVIDE A TRANSLATION OF A GERMAN SUMMARY OF A 30-MONTH STUDY OF SODIUM BICHROMATE AND CALCIUM CHROMATE ADMINISTERED AT VARYING DOSES VIA INTERTRACHEAL INSTILLATION TO S-D RATS. WITH REGARD TO CARCINOGENICITY, THE SUMMARY STATED THAT "A CARCINOGENIC EFFECT WAS ONLY PRODUCED BY BOTH CHROMATES IN THE LUNG" AND THAT "THERE WAS NO DIFFERENCE IN INTENSITY OF EFFECT BETWEEN THE SODIUM DICHROMATE AND THE DISSOLVED CALCIUM CHROMATE."

REPORT 03 : IARC MONOGRAPH 23 1980
CHROMIUM (VI) COMPOUNDS CAUSE MUTATIONS AND ALLIED EFFECTS IN A VERY WIDE RANGE OF PROKARYOTIC AND EUKARYOTIC SYSTEMS BOTH IN VITRO AND IN VIVO.

REPT'D USE : MFR OF CHROMIC ACID, PIGMENTS, CORROSION INHIBITORS, CLEANING AGENTS PAINTS/VARNISHES AEROSOLS AND RUST REMOVERS

CAS NO : **148-18-5**

CHEM NAME : ***** SODIUM DIETHYLDITHIOCARBAMATE**

REPORT 01 : NCI CARC TRS 172 DHEW PUB NO NIH 79-1728
UNDER THE CONDITIONS OF THIS BIOASSAY, SODIUM

DIETHYLDITHIOCARBAMATE WAS NOT CARCINOGENIC FOR F344 RATS OR B6C3F1 MICE (FEEDING STUDY).

REPORT 02 : IARC MONOGRAPH 12 1976
SODIUM DIETHYLDITHIOCARBAMATE HAS BEEN TESTED BY ORAL ADMINISTRATION AND BY SINGLE SUBCUTANEOUS INJECTION IN TWO STRAINS OF MICE. ALTHOUGH IT PRODUCED AN INCREASE IN THE INCIDENCE OF LIVER CELL TUMORS IN MALES OF ONE STRAIN AND A SLIGHT INCREASE OF LUNG TUMORS IN MALES OF THE OTHER STRAIN FOLLOWING ITS ORAL ADMINISTRATION, THE AVAILABLE DATA ARE INSUFFICIENT TO MAKE AN EVALUATION OF THE CARCINOGENICITY OF THIS COMPOUND.

REPORT 03 : NTP TECHNICAL BULLETIN 7 APR 82
SODIUM DIETHYLDITHIOCARBARMATE WAS NEGATIVE IN THE SALMONELLA TYPHIMURIUM MUTAGENICITY ASSAY.

REPORT 04 : IARC MONOGRAPH 12 1976
SODIUM DIETHYLDITHIOCARBAMATE PRODUCED A SMALL BUT SIGNIFICANT INCREASE IN THE NUMBER OF CHROMOSOME BREAKAGES AND ABERRATIONS IN V. FABA.
REPT'D USE : LATEX RUBBER ACCELERATOR

CAS NO : **7681-49-4**
CHEM NAME : ***** SODIUM FLUORIDE**

REPORT 01 : IARC MONOGRAPH 27 APR 82
SODIUM FLUORIDE WAS TESTED IN THREE EXPERIMENTS IN THREE DIFFERENT STRAINS OF MICE BY ORAL ADMINISTRATION. THE AVAILABLE DATA ARE INSUFFICIENT FOR EVALUATION.

REPORT 02 : IARC MONOGRAPH 27 APR 82
SODIUM FLUORIDE WAS NOT MUTAGENIC TO SALMONELLA OR DROSOPHILA AND DID NOT INDUCE GENE CONVERSION IN SACCHAROMYCES.

REPORT 03 : NTP CARC TEST PROGRAM JUL 82
IN A CURRENT NTP BIOASSAY SODIUM FLUORIDE IS BEING ADMINSITERED IN THE WATER OF F344 RATS AND B6C3F1 MICE. THE CHRONIC EXPOSURE PHASE IS IN PROGRESS.
REPT'D USE : FLUORIDATION AGENT, GLASS FORSTING AGENT, GLUES/ADHESIVES, TOOTHPASTES, PEDULICIDE, ACARICIDE, VITREOUS ENAMELS AND ANTIHELMINTIC

CAS NO : **1681-52-9**
CHEM NAME : *****SODIUM HYPOCHLORITE**

REPORT 01 : NTP DRAFT TR 392 NIH PUB. 91-2847 NOV 90
UNDER 2-YEAR STUDIES NO EVIDENCE OF CARCINOGENIC ACTIVITY OF CHLORINATED DRINKING WATER FOR MALE F344/N RATS. EQUIVOCAL EVIDENCE OF CARCINOGENIC ACTIVITY OF CHLORINATED DRINKING WATER FOR FEMALE F344/N RATS, INDICATED BY INCREASE IN INCIDENCE OF MONONUCLEAR CELL LEUKEMIA. NO EVIDENCE OF CARCINOGENIC ACTIVITY OF CHLORINATED DRINKING WATER FOR MALE OR FEMALE B6C341 MICE.
REPT'D USE : DISINFECTANT, BLEACHING, INTERMEDIATE, ORGANIC CHEMICALS, WATER PURIFICATION MEDICINE, FUNGICIDES AND GERMICIDE

CAS NO : **137-20-2**
CHEM NAME : ***** SODIUM N-METHYL-N-OLEOYLTAURINE**

REPORT 01 : DECISION NOT TO TEST, 50FR4617 8 NOV 85
EPA'S RESPONSE TO ITC'S RECOMMENDATION THAT EPA CONSIDER HEALTH EFFECTS TESTING OF SMOT UNDER TSCA 4(A). EPA NOT INITIATING RULEMAKING NOW BECAUSE LOW EXPOSURE AND ONLY SMALL AMOUNTS RELEASED INTO THE ATMOSPHERE. EXISTING INFORMATION DOES NOT SUGGEST POTENTIAL UNREASONABLE RISK OF EXPOSURE.
REPT'D USE : WASHING TEXTILES BEFORE DYEING, PHOTOGRAPHIC PAPER PRODUCTION AND PESTICIDE FORMULATIONS

CAS NO : **8030-30-6**
CHEM NAME : ***** SOLVENT NAPHTHA (HEAVY AROMATIC SOLVENT NAPHTHA)**

CAS NO : **CLASS-05**
CHEM NAME : ***** SOOT**

REPORT 01 : IARC SUPPLEMENT 1 SEP 79
SOOTS, COAL-TARS, CREOSOTE OILS, SHALE OILS, AND CUTTING OILS ARE CARCINOGENIC IN EXPERIMENTAL ANIMALS AFTER SKIN PAINTING OR SUBCUTANEOUS INJECTIONS. OCCUPATIONAL EXPOSURE TO THESE CHEMICALS CAUSES CANCER AT SEVERAL SITES, INCLUDING SKIN, LUNG, BLADDER AND GASTROINTESTINAL TRACT. EFFECT MAY BE DUE TO POLYCYCLIC AROMATIC HYDROCARBONS IN THESE MATERIALS.

REPORT 02 : IARC MONOGRAPH 35 JAN 85
THERE IS SUFFICIENT EVIDENCE FOR CARCINOGENICITY OF COAL-SOOT EXTRACT AND OF OIL-SHALE SOOT EXTRACT IN ANI-

MALS, BUT LIMITED EVIDENCE FOR CARCINOGENICITY OF IMPLANTED WOOD SOOT AND SOOT FROM HEATING OIL; AND INADEQUATE EVIDENCE OF COAL SOOT, EXTRACT OF WOOD SOOT AND OF EXTRACT OF FUEL-OIL SOOT. THERE IS SUFFICIENT EVIDENCE THAT SOOT IS CARCINOGENIC TO HUMANS. THE CARCINOGENICITY IS DEMONSTRATED BY NUMEROUS REPORTS DATING BACK OVER 200 YEARS, OF SKIN CANCER, PARTICULARLY OF THE SCROTUM, AMONG CHIMNEY-SWEEPS.

REPORT 03 : IARC MONOGRAPH 35 JAN 85
EXTRACTS OF SOOT SAMPLES FROM DOMESTIC SOURCES WERE MUTAGENIC IN SALMONELLA TYPHIMURIUM. EXTRACTS OF AN EXPERIMENTALLY DERIVED SOOT WERE MUTAGENIC IN FORWARD MUTATION ASSAYS IN S. TYPHIMURIUM AND IN CULTURED HUMAN LYMPHOBLASTS.

REPORT 06 : MULTI-AGENCY CROSS CHECK
Y HHS MAGUIRE
N OSHA CARC
Y EPA CAG
N EPA RCRA
N EPA CLEAN AIR
N EPA CLEAN WATER

REPT'D USE : COMBUSTION BY-PRODUCT AND FERTILIZER

CAS NO : **52-01-7**
CHEM NAME : ***** SPIRONOLACTONE**

REPORT 01 : IARC MONOGRAPH 24 1980
SPIRONOLACTONE WAS TESTED BY ORAL ADMINISTRATION IN TWO EXPERIMENTS IN RATS. AN INCREASED INCIDENCE OF THYROID AND TESTICULAR TUMORS WAS REPORTED IN ONE EXPERIMENT BUT NOT IN ANOTHER EXPERIMENT OF LONGER DURATION WITH LOWER DOSES. EPIDEMIOLOGIC STUDIES HAVE NOT CONFIRMED THE SUSPICION RAISED BY CASE REPORTS THAT SPIRONOLACTONE MAY CAUSE BREAST CANCER IN HUMANS. THE DATA ARE INSUFFICIENT, HOWEVER, TO PERMIT CONFIDENT EXCLUSION OF SUCH AN EFFECT.

REPT'D USE : ANTIHYPERTENSIVE AGENT AND TREATMENT OF HEPATIC CIRRHOSIS

CAS NO : **7772-99-8**
CHEM NAME : ***** STANNOUS CHLORIDE**

REPORT 01 : NTP CARC TRS 231 NIH PUB NO. 82-1787
UNDER THE CONDITIONS OF THIS BIOASSAY, DIETARY ADMINISTRATION OF STANNOUS CHLORIDE WAS JUDGED NOT TO BE CARCINOGENIC FOR MALE OR FEMALE F344/N RATS OR B6C3F1/N

MICE, ALTHOUGH C-CELL TUMORS OF THE THYROID GLAND IN MALE RATS MAY HAVE BEEN ASSOCIATED WITH THE ADMINISTRATION OF THE CHEMICAL.

REPT'D USE : FOOD PRESERVATIVE, COLOR/PERFUME STABILIZER, REDUCING AGENT IN TIN PLATING, TEXTILE MORDANT, SILVERING AGENT FOR GLASS, SOLDERING FLUXES, ANTI-SLUDGE ADDITIVE TO OIL AND SWIMMING POOL CHEMICALS

CAS NO : **10048-13-2**
CHEM NAME : ***** STERIGMATOCYSTIN**

REPORT 01 : IARC MONOGRAPH 10 1976
STERIGMATOCYSTIN IS CARCINOGENIC IN MICE AND RATS FOLLOWING ITS ORAL ADMINISTRATION; IT PRODUCED LUNG TUMORS IN MICE AND LIVER TUMORS IN RATS. IN RATS, IT ALSO PRODUCED SKIN AND LIVER TUMORS FOLLOWING ITS APPLICATION TO THE SKIN AND SARCOMAS AT THE SITE OF ITS SUBCUTANEOUS INJECTION. IN A 1979 SUPPLEMENT, IARC DETERMINED THAT THERE IS SUFFICIENT EVIDENCE FOR THE CARCINOGENICITY OF STERIGMATOCYSTIN IN ANIMALS.

REPORT 02 : IARC MONOGRAPH 10 1976
STERIGMATOCYSTIN INDUCED FRAMESHIFT MUTATIONS IN SALMONELLA.
REPT'D USE : NATURAL PRODUCT OF SELECTED FUNGI

CAS NO : **8052-41-3**
CHEM NAME : ***** STODDARD SOLVENT**

REPORT 01 : EPA TSCA 8(E) 8EHQ-0584-0517 JUN 84
SOLVENT-CUT BACK RUST PREVENTIVE (STODDARD SOLVENT - 90.9 VOL%/2-BUTO XYETHANOL - 2.5 VOL%/CALCIUM PETROLEUM SULFONATE - 6.6 VOL%) CARCINOGENIC IN A SKIN PAINTING STUDY USING C3H/HEJ MALE MICE. CPD INDUCED SQUAMOUS CELL CARCINOMAS IN 12% OF MICE TESTED. NIOSH RECOMMENDED STANDARD TWA 350 MG/CUBIC METER FOR PETROLEUM SOLVENTS (7/77); SEE 68477-31-6.
REPT'D USE : SOLVENT

CAS NO : **3810-74-0**
CHEM NAME : ***** STREPTOMYCIN SULFATE (2:3)**

REPORT 01 : NTP TECHNICAL BULLETIN 6/8 JAN 82, JUL 82
STREPTOMYCIN SULFATE WAS POSITIVE IN THE SALMONELLA ASSAY, BUT NEGATIVE IN DROSOPHILA WHEN TESTED FOR SEX-LINKED RECESSIVE LETHALS.
REPT'D USE : NO AVAILABLE PRODUCTION OR USE DATA

CAS NO : **18883-66-4**
CHEM NAME : ***** STREPTOZOTOCIN**

REPORT 01 : IARC MONOGRAPH 17 1978
STREPTOZOTOCIN IS CARCINOGENIC IN MICE, RATS AND SYRIAN GOLDEN AND CHINESE HAMSTERS FOLLOWING ITS INTRAVENOUS OR INTRAPERITONEAL ADMINISTRATION. IT PRODUCES BENIGN AND MALIGNANT TUMORS OF THE LIVER AND KIDNEY AND ISLET-CELL TUMORS OF THE PANCREAS. IT IS CARCINOGENIC AFTER ITS ADMINISTRATION IN SINGLE DOSES. IN A 1979 SUPPLEMENT, IARC DETERMINED THAT THERE IS SUFFICIENT EVIDENCE FOR THE CARCINOGENICITY OF STREPTOZOTOCIN IN ANIMALS.

REPORT 02 : NCI BIOASSAY PROGRAM CANCER VOL. 40 1977
UNDER THE CONDITIONS OF THE BIOASSAY, I.P. INJECTION OF STREPTOZOTOCIN WAS CARCINOGENIC TO SWISS-WEBSTER MALE AND FEMALE MICE, CAUSING LUNG AND RENAL TUMORS. IT WAS ALSO CARCINOGENIC TO MALE AND FEMALE SPRAGUE-DAWLEY (CD) RATS, CAUSING RENAL TUMORS. IN ADDITION, MALE RATS HAD PERITONEAL SARCOMAS AND FEMALE RATS HAD LIVER TUMORS.

REPORT 06 : MULTI-AGENCY CROSS CHECK
Y HHS MAGUIRE
N OSHA CARC
Y EPA CAG
N EPA RCRA
N EPA CLEAN AIR
N EPA CLEAN WATER
REPT'D USE : ANTIMICROBIAL AGENT AND ANTINEOPLASTIC AGENT

CAS NO : **100-42-5**
CHEM NAME : ***** STYRENE**

REPORT 01 : IARC MONOGRAPH 19 1979
STYRENE WAS TESTED BY ORAL ADMINISTRATION TO MOTHERS AND OFFSPRING OF TWO STRAINS OF MICE AND ONE STRAIN OF RATS. HIGH DOSES OF STYRENE INCREASED THE INCIDENCE OF LUNG TUMORS IN OFFSPRING OF ONE STRAIN OF MICE. A LOWER DOSE OF STYRENE SLIGHTLY INCREASED THE INCIDENCE OF LIVER CELL TUMORS IN MALE OFFSPRING OF THE OTHER STRAIN. IN RATS, THE TOTAL INCIDENCE OF TUMORS WAS NOT STATISTICALLY DIFFERENT IN STYRENE TREATED AND CONTROL ANIMALS. THE FINDING OF CHROMOSOMAL ABERRATIONS IN WORKERS INDICATES A NEED FOR EPIDEMIOLOGICAL INVESTIGATIONS.

REPORT 02 : NCI CARC TRS 170 DHEW PUB NO NIH 79-1726
UNDER THE CONDITIONS OF THIS BIOASSAY, THERE WAS NO CONVINCING EVIDENCE FOR THE CARCINOGENICITY OF A SOLU-

TION OF B-NITROSTYRENE AND STYRENE IN FISCHER 344 RATS OF IN B6C3F1 MICE. (THE COMPOUND WAS ADMINISTERED BY GAVAGE.)

REPORT 03 : NCI CARC TRS 185 DHEW PUB NO 79-1741
THE FINDINGS OF AN INCREASED INCIDENCE OF A COMBINATION OF ADENOMAS AND CARCINOMAS OF THE LUNG PROVIDED SUGGESTIVE EVIDENCE FOR THE CARCINOGENICITY OF STYRENE IN MALE B6C3F1 MICE. HOWEVER, IT IS CONCLUDED THAT, UNDER THE CONDITIONS OF THIS BIOASSAY, NO CONVINCING EVIDENCE FOR THE CARCINOGENICITY OF THE COMPOUND WAS OBTAINED IN FISCHER 344 RATS OR B6C3F1 MICE OF EITHER SEX (THE COMPOUND WAS ADMINISTERED BY GAVAGE).

REPORT 04 : CMA 2-YEAR AND REPRODUCTION STUDY AUG 80
IT IS CONCLUDED THAT INGESTION OF 125 PPM AND 250 PPM STYRENE IN THE DRINKING WATER UNDER THE CONDITIONS OF THIS STUDY PRODUCED NO DELETERIOUS DOSE-RELATED EFFECTS IN THE ALBINO RAT. THE DAILY STYRENE CONSUMPTION IS ESTIMATED TO HAVE BEEN, IN MG/KG BODY WEIGHT, 7.7 AND 14 FOR THE LOW AND HIGH DOSE MALES, AND 12 AND 21 FOR THE LOW AND HIGH DOSE FEMALES. IT IS ALSO CONCLUDED THAT NO ADVERSE EFFECTS ON REPRODUCTIVE CAPACITY IN THE RATS OF THIS STUDY WERE PRODUCED BY STYRENE ADMINISTERED IN THE DRINKING WATER.

REPORT 05 : IARC MONOGRAPH 19 1979
STYRENE INDUCED REVERSE MUTATIONS IN SALMONELLA TA1535 AND TA100 WITH METABOLIC ACTIVATION; IT WAS NOT MUTAGENIC TO TA1537, OR TA98. STYRENE DID NOT INDUCE FORWARD MUTATIONS IN SACCHAROMYCES, EVEN WITH METABOLIC ACTIVATION. AN INCREASE IN THE FREQUENCY OF CHROMOSOMAL ABERRATIONS IN BONE MARROW CELLS OF TREATED MALE MICE.

REPORT 07 : INTERIM REPORT OF SURVEY OF STYRENE WORKERS FEB 82
THE NIOSH IS CONDUCTING A MORTALITY STUDY OF WORKERS EXPOSED TO STYRENE AT TWO FACILITIES IN WASHINGTON STATE WHICH MANUFACTURE FIBERGLASS, REINFORCED PLASTIC BOATS. AS OF DEC 77, 75 PERCENT OF THE COHORT WAS ALIVE AND 3 PERCENT DECEASED. VITAL STATUS WAS UNKNOWN FOR 22 PERCENT. PRELIMINARY RESULTS INDICATE NO DIFFERENCE BETWEEN OBSERVED AND EXPECTED DEATHS DUE TO LUNG CANCER; HOWEVER THERE WERE ONLY FOUR DEATHS IN THIS CATEGORY COMPARED TO 1.7 EXPECTED. NIOSH IS CONTINUING TO ASCERTAIN VITAL STATUS FOR THE COHORT.

REPORT 08 : NIOSH CRITERIA DOCUMENT SEP 83
NIOSH RECOMMENDS A STANDARD OF 50 PPM AS A TWA CONCENTRATION TO PRESENT CNS DEPRESSION AND IRRITATION OF THE

EYES AND RESPERTORY TRACT. EXPOSURE SHOULD NOT EXCEED 100 PPM. THE OSHA STANDARD IS 100 PPM.

REPORT 09 : CMA REPORT LBI PROJECT 2612 AND 21175
STYRENE INCORPORATED IN DRINKING WATER OF RATS FOR 2 YEARS AT 250 PPM IN CONJUNCTION WITH A 3-GENERATION REPRODUCTION STUDY SHOWED NO TREATMENT-RELATED INCREASE IN TUMOR INCIDENCE OR OTHER EVIDENCE FOR CARCINOGENICITY. HISTOPATHOLOGICAL FINDINGS SHOWED NO EVIDENCE OF TOXIGENIC EFFECT ON SPRAGUE-DAWLEY RATS.

REPT'D USE : CHEMICAL INTERMEDIATE, PLASTIC MONOMER, CASTING PLASTICS AND AUTO BODY FILLERS

CAS NO : **9003-55-8**

CHEM NAME : ***** STYRENE-BUTADIENE COPOLYMERS**

REPORT 01 : IARC MONOGRAPH 19 1979
NO DATA ON THE CARCINOGENICITY OF STYRENE-BUTADIENE COPOLYMERS WERE AVAILABLE. RECENT EPIDEMIOLOGICAL INFORMATION ON STYRENE-BUTADIENE COPOLYMER WORKERS, WHICH INDICATES LYMPHATO-HEMATOPOIETIC MALIGNANCIES, CLEARLY REQUIRES ELUCIDATION BY FURTHER STUDIES.

REPORT 02 : NTP CARC TEST PROGRAM JUL 82
IN A CURRENT NTP BIOASSAY, STYRENE OXIDE WAS ADMINISTERED BY GAVAGE TO F344 RATS AND B6C3F1 MICE. THE CHRONIC EXPOSURE PHASE HAS BEEN COMPLETED AND HISTOPATHOLOGY IS IN PROGRESS.

REPT'D USE : TIRES, FOOTWEAR, CARPET BACKINGS, ADHESIVES AND MOLDED FOAM APPLICATIONS

CAS NO : **96-09-3**

CHEM NAME : ***** STYRENE OXIDE**

REPORT 01 : IARC MONOGRAPH 11 1976
STYRENE OXIDE HAS BEEN TESTED IN TWO LIMITED STUDIES IN MICE BY SKIN APPLICATION. NO SIGNIFICANT INCREASE IN THE INCIDENCE OF SKIN TUMORS WAS OBSERVED.

REPORT 02 : IARC MONOGRAPH 19 1979
STYRENE OXIDE WAS MUTAGENIC TO SALMONELLA TA1535 AND TA100 IN THE ABSENCE OF METABOLIC ACTIVATION. A DOSE-DEPENDENT INCREASE IN FORWARD MUTATIONS IN S. POMBE AND GENE CONVERSIONS IN SACCHAROMYCES HAVE BEEN NOTED.

REPORT 03 : NTP TECHNICAL BULLETIN 7 APR 82
STYRENE OXIDE IN RAT REPRODUCTIVE ASSAY SHOWED ANTIFER-

TILITY EFFECTS IN MALES WITHOUT DOMINANT LETHAL MUTATIONS, WAS POSITIVE IN A MOUSE SPERM HEAD MORPHOLOGY TEST; AND NEGATIVE IN DROSOPHILA SEX-LINKED RECESSIVE LETHAL, IN VITRO UNSCHEDULED DNA SYNTHESES IN HUMAN FIBROBLOSTS, AND IN RAT BONE MARROW CYTOLOGY TESTS.

REPORT 04 : NTP TECHNICAL BULLETIN 3 DEC 80
INHALATION EXPOSURE AT 100 PPM TO RATS FOR 7 HRS/DAY, 5 DAYS/WK FOR 3 WKS. PRIOR TO GESTATION AND DAYS 1-19 OF GESTATION, AND ON DAYS 1-24 OF GESTATION AT 15 PPM TO RABBITS INDUCED MATERNAL TOXICITY.

REPORT 05 : IARC MONOGRAPH 36 FEB 85
TESTED BY INTRAGASTRIC INTUBATION IN RATS OF ONE STRAIN AND INDUCED SQUAMOUS-CELL CARCINOMAS AND PAPILLOMAS OF FORESTOMACH. PRENATAL EXPOSURE FOLLOWED BY POSTNATAL ORAL ADMINISTRATION TO RATS OF ANOTHER STRAIN ALSO PRODUCED SQUAMOUS-CELL CARCINOMAS AND INCREASED THE INCIDENCE OF PAPILLOMAS OF THE FORESTOMACH. NO INCREASE IN INCIDENCE OF SKIN TUMORS WAS OBSERVED IN MICE OF TWO STRAINS FOLLOWING TOPICAL APPLICATION OF STYRENE OXIDE.

REPORT 07 : IARC MONOGRAPH 36 FEB 85
EXPOSURE OF RATS AND RABBITS BY INHALATION TO STYRENE OXIDE VAPOR AT MATERNALLY TOXIC DOSES DID NOT RESULT IN MALFORMATIONS; HOWEVER, THERE WAS AN INCREASE IN PREIMPLANTATION LOSSES AND OSSIFICATION DEFECTS IN RATS AND AN INCREASED RESORPTION FREQUENCY IN RABBITS.

REPORT 08 : IARC MONOGRAPH 36 FEB 85
STYRENE OXIDE MUTAGENIC TO BACTERIA, YEAST AND INSECTS; INDUCED CHROMOSOMAL ABERRATIONS AND MICRONUCLEI IN PLANTS. MUTAGENIC TO MAMMALIAN CELLS IN VITRO; INDUCED DNA DAMAGE IN MAMMALIAN CELLS BOTH IN VIVO AND IN VITRO, CHROMOSOMAL ABERRATIONS AND SISTER CHROMATID EXCHANGES IN VITRO. IN STUDIES IN MICE AND HAMSTERS IN VIVO, NO DOMINANT LETHAL MUTATIONS, CHROMOSOMAL ABERRATIONS, MICRONUCLEI OR SISTER CHROMATID EXCHANGES WERE INDUCED; HOWEVER, ONE STUDY IN MICE INDUCED CHROMOSOMAL ABERRATIONS.

REPT'D USE : DILUENT FOR EPOXY RESINS, REACTIVE PLASTICIZER AND CHEMICAL INTERMEDIATE

CAS NO : **UNKNOWN-30**
CHEM NAME : ***** SUBSTITUTED POLYGLYCIDYL BENZENEAMINE**

REPORT 01 : EPA STATUS REPORT 8EHQ-1283-05 00 DEC 83
COMPANY CLAIMS CHEMICAL NAME IS TSCA CBI. RESULTS FROM

IN-VITRO GENOTOXICITY TESTS WERE POSITIVE BOTH WITH AND WITHOUT METABOLIC ACTIVATION IN AMES S. TYPHIMURIUM ASSAY, AS L5178Y MOUSE LYMPHOMA MUTATION ASSAY AND A CHO CELL SCE ASSAY. ALSO REPORTED AS MUTAGENIC IN S. CEREVISIAE STRAIN D7 ONLY WITH METABOLIC ACTIVATION.

REPT'D USE : RESEARCH AND DEVELOPMENT

CAS NO : **108-30-5**

CHEM NAME : ***** SUCCINIC ANHYDRIDE**

REPORT 01 : NTP TR 373 NIH PUB 89-2828 JAN 90
UNDER 2-YEAR GAVAGE STUDIES, NO EVIDENCE OF CARCINOGENIC ACTIVITY OF SUCCINIC ANHYDRIDE FOR MALE OR FEMALE F344/N RATS GIVEN 50/100 MG/KG. NO EVIDENCE OF CARCINOGENIC ACTIVITY FOR MALE B6C3F1 MICE GIVEN 38 OR 75 MG/KG SUCCINIC ANHYDRIDE OR FOR FEMALE B6C3F1 MICE GIVEN 75 OR 150 MG/KG.

REPORT 02 : IARC MONOGRAPH 15 1977
SUCCINIC ANHYDRIDE DID NOT INDUCE REVERSE MUTATIONS IN SALMONELLA TA1535, TA1537, TA98 AND TA100, WITH OR WITHOUT METABOLIC ACTIVATION.

REPORT 03 : IARC MONOGRAPH 15 1977
IN THE ONLY STUDY AVAILABLE, WHICH INVOLVED FEW RATS, SUCCINIC ANHYDRIDE PRODUCED LOCAL SARCOMAS AFTER ITS SUBCUTANEOUS INJECTION. FURTHER STUDIES ARE REQUIRED BEFORE AN EVALUATION OF THE CARCINOGENICITY OF THIS COMPOUND CAN BE MADE.

REPT'D USE : INDUSTRIAL INTERMEDIATE FOR POLYMERS, PHARMACEUTICALS, DYES, PHOTOGRAPHIC CHEMICALS, FR CHEMICALS, CLEANING AGENTS, ADHESIVES, DEHYDRATING AGENT, FOOD ADDITIVE, MANUFACTURE OF POLYMERIC MATERIALS, PHARMACEUTICALS AND AGRICULTURAL/INDUSTRIAL CHEMICALS

CAS NO : **95-06-7**

CHEM NAME : ***** SULFALLATE**

REPORT 01 : IARC MONOGRAPH 30 JAN 83
SULFALLATE (TECHNICAL GRADE) TESTED FOR CARCINOGENICITY IN ONE STRAIN OF MICE AND IN ONE STRAIN OF RATS BY ADMINISTRATION IN DIET. CARCINOGENIC IN MICE, INDUCING MAMMARY GLAND TUMORS IN FEMALES AND LUNG TUMORS IN MALES. ALSO CARCINOGENIC IN RATS, INDUCING MAMMARY GLAND TUMORS IN FEMALES AND BENIGN AND MALIGNANT TUMORS OF FORESTOMACH IN MALES. SULFALLATE SHOULD BE REGARDED, FOR PRACTICAL PURPOSES, AS IF IT PRESENTED A CARCINOGENIC RISK TO HUMANS.

REPORT 02 : IARC MONOGRAPH 30 JAN 83
MUTAGENIC TO SALMONELLA TYPHIMURIUM IN PRESENCE OF EXOGENOUS METABOLIC ACTIVATION AND TO ASPERGILLUS AND STREPTOMYCES BUT NOT TO SACCHAROMYCES CERVISIAE WITH OR WITHOUT METABOLIC ACTIVATION. THERE IS SUFFICIENT EVIDENCE FOR THE MUTAGENICITY OF SULFALLATE IN BACTERIAL SYSTEMS BUT INSUFFICIENT EVIDENCE FOR ITS MUTAGENICITY IN MAMMALIAN CELLS OR IN MAMMALS.

REPORT 06 : MULTI-AGENCY CROSS CHECK
Y HHS MAGUIRE
N OSHA CARC
N EPA CAG
N EPA RCRA
N EPA CLEAN AIR
N EPA CLEAN WATER
REPT'D USE : HERBICIDE

CAS NO : **57-68-1**
CHEM NAME : ***** SULFAMETHAZINE**

REPORT 01 : NTP TECHNICAL BULLETIN 8 JUL 82
SULFAMETHAZINE ADMINISTERED ORALLY TO RATS INDUCED MALFORMATIONS AT DOSES NOT TOXIC TO THE MOTHER.
REPT'D USE : FOOD ADDITIVE

CAS NO : **723-46-6**
CHEM NAME : ***** SULFAMETHOXAZOLE**

REPORT 01 : IARC MONOGRAPH 24 1980
SULFAMETHOXAZOLE WAS TESTED IN ONE EXPERIMENT IN RATS BY ORAL ADMINISTRATION: IT PRODUCED THYROID TUMORS. IN ONE HYPOTHESIS-SEEKING EPIDEMIOLOGIC STUDY, AN ASSOCIATION BETWEEN SULFAMETHOXAZOLE USE AND NASOPHARYNGEAL AND CERVICAL CANCERS WAS NOTED; HOWEVER, THE EPA DATA WERE INSUFFICIENT.

REPORT 02 : NTP TECHNICAL BULLETIN 9 APR 83
SULFAMETHOXAZDLE WAS NEGATIVE IN ONE SALMONELLA TYPHIMURIUM MUTAGENICITY ASSAY.

REPORT 06 : MULTI-AGENCY CROSS CHECK
N HHS MAGUIRE
Y OSHA CARC
N EPA CAG
N EPA RCRA
N EPA CLEAN AIR
N EPA CLEAN WATER

REPT'D USE : ANTIBACTERIAL AGENT

CAS NO : **127-69-5**
CHEM NAME : ***** SULFISOXAZOLE**

REPORT 01 : NCI CARC TRS 138 DHEW PUB NIH 79-1393
UNDER THE CONDITIONS OF THIS BIOASSAY, SULFISOXAZOLE WAS NOT CARCINOGENIC FOR EITHER F344 RATS OR B6C3F1 MICE. SULFISOXAZOLE WAS GIVEN BY GAVAGE.

REPORT 02 : IARC MONOGRAPH 24 SEP 80
SULFAFURAZOLE WAS TESTED IN MICE AND RATS BY ORAL ADMINISTRATION: NO INCREASES IN TUMOR INCIDENCES WERE OBSERVED.

REPORT 03 : IARC MONOGRAPH 24 1980
WHEN MICE AND RATS WERE ADMINISTERED SULFATURAZOLE ORALLY ON DAYS 7-12 AND 9-14 OF PREGNANCY, RESPECTIVELY, A SIGNIFICANT INCREASE IN CLEFT PALATE AND SKELETAL DEFECTS WAS FOUND IN OFFSPRING OF BOTH SPECIES. IN TWO LARGE STUDIES OF WOMEN EXPOSED DURING PREGNANCY, NO INCREASE IN MALFORMATION RATE WAS OBSERVED IN THE OFF-SPRING.
REPT'D USE : ANTIBACTERIAL AGENT AND VETERINARY MEDICINE

CAS N0 : **77-79-2**
CHEM NAME : ***** 3-SULFOLENE**

REPORT 01 : NCI CARC TRS 102 DHEW PUB NO 78-1352
UNDER THE CONDITIONS OF THIS BIOASSAY, THERE WAS NO EVIDENCE FOR THE CARCINOGENICITY OF 3-SULFOLENE TO OSBORNE-MENDEL RATS OR B6C3F1 MICE. 3-SULFOLENE WAS ADMINISTERED BY GAVAGE.

REPORT 02 : NTP TECHNICAL BULLETIN 6 JAN 82
THE COMPOUND WAS NEGATIVE IN THE SALMONELLA TYPHIMURIUM MUTAGENICITY ASSAY IN ONE LABORATORY.
REPT'D USE : ORGANIC SYNTHESIS

CAS NO : **9080-79-9**
CHEM NAME : ***** SULFONATED POLYSTYRENE, SODIUM SALT**

REPORT 01 : 8EHQ-0485-0549 8(E) SUBMISSION MAY 85
RESULTS OF INHALATION STUDIES SUGGEST A POSSIBLE RISK OF LUNG FUNCTION IMPAIRMENT FOLLOWING INHALATION OF RESPIRABLE DRY POLYMER PARTICLES. IN THIS STUDY, SPRAGUE-DAWLEY RATS WERE EXPOSED TO A POLYMER DUST

AEROSOL WITH CONCENTRATIONS OF 1.2, 2.2, 3.0, OR 4.6 MG/L FOR 4 HOURS FOLLOWED BY A 14-DAY OBSERVATION PERIOD. THE LC50'5 WERE FOUND TO BE 2.6 MG/L AND 3.5 MG/L FOR MALE AND FEMALE RATS RESPECTIVELY. THE COMBINED SEX LC50 WAS CALCULATED TO BE 3.0 MG/L.

CAS NO : **2257-95-5**
CHEM NAME : ***** SYMPHYTINE**

REPORT 01 : IARC MONOGRAPH 31 JUL 83
TESTS IN MALE RATS BY INTRAPERITONEAL ADMINISTRATION NON-SIGNIFICANT INCREASE IN INCIDENCE OF LIVER TUMORS OBSERVED. WHEN LEAVES AND ROOTS OF SYMPHYTUM OFFICINALE L. TESTED IN RATS BY ADMINISTRATION IN DIET INCIDENCE OF LIVER TUMORS INCREASED.

REPORT 02 : IARC MONOGRAPH 31 JUL 83
SYMPHYTINE MUTAGENIC IN SALMONELLA TYPHIMURIUM AND IN MAMMALIAN CELLS IN VITRO.

REPT'D USE : FOUND IN SEVERAL PLANT SPECIES; USED AS FOODS AND HERBAL REMEDIES AND ANIMAL FEED

CAS NO : **14807-96-6**
CHEM NAME : ***** TALC**

REPORT 01 : IARC 42 1987
DIFFERENT GRADES TALC TESTED FOR CARCINOGENICITY IN MICE, RATS AND HAMSTERS. THE MAJORITY OF STUDIES INADEQUATE. TUMOR INCIDENCE NOT INCREASED FOLLOWING EITHER ADMINISTRATION OF SINGLE INTRAPLEURAL DOSE OR 4 INTRAPERITONEAL INJECTIONS TO RATS. SINGLE SC INJECTION IN MICE DID NOT PRODUCE LOCAL TUMORS. NO TUMORS PRODUCED IN RATS AFTER ADMINISTRATION IN DIET. CHARACTERIZATION OF TALC INSUFFICIENT TO TELL IF IT CONTAINED ASBESTIFORM FIBERS. IARC CONCLUDED THAT EVIDENCE INADEQUATE FOR DETERMINING CARCINOGENICITY OF TALC. TALC NOT MUTAGENIC TO S. TYPHIMURIUM OR S. CEREVISAE IN HOST-MEDIATED ASSAYS. NO CHROMOSOMAL ABERRATIONS OR DOMINANT LETHAL MUTATIONS SEEN.

REPORT 02 : IARC MONOGRAPH 42 1987
SUGGESTED ASSOCIATION BETWEEN EXPOSURE TO TALC CONTAINING ASBESTIFORM FIBERS AND MESOTHELIOMA. STUDIES OF MINERS AND MILLERS OF TALC CONTAINING ASBESTIFORM TREMOLITE AND ANTHOPHYLLITES SHOWED EXCESS OF LUNG CANCER AND 1 CASE OF MESOTHELIOMA. ONE COHORT STUDY REVEALED EXCESS MORTALITY FROM LUNG CANCER AND NONMALIGNANT RESPIRATORY DISEASE. CONCLUSION: INADEQUATE

EVIDENCE OF CARCINOGENICITY TO HUMANS OF TALC NOT CONTAINING ASBESTIFORM FIBERS, WHILE SUFFICIENT EVIDENCE FOR CARCINOGENICITY TO HUMANS OF TALC CONTAINING ASBESTIFORM FIBERS.

REPT'D USE : CERAMICS, PAPER, PLASTICS, BUILDING MATERIALS, PAINTS, COSMETICS, TOOTHPASTE AND SOAP FILLER, PHARMACEUTICALS, SHOE POLISH, LUBRICANT AND TEXTILES

CAS NO : **1401-55-4**
CHEM NAME : ***** TANNIC ACID**

REPORT 01 : NAS REPORT ON DIET NUTRITION AND CANCER 1982
NO ADEQUATE STUDIES HAVE BEEN CONDUCTED TO TEST THE CARCINOGENICITY OF ORALLY ADMINISTERED TANNINS. TANNIC ACID WAS FOUND NOT TO BE GENOTOXIC OR MUTAGENIC TO SACCHAROMYCES AND SALMONELLA WITHOUT ACTIVATION. TANNINS FROM VARIOUS SOURCES SUCH AS APPLE JUICE, GRAPE JUICE, WINE WERE FOUND TO BE STRONGLY CLASTOGENIC FOR CHINESE HAMSTER OVARY CELLS.

REPORT 02 : IARC MONOGRAPH 10 1976
TANNIC ACID IS CARCINOGENIC IN RATS FOLLOWING ITS SUBCUTANEOUS INJECTION; IT PRODUCED LIVER TUMORS. IN MICE, SUBCUTANEOUS INJECTION OF HYDROLYSABLE TANNINS PRODUCED LIVER TUMORS, AND THAT OF CONDENSED TANNINS PRODUCED BOTH LOCAL SARCOMAS AND LIVER TUMORS. NO ADEQUATE PUBLISHED STUDIES INVOLVING ORAL ADMINISTRATION OF TANNINS WERE AVAILABLE.

REPT'D USE : ALCOHOL DENATURANT, CLARIFYING AGENT IN BEER/WINE, CHEMICAL REAGENT, CHEMICAL INTERMEDIATE, LEATHER TANNING, TEXTILE MORDANT AND NATURAL OCCURRENCE IN FOODS

CAS NO : **39300-88-4**
CHEM NAME : ***** TARA GUM**

REPORT 01 : NTP BOARD OF SCIENTIFIC COUNSELORS FEB 81
THE BIOASSAY REPORT CONCLUSIONS FOR TARA GUM WERE REVIEWED BY THE PEER REVIEW PANEL AND APPROVED WITH MINOR MODIFICATIONS. UNDER CONDITIONS OF THIS BIOASSAY, TARA GUM ADMININISTERED IN THE FEED WAS NOT CARCINOGENIC FOR F344 RATS OR B6C3F1 MICE OF EITHER SEX.

REPT'D USE : SUBSTITUTE FOR LOCUST BEAN GUM IN COSMETICS

CAS N0 : **72-54-8**
CHEM NAME : ***** TDE**

REPORT 01 : NCI CARC TRS 131 DHEW PUB NO 78-1386
UNDER THE CONDITIONS OF THESE BIOASSAYS THERE WAS NO EVIDENCE FOR THE CARCINOGENICITY OF . . . TDE IN FEMALE OSBORNE-MENDEL RATS OR B6C3F1 MICE OF EITHER SEX. THE FINDINGS SUGGEST A POSSIBLE CARCINOGENIC EFFECT OF TDE IN MALE OSBORNE-MENDEL RATS, BASED ON THE INDUCTION OF COMBINED FOLLICULAR-CELL CARCINOMAS AND FOLLICULAR-CELL ADENOMAS OF THE THYROID. DDT AND P,P'-DDE RESULTS ALSO REPORTED.

REPORT 02 : IARC MONOGRAPH 5 1974
THE TWO DDT METABOLITES, P,P'-DDE AND TDE (DDD), WERE TESTED BY ORAL ADMINISTRATION TO MICE. THE LATTER PRODUCED A SIGNIFICANT INCREASE IN LUNG TUMORS, WHILE DDE WAS FOUND TO PRODUCE A HIGH INCIDENCE OF LIVER-CELL TUMORS.

REPORT 03 : EPA AMBIENT WATER QUALITY CRITERIA OCT 80
FOR METABOLITES OF DDT, THE CRITERION OF .001 UG/L IS ESTABLISHED TO PROTECT AQUATIC LIFE. A HUMAN HEALTH EFFECTS CRITERION, BASED ON CARCINOGENICITY, HAS BEEN DERIVED FOR DDT.

REPORT 06 : MULTI-AGENCY CROSS CHECK
N HHS MAGUIRE
Y OSHA CARC
N EPA CAG
N EPA RCRA
N EPA CLEAN AIR
Y EPA CLEAN WATER
REPT'D USE : INSECTICIDE

CAS NO : **100-21-0**
CHEM NAME : ***** TEREPHTHALIC ACID**

REPORT 01 : CIIT 24 MONTH TOXICOLOGY STUDY APR 81
TEREPHTHALIC ACID IS BEING EVALUATED FOR TOXICOLOGIC AND/OR CARCINOGENIC EFFECTS IN F344 RATS FOLLOWING DIETARY ADMINISTRATION AT LEVELS OF 1000, 142 AND 200 NG/KG/DAY. NECROPSY AT WEEK 78 REVEALED HYPERPLASIA AND/OR TRANSITIONAL CELL ADENOMAS OF THE URINARY BLADDER AND ADENOMAS OR ADENOCARCINOMAS OF THE UTERUS IN THE HIGH DOSE GROUP.
REPT'D USE : INTERMEDIATE IN MFR CF POLYESTERS, ANALYTICAL REAGENT AND POULTRY FEED ADDITIVE

CAS NO : **8001-50-1**
CHEM NAME : ***** TERPENE POLYCHLORINATE**

REPORT 01 : IARC MONOGRAPH 1974
TERPENE POLYCHLORINATES WERE TESTED IN A PRELIMINARY STUDY BY THE ORAL ROUTE IN TWO STRAINS OF MICE AND PRODUCED AN INCREASED INCIDENCE OF HEPATOMAS IN MALES OF ONE STRAIN.

REPORT 06 : MULTI-AGENCY CROSS CHECK
N HHS MAGUIRE
Y OSHA CARC
N EPA CAG
N EPA RCRA
N EPA CLEAN AIR
N EPA CLEAN WATER
REPT'D USE : INSECTICIDE MOTHPROOFING AGENT

CAS NO : **58-22-0**
CHEM NAME : ***** TESTOSTERONE**

REPORT 01 : IARC MONOGRAPH 21 1979
TESTOSTERONE PROPIONATE IMPLANTED SUBCUTANEOUSLY IN MICE INDUCED CERVICAL-UTERINE TUMORS, WHICH METASTASIZED IN SOME CASES; IN RATS, METASTASIZING PROSTATIC ADENOCARCINOMAS WERE INDUCED IN MALES. THERE IS SUFFICIENT EVIDENCE FOR THE CARCINOGENICITY OF TESTOSTERONE IN EXPERIMENTAL ANIMALS. IN THE ABSENCE OF ADEQUATE DATA IN HUMANS, IT IS REASONABLE, FOR PRACTICAL PURPOSES, TO REGARD TESTOSTERONE AS IF IT PRESENTED A CARCINOGENIC RISK TO HUMANS.

REPORT 02 : IARC MONOGRAPH 21 1979
NO RELEVANT MUTAGENICITY DATA ON TESTOSTERONE WERE AVAILABLE. TESTOSTERONE IS EMBRYOLETHAL IN PRE- AND POSTIMPLANTATION EMBRYOS AND CAUSES VIRILIZATION IN FEMALE OFFSPRING.
REPT'D USE : ANDROGENIC ANABOLIC HORMONE

CAS NO : **79-94-7**
CHEM NAME : *****TETRABROMOBISPHENOL A (TBBPA)**

REPORT 01 : EPA FR NOTICE 52 FR 25219 MAY 86
EPA ISSUING FINAL TEST RULE UNDER TSCA 4(A) REQUIRING MANUFACTURERS AND PROCESSORS TO PERFORM TESTING FOR CHEMICAL FATE AND ENVIRONMENTAL EFFECTS. THIS IS IN RESPONSE TO ITC, MAY 2, 1985 DESIGNATION OF TBBPA FOR PRIORITY CONSIDERATION FOR TESTING.

REPT'D USE : REACTIVE AND ADDITIVE FLAME RETARDANT

CAS NO : **14047-09-7**
CHEM NAME : ***** 3,3',4,4'-TETRACHLOROAZOBENZENE (TCAB)**

REPORT 01 : EPA TSCA 8(E) NOTICE 8EHQ-1281-0424 JAN 82
VELSICOL CHEMICAL SUBMITTED PREPUBLICATION RESULTS OF A CHICK EMBRYO STUDY OF TCAB. ACCORDING TO THE INFORMATION CONTAINED IN THE SUBMITTED ABSTRACT, TCAB APPEARS TO HAVE SOME DEGREE OF TERATOGENIC AND EMBRYOTOXIC ACTIVITY.
REPT'D USE : MFR OF PESTICIDE AND DYE INTERMEDIATES

CAS NO : **21232-47-3**
CHEM NAME : ***** 3,3',4,4'-TETRACHLOROAZOXYBENZENE (TCAOB)**

REPORT 01 : EPA TSCA 8(E) NOTICE 8E-Q-1282-04.24 JAN 82
VELSICOL CHEMICAL SUBMITTED PREPUBLICATION RESULTS OF A CHICK EMBRYO STUDY OF TCAOB. ACCORDING TO THE INFORMATION CONTAINED IN THE SUBMITTED ABSTRACT, TCAOB APPEARS TO HAVE SOME DEGREE OF TERATOGENIC AND EMBRYOTOXIC ACTIVITY.
REPT'D USE : MFR PESTICIDE AND DYE INTERMEDIATES

CAS NO : **95-94-3**
CHEM NAME : ***** 1,2,4,5-TETRACHLOROBENZENE (1,2,4,5-TCB)**

REPORT 01 : 51 CFR 24657 CHLORINATED BENZENES; JUL 86
FINAL TEST RULE
IN ACCORDANCE WITH 40 CFR, PART 798, EPA IS PROMOLGATING A TEST RULE FOR CHLORINATED BENZENES.
REPT'D USE : PESTICIDES

CAS NO : **15721-02-5**
CHEM NAME : ***** 2,2',5,5'-TETRACHLOROBENZIDINE**

REPORT 01 : IARC MONOGRAPH 27 APR 82
2,2'5,5'-TETRACHLOROBENZIDINE WAS TESTED FOR CARCINOGENICITY BY DIETARY ADMINISTRATION IN MICE AND RATS. THE AVAILABLE DATA WERE INSUFFICIENT FOR EVALUATION.

REPORT 02 : IARC MONOGRAPH 27 APR 82
2,2'5,5'-TETRACHLOROBENZIDINE INDUCED REVERSE MUTATIONS IN SALMONELLA STRAINS TA98 AND TA100 WITH METABOLIC ACTIVATION. NO DATA WERE AVAILABLE CONCERNING EFFECTS

ON REPRODUCTION OR PRENATAL TOXICITY.
REPT'D USE : PIGMENT INTERMEDIATE

CAS NO : **1746-01-6**
CHEM NAME : ***** 2,3,7,8-TETRACHLORODIBENZO-P-DIOXIN (TCDD)**

REPORT 01 : IARC SUPPLEMENT 1 1979
RATS FED 2,3,7,8-TETRACHLORODIBENZO-P-DIOXIN DEVELOPED A VARIETY OF MALIGNANT TUMORS, MAINLY OF THE RESPIRATORY TRACT, SEBACEOUS GLANDS, LIVER AND BILE DUCTS. A NUMBER OF CASES OF CANCER HAVE BEEN REPORTED IN WORKERS EXPOSED TO TCDD, BUT NO ADEQUATE EPIDEMIOLOGICAL STUDIES WERE AVAILABLE.

REPORT 02 : EPA RCRA HEALTH AND ENV. EFFECTS PROFILE OCT 80
TCDD HAS BEEN FOUND TO INDUCE HEPATOCELLULAR CARCINOMA AND TUMORS IN TWO RAT FEEDING STUDIES. TCDD HAS ALSO PRODUCED FETOTOXIC AND TERATOGENIC EFFECTS IN LAB ANIMALS. POSITIVE MUTAGENICITY OF TCDD HAS BEEN DEMONSTRATED IN THREE BACTERIAL BIOASSAY SYSTEMS.

REPORT 03 : NTP BIOASSAY IRS 201 NIH PUB.NO. 82-1757
UNDER THE CONDITIONS OF THIS BIOASSAY TCDD APPLIED TO THE SKIN WAS NOT CARCINOGENIC FOR MALE SWISS-WEBSTER MICE (THE INCREASE OF FIBROSARCOMAS IN THE INTEGUMENTARY SYSTEM MAY HAVE BEEN ASSOCIATED WITH THE SKIN APPLICATION OF TCDD). TCDD WAS CARCINOGENIC FOR FEMALE SWISS -WEBSTER MICE CAUSING FIBROSARCOMAS IN THE INTEGUMENTARY SYSTEM.

REPORT 04 : NTP BIOASSAY TRS 209 NIH PUB NO 82-1765
UNDER THE CONDITIONS OF THIS BIOASSAY, TCDD WAS CARCINOGENIC FOR OSBORNE-MENDEL RATS INDUCING FOLLICULAR-CELL THYROID ADENOMAS IN MALES AND NEOPLASTIC NODULES OF THE LIVER IN FEMALES. TCDD WAS ALSO CARCINOGENIC FOR B6C3F1 MICE INDUCING HEPATOCELLULAR CARCINOMAS IN MALES AND FEMALES AND FOLLICULAR-CELL THYROID ADENOMAS IN FEMALES (GAVAGE).

REPORT 05 : IARC MONOGRAPH 15 1977
TCDD IS EMBRYOTOXIC AND TERATOGENIC IN MICE IN REPEATED OR SINGLE DOSES OF AS LITTLE AS 1-10 UG/KG BW, CAUSING INCREASED FREQUENCIES OF CLEFT PALATE AND KIDNEY ABNORMALITIES. A DOSE-RESPONSE RELATIONSHIP HAS BEEN OBSERVED. IN SALMONELLA, FRAMESHIFT MUTATIONS IN STRAIN TA1532 BUT NOT BASE SUBSTITUTIONS IN STRAIN TA1530 HAVE BEEN OBSERVED. TCDD INCREASED THE REVERSION FREQUENCY TO STREPTOMYCIN INDEPENDENCE IN E. COLI.

REPORT 06 : MULTI-AGENCY CROSS CHECK
Y HHS MAGUIRE
N OSHA CARC
Y EPA CAG
N EPA RCRA
N EPA CLEAN AIR
N EPA CLEAN WATER

REPORT 07 : NTP TECHNICAL BULLETIN 6/8 JAN 82
TCDD WAS NEGATIVE IN THE SALMONELLA ASSAY AND IN DROSOPHILA WHEN TESTED FOR SEX-LINKED RECESSIVE LETHALS.

REPORT 08 : NIOSH INTELLIGENCE BULLETIN 40 JAN 84
NIOSH RECOMMENDS THAT TCDD BE REGARDED AS POTENTIAL OCCUPATIONAL CARCINOGEN, THAT OCCUPATIONAL EXPOSURE TO TCDD BE CONTROLLED, AND THAT DECONTAMINATION MEASURES BE USED FOR TCDD-CONTAMINATED WORK ENVIRONMENTS BECAUSE OF ITS CARCINOGENICITY IN RATS AND MICE. TCDD CAUSES A VARIETY OF SYSTEMS AND IMMULOGICAL EFFECTS. RESULTS OF MUTAGENICITY TESTS ARE INCONCLUSIVE.

REPORT 09 : NIOSH INTELLIGENCE BULLETIN 40 JAN 84
IN ANIMALS TCDD CAUSES VARIOUS SYSTEMIC EFFECTS AT WIDE RANGE OF EXPOSURE CONCENTRATIONS, INCLUDING TUMORIGENESIS, IMMUNOLOGICAL DYSFUNCTION, AND TERATOGENESIS. STUDIES OF HUMANS EXPOSED TO TCDD-CONTAMINATED MATERIALS SUGGEST IT CAUSES CHLORACNE, METABOLIC DISORDERS (PORPHYRIA), AND OTHER SYSTEMIC PROBLEMS. NIOSH RECOMMENDS TCDD BE REGARDED AS POTENTIAL OCCUPATIONAL CARCINOGEN AND OCCUPATIONAL EXPOSURE BE CONTROLLED BECAUSE TCDD IS CARCINOGENIC IN RATS AND MICE.

REPT'D USE : CONTAMINANT IN HERBICIDE

CAS NO : **630-20-6**

CHEM NAME : ***** 1,1,1,2-TETRACHLOROETHANE**

REPORT 01 : NIOSH CURRENT INTELLIGENCE BULLETIN 27 AUG 78
NCI BIOASSAY TESTING IN PROGRESS AS OF JULY 1978 BUT NO REPORT IS AVAILABLE. NO OSHA EXPOSURE STANDARD EXISTS. ALL CHLOROETHANE COMPOUNDS ARE KNOWN TO CAUSE CNS DEPRESSION IN LABORATORY ANIMALS. THIS COMPOUND IS AN EMBRYOTOXIN IN RABBITS AND RATS; CAUSES LIVER DYSFUNCTION IN RATS; AND IS A MUTAGEN IN RATS. NIOSH RECOMMENDS THAT THE COMPOUND SHOULD BE TREATED WITH CAUTION IN THE WORKPLACE BECAUSE OF ITS RELATION TO FOUR CARCINOGENIC CHLOROETHANE COMPOUNDS.

REPORT 02 : NTP BOARD OF SCIENTIFIC COUNSELORS DEC 81
UNDER THE CONDITIONS OF THIS BIOASSAY, TECHNICAL GRADE 1,1,1,2-TECE WAS NOT DEMONSTRATED CARCINOGENIC FOR F344 RATS (MAY BE ASSOCIATED WITH INCREASED LIVER TUMORS IN MALE RATS). THE COMPOUND CAUSED THE FORMATION OF HEPATOCELLULAR ADENOMAS IN B6C3F1 MICE BUT HEPATOCELLULAR CARCINOMAS IN B6C3F1 FEMALES WAS NOT INTERPRETABLE DUE TO POOR SURVIVAL OF HIGH DOSE MICE. 1,1,1,2-TECE SHOULD BE CONSIDERED CARCINOGENIC FOR B6C3F1 MICE BECAUSE OF THE INCREASED INCIDENCE OF LIVER TUMORS. (ADMINISTRATION BY GAVAGE- SEE CAVEATS IN BIOASSAY).

REPORT 03 : NTP TECHNICAL BULLETIN 7 APR 82
1,1,1,2-TETRACLOROETHANE WAS NEGATIVE IN THE SALMONELLA TYPHIMURIUM MUTAGENICITY ASSAY.

REPORT 04 : IARC VOL 41 1986
LIMITED EVIDENCE FOR CARCINOGENICITY OF CPD TO TEST ANIMALS. THE REPRODUCTIVE EFFECTS AND PRENATAL TOXICITY COULD NOT BE EVALUATED WITH THE AVAILABLE DATA. CPD NOT MUTAGENIC TO BACTERIA AND DOES NOT INDUCE TRANSFORMATION IN CULTURED MAMMALIAN CELLS. THE ORAL LD50 IN MICE IS SAID TO BE 1500 KG/KG. THE ORAL LD50 IN FEMALE RATS IS SAID TO BE 670 MG/KG AND IN MALE RATS, 780 MG/KG.

REPORT 06 : MULTI-AGENCY CROSS CHECK
N HHS MAGUIRE
N OSHA CARC
N EPA CAG
N EPA RCRA
N EPA CLEAN AIR
N EPA CLEAN WATER

REPT'D USE : INTERMEDIATE IN MFR OF TCE/PCE, SOLVENT, MFR OF INSECTICIDES/HERBICIDES, BLEACHES AND PAINTS/VARNISHES

CAS NO : **79-34-5**
CHEM NAME : ***** 1,1,2,2-TETRACHLOROETHANE**

REPORT 01 : NCI CARC TRS 27 DHEW PUB NO 78-827
UNDER THE CONDITIONS OF THIS STUDY, ORALLY ADMINISTERED 1,1,2,2-TETRACHLOROETHANE IS A LIVER CARCINOGEN IN B6C3F1 MICE OF BOTH SEXES. THE RESULTS DO NOT PROVIDE CONCLUSIVE EVIDENCE FOR THE CARCINOGENICITY OF 1,1,2,2-TETRACHLOROETHANE IN OSBORNE-MENDEL RATS.

REPORT 02 : IARC MONOGRAPH 20 1979
1,1,2,2-TETRACHLOROETHANE WAS TESTED IN ONE EXPERIMENT IN MICE AND IN ONE IN RATS BY ORAL ADMINISTRATION. IN MALE AND FEMALE MICE, IT PRODUCED HEPATOCELLULAR CARCINO-

MAS. ALTHOUGH A FEW HEPATOCELLULAR CARCINOMAS WERE OBSERVED IN MALE RATS, NO SIGNIFICANT INCREASE IN THE INCIDENCE OF TUMORS WAS OBSERVED IN ANIMALS OF EITHER SEX.

REPORT 03 : NIOSH CRITERIA DOCUMENT DEC 76
NIOSH RECOMMENDS A STANDARD OF 1 PPM TWA BASED ON LIVER, GI, AND CNS EFFECTS. THE PRESENT OSHA STANDARD IS 5 PPM TWA. WHILE THE OBSERVATIONS OF SMALL INCREASES OF POINT MUTATION FREQUENCY IN BACTERIA ARE CAUSE FOR CONCERN, THEY ARE INSIGNIFICANT TO ESTABLISH THE EXISTENCE OF ANY SIGNIFICANT RISK OF GENETIC HAZARDS TO THE HUMAN POPULATION EXPOSED TO TETRACHLOROETHANE.

REPORT 04 : EPA AMBIENT WATER QUALITY CRITERIA OCT 80
1,1,2,2-TETRACHLOROETHANE IS CONSIDERED TO BE CARCINOGENIC IN AT LEAST ONE RODENT SPECIES. AMBIENT WATER CONCENTRATIONS OF 1.7 UG/L .17 UG/L, .017 UG/L ARE ESTIMATED TO KEEP THE LIFETIME CANCER RISK BELOW 10 -5, 10 -6, AND 10 -7 RESPECTIVELY.

REPORT 06 : MULTI-AGENCY CROSS CHECK
N HHS MAGUIRE
N OSHA CARC
Y EPA CAG
Y EPA RCRA
N EPA CLEAN AIR
Y EPA CLEAN WATER

REPORT 07 : EPA RCRA HEALTH AND ENV. EFFECTS PROFILE OCT 80
AN INCREASED INCIDENCE OF HEPATOCELLULAR CARCINOMAS HAS BEEN SHOWN IN MICE FOLLOWING ORAL ADMINISTRATION OF 1,1,2,2-TETRACHLOROETHANE. MUTAGENIC EFFECTS HAVE BEEN REPORTED IN THE AMES SALMONELLA ASSAY AND IN E. COLI. THERE IS NO AVAILABLE EVIDENCE TO INDICATE THAT 1,1,2,2-TECE PRODUCES TERATOGENIC EFFECTS.

REPORT 08 : IARC MONOGRAPH 20 1979
1,1,2,2-TETRACHLORETHANE IS TERATOGENIC IN MICE. THE CHEMICAL WAS MUTAGENIC IN SALMONELLA TA1530 AND TA1535, SHOWING A LINEAR INCREASE IN REVERSION FREQUENCY WITH INCREASING CONCENTRATION. NEGATIVE RESULTS WERE OBTAINED IN STRAIN TA1538.

REPORT 09 : NTP TECHNICAL BULLETIN 6/7 JAN 82
1,1,2,2-TECE WAS NEGATIVE IN THE SALMONELLA ASSAY, THE RAT DOMINANT LETHAL ASSAY, THE DROSOPHILA SEX-LINKED RECESSIVE LETHAL ASSAY, AND IN VITRO UNSCHEDULED DNA SYNTHESIS IN HUMAN FIBROBLASTS. RESULTS IN THE MOUSE SPERM HEAD MORPHOLOGY ASSAY AND THE RAT BONE MARROW CYTOLOGY ASSAY WERE SUGGESTIVE BUT NOT CLEARLY POSITIVE.

REPT'D USE : INDUSTRIAL SOLVENT AND INSECTICIDE INTERMEDIATE

CAS NO : **2438-88-2**
CHEM NAME : ***** 2,3,5,6-TETRACHLORO-4-NITROANISOLE (TCNA)**

REPORT 01 : NCI CARC TRS 114 DHEW PUB NO 78-1369
UNDER THE CONDITIONS OF THIS BIOASSAY, DIETARY ADMINISTRATION OF 2,3,5,6-TETRACHLORO-4-NITROANISOLE WAS NOT CARCINOGENIC TO MALE OR FEMALE FISCHER 344 RATS OR B6C3F1 MICE OF EITHER SEX.
REPT'D USE : FUNGICIDE AND ACARICIDE

CAS NO : **22248-79-9**
CHEM NAME : ***** TETRACHLORVINPHOS**

REPORT 01 : NCI CARC TRS 33 DHEW PUB NO 78-833
UNDER THE CONDITIONS OF THIS BIOASSAY, THE ADMINISTRATION OF TECHNICAL-GRADE TETRACHLORVINPHOS IN OSBORNE-MENDEL RATS WAS ASSOCIATED WITH PROLIFERATIVE LESIONS OF THE C-CELLS OF THE THYROID AND CORTICAL ADENOMAS OF THE ADRENAL IN FEMALES. IN FEMALE B6C3F1 MICE, THE INCIDENCE OF NEOPLASTIC NODULE OF THE LIVER WAS ASSOCIATED WITH TREATMENT AND IN MALE MICE TETRACHLORVINPHOS WAS CARCINOGENIC, CAUSING HEPATOCELLULAR CARCINOMA OF THE LIVER (FEEDING STUDY).

REPORT 02 : IARC MONOGRAPH 30 JAN 83
TESTED FOR CARCINOGENICITY IN MICE AND RATS BY ADMINISTRATION IN DIET. IT PRODUCED HEPATOCELLULAR CARCINOMAS IN MALE MICE AND BENIGN AND MALIGNANT LIVER-CELL TUMORS IN FEMALE MICE. IN RATS, IT INCREASED INCIDENCES OF C-CELL ADENOMA OF THYROID AND OF CORTICAL ADENOMA OF ADRENAL IN FEMALES.

REPORT 03 : IARC MONOGRAPH 30 JAN 83
TETRACHLORVINPHOS NOT MUTAGENIC TO BACTERIA IN SYSTEMS STUDIED. IT WAS A WEAK INDUCER OF CHROMOSOMAL ABERRATIONS IN HUMAN LYMPHOCYTES IN VITRO. NO OVERALL EVALUATION OF THE MUTAGENICITY OF TETRACHLORVINPHOS COULD BE MADE.

REPORT 06 : MULTI-AGENCY CROSS CHECK
N HHS MAGUIRE
Y OSHA CARC
N EPA CAG
N EPA RCRA

N EPA CLEAN AIR
N EPA CLEAN WATER
REPT'D USE : INSECTICIDE

CAS NO : **64-75-5**
CHEM NAME : ***** TETRACYCLINE HYDROCHLORIDE**

REPORT 01 : NTP TECH REPORT TR344 NIH PUB. NO. 89-2600 AUG 89
UNDER CONDITIONS OF 2-YEAR FEED STUDY, NO EVIDENCE OF CARCINOGENIC ACTIVITY IN MALE AND FEMALE F344 RATS AND B6C3F1 MICE FED 12,500 OR 25,000 PPM IN DIET.
REPT'D USE : DRUG

CAS NO : **112-60-7**
CHEM NAME : ***** TETRAETHYLENE GLYCOL**

REPORT 01 : EPA STATUS REPORT 8EHQ-0987-0693 SEP 87
NEGATIVE IN S. TYPHIMURIUM ASSAY USING 5 STRAINS OF BACTERIA WITH AND WITHOUT ACTIVATION. NO STATISTICALLY SIGNIFICANT INCREASES IN GENE MUTATION ACTIVITY WITH OR WITHOUT METABOLIC ACTIVATION IN A FORWARD GENE MUTATION TEST IN CHINESE HAMSTER OVARY CELLS (HGPRT-LOCUS). A SISTER CHROMATID EXCHANGE TEST USING CHINESE HAMSTER OVARY CELLS SHOWED STATISTICALLY SIGNIFICANT INCREASES IN NUMBERS OF EXCHANGES COMPARED TO CONTROLS WITH AND WITHOUT METABOLIC ACTIVATION.
REPT'D USE : SOLVENT FOR NITROCELLULOSE, PLASTICIZER, LACQUERS AND COATING COMPOSITIONS

CAS NO : **17831-71-9**
CHEM NAME : ***** TETRAETHYLENE GLYCOL DIACRYLATE**

REPORT 01 : EPA TSCA 8(E) NOTICE 8EHQ-0981-0411 SEP 81
CELANESE CORP REPORTED PRELIMINARY RESULTS FROM A CHRONIC DERMAL APPLICATION STUDY IN MICE. SQUAMOUS CELL CARCINOMAS AND MICROSCOPIC PAPILLOMAS WERE REPORTED AFTER TREATMENTS TWICE WEEKLY FOR 18 MONTHS.
REPT'D USE : NO AVAILABLE PRODUCTION OR USE DATA

CAS NO : **112-57-2**
CHEM NAME : ***** TETRAETHYLENEPENTAMINE**

REPORT 01 : NTP TECHNICAL BULLETIN 9 APR 83
TETRAETHYLENEPENTAMINE WAS POSITIVE IN ONE SALMONELLA TYPHIMURIUM MUTAGENICITY ASSAY.

CAS NO : **78-00-2**
CHEM NAME : ***** TETRAETHYLLEAD (TEL)**

REPORT 01 : IARC MONOGRAPH 23 1980
WHEN TETRAETHYLLEAD WAS ADMINISTERED BY SUBCUTANEOUS INJECTION TO NEONATAL MICE, AN INCREASED INCIDENCE OF LYMPHOMAS OCCURRED IN FEMALE ANIMALS ONLY: ADDITIONAL STUDIES ARE REQUIRED BEFORE AN EVALUATION OF THE CARCINOGENICITY OF THIS COMPOUND CAN BE MADE.

REPORT 02 : NTP TECHNICAL BULLETIN NO. 3 DEC 80
THE COMPOUND WAS FOUND TO BE NEGATIVE IN THE SALMONELLA TYPHIMURIUM MUTAGENICITY ASSAY IN TWO LABORATORIES.
REPT'D USE : ANTIKNOCK AGENT IN GASOLINE AND INTERMEDIATE IN MFR OF PESTICIDES

CAS NO : **97-77-8**
CHEM NAME : ***** TETRAETHYLTHIURAM DISULFIDE (DISULFIRAM)**

REPORT 01 : NCI CARC TRS 166 DHEW PUB NO 79-1722
UNDER THE CONDITIONS OF THIS BIOASSAY, TETRAETHYLTHIURAM DISULFIDE WAS NOT CARCINOGENIC FOR F344 RATS OR B6C3F1 MICE OF EITHER SEX (FEEDING STUDY).

REPORT 02 : NIOSH CURRENT INTELLIGENCE BULLETIN 23 APR 78
NIOSH REPORTS THAT RATS EXPOSED TO 20 PPM ETHYLENE DIBROMIDE BY INHALATION (THE CURRENT 8-HOUR TWA OSHA EXPOSURE STANDARD) AND ALSO RECEIVING A DIET CONTAINING 0.05% DISULFIRAM ARE EXPERIENCING EXCEEDINGLY HIGH MORTALITY LEVELS AS WELL AS A HIGH INCIDENCE OF TUMORS. NIOSH RECOMMENDS THAT NO WORKER BE EXPOSED TO BOTH ETHYLENE DIBROMIDE AND DISULFIRAM.

REPORT 03 : IARC MONOGRAPH 12 1976
TEST RESULTS TO DATE ARE DERIVED FROM A STUDY IN WHICH AN IMPURE MATERIAL WAS TESTED. THE LIMITED DATA AVAILABLE DO NOT ALLOW AN EVALUATION OF THE CARCINOGENICITY OF DISULFIRAM TO BE MADE.
REPT'D USE : RUBBER ACCELERATOR/VULCANIZING AGENT, SEED DISINFECTANT AND ALCOHOL DENATURANT

CAS NO : **116-14-3**
CHEM NAME : ***** TETRAFLUOROETHYLENE**

REPORT 01 : IARC MONOGRAPH 19 1979
NO DATA ON THE CARCINOGENICITY OR MUTAGENICITY OF TETRAFLUOROETHYLENE WERE AVAILABLE TO THE WORKING GROUP.

REPORT 02 : EPA RESPONSE TO ITC RECOMMS (46FR53704) OCT 81
IN RESPONSE TO THE ITC RECOMMENDATIONS (45FR78432) THE EPA PUBLISHED AN ANPR PROPOSING A STRUCTURE-ACTIVITY CATEGORIZATION AND SELECTED TESTING REQUIREMENTS. INDUSTRY WILL CONDUCT ONCOGENICITY TESTING FOR VDF; RESULTS FROM VDF SHOULD APPLY TO VINYL FLUORIDE. THE EPA AND INDUSTRY WILL CONDUCT MUTAGENICITY TESTING AND WILL DECIDE ON FURTHER TIER TESING, AS APPROPRIATE. EXISTING TERATOGENICITY DATA ARE SUFFICIENT FOR VDF. PRODUCTION FOR SEVERAL FLUORALKENES IS INSUFFICIENT TO WARRANT TESTING.

REPORT 03 : PROPOSED TEST RULE, 50FR46133 NOV 85
DOCUMENT PROPOSES A RULE TO REQUIRE TIERED MUTAGENICITY BIOASSAYS AND DEPENDING ON THE OUTCOME OF THE MUTAGENICITY TESTING, CHRONIC ONCOGENICITY BIOASSAYS FOR TFE. THIS IS EPA'S RESPONSE TO ITC'S DESIGNATING FLUOROALKANES FOR PRIORITY TESTING CONSIDERATION IN ITS 7TH ANNUAL REPORT, PUBLISHED IN 45FR78432.

REPORT 07 : FLUOROALKENES; FINAL TEST RULE 52FR21516 JUN 87
EPA IS ISSUING A FINAL TEST RULE UNDER TSCA TO REQUIRE HEALTH EFFECTS TESTING OF CERTAIN FLUOROALKENES. TEST RULE BECOMES EFFECTIVE ON JULY 22. 1987.

REPT'D USE : CHEMICAL INTERMEDIATE AND PLASTIC MONOMER

CAS NO : **124-64-1**

CHEM NAME : ***** TETRAKIS(HYDROXYMETHYL)PHOSPHONIUM CHLORIDE (THPC)**

REPORT 01 : CPSC FLAM.E RETARDANTS MONOGRAPH JUL 82
THPC WAS NOT MUTAGENIC FOR SALMONELLA TA98 OR TA100. IN THE REC ASSAY IN B. SUBTILIS, THPC WAS ACTIVE WITH AND WITHOUT METABOLIC ACTIVATION. ABILITY OF THPC TO PRODUCE CHROMOSOME ABERRATIONS IN RAT BONE-MARROW CELLS WAS QUESTIONABLE. THPC PRODUCED A SIGNIFICANT INCREASE IN THE FREQUENCY OF CHROMOSOME BREAKS IN CHINESE HAMSTER CELLS.

REPT'D USE : FLAME RETARDANT

CAS NO : **55566-30-8**

CHEM NAME : ***** TETRAKIS(HYDROXYMETHYL)PHOSPHONIUM SULFATE (THPS)**

REPORT 01 : NTP TECH REPORT NTP TR 296 NI PUB #87-2552 FEB 87
UNDER CONDITIONS OF A 2-YEAR GAVAGE STUDY, NO EVIDENCE OF CARCINOGENICITY IN EITHER SEX OF F344/N RATS OR B6C3F1 MICE GIVEN 5 OR 10 MG/KG. NO EVIDENCE OF CARCINOGENICITY

IN EITHER SEX OF F344/N RATS GIVEN 3.75 OR 7.5 MG/KG, IN MALE B6C3F1 MICE GIVEN 7.5 OR 15 MG/KG, OR IN FEMALE B6C3F1 MICE GIVEN 15 OR 30 MG/KG.

REPT'D USE : TEXTILE FLAME RETARDANT

CAS NO : **140-66-9**
CHEM NAME : ***** 4-(1,1,3,3-TETRAMETHYLBUTYL)PHENOL (TMBP)**

REPORT 01 : DECISION TO ADOPT NEGOT. TESTING PROG. 49FR29449 JUL 84
NO PUBLIC COMMENTS RECEIVED REGARDING NOV 15, 1983 DECISION NOT TO TEST TMBP, SO EPA NOT PROPOSING A TSCA 4(A) RULE REQUIRING ENVIRONMENTAL EFFECTS TESTING.

CAS NO : **75-74-1**
CHEM NAME : ***** TETRAMETHYLLEAD (TML)**

REPORT 01 : IARC MONOGRAPH 1973
THE WORKING GROUP WAS NOT AWARE OF ANY ADEQUATE INHALATION STUDY ON TEL OR TML.

REPT'D USE : GASOLINE ADDITIVE

CAS NO : **2782-91-4**
CHEM NAME : ***** 1,1,3,3-TETRAMETHYL-2-THIOUREA**

REPORT 06 : MULTI-AGENCY CROSS CHECK
N HHS MAGUIRE
Y OSHA CARC
N EPA CAG
N EPA RCRA
N EPA CLEAN AIR
N EPA CLEAN WATER

REPT'D USE : NO AVAILABLE PRODUCTION OR USE DATA

CAS NO : **509-14-8**
CHEM NAME : ***** TETRANITROMETHANE**

REPORT 01 : NTP TECHNICAL BULLETIN 3 DEC 80
UNDER 2-YEAR INHALATION STUDIES CLEAR EVIDENCE OF CARCINOGENIC ACTIVITY OF TETRANITROMETHANE FOR MALE AND FEMALE F344/N RATS AND MALE AND FEMALE B6C3F1 MICE BASED ON INCREASED INCIDENCES OF ALVEOLAR/BRONCHIOLAR NEOPLASMS IN BOTH SPECIES AND SQUAMOUS CELL CARCINOMAS OF LUNG IN RATS.

REPORT 02 : NTP TECHNICAL BULLETIN NO. 3 DEC 80
TETRANITROMETHANE WAS POSITIVE IN THE SALMONELLA TYPHIMURIUM MUTAGENICITY ASSAY IN ONE LABORATORY.

REPORT 03 : NTP REPORT 386 NIH PUBLICATION NO 90-2841 MAR 90
UNDER 2-YEAR INHALATION STUDIES CLEAR EVIDENCE OF CARCINOGENIC ACTIVITY OF TETRANITROMETHANE FOR CARCINOGENIC ACTIVITY OF TETRANITROMETHANE FOR MALE AND FEMALE F344/N RATS AND MALE AND FEMALE B6C3F1 MICE, BASED ON INCREASED INCIDENCES OF ALVEOLAR/BRONCHIOLAR NEOPLASMS IN BOTH SPECIES AND SQUAMOUS CELL CARCINOMAS OF LUNG IN RATS. CHRONIC INFLAMMATION OF NASAL MUCOSA RELATED TO EXPOSURE IN RATS AND FEMALE MICE, AND HYPERPLASIA AND SQUAMOUS METAPLASIA OF RESPIRATORY EPITHELIUM INCREASED IN EXPOSED MALE RATS.

REPT'D USE : ROCKET PROPELLANT OXIDIZER, DIESEL FUEL ADDITIVE, TNT CONTAMINANT AND BIOCHEMICAL REAGENT

CAS NO : **83-67-0**
CHEM NAME : ***** THEOBROMINE**

REPORT 01 : IARC MONOGRAPHS 51 1991
INADEQUATE EVIDENCE FOR CARCINOGENICITY IN HUMANS OF THEOBROMINE. NO DATA ON THE CARCINOGENICITY IN EXPERIMENTAL ANIMALS. ORAL ADMINISTRATION OF HIGH DOSES OF THEOBROMINE TO RATS CAUSED SEVERE TESTICULAR ATROPHY, WHICH LOWER LEVELS FOR PROLONGED PERIODS HAD NO SIGNIFICANT ADVERSE EFFECTS ON TESTES. MICE, HAMSTERS AND DOGS LESS SENSITIVE THAN RATS OR RESISTANT TO EFFECT IN CAUSING TESTCULAR CHANGES. NO ADVERSE REPRODUCTIVE EFFECT IN THREE-GENERATION STUDY IN RATS GIVEN COCOA POWDER CONTAINING THEOBROMINE IN DIET. TERATOGENIC EFFECTS OBSERVED IN RABBITS AFTER GAVAGE BUT NOT AFTER DIETARY ADMINISTRATION; DEVELOPMENTAL TOXICITY AT LOWEST DOES LEVEL IN RABBITS.

REPORT 02 : IARC MONOGRAPH 51 1991
IN HUMAN CELLS IN VITRO, THEOBROMINE INDUCED SISTER CHROMATID EXCHANGE AND CHROMOSOMAL BREAKS. IN CULTURED MAMMALIAN CELLS, INDUCED GENE MUTATIONS AND SISTER CHROMATID EXCHANGE BUT NOT CHROMOSOMAL ABERRATIONS OR CELL TRANSFORMATION. IN PLANTS, DID NOT INDUCE CHROMOSOMAL ABERRATIONS. INDUCED GENE MUTATIONS IN LOWER EUKARYOTES AND BACTERIA BUT GAVE NEGATIVE RESULTS IN SALMONELLA/MAMMALIAN MICROSOME ASSAY.

REPT'D USE : PRINCIPAL ALKALOID OF CACAO BEAN, CAFFEINE, PHARMACEUTICAL PRODUCTS, CHOCOLATE BEVERAGES, CHOCOLATE-BASED FOODS, GREEN COFFEE BEANS, TEA AND MATE.

CAS NO : **58-55-9**
CHEM NAME : ***** THEOPHYLLINE**

REPORT 01 : IARC MONOGRAPHS 51 1991
INADEQUATE EVIDENCE FOR CARCINOGENICITY IN HUMANS OF THEOPHYLLINE. INADEQUATE EVIDENCE FOR THE CARCINOGENICITY IN EXPERIMENTAL ANIMALS; LIMITED DATA ON MOTHERS DURING PREGNANCY SHOWED NO EXCESS IN FREQUENCY OF MALFORMATIONS IN OFFSPRING. GIVEN BY GAVAGE AT HIGH DOSES DECREASED TESTICULAR WEIGHT IN RATS AND MICE, BUT NO CHANGE IN SEMEN CHARACTERISTICS. ADMINISTRATION IN DIET AT DOSE LEVELS MILDLY TOXIC TO ADULTS CASUED DECREASED NUMBERS OF LITTERS, DECREASED LIVE LITTER SIZE, INCREASED NUMBER OF RESORPTIONS, DECREASED NEONATAL WEIGHT. ABNORMAL SPERM IN RATS BUT NOT IN MICE AT HIGH DOSE LEVELS.

REPORT 02 : IARC MONOGRAPHS 51 1991
THEOPHYLLINE INDUCED SISTER CHROMATID EXCHANGE IN CHINESE HAMSTERS IN VIVO BUT NOT DOMINANT LETHAL MUTATIONS IN MICE OR CHROMOSOMAL ABERRATIONS IN BONE MARROW OF RATS. NEGATIVE RESULTS IN HOST-MEDIATED ASSAY WITH SALMONELLA TYPHIMURIUM IN MICE. IN CULTURED HUMAN CELLS, INDUCED SISTER CHROMATID EXCHANGE AND CHROMOSOMAL BREAKS BUT NOT MICRONUCLEI OR CHROMOSOMAL ABERRATIONS. INDUCED SISTER CHROMATID EXCHANGE AND CHROMOSOMAL ABERRATIONS BUT NOT MICRONUCLEI OR GENE MUTATION IN ANIMAL CELLS IN VITRO. RESULTS ON THE INDUCTION OF CHROMOSOMAL ABERRATIONS IN PLANTS EQUIVOCAL. IN LOWER EUKARYOTES, INDUCED GENE MUTATIONS. NEGATIVE RESULTS SALMONELLA/MAMMALIAM.

RFPT'D USE : BLACK TEA, GREEN COFFEE, COCOA COTYLEDON, DRIED MATE AND PHARMACEUTICAL

CAS NO : **62-55-5**
CHEM NAME : ***** THIOACETAMIDE (TAA)**

REPORT 01 : IARC MONOGRAPH 7 1974
THIOACETAMIDE IS CARCINOGENIC TO MICE AND RATS FOLLOWING ORAL ADMINISTRATION, THE ONLY ROUTE TESTED. IT INDUCED LIVER-CELL TUMORS IN MICE AND LIVER-CELL AND BILE DUCT TUMORS IN RATS. IN A 1979 SUPPLEMENT, IARC DETERMINED THAT THERE IS SUFFICIENT EVIDENCE FOR THE CARCINOGENICITY OF THIOACETAMIDE IN ANIMALS.

REPORT 02 : NTP TECHNICAL BULLETIN 8 JUL 82
THIOACETAMIDE WAS POSITIVE IN THE L5178Y MOUSE LYMPHOMA ASSAY.

REPORT 06 : MULTI-AGENCY CROSS CHECK
Y HHS MAGUIRE
N OSHA CARC
Y EPA CAG
N EPA RCRA
N EPA CLEAN AIR
N EPA CLEAN WATER
REPT'D USE : CHEMICAL REAGENT, ORGANIC SOLVENT, MOTOR FUEL STABILIZER AND VULCANIZATION ACCELERATOR

CAS NO : **139-65-1**
CHEM NAME : ***** 4,4'-THIODIANILINE**

REPORT 01 : NCI CARC TRS 47 DHEW PUB NO 78-847
UNDER THE CONDITIONS OF THIS BIOASSAY, 4,4'-THIODIANILINE WAS CARCINOGENIC FOR FISCHER 344 RATS, INDUCING TUMORS IN THE LIVER, THYROID, COLON, AND EAR CANAL OF MALE RATS, AND THE THYROID, UTERUS, AND EAR CANAL OF FEMALE RATS. 4,4'-THIODIANILINE WAS CARCINOGENIC FOR B6C3F1 MICE, INDUCING TUMORS IN THE LIVER AND THYROID OF BOTH MALES AND FEMALES (FEEDING STUDY).

REPORT 02 : IARC MONOGRAPH 16 1978
4,4'-THIODIANILINE IS CARCINOGENIC IN RATS AFTER ITS ORAL ADMINISTRATION; IT PRODUCED AN INCREASED INCIDENCE OF MAMMARY CARCINOMAS.

REPORT 03 : IARC MONOGRAPH 27 1982
4,4'-THIODIANILINE WAS TESTED IN MICE AND RATS BY DIETARY ADMINSTRATION. IT WAS CARCINOGENIC FOR ANIMALS OF BOTH SPECIES. THERE IS SUFFICIENT EVIDENCE FOR THE CARCINOGENICITY OF 4,4'-THIODIANILINE IN ANIMALS.

REPORT 04 : IARC MONOGRAPH 27 1982
4,4'-THIODIANILINE INDUCED REVERSE MUTATIONS IN SALMONELLA STRAINS TA98 AND TA100 WITH METABOLIC ACTIVATION. NO DATA CONCERNING EFFECTS ON REPRODUCTION OR PRENATAL TOXICITY WERE AVAILABLE.

REPORT 06 : MULTI-AGENCY CROSS CHECK
N HHS MAGUIRE
Y OSHA CARC
N EPA CAG
N EPA RCRA
N EPA CLEAN AIR
N EPA CLEAN WATER
REPT'D USE : DYE INTERMEDIATE FOR DISAZO DYES. NO KNOWN CURRENT PRODUCTION OR USE

CAS NO : **64039-27-6**
CHEM NAME : ***** BETA-THIOGUANINE DEOXYRIBOSIDE (B-TGDR)**

REPORT 01 : NCI CARC TRS NO 57 DHEW PUB NO NIH 78-1363
UNDER THE CONDITIONS OF THIS BIOASSAY, THE LOW SURVIVAL OF THE DOSED AND VEHICLE-CONTROL GROUPS OF MICE, AS WELL AS THE POSSIBLE PROCEDURAL PROBLEM THAT MAY HAVE AFFECTED THE INCIDENCES OF TUMORS IN THESE GROUPS, DOES NOT ALLOW A DETERMINATION TO BE MADE OF THE CARCINOGENIC POTENTIAL OF B-TGDR IN THIS SPECIES. B-TGDR IN THE VEHICLE OF 0.05% POLYSORBATE 80 WAS, HOWEVER, CARCINOGENIC IN RATS, PRODUCING CARCINOMAS OF THE EAR CANAL IN THE FEMALES AND POSSIBLY ALSO IN THE MALES. B-TGDR WAS GIVEN BY I.P. INJECTION.
REPT'D USE : EXPERIMENTAL ANTINEOPLASTIC AGENT

CAS NO : **52-24-4**
CHEM NAME : ***** TRIS(1-AZIRIDINYL) PHOSPHINE SULFIDE**

REPORT 01 : NCI CARC TRS 58 DHEW PUB NO 78-1308
UNDER THE CONDITIONS OF THIS BIOASSAY, THIO-TEPA WAS CARCINOGENIC IN BOTH SPRAGUE-DAWLEY RATS AND B6C3F1 MICE. IN THE RATS, THE CHEMICAL INDUCED SQUAMOUS-CELL CARCINOMA OF THE SKIN OR EAR CANAL IN BOTH MALES AND FEMALES, AND HEMATOPOIETIC NEOPLASMS IN THE MALES; IN THE MICE, IT INDUCED LYMPHOMA OR LYMPHOCYTIC LEUKEMIA IN BOTH SEXES AND SQUAMOUS-CELL CARCINOMA IN THE SKIN AND ASSOCIATED GLANDS OF MALES. THIO-TEPA WAS GIVEN BY ABDOMINAL INJECTION.

REPORT 02 : IARC SUPPLEMENT 1 1979
THIO-TEPA IS CARCINOGENIC IN MICE AND RATS AFTER ADMINISTRATION BY VARIOUS ROUTES, PRODUCING A VARIETY OF MALIGNANT TUMORS. THERE ARE SEVERAL CASE REPORTS AND EPIDEMIOLOGICAL STUDIES SUGGESTING THE DEVELOPMENT OF ACUTE NON-LYMPHOCYTIC LEUKEMIA IN PATIENTS TREATED WITH THIOTEPA. THERE IS SUFFICIENT EVIDENCE FOR THE CARCINOGENICITY OF THIO-TEPA IN ANIMALS. IN A 1979 SUPPLEMENT, IARC CONCLUDED THAT THIO-TEPA IS CARCINOGENIC FOR HUMANS.

REPORT 03 : IARC MONOGRAPH 9 1979
THIO-TEPA WAS TERATOGENIC IN MICE INJECTED I.P. WITH SINGLE DOSES ON VARIOUS DAYS OF GESTATION. THIO-TEPA INCREASED THE FREQUENCY OF INHERITED RECESSIVE LETHALS IN DROSOPHILA. TREATMENT OF MALE MICE WITH SINGLE I.P. INJECTIONS OF THIO-TEPA INDUCED A HIGH FREQUENCY OF DOMINANT LETHALS.

REPORT 06 : MULTI-AGENCY CROSS CHECK
Y HHS MAGUIRE
N OSHA CARC
Y EPA CAG
N EPA RCRA
N EPA CLEAN AIR
N EPA CLEAN WATER
REPT'D USE : CHEMOTHERAPEUTIC AGENT

CAS NO : **141-90-2**
CHEM NAME : ***** THIOURACIL**

REPORT 01 : IARC MONOGRAPH 7 1974
THIORACIL INCREASED THE INCIDENCE OF LIVER-CELL TUMORS IN MICE AND PRODUCED THYROID TUMORS IN SEVERAL STRAINS OF RATS FOLLOWING ORAL ADMINISTRATION, THE ONLY ROUTE TESTED.

REPORT 06 : MULTI-AGENCY CROSS CHECK
N HHS MAGUIRE
Y OSHA CARC
N EPA CAG
N EPA RCRA
N EPA CLEAN AIR
N EPA CLEAN WATER
REPT'D USE : MFR OF ANTI-THYROID MEDICATIONS

CAS NO : **62-56-6**
CHEM NAME : ***** THIOUREA**

REPORT 01 : NTP TECHNICAL BULLETIN 8 JUL 82
THIOUREA GAVE QUESTIONABLE RESULTS IN THE L5178Y MOUSE LYMPHOMA ASSAY.

REPORT 02 : IARC MONOGRAPH 7 1974
THIOUREA PRODUCED LIVER, THYROID AND ZYMBAL GLAND TUMORS IN RATS FOLLOWING ORAL ADMINISTRATION. INTRAPERITONEAL INJECTION FOLLOWED ORAL ADMINISTRATION ALSO LED TO THE FORMATION OF ZYMBAL GLAND TUMORS IN RATS. ORAL AND S.C. ADMINISTRATION TO MICE DID NOT PRODUCE THYROID TUMORS; HOWEVER, THE EXPERIMENT USING THE S.C. ROUTE WAS INADEQUATE. EVIDENCE FOR THE CARCINOGENICITY OF THIOUREA IN EXPERIMENTAL ANIMALS IS CONSIDERED SUFFICIENT.

REPORT 03 : NAS REPORT ON DIET, NUTRITION AND CANCER 1982
THIOUREA HAS BEEN SHOWN TO CAUSE THYROID TUMORS, HEPATIC ADENOMAS, AND EPIDERMOID CARCINOMAS OF ZYMBAL

GLANDS WHEN ADMINISTERED TO RATS. THIOUREA WAS NEGATIVE IN THE SALMONELLA ASSAY, BUT POSITIVE IN THE HOST-MEDIATED ASSAY. IT ALSO INDUCED TRANSFORMATION IN HAMSTER EMBRYO CELLS.

REPORT 04 : 4TH ANNUAL REPORT ON CARCINOGENS 1985
(MAGUIRE REPT)
SUFFICIENT EVIDENCE FOR CARCINOGENICITY OF THIOUREA IN TEST ANIMALS. THYROID, LIVER AND ZYMBALS GLAND TUMORS WERE PRODUCED IN RATS AFTER ADMINISTRATION IN DRINKING WATER. IP INJECTION + DRINKING WATER INDUCED ZYMBALS GLAND TUMORS IN RATS. ADMINISTRATION OF CPD IN DIET OR SC INJECTION DID NOT PRODUCE TUMORS IN MICE; HOWEVER, THIS TEST CONSIDERED INADEQUATE. RESULTS OF A FEEDING EXPERIMENT IN TROUT CAN NOT BE CONSIDERED CONCLUSIVE.

REPORT 06 : MULTI-AGENCY CROSS CHECK
Y HHS MAGUIRE
N OSHA CARC
Y EPA CAG
Y EPA RCRA
N EPA CLEAN AIR
N EPA CLEAN WATER
REPT'D USE : PHOTOGRAPHIC PAPERS/TONING AGENT, RUBBER ACCELERATOR, ANIMAL HIDE GLUES, FLAME RETARDANTS, CORROSION INHIBITOR, DRAIN CLEANER, FUNGICIDE, TARNISH REMOVERS, LAB REAGENT AND ANTITHYROID AGENT

CAS NO : **137-26-8**
CHEM NAME : ***** THIURAM**

REPORT 01 : IARC MONOGRAPH 12 1976
THIURAM HAS BEEN TESTED BY ORAL ADMINISTRATION AND BY SINGLE SUBCUTANEOUS INJECTION IN MICE. ALTHOUGH NO CARCINOGENIC EFFECT WAS OBSERVED IN THESE STUDIES, THE AVAILABLE DATA ARE INSUFFICIENT FOR AN EVALUATION OF THE CARCINOGENICITY OF THIS COMPOUND TO BE MADE.

REPORT 02 : IARC MONOGRAPH 12 1976
THIURAM INDUCED AND INCREASED NUMBER OF CHROMOSOME ABERRATIONS IN METAPHASES OF BONE MARROW CELLS OF TREATED MICE.
REPT'D USE : RUBBER ACCELERATOR, FUNGICIDE, BACTERIOSTAT AND ANTISEPTIC

CAS NO : **1314-20-1**
CHEM NAME : ***** THORIUM DIOXIDE**

REPORT 01 : 5TH DRAFT ANNUAL REPORT ON CARCINOGENS OCT 89
SUFFICIENT EVIDENCE FOR CARCINOGENICITY OF THORIUM DIOXIDE IN HUMANS. INTRAVASCULAR INJECTION INDUCED TUMORS OF LIVER, INCLUDING HEPATOCELLULAR AND CARCINOMAS, CHOLANGIOCELLAR, EXTRA-HEPATIC BILIARY SYSTEM COMMON HEPATIC DUCT, LIVER CELL, AND UNDIFFERENTIATED CARCINOMAS, SARCOMAS, HEMANGIOENDOTHELLIOMAS, RETICULUM CELL SARCOMAS, ADENOCARCINOMAS, AND HEPATOMAS, WITH LATENCY OF 21 TO 23 YEARS. INTRAVASCULAR INJECTION INDUCED TUMORS OF KIDNEY, INCLUDING CARCINOMAS OF RENAL PELVIS, WITH LATENCY OF 21 TO 25 YEARS; CARCINOMAS OF MAXILLARY SENSES, SPINDLE CELL SARCOMAS IN AND ASSOCIATED WITH RECENT STUDY SUGGESTS LATENCY OF 36 YEARS BETWEEN EXPOSURE AND ONSET OF ANGIOSARCOMAS.

REPORT 02 : DRAFT 5TH ANNUAL REPORT ON CARCINOGENS OCT 89
SUFFICIENT EVIDENCE FOR CARCINOGENICITY OF THORIUM DIOXIDE IN EXPERIMENTAL ANIMALS. ADMINISTERED BY INTRAVENOUS INJECTION INDUCED HEMANGIOENDOTHELIO-SARCOMAS AND RETICULOENDOTHELIOSARCOMAS OF LIVER, SPLEEN, AND LUNG, RHABDOMYOSARCOMAS, ALVEOLAR CELL CARCINOMAS, POLYMORPHOUS SARCOMAS OF LIVER, AND RETICULUM CELL SARCOMAS IN RABBITS; AND ADENOMAS OF LUNG, HEPATOMAS, HEMANGIOENDOTHELIO-SARCOMAS OF LIVER, LOCAL FIBER SARCOMAS, LYMPHOMAS, HEPATOCELLULAR CARINOMAS AND ADENOMAS, ANGIOMAS OF SPLEEN, AND SPINDLE CELL SARCOMAS IN MICE.

REPORT 06 : MULTI-AGENCY CROSS CHECK
Y HHS MAGUIRE
N OSHA CARC
N EPA CAG
N EPA RCRA
N EPA CLEAN AIR
N EPA CLEAN WATER
REPT'D USE : GAS MAINTLESS, RADIOPAQUE, X-RAY IMAGING, ELECTRODES AND ARC WELDING

CAS NO : **3864-99-1**
CHEM NAME : ***** TINUVIN 327**

REPORT 01 : EPA TSCA 8(E) 1088-0756 OCT 88
IN 90-DAY FEEDING STUDY INVOLVING TINUVIN 327 AT DOSE LEVELS OF 2,000 TO 50,000 PPM IN DAILY DIET OF RATS, "DISTINCT" LIVER DAMAGE EVIDENCED BY NECROTIC HEPATIC CELLS (AND FOCI OF NECROSIS IN SOME CASES) AND SIGNS OF ANEMIA

OBSERVED IN THE MALE RATS. THIS STUDY ALSO SHOWED PANCREAS OF THE MALE RATS HAD LESIONS. IN ANOTHER 90 DAY FEEDING STUDY DECREASED HEMOGLOBIN CONTENT AND PACKED CELL VOLUME, INCREASED LIVER, KIDNEY AND THYROID WEIGHTS, AND ISOLATED LIVER CELL NECROSIS WERE OBSERVED IN THE MALE RATS ONLY IN 100 AND 200 PPM GROUPS.

REPT'D USE : PLASTICS AND POLYMER COATINGS STABILIZER

CAS NO : **13463-67-7**

CHEM NAME : ***** TITANIUM DIOXIDE**

REPORT 01 : NCI CARC TRS 97 DHEW PUB NO 79-1347
UNDER THE CONDITIONS OF THIS BIOASSAY, TITANIUM DIOXIDE WAS NOT CARCINOGENIC BY THE ORAL ROUTE FOR FISCHER 344 RATS OR B6C3F1 MICE.

REPORT 02 : EPA TSCA 8(E) 8EHQ-1083-0497 OCT 83
DU PONT SUBMITTED SUMMARIZED PRELIMINARY FINDINGS FROM A 2 YEAR STUDY CONDUCTED AT ITS HASKELL LABORATORY: 100 MALE AND 100 FEMALE CHARLES RIVER-CD RATS EXPOSED TO TITANIUM DIOXIDE DUST LEVELS OF 0, 10, 50 OR 250 MG/M3 FOR 6 HOURS PER DAY, 5 DAYS PER WEEK FOR 104 WEEKS. RATS SACRIFICED AT THE END OF THE 104 WEEK EXPOSURE PERIOD SHOWED A STATISTICALLY SIGNIFICANT INCIDENCE OF MICROSCOPIC LUNG TUMORS IDENTIFIED AS SQUAMOUS CELL CARCINOMA (IN ABOUT 10% OF THE RATS) AND BRONCHIOALVEOLAR ADENOMA (IN ABOUT 15% OF THE RATS) AT THE 250 MG/M3 LEVEL.

REPT'D USE : PIGMENT PAINTS/VARNISHES, FOOD/PHARMACEUTICAL/COSMETIC; COLOR PROTECTIVE COATINGS FOR PAPER AND PLASTIC, INKS, CLEANING AGENTS, FR CHEMICALS, AEROSOLS, SPACE DEODORIZERS, PHOTOGRAPHIC CHEMICAL, SWIMMING POOL CHEMICALS, ADHESIVES/GLUES, FLOOR WAXES AND MODEL TOY FUELS

CAS NO : **1271-19-8**

CHEM NAME : ***** TITANIUM FERROCENE**

REPORT 01 : NTP TECHNICAL BULLETIN 3 DEC 80
THE COMPOUND WAS FOUND TO BE POSITIVE IN THE SALMONELLA TYPHIMURIUM MUTAGENICITY ASSAY IN ONE LABORATORY.

REPORT 02 : NTP DRAFT TR 399 NIH PUB. NO 91-2854 NOV 90
UNDER CONDITIONS OF 2 YEAR STUDIES EQUIVOCAL EVIDENCE OF CARCINOGENIC ACTIVITY OF TITANOCENE DICHLORIDE FOR MALE AND FEMALE F344/N RATS BASED ON MARGINAL INCREASE IN INCIDENCE OF SQUAMOUS CELL NEOPLASMS OF FORESTOMACH. NONNEOPLASTIC LESIONS ASSOCIATED WITH ADMINISTRATION FOR UP TO 2 YEARS INCLUDED EROSIONS AND INFLAMMATION OF

GASTRIC MUCOSA, HYPERPLASIA AND METAPLASIA OF FUNDIC GLANDS WITH FIBROSIS OF LAMINA PROPRIA IN GLANDULAR STOMACH, AND ACANTHOSIS (HYPERPLASIA) AND HYPERKERATOSIS OF FORESTOMACH EPITHELIUM.

REPT'D USE : CATALYST

CAS NO : **7550-45-0**
CHEM NAME : ***** TITANIUM TETRACHLORIDE**

REPORT 01 : EPA TSCA 8(E) STATUS REPORT 8EHQ-0984-0530 SEP 84
DU PONT DE NEMOURS SUBMITTED RESULTS FROM 2-YEAR INHALATION STUDY OF MALE AND FEMALE CHARLES RIVER-CD RATS EXPOSED TO TITANIUM TETRACHLORIDE VAPORS AT EXPOSURE LEVELS OF 0, 0.1, 1 OR 10 MG/M3 FOR 6 HOURS A DAY, 5 DAYS A WEEK. IN LUNGS, HISTOPATHOLOGICAL EXAMINATION REVEALED TUMORS AT 10 MG/M3 EXPOSURE LEVEL. TUMORS WERE KERATINIZED, CYSTIC SQUAMOUS CELL CARCINOMA AND SQUAMOUS CELL CARCINOMA, DIFFERENTIATED.

CAS NO : **CLASS-17**
CHEM NAME : ***** TOBACCO HABITS OTHER THAN SMOKING**

REPORT 01 : IARC MONOGRAPH 37 SEP 85
SUFFICIENT EVIDENCE THAT ORAL USE OF SOME COMMONLY USED SNUFFS IN N. AMERICA AND WESTERN EUROPE CARCINOGENIC. IN AGGREGATE SOME STUDIES SHOW ORAL USE OF SMOKELESS TOBACCO PRODUCTS AND OF TOBACCO MIXED WITH LIME (KHAINI) ARE CARCINOGENIC TO HUMANS. LIMITED EVIDENCE CHEWING TOBACCO COMMONLY USED IN THESE AREAS IS CARCINOGENIC. INADEQUATE EVIDENCE ORAL USE OF OTHER SMOKELESS TOBACCO PREPARATIONS (I.E. NASS, NASWAR, MISHRI, GUDAKHU AND SHAMMAH) OR NASAL USE OF SNUFF CARCINOGENIC TO HUMANS. ALSO INADEQUATE EVIDENCE TO EVALUATE CARCINOGENICITY OF CHEWING TOBACCO, SNUFF OR NASS TO EXPERIMENTAL ANIMALS.

REPORT 02 : IARC MONOGRAPH 37 SEP 85
ETHANOL EXTRACTS OF CHEWING TOBACCO (NICOTIANA TABACUM) INDUCE MUTATIONS IN SALMONELLA TYPHIMURIUM AND IN CHINESE HAMSTER V79 CELLS, AND MICRONUCLEI IN BONE-MARROW CELLS OF MICE. ETHYL ACETATE EXTRACTS INDUCE SISTER CHROMATID EXHANGES IN CULTURED HUMAN-LYMPHOCYTES AND IN HUMAN LYMPHOBLASTOID CELL LINE. ETHYL ACETATE AND ETHANOL EXTRACTS INDUCE TRANSFORMATION IN SYRIAN HAMSTER EMBRYO CELLS. EXTRACTS OF NASS AND KHAINI INDUCE CHROMOSOMAL ABERRATIONS IN CHINESE HAMSTER OVARY CELLS. SALIVA COLLECTED DURING CHEWING OF INDIAN TOBACCO INDUCED CHROMOSOMAL ABERRATIONS IN

CHINESE HAMSTER OVARY CELLS. MICRONUCLEATED CELLS FOUND IN ORAL-MUCOSA CELLS FROM KHAINI AND USERS.
REPT'D USE : TOBACCO CHEWING AND SNUFF TAKING

CAS NO : **1156-19-0**
CHEM NAME : ***** TOLAZAMIDE**

REPORT 01 : NCI CARC TRS 51 DHEW PUB NO 78-1301
UNDER THE CONDITIONS OF THIS BIOASSAY, TOLAZAMIDE WAS NOT CARCINOGENIC FOR FISCHER 344 RATS OR B6C3F1 MICE (FEEDING STUDY).
REPT'D USE : ORAL HYPOGLYCEMIC AGENT

CAS NO : **64-77-7**
CHEM NAME : ***** TOLBUTAMIDE**

REPORT 01 : NTP TECHNICAL BULLETIN 5 AUG 81
TOLBUTAMIDE WAS NEGATIVE IN THE SALMONELLA TYPHIMURIUM MUTAGENICITY ASSAY IN ONE LABORATORY.

REPORT 02 : NCI CARC TRS 31 DHEW PUB NO 77-831
UNDER THE CONDITIONS OF THIS BIOASSAY, TOLBUTAMIDE WAS NOT CARCINOGENIC FOR FISCHER 344 RATS OR B6C3F1 MICE (FEEDING STUDY).
REPT'D USE : HYPOGLYCEMIC AGENT

CAS NO : **119-93-7**
CHEM NAME : ***** O-TOLIDINE(3,3-DIMETHYLBENZIDINE)**

REPORT 01 : TSCA ITC RECOMMENDATIONS (44FR70664) DEC 79
IN A JANUARY 1981 RULING, THE SOUTHERN DISTRICT COURT OF NEW YORK ORDERED EPA TO PROPOSE TEST RULES, OR DECISIONS NOT TO REQUIRE TESTING, BY THE END OF 1983 FOR ALL CHEMICALS LISTED IN THE FIRST FIVE ITC REPORTS. THE SCHEDULE SUBMITTED BY THE EPA, AND ACCEPTED BY THE COURT WITHOUT CHANGE, SHOWS THAT A DECISION FOR TESTING OF HUMAN HEALTH EFFECTS OF O-TOLIDINE BASED DYES WILL BE MADE IN 1982.

REPORT 02 : NIOSH CRITERIA DOCUMENT AUG 78
OCCUPATIONAL EXPOSURE TO O-TOLIDINE SHALL BE CONTROLLED SO THAT EMPLOYEES ARE NOT EXPOSED AT A CONCENTRATION GREATER THAN 20 MICROGRAMS PER CUBIC METER OF AIR, DETERMINED AS A CEILING CONCENTRATION IN A 1-HOUR SAMPLING PERIOD. SKIN CONTACT WITH O-TOLIDINE SHALL BE AVOIDED. ANIMALS EXPOSED TO O-TOLIDINE HAVE DEVEL. CANCERS AT VARIOUS SITES AND RESULTS OF BACTERIAL TESTS INDICATE THAT

O-TOLIDINE MAY BE MUTAGENIC. DAMAGE TO MAMMALIAN DNA METABOLISM HAS ALSO BEEN REPORTED. HUMANS HAVE DEVELOPED NASAL IRRITATION FOLLOWING EXP. NO OSHA STANDARD.

REPORT 03 : IARC MONOGRAPH 1 1979
PURIFIED O-TOLIDINE IS A SYSTEMIC CARCINOGEN IN THE RAT WHEN GIVEN SUBCUTANEOUSLY. IN FEEDING EXPERIMENTS, THE COMMERCIAL PRODUCT FAILED TO PRODUCE TUMORS IN HAMSTERS. IN A 1979 SUPPLEMENT, IARC DETERMINED THAT THERE IS SUFFICIENT EVIDENCE FOR THE CARCINOGENICITY OF O-TOLIDINE IN ANIMALS.

REPORT 04 : NTP CARC TEST PROGRAM JUL 82
IN A CURRENT NTP BIOASSAY, O-TOLIDINE IS BEING ADMINISTERED IN THE WATER TO F344 RATS. THE PRECHRONIC TEST PHASE IS IN PROGRESS.

REPORT 05 : EPA RESPONSE TO ITC RECOMS (46FR55004) NOV 81
EPA HAS DECIDED NOT TO PURSUE TESTING PROPOSALS FOR HEALTH AND ENVIRONMENTAL EFFECTS TESTING OF O-TOLIDINE DYES AT THIS TIME BECAUSE TESTING PROGRAMS WHICH ARE EXPECTED TO SUPPLY DATA SUFFICIENT FOR EPA NEEDS ARE BEING CURRENTLY PLANNED AND CONDUCTED, AND REGULATORY EFFORTS ARE BEING PURSUED.

REPORT 06 : MULTI-AGENCY CROSS CHECK
Y HHS MAGUIRE
N OSHA CARC
Y EPA CAG
N EPA RCRA
N EPA CLEAN AIR
N EPA CLEAN WATER

REPT'D USE : DYE INTERMEDIATE, COLORIMETRIC REAGENT FOR CHLORINE, SWIMMING POOL CHEMICALS AND CURING AGENT FOR PV RESINS

CAS NO : **108-88-3**
CHEM NAME : ***** TOLUENE**

REPORT 01 : EPA RCRA HEALTH AND ENV. EFFECTS PROFILE OCT 80
THE AVAILABLE STUDIES THAT DESCRIBE THE CARCINOGENIC OR MUTAGENIC POTENTIAL OF TOLUENE ARE INADEQUATE FOR DRAWING ANY CONCLUSIONS ABOUT ITS CARCINOGENICITY. THREE STUDIES IN RATS INDICATE THAT TOLUENE DAMAGES CHROMOSOMES IN BONE MARROW CELLS.

REPORT 02 : NIOSH CRITERIA DOCUMENT JUL 73
NIOSH RECOMMENDS THAT OCCUPATIONAL EXPOSURE TO TOLUENE BE NO MORE THAN 100 PARTS PER MILLION OF AIR (TWA) FOR 8 HOURS WITH A CEILING OF 200 PARTS PER MILLION OF AIR IN A

SAMPLING TIME OF 10 MINUTES. CERTAIN GRADES OF TOLUENE MAY CONTAIN AS MUCH AS 25% BENZENE AND OTHER AROMATICS. THE MAJOR PROBLEM OF TOLUENE TOXICITY CONCERNS ITS NARCOTIC EFFECTS ON WORKERS BY CAUSING SYMPTOMS AND SIGNS SUCH AS MUSCULAR WEAKNESS, INCOORDINATION, AND MENTAL CONFUSION WHICH MAY POSE A RISK TO THE WORKER AND OTHERS. THE OSHA STANDARD IS 200 PPM TWA.

REPORT 03 : CIIT 24-MONTH EXECUTIVE SUMMARY OCT 80
UNDER THE CONDITIONS OF THIS BIOASSAY, CHRONIC INHALATION EXPOSURE OF TOLUENE, AT VAPOR CONCENTRATIONS OF 300, 100, AND 30 PPM TO F344 ALBINO RATS, RESULTED IN NO EXPOSURE-RELATED LESIONS. IT IS CONCLUDED THAT TOLUENE VAPOR IS NON-TOXIC AT CONCENTRATIONS AS HIGH AS 300 PPM FOR UP TO TWO YEARS OF EXPOSURE IN RATS.

REPORT 04 : NTP TECHNICAL BULLETIN 2 APR 80
TOLUENE WAS FOUND TO BE NEGATIVE IN THE SALMONELLA TYPHIMURIUM MUTAGENICITY ASSAY IN ONE LABORATORY.

REPORT 05 : ITC FIRST REPORT TO EPA (42FR55026) OCT 77
THE INTERAGENCY TESTING COMMITTEE RECOMMENDS THAT THE EPA ADMINSTRATOR CONSIDER THE DEVELOPMENT OF TEST RULES TO TEST TOLUENE FOR CARCINOGENICITY, TERATOGENICITY, AND OTHER CHRONIC EFFECTS; AND THAT AN EPIDEMIOLOGIC STUDY BE CONDUCTED.

REPORT 06 : MULTI-AGENCY CROSS CHECK
N HHS MAGUIRE
N OSHA CARC
N EPA CAG
Y EPA RCRA
Y EPA CLEAN AIR
Y EPA CLEAN WATER

REPORT 07 : EPA CAG ASSESSMENT JUN 80
THERE IS SUGGESTIVE EVIDENCE, BASED ON SKIN PAINTING EXPERIMENTS IN MICE, THAT TOLUENE HAS A WEAK PROMOTING EFFECT ON DMBA-INITIATED SKIN CARCINOGENESIS. THREE STUDIES IN RATS INDICATE THAT TOLUENE DAMAGES CHROMOSOMES IN BONE MARROW CELLS WHEN ADMINISTERED BY INHALATION OR INJECTION. THE AVAILABLE STUDIES THAT DESCRIBE THE CARCINOGENIC OR MUTAGENIC POTENTIAL OF TOLUENE ARE INADEQUATE FOR DRAWING ANY CONCLUSIONS ABOUT THE CARCINOGENICITY OF TOLUENE.

REPORT 08 : NTP TR 371 NIH PUB. 89-2826 FEB 90
UNDER 2-YEAR OLD INHALATION STUDIES NO EVIDENCE OF CARCINOGENIC ACTIVITY FOR MALE OR FEMALE F344/N RATS EXPOSED TO TOLUENE AT CONCENTRATIONS OF 600 OR 1,200 PPM.

NO EVIDENCE OF CARCINOGENIC ACTIVITY FOR MALE OR FEMALE B6C3F1 MICE EXPOSED BY INHALATION TO TOLUENE AT CONCENTRATIONS OF 120, 600, OR 1,200 PPM FOR 2 YEARS. MALE AND FEMALE MICE MIGHT HAVE BEEN ABLE TO ENDURE SOMEWHAT HIGHER EXPOSURE CONCENTRATIONS WITHOUT COMPROMISING HEALTH OR LONGEVITY.

REPORT 09 : EPA FYI-AX- 0885-0294 AUG 85
API REPORTS PETROLEUM DERIVED HYDROCARBON WITH TOLUENE ADMINISTERED TO RATS OF BOTH SEXES AT 100,500 AND 2000 PPM VIA INHALATION; AT 2000 PPM EXPOSURE TO DAMS, SIGNIFICANT INHIBITION OF GROWTH OF OFFSPRING OVER BOTH GENERATIONS OBSERVED. NOTED BOTH IN UTERO AS EVIDENCED BY DECREASED INTACT UTERINE WEIGHTS AND FETAL BODY WEIGHT ON GESTATION DAY 21 IN 2000 PPM DOSE GROUP AT CESAREAN SECTION AND POSTPARTUM IN DECREASED PUP WEIGHTS THROUGHOUT LACTATION IN PUPS OF FOUR GROUPS. EXPOSURE AT THIS LEVEL TO MALE PARENTS ONLY IN TWO GROUPS DID NOT ELICIT SIMILAR FINDINGS AND CONSISTENT FINDINGS NOT EVIDENT AT 500 PPM OR LESS. THUS, 500 PPM CONCENTRATION CONSIDERED A "NO-EFFECT" LEVEL.

REPORT 10 : HEALTH ASSESS. DOC. EPA-600/8- 82-008F0/8-82- AUG 83
DOSE-RESPONSE RELATIONSHIPS FOR ACUTE EFFECTS IN HUMANS: AT 100-300 PPM, SIGNS OF INCOORDINATION, LACRIMATION, IRRITATION TO MUCOUS MEMBRANES; AT 300-800 PPM, GROSS SIGNS OF INCOORDINATION DURING EXPOSURES UP TO 8 HRS; AT 1500 PPM, PROBABLY NOT LETHAL FOR UP TO >4000 PPM, PROBABLY CAUSES RAPID IMPAIRMENT OF REACTION TIME AND COORDINATION AND MIGHT LEAD TO DEATH; AT 10000-30000 PPM, NARCOSIS WITHIN 30 MINUTES; LONGER MAY BE LETHAL.

REPT'D USE : FUEL COMPONENT, CHEMICAL INTERMEDIATE, SOLVENT, DYE INTERMEDIATE, ADHESIVES, LABORATORY AGENT, CLEANING AGENTS, AEROSOLS, PAINTS/VARNISHES, CHEMISTRY TOY SETS, FR CHEMICALS, PYROTECHNICS, AND SWIMMING POOL CHEMICALS

CAS NO : **15481-70-6**
CHEM NAME : ***** 2,6-TOLUENEDIAMINE DIHYDROCHLORIDE**

REPORT 01 : NCI/NTP CARC TRS 200 DHEW PUB. NO. 80-20
THE COMPOUND WAS NOT CARCINOGENIC FOR F344 RATS AND B6C3F1 MICE UNDER THE TEST CONDITIONS (FEEDING STUDY).

REPT'D USE : DYE INTERMEDIATE AND INTERMEDIATE IN MFR TDI FOR PU FOAMS. NO LISTING ON TSCA INVENTORY

CAS NO : **6369-59-1**
CHEM NAME : ***** 2,5-TOLUENEDIAMINE SULFATE**

REPORT 01 : NCI CARC TRS 126 DHEW PUB NO 78-1381
UNDER THE CONDITIONS OF THIS BIOASSAY, SUFFICIENT EVIDENCE WAS NOT OBTAINED TO DEMONSTRATE THE CARCINOGENICITY OF 2,5-TOLUENEDIAMINE SULFATE IN EITHER FISCHER 344 RATS OR B6C3F1 MICE (FEEDING STUDY).

REPORT 02 : NTP TECHNICAL BULLETIN 3 DEC 80
TERATOLOGY ASSAYS SHOWED THAT ORAL ADMINISTRATION OF 2,5-TOLUENEDIAMINE SULFATE WAS NOT ASSOCIATED WITH PRENATAL DEATH OR INCREASED MALFORMATIONS TO MICE AT DOSES TOXIC TO THE MOTHER.

REPORT 03 : EPA RESPONSE TO ITC RECOMMEND'N (47FR973) JAN 81
IN RESPONSE TO THE ITC RECOMMENDATIONS (45FR35897) THE EPA PROPOSES AN ADVANCED NOTICE REQUESTING COMMENTS CONCERNING A PROPOSAL TO CONDUCT MUTAGENICITY AND CARCINOGENICITY TESTING FOR 11 SELECTED PHENYLENEDIAMINES, AND TERATOGENICITY, REPRODUCTIVE EFFECTS AND OTHER CHRONIC EFFECTS TESTING, AS WELL AS AN EPIDEMIOLOGIC STUDY, FOR 12 PHENYLENEDIAMINES. THE EPA WILL CONSIDER ADDITIONAL TESTING FOR AN ADDITIONAL 37 PHENYLENEDIAMINES, AS APPROPRIATE.
REPT'D USE : DYE INTERMEDIATE, LABORATORY AGENT, HAIR DYE FORMULATIONS, FUR DYE, ANTIFREEZE AND NAIL POLISH

CAS NO : **26471-62-5**
CHEM NAME : ***** TOLUENE DIISOCYANATE (TDI)**

REPORT 01 : NTP CARC TEST PROGRAM DRAFT REPORT 82-2 JUL 82
NO. 82-2507
IN A BIOASSAY, 2,4-TDI MIXED WITH 2,6-TDI WAS ADMINISTERED BY GAVAGE TO F344 RATS AND B6C3F1 MICE. THE STUDY HAS BEEN COMPLETED AND THE REPORT IS IN PEER REVIEW.

REPORT 02 : NIOSH CRITERIA DOCUMENT
NIOSH RECOMMENDS THAT EXPOSURE CONCENTRATION LIMITS EQUIVALENT TO A CEILING CONCENTRATION OF 20 PPB AND A TWA CONCENTRATION OF 5 PPB BE OBSERVED FOR DIISOCYANATES. FOR TOLUENE DIISOCYANATE, THE RECOMMENDED TWA IS 35 MICROGRAMS/CU M FOR UP TO A 10-HOUR WORKSHIFT, 40-HOUR WORK WEEK WITH A CEILING CONCENTRATION OF 140 MICROGRAMS/CU M FOR ANY 10-MINUTE SAMPLING PERIOD BASED ON AIRWAY EFFECTS. THE PRESENT OSHA STANDARD IS 0.02 PPM CEILING.

REPORT 03 : NTP TR 251, NIH PUB.#86-2507 AUG 86
IN STUDY, COMMERCIAL GRADE 2,4 (80%)- AND 2,6 (20%)-TDI WAS CARCINOGENIC FOR F344/N RATS, CAUSING SUBCUTANEOUS FIBROMAS AND FIBROSARCOMAS IN MALES AND FEMALES, PANCREATIC ACINAR CELL ADENOMAS IN MALES, PANCREATIC ISLET CELL ADENOMAS, NEOPLASTIC NODULES OF LIVER AND MAMMARY GLAND FIBROADENOMAS IN FEMALES. TDI WAS NOT CARCINOGEGENIC FOR MALE MICE. TDI WAS CARCINOGENIC FOR FEMALE B6CEF1 MICE. THIS IS THE FINAL REPORT.

REPORT 05 : NTP TR 251 NIH PUB # 86-2507 AUG 86
COMM. GRADE TDI IN CORN OIL CARCINOGENIC FOR F344/N MALE AND FEMALE RATS. IT WAS NOT CARCINOGENIC FOR MALE MICE, BUT WAS CARCINOGENIC FOR FEMALE B6C3F1 MICE.TESTING DONE VIA GAVAGE.

REPORT 07 : NIOSH INTELLIGENCE BULLETIN 53 DEC 89
STUDIES IN ANIMALS DEMONSTRATED TOLUENE DIISOCYANATE (TDI) A CARCINOGEN. WHEN RATS AND MICE EXPOSED ORALLY TO COMMERCIAL-GRADE TDI (80:20 MIXTURE OF 2,4-AND 2.6-TDI), TUMORS INDUCED SYSTEMIC NATURE OF TDI CARCINOGENICITY DEMONSTRATED BY APPEARANCE OF TUMORS AT MULTIPLE SITES (PANCREAS, LIVER, SKIN, MAMMARY GLANDS, AND CIRCULATORY SYSTEM). RARE BRAIN TUMORS FOUND IN RATS EXPOSED TO TDI (TWO GLIOMAS AND ONE PINEALOMA). EXPERIMENTAL STUDIES DEMONSTRATED 2,4-TOLUENEDIAMINE (TDA), A HYDROLYSIS PRODUCT OF 2,4-TDI, A CARCINOGEN. WHEN RATS AND MICE EXPOSED ORALLY TO TDA, TUMORS INDUCED IN LIVERS, SKIN AND MAMMARY GLANDS.

REPT'D USE : COMMERCIAL MIXTURE OF 2,4 -AND 2,6 - ISOMERS

CAS NO : **584-84-9**

CHEM NAME : ***** 2,4-TOLUENE DIISOCYANATE (2,4-TDI)**

REPORT 01 : NTP CARC TEST PROGRAM JUL 82
IN A CURRENT BIOASSAY, 2,4-TDI, AS PART OF A MIXTURE WITH 2,6-TDI, WAS ADMINISTERED BY GAVAGE TO F344 RATS AND B6C3F1 MICE. THE STUDY HAS BEEN COMPLETED AND THE REPORT IS IN PEER REVIEW.

REPORT 02 : NIOSH CRITERIA DOCUMENT SEP 78
NIOSH RECOMMENDS THAT EXPOSURE CONCENTRATION LIMITS EQUIVALENT TO A CEILING CONCENTRATION OF 20 PPB AND A TWA CONCENTRATION OF 5 PPB BE OBSERVED FOR DIISO-CYANATES. FOR TOLUENE DIISOCYANATE, THE RECOMMENDED TWA IS 35 MICROGRAMS/CU M FOR UP TO A 10-HOUR WORKSHIFT,

40-HOUR WORK WEEK WITH A CEILING CONCENTRATION OF 140 MICROGRAMS/CU M FOR ANY 10 MINUTE SAMPLING PERIOD BASED ON AIRWAY EFFECTS. THE PRESENT OSHA STANDARD IS 0.02 PPM CEILING.

REPORT 03 : IARC MONOGRAPH 19 1979
NO DATA ON THE CARCINOGENICITY OR MUTAGENICITY OF 2,4- AND 2,6-TOLUENE DIISOCYANATES, 1,5-NAPHTHALENE DIISOCYANATE, 4,4'-METHYLENEDIPHENYL DIISOCYANATE OR POLYMETHYLENE POLYPHENYL ISOCYANATE WERE AVAILABLE.

REPORT 04 : EPA RCRA HEALTH AND ENV. EFFECTS PROFILE OCT 80
TDI AFTER BEING TESTED IN AN NTP BIOASSAY WAS FOUND NOT TO BE CARCINOGENIC. ADDITIONALLY, TDI DID NOT SHOW MUTAGENIC ACTIVITY ON TESTING OF SALMONELLA WITH AND WITHOUT A MAMMALIAN LIVER MICROSOME ACTIVATING SYSTEM.

REPORT 05 : IARC MONOGRAPH 19 1979
NO DATA ON THE EMBRYOTOXICITY, TERATOGENICITY, METABOLISM OR MUTAGENICITY OF 2,4-TOLUENE DIISOCYANATE WERE AVAILABLE.

REPORT 06 : MULTI-AGENCY CROSS CHECK 1985
Y HHS MAGUIRE
N OSHA CARC
N EPA CAG
Y EPA RCRA
N EPA CLEAN AIR
N EPA CLEAN WATER

REPORT 07 : IARC MONOGRAPH 39, 1986 JUN 85
INTRAGASTRIC DOSING INDUCED INCREASED INCIDENCE OF SUBCUT. FIBROMAS, NEOPLASMS AND ADENOMAS IN MALE MICE. FEMALE MICE IN SAME TEST SHOWED INCREASE IN HEMANGIOMAS AND ADENOMAS. INHALATION TESTS SHOWED NO TREATMENT-RELATED TUMORS. NO DATA AVAILABLE TO EVALUATE REPRODUCTIVE EFFECTS OR PRENATAL TOXICITY TO HUMANS OR ANIMALS. DATA INADEQUATE TO EVALUATE MUTAGENICITY, BUT METABOLITES MUTAGENIC TO S. TYPHIMURIUM AND D. MELANOGASTER WITH ACTIVATION. DNA DAMAGE AND REPAIR MUTATIONS IN CULTURED MAMMALIAN CELLS ATTRIBUTED TO THE METABOLITE. RESPIRATORY IRRITANT AND SENSITIZER IN HUMANS. IARC: SUFFICIENT EVIDENCE FOR CARCINOGENICITY IN ANIMALS, INADEQUATE FOR HUMANS.

REPT'D USE : INTERMEDIATE IN PRODUCTION OF URETHANE; USUALLY FOUND IN MIXTURE WITH 2,6 ISOMER; AND DYE INTERMEDIATE

CAS NO : **91-08-7**
CHEM NAME : ***** 2,6-TOLUENE DIISOCYANATE (2,6-TDI)**

REPORT 01 : IARC MONOGRAPH 19 1979
NO DATA ON THE CARCINOGENICITY OR MUTAGENICITY OF 2,4- AND 2,6-TOLUENE DIISOCYANATES, 1,5-NAPHTHALENE DIISOCYANATE, 4,4'-METHYLENEDIPHENYL DIISOCYANATE OR POLYMETHYLENE POLYPHENYL ISOCYANATE WERE AVAILABLE.

REPORT 02 : EPA RCRA HEALTH AND ENV. EFFECTS PROFILE OCT 80
TDI AFTER BEING TESTED IN AN NTP BIOASSAY WAS FOUND NOT TO BE CARCINOGENIC. ADDITIONALLY, TDI DID NOT SHOW MUTAGENIC ACTIVITY ON TESTING OF SALMONELLA WITH AND WITHOUT A MAMMALIAN LIVER MICROSOME ACTIVATING SYSTEM.

REPORT 03 : NTP CARC TEST PROGRAM DRAFTR EPORT 82-2 JUL 82
NO. 8 2-2507
IN A CURRENT BIOASSAY, 2,4-TDI WAS ADMINISTERED BY GAVAGE TO F344 RATS AND B6C3F1 MICE AS PART OF A MIXTURE WITH 2,6-TDI. THIS STUDY HAS BEEN COMPLETED AND THE REPORT IS IN PEER REVIEW.

REPORT 04 : NIOSH CRITERIA DOCUMENT SEP 78
NIOSH RECOMMENDS THAT EXPOSURE CONCENTRATION LIMITS EQUIVALENT TO A CEILING CONCENTRATION OF 20 PPB AND A TWA CONCENTRATION OF 5 PPB BE OBSERVED FOR DIISOCYANATES. FOR TOLUENE DIISOCYANATE, THE RECOMMENDED TWA IS 35 MICROGRAMS/CU M FOR UP TO A 10-HOUR WORKSHIFT, 40-HOUR WORK WEEK WITH A CEILING CONCENTRATION OF 140 MICROGRAMS/CU M FOR ANY 10-MINUTE SAMPLING PERIOD BASED ON AIRWAY EFFECTS. THE PRESENT OSHA STANDARD IS 0.02 PPM CEILING.

REPORT 05 : IARC MONOGRAPH 39, 1986 JUN 85
MIXTURES OF 2,4- AND 2,6-TDI ADMINISTERED TO MICE AND RATS VIA INTRAGASTRIC EXPOSURE AND INHALATION. INTRAGASTRIC ROUTE INCREASED INCIDENCE OF SC FIBROMAS AND FIBROSARCOMAS IN MALE RATS PLUS INVOLVEMENT OF PANCREAS, LIVER AND MAMMARY GLANDS IN RATS. IN MICE, BLOOD DYSCRASIAS AND HEPATOCELLULAR ADENOMAS SEEN. NO TUMORS SEEN IN ANIMALS EXPOSED BY INHALATION. NO DATA AVAILABLE TO EVALUATE REPRODUCTIVE EFFECTS OR PRENATAL TOXICITY IN HUMANS OR ANIMALS. ONE CASE REPORT OF ADENOCARCINOMA OF LUNG INADEQUATE TO EVALUATE CARCINOGENICITY OF TDI TO HUMANS. IARC SAYS EVIDENCE FOR CARCINOGENICITY OF TDI'S TO ANIMALS SUFFICIENT TO EVALUATE, BUT EVIDENCE INADEQUATE FOR HUMANS.

REPORT 06 : MULTI-AGENCY CROSS CHECK 1985
Y HHS MAGUIRE
N OSHA CARC
N EPA CAG
N EPA RCRA
N EPA CLEAN AIR
N EPA CLEAN WATER
REPT'D USE : INTERMEDIATE FOR POLYURETHANE FOAMS, COATINGS, ELASTOMERS AND ADHESIVES

CAS NO : **88-19-7**
CHEM NAME : ***** ORTHO-TOLUENESULPHONAMIDE**

REPORT 01 : IARC MONOGRAPH 22 1980
ORTHO-TOLUENESULPHONAMIDE WAS TESTED BY ORAL ADMINISTRATION IN A TWO GENERATION STUDY: NO INCREASE IN BLADDER TUMOR INCIDENCE WAS NOTED IN ANIMALS OF EITHER GENERATION. IN ONE OF TWO SINGLE-GENERATION STUDIES IN RATS, BENIGN AND MALIGNANT BLADDER TUMORS WERE FOUND. THERE IS LIMITED EVIDENCE THAT ORTHO-TOLUENESULPHONAMIDE IS CARCINOGENIC WHEN GIVEN ORALLY TO RATS; BUT THE AVAILABLE DATA SUGGEST THAT IMPURITIES AT THE LEVELS NORMALLY FOUND IN COMMERCIAL SACCHARIN DO NOT CONTRIBUTE TO THE CARCINOGENICITY OF SACCHARIN.

REPORT 02 : IARC MONOGRAPH 22 1980
THERE IS NO EVIDENCE THAT O-TOLUENESULPHONAMIDE IS MUTAGENIC, ALTHOUGH IMPURITIES EXTRACTED FROM SOME LOTS OF SACCHARIN WERE MUTAGENIC IN SALMONELLA.
REPT'D USE : INTERMEDIATE IN MFR OF SACCHARIN

CAS NO : **95-53-4**
CHEM NAME : ***** O-TOLUIDINE**

REPORT 01 : IARC MONOGRAPH 27 1982
O-TOLUIDINE WAS TESTED IN SEVERAL EXPERIMENTS, BUT ONLY TWO EXPERIMENTS IN MICE AND TWO EXPERIMENTS IN RATS BY ORAL ADMINISTRATION WERE ADEQUATE. IT PRODUCED HEPATOCELLULAR CARCINOMAS AND HEMANGIOSARCOMAS IN MICE AND SARCOMAS, PAPILLOMAS, MESOTHELIOMAS IN RATS. THERE IS SUFFICIENT EVIDENCE FOR THE CARCINOGENICIY OF O-TOLUIDINE.HCL IN ANIMALS.

REPORT 02 : IARC MONOGRAPH 27 1982
O-TOLUIDINE GAVE POSITIVE RESULTS IN E. COLI DNA REPAIR AND PHAGE INDUCTION ASSAYS, BUT WAS NOT MUTAGENIC IN SALMONELLA UNLESS TESTED WITH BOTH NONHARMAN AND A METABOLIC ACTIVATION SYSTEM. URINE CONCENTRATES FROM

RATS TREATED WITH O-TOLUIDINE WERE MUTAGENIC FOR SALMONELLA WITH METABOLIC ACTIVATION.

REPORT 04 : EPA/TSCA 8(E) FY1-AX 0188- 0266 JAN 88
OCAW/AFL-C10 HAS DISCOVERED EIGHT CASES OF BLADDER CANCER AMONG 450 WORKERS EXPOSED TO O-TOLUIDINE AT THE GOODYEAR PLANT IN NIAGARA FALLS, N.Y. ALL VICTIMS WORKED IN THE SAME AREA OF THE PLANT. NOT REGULATED AS A CARCINOGEN BY OSHA/EPA.

REPORT 06 : MULTI-AGENCY CROSS CHECK
Y HHS MAGUIRE
Y OSHA CARC
N EPA CAG
N EPA RCRA
N EPA CLEAN AIR
N EPA CLEAN WATER

REPT'D USE : DYE INTERMEDIATE, PHOTOGRAPHIC, CHEMICAL, RUBBER ANTIOXIDANT AND LABORATORY AGENT

CAS NO : **636-21-5**
CHEM NAME : ***** O-TOLUIDINE HYDROCHLORIDE**

REPORT 01 : NCI CARC TRS 153 DHEW PUB NO 79-1709
UNDER THE CONDITIONS OF THIS BIOASSAY, O-TOLUIDINE HYDROCHLORIDE WAS CARCINOGENIC IN BOTH MALE AND FEMALE F344 RATS AND B6C3F1 MICE, PRODUCING A SIGNIFICANT INCREASED INCIDENCE OF ONE OR MORE TYPES OF NEOPLASMS (FEEDING STUDY).

REPORT 02 : IARC SUPPLEMENT 1 1979
O-TOLUIDINE IS CARCINOGENIC IN MICE AND RATS FOLLOWING ITS ORAL ADMINISTRATION AS THE HYDROCHLORIDE. AN EPIDEMIOLOGIC STUDY APPEARED TO INVOLVE OCCUPATIONAL EXPOSURE TO TOLUIDINES; IT WAS SUGGESTIVE OF A CARCINOGENIC EFFECT, BUT WAS INSUFFICIENT TO ALLOW EVALUATION.

REPORT 03 : IARC MONOGRAPH 27 1982
O-TOLUIDINE HCL WAS TESTED IN SEVERAL EXPERIMENTS, BUT ONLY TWO EXPERIMENTS IN MICE AND TWO EXPERIMENTS IN RATS BY ORAL ADMINISTRATION WERE ADEQUATE. IT PRODUCED HEPATOCELLULAR CARCINOMAS AND HEMANGIOSARCOMAS IN MICE AND SARCOMAS, PAPILLOMAS, MESOTHELIOMAS IN RATS. THERE IS SUFFICIENT EVIDENCE FOR THE CARCINOGENICITY OF O-TOLUIDINE HCL IN ANIMALS.

REPORT 04 : IARC MONOGRAPH 27 1982
O-TOLUIDINE GAVE POSITIVE RESULTS IN E COLI DNA REPAIR AND PHAGE INDUCTION ASSAY, BUT WAS NOT MUTAGENIC IN SALMO-

NELLA UNLESS TESTED WITH BOTH NONHARMAN AND A METABOLIC ACTIVATION SYSTEM. URINE CONCENTRATES FROM RATS TREATED WITH O-TOLUIDINE WERE MUTAGENIC FOR SALMONELLA WITH METABOLIC ACTIVATION.

REPORT 05 : NTP TECHNICAL BULLETIN 5 AUG 81
O-TOLUIDINE HCL WAS NEGATIVE IN THE SALMONELLA TYPHIMURIUM MUTAGENICITY ASSAY IN ONE LABORATORY.

REPORT 06 : MULTI-AGENCY CROSS CHECK
Y HHS MAGUIRE
N OSHA CARC
Y EPA CAG
N EPA RCRA
N EPA CLEAN AIR
N EPA CLEAN WATER

REPT'D USE : DYE INTERMEDIATE, RUBBER ANTIOXIDANT, PHOTOGRAPHIC CHEMICALS AND CHEMICAL INDICATOR; NO KNOWN PRODUCTION/INFORMATION SINCE 1980

CAS NO : **8001-35-2**
CHEM NAME : ***** TOXAPHENE**

REPORT 01 : NCI CARC TRS 37 DHEW PUB NO 79-837
UNDER THE CONDITIONS OF THIS BIOASSAY, TOXAPHENE WAS CARCINOGENIC IN MALE AND FEMALE B6C3F1 MICE, CAUSING INCREASED INCIDENCES OF HEPATOCELLULAR CARCINOMAS. THE TEST RESULTS ALSO SUGGEST CARCINOGENICITY OF TOXAPHENE FOR THE THYROID OF MALE AND FEMALE OSBORNE-MENDEL RATS (FEEDING STUDY).

REPORT 02 : IARC MONOGRAPH 20 1979
TOXAPHENE WAS TESTED IN ONE EXPERIMENT IN MICE AND ONE IN RATS BY ORAL ADMINISTRATION: AN INCREASED INCIDENCE OF HEPATOCELLULAR CARCINOMAS WAS OBSERVED IN MALE AND FEMALE MICE, AND AN INCREASED INCIDENCE OF THYROID TUMORS WAS OBSERVED IN MALE AND FEMALE RATS. IN A 1979 SUPPLEMENT, IARC DETERMINED THAT THERE IS SUFFICIENT EVIDENCE FOR THE CARCINOGENICITY OF TOXAPHENE IN ANIMALS.

REPORT 03 : EPA RCRA HEALTH AND ENV. EFFECTS PROFILE OCT 80
TOXAPHENE HAS NOT PRODUCED TERATOGENIC EFFECTS IN LABORATORY ANIMALS, BUT HAS BEEN FOUND TO BE MUTAGENIC IN TWO STRAINS OF SALMONELLA WITH METABOLIC ACTIVATION. AN NCI 1979 STUDY FOUND THAT TOXAPHENE SIGNIFICANTLY INCREASED THE INCIDENCES OF HEPATOCELLULAR CARCINOMAS IN MICE.

REPORT 04 : EPA AMBIENT WATER QUALITY CRITERIA OCT 80
TOXAPHENE IS SUSPECTED OF BEING A HUMAN CARCINOGEN. AMBIENT WATER CONCENTRATIONS OF 7.1 NG/L, .71 NG/L, AND .07 NG/L ARE ESTIMATED TO RESULT IN A RESPECTIVE 10 -5, 10 -6, AND 10 -7 INCREASE IN CANCER RISK OVER A LIFETIME.

REPORT 05 : NAS REPORT ON DIET, NUTRITION AND CANCER 1982
ON THE BASIS OF STUDIES IN ANIMALS, AND IN THE ABSENCE OF ADEQUATE DATA FROM EPA STUDIES, IT APPEARS THAT TOXAPHENE PRESENTS A CARCINOCENIC RISK TO HUMANS. HOWEVER, IT IS REASONALBE TO ASSUME THAT AMOUNTS PRESENT IN THE AVERAGE U.S. DIET DO NOT MAKE A MAJOR CONTRIBUTION TO THE OVERALL RISK OF CANCER FOR HUMANS.

REPORT 06 : MULTI-AGENCY CROSS CHECK
Y HHS MAGUIRE
N OSHA CARC
Y EPA CAG
Y EPA RCRA
N EPA CLEAN AIR
Y EPA CLEAN WATER

REPORT 07 : IARC MONOGRAPH 20 1979
TOXAPHENE IS MUTAGENIC IN THE SALMONELLA ASSAY WITHOUT REQUIRING LIVER HOMOGENATE FOR ACTIVITY. THE CHEMICAL DID NOT INDUCE DOMINANT LETHALS IN MICE.

REPORT 08 : NTP TECHNICAL BULLETIN 9 APR 83
TOXAPHENE WAS POSITIVE IN ONE SALMONELLA TYPHIMURIUM MUTAGENICITY ASSAY.
REPT'D USE : INSECTICIDE

CAS NO : **82244-86-8**
CHEM NAME : ***** TRANS-A-CHLORO-3-OXYAMINOPROPENE HYDROCHLORIDE**

REPORT 01 : STATUS REPORT 8EHQ-1086-0637 NOV 86
PRELIMINARY EVALUATION OF DATA FROM A 28-DAY DERMAL TOXICITY STUDY OF THIS CPD INDICATES THAT IT PRODUCED METHEMOGLOBINEMIA, MACROCYTIC ANEMIA AND HISTOPATHOLOGICAL CHANGES IN SPLEEN OF BOTH SEXES OF RATS. ANIMALS WERE TREATED DERMALLY FOR 6 HRS/DAY, 7 DAYS/WK FOR FOUR WEEKS WITH A 1.0% SOL'N OF 5.0 MG/KG, 20.0 MG/KG OR 80.0 MG/KG OF AN AQUEOUS SOLUTION CONTAINING 47.4% OF CPD. CPD-RELATED EFFECTS SEEN AT ALL LEVELS.
REPT'D USE : RESEARCH CPD

CAS NO : **3234-02-4**
CHEM NAME : ***** TRANS-2,3-DIBROMO-2-BUTENE-1,4-DIOL**

REPORT 01 : NTP TECHNICAL BULLETIN 9 APR 83
TRANS-2,3-DIBROMO-2-BUTENE-1,4-DIOL WAS POSITIVE IN ONE SALMONELLA TYPHIMURIUM MUTAGENICITY ASSAY.
REPT'D USE : UNKNOWN

CAS NO : **764-41-0**
CHEM NAME : ***** TRANS-1,4-DICHLOROBUTENE**

REPORT 01 : IARC MONOGRAPH 15 AUG 77
WHEN INJECTED SUBCUTANEOUSLY OR INTRAPERITONEALLY 1,4-DICHLOROBUTENE PRODUCED LOW INCIDENCES OF LOCAL SARCOMAS. THE AVAILABLE DATA DO NOT ALLOW AN EVALUATION OF THE CARCINOGENICITY OF THIS COMPOUND TO BE MADE.

REPORT 02 : IARC MONOGRAPH 15 AUG 77
NO DATA ON THE CHRONIC TOXICITY, EMBRYOTOXICITY OR TERATOGENICITY OF 1,4-DICHLOROBUTENE WERE AVAILABLE. THE CHEMCIAL CAUSED REVERSE MUTATIONS IN SALMONELLA TA100 METABOLIC ACTIVATION ENHANCED THE MUTAGENIC EFFECT. THE CHEMICAL WAS ALSO REPORTED TO BE MUTGENIC IN E. COLI.

REPORT 03 : TSCA SECT 8(E) STATUS REPORT SEP 85
MALE RATS WERE EXPOSED TO 0 PPM, 0.1 PPM, 0.3 PPM OR 1.0 PPM 1,4-DCB FOR 19 MONTHS. STATISTICALLY SIGNIFICANT INCREASES IN INCIDENCE OF BENIGN NASAL TUMORS SEEN IN ALL GROUPS AND MALIGNANT NASAL TUMORS WERE SEEN IN THE 1.0 PPM GROUP. WHEN INCIDENCES OF BENIGN AND MALIGNANT TUMORS COMBINED, STATISTICALLY SIGNIFICANT INCREASES SEEN IN THE 0.3 PPM AND 1.0 PPM GROUPS. NO OTHER COMPOUND-RELATED TUMORS WERE OBSERVED.
REPT'D USE : INTERMEDIATE IN MFR OF CHLOROPRENE AND HEXAMETHYLENEDIAMINE

CAS NO : **25962-77-0**
CHEM NAME : ***** TRANS-2-((DIMETHYLAMINO)METHYLAMINO)-5-(2-(5-NITRO-2-FURYL)VINYL)-1,3**

REPORT 01 : IARC MONOGRAPH 1974
THIS COMPOUND IS CARCINOGENIC IN RATS FOLLOWING ORAL ADMINISTRATION, THE ONLY SPECIES AND ROUTE TESTED. IT PRODUCED MAMMARY AND INTESTINAL TRACT CARCINOMAS. IN

A 1979 SUPPLEMENT, IARC DETERMINED THAT THERE IS SUFFICIENT EVIDENCE FOR THE CARCINOGENICITY OF THIS COMPOUND IN ANIMALS.

REPT'D USE : PHARMACEUTICAL

CAS NO : **14567-73-8**

CHEM NAME : ***** TREMOLITE**

REPORT 01 : UNDER FEED STUDIES, NON-FIBROUS TREMOLITE NOT OVERTLY TOXIC OR CARCINOGENIC FOR MALE OR FEMALE F344/N RATS, FOLLOWING LIFETIME INGESTION OF DIET CONTAINING 1% TREMOLITE.

CAS NO : **299-75-2**

CHEM NAME : ***** TREOSULPHAN**

REPORT 01 : IARC MONOGRAPH 26 MAY 81
NO DATA ON THE CARCINOGENICITY OR TERATOGENICITY OF TREOSULPHAN IN EXPERIMENTAL ANIMALS WERE AVAILABLE. THE ONLY AVAILABLE EPIDEMIOLOCICAL STUDY INDICATES THAT USE OF THE DRUG IS FOLLOWED BY AN INCREASED RISK OF ACUTE MONOLYMPHOCYTIC LEUKEMIA. THERE IS SUFFICIENT EVIDENCE FOR THE CARCINOGENICITY OF TREOSULPHAN IN HUMANS.

REPORT 02 : IARC MONOGRAPH 26 MAY 81
NO DATA ON THE TERATOGENICITY OF TREOSULPHAN WERE AVAILABLE. THE ONLY AVAILABLE STUDIES ON CHROMOSOMAL EFFECTS SHOWED THAT THE DRUG PRODUCES CHROMOSOMAL ABERRATIONS IN PLANTS.

REPT'D USE : ANTINEOPLASTIC AGENT

CAS NO : **959-52-4**

CHEM NAME : ***** 1,3,5-TRIACRYLOYLHEXA-HYDRO-S-TRIAZINE**

REPORT 01 : EPA TSCA 8(E) 8EHQ-0384-0506 S APR 84
DID NOT PRODUCE ACTIVITY IN-VITRO, CONSISTENT WITH THAT OF A CHEMICAL MUTAGEN IN A FORWARD POINT MUTATION ASSAY IN CHO CELLS, IN AN SCE TEST IN CHO CELLS OR IN A TEST FOR UDS. PRESENCE OF CHROMOSOMAL ABERRATIONS IN CHO CELLS, IN A CONCENTRATION-RELATED MANNER, CONFIRMED THAT CLASTOGENIC EFFECT WAS AN IN-VITRO CLASTOGENIC EFFECT. ADDITIONAL CLASTOGENIC POTENTIAL DEMONSTRATED IN-VIVO BY MICRONUCLEUS TEST. FINDINGS SHOW THAT CPD IS A CLASTOGENIC MATERIAL BUT DEVOID OF MUTAGENIC ACTIVITY.

CAS NO : **75-25-2**
CHEM NAME : ***** TRIBROMOMETHANE**

REPORT 01 : NTP TR 350 NIH PUB. 89-2805 MAY 89
TWO YEAR GAVAGE STUDIES SHOW EVIDENCE OF CARCINOGENIC ACTIVITY OF TRIBROMOMETHANE FOR MALE AND FEMALE F 344/N RATS BASED ON UNCOMMON NEOPLASMS OF THE LARGE INTESTINE. THE DETECTION OF CARCINOGENIC RESPONSES WAS LOWERED BY REDUCED SURVIVAL OF RATS GIVEN 200 MG/KG TRIBROMOMETHANE. CHEMICALLY RELATED NONNEOPLASTIC LESIONS INCLUDED FATTY CHANGE AND ACTIVE CHRONIC INFLAMMATION OF THE LIVER IN MALE AND FEMALE RATS, MINIMAL NECROSIS OF THE LIVER IN MALE RATS, AND MIXED CELL FOCI OF THE LIVER IN FEMALE RATS. NO EVIDENCE OF CARCINOGENIC ACTIVITY FOR MALE B6C3F1 MICE GIVEN 50 OR 100 MG/KG TRIBROMOMETHANE OR FOR FFMALE B6C3F1 MICE GIVEN 100 OR 200 MG/KG.

REPORT 02 : EPA AMBIENT WATER QUALITY CRITERIA OCT 80
IT IS NOT PRESENTLY POSSIBLE TO DERIVE A VALID WATER QUALITY CRITERION FOR TRIBROMOMETHANE. HOWEVER, FOR THE MAXIMUM PROTECTION OF HUMAN HEALTH FROM THE POTENTIAL CARCINOGENIC EFFECTS FROM TRIBROMOMETHANE, CONCENTRATIONS OF 1.9 UG/L, .19 UG/L AND .019 UG/LARE ESTIMATED TO RESULT IN AN INCREMENTAL INCREASE OF CANCER RISK OVER THE LIFETIME OF 10 -5, 10 -6, AND 10 -7 RESPECTIVELY.

REPORT 03 : EPA AMBIENT WATER QUALITY CRITERIA OCT 80
BROMOFORM GAVE POSITIVE RESULTS IN MUTAGENICITY TESTS WITH SALMONELLA. IN A SHORT TERM IN VIVO ONCOGENICITY ASSAY IT CAUSED A SIGNIFICANT INCREASE IN TUMOR INCIDENCE AT ONE DOSE LEVEL.

REPORT 06 : MULTI-AGENCY CROSS CHECK
N HHS MAGUIRE
N OSHA CARC
N EPA CAG
N EPA RCRA
N EPA CLEAN AIR
Y EPA CLEAN WATER

REPT'D USE : CHEMICAL INTERMEDIATE, SOLVENT, FLAME RETARDANTS AND GAUGE FLUIDS

CAS NO : **126-73-8**
CHEM NAME : ***** TRIBUTYL PHOSPHATE (PHOSPHORIC ACID, TRIBUTYL ESTER)**

REPORT 01 : EPA TSCA FEDERAL REGISTER NOTICE 51 FR 18368 MAY 86
IN RESPONSE TO ITC'S RECOMMENDATION TO GIVE THIS CHEMICAL

PRIORITY ATTENTION, EPA IS ADDING IT TO 2 MODEL INFORMATION GATHERING RULES: TSCA 8(A) PRELIMINARY ASSESSMENT INFORMATION RULE(PAIR) AND TSCA 8(D) HEALTH AND SAFETY DATA REPORTING RULE. THIS RULE SHALL BECOME EFFECTIVE ON JUNE 18, 1986.

REPORT 02 : 19TH REPORT OF TSCA ITC, NOV. 1986 NOV 86
REPORT ADDS TBP TO SECT 4(E) PRIORITY LIST, DESIGNATED FOR RESPONSE IN 12 MONTHS. AREAS FOR TESTING INCLUDE CHRONIC TOXICITY - ONCOGENIC, NEUROTOXIC, RENAL, REPRODUCTIVE AND DEVELOPMENTAL; CHEMICAL FATE AND PERSISTENCE IN SOILS AND SEDIMENTS; CHRONIC EFFECTS ON AQUATIC AND TERRESTRIAL PLANTS, DAPHNIDS OR OTHER AQUATIC INVERTEBRATES; ACUTE AND CHRONIC EFFECTS ON BENTHIC ORGANISMS AND SOIL INVERTEBRATES FOUND TO BE PERSISTENT UNDER ANAEROBIC CONDITIONS.

REPORT 03 : EPA TSCA FEDERAL REGISTER NOTICE 52 FR 43346 NOV 87
EPA PPROPOSING THAT MANUFACTURERS AND PROCESSORS OF TBP BE REQUIRED UNDER TSCA 4 TO DO TESTING FOR NEUROTOXICITY, DEVELOPMENTAL TOXICITY, REPRODUCTIVE TOXICITY, MUTAGENICITY, ONCOGENICITY, DERMAL SENSITIZATION, ORAL/DERMAL PHARMACOKINETICS, ENVIRONMENTAL EFFECTS, AND CHEMICAL FATE. RULE PROPOSED IN RESPONSE TO ITC'S DESIGNATING TBP FOR PRIORITY CONSIDERATION FOR CHEMICAL FATE, HEALTH EFFECTS AND ENVIRONMENTAL EFFECTS TESTING.

CAS NO : **52-68-6**
CHEM NAME : ***** TRICHLORFON**

REPORT 01 : IARC MONOGRAPH 30 JAN 83
TESTED FOR CARCINOGENICITY BY ORAL ADMINSTRATION IN MICE AND RATS, BY INTRAPERITONEAL ADMINISTRATION IN MICE, RATS AND HAMSTERS, BY DERMAL APPLICATION IN MICE AND BY INTRAMUSCULAR ADMINISTRATION IN RATS. STUDIES IN MICE AND TWO IN RATS CONSIDERED INADEQUATE FOR EVALUATION. OTHER STUDIES IN RATS AND HAMSTERS NOT INDICATIVE OF CARCINOGENIC EFFECT.

REPORT 02 : IARC MONOGRAPH 30 JAN 83
TRICHLORFON TERATOGENIC TO MICE, RATS AND HAMSTERS WHEN GIVEN AT DOSE LEVELS WHICH ALSO CAUSED REDUCTIONS IN MATERNAL FOOD INTAKE AND MATERNAL BODY WEIGHT GAIN.

REPORT 03 : IARC MONOGRAPH 30 JAN 83
SUFFICIENT EVIDENCE THAT TRICHLORFON OR ITS DEGRADATION PRODUCTS IS MUTAGENIC IN BACTERIA, MAMMALIAN CELLS AND MICE. INCONSISTENT RESULTS HAVE BEEN REPORTED WITH MANY

SYSTEMS, HOWEVER, AND INSTABILITY OF TRICHLORFON MAY BE IMPORTANT CONSIDERATION IN INTERPRETING DATA.
REPT'D USE : INSECTICIDE; NOT PRODUCED COMMERCIALLY IN U.S.

CAS NO : **634-93-5**
CHEM NAME : ***** 2,4,6-TRICHLOROANILINE**

REPORT 01 : ITC FOURTH REPORT TO EPA (44FR31866) JUN 79
THE INTERAGENCY TESTING COMMITTEE RECOMMENDS THAT THE EPA ADMINISTRATOR CONSIDER THE DEVELOPMENT OF TEST RULES TO TEST ANILINE, AND BROMOCHLORO AND NITROANILINES FORCARCINOGENICITY,MUTAGENICITY,TERATOGENICITY/REPRODUCTIVE EFFECTS AND NEUROLOGICAL EFFECTS; AND THAT AN EPIDEMIOLOGIC STUDY BE CONDUCTED.

REPORT 02 : EPA TSCA FEDERAL REGISTER NOTICE 49FR198 JAN 84
EPA PUBLISHING ANPR TO REQUIRE TESTING OF THE ANILINES FOR HEALTH AND ENVIRONMENTAL EFFECTS. THIS IS IN RESPONSE TO ITC'S DESIGNATION OF THE ANILINES FOR TESTING.

REPORT 06 : MULTI-AGENCY CROSS CHECK
N HHS MAGUIRE
Y OSHA CARC
N EPA CAG
N EPA RCRA
N EPA CLEAN AIR
N EPA CLEAN WATER
REPT'D USE : LABORATORY AGENT

CAS NO : **87-61-6**
CHEM NAME : ***** 1,2,3-TRICHLOROBENZENE**

REPORT 01 : EPA TSCA FEDERAL REGISTER NOTICE 45FR11728 APR 86
PROMULGATES EPA'S DECISION TO REQUIRE MFRS AND PROCESSORS OF 1,2,3-TRICHLOROBENZENE TO CONDUCT ENVIRONMENTAL EFFECTS TESTING IN REPONSE TO ITC'S RECOMMENDATION.

REPORT 02 : EPA TSCA FEDERAL REGISTER NOTICE 52FR24460 JUL 87
EPA ISSUING FINAL RULE SPECIFYING FINAL TEST STANDARDS AND REPORTING REQUIREMENTS FOR ENVIRONMENTAL EFFECTS TESTING.

CAS NO : **120-82-1**
CHEM NAME : ***** 1,2,4-TRICHLOROBENZENE**

REPORT 01 : EPA TSCA 4(E) TEST RULE (51FR11728) APR 86
PROMULGATES EPA'S DECISION TO REQUIRE MFR'S AND PROCES-

SORS OF TRCHLOROBENZENE TO CONDUCT CHEMICAL FATE AND ENVIRONMENTAL EFFECTS TESTING IN RESPONSE TO ITC'S RECOMMENDATION.

REPORT 02: : EPA TSCA REDERAL REGISTER NOTICE 51 CFR 24658 JUL 86
EPA PROMULGATES RULE REQUIRING TEST OF CHLORINATED BENZENES ACCORDING TO TSCA HEALTH EFFECTS TESTING GUIDELINES IN 40 CFR PART 798.

REPORT 03 : EPA TSCA FEDERAL REGISTER NOTICE 52 FR 24460 JUL 87
EPA ISSUING FINAL TEST STANDARDS AND REPORTING REQUIREMENTS FOR ENVIRONMENTAL EFFECTS TESTING.

CAS NO : **71-55-6**
CHEM NAME : ***** 1,1,1-TRICHLOROETHANE**

REPORT 01 : EPA TSCA FEDERAL REGISTER NOTICE 50 FR 51683 DEC 85
IN RESPONSE TO ITC'S DESIGNATION OF TCE FOR PRIORITY TESTING, EPA IS PROMULGATING FINAL TEST STANDARDS AND REPORTING REQUIREMENTS FOR TCE.

REPORT 02 : NIOSH CRITERIA DOCUMENT JUL 76
NIOSH RECOMMENDS A STANDARD OF 200 PPM (TWA) BASED ON CNS RESPONSES TO ACUTE EXPOSURE AND CARDIOVASCULAR AND RESPIRATORY EFFECTS ASSOCIATED WITH CHRONIC EXPOSURES. THE PRESENT OSHA STANDARD IS 350 PPM TWA.

REPORT 03 : NIOSH CURRENT INTELLIGENCE BULLETIN 27 AUG 78
NCI BIOASSAY IN PROGRESS AS OF JULY 1978 BECAUSE THE PREVIOUS TESTS WERE INCONCLUSIVE. OSHA EXPOSURE STANDARD IS 350 PPM TWA WHEREAS 350 PPM CEILING IS THE NIOSH RECOMMENDED STANDARD. ALL CHLOROETHANE COMPOUNDS ARE KNOWN TO CAUSE CNS DEPRESSION IN LAB ANIMALS. THIS COMPOUND INDUCES RESPIRATORY FAILURE, CARDIAC ARRYTHMIAS, CARDIAC FAILURE, PULMONARY CONGESTION AND DEATH IN TEST ANIMALS. NIOSH RECOMMENDS THAT THE COMPOUND BE TREATED IN THE WORKPLACE WITH CAUTION BECAUSE OF ITS STRUCTURAL SIMILARITY TO FOUR CARCINOGENIC CHLOROETHANE CMPDS.

REPORT 04 : TSCA INTER- AGENCY TESTING COMMITTEE APR 78
IN A JAN 1981 RULING, THE SOUTHERN DISTRICT COURT OF NEW YORK ORDERED EPA TO PROPOSE TEST RULES, OR DECISIONS NOT TO REQUIRE TESTING, BY THE END OF 1983 FOR ALL CHEMICALS LISTED IN THE FIRST FIVE ITC REPORTS. THE SCHEDULE SUBMITTED BY EPA, AND ACCEPTED BY THE COURT WITHOUT CHANGE, SHOWS THAT A DECISION FOR TESTING OF CARCINOGENICITY, MUTAGENICITY, TERATOGENICITY, AND EPA STUDIES OF 1,1,1-TCE WILL BE MADE IN 1981.

REPORT 05 : NTP DRAFT REPORT 82-91 NIH PUB 82-2518 JAN 83
1,1,1-TRICHLOROETANE WAS NOT CARCINOGENIC FOR MALE OR FEMALE F344/N RATS; FEMALE RATS MAY NOT HAVE BEEN ADEQUATELY TESTED BECAUSE HIGH DOSE WAS TOXIC AND THERE WAS LARGE NUMBER OF ACCIDENTAL DEATHS. ASSOCIATION BETWEEN ADMINISTRATION OF 1,1,1-TRICHLOROETHANE AND INCREASED INCIDENCES OF HEPATOCELLULAR CARCINOMAS IN MALE B6C3F1 MICE WAS CONSIDERED EQUIVOCAL. 1,1,1-TRICHLOROETHANE WAS CARCINOGENIC FOR FEMALE B6C3F1 MICE, CAUSING INCREASED INCIDENCE OF HEPATOCELLULAR CARCINOMAS (GAVAGE STUDY).

REPORT 06 : MULTI-AGENCY CROSS CHECK
N HHS MAGUIRE
N OSHA CARC
N EPA CAG
Y EPA RCRA
Y EPA CLEAN AIR
Y EPA CLEAN WATER

REPORT 07 : EPA RCRA HEALTH AND ENV. EFFECTS PROFILE OCT 80
RESULTS OF AN NCI BIOASSAY OF METHYL CHLOROFORM WERE INCONCLUSIVE DUE TO EXPERIMENTAL PROBLEMS. IN VITRO STUDIES HAVE INDICATED THAT 1,1,1-TCE IS SLIGHTLY MUTAGENIC WITH OR WITHOUT ACTIVATION, AND CAN CAUSE MAMMALIAN CELL TRANSFORMATION. STUDIES OF TERATOGENIC POTENTIAL OF 1,1,1-TCE ARE SUGGESTIVE; HOWEVER MORE STUDIES ARE NEEDED TO MAKE A CONCLUSIVE STATEMENT.

REPORT 08 : IARC MONOGRAPH 20 OCT 79
1,1,1-TRICHLOROETHANE WAS TESTED IN ONE EXPERIMENT IN MICE AND IN ONE IN RATS BY ORAL ADMINISTRATION AND IN ONE EXPERIMENT BY INHALATION EXPOSURE IN RATS. ALTHOUGH A FEW LIVER TUMORS WERE OBSERVED IN MALE MICE, THESE EXPERIMENTS WERE CONSIDERED TO BE INADEQUATE. THE AVAILABLE DATA DO NOT PERMIT AN EVALUATION OF THE CARCINOGENICITY OF 1,1,1-TRICHLOROETHANE TO BE MADE.

REPORT 09 : IARC MONOGRAPH 20 1979
1,1,1-TRICHLOROETHANE WAS MUTAGENIC IN SALMONELLA TA100, WITH OR WITHOUT METABOLIC ACTIVATION.

REPORT 10 : EPA RESPONSE TO ITC RECOMMS 46FR30300 JUN 81
IN RESPONSE TO THE ITC RECOMMENDATIONS (43FR16684) THE EPA PROPOSES THAT MANUFACTURERS AND PROCESSORS OF 1,1,1-TRICHLOROETHANE CONDUCT TESTING FOR TERATOGENICITY. AVAILABLE REPRODUCTIVE EFFECTS DATA ARE SUFFICIENT. THE NTP IS CONDUCTING AN ONCOGENICITY STUDY. THE EPA WILL CONDUCT MUTAGENICITY TESTING AND DECIDE

IF FURTHER TIER TESTING IS APPROPRIATE. NO STUDY POPULATION FOR AN EPIDEMIOLOGIC STUDY COULD BE IDENTIFIED.

REPORT 11 : NCI CARC TRS 3 DHEW PUB NO 77-803
IT WOULD BE INAPPRORIATE TO MAKE AN ASSESSMENT OF CARCINOGENICITY OF 1,1,1-TRICHLOROETHANE ON THE BASIS OF THIS TEST, BECAUSE OF THE ABBREVIATED LIFE SPANS OF BOTH THE RATS AND THE MICE. THE TECHNICAL GRADE OF THE CHEMICAL WAS GIVEN BY GAVAGE.

REPORT 12 : EPA TSCA 8(E) 1084-0535 OCT 84
RESULTS OF SEVERAL ACUTE INHALATION TOXICITY STUDIES IN RATS AND QUAIL INDICATE MIXTURES OF 1,1,1-TRICHLOROETHANE AND PERFLUOROALKYL RESINS ARE MORE TOXIC THAN 1,1,1-TRICHLETHANE ALONE. THE LC50 FOR 1,1,1-TCE ALONE WAS DETERMINED TO BE 14,000 PPM FOR BOTH RATS AND QUAIL.

REPORT 13 : EPA TSCA FEDERAL REGISTER NOTICE 51FR51683 DEC 85
EPA HAS REVIEWED PUBLIC COMMENTS ON INDUSTRY'S PROPOSED PROTOCOLS AND SCHEDULES FOR TEST STANDARDS AND REPORTING REQUIREMENTS AND HAS DECIDED TO PROMULGATE A FINAL RULE THAT INCORPORATES THESE PROTOCOLS AND SCHEDULES FOR TESTING TCE FOR DEVELOPMENTAL TOXICITY.

REPORT 14 : PROPOSED RULE MAKING JUN 84
NAT'L DRINKING WATER REGS 49FR24330
UNDER SAFE DRINKING WATER ACT, EPA PROPOSES TO ESTABLISH A LEVEL OF 0.20 MG/L FOR 1,1,1-TCE.

REPORT 15 : ANPR EPA SAFE DRINKING WATER STDS -49 FR 24330 JUN 84
EPA PROPOSES TO ESTABLISH RECOMMENDED MAXIMUM CONTAMINAT LEVELS FOR 1,1,1-TCE UNDER SAFE DRINKING WATER ACT.

REPT'D USE : SOLVENT, CLEANING AGENTS, PHOTOGRAPHIC CHEMICAL, ADHESIVES, FLAME RETARDANT, AEROSOLS, DEODORIZING AGENTS, METAL DEGREASING AGENT, DRAIN CLEANERS AND DRY CLEANING AGENT

CAS NO : **79-00-5**
CHEM NAME : ***** 1,1,2-TRICHLOROETHANE**

REPORT 01 : NCI CARC TRS 74 DHEW PUB NO 78-1324
THE RESULTS OF THIS STUDY DO NOT PROVIDE CONVINCING EVIDENCE FOR THE CARCINOGENICITY OF 1,1,2-TRICHLOROETHANE IN OSBORNE-MENDEL RATS. UNDER THE CONDITIONS OF THIS BIOASSAY 1,1,2-TRICHLOROETHANE IS CARCINOGENIC IN B6C3F1 MICE, CAUSING HEPATOCELLULAR CARCINOMAS AND ADRENAL PHEOCHROMOCYTOMAS (STOMACH INTUBATION STUDY).

REPORT 02 : NIOSH CURRENT INTELLIGENCE BULLETIN 27 AUG 78
NCI BIOASSAY RESULTS INDICATE THAT 1,1,2-TRICHLOROETHANE IS CARCINOGENIC IN MICE, INDUCING LIVER CANCER IN BOTH SEXES AND INCREASED ADRENAL PHEOCHROMOCYTOMA, A TUMOR WHICH GIVES RISE TO HIGH BLOOD PRESSURE AND HYPERGLYCEMIA. NO CONCLUSIVE EVIDENCE OF CANCER IN RATS WAS FOUND. OSHA EXPOSURE STANDARD IS 10 PPM BUT NIOSH HAS NO RECOMMENDED STANDARD. THE COMPOUND INDUCES LIVER AND KIDNEY INJURY IN DOGS AND GUINEA PIGS. NIOSH RECOMMENDS THAT OCCUPATIONAL EXPOSURE TO THIS COMPOUND BE MINIMIZED.

REPORT 03 : IARC MONOGRAPH 20 1979
1,1,2-TRICHLOROETHANE WAS TESTED IN ONE EXPERIMENT IN MICE AND IN ONE IN RATS BY ORAL ADMINISTRATION. IT PRODUCED HEPATOCELLULAR CARCINOMAS AND ADRENAL PHEOCHROMOCYTOMAS IN MICE OF BOTH SEXES. THE STUDY IN RATS GAVE INCONCLUSIVE RESULTS.

REPORT 04 : EPA AMBIENT WATER QUALITY CRITERIA OCT 80
1,1,2-TRICHLOROETHANE IS CONSIDERED TO BE CARCINOGENIC IN AT LEAST ONE RODENT SPECIES. AMBIENT WATER CONCENTRATIONS OF 6 UG/L, .6 UG/L, AND .06 UG/L ARE ESTIMATED TO KEEP THE LIFETIME RISK BELOW 10 -5, 10 -6. AND 10 -7 RESPECTIVELY.

REPORT 05 : EPA RCRA HEALTH AND ENV. EFFECTS PROFILE OCT 80
RESULTS OF AN NTP BIOASSAY INDICATE THAT ORAL ADMINISTRATION OF 1,1,2-TRICHLOROETHANE PRODUCES AN INCREASE OF SEVERAL TUMOR TYPES IN RATS AND MICE.

REPORT 06 : MULTI-AGENCY CROSS CHECK
N HHS MAGUIRE
N OSHA CARC
Y EPA CAG
Y EPA RCRA
N EPA CLEAN AIR
Y EPA CLEAN WATER

REPORT 07 : IARC MONOGRAPH 20 1979
1,1,2-TRICHLOROETHANE WAS NOT MUTAGENIC IN A PLATE ASSAY WITH SALMONELLA TA1535, WITH OR WITHOUT METABOLIC ACTIVATION. NO DATA ON THE EMBRYOTOXICITY OR TERATOGENICITY OF 1,1,2-TRICHLOROETHANE WERE AVAILABLE.

REPORT 08 : EPA TSCA 8(E) 0283-0471 AND 0383-0471 MAR 83
FERRO CORP REPORTED ITS INDUSTRIAL LUBRICANT ADDITIVE PRODUCT (MAJOR COMPONENT BIS(2-CHLOROOCTYL) DISULFIDE) EXHIBITED MODERATE TO EXTREME DERMAL SENSITIZATION IN GUINEA PIG. CONTACT DERMATITIS REPORTED AMONG METALWORKERS USING PRODUCT WITH 1,1,1-TRICHLOROETHANE, STODDARD SOLVENT, AND IN MOST CASES KEROSENE. PRODUCT

MUTAGENIC IN SACCHAROMYCES CEREVISIAL WITHOUT ACTIVATION, AND CAUSED SOME MUTANTS WITH ACTIVATION.

REPT'D USE : INDUSTRIAL SOLVENT

CAS NO : **79-01-6**

CHEM NAME : ***** TRICHLOROETHYLENE (TCE)**

REPORT 01 : ADDENDUM TO EPA HEALTH ASSESS JUN 87
DHEW PUB NO 76-802
THE RESULTS OF THIS CARCINOGENESIS TEST OF TRICHLOROETHYLENE CLEARLY INDICATE THAT TRICHLOROETHYLENE INDUCED A HEPATOCELLULAR CARCINOMA RESPONSE IN MICE. WHILE THE ABSENCE OF A SIMILAR EFFECT IN RATS APPEARS MOST LIKELY ATTRIBUTABLE TO A DIFFERENCE IN SENSITIVITY BETWEEN THE OSBORNE-MENDEL RAT AND THE B6C3F1 MOUSE, THE EARLY MORTALITY OF RATS DUE TO TOXICITY MUST ALSO BE CONSIDERED, INDUSTRIAL GRADE TCE WAS GIVEN BY GAVAGE.

REPORT 02 : NIOSH CURRENT INTELLIGENCE BULLETIN 2 JUN 75
PRELIMINARY EVALUATION OF THE CARCINOGENIC ACTIVITY OF TCE IN LABORATORY RODENTS BY THE NCI INDICATES THAT THIS MATERIAL IS A POTENT LIVER CARCINOGEN.

REPORT 03 : EPA AMBIENT WATER QUALITY CRITERIA OCT 80
TRICHLOROETHYLENE IS SUSPECTED OF BEING A HUMAN CARCINOGEN. AMBIENT WATER CONCENTRATIONS OF 27 UG/L, 2.7 UG/L AND .27 UG/L ARE ESTIMATED TO RESULT IN AN INCREASED CANCER RISK OVER A LIFETIME OF 10 -5, 10 -6, AND 10 -7 RESPECTIVELY.

REPORT 04 : IARC MONOGRAPH 20 1979
TRICHLOROETHYLENE WAS TESTED IN ONE EXPERIMENT IN MICE AND IN ONE IN RATS BY ORAL ADMINISTRATION. IN MICE, IT PRODUCED HEPATOCELLULAR CARCINOMAS AND LUNG TUMORS IN BOTH MALES AND FEMALES. THE EXPERIMENT IN RATS WAS CONSIDERED TO BE INADEQUATE.

REPORT 05 : NIOSH SPECIAL HAZARD REVIEW FEB 78
NIOSH RECOMMENDS A STANDARD OF 25 PPM TWA BASED ON CARCINOGENIC AND CNS DEPRESSANT EFFECTS. THE PRESENT OSHA STANDARD IS 100 PPM TWA.

REPORT 06 : MULTI-AGENCY CROSS CHECK
N HHS MAGUIRE
N OSHA CARC
Y EPA CAG
Y EPA RCRA
Y EPA CLEAN AIR
Y EPA CLEAN WATER

REPORT 07 : NTP TR 243 NIH PUB. 90- 1799 MAY 90
UNDER CONDITIONS OF THIS BIOASSAY, TRICHLOROETHYLENE (EPICHLORODYDRIN-FREE) CARCINOGENIC IN MALE F344/N RATS GIVEN 500 AND 1,000 MG/KG (EPICHLOROHYDRIN-FREE), CARCINOGENIC IN MALE F344/N RATS GIVEN 500 AND INDUCING INCREASE IN RENAL TUBULAR ADENOMAS AND ADENCARCINOMAS. FINDINGS MAY BE CONFOUNDED BECAUSE BOTH DOSE LEVELS TO TCE EXCEEDED MTD FOR MALE RATS. CARCINOGENIC FOR BOTH SEXES OF B6C3F1/N MICE GIVEN 1,000 MG/KG, CAUSING INCREASED INCIDENCES OF HEPATOCELLULAR CARCINOMAS AND OF HEPATOCELLULAR CARCINOGENIC FOR FEMALE F344/N RATS.

REPORT 09 : EPA RCRA HEALTH AND ENV. EFFECTS PROFILE OCT 80
TRICHLOROETHYLENE HAS NOT BEEN SHOWN TO BE A TERATOGEN, AND THE DATA SUGGESTING MUTAGENICITY AND CARCINOGENICITY ARE WEAK. THE STUDIES OF MUTAGENICITY AND CARCINOGENICITY HAVE BEEN COMPLICATED BY THE PRESENCE OF CONTAMINANTS WITH KNOWN CARCINOGENIC AND MUTAGENIC ACTIVITY. HOWEVER, THE EPA CAG HAS DETERMINED THAT TRICHLOROETHYLENE IS CARCINOGENIC.

REPORT 10 : IARC MONOGRAPH 20 1979
TCE WAS MUTAGENIC IN E. COLI K12 AND IN SALMONELLA TA100, IN THE PRESENCE OF METABOLIC ACTIVATION. IN ANOTHER ASSAY WITH SALMONELLA TA100, THE PURIFIED CHEMICAL WAS SHOWN NONMUTAGENIC. IN SACCHAROMYCES, TCE INDUCED REVERSEMUTATIONS, WITH METABOLIC ACTIVATION. THE EXTENT TO WHICH POSITIVE MUTAGENIC RESULTS REPORTED WITH TCE ARE DUE TO IMPURITIES, SUCH AS EPICHLOROHYDRIN AND 1,2-EPOXYBUTANE, COULD NOT BE DETERMINED.

REPORT 11 : NTP TECHNICAL BULLETIN 3 DEC 80
TERATOLOGY ASSAYS SHOW THAT INHALATION OF TCE IN RATS AND RABBITS IS NOT ASSOCIATED WITH PRENATAL DEATH OR INCREASED MALFORMATIONS AT DOSES TOXIC TO THE MOTHER.

REPORT 12 : NTP TECHNICAL BULLETIN 9 APR 83
TRICHLOROETHYLENE WAS NEGATIVE IN ONE SALMONELLA TYPHIMURIUM MUTAGENICITY ASSAY.

REPORT 13 : NTP TR 273 NIH PUB #88-2520 APR 88
IN 2 YEAR GAVAGE STUDIES OF TRICHLOROETHYLENE IN MALE AND FEMALE ACI, AUGUST, MARSHALL, AND OSBORNE-MENDEL RATS, TRICHLOROETHYLENE ADMINISTRATION CAUSED RENAL TUBULAR CELL CYTOMEGALY AND TOXIC NEPHROPATHY. THESE ARE CONSIDERED TO BE INADEQUATE STUDIES OF CARCINOGENIC ACTIVITY BECAUSE OF CHEMICALLY INDUCED TOXICITY, REDUCED SURVIVAL, AND DEFICIENCIES IN CONDUCT OF THE STUDIES. DESPITE THESE LIMITATIONS, TUBULAR CELL NEOPLASMS OF THE KIDNEY OVSERVED, AND INTERSTITIAL CELL NEOPLASMS OF TEST IS OBSERVED IN MARSHALL RATS.

REPORT 14 : EPA SAFE DRINKING WATER ANPR FR NOTICE 49FR24330 DEC 86
PROPOSED RULE UNDER THE SAFE DRINKING WATER ACT WATER ESTABLISHES RECOMMENDED MAXIMUM CONTAMINANT LEVELS IN DRINKING WATER.

REPT'D USE : SOLVENT, CHEMICAL INTERMEDIATE, ADHESIVES, PAINSTS/VARNISHES, SPOT REMOVERS, CLEANING AGENTS, AEROSOL ENGINE CLEANERS, FIRE EXTINGUISHERS SOLVENT AND ANESTHETIC EXTRACTANT

CAS NO : **75-69-4**
CHEM NAME : ***** TRICHLOROFLUOROMETHANE**

REPORT 01 : NCI CARC TRS 106 DHEW PUB NO 78-1356
THE RESULTS OF THE BIOASSAY OF TRICHLOROFLUOROMETHANE IN OSBORNE-MENDEL RATS FOR POSSIBLE CARCINOGENICITY ARE NOT CONCLUSIVE BECAUSE INADEQUATE NUMBERS OF RATS SURVIVED LONG ENOUGH TO BE AT RISK FROM LATE-DEVELOPING TUMORS. UNDER THE CONDITIONS OF THIS BIOASSAY, TRICHLOROFLUOROMETHANE WAS NOT CARCINOGENIC TO MALE OR FEMALE B6C3F1 MICE. TRICHLOROFLUOROMETHANE WAS GIVEN BY GAVAGE.

REPORT 02 : EPA AMBIENT WATER QUALITY CRITERIA OCT 80
A CRITERION, BASED ON A NO OBSERVED ADVERSE EFFECT LEVEL IN RATS, AND CALCULATED ON AN ACCEPTABLE DAILY HUMAN INTAKE OF 64.5G AND AN UNCERTAINTY FACTOR OF 1000, IS 32.3MG/L. HOWEVER, FOR THE MAXIMUM PROTECTION OF HUMAN HEALTH FROM THE POTENTIAL CARCINOGENIC EFFECTS FROM TCFM, CONCENTRATIONS OF 1.9 UG/L, .19 UG/L AND .019 UG/L ARE ESTIMATED TO RESULT IN AN INCREMENTAL INCREASE OF CANCER RISK OVER THE LIFETIME OF 10 -5, 10 -6 AND 10 -7 RESPECTIVELY.

REPORT 03 : EPA RCRA HEALTH AND ENV. EFFECTS PROFILE OCT 80
FULLY HALOGENATED CHLOROFLUORALKANES, INCLUDING TRICHLOROFLUOROMETHANE (F-11) ARE ALLEGED TO DEPLETE THE STRATOSPHERIC OZONE, LEADING TO VARIOUS ADVERSE EFFECTS. F-11 WAS NOT FOUND CARCINOGENIC IN A BIOASSAY IN MICE, AND WAS NEGATIVE IN MUTAGENICITY TESTS IN SALMONELLA.

REPORT 04 : NTP TECHNICAL BULLETIN 9 APR 83
TRICHLOROFLUOROMETHANE WAS NEGATIVE IN ONE SALMONELLA TYPHIMURIUM MUTAGENICITY ASSAY.

REPORT 06 : MULTI-AGENCY CROSS CHECK
N HHS MAGUIRE
N OSHA CARC
N EPA CAG
Y EPA RCRA

N EPA CLEAN AIR
Y EPA CLEAN WATER
REPT'D USE : AEROSOL PROPELLANT, REFRIGERANT, SOLVENT, CLEANING AGENTS, SPACE DEODORANTS, FIRE EXTINGUISHERS AND PLASTIC MONOMER

CAS NO : **95-95-4**
CHEM NAME : ***** 2,4,5-TRICHLOROPHENOL**

REPORT 01 : IARC MONOGRAPH 20 1979
2,4,5- AND 2,4,6-TRICHLOROPHENOLS WERE TESTED BY S.C. INJECTION IN TWO STRAINS OF MICE. THE STUDY WAS CONSIDERED INADEQUATE. NO DATA ON THE EMBRYOTOXICITY, TERATOGENICITY OR MUTAGENICITY OF TRICHLOROPHENOLS WERE AVAILABLE.

REPORT 02 : NTP TECHNICAL BULLETIN 9 APR 83
2,4,5-TRICHLOROPHENOL WAS NEGATIVE IN ONE SALMONELLA TYPHIMURIUM MUTAGENICITY ASSAY.

REPORT 06 : MULTI-AGENCY CROSS CHECK
N HHS MAGUIRE
N OSHA CARC
N EPA CAG
Y EPA RCRA
N EPA CLEAN AIR
N EPA CLEAN WATER
REPT'D USE : HERBICIDE INTERMEDIATE AND FUNGICIDE

CAS NO : **88-06-2**
CHEM NAME : ***** 2,4,6-TRICHLOROPHENOL**

REPORT 01 : NCI CARC TRS 155 DHEW PUB NO 79-1711
UNDER THE CONDITIONS OF THIS BIOASSAY, 2,4,6-TRICHLOROPHENOL WAS CARCINOGENIC IN MALE F344 RATS, INDUCING LYMPHOMAS OR LEUKEMIAS. THE TEST CHEMICAL WAS ALSO CARCINOGENIC IN BOTH SEXES OF B6C3F1 MICE, INDUCING HEPATOCELLULAR CARCINOMAS OR ADENOMAS (FEEDING STUDY).

REPORT 02 : IARC MONOGRAPH 20 1979
2,4,6-TRICHLOROPHENOL WAS TESTED IN ONE EXPERIMENT IN TWO STRAINS OF MICE BY ORAL ADMINISTRATION, AND 2,4,5- AND 2,4,6-TRICHLOROPHENOLS WERE TESTED BY SUBCUTANEOUS INJECTION IN TWO STRAINS OF MICE. NO DATA ON THE EMBRYOTOXICITY, TERATOGENICITY OR MUTAGENICITY OF TRICHLOROPHENOLS WERE AVAILABLE. BOTH EXPERIMENTS WERE CONSIDERED INADEQUATE.

REPORT 03 : NTP TECHNICAL BULLETIN 3 DEC 80
THE COMPOUND WAS FOUND TO BE NEGATIVE IN THE SALMONELLA TYPHIMURIUM MUTAGENICITY ASSAY IN ONE LABORATORY.

REPORT 04 : EPA AMBIENT WATER QUALITY CRITERIA OCT 80
2,4,6-TRICHLOROPHENOL IS SUSPECTED OF BEING A HUMAN CARCINOGEN. AMBIENT WATER CONCENTRATIONS OF 12 UG/L, 1.2 UG/L, AND .12 UG/L ARE ESTIMATED TO KEEP THE LIFETIME CANCER RISK BELOW 10 -5, 10 -6, AND 10 -7 RESPECTIVELY.

REPORT 05 : EPA RCRA HEALTH AND ENV. EFFECTS PROFILE OCT 80
IN A 1979 STUDY, THE NCI CONCLUDED THAT 2,4,6-TRICHLOROPHENOL IS CARCINOGENIC IN RATS AND MICE. EPA'S CARCINOGEN ASSESSMENT GROUP HAS DETERMINED THAT THERE IS SUBSTANTIAL EVIDENCE THAT 2,4,6TRICHLOROPHENOL IS CARCINOGENIC IN MAN.

REPORT 06 : MULTI-AGENCY CROSS CHECK
Y HHS MAGUIRE
N OSHA CARC
N EPA CAG
Y EPA RCRA
N EPA CLEAN AIR
Y EPA CLEAN WATER

REPT'D USE : WOOD PRESERVATIVES AND DISINFECTANT FOR URINALS; TOILET BOWLS AND BATHROOMS

CAS N0 : **93-76-5**
CHEM NAME : ***** 2,4,5-TRICHLOROPHENOXYACETIC ACID (2,4,5-T)**

REPORT 01 : IARC MONOGRAPH 15 1977
2,4,5-T WAS TESTED IN MICE IN THREE STUDIES BY ORAL AND S.C. ADMINISTRATION. THE CARCINOGENICITY OF THIS COMPOUND COULD NOT BE EVALUATED ON THE BASIS OF THE AVAILABLE DATA. THE RESULTS OF THE SINGLE COHORT STUDY OF A SMALL NUMBER OF WORKERS EXPOSED TO VARIOUS HERBICIDES, INCLUDING 2.4-D, 2,4,5-T, AND 3-AMINO-1,2,4-TRIAZOLE, ARE NOT SUFFICIENT TO EVALUATE THE CARCINOGENICITY OF 2,4,5-T.

REPORT 02 : IARC MONOGRAPH 15 1977
NO CLEARCUT TERATOGENIC EFFECTS HAVE BEEN REPORTED FOR 2,4,5-T IN RATS, RABBITS, SHEEP OR MONKEYS. 2,4,5-T IS NOT MUTAGENIC IN SALMONELLA TA1535, TA1537 OR TA1538, E. COLI, OR SACCHAROMYCES. 2,4,5-T DID NOT INCREASE DOMINANT LETHAL MUTATIONS IN MICE.

REPORT 03 : NTP TECHNICAL BULLETIN 3/8 DEC 81 , JUL 82
2,4,5-T WAS NEGATIVE IN THE SALMONELLA ASSAY AND IN DROSOPHILA WHEN TESTED FOR SEX-LINKED RECESSIVE LETHALS.

REPORT 06 : MULTI-AGENCY CROSS CHECK
N HHS MAGUIRE
N OSHA CARC
N EPA CAG
Y EPA RCRA
N EPA CLEAN AIR
N EPA CLEAN WATER

REPT'D USE : HERBICIDE

CAS NO : **96-18-4**
CHEM NAME : ***** 1,2,3-TRICHLOROPROPANE**

REPORT 01 : NTP CARC TEST PROGRAM JUL 82
IN A CURRENT NTP BIOASSAY, 1,2,3-TRICHLOROPROPANE IS BEING ADMINISTERED BY GAVAGE TO F344 RATS AND B6C3F1 MICE. THE PRECHRONIC TEST PHASE IS IN PROGRESS.

REPORT 02 : NTP TECHNICAL BULLETIN 6 JAN 82
THE COMPOUND WAS POSITIVE IN THE SALMONELLA TYPHIMURIUM MUTAGENICITY ASSAY IN ONE LABORATORY.

REPT'D USE : SOLVENT, DEGREASING AGENT AND PAINT/VARNISH REMOVERS

CAS NO : **817-09-4**
CHEM NAME : ***** TRICHLOROTRIETHYLAMINE HYDROCHLORIDE**

REPORT 01 : IARC MONOGRAPH 9 1975
ALTHOUGH THERE IS AN INDICATION IN ONE STUDY THAT TRICHLOROTRIETHYLAMINE HYDROCHLORIDE IS CARCINOGENIC IN MICE AND RATS, THE DATA ARE INSUFFICIENT TO ALLOW AN EVALUATION OF THE CARCINOGENICITY OF THIS CHEMICAL.

REPORT 02 : IARC MONOGRAPH 9 1975
TRICHLOROTRIETHYLAMINE HYDROCHLORIDE INDUCES POINT MUTATIONS IN S. POMBE AND CHROMOSOME ABERRATIONS IN CARCINOMA CELLS.

REPT'D USE : ANTINEOPLASTIC AGENT AND FIXING AGENT FOR TEXTILE DYES; NO LISTING ON TSCA INVENTORY

CAS NO : **76-13-1**
CHEM NAME : ***** 1,1,2-TRICHLORO-1,2,2-TRIFLUOROETHANE (F113)**

REPORT 01 : EPA RCRA HEALTH AND ENV. EFFECTS PROFILE OCT 80
FULLY HALOGENATED CHLOROFLUORALKANES, INCLUDING DICHLORODIFLUOROMETHANE (F-12) ARE ALLEGED TO DEPLETE THE STRATOSPHERIC OZONE, LEADING TO VARIOUS ADVERSE EFFECTS. F-12 HAS TESTED NEGATIVE IN SALMONELLA ASSAYS, BUT POSITIVE IN NEUROSPORA.

REPORT 02 : CPSC SCHOOL LAB CHEMICAL SURVEY JAN 82
THE CHEMICAL IS REPORTED IN ONE MAJOR TEXT TO BE AN EXPERIMENTAL CARCINOGEN. SUBSEQUENT CPSC STAFF REVIEW SHOWS THERE IS INSUFFICIENT EVIDENCE FOR THE CARCINOGENICITY OF THIS CHEMICAL IN ANIMALS OR MAN. THE CHEMICAL HAS SHOWN NO MUTAGENIC ACTIVITY IN SALMONELLA OR DOMINANT LETHAL ASSAY IN MICE. TERATOLOGY AND REPRODUCTIVE STUDIES ARE REPORTED NEGATIVE. IN A RECENT BIOASSAY, DU PONT EXPOSED RATS BY INHALATION TO 20000, 10000, AND PPM OF CFC-113. THE CHRONIC EXPOSURE PHASE HAS BEEN COMPLETED AND HISTOPATHOLOGY WILL BE COMPLETED DURING 1983.

REPT'D USE : DRY CLEANING SOLVENT, REFRIGERANT, AEROSOL, PROPELLANT AND FIRE EXTINGUISHER

CAS NO : **21259-20-1**
CHEM NAME : ***** T2-TRICHOTHECENE**

REPORT 01 : IARC MONOGRAPH 31 JUL 83
TESTS IN RATS BY INTRAGASTRIC ADMINISTRATION, IN TROUT IN DIET AND IN MICE BY SKIN PAINTING STUDIES TO INVESTIGATE INITIATING AND PROMOTING ACTIVITY; ALL INADEQUATE FOR EVALUATION.

REPORT 02 : IARC MONOGRAPH 31 JUL 83
T2-TRICHOTHECENE WAS NOT MUTAGENIC TO BACTERIA OR YEAST. INDUCED CHROMOSOMAL LOSS, NON-DISJUNCTION OF SEX CHROMOSOMES AND A SLIGHT INCREASE IN RECESSIVE LETHAL MUTATIONS IN DROSOPHILA MELANOGASTER. ITS 4-0-DEACYLATED METABOLITE INDUCED UNSCHEDULED DNA SYNTHESIS IN HUMAN FIBROBLASTS IN VITRO. CHROMATID ABERRATIONS BUT NOT MICRONUCLEI INDUCED IN HAMSTERS EXPOSED IN VIVO. T2-TRICHOTHECENE IS EMBRYOLETHAL AND TERATOGENIC TO MICE, INDUCING TAIL AND LIMB ANOMALIES AND EXENCEPHALY.

REPT'D USE : NATURALLY OCCURRING MYCOTOXIN

CAS NO : **1330-78-5**
CHEM NAME : ***** TRICRESYL PHOSPHATE (TCP)**

REPORT 01 : EPA TSCA 8EHQ 0788-0744 SEP 88
THE MOBIL RESEARCH AND DEVELOPMENT CORPORATION RESEARCH SHOWED THAT REPEATED SKIN APPLICATIONS OF THE GENERIC JET ENGINE OIL CONTAINING 3% TCP (ONE DERMAL APPLICATION/DAY, 5 DAYS/WEEK, FOR 90 DAYS) TO MALE AND FEMALE SPRAGUE-DAWLEY RATS DECREASED THE ACTIVITIES OF BOTH SERUM AND ERYTHROCYTE CHOLINESTERASE. A FOLLOW-UP STUDY, DESIGNED TO IDENTIFY THE COMPONENT CAUSING

		CHOLINESTERASE INHIBITION, SHOWED THAT THE TCP ADDITIVE WAS ENTIRELY RESPONSIBLE. AN ADDITIONAL ACUTE STUDY, PERFORMED IN MALE LONG-EVANS RATS, SHOWED THAT SINGLE ORAL OR DERMAL DOSES OF TCP OR TRIORTHO-CRESYL PHOSPHATE INHIBITED SERUM CHOLINESTERASE AND BRAIN NEUROPATHY.
REPT'D USE	:	JET ENGINE OIL ANTI-WEAR AGENT, MENERAL OIL LUBRICANT, FIRE RESISTANT HYDRAULIC FLUIDS, PLASTICIZER, PLASTICS, FIRE RETARDANT, AIR FILTER MEDIUM, SOLVENT AND WATERPROOFING

CAS NO : **15468-32-3**
CHEM NAME : ***** TRIDYMITE**

REPORT 01	:	NIOSH CRITERIA DOCUMENT NOV 74 NIOSH RECOMMENDS A STANDARD OF 50 UG/CU M TWA BASED ON SILICOSIS. THE PRESENT OSHA STANDARD IS 10 MG/CU TWA.
REPT'D USE	:	NO LISTING ON TSCA INVENTORY

CAS NO : **102-76-6**
CHEM NAME : ***** TRIETHANOLAMINE**

REPORT 01	:	EPA TSCA FYI-0638 TRIETHANOLAMINE (TEA) UNIFORMALY RADIO LABELED WITH 14C WAS ADMINISTERED TO C3HEJ MICE INTRAVENOUSLY (1 MG/KG; IN WATER) OR DERMALLY, (1000 MG/KG; IN ACETONE; 2000 MG/KG; UNDILUTED OR IN WATER) AND TO FISCHER 344 RATS DERMALLY (1000 MG/KG; UNDILUTED). RESULTS INDICATE THAT TEA WAS EXTENSIVELY AND RAPIDLY ABSORBED FOLLOWING DERMAL APPLICATION TO C3H/HEJ MICE. THE DATA SUGGEST A MUCH SLOWER RATE OF DERMAL ABSORPTION OF TEA FOR RATS.
REPT'D USE	:	COSMETIC

CAS NO : **1680-21-3**
CHEM NAME : ***** TRIETHYLENE GLYCOL DIACRYLATE**

REPORT 01	:	EPA TSCA 8(E) NOTICE 8EHQ-0981-0410 SEP 81 CELANESE PROVIDED PRELIMINARY RESULTS FROM A CHRONIC DERMAL APPLICATION STUDY OF TRIETHYLENE GLYCOL DIACRYLATE IN MICE. SQUAMOUS CELL CARCINOMAS AND MICROSCOPIC PAPILLOMAS WERE REPORTED AFTER TREATMENTS TWICE WEEKLY FOR 18 MONTHS.
REPT'D USE	:	NO AVAILABLE PRODUCTION OR USE DATA

CAS NO : **1954-28-5**
CHEM NAME : ***** TRIETHYLENE GLYCOL DIGLYCIDYL ETHER**

REPORT 01 : IARC MONOGRAPH 11 1976
TRIETHYLENE GLYCOL DIGLYCIDYL ETHER WAS CARCINOGENIC IN A LIMITED STUDY IN MICE BY INTRAPERITONEAL INJECTION: IT CAUSED AN INCREASE IN THE INCIDENCE OF PULMONARY TUMORS.
REPT'D USE : LABORATORY AGENT; NO LISTING ON TSCA INVENTORY

CAS NO : **143-22-6**
CHEM NAME : ***** TRIETHYLENE GLYCOL MONOBUTYL ETHER**

REPORT 01 : EPA TSCA FEDERAL REGISTER NOTICE 51 FR 17883 MAY 86
IN RESPONSE TO ITC MAY 21, 1985 RECOMMENDATION (50 FR 20930) TESTING UNDER TSCA 4(A) TO REQUIRE SUBCHRONIC, DEVELOPMENTAL, NEUROTOXIC, MUTAGENIC, REPRODUCTIVE AND ONCOGENIC TOXICITY TESTING.
REPT'D USE : BRAKE FLUID

CAS NO : **112-50-5**
CHEM NAME : ***** TRIETHYLENE GLYCOL MONOETHYL ETHER**

REPORT 01 : EPA FEDERAL REGISTER NOTICE 51 FR 17883 MAY 86
EPA RESPONSE TO ITC MAY 21, 1985 (50 FR 20930) RECOMMENDATION FOR TESTING PROPOSES UNDER TSCA 4(A) TO REQUIRE SUBCHRONIC, DEVELOPMENTAL NEUROTOXIC, MUTAGENIC, REPORDUCTIVE AND ONCOGENIC TOXICITY TESTING.
REPT'D USE : BRAKE FLUID

CAS NO : **112-35-6**
CHEM NAME : ***** TRIETHYLENE GLYCOL MONOMETHYL ETHER**

REPORT 01 : EPA FR NOTICE 51 FR 17883 MAY 86
IN RESPONSE TO ITC MAY 21, 1985 (50 FR 20930) RECOMMENDATION, EPA PROPOSES TESTING UNDER TSCA 4(A) TO REQUIRE SUBCHRONIC, DEVELOPMENTAL NEUROTOXIC, MUTAGENIC, REPRODUCTIVE AND ONCOGENIC TOXICITY TESTING.
REPT'D USE : BRAKE FLUID

CAS NO : **112-24-3**
CHEM NAME : ***** TRIETHYLENE TETRAMINE (TETA)**

REPORT 01 : CHEMICAL HAZARD INFORMATION PROFILE SEP 82
TETA MUTAGENIC IN S. TYPHIMURIUM TEST WITHOUT ACTIVATION IN STRAINS TA1535, TA1538, TA98 AND TA100, BUT NEGATIVE IN TA1537. SHOWN TO HAVE DNA-DAMAGING CAPABILITY IN AN E. COLI DIFFERENTIAL REPAIR ASSAY. ADMINISTRATION OF TETA TO PREGNANT ANIMALS RESULTED IN DAMAGE OR DEATH TO OFF-SPRING.
REPT'D USE : PLASTICIZER, COSMETICS, CHELATING AGENT AND CHEMICAL INTERMEDIATE

CAS NO : **78-40-0**
CHEM NAME : ***** TRIETHYL PHOSPHATE**

REPORT 01 : CPSC FLAME RETARDANTS MONOGRAPH JUL 82
TRIETHYL PHOSPHATE WAS MUTAGENIC IN S. TYPHIMURIUM AND IN PSEUDOMONAS. TRIETHYL PHOSPHATE FAILED TO PRODUCE MUTATIONS IN S. TYPHIMURIUM TA98, TA100, TA1535, TA1537 AND TA1538, WITH OR WITHOUT METABOLIC ACTIVATION. NO MUTAGENIC EFFECTS WERE OBSERVED IN A DOMINANT LETHAL STUDY IN SWISS MICE.

CAS NO : **1582-09-8**
CHEM NAME : ***** TRIFLURALIN**

REPORT 02 : NCI CARC TRS 34 DHEW PUB NO 78-834
EVALUATION OF THE RESULTS OF THIS BIOASSAY INDICATES THAT TECHNICAL GRADE TRIFLURALIN IS A CARCINOGEN IN FEMALE B6C3F1 MICE, BEING ASSOCIATED IN THESE ANIMALS WITH AN ELEVATED INCIDENCE OF HEPATOCELLULAR CARCINOMAS, ALVEOLAR/BRONCHIOLAR ADENOMAS AND SQUAMOUS-CELL CARCINOMAS OF THE FORESTOMACH. SUFFICIENT EVIDENCE WAS NOT PROVIDED FOR THE CARCINOGENICITY OR TUMORIGENICITY OF TRIFLURALIN IN MALE B6C3F1 MICE OR IN OSBORNE-MENDEL RATS OF EITHER SEX (FEEDING STUDY).

REPORT 06 : MULTI-AGENCY CROSS CHECK
N HHS MAGUIRE
Y OSHA CARC
N EPA CAG
N EPA RCRA
N EPA CLEAN AIR
N EPA CLEAN WATER
REPT'D USE : PESTICIDE AND HERBICIDE

CAS NO : **1421-63-2**
CHEM NAME : ***** 2,4,5-TRIHYDROXYBUTYROPHENONE (THBP)**

REPORT 02 : NTPTECHNICAL BULLETIN 6 JAN 82
THE COMPOUND WAS POSITIVE IN THE S. TYPHIMURIUM MUTAGENICITY ASSAY IN ONE LABORATORY.
REPT'D USE : ANTIOXIDANT FOR POLYOLEFIN PLASTICS AND FOOD ADDITIVE

CAS NO : **137-17-7**
CHEM NAME : ***** 2,4,5-TRIMETHYLANILINE**

REPORT 01 : NCI CARC TRS 160 DHEW PUB NO 79-1716
UNDER THE CONDITIONS OF THIS BIOASSAY, 2,4,5-TRIMETHYLANILINE WAS CARCINOGENIC FOR MALE AND FEMALE F344 RATS AND FEMALE B6C3F1 MICE, INDUCING HEPATOCELLULAR CARCINOMAS OR NEOPLASTIC NODULES IN THE RATS OF EACH SEX, ALVEOLAR/BRONCHIOLAR CARCINOMAS IN THE FEMALE RATS, AND HEPATOCELLULAR CARCINOMAS IN THE FEMALE MICE (FEEDING STUDY).

REPORT 02 : NTP TECHNICAL BULLETIN 5/8 AUG 81, JUL 82
2,4,5-TMA WAS POSITIVE IN THE SALMONELLA ASSAY, BUT NEGATIVE IN DROSOPHILA WHEN TESTED FOR SEX-LINKED RECESSIVE LETHALS.

REPORT 03 : IARC MONOGRAPH 27 1982
2,4,5-TRIMETHYLANILINE AND ITS HCL WERE TESTED IN TWO EXPERIMENTS IN MICE AND TWO EXPERIMENTS IN RATS BY DIETARY ADMINISTRATION. ONE EXPERIMENT IN MICE WAS INADEQUATE FOR EVALUATION. IN THE OTHER, 2,4,5-TMA PRODUCED HEPATOCELLULAR CARCINOMAS IN FEMALE MICE. IN ONE STUDY IN RATS THERE WAS NO SIGNIFICANT INCREASE IN TUMORS. IN THE OTHER, IT PRODUCED LIVER CARCINOMAS AND LUNG ADENOMAS. THERE IS LIMITED EVIDENCE FOR THE CARCINOGENICITY OF 2,4,5-TMA IN ANIMALS.

REPORT 04 : IARC MONOGRAPH 27 1982
2,4,5-TRIMETHYLANILINE WAS MUTAGENIC TO SALMONELLA WITH METABOLIC ACTIVATION. NO DATA CONCERNING EFFECTS ON REPRODUCTION OR PRENATAL TOXICITY WERE AVAILABLE.
REPT'D USE : DYE INTERMEDIATE FOR FOOD COLOR; NO KNOWN CURRENT PRODUCTION

CAS NO : **88-05-1**
CHEM NAME : ***** 2,4,6-TRIMETHYLANILINE**

REPORT 01 : IARC MONOGRAPH 27 1982
2,4,6-TRIMETHYLANILINE WAS TESTED IN ONE EXPERIMENT IN

MICE AND ONE EXPERIMENT IN MALE RATS. THE EXPERIMENTS WERE CONSIDERED INADEQUATE FOR EVALUATION.

REPORT 02 : IARC MONOGRAPH 27 1982
THE AVAILABLE DATA WERE INADEQUATE TO EVALUATE THE MUTAGENICITY OF 2,4,6-TRIMETHYLANILINE. NO DATA WERE AVAILABLE CONCERNING EFFECTS ON REPRODUCTION OR PRENATAL TOXICITY.

REPORT 06 : MULTI-AGENCY CROSS CHECK
N HHS MAGUIRE
Y OSHA CARC
N EPA CAG
N EPA RCRA
N EPA CLEAN AIR
N EPA CLEAN WATER
REPT'D USE : ORGANIC SYNTHESIS

CAS NO : **526-73-8**
CHEM NAME : ***** 1,2,3-TRIMETHYL BENZENE**

REPORT 01 : EPA TSCA FEDERAL REGISTER NOTICE 52FR2522 JAN 87
EPA IS ISSUING FINAL TEST RULE UNDER TSCA 4(A) THAT SPECIFIES TEST STANDARDS AND REPORTING REQUIREMENTS FOR TESTING OF THE C9 FRACTION OF AROMATIC HYDROCARBONS.

CAS NO : **95-63-6**
CHEM NAME : ***** 1,2,4-TRIMETHYL BENZENE**

REPORT 01 : EPA TSCA FR NOTICE 52FR2522 JAN 87
EPA IS ISSUING A FINAL TEST RULE UNDER TSCA 4(A) THAT SPECIFIES TEST STANDARDS AND REPORTING REQUIREMENTS FOR TESTING OF THE C9 FRACTION OF AROMATIC HYDROCARBONS.

CAS NO : **25551-13-7**
CHEM NAME : ***** TRIMETHYL BENZENE (MIXED ISOMERS)**

REPORT 01 : EPA TSCA FEDERAL REGISTER NOTICE 52FR2522 JAN 87
EPA IS ISSUING A FINAL TEST RULE UNDER TSCA 4(A) THAT SPECIFIES TEST STANDARDS AND REPORTING REQUIREMENT FOR TESTING C9 FRACTIONS OF AROMATIC HYDROCARBONS.

CAS NO : **540-84-1**
CHEM NAME : ***** 2,2,4-TRIMETHYLPENTANE**

REPORT 01 : FTI-0187 -0815 JAN 87
DERMAL ABSORPTION FOLLOWING BRIEF EXPOSURE IS LESS BY ONE HALF THAT OF BENZENE. THIS MAY BE DUE TO THE GREATER VOLATILITY OF ISOOCTANE THAN BENZENE. EXCRETION OF CPD WAS SLOW FOLLOWING IV ADMINISTRATION. PRESUMABLY CPD IS STORED IN FAT, CAUSING THIS LONG EXCRETION TIME.
REPT'D USE : MOTOR FUEL

CAS NO : **512-56-1**
CHEM NAME : ***** TRIMETHYLPHOSPHATE**

REPORT 01 : NCI CARC TRS 81 DHEW PUB NO 78-1331
UNDER THE CONDITIONS OF THIS BIOASSAY, TRIMETHYLPHOSPHATE WAS CARCINOGENIC IN FEMALE B6C3F1 MICE, INDUCING ADENOCARCINOMAS OF THE UTERUS/ENDOMETRIUM. TRIMETHYLPHOSPHATE WAS ASSOCIATED WITH THE INDUCTION OF BENIGN FIBROMAS OF THE SUBCUTANEOUS TISSUE IN MALE FISCHER 344 RATS. NO EVIDENCE OF CARCINOGENICITY OF THE COMPOUND WAS OBTAINED IN FEMALE RATS OR IN MALE MICE. TRIMETHYLPHOSPHATE WAS GIVEN BY GAVAGE.

REPORT 02 : CPSC FLAME RETARDANTS MONOGRAPH JUL 82
TRIMETHYL PHOSPHATE HAS SHOWN MUTAGENIC ACTIVITY IN VARIOUS BACTERIAL ASSAY SYSTEMS. THE CHEMICAL SHOWED HIGH MUTAGENIC ACTIVITY WHEN ADMINISTERED TO MICE IN A DOMINANT LETHAL STUDY. CHROMOSOME BREAKS IN HUMAN LYMPHOCYTES HAVE BEEN OBSERVED AFTER TREATMENT IN VITRO. CHROMOSOME ABERRATIONS WERE OBSERVED IN BONE-MARROW CELLS AFTER ORAL AND I.P. ADMINISTRATION TO RATS AND MICE.

REPORT 06 : MULTI-AGENCY CROSS CHECK
N HHS MAGUIRE
Y OSHA CARC
N EPA CAG
N EPA RCRA
N EPA CLEAN AIR
N EPA CLEAN WATER
REPT'D USE : INDUSTRIAL SOLVENT AND FR CHEMICAL

CAS NO : **3902-71-4**
CHEM NAME : ***** 4,5',8-TRIMETHYLPSORALEN**

REPORT 01 : IARC VOL 40 1986
THERE IS INADEQUATE EVIDENCE FOR CARCINOGENICITY OF CPD

IN COMBINATION WITH ULTRAVIOLET A (UVA) RADIATION TO EXPERIMENTAL ANIMALS. NO DATA AVAILABLE TO EVALUATE IT IN ABSENCE OF UVA RADIATION. THERE IS INADEQUATE EVIDENCE FOR CARCINOGENICITY OF CPD TO HUMANS. NO DATA AVAILABLE TO WORKING GROUP CONCERNING MUTAGENICITY AND CHROMOSOMAL EFFECTS. NO SKIN TUMOR DESCRIBED IN ONE INCOMPLETELY REPORTED STUDY IN MICE (CPD + UVA RAD.). EVIDENCE INADEQUATE TO DETERMINE GENETIC ACTIVITY IN SHORT-TERM TESTS IN THE DARK, BUT SUFFICIENT IN COMBINATION WITH UVA.

REPT'D USE : CHEMO-THERAPEUTIC AGENT

CAS NO : **2489-77-2**

CHEM NAME : ***** 1,1,3-TRIMETHYL-2-THIOUREA**

REPORT 01 : NCI CARC TRS 129 DHEW PUB NO 79-1384 1979
UNDER THE CONDITIONS OF THIS BIOASSAY, DIETARY ADMINISTRATION OF TRIMETHYLTHIOUREA WAS CARCINOGENIC IN FEMALE FISCHER 344 RATS, INDUCING FOLLICULAR-CELL CARCINOMAS OF THE THYROID. THERE WAS NOT SUFFICIENT EVIDENCE FOR THE CARCINOGENICITY OF THE COMPOUND IN MALE FISCHER 344 RATS OR IN B6C3F1 MICE OF EITHER SEX.

REPORT 06 : MULTI-AGENCY CROSS CHECK
N HHS MAGUIRE
Y OSHA CARC
N EPA CAG
N EPA RCRA
N EPA CLEAN AIR
N EPA CLEAN WATER

REPT'D USE : RUBBER ACCELERATOR AND ADHESIVES

CAS NO : **129-79-3**

CHEM NAME : ***** 2,4,7-TRINITROFLUORENONE (TNF)**

REPORT 01 : EPA SECTION 8E NOTIFICATON FROM IBM CORP. 8EHQ-0480-0339
IBM USES TNF AS A COMPONENT IN CHARGE-TRANSFER COMPLEXES FOR ELECTROPHOTOGRAPHIC COPIERS AND PRINTERS. IBM REPORTS THAT TNF IS A DIRECT-ACTING MUTAGEN (DOES NOT REQUIRE METABOLIC ACTIVATION) IN THE AMES SALMONELLA/MICROSOMAL MUTATION ASSAY (HIST/HIS-), A CHINESE HAMSTER OVARY CELL SISTER CHROMATID EXCHANGE ASSAY, AND A MAMMALIAN MUTATION ASSAY WITH L5178Y MOUSE CELLS. A 1962 STUDY INDICATES THAT TNF INDUCED MAMMARY GLAND TUMORS FOLLOWING A SINGLE ORAL ADMINISTRATION TO RATS. IBM CONTENDS THAT TNF POSES NO HEALTH TO PEOPLE.

REPORT 02 : IBM LETTER TO EPA 6/3/80
IBM CONTENDS THAT THE NATURE OF THE ELECTROPHOTOGRAPHIC PROCESSES AND THE SMALLER AMOUNTS OF TNF IN THE MACHINES COMPRISING APPROXIMATELY 98.5% OF THE SUBJECT MACHINES LEADS IBM TO BELIEVE THAT THEY PRESENT LESS HUMAN EXPOSURE THAN IDENTIFIED IN THE WORST-CASE ANALYSIS.

REPORT 03 : EPA TSCA 8(E) NOTICE 8EHQ-0480-0339 APR 80
IBM REPORTS THAT TNF PRODUCED POSITIVE RESULTS WITHOUT METABOLIC ACTIVATION IN THE SALMONELLA ASSAY, CHINESE HAMSTER OVARY CELL SITER CHROMATID EXCHANGE ASSAY, AND A MAMMALIAN MUTATION ASSAY WITH L5178Y MOUSE CELLS. A 1962 STUDY INDICATES THAT TNF INDUCED MAMMARY GLAND TUMORS FOLLOWING A SINGLE ORAL ADMINISTRATION TO RATS.

REPORT 06 : MULTI-AGENCY CROSS CHECK
N HHS MAGUIRE
Y OSHA CARC
N EPA CAG
N EPA RCRA
N EPA CLEAN AIR
N EPA CLEAN WATER

REPT'D USE : CHARGE-TRANSFER COMPLEX AND ELECTROPHOTOGRAPHIC COPIERS

CAS NO : **217-59-4**
CHEM NAME : ***** TRIPHENYLENE**

REPORT 01 : IARC MONOGRAPH 32 DEC 83
TESTED FOR CARCINOGENICITY IN ONE STUDY BY SKIN APPLICATION TO MALE MICE. NO INCREASE IN INCIDENCE OF SKIN TUMORS OBSERVED. TRIPHENYLENE MUTAGENIC TO SALMONELLA TYPHIMURIUM IN PRESENCE OF EXOGENOUS METABOLIC SYSTEM.

REPT'D USE : INCOMPLETE COMBUSTION PRODUCT AND CIGARETTE SMOKE

CAS NO : **603-35-0**
CHEM NAME : ***** TRIPHENYLPHOSPHINE (TPP)**

REPORT 02 : NTP TECHNICAL BULLETIN 2 APR 80
THE COMPOUND WAS FOUND TO BE NEGATIVE IN THE SALMONELLA TYPHIMURIUM MUTAGENICITY ASSAY IN ONE LABORATORY.

REPT'D USE : ORGANIC SYNTHESIS

CAS NO : **101-02-0** .
CHEM NAME : ***** TRIPHENYL PHOSPHITE**

REPORT 01 : EPA TSCA 8(E) NOTICE 8EHQ-0682-0451 JUN 82
BORG-WARNER REPORTED THAT IT HAD BEEN VERBALLY INFORMED BY A LABORATORY IN ENGLAND THAT TRIPHENYL PHOSPHITE, DERMALLY APPLIED AT DOSES OF 500 AND 5000 MG/KG, HAD BEEN FOUND TO PRODUCE NEURO-TOXIC EFFECTS IN CHICKENS.

REPORT 02 : EPA TESC 8(E) NOTICE 8EHQ-0682-0451 JUL 82
BORG-WARNER REPORTED IT HAD BEEN VERBALLY INFORMED BY A LABORATORY IN UK THAT TRIPHENYL PHOSPHITE, DERMALLY APPLIED AT DOSES OF 500 AND 5000 MK/KG, HAD BEEN FOUND TO PRODUCE NEUROTOXIC EFFECTS IN CHICKENS. NO EFFECTS HAD BEEN OBSERVED THUS FAR AT THE 50 MG/KG DOSE LEVEL.
REPT'D USE : ADHESIVES/GLUES, CAULKING/SPACKLING COMPOUNDS, ANTIOXIDANT, POLYMER STABILIZER, CHEMICAL INTERMEDATE, EPOXY RESIN SOLVENT AND STABILIZER IN ALKYD RESIN

CAS NO : **76-87-9**
CHEM NAME : ***** TRIPHENYLTIN HYDROXIDE**

REPORT 01 : NCI CARC TRS 139 DHEW PUB NO 78-1394
UNDER THE CONDITIONS OF THIS BIOASSAY, THERE WAS NO EVIDENCE FOR THE CARCINOGENICITY OF TRIPHENYLTIN HYDROXIDE TO FISCHER 344 RATS OR TO B6C3F1 MICE (FEEDING STUDY).
REPT'D USE : FUNGICIDE AND INSECT CHEMOSTERILANT

CAS NO : **68-76-8**
CHEM NAME : ***** TRIS(AZIRIDINYL)-P-BENZOQUINONE (TRIAZIQUINONE)**

REPORT 01 : IARC MONOGRAPH 9 1975
TRIS(AZIRIDINYL)-PARA-BENZOQUINONE IS CARCINOGENIC IN RATS FOLLOWING ITS INTRAVENOUS INJECTION AND ALSO ITS INTRAVENOUS FOLLOWED BY INTRAPERITONEAL ADMINISTRATION, PRODUCING A VARIETY OF MALIGNANT TUMORS.

REPORT 02 : IARC MONOGRAPH 9 1975
IN DROSOPHILA INJECTED WITH TRIAZIQUONE, THE FREQUENCY OF SEX CHROMOSOME LOSS AND OF THE FORMATION OF SEX-LINKED RECESSIVES INCREASED SIGNIFICANTLY. I.P. INJECTION OF TRIAZIQUONE INTO MALE MICE RESULTED IN AN INCREASED FREQUENCY OF DOMINANT LETHALS. HUMAN LEUCOCYTES TREATED WITH TRIAZIQUONE IN VITOR SHOWED A SIGNIFICANT

INCREASE IN THE FREQUENCY OF SEVERAL DIFFERENT TYPES OF ABERRATIONS.
REPT'D USE : ANTINEOPLASTIC AGENT AND INSECT CHEMOSTERILANT

CAS NO : **545-55-1**
CHEM NAME : ***** TRIS(1-AZIRIDINYL)PHOSPHINE OXIDE (TEPA)**

REPORT 01 : IARC MONOGRAPH 9 1975
TRIS(l-AZIRIDINYL)PHOSPHINE OXIDE PRODUCED A LOW INCIDENCE OF BENIGN AND MALIGNANT TUMORS IN RATS FOLLOWING ITS ORAL ADMINISTRATION, THE ONLY SPECIES AND ROUTE TESTED. THE AVAILABLE DATA ARE INSUFFICIENT FOR EVALUATION OF THE CARCINOGENICITY OF THIS COMPOUND.

REPORT 02 : IARC MONOGRAPH 9 1975
TEPA INDUCES BACK-MUTATIONS IN S. POMBE, DOMINANT LETHALS IN M. DOMESTICA, AND DOMINANT LETHALS, SEX LINKED RECESSIVE LETHALS AND CHROMOSOME TRANSLOCATIONS IN DROSOPHILA. FREQUENCIES OF DOMINANT LETHAL WERE SIGNIFICANTLY INCREASED IN THE PROGENY OF MICE GIVEN I.P. INJECTIONS OF TEPA.
REPT'D USE : PHOTOEMULSION HARDENER, CHEMOSTERILANT, ACARICIDE, TEXTILE DYEING/FINISHING AGENT, POLYMER STABILIZER, INSECTICIDE, ADHESION MODIFIER AND ANTINEOPLASTIC AGENT. NO KNOWN CURRENT PRODUCTION; NO LISTING ON TSCA INVENTORY

CAS NO : **51-18-3**
CHEM NAME : ***** 2,4,6-TRIS(1-AZIRIDINYL)-S-TRIAZINE (TEM)**

REPORT 01 : IARC MONOGRAPH 9 1975
2,4,6-TRIS(1-AZIRIDINYL)-S-TRIAZINE IS CARCINOGENIC IN MICE FOLLOWING ITS INTRAPERITONEAL INJECTION, PRODUCING AN INCREASE IN THE INCIDENCE OF OVARIAN, THYMIC AND LUNG TUMORS. IT ALSO ACTED AS AN INITIATOR OF SKIN CARCINOGENESIS. IN RATS, IT PRODUCED SARCOMAS AT THE SITE OF ITS SUBCUTANEOUS INJECTION.

REPORT 02 : IARC MONOGRAPH 9 1975
TEM INDUCES POINT MUTATIONS IN NEUROSPORA AND CAUSES REVERSIONS OF BASE-PAIR SUBSTITUTION MUTATIONS IN BOTH SALMONELLA AND SACCHAROMYCES. TEM INDUCES DOMINANT LETHALS IN MALE MICE AFTER SINGLE I.P. INJECTIONS, AND INDUCED CHROMOSOME ABERRATIONS IN CULTURED HUMAN LEUCOCYTES.
REPT'D USE : ADHESIVES, ANTINEOPLASTIC AGENT, INTERMEDIATE IN MFR OF TEXTILE FINISHING AGENTS AND CHEMOSTERILANT

CAS NO : **115-96-8**
CHEM NAME : ***** TRIS(2-CHLOROETHYL)PHOSPHATE**

REPORT 01 : CPSC FLAME RETARDANTS MONOGRAPH JUL 82
NO MUTAGENIC ACTIVITY WAS DETECTED FOR TRIS(2-CHLOROETHYL)PHOSPHATE IN SALMONELLA TA1535, TA1537, TA1538, AND TA100, WITH OR WITHOUT METABOLIC ACTIVATION. A SECOND STUDY FOUND WEAK MUTAGENIC ACTIVITY IN TA100 AND TA1535, WHICH INCREASED WITH METABOLIC ACTIVATION.

REPORT 02 : NTP DRAFT TR 391 NIH PUB 90-2846 APR 90
UNDER 2-YEAR GAVAGE STUDIES, CLEAR EVIDENCE OF CARCINOGENIC ACTIVITY FOR MALE AND FEMALE F344/N RATS RECEIVING TRIS(2-CHLOROETHYL)PHOSPHATE SHOWN BY RENAL TUBULAR ADENOMAS. NO EVIDENCE OF CARCINOGENIC ACTIVITY FOR MALE B6C3F1 MICE RECEIVING 175 OR 350 MG/KG EQUIVOCAL EVIDENCE OF CARCINOGENIC ACTIVITY FOR FEMALE B6C3F1 MICE, SHOWN BY HARDERIAN GLAND ADENOMAS. RENAL TUBULAR CELL HYPERPLASIA IN RATS, GLIOSIS, HEMORRHAGE, PIGMENTATION AND MINERALIZATION IN BRAINS OF FEMALE RATS; AND KARYOMERGALY OF KIDNEY TUBULAR EPITHELIA CELLS OF MICE CHEMICAL RELATED.

REPT'D USE : FLAME-RETARDANT PLASTICIZER, POLYURETHANE, POLYISOCYAN URATE FOAMS, CARPET BACKINGS, PAINTS, LACQUERS, EPOXY RESINS, ADHESIVES, URETHANE COATINGS, ACRYLIC SHEETS AND PARTICLE BOARD

CAS NO : **38571-73-2**
CHEM NAME : ***** 1,2,3-TRIS(CHLOROMETHOXY)-PROPANE**

REPORT 01 : IARC MONOGRAPH 15 1977
1,2,3-TRIS(CHLOROMETHOXY)PROPANE IS CARCINOGENIC IN MICE, THE ONLY SPECIES TESTED, FOLLOWING ITS SUBCUTANEOUS OR INTRAPERITONEAL ADMINISTRATION; IT PRODUCED MALIGNANT TUMORS AT THE SITES OF ADMINISTRATION. SKIN PAPILLOMAS AND A LOW INCIDENCE OF SKIN CARCINOMAS WERE OBSERVED IN SKIN-PAINTING STUDIES IN MICE.

REPT'D USE : EXPERIMENTAL CURING AGENT FOR EPOXY RESINS. NO LISTING ON TSCA INVENTORY

CAS NO : **126-72-7**
CHEM NAME : ***** TRIS (2,3-DIBROMPROPYL) PHOSPHATE (TRIS)**

REPORT 01 : NCI CARC TRS 76 DHEW PUB NO 78-1326
UNDER THE CONDITIONS OF THIS STUDY, ORALLY ADMINISTERED TRIS WAS CARCINOGENIC TO B6C3F1 MICE, CAUSING INCREASED INCIDENCE OF TUMORS OF THE LIVERS, LUNGS, AND STOMACHS OF FEMALE MICE AND IN KIDNEYS, LUNGS AND STOMACHS OF MALE

MICE. TBP WAS ALSO CARCINOGENIC IN FISCHER 344 RATS CAUSING AN INCREASED INCIDENCE OF KIDNEY TUMORS IN BOTH SEXES.

REPORT 02 : IARC MONOGRAPH 20 1979
TRIS IS MUTAGENIC IN SALMONELLA TA100 AND TA153, BUT NOT IN TA1537 AND TA1538, INDICATING THAT MUTATIONS INDUCED ARE OF THE BASE-PAIR SUBSTITUTION TYPE. THE CHEMICAL WAS ALSO MUTAGENIC TO DROSOPHILA INDUCING SEX-LINKED RECESSIVE LETHALS IN MALE GERM CELL STAGES. A SIGNIFICANT DOSE-DEPENDENT INCREASE IN SISTER CHROMATID EXCHANGES WAS OBSERVED IN CHINESE HAMSTER V79 CELLS TREATED WITH TRIS.

REPORT 03 : IARC MONOGRAPH 20 1979
IN MICE, FOLLOWING ORAL ADMINISTRATION, TRIS PRODUCED TUMORS OF THE FORESTOMACH AND LUNG IN ANIMALS OF BOTH SEXES, BENIGN AND MALIGNANT LIVER TUMORS IN FEMALES AND BENIGN AND MALIGNANT TUMORS OF THE KIDNEY IN MALES. IN RATS, IT PRODUCED BENIGN AND MALIGNANT TUMORS OF THE KIDNEY IN MALES AND BENIGN KIDNEY TUMORS IN FEMALES. AFTER SKIN APPLICATION OT FEMALE MICE, IT PRODUCED TUMORS OF THE SKIN, LUNG, FORESTOMACH AND ORAL CAVITY. IN A 1979 SUPPLEMENT, IARC DETERMINED THERE IS SUFFICIENT EVIDENCE FOR THE CARCINOGENICITY OF TRIS IN ANIMALS.

REPORT 04 : VELSICOL/TABER- SHAW OCCUP. MEDICINE ASSOC. OCT 80
NO OVERALL OR CAUSE-SPECIFIC MORTALITY EXCESS WAS DETECTED AMONG 628 PERSONS POTENTIALLY EXPOSED TO TRIS ON A ROUTINE OR NON-ROUTINE BASIS. HOWEVER, 94 PERCENT OF THE COHORT WAS STILL ALIVE AS OF DECEMBER 31, 1976.

REPORT 05 : EPA TSCA 8(E) NOTICE 8EHQ-0478-0128
RESULTS OF STERILITY RETESTING OF EMPLOYEES EXPOSED TO TRIS SHOWS THAT OF SIX WORKERS EXPOSED, TWO WHO HAD LOW SPERM COUNTS AT THE FIRST EXAMINATION HAD NORMAL COUNTS ON THE SECOND. TWO WORKERS DID NOT HAVE DEFECTIVE COUNTS ON THE INITIAL EXAMINATION, BUT ONE OF THESE HAD MARKEDLY HIGHER COUNTS ON THE SECOND EXAMINATION. SUBMITTER CONTENDS THAT DBCP CONTAMINATION IS RESPONSIBLE FOR THE OBSERVED EFFECT.

REPORT 06 : MULTI-AGENCY CROSS CHECK
Y HHS MAGUIRE
N OSHA CARC
Y EPA CAG
N EPA RCRA
N EPA CLEAN AIR
N EPA CLEAN WATER

REPT'D USE : FLAME RETARDANT; NOT PRODUCED OR IMPORTED

CAS NO : **13674-87-8**
CHEM NAME : ***** TRIS(1,3-DICHLOROISOPROPYL)PHOSPHATE (FYROL FR2)**

REPORT 01 : EPA TSCA 8(E) NOTICE 8EHQ-1280-0401 DEC 80
IN A STUDY SPONSORED BY STAUFFER CHEMICAL CO., RATS WERE FED 2,5,20, AND 80 MG/KG TDCP/DAY FOR TWO YEARS. STAUFFER REPORTED NEOPLASTIC CHANGES IN THE LIVER, BENIGN TESTICULAR INTERSTITIAL CELL TUMORS, AND A STATISTICALLY SIGNIFICANT INCIDENCE OF HEPATOCELLULAR CARCINOMAS AT THE 80 MG/KG DOSE ONLY IN TEST ANIMALS. THE CONTROL MALE RATS ALSO DEMONSTRATED SUBSTANTIAL INTERSTITIAL CELL TUMORS.

REPORT 03 : NTP TECHNICAL BULLETIN 9 APR 83
TRIS (1,3-DICHLORO-2-PROPYL) PHOSPHATE WAS POSITIVE IN ONE SALMONELLA TYPHIMURIUM MUTAGENICITY ASSAY.
REPT'D USE : FLAME RETARDANT IN POLYURETHANE FOAMS, MATTRESSES AND TEXTILES

CAS NO : **78-43-3**
CHEM NAME : ***** TRIS(2,3-DICHLOROPROPYL)PHOSPHATE**

REPORT 01 : CPSC FLAME RETARDANTS MONOGRAPH JUL 82
TRIS(2,3-DICHLOROPROPYL)PHOSPHATE SHOWED HIGH MUTAGENIC ACTIVITY IN SALMONELLA TA1535 AND TA100 ASSAYS, WHICH INCREASED WITH METABOLIC ACTIVATION. NO ACTIVITY WAS DETECTED IN TA98, TA1537 AND TA1538. PRODUCED CHROMOSOMAL ABERRATIONS IN HAMSTER LUNG FIBROBLAST CELLS, BUT RESULTS FOR RAT BONE MARROW CELLS WERE INDECISIVE. THE CHEMICAL DID NOT PRODUCE CHROMOSOMAL ABERRATIONS OR SISTER CHROMATID EXCHANGES IN HUMAN EMBRYO FIBROBLASTS.
REPT'D USE : FLAME RETARDANT

CAS NO : **78-42-2**
CHEM NAME : ***** TRIS(2-ETHYLHEXYL)PHOSPHATE**

REPORT 01 : NTP TECHNICAL REPORT 83-274 NIH PUB 84-2530 AUG 84
THERE WAS EQUIVOCAL EVIDENCE OF CARCINOGENICITY IN MALE F344/N RATS RECEIVING 2000 AND 4000 MG/KG TRIS (2-ETHYLHEXYL)-PHOSPHATE, EVIDENCE BY INCREASED PHEOCHROMOCYTOMAS OF ADRENAL GLAND NO EVIDENCE OF CARCINOGENICITY IN FEMALE F344/N RATS OR IN MALE B6C3F1 MICE; SOME EVIDENCE OF CARCINOGENICITY IN FEMALE B6C3F1 MICE THAT RECEIVED 1000 MG/KG SHOWN BY INCREASED HEPATOCELLULAR CARCINOMA. ASSOCIATED WITH INCREASED INCIDENCES

OF FOLLICULAR CELL HYPERPLASIAS OF THYROID GLAND IN MALE AND FEMALE B6C3F1 MICE.
REPT'D USE : FIRE RETARDANT AND PLASTICIZER

CAS NO : **3319-31-1**
CHEM NAME : ***** TRIS(2-ETHYLHEXYL)TRIMELLITATE**

REPORT 01 : RESPONSE TO ITC 48FR51842 NOV 83
IN RESPONSE TO ITC'S DESIGNATION OF CPD FOR HEALTH AND ENVIRONMENTAL EFFECTS TESTING CONSIDERATION, EPA CONCLUDED THAT THE CMA'S TESTING PROGRAM COMBINED WITH RELATED DATA FROM PHTHALATE ESTERS PROGRAM FROM CMA AND NTP WOULD SUPPLY INFORMATION TO ENABLE EPA TO REASONABLY PREDICT TOXICITY OF CPD. THEREFORE, EPA NOT INITIATING RULEMAKING NOW UNDER TSCA 4(A).
REPT'D USE : PVC PLASTICIZER

CAS NO : **57-39-6**
CHEM NAME : ***** TRIS(2-METHYL-1-AZIRIDINYL)PHOSPHINE OXIDE**

REPORT 01 : IARC MONOGRAPH 9 1975
TRIS(2-METHYL-1-AZIRIDINYL)PHOSPHINE OXIDE INDUCED LEUKEMIA IN RATS FOLLOWING ITS ORAL ADMINISTRATION. THE DATA ARE INSUFFICIENT FOR EVALUATION OF ITS CARCINOGENICITY.

REPORT 02 : IARC MONOGRAPH 9 1975
TREATMENT OF MALE MICE WITH METEPA RESULTED IN A HIGH FREQUENCY OF DOMINANT LETHALS.
REPT'D USE : POLYMER CROSS LINKING AGENT FOR PAINTS/ADHESIVES, EXPERIMENTAL AMD ANTINEOPLASTIC AGENT

CAS NO : **150-38-9**
CHEM NAME : ***** TRISODIUM ETHYLENEDIAMINETETRAACETATE TRIHYDRATE (EDTA)**

REPORT 01 : NCI CARC TRS 11 DHEW PUB NO 77-811 1977
NO COMPOUND-RELATED SIGNS OF CLINICAL TOXICITY WERE NOTED. ALTHOUGH A VARIETY OF TUMORS OCCURRED AMONG TEST AND CONTROL ANIMALS OF BOTH SPECIES, NO TUMORS WERE RELATED TO TREATMENT. SINCE SURVIVAL WAS SATISFACTORY AND SHOWED NO CONSISTENT VARIATION AMONG TEST AND CONTROL GROUPS, THE ABSENCE OF TREATMENT-RELATED TUMORS COULD NOT BE ATTRIBUTED TO EARLY MORTALITY (FEEDING STUDY).
REPT'D USE : CHELATING AGENT, FOOD SUBSTANCE, COSMETICS/PHARMACEUTICALS AND TEXTILE PROCESSING/DYEING AID

CAS NO : **62450-06-0**
CHEM NAME : ***** TRP-P-1(3-AMINO-1,4-DIMETHYL-5H-PYRIDO-(4,3-B)INDOLE) AND ITS ACETATE**

REPORT 01 : IARC MONOGRAPH 31 JUL 83
TESTED IN MICE IN DIET, BY SKIN APPLICATION AND BY BLADDER IMPLANTATION AND IN RATS AND HAMSTERS BY SUBCUTANEOUS ADMINISTRATION. HEPATIC TUMORS INDUCED BY ORAL ADMINISTRATION IN BOTH MALE AND FEMALE MICE, ALTHOUGH INCREASED INCIDENCE SIGNIFICANT ONLY IN FEMALES. IN FEMALE RATS SIGNIFICANT INCREASE IN INCIDENCE OF FIBROSARCOMAS AT SITE OF INJECTION; IN HAMSTERS INCREASED INCIDENCE OF SUBCUTANEOUS SARCOMAS. BLADDER CARCINOMAS OBSERVED IN FEMALE MICE AFTER IMPLANTATION OF PARAFFIN PELLETS CONTAINING TRP-P-1. STUDY IN MICE BY SKINPAINTING INADEQUATE.

REPORT 02 : IARC MONOGRAPH 31 JUL 83
TRP-P-1 MUTAGENIC TO BACTERIA FOLLOWING METABOLIC ACTIVATION. INDUCED CHROMOSOMAL ANOMALIES IN MAMMALIAN CELLS, INCULDING HUMAN CELLS, IN VITRO. CAUSED MORPHOLOGICAL TRANSFORMATION OF HAMSTER EMBRYO CELLS.
REPT'D USE : FOUND IN CHARRED FRACTION OF COOKED FISH AND MEAT

CAS NO : **62450-07-1**
CHEM NAME : ***** TRP-P-2(3-AMINO-1-METHYL-5H-PYRIDO(4,3-B) INDOLE) AND ITS ACETATE**

REPORT 01 : IARC MONOGRAPH 31 JUL 83
TESTED IN MICE AND RATS IN DIET, IN MICE BY SKIN APPLICATION AND BLADDER IMPLANTATION AND IN RATS AND HAMSTERS BY SUBCUTANEOUS ADMINISTRATION. INCREASED INCIDENCE OF LIVER TUMORS OBSERVED IN FEMALE MICE AND FEMALE RATS FOLLOWING ORAL ADMINISTRATION. BLADDER CARCINOMAS OBSERVED IN FEMALE MICE AFTER IMPLANTATION OF TRP-P-2 IN PARAFFIN PELLETS. RESULTS OF SUBCUTANEOUS INJECTION AND SKIN APPLICATION INADEQUATE FOR EVALUATION.

REPORT 02 : IARC MONOGRAPH 31 JUL 83
TRP-P-2 MUTAGENIC TO SALMONELLA TYPHIMURIUM FOLLOWING METABOLIC ACTIVATION. INDUCED CHROMOSOMAL ANOMALIES IN MAMMALIAN CELLS, INCULDING HUMAN CELLS, IN VITRO. CAUSED NEOPLASTIC TRANSFORMATION IN VITRO IN SYRIAN GOLDEN HAMSTER CELLS.
REPT'D USE : FOUND IN CHARRED FRACTION OF COOKED FISH

CAS NO : **73-22-3**
CHEM NAME : ***** L-TRYPTOPHAN**

REPORT 01 : NCI CARC TRS 71 DHEW PUB NO 78-1321
UNDER THE CONDITIONS OF THIS BIOASSAY, L-TRYPTOPHAN WAS NOT CARCINOGENIC FOR FISCHER 344 RATS OR B6C3FI MICE (FEEDING STUDY).

REPORT 02 : NAS REPORT ON DIET NUTRITION AND CANCER 1982
DOGS FED HIGH LEVELS OF TRYPTOPHAN DEVELOPED HYPERPLASIA OF THE URINARY BLADDER. TRYPTOPHAN AND ITS METABOLITES WERE NOT MUTAGENIC IN THE SALMONELLA ASSAY. ATTEMPTS TO RELATE THE DEVELOPMENT OF TUMORS IN THE URINARY BLADDER OF HUMANS TO ABNORMALITIES IN THE METABOLISM OF TRYPTOPHAN HAVE NOT BEEN DEFINITIVE.
REPT'D USE : ENDOGENOUS CHEMICAL, NATURAL FOOD CONSTITUENT AND NUTRITIONAL SUPPLEMENT

CAS NO : **9038-95-3**
CHEM NAME : ***** UCON 50-HB POLYETHERS**

REPORT 01 : STATUS REPORT 8EHQ-1086-0635 NOV 86
UNION CARBIDE CORP SUBMITTED FOLLOWING RESULTS OF A 14-WEEK AEROSOL INHALATION STUDY IN FISCHER 344 RATS. RATS WERE EXPOSED TO 0, 0.3, 1.1, OR 5.2 MG/M3 OF TEST MATERIAL FOR 6 HRS/DAY, 5 DAYS/WK FOR 14 WEEKS. PRINCIPAL EFFECTS WERE NOTED IN LUNGS OF EXPOSED ANIMALS. THE INJURY TO LUNG TISSUE SEEN IN THIS STUDY WAS A SLIGHT TO MODERATE DEGREE OF SEVERITY. SOME SIGNS OF RECOVERY WERE NOTED.
REPT'D USE : TEXTILE FINISH

CAS NO : **66-75-1**
CHEM NAME : ***** URACIL MUSTARD**

REPORT 01 : IARC MONOGRAPH 9 1975
URACIL MUSTARD IS CARCINOGENIC IN MICE AND RATS FOLLOWING ITS INTRAPERITONEAL INJECTION, PRODUCING A DOSE-RELATED INCREASE IN THE INCIDENCE OF LUNG TUMORS IN MICE AND A VARIETY OF TUMORS IN BOTH MICE AND RATS. IN A 1979 SUPPLEMENT, IARC DETERMINED THAT THERE IS SUFFICIENT EVIDENCE FOR THE CARCINOGENICITY OF URACIL MUSTARD IN ANIMALS.

REPORT 02 : NCI BIOASSAY PROGRAM, CANCER VOL. 40 1977
UNDER THE CONDITIONS OF THE BIOASSAY, I.P. INJECTION OF URACIL MUSTARD WAS CARCINOGENIC TO SWISS-WEBSTER MALE AND FEMALE MICE, CAUSING LUNG TUMORS. IN ADDITION, THERE WERE MALIGNANT LYMPHOMAS AND OVARIAN TUMORS IN FEMALES AND LIVER TUMORS IN MALES. IT WAS ALSO CARCINO-

GENIC TO MALE AND FEMALE SPRAGUE-DAWLEY (CD) RATS, CAUSING PERITONEAL SARCOMAS AND LYMPHOMAS. IN ADDITION, MALE RATS HAD PANCREATIC TUMORS AND FEMALE RATS HAD OVARIAN TUMORS AND BREAST CARCINOMAS.

REPORT 06 : MULTI-AGENCY CROSS CHECK
N HHS MAGUIRE
N OSHA CARC
Y EPA CAG
N EPA RCRA
N EPA CLEAN AIR
N EPA CLEAN WATER
REPT'D USE : ANTINEOPLASTIC AGENT

CAS NO : **1000-82-4**
CHEM NAME : ***** UREA-FORMALDEHYDE RESINS**

REPORT 01 : EPA RESPONSE TO ITC, 49FR21371 MAY 84
THIS IS EPA'S RESPONSE TO ITC'S DESIGNATION OF UREA-FORMALDEHYDE RESINS ("METHYLOLUREA") FOR PRIORITY CONSIDERATION FOR HEALTH EFFECTS TESTING. EPA CONCLUDES TENTATIVELY THAT SUCH TESTING IS CALLED FOR UNDER TSCA 4(A).
REPT'D USE : ADHESIVES (80-85%), THERMOSETTING PLASTICS, SLOW-RELEASE FERTILIZER, FABRIC FINISHES, PAINT ADDITIVES AND PAPER FINISHES.

CAS NO : **865-21-4**
CHEM NAME : ***** VINBLASTINE**

REPORT 01 : NCI BIOASSAY PROGRAM, CANCER VOL. 40 1977
UNDER THE CONDITIONS OF THE BIOASSAY, I.P. INJECTION OF VINBLASTINE WAS CARCINOGENIC TO MALE SPRAGUE-DAWLEY (CD) RATS, CAUSING LYMPHOSARCOMAS, AND PITUITARY TUMORS. IT WAS NOT CARCINOGENIC TO FEMALE RATS AND MALE AND FEMALE SWISS-WEBSTER MICE.

REPORT 02 : IARC MONOGRAPH 26 1981
VINBLASTINE SULPHATE WAS TESTED IN THREE STUDIES, TWO BY IP INJECTION IN MICE AND RATS, AND ONE BY IV INJECTION IN RATS. NO EVIDENCE OF CARCINOGENICITY WAS FOUND, BUT THE DRUG HAS NOT BEEN ADEQUATELY TESTED AT HIGH DOSES.

REPORT 03 : IARC MONOGRAPH 26 1981
VINBLASTINE SULPHATE CAN INDUCE TERATOGENIC EFFECTS IN SEVERAL ANIMAL SPECIES AND EMBRYOLETHALITY AT DOSES NONTOXIC TO THE MOTHER. ON THE BASIS OF AVAILABLE DATA, THIS COMPOUND CANNOT BE CONSIDERED TO BE MUTAGENIC.
REPT'D USE : ANTIMITOTIC ANTINEOPLASTIC AGENT

CAS NO : **57-22-7**
CHEM NAME : ***** VINCRISTINE**

REPORT 01 : NCI BIOASSAY PROGRAM, CANCER VOL. 40 1977
UNDER THE CONDITIONS OF THE BIOASSAY, I.P. INJECTION OF VINCRISTINE WAS NOT CARCINOGENIC TO SWISS-WEBSTER MICE OR SPRAGUE-DAWLEY (CD) RATS OF EITHER SEX.

REPORT 02 : IARC MONOGRAPH 26 1981
VINCRISTINE SULPHATE CAN INDUCE TERATOGENIC EFFECTS IN SEVERAL ANIMAL SPECIES, AND IT INDUCED EMBRYOLETHALITY AT DOSES NONTOXIC TO THE MOTHER. THERE IS NO EVIDENCE TO SUGGEST THAT THIS COMPOUND IS MUTAGENIC.

REPORT 03 : IARC MONOGRAPH 26 1981
VINCRISTINE SULPHATE WAS TESTED IN MICE AND RATS BY IP INJECTION. IN THESE LIMITED STUDIES NO EVIDENCE OF CARCINOGENICITY WAS FOUND. THE REASON IS INADEQUATE EVIDENCE FOR THE CARCINOGENICITY OF VINCRISTINE SULPHATE.
REPT'D USE : ANTINEOPLASTIC AGENT

CAS NO : **108-05-4**
CHEM NAME : ***** VINYL ACETATE**

REPORT 01 : IARC MONOGRAPH 19
IN ONLY ONE STUDY AVAILABLE, VINYL ACETATE WAS TESTED IN RATS BY INHALATION EXPOSURE; IT PRODUCED NO EVIDENCE OF CARCINOGENICITY.

REPORT 02 : IARC MONOGRAPH 19 1979
EXPOSURE OF SALMONELLA TA100 AND TA1530 TO VAPORS OF VINYL ACETATE, WITH OR WITHOUT METABOLIC ACTIVATION, CAUSED NO MUTAGENIC EFFECTS. NO DATA ON THE EMBRYOTOXICITY, TERATOGENICITY OR METABOLISM OF VINYL ACETATE WERE AVAILABLE.

REPORT 03 : EXP. PATH. LAB./NCI SEP 82
MICROSCOPIC EXAMINATION OF TISSUES FROM MALE AND FEMALE F344 RATS TREATED WITH 1000 MG/L AND 2500 MG/L VINYL ACETATE IN DRINKING WATER INDICATE INCIDENCE OF NEOPLASTIC NODULES OF LIVER OF BOTH SEXES AND NEOPLASMS OF UTERUS WAS GREATER THAN IN CONTROL RATS. THE INCIDENCE OF PROLIFERATIVE C-CELL LESIONS OF THE THYROID GLAND IN TREATED FEMALE RATS WAS INCREASED WHEN COMPARED TO CONTROL FEMALE RATS.

REPORT 04 : EPA TSCA 8(E) 0185-0543 JAN 85
SOCIETY OF PLASTICS INDUSTRY REPORTED RATS DOSED ORALLY AT 0, 200, 1000 AND 5000 PPM VINYL ACETATE IN WATER SHOWED A DECREASE IN THE NUMBER OF LITTERS PRODUCED AT 5000 PPM. ALSO IN SECOND GENERATION THERE WAS DECREASE IN FERTILITY AT 5000 PPM.

REPORT 05 : IARC MONOGRAPH 39, 1986 JUN 85
VA IN DRINKING WATER TO RATS LED TO INCREASED INCIDENCE OF UTERINE ADENOCARCINOMAS; HOWEVER, LESIONS COULD NOT BE DIRECTLY LINKED TO CPD. VA NOT MUTAGENIC TO S. TYPHIMURIUM WITH OR WITHOUT ACTIVATION. IT INDUCED SCE AND CHROMOSOMAL ABERRATIONS IN-VITRO IN HUMAN LYMPHOCYTES AND WHOLE-BLOOD CULTURES AND IN CHO CELLS. NO DATA TO EVALUATE REPRODUCTIVE EFFECTS OR PRENATAL TOXICITY IN ANIMALS OR HUMANS. INCREASED INCIDENCE OF LUNG CANCER IN SYNTHETIC CHEMICAL PLANT NOT SPECIFICALLY ASSOCIATED WITH CPD. IARC SAYS THAT DATA IS INADEQUATE FOR CARCINOGENICITY OF CPD TO ANIMALS OR HUMANS. LACK OF EPI DATA FORESTALLS EVALUATION OF CARCINOGENICITY IN HUMANS.

REPORT 07 : STATUS REPORT 8EHQ-0187-0650 JAN 87
IN AN INHALATION STUDY, SPRAGUE-DAWLEY RATS WERE EXPOSED TO 0, 50, 200 AND 600 PPM OF CPD FOR 6 HRS/DAY, 5 DAYS/WEEK FOR TWO YEARS. PRELIM. PATHOLOGY REPORT SHOWS SQUAMOUS CELL CARCINOMAS SEEN IN 2/77 ANIMALS AND PAPILLOMAS IN 8/77 RATS. NO OTHER TUMORS SEEN IN OTHER EXPOSURE ANIMALS OR THE CONTROL ANIMALS.

REPORT 08 : STATUS REPORT 8EHQ-1086-0642 OCT 86
GROUPS OF 90 MICE WERE EXPOSED TO CONCENTRATIONS OF CPD OF 0, 50, 200 AND 600 PPM FOR 6 HRS/DAY, 5 DAYS/WK FOR 2 YRS. ONLY ADVERSE EFFECTS SEEN IN RESPIRATORY TRACT. MICE EXPOSED TO 600 PPM LEVEL SHOWED BRONCHIOLAR EPITHELIAL LESIONS IN LUNG, PLUS ONE MOUSE WITH A SQUAMOUS CELL NODULE IN A TERMINAL BRONCHIAL AIRWAY AND ANOTHER MOUSE WITH ONE SQUAMOUS CELL CARCINOMA IN A MAJOR AIRWAY. NO SUCH TUMORS SEEN IN CONTROL ANIMALS OR AT OTHER CONCENTRATIONS. 50 PPM SEEMED TO BE A NO OBSERVABLE EFFECT LEVEL.

REPT'D USE : PLASTIC MONOMER, PAINT, ADHESIVES AND AEROSOLS

CAS NO : **593-60-2**
CHEM NAME : ***** VINYL BROMIDE**

REPORT 02 : IARC MONOGRAPH 19 1979
EXPOSURE OF SALMONELLA TA1530 OR TA100 TO VAPORS OF VINYL BROMIDE CAUSED MUTAGENIC EFFECTS. THE PRESENCE OF META-

BOLIC ACTIVATION SYSTEM ENHANCED THE MUTAGENICITY. NO DATA ON THE EMBRYOTOXICITY OR TERATOGENICITY OF VINYL BROMIDE WERE AVAILABLE.

REPORT 03 : EPA TSCA 8(E) NOTICE 8EHQ-0479-0281 APR 79
REPORT OF THE FINAL HISTOPATHOLOGIC RESULTS FROM A 2 YEAR INHALATION STUDY OF VINYL BROMIDE IN RATS. FINAL RESULTS SHOW AN INCREASED INCIDENCE OF ANGIOSARCOMAS AT ALL LEVELS - 1235, 247, 52, 10 PPM VINYL BROMIDE TESTED. VINYL BROMIDE APPEARS TO BE MORE POTENT THAN VINYL CHLORIDE IN THE INDUCTION OF ANGIOSARCOMAS.

REPORT 04 : IARC MONOGRAPH 19 1979
VINYL BROMIDE WAS TESTED IN MICE BY SKIN APPLICATION AND BY S.C. INJECTION. NO LOCAL TUMORS WERE PRODUCED. A COMMERCIAL SUSPENSION OF POLYMERIZED VINYL BROMIDE PRODUCED NO SKIN TUMORS IN MICE WHEN TESTED BY SKIN APPLICATION, BUT LOCAL SARCOMAS WERE PRODUCED FOLLOWING ITS SC INJECTION.

REPORT 05 : NIOSH CRITERIA DOCUMENT ON VINYL HALIDES SEP 78
NIOSH RECOMMENDS CONTROL AS PROMULGATED FOR VINYL CHLORIDE IN 29 CFR 1910 BASED ON POTENTIAL CARCINOGENICITY. THERE IS NO PRESENT OSHA STANDARD.

REPORT 06 : MULTI-AGENCY CROSS CHECK
N HHS MAGUIRE
Y OSHA CARC
N EPA CAG
N EPA RCRA
N EPA CLEAN AIR
N EPA CLEAN WATER

REPORT 07 : NIOSH CURRENT INTELLIGENCE BULLETIN 28 SEP 78
THE NIOSH AND OSHA JOINTLY RECOMMEND THAT VINYL BROMIDE BE CONSIDERED IN THE WORKPLACE AS POTENTIAL CARCINOGENS TO HUMANS. RECOMMENDATION IS BASED ON THE RESULTS OF RECENT STUDIES INDICATING THAT EXPOSURE TO VBR CAUSES ANGIOSARCOMA OF THE LIVER AND OTHER CANCER IN LABORATORY ANIMALS.

REPORT 08 : IARC MONOGRAPH 39, 1986 JUN 85
VB VIA INHALATION PRODUCED LIVER ANGIOSARCOMAS AND ZYMBAL GLAND CARCINOMAS AS WELL AS INCREASED INCIDENCE OF NEOPLASTIC NODULES AND HEPATOCELLULAR CARCINCOMAS IN RATS. VB MUTAGENIC TO S. TYPHIMURIUM WITH OR WITHOUT ACTIVATION. NO DATA AVAILABLE TO EVALUATE ANIMAL OR HUMAN REPRODUCTIVE EFFECTS OR PRENATAL TOXICITY. NO EPA OR CASE STUDIES TO EVALUATE CARCINOGENICITY OF VB IN HUMANS. IARC SAYS THERE IS SUFFICIENT EVIDENCE FOR

CARCINOGENICITY OF CPD TO EXPERIMENTAL ANIMALS, BUT NO DATA ON HUMANS IS AVAILABLE.

REPT'D USE : FR CHEMICAL, COMONOMER IN MODACRYLIC FIBERS, IMTERMEDIATE IN ORGANIC SYNTHESIS, PHARMACEUTICALS AND FUMIGANTS

CAS NO : **75-01-4**

CHEM NAME : ***** VINYL CHLORIDE MONOMER (VCM)**

REPORT 01 : EPA RCRA HEALTH AND ENV. CRITERIA 10/80 OCT 80

ANIMAL STUDIES INDICATE THAT VINYL CHLORIDE IS NOT TERATOGENIC, BUT IS MUTAGENIC IN SEVERAL BIOLOGICAL TEST SYSTEMS. VCM HAS BEEN FOUND TO BE CARCINOGENIC IN LAB ANIMALS AND HAS BEEN POSITIVELY ASSOCIATED WITH ANGIOSARCOMA OF THE LIVER IN HUMANS.

REPORT 02 : EPA STATUS REPORT APR 78

VINYL CHLORIDE IS AN ESTABLISHED CARCINOGEN FOR ANIMALS AND HUMANS. THEREFORE, INTERMEDIATES THAT CONTAIN VC AS AN IMPURITY OR INTERMEDIATES THAT CAN GIVE RISE TO VC DURING PROCESSING ARE OF CONCERN AS POLLUTANTS. OF EQUAL CONCERN IS THE REPORT THAT THE FINISHED PRODUCTS CONTAIN RESIDUAL VC AND ARE SOURCES OF ENVIRONMENTAL POLLUTION.

REPORT 03 : IARC MONOGRAPH 19 1979

VINYL CHLORIDE WAS TESTED IN RATS BY ORAL, S.C. AND INTRAPERITONEAL ADMINISTRATION; IN MICE, RATS AND HAMSTERS BY INHALATION EXPOSURE. AFTER ORAL AND INHALATION EXPOSURE, VC WAS CARCINOGENIC IN ALL THREE SPECIES, PRODUCING TUMORS AT DIFFERENT SITES. VC WAS CARCINOGENIC IN RATS FOLLOWING PRENATAL EXPOSURE. SEVERAL INDEPENDENT BUT MUTUALLY CONFIRMATORY STUDIES HAVE SHOWN THAT EXPOSURE TO VINYL CHLORIDE RESULTS IN AN INCREASED CARCINOGENIC RISK IN HUMANS, INVOLVING THE LIVER, BRAIN, LUNG AND HAEMO-LYMPHOPOIETIC SYSTEM.

REPORT 04 : NIOSH CRITERIA DOCUMENT MAR 74

NIOSH RECOMMENDS A 1 PPM CEILING BASED ON LIVER CANCER. THE PRESENT OSHA STANDARD IS 1 PPM TWA (29CFR1910.1017).

REPORT 05 : NIOSH CURRENT INTELLIGENCE BULLETIN 28 SEP 78

VINYL CHLORIDE IS KNOWN TO CAUSE ANGIOSARCOMA OF THE LIVER AND CANCERS OF OTHER SITES IN LABORATORY ANIMALS AND IN HUMANS. VINYL CHLORIDE IS THE ONLY VINYL HALIDE FOR WHICH AN OSHA EXPOSURE STANDARD CURRENTLY EXISTS.

REPORT 06 : MULTI-AGENCY CROSS CHECK
Y HHS MAGUIRE
Y OSHA CARC
Y EPA CAG
Y EPA RCRA
Y EPA CLEAN AIR
Y EPA CLEAN WATER

REPORT 07 : EPA AMBIENT WATER QUALITY CRITERIA OCT 80
VINYL CHLORIDE IS A WELL-KNOWN HUMAN AND ANIMAL CARCINOGEN. AMBIENT WATER CONCENTRATIONS OF 20 UG/L, 2.0 UG/L, AND .2 UG/L ARE ESTIMATED TO RESULT IN A RESPECTIVE 10 -5, 10 -6, AND 10 -7 INCREASE IN CANCER RISK OVER A LIFETIME.

REPORT 08 : NAS REPORT ON DIET, NUTRITION AND CANCER 1982
OCCUPATIONAL EXPOSURE TO VINYL CHLORIDE IS ASSOCIATED WITH INCREASED INCIDENCE OF CANCER, BUT THIS EVIDENCE HAS BEEN DERIVED FROM STUDIES OF GROUPS OCCUPATIONALLY EXPOSED TO HIGH DOSES OF VCM. SIMILAR CARCINOGENIC EFFECTS WERE DEMONSTRATED IN RATS THAT INGESTED OR INHALED LARGE AMOUNTS OF VCM. THESE RESULTS WERE LATER CONFIRMED IN MICE AND HAMSTERS.

REPORT 09 : EPA NAT'L DRINK ING WATER STDS - ANPR 49FR2433 0 JUN 84
EPA PROPOSES TO REGULATE VC IN DRINKING WATER AT ZERO LEVEL IN COMPLIANCE WITH THE SAFE DRINKING WATER ACT.

REPORT 10 : STATUS REPORT 8EHQ-0986-0629 OCT 86
UNION CARBIDE CORP DID AN EPA STUDY OF 10,173 MEN WHO WORKED FOR AT LEAST ONE YEAR IN JOBS EXPOSED TO VC. THE STUDY CONFIRMED THAT THE VC WIORKERS HAD SIGNIFICANTLY HIGHER MORTALITY FROM ANGIOSARCOMAS, CANCER OF THE LIVER AND BILIARY TRACT AND CANCER OF THE BRAIN AND OTHER PARTS OF CNS. THERE WAS NO EXCESS OF EITHER RESPIRATORY CANCER OR LYMPHATIC OR HEMATOPOIETIC CANCER.

REPORT 11 : CPSC FR NOTICE VOL. 40, NO. 173, P. 11170 SEPT 75
ON AUGUST 21, 1941 CPSC PUBLISHED FR NOTICE 39 FR 30112-3014 UNDER AUTHORITY OF FHSA BANNING SELF-PRESSURIZED PRODUCTS CONTAINING VINYL CHLORIDE MONOMER AS INGREDIENT OR PROPELLANT, EFFECTIVE OCTOBER 7, 1974.

REPT'D USE : CHEMICAL INTERMEDIATE, PLASTIC MONOMER, PLASTICS ADHESIVE, REFRIGERANT AND FLAME RETARDANT

CAS NO : **9003-22-9**
CHEM NAME : ***** VINYL CHLORIDE-VINYL ACETATE COPOLYMERS**

REPORT 01 : IARC MONOGRAPH 19 1979
VINYL CHLORIDE-VINYL ACETATE COPOLYMERS WERE TESTED IN RATS BY SUBCUTANEOUS IMPLANTATION AS FILMS OR POWDER; LOCAL SARCOMAS WERE INDUCED FOLLOWING IMPLANTATION OF FILMS.

REPORT 02 : IARC MONOGRAPH 19 1979
NO DATA ON THE EMBRYOTOXICITY, TERATOGENICITY, METABOLISM OR MUTGENICITY OF VINYL ACETATE-VINYL CHLORIDE COPOLYMERS WERE AVAILABLE.

REPT'D USE : RECORDS, VINYL ASBESTOS FLOORING, PAINTS/VARNISHES, ADHESIVES, CAULKING/SPACKLING COMPOUNDS, AS CHEMICAL AND FLAME RETARDANT

CAS NO : **100-40-3**
CHEM NAME : ***** 4-VINYLCYCLOHEXENE**

REPORT 01 : IARC MONOGRAPH 11 1976
4-VINYLCYCLOHEXENE HAS BEEN TESTED ONLY IN MICE BY SKIN APPLICATION. THE DATA DO NOT ALLOW AN EVALUATION.

REPORT 02 : NTP TECHNICAL BULLETIN 6 JAN 82
THE COMPOUND WAS NEGATIVE IN THE SALMONELLLA TYPHIMURIUM MUTAGENICITY ASSAY IN ONE LABORATORY.

REPORT 03 : NTP TR 303 NIH PUB.# 86-2559 AUG 86
GAVAGE STUDY IN MALE AND FEMALE F344/N RATS CONSIDERED INADEQUATE STUDIES OF CARCINOGENICITY BECAUSE OF EXTENSIVE AND EARLY MORTALITY AT THE HIGH DOSE OR AT BOTH DOSES AND THE LACK OF CONCLUSIVE EVIDENCE OF CARCINOGENIC EFFECT. THERE WAS CLEAR EVIDENCE OF CARINOGENICITY IN FEMALE MICE AS SHOWN BY INCREASED INCIDENCE OF UNCOMMON OVARIAN NEOPLASMS AT BOTH DOSES. ALSO, INCREASED INCIDENCE OF ADRENAL GLAND ADENOMAS IN HIGH DOSE MICE MAY HAVE BEEN RELATED TO ADMINISTRATION OF 4-VCH.

REPORT 04 : IARC MONOGRAPH 39, 1986 JUN 85
4-VCH TFSTED IN MICE BY GASTRIC INTUBATION IN CORN OIL SHOWED INCREASED INCIDENCE OF OVARIAN TUMORS AND HIGH MORTALITY IN MALES. NO CLEAR EVIDENCE OF CARCINOGENICITY IN RATS - STUDY COMPROMISED BY HIGH MORTALITY. NO DATA AVAILABLE TO EVALUATE REPRODUCTIVE EFFECTS OR PRENATAL TOXICITY IN HUMANS OR ANIMALS. 4-VCH NOT MUTAGENIC TO S. TYPHIMURIUM WITH OR WITHOUT ACTIVATION. NO EPA STUDIES OR CASE REPORTS AVAILABLE TO EVALUATE

CARCINOGENICITY OF 4-VCH TO HUMANS. LIMITED EVIDENCE OF CARCINOGENICITY OF 4-VCH TO ANIMALS. NO DATA ON HUMANS IS AVAILABLE FOR EVALUATION OF CARCINOGENIC RISK.

REPT'D USE : CHEMICAL INTERMEDIATE IN PRODUCTION OF VINYLCYCLOHEXENE DIEPOXIDE, DIMER OF 1,3-BUTADIENE, BY-PRODUCT OF RUBBER PRODUCTION, FLAME RETARDANT, THIOCYANATE INSECTICIDES AND ANTIOXIDANT

CAS NO : **106-87-6**

CHEM NAME : ***** 4-VINYL-3-CYCLOHEXENE DIOXIDE (1-EPOXYETHYL-3,4-EPOXY CYCLOHEXANE)**

REPORT 01 : IARC MONOGRAPH 11 1976
1-VINYL-3-CYCLOHEXENE DIOXIDE IS CARCINOGENIC IN MICE BY SKIN APPLICATION; IT PRODUCED SQUAMOUS-CELL SKIN CARCINOMAS.

REPORT 02 : NTP TR362 NIH 90-2817 NOV 89
UNDER 2-YEAR DERMAL STUDIES CLEAR EVIDENCE OF CARCINOGENIC ACTIVITY OF 4-VINYL-1-CYCLOHEXENE DIEPOXIDE FOR MALE AND FEMALE F344/N RATS SHOWN BY SQUAMOUS CELL AND BASAL CELL NEOPLASMS OF SKIN. CLEAR EVIDENCE OF CARCINOGENIC ACTIVITY FOR B6C3F1 MICE SHOWN BY SQUAMOUS CELL CARCINOMAS OF SKIN IN MALES AND SQUAMOUS CELL CARCINOMAS OF SKIN AND OVARIAN NEOPLASMS IN FEMALES; INCREASED INCIDENCES OF LUNG NEOPLASMS IN FEMALES MAY ALSO HAVE BEEN RELATED TO CHEMICAL APPLICATION.

REPT'D USE : REACTIVE DILUENT FOR DIEPOXIDES AND EPOXY RESINS

CAS NO : **75-02-5**

CHEM NAME : ***** VINYL FLUORIDE**

REPORT 01 : NIOSH CURRENT INTELLIGENCE BULLETIN 28 SEP 78
THERE IS A LACK OF INFORMATION REGARDING THE CARCINOGENICITY OF VINYL FLUORIDE. HOWEVER, IT HAS BEEN SHOWN TO BE MUTAGENIC IN BACTERIAL SYSTEMS. THIS EVIDENCE FOR MUTAGENICITY IS CAUSE FOR CONCERN.

REPORT 02 : EPA TSCA FEDERAL REGISTER NOTICE 50FR46133 NOV 85
DOCUMENT PROPOSES RULE TO REQUIRE TIERED MUTAGENICITY TESTING FOR VF. THIS IS IN RESPONSE TO ITC'S DESIGNATION OF FLUORALKANES FOR PRIORITY TESTING CONSIDERATION IN ITS 7TH ANNUAL REPORT PUBLISHED IN 45FR78432.

REPORT 05 : IARC MONOGRAPH 39 JUN 85
NO DATA AVAILABLE TO EVALUATE CARCINOGENICITY, REPRODUCTIVE EFFECTS OR PRENATAL TOXICITY OF VF TO HUMANS OR EXPERIMENTAL ANIMALS OR THE ACTIVITY OF VF IN SHORT-TERM

TESTS. NO CASE REPORTS OR EPA STUDIES AVAILABLE TO EVALUATE CARCINOGENICITY OF VF IN HUMANS.

REPORT 07 : EPA TSCA FEDERAL REGISTER NOTICE 52 FR 21516 JUN 87
EPA IS ISSUING FINAL TEST RULE TO REQUIRE TESTING FOR CERTAIN HEALTH EFFECTS UNDER TSCA. THIS RULE BECOMES EFFECTIVE ON JULY 22, 1987.

REPT'D USE : INTERMIATE IN MFR OF POLY VINYL FLUORIDE

CAS NO : **75-35-4**

CHEM NAME : ***** VINYLIDENE CHLORIDE (1,1-DICHLOROETHYLENE)**

REPORT 01 : NTP TECHNICAL BULLETIN 7 APR 82
VINYLIDENE CHLORIDE WAS NEGATIVE IN THE SALMONELLA TYPHIMURIUM MUTAGENICITY ASSAY.

REPORT 02 : EPA AMBIENT WATER QUALITY CRITERIA OCT 80
VINYLIDENE CHLORIDE IS SUSPECTED OF BEING A HUMAN CARCINOGEN. AMBIENT WATER CONCENTRATIONS OF .33 UG/L, .03 UG/L AND .003 UG/L ARE ESTIMATED TO KEEP THE LIFETIME CANCER RISK BELOW 10-5, 10-6, AND 10-7 RESPECTIVELY.

REPORT 03 : IARC MONOGRAPH 19 1979
VINYLIDENE CHLORIDE WAS TESTED IN RATS BY ORAL ADMINISTRATION AND IN MICE, RATS AND HAMSTERS BY INHALATION EXPOSURE. WHEN GIVEN BY INHALATION IN RATS AND MICE, IT INDUCED MALIGNANT TUMORS, INCLUDING ANGIOSARCOMAS. THE PRELIMINARY RESULTS OF ANOTHER INHALATION STUDY IN RATS AND MICE INDICATE THE INDUCTION OF MALIGNANT TUMORS OF THE KIDNEY IN MICE, MOSTLY MALES, AND AN INCREASED INCIDENCE OF MAMMARY TUMORS IN RATS.

REPORT 04 : EPA RCRA HEALTH AND ENV. EFFECTS PROFILE OCT 80
VINYLIDENE CHLORIDE HAS BEEN SHOWN TO BE A MUTAGEN IN BACTERIAL SYSTEMS AND A CARCINOGEN IN MICE. BOTH KIDNEY ADENOCARCINOMAS AND MAMMARY ADENOCARCINOMAS WERE PRODUCED AFTER EXPOSURE TO VINYLIDENE CHLORIDE (1,1-DCE) BY INHALATION. NO TERATOGENIC EFFECTS HAVE BEEN OBSERVED.

REPORT 05 : NIOSH CURRENT INTELLIGENCE BULLETIN NO. 28 SEP 78
NIOSH AND OSHA JOINTLY RECOMMEND THAT VINYLIDENE CHLORIDE BE CONSIDERED IN THE WORKPLACE AS A POTENTIAL CARCINOGEN TO HUMANS. RECOMMENDATION IS BASED ON THE RESULTS OF RECENT STUDIES INDICATING THAT EXPOSURE TO THIS CHEMICAL CAUSES ANGIOSARCOMA OF THE LIVER AND OTHER CANCERS IN LABORATORY ANIMALS.

REPORT 06 : MULTI-AGENCY CROSS CHECK
N HHS MAGUIRE
N OSHA CARC
Y EPA CAG
Y EPA RCRA
N EPA CLEAN AIR
Y EPA CLEAN WATER

REPORT 07 : NIOSH CRITERIA DOCUMENT ON VINYL HALIDES SEP 78
NIOSH RECOMMENDS CONTROL AS PROMULGATED FOR VINYL CHLORIDE IN 29 CFR 1910.1017 BASED ON POTENTIAL CARCINOGENICITY. THERE IS NO PRESENT OSHA STANDARD.

REPORT 08 : NTP CARC TRS 228 NIH PUB NO. 82-1784
UNDER THE CONDITIONS OF THIS BIOASSAY, VINYLIDENE CHLORIDE ADMINISTERED BY GAVAGE WAS NOT CARCINOGENIC FOR F344/N RATS OR B6C3F1/N MICE OF EITHER SEX. HOWEVER, BECAUSE OF LIMITATIONS IN THE DOSAGE ADMINISTERED, AND BECAUSE PREVIOUS STUDIES HAVE REPORTED VINYLIDENE CHLORIDE TO BE CARCINOGENIC, THIS STUDY CANNOT BE TAKEN AS PROOF THAT THE CHEMICAL IS NOT A CARCINOGEN.

REPORT 09 : IARC MONOGRAPH 19 1979
VINYLIDENE CHLORIDE PRODUCED REVERSE MUTATIONS IN SALMONELLA TA100 AND TA1530 IN THE PRESENCE OF METABOLIC ACTIVATION. VINYLIDENE CHLORIDE WAS NOT MUTAGENIC IN THE DOMINANT LETHAL TEST IN MALE CD.1 MICE.

REPORT 10 : CMA INHALATION ONCOGENICITY STUDY SEP 82
DS-D RATS WERE EXPOSED TO VDC AT 75, 25, AND 0 PPM FOR 6 H/D 5 D/WK FOR 18 MONTHS. ALTHOUGH THE INCIDENCE OF SEVERAL TUMORS OR TUMOR TYPES WAS FOUND, AT THE 24 MONTH SACRIFICE, TO BE STATISTICALLY INCREASED OR DECREASED IN VDC-EXPOSED RATS WHEN COMPARED TO CONTROL GROUPS, NONE OF THESE DIFFERENCES WAS JUDGED TO BE ATTRIBUTABLE TO VDC EXPOSURE.

REPORT 11 : IARC MONOGRAPH 39, 1986 JUN 85
VC TESTED FOR CARCINOGENICITY IN MICE ORALLY GAVE NEGATIVE RESULTS BUT MICE TESTED VIA INHALATION SHOWED INCREASED INCIDENCE OF KIDNEY ADENOCARCINOMAS IN MALES AND MAMMARY CARCINOMAS IN FEMALES. INHALATION EXPOSURE OF PREGNANT RABBITS AND RATS INDUCED EMBRYOTOXICITY/MALFUNCTIONS IN PRESENCE OF MATERNAL TOXICITY. VC MUTAGENIC TO BACTERIA AND YEAST WITH ACTIVATION. IT INDUCED UNSCHEDULED DNA SYNTHESIS IN RAT HEPATOCYTES. IT INDUCED DNA REPAIR IN MICE EXPOSED BY INHALATION, BUT DID NOT INDUCE DOMINANT LETHAL MUTATIONS. NO DATA AVAILABLE TO EVALUATE REPRODUCTIVE EFFECTS OR PRENATAL TOXICITY TO ANIMALS. THERE IS INADEQUATE EVIDENCE FOR CARCINOGENICITY IN HUMANS.

REPORT 12 : EPA SAFE DRINKING WATER FR NOTICE JUN 84
THIS PROPOSED RULE MAKING UNDER THE SAFE DRINKING WATER ACT ESTABLISHES RECOMMENDED MAXIMUM CONTAMINANT LEVELS (RLMCLS) IN DRINKING WATER.

REPT'D USE : PLASTIC MONOMER, FOOD PACKAGING FILM, SOLVENT RESISTANT LACQUER RESINS, BARRIER COATINGS ON PAPER PRODUCTS AND PLASTIC FILMS AND FLAME RETARDANTS

CAS NO : **9011-067**
CHEM NAME : ***** VINYLIDENE CHLORIDE-VINYL CHLORIDE COPOLYMERS**

REPORT 01 : IARC MONOGRAPH NO. 19 2/78
NO DATA ON THE CARCINOGENICITY OF VINYLIDENE CHLORIDE-VINYL CHLORIDE COPOLYMERS WERE AVAILABLE TO THE WORKING GROUP.

REPORT 02 : IARC MONOGRAPH 19 1979
NO DATA ON TEH EMBRYO-TOXICITY, TERATOGENICITY, METABOLISM OR MUTAGENICITY OF VINYLIDENE CHLORIDE-VINYL CHLORIDE COPOLYMERS WERE AVAILABLE.

REPT'D USE : FOOD PACKAGING; AUDIO, VIDEO AND COMPUTER TAPES, PAINTS, NON-WOVEN FABRICS, AUTO SEAT COVERS AND OUTDOOR FURNITURE

CAS NO : **75-38-7**
CHEM NAME : ***** VINYLIDENE FLUORIDE**

REPORT 01 : EPA TSCA 8(E) NOTICE 8EHQ-0480-0337 APR 80
PENNWALT CORP SUBMITTED RESULTS FROM A LONG-TERM CARCINOGENICITY BIOASSAY ON VDF. PUBLISHED RESULTS INDICATE THAT VDF HAD APPARENTLY PRODUCED AN INCREASED INCIDENCE OF LIPOMAS AND LIPOSARCOMAS AT THE HIGHEST DOSE - 8.25 MG/KG - ADMINISTERED ORALLY TO RATS.

REPORT 02 : NIOSH CURRENT INTELLIGENCE BULLETIN NO. 28 SEP 78
THERE IS A LACK OF INFORMATION REGARDING THE CARCINOGENICITY OF VINYLIDINE FLUORIDE. HOWEVER, IT HAS BEEN SHOWN TO BE MUTAGENIC IN BACTERIAL SYSTEMS.

REPORT 03 : EPA TSCA FEDERAL REGISTER NOTICE 46FR53704 OCT 81
IN RESPONSE TO THE ITC RECOMMENDATIONS (45FR78432) THE EPA PUBLISHED AN ANPR PROPOSING A STRUCTURE-ACTIVITY CATEGORIZATION AND SELECTED TESTING REQUIREMENTS. INDUSTRY WILL CONDUCT ONCOGENICITY TESTING FOR VDF; RESULTS FROM VDF SHOULD APPLY TO VINYL FLOURIDE. THE EPA AND INDUSTRY WILL CONDUCT MUTAGENICITY TESTING AND DECIDE ON FURTHER TIER TESTING AS APPROPRIATE. EXISTING TERATOGENICITY DATA

ARE SUFFICIENT FOR VDF. PRODUCTION FOR SEVERAL FLUORO ALKENES IS INSUFFICIENT TO WARRANT TESTING.

REPORT 04 : EPA TSCA FEDERAL REGISTER NOTICE 50FR46133 NOV 85
DOCUMENT PROPOSES A RULE TO REQUIRE REPRODUCTIVE EFFECTS TESTING, CHRONIC ONCOGENICITY BIOASSAYS AND TIERED MUTAGENICITY TESTING FOR CPD. THIS IS EPA'S RESPONSE TO ITC'S DESIGNATING FLUORALKANES FOR PRIORITY TESTING CONSIDERATION REPORTED IN ITS 7TH ANNUAL REPORT, PUBLISHED IN 45FR78432.

REPORT 08 : IARC MONOGRAPH 39 JUN 85
LIMITED STUDY IN ONE STRAIN. RATS ORALLY SHOWED SMALL NUMBER OF LIPOSARCOMAS. VF GAVE EQUIVOCAL RESULTS FOR MUTAGENICITY TO S. TYPHIMURIUM ACTIVATION. NO DATA AVAILABLE TO EVALUATE REPRODUCTIVE EFFECTS OR PRENATAL TOXICITY IN ANIMALS OR HUMANS. NO CASE REPORTS OR EPA STUDIES FOUND FOR EVALUATION OF CARCINOGENICITY DATA FOR HUMANS. THERE WAS INADEQUATE EVIDENCE FOR CARCINOGENICITY OF CPD IN EXPERIMENTAL ANIMALS AND NO DATA FOR HUMANS.

REPORT 09 : EPA TSCA FEDERAL REGISTER NOTICE 52 FR 21516 JUN 87
EPA IS ISSUING A FINAL TEST RULE UNDER TSCA TO REQUIRE TESTING FOR CERTAIN HEALTH EFFECTS. THIS RULE BECOMES EFFECTIVE ON JULY 22, 1987

REPT'D USE : POLYMER CHEMICAL INTERMEDIATE

CAS NO : **88-12-0**

CHEM NAME : ***** N-VINYL-2-PYRROLIDONE (1-ETHENYL-2-PYRROLIDINONE)**

REPORT 01 : IARC MONOGRAPH 19 1979
NO DATA ON THE CARCINOGENICITY OR MUTAGENICITY OF N-VINYL- 2-PYRROLIDONE WERE AVAILABLE.

REPORT 02 : EPA TSCA 8(E) 0785-0561S JUL 85
BASF WYANDOTTE REPORTED 1-ETHENYL-2-PYRROLIDINONE FOUND NEGATIVE IN VITRO AMES SALMONELLA TYPHIMIRIUM MUTAGENICITY ASSAY WITH AND WITHOUT ACTIVATION, MOUSE LYMPHOMA CELL FORWARD MUTATION ASSAY WITH AND WITHOUT ACTIVATION, BABLB/3T3 MOUSE CELL TRANSFORMATION ASSAY AND UNSCHEDULED DNA SYNTHESIS ASSAY IN RAT HEPATOCYTES.

REPORT 03 : EPA TSCA 8(E) 0785-0561S JUL 85
BASF WYANDOTTE REPORTS 1-ETHENYL-2-PYRROLIDINONE HEPATOTOXIC AND WEAK HEPATIC CANCINOGEN IN RATS, BASED ON 1980, 3-MONTH INHALATION STUDY (DOSES AT 0, 5, 15, 45 AND

		120 PPM), AND MARCH 1983 STUDY OF FEMALE RATS (EXPOSED TO 45 PPM FOR 3-MONTHS). A 3-MONTH INHALATION, DRINKING WATER TEST IN HAMSTERS, AND A 3-MONTH GAVAGE STUDY IN RATS CONFIRMED HEPATOXICITY TO RATS BUT NOT TOXIC TO HAMSTER.
REPT'D USE	:	INTERMEDIATE FOR POLYVINYLPYROLLIDONE

CAS NO	:	**25013-15-4**
CHEM NAME	:	***** VINYL TOLUENE**
REPORT 01	:	NTP DRAFT TR 375 NIH PUB NO. 90-2830 NOV 89 UNDER CONDITIONS OF 2 YEAR INHALATION STUDIES NO EVIDENCE OF CARCINOGENIC ACTIVITY FOR MALE OR FEMALE F344/N RATS EXPOSED TO 100 OR 300 PPM VINYL TOLUENE AND NO EVIDENCE OF CARCINOGENIC ACTIVITY FOR MALE OR FEMALE B6C3F1 MICE EXPOSED TO 10 OR 25 PPM. THERE WAS EVIDENCE OF CHEMICAL-RELATED TOXICITY TO THE NASAL PASSAGE IN BOTH RATS AND MICE.
REPT'D USE	:	INSECTICIDES, POLYESTER RESINS, ALKYD COATING RESINS, DRYING OILS CHEMICAL AND INTERMEDIATE

CAS NO	:	**68-26-8**
CHEM NAME	:	***** VITAMIN A**
REPORT 01	:	NAS REPORT ON DIET, NUTRITION AND CANCER 1982 THE COMMITTEE CONCLUDES THAT THE LABORATORY EVIDENCE SHOWS THAT VITAMIN A IS LABLE TO SUPPRESS CHEMICALLY-INDUCED TUMORS. THE EPA EVIDENCE IS SUFFICIENT TO SUGGEST THAT FOODS RICH IN CAROTENES OR VITAMIN A ARE ASSOCIATED WITH A REDUCED RISK OF CANCER.
REPT'D USE	:	VITAMIN

CAS NO	:	**50-81-7**
CHEM NAME	:	***** VITAMIN C (ASCORBIC ACID)**
REPORT 01	:	NAS REPORT NO DIET, NUTRITION AND CANCER 1982 THE LIMITED EVIDENCE SUGGESTS THAT VITAMIC C CAN INHIBIT THE FORMATION OF SOME CARCINOGENS AND THAT THE COMSUMPTION OF VITAMIN C-CONTAINING FOODS IS ASSOCIATED WITH A LOWER RISK OF CANCERS OF THE STOMACH AND ESOPHAGUS.
REPORT 02	:	NTP TR 247 NIH PUB 83-2503 UNDER THE CONDITIONS OF THIS BIOASSAY, L-ASCORBIC ACID WAS NOT CARCINOGENIC FOR MALE AND FEMALE F344/N RATS OR MALE AND FEMALE B6C3F1 MICE.
REPT'D USE	:	VITAMIN

CAS NO : **1406-66-2**
CHEM NAME : ***** VITAMIN E (A-TOCOPHEROL)**

REPORT 01 : NAS REPORT ON DIET, NUTRITION AND CANCER 1982
VITAMIN E INHIBITS THE FORMATION OF NITROSOAMINES IN VIVO AND IN VITRO. HOWEVER, THERE ARE NO REPORTS ON THE EFFECT OF THIS VITAMIN ON NITROSOAMINE-INDUCED NEOPLASIA. THERE IS LIMITED EVIDENCE SUGGESTING THAT VITAMIN E MAY INHIBIT TUMORIGENESIS IN SEVERAL MODEL ANIMAL SYSTEMS.
REPT'D USE : VITAMIN

CAS NO : **13983-17-0**
CHEM NAME : ***** WOLLASTONITE**

REPORT 01 : IARC MONOGRAPH 42 1987
LIMITED EVIDENCE FOR CARCINOGENICITY TO EXPERIMENTAL ANIMALS; AND INADEQUATE EVIDENCE FOR CARCINOGENICITY TO HUMANS. 4 DIFFERENT PARTICLE SIZES TESTED FOR CARCINOGENICITY IN RATS BY INTRAPLEURAL IMPLANTS. SIGNIFICANTLY INCREASED INCIDENCE OF PLEURAL SARCOMAS SEEN WITH 2 GRADES AND NONSTATISTICALLY SIGNIFICANT INCREASE WITH 3RD. ALL CONTAINED FIBERS>4UM LONG X <0.5 UM IN DIA. PLEURAL SARCOMAS NOT SEEN AFTER IMPLANTATION OF GRADE CONTAINING RELATIVELY FEW FIBERS. MILD PARENCHYMAL AND PLEURAL CHANGES SEEN IN LUNGS OF WOLLASTONITE WORKERS. SMALL COHORT STUDY OF WORKERS IN QUARRY SHOWED EXPECTED NUMBER OF DEATHS FROM CANCER.
REPT'D USE : CERAMICS, PAINTS, PLASTICS, RUBBER, ABRASIVES, METALLURGY, EPOXY RESINS, TILES, BRAKE LININGS AND CEMENTS

CAS NO : **1330-20-7**
CHEM NAME : ***** XYLENE (MIXED ISOMERS)**

REPORT 01 : NIOSH CRITERIA DOCUMENT MAY 75
NIOSH RECOMMENDS A STANDARD OF 100 PPM TWA BASED ON CNS DEPRESSANT EFFECTS AND AIRWAY IRRITATION. THE PRESENT OSHA STANDARD IS 100 PPM TWA.

REPORT 02 : NTP TEST REGIMEN
TERATOLOGY ASSAYS SHOWED THAT ORAL ADMINISTRATION OF XYLENE (MIXED ISOMERS) CAUSED INCREASED MALFORMATIONS IN MICE. IN A CURRENT NTP BIOASSAY, XYLENE IS BEING ADMINISTERED TO F344 RATS AND B6C3F1 MICE BY GAVAGE. THE CHRONIC TESTING PHASE BEGAN JUNE 80.

REPORT 03 : TSCA EPA INTERAGENCY TESTING COMM. OCT 77
IN A JAN 1981 RULING, THE SOUTHERN DISTRICT COURT OF NEW YORK ORDERED EPA TO PROPOSE TEST RULES, OR DECISIONS NOT

TO REQUIRE TESTING, BY THE END OF 1983 FOR ALL CHEMICALS LISTED IN THE FIRST FIVE ITC REPORTS. THE SCHEDULE SUBMITTED BY THE EPA, AND ACCEPTED BY THE COURT WITHOUT CHANGE, SHOWS THAT A DECISION FOR TESTING OF MUTAGENICITY, TERATOGENICITY, AND EPA STUDIES OF XYLENES MAY BE MADE IN 1982.

REPT'D USE : CLEANING AGENT, PAINT, FLAME RETARDANT, AEROSOL, CAUSTIC ADHESIVE AND POLYESTER

CAS NO : **95-68-1**

CHEM NAME : ***** 2,4-XYLIDINE**

REPORT 01 : IARC MONOGRAPH 16 1977

DATA REPORTED IN AN ABSTRACT INDICATE THAT 2,4-XYLIDENE PRODUCED AN INCREASED INCIDENCE OF SUBCUTANEOUS AND LIVER TUMORS IN RATS FOLLOWING ITS ORAL ADMINISTRATION. THE INADEQUACY OF REPORTING DOES NOT ALLOW AN EVALUATION OF THIS COMPOUND TO BE MADE.

REPT'D USE : DYE INTERMEDIATE

CAS NO : **95-78-3**

CHEM NAME : ***** 2,5-XYLIDINE**

REPORT 01 : IARC MONOGRAPH 16 1977

DATA REPORTED IN AN ABSTRACT INDICATE THAT 2,5-XYLIDINE PRODUCED AN INCREASED INCIDENCE OF SUBCUTANEOUS AND LIVER TUMORS IN RATS FOLLOWING ITS ORAL ADMINISTRATION. THE INADEQUACY OF THE REPORTING DOES NOT ALLOW AN EVALUATION OF THE CARCINOGENICITY OF THIS COMPOUND TO BE MADE.

REPT'D USE : DYE INTERMEDIATE

CAS NO : **87-62-7**

CHEM NAME : ***** 2,6-XYLIDINE**

REPORT 01 : NTP REPORT NO. 278; NIH PUBLICATION NO. 90-2534 JAN 90

IN GAVAGE AND FEED STUDIES, 2,6-XYLIDINE CLEARLY CARCINOGENICY FOR MALE AND FEMALE CHARLES RIVER CD RATS, GIVEN DOSES OF 0, 300, 100 OR 3,000 PPM CAUSING SIGNIFICANT INCREASES IN INCIDENCES OF ADENOMAS AND CARCINOMAS OF NASAL CAVITY. RHABDOMYOSARCOMAS, RARE TUMORS OF NASAL CAVITY OBSERVED IN DOSED RATS OF EACH SEX. INCREASED INCIDENCES OF SUBCUTANEOUS FIBROMAS AND FIBROSARCOMAS IN MALE AND FEMALE RATS AND OF NEOPLASTIC NODULES OF LIVER IN FEMALE RATS MAY HAVE BEEN RELATED TO ADMINISTRATION OF 2,6-XYLIDINE.

REPT'D USE : CHEMICAL INTERMEDIATE IN DYES PRODUCTION, AND C.I. ACID ORANGE 134, C.I. DIRECT RED 150 AND COSMETICS.

CAS NO : **17924-92-4**
CHEM NAME : ***** ZEARALENONE**

REPORT 01 : NTP BOARD OF SCIENTIFIC COUNSELORS DEC 81
UNDER THE CONDITIONS OF THIS BIOASSAY, ZEARALENONE ADMINISTERED IN THE FEED WAS ASSOCIATED WITH AN INCREASED INCIDENCE OF HEPATOCELLULAR ADENOMAS IN FEMALE B6C3F1 MICE AND AN INCREASED INCIDENCE OF PITUITARY ADENOMAS IN MALE AND FEMALE B6C3F1 MICE. ZEARALENONE WAS NOT CARCINOGENIC FOR F344 RATS OF EITHER SEX. THERE IS INDIRECT EVIDENCE THAT ZEARALENONE IS A CARCINOGEN FOR MALE AND FEMALE B6C3F1 MICE.

REPORT 02 : IARC MONOGRAPH 31 JUL 83
IN TESTS IN MICE AND RATS DIET INCREASED INCIDENCES OF HEPATOCELLULAR ADENOMAS IN FEMALE MICE AND OF PITUITARY ADENOMAS IN MICE OF BOTH SEXES OBSERVED. NO EVIDENCE OF CARCINOGENIC EFFECT IN RATS.

REPORT 03 : IARC MONOGRAPH 31 JUL 83
ZEARALENONE POSITIVE IN BACTERIAL REPAIR TEST BUT NOT MUTAGENIC IN BACTERIA OR YEAST. ZEARALENONE INDUCES MUTIPLE REPRODUCTIVE DEFICIENCIES, SUCH AS INFERTILITY AND SMALL LITTERS, IN RATS AND PIGS. DATA INADEQUATE TO ASSESS TERATOGENICITY TO EXPERIMENTAL ANIMALS.
REPT'D USE : INTERMEDIATE FOR MANUFACTURE OF ZEARALANOL

CAS NO : **315-18-4**
CHEM NAME : ***** ZECTRAN**

REPORT 02 : NCI CARC TRS NO 147 DHEW PUB NO NIH 78-1703 1978
UNDER THE CONDITIONS OF THIS BIOASSAY, SUFFICIENT EVIDENCE WAS NOT OBTAINED FOR THE CARCINOGENICITY OF MEXACARBATE FOR OSBORNE-MENDEL RATS OR B6C3F1 MICE (FEEDING STUDY).

REPORT 03 : IARC MONOGRAPH 12 1976
ZECTRAN HAS BEEN TESTED BY ORAL ADMINISTRATION AND BY SINGLE SUBCUTANEOUS INJECTION IN TWO STRAINS OF MICE. FOLLOWING ITS ORAL ADMINISTRATION, AN INCREASED INCIDENCE OF LUNG ADENOMAS WAS OBSERVED IN BOTH SEXES OF ONE STRAIN COMBINED, TOGETHER WITH A SLIGHT INCREASE IN THE NUMBER OF LIVER-CELL TUMORS IN MALES OF THE SAME STRAIN. THE AVAILABLE DATA DO NOT ALLOW AN EVALUATION

OF THE CARCINOGENICITY OF ZECTRAN TO BE MADE.
REPT'D USE : PESTICIDE

CAS NO : **7440-66-6**
CHEM NAME : ***** ZINC**

REPORT 01 : EPA HEALTH ASSESSMENT 600/8-87/022F JUL 87
ZINC TOXIC TO AQUATIC ORGANISMS, WILDLIFE AT HIGH EXPOSURE LEVELS. WORKER EXPOSURE TO ZINC OXIDE, CAN CAUSE ACUTE "METAL FUME FEVER", CHRONIC RESPIRATORY INFECTIONS, DERMATITIS, CONJUNCTIVITIS, GASTRITIS, BRONCHITIS, DIFFUSE PNEUMOSCLEROSIS, HYPOCALCEMIA, AND EPIGASTRIC PAIN. EXPOSURE TO ZINC CHLORIDE SMOKE BOMBS RESULTED IN PULMONARY DISEASE/DEATH. ADMINISTRATION OF 150 MG ELEMENTAL ZINC SHOWED ADVERSE EFFECTS ON IMMUNOLOGIC AND CARDIOVASCULAR SYSTEMS. SOME DATA INDICATES ZINC INVOLVED IN TUMOR GROWTH PROMOTION OR INHIBITION. CARCINOGENIC RESPONSE FOLLOWED INJECTION OF ZINC SALTS INTO TESTES OF FOWL AND RATS.

REPORT 02 : EPA HEALTH ASSESSMENT 600/8/87/022F JUL 87
ZINC OXIDE CAUSE CHROMOSOMAL DAMAGE IN INCREASED FREQUENCY OF HYPERDIPLOID CELLS IN BONE MARROW OF NONINBRED WHITE RATS AT 0.1 AND 0.5 MG/M3. INCREASE IN FREQUENCY OF STRUCTURAL ABERRATIONS OF CHROMOSOMES AND HYPERDIPLOID CELLS SEEN WHEN HUMAN LYMPHOCYTES AT G STAGE OF CELL CYCLE WERE EXPOSED TO ZINC ACETATE AT CONCENTRATIONS OF 7.0 TO 20.0 UG/ML. ZINC OXIDE NOT MUTAGENIC AT LEVELS OF 100 TO 5,000 UG/PLATE IN THE SALMONELLA REVERSION ASSAY. INGESTION OF HIGH LEVELS OF ZINC MAY HAVE ADVERSE IMPACT ON REPRODUCTION.
REPT'D USE : GALVANIZING OF IRON AND STEEL, RUBBER VULCANIZATION AND MEDICINE

CAS NO : **7646-85-7**
CHEM NAME : ***** ZINC CHLORIDE**

REPORT 01 : MICROBIOLOGICAL ASSOC'S FINAL REPORT TO CMA OCT 84
IN THE L5178Y TK+/- MOUSE LYMPHOMA MUTAGENESIS ASSAY IN THE PRESENCE OF METABOLIC ACTIVATION, ZINC CHLORIDE GAVE A POSITIVE RESPONSE.

CAS NO : **15930-94-6**
CHEM NAME : ***** ZINC CHROMATE HYDROXIDE**

REPORT 01 : IARC MONOGRAPH 23 1980
THERE IS SUFFICIENT EVIDENCE OF RESPIRATORY CARCINOGE-

NICITY IN MEN OCCUPATIONALLY EXPOSED DURING CHROMATE PRODUCTION. CHROMIUM (VI) COMPOUNDS CAUSE MUTATIONS AND ALLIED EFFECTS IN A VERY WIDE RANGE OF PROKARYOTIC AND EUKARYOTIC SYSTEMS BOTH IN VITRO AND IN VIVO.

REPT'D USE : PAINTS/VARNISHES, OIL COLORS AND AEROSOLS

CAS NO : **5823-20-1**

CHEM NAME : ***** ZINC DIALKYLDITHIOPHOSPHATE**

REPORT 01 : MCIROBIOLOGICAL ASSOC'S FINAL REPORT FOR CMA OCT 84
UNDER CONDITIONS OF ASSAY, DATA INDICATE THAT 9 ZDDP'S INDUCE MORPHOLOGICAL TRANSFORMATION IN BALB/3T3 CLONE A31-1 CELLS IN PRESENCE OF EXOGENOUS SOURCE OF METABOLIC ACTIVATION. IN L51784Y TK+/- MOUSE LYMPHOMA MUTAGENESIS ASSAY, THEY GAVE NEGATIVE RESPONSES BOTH WITH AND WITHOUT METABOLIC ACTIVATION.

CAS NO : **557-07-3**

CHEM NAME : ***** ZINC OLEATE**

REPORT 01 : MICROBIOLOGICAL ASSOC'S FINAL REPORT TO CMA OCT 84
IN THE MOUSE LYMPHOMA MUTAGENESIS ASSAY, ZINC OLEATE GAVE A NEGATIVE RESPONSE IN PRESENCE OF METABOLIC ACTIVATION (L5178Y TK+/- MOUSE LYMPHOMA ASSAY).

CAS NO : **1314-13-2**

CHEM NAME : ***** ZINC OXIDE**

REPORT 01 : EPA HEALTH ASSESSMENT 600/8-87/022F JUL 87
ZINC TOXIC TO AQUATIC ORGANISMS, WILDLIFE AT HIGH EXPOSURE LEVELS. WORKER EXPOSURE TO ZINC OXIDE CAN CAUSE ACUTE "METAL FUME FEVER", CHRONIC RESPIRATORY INFECTIONS, DERMATITIS, CONJUNCTIVITIS, GASTRITIS, BRONCHITIS, DIFFUSE PNEUMOSCLEROSIS, HYPOCALCEMIA, AND EPIGASTRIC PAIN. EXPOSURE TO ZINC CHLORIDE SMOKE BOMBS RESULTED IN PULMONARY DISEASE/DEATH. ADMINISTRATION OF 150 MG ELEMENTAL ZINC SHOWED ADVERSE EFFECTS ON IMMUNOLOGIC AND CARDIOVASCULAR SYSTEMS. SOME DATA INDICATES ZINC INVOLVED IN TUMOR GROWTH PROMOTION OR INHIBITION. CARCINOGENIC RESPONSE FOLLOW INJECTION OF ZINC SALTS INTO TESTES OF FOWL AND RATS.

REPORT 02 : EPA HEALTH ASSESSMENT 600/8-87/022F JUL 87
ZINC OXIDE CAUSE CHROMOSOMAL DAMAGE IN INCREASED FREQUENCY OF HYPERDIPLOID CELLS IN BONE MARROW OF NONINBRED WHITE RATS AT 0.1 AND 0.5 MG/M3. INCREASES IN FREQUENCY OF STRUCTURAL ABERRATIONS OF CHROMOSOMES

		AND HYPERIDIPLOID CELLS SEEN WHEN HUMAN LYMPHOCYTES AT G STAGE OF CELL CYCLE WERE EXPOSED TO ZINC ACETATE AT CONCENTRATIONS OF 7.0 TO 20.0 KG/ML. ZINC OXIDE NOT MUTAGENIC AT LEVELS OF 100 TO 5,000 KG/PLATE IN SALMONELLA REVERSION ASSAY. INGESTION OF HIGH LEVELS OF ZINC MAY HAVE ADVERSE IMPACT ON REPRODUCTION.
REPT'D USE	:	GALVANIZING OF IRON AND STEEL, RUBBER VULCANIZATION AND MEDICINE

CAS NO : **12433-50-0**
CHEM NAME : ***** ZINC POTASSIUM CHROMATE**

REPORT 01	:	IARC MONOGRAPH 23 1980 EXPERIMENTS IN MICE AND RATS EXPOSED TO ZINC YELLOW (VI) BY INTRATRACHEAL INJECTION AND I.M. AND IP IMPLANTATION WERE INADEQUATE TO EVALUATE CARCINOGENICITY. CHROMIUM (VI) CPDS CAUSE MUTATIONS AND ALLIED EFFECTS IN A VERY WIDE RANGE OF PROKARYOTIC AND EUKARYOTIC SYSTEMS, BOTH IN VIVO AND IN VITRO.
REPT'D USE	:	COLORING AGENT IN LINOLEUM, RUBBER, ARTISTS' OIL PAINTS, PLASTICS, AND ANTI-CORROSION PAINT

CAS NO : **12122-67-7**
CHEM NAME : ***** ZINEB**

REPORT 01	:	IARC MONOGRAPH 12 1976 ZINEB PRODUCED AN INCREASED INCIDENCE OF LUNG TUMORS AFTER ITS ORAL ADMINISTRATION TO MICE OF ONE STRAIN. SYSTEMIC RETICULUM-CELL SARCOMAS WERE OBSERVED IN MICE AND A VARIETY OF SARCOMAS IN RATS AFTER ITS SUBCUTANEOUS INJECTION. NO INCREASES IN TUMOR INCIDENCES WERE OBSERVED IN TWO OTHER STRAINS OF MICE AND IN TWO LIMITED STUDIES IN RATS FOLLOWING ORAL ADMINISTRATION. THE AVAILABLE DATA DO NOT ALLOW AN EVALUATION OF THE CARCINOGENICITY OF ZINEB TO BE MADE.
REPT'D USE	:	INSECTICIDE AND FUNGICIDE

CAS NO : **137-30-4**
CHEM NAME : ***** ZIRAM**

REPORT 01	:	IARC MONOGRAPH 12 1976 ORAL ADMINISTRATION OF ZIRAM INCREASED THE INCIDENCE OF LIVER AND SUBCUTANEOUS TUMORS IN ONE STUDY IN RATS. WITH A LIMITED NUMBER OF SURVIVING ANIMALS; ITS SUBCUTANEOUS ADMINISTRATION INDUCED A LOWER YIELD OF THE SAME TYPES OF TUMORS. IN ONE STUDY BY ORAL ADMINISTRATION IN TWO STRAINS OF MICE, NO CARCINOGENIC EFFECT WAS OBSERVED.

THE LIMITED DATA AVAILABLE DO NOT ALLOW AN EVALUATION OF THE CARCINOGENICITY OF ZIRAM TO BE MADE.

REPORT 03 : NTP TECHNICAL BULLETIN 6 JAN 82
THE COMPOUND WAS POSITIVE IN THE SALMONELLA TYPHIMURIUM MUTAGENICITY ASSAY IN ONE LABORATORY.

REPORT 04 : IARC MONOGRAPH 12 1976
NO INCREASE IN THE NUMBER OF RECESSIVE LETHALS IN DROSOPHILA WAS OBTAINED WITH ZIRAM. NO SIGNIFICANT INCREASE IN GENE CONVERSION WAS NOTED IN SACCHAROMYCES. METABOLIC ACTIVATION WAS NOT USED IN THESE TESTS. AN INCREASED INCIDENCE OF CHROMOSOME ABERRATIONS OF BONE-MARROW CELLS WAS FOUND IN TREATED MICE.

REPORT 05 : NTP TR 238 NIH PUB 83-1794 1983
UNDER THE CONDITIONS OF THESE STUDIES, ZIRAM WAS CARCINOGENIC FOR MALE F344/N RATS, CAUSING INCREASED INCIDENCES OF C-CELL CARCINOMAS OF THE THYROID GLAND. ZIRAM WAS NOT CARCINOGENIC FOR EITHER FEMALE F344/N RATS OF FOR MALE B6C3F1 MICE. INCREASED INCIDENCES OF ALVEOLAR/BRONCHIOLAR ADENOMAS OR CARCINOMAS OCCURRED IN FEMALE B6C3F1 MICE. HOWEVER, THE INTERPRETAION OF THIS INCREASE IN LUNG TUMORS IS COMPLICATED BY AN INTERCURRENT SENDAI VIRUS INFECTION.

REPT'D USE : FUNGICIDE, RUBBER ACCELERATOR AND ADHESIVES

INDEX A

ALPHABETICAL LISTING OF CHEMICALS

CHEMICAL NAME	CAS NO.
ACETOPHENONE (ACP)	96-86-2
ACETALDEHYDE	75-07-0
ACETAMIDE	60-35-5
ACETAMINOPHEN	103-90-2
ACETIN	26446-35-5
ACETOHEXAMIDE	968-81-0
ACETONITRILE	75-05-8
P-ACETOPHENETIDIDE CAFFEINE ASPIRIN MIXTURE	8003-03-0
2-ACETYLAMINOFLUORENE (2-AAF)	53-96-3
1-ACETYL-2-PICOLINOYL HYDRAZINE (1A2PH)	17433-31-7
ACETYL-O-TOLUIDINE	120-66-1
ACRIFLAVINIUM CHLORIDE	8048-52-0
ACROLEIN	107-02-8
ACRONYCINE	7008-42-6
ACRYLAMIDE	79-06-1
ACRYLIC ACID	79-10-7
ACRYLONITRILE (AN)	107-13-1
ACRYLONITRILE-BUTADIENE-STYRENE COPOLYMER RESINS	9003-56-9
ACTINOMYCIN-D	50-76-0
ADRIAMYCIN	23214-92-8
AF-2(2-(2-FURYL)-3-(5-NITRO-2-FURYL)ACRYLAMIDE)	3688-53-7
AFLATOXIN	1402-68-2
AGAR-AGAR	9002-18-0
AGARITINE(L-GLUTAMIC ACID-5-(2-(4-HYDROXYMETHYL)-PHEN	2757-90-6

CHEMICAL NAME	CAS NO.
ALDICARB	116-06-3
ALDRIN	309-00-2
ALKYL EPOXIDES	CLASS-20
ALKYL PHTHALATES	CLASS-02
ALLYL CHLORIDE	107-05-1
ALLYL GLYCIDYL ETHER (AGE)	106-92-3
ALLYL ISOTHIOCYANATE (ATC)	57-06-7
ALLYL ISOVALERATE	2835-39-4
ALLYLAMINE (2-PROPINE-1-AMINE)	107-11-7
ALUMINUM	7429-90-5
5-AMINOACENAPHTHENE	4657-93-6
9-AMINOACRIDINE HC1 H20	52417-22-8
2-AMINOANTHRACENE	613-13-8
2-AMINOANTHRAQUINONE	117-79-3
PARA-AMINOBENZOIC ACID	150-13-0
4-AMINOBIPHENYL	92-67-1
1-AMINO-2,4-DIBROMO ANTHRAQUINONE	81-49-2
7-AMINO-2,2-DIMETHYL-2,3-DIHYDROBENZOFURAN	88296-46-4
3-AMINO-4-ETHOXYACETANILIDE	17026-81-2
3-AMINO-9-ETHYLCARBAZOLE	132-32-1
3-AMINO-9-ETHYLCARBAZOLE HYDROCHLORIDE	6109-97-3
2-AMINO-5-(5-NITRO-2-FURYL)-1,3,4-THIADIAZOLE	712-68-5
2-AMINO-NITROPHENOL	121-88-0
2-AMINO-4-NITROPHENOL	99-57-0
4-AMINO-2-NITROPHENOL	119-34-6
2-AMINO-5-NITROTHIAZOLE	121-66-4
O-AMINOPHENOL	95-55-6
3-AMINO-S-TRIAZOLE (AMITROLE)	61-82-5
11-AMINOUNDECANOIC ACID	2432-99-7
AMPICILLIN TRIHYDRATE	7177-48-2
ANGELICIN AND SOME SYNTHETIC DERIVATIVES	523-50-2
ANILAZINE	101-05-3

CHEMICAL NAME	CAS NO.
ANILINE	62-53-3
ANILINE HYDROCHLORIDE	142-04-1
O-ANISIDINE	90-04-0
O-ANISIDINE HYDROCHLORIDE	134-29-2
P-ANISIDINE HYDROCHLORIDE	20265-97-8
ANTHANTHRENE	191-26-4
ANTHRACENE	120-12-7
ANTHRANILIC ACID	118-92-3
ANTHRAQUINONE	84-65-1
ANTIMONY	7440-36-0
ANTIMONY SULFIDE (PENTASULFIDE)	1315-04-4
ANTIMONY TRIOXIDE	1309-64-4
ANTIMONY TRISULFIDE	1345-04-6
APHOLATE	52-46-0
ARABINOSYLCYTOSINE	147-94-4
ARAMITE	140-57-8
ARISTOLOCHIC ACID	10130-99-5
AROCHLOR 1254	11097-69-1
C9 AROMATIC HYDROCARBONS (NAPTHA)	UNKNOWN-04
AROMATIC PETROLEUM PITCH	68187-58-6
ARSANILIC ACID	98-50-0
ARSENIC	7440-38-2
ARSENIC PENTOXIDE	1303-28-2
ARSENIC SULFIDE	1303-33-9
ARSENIC TRIOXIDE	1327-53-3
ARSINE	7784-42-1
ARYL PHOSPHATES	CLASS-01
ASBESTOS	1332-21-4
ATTAPULGITE	12174-11-7
AURAMINE HYDROCHLORIDE	2465-27-2
AUROTHIOGLUCOSE	12192-57-3
5-AZACYTIDINE	320-67-2

CHEMICAL NAME	CAS NO.
AZASERINE	115-02-6
AZATHIOPRINE	446-86-6
AZINPHOS-METHYL	86-50-0
AZIRIDINE	151-56-4
2-(1-AZIRIDINYL)-ETHANOL	1072-52-2
AZIRIDYL BENZOQUINONE	800-24-8
AZOBENZENE	103-33-3
AZODICARBONAMIDE	123-77-3
BENZ(A)ACRIDINE	225-11-6
BENZ(C)ACRIDINE	225-51-4
BENZAL CHLORIDE	98-87-3
BENZALDEHYDE	100-52-7
BENZAMINE, 4,4-(1,4-PHENYLENE BIS (1-METHYLETHYLIDENE	2716-12-3
BENZ(A)ANTHRACENE (BA)	56-55-3
BENZENE	71-43-2
BENZIDINE	92-87-5
BENZIDINE CONGENER DYES	CLASS-09
BENZIDINE DIHYDROCHLORIDE	531-85-1
BENZO(B)FLUORANTHENE	205-99-2
BENZO(GHI)FLUORANTHENE	203-12-3
BENZO(J)FLUORANTHENE	205-82-3
BENZO(K)FLUORANTHENE	207-08-9
BENZO(A)FLUORENE	238-84-6
BENZO(B)FLUORENE	243-17-4
BENZO(C)FLUORENE	205-12-9
BENZOFURAN	271-89-6
BENZOFURAZAN-1-OXIDE	480-96-6
BENZOIC ACID, HYDRAZIDE	613-94-5
BENZOIN	119-53-9
BENZO(GHI)PERYLENE	191-24-2
BENZO(C)PHENANTHRENE	195-19-7
BENZO(A)PYRENE	50-32-8

CHEMICAL NAME	CAS NO.
BENZO(E)PYRENE	192-97-2
1H-BENZOTRIAZOLE	95-14-7
BENZOTRICHLORIDE (BTC)	98-07-7
BENZOYL CHLORIDE (BC)	98-88-4
BENZOYL PEROXIDE	94-36-0
BENZYL ACETATE	140-11-4
BENZYL ALCOHOL	100-51-6
BENZYL CHLORIDE	100-44-7
BERTRANDITE	12161-82-9
BERYLLIUM	7440-41-7
BERYLLIUM ACETATE	543-81-7
BERYLLIUM-ALUMINUM ALLOY	12770-50-2
BERYLLIUM CARBONATE	13106-47-3
BERYLLIUM CHLORIDE	7787-47-5
BERYLLIUM-COPPER ALLOY	11133-98-5
BERYLLIUM-COPPER-COBALT ALLOY	55158-44-6
BERYLLIUM FLUORIDE	7787-49-7
BERYLLIUM HYDROXIDE	13327-32-7
BERYLLIUM-NICKEL ALLOY	37227-61-5
BERYLLIUM OXIDE	1304-56-9
BERYLLIUM PHOSPHATE	13598-15-7
BERYLLIUM SILICATE	15191-85-2
BERYLLIUM SULFATE TETRAHYDRATE	7787-56-6
BERYLLIUM ZINC SILICATE	39413-47-3
BERYL ORE	1302-52-9
BETEL-QUID AND ARECA NUT CHEWING	CLASS-03
BIPHENYL	92-52-4
2-BIPHENYLAMINE HYDROCHLORIDE	90-41-5
3,3',4,4'-BIPHENYL-TETRAMINE	7411-49-6
BIS(1-AZIRIDINYL)-MORPHOLINOPHOSPHINE SULFIDE	2168-68-5
BISCHLORDETHYL NITROSOUREA (BCNU)	154-93-8
BIS(2-CHLOROETHYL)ETHER (BCEE)	111-44-4

CHEMICAL NAME	CAS NO.
BIS(2-CHLOROISOPROPYL)ETHER (BCIE)	39638-32-9
1,2-BIS(CHLOROMETHOXY)ETHANE (BCMXE)	13483-18-6
1,4-BIS(CHLOROMETHOXYMETHYL)-BENZENE	56894-91-8
BIS(CHLOROMETHYL) ETHER (BCME)	542-88-1
BIS(2-CHLORO-1-METHYLETHYL) ETHER (BCMEE)	108-60-1
BIS(2,3-DIBROMPROPYL ETHER) OF TETRABROMBISPHENOL A	21850-44-2
BIS(2-DIMETHYLAMINOETHYL)ETHER (DMAEE) NIAX CATALYST	3033-62-3
BIS(2,4-DI-TERT-BUTYLPHENYL)PENTAERYTHRITOL DIPHOSPHI	26741-53-7
BIS(2,2-EPOXYCYCLOPENTENYL) ETHER	2386-90-5
BIS(2-ETHYLHEXYL)-TEREPHTHALATE	6422-86-2
BIS(2-METHOXYETHYL)ETHER	111-96-6
BISPHENOL A	80-05-7
BISPHENOL A DIGLYCIDYL ETHER	1675-54-3
BIS(TRICHLOROMETHYL)SULFONE	3064-70-8
BITUMENS (FROM PETROLEUM REFINING)	8052-42-4
BLEOMYCIN	11056-06-7
BOOT AND SHOE MANUFACTURE AND REPAIR	CLASS-36
BORON TRIFLUORIDE	7637-07-2
BRACKEN FERN	UNKNOWN-31
BRILLIANT RED HISALONIC	UNKNOWN-18
BROMOACETIC ACID, ETHYL ESTER	105-36-2
BROMODICHLOROETHANE	75-27-4
2-BROMO-4,6-DINITROANLINE	1817-73-8
BROMOETHANE	74-96-4
BROMOMETHANE (BM) (METHYL BROMIDE)	74-83-9
7-BROMOMETHYL-12-METHYL-BENZ(A)ANTHRACENE	16238-56-5
1,3-BUTADIENE	106-99-0
1,4-BUTANEDIOL DIMETHANESULPHONATE	55-98-1
2-BUTOXYETHANOL (EGMBE)	111-76-2
2(2-BUTOXYETHOXY) ETHYL ACETATE (DGBA)	124-17-4
N-BUTYL ACETATE	123-86-4
N-BUTYL ACRYLATE	141-32-2

CHEMICAL NAME	CAS NO.
T-BUTYL ALCOHOL (T-BUTANOL)	75-65-0
BUTYLATED HYDROXYANISOLE (BHA)	25013-16-5
BUTYLATED HYDROXYTOLUENE (BHT)	128-37-0
P-TERT-BUTYLBENZALDEHYDE (P-TBB)	939-97-9
P-TERT-BUTYLBENZOIC ACID (P-TBBA)	98-73-7
BUTYL BENZYL PHTHALATE	85-68-7
N-BUTYL CHLORIDE	109-69-3
BUTYL GLYCIDYL ETHER (BGE)	2426-08-6
T-BUTYL HYDROPEROXIDE (TBHP)	75-91-2
1-BUTYLMETACRYLATE/STYRENE RANDOM COPOLYMER	25213-39-2
P-TERT-BUTYLTOLUENE (P-TBT)	98-51-1
BETA-BUTYROLACTONE	3068-88-0
GAMMA-BUTYROLACTONE	96-48-0
CADMIUM	7440-43-9
CADMIUM CHLORIDE	10108-64-2
CADMIUM OXIDE	1306-19-0
CADMIUM SULFATE	10124-36-4
CADMIUM SULFIDE	1306-23-6
CAFFEINE	58-08-2
CALCIUM ARSENATE	7778-44-1
CALCIUM CHROMATE	8012-75-7
CALCIUM CYANAMIDE	156-62-7
CALCIUM CYCLAMATE	139-06-0
CALCIUM DIALKYLTHIOPHOSPHATES	UNKNOWN-35
CALCIUM NAPHTHENATE	61789-36-4
CALCIUM SODIUM METAPHOSPHATE FIBERS	23209-59-8
CALIBAN (F/R-P 44)	61345-53-7
CAMPHOR	76-22-2
CANTHARIDIN	56-25-7
CAPROLACTAM	105-60-2
CAPTAN	133-06-2
CARBARYL	63-25-2

CHEMICAL NAME	CAS NO.
CARBAZOLE	86-74-8
3-CARBETHOXYPSORALEN	20073-24-9
CARBON BLACKS	1333-86-4
CARBON DISULFIDE	75-15-0
CARBON FIBERS	UNKNOWN-08
CARBON MONOXIDE	630-08-0
CARBON TETRACHLORIDE	56-23-5
CARBROMAL	77-65-6
CARPENTRY AND JOINERY	UNKNOWN-24
CARRAGEENANS	9000-07-1
D-CARVONE	2244-16-8
CATECHOL	120-80-9
CHLORAL HYDRATE	302-17-0
CHLORAMBEN	133-90-4
CHLORAMBUCIL	305-03-3
CHLORAMINE	10599-90-3
CHLORAMPHENICOL	56-75-7
CHLORDANE	57-74-9
CHLORDECONE	143-50-0
CHLORDIMEFORM	6164-98-3
CHLORENDIC ACID	115-28-6
CHLORINATED NAPHTHALENES (CNS)	CLASS 11
CHLORINATED PARAFFIN	CLASS 44
CHLORINATED TERPHENYL	61788-33-8
CHLORINATED TRISODIUM PHOSPHATE	56802-99-4
CHLORINE	7782-50-5
CHLORMADINONE ACETATE	302-22-7
CHLORNAPHAZINE	494-03-1
CHLOROACETALDEHYDE	107-20-0
CHLOROACETONE	78-95-5
2-CHLOROACETOPHENONE	532-27-4
4'-(CHLOROACETYL)-ACETANILIDE	140-49-8

CHEMICAL NAME	CAS NO.
P-CHLOROANILINE	106-47-8
CHLOROBENZENE	108-90-7
CHLOROBENZILATE	510-15-6
P-CHLOROBENZOTRICHLORIDE (PCBTC)	5216-25-1
4-CHLOROBENZOTRIFLUORIDE	98-56-6
CHLORODIBROMOMETHANE	124-48-1
CHLORODIFLUOROMETHANE	75-45-6
CHLOROETHANE	75-00-3
2-CHLOROETHANOL (ETHYLENE CHLOROHYDRIN)	107-07-3
(2-CHLOROETHYL)TRIMETHYLAMMONIUM CHLORIDE	999-81-5
1-(2-CHLOROETHYL)-3-(4-METHYLCYCLOHEXYL)-1-NITROSOUREA	13909-09-6
CHLOROFLUOROCARBON 22	UNKNOWN-19
CHLOROFLUOROCARBONS	CLASS-08
CHLOROFLUOROMETHANE	593-70-4
CHLOROFORM	67-66-3
3-CHLORO-2-HYDROXYPROPYLTRIMETHYL AMMONIUM CHLORIDE	3327-22-8
CHLOROMETHANE	74-87-3
2-CHLORO-1-METHYLETHYL(2-CHLOROPROPYL) ETHER	UNKNOWN-02
CHLOROMETHYL METHYL ETHER (CMME)	107-30-2
3-CHLORO-2-METHYLPROPENE	563-47-3
2-(CHLOROMETHYL)PYRIDINE HYDROCHLORIDE	6959-47-3
3-(CHLOROMETHYL)PYRIDINE HYDROCHLORIDE	6959-48-4
4-CHLORO-2-NITROANILINE	89-63-4
1-CHLORO-4-NITROBENZENE	100-00-5
2-CHLORONITROBENZENE	88-73-3
CHLOROPHENOLS	CLASS-29
CHLOROPHENOXY HERBICIDES	CLASS-30
4-CHLORO-M-PHENYLENEDIAMINE	5131-60-2
4-CHLORO-O-PHENYLENEDIAMINE	95-83-0
2-CHLORO-P-PHENYLENEDIAMINE SULFATE	61702-44-1
CHLOROPICRIN	76-06-2
CHLOROPRENE	126-99-8

CHEMICAL NAME	CAS NO.
CHLOROPROPHAM	101-21-3
CHLOROQUINE	54-05-7
CHLOROTHALONIL	1897-45-6
2-CHLOROTOLUENE	95-49-8
3-CHLORO-P-TOLUIDINE	95-74-9
5-CHLORO-O-TOLUIDINE	95-79-4
4-CHLORO-O-TOLUIDINE HYDROCHLORIDE	3165-93-3
CHLOROTRIANISENE	569-57-3
2-CHLORO-1,1,1-TRIFLUOROETHANE	75-88-7
CHLORPHENIRAMINE MALEATE	113-92-8
CHLORPROPAMIDE	94-20-2
CHOLESTEROL	57-88-5
CHROMIC ACETATE	1066-30-4
CHROMITE ORE	1308-31-2
CHROMIUM (ELEMENTAL)	7440-47-3
CHROMIUM TRIOXIDE	1333-82-0
CHRYSENE	218-01-9
CHRYSOTILE ASBESTOS	12001-29-5
C.I. ACID BLUE 1 (BLUE VRS)	129-17-9
C.I. ACID BLUE 9 (BRILLIANT BLUE FCF)	2650-18-2
C.I. ACID BLUE 43	2150-60-9
C.I. ACID GREEN 3 (GUINEA GREEN B)	4680-78-8
C.I. ACID GREEN 5 (LIGHT GREEN SF)	5141-20-8
C.I. ACID ORANGE 3	6373-74-6
C.I. ACID ORANGE 10 (ORANGE G)	1936-15-8
C.I. ACID ORANGE 12	1934-20-9
C.I. ACID ORANGE 20 (ORANGE 1)	523-44-4
C.I. ACID RED 1	3734-67-6
C.I. ACID RED 2 (METHYL RED)	493-52-7
C.I. ACID RED 14 (CARMOISINE)	3567-69-9
C.I. ACID RED 27 (AMARANTH)	915-67-3
C.I. ACID VIOLET 49 (BENZYL VIOLET 4B)	1694-09-3

CHEMICAL NAME	CAS NO.
C.I. ACID YELLOW 23 (TARTRAZINE)	1934-21-0
C.I. BASIC ORANGE 2 (CHRYSOIDINE)	532-82-1
C.I. BASIC RED 1 (RHODAMINE 6G)	989-38-8
C.I. BASIC RED 9 MONOMYDROCHLORIDE	569-61-9
C.I. BASIC VIOLET 10 (RHODAMINE B)	81-88-9
C.I. BASIC VIOLET 14 (MAGENTA)	632-99-5
C.I. BASIC YELLOW 2 (AURAMINE)	492-80-8
C.I. 37085 (P-CHLORO-O-TOLUIDINE)	95-69-2
C.I. 77305 (CHROMIUM SULFATE)	10101-53-8
C.I. DIRECT BLACK 38	1937-37-7
C.I. DIRECT BLUE 6	2602-46-2
C. I. DIRECT BLUE 14 (TRYPAN BLUE)	72-57-1
C. I. DIRECT BLUE 15	2429-74-5
C.I. DIRECT BLUE 53 (EVANS BLUE)	314-13-6
C.I. DIRECT BROWN 95	16071-86-6
C.I. DIRECT BROWN 191	10190-66-6
C.I. DIRECT YELLOW 28	8005-72-9
C.I. DISPERSE BLUE 1	2475-45-8
C.I. DISPERSE BLUE 7	3179-90-6
C.I. DISPERSE BLUE 79	3956-55-6
C.I. DISPERSE BLUE 79:1	3618-72-2
C.I. DISPERSE BLUE 291	UNKNOWN-10
C.I. DISPERSE ORANGE 11	82-28-0
C.I. DISPERSE RED 9	12226-87-8
C.I. DISPERSE RED 15	116-85-8
C.I. DISPERSE YELLOW NO. 3	2832-40-8
C.I. FOOD GREEN 3 (FAST GREEN FCF)	2353-45-9
C.I. FOOD RED 1 (PONCEAU SX)	4548-53-2
C.I. FOOD RED 5 (PONCEAU MX)	3761-53-3
C.I. FOOD RED 6 (PONCEAU 3R)	3564-09-8
C.I. FOOD RED 17 (FD&C RED 40)	25956-17-6
C.I. FOOD YELLOW 3 (SUNSET YELLOW FCF)	2783-94-0

CHEMICAL NAME	CAS NO.
C.I. MORDANT BLACK	2538-85-4
C.I. MORDANT ORANGE 1	2243-76-7
C.I. MORDANT RED 9	1836-22-2
C.I. MORDANT RED 11	72-48-0
C.I. NATURAL RED 8 (PURPURIN)	81-54-9
C.I. PIGMENT GREEN 17 (CHROMIC OXIDE)	1308-38-9
C.I. PIGMENT ORANGE 5	3468-63-1
C.I. PIGMENT RED 1	6410-10-2
C.I. PIGMENT RED 53, BARIUM SALT	5160-02-1
C.I. PIGMENT YELLOW 12 (DIARYLANILIDE YELLOW)	6358-85-6
C.I. PIGMENT YELLOW 31 (BARIUM CHROMATE)	10294-40-3
C.I. PIGMENT YELLOW 32 (STRONTIUM CHROMATE)	7789-06-2
C.I. PIGMENT YELLOW 33 (CALCIUM CHROMATE)	13765-19-0
C.I. PIGMENT YELLOW 34 (LEAD CHROMATE)	7758-97-6
C.I. PIGMENT YELLOW 36 (ZINC CHROMATE)	13530-65-9
C.I. 58050 (QUINIZARIN)	81-64-1
C.I. REACTIVE BLACK 5	12225-25-1
C.I. REACTIVE BLUE 19	2580-78-1
C.I. REACTIVE BLUE 36	12236-89-4
C.I. REACTIVE BLUE 50	12225-61-5
C.I. REACTIVE RED 22	19526-81-9
C.I. REACTIVE RED 78	12270-86-9
C.I. REACTIVE RED 83	61969-26-4
C.I. REACTIVE RED 84	61969-27-5
C.I. REACTIVE YELLOW 14	18976-74-4
C.I. REACTIVE YELLOW 31	12237-13-7
C.I. REACTIVE YELLOW 39	12226-61-8
C.I. SOLVENT BROWN 1 (SUDAN BROWN RR)	6416-57-5
C.I. SOLVENT ORANGE 1	2051-85-1
C.I. SOLVENT ORANGE 2 (OIL ORANGE SS)	2646-17-5
C.I. SOLVENT ORANGE 7 (SUDAN II)	3118-97-6
C.I. SOLVENT ORANGE 15 (ACRIDINE ORANGE)	494-38-2

CHEMICAL NAME	CAS NO.
C.I. SOLVENT RED 19 (SUDAN RED 7B)	6368-72-5
C.I. SOLVENT RED 23 (SUDAN III)	85-86-9
C.I. SOLVENT RED 24 (SCARLET RED)	85-83-6
C.I. SOLVENT RED 80 (CITRUS RED NO. 2)	6358-53-8
C.I. SOLVENT YELLOW 1 (P-AMINOAZOBENZENE)	60-09-3
C.I. SOLVENT YELLOW 2 (P-DIMETHYLAMINOAZOBENZENE)	60-11-7
C.I. SOLVENT YELLOW 3 (O-AMINOAZOTOLUENE)	97-56-3
C.I. SOLVENT YELLOW 5 (YELLOW AB)	85-84-7
C.I. SOLVENT YELLOW 6 (YELLOW OB)	131-79-3
C.I. SOLVENT YELLOW 7 (4-HYDROXYAZOBENZENE)	1689-82-3
C.I. SOLVENT YELLOW 14 (SUDAN 1)	842-07-9
C.I. VAT BLUE 4	81-77-6
C.I. VAT GREEN 9	128-60-9
C.I. VAT YELLOW 4	128-66-5
CINNAMYL ANTHRANILATE	87-29-6
CISPLATIN	15663-27-1
CITRININ	518-75-2
CLOFIBRATE	637-07-0
CLOMIPHENE CITRATE	50-41-9
CLONITRALID	1420-04-8
COAL-TARS AND DERIVED PRODUCTS	8007-45-2
COBALT-CHROMIUM ALLOY	11114-92-4
COBALT NAPHTHENATE	61789-51-3
COFFEE	UNKNOWN-15
COKE OVEN EMISSIONS (POLYCYCLIC ORGANIC MATTER)	CLASS-07
COLCHICINE	64-86-8
CONJUGATED ESTROGENS	CLASS-10
COPPER 8-HYDROXYQUINOLINE	10380-28-6
CORONENE	191-07-1
COUMAPHOS	56-72-4
COUMARIN	91-64-5
CREOSOTE	8001-58-9

CHEMICAL NAME	CAS NO.
M-CRESIDINE	102-50-1
P-CRESIDINE	120-71-8
M-CRESOL	108-39-4
MIXED CRESOLS	1319-77-3
O-CRESOL	95-48-7
P-CRESOL	106-44-5
CRISTOBALITE	14464-46-1
CROCIDOLITE ASBESTOS	12001-28-4
CROTONALDEHYDE	4170-30-3
CUMENE (1-METHYLETHYL)BENZENE)	98-82-8
CUPFERRON	135-20-6
CYANURIC ACID	108-80-5
CYCASIN	14901-08-7
CYCLOCHLOROTINE	12663-46-6
CYCLOHEXANE	110-82-7
CYCLOHEXANONE	108-94-1
CYCLOHEXYLAMINE (CHA)	108-91-8
M-(CYCLOHEXYLTHIO) PHTHALIMIDE	17796-82-6
CYCLOPENTA(CD)PYRENE	27208-37-3
CYCLOPHOSPHAMIDE	50-18-0
CYTEMBENA	21739-91-3
CYTOXAL ALCOHOL	4465-94-5
2,4-D	94-75-7
DACARBAZINE	4342-03-4
DAMINOZIDE	1596-84-5
DAPSONE	80-08-0
DAUNOMYCIN	20830-81-3
O,P'-DDD	53-19-0
P,P'-DDE	72-55-9
DDT	50-29-3
DDT (TECHNICAL GRADE-MIXTURE OF ISOMERS)	8017-34-3
DECABROMOBIPHENYL	13654-09-6

CHEMICAL NAME	CAS NO.
DECABROMODIPHENYL OXIDE	1163-19-5
DECALIN	91-17-8
DIACETONE ALCOHOL (DA)	123-42-2
DIACETYLAMINOAZOTOLUENE	83-63-6
N,N-DIACETYLBENZIDINE	613-35-4
DIALLATE	2303-16-4
DIALLYL DIGLYCOL CARBONATE (CR39 MONOMER)	142-22-1
DIALLYL PHTHALATE	131-17-9
2,4-DIAMINOANISULE SULFATE	39156-41-7
2,4-DIAMINOPHENOL:2HCL	137-09-7
2,6-DIAMINO-3-(PHENYLAZO)PYRIDINE	94-78-0
2,4-DIAMINOTOLUENE	95-80-7
2,5-DIAMINOTOLUENE	95-70-5
O-DIANISIDINE (3,3'-DIMETHOXYBENZIDINE)	119-90-4
DIAZEPAM	439-14-5
DIAZINON	333-41-5
DIAZOMETHANE	334-88-3
DIBENZ(A,H)ACRIDINE	226-36-8
DIBENZ(A,J)ACRIDINE	224-42-0
DIBENZ(A,C)ANTHRACENE	215-58-7
DIBENZ(A,H)ANTHRACENE (DBA)	53-70-3
DIBENZ(A,J)ANTHRACENE	224-41-9
7H-DIBENZO(C,G)CARBAZOLE	194-59-2
DIBENZO-P-DIOXIN	262-12-4
DIBENZO(A,E)FLUORANTHENE	5385-75-1
DIBENZO(H,RST)PENTAPHENE	194-47-2
DIBENZO(A,E)PYRENE	192-65-4
DIBENZO(A,H)PYRENE	189-64-0
DIBENZO(A,I)PYRENE	189-55-9
DIBENZO(A,L)PYRENE	191-30-0
1,2-DIBROMO-3-CHLOROPROPANE, DBCP	96-12-8
1,2-DIBROMO-4-(1,2-DIBROMOETHYL)CYCLOHEXANE	3322-93-8

CHEMICAL NAME	CAS NO.
DIBROMODULCITOL	10318-26-0
1,2-DIBROMOETHANE (EDB)	106-93-4
DIBROMOMANNITOL	488-41-5
2,3-DIBROMO-1-PROPANOL	96-13-9
2,3-DIBROMOPROPYL ACRYLATE	19660-16-3
2,3-DIBROMOPROPYL METHACRYLATE	3066-70-4
DI-N-BUTYL PHTHALATE (PHTHALIC ACID, DIBUTYL ESTER) (DBP)	84-74-2
DIBUTYLTIN DIACETATE	1067-33-0
1,2-DICHLORO-1,1-DIFLUOROETHANE	1649-08-7
DICHLOROACETYLENE	7572-29-4
1,2-DICHLOROBENZENE (ODCB)	95-50-1
1,4-DICHLOROBENZENE (PDCB)	106-46-7
3,3'-DICHLOROBENZIDINE (DCB)	91-94-1
3,4'-DICHLOROBENZOTRIFLUORIDE	328-84-7
3,3'-DICHLORO-4,4'-DIAMINODIPHENYL ETHER	28434-86-8
2,7-DICHLORODIBENZO-P-DIOXIN (DCDD)	33857-26-0
DICHLORODIFLUOROMETHANE (CFC-12)	75-71-8
1,1-DICHLOROETHANE	75-34-3
1,2-DICHLOROETHANE (EDC)	107-06-2
1,2-DICHLOROETHYLENE	540-59-0
DICHLOROMETHOTREXATE	528-74-5
2,4-DICHLORONITROBENZENE	611-06-3
3,4-DICHLORONITROBENZENE	99-54-7
2,4-DICHLOROPHENOL	120-83-2
2,6-DICHLORO-P-PHENYLENEDIAMINE	609-20-1
1,2-DICHLOROPROPANE	78-87-5
1,3-DICHLOROPROPENE	542-75-6
DICHLORVOS	62-73-7
DICOFOL	115-32-2
N,N'-DICYCLOHEXYLTHIOUREA	1212-29-9
DIELDRIN	60-57-1
DIENOESTROL	84-17-3

CHEMICAL NAME	CAS NO.
DIEPOXYBUTANE	1464-53-5
1,2,3,4-DIEPOXYBUTANE, DL	564-00-1
DI(2,3-EPOXYPROPYL) ETHER (DGE)	2238-07-5
DIESEL FUELS	68334-30-5
2,2'-DIETHOXYACETOPHENONE (DEAP)	6175-45-7
DIETHYLCARBAMOYL CHLORIDE (DECC)	88-10-8
DIETHYLENE GLYCOL BUTYL ETHER (DGBE)	112-34-5
DIETHYLENETRIAMINE (DETA)	111-40-0
DI(2-ETHYLHEXYL)ADIPATE	103-23-1
DI(2-ETHYLHEXYL)PHTHALATE (DEHP)	117-81-7
1,2-DIETHYLHYDRAZINE (SDEH)	1615-80-1
DIETHYL PHTHALATE	84-66-2
DIETHYLSTILBESTROL	56-53-1
DIETHYLSTILBESTROL DIPROPRIONATE	130-80-3
DIETHYL SULFATE	64-67-5
N,N'-DIETHYLTHIOUREA	105-55-5
DIETHYL TOLUENE DIAMINE	68479-98-1
DIGLYCIDYL RESORCINOL ETHER (DGRE)	101-90-6
DIHYDROSAFROLE	94-58-6
1,5-DIHYDROXYANTHRAQUINONE	117-12-4
1,8-DIHYDROXYANTHRAQUINONE (CHRYSAZIN)	117-10-2
2,2-DIHYDROXY-4-METHOXYBENZOPHENONE	131-53-3
DIHYDROXYMETHYLFURATRIZINE	794-93-4
DIISOPROPYL SULFATE (DIPS)	2973-10-6
DIISOPROPYLAMINE	108-18-9
DIMETHISTERONE	79-64-1
DIMETHOATE	60-51-5
DIMETHOXANE	828-00-2
2,4-DIMETHOXYANILINE HYDROCHLORIDE	54150-69-5
3,3'-DIMETHOXYBENZIDINE DIHYDROCHLORIDE	20325-40-0
3,3'-DIMETHOXYBENZIDINE-4,4'-DIISOCYANATE	91-93-0
DIMETHOXYETHYL PHTHALATE	117-82-8

CHEMICAL NAME	CAS NO.
DIMETHYL ADIPATE (DMA)	627-93-0
DIMETHYL AMINE (DMA)	124-40-3
N,N-DIMETHYLANILINE	121-69-7
DIMETHYLARSINIC ACID	75-60-5
7,12-DIMETHYL-BENZO(A)ANTHRACENE	57-97-6
DIMETHYLCARBAMOYL CHLORIDE (DMCC)	79-44-7
DIMETHYL GLUTARATE (DMG)	1119-40-0
1,1-DIMETHYLHYDRAZINE (UDMH)	57-14-7
1,2-DIMETHYLHYDRAZINE (SDMH)	540-73-8
1,2-DIMETHYLHYDRAZINE DIHYDROCHLORIDE	306-37-6
DIMETHYL HYDROGEN PHOSPHITE	868-85-9
DIMETHYL N-METHYLOLPHOSPHONOPROPIONAMIDE (DMMP)	20120-33-6
DIMETHYL METHYLPHOSPHONATE	756-79-6
2,6-DIMETHYL MORPHOLINE	141-91-3
N,N-DIMETHYL-P-NITROSOANILINE	138-89-6
1,4-DIMETHYLPHENANTHRENE	22349-59-3
DIMETHYL PHTHALATE	131-11-3
DIMETHYL SULFATE	77-78-1
DIMETHYL SULFOXIDE	67-68-5
DIMETHYL TEREPHTHALATE	120-61-6
N,N-DIMETHYL-P-(M-TOLYLAZO)-ANILINE	55-80-1
DIMETHYLVINYL CHLORIDE	513-37-1
2,4-DINITROANILINE	97-02-9
2,4-DINITRO-1-CHLOROBENZENE	97-00-7
4,6-DINITRO-O-CRESOL	534-52-1
2,4-DINITROPHENOL	51-28-5
1,8-DINITROPYRENE	42397-65-9
DINITROSOPENTAMETHYLENETETRAMINE (DNPMT)	101-25-7
1,4-DINITROSOPIPERAZINE	140-79-4
2,4-DINITROTOLUENE (DNT)	121-14-2
2,6-DINITROTOLUENE	606-20-2
DINONYL PHTHALATE ESTER (1,2-BENZENEDICARBOXYLIC ACID)	84-76-4

CHEMICAL NAME	CAS NO.
DI-N-OCTYL PHTHALATE (DOP)	117-84-0
1,3-DIOXANE	505-22-6
1,4-DIOXANE	123-91-1
DIOXATHION	78-34-2
1,3-DIOXOLANE	646-06-0
DIPHENHYDRAMINE HYDROCHLORIDE	147-24-0
2,4'-DIPHENYLDIAMINE	492-17-1
1,3-DIPENYLGUANIDINE (DPC)	102-06-7
5,5'-DIPHENYLHYDANTOIN (PHENYTOIN)	57-41-0
1,2-DIPHENYLHYDRAZINE (HYDRAZOBENZENE)	122-66-7
1,1-DIPHENYLHYDRAZINE HYDROCHLORIDE	530-47-2
4,4'-DIPHENYLMETHANE DIISOCYANATE (MDI)	101-68-8
DIPROPYLENEGLYCOL DIACRYLATE	57472-68-1
2,6-DI-TERT-BUTYLPHENOL	128-39-2
2,5-DITHIOBIUREA	142-46-1
DITHRANOL	480-22-8
DL-AMPHETAMINE SULFATE	60-13-9
DODECYLPHENOL	27193-86-8
EMETINE	483-18-1
ENDOSULFAN	115-29-7
ENDRIN	72-20-8
EOSIN	15086-94-9
EOSIN DISODIUM SALT	17372-87-1
EPHEDRINE SULFATE	134-72-5
EPIBROMOHYDRIN (1,2-EPOXY-3-BROMOPROPANE)	3132-64-7
EPICHLOROHYDRIN	106-89-8
1,2-EPOXYBUTANE	106-88-7
BETA-(3,4-EPOXYCYCLOHEXYL)ETHYLTRIMETHOXY SILANE	3388-04-3
3,4-EPOXY-6-METHYLCYCLOHEXYLMETHYL-3,4-EPOXY-6-METHYL	141-37-7
2,3-EPOXYPROPYLTRIMETHYLAMMONIUMCHLORIDE	3033-77-0
EPOXY RESIN	25068-38-6
CIS-9,10-EPOXYSTEARIC ACID	2443-39-2

CHEMICAL NAME	CAS NO.
1,2-EPOXY-3,3,3-TRICHLOROPROPANE	3083-23-6
ERIONITE	66733-21-9
ERYTHROMYCIN STEARATE	643-22-1
ESTRADIOL - 17B	50-28-2
ESTRADIOL MUSTARD	22966-79-6
ESTRONE	53-16-7
1,2-ETHANEDITHIOL	26914-40-9
1,2-ETHANEDIYL-BIS (OXY-2, 1-ETHANEDIYL) ESTER	36443-68-2
ETHANOLAMINE	141-43-5
ETHINYLESTRADIOL	57-63-6
ETHIONAMIDE	536-33-4
ETHIONINE	13073-35-3
DL-ETHIONINE	67-21-0
2-ETHOXYETHANOL ACETATE	111-15-9
P-ETHOXYPHENYLUREA	150-69-6
ETHYL ACRYLATE (ACRYLIC ACID, ETHYL ESTER)	140-88-5
ETHYL AURAMINE NITRATE	43130-12-7
ETHYLBENZENE	100-41-4
ETHYL CARBAMATE (URETHANE)	51-79-6
P,P'-ETHYL-DDD	72-56-0
ETHYLENE	74-85-1
ETHYLENE BIS DITHIOCARBAMATE	111-54-6
ETHYLENE BIS(OXYETHYLENE) DIACETATE	111-21-7
ETHYLENEBIS TRIS (2-CYANOETHYL) PHOSPHONIUM BROMIDE	10310-38-0
ETHYLENEDIAMINE	107-15-3
ETHYLENEDIAMINETETRA(METHYL PHOSPHONIC ACID)	1429-50-1
ETHYLENEDIAMINETETRA(METHYL PHOSPHONIC ACID) HEXASODI	15142-96-8
ETHYLENEDIAMINETETRA(METHYL PHOSPHONIC ACID) TETRAPOT	68188-96-5
ETHYLENE GLYCOL MONOMETHYL ETHER (2-ETHOXYETHANOL) (2	110-80-5
ETHYLENE OXIDE (ETO)	75-21-8
ETHYLENE SULFIDE	420-12-2
ETHYLENE THIOUREA (ETU)	96-45-7

CHEMICAL NAME	CAS NO.
2-ETHYLHEXANOIC ACID	149-57-5
2-ETHYLHEXANOL	104-76-7
ETHYL METHACRYLATE	97-63-2
ETHYL METHANESULFONATE (EMS)	62-50-0
5-ETHYL-5-METHYL-1,3-BIS-(OXIRANYLMETHYL)-2,4-IMIDAZOLIDIN	15336-82-0
ETHYLPHOSPHONOTHIOIC DICHLORIDE (EPTD)	993-43-1
ETHYL SELENAC	12367-47-4
ETHYL TELLURAC	20941-65-5
O-ETHYL TOLUENE	611-14-3
ETHYLTOLUENES	25550-14-5
ETHYNODIOL DIACETATE	297-76-7
EUGENOL	97-53-0
FENTHION	55-38-9
FERBAM	14484-64-1
FERRIC OXIDE	1309-37-1
FERROCENE	102-54-5
FLAGYL	443-48-1
FLUOMETURON	2164-17-2
FLUORANTHENE	206-44-0
FLUORENE	86-73-7
FLUOROALKANES	CLASS-21
2-FLUOROBENZOYL CHLORIDE	393-52-2
1-FLUORO-2,4-DINITROBENZENE	70-34-8
FLUOROSILICON FLUIDS AND GREASES	CLASS-27
5-FLUOROURACIL	51-21-8
FORMALDEHYDE	50-00-0
FORMAMIDE	75-12-7
FORMULATED FENAMINOSULF	140-56-7
2-(2-FORMYLHYDRAZINO)-4-(5-NITRO-2-FURYL)THIAZOLE	3570-75-0
NO. 2 FUEL OIL	68476-30-2
FUEL OILS (HEATING OILS)	CLASS-33
FURAZLIDONE	67-45-8

CHEMICAL NAME	CAS NO.
FURFURAL	98-01-1
FURFURYL ACETATE	623-17-6
FURNACE CARBON BLACK	7440-44-0
FURNITURE AND CABINET-MAKING INDUSTRY	UNKNOWN-25
FUROSEMIDE	54-31-9
FURSARENON-X	23255-69-8
GASOLINE	8006-61-9
GLU-P-1 (2-AMINO-6-METHYLPYRIDO)1,2-A:3',2'-D(IMIDAZOLE)	67730-11-4
GLU-P-2(2-AMINOPYRIDO)1,2-A:3'2-D(IMADAZOLE)	67730-10-3
GLUTERALDEHYDE	111-30-8
GLYCIDALDEHYDE	765-34-4
GLYCIDOL	556-52-5
GLYCIDOL AND DERIVATIVES (ETHERS, ESTERS)	CLASS-12
GLYCIDYL ACRYLATE (GA)	106-90-1
GLYCIDYL METHACRYLATE (GMA)	106-91-2
GLYCIDYL OLEATE	5431-33-4
GLYCIDYL STEARATE	7460-84-6
GOSSYPOL	303-45-7
GRISEOFULVIN	126-07-8
GUANAZOLE	1455-77-2
GUAR GUM	9000-30-0
GUM ARABIC	9000-01-5
GYROMITRIN (ACETALDEHYDE FORMYLMETHYLHYDRAZONE)	16568-02-8
HAIR SERVICE OCCUPATION AND USERS OF HAIR DYES	UNKNOWN-21
HALAZONE	80-13-7
HAMAMELIS WATER	68916-39-2
HC BLUE NO. 1	2784-94-3
HC BLUE NO. 2	33229-34-4
HC RED NO. 3	2871-01-4
HC YELLOW 4	52551-67-4
HEMATITE MINING	UNKNOWN-05
HEPTACHLOR	76-44-8

CHEMICAL NAME	CAS NO.
HEPTACHLOR EPOXIDE	1024-57-3
HEXABROMOBIPHENYL (PURE)	59080-40-9
HEXABROMOBIPHENYL (TECH)	36355-01-8
HEXACHLOROBENZENE	118-74-1
2,2',3,3',5,5'-HEXACHLOROBIPHENYL	35694-04-3
HEXACHLORO-1,3-BUTADIENE	87-68-3
ALPHA-HEXACHLOROCYCLOHEXANE	319-84-6
BETA-HEXACHLOROCYCLOHEXANE	319-85-7
1,2,3,4,5,6-HEXACHLOROCYCLOHEXANE (HCH)	608-73-1
HEXACHLOROCYCLOPENTADIENE	77-47-4
1,2,3,6,7,8-HEXACHLORODIBENZO-P-DIOXIN (HCDD)	57653-85-7
1,2,3,7,8,9-HEXACHLORODIBENZO-P-DIOXIN (HCDD)	19408-74-3
HEXACHLOROETHANE	67-72-1
1,2,3,4,7,7-HEXACHLORONORBORNADIENE	3389-71-7
HEXACHLOROPHENE	70-30-4
HEXAFLUOROPROPENE	116-15-4
HEXAFLURORPROPYLENE OXIDE	428-59-1
HEXAMETHYLPHOSPHORAMIDE (HMPA)	680-31-9
HEXANE	110-54-3
4-HEXYLRESORCINOL	136-77-6
HYCANTHONE MESYLATE	23255-93-8
HYDRALAZINE	86-54-4
HYDRALAZINE HYDROCHLORIDE	304-20-1
HYDRAZINE	302-01-2
HYDRAZINE MONOHYDRATE	7803-57-8
HYDRAZINE SULFATE (1:1)	10034-93-2
2-HYDRAZINOETHANOL	109-84-2
HYDROCHLOROTHIAZIDE	58-93-5
HYDRODESULFURIZED VACUUM GAS OIL	64742-86-5
HYDROGEN FLUORIDE	7664-39-3
HYDROGEN PEROXIDE	7722-84-1
HYDROQUINONE	123-31-9

CHEMICAL NAME	CAS NO.
1-(2-HYDROXYETHYL)-2-PYRROLIDONE	3445-11-2
HYDROXYLAMINE	7803-49-8
HYDROXYLAMINE HYDROCHLORIDE	5470-11-1
HYDROXYLAMINE SULFATE (1:1) (SALT)	10046-00-1
HYDROXYLAMINE SULFATE (2:1) SALT	10039-54-0
17-ALPHA-HYDROXYPROGESTERONE CAPROATE	630-56-8
8-HYDROXYQUINOLINE	148-24-3
HYDROXYSENKIRKINE	26782-43-4
HYDROXYUREA	127-07-1
ICRF - 159	21416-87-5
IMIDAZOLE MUSTARD	5034-77-5
INDENO(1,2,3-CD)PYRENE (IP)	193-39-5
INTRACRON BLUE 3G	75432-59-6
INTRACRON ORANGE G	UNKNOWN-20
IODINATED GLYCEROL	5634-39-9
IODOFORM	75-47-8
IPD	3458-22-8
IQ(2-AMINO-3-METHYLIMIDAZO)(4,5-F QUINOLINE)	76180-96-6
IRON DEXTRAN	9004-66-4
IRON DEXTRIN	8050-93-9
ISATIDINE	15503-86-3
ISOAMYL ALCOHOL	71-41-0
ISOBUTYL ALCOHOL	78-83-1
ISOBUTYL NITRITE	542-56-3
ISONICOTINIC ACID HYDRAZIDE (INH)	54-85-3
ISOPHORONE (3,5,5-TRIMETHYL-2-CYCLOHEXEN-1-ONE)	78-59-1
ISOPHOSPHAMIDE	3778-73-2
ISOPRENE	78-79-5
ISOPROPYL ALCOHOL (ISOPROPANOL)	67-63-0
ISOPROPYL GLYCIDYL ETHER	4016-14-2
ISOPROTERENOL HYDROCHLORIDE	51-30-9
ISOSAFROLE	120-58-1

CHEMICAL NAME	CAS NO.
JACOBINE	471-14-7
JET FUEL	CLASS-32
KAEMPFEROL	520-18-3
KEROSENE	8008-20-6
L-EPINEPHRINE HYDROCHLORIDE	55-31-2
LASIOCARPINE	303-34-4
LAUROYL PEROXIDE	105-74-8
LEAD	7439-92-1
LEAD ACETATE	301-04-2
LEAD ARSENATE	7784-40-9
LEAD CARBONATE	598-63-0
LEAD CHLORIDE	7758-95-4
LEAD CHROMATE (VI) OXIDE	1344-38-3
LEAD DIMETHYLDITHIOCARBAMATE	19010-66-3
LEAD NAPHTHENATE	50825-29-1
LEAD NITRATE	10099-74-8
LEAD OXIDE	1317-36-8
LEAD PHOSPHATE	7446-27-7
LEAD SUBACETATE	1335-32-6
LEATHER GOODS MANUFACTURE	UNKNOWN-16
LEATHER TANNING AND PROCESSING INDUSTRIES	UNKNOWN-27
D-LIMONENE	5989-27-5
LINDANE (GAMMA-HEXACHLOROCYCLOHEXANE)	58-89-9
LITHIUM	7439-93-2
LITHOCHOLIC ACID	434-13-9
LOCUST BEAN GUM	9000-40-2
LOMUSTINE1-(2-CHLOROETHYL)-3-CYCLOHEXYL-1-NITROSOUREA	13010-47-4
LUMBER AND SAWMILL INDUSTRIES	UNKNOWN-23
LUTEOSKYRIN	21884-44-6
LYNOESTRENOL	52-76-6
MALAOXON	1634-78-2
MALATHION	121-75-5

CHEMICAL NAME	CAS NO.
MALEIC ANHYDRIDE	108-31-6
MALEIC HYDRAZIDE	123-33-1
MALONALDEHYDE	542-78-9
MALONONITRILE	4341-85-9
MANEB	12427-38-2
D-MANNITOL	69-65-8
MANNOMUSTINE (DIHYDROCHLORIDE)	551-74-6
MANUFACTURE OF AURAMINE (C. I. BASIC YELLOW 2)	UNKNOWN-03
MANUFACTURE OF CHROMATE	UNKNOWN-01
MATE	CLASS-38
MCPA	94-74-6
MEA-A-C(2-AMINO-3-METHYL-9H-PYRIDO(2,3-B)INDOLE	26148-68-5
MEDROXYPROGESTERONE ACETATE	71-58-9
MEGESTROL ACETATE	595-33-5
MEIQ (2-AMINO-3,4-DIMETHYLIMIDAZO(4,5-F)QUINOLINE (MEIQ)	77094-11-2
MEIQX (2-AMINO-3,4-DIMETHYLIMIDAZO(4,5-F)QUINOXALINE)	77500-04-0
MELAMINE	108-78-1
MELPHALAN	148-82-3
DL-MENTHOL	89-78-1
2-MERCAPTOBENZOTHIAZOLE	149-30-4
6-MERCAPTOPURINE	50-44-2
6-MERCAPTOPURINE-RIBOSIDE	574-25-4
MERCURY	7439-97-6
MERPHALAN	531-76-0
MESITYLENE	108-67-8
MESITYL OXIDE	141-97-7
MESTRANOL	72-33-3
METAL-WORKING FLUIDS	CLASS-06
METHALLY-2-NITROPHENYL ETHER	13414-54-5
METHANEARSONIC ACID.NA	2163-80-6
METHANEARSONIC ACID.2NA	144-21-8
METHANOL	67-56-1

CHEMICAL NAME	CAS NO.
METHAPYRILENE	91-80-5
METHOTREXATE	59-05-2
METHOXYCHLOR	72-43-5
2-METHOXYETHANOL (2-ME)	109-86-4
2-METHOXYETHANOL ACETATE	110-49-6
2-METHOXYETHYL PALMITATE	111-07-9
METHOXYPSORALEN	298-81-7
5-METHOXYPSORALEN	484-20-8
METHYCYCLOPENTANE (MCP)	96-37-7
METHYL ACRYLATE	96-33-3
4-METHYL-P-AMINOPHENOL SULFATE	55-55-0
METHYLAZOXYMETHANOL ACETATE	592-62-1
A-METHYLBENZYL ALCOHOL	98-85-1
METHYL TERT-BUTYL ETHER	1634-04-4
METHYL-N-BUTYL KETONE (MBK)	591-78-6
METHYL CARBAMATE	598-55-0
3-METHYLCHOLANTHRENE	56-49-5
METHYLCHRYSENES	CLASS-13
METHYL-2-CYANOACRYLATE	137-05-3
ALPHA-METHYLDOPA SESQUIHYDRATE	41372-08-1
4,4'-METHYLENEBIS(2-CHLOROANILINE) (MBOCA)	101-14-4
4,4'-METHYLENEBIS(N,N-DIMETHYL)ANILINE	101-61-1
4,4'-METHYLENEBIS(2-METHYLANILINE)	1807-55-2
METHYLENE CHLORIDE	75-09-2
4,4'-METHYLENEDIANILINE (MDA)	101-77-9
METHYL ETHYL KETONE	78-93-3
METHYL ETHYL KETOXIME	96-29-7
2-METHYLFLUORANTHENE	33543-31-6
3-METHYLFLUORANTHENE	1706-10-0
METHYLHYDRAZINE	60-34-4
METHYLHYDRAZINE SULFATE (1:1)	302-15-8
METHYL IODIDE	74-88-4

CHEMICAL NAME	CAS NO.
METHYL ISOBUTYL KETONE	108-10-1
METHYL METHACRYLATE	80-62-6
METHYL METHANESULFONATE (MMS)	66-27-3
2-METHYL-1-NITROANTHRAQUINONE (MNA)	129-15-7
N-METHYL-N'-NITRO-N-NITROSOGUANIDINE (MNNG)	70-25-7
4-(METHYLNITROSAMINO)-4-(3-PYRIDYL) BUTANAL (NNA)	64091-90-3
4-(METHYLNITROSAMINO)-1-(3-PYRIDYL)-1-BUTANONE (NNK)	64091-91-4
N-METHYL-N-NITROSOANILINE	614-00-6
METHYL PARATHION	298-00-0
1-METHYLPHENANTHRENE	832-69-9
METHYL SELENAC	144-34-3
METHYLTHIOURACIL	56-04-2
MICA DUST	CLASS-26
MICHLER'S KETONE	90-94-8
MIREX	2385-85-5
MITOMYCIN C	50-07-7
MONOCHLOROACETIC ACID	79-11-8
MONOCROTALINE	315-22-0
MONOLINURON	1746-81-2
MONURON	150-68-5
5-(MORPHOLINOMETHYL)-3-((5-NITROFURFURYLIDENE)-AMINOL)-2-	139-91-3
MORPHOLINOPHOSPHORAMIDATE	597-25-1
MUSTARD GAS	505-60-2
N-METHYLOLACRYLAMIDE	924-42-5
N-METHYLPYRROLIDONE (NMP)	872-50-4
NAFENOPIN	3771-19-5
NALDIXIC	389-08-2
NAPHTHALENE	91-20-3
1,5-NAPHTHALENEDIAMINE	2243-62-1
1,5-NAPHTHALENE DIISOCYANATE	3173-72-6
NAPHTHENIC DISTILLATE	64741-52-2
1,4-NAPHTHOQUINONE	130-15-4

CHEMICAL NAME	CAS NO.
1-NAPHTHYLAMINE	134-32-7
2-NAPHTHYLAMINE	91-59-8
N-(-1-NAPHTHYL)ETHYLENEDIAMINE DIHYDROCHLORIDE	1465-25-4
1-NAPHTHYLTHIOUREA	86-88-4
NICKEL (INCL CERTAIN NI CPDS)	7440-02-0
NICKEL-BIS(ETHYL(3,5-DI-T-BUTYL-4-HYDROXYBENZYL)PHOSP	30947-30-9
NICKEL CARBONATE	3333-67-3
NICKEL CARBONYL	13463-39-3
NICKELOCENE	1271-28-9
NICKEL OXIDE	1313-99-1
NICKEL REFINING	UNKNOWN-06
NICKEL SUBSULFIDE	12035-72-2
NIRIDAZOLE	61-57-4
NITHIAZIDE	139-94-6
NITRILOTRIACETIC ACID (NTA)	139-13-9
NITRILOTRIACETIC ACID, TRISODIUM SALT, MONOHYDRATE	18662-53-8
5-NITROACENAPHTHENE	602-87-9
5-NITRO-P-ACETOPHENETIDE	1777-84-0
P-NITROANILINE	100-01-6
5-NITRO-O-ANISIDINE	99-59-2
O-NITROANISOLE	91-23-6
9-NITROANTHRACENE	602-60-8
4-NITROANTHRANILIC ACID	619-17-0
NITROBENZENE	98-95-3
6-NITROBENZIMIDAZOLE	94-52-0
6-NITROBENZO(A)PYRENE	63041-90-7
P-NITROBENZOYL CHLORIDE	122-04-3
P-NITROBENZYL CHLORIDE	100-14-1
4-NITROBIPHENYL	92-93-3
7-NITRO-2,2-DIMETHYL-2,3-DIHYDROBENZOFURAN	13414-55-6
NITROFEN	1836-75-5
3-NITROFLUORANTHENE	892-21-7

CHEMICAL NAME	CAS NO.
2-NITROFLUORENE	607-57-8
NITROFURANTOIN	67-20-9
NITROFURAZONE	59-87-0
1-((5-NITROFURFURYLIDENE)AMINO)-2-IMIDAZOLIDINONE	555-84-0
N-(4-(5-NITRO-2-FURYL)-2-THIAZOLYL) ACETAMIDE	531-82-8
NITROGEN DIOXIDE	10102-44-0
NITROGEN MUSTARD	51-75-2
NITROGEN MUSTARD HYDROCHLORIDE	55-86-7
NITROGEN MUSTARD N-OXIDE	126-85-2
NITROGEN MUSTARD N-OXIDE HYDROCHLORIDE	302-70-5
1-NITRONAPHTHALENE	86-57-7
2-NITRONAPHTHYLENE	581-89-5
P-NITROPHENOL	100-02-7
2-NITRO-P-PHENYLENEDIAMINE	5307-14-2
4-NITRO-O-PHENYLENEDIAMINE	99-56-9
2-NITROPROPANE	79-46-9
3-NITROPROPIONIC ACID	504-88-1
1-NITROPYRENE	5522-43-0
8-NITROQUINOLINE	607-35-2
4-NITROQUINOLINE-1-OXIDE	56-57-5
N-NITROSAMINES DERIVED FROM ARECA-NUT ALKALOIDS	CLASS-18
N'-NITROSOANABASINE (NAB)	37620-20-5
N'-NITROSOANATABINE	71267-22-6
N-NITROSODI-N-BUTYLAMINE (DBNA)	924-16-3
N-NITROSODIETHANOLAMINE	1116-54-7
N-NITROSODIETHYLAMINE (NDEA)	55-18-5
N-NITROSODIMETHYLAMINE (NDMA)	62-75-9
N-NITROSODIPHENYLAMINE	86-30-6
P-NITROSODIPHENYLAMINE	156-10-5
N-NITROSODI-N-PROPYLAMINE	621-64-7
N-NITROSOFOLIC ACID	29291-35-8
N-NITROSOHYDROXYPROLINE (NHPRO)	30310-80-6

CHEMICAL NAME	CAS NO.
N-NITROSOMETHYLETHYLAMINE	10595-95-6
N-NITROSO-N-METHYLUREA (NMU)	684-93-5
N-NITROSO-N-METHYLURETHANE (NMUT)	615-53-2
N-NITROSOMETHYLVINYLAMINE	4549-40-0
N-NITROSOMORPHOLINE	59-89-2
N-NITROSO-N-ETHYLUREA	759-73-9
N-NITROSONORNICOTINE	16543-55-8
N-NITROSOPIPERIDINE	100-75-4
N-NITROSOPROLINE (NPRO)	7519-36-0
N-NITROSOPYRROLIDINE (NPYR)	930-55-2
N-NITROSOSARCOSINE	13256-22-9
5-NITRO-O-TOLUIDINE	99-55-8
NITROUS OXIDE (NITROGEN OXIDE)	10024-97-2
NITROVIN	804-36-4
NORETHISTERONE	68-22-4
NORETHISTERONE ACETATE	51-98-9
NORETHYNODREL	68-23-5
NORGESTREL	797-63-7
NYLON 6	25038-54-4
O-CHLOROBENZALMALONONITRILE (CS2)	2698-41-1
OCHRATOXIN A	303-47-9
OCTABROMOBIPHENYL	27858-07-7
OCTAMETHYLCYCLOTETRASILOXANE	556-67-2
1-OCTANOL	111-87-5
OESTRIOL	50-27-1
OLEYLAMINE (9-OCTADECENYLAMINE, ODA)	112-90-3
ORANGE HISALONIC	UNKNOWN-22
ORGANOTIN COMPOUNDS	CLASS-16
7-OXABICYCLO(4.1.0)HEPTANE-3-CARBOXYLIC ACID	2386-87-0
OXAZEPAM	604-75-1
N-(4-(OXIRANYLMETHOXY)PHENYL)-N-(OXIRANYLMETHYL)-OXI	5026-74-4
4,4'-OXYDIANILINE	101-80-4

CHEMICAL NAME	CAS NO.
N-OXYDIETHYLENE-2-BENZOTHIAZYL SULFENAMIDE	102-77-2
N-OXYDIETHYLENE THIOCARBAMYL-N-OXYDIETHYLENE	13752-51-7
OXYMETHOLONE	434-07-1
OXYPHENBUTAZONE	129-20-4
OXYTETRACYCLINE HYDROCHLORIDE	2058-46-0
PARA-CHLOROANILINE HYDROCHLORIDE	20265-96-7
PARASORBIC ACID	10048-32-5
PARATHION	56-38-2
PATULIN	149-29-1
PENICILLIC ACID	90-65-3
PENICILLIN VK	132-98-9
PENTABROMOETHYLBENZENE (PEB)	85-22-3
PENTABROMOPHENYL ALLYL ETHER	3555-11-1
PENTACHLOROBENZENE	608-93-5
PENTACHLOROETHANE	76-01-7
PENTACHLORONITROBENZENE (PCNB)	82-68-8
PENTACHLOROPHENOL	87-86-5
PENTAERYTHRITOL MONOOLEATE TRIACRYLATE	UNKNOWN-09
PENTAERYTHRITOL TETRANNITRATE (PETN,NF)	78-11-5
PENTAERYTHRITOL-TRIS-(B-(N-AZIRIDINYL) PROPIONATE)	57116-45-7
PENTAERYTHRITOL TRIACRYLATE (PETA)	3524-68-3
N-PENTANOIC ACID (N-VALERIC ACID)	109-52-4
PERCHLOROETHYLENE	127-18-4
PERFLUOROALKYL RESINS	68298-78-2
PERLITE	12427-27-9
PEROXYACETIC ACID	79-21-0
PERYLENE	198-55-0
PETASITENINE	60102-37-6
PETROLEUM DISTILLATES	68477-31-6
PETROLEUM MINERAL AND CRUDE OILS	8002-05-9
PETROLEUM NAPHTHA, HEAVY CATALYTIC REFORMED	64741-68-0
PETROLEUM NAPHTHA SOLVENT	64742-95-6

CHEMICAL NAME	CAS NO.
PETROLEUM REFINING OCCUPATIONAL EXPOSURES	CLASS-23
PHENACETIN	62-44-2
PHENANTHRENE	85-01-8
PHENAZOPYRIDINE HYDROCHLORIDE	136-40-3
PHENELZINE SULFATE	156-51-4
PHENESTERIN	3546-10-9
PHENFORMIN HYDROCHLORIDE	114-86-3
PHENICARBAZIDE	103-03-7
PHENOBARBITAL SODIUM	57-30-7
PHENOBARBITONE	50-06-6
PHENOL	108-95-2
PHENOXYBENZAMINE	59-96-1
PHENOXYBENZAMINE HCL	63-92-3
2-PHENOXYETHANOL	122-99-6
4'-PHENYLACETANILIDE	4075-79-0
PHENYLBUTAZONE	50-33-9
4-PHENYLCYCLOHEXENE (4-PCH)	4994-16-5
PHENYLENEDIAMINES CATEGORY	CLASS-22
M-PHENYLENEDIAMINE	108-45-2
M-PHENYLENEDIAMINE HYDROCHLORIDE	541-69-5
M-PHENYLENEDIAMINE.H2SO4	54-17-8
O-PHENYLENEDIAMINE	95-54-5
O-PHENYLENEDIAMINE DIHYDROCHLORIDE	615-28-1
P-PHENYLENEDIAMINE	106-50-3
P-PHENYLENEDIAMINE DIHYDROCHLORIDE	624-18-0
P-PHENYLENEDIAMINE.H2SO4 SALT	1624-57-75
PHENYLEPHRINE HYDROCHLORIDE	61-76-7
2-PHENYLETHANOL (2-PE)	60-12-8
PHENYL GLYCIDYL ETHER	122-60-1
PHENYLHYDRAZINE MONOHYDROCHLORIDE	59-88-1
N-PHENYLHYDROXYLAMINE	100-65-2
1-PHENYL-3-METHYL-5-PYRAZOLONE	89-25-8

CHEMICAL NAME	CAS NO.
N-PHENYL-2-NAPHTHYLAMINE (PBNA)	135-88-6
O-PHENYLPHENOL	90-43-7
N-PHENYL-P-PHENYLENEDIAMINE	101-54-2
1-PHENYL-3-PYRAZOLIDINONE (PHENIDONE)	92-43-3
1-PHENYL-2-THIOUREA	103-85-5
PHENYTOIN SODIUM	630-93-3
PHOSPHAMIDON	13171-21-6
PHOSPHORUS	7723-14-0
PHOSPHORUS ACID, ISODECYL DIPHENYL ESTER	26544-23-0
PHOTODIELDRIN	13366-73-9
PHTHALAMIDE	88-96-0
PHTHALATE ESTERS	CLASS-24
PHTHALIC ANHYDRIDE	85-44-9
PICLORAM	1918-02-1
PICRIC ACID	88-89-1
PIPERONYL BUTOXIDE	51-03-6
PIPERONYL SULFOXIDE	120-62-7
PIVALOLACTONE	1955-45-9
POLYACRYLIC ACID	9003-01-4
POLYBROMINATED BIPHENYL MIXTURE (PBB)	67774-32-7
POLYCHLORINATED BIPHENYLS (PCB'S)	1336-36-3
POLYCHLORINATED DIBENZOFURAN (PCDB)	CLASS-19
POLYCHLOROPRENE	9010-98-4
POLYETHYLENE	9002-88-4
POLYETHYLENE GLYCOL SORBITAN MONOLAURATE	9005-64-5
POLYHALOGENATED DIBENZO-P-DIOXINS (HDD)	UNKNOWN-32
POLYHALOGENATED DIBENZOFURAN (HDF) (2,3,7,8-TCDF)	UNKNOWN-29
POLYMETHYLENE POLYPHENYL ISOCYANATE	9016-87-9
POLYMETHYL METHACRYLATE	9011-14-7
POLYPROPYLENE	9003-07-0
POLYSTYRENE	9003-53-6
POLYTETRAFLUOROETHYLENE (PTFE)	9002-84-0

CHEMICAL NAME	CAS NO.
POLYURETHANE	9009-54-5
POLYVINYL ACETATE	9003-20-7
POLYVINYL ALCOHOL	9002-89-5
POLYVINYL CHLORIDE	9002-86-2
POLYVINYL PYRROLIDONE	9003-39-8
POTASSIUM ARSENATE	7784-41-0
POTASSIUM ARSENITE	13464-35-2
POTASSIUM AZIDE	20762-60-1
POTASSIUM BIS(S-HYDROXYETHYL)-DITHIOCARBAMATE	23746-34-1
POTASSIUM BROMATE	7758-01-2
POTASSIUM CHLORIDE	7447-40-7
POTASSIUM CHROMATE	7789-00-6
POTASSIUM DICHROMATE	7778-50-9
PREDNISONE	53-03-2
PROBENECID	57-66-9
PROCARBAZINE HYDROCHLORIDE	366-70-1
PROFLAVINE HEMISULPHATE	1811-28-5
PROFLAVINE MONOHYDROCHLORIDE HEMIHYDRATE	952-23-8
PROGESTERONE	57-83-0
PRONAMIDE	23950-58-5
PRONETALOL HYDROCHLORIDE	51-02-5
PROPANEDIOIC ACID COMPOUND	63843-89-0
1,3-PROPANE SULTONE	1120-71-4
PROPHAM	122-42-9
B-PROPIOLACTONE	57-57-8
PROPIONIC ACID	79-09-4
1-PROPOXY-2-PROPANOL	1569-01-3
N-PROPYL ALCOHOL	71-23-8
N-PROPYL CARBAMATE	627-12-3
PROPYL GALLATE	121-79-9
PROPYLENE	115-07-1
PROPYLENE BASED GLYCOL ETHERS	CLASS-31

CHEMICAL NAME	CAS NO.
PROPYLENEIMINE (2-METHYLAZIRIDINE)	75-55-8
PROPYLENE OXIDE (1,2-EPOXYPROPANE)	75-56-9
PROPYLTHIOURACIL	51-52-5
PULP AND PAPER INDUSTRY	UNKNOWN-26
PYRAZINAMIDE	98-96-4
PYRAZOLE	288-13-1
PYRENE	129-00-0
PYRIDINE	110-86-1
PYRIDO(3,4-C)PSORALEN AND 7-METHYLPYRIDO(3,4-C)PSORALEN	85878-62-2
PYRIMETHAMINE	58-14-0
PYROGALLOL	87-66-1
QUARTZ	14808-60 7
QUERCETIN	117-39-5
4-QUINOLINAMINE-N-HYDROXY-1-OXIDE	4637-56-3
QUINOLINE	91-22-5
P-QUINONE	106-51-4
P-QUINONE DIOXIME	105-11-3
RESERPINE	50-55-5
RESORCINOL	108-46-3
RETROSINE	480-54-6
RHU LIGHT VACUUM GAS OIL	64741-75-9
RHU MIDDLE DISTILLATE	64741-76-0
RIDDELLIINE	23246-96-0
RIFAMPICIN	13292-46-1
ROTENONE	83-79-4
ROXARSONE	121-19-7
RUBBER INDUSTRY	UNKNOWN-17
RUGULOSIN	23537-16-8
SACCHARATED IRON	8047-67-4
SACCHARIN	81-07-2
SACCHARIN, SODIUM	128-44-9
SAFROLE	94-59-7

CHEMICAL NAME	CAS NO.
SELENIUM	7782-49-2
SELENIUM SULFIDE	7446-34-6
SEMICARBAZIDE	57-56-7
SEMICARBAZIDE HYDROCHLORIDE	563-41-7
SENECIPHYLLINE	480-81-9
SENKIRKINE	2318-18-5
SEPIOLITE	18307-23-8
SHALE-OILS	68308-34-9
SILICA	7631-86-9
SODIUM ARSENATE	7631-89-2
SODIUM ARSENITE	7784-46-5
SODIUM AZIDE	26628-22-8
SODIUM CACODYLATE	124-65-2
SODIUM CHROMATE	7775-11-3
SODIUM CYCLAMATE	139-05-9
SODIUM DICHROMATE (VI)	10588-01-9
SODIUM DIETHYLDITHIOCARBAMATE	148-18-5
SODIUM FLUORIDE	7681-49-4
SODIUM HYPOCHLORITE	1681-52-9
SODIUM N-METHYL-N-OLEOYLTAURINE	137-20-2
SOLVENT NAPHTHA (HEAVY AROMATIC SOLVENT NAPHTHA)	8030-30-6
SOOT	CLASS-05
SPIRONOLACTONE	52-01-7
STANNOUS CHLORIDE	7772-99-8
STERIGMATOCYSTIN	10048-13-2
STODDARD SOLVENT	8052-41-3
STREPTOMYCIN SULFATE (2:3)	3810-74-0
STREPTOZOTOCIN	18883-66-4
STYRENE	100-42-5
STYRENE-BUTADIENE COPOLYMERS	9003-55-8
STYRENE OXIDE	96-09-3
SUBSTITUTED POLYGLYCIDYL BENZENEAMINE	UNKNOWN-30

CHEMICAL NAME	CAS NO.
SUCCINIC ANHYDRIDE	108-30-5
SULFALLATE	95-06-7
SULFAMETHAZINE	57-68-1
SULFAMETHOXAZOLE	723-46-6
SULFISOXAZOLE	127-69-5
3-SULFOLENE	77-79-2
SULFONATED POLYSTYRENE, SODIUM SALT	9080-79-9
SYMPHYTINE	2257-95-5
TALC	14807-96-6
TANNIC ACID	1401-55-4
TARA GUM	39300-88-4
TDE	72-54-8
TEREPHTHALIC ACID	100-21-0
TERPENE POLYCHLORINATE	8001-50-1
TESTOSTERONE	58-22-0
TETRABROMOBISPHENOL A (TBBPA)	79-94-7
3,3',4,4'-TETRACHLOROAZOBENZENE (TCAB)	14047-09-7
3,3',4,4'-TETRACHLOROAZOXYBENZENE (TCAOB)	21232-47-3
1,2,4,5-TETRACHLOROBENZENE (1,2,4,5-TCB)	95-94-3
2,2',5,5'-TETRACHLOROBENZIDINE	15721-02-5
2,3,7,8-TETRACHLORODIBENZO-P-DIOXIN (TCDD)	1746-01-6
1,1,1,2-TETRACHLOROETHANE	630-20-6
1,1,2,2-TETRACHLOROETHANE	79-34-5
2,3,5,6-TETRACHLORO-4-NITROANISOLE (TCNA)	2438-88-2
TETRACHLORVINPHOS	22248-79-9
TETRACYCLINE HYDROCHLORIDE	64-75-5
TETRAETHYLENE GLYCOL	112-60-7
TETRAETHYLENE GLYCOL DIACRYLATE	17831-71-9
TETRAETHYLENEPENTAMINE	112-57-2
TETRAETHYLLEAD (TEL)	78-00-2
TETRAETHYLTHIURAM DISULFIDE (DISULFIRAM)	97-77-8
TETRAFLUOROETHYLENE	116-14-3

CHEMICAL NAME	CAS NO.
TETRAKIS(HYDROXYMETHYL)PHOSPHONIUM CHLORIDE (THPC)	124-64-1
TETRAKIS(HYDROXYMETHYL)PHOSPHONIUM SULFATE (THPS)	55566-30-8
4-(1,1,3,3-TETRAMETHYLBUTYL)PHENOL (TMBP)	140-66-9
TETRAMETHYLLEAD (TML)	75-74-1
1,1,3,3-TETRAMETHYL-2-THIOUREA	2782-91-4
TETRANITROMETHANE	509-14-8
THEOBROMINE	83-67-0
THEOPHYLLINE	58-55-9
THIOACETAMIDE (TAA)	62-55-5
4,4'-THIODIANILINE	139-65-1
BETA-THIOGUANINE DEOXYRIBOSIDE (B-TGDR)	64039-27-6
TRIS(1-AZIRIDINYL) PHOSPHINE SULFIDE	52-24-4
THIOURACIL	141-90-2
THIOUREA	62-56-6
THIURAM	137-26-8
THORIUM DIOXIDE	1314-20-1
TINUVIN 327	3864-99-1
TITANIUM DIOXIDE	13463-67-7
TITANIUM FERROCENE	1271-19-8
TITANIUM TETRACHLORIDE	7550-45-0
TOBACCO HABITS OTHER THAN SMOKING	CLASS-17
TOLAZAMIDE	1156-19-0
TOLBUTAMIDE	64-77-7
O-TOLIDINE(3,3-DIMETHYLBENZIDINE)	119-93-7
TOLUENE	108-88-3
2,6-TOLUENEDIAMINE DIHYDROCHLORIDE	15481-70-6
2,5-TOLUENEDIAMINE SULFATE	6369-59-1
TOLUENE DIISOCYANATE (TDI)	26471-62-5
2,4-TOLUENE DIISOCYANATE (2,4-TDI)	584-84-9
2,6-TOLUENE DIISOCYANATE (2,6-TDI)	91-08-7
ORTHO-TOLUENESULPHONAMIDE	88-19-7
O-TOLUIDINE	95-53-4

CHEMICAL NAME	CAS NO.
O-TOLUIDINE HYDROCHLORIDE	636-21-5
TOXAPHENE	8001-35-2
TRANS-A-CHLORO-3-OXYAMINOPROPENE HYDROCHLORIDE	82244-86-8
TRANS-2,3-DIBROMO-2-BUTENE-1,4-DIOL	3234-02-4
TRANS-1,4-DICHLOROBUTENE	764-41-0
TRANS-2-((DIMETHYLAMINO)METHYLAMINO)-5-(2-(5-NITRO-2-	25962-77-0
TREMOLITE	14567-73-8
TREOSULPHAN	299-75-2
1,3,5-TRIACRYLOYLHEXA-HYDRO-S-TRIAZINE	959-52-4
TRIBROMOMETHANE	75-25-2
TRIBUTYL PHOSPHATE (PHOSPHORIC ACID, TRIBUTYL ESTER)	126-73-8
TRICHLORFON	52-68-6
2,4,6-TRICHLOROANILINE	634-93-5
1,2,3-TRICHLOROBENZENE	87-61-6
1,2,4-TRICHLOROBENZENE	120-82-1
1,1,1-TRICHLOROETHANE	71-55-6
1,1,2-TRICHLOROETHANE	79-00-5
TRICHLOROETHYLENE (TCE)	79-01-6
TRICHLOROFLUOROMETHANE	75-69-4
2,4,5-TRICHLOROPHENOL	95-95-4
2,4,6-TRICHLOROPHENOL	88-06-2
2,4,5-TRICHLOROPHENOXYACETIC ACID (2,4,5-T)	93-76-5
1,2,3-TRICHLOROPROPANE	96-18-4
TRICHLOROTRIETHYLAMINE HYDROCHLORIDE	817-09-4
1,1,2-TRICHLORO-1,2,2-TRIFLUOROETHANE (F113)	76-13-1
T2-TRICHOTHECENE	21259-20-1
TRICRESYL PHOSPHATE (TCP)	1330-78-5
TRIDYMITE	15468-32-3
TRIETHANOLAMINE	102-76-6
TRIETHYLENE GLYCOL DIACRYLATE	1680-21-3
TRIETHYLENE GLYCOL DIGLYCIDYL ETHER	1954-28-5
TRIETHYLENE GLYCOL MONOBUTYL ETHER	143-22-6

CHEMICAL NAME	CAS NO.
TRIETHYLENE GLYCOL MONOETHYL ETHER	112-50-5
TRIETHYLENE GLYCOL MONOMETHYL ETHER	112-35-6
TRIETHYLENE TETRAMINE (TETA)	112-24-3
TRIETHYL PHOSPHATE	78-40-0
TRIFLURALIN	1582-09-8
2,4,5-TRIHYDROXYBUTYROPHENONE (THBP)	1421-63-2
2,4,5-TRIMETHYLANILINE	137-17-7
2,4,6-TRIMETHYLANILINE	88-05-1
1,2,3-TRIMETHYL BENZENE	526-73-8
1,2,4-TRIMETHYL BENZENE	95-63-6
TRIMETHYL BENZENE (MIXED ISOMERS)	25551-13-7
2,2,4-TRIMETHYLPENTANE	540-84-1
TRIMETHYLPHOSPHATE	512-56-1
4,5',8-TRIMETHYLPSORALEN	3902-71-4
1,1,3-TRIMETHYL-2-THIOUREA	2489-77-2
2,4,7-TRINITROFLUORENONE (TNF)	129-79-3
TRIPHENYLENE	217-59-4
TRIPHENYLPHOSPHINE (TPP)	603-35-0
TRIPHENYL PHOSPHITE	101-02-0
TRIPHENYLTIN HYDROXIDE	76-87-9
TRIS(AZIRIDINYL)-P-BENZOQUINONE (TRIAZIQUINONE)	68-76-8
TRIS(1-AZIRIDINYL)PHOSPHINE OXIDE (TEPA)	545-55-1
2,4,6-TRIS(1-AZIRIDINYL)-S-TRIAZINE (TEM)	51-18-3
TRIS(2-CHLOROETHYL)PHOSPHATE	115-96-8
1,2,3-TRIS(CHLOROMETHOXY)-PROPANE	38571-73-2
TRIS (2,3-DIBROMOPROPYL) PHOSPHATE (TRIS)	126-72-7
TRIS(1,3-DICHLOROISOPROPYL)PHOSPHATE (FYROL FR2)	13674-87-8
TRIS(2,3-DICHLOROPROPYL)PHOSPHATE	78-43-3
TRIS(2-ETHYLHEXYL)PHOSPHATE	78-42-2
TRIS(2-ETHYLHEXYL)TRIMELLITATE	3319-31-1
TRIS(2-METHYL-1-AZIRIDINYL)PHOSPHINE OXIDE	57-39-6
TRISODIUM ETHYLENEDIAMINETETRAACETATE TRIHYDRATE (EDT	150-38-9

CHEMICAL NAME	CAS NO.
TRP-P-1(3-AMINO-1,4-DIMETHYL-5H-PYRIDO(4,3-B)INDOLE)	62450-06-0
TRP-P-2(3-AMINO-1-METHYL-5H-PYRIDO(4,3-B)INDOLE) AND	62450-07-1
L-TRYPTOPHAN	73-22-3
UCON 50-HB POLYETHERS	9038-95-3
URACIL MUSTARD	66-75-1
UREA-FORMALDEHYDE RESINS	1000-82-4
VINBLASTINE	865-21-4
VINCRISTINE	57-22-7
VINYL ACETATE	108-05-4
VINYL BROMIDE	593-60-2
VINYL CHLORIDE MONOMER (VCM)	75-01-4
VINYL CHLORIDE-VINYL ACETATE COPOLYMERS	9003-22-9
4-VINYLCYCLOHEXANE	100-40-3
4-VINYL-3-CYCLOHEXENE DIOXIDE (1-EPOXYETHYL-3,4-EPOXY CYC	106-87-6
VINYL FLUORIDE	75-02-5
VINYLIDENE CHLORIDE (1,1-DICHLOROETHYLENE)	75-35-4
VINYLIDENE CHLORIDE-VINYL CHLORIDE COPOLYMERS	9011-06-7
VINYLIDENE FLUORIDE	75-38-7
N-VINYL-2-PYRROLIDONE (1-ETHENYL-2-PYRROLIDINONE)	88-12-0
VINYL TOLUENE	25013-15-4
VITAMIN A	68-26-8
VITAMIN C (ASCORBIC ACID)	50-81-7
VITAMIN E (A-TOCOPHEROL)	1406-66-2
WOLLASTONITE	13983-17-0
XYLENE (MIXED ISOMERS)	1330-20-7
2,4-XYLIDINE	95-68-1
2,5-XYLIDINE	95-78-3
2,6-XYLIDINE	87-62-7
ZEARALENONE	17924-92-4
ZECTRAN	315-18-4
ZINC	7440-66-6
ZINC CHLORIDE	7646-85-7

CHEMICAL NAME	CAS NO.
ZINC CHROMATE HYDROXIDE	15930-94-6
ZINC DIALKYLDITHIOPHOSPHATE	5823-20-1
ZINC OLEATE	557-07-3
ZINC OXIDE	1314-13-2
ZINC POTASSIUM CHROMATE	12433-50-0
ZINEB	12122-67-7
ZIRAM	137-30-4

INDEX B

LISTING OF CHEMICALS BY CAS NUMBER

CHEMICAL NAME	CAS NO.
FORMALDEHYDE	50-00-0
PHENOBARBITONE	50-06-6
MITOMYCIN C	50-07-7
CYCLOPHOSPHAMIDE	50-18-0
OESTRIOL	50-27-1
ESTRADIOL - 17B	50-28-2
DDT	50-29-3
BENZO(A)PYRENE	50-32-8
PHENYLBUTAZONE	50-33-9
CLOMIPHENE CITRATE	50-41-9
6-MERCAPTOPURINE	50-44-2
RESERPINE	50-55-5
ACTINOMYCIN-D	50-76-0
VITAMIN C (ASCORBIC ACID)	50-81-7
PRONETALOL HYDROCHLORIDE	51-02-5
PIPERONYL BUTOXIDE	51-03-6
2,4,6-TRIS(1-AZIRIDINYL)-S-TRIAZINE (TEM)	51-18-3
5-FLUOROURACIL	51-21-8
2,4-DINITROPHENOL	51-28-5
ISOPROTERENOL HYDROCHLORIDE	51-30-9
PROPYLTHIOURACIL	51-52-5
NITROGEN MUSTARD	51-75-2
ETHYL CARBAMATE (URETHANE)	51-79-6
NORETHISTERONE ACETATE	51-98-9

CHEMICAL NAME	CAS NO.
SPIRONOLACTONE	52-01-7
TRIS(1-AZIRIDINYL) PHOSPHINE SULFIDE	52-24-4
APHOLATE	52-46-0
TRICHLORFON	52-68-6
LYNOESTRENOL	52-76-6
PREDNISONE	53-03-2
ESTRONE	53-16-7
O,P'-DDD	53-19-0
DIBENZ(A,H)ANTHRACENE (DBA)	53-70-3
2-ACETYLAMINOFLUORENE(2-AAF)	53-96-3
CHLOROQUINE	54-05-7
M-PHENYLENEDIAMINE H2S04	54-17-8
FUROSEMIDE	54-31-9
ISONICOTINIC ACID HYDRAZIDE (INH)	54-85-3
N-NITROSODIETHYLAMINE (NDEA)	55-18-5
L-EPINEPHRINE HYDROCHLORIDE	55-31-2
FENTHION	55-38-9
4-METHYL-P-AMINOPHENOL SULFATE	55-55-0
N,N-DIMETHYL-P-(M-TOLYLAZO)-ANILINE	55-80-1
NITROGEN MUSTARD HYDROCHLORIDE	55-86-7
1,4-BUTANEDIOL DIMETHANESULPHONATE	55-98-1
METHYLTHIOURACIL	56-04-2
CARBON TETRACHLORIDE	56-23-5
CANTHARIDIN	56-25-7
PARATHION	56-38-2
3-METHYLCHOLANTHRENE	56-49-5
DIETHYLSTILBESTROL	56-53-1
BENZ(A)ANTHRACENE (BA)	56-55-3
4-NITROQUINOLINE-1-OXIDE	56-57-5
COUMAPHOS	56-72-4
CHLORAMPHENICOL	56-75-7
ALLYL ISOTHIOCYANATE (ATC)	57-06-7

CHEMICAL NAME	CAS NO.
1,1-DIMETHYLHYDRAZINE (UDMH)	57-14-7
VINCRISTINE	57-22-7
PHENOBARBITAL SODIUM	57-30-7
TRIS(2-METHYL-1-AZIRIDINYL)PHOSPHINE OXIDE	57-39-6
5,5'-DIPHENYLHYDANTOIN (PHENYTOIN)	57-41-0
SEMICARBAZIDE	57-56-7
B-PROPIOLACTONE	57-57-8
ETHINYLESTRADIOL	57-63-6
PROBENECID	57-66-9
SULFAMETHAZINE	57-68-1
CHLORDANE	57-74-9
PROGESTERONE	57-83-0
CHLOLESTEROL	57-88-5
7,12-DIMETHYL-BENZO(A)ANTHRACENE	57-97-6
CAFFEINE	58-08-2
PYRIMETHAMINE	58-14-0
TESTOSTERONE	58-22-0
THEOPHYLLINE	58-55-9
LINDANE (GAMMA-HEXACHLOROCYCLOHEXANE)	58-89-9
HYDROCHLOROTHIAZIDE	58-93-5
METHOTREXATE	59-05-2
NITROFURAZONE	59-87-0
PHENYLHYDRAZINE MONOHYDROCHLORIDE	59-88-1
N-NITROSOMORPHOLINE	59-89-2
PHENOXYBENZAMINE	59-96-1
C. I. SOLVENT YELLOW 1 (P-AMINOAZOBENZENE)	60-09-3
C. I. SOLVENT YELLOW 2 (P-DIMETHYLAMINOAZOBENZENE)	60-11-7
2-PHENYLETHANOL (2-PE)	60-12-8
DL-AMPHETAMINE SULFATE	60-13-9
METHYLHYDRAZINE	60-34-4
ACETAMIDE	60-35-5
DIMETHOATE	60-51-5

CHEMICAL NAME	CAS NO.
DIELDRIN	60-57-1
NIRIDAZOLE	61-57-4
PHENYLEPHRINE HYDROCHLORIDE	61-76-7
3-AMINO-S-TRIAZOLE (AMITROLE)	61-82-5
PHENACETIN	62-44-2
ETHYL METHANESULFONATE (EMS)	62-50-0
ANILINE	62-53-3
THIOACETAMIDE (TAA)	62-55-5
THIOUREA	62-56-6
DICHLORVOS	62-73-7
N-NITROSODIMETHYLAMINE (NDMA)	62-75-9
CARBARYL	63-25-2
PHENOXYBENZAMINE HCL	63-92-3
DIETHYL SULFATE	64-67-5
TETRACYCLINE HYDROCHLORIDE	64-75-5
TOLBUTAMIDE	64-77-7
COLCHICINE	64-86-8
METHYL METHANESULFONATE (MMS)	66-27-3
URACIL MUSTARD	66-75-1
NITROFURANTOIN	67-20-9
DL-ETHIONINE	67-21-0
FURAZLIDONE	67-45-8
METHANOL	67-56-1
ISOPROPYL ALCOHOL (ISOPROPANOL)	67-63-0
CHLOROFORM	67-66-3
DIMETHYL SULFOXIDE	67-68-5
HEXACHLOROETHANE	67-72-1
NORETHISTERONE	68-22-4
NORETHYNODREL	68-23-5
VITAMIN A	68-26-8
TRIS(AZIRIDINYL)-P-BENZOQUINONE (TRIAZIQUINONE)	68-76-8
D-MANNITOL	69-65-8

CHEMICAL NAME	CAS NO.
N-METHYL-N'-NITRO-N-NITROSOGUANIDINE (MNNG)	70-25-7
HEXACHLOROPHENE	70-30-4
1-FLUORO-2,4-DINITROBENZENE	70-34-8
N-PROPYL ALCOHOL	71-23-8
ISOAMYL ALCOHOL	71-41-0
BENZENE	71-43-2
1,1,1-TRICHLOROETHANE	71-55-6
MEDROXYPROGESTERONE ACETATE	71-58-9
ENDRIN	72-20-8
MESTRANOL	72-33-3
METHOXYCHLOR	72-43-5
C. I. MORDANT RED 11	72-48-0
TDE	72-54-8
P,P'-DDE	72-55-9
P,P'-ETHYL-DDD	72-56-0
C. I. DIRECT BLUE 14 (TRYPAN BLUE)	72-57-1
L-TRYPTOPHAN	73-22-3
BROMOMETHANE (BM) (METHYL BROMIDE)	74-83-9
ETHYLENE	74-85-1
CHLOROMETHANE	74-87-3
METHYL IODIDE	74-88-4
BROMOETHANE	74-96-4
CHLOROETHANE	75-00-3
VINYL CHLORIDE MONOMER (VCM)	75-01-4
VINYL FLUORIDE	75-02-5
ACETONITRILE	75-05-8
ACETALDEHYDE	75-07-0
METHYLENE CHLORIDE	75-09-2
FORMAMIDE	75-12-7
CARBON DISULFIDE	75-15-0
ETHYLENE OXIDE (ETO)	75-21-8
TRIBROMOMETHANE	75-25-2

CHEMICAL NAME	CAS NO.
BROMODICHLOROETHANE	75-27-4
1,1-DICHLOROETHANE	75-34-3
VINYLIDENE CHLORIDE (1,1-DICHLOROETHYLENE)	75-35-4
VINYLIDENE FLUORIDE	75-38-7
CHLORODIFLUOROMETHANE	75-45-6
IODOFORM	75-47-8
PROPYLENEIMINE (2-METHYLAZIRIDINE)	75-55-8
PROPYLENE OXIDE (1,2-EPOXYPROPANE)	75-56-9
DIMETHYLARSINIC ACID	75-60-5
T-BUTYL ALCOHOL (T-BUTANOL)	75-65-0
TRICHLOROFLUOROMETHANE	75-69-4
DICHLORODIFLUOROMETHANE (CFC-12)	75-71-8
TETRAMETHYLLEAD (TML)	75-74-1
2-CHLORO-1,1,1-TRIFLUOROETHANE	75-88-7
T-BUTYL HYDROPEROXIDE (TBHP)	75-91-2
PENTACHLOROETHANE	76-01-7
CHLOROPICRIN	76-06-2
1,1,2-TRICHLORO-1,2,2-TRIFLUOROETHANE (F113)	76-13-1
CAMPHOR	76-22-2
HEPTACHLOR	76-44-8
TRIPHENYLTIN HYDROXIDE	76-87-9
HEXACHLOROCYCLOPENTADIENE	77-47-4
CARBROMAL	77-65-6
DIMETHYL SULFATE	77-78-1
3-SULFOLENE	77-79-2
TETRAETHYLLEAD (TEL)	78-00-2
PENTAERYTHRITOL TETRANNITRATE (PETN,NF)	78-11-5
DIOXATHION	78-34-2
TRIETHYL PHOSPHATE	78-40-0
TRIS(2-ETHYLHEXYL)PHOSPHATE	78-42-2
TRIS(2,3-DICHLOROPROPYL)PHOSPHATE	78-43-3
ISOPHORONE (3,5,5-TRIMETHYL-2-CYCLOHEXEN-1-ONE)	78-59-1

CHEMICAL NAME	CAS NO.
ISOPRENE	78-79-5
ISOBUTYL ALCOHOL	78-83-1
1,2-DICHLOROPROPANE	78-87-5
METHYL ETHYL KETONE	78-93-3
CHLOROACETONE	78-95-5
1,1,2-TRICHLOROETHANE	79-00-5
TRICHLOROETHYLENE (TCE)	79-01-6
ACRYLAMIDE	79-06-1
PROPIONIC ACID	79-09-4
ACRYLIC ACID	79-10-7
MONOCHLOROACETIC ACID	79-11-8
PEROXYACETIC ACID	79-21-0
1,1,2,2-TETRACHLOROETHANE	79-34-5
DIMETHYLCARBAMOYL CHLORIDE (DMCC)	79-44-7
2-NITROPROPANE	79-46-9
DIMETHISTERONE	79-64-1
TETRABROMOBISPHENOL A (TBBPA)	79-94-7
BISPHENOL A	80-05-7
DAPSONE	80-08-0
HALAZONE	80-13-7
METHYL METHACRYLATE	80-62-6
SACCHARIN	81-07-2
1-AMINO-2,4-DIBROMO ANTHRAQUINONE	81-49-2
C. I. NATURAL RED 8 (PURPURIN)	81-54-9
C. I. 58050 (QUINIZARIN)	81-64-1
C. I. VAT BLUE 4	81-77-6
C. I. BASIC VIOLET 10 (RHODAMINE B)	81-88-9
C. I. DISPERSE ORANGE 11	82-28-0
PENTACHLORONITROBENZENE (PCNB)	82-68-8
DIACETYLAMINOAZOTOLUENE	83-63-6
THEOBROMINE	83-67-0
ROTENONE	83-79-4

CHEMICAL NAME	CAS NO.
DIENOESTROL	84-17-3
ANTHRAQUINONE	84-65-1
DIETHYL PHTHALATE	84-66-2
DI-N-BUTYL PHTHALATE (PHTHALIC ACID, DIBUTYL ESTER) (DBP)	84-74-2
DINONYL PHTHALATE ESTER (1,2-BENZENEDICARBOXYLIC ACID)	84-76-4
PHENANTHRENE	85-01-8
PENTABROMOETHYLBENZENE (PEB)	85-22-3
PHTHALIC ANHYDRIDE	85-44-9
BUTYL BENZYL PHTHALATE	85-68-7
C. I. SOLVENT RED 24 (SCARLET RED)	85-83-6
C. I. SOLVENT YELLOW 5 (YELLOW AB)	85-84-7
C. I. SOLVENT RED 23 (SUDAN III)	85-86-9
N-NITROSODIPHENYLAMINE	86-30-6
AZINPHOS-METHYL	86-50-0
HYDRALAZINE	86-54-4
1-NITRONAPHTHALENE	86-57-7
FLUORENE	86-73-7
CARBAZOLE	86-74-8
1-NAPHTHYLTHIOUREA	86-88-4
CINNAMYL ANTHRANILATE	87-29-6
1,2,3-TRICHLOROBENZENE	87-61-6
2,6-XYLIDINE	87-62-7
PYROGALLOL	87-66-1
HEXACHLORO-1,3-BUTADIENE	87-68-3
PENTACHLOROPHENOL	87-86-5
2,4,6-TRIMETHYLANILINE	88-05-1
2,4,6-TRICHLOROPHENOL	88-06-2
DIETHYLCARBAMOYL CHLORIDE (DECC)	88-10-8
N-VINYL-2-PYRROLIDONE (1-ETHENYL-2-PYRROLIDINONE)	88-12-0
ORTHO-TOLUENESULPHONAMIDE	88-19-7
2-CHLORONITROBENZENE	88-73-3
PICRIC ACID	88-89-1

CHEMICAL NAME	CAS NO.
PHTHALAMIDE	88-96-0
1-PHENYL-3-METHYL-5-PYRAZOLONE	89-25-8
4-CHLORO-2-NITROANILINE	89-63-4
DL-MENTHOL	89-78-1
O-ANISIDINE	90-04-0
2-BIPHENYLAMINE HYDROCHLORIDE	90-41-5
O-PHENYLPHENOL	90-43-7
PENICILLIC ACID	90-65-3
MICHLER'S KETONE	90-94-8
2,6-TOLUENE DIISOCYANATE (2,6-TDI)	91-08-7
DECALIN	91-17-8
NAPHTHALENE	91-20-3
QUINOLINE	91-22-5
O-NITROANISOLE	91-23-6
2-NAPHTHYLAMINE	91-59-8
COUMARIN	91-64-5
METHAPYRILENE	91-80-5
3,3'-DIMETHOXYBENZIDINE-4,4'-DIISOCYANATE	91-93-0
3,3'-DICHLOROBENZIDINE (DCB)	91-94-1
1-PHENYL-3-PYRAZOLIDINONE (PHENIDONE)	92-43-3
BIPHENYL	92-52-4
4-AMINOBIPHENYL	92-67-1
BENZIDINE	92-87-5
4-NITROBIPHENYL	92-93-3
2,4,5-TRICHLOROPHENOXYACETIC ACID (2,4,5-T)	93-76-5
CHLORPROPAMIDE	94-20-2
BENZOYL PEROXIDE	94-36-0
6-NITROBENZIMIDAZOLE	94-52-0
DIHYDROSAFROLE	94-58-6
SAFROLE	94-59-7
MCPA	94-74-6
2,4-D	94-75-7

CHEMICAL NAME	CAS NO.
2,6-DIAMINO-3-(PHENYLAZO)PYRIDINE	94-78-0
SULFALLATE	95-06-7
1H-BENZOTRIAZOLE	95-14-7
O-CRESOL	95-48-7
2-CHLOROTOLUENE	95-49-8
1,2-DICHLOROBENZENE (ODCB)	95-50-1
O-TOLUIDINE	95-53-4
O-PHENYLENEDIAMINE	95-54-5
O-AMINOPHENOL	95-55-6
1,2,4-TRIMETHYL BENZENE	95-63-6
2,4-XYLIDINE	95-68-1
C. I. 37085 (P-CHLORO-O-TOLUIDINE)	95-69-2
2,5-DIAMINOTOLUENE	95-70-5
3-CHLORO-P-TOLUIDINE	95-74-9
2,5-XYLIDINE	95-78-3
5-CHLORO-O-TOLUIDINE	95-79-4
2,4-DIAMINOTOLUENE	95-80-7
4-CHLORO-O-PHENYLENEDIAMINE	95-83-0
1,2,4,5-TETRACHLOROBENZENE (1,2,4,5-TCB)	95-94-3
2,4,5-TRICHLOROPHENOL	95-95-4
STYRENE OXIDE	96-09-3
1,2-DIBROMO-3-CHLOROPROPANE, DBCP	96-12-8
2,3-DIBROMO-1-PROPANOL	96-13-9
1,2,3-TRICHLOROPROPANE	96-18-4
METHYL ETHYL KETOXIME	96-29-7
METHYL ACRYLATE	96-33-3
METHYCYCLOPENTANE (MCP)	96-37-7
ETHYLENE THIOUREA (ETU)	96-45-7
GAMMA-BUTYROLACTONE	96-48-0
ACETOPHENONE (ACP)	96-86-2
2,4-DINITRO-1-CHLOROBENZENE	97-00-7
2,4-DINITROANILINE	97-02-9

CHEMICAL NAME	CAS NO.
EUGENOL	97-53-0
C. I. SOLVENT YELLOW 3 (O-AMINOAZOTOLUENE)	97-56-3
ETHYL METHACRYLATE	97-63-2
TETRAETHYLTHIURAM DISULFIDE (DISULFIRAM)	97-77-8
FURFURAL	98-01-1
BENZOTRICHLORIDE (BTC)	98-07-7
ARSANILIC ACID	98-50-0
P-TERT-BUTYLTOLUENE (P-TBT)	98-51-1
4-CHLOROBENZOTRIFLUORIDE	98-56-6
P-TERT-BUTYLBENZOIC ACID (P-TBBA)	98-73-7
CUMENE (1-METHYLETHYL)BENZENE)	98-82-8
A-METHYLBENZYL ALCOHOL	98-85-1
BENZAL CHLORIDE	98-87-3
BENZOYL CHLORIDE (BC)	98-88-4
NITROBENZENE	98-95-3
PYRAZINAMIDE	98-96-4
3,4-DICHLORONITROBENZENE	99-54-7
5-NITRO-O-TOLUIDINE	99-55-8
4-NITRO-O-PHENYLENEDIAMINE	99-56-9
2-AMINO-4-NITROPHENOL	99-57-0
5-NITRO-O-ANISIDINE	99-59-2
1-CHLORO-4-NITROBENZENE	100-00-5
P-NITROANILINE	100-01-6
P-NITROPHENOL	100-02-7
P-NITROBENZYL CHLORIDE	100-14-1
TEREPHTHALIC ACID	100-21-0
4-VINYLCYCLOHEXENE	100-40-3
ETHYLBENZENE	100-41-4
STYRENE	100-42-5
BENZYL CHLORIDE	100-44-7
BENZYL ALCOHOL	100-51-6
BENZALDEHYDE	100-52-7

CHEMICAL NAME	CAS NO.
N-PHENYLHYDROXYLAMINE	100-65-2
N-NITROSOPIPERIDINE	100-75-4
TRIPHENYL PHOSPHITE	101-02-0
ANILAZINE	101-05-3
4,4'-METHYLENEBIS(2-CHLOROANILINE) (MBOCA)	101-14-4
CHLOROPROPHAM	101-21-3
DINITROSOPENTAMETHYLENETETRAMINE (DNPMT)	101-25-7
N-PHENYL-P-PHENYLENEDIAMINE	101-54-2
4,4'-METHYLENEBIS(N,N-DIMETHYL)ANILINE	101-61-1
4,4'-DIPHENYLMETHANE DIISOCYANATE (MDI)	101-68-8
4,4'-METHYLENEDIANILINE (MDA)	101-77-9
4,4'-OXYDIANILINE	101-80-4
DIGLYCIDYL RESORCINOL ETHER (DGRE)	101-90-6
1,3-DIPHENYLGUANIDINE (DPC)	102-06-7
M-CRESIDINE	102-50-1
FERROCENE	102-54-5
TRIETHANOLAMINE	102-76-6
N-OXYDIETHYLENE-2-BENZOTHIAZYL SULFENAMIDE	102-77-2
PHENICARBAZIDE	103-03-7
DI(2-ETHYLHEXYL)ADIPATE	103-23-1
AZOBENZENE	103-33-3
1-PHENYL-2-THIOUREA	103-85-5
ACETAMINOPHEN	103-90-2
2-ETHYLHEXANOL	104-76-7
P-QUINONE DIOXIME	105-11-3
BROMOACETIC ACID, ETHYL ESTER	105-36-2
N,N'-DIETHYLTHIOUREA	105-55-5
CAPROLACTAM	105-60-2
LAUROYL PEROXIDE	105-74-8
P-CRESOL	106-44-5
1,4-DICHLOROBENZENE (PDCB)	106-46-7
P-CHLOROANILINE	106-47-8

CHEMICAL NAME	CAS NO.
P-PHENYLENEDIAMINE	106-50-3
P-QUINONE	106-51-4
4-VINYL-3-CYCLOHEXENE DIOXIDE (1-EPOXYETHYL-3,4-EPOXY CYC	106-87-6
1,2-EPOXYBUTANE	106-88-7
EPICHLOROHYDRIN	106-89-8
GLYCIDYL ACRYLATE (GA)	106-90-1
GLYCIDYL METHACRYLATE (GMA)	106-91-2
ALLYL GLYCIDYL ETHER (AGE)	106-92-3
1,2-DIBROMOETHANE (EDB)	106-93-4
1,3-BUTADIENE	106-99-0
ACROLEIN	107-02-8
ALLYL CHLORIDE	107-05-1
1,2-DICHLOROETHANE (EDC)	107-06-2
2-CHLOROETHANOL (ETHYLENE CHLOROHYDRIN)	107-07-3
ALLYLAMINE (2-PROPINE-1-AMINE)	107-11-7
ACRYLONITRILE (AN)	107-13-1
ETHYLENEDIAMINE	107-15-3
CHLOROACETALDEHYDE	107-20-0
CHLOROMETHYL METHYL ETHER (CMME)	107-30-2
VINYL ACETATE	108-05-4
METHYL ISOBUTYL KETONE	108-10-1
DIISOPROPYLAMINE	108-18-9
SUCCINIC ANHYDRIDE	108-30-5
MALEIC ANHYDRIDE	108-31-6
M-CRESOL	108-39-4
M-PHENYLENEDIAMINE	108-45-2
RESORCINOL	108-46-3
BIS(2-CHLORO-1-METHYLETHYL) ETHER (BCMEE)	108-60-1
MESITYLENE	108-67-8
MELAMINE	108-78-1
CYANURIC ACID	108-80-5
TOLUENE	108-88-3

CHEMICAL NAME	CAS NO.
CHLOROBENZENE	108-90-7
CYCLOHEXYLAMINE (CHA)	108-91-8
CYCLOHEXANONE	108-94-1
PHENOL	108-95-2
N-PENTANOIC ACID (N-VALERIC ACID)	109-52-4
N-BUTYL CHLORIDE	109-69-3
2-HYDRAZINOETHANOL	109-84-2
2-METHOXYETHANOL (2-ME)	109-86-4
2-METHOXYETHANOL ACETATE	110-49-6
HEXANE	110-54-3
ETHYLENE GLYCOL MONOMETHYL ETHER (2-ETHOXYETHANOL) (2	110-80-5
CYCLOHEXANE	110-82-7
PYRIDINE	110-86-1
2-METHOXYETHYL PALMITATE	111-07-9
2-ETHOXYETHANOL ACETATE	111-15-9
ETHYLENE BIS(OXYETHYLENE) DIACETATE	111-21-7
GLUTERALDEHYDE	111-30-8
DIETHYLENETRIAMINE (DETA)	111-40-0
BIS(2-CHLOROETHYL)ETHER (BCEE)	111-44-4
ETHYLENE BIS DITHIOCARBAMATE	111-54-6
2-BUTOXYETHANOL (EGMBE)	111-76-2
1-OCTANOL	111-87-5
BIS(2-METHOXYETHYL)ETHER	111-96-6
TRIETHYLENE TETRAMINE (TETA)	112-24-3
DIETHYLENE GLYCOL BUTYL ETHER (DGBE)	112-34-5
TRIETHYLENE GLYCOL MONOMETHYL ETHER	112-35-6
TRIETHYLENE GLYCOL MONOETHYL ETHER	112-50-5
TETRAETHYLENEPENTAMINE	112-57-2
TETRAETHYLENE GLYCOL	112-60-7
OLEYLAMINE (9-OCTADECENYLAMINE, ODA)	112-90-3
CHLORPHENIRAMINE MALEATE	113-92-8
PHENFORMIN HYDROCHLORIDE	114-86-3

CHEMICAL NAME	CAS NO.
AZASERINE	115-02-6
PROPYLENE	115-07-1
CHLORENDIC ACID	115-28-6
ENDOSULFAN	115-29-7
DICOFOL	115-32-2
TRIS(2-CHLOROETHYL)PHOSPHATE	115-96-8
ALDICARB	116-06-3
TETRAFLUOROETHYLENE	116-14-3
HEXAFLUOROPROPENE	116-15-4
C. I. DISPERSE RED 15	116-85-8
1,8-DIHYDROXYANTHRAQUINONE (CHRYSAZIN)	117-10-2
1,5-DIHYDROXYANTHRAQUINONE	117-12-4
QUERCETIN	117-39-5
2-AMINOANTHRAQUINONE	117-79-3
DI(2-ETHYLHEXYL)PHTHALATE (DEHP)	117-81-7
DIMETHOXYETHYL PHTHALATE	117-82-8
DI-N-OCTYL PHTHALATE (DOP)	117-84-0
HEXACHLOROBENZENE	118-74-1
ANTHRANILIC ACID	118-92-3
4-AMINO-2-NITROPHENOL	119-34-6
BENZOIN	119-53-9
O-DIANISIDINE (3,3'-DIMETHOXYBENZIDINE)	119-90-4
O-TOLIDINE(3,3-DIMETHYLBENZIDINE)	119-93-7
ANTHRACENE	120-12-7
ISOSAFROLE	120-58-1
DIMETHYL TEREPHTHALATE	120-61-6
PIPERONYL SULFOXIDE	120-62-7
ACETYL-O-TOLUIDINE	120-66-1
P-CRESIDINE	120-71-8
CATECHOL	120-80-9
1,2,4-TRICHLOROBENZENE	120-82-1
2,4-DICHLOROPHENOL	120-83-2

CHEMICAL NAME	CAS NO.
2,4-DINITROTOLUENE (DNT)	121-14-2
ROXARSONE	121-19-7
2-AMINO-5-NITROTHIAZOLE	121-66-4
N,N-DIMETHYLANILINE	121-69-7
MALATHION	121-75-5
PROPYL GALLATE	121-79-9
2-AMINO-NITROPHENOL	121-88-0
P-NITROBENZOYL CHLORIDE	122-04-3
PROPHAM	122-42-9
PHENYL GLYCIDYL ETHER	122-60-1
1,2-DIPHENYLHYDRAZINE (HYDRAZOBENZENE)	122-66-7
2-PHENOXYETHANOL	122-99-6
HYDROQUINONE	123-31-9
MALEIC HYDRAZIDE	123-33-1
DIACETONE ALCOHOL (DA)	123-42-2
AZODICARBONAMIDE	123-77-3
N-BUTYL ACETATE	123-86-4
1,4-DIOXANE	123-91-1
2-(2-BUTOXYETHOXY) ETHYL ACETATE (DGBA)	124-17-4
DIMETHYL AMINE (DMA)	124-40-3
CHLORODIBROMOMETHANE	124-48-1
TETRAKIS(HYDROXYMETHYL)PHOSPHONIUM CHLORIDE (THPC)	124-64-1
SODIUM CACODYLATE	124-65-2
GRISEOFULVIN	126-07-8
TRIS (2,3-DIBROMOPROPYL) PHOSPHATE (TRIS)	126-72-7
TRIBUTYL PHOSPHATE (PHOSPHORIC ACID, TRIBUTYL ESTER)	126-73-8
NITROGEN MUSTARD N-OXIDE	126-85-2
CHLOROPRENE	126-99-8
HYDROXYUREA	127-07-1
PERCHLOROETHYLENE	127-18-4
SULFISOXAZOLE	127-69-5
BUTYLATED HYDROXYTOLUENE (BHT)	128-37-0

CHEMICAL NAME	CAS NO.
2,6-DI-TERT-BUTYLPHENOL	128-39-2
SACCHARIN, SODIUM	128-44-9
C. I. VAT GREEN 9	128-60-9
C. I. VAT YELLOW 4	128-66-5
PYRENE	129-00-0
2-METHYL-1-NITROANTHRAQUINONE (MNA)	129-15-7
C. I. ACID BLUE 1 (BLUE VRS)	129-17-9
OXYPHENBUTAZONE	129-20-4
2,4,7-TRINITROFLUORENONE (TNF)	129-79-3
1,4-NAPHTHOQUINONE	130-15-4
DIETHYLSTILBESTROL DIPROPRIONATE	130-80-3
DIEMTHYL PHTHALATE	131-11-3
DIALLYL PHTHALATE	131-17-9
2,2,-DIHYDROXY-4-METHOXYBENZOPHENONE	131-53-3
C. I. SOLVENT YELLOW 6 (YELLOW OB)	131-79-3
3-AMINO-9-ETHYLCARBAZOLE	132-32-1
PENICILLIN VK	132-98-9
CAPTAN	133-06-2
CHLORAMBEN	133-90-4
O-ANISIDINE HYDROCHLORIDE	134-29-2
1-NAPHTHYLAMINE	134-32-7
EPHEDRINE SULFATE	134-72-5
CUPFERRON	135-20-6
N-PHENYL-2-NAPHTHYLAMINE (PBNA)	135-88-6
PHENAZOPYRIDINE HYDROCHLORIDE	136-40-3
4 HEXYLRESORCINOL	136-77-6
METHYL-2-CYANOACRYLATE	137-05-3
2,4-DIAMINOPHENOL:2HCL	137-09-7
2,4,5-TRIMETHYLANILINE	137-17-7
SODIUM N-METHYL-N-OLEOYLTAURINE	137-20-2
THIURAM	137-26-8
ZIRAM	137-30-4

CHEMICAL NAME	CAS NO.
N,N-DIMETHYL-P-NITROSOANILINE	138-89-6
SODIUM CYCLAMATE	139-05-9
CALCIUM CYCLAMATE	139-06-0
NITRILOTRIACETIC ACID (NTA)	139-13-9
4,4'-THIODIANILINE	139-65-1
5-(MORPHOLINOMETHYL)-3-((5-NITROFURFURYLIDENE)-AMINOL)-2-	139-91-3
NITHIAZIDE	139-94-6
BENZYL ACETATE	140-11-4
4'-(CHLOROACETYL)-ACETANILIDE	140-49-8
FORMULATED FENAMINOSULF	140-56-7
ARAMITE	140-57-8
4-(1,1,3,3-TETRAMETHYLBUTYL)PHENOL (TMBP)	140-66-9
1,4-DINITROSOPIPERAZINE	140-79-4
ETHYL ACRYLATE (ACRYLIC ACID, ETHYL ESTER)	140-88-5
N-BUTYL ACRYLATE	141-32-2
3,4-EPOXY-6-METHYLCYCLOHEXYLMETHYL-3,4-EPOXY-6-METHYL	141-37-7
ETHANOLAMINE	141-43-5
THIOURACIL	141-90-2
2,6-DIMETHYL MORPHOLINE	141-91-3
MESITYL OXIDE	141-97-7
ANILINE HYDROCHLORIDE	142-04-1
DIALLYL DIGLYCOL CARBONATE (CR39 MONOMER)	142-22-1
2,5-DITHIOBIUREA	142-46-1
TRIETHYLENE GLYCOL MONOBUTYL ETHER	143-22-6
CHLORDECONE	143-50-0
METHANEARSONIC ACID.2NA	144-21-8
METHYL SELENAC	144-34-3
DIPHENHYDRAMINE HYDROCHLORIDE	147-24-0
ARABINOSYLCYTOSINE	147-94-4
SODIUM DIETHYLDITHIOCARBAMATE	148-18-5
8-HYDROXYQUINOLINE	148-24-3
MELPHALAN	148-82-3

CHEMICAL NAME	CAS NO.
PATULIN	149-29-1
2-MERCAPTOBENZOTHIAZOLE	149-30-4
2-ETHYLHEXANOIC ACID	149-57-5
PARA-AMINOBENZOIC ACID	150-13-0
TRISODIUM ETHYLENEDIAMINETETRAACETATE TRIHYDRATE (EDT	150-38-9
MONURON	150-68-5
P-ETHOXYPHENYLUREA	150-69-6
AZIRIDINE	151-56-4
BISCHLORDETHYL NITROSOUREA (BCNU)	154-93-8
P-NITROSODIPHENYLAMINE	156-10-5
PHENELZINE SULFATE	156-51-4
CALCIUM CYANAMIDE	156-62-7
DIBENZO(A,I)PYRENE	189-55-9
DIBENZO(A,H)PYRENE	189-64-0
CORONENE	191-07-1
BENZO(GHI)PERYLENE	191-24-2
ANTHANTHRENE	191-26-4
DIBENZO(A,L)PYRENE	191-30-0
DIBENZO(A,E)PYRENE	192-65-4
BENZO(E)PYRENE	192-97-2
INDENO(1,2,3-CD)PYRENE (IP)	193-39-5
DIBENZO(H,RST)PENTAPHENE	194-47-2
7H-DIBENZO(C,G)CARBAZOLE	194-59-2
BENZO(C)PHENANTHRENE	195-19-7
PERYLENE	198-55-0
BENZO(GHI)FLUORANTHENE	203-12-3
BENZO(C)FLUORENE	205-12-9
BENZO(J)FLUORANTHENE	205-82-3
BENZO(B)FLUORANTHENE	205-99-2
FLUORANTHENE	206-44-0
BENZO(K)FLUORANTHENE	207-08-9
DIBENZ(A,C)ANTHRACENE	215-58-7

CHEMICAL NAME	CAS NO.
TRIPHENYLENE	217-59-4
CHRYSENE	218-01-9
DIBENZ(A,J)ANTHRACENE	224-41-9
DIBENZ(A,J)ACRIDINE	224-42-0
BENZ(A)ACRIDINE	225-11-6
BENZ(C)ACRIDINE	225-51-4
DIBENZ(A,H)ACRIDINE	226-36-8
BENZO(A)FLUORENE	238-84-6
BENZO(B)FLUORENE	243-17-4
DIBENZO-P-DIOXIN	262-12-4
BENZOFURAN	271-89-6
PYRAZOLE	288-13-1
ETHYNODIOL DIACETATE	297-76-7
METHYL PARATHION	298-00-0
METHOXYPSORALEN	298-81-7
TREOSULPHAN	299-75-2
LEAD ACETATE	301-04-2
HYDRAZINE	302-01-2
METHYLHYDRAZINE SULFATE (1:1)	302-15-8
CHLORAL HYDRATE	302-17-0
CHLORMADINONE ACETATE	302-22-7
NITROGEN MUSTARD N-OXIDE HYDROCHLORIDE	302-70-5
LASIOCARPINE	303-34-4
GOSSYPOL	303-45-7
OCHRATOXIN A	303-47-9
HYDRALAZINE HYDROCHLORIDE	304-20-1
CHLORAMBUCIL	305-03-3
1,2-DIMETHYLHYDRAZINE DIHYDROCHLORIDE	306-37-6
ALDRIN	309-00-2
C. I. DIRECT BLUE 53 (EVANS BLUE)	314-13-6
ZECTRAN	315-18-4
MONOCROTALINE	315-22-0

CHEMICAL NAME	CAS NO.
ALPHA-HEXACHLOROCYCLOHEXANE	319-84-6
BETA-HEXACHLOROCYCLOHEXANE	319-85-7
5-AZACYTIDINE	320-67-2
3,4-DICHLOROBENZOTRIFLUORIDE	328-84-7
DIAZINON	333-41-5
DIAZOMETHANE	334-88-3
PROCARBAZINE HYDROCHLORIDE	366-70-1
NALDIXIC	389-08-2
2-FLUROBENZOYL CHLORIDE	393-52-2
ETHYLENE SULFIDE	420-12-2
HEXAFLURORPROPYLENE OXIDE	428-59-1
OXYMETHOLONE	434-07-1
LITHOCHOLIC ACID	434-13-9
DIAZEPAM	439-14-5
FLAGYL	443-48-1
AZATHIOPRINE	446-86-6
JACOBINE	471-14-7
DITHRANOL	480-22-8
RETROSINE	480-54-6
SENECIPHYLLINE	480-81-9
BENZOFURAZAN-1-OXIDE	480-96-6
EMETINE	483-18-1
5-METHOXYPSORALEN	484-20-8
DIBROMOMANNITOL	488-41-5
2,4'-DIPHENYLDIAMINE	492-17-1
C. I. BASIC YELLOW 2 (AURAMINE)	492-80-8
C. I. ACID RED 2 (METHYL RED)	493-52-7
CHLORNAPHAZINE	494-03-1
C. I. SOLVENT ORANGE 15 (ACRIDINE ORANGE)	494-38-2
3-NITROPROPIONIC ACID	504-88-1
1,3-DIOXANE	505-22-6
MUSTARD GAS	505-60-2

CHEMICAL NAME	CAS NO.
TETRANITROMETHANE	509-14-8
CHLOROBENZILATE	510-15-6
TRIMETHYLPHOSPHATE	512-56-1
DIMETHYLVINYL CHLORIDE	513-37-1
CITRININ	518-75-2
KAEMPFEROL	520-18-3
C. I. ACID ORANGE 20 (ORANGE 1)	523-44-4
ANGELICIN AND SOME SYNTHETIC DERIVATIVES	523-50-2
1,2,3-TRIMETHYL BENZENE	526-73-8
DICHLOROMETHOTREXATE	528-74-5
1,1-DIPHENYLHYDRAZINE HYDROCHLORIDE	530-47-2
MERPHALAN	531-76-0
N-(4-(5-NITRO-2-FURYL)-2-THIAZOLYL) ACETAMIDE	531-82-8
BENZIDINE DIHYDROCHLORIDE	531-85-1
2-CHLOROACETOPHENONE	532-27-4
C. I. BASIC ORANGE 2 (CHRYSOIDINE)	532-82-1
4,6-DINITRO-O-CRESOL	534-52-1
ETHIONAMIDE	536-33-4
1,2-DICHLOROETHYLENE	540-59-0
1,2-DIMETHYLHYDRAZINE (SDMH)	540-73-8
2,2,4-TRIMETHYLPENTANE	540-84-1
M-PHENYLENEDIAMINE HYDROCHLORIDE	541-69-5
ISOBUTYL NITRITE	542-56-3
1,3-DICHLOROPROPENE	542-75-6
MALONALDEHYDE	542-78-9
BIS(CHLOROMETHYL) ETHER (BCME)	542-88-1
BERYLLIUM ACETATE	543-81-7
TRIS(1-AZIRIDINYL)PHOSPHINE OXIDE (TEPA)	545-55-1
MANNOMUSTINE (DIHYDROCHLORIDE)	551-74-6
1-((5-NITROFURFURYLIDENE)AMINO)-2-IMIDAZOLIDINONE	555-84-0
GLYCIDOL	556-52-5
OCTAMETHYLCYCLOTETRASILOXANE	556-67-2

CHEMICAL NAME	CAS NO.
ZINC OLEATE	557-07-3
SEMICARBAZIDE (HYDROCHLORIDE)	563-41-7
3-CHLORO-2-METHYLPROPENE	563-47-3
1,2,3,4-DIEPOXYBUTANE, DL	564-00-1
CHLOROTRIANISENE	569-57-3
C. I. BASIC RED 9 MONOHYDROCHLORIDE	569-61-9
6-MERCAPTOPURINE-RIBOSIDE	574-25-4
2-NITRONAPHTHYLENE	581-89-5
2,4-TOLUENE DIISOCYANATE (2,4-TDI)	584-84-9
METHYL-N-BUTYL KETONE (MBK)	591-78-6
METHYLAZOXYMETHANOL ACETATE	592-62-1
VINYL BROMIDE	593-60-2
CHLOROFLUOROMETHANE	593-70-4
MEGESTROL ACETATE	595-33-5
MORPHOLINOPHOSPHORAMIDATE	597-25-1
METHYL CARBAMATE	598-55-0
LEAD CARBONATE	598-63-0
9-NITROANTHRACENE	602-60-8
5-NITROACENAPHTHENE	602-87-9
TRIPHENYLPHOSPHINE (TPP)	603-35-0
OXAZEPAM	604-75-1
2,6-DINITROTOLUENE	606-20-2
8-NITROQUINOLINE	607-35-2
2-NITROFLUORENE	607-57-8
1,2,3,4,5,6-HEXACHLOROCYCLOHEXANE (HCH)	608-73-1
PENTACHLOROBENZENE	608-93-5
2,6-DICHLORO-P-PHENYLENEDIAMINE	609-20-1
2,4-DICHLORONITROBENZENE	611-06-3
O-ETHYL TOLUENE	611-14-3
2-AMINOANTHRACENE	613-13-8
N,N-DIACETYLBENZIDINE	613-35-4
BENZOIC ACID, HYDRAZIDE	613-94-5

CHEMICAL NAME	CAS NO.
N-METHYL-N-NITROSOANILINE	614-00-6
O-PHENYLENEDIAMINE DIHYDROCHLORIDE	615-28-1
N-NITROSO-N-METHYLURETHANE (NMUT)	615-53-2
4-NITROANTHRANILIC ACID	619-17-0
N-NITROSODI-N-PROPYLAMINE	621-64-7
FURFURYL ACETATE	623-17-6
P-PHENYLENEDIAMINE DIHYDROCHLORIDE	624-18-0
N-PROPYL CARBAMATE	627-12-3
DIMETHYL ADIPATE (DMA)	627-93-0
CARBON MONOXIDE	630-08-0
1,1,1,2-TETRACHLOROETHANE	630-20-6
17-ALPHA-HYDROXYPROGESTERONE CAPROATE	630-56-8
PHENYTOIN SODIUM	630-93-3
C. I. BASIC VIOLET 14 (MAGENTA)	632-99-5
2,4,6-TRICHLOROANILINE	634-93-5
O-TOLUIDINE HYDROCHLORIDE	636-21-5
CLOFIBRATE	637-07-0
ERYTHROMYCIN STEARATE	643-22-1
1,3-DIOXOLANE	646-06-0
HEXAMETHYLPHOSPHORAMIDE (HMPA)	680-31-9
N-NITROSO-N-METHYLUREA (NMU)	684-93-5
2-AMINO-5-(5-NITRO-2-FURYL)-1,3,4-THIADIAZOLE	712-68-5
SULFAMETHOXAZOLE	723-46-6
DIMETHYL METHYLPHOSPHONATE	756-79-6
N-NITROSO-N-ETHYLUREA	759-73-9
TRANS-1,4-DICHLOROBUTENE	764-41-0
GLYCIDALDEHYDE	765-34-4
DIHYDROXYMETHYLFURATRIZINE	794-93-4
NORGESTREL	797-63-7
AZIRIDYL BENZOQUINONE	800-24-8
NITROVIN	804-36-4
TRICHLOROTRIETHYLAMINE HYDROCHLORIDE	817-09-4

CHEMICAL NAME	CAS NO.
DIMETHOXANE	828-00-2
1-METHYLPHENANTHRENE	832-69-9
C. I. SOLVENT YELLOW 14 (SUDAN I)	842-07-9
VINBLASTINE	865-21-4
DIMETHYL HYDROGEN PHOSPHITE	868-85-9
N-METHYLPYRROLIDONE (NMP)	872-50-4
3-NITROFLUORANTHENE	892-21-7
C. I. ACID RED 27 (AMARANTH)	915-67-3
N-NITROSODI-N-BUTYLAMINE (DBNA)	924-16-3
N-METHYLOLACRYLAMIDE	924-42-5
N-NITROSOPYRROLIDINE (NPYR)	930-55-2
P-TERT-BUTYLBENZALDEHYDE (P-TBB)	939-97-9
PROFLAVINE MONOHYDROCHLORIDE HEMIHYDRATE	952-23-8
1,3,5-TRIACRYLOYLHEXA-HYDRO-S-TRIAZINE	959-52-4
ACETOHEXAMIDE	968-81-0
C. I. BASIC RED 1 (RHODAMINE 6G)	989-38-8
ETHYLPHOSPHONOTHIOIC DICHLORIDE	993-43-1
(2-CHLOROETHYL)TRIMETHYLAMMONIUM CHLORIDE	999-81-5
UREA-FORMALDEHYDE RESINS	1000-82-4
HEPTACHLOR EPOXIDE	1024-57-3
CHROMIC ACETATE	1066-30-4
DIBUTYLTIN DIACETATE	1067-33-0
2-(1-AZIRIDINYL)-ETHANOL	1072-52-2
N-NITROSODIETHANOLAMINE	1116-54-7
DIMETHYL GLUTARATE (DMG)	1119-40-0
1,3-PROPANE SULTONE	1120-71-4
TOLAZAMIDE	1156-19-0
DECARBROMODIPHENYL OXIDE	1163-19-5
N,N'-DICYCLOHEXYLTHIOUREA	1212-29-9
TITANIUM FERROCENE	1271-19-8
NICKELOCENE	1271-28-9
BERYL ORE	1302-52-9

CHEMICAL NAME	CAS NO.
ARSENIC PENTOXIDE	1303-28-2
ARSENIC SULFIDE	1303-33-9
BERYLLIUM OXIDE	1304-56-9
CADMIUM OXIDE	1306-19-0
CADMIUM SULFIDE	1306-23-6
CHROMITE ORE	1308-31-2
C. I. PIGMENT GREEN 17 (CHROMIC OXIDE)	1308-38-9
FERRIC OXIDE	1309-37-1
ANTIMONY TRIOXIDE	1309-64-4
NICKEL OXIDE	1313-99-1
ZINC OXIDE	1314-13-2
THORIUM DIOXIDE	1314-20-1
ANTIMONY SULFIDE (PENTASULFIDE)	1315-04-4
LEAD OXIDE	1317-36-8
MIXED CRESOLS	1319-77-3
ARSENIC TRIOXIDE	1327-53-3
XYLENE (MIXED ISOMERS)	1330-20-7
TRICRESYL PHOSPHATE (TCP)	1330-78-5
ASBESTOS	1332-21-4
CHROMIUM TRIOXIDE	1333-82-0
CARBON BLACKS	1333-86-4
LEAD SUBACETATE	1335-32-6
POLYCHLORINATED BIPHENYLS (PCB'S)	1336-36-3
LEAD CHROMATE (VI) OXIDE	1344-38-3
ANTIMONY TRISULFIDE	1345-04-6
TANNIC ACID	1401-55-4
AFLATOXIN	1402-68-2
VITAMIN E (A-TOCOPHEROL)	1406-66-2
CLONITRALID	1420-04-8
2,4,5-TRIHYDROXYBUTYROPHENONE (THBP)	1421-63-2
ETHYLENEDIAMINETETRA(METHYL PHOSPHONIC ACID)	1429-50-1
GUANAZOLE	1455-77-2

CHEMICAL NAME	CAS NO.
DIEPOXYBUTANE	1464-53-5
N-(1-NAPHTHYL)ETHYLENEDIAMINE DIHYDROCHLORIDE	1465-25-4
1-PROPOXY-2-PROPANOL	1569-01-3
TRIFLURALIN	1582-09-8
DAMINOZIDE	1596-84-5
1,2-DIETHYLHYDRAZINE (SDEH)	1615-80-1
P-PHENYLENE DIAMINE.H2SO4 SALT	1624-57-75
METHYL TERT-BUTYL ETHER	1634-04-4
MALAOXON	1634-78-2
1,2-DICHLORO-1,1-DIFLUOROETHANE	1649-08-7
BISPHENOL A DIGLYCIDYL ETHER	1675-54-3
TRIETHYLENE GLYCOL DIACRYLATE	1680-21-3
SODIUM HYPOCHLORITE	1681-52-9
C. I. SOLVENT YELLOW 7 (4-HYDROXYAZOBENZENE)	1689-82-3
C. I. ACID VIOLET 49 (BENZYL VIOLET 4B)	1694-09-3
3-METHYLFLUORANTHENE	1706-10-0
2,3,7,8-TETRACHLORODIBENZO-P-DIOXIN (TCDD)	1746-01-6
MONOLINURON	1746-81-2
5-NITRO-P-ACETOPHENETIDE	1777-84-0
4,4'-METHYLENEBIS(2-METHYLANILINE)	1807-55-2
PROFLAVINE HEMISULPHATE	1811-28-5
2-BROMO-4,6-DINITROANLINE	1817-73-8
C. I. MORDANT RED 9	1836-22-2
NITROFEN	1836-75-5
CHLOROTHALONIL	1897-45-6
PICLORAM	1918-02-1
C. I. ACID ORANGE 12	1934-20-9
C. I. ACID YELLOW 23 (TARTRAZINE)	1934-21-0
C. I. ACID ORANGE 10 (ORANGE G)	1936-15-8
C. I. DIRECT BLACK 38	1937-37-7
TRIETHYLENE GLYCOL DIGLYCIDYL ETHER	1954-28-5
PIVALOLACTONE	1955-45-9

CHEMICAL NAME	CAS NO.
C. I. SOLVENT ORANGE 1	2051-85-1
OXYTETRACYCLINE HYDROCHLORIDE	2058-46-0
C. I. ACID BLUE 43	2150-60-9
METHANEARSONIC ACID.NA	2163-80-6
FLUOMETURON	2164-17-2
BIS(1-AZIRIDINYL)-MORPHOLINOPHOSPHINE SULFIDE	2168-68-5
DI(2,3-EPOXYPROPYL) ETHER (DGE)	2238-07-5
1,5-NAPHTHALENEDIAMINE	2243-62-1
C. I. MORDANT ORANGE 1	2243-76-7
D-CARVONE	2244-16-8
SYMPHYTINE	2257-95-5
DIALLATE	2303-16-4
SENKIRKINE	2318-18-5
C. I. FOOD GREEN 3 (FAST GREEN FCF)	2353-45-9
MIREX	2385-85-5
7-OXABICYCLO(4.1.O)HEPTANE-3-CARBOXYLIC ACID	2386-87-0
BIS(2,2-EPOXYCYCLOPENTENYL) ETHER	2386-90-5
BUTYL GLYCIDYL ETHER (BGE)	2426-08-6
C. I. DIRECT BLUE 15	2429-74-5
11-AMINOUNDECANOIC ACID	2432-99-7
2,3,5,6-TETRACHLORO-4-NITROANISOLE (TCNA)	2438-88-2
CIS-9,10-EPOXYSTEARIC ACID	2443-39-2
AURAMINE HYDROCHLORIDE	2465-27-2
C. I. DISPERSE BLUE 1	2475-45-8
1,1,3-TRIMETHYL-2-THIOUREA	2489-77-2
C. I. MORDANT BLACK	2538-85-4
C. I. REACTIVE BLUE 19	2580-78-1
C. I. DIRECT BLUE 6	2602-46-2
C. I. SOLVENT ORANGE 2 (OIL ORANGE SS)	2646-17-5
C. I. ACID BLUE 9 (BRILLIANT BLUE FCF)	2650-18-2
O-CHLOROBENZALMALONONITRILE (CS2)	2698-41-1
BENZAMINE, 4,4-(1,4-PHENYLENE BIS (1-METHYLETHYLIDENE	2716-12-3

CHEMICAL NAME	CAS NO.
AGARITINE(L-GLUTAMIC ACID-5-(2-(4-HYDROXYMETHYL)-PHEN	2757-90-6
1,1,3,3-TETRAMETHYL-2-THIOUREA	2782-91-4
C. I. FOOD YELLOW 3 (SUNSET YELLOW FCF)	2783-94-0
HC BLUE NO. 1	2784-94-3
C. I. DISPERSE YELLOW NO. 3	2832-40-8
ALLYL ISOVALERATE	2835-39-4
HC RED NO. 3	2871-01-4
DIISOPROPYL SULFATE (DIPS)	2973-10-6
BIS(2-DIMETHYLAMINOETHYL)ETHER (DMAEE) NIAX CATALYST	3033-62-3
2,3-EPOXYPROPYLTRIMETHYLAMMONIUMCHLORIDE	3033-77-0
BIS(TRICHLOROMETHYL)SULFONE	3064-70-8
2,3-DIBROMOPROPYL METHACRYLATE	3066-70-4
BETA-BUTYROLACTONE	3068-88-0
1,2-EPOXY-3,3,3-TRICHLOROPROPANE	3083-23-6
C. I. SOLVENT ORANGE 7 (SUDAN II)	3118-97-6
EPIBROMOHYDRIN (1,2-EPOXY-3-BROMOPROPANE)	3132-64-7
4-CHLORO-O-TOLUIDINE HYDROCHLORIDE	3165-93-3
1,5-NAPHTHALENE DIISOCYANATE	3173-72-6
C. I. DISPERSE BLUE 7	3179-90-6
TRANS-2,3-DIBROMO-2-BUTENE-1,4-DIOL	3234-02-4
TRIS(2-ETHYLHEXYL)TRIMELLITATE	3319-31-1
1,2-DIBROMO-4-(1,2-DIBROMOETHYL)CYCLOHEXANE	3322-93-8
3-CHLORO-2-HYDROXYPROPYLTRIMETHYL AMMONIUM CHLORIDE	3327-22-8
NICKEL CARBONATE	3333-67-3
BETA-(3,4-EPOXYCYCLOHEXYL)-ETHYLTRIMETHOXY SILANE	3388-04-3
1,2,3,4,7,7-HEXACHLORONORBORNADIENE	3389-71-7
1-(2-HYDROXYETHYL)-2-PYRROLIDONE	3445-11-2
IPD	3458-22-8
C. I. PIGMENT ORANGE 5	3468-63-1
PENTAERYTHRITOL TRIACRYLATE (PETA)	3524-68-3
PHENESTERIN	3546-10-9
PENTABROMOPHENYL ALLYL ETHER	3555-11-1

CHEMICAL NAME	CAS NO.
C. I. FOOD RED 6 (PONCEAU 3R)	3564-09-8
C. I. ACID RED 14 (CARMOISINE)	3567-69-9
2-(2-FORMYLHYDRAZINO)-4-(5-NITRO-2-FURYL)THIAZOLE	3570-75-0
C.I. DISPERSE BLUE 79:1	3618-72-2
AF-2(2-(2-FURYL)-3-(5-NITRO-2-FURYL)ACRYLAMIDE)	3688-53-7
C. I. ACID RED 1	3734-67-6
C. I. FOOD RED 5 (PONCEAU MX)	3761-53-3
NAFENOPIN	3771-19-5
ISOPHOSPHAMIDE	3778-73-2
STREPTOMYCIN SULFATE (2:3)	3810-74-0
TINUVIN 327	3864-99-1
4,5',8-TRIMETHYLPSORALEN	3902-71-4
C. I. DISPERSE BLUE 79	3956-55-6
ISOPROPYL GLYCIDYL ETHER	4016-14-2
4'-PHENYLACETANILIDE	4075-79-0
CROTONALDEHYDE	4170-30-3
MALONONITRILE	4341-85-9
DACARBAZINE	4342-03-4
CYTOXAL ALCOHOL	4465-94-5
C. I. FOOD RED 1 (PONCEAU SX)	4548-53-2
N-NITROSOMETHYLVINYLAMINE	4549-40-0
4-QUINOLINAMINE-N-HYDROXY-1-OXIDE	4637-56-3
5-AMINOACENAPHTHENE	4657-93-6
C. I. ACID GREEN 3 (GUINEA GREEN B)	4680-78-8
4-PHENYLCYCLOHEXENE (4-PCH)	4994-16-5
N-(4-(OXIRANYLMETHOXY)PHENYL)-N-(OXIRANYLMETHYL)-OXI	5026-74-4
IMIDAZOLE MUSTARD	5034-77-5
4-CHLORO-M-PHENYLENEDIAMINE	5131-60-2
C. I. ACID GREEN 5 (LIGHT GREEN SF)	5141-20-8
C. I. PIGMENT RED 53, BARIUM SALT	5160-02-1
P-CHLOROBENZOTRICHLORIDE (PCBTC)	5216-25-1
2-NITRO-P-PHENYLENEDIAMINE	5307-14-2

CHEMICAL NAME	CAS NO.
DIBENZO(A,E)FLUORANTHENE	5385-75-1
GLYCIDYL OLEATE	5431-33-4
HYDROXYLAMINE HYDROCHLORIDE	5470-11-1
1-NITROPYRENE	5522-43-0
IODINATED GLYCEROL	5634-39-9
ZINC DIALKYLDITHIOPHOSPHATE	5823-20-1
D-LIMONENE	5989-27-5
3-AMINO-9-ETHYLCARBAZOLE HYDROCHLORIDE	6109-97-3
CHLORDIMEFORM	6164-98-3
2,2'-DIETHOXYACETOPHENONE (DEAP)	6175-45-7
C. I. SOLVENT RED 80 (CITRUS RED NO. 2)	6358-53-8
C. I. PIGMENT YELLOW 12 (DIARYLANILIDE YELLOW)	6358-85-6
C. I. SOLVENT RED 19 (SUDAN RED 7B)	6368-72-5
2,5-TOLUENEDIAMINE SULFATE	6369-59-1
C. I. ACID ORANGE 3	6373-74-6
C. I. PIGMENT RED 1	6410-10-2
C. I. SOLVENT BROWN 1 (SUDAN BROWN RR)	6416-57-5
BIS(2-ETHYLHEXYL)-TEREPHTHALATE	6422-86-2
2-(CHLOROMETHYL)PYRIDINE HYDROCHLORIDE	6959-47-3
3-(CHLOROMETHYL)PYRIDINE HYDROCHLORIDE	6959-48-4
ACRONYCINE	7008-42-6
AMPICILLIN TRIHYDRATE	7177-48-2
3,3',4,4'-BIPHENYL-TETRAMINE	7411-49-6
ALUMINUM	7429-90-5
LEAD	7439-92-1
LITHIUM	7439-93-2
MERCURY	7439-97-6
NICKEL (INCL CERTAIN NI CPDS)	7440-02-0
ANTIMONY	7440-36-0
ARSENIC	7440-38-2
BERYLLIUM	7440-41-7
CADMIUM	7440-43-9

CHEMICAL NAME	CAS NO.
FURNACE CARBON BLACK	7440-44-0
CHROMIUM (ELEMENTAL)	7440-47-3
ZINC	7440-66-6
LEAD PHOSPHATE	7446-27-7
SELENIUM SULFIDE	7446-34-6
POTASSIUM CHLORIDE	7447-40-7
GLYCIDYL STEARATE	7460-84-6
N-NITROSOPROLINE (NPRO)	7519-36-0
TITANIUM TETRACHLORIDE	7550-45-0
DICHLOROACETYLENE	7572-29-4
SILICA	7631-86-9
SODIUM ARSENATE	7631-89-2
BORON TRIFLUORIDE	7637-07-2
ZINC CHLORIDE	7646-85-7
HYDROGEN FLUORIDE	7664-39-3
SODIUM FLUORIDE	7681-49-4
HYDROGEN PEROXIDE	7722-84-1
PHOSPHORUS	7723-14-0
POTASSIUM BROMATE	7758-01-2
LEAD CHLORIDE	7758-95-4
C. I. PIGMENT YELLOW 34 (LEAD CHROMATE)	7758-97-6
STANNOUS CHLORIDE	7772-99-8
SODIUM CHROMATE	7775-11-3
CALCIUM ARSENATE	7778-44-1
POTASSIUM DICHROMATE	7778-50-9
SELENIUM	7782-49-2
CHLORINE	7782-50-5
LEAD ARSENATE	7784-40-9
POTASSIUM ARSENATE	7784-41-0
ARSINE	7784-42-1
SODIUM ARSENITE	7784-46-5
BERYLLIUM CHLORIDE	7787-47-5

CHEMICAL NAME	CAS NO.
BERYLLIUM FLUORIDE	7787-49-7
BERYLLIUM SULFATE TETRAHYDRATE	7787-56-6
POTASSIUM CHROMATE	7789-00-6
C. I. PIGMENT YELLOW 32 (STRONTIUM CHROMATE)	7789-06-2
HYDROXYLAMINE	7803-49-8
HYDRAZINE MONOHYDRATE	7803-57-8
TOXAPHENE	8001-35-2
TERPENE POLYCHLORINATE	8001-50-1
CREOSOTE	8001-58-9
PETROLEUM MINERAL AND CRUDE OILS	8002-05-9
P-ACETOPHENETIDIDE CAFFEINE ASPIRIN MIXTURE	8003-03-0
C. I. DIRECT YELLOW 28	8005-72-9
GASOLINE	8006-61-9
COAL-TARS AND DERIVED PRODUCTS	8007-45-2
KEROSENE	8008-20-6
CALCIUM CHROMATE	8012-75-7
DDT (TECHNICAL GRADE-MIXTURE OF ISOMERS)	8017-34-3
SOLVENT NAPHTHA (HEAVY AROMATIC SOLVENT NAPHTHA)	8030-30-6
SACCHARATED IRON	8047-67-4
ACRIFLAVINIUM CHLORIDE	8048-52-0
IRON DEXTRIN	8050-93-9
STODDARD SOLVENT	8052-41-3
BITUMENS (FROM PETROLEUM REFINING)	8052-42-4
GUM ARABIC	9000-01-5
CARRAGEENANS	9000-07-1
GUAR GUM	9000-30-0
LOCUST BEAN GUM	9000-40-2
AGAR-AGAR	9002-18-0
POLYTETRAFLUOROETHYLENE (PTFE)	9002-84-0
POLYVINYL CHLORIDE	9002-86-2
POLYETHYLENE	9002-88-4
POLYVINYL ALCOHOL	9002-89-5

CHEMICAL NAME	CAS NO.
POLYACRYLIC ACID	9003-01-4
POLYPROPYLENE	9003-07-0
POLYVINYL ACETATE	9003-20-7
VINYL CHLORIDE-VINYL ACETATE COPOLYMERS	9003-22-9
POLYVINYL PYRROLIDONE	9003-39-8
POLYSTYRENE	9003-53-6
STYRENE-BUTADIENE COPOLYMERS	9003-55-8
ACRYLONITRILE-BUTADIENE-STYRENE COPOLYMER RESINS	9003-56-9
IRON DEXTRAN	9004-66-4
POLYETHYLENE GLYCOL SORBITAN MONOLAURATE	9005-64-5
POLYURETHANE	9009-54-5
POLYCHLOROPRENE	9010-98-4
VINYLIDENE CHLORIDE-VINYL CHLORIDE COPOLYMERS	9011-06-7
POLYMETHYL METHACRYLATE	9011-14-7
POLYMETHYLENE POLYPHENYL ISOCYANATE	9016-87-9
UCON 50-HB POLYETHERS	9038-95-3
SULFONATED POLYSTYRENE, SODIUM SALT	9080-79-9
NITROUS OXIDE (NITROGEN OXIDE)	10024-97-2
HYDRAZINE SULFATE (1:1)	10034-93-2
HYDROXYLAMINE SULFATE (2:1) SALT	10039-54-0
HYDROXYLAMINE SULFATE (1:1) (SALT)	10046-00-1
STERIGMATOCYSTIN	10048-13-2
PARASORBIC ACID	10048-32-5
LEAD NITRATE	10099-74-8
C. I. 77305 (CHROMIUM SULFATE)	10101-53-8
NITROGEN DIOXIDE	10102-44-0
CADMIUM CHLORIDE	10108-64-2
CADMIUM SULFATE	10124-36-4
ARISTOLOCHIC ACID	10130-99-5
C. I. DIRECT BROWN 191	10190-66-6
C. I. PIGMENT YELLOW 31 (BARIUM CHROMATE)	10294-40-3
ETHYLENEBIS TRIS (2-CYANOETHYL) PHOSPHONIUM BROMIDE	10310-38-0

CHEMICAL NAME	CAS NO.
DIBROMODULCITOL	10318-26-0
COPPER 8-HYDROXYQUINOLINE	10380-28-6
SODIUM DICHROMATE (VI)	10588-01-9
N-NITROSOMETHYLETHYLAMINE	10595-95-6
CHLORAMINE	10599-90-3
BLEOMYCIN	11056-06-7
AROCHLOR 1254	11097-69-1
COBALT-CHROMIUM ALLOY	11114-92-4
BERYLLIUM-COPPER ALLOY	11133-98-5
CROCIDOLITE ASBESTOS	12001-28-4
CHRYSOTILE ASBESTOS	12001-29-5
NICKEL SUBSULFIDE	12035-72-2
ZINEB	12122-67-7
BERTRANDITE	12161-82-9
ATTAPULGITE	12174-11-7
AUROTHIOGLUCOSE	12192-57-3
C. I. REACTIVE BLACK 5	12225-25-1
C. I. REACTIVE BLUE 50	12225-61-5
C. I. REACTIVE YELLOW 39	12226-61-8
C. I. DISPERSE RED 9	12226-87-8
C. I. REACTIVE BLUE 36	12236-89-4
C. I. REACTIVE YELLOW 31	12237-13-7
C. I. REACTIVE RED 78	12270-86-9
ETHYL SELENAC	12367-47-4
PERLITE	12427-27-9
MANEB	12427-38-2
ZINC POTASSIUM CHROMATE	12433-50-0
CYCLOCHLOROTINE	12663-46-6
BERYLLIUM-ALUMINUM ALLOY	12770-50-2
LOMUSTINE1-(2-CHLOROETHYL)-3-CYCLOHEXYL-1-NITROSOUREA	13010-47-4
ETHIONINE	13073-35-3
BERYLLIUM CARBONATE	13106-47-3

CHEMICAL NAME	CAS NO.
PHOSPHAMIDON	13171-21-6
N-NITROSOSARCOSINE	13256-22-9
RIFAMPICIN	13292-46-1
BERYLLIUM HYDROXIDE	13327-32-7
PHOTODIELDRIN	13366-73-9
METHALLY-2-NITROPHENYL ETHER	13414-54-5
7-NITRO-2,2-DIMETHYL-2,3-DIHYDROBENZOFURAN	13414-55-6
NICKEL CARBONYL	13463-39-3
TITANIUM DIOXIDE	13463-67-7
POTASSIUM ARSENITE	13464-35-2
1,2-BIS(CHLOROMETHOXY)ETHANE (BCMXE)	13483-18-6
C. I. PIGMENT YELLOW 36 (ZINC CHROMATE)	13530-65-9
BERYLLIUM PHOSPHATE	13598-15-7
DECABROMOBIPHENYL	13654-09-6
TRIS(1,3-DICHLOROISOPROPYL)PHOSPHATE (FYROL FR2)	13674-87-8
N-OXYDIETHYLENE THIOCARBAMYL-N-OXYDIETHYLENE	13752-51-7
C. I. PIGMENT YELLOW 33 (CALCIUM CHROMATE)	13765-19-0
1-(2-CHLOROETHYL)-3-(4-METHYLCYCLOHEXYL)-1-NITROSOUREA	13909-09-6
WOLLASTONITE	13983-17-0
3,3',4,4'-TETRACHLOROAZOBENZENE (TCAB)	14047-09-7
CRISTOBALITE	14464-46-1
FERBAM	14484-64-1
TREMOLITE	14567-73-8
TALC	14807-96-6
QUARTZ	14808-60-7
CYCASIN	14901-08-7
EOSIN	15086-94-9
ETHYLENEDIAMINETETRA(METHYL PHOSPHONIC ACID) HEXASODI	15142-96-8
BERYLLIUM SILICATE	15191-85-2
5-ETHYL-5-METHYL-1,3-BIS-(OXIRANYLMETHYL)-2,4-IMIDAZO	15336-82-0
TRIDYMITE	15468-32-3
2,6-TOLUENEDIAMINE DIHYDROCHLORIDE	15481-70-6

CHEMICAL NAME	CAS NO.
ISATIDINE	15503-86-3
CISPLATIN	15663-27-1
2,2',5,5'-TETRACHLOROBENZIDINE	15721-02-5
ZINC CHROMATE HYDROXIDE	15930-94-6
C. I. DIRECT BROWN 95	16071-86-6
7-BROMOMETHYL-12-METHYL-BENZ(A)ANTHRACENE	16238-56-5
N-NITROSONORNICOTINE	16543-55-8
GYROMITRIN (ACETALDEHYDE FORMYLMETHYLHYDRAZONE)	16568-02-8
3-AMINO-4-ETHOXYACETANILIDE	17026-81-2
EOSIN DISODIUM SALT	17372-87-1
1-ACETYL-2-PICOLINOYL HYDRAZINE (1A2PH)	17433-31-7
M-(CYCLOHEXYLTHIO) PHTHALIMIDE	17796-82-6
TETRAETHYLENE GLYCOL DIACRYLATE	17831-71-9
ZEARALENONE	17924-92-4
SEPIOLITE	18307-23-8
NITRILOTRIACETIC ACID, TRISODIUM SALT, MONOHYDRATE	18662-53-8
STREPTOZOTOCIN	18883-66-4
C. I. REACTIVE YELLOW 14	18976-74-4
LEAD DIMETHYLDITHIOCARBAMATE	19010-66-3
1,2,3,7,8,9-HEXACHLORODIBENZO-P-DIOXIN (HCDD)	19408-74-3
C. I. REACTIVE RED 22	19526-81-9
2,3-DIBROMOPROPYL ACRYLATE	19660-16-3
3-CARBETHOXYPSORALEN	20073-24-9
DIMETHYL N-METHYLOLPHOSPHONOPROPIONAMIDE (DMMP)	20120-33-6
PARA-CHLOROANILINE HYDROCHLORIDE	20265-96-7
P-ANISIDINE HYDROCHLORIDE	20265-97-8
3,3'-DIMETHOXYBENZIDINE DIHYDROCHLORIDE	20325-40-0
POTASSIUM AZIDE	20762-60-1
DAUNOMYCIN	20830-81-3
ETHYL TELLURAC	20941-65-5
3,3',4,4'-TETRACHLOROAZOXYBENZENE (TCAOB)	21232-47-3
T2-TRICHOTHECENE	21259-20-1

CHEMICAL NAME	CAS NO.
ICRF - 159	21416-87-5
CYTEMBENA	21739-91-3
BIS(2,3-DIBROMPROPYL ETHER) OF TETRABROMBISPHENOL A	21850-44-2
LUTEOSKYRIN	21884-44-6
TETRACHLORVINPHOS	22248-79-9
1,4-DIMETHYLPHENANTHRENE	22349-59-3
ESTRADIOL MUSTARD	22966-79-6
CALCIUM SODIUM METAPHOSPHATE FIBERS	23209-59-8
ADRIAMYCIN	23214-92-8
RIDDELLIINE	23246-96-0
FURSARENON-X	23255-69-8
HYCANTHONE MESYLATE	23255-93-8
RUGULOSIN	23537-16-8
POTASSIUM BIS(S-HYDROXYETHYL)-DITHIOCARBAMATE	23746-34-1
PRONAMIDE	23950-58-5
VINYL TOLUENE	25013-15-4
BUTYLATED HYDROXYANISOLE (BHA)	25013-16-5
NYLON 6	25038-54-4
EPOXY RESIN	25068-38-6
1-BUTYLMETACRYLATE/STYRENE RANDOM COPOLYMER	25213-39-2
ETHYLTOLUENES	25550-14-5
TRIMETHYL BENZENE (MIXED ISOMERS)	25551-13-7
C. I. FOOD RED 17 (FD&C RED 40)	25956-17-6
TRANS-2-((DIMETHYLAMINO)METHYLAMINO)-5-(2-(5-NITRO-2-	25962-77-0
MEA-A-C(2-AMINO-3-METHYL-9H-PYRIDO(2,3-B)INDOLE	26148-68-5
ACETIN	26446-35-5
TOLUENE DIISOCYANATE (TDI)	26471-62-5
PHOSPHORUS ACID, ISODECYL DIPHENYL ESTER	26544-23-0
SODIUM AZIDE	26628-22-8
BIS(2,4-DI-TERT-BUTYLPHENYL)PENTAERYTHRITOL DIPHOSPHI	26741-53-7
HYDROXYSENKIRKINE	26782-43-4
1,2-ETHANEDITHIOL	26914-40-9

CHEMICAL NAME	CAS NO.
DODECYLPHENOL	27193-86-8
CYCLOPENTA(CD)PYRENE	27208-37-3
OCTABROMOBIPHENYL	27858-07-7
3,3'-DICHLORO-4,4'-DIAMINODIPHENYL ETHER	28434-86-8
N-NITROSOFOLIC ACID	29291-35-8
N-NITROSOYDROXYPROLINE (NHPRO)	30310-80-6
NICKEL-BIS(ETHYL(3,5-DI-T-BUTYL-4-HYDROXYBENZYL)PHOSP	30947-30-9
HC BLUE NO. 2	33229-34-4
2-METHYLFLUORANTHENE	33543-31-6
2,7-DICHLORODIBENZO-P-DIOXIN (DCDD)	33857-26-0
2,2',3,3',5,5'-HEXACHLOROBIPHENYL	35694-04-3
HEXABROMOBIPHENYL (TECH)	36355-01-8
1,2-ETHANEDIYL-BIS (OXY-2, 1-ETHANEDIYL) ESTER	36443-68-2
BERYLLIUM-NICKEL ALLOY	37227-61-5
N'-NITROSOANABASINE (NAB)	37620-20-5
1,2,3-TRIS(CHLOROMETHOXY)-PROPANE	38571-73-2
2,4-DIAMINOANISOLE SULFATE	39156-41-7
TARA GUM	39300-88-4
BERYLLIUM ZINC SILICATE	39413-47-3
BIS(2-CHLOROISOPROPYL)ETHER (BCIE)	39638-32-9
ALPHA-METHYLDOPA SESQUIHYDRATE	41372-08-1
1,8-DINITROPYRENE	42397-65-9
ETHYL AURAMINE NITRATE	43130-12-7
LEAD NAPHTHENATE	50825-29-1
9-AMINOACRIDINE HC1 H20	52417-22-8
HC YELLOW 4	52551-67-4
2,4-DIMETHOXYANILINE HYDROCHLORIDE	54150-69-5
BERYLLIUM-COPPER-COBALT ALLOY	55158-44-6
TETRAKIS(HYDROXYMETHYL)PHOSPHONIUM SULFATE (THPS)	55566-30-8
CHLORINATED TRISODIUM PHOSPHATE	56802-99-4
1,4-BIS(CHLOROMETHOXYMETHYL)-BENZENE	56894-91-8
PENTAERYTHRITOL-TRIS-(B-(N-AZIRIDINYL) PROPIONATE)	57116-45-7

CHEMICAL NAME	CAS NO.
DIPROPYLENEGLYCOL DIACRYLATE	57472-68-1
1,2,3,6,7,8-HEXACHLORODIBENZO-P-DIOXIN (HCDD)	57653-85-7
HEXABROMOBIPHENYL (PURE)	59080-40-9
PETASITENINE	60102-37-6
CALIBAN (F/R-P 44)	61345-53-7
2-CHLORO-P-PHENYLENEDIAMINE SULFATE	61702-44-1
CHLORINATED TERPHENYL	61788-33-8
CALCIUM NAPHTHENATE	61789-36-4
COBALT NAPHTHENATE	61789-51-3
C. I. REACTIVE RED 83	61969-26-4
C. I. REACTIVE RED 84	61969-27-5
TRP-P-1(3-AMINO-1,4-DIMETHYL-5H-PYRIDO(4,3-B)INDOLE)	62450-06-0
TRP-P-2(3-AMINO-1-METHYL-5H-PYRIDO(4,3-B)INDOLE) AND	62450-07-1
6-NITROBENZO(A)PYRENE	63041-90-7
PROPANEDIOIC ACID COMPOUND	63843-89-0
BETA-THIOGUANINE DEOXYRIBOSIDE (B-TGDR)	64039-27-6
4-(METHYLNITROSAMINO)-4-(3-PYRIDYL)BUTANAL (NNA)	64091-90-3
4-(METHYLNITROSAMINO)-1-(3-PYRIDYL)-1-BUTANONE (NNK)	64091-91-4
NAPHTHENIC DISTILLATE	64741-52-2
PETROLEUM NAPHTHA, HEAVY CATALYTIC REFORMED	64741-68-0
RHU LIGHT VACUUM GAS OIL	64741-75-9
RHU MIDDLE DISTILLATE	64741-76-0
HYDRODESULFURIZED VACUUM GAS OIL	64742-86-5
PETROLEUM NAPHTHA SOLVENT	64742-95-6
ERIONITE	66733-21-9
GLU-P-2(2-AMINOPYRIDO)1,2-A:3'2-D(IMADAZOLE)	67730-10-3
GLU-P-1 (2-AMINO-6-METHYLPYRIDO)1,2-A:3',2'-D(IMIDAZOLE	67730-11-4
POLYBROMINATED BIPHENYL MIXTURE (PBB)	67774-32-7
AROMATIC PETROLEUM PITCH	68187-58-6
ETHYLENEDIAMINETETRA(METHYL PHOSPHONIC ACID) TETRAPOT	68188-96-5
PERFLUOROALKYL RESINS	68298-78-2
SHALE-OILS	68308-34-9

CHEMICAL NAME	CAS NO.
DIESEL FUELS	68334-30-5
NO. 2 FUEL OIL	68476-30-2
PETROLEUM DISTILLATES	68477-31-6
DIETHYL TOLUENE DIAMINE	68479-98-1
HAMAMELIS WATER	68916-39-2
N'-NITROSOANATABINE	71267-22-6
INTRACRON BLUE 3G	75432-59-6
IQ(2-AMINO-3-METHYLIMIDAZO (4,5-F QUINOLINE)	76180-96-6
MEIQ (2-AMINO-3,4-DIMETHYLIMIDAZO(4,5-F)QUINOLINE (MEIQ)	77094-11-2
MEIQX (2-AMINO-3,4-DIMETHYLIMIDAZO(4,5-F)QUINOXALINE)	77500-04-0
TRANS-A-CHLORO-3-OXYAMINOPROPENE HYDROCHLORIDE	82244-86-8
PYRIDO(3,4-C)PSORALEN AND 7-METHYLPYRIDO(3,4-C)PSORALEN	85878-62-2
7-AMINO-2,2-DIMETHYL-2,3-DIHYDROBENZOFURAN	88296-46-4
ARYL PHOSPHATES	CLASS-01
ALKYL PHTHALATES	CLASS-02
BETEL-QUID AND ARECA NUT CHEWING	CLASS-03
SOOT	CLASS-05
METAL-WORKING FLUIDS	CLASS-06
COKE OVEN EMISSIONS (POLYCYCLIC ORGANIC MATTER)	CLASS-07
CHLOROFLUOROCARBONS	CLASS-08
BENZIDINE CONGENER DYES	CLASS-09
CONJUGATED ESTROGENS	CLASS-10
CHLORINATED NAPHTHALENES (CNS)	CLASS-11
GLYCIDOL AND DERIVATIVES (ETHERS, ESTERS)	CLASS-12
METHYLCHRYSENES	CLASS-13
ORGANOTIN COMPOUNDS	CLASS-16
TOBACCO HABITS OTHER THAN SMOKING	CLASS-17
N-NITROSAMINES DERIVED FROM ARECA-NUT ALKALOIDS	CLASS-18
POLYCHLORINATED DIBENZOFURAN (PCDB)	CLASS-19
ALKYL EPOXIDES	CLASS-20
FLUOROALKANES	CLASS-21
PHENYLENEDIAMINES CATEGORY	CLASS-22

CHEMICAL NAME	CAS NO.
PETROLEUM REFINING OCCUPATIONAL EXPOSURES	CLASS-23
PHTHALATE ESTERS	CLASS-24
MICA DUST	CLASS-26
FLUOROSILICON FLUIDS AND GREASES	CLASS-27
CHLOROPHENOLS	CLASS-29
CHLOROPHENOXY HERBICIDES	CLASS-30
PROPYLENE BASED GLYCOL ETHERS	CLASS-31
JET FUEL	CLASS-32
FUEL OILS (HEATING OILS)	CLASS-33
BOOT AND SHOE MANUFACTURE AND REPAIR	CLASS-36
MATE	CLASS-38
CHLORINATED PARAFFIN	CLASS-44
MANUFACTURE OF CHROMATE	UNKNOWN-01
2-CHLORO-1-METHYLETHYL(2-CHLOROPROPYL) ETHER	UNKNOWN-02
MANUFACTURE OF AURAMINE (C. I. BASIC YELLOW 2)	UNKNOWN-03
C9 AROMATIC HYDROCARBONS (NAPTHA)	UNKNOWN-04
HEMATITE MINING	UNKNOWN-05
NICKEL REFINING	UNKNOWN-06
CARBON FIBERS	UNKNOWN-08
PENTAERYTHRITOL MONOOLEATE TRIACRYLATE	UNKNOWN-09
C. I. DISPERSE BLUE 291	UNKNOWN-10
COFFEE	UNKNOWN-15
LEATHER GOODS MANUFACTURE	UNKNOWN-16
RUBBER INDUSTRY	UNKNOWN-17
BRILLIANT RED HISALONIC	UNKNOWN-18
CHLOROFLUOROCARBON 22	UNKNOWN-19
INTRACRON ORANGE G	UNKNOWN-20
HAIR SERVICE OCCUPATION AND USERS OF HAIR DYES	UNKNOWN-21
ORANGE HISALONIC	UNKNOWN-22
LUMBER AND SAWMILL INDUSTRIES	UNKNOWN-23
CARPENTRY AND JOINERY	UNKNOWN-24
FURNITURE AND CABINET-MAKING INDUSTRY	UNKNOWN-25

CHEMICAL NAME	CAS NO.
PULP AND PAPER INDUSTRY	UNKNOWN-26
LEATHER TANNING AND PROCESSING INDUSTRIES	UNKNOWN-27
POLYHALOGENATED DIBENZOFURAN (HDF) (2,3,7,8-TCDF)	UNKNOWN-29
SUBSTITUTED POLYGLYCIDYL BENZENEAMINE	UNKNOWN-30
BRACKEN FERN	UNKNOWN-31
POLYHALOGENATED DIBENZO-P-DIOXINS (HDD)	UNKNOWN-32
CALCIUM DIALKYLTHIOPHOSPHATES	UNKNOWN-35